Experimental Design for Formulation

ASA-SIAM Series on Statistics and Applied Probability

The ASA-SIAM Series on Statistics and Applied Probability is published jointly by the American Statistical Association and the Society for Industrial and Applied Mathematics. The series consists of a broad spectrum of books on topics in statistics and applied probability. The purpose of the series is to provide inexpensive, quality publications of interest to the intersecting membership of the two societies.

Gan, G., Ma, C., and Wu, J., *Data Clustering: Theory, Algorithms, and Applications*

Hubert, L., Arabie, P., and Meulman, J., *The Structural Representation of Proximity Matrices with MATLAB*

Nelson, P. R., Wludyka, P. S., and Copeland, K. A. F., *The Analysis of Means: A Graphical Method for Comparing Means, Rates, and Proportions*

Burdick, R. K., Borror, C. M., and Montgomery, D. C., *Design and Analysis of Gauge R&R Studies: Making Decisions with Confidence Intervals in Random and Mixed ANOVA Models*

Albert, J., Bennett, J., and Cochran, J. J., eds., *Anthology of Statistics in Sports*

Smith, W. F., *Experimental Design for Formulation*

Baglivo, J. A., *Mathematica Laboratories for Mathematical Statistics: Emphasizing Simulation and Computer Intensive Methods*

Lee, H. K. H., *Bayesian Nonparametrics via Neural Networks*

O'Gorman, T. W., *Applied Adaptive Statistical Methods: Tests of Significance and Confidence Intervals*

Ross, T. J., Booker, J. M., and Parkinson, W. J., eds., *Fuzzy Logic and Probability Applications: Bridging the Gap*

Nelson, W. B., *Recurrent Events Data Analysis for Product Repairs, Disease Recurrences, and Other Applications*

Mason, R. L. and Young, J. C., *Multivariate Statistical Process Control with Industrial Applications*

Smith, P. L., *A Primer for Sampling Solids, Liquids, and Gases: Based on the Seven Sampling Errors of Pierre Gy*

Meyer, M. A. and Booker, J. M., *Eliciting and Analyzing Expert Judgment: A Practical Guide*

Latouche, G. and Ramaswami, V., *Introduction to Matrix Analytic Methods in Stochastic Modeling*

Peck, R., Haugh, L., and Goodman, A., *Statistical Case Studies: A Collaboration Between Academe and Industry, Student Edition*

Peck, R., Haugh, L., and Goodman, A., *Statistical Case Studies: A Collaboration Between Academe and Industry*

Barlow, R., *Engineering Reliability*

Czitrom, V. and Spagon, P. D., *Statistical Case Studies for Industrial Process Improvement*

Experimental Design for Formulation

Wendell F. Smith

Pittsford, New York

Society for Industrial and Applied Mathematics
Philadelphia, Pennsylvania

American Statistical Association
Alexandria, Virginia

The correct bibliographic citation for this book is as follows: Smith, Wendell F., *Experimental Design for Formulation*, ASA-SIAM Series on Statistics and Applied Probability, SIAM, Philadelphia, ASA, Alexandria, VA, 2005.

10 9 8 7 6 5 4 3 2

Library of Congress Cataloging-in-Publication Data

Smith, Wendell F. (Wendell Franklyn), 1931-
Experimental design for formulation / Wendell F. Smith.
p. cm. – (ASA-SIAM series on statistics and applied probability)
Includes bibliographical references and index.
ISBN 978-0-898715-80-4 (pbk.)
1. Experimental design. I. Title. II. Series.

QA279.S64 2005
519.5'7–dc22 2004065317

Contents

List of Figures

List of Tables

Preface

Few things impact our everyday lives more than those products that are manufactured by mixing ingredients together. From the time we arise in the morning until we retire at night, we depend on *formulated products*. Some examples are

Adhesives	Herbicides
Beverages	Hydrogels
Biological solutions	Inks
Cements	Paints
Ceramic glazes	Personal care products
Cleaning agents	Pesticides
Combination vaccines	Petroleum products
Cosmetics	Pharmaceuticals
Construction materials	Photoconductors
Dyes	Photoresists
Fiber finishes	Polymer additives
Floor coverings	Polymers
Floor finishes	Powder coatings
Foams	Protective coatings
Food ingredients	Rubber
Froth flotation reagents	Sealants
Gasketing materials	UV curable coatings
Glasses	Water treatment chemicals

At the time of this writing there is only one book in the English language dedicated to experimental design for formulation [29]. The probable reason for this is because the subject is a specialized subset of experimental design in general. As a consequence the topic is either relegated to a single chapter in books on regression or the design of experiments [49, 79, 107] or simply to sections within chapters [93, 94, 102].

This text has evolved from a short course that I have taught since 1995 for the American Chemical Society. Because the sponsoring organization for the course is the American Chemical Society, the majority of students have been chemists. However, this book is intended for a much broader audience that would include students and researchers in the physical sciences, engineering disciplines, or statistics.

The book is intended to provide a practical step-by-step guide to the design and analysis of experiments involving formulations. It contains many examples selected from a wide variety of fields along with output from several popular computing packages. Formulas underlying the computer output will be explained, and proper interpretation of the output will be emphasized. There is more than enough material in this book for it to be used as the supporting text for a course at the senior undergraduate or beginning graduate level. With selective abridgement, the text is also suitable for a two- or three-day short course.

The prerequisites for this book are relatively modest. Previous exposure to a first course in statistics and an introductory course on experimental design will be assumed. The reader should be comfortable with hypothesis tests, confidence intervals, the normal, t, and F distributions, and factorial, fractional factorial, and central composite designs. It is also assumed that the reader has some knowledge of matrix algebra. The matrix formulation of ordinary least squares (OLS) is well covered in texts on linear regression and will not be repeated here. At the same time, we will draw generously on the results of this approach to OLS. Statistical proofs are largely absent, as they can be found in texts on linear regression or experimental design.

Topics are ordered according to a sequence of steps one normally follows in a designed experiment — hypothesizing a model, designing an experiment to support the model, collecting data, and finally fitting a model and interpreting the results. The four major parts of the text are as follows:

- **Preliminaries** (Chapters 1–3). This section covers topics that one needs to know before beginning to design a mixture experiment. This includes a description of mixture space and an explanation of the model types commonly used in a mixture setting.

- **Design** (Chapters 4–7). These chapters cover the design of mixture experiments, design evaluation, and the modification of designs following evaluation. In addition, the blocking of mixture experiments is discussed in this section.

- **Analysis** (Chapters 8–12). This section covers model fitting, model evaluation, and the modification of models following evaluation. Other topics include the concept of an *effect* in a mixture setting and elementary optimization methods.

- **Special Topics** (Chapters 13 and 14). It is sometimes of interest to combine mixture and nonmixture variables (often called *process variables*) in a designed experiment. This topic is covered in Chapter 13. Chapter 14 explains the concept of collinearity and the possible problems that can result from its presence.

The beginning student in this area need not embark on a complete read-through. To get started as quickly as possible designing and analyzing mixture experiments, a beginning practitioner should be comfortable with the material in Chapters 1–6, 8, 9, 10 (Sections 1 and 2), 11 (Sections 1–4), and 12 (Sections 1 and 2). In addition, one could save the material on robust regression (Section 10.1, pages 212–218) for a later reading.

This book would not have been written were it not for my friend Henry Altland, retired from Corning Incorporated. "Hank" Altland has been encouraging me for years to write a book on this subject, and so in 2002 I decided to undertake the project. For the past two years he has provided timely feedback and assistance in the preparation of the manuscript, for which I am most grateful.

In 1984 I took an American Chemical Society short course titled *Sequential Simplex Optimization*. The course changed my professional life from a focus on photochemistry to a focus on the statistical design and analysis of mixture experiments. Stanley N. Deming,

Professor (now Professor Emeritus) of Analytical Chemistry, University of Houston, was one of the teachers of that course, and we have continued a personal and professional relationship over the past 20 years. Stan provided extremely valuable critiques of the manuscript, which led to significant improvements in both clarity and content.

When I realized the important role that formulation plays in the development of color photographic products, I arranged to have John A. Cornell, Professor (now Professor Emeritus) of Statistics, University of Florida, come to Eastman Kodak Company as a consultant. For several years I served as John's host while John served as my mentor, and as a result much of what I learned I owe to him. John provided an extremely valuable critique of the manuscript that led to several improvements.

Patrick Whitcomb, Principal, Stat-Ease, Inc., also provided helpful feedback in the preparation of this book. Pat founded Stat-Ease, Inc. in 1985, and since that time Stat-Ease's product Design-Expert has enjoyed an ever-widening user base. Pat teaches several short courses on the design of experiments and has had considerable experience in the design and analysis of mixture experiments. Several of the examples in this book were checked by Pat using Design-Expert Version 7.

Finally, I would like to thank Linda Thiel, Acquisitions Editor, Society for Industrial and Applied Mathematics (SIAM). Linda encouraged me to submit a preliminary manuscript to SIAM for review and has provided advice and help at several points during the preparation of the final manuscript.

Part I

Preliminaries

Chapter 1

Introduction

1.1 The Experimental Design Process

Many scientific experiments can be broken down into three stages: the planning of the experiment, the implementation of the experiment, and the analysis and interpretation of the data that are collected. When the planning stage involves the statistical design of experiments (DOE), when the implementation stage entails a randomization scheme and possible blocking, and when the interpretation stage utilizes the statistical analysis of data, then these three stages comprise the **experimental design process**. While the phrase "experimental design process" tends to put the emphasis on the design (planning) stage, one should keep in mind that it encompasses the planning, implementation, and analysis of an experiment using valid statistical principles. Aspects of all three stages will be addressed in this book.

In the course of planning an experiment, one must decide what conditions are to be varied (the treatments or factors) and what response or responses are to be measured. One then hypothesizes that any responses to be observed and measured are functionally related to the levels (or values) of a factor or factors. This can be represented analytically as

$$\text{Response(s)} = f(\text{factor levels}) \tag{1.1}$$

where f(factor levels) is to be read "some mathematical function of the levels of a factor or factors".

A **formulation** is nothing more than a **mixture**, being composed of two or more **components**. Component proportions are not independent of one another — if the proportion of one component is increased, then the proportion of one or more of the other components must decrease if the total weight (or amount) of the mixture remains the same. The proportions of mixture components could be thought of as factor levels, although the word *factors* is usually reserved for nonmixture variables that are often (although not necessarily) independent of one another. Factors are sometimes called "process variables". Examples are time, temperature, pressure, coating speed, etc.

In a mixture setting, then, the responses to be observed and measured are functionally related to the component proportions, and we can rewrite Eq. 1.1 as

$$\text{Response(s)} = f(\text{component proportions}) \tag{1.2}$$

The responses that we will be primarily concerned with in this book will be *measured* responses — those that can be arranged on a continuous scale from smallest to largest, can be negative or positive, and have a consistent unit of measurement (that is, a difference of one unit has the same meaning wherever the difference occurs) [91]. Responses that are counts (taking only nonnegative *integer* values) or that are dichotomous (taking one or the other of two values) require special regression models (such as the generalized linear model [108]) and will not be considered in this book.

We will, however, consider cases where the response (the *left* side of Eq. 1.2) is a proportion — the quotient obtained when the magnitude of a part is divided by the magnitude of the whole. For example, if the part were the number of successes (or failures) and the whole were the number of attempts, then the response would be the proportion of successes (or failures) out of the number of attempts. An example of such an experimental setting is the proportion out of 30 rats exhibiting DMBA (7,12-dimethylbenz(a)anthracene)-induced mammary gland tumors as a function of the relative proportions of fat, carbohydrate, and fiber in an isocaloric diet [17]. Although the outcome on a "per-rat" basis is dichotomous (either tumor or no tumor), when the number of rats exhibiting tumors is divided by the number of rats in a group (30), the result is a proportion. In this case the left and right sides of Eq. 1.2 are in units of proportion.

Another example of an experimental setting that will be covered in this book is the dependence of textile hydrophilicity (the response) on the method of finishing [16]. Textile hydrophilicity was measured as a function of the relative proportions of three fabric softeners as well as the total amount of the softeners and the amount of a resin. Because *softener amount* and *resin amount* are factors (nonmixture variables), the model equation might be of the form

$$\text{Response} = f(\text{component proportions}) + f(\text{factor levels})$$

or perhaps of the form

$$\begin{aligned}\text{Response} = {} & f(\text{component proportions}) + f(\text{factor levels}) \\ & + f(\text{component proportions} \times \text{factor levels})\end{aligned}$$

Experiments in which component proportions and factor levels are combined and varied are called **mixture-process variable** (MPV) experiments [30, 32, 63, 75]. A unique process variable is the *total amount* of a mixture. Experiments in which the total amount of a mixture is varied as well as the relative proportions of the mixture components are called **mixture-amount** (MA) experiments [128, 129]. The textile hydrophilicity example is really a MA–MPV experiment, because the amount of the softeners (the mixture components) is varied as well as the resin level. While most of the examples in this book are mixture experiments, MA and MPV designs and models will be discussed in Chapter 13.

Although not the subject of this book, readers should be aware that there are experimental settings that are modeled by equations of the form

$$\text{Mixture composition} = f(\text{factor levels}) \tag{1.3}$$

In these models the mixture compositions are on the *left* side of the equal sign because the compositions are the response. An example of this would be an experiment to determine the relative proportions of sand, silt, and clay in sediments as a function of water depth (the factor) in a lake. Experiments modeled by equations of the form of Eq. 1.3 are the subject of the text *The Statistical Analysis of Compositional Data* by Aitchison [1].

There are four general goals to be achieved using model equations of the form of Eq. 1.2:

1. Use the model to gain insight as to why the mixture compositions behave as they do. (E.g., Is there synergistic or antagonistic blending among the components?) The model is used as a tool for understanding.

2. Use the model to determine the mixture composition(s) where the response is near a maximum, a minimum, or a target value. The model is used as a predictive tool.

3. Use the model to determine the mixture composition(s) where the effect of mixing measurement error is minimal. Mixing measurement error arises through imprecise measurement of the amounts of the mixture components. Such errors lead to actual mixture proportions that are different from the aim mixture proportions. If the formulation is going to be manufactured, this could be an important consideration.

4. Use the model to determine the mixture composition(s) where the effects of external uncontrollable variables, such as temperature and humidity, are minimized. As with item 3, if the formulation will be manufactured, then this may also be an important consideration.

The first three goals are addressed in this book, and leading references to the fourth are given in Section 13.5.

Model equation 1.2 implies that the response to be observed and measured is functionally related to the composition of the formulations by a model. Generally one does not have enough knowledge about the system to write a theoretical model, and so we fall back on an empirical model that we hope will be locally satisfactory. The empirical models most commonly used are polynomial functions of graduating degree (such as linear, quadratic, cubic, and quartic). See Box, Hunter, and Hunter, Chapter 9 of [13] for a discussion of empirical vs. theoretical models and Box and Draper, Chapter 12 of [12] for a discussion of the links between the two.

The reader may wonder if the cart has been put before the horse because we are talking about models before we have even considered the design of the experiment. The reason for this is that to properly design an experiment, one must have some idea of the model that the design is intended to support. By *support* it is meant that there are enough experiments to adequately fit the model plus additional experiments to provide some measure of experimental error. A measure of experimental error is needed to make inferences about the statistical significance of the model as well as of the terms in the model at the analysis stage. It is for these reasons that models for a mixture setting are described in one of the earlier chapters of this book.

At the beginning of an investigation, and without prior subject-matter knowledge, one would have no idea about the functional relationship between the response and the mixture variables. What is often done at this point is to hypothesize a **linear polynomial model**, sometimes called a **screening model**. Fitting the data to a screening model helps to sort out which components of the mixture have an effect on the response and which do not. Components that have no effect on the response can be held constant in a future experiment, thus reducing the number of variables.

Second-degree and higher-order polynomial models — commonly called **response-surface models** (even though linear models also generate response surfaces) — require more experiments than linear polynomial models. Second-order response surfaces have **stationary points**: tops of "mountains", bottoms of "valleys", and "saddle" points. For this reason optimization is usually carried out using response-surface models. One might hypothesize a second-degree (quadratic) model at the outset of an investigation because of prior subject-matter knowledge or possibly because experimentation is inexpensive. Subject-matter knowledge could have arisen from having run a screening experiment and discovering that there was **lack of fit**. A test for lack of fit, explained in Chapter 8, requires that a design include **replicates**, multiple experiments carried out at the same set of conditions.

How the experiment is to be conducted must also be thought about at the design stage because the way that the experiment is carried out determines the calculations required to make meaningful statistical inferences. **Randomization** and **blocking** provide two strategies for handling unwanted variability. Unplanned systematic variability will lead to distortion (bias) of the estimates in a fitted model. Randomization does not remove this systematic variability, but it converts it into random, or chance-like, variability. Variability arising from unwanted step changes, such as day-to-day or batch-to-batch variations, can be minimized by blocking. Blocking transforms nuisance variables that are known or suspected of undergoing discrete changes into factors of the design. Randomization and blocking are not mutually exclusive and are often used together.

At the analysis stage, **linear regression** using ordinary least squares will be used to fit models to data. On fitting a model it is not unusual to conclude that it is either overspecified or underspecified. If the former, then there are terms in the model that have no statistical significance. In this case they may be eliminated, leading to a simpler, more parsimonious model. On the other hand, a lack-of-fit test may suggest that the model is underspecified, in which case additional, usually higher-order, terms are needed. To support the additional terms the design may need augmentation, in which case new data will need to be collected and the results reanalyzed. Thus the sequence "plan $\rightarrow$ conduct $\rightarrow$ analyze" is often an **iterative process**.

1.2 Resources

At the time of this writing, the only other book in the English language dedicated to the subject matter in this book is *Experiments with Mixtures* by John Cornell [29]. A few texts, such as Draper and Smith [49], Khuri and Cornell [79], and Myers and Montgomery [107], do contain chapters introducing and discussing design and analysis of mixture experiments. Cornell's excellent third edition thoroughly covers the literature up to 2002. Because of its thorough coverage the book is highly recommended to practitioners who need a single reference with virtually complete coverage of the literature. *Experimental Design for Formulation* does not purport to cover all of the subjects in Cornell's treatise. Its purpose is to provide the industrial scientist or engineer, who may have limited knowledge of statistics, with the basic tools needed to put these methods into practice.

It is the author's belief that anyone who aspires to become a practitioner of these methods should have at least one book on linear regression analysis within arm's reach. Books on linear regression have proliferated in recent years, and many excellent texts are available. Some well-known texts are Draper and Smith [49], Montgomery, Peck, and Vining [100], and Myers [104]. If one is completely new to model fitting, then Lunneborg's *Modeling Experimental and Observational Data* [91] provides an excellent introduction.

Inevitably one will need software to implement the methods in this book. There is a plethora of products to choose from. Examples of products with DOE functionality are JMP, MINITAB, SAS (including the ADX Interface), S-PLUS, STATGRAPHICS *Plus*, and STATISTICA.[1] Design-Expert and ECHIP are examples of dedicated DOE products with mixture and mixture-process variable capabilities. The software package MIXSOFT is a collection of FORTRAN routines for the statistical design and analysis of mixture experiments and other experiments with constrained experimental spaces. For the advanced practitioner, products such as GAUSS, MATLAB, and SAS/IML are tools for doing statistical computations using matrices and vectors. Information about most of these products can be found on the Web, and in some cases product reviews are available. Only a subset of these products will be cited in the text.

Arc is a computer program written in the *Xlisp-Stat* language that is designed to be used with the book *Applied Regression Including Computing and Graphics* by Cook and Weisberg [24]. Both *Arc* and *Xlisp-Stat* can be downloaded for free from

http://www.stat.umn.edu/arc/

To use *Arc*, you do not need to know how to program in *Xlisp-Stat*. A useful feature of *Arc* is a dialog window for calculating probabilities (p values) from the values of t, χ^2, and F tests.

A Web site that can be very useful is StatLib. This is a system for distributing statistical software, datasets, and information by electronic mail, FTP, and WWW. It also provides links to other Web sites of interest to the statistics community. The site is maintained by the Department of Statistics at Carnegie Mellon University; the URL is

http://lib.stat.cmu.edu/

A representative overview of scientific journals with articles on mixture experiments can be obtained by perusing the Bibliography. The greatest percentage of these have appeared in *Technometrics* and the *Journal of Quality Technology*. Taken together these two journals constitute approximately 50% of the journal entries in the Bibliography. Much of the early work in this area appeared in *Technometrics*, with the lion's share of present-day work appearing in the *Journal of Quality Technology*. "41 Years of Technometrics" is a set of four CDs that includes a fully searchable archive of all *Technometrics* articles from 1959 to 2000 in PDF format. This is available from the American Statistical Association at

http://www.amstat.org/publications/

[1] All software product names mentioned in this book are trademarked.

A useful overview that readers should find helpful is Piepel and Cornell, "Mixture experiment approaches: Examples, discussion, and recommendations" [130]. The same authors have compiled *A Catalog of Mixture Experiment Examples* [131]. The Catalog contains a comprehensive bibliography with over 400 entries plus numerous tables. Between the tables and the bibliography one can quickly track down published articles based on a variety of search criteria, such as the number of mixture components, the types of constraints, the number of design points, the fitted model(s), etc. The Catalog is available as a Word file via email from greg.piepel@pnl.gov.

Chapter 2

Mixture Space

The first subject that needs to be addressed is to define the experimental space within which one is going to carry out experiments. This space, known as the **mixture simplex**, is defined by the following two constraints:

1. The **summation** or **equality** constraint:

$$\sum_{i=1}^{q} X_i = 1 \tag{2.1}$$

2. The **nonnegativity** constraint:

$$X_i \geq 0 \quad i = 1, 2, \ldots, q \tag{2.2}$$

The symbol q is used throughout the mixture literature to represent the number of mixture components. The symbol X_i is used to symbolize component i as well as its proportion in a mixture. The latter may be expressed in units of proportion by weight, volume, or mole fraction. Most often, however, the units are proportions by weight.

One could, of course, express the total in terms of ounces, pounds, grams, etc., as long as the total always adds up to the same amount. For example, if one were formulating a cake recipe, the amounts of the individual components (flour, sugar, butter, etc.) could be expressed in terms of ounces as long as the total always adds up, say, to 16 ounces. Design-Expert allows one to express a formulation in terms of **actuals** (e.g., ounces, pounds, grams) or in terms of **reals** (component proportions). Although the words "actual" and "real" are synonyms, it is convenient to adopt this nomenclature to distinguish between the two methods of expressing the composition of a mixture.

Let us begin by using the mixture constraints to define the 2-simplex, that is, the simplex for $q = 2$ components. If we were working with two **factors** (nonmixture variables) instead of two mixture **components**, we could define our experimental space in terms of a set of x,y axes (see Fig. 2.1). Along the horizontal X_1 axis, the point where $X_1 = 1.0$ (labeled a) may be viewed as a "mixture" in which the proportions of X_1 and X_2 are 1.0

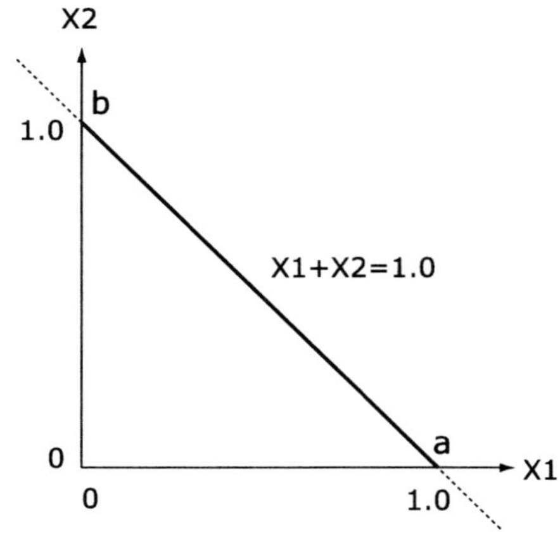

Figure 2.1. *A two-component simplex.*

and 0, respectively. Similarly, along the vertical X_2 axis, the point where $X_2 = 1.0$ (labeled b) can be viewed as another "mixture" where the proportions of X_1 and X_2 are 0 and 1.0, respectively. If we pass a line through these two points, with no restrictions on the length of the line, then anywhere along this line it will be true that $X_1 + X_2 = 1.0$ and the equality constraint will be satisfied. However, this will be true even when X_1 or X_2 takes on negative values (dashed lines). Applying the nonnegativity constraint restricts the line to that which is bolder in Fig. 2.1.

The steps that we have taken to define the 2-simplex can be summarized as follows:

- The summation constraint restricted the two-dimensional factor space to a one-dimensional line.
- The equality part of the nonnegativity constraint defined two bounding lines.
- The two constraints reduced the factor space to a one-dimensional simplex.

Note that we have gone from a *two-dimensional* factor space to a *one-dimensional* mixture space, and as a result we have lost one degree of freedom. In general we shall see that mixture space for a q-component mixture is always a $q - 1$ dimensional simplex. As we are capable of illustrating in three dimensions, this means that we can also illustrate three- and four-component simplexes.

Moving on to the 3-simplex, consider a set of orthogonal x, y, z axes labeled X_1, X_2, and X_3 (Fig. 2.2). If we tick off points a, b, and c on these axes where $X_1 = X_2 = X_3 = 1$, and then pass a plane through these points, we will have an unbounded plane. If we now apply the nonnegativity constraint, then this plane becomes bounded by three additional planes — the X_1X_2, X_2X_3, and X_1X_3 planes — leading to the three edges ab, bc, and ac, respectively. The resulting 3-simplex is an equilateral triangle.

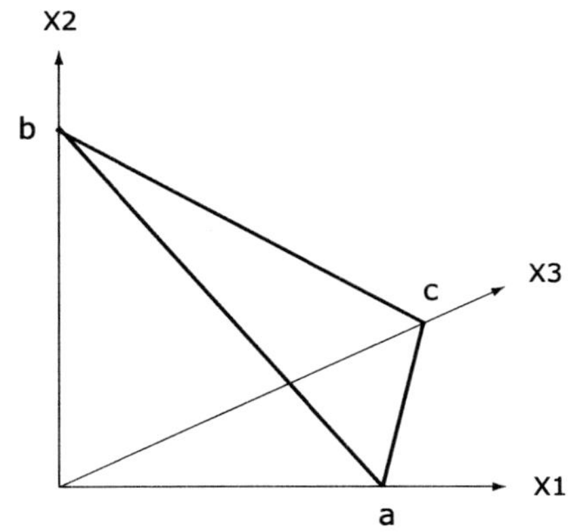

Figure 2.2. *A three-component simplex.*

Because the 3-simplex is so easy to draw, we will use three-component mixtures for many of the examples in this book.

For $q \geq 3$, we can generalize the steps that are taken to define a simplex.

- The summation constraint restricts the factor space to a $q-1$ dimensional plane ($q=3$) or hyperplane ($q>3$).
- The equality part of the nonnegativity constraint creates q bounding planes ($q=3$) or bounding hyperplanes ($q>3$) of dimension $q-1$.
- The two constraints reduce the factor space to a regular $q-1$ dimensional simplex.

In the case of four components, we cannot really conceptualize four-dimensional factor space. However, if we could and if we applied the stepwise procedure, our resulting 4-simplex would look like Fig. 2.3, i.e., a tetrahedron. As far as illustrations go, this is all we can do.

We can infer from Figs. 2.1–2.3 a few properties of higher-order simplexes.

Property 2.1 Simplexes are modular — all of the boundaries are simplexes. For example, the three-dimensional tetrahedron is bounded by four two-dimensional simplexes (triangles), six one-dimensional simplexes (edges), and four zero-dimensional simplexes (vertices), for a total of 14.

Property 2.2 Because each vertex is connected to every other vertex, the number of d-dimensional simplexes in a q-component simplex is

$$N_d = \binom{q}{d+1} = \frac{q!}{(d+1)!(q-d-1)!}$$

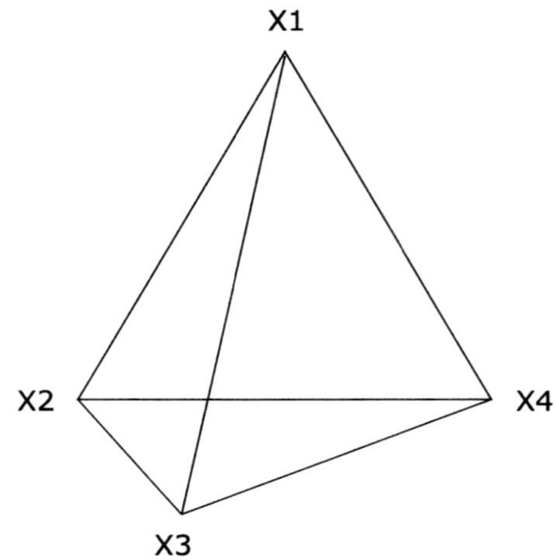

Figure 2.3. *A four-component simplex.*

For example, the total number of one-dimensional edges ($d = 1$) in a q-component simplex is

$$N_1 = \binom{q}{2} = \frac{q!}{2!(q-2)!} = \frac{q(q-1)}{2}$$

This property leads directly to Property 2.3.

Property 2.3 The total number of simplexes bounding any simplex is

$$N_{total} = \binom{q}{1} + \binom{q}{2} + \binom{q}{3} + \cdots + \binom{q}{q-1}$$
$$= 2^q - 2$$

For example, a 10-component simplex is 9-dimensional and is bounded by $2^{10} - 2 = 1022$ simplexes of lower dimensions. The number of simplexes of each dimension is tabulated in Table 2.1.

Each vertex of a simplex represents a pure component. Binary blends occur on the one-dimensional edges. Ternary blends occur on the two-dimensional faces or constraint planes (the triangle in Fig. 2.2 and the faces of the tetrahedron in Fig. 2.3). Four-component blends are located within the tetrahedron.

It is fair to ask why one might want to count the number of edges in a simplex. One reason is that the edges are good places to locate design points if one is interested in detecting binary blending behavior. For example, if one were interested in detecting quadratic blending of two components, then the vertices and the midpoints of the edges would be good places to locate design points. If one were interested in detecting cubic blending of binary blends, then the vertices and the 1/3 – 2/3 blends would be good places to locate design points. Thus the total number of design points needed to detect binary

Table 2.1. *Lower-dimensional simplexes contained within a 10-component simplex*

Type	Dimension	Number
Vertices	0	10
Edges	1	45
Constraint planes	2	120
	3	210
	4	252
	5	210
	6	120
	7	45
	8	10
Total		1022

blending is equal to the number of vertices plus some multiple of the number of edges. Similar arguments would apply for detecting and measuring ternary blending. These arguments assume that one is able to explore the whole simplex, which will not always be the case. Further consideration will be given to design points and their location in Chapters 4 and 5.

A word should be said about the **simplex coordinate system** that is customarily used with mixture simplexes [29]. One-hundred percent of a component is always located at a vertex. Zero percent of a component is always located on the opposite $q-2$ dimensional subsimplex. For two components, 100% of component X_1 is located at one end of the 2-simplex (a vertex). Zero percent of component X_1 (or 100% of X_2) is located on the $q-2=0$ dimensional subsimplex that does not contain X_1, which is the opposite end of the 2-simplex (a vertex).

In the case of three components (Fig. 2.4), 100% of component X_1 is located at the top vertex. Zero percent of component X_1 is located on the opposite $q-2=1$ dimensional subsimplex, which is the bottom edge of the triangle. Proceeding from this edge, the horizontal dotted lines represent increasing amounts (in increments of 0.2) of component X_1. The proportions of X_1 are indicated along the right side of the triangle. In a similar fashion, the proportions of X_2 are indicated along the left side of the triangle and correspond to the dotted lines sloping downward to the right. The proportions of X_3 are indicated along the bottom of the triangle and correspond to the dotted lines sloping upward to the right. At *any* point in the plane, the proportions sum to one.

In the 4-simplex illustrated in Fig. 2.3, 100% of component X_1 is located at the top vertex. Zero percent of component X_1 is located on the opposite $q-2=2$ dimensional subsimplex, the triangular base of the tetrahedron. The coordinate lines in the 3-simplex are replaced by coordinate planes in the 4-simplex. In a 5- or higher-simplex, the planes become hyperplanes.

It was stated above that the simplex coordinate system is the system that is most often used. It should be pointed out that one could pass a set of orthogonal axes through a simplex and represent the component proportions in terms of coordinates based on the set of

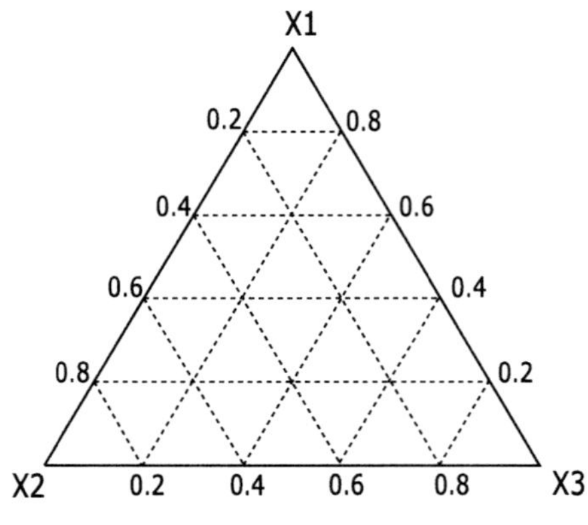

Figure 2.4. *Simplex coordinate system for a 3-simplex.*

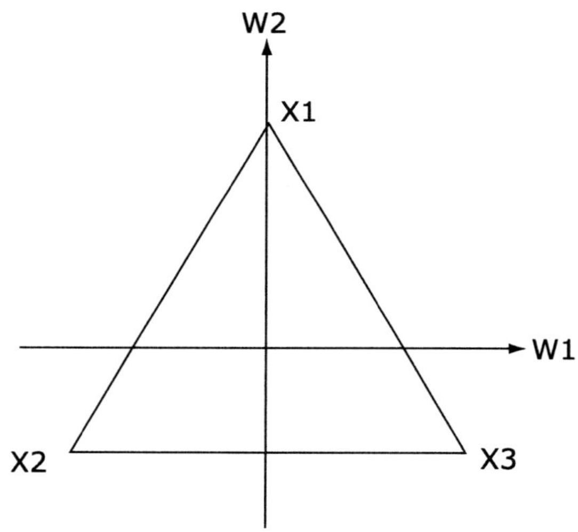

Figure 2.5. *A set of coordinate axes for mixture-related variables.*

orthogonal axes. This is illustrated for the case where $q = 3$ in Fig. 2.5. Here the origin of the orthogonal x, y axes, labeled W_1 and W_2, is located at the overall centroid of the simplex (where $(X_1, X_2, X_3) = (1/3, 1/3, 1/3)$). The origin could have been located elsewhere, for example at a vertex. In the general case, the W_i, $i = 1, \ldots, (q - 1)$, will be linear combinations of the X_i, $i = 1, \ldots, q$. Cornell [25] refers to the W_i as **mixture-related variables** (MRVs). Designing with MRVs is beyond the intended scope of this book. For details see Cornell's text [29].

Chapter 3

Models for a Mixture Setting

Many types of regression models can be used in a mixture setting, but we shall focus in this chapter on **Scheffé canonical polynomials**. According to Webster, canonical means "reduced to the simplest or clearest schema possible." Scheffé polynomials are by far the most commonly encountered mixture model forms in technical articles and books as well as in software packages that have mixture capabilities. Certain reparameterized forms of the Scheffé models can be useful, and these will be discussed as well. The models discussed in this chapter certainly do not embrace all of the models that can be used in a mixture setting, but they do cover those that are most frequently used. Cornell and Gorman [33] discuss other mixture model forms.

First, a word should be said about the meaning of *linear* in the context of regression models, as it has two meanings. The term "linear model", as opposed to a "nonlinear model", is one that can be written in the form

$$E(Y) = \alpha_0 Z_0 + \alpha_1 Z_1 + \cdots + \alpha_p Z_p \tag{3.1}$$

where $E(Y)$ is the expected value of Y, the α s are called **parameters** or **coefficients**, and the Z_i, $i = 0, 1, \ldots, p$, are **predictor variables**, **regressor variables**, or simply **regressors**. In the sense used here, a linear model is said to be *linear in the parameters*. For example, the following are linear regression models:

$$E(Y) = \alpha \sin(\theta) \tag{3.2}$$

$$E(Y) = \alpha + \beta \ln(x) \tag{3.3}$$

$$E(Y) = \alpha + \beta x + \gamma x^2 \tag{3.4}$$

whereas the regression models

$$E(Y) = x^{\alpha}$$

$$E(Y) = \exp(\alpha + \beta x)$$

$$E(Y) = \frac{1}{1 + exp[-(\alpha + \beta x + \gamma x^2)]}$$

are nonlinear because they cannot be written in the form of Eq. 3.1. In all six expressions, the αs, βs, and γs are the parameters.

In the case of the three linear models, Eqs. 3.2–3.4, and with reference to Eq. 3.1,

In Eq. 3.2: $\alpha_0 \equiv \alpha;\ Z_0 \equiv \sin(\theta)$
In Eq. 3.3: $\alpha_0 \equiv \alpha;\ Z_0 \equiv 1;\ \alpha_1 \equiv \beta;\ Z_1 \equiv \ln(x)$
In Eq. 3.4: $\alpha_0 \equiv \alpha;\ Z_0 \equiv 1;\ \alpha_1 \equiv \beta;\ Z_1 \equiv x;\ \alpha_2 \equiv \gamma;\ Z_2 \equiv x^2$

Thus the Zs can be algebraic (as in Eq. 3.4) or transcendental (as in Eqs. 3.2 and 3.3).

Particularly relevant to our purposes is Eq. 3.4, which is an example of a **polynomial model**. A polynomial model is of the form

$$E(Y) = \alpha_0 + \alpha_1 x + \alpha_2 x^2 + \cdots + \alpha_p x^p$$

where the exponents are whole numbers (i.e., members of the set $\{\ 0, 1, 2, 3, \ldots\}$). The value of the highest power in a polynomial model is called the **order** or **degree** of the model. Equation 3.4, for example, is a second-order or second-degree polynomial. Polynomials of degree one are called linear polynomials, of degree 2 quadratic, of degree 3 cubic, and of degree 4 quartic. Thus the adjective "linear" is used here in a sense that differs from its use to describe a linear vs. nonlinear model. *All* of the models used in this text will be linear in the parameters, so when the adjective "linear" is used it should be taken to mean "first-order".

More generally, the degree or order of a polynomial model is usually taken as equal to the largest *sum* of exponents appearing in any term. For example, consider the polynomial model

$$\begin{aligned} E(Y) = {} & \alpha_0 + \alpha_1 x_1 + \alpha_2 x_2 + \alpha_{11} x_1^2 + \alpha_{22} x_2^2 + \alpha_{12} x_1 x_2 \\ & + \alpha_{112} x_1^2 x_2 + \alpha_{122} x_1 x_2^2 + \alpha_{111} x_1^3 + \alpha_{222} x_2^3 \end{aligned} \tag{3.5}$$

This is a third-degree polynomial because the exponents of the terms in the second line of Eq. 3.5 (the terms of highest order) sum to three. The terms containing $x_1^2 x_2$ and $x_1 x_2^2$ are *mixed* third-order terms.

3.1 Model Assumptions

For ease in speech and writing, we often make reference to "models" without being specific about what we mean by a model. To be precise, a **model** is composed of two parts, the *model equation* and any *assumptions* that need to be made about the terms in the model equation [150]. Looking ahead to Chapter 8, we shall eventually be fitting response data to regression models using the method of **ordinary least squares** (OLS). In so doing, certain assumptions will be made, and these assumptions will be part of the OLS regression models.

The situation will be illustrated using a simple linear regression "model", for which we can write

$$E(Y) = \alpha_0 + \alpha_1 X \tag{3.6}$$

where $E(Y)$ is the expected value of Y conditional on a specific value of X (sometimes written $E(Y|X)$ or simply η). $E(Y)$ is equal to the average, or mean, value expected after an infinite number of samplings at the specific value of X. Although this expression was loosely referred to as a "model", it is more precisely an **expectation function**. For the ith observation, the expected value $E(Y_i)$ is, of course, not the same as a single observed value, Y_i. The difference between the two is given by

$$\begin{aligned}\epsilon_i &= Y_i - E(Y_i) \\ &= Y_i - \alpha_0 - \alpha_1 X\end{aligned}$$

which on rearrangement leads to

$$Y_i = \alpha_0 + \alpha_1 X + \epsilon_i$$

As the model is applicable to all observations, we can drop the subscript i and write

$$Y = \alpha_0 + \alpha_1 X + \epsilon$$

Generalizing, this can be written

$$Y = f(\alpha, X) + \epsilon \tag{3.7}$$

which we now take as our **model equation** [148, 150]. The last term represents a disturbance and is a recognition of the fact that our observations will not fit the model exactly. Although each X is assumed to be measured without error, ϵ is a random variable that has some **assumed probability distribution**. The distributional properties of ϵ are assumed to be passed to the Ys. The observed response is thus also a random variable with a probability distribution at each specific value of X.

When model 3.6 is fitted to data, we write the fitted model as

$$\hat{Y} = a_0 + a_1 X$$

where the symbol $\hat{Y}$ is read "Y hat" to symbolize the fitted value of Y, and a_0 and a_1 are the **least-squares estimators** of the intercept and slope, respectively. Note that the α s in Eq. 3.6 have been replaced with a s. In OLS, the estimators are *linear combinations* of the observations, Y_i. As a result, the estimators are also random variables and also have a probability distribution.

The differences between the observed values Y_i and the corresponding fitted values $\hat{Y}_i$ are called **residuals**. A residual is symbolized by a lower-case Roman e, to distinguish it from the conceptual errors, which are symbolized by a lower-case Greek epsilon, ϵ. Thus we have

$$\begin{aligned}e_i &= Y_i - \hat{Y}_i \\ &= Y_i - a_0 - a_1 X\end{aligned}$$

To complete the description of our model(s), we need to specify the assumptions about the ϵ_i s. These assumptions are embodied in the **Gauss–Markov conditions**, which are

$$E(\epsilon_i) = 0 \qquad \text{for all } i \tag{3.8}$$

$$\text{var}(\epsilon_i) = \sigma^2 \qquad \text{for all } i \tag{3.9}$$

$$\text{cov}(\epsilon_i, \epsilon_j) = 0 \qquad \text{for all } i \neq j \tag{3.10}$$

Condition 3.8 implies that no necessary explanatory variables have been left out of the model, the result of which would lead to a systematic bias in the disturbances. Condition 3.9 implies that all of the observed responses (Y_i) are equally unreliable and is known as the **homogeneous variance assumption**. This assumption is usually checked after a model is fitted to the data. When this condition is violated, the problem is handled by a **variance stabilizing transformation** of the response, or in some cases by **weighted least squares**. Condition 3.10 means the disturbances are pairwise **independent** of one another. Least-squares fitting of models when the disturbances are not independent of one another is handled by **generalized least squares**. Weighted least squares will be discussed in Chapter 10 in the context of robust regression, but generalized least squares will not be discussed at all. The reader is referred to texts on linear regression, such as Draper and Smith [49], Montgomery, Peck, and Vining [100], or Myers [105], for information about generalized least squares.

Taken together, the Gauss–Markov conditions imply that the disturbances are *independently and identically distributed*, often abbreviated $\epsilon_i \sim$ i.i.d. An important result about the quality of the least-squares estimators is the **Gauss–Markov theorem**, which states the following:

> *Under the G-M conditions, the least-squares estimators are unbiased and have minimum variance among all unbiased linear estimators.*

The estimators are called *best linear unbiased estimators* (BLUE) because

- "best" implies minimum variance (precise),
- "linear" means the estimated coefficients are linear combinations of the Y_i, and
- "unbiased" means $E(a_j) = \alpha_j$.

Note that the Gauss–Markov conditions do not say anything about the ϵ_i being normally distributed. The assumption of normality is in fact not required for least-squares estimation *unless* one cares to engage in hypothesis testing and confidence-interval estimation, which is most often the case. Like the homogeneous variance assumption, the normality assumption is usually checked after the model is fit. Thus the complete assumptions are that the ϵ_i are normally and independently distributed with mean 0 and variance σ^2, abbreviated $\epsilon_i \sim \text{NID}(0, \sigma^2)$.

In the sections to follow, the reader will encounter a variety of expectation functions, but in the interest of simplicity, these shall be referred to as "models" or "model functions". Despite their variety, *all* of the models can be succinctly represented by the **general linear model**

$$\mathbf{Y} = \mathbf{X}\boldsymbol{\beta} + \boldsymbol{\epsilon} \tag{3.11}$$

Let n be the number of observations (i.e., mixtures, formulations, experiments, ...) and p be the number of parameters in the model. $\mathbf{Y}$ is then a $n \times 1$ vector of observed responses, $\mathbf{X}$ is a $n \times p$ matrix of regressor variables, $\boldsymbol{\beta}$ is a $p \times 1$ vector of coefficients, and ϵ is a $n \times 1$ vector of disturbances.

3.2 Linear Models

Assume that the following disturbance-free data have been collected for a two-component mixture:

X_1	X_2	Y
1.0	0.0	10
0.8	0.2	12
0.6	0.4	14
0.4	0.6	16
0.2	0.8	18
0.0	1.0	20

Fig. 3.1 displays the response, Y, as a function of the composition of the mixture. The plotted line is the first-order (linear) response surface for the two-component mixture.

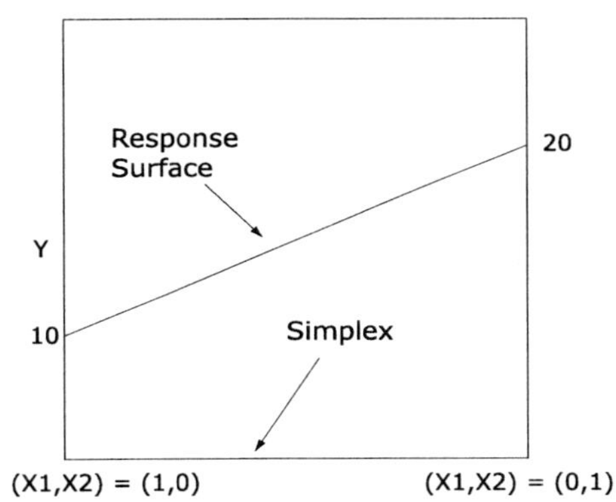

Figure 3.1. *A two-component linear response surface.*

Let us conjecture that this surface can be described by the linear regression function Eq. 3.12:

$$E(Y) = \alpha_0 + \alpha_1 X_1 + \alpha_2 X_2 \tag{3.12}$$

A model that will fit this data is

$$\hat{Y} = 5 + 5X_1 + 15X_2 \tag{3.13}$$

That this model fits the data is easily verified by substituting values for X_1 and X_2 into Eq. 3.13. Unfortunately, this equation is not unique. Another model that will fit the data equally well is

$$\hat{Y} = 15 - 5X_1 + 5X_2 \tag{3.14}$$

With a little reflection it becomes apparent that there are an infinite number of models of the form of Eq. 3.12 that will fit this data. What we have encountered is a situation where the regression function is **overparameterized** — there are more parameters than can be estimated uniquely.

For this particular example, the reason for the problem is that there are actually *three* explanatory variables in Eq. 3.12. Two of these are explicit, while one is implicit. We could rewrite Eq. 3.12 as

$$E(Y) = \alpha_0(1) + \alpha_1 X_1 + \alpha_2 X_2$$

where 1 is an implied (albeit constant) regressor. If we note that $X_1 + X_2 = 1$ — that is, there is an exact **linear dependency** between the regressors — then we have an overparameterized model. To correct the problem we can carry out the following algebra:

$$\begin{aligned} E(Y) &= \alpha_0(1) + \alpha_1 X_1 + \alpha_2 X_2 \\ &= \alpha_0(X_1 + X_2) + \alpha_1 X_1 + \alpha_2 X_2 \\ &= (\alpha_0 + \alpha_1)X_1 + (\alpha_0 + \alpha_2)X_2 \\ &= \beta_1 X_1 + \beta_2 X_2 \end{aligned} \tag{3.15}$$

The parameters of Eq. 3.15 may be written in terms of the parameters of Eq. 3.12 as

$$\begin{aligned} \beta_1 &= \alpha_0 + \alpha_1 \\ \beta_2 &= \alpha_0 + \alpha_2 \end{aligned}$$

As a specific example, consider the fitted model Eq. 3.13. The necessary algebraic transformations would be the following:

$$\begin{aligned} \hat{Y} &= 5 + 5X_1 + 15X_2 \\ &= 5(1) + 5X_1 + 15X_2 \\ &= 5(X_1 + X_2) + 5X_1 + 15X_2 \\ &= (5 + 5)X_1 + (5 + 15)X_2 \\ &= 10X_1 + 20X_2 \end{aligned} \tag{3.16}$$

The reader may care to show that Eq. 3.14 may be transformed to Eq. 3.16 as well.

Generalizing, we can state the following: **A first-order polynomial in *q* factors has *q + 1* terms, of which one term must be deleted to obtain a full-rank mixture model.** A full-rank model, whether it is a mixture or a nonmixture model, is one in which there are no linear dependencies among the regressors, and as a result the parameter estimates will be unique. Note that the bold-face statement specifies only that one term must be deleted, not that the intercept must specifically be deleted. This is discussed further in the subsection to follow.

The model function Eq. 3.15 is referred to in the mixture-experiment literature as a linear **Scheffé polynomial**. The general form of a linear Scheffé polynomial may be written

$$E(Y) = \sum_{i=1}^{q} \beta_i X_i \tag{3.17}$$

The number of terms in a linear Scheffé polynomial is the same as the number of components in the mixture, namely q.

As a second example, consider the linear Scheffé model of the form

$$\hat{Y} = 0X_1 + 10X_2 + 20X_3$$

The response surface for this model is shown in Fig. 3.2 (left). It is important to realize that linear coefficients in Scheffé models are **estimates of the response at each vertex** — *not estimates of the effects of the components.* This important distinction is not helped by some software products that label coefficients in linear Scheffé models "effects". To see the difference, consider the surface on the right in Fig. 3.2, for which the model is

$$\hat{Y} = 10X_1 + 20X_2 + 30X_3$$

For the same mixture composition, a response in the right illustration is displaced from a response in the left illustration by +10. This is true for any composition, not just the vertices.

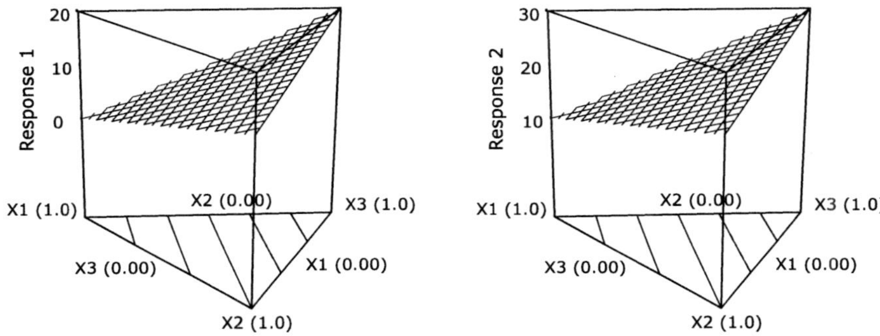

Figure 3.2. *Two three-component linear response surfaces.*

The *effect* of a component is manifested by a gradient (slope) in some specified direction. Whatever direction one might choose in Fig. 3.2, slopes and therefore effects are identical for the two surfaces.

3.2.1 Intercept Forms

Previously it was stated that we must delete one term from a q-factor linear regression model to obtain a full-rank mixture model. For the specific case of model function Eq. 3.12, page 19, either the $\alpha_1 X_1$ term or the $\alpha_2 X_2$ term could have been selected for deletion.

To see this, make the substitution

$$X_1 = 1 - X_2$$

in Eq. 3.12. Algebraic rearrangement will then lead to an intercept form for the mixture model.

$$\begin{aligned} E(Y) &= \alpha_0 + \alpha_1 X_1 + \alpha_2 X_2 \\ &= \alpha_0 + \alpha_1(1 - X_2) + \alpha_2 X_2 \\ &= (\alpha_0 + \alpha_1) + (\alpha_2 - \alpha_1)X_2 \\ &= \gamma_0 + \gamma_2 X_2 \end{aligned} \tag{3.18}$$

The parameters in Eq. 3.18 have the following meaning:

$$\gamma_0 = \alpha_0 + \alpha_1$$
$$\gamma_2 = \alpha_2 - \alpha_1$$

Alternatively, had we made the substitution

$$X_2 = 1 - X_1$$

in Eq. 3.12, then the reparameterized model function would have become

$$E(Y) = \gamma_0 + \gamma_1 X_1 \tag{3.19}$$

In this case the parameters have the meaning

$$\gamma_0 = \alpha_0 + \alpha_2$$
$$\gamma_1 = \alpha_1 - \alpha_2$$

Regression functions Eqs. 3.18 and 3.19 are called **intercept mixture models**. The intercept (γ_0) in these models is the estimated response at the vertex of the mixture variable that has been algebraically eliminated from the model.

Another way to think about the intercept forms of mixture models is to derive them starting from the full-rank Scheffé model Eq. 3.15. For example, model function Eq. 3.18 may be derived as follows:

$$\begin{aligned} E(Y) &= \beta_1 X_1 + \beta_2 X_2 \\ &= \beta_1(1 - X_2) + \beta_2 X_2 \\ &= \beta_1 + (\beta_2 - \beta_1) X_2 \\ &= \gamma_0 + \gamma_2 X_2 \end{aligned}$$

Using this type of algebra, and starting with the Scheffé model Eq. 3.16, page 20, one can derive the equivalent models

$$\hat{Y} = 10 + 10X_2 \tag{3.20}$$
$$\hat{Y} = 20 - 10X_1 \tag{3.21}$$

"Equivalent" means that models Eqs. 3.16, 3.20, and 3.21 will each reproduce the response surface in Fig. 3.1. However, the meanings of the parameter estimates in the three models are not equivalent. Let us write Eq. 3.20 and Eq. 3.21 as

$$\hat{Y} = g_0 + g_i X_i$$

where g_0 and g_i are least-squares estimates of γ_0 and γ_i, $i = 1, 2$. The value of g_0 estimates the response at the X_1 vertex when $i = 2$ and at the X_2 vertex when $i = 1$. The value of g_i, $i = 1, 2$, estimates the *difference* between the response at the vertex of the mixture variable missing from the model and at the ith vertex.

3.3 Quadratic Models

In a manner that is similar to Section 3.2, assume that the following disturbance-free data have been collected for a two-component mixture:

X_1	X_2	Y
1.0	0.0	10.00
0.8	0.2	12.64
0.6	0.4	14.96
0.4	0.6	16.96
0.2	0.8	18.64
0.0	1.0	20.00

Fig. 3.3 displays the response, Y, as a function of the composition of the mixture.

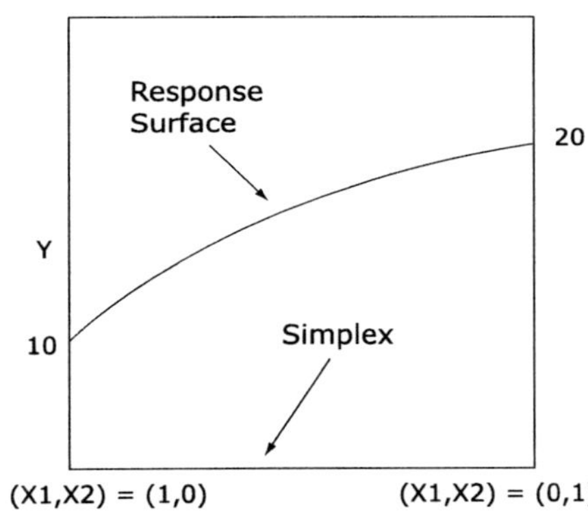

Figure 3.3. *A two-component quadratic response surface.*

The data in the table can be fitted to the quadratic response function Eq. 3.22.

$$E(Y) = \alpha_0 + \alpha_1 X_1 + \alpha_2 X_2 + \alpha_{12} X_1 X_2 + \alpha_{11} X_1^2 + \alpha_{22} X_2^2 \tag{3.22}$$

A model that fits the data is

$$\hat{Y} = 17 - 5X_1 + 4X_2 + X_1 X_2 - 2X_1^2 - X_2^2 \tag{3.23}$$

Another model that fits the data equally well is

$$\hat{Y} = 16 - 5X_1 + 5X_2 + 2X_1 X_2 - X_1^2 - X_2^2 \tag{3.24}$$

In fact, there are an infinite number of models of the form Eq. 3.22 that will fit this data.

As in the linear case, the reason for the problem is that model Eq. 3.22 is overparameterized for the two-component mixture setting. To see this, it is convenient to augment the

X data in the previous table as follows:

(1)	X_1	X_2	X_1X_2	X_1^2	X_2^2
1.0	1.0	0.0	0.00	1.00	0.00
1.0	0.8	0.2	0.16	0.64	0.04
1.0	0.6	0.4	0.24	0.36	0.16
1.0	0.4	0.6	0.24	0.16	0.36
1.0	0.2	0.8	0.16	0.04	0.64
1.0	0.0	1.0	0.00	0.00	1.00

(1) is the implied regressor for the intercept. Inspection of the columns reveals that there are three linear dependencies:

$$\begin{aligned} 1 &= X_1 + X_2 \\ X_1^2 &= X_1 - X_1X_2 \\ X_2^2 &= X_2 - X_1X_2 \end{aligned}$$

If substitutions are made for the three terms on the left in model 3.23, then an equivalent full-rank mixture model may be derived as follows:

$$\begin{aligned} \hat{Y} &= 17(1) - 5X_1 + 4X_2 + X_1X_2 - 2X_1^2 - X_2^2 \\ &= 17(X_1 + X_2) - 5X_1 + 4X_2 + X_1X_2 - 2(X_1 - X_1X_2) - (X_2 - X_1X_2) \\ &= (17 - 5 - 2)X_1 + (1 + 2 + 1)X_1X_2 + (17 + 4 - 1)X_2 \\ &= 10X_1 + 4X_1X_2 + 20X_2 \end{aligned} \tag{3.25}$$

In the same fashion, model 3.24 may be transformed into Eq. 3.25 as well.

Let us now generalize the situation. The general form of a q-factor quadratic regression model is

$$E(Y) = \alpha_0 + \sum_{i=1}^{q} \alpha_i X_i + \sum_{i=1}^{q-1} \sum_{j=i+1}^{q} \alpha_{ij} X_i X_j + \sum_{i=1}^{q} \alpha_{ii} X_i^2$$

To reparameterize these models to full-rank mixture models, we make the following substitutions. For the intercept we write

$$\alpha_0 = \alpha_0(1) = \alpha_0 \sum_{i=1}^{q} X_i = \sum_{i=1}^{q} \alpha_0 X_i \tag{3.26}$$

which has the effect of converting the intercept into a sum of first-order terms. Second, wherever there is a squared term, we write

$$\alpha_{ii} X_i^2 = \alpha_{ii} X_i \cdot \left(1 - \sum_{j \neq i}^{q} X_j\right) = \alpha_{ii} X_i - \alpha_{ii} X_i \sum_{j \neq i}^{q} X_j \tag{3.27}$$

which has the effect of converting all squared terms into a single linear term plus a string of crossproduct terms. Collecting the linear and crossproduct terms leads to the general form of the quadratic Scheffé polynomial.

$$E(Y) = \sum_{i=1}^{q} \beta_i X_i + \sum_{i=1}^{q-1} \sum_{j=i+1}^{q} \beta_{ij} X_i X_j \tag{3.28}$$

Although the terms $\beta_{ij} X_i X_j$ look like interaction terms, they are referred to as quadratic blending terms in the mixture-experiment literature, and the coefficients, β_{ij}, are referred to as quadratic or nonlinear blending coefficients. When $\beta_{ij} > 0$ and a high response value is desirable, we say that the blending between components i and j is *synergistic*; otherwise we say that it is *antagonistic*. When $\beta_{ij} < 0$ and a low response value is desirable, we say that the blending between components i and j is synergistic, and otherwise we say that it is antagonistic.

The number of terms in a quadratic Scheffé polynomial can be calculated as follows:

$$\begin{aligned} p &= q \text{ linear terms} + \binom{q}{2} \text{ cross-product terms} \\ &= q + \frac{q!}{2!(q-2)!} \\ &= q + \frac{q(q-1)}{2} \\ &= \frac{q(q+1)}{2} \end{aligned}$$

Thus the *minimum* number of design points needed to support a quadratic Scheffé polynomial is $q(q+1)/2$. Because quadratic models are so common in a mixture setting, it is worth committing this formula to memory.

We now state the following: **A second-order polynomial in q factors has $(q + 1)(q + 2)/2$ terms, of which $q + 1$ must be deleted to obtain a full-rank mixture model.** A sufficient but not a necessary condition is to delete the constant term and the q pure quadratic terms. Deletion of other terms can lead to intercept forms of the quadratic model, and these are now discussed.

3.3.1 Intercept Forms

Referring again to the two-factor regression function Eq. 3.22, page 23, this function has $[(2+1)(2+2)]/2 = 6$ terms, of which $2 + 1 = 3$ must be deleted to obtain a full-rank mixture model. If we retain the intercept, then we may retain two more regressors from the remaining five. We cannot choose X_1 and X_2 together, as this will lead to an exact dependency and will not model the quadratic curvature, and so we have a choice of nine possible models, all of which will be full rank. As q gets larger and larger, the number of possible intercept mixture models increases very rapidly. However, there are a limited number of *useful* intercept models, two of which we shall focus on here.

The first is what is known as a **slack-variable model**. For the particular case where $q = 2$, the quadratic slack-variable model function takes the form

$$E(Y) = \beta_0 + \beta_i X_i + \beta_{ii} X_i^2$$

When $i = 1$ we say that X_2 is slack, and when $i = 2$ we say that X_1 is slack. The slack-variable forms of model 3.25 are

$$\hat{Y} = 20 - 6X_1 - 4X_1^2 \quad X_2 \text{ slack}$$
$$\hat{Y} = 10 + 14X_2 - 4X_2^2 \quad X_1 \text{ slack}$$

The first equation (X_2 slack) can be derived starting from the full-rank Scheffé model 3.25 as follows:

$$\begin{aligned}\hat{Y} &= 10X_1 + 4X_1X_2 + 20X_2 \\ &= 10X_1 + 4X_1(1 - X_1) + 20(1 - X_1) \\ &= 20 + (10 + 4 - 20)X_1 - 4X_1^2 \\ &= 20 - 6X_1 - 4X_1^2\end{aligned}$$

The model for X_1 slack is derived in a similar manner.

For $q > 2$ there will be additional terms in the model function of the form

$$\sum_{i=1}^{q-2} \sum_{j=i+1}^{q-1} X_i X_j$$

with terms in the qth (slack) component absent. Thus the general form of a quadratic slack-variable model function is

$$E(Y) = \beta_0 + \sum_{i=1}^{q-1} \beta_i X_i + \sum_{i=1}^{q-2} \sum_{j=i+1}^{q-1} \beta_{ij} X_i X_j + \sum_{i=1}^{q-1} \beta_{ii} X_i^2 \tag{3.29}$$

The number of terms contributed by each type of term is

$$(1) + (q - 1) + \left(\frac{(q - 1)(q - 2)}{2}\right) + (q - 1)$$

which, when added together, lead to a total of $q(q + 1)/2$ terms, the same as the quadratic Scheffé model function.

The slack-variable model is occasionally used in situations where one of the mixture components is present in very large amount while the other $q - 1$ components are present in much smaller amounts. Varying the proportions of the components present in small amounts will have little effect on the proportion of the component present in large amount, and so the latter is viewed as taking up the slack.

The second quadratic intercept model function that occasionally is useful is the following:

$$E(Y) = \beta_0 + \sum_{i=1}^{q-1} \beta_i X_i + \sum_{i=1}^{q-1} \sum_{j=i+1}^{q} \beta_{ij} X_i X_j \tag{3.30}$$

Inspection of this function reveals that it differs from the quadratic Scheffé function Eq. 3.28, page 25, in that one of the linear terms in Eq. 3.28 has been replaced by an intercept. This procedure whereby one of the linear terms is replaced with an intercept can be applied to Scheffé polynomials of any order (linear, quadratic, or higher order). Casting model Eq. 3.25, page 24, in this form leads to the equations

$$\hat{Y} = 20 - 10X_1 + 4X_1X_2$$
$$\hat{Y} = 10 + 10X_2 + 4X_1X_2$$

The first equation is derived as follows:

$$\begin{aligned}\hat{Y} &= 10X_1 + 20X_2 + 4X_1X_2 \\ &= 10X_1 + 20(1 - X_1) + 4X_1X_2 \\ &= 20 - 10X_1 + 4X_1X_2\end{aligned}$$

In these intercept forms *the meaning of the linear coefficients differs from their meaning in the Scheffé model form*, but the higher-order terms retain their meaning. These models are useful when regression software does not output correct regression statistics for Scheffé models. The reasons for this are discussed in Chapter 8.

3.4 Cubic and Quartic Scheffé Models

Using an approach similar to that described for the linear and quadratic Scheffé models, one can derive higher-order Scheffé models. The terms in a cubic Scheffé model can be exemplified by the model for $q = 3$.

$$\begin{aligned}E(Y) = {} & \beta_1X_1 + \beta_2X_2 + \beta_3X_3 + \beta_{12}X_1X_2 + \beta_{13}X_1X_3 + \beta_{23}X_2X_3 \\ & + \gamma_{12}X_1X_2(X_1 - X_2) + \gamma_{13}X_1X_3(X_1 - X_3) + \gamma_{23}X_2X_3(X_2 - X_3) \\ & + \beta_{123}X_1X_2X_3\end{aligned}$$

Just as the quadratic Scheffé model can be viewed as an augmented linear model, the cubic model can be viewed as an augmented quadratic model. The cubic terms are of two types. The $X_iX_j(X_i - X_j)$ terms model cubic blending of binaries. The coefficients of these terms are symbolized by γ_{ij}s to distinguish them from the coefficients of the quadratic terms, the β_{ij}s. The $X_iX_jX_k$ terms model cubic blending of ternaries. Although when $q = 3$ there are many more terms of the type $X_iX_j(X_i - X_j)$ than of the type $X_iX_jX_k$, beyond $q = 5$ the reverse is true. For example, when $q = 8$ there are 56 terms of the type $X_iX_jX_k$ but "only" 28 terms of the type $X_iX_j(X_i - X_j)$.

In presenting the general forms for cubic and quartic Scheffé polynomial, the following "shorthand" notation will be used:

$$\sum_{i<j}^{q}\sum \equiv \sum_{i=1}^{q-1}\sum_{j=i+1}^{q}$$

$$\sum\sum_{i<j<k}^{q}\sum \equiv \sum_{i=1}^{q-2}\sum_{j=i+1}^{q-1}\sum_{k=j+1}^{q}$$

$$\sum\sum\sum_{i<j<k<l}^{q}\sum \equiv \sum_{i=1}^{q-3}\sum_{j=i+1}^{q-2}\sum_{k=j+1}^{q-1}\sum_{l=k+1}^{q}$$

The general form for a cubic Scheffé polynomial is then

$$E(Y) = \sum_{i}^{q} \beta_i X_i + \sum_{i<j}\sum^{q} \beta_{ij} X_i X_j + \sum_{i<j}\sum^{q} \gamma_{ij} X_i X_j (X_i - X_j)$$
$$+ \sum\sum_{i<j<k}\sum^{q} \beta_{ijk} X_i X_j X_k \tag{3.31}$$

while the general form of the quartic Scheffé polynomial is

$$E(Y) = \sum_{i}^{q} \beta_i X_i + \sum_{i<j}\sum^{q} \beta_{ij} X_i X_j + \sum_{i<j}\sum^{q} \gamma_{ij} X_i X_j (X_i - X_j)$$
$$+ \sum_{i<j}\sum^{q} \delta_{ij} X_i X_j (X_i - X_j)^2 + \sum\sum_{i<j<k}\sum^{q} \beta_{iijk} X_i^2 X_j X_k$$
$$+ \sum\sum_{i<j<k}\sum^{q} \beta_{ijjk} X_i X_j^2 X_k + \sum\sum_{i<j<k}\sum^{q} \beta_{ijkk} X_i X_j X_k^2$$
$$+ \sum\sum\sum_{i<j<k<l}\sum^{q} \beta_{ijkl} X_i X_j X_k X_l \tag{3.32}$$

Terms of the form $X_1^2 X_j X_k$, $X_i X_j^2 X_k$, and $X_i X_j X_k^2$ can be included only when $q \geq 3$, while terms of the form $X_i X_j X_k X_l$ can be included only when $q \geq 4$.

Recall that the number of terms in a quadratic Scheffé polynomial is $q(q+1)/2$. This can be expressed as

$$p_{q,2} = \frac{q(q+1)}{2!}$$

It can be shown that the number of terms in a cubic Scheffé polynomial is

$$p_{q,3} = \frac{q(q+1)(q+2)}{3!}$$

and that the number of terms in a quartic Scheffé polynomial is

$$p_{q,4} = \frac{q(q+1)(q+2)(q+3)}{4!}$$

Clearly, a pattern is evolving. This pattern can be summarized by the following combinatorial expression:

$$p_{q,m} = \binom{q+m-1}{m} = \frac{(q+m-1)!}{m!(q-1)!} \tag{3.33}$$

Equation 3.33 can be used to calculate the number of terms in a Scheffé polynomial of any order m and for any q.

3.4.1 Special Forms

A motivation for truncated forms of the cubic and quartic polynomial functions arises because, as q increases even moderately, the number of terms that must be supported by a design increases dramatically. This can be seen in Table 3.1.

Table 3.1. *Number of terms in some Scheffé polynomials*

q	Linear	Quadratic	Cubic	Quartic
2	2	3	4	5
3	3	6	10	15
4	4	10	20	35
5	5	15	35	70
6	6	21	56	126
7	7	28	84	210
8	8	36	120	330

Truncated forms of the cubic and quartic models are called **special cubic** and **special quartic** models. Being truncated, these polynomials will not model as much complex curvature in a response surface as the full model forms.

The general form of the special cubic polynomial is

$$E(Y) = \sum_{i}^{q} \beta_i X_i + \sum\sum_{i<j}^{q} \beta_{ij} X_i X_j + \sum\sum\sum_{i<j<k}^{q} \beta_{ijk} X_i X_j X_k \tag{3.34}$$

and that for the special quartic polynomial is

$$\begin{aligned} E(Y) = & \sum_{i}^{q} \beta_i X_i + \sum\sum_{i<j}^{q} \beta_{ij} X_i X_j + \sum\sum\sum_{i<j<k}^{q} \beta_{iijk} X_i^2 X_j X_k \\ & + \sum\sum\sum_{i<j<k}^{q} \beta_{ijjk} X_i X_j^2 X_k + \sum\sum\sum_{i<j<k}^{q} \beta_{ijkk} X_i X_j X_k^2 \end{aligned} \tag{3.35}$$

For the specific case where $q = 3$ these polynomials take the forms

$$\begin{aligned} E(y) = & \beta_1 X_1 + \beta_2 X_2 + \beta_3 X_3 + \beta_{12} X_1 X_2 + \beta_{13} X_1 X_3 + \beta_{23} X_2 X_3 \\ & + \beta_{123} X_1 X_2 X_3 \end{aligned}$$

$$\begin{aligned} E(y) = & \beta_1 X_1 + \beta_2 X_2 + \beta_3 X_3 + \beta_{12} X_1 X_2 + \beta_{13} X_1 X_3 + \beta_{23} X_2 X_3 \\ & + \beta_{1123} X_1^2 X_2 X_3 + \beta_{1223} X_1 X_2^2 X_3 + \beta_{1233} X_1 X_2 X_3^2 \end{aligned}$$

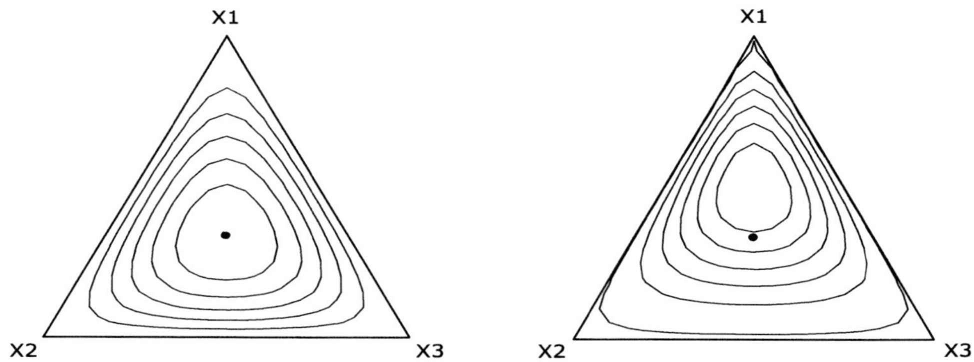

Figure 3.4. *Curvature modeled by* $X_1X_2X_3$ *(left) and* $X_1^2X_2X_3$ *(right).*

The cubic term in the special cubic polynomial and the three quartic terms in the special quartic polynomial model are useful for modeling curvature of a response surface in the interior of the triangle. As illustrated in Fig. 3.4, a term such as $X_iX_jX_k$ models peaks or valleys that are symmetrically located with respect to the centroid of the $X_iX_jX_k$ simplex. Terms such as $X_i^2X_jX_k$ also model peaks and valleys, but these are offset from the centroid along the X_i component axis (cf. page 53).

The number of terms in some special cubic and special quartic polynomials are summarized in Table 3.2. The number of terms in special cubic models is $q(q^2+5)/6$, while the number of terms in special quartic models is $q(q^2-2q+3)/2$. Although the special models contain fewer terms than the full models, we are still confronted with a large number of terms for moderately large q.

Table 3.2. *Number of terms in some special Scheffé polynomials*

q	Special Cubic	Speial Quartic
2	.	.
3	7	9
4	14	22
5	25	45
6	41	81
7	63	133
8	92	204

For further discussions of Scheffé models and examples, see Scheffé [146], Gorman and Hinman [59], and several papers by Lambrakis [87, 88, 89]. Lambrakis [88] presents an equation for the general form of a Scheffé model of any order and for any number of mixture components.

3.5 Choosing a Model

Assume for the moment that we begin an investigation without the complications of including process variables or mixture amount, and that we desire a design that will adequately support a Scheffé model. But where does one begin? Linear, quadratic, higher order? At the beginning of an investigation we have no knowledge of the functional relationship between the response and the mixture variables — or even that one exists. Perhaps, for example, the best "model" may turn out to be simply the average response. On the other hand, if an investigator is entering a research area in which scientists have built up experience about which model is most apt to describe the data, then he or she may well benefit from **subject-matter knowledge**. There is no substitute for knowing how similar experiments have turned out.

When there is no preexisting information, what is often done is to hypothesize a linear polynomial model, sometimes called a screening model. There are at least two reasons for this. First, experimentation is usually expensive. Supporting a first-order model requires fewer observations than supporting a second- or higher-order model. With a properly designed experiment, a formal statistical test may be performed to check the adequacy of the linear model. If the fitted model is not adequate, then we need to consider augmenting the model and possibly the design.

A second reason for beginning with a linear model is that such a model can help us determine if there are components that have *no effect* on the response or if there are components that have the *same effect* on the response. Components that have no effect can be disregarded in the analysis (after renormalization of the proportions of the components that do have an effect), leading to a more parsimonious model (and perhaps a more parsimonious formulation). Furthermore, in future experiments the proportion of a component with no effect could be held constant, thus reducing the number of variables. Components that have the same effect can be combined in the analysis, again leading to a more parsimonious model.

Second-degree and higher-order models are commonly called **response-surface models**. Optimization is usually carried out using response-surface models. One might hypothesize a second-order model at the beginning of an investigation because of subject-matter knowledge or perhaps because experimentation is inexpensive. A higher-degree model can always be reduced to a lower-degree model, provided certain statistical criteria are met.

Once a decision has been made about the order of the model, the next step is to develop a suitable design to support the model. Chapters 4 and 5 discuss designing in a mixture setting, while Chapter 6 discusses design evaluation — an exercise worth engaging in before starting what might be an expensive experimental program.

Part II

Design

Chapter 4

Designs for Simplex-Shaped Regions

This chapter and the next divide the subject of design into two parts, depending on the shape of the design region. Before discussing designs, then, we need first to consider those conditions that determine the possible shapes that design regions may assume in a mixture setting.

4.1 Constraints and Subspaces

The shape of a design region in a mixture setting is determined by the constraints that are imposed on the component proportions. Mixture constraints can be divided into two broad categories: **single-component** and **multicomponent** constraints. Single-component constraints are of the form

$$L_i \le X_i \le U_i \tag{4.1}$$

where L_i and U_i are lower and upper bounds, respectively, on the proportion X_i of component i. Equation 4.1 has two single-component constraints, a lower- and an upper-bound constraint. Multicomponent constraints are of the form

$$L_k \le a_{k1}X_1 + a_{k2}X_2 + \cdots + a_{kq}X_q \le U_k \tag{4.2}$$

where L_k and U_k are lower and upper bounds, respectively, for the kth two-sided constraint, and where some of the a_{ki} may be zero.

Ratio constraints are also common in formulation work. Consider, for example, the ratio constraint

$$\frac{X_3 + X_4 + X_5}{X_1 + X_2} \le 1.0$$

With a little algebra this can be rewritten as the multicomponent constraint:

$$0 \le X_1 + X_2 - X_3 - X_4 - X_5$$

In terms of Eq. 4.2, $L_k = 0$, $a_{k1} = a_{k2} = 1$, and $a_{k3} = a_{k4} = a_{k5} = -1$.

The set of single-component constraints

$$0 \leq X_i \leq 1 \qquad i = 1, 2, \ldots, q$$

might be considered the trivial case. These constraints, in combination with the summation constraint Eq. 2.1, page 9, lead to q-simplexes.

Consider the following modified constraint on X_1:

$$0.1 \leq X_1 \tag{4.3}$$

Here we have a nonzero lower bound on component X_1 only. The cases for $q = 3$ and $q = 4$ are illustrated in Fig. 4.1.

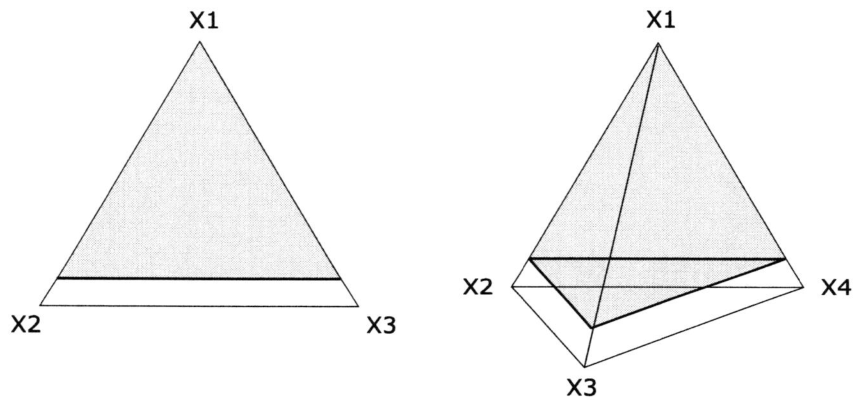

Figure 4.1. *A single lower bound in a 3- and a 4-simplex.*

In the 3-simplex the shaded constrained region lies above the line at $X_1 = 0.1$; in the 4-simplex, it lies above the triangle (3-simplex) at $X_1 = 0.1$. In both cases the constrained region is still simplex-shaped.

Let us now add a second nonzero lower bound.

$$\begin{aligned} 0.1 &\leq X_1 \\ 0.2 &\leq X_2 \end{aligned} \tag{4.4}$$

Fig. 4.2 illustrates the result. Again, the shaded constrained regions are still shaped like a simplex.

Finally, let us add a third nonzero lower bound.

$$\begin{aligned} 0.1 &\leq X_1 \\ 0.2 &\leq X_2 \\ 0.3 &\leq X_3 \end{aligned} \tag{4.5}$$

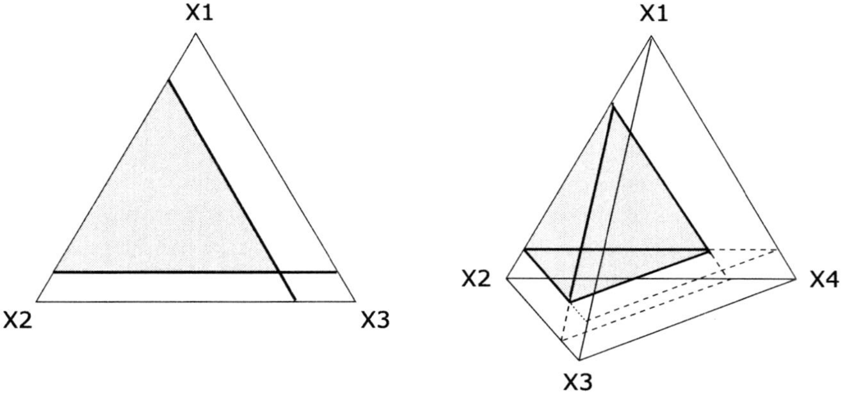

Figure 4.2. *Two lower bounds in a 3- and a 4-simplex.*

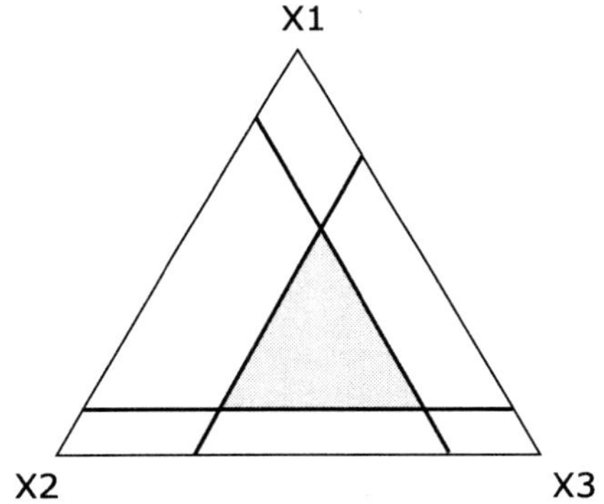

Figure 4.3. *Three lower bounds in a 3-simplex.*

Fig. 4.3 illustrates the result for $q = 3$. (The diagram for $q = 4$ is messy and is not included.) Again the constrained region is simplex-shaped.

From these examples, the following property is inferred.

Property 4.1 When there are only lower bounds on component proportions, the subregion of interest is always shaped like a simplex.

As a consequence, the simplex designs discussed in this chapter apply not only to the full simplex *but also to mixtures that have only lower-bound constraints.*

Let us take a closer look at the bound in Eq. 4.3 and Fig. 4.1. If the *minimum* proportion of component X_1 is 0.1, then the *maximum* proportion of any other components cannot exceed 0.9. Any proportion X_i, $i = 2, 3, \ldots, q$, greater than 0.9 is unattainable. When $q = 3$ (for example), upper bounds of 1.0 on X_2 and X_3 are said to be **inconsistent** [119, 123].

When Eq. 4.3 applies and $q = 3$, the complete set of constraints would be written

$$0.1 \leq X_1 \leq 1.0$$
$$0.0 \leq X_2 \leq 0.9$$
$$0.0 \leq X_3 \leq 0.9$$

The adjusted bounds on X_2 and X_3 are referred to as **implied constraints**.

The set of lower bounds in Eqs. 4.4 leads to q implied constraints no matter what the value of q. U_1 must be adjusted to 0.8 to account for $L_2 = 0.2$, U_2 must be adjusted to 0.9 to account for $L_1 = 0.1$, and U_i, $i = 3, \ldots, q$, must be adjusted to 0.7 to account for $L_1 + L_2 = 0.3$. Cf. Fig. 4.2.

When Eqs. 4.5 apply and $q = 3$, the complete set of constraints would be written

$$0.1 \leq X_1 \leq 0.5$$
$$0.2 \leq X_2 \leq 0.6$$
$$0.3 \leq X_3 \leq 0.7$$

U_1 is adjusted to 0.5 to account for $L_2 + L_3 = 0.5$, U_2 is adjusted to 0.6 to account for $L_1 + L_3 = 0.4$, and U_3 is adjusted to 0.7 to account for $L_1 + L_2 = 0.3$. Cf. Fig. 4.3. When Eqs. 4.5 apply and $q > 3$, then $U_i = 0.4$ for $i = 4, \ldots, q$ because $L_1 + L_2 + L_3 = 0.6$.

These considerations lead us to Property 4.2 [119].

Property 4.2 Inconsistent constraints occur whenever

$$U_i + \sum_{j \neq i}^{q} L_j > 1 \tag{4.6}$$

or whenever

$$L_i + \sum_{j \neq i}^{q} U_j < 1 \tag{4.7}$$

Property 4.2 implies that U_i and L_i are expressed in terms of component proportions (the reals; p. 9). If U_i and L_i are expressed in terms of actuals (such as grams, ounces, or pounds), then Eq. 4.6 should be reexpressed as

$$U_i + \sum_{j \neq i}^{q} L_j > Total \tag{4.8}$$

and Eq. 4.7 as

$$L_i + \sum_{j \neq i}^{q} U_j < Total \tag{4.9}$$

Note that in the case of Eqs. 4.6 and 4.8, consistency can be achieved by lowering either U_i or the L_j, $j \neq i$, and that in the case of Eqs. 4.7 and 4.9, consistency can be achieved by raising L_i or the U_j, $j \neq i$. Software characteristically lowers U_i in the former case and raises L_i in the latter case.

When inconsistent constraints are entered in PC computing packages, range adjustments are automatically made so that the greater-than symbol in Eqs. 4.6 and 4.8 and/or the less-than symbol in Eqs. 4.7 and 4.9 become equal signs. Some software packages will print a message when these types of range adjustments are made. Design-Expert, for example, prints the message

"Ranges will be adjusted to meet constraints"

whereas MINITAB prints the message

"Bounds were adjusted to accommodate specified constraints"

Inconsistent constraints are not always obvious and can lull the unwary into believing that he or she is exploring a broader range than is actually the case. Here is an example from the food industry. Soo, Sander, and Kess [161] studied the effects of composition and processing on the textural quality of cooked shrimp patties. The mixture components were shrimp, isolated soy protein (ISP), sodium chloride, and sodium tritolylphosphate (STP). The following ranges were specified by the authors:

$$\begin{array}{ccccc} 0.85 & \leq & \text{Shrimp} & \leq & 1.0 \\ 0.05 & \leq & \text{ISP} & \leq & 0.10 \\ 0.01 & \leq & \text{NaCl} & \leq & 0.03 \\ 0.001 & \leq & \text{STP} & \leq & 0.005 \end{array}$$

The effective range of one component is $\sim$ 50% smaller than its stated range. Identifying the inconsistent constraint(s) will elevate one's appreciation for the fact that software packages usually take care of this automatically.

What shapes might we expect when there are only upper bounds? Let us begin as before with a simple case where there is only one upper bound that is less than 1.0 and $q = 3$:

$$0.0 \leq X_1 \leq 0.1$$
$$0.0 \leq X_2 \leq 1.0$$
$$0.0 \leq X_3 \leq 1.0$$

These constraints are consistent because neither Eq. 4.6 nor 4.7 is violated. The situation is illustrated in Fig. 4.4 (left). This figure is the same as Fig. 4.1 (left), but now the constraint is interpreted as an *upper-bound* rather than a lower-bound constraint. As a result, the shaded constrained region lies below the line at $X_1 = 0.1$ and is shaped like a trapezoid. Within the constrained region, X_2 and X_3 are free to range between 0.0 and 1.0.

Adding a second upper constraint that is less than 1.0, such as

$$0.0 \leq X_1 \leq 0.1$$
$$0.0 \leq X_2 \leq 0.2$$
$$0.7 \leq X_3 \leq 1.0$$

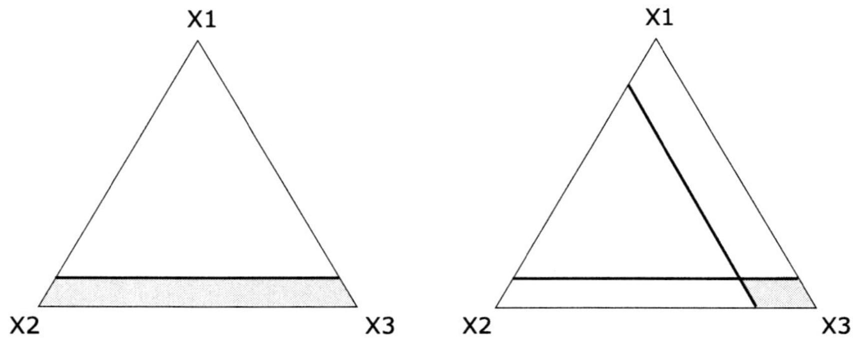

Figure 4.4. *One and two upper bounds in a 3-simplex.*

leads to a subspace that is now a parallelogram (shaded corner of the triangle in Fig. 4.4 (right)). The lower bound on X_3 has been adjusted to maintain consistency, and it is easy to see from the figure that X_3 is no longer free to range to 0.0. Clearly, *we cannot state a simple property about the shape of the design region when there are upper bounds.* The shape will depend on the nature of the particular bound(s).

Before adding a third upper constraint and continuing with this example, it is useful to digress for a moment. Assume that there are the same two upper-bound constraints on X_1 and X_2, but now $q = 4$. In this case it may come as a bit of a surprise that the following constraints are consistent because neither Eqs. 4.6 nor 4.7 is violated:

$$\begin{aligned} 0.0 &\le X_1 \le 0.1 \\ 0.0 &\le X_2 \le 0.2 \\ 0.0 &\le X_3 \le 1.0 \\ 0.0 &\le X_4 \le 1.0 \end{aligned}$$

This result is general no matter what the upper bounds on X_1 and X_2 and is true as well if we replace X_4 with $X_i, i = 4, 5, \ldots, q$.

Continuing with the three-component example, let us add a third upper bound that is less than 1.0:

$$\begin{aligned} 0.0 &\le X_1 \le 0.1 \\ 0.0 &\le X_2 \le 0.2 \\ 0.0 &\le X_3 \le 0.3 \end{aligned}$$

where the lower bound on X_3 has been reset to 0.0. The picture now looks like Fig. 4.5. In the center triangle, all three constraints are violated. In each of the three trapezoidal regions, two constraints are violated. And in each of the three parallelograms, one of the three constraints is violated. There is no shading because nowhere in the large triangle are all three constraints satisfied simultaneously. Thus we have an empty constrained region and the constraints are again said to be inconsistent [119, 123]. This leads to Property 4.3.

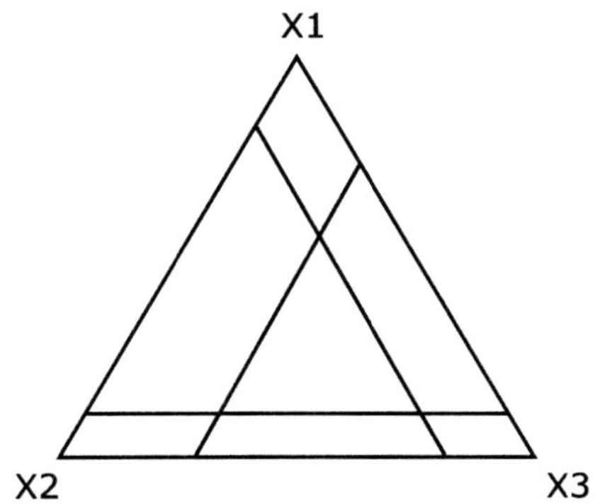

Figure 4.5. *Three upper bounds in a 3-simplex.*

Property 4.3 Upper bounds are inconsistent whenever

$$\sum_{i}^{q} U_i \leq 1 \tag{4.10}$$

and lower bounds are inconsistent whenever

$$\sum_{i}^{q} L_i \geq 1 \tag{4.11}$$

If, for example, the bounds

$$X_1 \leq 0.1$$
$$X_2 \leq 0.2$$
$$X_3 \leq 0.3$$
$$X_4 \leq 0.4$$

were specified for a $q = 4$ design, then the equality part of Eq. 4.10 would apply. The constrained region would not be a region at all but rather the single mixture $(X_1, X_2, X_3, X_4) = (0.1, 0.2, 0.3, 0.4)$.

Had the U_i in Eq. 4.10 or the L_i in Eq. 4.11 been expressed in terms of actuals, then Eq. 4.10 would become

$$\sum_{i}^{q} U_i \leq Total \tag{4.12}$$

and Eq. 4.11 would become

$$\sum_{i}^{q} L_i \geq Total \tag{4.13}$$

Software packages handle the inconsistencies exemplified by Property 4.3 in different ways. JMP outputs an empty data table but no message. Design-Expert and MINITAB print error messages only. For example, if Eq. 4.10 or 4.12 applies, Design-Expert prints the message

"Sum of maximums is less than or equal to the total"

whereas if Eq. 4.11 or 4.13 applies, then the word "maximums" is replaced by the word "minimums" and "less than" by "greater than". Again, if Eq. 4.10 or Eq. 4.12 applies, MINITAB prints the message

"The total for high values ([total printed]) is less or equal to the total for the mixture ([total printed]). Increase the high values or decrease the mixture total"

and if Eq. 4.11 or 4.13 applies, then the message is

"The total for the mixture ([total printed]) is less than or equal to the sum of the lower bounds ([sum printed]). Increase the mixture total"

The implications of these error messages are that (a) the burden of adjustment is on the user, and (b) adjustments may be made to either the relevant bounds *or* the total so that the $\leq$ sign in Eq. 4.10 becomes a $>$ sign, or the $\geq$ sign in Eq. 4.11 becomes a $<$ sign. If the user chooses to adjust the total, then the total no longer is equal to 1.0, and one is automatically designing in the actuals rather than the reals.

To illustrate, assume the following constraints are specified in MINITAB:

$$0.3 \leq X_1 \leq 1.0$$
$$0.4 \leq X_2 \leq 1.0$$
$$0.5 \leq X_3 \leq 1.0$$

Because the sum of the lower bounds (1.2) exceeds 1.0 and Eq. 4.11 is violated, an error message is printed.

"The total for the mixture (1.000000) is less than or equal to the sum of the lower bounds (1.200000). Increase the mixture total"

One way to correct the situation is to increase the default total (1.0) to something that is greater than 1.2, such as 1.25, in which case one is specifying the constraints in terms of the actuals. MINITAB will output a design and print the following tables in the session window:

	Amount		Proportion		Pseudocomponent	
Comp	Lower	Upper	Lower	Upper	Lower	Upper
A	0.30000	0.35000	0.24000	0.28000	0.00000	1.00000
B	0.40000	0.45000	0.32000	0.36000	0.00000	1.00000
C	0.50000	0.55000	0.40000	0.44000	0.00000	1.00000

The upper bounds in the Amount columns are adjusted to be consistent with a total amount of 1.25. For example, the upper bound on X_1 (0.35), when added to the lower bounds for X_2 and X_3, is equal to 1.25. The numbers in the Proportion columns are simply normalized values of the numbers in the Amount table. In this particular example, the numbers in the Amount column have been divided by 1.25 to give the numbers in the Proportion columns. In terms of proportions, the upper bound on X_1 (0.28), when added to the lower bounds for X_2 and X_3, is equal to 1.00. Pseudocomponents are explained in the last section of this chapter.

Under what circumstances might upper bounds lead to simplex-shaped design regions? Fig. 4.6 shows an example for $q = 3$ and the set of constraints

$$0 \leq X_1 \leq 0.5$$
$$0 \leq X_2 \leq 0.5$$
$$0 \leq X_3 \leq 0.5$$

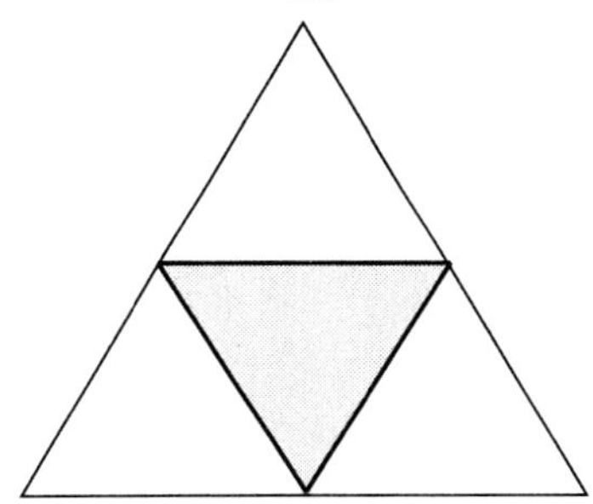

Figure 4.6. *Three upper bounds in a 3-simplex.*

The inverted subsimplex has been called a U-simplex, in contrast to the constrained region in Fig. 4.3, page 37, which is sometimes called an L-simplex. If any one of the three upper bounds were increased beyond 0.5, the constrained region would no longer be simplex-shaped. This leads to Property 4.4 [36].

Property 4.4 If there are *only* upper-bound constraints, then a U-simplex will lie within the full simplex whenever

$$\sum_{i=1}^{q} U_i - U_{min} \leq 1 \tag{4.14}$$

As additional examples, consider the three sets of constraints:

$0.2 \leq X_1 \leq 0.5$	$X_1 \leq 0.6$	$X_1 \leq 0.3$
$0.0 \leq X_2 \leq 0.5$	$X_2 \leq 0.5$	$X_2 \leq 0.4$
$0.0 \leq X_3 \leq 0.5$	$X_3 \leq 0.5$	$X_3 \leq 0.5$

The set on the left does not lead to a simplex-shaped subregion because there are not *only* lower-bound or *only* upper-bound constraints. The set in the middle has only upper-bound constraints, but Eq. 4.14 is violated. The set on the right has only upper-bound constraints and does not violate Eq. 4.14, and therefore will lead to a simplex-shaped subregion. Software will adjust the last set so that the bounds are consistent:

$$\begin{aligned} 0.1 \le X_1 \le 0.3 \\ 0.2 \le X_2 \le 0.4 \\ 0.3 \le X_3 \le 0.5 \end{aligned} \tag{4.15}$$

The small inverted triangle in Fig. 4.7 shows the constrained region defined by the constraints in Eqs. 4.15. The sides of this triangle are delineated by the upper-bound constraints, while the vertices are defined by the lower-bound constraints. The compositions of vertices 1, 2, and 3 are $(X_1, X_2, X_3) = (0.1, 0.4, 0.5)$, $(0.3, 0.2, 0.5)$, and $(0.3, 0.4, 0.3)$, respectively.

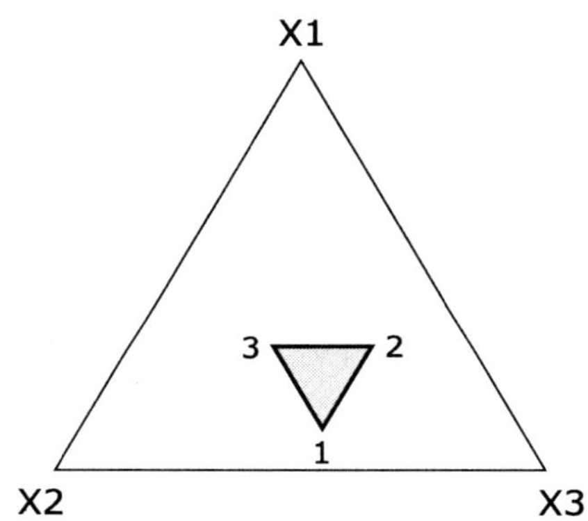

Figure 4.7. *Constrained region defined by Eqs.* 4.15.

Figure 4.8 shows an example of a $q = 4$ U-simplex. In this case the upper bounds have been set to

$$U_i \le \frac{1}{3} \qquad i = 1, 2, 3, 4$$

The vertices numbered 1–4 are located at the centroids of each of the four triangular constraint planes and have 0% of components X_1, X_2, X_3, and X_4, respectively. The design space is therefore an inverted tetrahedron.

Although sets of constraints that satisfy Property 4.4 lead to simplex-shaped design regions and consequently designs in this chapter apply, some computing packages do not recognize this and defer to the procedures discussed in the next chapter. Exceptions include Design-Expert Version 7 and MIXSOFT. More will be said about this in Section 4.5.

Piepel [119] also discusses checking constraints on linear combinations of components, such as Eq. 4.2. The procedures are much more complex and are beyond the intended scope of this text. The interested reader is referred to the discussion by Piepel.

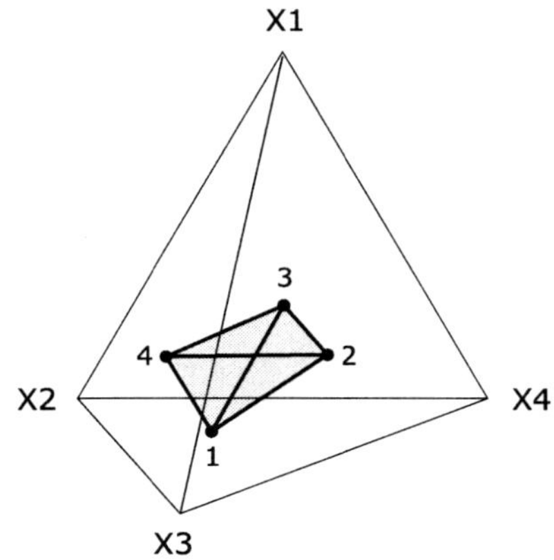

Figure 4.8. *Constrained region defined by* $U_i \leq 1/3$, $i = 1, 2, 3, 4$.

4.2 Some Design Considerations

Box and Draper [11, 12] list several properties of a good experimental design, many of which have been discussed by other authors (see, for example, Atkinson and Donev [3], Myers [106], and Myers and Montgomery [107]). These properties can be grouped according to the design or analysis stage where they are implemented, checked, or modified. Some can be implemented at the design stage — before any data are collected — but others cannot be checked and possibly adjusted until after data are collected and an analysis is performed. Consequently the list below has been divided into two broad groups, depending on the stage (design or analysis) at which the property is checked. Furthermore, the design stage has been divided into two subgroups, reflecting the order of implementation in most statistical software packages.

1. Design stage.
   (a) Initial goals:
      i. Generate a satisfactory distribution of information throughout the region of interest.
      ii. Provide sufficient design points to allow a test for model lack of fit.
      iii. Provide replicate design points to allow an estimate of pure experimental error.
      iv. Allow experiments to be performed in blocks.
      v. Allow designs to be built up sequentially.
      vi. Be cost-effective — we do not have an infinite amount of time or money.
   (b) Design evaluation stage:
      i. Check for the presence of high influence points.
      ii. Is the design robust to the presence of outliers in the data?
      iii. Is there a good distribution (however that may be defined) of prediction variances?

2. Analysis checks and modifications:

   (a) Homogeneous variance assumption.

   (b) Normally distributed residuals.

   (c) Outliers.

   (d) Transformation of the response, when necessary.

   (e) Augmentation of the design and model, with blocking, when necessary.

The aims under 1(a) and 1(b) are the subject of this chapter and Chapters 5–7; those under 2 are covered in Chapter 9.

To clarify points i–iii under 1(a), assume that one wants to design an experiment to support the linear Scheffé model

$$Y = \beta_1 X_1 + \beta_2 X_2 \tag{4.16}$$

which contains only two unknown parameters. Table 4.1 gives several designs that one might consider. The minimum number of design points required to support a polynomial model is equal to the number of unknown parameters in the model. Thus to support model 4.16 requires only two design points, and design a is adequate. The idea of fitting a two-term model to two design points is not ideal, and it would be much better to include additional points, such as a 50:50 blend (design b) or perhaps a 50:50 blend plus two additional design points midway between the 50:50 blend and the vertices (design c).

Table 4.1. *Some designs to support the model* $Y = \beta_1 Y_1 + \beta_2 X_2$

	Number of design pts. of composition†					Degrees of freedom for		
Design	1,0	0.75,0.25	0.5,0.5	0.25,0.75	0,1	Residuals	LOF	PE
a	1				1	0	0	0
b	1		1		1	1	1	0
c	1	1	1	1	1	3	3	0
d	2		1		2	3	1	2
e	2		2		2	4	1	3
f	2	1	2	1	2	6	3	3

† Compositions are expressed as X_1, X_2

For every data point that we collect, we "earn" a **degree of freedom**. For every parameter that we estimate, we "spend" a degree of freedom. If we earn more degrees of freedom than we spend estimating the model parameters, then we have degrees of freedom left over, which we call residual degrees of freedom. Thus in designs a–f we earn 2, 3, 5, 5, 6, and 8 degrees of freedom. In each case, however, we spend only two of these estimating the parameters, β_1 and β_2, in model 4.16. As a result we have residual degrees of freedom as given in column 7 of Table 4.1. Degrees of freedom play an important role when fitting models to data and will be discussed in detail in Chapters 8 and 9.

Designs c and d both have three residual degrees of freedom. However, there is clearly a difference between these two designs. Design c has five *discrete* design points, whereas design d has only three. As a result we can subclassify the residual degrees of freedom into two subcategories, those for **lack of fit** (LOF) and those for **pure experimental error** (PE). The residual degrees of freedom in design c are all lack-of-fit degrees of freedom because we have five discrete design points, and $5 - 2 = 3$. In design d we have one lack-of-fit degree of freedom because we have three discrete design points, and $3 - 2 = 1$; the remaining residual degrees of freedom would be classified as pure-error degrees of freedom, because they arise from replication. To obtain an estimate of pure error, replicates *must* be included in a design. Columns 8 and 9 of Table 4.1 give the breakdown of the residual degrees of freedom into lack of fit and pure error for the six designs.

Having degrees of freedom for lack of fit *and* pure error is desirable. This is because it allows for a formal statistical test, called a **lack-of-fit test**, to be carried out that provides information about the adequacy of the model. If lack of fit is statistically significant, then one needs to consider a higher-order model. Details of this test are reserved for discussion in the chapters on analysis.[1]

4.3 Three Designs

In this section, three designs for simplex-shaped design regions are presented. The first two, the simplex lattice and simplex centroid designs, were both introduced by Scheffé in the 1950s [146] and 1960s [147]. Both designs are available in Design-Expert, JMP, MINITAB, and MIXSOFT. The third design, the simplex screening design, was introduced about 20 years later by Snee and Marquardt [159]. Screening designs are offered by Design-Expert and JMP. Despite differences in software packages, these designs may be created in most packages with a minimum of fuss.

4.3.1 Simplex Lattice Designs

A simplex lattice design always has a descriptor of the form $\{q, m\}$. The q within the curly braces has the usual meaning, i.e., the number of mixture components. The m within the curly braces describes the order of the model that is supported by the design. Thus a {4,2} simplex lattice design will support a $q = 4$ second-order Scheffé mixture model.

The treatment combinations for a $\{q, m\}$ simplex lattice design consist of *all* mixtures whose proportions are members of the set

$$\left(\frac{0}{m}, \frac{1}{m}, \dots, \frac{m}{m}\right) \tag{4.17}$$

[1] As a rule of thumb, for a mixture setting Design-Expert [163] recommends LOF and PE degrees of freedom each equal to the number of components plus one ($q + 1$) up to a maximum of five each.

To illustrate, the {4,2} simplex lattice design would consist of the following design points:

X_1	X_2	X_3	X_4
2/2	0/2	0/2	0/2
0/2	2/2	0/2	0/2
0/2	0/2	2/2	0/2
0/2	0/2	0/2	2/2
1/2	1/2	0/2	0/2
1/2	0/2	1/2	0/2
1/2	0/2	0/2	1/2
0/2	1/2	1/2	0/2
0/2	1/2	0/2	1/2
0/2	0/2	1/2	1/2

or equivalently

X_1	X_2	X_3	X_4
1	0	0	0
0	1	0	0
0	0	1	0
0	0	0	1
1/2	1/2	0	0
1/2	0	1/2	0
1/2	0	0	1/2
0	1/2	1/2	0
0	1/2	0	1/2
0	0	1/2	1/2

There are two categories of points in this design — the vertices and the midpoints of the edges (edge centroids). The design is displayed on the left in Fig. 4.9.

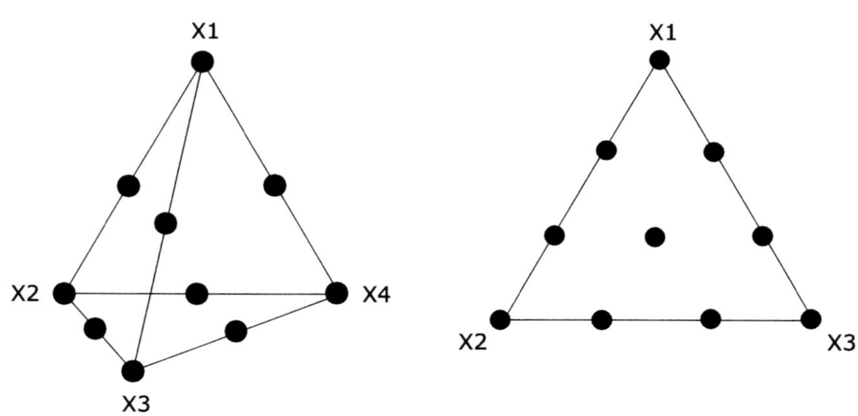

Figure 4.9. {4, 2} *and* {3, 3} *simplex lattice designs.*

As a second example, the design points for the {3,3} simplex lattice design are tabulated as follows:

X_1	X_2	X_3
1	0	0
0	1	0
0	0	1
1/3	2/3	0
2/3	1/3	0
1/3	0	2/3
2/3	0	1/3
0	1/3	2/3
0	2/3	1/3
1/3	1/3	1/3

This design is displayed on the right in Fig. 4.9. Both designs in Fig. 4.9 are 10-point designs. The number of design points in a $\{q, m\}$ simplex lattice design is equal to the number of terms in a q-component Scheffé model of degree m (cf. Eq. 3.33, page 28). Because of this, Scheffé models of degree m are sometimes referred to as $\{q, m\}$ Scheffé models when they are associated with the corresponding $\{q, m\}$ lattice designs [87, 88, 146]. If the design on the left in Fig. 4.9 were used to support the {4,2} Scheffé model, then the design is said to be **saturated** — there are no degrees of freedom for estimating lack of fit. The same can be said about the design on the right if it were used to support the {3,3} Scheffé model.

Note that the {3,3} design has one **complete mixture** [29] — a formulation in which all of the components are present — whereas the {4,2} design does not. Whenever $m < q$, a simplex lattice design will consist of mixtures containing up to m components, and consequently there will be no complete mixtures; if $m = q$, there will be one complete mixture; and if $m > q$, there will be more than one complete mixture. In a {3,4} lattice design, for example, there will be three complete mixtures. Table 4.2 shows, for $3 \leq q \leq 6$ and $2 \leq m \leq 4$, the number of different types of design points in several $\{q, m\}$ lattice designs. Underlined values are complete mixtures.

Note that among the 12 designs in Table 4.2, there are few complete mixtures and that the design points tend to "pile up" on the lower-dimensional subsimplexes. Thus the distribution of information is weighted towards the boundaries of the simplex [26]. For this reason, plus the fact that the designs are saturated designs when used to support the corresponding $\{q, m\}$ Scheffé models, the designs are often augmented with $q+1$ additional complete blends. One of these blends is always the overall centroid, while the remaining q are **axial check blends** — design points located halfway between the vertices and the overall centroid. Design-Expert and MINITAB provide this feature as an option. Fig. 4.10 displays the {3,2} lattice design (filled circles) augmented by the center point and axial check blends (open circles). Augmenting $\{q, 2\}$ lattice designs in this manner leads to designs that are sometimes called **simplex response-surface** designs. Such designs support quadratic Scheffé models with $q + 1$ additional degrees of freedom for lack of fit.

It is still necessary to replicate design points to obtain degrees of freedom for pure error. Software handles this in different ways. JMP replicates the entire design. Design-Expert picks the highest leverage points to replicate, and as the vertices often are high

Table 4.2. *Simplex lattice designs. Point types*

q	m	Blends pure	2	3	4	Total blends
3	2	3	3	·	·	6
	3	3	6	1	·	10
	4	3	9	3	·	15
4	2	4	6	·	·	10
	3	4	12	4	·	20
	4	4	18	12	1	35
5	2	5	10	·	·	15
	3	5	20	10	·	35
	4	5	30	30	5	70
6	2	6	15	·	·	21
	3	6	30	20	·	56
	4	6	45	60	15	126

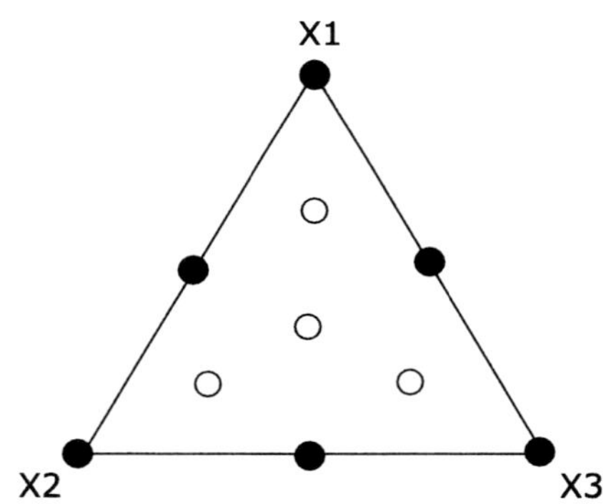

Figure 4.10. *Augmented* $\{3, 2\}$ *simplex lattice design.*

leverage points, they are usually selected.[2] MINITAB allows the user to choose points to be replicated. In all cases the resulting design can be further modified by the user.

4.3.2 Simplex Centroid Designs

For given q, there is only one simplex centroid design. The design consists of all mixtures located at the centroid of each simplex contained within a q-component simplex — all the pure "blends" $(1,\ 0,\ 0,\ \ldots,\ 0)$, all the binary blends $(1/2,\ 1/2,\ 0,\ \ldots,\ 0)$, all the ternary blends $(1/3,\ 1/3,\ 1/3,\ 0,\ \ldots), \ldots,$ plus the overall centroid $(1/q,\ 1/q,\ \ldots,\ 1/q)$. As the number of vertices can be viewed as q components taken one at a time, the number of

[2]Leverage is discussed in Section 6.2.

binary blends as q components taken two at a time, and so on, the number of points in a full simplex centroid design is then

$$\binom{q}{1}+\binom{q}{2}+\binom{q}{3}+\cdots+\binom{q}{q}=2^q-1 \tag{4.18}$$

The number of design points in a simplex centroid design increases rapidly with q. Assume, for example, that one desires to support a $q = 6$ quadratic Scheffé polynomial, which has 21 terms. If one used a simplex centroid design to support this model, there would be $63 - 21 = 42$ degrees of freedom for lack of fit. One might think that it would be a better idea to use this design to support a $q = 6$ cubic Scheffé model, which has 56 terms. Unfortunately, this model has terms of the form $X_iX_j(X_i - X_j)$, and these are equal to zero in a simplex centroid design because either X_iX_j or $(X_i - X_j)$ will always be equal to zero. Problems also arise with terms such as $X_iX_j(X_i - X_j)^2$ and $X_i^2X_jX_k$. For these reasons, the following **special polynomials**, which have the same number of terms as the simplex centroid designs, are used with these designs:

$$\begin{aligned} E(y) = &\sum_{1\le i\le q} \beta_i X_i + \sum_{1\le i\le j\le q} \beta_{ij} X_i X_j \\ &+ \sum_{1\le i\le j\le k\le q} \beta_{ijk} X_i X_j X_k + \cdots + \beta_{12\ldots q} X_1 X_2 \cdots X_q \end{aligned} \tag{4.19}$$

As pointed out by Piepel [123], there may be instances where one may care to fit a truncated form of this model, in which case it would be desirable to generate a p-level fraction of the full design. Piepel proposed truncating the series 4.18 at the $\binom{q}{p}$ term but always including the overall centroid. For example, if one wanted a fractional design plan to support terms in model 4.19 up through X_iX_j, then the number of points in the $q = 6$ simplex centroid design that one may choose to include could be

$$\binom{6}{1}+\binom{6}{2}+\binom{6}{6}=22$$

or perhaps

$$\binom{6}{1}+\binom{6}{2}+\binom{6}{3}+\binom{6}{6}=42$$

Neither design is ideal, as the 22-point design suffers from having only one degree of freedom for lack of fit, whereas the 42-point design has an excess of lack-of-fit degrees of freedom. However, with the addition of replicates and axial check blends to the 22-point design, one could perform a formal lack-of-fit test and if necessary augment the special polynomial model up through terms in $X_iX_jX_k$. This would, of course, require the addition of the $\binom{6}{3} = 20$ additional ternary blends, which are the *two-dimensional* centroids.

A p-level fraction of a q-component simplex centroid design is called a $\{q, p\}$ simplex centroid design. The 22- and 42-point designs above are {6,2} and {6,3} designs, respectively. One must be careful to distinguish the meaning of the second number in the curly braces, as it has a different meaning with the centroid designs than it does with the lattice designs. A $\{q, 3\}$ centroid design, for example, has edge *centroids*, whereas a $\{q, 3\}$ lattice design has design points on the edges located at the 1/3,2/3 and 2/3,1/3 blends of the components that define the edges. Both designs include the vertices and the two-dimensional centroids.

Fig. 4.11 shows pictures of the full (i.e., {3,3}) simplex centroid design and a 2-level fraction of the centroid design for four components. The overall centroid in the {4,2} design is indicated by a filled square. This design has 11 design points, equal to the sum of terms in the series

$$\binom{4}{1} + \binom{4}{2} + \binom{4}{4}$$

The full simplex centroid design for four components would have $2^4 - 1 = 15$ runs. The additional four runs would be located at the centroids of the two-dimensional constraint planes (triangles).

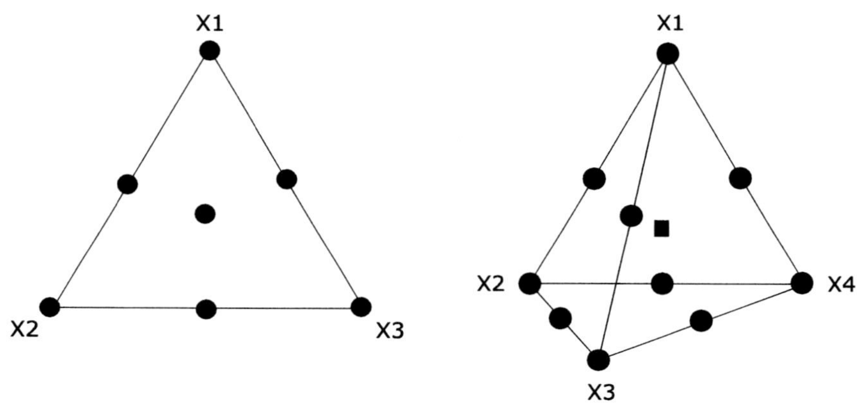

Figure 4.11. $\{3, 3\}$ *and* $\{4, 2\}$ *simplex centroid designs.*

Neither Design-Expert nor MINITAB has an option for specifying a p-level fraction of a simplex centroid design. One must generate the full design composed of $2^q - 1$ formulations and then delete unwanted observations from the data table. JMP and MIXSOFT, on the other hand, provide an option to choose a p-level fraction of the design. Design-Expert and MINITAB have options for augmenting the designs with axial check blends, while JMP and MIXSOFT do not.

4.3.3 Simplex-Screening Designs

In the initial stages of an experimental program, and in the absence of subject-matter knowledge, simplex-screening designs deserve serious consideration by the formulator. As will be clear with the example to be presented, a single two-dimensional plot of the observed responses can provide an initial visual indication of the effects of the components on the response. This is before a model is fit to the data.

For a given q there is only one screening design, and this consists of the set of points in Table 4.3. End points are the q blends with $q - 1$ components present at $100/(q - 1)\%$.

Thus when $q = 3$, the end points are the three blends with two components present at 50%. See Fig. 4.12 (left) for a picture of this design. When $q = 4$, the end points are the four blends with three components present at $33\frac{1}{3}\%$ — the centroids of the four triangular constraint planes. Fig. 4.12 (right) shows four of the 13 design points — those that are on the X_1 axis. The design point at the bottom of the tetrahedron, which is one of the four end points, is the centroid of the (X_2, X_3, X_4) constraint plane, where $X_1 = 0$. The end points are sometimes called **constraint-plane centroids** [154] even though when $q > 4$ the planes are really hyperplanes.

Table 4.3. *Simplex-screening designs. Point types*

Type	Number
Vertices	q
Overall centroid	1
Axial check blends	q
End points	q
Total	$3q + 1$

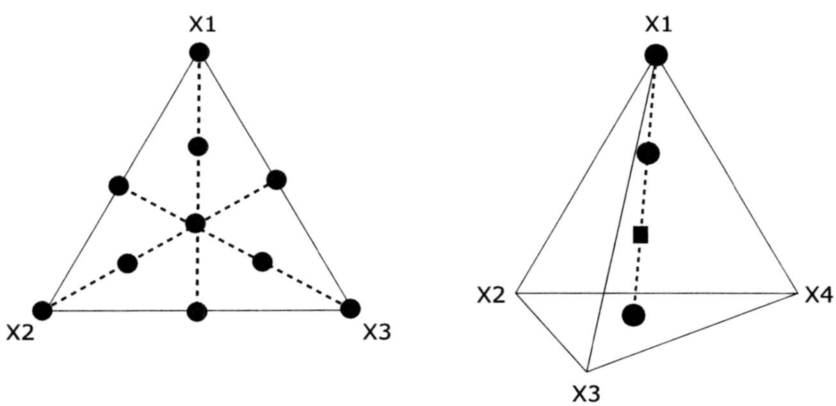

Figure 4.12. *Full ($q = 3$) and partial ($q = 4$) simplex-screening designs.*

Two comments need to be made about these designs, one specifically directed at the $q = 3$ screening design and the other toward these designs in general. The design on the left in Fig. 4.12 is the same design as the augmented {3,2} simplex lattice design (Fig. 4.10). Thus, when $q = 3$, the simplex-screening design will support the $q = 3$ Scheffé quadratic model, despite the fact that screening designs were developed for the purpose of supporting linear, not quadratic, models. The case for $q = 3$ is a special (and very versatile) case. Simplex-screening designs where $q \geq 4$ do not support quadratic Scheffé models.

The second comment is that the design points in these designs fall on the component axes, which are indicated in Fig. 4.12 by dotted lines. For this reason, these designs are also referred to as **axial designs** [29]. The **component axis** of a component, X_i, is the locus of compositions extending from $X_i = 1$, $X_j = 0$, $j \neq i$ to the composition $X_i = 0$, $X_j = 1/(q-1)$ for all $j \neq i$. Along the X_i component axis, the proportions of the other $q-1$ components (X_j, $j \neq i$) are changing but remain equal to one another. For example, in the $q = 3$ case, along the X_1 axis the ratio of X_2 to X_3 is always 1:1; in the $q = 4$ case, along the X_1 axis components X_2, X_3, and X_4 are always in a ratio of 1:1:1. One can envision the component axes as 2-simplexes, in which at one vertex (end) we have pure component X_i, and at the other vertex, we have a mixture of the X_j, $j \neq i$, in equal amounts to one another and, of course, $X_i = 0$.

To illustrate how useful these designs can be, an example is presented from the pharmaceutical literature. Belloto et al. [4] were interested in studying the solubility of the drug Diazepam in mixtures of ethanol, propylene glycol, and water. Using a modest number of measurements, they hoped to develop a good working model that would predict the solubility of Diazepam in any mixture of the solvents within the 3-simplex. They chose a simplex-screening design, and their results are given in Table 4.4.

Table 4.4. *Diazepam solubility experiment*

	Volume fraction				
ID	Ethanol	Glycol	Water	Solubility (mg/ml)	ln(Solubility)
1	0.50	0.50	0	27.0	3.30
2	0.50	0	0.50	6.02	1.80
3	0	0.50	0.50	0.610	−0.494
4	0.33	0.33	0.33	9.52	2.25
5	0.66	0.17	0.17	28.0	3.33
6	0.16	0.66	0.17	13.0	2.56
7	0.17	0.17	0.66	0.408	−0.896
8	1.00	0	0	27.8	3.32
9	0	1.00	0	7.42	2.00
10	0	0	1.00	0.0479	−3.04

A simplex-screening plot [159] of the Diazepam data is displayed in Fig. 4.13. (Numbers near the data points are IDs in Table 4.4.) The measured solubilities vary over nearly three orders of magnitude. Because of this the figure shows the natural log of the solubility, rather than the solubility, plotted against the proportions along the component axes. An important feature of such a plot is that no matter how many components are present, the observed responses can always be displayed in one two-dimensional plot per response.

The screening plot allows preliminary inferences to be made about the behavior of the components before a model is fit to the data. In this particular example, one might note that propylene glycol plays an intermediate role, while water and ethanol are the big actors controlling the solubility of Diazepam. Furthermore, the plot for ethanol appears definitely curvilinear, suggesting that a quadratic or reduced quadratic model may fit the logged data better than a linear model. This turns out to be the case. Since the design supports the quadratic model with degrees of freedom to spare, one could just go ahead and

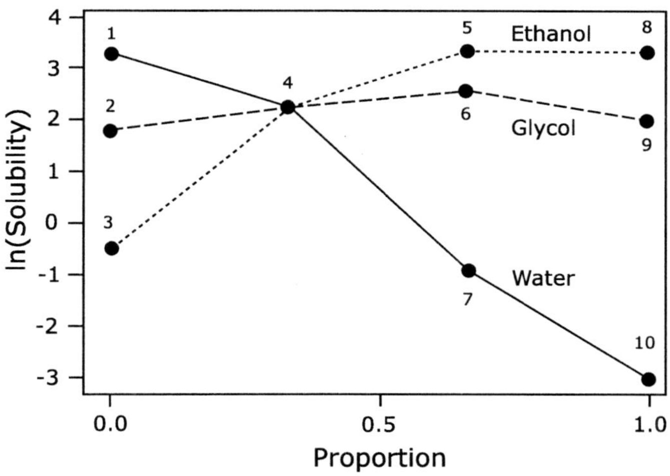

Figure 4.13. *Diazepam solubility experiment. Screening plot.*

fit the quadratic model and then examine the nonlinear blending coefficients to see if any or all are significant.

Design-Expert offers screening designs as an option for $q \geq 6$. Screening designs for $q < 6$ can be created using the `User Defined` option. MINITAB does not offer screening designs per se, but they can be created using the `Extreme Vertices` design option. This will create more design points than needed, but the unwanted points (which are identified by point type) can be deleted from the worksheet. JMP offers screening designs under the name **ABCD designs**.

4.4 Designs for Three Components

A Catalog of Mixture Experiment Examples, compiled and maintained by Piepel and Cornell [131], contains a table of component-proportion design and analysis examples. It is quite clear from the table that the preponderance of examples reported in the literature are for $q = 3$ mixtures. Recognizing the importance of designs for three-component systems, Cornell [26] reported the results of a study comparing the pros and cons of two 10-point designs for three-component mixtures.

The designs studied were the {3,3} simplex lattice design (Fig. 4.14, left) and the augmented {3,2} simplex lattice design (Fig. 4.14, right). The latter is, of course, also the simplex-screening design and can be viewed equally well as an augmented simplex centroid design. Both are 10-point designs. Following Cornell, let us refer to the design on the left as A and the design on the right as B.

Both designs would support a sequential model-building process from linear to quadratic to special cubic. Beyond special cubic, however, things change. Design A will support the additional terms

$$\gamma_{12}X_1X_2(X_1 - X_2) + \gamma_{13}X_1X_3(X_1 - X_3) + \gamma_{23}X_2X_3(X_2 - X_3)$$

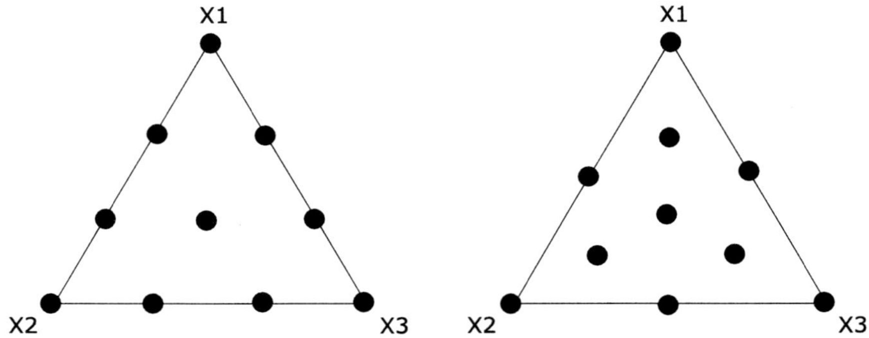

Figure 4.14. {3, 3} *and augmented* {3, 2} *simplex lattice designs.*

leading to the 10-term full cubic model. When this model is fit, however, the design is saturated, and so there are no degrees of freedom for lack of fit. Design B will not support all of the cubic terms because of the linear dependency

$$X_1X_2(X_1 - X_2) = X_1X_3(X_1 - X_3) - X_2X_3(X_2 - X_3)$$

Design B *would* support a model containing the linear, quadratic, special cubic, and two of the three cubic terms (nine terms total). But the question arises as to which two of the three cubic terms to choose. One does not know in advance which of the three cubic terms might not be needed.

Because of the different distribution of information in design B, and in particular because of the presence of design points inside the triangle and the multiple levels of each component, design B will support the nine-term special quartic model:

$$E(Y) = \sum_{i}^{q} \beta_i X_i + \sum_{i<j}\sum^{q} \beta_{ij} X_i X_j + \sum\sum_{i<j<k}\sum^{q} \beta_{iijk} X_i^2 X_j X_k$$
$$+ \sum\sum_{i<j<k}\sum^{q} \beta_{ijjk} X_i X_j^2 X_k + \sum\sum_{i<j<k}\sum^{q} \beta_{ijkk} X_i X_j X_k^2$$

Design A will not support this model because, for every design point, the dependencies

$$X_1^2X_2X_3 = X_1X_2^2X_3 = X_1X_2X_3^2$$

apply.

One of the initial goals of a good design is to generate a satisfactory distribution of information throughout the design region (cf. page 45). Exactly what is "satisfactory" depends to some extent on the goal of the experiment. If the goal is to focus on the blending properties of the binary blends defining the boundaries of the 3-simplex, then design A would be the design of choice. On the other hand, if the goal of the experiment is to learn as much as possible about the blending properties of complete mixtures, then design B would be the design of choice. Design B has the additional advantage in that it will support the linear, quadratic, special cubic, and special quartic models with one or more degrees of freedom for lack of fit.

Cornell discusses several other aspects of these two designs, and the reader is referred to the original article [26] or Cornell's text [29] for further details.

4.5 Coding Mixture Variables

Consider the following hypothetical constraints for a three-component mixture experiment:

$$0.12 \leq X_1$$
$$0.22 \leq X_2$$
$$0.32 \leq X_3$$

Because there are only lower bounds on the component proportions, we know that this must lead to a simplex-shaped design region similar to that in Fig. 4.3. To be more precise, we should also include the implied upper bounds:

$$0.12 \leq X_1 \leq 0.46$$
$$0.22 \leq X_2 \leq 0.56$$
$$0.32 \leq X_3 \leq 0.66$$

The component proportions for a simplex-screening design based on these constraints are displayed in the first three columns of Table 4.5. The reader is invited to study these columns and try to identify the vertices, axial check blends, overall centroid, and the end points.

Table 4.5. *Hypothetical simplex-screening design*

X_1	X_2	X_3	X_1^*	X_2^*	X_3^*
0.177	0.277	0.547	0.166	0.166	0.667
0.177	0.447	0.377	0.166	0.667	0.166
0.347	0.277	0.377	0.667	0.166	0.166
0.233	0.333	0.433	0.333	0.333	0.333
0.460	0.220	0.320	1.000	0.000	0.000
0.120	0.560	0.320	0.000	1.000	0.000
0.120	0.220	0.660	0.000	0.000	1.000
0.290	0.220	0.490	0.500	0.000	0.500
0.290	0.390	0.320	0.500	0.500	0.000
0.120	0.390	0.490	0.000	0.500	0.500

The last three columns in Table 4.5, the X_i^* s, are *coded* values of the X_i s. As the subregion is simplex-shaped, there is no reason why we cannot represent the design points in terms of the subsimplex rather than in terms of the full simplex. It is clearly much easier to identify the point types when the component proportions are expressed in terms of the coded variables rather than in terms of the uncoded variables. This is one reason why coded variables are often used. For example, if the lower bounds in this example were instead 0.107, 0.195, and 0.314, the uncoded X_i would be even more complicated, but the coded values would be the same.

The coded components, X_i^*, in Table 4.5 are known as **pseudocomponents**. Pseudocomponents were first introduced by Kurotori [85] and are discussed extensively by Crosier [36, 37].

Let us define the lower-bound constraints on the X_i as L_i, $i = 1, 2, \ldots, q$, where some of the L_i may be equal to zero. For the three-component example above, $L_1 = 0.12$, $L_2 = 0.22$, and $L_3 = 0.32$. Define L as

$$L = \sum_i^q L_i \qquad i = 1, 2, \ldots, q \tag{4.20}$$

For this example, $L = 0.12 + 0.22 + 0.32 = 0.66$. Pseudocomponent proportions are calculated from the expression

$$X_i^* = \frac{X_i - L_i}{1 - L} \tag{4.21}$$

The pseudocomponent proportions, X_i^*, in Table 4.5 are consequently calculated from the formulas

$$X_1^* = \frac{X_1 - 0.12}{0.34}, \qquad X_2^* = \frac{X_2 - 0.22}{0.34}, \qquad X_3^* = \frac{X_3 - 0.32}{0.34}$$

Calculating the X_i from the X_i^* is simply a matter of algebraically rearranging Eq. 4.21.

The pseudocomponent transformations discussed to this point are more correctly referred to as L-pseudocomponent transformations, where the "L" stands for "lower". For inverted simplexes, such as those illustrated in Figs. 4.7 and 4.8, page 45, one could use what is called the U-pseudocomponent transformation, where "U" refers to "upper" [36]. In place of Eq. 4.21 one would use the expression

$$X_i^{**} = \frac{U_i - X_i}{U - 1} \tag{4.22}$$

where X^{**} is the upper-pseudocomponent proportion, U_i is the upper bound for the ith component, and U is defined

$$U = \sum_i^q U_i \qquad i = 1, 2, \ldots, q \tag{4.23}$$

To illustrate, consider Fig. 4.15, which is the same as Fig. 4.7 except that the L-pseudocomponent simplex has been added (dashed triangle). The small shaded triangle is the constrained region for the set of constraints in Eqs. 4.15, page 44. It is also the U-pseudocomponent simplex. The size of the L-pseudocomponent simplex is $1 - L$ (the denominator in Eq. 4.21), where by size we mean the range of the X_i^* *in terms of the reals*. The size of the U-pseudocomponent simplex is $U - 1$ (the denominator in Eq. 4.22), where

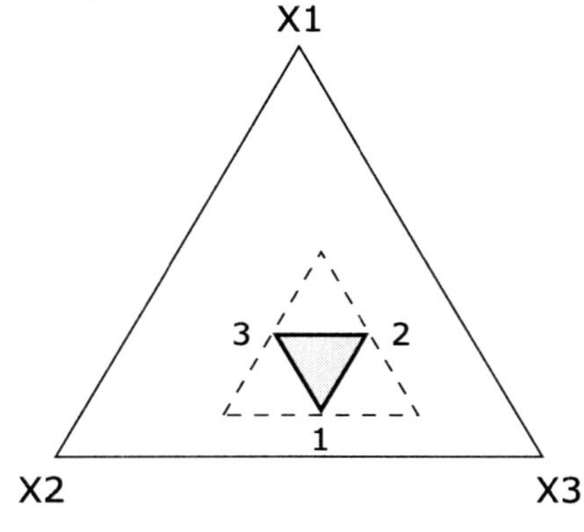

Figure 4.15. *Constrained region defined by Eqs.* 4.15.

by size we mean the range of the X_i^{**} *in terms of the reals*. In this example, $1 - L = 0.4$ and $U - 1 = 0.2$, and since $0.2 < 0.4$, the U-pseudocomponent simplex is half the size of the L-pseudocomponent simplex. The compositions of the numbered vertices in terms of the reals, lower-, and upper-pseudocomponent proportions are

	Reals			Lower pseudos			Upper pseudos		
Vertex	X_1	X_2	X_3	X_1^*	X_2^*	X_3^*	X_1^{**}	X_2^{**}	X_3^{**}
1	0.1	0.4	0.5	0.0	0.5	0.5	1	0	0
2	0.3	0.2	0.5	0.5	0.0	0.5	0	1	0
3	0.3	0.4	0.3	0.5	0.5	0.0	0	0	1

When using the U-pseudocomponent transformation, one must keep in mind that the U-pseudocomponents have effects that are opposite those of the real components.

Pseudocomponent transformations are not restricted to situations where there are only lower-bound constraints. These transformations are also applicable to situations where there are both lower- and upper-bound constraints that may lead to subregions of the simplex that are not shaped like a simplex. Fig. 4.16 shows an example of such a subregion. The triangle drawn with dashed lines is the L-pseudocomponent simplex. In all cases a pseudocomponent simplex, whether lower or upper, is the *smallest simplex of that type (L- or U-) that will include all of the observations.*

It is important to note in Fig. 4.16 that the vertices of the pseudocomponent simplex lie outside the constrained region. Thus whenever a Scheffé model such as

$$E(Y) = \beta_1^* X_1^* + \beta_2^* X_2^* + \beta_3^* X_3^*$$

is fit to data collected on the boundary and perhaps the interior of a constrained region, the estimates of the parameters, the b_i^* s, are applicable *only* to the constrained region. They should not be used to predict the response at a vertex of the pseudocomponent simplex unless the particular vertex is part of the constrained region; otherwise one is extrapolating.

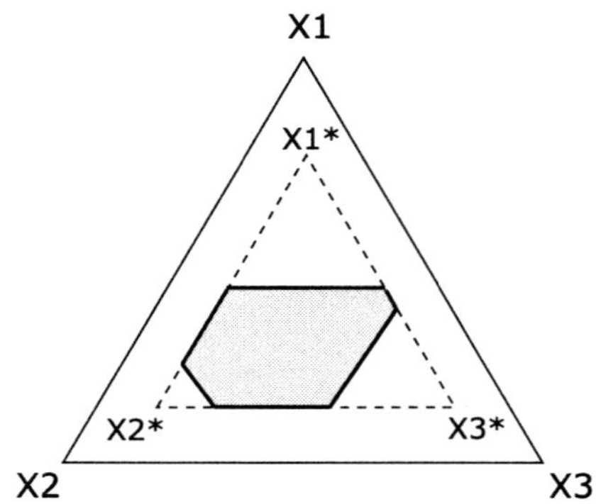

Figure 4.16. *A pseudocomponent simplex.*

Thus, we have three **metrics** that can be used in a mixture setting: (i) *actuals* (grams, ounces, pounds, etc.); (ii) *reals* (component proportions); and (iii) *pseudos* (pseudocomponent proportions). Unless otherwise stated, further references to pseudocomponents shall imply L-pseudocomponents. Statistical software packages handle model fitting in these metrics in various ways.

- In Design-Expert Version 7, if model fitting is done in terms of the pseudos, then the model is also back transformed to reals and actuals. If the model is fitted in the reals, then the model is back transformed to the actuals but not to the pseudos. If the model is fitted in the actuals, then the model is back transformed to the reals but not to the pseudos.

  If constraints are entered in terms of component proportions, then actuals and reals are the same. If there are no constraints on the component proportions (other than the equality and nonnegativity constraints, page 9), then reals and pseudos are the same. If both conditions hold, then actuals = reals = pseudos.

- With MINITAB, one enters the constraints in terms of actuals or reals. Model fitting is done in the reals or the pseudos. Pseudos are not automatically displayed in MINITAB's worksheet but can be displayed with a simple point-and-click.

- Component constraints in JMP are specified in terms of reals, and the data table is constructed in terms of reals. To display pseudos in the data table one must use JMP's calculator and Eq. 4.21 or Eq. 4.22, page 58. Models can then be fit in terms of either the reals or the pseudos.

Chapter 5

Designs for Non-Simplex-Shaped Regions

In Chapter 4 we saw that when there are upper-bound constraints, in most cases the design region will not be shaped like a simplex. For example, the constraints exhibited in Fig. 4.4, page 40, led to subspaces shaped like a trapezoid and a parallelogram. Fig. 5.1 displays the same constrained region as illustrated in Fig. 4.16, page 60, except that in this case lower and upper bounds are indicated. When $q = 3$, lower and upper bounds lead in many cases to irregular polygons. When $q = 4$, the irregular polygons become irregular polyhedrons, and when $q > 4$, irregular polyhedrons become irregular hyperpolyhedrons.[1] Thus we can say that this chapter is about designs for irregularly shaped regions. Most of the methods used to develop designs for such regions fall under the heading of **computer-aided experimental design**.

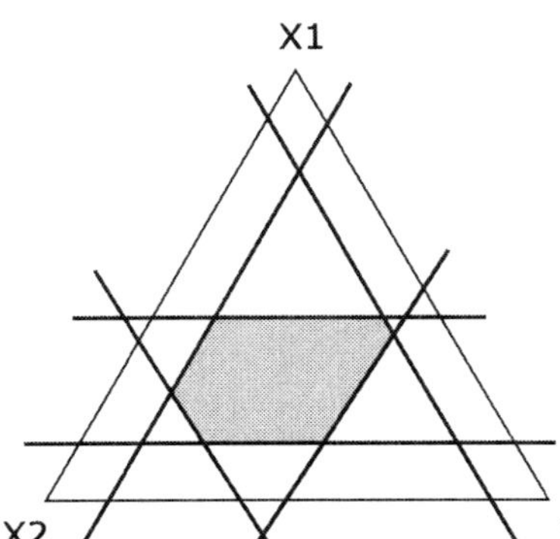

Figure 5.1. *Irregular polygonal-shaped subregion.*

The boundaries of the subspace in Fig. 5.1, which are one-dimensional edges, are parallel to the sides of the simplex. When there are only single-component constraints, the $(q - 2)$-dimensional boundaries of a constrained region will always be parallel to the

[1]More precisely, constrained design regions within a simplex are *convex polytopes*. A polytope is a generalization of a polygon or polyhedron to any number of spatial dimensions. A polytope is convex if all line segments connecting any pair of points lie completely within the polytope.

$(q-2)$-dimensional boundaries of the simplex. This is not necessarily the case when there are multicomponent constraints. Consider, for example, the two ratio constraints

$$\frac{X_1}{X_3} \leq 2$$

$$\frac{3X_1 + X_2}{X_3} \geq 1$$

which can be rewritten as the multicomponent constraints

$$0 \leq -X_1 + 2X_3$$
$$0 \leq 3X_1 + X_2 - X_3$$

The first ratio constraint leads to the boundary nearest vertex X_1 in Fig. 5.2, while the second ratio constraint leads to the boundary nearest vertex X_3 in the figure. The shaded subspace is again an irregular polygon, but in this case two of the boundaries are not parallel to the sides of the simplex. More will be said about multicomponent constraints later in this chapter.

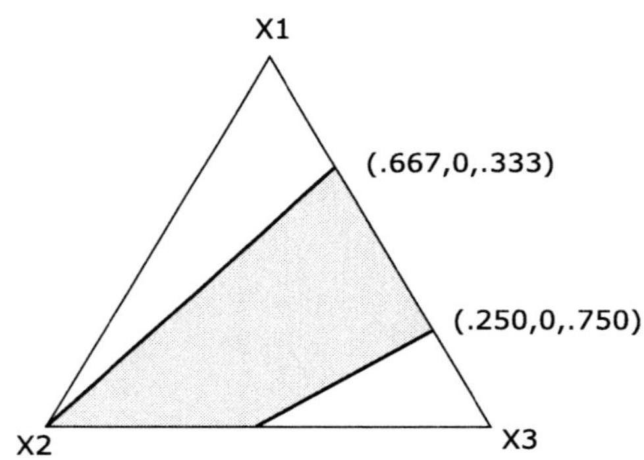

Figure 5.2. *Irregular polygonal-shaped subregion.*

Designs such as the simplex lattice, simplex centroid, and simplex screening designs do not apply to irregularly shaped design regions (although a method for mapping these designs into irregularly shaped regions has been described [145]). Instead of "prepackaged designs", design-of-experiments (DOE) software uses "prepackaged algorithms". In the sections to follow, we first look at the overall strategy used to create designs for irregularly shaped regions. Next, some of the algorithms that have evolved to create these designs will be briefly described. The remaining sections go into the specifics of design implementation.

5.1 Strategy Overview

One thing that all irregularly shaped design regions have in common are vertices. As a first step, we could calculate (by a method yet to be described) the composition of all of the vertices of a constrained region and consider these as **candidates points** for a design.

Certain pairs of vertices will share the same one-dimensional edge, and so as a next step we could calculate the centroids of the edges and add these to the list of candidate points. The composition of the edge centroids would be determined by averaging the composition of the vertices that share the same edge. Continuing in this manner, we could calculate the centroids of the two-dimensional constraint planes, present whenever $q \geq 4$. This would require identifying those vertices that share the same two-dimensional constraint plane and averaging their composition. The process could continue up to the point where we average all of the vertices and determine the composition of the $(q-1)$-dimensional overall centroid, which would in turn be added to the list of candidates. We would now have a list composed of the vertices, the one-dimensional centroids, the two-dimensional centroids, ..., the $(q-1)$-dimensional centroid, some or all of which could be included in a **candidate list**. The centroids that are calculated by averaging vertices are called **averaged-extreme-vertices** centroids, or **AEV centroids**.

As one might suspect, as q gets larger and larger, the list of candidate points grows rather rapidly. Crosier [37] has published formulas for counting the maximum possible number of boundaries in constrained mixture regions when there are only single-component constraints. Table 5.1 is based on Crosier's work and gives the maximum possible number of vertices and edges for $q \leq 12$. Fig. 5.1 exemplifies the case where $q = 3$. Crosier's formula for calculating the number of vertices for a mixture region defined by single-component constraints is implemented in MIXSOFT.

Table 5.1. *Maximum possible number of vertices and edge centroids in constrained regions defined by single-component constraints*

q	Vertices	Edge Centroids
3	6	6
4	12	18
5	30	60
6	60	150
7	140	420
8	280	980
9	630	2520
10	1260	5670
11	2772	13860
12	5544	30492

Do real-life situations arise where the number of possible vertices is in the hundreds or thousands? The answer is "yes". For example, Piepel and coworkers have spent many years formulating nonleachable glasses for the purpose of burying radioactive nuclear waste. Glasses are composed of many metal oxides. The formulations investigated by Piepel et al. have been composed of as many as 21 components. One recent study involved a nine-component glass formulation [133, 135]. Based on the published constraints, the number of vertices in the design space was 509, about 80% of the maximum possible. In another study [132, 135], a 12-component formulation was under investigation. Although the constraints were not explicitly stated, the constraints implied by the 35-observation data set would lead to the conclusion that the design region had more than 4000 vertices.

As the candidate list grows rapidly with q, the need arises for a means to reduce the list to a reasonable number of design points. Much research was done in this area during the 1960s and 1970s, and as a result several algorithms evolved, some of which are now incorporated into statistical computing packages.

Thus the steps involved in creating designs for irregularly shaped design regions can be summarized as follows:

1. Create a candidate list of design points.
   (a) Start the list by calculating the composition of the extreme vertices.
   (b) Calculate the composition of the various dimensional centroids and add the desired ones to the list.
   (c) Calculate the composition of any other design points of interest and add these to the list.
2. Use an algorithm to reduce the candidate list to a reasonable number of design points.

In the case of screening designs, where the intention is to support a first-degree model, it is possible that the number of extreme vertices calculated in step 1(a) may be adequate. Before the ubiquity of desktop computers, and because of the complexity of calculating the composition of higher-dimension centroids, the process sometimes ended here. In some instances — particularly when there is a large number of mixture components and several constraints — the number of extreme vertices may be impractically large (cf. Table 5.1), in which case step 2 is imperative. In other cases, the experimenter may prefer or even require more candidates than provided by step 1(a). This would be the case, for example, when designing to support a higher-degree model (quadratic, cubic, etc.). In these cases, step 1(b) and possibly 1(c) would be implemented.

Steps 1(b) and 1(c) are often software dependent. MIXSOFT will calculate all of the centroids of any dimension from zero (the vertices) up to $(q - 1)$ (the overall centroid). In JMP and MINITAB, the user specifies the "degree" of the design. Choosing "1" gives the vertices (zero-dimensional centroids), "2" adds the edge centroids (one-dimensional centroids), ..., up to "q", which adds the overall ($(q - 1)$-dimensional) centroid. In MINITAB edge centroids are called "double blends", two-dimensional centroids are called "triple blends", etc. The user must keep in mind that if the constrained region is completely embedded within the simplex, all of the components may have nonzero proportions no matter what the description of the blend.

MINITAB has options for including the overall centroid and/or axial check blends no matter what the degree of the model. The composition of axial check blends are determined by averaging the composition of the overall centroid with each vertex, and thus the number of these is equal to the number of extreme vertices. Unwanted design points can always be stripped out of the worksheet.

Design-Expert offers a variety of candidate points, all or some of which can be included in the candidate list. In addition to the vertices, edge centroids, axial check blends, and overall centroid, the following points are offered as potential candidates:

- Thirds of edges.
- Triple blends. These are averages of three adjacent vertices, and thus are defined differently from triple blends in MINITAB.
- Constraint-plane centroids (CPCs). Design-Expert defines these as the $(q - 2)$-dimensional centroids (which, for $q > 4$, will be the centroids of polyhedrons or hyperpolyhedrons rather than "planes"). If the design region is simplex-shaped, then the CPCs are the same as the axial end points, and there will be q of these. Otherwise the number will depend on the shape of the design region.
- Interior blends. These lie midway between the overall centroid and the edge centroids and also midway between the overall centroid and the CPCs. The number of these is equal to the sum of the number of edge centroids and CPCs.

In the Design Study at the end of this chapter, Fig. 5.7, page 88, illustrates the experimental subspace for a four-component mixture experiment that is described in the literature. The reader may find it helpful at this point to read the first few paragraphs of the Design Study, where the origin of the different types of candidate points is explained in somewhat more detail.

5.2 Algorithm Overview

Table 5.2 is a summary of some of the algorithms that have evolved for computer-aided design of experiments. Motivation for developing these arose from the need to develop designs for irregularly shaped experimental regions. These can arise in nonformulation settings as well as in mixture settings. See Snee [157] for several interesting examples in both areas. The McLean–Anderson, XVERT, and CONSIM algorithms were developed specifically for mixture settings. Although the others were not, they nonetheless have found much use in the mixture setting. Those that are capitalized are acronyms: CADEX (computer-aided design of experiments), DETMAX (determinant maximization), XVERT (extreme vertices), and CONSIM (constrained simplex).

Table 5.2. *Point-generation and point-selection algorithms*

Year	Name	Author(s)	Reference
1966	McLean–Anderson	McLean and Anderson	[93]
1969	CADEX	Kennard and Stone	[78]
1971	Dykstra	Dykstra	[50]
1972	Wynn–Mitchell	Wynn; Mitchell and Miller	[99, 175]
1972	Fedorov	Fedorov	[51]
1974	DETMAX	Mitchell	[97]
1974	XVERT	Snee and Marquardt	[158]
1979	CONSIM	Snee	[156]

The **McLean–Anderson** (M-A) and **XVERT** algorithms are for mixture settings where there are *only single-component constraints.* The M-A algorithm produces a total of $q2^{q-1}$ *potential* vertices, many of which may fall outside the constraints and therefore are not vertices at all. A drawback to the procedure was that just to calculate the extreme vertices, the number of potential design points that had to be calculated became quite large when $q \geq 5$. McLean and Anderson also suggested that the centroids of some or all of the d-dimensional bounding faces be added to the list of design points, an exercise that is quite labor intensive without the assistance of a computer. Although mainframe computers were available when McLean and Anderson published their procedure, the convenience of the desktop PC was still several years away.

XVERT was developed as an improved version of the M-A algorithm. It is more efficient and more versatile because it can be used to generate all or only a fraction of the vertices. It is the algorithm that is implemented by most computer packages when only single-component constraints are specified. Because it forms the basis of methods for generating only a fraction of the extreme vertices, Section 5.3.1 is devoted to a description of this algorithm. An abbreviated version of the XVERT algorithm, called XVERT1, has been published [115].

The **CONSIM** algorithm was developed because of the need to handle situations where there are multicomponent constraints, and consequently it is the algorithm implemented by statistical computing packages for such situations. While the XVERT algorithm could be viewed as a "back-of-the-envelope" calculation (although for $q \geq 5$ large envelopes are needed), this is definitely not the case with the CONSIM algorithm. It is a rather complicated algorithm that must be run on a computer. Like the XVERT algorithm, it is important enough in formulation work that a short section (Section 5.3.2) will be devoted to it.

The remaining algorithms — CADEX, Dykstra, Wynn–Mitchell, Fedorov, and DETMAX — were originally developed for nonmixture settings, but they are applicable as well to mixture settings. **CADEX** is not a point-generation algorithm in the sense of the M-A and XVERT algorithms but is rather a distance-based algorithm for selecting design points from a candidate list. The aim of CADEX is to arrive at a set of design points that are spaced as uniformly as possible over the design region. Each of the N candidates can be viewed as a single point in $(q-1)$-dimensional space. In a mixture setting, the first candidate point selected is that point with the smallest Euclidean distance from a pure component. Each additional point is picked from among those remaining by choosing the one whose minimum Euclidean distance to other points already in the design is as large as possible. No model is assumed, so design optimality (discussed in Section 5.4.2) is not a criterion. Both Design-Expert and MINITAB support distance-based designs.

Starting with a candidate list, sequential methods for constructing designs are often attributed to Dykstra [50], although Wynn [174] had previously published a similar method. The **Dykstra** and **Wynn** methods were originally developed for the *augmentation* of existing experimental data. Starting with an existing experimental design, the algorithm would choose additional design points one at a time such that the volume of the confidence region for the regression coefficients (the **generalized variance**) was minimized. The method has been adapted for constructing designs from scratch, i.e., when there is only a candidate list and no starting design (see, for example, [3] and [97]). The method is sometimes used to construct the initial designs for the other methods (Wynn–Mitchell, Fedorov, and DETMAX). MINITAB and MIXSOFT have an option for creating designs using the sequential method.

The **Wynn–Mitchell** method starts with an initial design of the requested size, often constructed using the sequential search procedure. In each iteration, one point is added to the design, and then one point is dropped from the resulting design. The added point is that point that contributes the most to some chosen objective function, and the point that is dropped from the $(n+1)$-point design is that point that contributes the least. The algorithm stops when there is no further improvement in the design. Van Schalkwyk [165] described a similar algorithm carried out in reverse, i.e., a point is first dropped, and then a point is added; cf. Johnson and Nachtsheim [76] for discussion.

The **DETMAX** algorithm, also known as the **excursion** algorithm, is the best known and most widely used algorithm, being implemented in most DOE computing packages. First, an initial design is constructed, possibly randomly or perhaps by the sequential procedure. When single-point exchanges (the Wynn–Mitchell method) no longer improve the chosen objective function, then the algorithm undertakes *excursions*. The algorithm adds or subtracts more than one point at a time, so that during the search procedure the number of points in the design varies between $n+k$ and $n-k$, where n is the requested design size and k is the maximum allowable excursion.

The **Fedorov** algorithm is the most computer intensive and therefore the slowest. The method is based on *simultaneous switching*, adding and deleting points at the same time. At each iteration the algorithm evaluates all possible exchanges of pairs of points, one point from the design and one point from the set of candidate points. Thus for each iteration, $N_{design\ pts} \times N_{candidate\ pts}$ comparisons are made, and this is what makes the algorithm so slow. Cook and Nachtsheim [21] have published a modified procedure, known as the **modified Fedorov** exchange algorithm, that speeds up the process. MINITAB supports the Fedorov algorithm.

For a comparison of these and other algorithms for constructing optimal designs, see Atkinson and Donev [3] and Cook and Nachtsheim [21].

5.3 Creating a Candidate List

5.3.1 XVERT

In the 1970s, before the PC became popular, one would learn to do the XVERT algorithm by hand because, without access to a mainframe computer, one might very well *have* to do it by hand. Today there is no longer a reason to do it by hand because any statistical computing package that has mixtures capability will have the algorithm built in. Nonetheless, it is important to understand how the algorithm is implemented because it forms the basis for one method of generating a subset of the extreme vertices. This method is explained in Section 5.4.1.

XVERT will be illustrated with a five-component alloy example [123]. The constraints on the component proportions are

$$0.03 \leq \text{Cu} \leq 0.10$$
$$0.00 \leq \text{Al} \leq 0.15$$
$$0.00 \leq \text{Cr} \leq 0.15$$
$$0.10 \leq \text{Fe} \leq 0.30$$
$$0.35 \leq \text{Ni} \leq 0.65$$

Crosier's formula [37] calculates 22 vertices, so we shall use XVERT to find their composition. Bear in mind that this can be done today simply by entering the constraints into statistical computing packages that have mixtures capabilities.

The steps in the procedure are as follows:

1. Rank the components in order of increasing range ($U_i - L_i$), checking that the constraints are consistent. This has already been done in the list of constraints, the ranges being in the order Cu $<$ Al $=$ Cr $<$ Fe $<$ Ni. The column headings in Table 5.3 are arranged from left to right in this order. The reason for this is that the component with the largest range (Ni in this case) will eventually be used to take up the "slack".

2. Form a classical 2^{q-1} factorial design from the $q - 1$ components with the smallest ranges. This is done in the top panel, columns 2–5 (Cu–Fe) in the table. The proportion of the qth component (Ni in this example) is calculated by subtraction, thus taking up the "slack". The points generated in this manner are referred to as **base points**. A point is an extreme vertex if the proportion of the qth component falls within its bounds. Therefore, if $0.35 \leq \text{Ni} \leq 0.65$, then the point is an extreme vertex. Checked points are those that pass the test and are called **core points**. However, we have only found 10 of the 22 vertices.

3. For those points for which X_q is outside its limits (points 1–3, 5, 9, and 16 in this example), adjust X_q to U_q or L_q, whichever is closest. In this example, points 1–3, 5, and 9 have Ni levels > 0.65, and so the Ni level for each of these points should be adjusted *downward* to 0.65. Point 16 has a Ni level < 0.35, and so the Ni level for this point should be adjusted *upward* to 0.35.

4. For each adjusted observation (row), generate additional points by adjusting the level of one component at a time so that $\sum_i^q X_i = 1.0$. This will generate $q - 1$ additional points.

The procedure for steps 3 and 4 is illustrated for base points 2 and 16 in the lower two panels of Table 5.3. Points 2a–2d have had the Ni level adjusted *downward* from 0.80 to 0.65. The levels of Fe, Cr, Al, and Cu in 2a–2d, respectively, have each been adjusted *upward* by 0.15 to maintain the equality constraint. This procedure has identified three more vertices, but two of these are duplicates. Point 2d violates the upper constraint for Cu and is of no further interest. Snee and Marquardt [158] refer to this group as a **candidate subgroup**.

Point 16 has a Ni level of 0.30 which is < 0.35, and so its level must be adjusted *upward*. Adjustments in this subgroup are displayed in the bottom panel of Table 5.3.

Vertices and duplicate vertices in all six candidate subgroups are summarized in the tabular summary below. Subgroup IDs that are not in the table (such as 2d) are not vertices because of constraint violations. In summary, then, XVERT identifies 10 vertices in the core group and 12 vertices in the candidate subgroups, for a total of 22.

Subgroup	Vertices	Duplicates
1	none	none
2	2a	2b(6), 2c(4)
3	3a, 3b	3d(4)
5	5a, 5c	5d(6)
9	9b, 9c, 9d	none
16	16a, 16b, 16c, 16d	none

Table 5.3. *Alloy example. XVERT design*

ID	Cu	Al	Cr	Fe	Ni	
1	.03	.00	.00	.10	.87	
2	.10	.00	.00	.10	.80	
3	.03	.15	.00	.10	.72	
4	.10	.15	.00	.10	.65	√
5	.03	.00	.15	.10	.72	
6	.10	.00	.15	.10	.65	√
7	.03	.15	.15	.10	.57	√
8	.10	.15	.15	.10	.50	√
9	.03	.00	.00	.30	.67	
10	.10	.00	.00	.30	.60	√
11	.03	.15	.00	.30	.52	√
12	.10	.15	.00	.30	.45	√
13	.03	.00	.15	.30	.52	√
14	.10	.00	.15	.30	.45	√
15	.03	.15	.15	.30	.37	√
16	.10	.15	.15	.30	.30	
2	.10	.00	.00	.10	.80	
2a	.10	.00	.00	<u>.25</u>	.65	√
2b	.10	.00	<u>.15</u>	.10	.65	duplicate of 6
2c	.10	<u>.15</u>	.00	.10	.65	duplicate Of 4
2d	<u>.25</u>	.00	.00	.10	.65	
16	.10	.15	.15	.30	.30	
16a	.10	.15	.15	<u>.25</u>	.35	√
16b	.10	.15	<u>.10</u>	.30	.35	√
16c	.10	<u>.10</u>	.15	.30	.35	√
16d	<u>.05</u>	.15	.15	.30	.35	√

Once the composition of the vertices is determined, it is a matter of averaging to obtain the AEV centroids, overall centroid, interior blends, and so forth. To calculate d-dimensional AEV centroids, one must first identify the vertices that define the various d-dimensional faces. The minimum number of points required to define a d-dimensional face is $d+1$. For example, an edge requires two points, a two-dimensional face requires at least three points, a three-dimensional "face" requires at least four points, etc. To identify vertices that define a d-dimensional face requires identifying those for which each of $q-d-1$ proportions are constant. Once identified, the compositions are averaged to get the composition of the AEV centroid.

Again, an example based on the vertices for the alloy formulation should help to clarify the procedure.

- To identify an edge, one needs to find pairs of extreme vertices for which $q-d-1=3$ proportions are constant. Vertices 2a and 10 are the only two among the set of 22 for which Cu $= 0.10$, Al $= 0.00$, and Cr $= 0.00$. Although the compositions of vertices in subgroups 3, 5, and 9 are not reproduced here, there are no vertices in these subgroups that have this particular combination.

- To identify a two-dimensional face requires finding sets of vertices of size three or greater that have $q - d - 1 = 2$ proportions in common. One such set is composed of vertices 4, 8, 12, 16a, and 16b, for which Cu = 0.10 and Al = 0.15. There are no vertices in subgroups 3, 5, or 9 with this particular combination.

- To identify a three-dimensional face requires finding sets of vertices of size four or greater that have $q - d - 1 = 1$ proportion in common. One such set is composed of vertices 4, 6, 8, 10, 12, 14, 2a, 16a, 16b, and 16c, for which Cu = 0.1. Again, there are no vertices in subgroups 3, 5, or 9 (other than duplicates) that have Cu = 0.1.

In this particular example, there are 46 edge centroids, 34 two-dimensional centroids, and 10 three-dimensional centroids. More examples of calculating centroids are given in the Design Study at the end of this chapter.

5.3.2 CONSIM

As indicated earlier, the CONSIM algorithm was developed to handle situations where there are multicomponent constraints. One might wonder how important such constraints really are, and so three examples will be given where multicomponent constraints were useful or where they can be useful.

- Situations where two or more components play a similar role can often benefit from a multicomponent constraint. For example, in formulating a color photographic dispersion for color paper, one prepares a dispersion consisting of an oil phase dispersed in an aqueous gelatin matrix. Commonly the oil phase is composed of three primary classes of compounds: (i) a coupler, which is the precursor to the dye; (ii) an oily coupler solvent such as tricresyl phosphate; and (iii) stabilizers to stabilize the dye against heat and light. A typical set of constraints might be

$$\begin{aligned} 0.3 \leq &\ \text{Coupler} \leq 0.70 \\ 0.0 \leq &\ \text{Solvent A} \leq 0.35 \\ 0.0 \leq &\ \text{Solvent B} \leq 0.35 \\ 0.0 \leq &\ \text{Stabilizer A} \leq 0.35 \\ 0.0 \leq &\ \text{Stabilizer B} \leq 0.35 \end{aligned}$$

The problem with this set of constraints is that either the solvents or the stabilizers could go to zero, which would not be satisfactory. It is necessary that at least *some* solvent and *some* stabilizer be present in the dispersion. This can be taken care of by multicomponent constraints such as the following:

$$\begin{aligned} 0.15 \leq &\ \text{Solvent A + Solvent B} \leq 0.35 \\ 0.15 \leq &\ \text{Stabilizer A + Stabilizer B} \leq 0.35 \end{aligned}$$

The multicomponent constraints add the requirement that although solvent A or B or stabilizer A or B may go to zero, a certain minimum amount of solvent and stabilizer must be present. Also, although either solvent or either stabilizer may be present up to a level of 0.35, the total amount of solvent and the total amount of stabilizer cannot exceed 0.35.

- In some areas of research and manufacturing it is common practice to express the relative amounts of components in terms of ratios. Such is the case with color photographic dispersions, where the relative amounts of coupler and coupler solvent are traditionally expressed as the "coupler to coupler-solvent ratio". The ratio is usually fixed, but if one wanted to incorporate the ratio as a factor in a mixture experiment, it would be necessary to reexpress the range of ratios as multicomponent constraints. Many software packages can design in terms of both ratios and component proportions, but the two metrics cannot be mixed.

  Continuing with the previous example, constraints on the coupler to coupler-solvent ratio would be expressed as

$$L \leq \frac{\text{Coupler}}{\text{Solvent A} + \text{Solvent B}} \leq U$$

  where L and U are the upper and lower bounds on the ratios. These constraints can be reexpressed as the multicomponent constraints

$$\begin{aligned} L(\text{Solvent A}) + L(\text{Solvent B}) - \text{Coupler} &\leq 0 \\ 0 &\leq U(\text{Solvent A}) + U(\text{Solvent B}) - \text{Coupler} \end{aligned}$$

  For example, if one wanted to investigate coupler to coupler-solvent ratios in the range $L = 1$ to $U = 4$, then the following multicomponent constraints would be specified:

$$\begin{aligned} (\text{Solvent A}) + (\text{Solvent B}) - \text{Coupler} &\leq 0 \\ 0 &\leq 4(\text{Solvent A}) + 4(\text{Solvent B}) - \text{Coupler} \end{aligned}$$

  Combining these two constraints with the single and multiple constraints in the previous example, the CONSIM algorithm would find a design space with 24 vertices, 48 edges, 34 two-dimensional constraint planes, and 10 three-dimensional constraint polyhedrons.

- Cases in which two or more components contain an active ingredient can also benefit from a multicomponent constraint. As an example, iron ore sinter is a type of material that is used in blast furnaces to produce molten iron. It is a combination of several materials that are fused together through a high-temperature process called sintering. Koons [82] describes an experiment designed to identify the source of an environmental problem called "blue haze", which occurred in the vicinity of a sinter plant's off-gas stacks. This was an eight-component experiment consisting of the following components:

X_i	Class: Name	X_i	Class: Name
A	Earthy hematite	E	Mill scale
B	Specular hematite	F	Dolomite
C	Flue dust	G	Limestone
D	BOF slag	H	Coke

It was important that the total iron be maintained at or above a proportion of 0.46. It was known by analysis that components A–E contained, respectively, 60%, 60%, 35%, 20%, and 70% iron. Consequently the following multicomponent constraint was specified in the design of the experiment:

$$0.46 \leq 0.60A + 0.60B + 0.35C + 0.20D + 0.70E$$

In addition, it was important to maintain the total carbon level between 0.043 and 0.085. Components C and H contain 17% and 85% carbon, respectively. Consequently the following additional multicomponent constraint was included:

$$0.043 \leq 0.17C + 0.85H \leq 0.085$$

The mathematical details of the CONSIM algorithm are beyond the intended scope of this book, but the procedure can be described as follows:

1. Introduce a single constraint.
2. Determine which vertices lie inside and which lie outside the constraint.
3. Form new points (vertices) by taking a suitable linear combination of each vertex outside the constraint with a vertex that is inside the constraint and that lies on the same one-dimensional edge. The new point (vertex) will be an intersection of the constraint plane and the one-dimensional edge.
4. After all the new vertices have been generated, delete those outside the constraint.
5. Return to step 1 and introduce the next constraint.

Whenever there are multicomponent constraints the CONSIM algorithm is implemented by most statistical software packages with mixtures capabilities. Most, if not all of these programs, derive from a FORTRAN program called CONVRT published by Piepel [120]. A companion program, called CONAEV, was included in the same article and is used to calculate AEV centroids of any requested dimension(s) (cf. page 63). Improved versions of these two programs (MCCVRT and AEVC, respectively) have been incorporated into MIXSOFT.

5.4 Choosing Design Points

The previous section was devoted to methods for creating a candidate list, and so now we consider methods for selecting points from such a list to achieve a "good" design of reasonable size that will support the intended model. It is not always necessary to include in the candidate list all of the vertices, the d-dimensional centroids ($1 \leq d \leq q-1$), axial check blends, interior blends, and so forth. Depending on the number of mixture components and the particular constraints on the component proportions, the number of points one could consider for inclusion in a candidate list can become quite large.

To illustrate this concept, consider the eight-component iron ore sinter example (page 71). The experimenters specified 17 constraints, of which 11 were single-component and six were multiple-component. Based on these constraints, Design-Expert calculates 4425 points for possible inclusion in a candidate list. Depending on the degree of the model one cares to support, Design-Expert may recommend a subset of these points for inclusion in a candidate list. Table 5.4 illustrates the default candidate lists for linear, quadratic, special cubic, and cubic models. See page 65 for a description of Design-Expert's design points.

Table 5.4. *Iron ore sinter example. Default candidate lists in Design-Expert*

	Model†			
Point type	L	Q	SC	C
Vertices	184	184	184	184
Edge centroids		644	644	644
Thirds of edges				1288
Triple blends			1448	1448
Constraint plane centroids		16	16	16
Axial check blends	184	184	184	184
Interior blends		660	660	660
Overall centroid	1	1	1	1
TOTAL	369	1689	3137	4425

† L = linear, Q = quadratic, SC = special cubic, C = cubic

As the degree of the model increases, the number of points in a candidate list will increase. For $q = 8$, the linear, quadratic, special cubic, and cubic models have 8, 36, 92, and 120 terms, respectively. In this particular example there are enough unique points in the 369-point linear candidate list to support the 120-term cubic model. All 369 points in this list have four or more components with nonzero proportions, and 190 of the 396 are complete mixtures (i.e., have eight components with nonzero component proportions). It would appear as though a candidate list composed of only the 184 vertices should be sufficient from which to choose 120 points to support the cubic model. However, one gets into the same problem as described for design B in Section 4.4, page 55, i.e., linear dependencies among the columns of the $\mathbf{X}$ matrix.

Methods used to reduce candidate lists can be grouped into two broad categories. The first is applicable to screening designs, where the focus is on fitting a first-degree model. It is based on the XVERT algorithm and is described in Section 5.4.1. The second method is based on **design optimality** and is used for both screening and response-surface designs. This method is described in Section 5.4.2.

5.4.1 Designs Based on Classical Two-Level Screening Designs

If one is interested in a screening design to support a first-degree model, it may very well be sufficient to limit the candidate list to the extreme vertices. To illustrate, if $q = 8$, we need eight points minimum plus perhaps an additional five for lack of fit (ignoring replication for the moment), for a total of 13 runs. A candidate list of 184 vertices (as in the iron ore sinter example, Table 5.4) would be adequate. By default, Design-Expert always includes the axial check blends plus the overall centroid in a candidate list, providing more than enough points from which to choose a subset to support a first-degree model.

One approach to choosing a subset would be to begin the XVERT algorithm with a smaller list of base points. In an eight-component experiment, for example, rather than start the algorithm with $2^7 = 128$ base points, one could instead begin with $2^{7-3} = 16$ points. Generalizing, the number of design points in the final design could be reduced by starting the algorithm with a fractional factorial design or $q - 1$ columns of a Plackett–Burman design [136]. This approach was suggested by Snee and Marquardt in their article describing the XVERT algorithm [158]. A slightly modified version of this approach (explained below) is implemented in MIXSOFT. This is the *only* software package that supports the procedure(s) described in this section.

If one elects to start the algorithm with a $2^{(q-1)-1}$ fractional factorial design (for example), one has some options. (i) The half-fraction might be chosen in a systematic way (yet to be defined), or the half-fraction might be chosen randomly. (ii) In the likely event that there are candidate subgroups (page 68), one could select one or all possible vertices from the subgroups. There are no rules for these choices — they are left to the discretion of the experimenter.

To illustrate use of the five-component alloy example, let us code the low and the high levels of Cu, Al, Cr, and Fe in the table on page 69 as −1 and +1, respectively. In addition, let us add a column labeled "Cu × Al × Cr × Fe" which gives the product of the coded levels of Cu, Al, Cr, and Fe. This results in the following table:

ID	Cu	Al	Cr	Fe	Cu × Al × Cr × Fe
1	−1	−1	−1	−1	+1
2	+1	−1	−1	−1	−1
3	−1	+1	−1	−1	−1
4	+1	+1	−1	−1	+1
5	−1	−1	+1	−1	−1
6	+1	−1	+1	−1	+1
7	−1	+1	+1	−1	+1
8	+1	+1	+1	−1	−1
9	−1	−1	−1	+1	−1
10	+1	−1	−1	+1	+1
11	−1	+1	−1	+1	+1
12	+1	+1	−1	+1	−1
13	−1	−1	+1	+1	+1
14	+1	−1	+1	+1	−1
15	−1	+1	+1	+1	−1
16	+1	+1	+1	+1	+1

The last column contains eight plus signs and eight minus signs. On this basis, a systematic approach to choosing a $2^{(5-1)-1} = 2^3$ design can be implemented by dividing the 16 possible treatment combinations into two half-fractions of 8, one set for which the product is equal to +1 (IDs 1, 4, 6, 7, 10, 11, 13, and 16) and the other set for which the product is equal to −1 (IDs 2, 3, 5, 8, 9, 12, 14, and 15). One would then select one or the other half-fraction. In a random approach, one would disregard the signs and randomly select 8 of the 16 treatment combinations.

Depending on the choices made for (i) and (ii), the number of final design points will vary. For the alloy case, for example, the range of possible design points that could be found from the chosen set of base points is as follows:

	(ii) Vertices per subgroup	
(i) Choice of half-fraction	One	All possible
Systematic	7–8	10–14
Random	7–8	7–17

By way of example, the 17 in the Random/All Possible case could come about if, by chance, the following set of base points were selected:

		Candidate subgroup			
Base pt.	Core pt.	a	b	c	d
2		√	√	√	
3		√	√		$(= 2c)$
5		√		√	$(= 2b)$
7	√				
8	√				
9			√	√	√
10	√				
16		√	√	√	√

Adding the number of check marks gives 17. The 10 in the Systematic/All Possible case arises from the fraction for which Cu × Al × Cr × Fe = +1, while the 14 arises from the fraction for which Cu × Al × Cr × Fe = −1. The reader can verify the number of design points in these half-fractions by constructing charts such as the one above.

The value 7 in the table comes about because base point 1 is not a core point (vertex), and there are no vertices in #1's candidate subgroup. To ensure that a fractional factorial will lead to the expected number of design points (i.e., 8, 16, 32, ...) whenever one vertex per base point is to be generated, a modified version of XVERT has been incorporated into MIXSOFT [121, 122]. The modification is illustrated by taking base point 1.

ID	Cu	Al	Cr	Fe	Ni	
1	.03	.00	.00	.10	.87	
1a′	.03	.00	<u>.02</u>	<u>.30</u>	.65	√

Here, an adjustment of +0.22 cannot be made to Fe, as this will result in a proportion (0.32) that exceeds the upper bound for Fe. Instead, as much of the adjustment as possible (+0.20) is made to Fe, and the remainder (+0.02) is made to Cr. If the adjustment to Cr had been too large, then as much as possible would be made to Cr and, continuing to work from right to left, the remainder to Al. As a result, at most only one component will not be at its upper or lower bound, and the resulting point will be an extreme vertex. Vertex 1a′ is in fact a duplicate of vertex 9b.

A modification of this procedure for generating one or all possible vertices from a fractional design plan has been developed by Piepel [121, 122] and is implemented in

MIXSOFT. The resulting designs are called **Modified XVERT mixture screening designs**, or MXMSDs. The reader is encouraged to read the original articles before using these designs.

5.4.2 *D*-Optimality Criterion

In this section we focus on the relationship between an experimental design and the confidence region for the coefficient estimates in a fitted model. This is not to imply that prediction variances are not equally important. The emphasis on precision in the coefficient estimates arises because most commonly available software defaults to the D-optimality criterion which is parameter-based rather than prediction-based. Software products that support prediction-based criteria are ACED (Algorithms for the Construction of Experimental Designs) [168, 169, 170] and Gosset [62]. Discussion of methods for minimizing prediction variances using commonly available software will be deferred to the design evaluation stage (Chapter 6).

In addition, choosing replicates for the purpose of estimating pure experimental error is an important part of the design process. Design approaches such as D-optimality may or may not select replicates. In cases where the software or the particular approach does not select replicates, they must be selected and added in a separate step. Choosing which points to replicate is related to minimizing prediction variances, and so this too will be relegated to Chapter 6. Designs in the remainder of this chapter will focus on points to support the model plus lack-of-fit points.

A two-component mixture experiment will be used as an example because the relationships can be illustrated graphically. With more than two components, graphical methods must be replaced by analytical methods, but the concepts remain the same.

Assume that our aim is to design an experiment that will support the $q = 2$ linear Scheffé model

$$E(Y) = \beta_1 X_1 + \beta_2 X_2$$

The fitted model will be

$$\hat{Y} = b_1 X_1 + b_2 X_2$$

where b_1 and b_2 are sample-based estimates of β_1 and β_2. In all cases to be discussed in this book, models will be fit using OLS, and so the estimates b_1 and b_2 of β_1 and β_2 are least-squares estimates.

Table 5.5 displays three possible six-point designs. Keep in mind that these are designs, not candidate lists. Design A is composed of two replicates, each of the axial check blends and the overall centroid. In design B, the axial check blends have been moved to the boundary of the simplex, where they become vertices. In design C, one replicate of the overall centroid in B has been moved to one boundary (vertex), and the other replicate has been moved to the other boundary.

Fig. 5.3 displays **joint confidence regions** for b_1 and b_2 on the assumption that $b_1 = 10$, $b_2 = 20$, and $s = 0.25$. The largest ellipse in Fig. 5.3 is a 95% joint confidence region for b_1 and b_2 based on design A. The middle-sized ellipse is the corresponding region based on design B, while the smallest ellipse is for design C. Thus, as design points are moved to the boundaries of the design region, the joint confidence region gets smaller and smaller, and estimates of β_1 and β_2 have more and more precision. For designs where

Table 5.5. *Designs A, B, and C. Three two-component, six-point designs*

Design					
A		B		C	
0.75	0.25	1.0	0.0	1.0	0.0
0.75	0.25	1.0	0.0	1.0	0.0
0.50	0.50	0.5	0.5	1.0	0.0
0.50	0.50	0.5	0.5	0.0	1.0
0.25	0.75	0.0	1.0	0.0	1.0
0.25	0.75	0.0	1.0	0.0	1.0

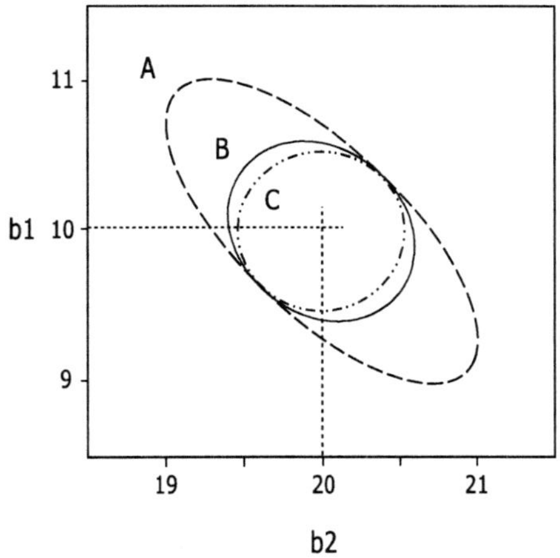

Figure 5.3. *Designs A, B , and C. Joint 95% confidence regions for b_1 and b_2.*

more than two coefficients are estimated, the ellipses of Fig. 5.3 would be replaced by ellipsoids. The method for calculating confidence ellipsoids is beyond the scope of this book but can be found in most books on regression analysis. (See, for example, Draper and Smith [49], Montgomery, Peck, and Vining [100], Myers [105], or Rawlings, Pantula, and Dickey [143].)

A design that minimizes the area of the confidence ellipse (or volume of a confidence ellipsoid when $p > 2$) is said to be ***D*-optimal**. *D*-optimality is one of many optimality criteria that were alluded to in Section 5.2. While *D*-optimality is undoubtedly the most popular optimality criterion, there are many others. A few of these are discussed in Section 5.4.3 and in Chapter 6.

Most practical formulation problems involve more than two mixture components, and therefore an analytical measure of the volume of the confidence ellipsoid is needed. Design-Expert, for example, has a section titled **Measures Derived from the $\mathbf{X'X}^{-1}$ Matrix**. It is in this section that information is found about various optimality criteria, including *D*-optimality. It turns out that the $\mathbf{X'X}$ matrix, which is also known as the **information matrix**, plays a *central role in ordinary least squares*, and if one is to benefit from the information provided by the various DOE software packages, then one needs to understand some of the properties of the $\mathbf{X}'\mathbf{X}$ matrix.

Design A in Table 5.5 is put in the form of a **design matrix**:

$$\begin{bmatrix} 0.75 & 0.25 \\ 0.75 & 0.25 \\ 0.50 & 0.50 \\ 0.50 & 0.50 \\ 0.25 & 0.75 \\ 0.25 & 0.75 \end{bmatrix}$$

This matrix is called a design matrix for obvious reasons—each row of the matrix represents an observation, while the columns give the component proportions. Because we are focusing on fitting a linear model, the design matrix is also the **model matrix**, commonly referred to as the **X matrix**. Had we been designing to support a quadratic model, then the design matrix would be augmented with one additional column corresponding to the X_1X_2 products, and the **X** matrix would be a three-column matrix. For the cubic model, the **X** matrix would be a four-column matrix.

The volume of the confidence region for the regression coefficients (the **generalized variance**) is *inversely proportional to the square root of the determinant of* $\mathbf{X}'\mathbf{X}$ [3, 12, 141]. Thus one needs to first calculate $\mathbf{X}'\mathbf{X}$ and then calculate its determinant. The **X** matrix is of dimension $n \times p$, where n is equal to the number of observations and p is equal to the number of coefficients in the model. The transposed **X** matrix ($\mathbf{X}'$) is therefore of dimension $p \times n$. As a result, the $\mathbf{X}'\mathbf{X}$ is always a square matrix of dimension $p \times p$.

For design A, then,

$$\mathbf{X}'\mathbf{X} = \begin{bmatrix} 0.75 & 0.75 & 0.50 & 0.50 & 0.25 & 0.25 \\ 0.25 & 0.25 & 0.50 & 0.50 & 0.75 & 0.75 \end{bmatrix} \begin{bmatrix} 0.75 & 0.25 \\ 0.75 & 0.25 \\ 0.50 & 0.50 \\ 0.50 & 0.50 \\ 0.25 & 0.75 \\ 0.25 & 0.75 \end{bmatrix}$$

$$= \begin{bmatrix} 1.75 & 1.25 \\ 1.25 & 1.75 \end{bmatrix}$$

A determinant is defined only for square matrices and is a scalar quantity. For a 2×2 matrix it is calculated as follows:

$$|\mathbf{A}| = \begin{vmatrix} a & b \\ c & d \end{vmatrix} = ad - bc \tag{5.1}$$

Thus for design A we have

$$|\mathbf{X}'\mathbf{X}| = \begin{vmatrix} 1.75 & 1.25 \\ 1.25 & 1.75 \end{vmatrix} = (1.75)^2 - (1.25)^2 = 1.5$$

Table 5.6 summarizes the $\mathbf{X}'\mathbf{X}$ matrices and their determinants for designs A, B, and C. Based on the determinants, the relative areas of ellipses A, B, and C in Fig. 5.3 are $1/\sqrt{1.5} : 1/\sqrt{6} : 1/\sqrt{9} \equiv 1.0 : 0.50 : 0.41$, respectively.

Table 5.6. *Designs A, B, and C.* $\mathbf{X'X}$ *and* $|\mathbf{X'X}|$ *matrices*

Design	$\mathbf{X'X}$	$\lvert\mathbf{X'X}\rvert$
A	$\begin{bmatrix} 1.75 & 1.25 \\ 1.25 & 1.75 \end{bmatrix}$	1.5
B	$\begin{bmatrix} 2.5 & 0.5 \\ 0.5 & 2.5 \end{bmatrix}$	6.0
C	$\begin{bmatrix} 3 & 0 \\ 0 & 3 \end{bmatrix}$	9.0

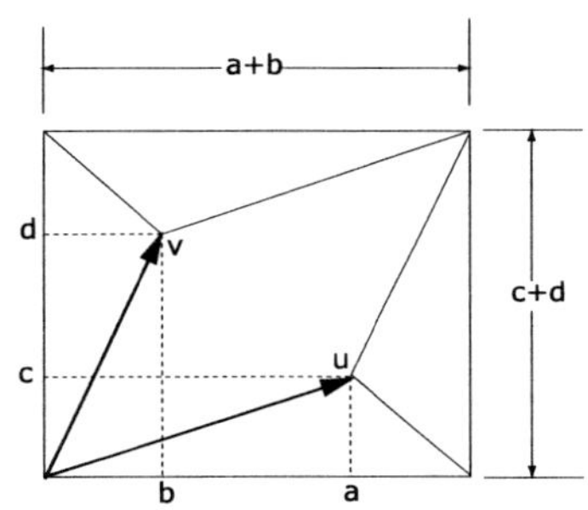

Figure 5.4. *Geometry of a determinant.*

As determinants play an important role in design optimality, it is worth devoting a few paragraphs to the geometric interpretation of a determinant. Equation 5.1 can be written $|A| = |\mathbf{u}\ \mathbf{v}|$, where $\mathbf{u} = [\,a\ \ c\,]'$ and $\mathbf{v} = [\,b\ \ d\,]'$. The parallelogram in Fig. 5.4 is said to be *spanned* by the vectors **u** and **v**. The determinant of $\mathbf{A} = |\mathbf{u}\ \mathbf{v}|$ is equal to the area of the parallelogram. This can be obtained by subtracting the area of the four triangles from the area of the rectangle. The areas of the top and bottom triangles are each equal to $\frac{1}{2}(a+b)c$, while the areas of the left and right triangles are each equal to $\frac{1}{2}(c+d)b$. The area of the parallelogram is therefore

$$(a+b)(c+d) - (a+b)c - (c+d)b = ad - bc$$

The determinant of a 3×3 matrix would be the volume of a parallelepiped spanned by three column vectors, **u**, **v**, and **w**. Illustrating the geometry would require a complicated three-dimensional picture, and so we will not attempt this. The concepts, however, are the same.

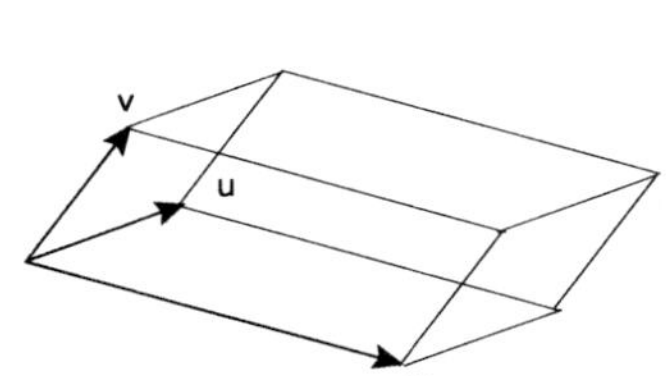

Referring back to Table 5.6, the increasing magnitude of $|\mathbf{X}'\mathbf{X}|$ as one goes from design A to B to C means that the areas of the parallelograms spanned by the columns of the respective $\mathbf{X}'\mathbf{X}$ matrices are increasing. Because the generalized variance is inversely proportion to $\sqrt{|\mathbf{X}'\mathbf{X}|}$, the generalized variances decrease as the areas of the parallelograms increase.

Depending on the software that one is using, the D-optimality criterion is reported in different ways. MINITAB and MIXSOFT, for example, simply report the determinant of $\mathbf{X}'\mathbf{X}$. JMP does not support two-component mixture experiments, but for $q \geq 3$ JMP reports what it refers to as the D efficiency [98], defined as

$$D \text{ efficiency} = 100 \cdot \frac{|\mathbf{X}'\mathbf{X}|^{1/p}}{n}$$

where n is the number of observations and p is the number of coefficients. Design-Expert reports the inverse of this multiplied by 100 and calls it the **scaled D-optimality criterion**. The D efficiency as reported by JMP (larger is better) and the scaled D-optimality criterion as reported by Design-Expert (smaller is better) permit comparison of designs of different size. For example, if there were no replicates in design A (Table 5.5, page 77) so that $n = 3$, then $|\mathbf{X}'\mathbf{X}| = 0.375$, and, relative to ellipse A in Fig. 5.3, the area of the confidence ellipsoid would be doubled.[2] For design A,

$$D \text{ efficiency} = 100 \cdot \frac{(1.5)^{1/2}}{6} = 20.41$$

while for the three-point design

$$D \text{ efficiency} = 100 \cdot \frac{(0.375)^{1/2}}{3} = 20.41$$

Considering the relative areas of the confidence ellipses for the two designs, parameter estimates based on the six-point design (A) will be more precise than estimates based on the three-point design. On a *per-point* basis, however, the six-point design is no more efficient than the three-point design. In effect, one pays a penalty for doing more experiments.

To implement a search for an optimal design in a software product, one must do the following:

1. Provide a candidate list. This is usually created within the software, but with some products (such as JMP, MINITAB, and MIXSOFT) a list may be created outside the software and then imported.

2. Specify a model. This is one of the criticisms of design optimality. In many if not most situations, the form of the model is not known a priori. A design that is optimal for a screening design will not be optimal for a response-surface design, even if there is an adequate number of design points.

3. Specify the desired number of design points.

4. Specify the optimality criterion. In most cases this will be the D-optimality criterion because this is what most popular computing packages default to.

[2]The relative areas of the confidence ellipses of the three-point design vs the six-point design would be $1/\sqrt{0.375} : 1/\sqrt{1.5} \equiv 2 : 1$.

Algorithms such as DETMAX that search a candidate list for designs based on D-optimality will tend to find designs that have most of the design points located at the boundaries of the design region, since these are the points that maximize $|\mathbf{X}'\mathbf{X}|$. Consider, for example, a simple case where the five-point candidate list in Table 5.7 is supplied to software, and a six-point design to support a linear Scheffé model is requested.

Table 5.7. *Six-point designs based on the D-criterion*

Candidates		Number of points selected by		
X_1	X_2	MINITAB	MIXSOFT	Design-Expert
1.00	0.00	3	3	2
0.75	0.25	0	0	1
0.50	0.50	0	0	1
0.25	0.75	0	0	1
0.00	1.00	3	3	1
$\|\mathbf{X}'\mathbf{X}\|$		9	9	5

MINITAB and MIXSOFT choose designs based strictly on the D-optimality criterion, and so by placing half of the design points at each of the two boundaries, $|\mathbf{X}'\mathbf{X}|$ is maximized. Design-Expert, on the other hand, has certain options. The user can divide the desired number of design points into points to support the model (called *model points*), lack-of-fit points, and replicates. Model points are picked first based on the D-optimality criterion. Lack-of-fit points are selected next using a distance-based algorithm, which tends to fill in the "holes" in the design. Finally, replicates are picked using the D-optimality criterion. Defaults for the number of lack-of-fit and replicate points are $q + 1$ to a maximum of five, but this can be changed.

Requesting six model points in Design-Expert (in spite of the fact that the $q = 2$ linear Scheffé model is only a two-term model) leads to the same design as that from MINITAB and MIXSOFT. All points are picked based on the D-optimality criterion, and so half of the points are placed at each of the two boundaries. Requesting two model points, three lack-of-fit points, and one replicate leads to the design in Table 5.7. In this case Design-Expert presumably first picks the two vertices using the D-optimality criterion, then picks three interior points using a distance-based criterion, and finally, for the replicate, it chooses a boundary (vertex) point using the D-optimality criterion again. Had the center point or an axial point been replicated instead, then $|\mathbf{X}'\mathbf{X}|$ would have equaled 3.75 or 4.06, respectively, both of which are less than 5.

Because the D-optimality criterion tends to pick boundary points, software often (but not always) will not pick center points for a design. The absence of a center point or points can lead to two potential problems: (1) Prediction *variances* in the interior of the design may be inflated, leading to imprecision in prediction. (2) If the model is underspecified (that is, the true underlying surface is of higher order than the assumed model), then inaccurate estimates (*biases*) in predicted values, coefficients, and s^2 may result. Designs that take potential bias into account tend not to have design points at the extremes of a design region [12, 107]. Including center points as well as placing some design points in the interior of a design can help to protect the fitted model from these potential problems. Piepel, Anderson, and Redgate [126] recommend forcing a center point into variance-optimal designs if the software package does not include one.

Software-to-software differences in the search procedure for "optimal designs" do exist. In addition, repeated "tries" within a product may lead to somewhat different designs. Table 5.8 compares four 20-point designs based on the alloy constraints. The points were chosen using the D-criterion to support the 15-term quadratic Scheffé model. The candidate list consisted of 22 vertices, 46 edge centroids, 10 constraint-plane centroids, and the overall centroid — 79 points total. Although there are also 34 two-dimensional centroids, for the purpose of software comparison these were not included in the candidate list because Design-Expert does not include these. Both the D-efficiency and the scaled D-optimality criterion (page 80) depend on $|X'X|^{1/p}$, where p is the number of parameters in the model. As can be seen from the last row of the table, either of these criteria would be nearly the same for all four designs.

Table 5.8. *Alloy example. D-optimal designs*

Point type†	Design point types from			
(dimension)	MINITAB	MIXSOFT	Design-Expert	JMP
V (0)	15	14	15	15
EC (1)	4	5	4	0
CPC (3)	1	1	1	0
OC (4)	0	0	0	0
Other	0	0	0	5
$10^{46} \cdot \lvert\mathbf{X'X}\rvert$	956	1017	957	1274
$10^{6} \cdot \lvert\mathbf{X'X}\rvert^{1/15}$	1355	1361	1355	1382

†V = vertex; EC = edge centroid; CPC = constraint-plane centroid; OC = overall centroid

Beginning with JMP Release 4 and Design-Expert Version 7, the default method for custom designs uses a candidate-set-free approach, although the criterion remains D-optimality. The method used is a **coordinate-exchange algorithm** [95] (as opposed to a point-exchange algorithm) modified for use in a mixture setting [127].[3] Although a candidate set is not used, the algorithm will often find vertices and possibly face centroids because of the search procedure that is used. The coordinate-exchange algorithm is a two-step process. The first step involves the creation of a nonoptimal starting design with the desired number of design points. The second step is the coordinate-exchange stage, in which the starting design is optimized.

In the first step, random points are generated within the whole mixture simplex. If a point falls within the constrained region, then it is a feasible point and is added to the starting design. If it does not fall within the constrained region, then it is projected onto the nearest constraint boundary. If all constraints are satisfied, then it is added to the starting

[3]Point exchange is still available by a radio button in Design-Expert Version 7.

design; otherwise it is projected once again onto the next nearest constraint boundary. The process of projection continues until all constraints are satisfied.

Fig. 5.5 shows a 3-simplex with a hexagon-shaped constrained region. The black dot within the hexagon is a hypothetical starting design point. Lines drawn through this point to each of the three vertices are known as **Cox-effect directions** or sometimes simply Cox directions [29, 118]. Along the X_1 Cox direction (for example), the proportion of X_1 will vary, but the *relative* proportions of X_2 and X_3 will remain constant and equal to the relative proportions at the end point of the X_1 direction (where $X_2{:}X_3 \approx 2 : 1$). Generalizing to $q > 3$, along the Cox-effect direction of any component, the pairwise ratios of the $q - 1$ remaining components will remain constant. There will be much more to say about Cox-effect directions in later chapters.

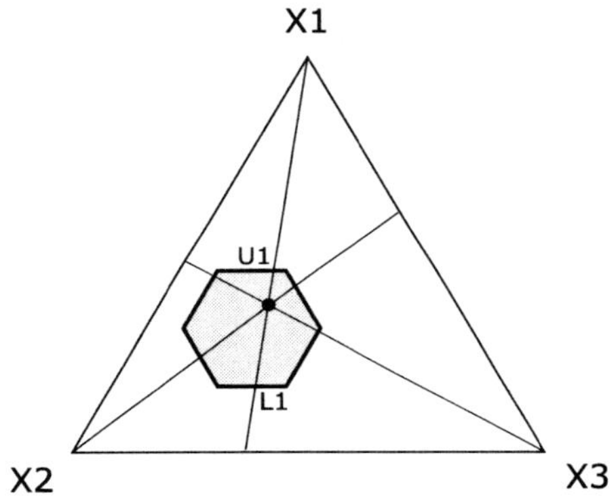

Figure 5.5. *Cox directions in a 3-simplex.*

In the second stage of the coordinate-exchange algorithm, the *first component* of the *first design point* in the starting design is varied along its Cox direction. Using X_1 in Fig. 5.5 as an example, the interval L_1–U_1 is divided into n subintervals (by default $n = 20$ in JMP, but this can be controlled). This leads to $n + 1$ additional points, and for each point the objective criterion (D-optimality in this case) is calculated. If there is an improvement in the criterion for one or more points, then the composition of the first design point is changed to the best point. Then the *second component* of the best point is varied along its Cox direction in the same manner. The process continues until *all* of the components in *all* of the starting design points have been considered for exchange. The whole process is then repeated until a complete pass through the list of design points results in no exchanges. It is not unusual for the algorithm to find vertices and edge centroids as optimal design points.

The JMP design in Table 5.8 was created in JMP Release 5 using the coordinate-exchange algorithm.[4] Points labeled "Other" are not face centroids, and thus would be absent from a candidate list. Designs can also be created in JMP Release 5 using the traditional point-exchange algorithm, but the method is not straightforward. Candidate points are first created using the `Mixture Design` dialog. Then, using the `Custom Design` dialog, the components are entered as covariates rather than as mixture variables. By specifying the

[4]The author is indebted to Bradley Jones, SAS Institute, for this particular design.

mixture variables as covariates, the component proportions become unchangeable values, and the design is then forced to be built around these.

5.4.3 *A*-Optimality Criterion

Like D-optimality, other optimality criteria are usually identified by a letter of the alphabet, and indeed there are nearly as many criteria as there are letters in the alphabet. Computer-aided experimental design is often said to use **alphabetic optimality** to find designs.

Another popular parameter-oriented criterion is ***A*-optimality**. Unlike D-optimality, A-optimality deals only with the variances of the coefficient estimates, ignoring the covariances among the estimates. To understand this, the *inverses* of the information matrices, $(\mathbf{X}'\mathbf{X})^{-1}$, for designs A, B, and C are tabulated in Table 5.9. An inverse is like the reciprocal of a scalar. Premultiplying a matrix by its inverse leads to an identity matrix, **I**. For example, for design A,

$$\begin{bmatrix} 1.1667 & -0.8333 \\ -0.8333 & 1.1667 \end{bmatrix}\begin{bmatrix} 1.75 & 1.25 \\ 1.25 & 1.75 \end{bmatrix} = \begin{bmatrix} 1 & 0 \\ 0 & 1 \end{bmatrix} = \mathbf{I}_2$$

Table 5.9. *Designs A, B, and C.* $(\mathbf{X}'\mathbf{X})^{-1}$ *matrix and* $\mathrm{tr}(\mathbf{X}'\mathbf{X})^{-1}$

Design	$(\mathbf{X}'\mathbf{X})^{-1}$	$\mathrm{tr}(\mathbf{X}'\mathbf{X})^{-1}$
A	$\begin{bmatrix} 1.1667 & -0.8333 \\ -0.8333 & 1.1667 \end{bmatrix}$	2.333
B	$\begin{bmatrix} 0.4167 & -0.0833 \\ -0.0833 & 0.4167 \end{bmatrix}$	0.8333
C	$\begin{bmatrix} 0.3333 & 0 \\ 0 & 0.3333 \end{bmatrix}$	0.6667

The matrix $\sigma^2 \cdot (\mathbf{X}'\mathbf{X})^{-1}$ is known as the **variance-covariance matrix** and like the $\mathbf{X}'\mathbf{X}$ matrix plays a major role in least squares. It acquires its name from the fact that the diagonal elements of $(\mathbf{X}'\mathbf{X})^{-1}$ are proportional to the variances of the coefficient estimates, and the off-diagonal elements are proportional to the covariances. In matrix notation, we write

$$(\mathrm{var}(\mathbf{b}))_{p\times p} = s^2 \cdot (\mathbf{X}'\mathbf{X})^{-1}_{p\times p} \tag{5.2}$$

where $\mathbf{b}$ is a $p \times 1$ vector of coefficient estimates, var(**b**) is a $p \times p$ matrix of variances and covariances, and s^2 is equal to the experimental mean square error. The variances of the coefficient estimates have both a *deterministic* component (a diagonal element of $(\mathbf{X}'\mathbf{X})^{-1}$) and a *stochastic* component (s^2). Apart from s^2, then, the *individual* variances of the coefficient estimates are equal to the diagonal elements of the $(\mathbf{X}'\mathbf{X})^{-1}$ matrix. Because these elements will be referred to time and again, they will be symbolized by c_{ii} and the off-diagonal elements by c_{ij}.

The standard errors of coefficient estimates are equal to the square roots of the diagonal elements, $\sqrt{c_{ii}}$, multiplied by s. Another way of describing the situation is to say that the $\sqrt{c_{ii}}$ are the standard errors of the coefficients estimates referenced to a **basis standard deviation** of 1.0. In the unlikely event that s, the estimate of σ, turns out to be exactly equal to 1.0, then the $\sqrt{c_{ii}}$ will be exactly equal to the standard errors of the coefficient estimates. Whatever the value of s, the standard errors of the coefficient estimates will be the product of the scalar, s, and the $\sqrt{c_{ii}}$ values.

As seen in Table 5.9 for designs A, B, and C, the variances of the two coefficient estimates for each linear model are equal to one another and, apart from s^2, are equal to 1.1667, 0.4167, and 0.3333, respectively. Thus, the relative precision of the coefficient estimates is *known in advance* — before the experiment is carried out. Once the data are collected and the results fitted to the model using OLS, one will have an estimate of σ^2, which is s^2.

A design is said to be A-optimal if it minimizes the trace of the $(\mathbf{X}'\mathbf{X})^{-1}$ matrix, symbolized $\mathrm{tr}(\mathbf{X}'\mathbf{X})^{-1}$. The trace is equal to the sum of the diagonal elements. Dividing the trace by p, the number of coefficients in the model, gives the *average variance* of the coefficients (which is where the letter designation A comes from). Software products such as Design-Expert, JMP, MINITAB, and MIXSOFT report the trace of the $(\mathbf{X}'\mathbf{X})^{-1}$ matrix (or related functions) but do not use the A-optimality criterion to choose a design.

Maximizing $|\mathbf{X}'\mathbf{X}|$ for D-optimality is equivalent to minimizing $|(\mathbf{X}'\mathbf{X})^{-1}|$ because

$$|(\mathbf{X}'\mathbf{X})^{-1}| = \frac{1}{|\mathbf{X}'\mathbf{X}|}$$

There is an interesting relationship between the determinant and the trace of the $(\mathbf{X}'\mathbf{X})^{-1}$ matrix. It turns out that these are given by

$$|(\mathbf{X}'\mathbf{X})^{-1}| = \prod_{i=1}^{p} \lambda_i$$

$$\mathrm{tr}(\mathbf{X}'\mathbf{X})^{-1} = \sum_{i=1}^{p} \lambda_i$$

where the λ_i are the eigenvalues of the $(\mathbf{X}'\mathbf{X})^{-1}$ matrix [3]. The eigenvalues of the $(\mathbf{X}'\mathbf{X})^{-1}$ matrix are *variances* of specific *linear combinations* of the columns of the $\mathbf{X}$ matrix, called principal components. The determinant of $(\mathbf{X}'\mathbf{X})^{-1}$ is then a product of p variances, and as a consequence, working with the pth root of $|(\mathbf{X}'\mathbf{X})^{-1}|$ reduces the determinant to units of variance. This is the basis for JMP's D efficiency and Design-Expert's scaled D-optimality criterion using the pth root of the determinant. $|(\mathbf{X}'\mathbf{X})^{-1}|^{1/p}$ is thus equal to the geometric mean of p variances, while $\mathrm{tr}(\mathbf{X}'\mathbf{X})^{-1}/p$ is equal to the arithmetic mean of the same variances. The trace, of course, deals only with the diagonal elements of the $(\mathbf{X}'\mathbf{X})^{-1}$ matrix, while the determinant deals with both the diagonal and off-diagonal elements [149].

Another way to look at the difference between the A and D criteria is shown in Fig. 5.6. The ellipse in this figure is the same as that shown for design A in Fig. 5.3. The horizontal lines at $b_1 = 9.25$ and 10.75 delineate the 95% confidence interval for b_1; the vertical lines at $b_2 = 19.25$ and 20.75 delineate the 95% confidence interval for b_2. The calculation of confidence intervals for parameter estimates is explained in Chapter 8, page 171.

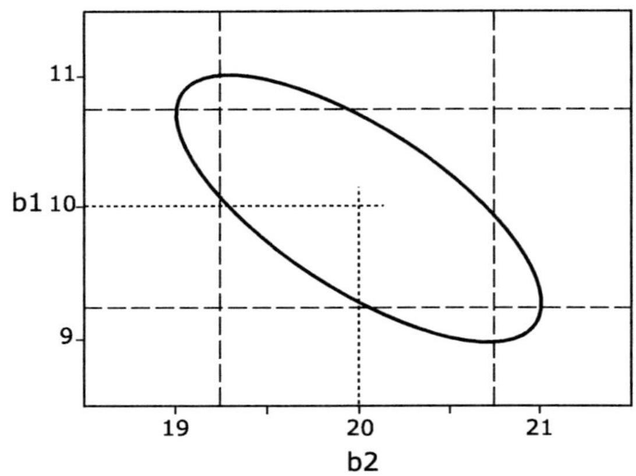

Figure 5.6. *Univariate and joint 95% confidence regions for b_1 and b_2. Design A.*

It is important to understand the difference in meaning between the *univariate confidence intervals* (related to A-optimality) and the *multivariate confidence region* (related to D-optimality). An A-optimal design is one that minimizes the average of the p confidence intervals. In the example illustrated in Fig. 5.6, an A-optimal design would minimize the mean of the two confidence intervals, one for b_1 and the other for b_2. A D-optimal design, on the other hand, is one that minimizes the volume of the confidence ellipsoid. For the $p = 2$ example illustrated here, this is the confidence ellipse in Fig. 5.6.

Confidence intervals and regions are a property of the method used to calculate them. This means that if we repeatedly went back into the laboratory and resampled (collected new data) and, for each sample, calculated a new confidence interval for a particular coefficient, then 95% (in this particular example) of such computed intervals would contain the true value of that coefficient. There is still a 5% chance that a computed 95% confidence interval would not contain the true value of the coefficient. This does not mean, however, that 95% of the rectangles bordered by the confidence intervals in Fig. 5.6 will *simultaneously* contain both b_1 and b_2. In this case we must say that 95% of the calculated joint confidence *regions* (the ellipses) will contain β_1 and β_2, the true values of b_1 and b_2. Thus the upper right and lower left corners of the rectangle bordered by the confidence intervals are excluded from the joint confidence region.

For further discussion of some of the pros and cons of optimal designs in general, see Box [8] and Myers and Montgomery [107]. For an evaluation of several optimal-design approaches in a mixture setting, see Piepel, Anderson, and Redgate [126].

In Section 4.2, page 45, several properties of a good experimental design were listed. In this chapter and the preceding, we have focused on items i, ii, and iv under 1(a). We have ignored several important concepts such as replication to estimate pure experimental error and minimizing prediction variances. With respect to replication, some software products such as Design-Expert will choose replicate points "on demand", that is, in the design-creation stage. Other products may or may not add replicates, and therefore the analyst is confronted with making decisions. These points and the others listed under 1(b) (page 45) will be covered in the next chapter.

Design Study

Heinsman and Montgomery (H&M) describe the design and analysis of a four-component mixture related to the manufacture of a soap product [68]. The formulation was a surfactant package composed of two nonionic surfactants, an anionic surfactant, and a zwitterionic surfactant. Although it was not explicitly stated, it is assumed that the surfactant package constituted a fixed small proportion of the soap product and that the relative proportions of the other ingredients in the product were held constant. The most important response was considered to be the life of the product measured in lather units. Three other responses, measures of the grease-cutting capability of the product, were also of interest. For now, however, we will not concern ourselves with the responses but rather focus on the design of the experiment.

The lower and upper bounds on the component proportions were

Nonionic surfactant A	$0.5 \le A \le 1.0$
Nonionic surfactant B	$0.0 \le B \le 0.5$
Anionic surfactant	$0.0 \le C \le 0.5$
Zwitterionic surfactant	$0.0 \le D \le 0.05$

With these constraints, Design-Expert outputs the *candidate* points in Table 5.10.

Table 5.10. *Surfactant experiment. Candidate points*

Point type	Number	Dimension
Vertices	6	0
Centers of edges	9	1
Thirds of edges	18	1
Triple blends	12	2
Constraint-plane centroids	5	2
Axial check blends	6	3
Interior blends	14	3
Overall centroid	1	3
Total candidate points	71	

The design region for the experiment is illustrated in Fig. 5.7. The numbering of the vertices in the enlarged view is explained below. Dimensions of the design points in Table 5.10 are *with respect to the design region*.[5] The number of vertices and edge centroids

[5] For example, points on the edge defined by vertices 5 and 6a lie on a two-dimensional constraint plane (triangle) with respect to the tetrahedron but on a one-dimensional edge with respect to the constrained design region.

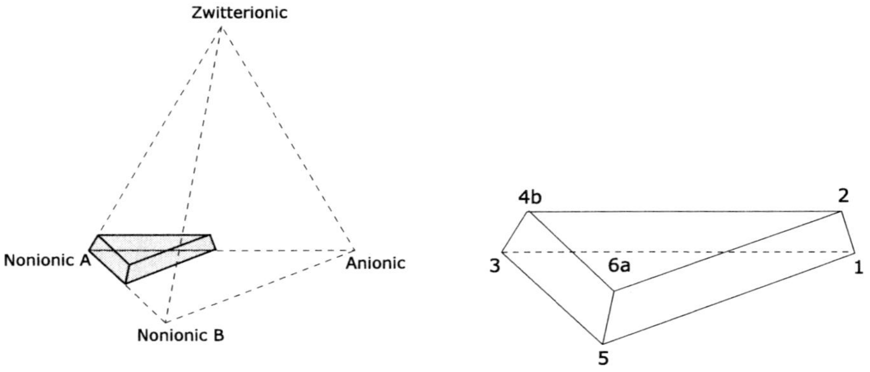

Figure 5.7. *Surfactant package design.*

should be apparent from the drawing. Thirds of edges are located at the 1/3,2/3 and 2/3,1/3 points of each edge, and so there are $2 \times 9 = 18$ of these. Each of the three faces defined by four vertices will contribute four triple blends (cf. page 65 for descriptions of point types), for a total of 12. This is because each of these faces has four sets of three adjacent vertices. For example, the sets on the front right face are 1-2-6a, 2-6a-5, 6a-5-1, and 5-1-2. The compositions of each of these sets is averaged to give a triple blend. Triple blends are two-dimensional because they are located on the faces of the design region.

In addition, each of the two faces defined by three vertices will contribute a triple blend, leading to a total of 14. Because the latter two are also CPCs, they are counted as CPCs rather than triple blends. The five CPCs are the AEV centroids of the five faces. As there are six vertices, there are also six axial check blends because these fall halfway between the overall centroid and each of the vertices. The interior blends lie midway between the overall centroid and the edge centroids as well as midway between the overall centroid and the CPCs, and so there are $9 + 5 = 14$ of these.

H&M assumed that the 10-term quadratic Scheffé model would be adequate for their study. Their design was constructed by forcing the six axial check blends and the overall centroid into the design and then choosing the remaining points using the D-optimality criterion. The rationale was that this would give a more uniform spacing of design points over the design region than would the pure D-optimality approach. Their final design consisted of 6 vertices, 5 edge centroids, 1 replicated constraint-plane centroid, 6 axial check blends, and the overall centroid, for a total of 20 points. A weighted average of the dimensions of the design points is

$$\begin{aligned}\overline{dim} &= \frac{6(0) + 5(1) + 2(2) + 7(3)}{20} \\ &= 1.5\end{aligned}$$

Following H&M, let us create and critique some 20-run designs capable of supporting a quadratic (10-term) Scheffé model. To reinforce the methods discussed in this chapter, we begin with a "longhand" implementation of the XVERT algorithm (Sections 5.3.1 and 5.4.1). This, of course, can be done on a computer, but it is instructive to see how the higher-dimensional centroids are calculated.

ID	*D*	*A*	*B*	*C*	
1	.00	.50	.00	.50	√
2	.05	.50	.00	.45	√
3	.00	1.00	.00	.00	√
4	.05	1.00	.00	−.05	
5	.00	.50	.50	.00	√
6	.05	.50	.50	−.05	
7	.00	1.00	.50	−.50	
8	.05	1.00	.50	−.55	
4	.05	1.00	.00	−.05	
4a	.05	1.00	−.05	.00	
4b	.05	.95	.00	.00	√
4c	.00	1.00	.00	.00	duplicate of 3
6	.05	.50	.50	−.05	
6a	.05	.50	.45	.00	√
6b	.05	.45	.50	.00	
6c	.00	.50	.50	.00	duplicate of 5
7	.00	1.00	.50	−.50	
7a	.00	1.00	.00	.00	duplicate of 3
7b	.00	.50	.50	.00	duplicate of 5
7c	−.50	1.00	.50	.00	
8	.05	1.00	.50	−.55	
8a	.05	1.00	−.05	.00	
8b	.05	.45	.50	.00	
8c	−.50	1.00	.50	.00	

A 2^{4-1} factorial design is used as a template to calculate eight base points (IDs 1–8 in the table), of which four (1, 2, 3, and 5) are core points. (Note that the order has been rearranged from A, B, C, D to D, A, B, C so that one of the components with the largest range takes up the "slack".) The remaining vertices are found after examining the subgroups for 4, 6, 7, and 8. The numbering of the vertices in Fig. 5.7 corresponds to the IDs of the six identified vertices.

As we are seeking a design with more than six points, we need to calculate the composition of the edge centroids. We know in advance that there will be nine edge centroids, which when combined with the six vertices will provide us with a total of only 15 design points, five short of our goal. If we include the five CPCs, however, we will have just the right number. To calculate the composition of the edge centroids, we must average pairs of extreme vertices for which there are $q - d - 1 = 2$ proportions in common (cf. page 69). To calculate the composition of the CPCs requires averaging sets of vertices of size three or greater that have $q - d - 1 = 1$ proportion in common. The table below shows the sets of vertices that need to be averaged to calculate the edge centroids and the CPCs.

Point type	ID	Extreme vertex 1	2	3	4b	5	6a
Edge centroids	1	√	√				
	2	√		√			
	3	√				√	
	4		√		√		
	5		√				√
	6			√	√		
	7			√		√	
	8				√		√
	9					√	√
CPCs	1	√	√	√	√		
	2	√	√			√	√
	3	√		√		√	
	4		√		√		√
	5			√	√	√	√

To illustrate, vertices 1 and 2 have the same proportions of A and B. This means they lie on the same edge, and so they are averaged to find an edge centroid. Similarly, vertices 1 and 3 have the same proportions of B and D, and so these would also be averaged. In the case of the CPCs, vertices 1, 2, 3, and 4b have the same proportion of component B, and so these define a two-dimensional constraint plane and are averaged to find the CPC. Note that there are three CPCs defined by four vertices, and two that are defined by three vertices, in agreement with Fig. 5.7.

The final design has 6 vertices, 9 edge centroids, and 5 constraint-plane centroids. A weighted average of the dimensions of the design points would be

$$\overline{dim} = \frac{6(0) + 9(1) + 5(2)}{20} = 0.95$$

Compared to the H&M design ($\overline{dim} = 1.5$), the design points are more heavily weighted towards the lower-dimensional design points, located near the boundaries of the design region. As H&M forced the overall centroid and the axial check blends into their design, their design has a better sampling of the interior of the design region where the higher-dimensional centroids are located. In addition, H&M have one replicate, which provides them with one degree of freedom for pure error. While this is far from ideal, it is better than zero degrees of freedom for pure error. To improve our design, we would have to consider deleting one or more points and replacing the deleted points with replicates. A method for choosing replicates is discussed in the next chapter.

How about the variance properties of the two designs? The scaled D-optimality criterion ($n \div |\mathbf{X}'\mathbf{X}|^{1/p}$, page 80) for the H&M design is 396, whereas for the XVERT-derived design it is 314. The trace of the variance-covariance matrix ($\text{tr}(\mathbf{X}'\mathbf{X})^{-1}$) for the H&M design is 2.79×10^5, while for the XVERT-derived design it is 1.47×10^5. (In both cases, the $\mathbf{X}$ matrix is expressed in terms of pseudocomponent proportions.) Thus on the basis of both the generalized variance (as evaluated by the scaled D-optimality criterion) and the average variance (based on the trace of the covariance matrix), the XVERT-based design is better. This is a reflection of the fact that boundary points are more effective in improving the precision of parameter estimates than interior points. The risk, however, is that if the assumed model is incorrect, and the true response surface is of higher order than quadratic, then predictions in the interior of the design region could be severely biased. This is why Piepel, Anderson, and Redgate [126] recommend forcing a center point into variance-optimal designs if the software package does not include one.

Let us now see how a software package such as Design-Expert might handle this problem. While it is not necessary to include all 71 design points in the candidate list (Table 5.10), there are so few that we shall use them all. In Design-Expert one specifies in advance the number of points to support the model, the desired number of lack-of-fit points, and the desired number of replicates. For notational purposes, let us designate these a, b, and c. The distinction is important because the points are picked in different ways (cf. page 81). Points to support the model are selected first and chosen on the basis of the D-optimality criterion. Lack-of-fit points (which also support the model) are picked next using a distance-based algorithm. Finally, replicates are picked using the D-optimality criterion.

The default value for a is always p, but a may be set greater than p. In so doing, one is guaranteed that the points will be picked by the D-optimality criterion, but one cannot control whether the additional points will be lack-of-fit or replicate points. Specifying b guarantees at least b lack-of-fit points, and these will be picked by a distance-based, rather than D-optimality, criterion. Specifying c guarantees at least c replicates picked by the D-optimality criterion.

When designing to support a $q = 4$ quadratic Scheffé model in Design-Expert, the default values for a, b, and c are 10, 5, and 5. Fortuitously, this is exactly what we need for a 20-run design. Accepting the default leads to a design consisting of 6 vertices (of which 5 are replicated), 3 edge centroids, 1 third of an edge, 2 CPCs, and 3 axial check blends. Note that the overall centroid was not picked. The weighted average of the dimensions of the design points is

$$\begin{aligned}\overline{dim} &= \frac{11(0) + 4(1) + 2(2) + 3(3)\}}{20} \\ &= 0.85\end{aligned}$$

The weighted average in this design is similar to that in the XVERT-based design, 0.95. The scaled D-optimality criterion is 345, less than that of the H&M design (396) but greater than the XVERT-based design (314). The trace is equal to 4.83×10^5, larger than either of the other two designs (2.79×10^5 for the H&M design, 1.47×10^5 for the XVERT-based design). However, this design has a strong point in its favor, and that is the five degrees of freedom for pure error. The XVERT-based design has no degrees of freedom for pure error, and so it would be impossible to estimate whether or not a linear model is adequate. In the H&M design, the tabled F statistic for testing lack of fit is $F_{.05;9,1} = 240.5$, leading to a very insensitive test. In the Design-Expert design, $F_{0.05;5,5} = 5.05$, and this would provide for a much more sensitive test. Much more will be said about lack of fit in Chapter 8.

As pointed out previously, the H&M design has the advantage of better sampling in the interior of the design space. In the event that the true underlying surface is of higher order than quadratic, this will provide better protection against prediction bias than will the Design-Expert design. On the other hand, the Design-Expert design has a clear advantage in terms of a more powerful test for lack of fit. We might consider incorporating both features into a single design, i.e., force a center point and the six axial check blends into the design and in addition include five replicates.

Forcing the seven interior points into the design and setting a-b-c = 8-0-5 will result in the remaining 13 points being selected by the D-optimality criterion, of which five will be replicates. The resulting design consists of 5 vertices (3 of which are replicated), 3 edge centroids (2 of which are replicated), plus the 6 axial check blends and the overall centroid. The weighted average of the dimensions of the design points is

$$\begin{aligned}\overline{dim} &= \frac{8(0)+5(1)+7(3)}{20}\\ &= 1.30\end{aligned}$$

Relative to the previous design, we now have better sampling of the interior of the design space. This is at the expense of variance minimization, as the scaled D-optimality criterion is now 443 and the trace is 5.88×10^5. Nonetheless, considering that we have five degrees of freedom for pure error, overall this is a good design.

Looking back at the constraints on the component proportions (page 87), we note that the range of component D (the zwitterionic surfactant) is considerably less than the range of the other three surfactants. Assuming it would be feasible, we might investigate the consequences of changing the constraints on D from $0 \leq D \leq 0.05$ to $0 \leq D \leq 0.10$. Carrying out the previous procedure (i.e., forced center point and axial check blends plus five replicates) would lead to a design with the same $\overline{dim}$ but with significantly smaller values of the scaled D-optimality criterion and the trace (232 and 0.332×10^5, respectively). Clearly, if expanding the range of D is practical, there is much to be gained in terms of variance reduction.

Table 5.11 displays $\sqrt{c_{ii}}$ values (for coefficients expressed in the pseudocomponent metric) for the designs discussed in the previous two paragraphs differing only in the range of D. (Chapter 6 explains how to output estimates of $\sqrt{c_{ii}}$ values, as most popular software

Table 5.11. *Surfactant experiment.* $\sqrt{c_{ii}}$ *values for two designs*

	Range of D	
Coefficient	0 – 0.05	0 – 0.10
A	0.698	0.698
B	0.698	0.699
C	0.697	0.699
D	359.	79.
AB	3.42	4.38
AC	4.35	3.44
AD	396.	89.5
BC	3.43	3.44
BD	396.	97.4
CD	382.	97.3

packages do not provide these numbers.) Within either design, a clear pattern emerges: *all terms that contain D have inflated standard errors.* Clearly, coefficient estimates for terms containing D are much less precise than those for terms not containing D. The source of the problem is the relatively small range of D compared to the range of the other components. This disparity is not removed by the pseudocomponent transformation. Had the pseudocomponent transformation removed the disparity in the ranges, then this problem would not have appeared in the pseudocomponent metric. More will be said about this in Chapter 6 and considerably more in Chapter 14.

Chapter 6

Design Evaluation

Among the reasons for fitting models to data, two of the most important are using the model for interpretive purposes so that one can better understand the system and using the model for prediction and optimization purposes. One or the other may take precedence, depending on the situation. In either case, before proceeding to the data collection and analysis stages, it would be well to take the time to see if the model will fulfill its intended use.

This chapter will explain some strategies that the analyst can use to evaluate a design that has been produced using the methods described in the previous chapter. In many cases design problems can be minimized or possibly even eliminated. At the very least, however, the analyst should be able to proceed armed with an understanding of any problems that might be encountered at the analysis stage.

In light of the reasons mentioned above for fitting models to data, this chapter is divided into two principal sections. The first section continues the discussion of topics related to coefficient estimation. Precise coefficient estimation is important if one is going to use the model for interpretive purposes. The second section focuses on prediction variances and replication, neither of which was covered in the last chapter.

6.1 Properties of the Least-Squares Estimators

In the surfactant Design Study at the end of the previous chapter, $\sqrt{c_{ii}}$ values for a quadratic Scheffé model were presented in Table 5.11 for two designs. Although Design-Expert automatically outputs these numbers, many other software products, such as JMP and MINITAB, do not. It is possible to output these numbers in JMP using JMP's scripting language or in MINITAB using MINITAB macros. However, the following procedure is simple and will give very good approximations to the numbers. It is *recommended* that the analyst carry out this procedure. The method is simply to create a normally distributed dummy response with a mean of any value but with a standard deviation of ~ 1.0. Most general-purpose statistical packages have normal random number generators that can implement this. One then simply fits a model to the data and examines the standard errors of the coefficient estimates. The results should be reasonable approximations of the square roots of the diagonal elements of the $(\mathbf{X}'\mathbf{X})^{-1}$ matrix.

An example is shown below for the H&M surfactant design described in the Design Study in the previous chapter. The column labeled "Basis $\sigma = 1.0$" lists the squares roots of the diagonal elements of the $(\mathbf{X}'\mathbf{X})^{-1}$ matrix (the $\sqrt{c_{ii}}$) for a quadratic model expressed in pseudocomponents. The column labeled "$s = 1.0243$" contains the standard errors of the coefficient estimates after fitting the model to a response that was simulated using a normal random number generator. The fitted model had a root mean square error (s) of 1.0243, and so the numbers in column three are equal within rounding error to 1.0243 × the numbers in column two. Although the numbers in the third column differ somewhat from the numbers in the second column, the message is clear.

	Standard errors	
Coefficient	Basis $\sigma = 1.0$	$s = 1.0243$
A	0.914	0.936
B	0.912	0.934
C	0.912	0.934
D	243.	249.
AB	3.62	3.70
AC	3.62	3.70
AD	272.	279.
BC	4.16	4.26
BD	270.	277.
CD	270.	277.

Table 6.1 displays $\sqrt{c_{ii}}$ values for a quadratic model (pseudocomponent metric) for an alloy design. The design was a 25-run design based on the D criterion and on the constraints listed on page 67. The design included five forced replicates. Lower and upper bounds on the component proportions in the reals (L_i, U_i) and pseudos (L_i^*, U_i^*) are summarized beneath Table 6.1. Again, inspection of the standard errors and the ranges reveals that the standard errors of the linear terms increases as the ranges decrease. There is also a clear pattern of the standard errors of the estimates for the quadratic terms.

What should an experimenter expect for $\sqrt{c_{ii}}$ values when there is little or no disparity in the ranges of the components? One approach to answering this question is to create a symmetrical design capable of supporting Scheffé models from linear to quartic and examine the $\sqrt{c_{ii}}$ values. A simplex-lattice design is symmetrical, and so all linear terms will have the same standard error, as will all quadratic terms, etc. Therefore, we need only to list standard errors of the term types.

Table 6.2 summarizes $\sqrt{c_{ii}}$ values (rounded) for model terms in linear through quartic Scheffé models. The $(\mathbf{X}'\mathbf{X})^{-1}$ matrices were based on two designs: (a) a 35-point {4,4} simplex-lattice design augmented with four axial check blends (39 points total); (b) a 56-point {4,5} simplex lattice design augmented with four axial check blends and a center point (61 points total). Two trends are clear. First, as the order of a term increases in a model, the standard error of the coefficient estimates tends to increase. Estimates for higher-order terms usually have lower precision than estimates for lower-order terms. Second, as the model order increases, coefficient estimates for terms of a given order tend to have slightly less precision. More will be said about precision of coefficient estimates in the chapter on collinearity, where variance inflation factors will be discussed.

Table 6.1. *Alloy example. $\sqrt{c_{ii}}$ values for a quadratic model, pseudocomponent metric*

Coefficient	Std. error
A	159.
B	21.9
C	21.2
D	10.3
E	3.51
AB	178.
AC	177.
AD	192.
AE	180.
BC	38.1
BD	36.1
BE	33.7
CD	35.6
CE	31.8
DE	16.4

	Reals			Pseudos	
Component	L_i	U_i	Range	L_i^*	U_i^*
A (Cu)	0.03	0.10	0.07	0.0	0.13462
B (Al)	0.00	0.15	0.15	0.0	0.28846
C (Cr)	0.00	0.15	0.15	0.0	0.28846
D (Fe)	0.10	0.30	0.20	0.0	0.38462
E (Ni)	0.35	0.65	0.30	0.0	0.57692

When standard errors of certain coefficient estimates are inflated because of a small range of a component relative to the others, the remedy seems clear — increase its range. Unfortunately, this may be easier said than done. In many instances this may not be practical and may very well be impossible, in which case one must live with the inflated variances. However, it is better to live with inflated variances knowing that they are there than it is to remain oblivious to their existence.

What problems might one be on the lookout for when there are known to be inflated variances of coefficient estimates? An obvious problem is imprecision in the coefficient estimates. If one is interested in interpreting the model, then one would like estimates that are as precise as possible. Small ranges tend to defeat this goal.

Second, at the analysis stage one is ultimately going to compare the magnitude of certain coefficient estimates against their standard errors to make inferences about significance. When standard errors are inflated, one is apt to conclude that a possibly significant coefficient estimate may not be significant because it is not sufficiently greater than its standard error. The result is called a **Type II error** (or more informally, a "miss" [19]). In a regression setting this means that one has concluded that a coefficient is not significant when

Table 6.2. *$\sqrt{c_{ii}}$ values for some Scheffé models*

Term type	Model†					
	L	Q	SC	C	SQr	Qr
	augmented {4,4} simplex lattice					
X_i	0.5	0.8	0.8	1.0	0.8	1.0
X_iX_j	·	3.3	4.0	4.1	4.0	4.8
$X_iX_jX_k$	·	·	24.	24.	·	·
$X_iX_j(X_i - X_j)$	·	·	·	7.2	·	8.4
$X_iX_j(X_i - X_j)^2$	·	·	·	·	·	21.
$X_i^2X_jX_k$	·	·	·	·	99.	110.
$X_iX_jX_kX_l$	·	·	·	·	·	260.
	augmented {4,5} simplex lattice					
X_i	0.4	0.8	0.8	0.9	0.8	1.0
X_iX_j	·	2.8	3.6	3.8	3.7	4.2
$X_iX_jX_k$	·	·	20.	20.	·	·
$X_iX_j(X_i - X_j)$	·	·	·	6.1	·	7.6
$X_iX_j(X_i - X_j)^2$	·	·	·	·	·	15.
$X_i^2X_jX_k$	·	·	·	·	71.	82.
$X_iX_jX_kX_l$	·	·	·	·	·	164.

† L = linear, Q = quadratic, SC = special cubic, C = cubic, SQr = special quartic, Qr = quartic model

in fact it may be significant. Thus, one should exercise caution when making inferences about coefficient estimates whose standard errors are known to be inflated.

When a peculiar pattern of standard errors of regression coefficients is discovered, as in Table 5.11, page 92, then it is worth checking for a third potential problem. This will be illustrated with the surfactant designs, but it is convenient to first give some background using the simpler designs in Table 5.5, page 77. The variance-covariance matrices (apart from σ^2) for designs A, B, and C are given in Table 6.3 (column headed $(\mathbf{X}'\mathbf{X})^{-1}$). From the elements of a variance-covariance matrix, it is a relatively easy matter to calculate the elements of a **correlation matrix of regression coefficients**. The diagonal elements of such a matrix are all 1s, and the off-diagonal elements are calculated from the elements of the covariance matrix using the formula

$$\rho_{ij} = \frac{c_{ij}}{\sqrt{c_{ii}c_{jj}}} \tag{6.1}$$

Many regression packages will output a correlation matrix of the explanatory variables (the *regressors*). Such a matrix should not be confused with the correlation matrix of regression *coefficients*, as discussed above.

Applying this simple formula to the covariance matrices in Table 6.3 leads to the correlation matrices of regression coefficients for the three designs (column headed **R**). The off-diagonal elements in the correlation matrix for design A mean that b_1 and b_2 are

Table 6.3. *Covariance matrices (apart from σ^2) and correlation matrices. Designs A, B, and C (Table 5.5, page 77)*

Design	$(\mathbf{X}'\mathbf{X})^{-1}$	$\mathbf{R}$
A	$\begin{bmatrix} 1.1667 & -0.8333 \\ -0.8333 & 1.1667 \end{bmatrix}$	$\begin{bmatrix} 1.0000 & -0.7142 \\ -0.7142 & 1.0000 \end{bmatrix}$
B	$\begin{bmatrix} 0.4167 & -0.0833 \\ -0.0833 & 0.4167 \end{bmatrix}$	$\begin{bmatrix} 1.0000 & -0.1999 \\ -0.1999 & 1.0000 \end{bmatrix}$
C	$\begin{bmatrix} 0.3333 & 0 \\ 0 & 0.3333 \end{bmatrix}$	$\begin{bmatrix} 1.0000 & 0 \\ 0 & 1.0000 \end{bmatrix}$

negatively correlated with one another. This means that with repeated sampling, when estimates of b_1 increase, then estimates of b_2 will tend to decrease, and vice versa. This can be inferred from the tipped sausage-shaped ellipse for design A in Fig. 5.3, page 77.

The off-diagonal elements in the correlation matrix for design C are equal to zero, which means that b_1 and b_2 are independent, or **orthogonal** to one another. Because the variances for these two coefficients are equal to one another (as evidenced by $c_{11} = c_{22}$ in the covariance matrix), the joint confidence region is a circle. Although it may not appear to be a circle in Fig. 5.3, this is an illusion because the circle is surrounded by two ellipses. If $c_{11} \neq c_{22}$ but $c_{12} = 0$, the joint confidence region would be elliptical, but it would not be tipped; the principal axis would be parallel to the x or y axis.

Returning to the surfactant designs, Table 6.4 displays the correlation matrix of regression coefficients for the H&M surfactant design. Because the correlation matrix is always symmetrical about the main diagonal, it is only necessary to show the lower (or upper) triangular part of the matrix. Note the signs and magnitudes of the correlations between

Table 6.4. *Correlation matrix of regression coefficients. Surfactant experiment*

	A	B	C	D	AB	AC	AD	BC	BD	CD
A	1.000									Symmetric
B	−0.088	1.000								
C	−0.088	−0.033	1.000							
D	0.068	−0.015	−0.015	1.000						
AB	−0.343	−0.305	0.090	0.241	1.000					
AC	−0.343	0.090	−0.305	0.241	0.093	1.000				
AD	−0.096	0.021	0.021	−0.999	−0.240	−0.240	1.000			
BC	0.151	−0.363	−0.363	0.358	0.010	0.010	−0.363	1.000		
BD	−0.061	−0.014	0.018	−0.999	−0.244	−0.245	0.997	−0.353	1.000	
CD	−0.061	0.018	−0.014	−0.999	−0.245	−0.244	0.997	−0.353	0.998	1.000

pairs of terms containing D. For example, D is highly correlated ($\rho_{ij} = -0.999$) with all quadratic terms that contain D. Because of the negative correlations, whenever coefficient estimates for D go up, estimates for AD, BD, and CD will tend to go down. Coefficients for AD, BD, and CD are positively correlated ($\rho_{ij} = 0.997 - 0.998$) with one another, so when an estimate of one goes up, estimates of the other two will also tend to go up. If one is trying to use the model for interpretive purposes, these correlations will tend to obfuscate clear interpretations. These problems originate, of course, from the small range of D relative to the other three components.

6.2 Leverage

The important role played by the $(\mathbf{X}'\mathbf{X})^{-1}$ matrix in OLS should be clear by now. Much was said in Chapter 5 and the first half of this chapter about its importance in determining the precision of coefficient estimates in regression models. There is another matrix, known as the **hat matrix**, that plays an equally important role in OLS, particularly with respect to precision in prediction. The hat matrix, which we symbolize $\mathbf{H}$, gets its name because it "puts the hats" on the Ys.

$$\hat{\mathbf{Y}} = \mathbf{HY} \tag{6.2}$$

In Eq. 6.2, $\mathbf{Y}$ is a $n \times 1$ vector of *observed* values of Y (the response), whereas $\hat{\mathbf{Y}}$ is a $n \times 1$ vector of *fitted* values of Y. This means that $\mathbf{H}$ is a square matrix of dimensions $n \times n$.

The hat matrix and its derivation is discussed in detail in most books on regression analysis, and indeed an entire article has been devoted to its properties [71]. Without concerning ourselves with the details of its derivation, we simply state that this matrix is given by

$$\mathbf{H} = \mathbf{X}(\mathbf{X}'\mathbf{X})^{-1}\mathbf{X}' \tag{6.3}$$

The dimensions of $\mathbf{X}$, $(\mathbf{X}'\mathbf{X})^{-1}$, and $\mathbf{X}'$ are, respectively, $(n \times p)$, $(p \times p)$, and $(p \times n)$, and so multiplying these together leads to $(n \times n)$, the dimensions of the hat matrix. The important point here is that the hat matrix depends only on the columns of $\mathbf{X}$, and therefore it can be calculated in advance of doing the experiment. Any insights that might be derived from the elements of the hat matrix can therefore be determined before doing the experiment.

It is helpful to look at a specific example of a hat matrix. An $n \times n$ matrix can be quite large, so let us look at a small example. To define a hat matrix, we must define $\mathbf{X}$, which in turn means that we must specify a model and a design to support the model. Let us assume, then, that we would like to design an experiment to support the linear Scheffé model

$$E(Y) = \beta_1 X_1 + \beta_2 X_2$$

A minimal design would be a {2,2} simplex lattice design. This would support the model with one degree of freedom left over for lack of fit. Obviously replicates are desirable, but we shall keep the design as small as possible. The design is

X_1	X_2
1.0	0.0
0.5	0.5
0.0	1.0

The hat matrix for this small design and model is given by

$$\mathbf{H}_1 = \begin{bmatrix} 1 & 0 \\ .5 & .5 \\ 0 & 1 \end{bmatrix} \begin{bmatrix} .8333 & -.1667 \\ -.1667 & .8333 \end{bmatrix} \begin{bmatrix} 1 & .5 & 0 \\ 0 & .5 & 1 \end{bmatrix} \tag{6.4}$$

$$= \begin{bmatrix} .8333 & .3333 & -.1667 \\ .3333 & .3333 & .3333 \\ -.1667 & .3333 & .8333 \end{bmatrix}$$

The subscript "1" attached to the symbol **H** is simply a label to distinguish it from other hat matrices that will be developed shortly. Equation 6.2 for this example becomes

$$\begin{bmatrix} \hat{Y}_1 \\ \hat{Y}_2 \\ \hat{Y}_3 \end{bmatrix} = \begin{bmatrix} .8333 & .3333 & -.1667 \\ .3333 & .3333 & .3333 \\ -.1667 & .3333 & .8333 \end{bmatrix} \begin{bmatrix} Y_1 \\ Y_2 \\ Y_3 \end{bmatrix}$$

By the properties of matrix multiplication, the fitted value for Y_1, which is $\hat{Y}_1$, is then

$$\hat{Y}_1 = 0.8333\, Y_1 + 0.3333\, Y_2 - 0.1667\, Y_3 \tag{6.5}$$

with obvious extensions to the other $\hat{Y}_i$, $i = 2, 3$. In other words, the $\hat{Y}_i$ s are *weighted sums (linear combinations) of the* Y_i*s*. Of the n weights, one of these will always be the *self weight*. In the case of $\hat{Y}_1$, the self weight is 0.8333; for $\hat{Y}_2$, it is 0.3333; and for $\hat{Y}_3$, it is 0.8333. The self weights are called **leverages** and are the diagonal elements of the hat matrix. These are symbolized h_{ii}, while the off-diagonal elements of a hat matrix are symbolized h_{ij}. The h_{ii} shall figure prominently in many influence diagnostics to be discussed in later chapters.

An implication of Eq. 6.5 is that the observed value for Y_1 plays a greater role than the observed values for Y_2 and Y_3 in determining the fitted value, $\hat{Y}_1$. Rather than Y_1 playing a dominant role, a preferable situation would be to have the Y_i, $i = 1, 2, 3$, play equal roles. Such is the case for $\hat{Y}_2$, where the weights are all 0.3333.

Consider the case where we use the same design to support the $q = 2$ quadratic Scheffé model

$$E(Y) = \beta_1 X_1 + \beta_2 X_2 + \beta_{12} X_1 X_2$$

This is a three-term model, and since we have only a three-observation design, the design is now saturated. The **X** matrix in this case will be

$$\mathbf{X} = \begin{bmatrix} 1 & 0 & 0 \\ .5 & .5 & .25 \\ 0 & 1 & 0 \end{bmatrix}$$

where the third column is the product $X_1 X_2$. The hat matrix turns out to be

$$\mathbf{H}_2 = \begin{bmatrix} 1 & 0 & 0 \\ 0 & 1 & 0 \\ 0 & 0 & 1 \end{bmatrix}$$

The implications of $\mathbf{H}_2$ are that $\hat{Y}_1 = Y_1$, $\hat{Y}_2 = Y_2$, and $\hat{Y}_3 = Y_3$. In other words, there is an exact fit.

Before summarizing some properties of hat matrices in general, let us first consider two more cases. If we replicate the above three-observation design, we have design B (Table 5.5, page 77).

X_1	X_2
1.0	0.0
1.0	0.0
0.5	0.5
0.5	0.5
0.0	1.0
0.0	1.0

Let us use this design to again support linear and quadratic Scheffé models. This will lead to hat matrices $\mathbf{H}_3$ and $\mathbf{H}_4$, respectively.

$$\mathbf{H}_3 = \begin{bmatrix} .4167 & .4167 & .1667 & .1667 & -.0833 & -.0833 \\ .4167 & .4167 & .1667 & .1667 & -.0833 & -.0833 \\ .1667 & .1667 & .1667 & .1667 & .1667 & .1667 \\ .1667 & .1667 & .1667 & .1667 & .1667 & .1667 \\ -.0833 & -.0833 & .1667 & .1667 & .4167 & .4167 \\ -.0833 & -.0833 & .1667 & .1667 & .4167 & .4167 \end{bmatrix}$$

$$\mathbf{H}_4 = \begin{bmatrix} .5 & .5 & 0 & 0 & 0 & 0 \\ .5 & .5 & 0 & 0 & 0 & 0 \\ 0 & 0 & .5 & .5 & 0 & 0 \\ 0 & 0 & .5 & .5 & 0 & 0 \\ 0 & 0 & 0 & 0 & .5 & .5 \\ 0 & 0 & 0 & 0 & .5 & .5 \end{bmatrix}$$

Table 6.5 summarizes the leverages for the four combinations of two designs and two models.

It will be convenient in later discussions to have a shorthand (matrix) representation for the leverages. Referring to Eq. 6.4, we can write the h_{ii}, $i = 1, 2, 3$, as follows:

$$h_{11} = \begin{bmatrix} 1 & 0 \end{bmatrix} \begin{bmatrix} .8333 & -.1667 \\ -.1667 & .8333 \end{bmatrix} \begin{bmatrix} 1 \\ 0 \end{bmatrix}$$

$$h_{22} = \begin{bmatrix} .5 & .5 \end{bmatrix} \begin{bmatrix} .8333 & -.1667 \\ -.1667 & .8333 \end{bmatrix} \begin{bmatrix} .5 \\ .5 \end{bmatrix}$$

$$h_{33} = \begin{bmatrix} 0 & 1 \end{bmatrix} \begin{bmatrix} .8333 & -.1667 \\ -.1667 & .8333 \end{bmatrix} \begin{bmatrix} 0 \\ 1 \end{bmatrix}$$

each of which can be represented in matrix notation as

$$h_{ii} = \mathbf{x}_i'(\mathbf{X}'\mathbf{X})^{-1}\mathbf{x}_i \tag{6.6}$$

Table 6.5. *Leverages for two designs and two models*

ID	X_1	X_2	h_{ii} in $\mathbf{H}_1$	$\mathbf{H}_2$	$\mathbf{H}_3$	$\mathbf{H}_4$
1	1.0	0.0	0.8333	1.0	0.4167	0.5
2	1.0	0.0			0.4167	0.5
3	0.5	0.5	0.3333	1.0	0.1667	0.5
4	0.5	0.5			0.1667	0.5
5	0.0	1.0	0.8333	1.0	0.4167	0.5
6	0.0	1.0			0.4167	0.5
Scheffé model:			linear	quadratic	linear	quadratic
# Parameters (p):			2	3	2	3
# Design pts (n):			3	3	6	6

Note that the primes (′) are reversed in Eq. 6.6 compared to Eq. 6.3. This is because vectors are commonly represented as column vectors, and so $\mathbf{x}_i'$ is a *row* of $\mathbf{X}$.

We now summarize, without proofs, several important properties of the hat matrix. Proofs can be found in many books on regression analysis (see, for example, Montgomery, Peck, and Vining [100] or Myers [105]).

Property 6.1 The leverage value for a design point always lies in the interval

$$\frac{1}{n} \leq h_{ii} \leq \frac{1}{r_i}$$

where n is the total number of observations and r_i is the number of replicates of the design point. This means that the maximum leverage of a design point is 1.0, and that its leverage can be reduced by replication — as r_i gets larger, $1/r_i$ gets smaller. This effect can be seen by comparing the leverages for IDs 1, 3, and 5 in $\mathbf{H}_1$ vs. $\mathbf{H}_3$ and in $\mathbf{H}_2$ vs. $\mathbf{H}_4$ (Table 6.5).

A method for choosing replicate design points that makes use of leverage has been described by Montgomery and Voth [101]. The method amounts to choosing for replication those design points that have the highest leverage. To know which points have high leverage, one needs of course a table of leverage values. Because the leverage values are independent of the response, there is no reason to wait until data are collected and a model is fit to examine them. By then, it may be too late. It is *recommended*, therefore, that for those software packages that do not output leverages in advance of collecting data and fitting a model, that one generate a dummy response of random data and fit a model to the data. One can then examine the leverage values and make decisions about replication.

The method for choosing replicates described in the previous paragraph appears to be in possible conflict with the statements made on pages 81 and 91 that Design-Expert picks replicates using the D-optimality criterion. Actually there is no conflict, the reason being that the design point already in the design that has the maximum leverage is the point that when replicated will maximize the determinant of the (augmented) information matrix, $|\mathbf{X}'\mathbf{X}|$.

Property 6.2 Leverage is a standardized measure of the distance of the ith design point from the data center in the Xs. A large leverage indicates that the ith observation is remote from the center of the n observations. For example, Table 6.6 displays leverage values for the design points in a four-component simplex centroid design. As the dimensionality of a centroid increases, the centroid moves closer and closer to the center of the design, and leverages will tend to decrease.

Table 6.6. *Leverages for a four-component simplex centroid design. Linear and quadratic models*

	Leverage	
Point type (dimension)	Linear model	Quadratic model
Vertex (0)	0.5322	0.9772
Edge centroid (1)	.2218	.7645
Constraint-plane centroid (2)	.1184	.3358
Overall centroid (3)	.0667	.1615

Property 6.3 A response surface will be pulled toward observations with high leverage values. Fitted values and observed values will be close. When $h_{ii} = 1$, the fitted value will be equal to the observed value, and the response surface will pass through the point.

Although leverage is classified as an **influence diagnostic**, the fact that a point may have high leverage is not a guarantee that the point is going to significantly influence the regression coefficients. A point that is remote from the centroid of the **X** space will assuredly have high leverage, but if it follows the general trend of the data, then it may not be influential on the regression coefficients.

Property 6.4 Apart from σ^2, *the variances of fitted values are equal to the leverage values.*

$$\begin{aligned}\operatorname{var}(\hat{Y}_i \,|\, \mathbf{x}_i) &= \sigma^2 \cdot h_{ii} \\ &= \sigma^2 \cdot \mathbf{x}_i'(\mathbf{X}'\mathbf{X})^{-1}\mathbf{x}_i \end{aligned} \tag{6.7}$$

The matrix representation of this equation is

$$(\operatorname{var}(\hat{\mathbf{Y}}))_{n\times n} = \sigma^2 \cdot \mathbf{H}_{n\times n}$$

where $\hat{\mathbf{Y}}$ is a $n \times 1$ vector of fitted values and $\operatorname{var}(\hat{\mathbf{Y}})$ is a $n \times n$ matrix of variances and covariances. The variances of fitted values have both a *deterministic* component (a diagonal element of **H**) and a *stochastic* component (σ^2). Fitted values at high leverage points will have low precision. Thus replication will both decrease the self weight of a datum and increase the precision of the fitted value.

An important extension of Eq. 6.7, applicable to points $\mathbf{x}_0$ that are not necessarily design points (such as those in a candidate list), is the following:

$$\begin{aligned} \mathrm{var}(\hat{Y}_i \,|\, \mathbf{x}_0) &= \sigma^2 \cdot \mathbf{x}_0'(\mathbf{X}'\mathbf{X})^{-1}\mathbf{x}_0 \\ &= \sigma^2 \cdot h_{00} \end{aligned} \tag{6.8}$$

The symbol h_{00} will be used to distinguish *any* point in the design region, while the symbol h_{ii} will be reserved for the diagonal elements of the hat matrix. At any point $\mathbf{x} = \mathbf{x}_0$ in the design region, then, the **standard error of prediction** is given by

$$\begin{aligned} \mathrm{s.e.}(\hat{Y}_i \,|\, \mathbf{x}_0) &= s \cdot \sqrt{\mathbf{x}_0'(\mathbf{X}'\mathbf{X})^{-1}\mathbf{x}_0} \\ &= s \cdot \sqrt{h_{00}} \end{aligned}$$

Another extension applicable to points that are not in a design (such as those in a candidate list), is the following [3]:

$$|\mathbf{X}_{n+1}'\mathbf{X}_{n+1}| = |\mathbf{X}_n'\mathbf{X}_n| \cdot \{1 + \mathbf{x}_0'(\mathbf{X}_\mathbf{n}'\mathbf{X}_\mathbf{n})^{-1}\mathbf{x}_0\} \tag{6.9}$$

The left side of this equation is the determinant of the information matrix of an augmented $(n+1)$-point design. The right side is the product of the determinant of the information matrix of an existing, unaugmented n-point design and the quantity within the curly braces, which is a scalar quantity. Thus, the point $\mathbf{x}_0$ in a candidate list that will maximize $|\mathbf{X}_{n+1}'\mathbf{X}_{n+1})|$ is that point which has the largest value of $\mathbf{x}_0'(\mathbf{X}_n'\mathbf{X}_n)^{-1}\mathbf{x}_0$, *which is the point that has the largest prediction variance.* Equation 6.9 is the basis for the sequential construction of D-optimal designs.

Property 6.5 The fifth and last property is

$$\sum_{i=1}^{n} h_{ii} = p$$

where p is the number of model parameters. The matrix equivalent of this expression is $\mathrm{tr}(\mathbf{H}) = p$, where tr stands for the trace. Apart from σ^2, then, *the sum of the prediction variances over the design points is equal to the number of parameters in the model.* For example, if the leverages in Table 6.5 in columns $\mathbf{H}_1$–$\mathbf{H}_4$ are added together, the sums will be equal to 2 ($\mathbf{H}_1$ and $\mathbf{H}_3$) or 3 ($\mathbf{H}_2$ and $\mathbf{H}_4$), corresponding to the number of coefficients in the linear and quadratic models, respectively. Also, in Table 6.6, page 104, after making allowance for the fact that the simplex centroid design has four vertices, six edge centroids, four constraint plane centroids, and one overall centroid, the sum of the leverages will be equal (within rounding error) to 4 and 10 for the linear and quadratic models, respectively.

The implications of this property are most interesting. *For a given model,* the total prediction variance over the design points is a constant regardless of the number of data points. This means that by increasing the number of design points, the total prediction variance will be more evenly spread out, leading to a lower average prediction variance over the design points. This property also lends some credibility to the choice of parsimonious models in the model building process. Caution is in order here, however. While decreasing the number of terms in a model will reduce p, at some point s^2 will become biased because of lack of fit and will start to increase. Thus an underfitted model can lead to inflated prediction variances.

As there are always n leverages, the average prediction variance over the design points is equal to p/n. This provides a norm against which leverages can be gauged and led Belsley, Kuh, and Welsch [6] to suggest that a value that exceeds $2p/n$ be considered a high leverage value.

A three-component example will be used to illustrate some of these properties. The example is a mixture experiment involving a poultry-feed blend [160]. The constraints on the component proportions were

$$\begin{array}{ccccc} 0.3 & \leq & \text{Maize} & \leq & 0.8 \\ 0.0 & \leq & \text{Fish} & \leq & 0.3 \\ 0.0 & \leq & \text{Soybean} & \leq & 0.5 \end{array}$$

Fig. 6.1 illustrates the constrained region in the context of the full three-component simplex.

The 10 solid circles in the figure are the points selected by Design-Expert to support a quadratic Scheffé model. The points were selected from a 25-point candidate list consisting of six vertices, six edge centroids, six axial check blends, six interior points (points that lie

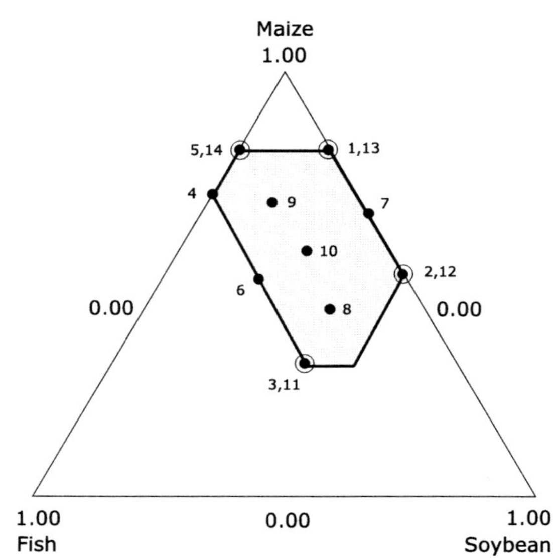

Figure 6.1. *Design points. Poultry-feed example* 1.

midway between the overall centroid and the edge centroids), and the overall centroid. The design points were chosen by specifying a-b-c = 6-4-0, where a is the number of points chosen by the D-optimality criterion to support the model, b is the number of lack-of-fit points selected by a distance-based criterion (which also support the model), and c is the number of replicates. The default values for a-b-c are 6-4-4, but we purposely chose 6-4-0 so that we can see how replicates are picked on the basis of leverage.

Table 6.7. *Leverages. Poultry-feed example* 1

Point		Design Size				
ID	Type[†]	10 pts	11 pts	12 pts	13 pts	14 pts
1	V	.7688	.7638	.7592	.4316	.4282
2	V	.8288	.8287	.4532	.4517	.4517
3	V	.8806	.4682	.4682	.4667	.4665
4	V	.6899	.6895	.6830	.6787	.6421
5	V	.7003	.6996	.6995	.6896	.4081
6	EC	.4016	.3955	.3574	.2896	.2716
7	EC	.5913	.5814	.5723	.5717	.5344
8	Ax	.4198	.3999	.3882	.3757	.3712
9	Ax	.3083	.2998	.2928	.2907	.2675
10	OC	.4106	.4055	.4048	.4040	.4040
11	V		.4682	.4682	.4667	.4665
12	V			.4532	.4517	.4517
13	V				.4316	.4282
14	V					.4081

[†]V = vertex, EC = edge centroid, Ax = axial check blend
OC = overall centroid

Leverages for the 10-point design are listed in Table 6.7 in the column labeled "10 pts". The ID value in the first column identifies the numbered points in Fig. 6.1. The point with the maximum leverage in the 10-point design is point 3 (a vertex), and so we replicate this point. In the column labeled "11 pts", points 3 and 11 are replicates. The point with the maximum leverage in the 11-point design is point 2 (a vertex), and so this point is replicated. Points 2 and 12 are now replicates. And so on. Replicated points in Fig. 6.1 are indicated by open circles surrounding solid circles. In going from the 10-point design to the 14-point design, the maximum leverage has decreased from 0.8806 to 0.6421. The sum of the leverages, however, has remained constant and equal to six, the number of terms in the quadratic Scheffé model. The 14-point design is the same design as that created by Design-Expert had we accepted the default a-b-c = 6-4-4.

As points are replicated and leverages drop, the prediction variances over the design points drop correspondingly. Clearly, it is desirable to have a design that minimizes prediction variances. A goal might be to seek a design that minimizes the *average* prediction variance, while another might be to seek a design that minimizes the *maximum* prediction variance. These goals are the basis of two alphabetic optimality criteria.

A ***V*-optimal design** is one that minimizes the average prediction variance (APV) over the design region. A ***G*-optimal design** is one that minimizes the maximum prediction variance (MPV) over the design region. In practice, both the average and maximum prediction variances are determined over sets of points that are software dependent. Some products evaluate APV and MPV over the design points, while others evaluate these over the candidate points. Keep in mind that estimating APV over the design points is not a particularly useful criterion, since, for a given model (and therefore p), all n-point designs will have the same APV.

A design criterion related to G-optimality is ***G* efficiency**, defined as

$$G \text{ efficiency} = 100 \cdot \left(\frac{p/n}{\text{MPV}} \right) \tag{6.10}$$

where p/n is the APV (apart from s^2) *over the design points* (Property 6.5) and the method used to evaluate MPV depends on the software one is using. Snee [154] suggested that from a practical point of view, a G-efficiency $\geq 50\%$ is a reasonable goal for a "good" design.

Table 6.8 summarizes APVs, MPVs, and G efficiencies for the designs in Table 6.7 as a function of four point sets — the design points, the candidate points, a calculated grid of points, and sets of random points. In all cases, G efficiencies were calculated using

Table 6.8. *Prediction-oriented criteria. Poultry-feed example* 1

	Design Size				
Criterion†	10 pts	11 pts	12 pts	13 pts	14 pts
			Over design points		
MPV	0.881	0.829	0.759	0.690	0.642
APV	0.600	0.545	0.500	0.462	0.429
G eff	68.1	65.8	65.8	66.9	66.7
SDC	105.7	104.6	103.2	101.8	100.4
			Over candidate points		
MPV	1.633	1.335	1.232	1.210	1.085
APV	0.569	0.519	0.482	0.455	0.418
G eff	36.7	40.9	40.6	38.2	39.5
			Over grid of points		
MPV	1.633	1.335	1.232	1.209	1.085
APV	0.458	0.430	0.402	0.380	0.357
G eff	36.7	40.9	40.6	38.2	39.5
			Over random points		
MPV	1.627	1.329	1.222	1.202	1.079
APV	0.445	0.419	0.392	0.372	0.350
G eff	36.9	41.0	40.9	38.4	39.7

†MPV = maximum prediction variance, APV = average prediction variance, G eff = G efficiency, SDC = scaled D-optimality criterion (pseudocomponent metric)

Eq. 6.10. The numerator is the APV in the design-point set ($6/n$ because $p = 6$ in a $q = 3$ quadratic Scheffé model, and n is the number of design points), while the denominator is the MPV in the set chosen to evaluate MPV. The third set, based on a grid of points, utilizes the grid described by Piepel, Anderson, and Redgate [126] in their study of response surface designs for irregularly shaped constrained regions. The grid consists of 1316 points formed by varying the proportions of the three components by 0.01 between their lower and upper bounds. The fourth set, based on random points, uses a method described by Borkowski [7]. The points consisted of 200 sets of 900 randomly generated points within the design region. For each set, APV and MPV were determined. The APVs reported in Table 6.8 for the random-point set are the means of the 200 sets. The MPVs are the maximum value of each set of 200.

To calculate design criteria over the candidate points requires MIXSOFT, ACED, or software capable of doing matrix calculations. To calculate the criteria over a grid of points or over randomly generated points requires software capable of doing matrix calculations. Although the reader may not have the necessary software to carry out these calculations, Table 6.8 carries a message. In this example, APVs calculated over the design points tend to somewhat overestimate average prediction variances and to severely underestimate maximum prediction variances compared to estimates made with a larger sampling of the design space. As a result, G efficiencies based on the design-point set are misleading. G efficiencies based on the candidate set are much better estimates of the true performance of the model over the design region. The bottom line is that APVs, MPVs, and G efficiencies based solely on design points may not always be reliable estimates of the performance of a model over the design region.

A second example using the poultry-feed constraints will illustrate a case where design criteria estimated over the design points *do* provide reasonably good estimates of the performance over the design region. In MINITAB, a candidate set was generated consisting of six vertices, six edge centroids, six axial check blends, plus the overall centroid, for a total of 19 points. (The 25-point Design-Expert candidate set included six interior points, but none of these were picked by the point selection algorithm.) One has a choice in MINITAB of selecting design points by the D-optimality criterion or by a distance-based criterion, but the two cannot be mixed. Using the D-optimality criterion, the design illustrated in Fig. 6.2, points 1–10, was generated. As with Fig. 6.1, the design is illustrated in the context of the full simplex.

Points 9 and 10, which are replicates of points 1 and 2, respectively, were chosen by MINITAB based on the D-optimality criterion. Points 11–14 in the figure are replicates picked by the "one-at-a-time method" used previously. The weighted average of the dimensions of the design points ($\overline{dim}$) in the 14-point design is 0.357. In the first example, the weighted average was 0.571, reflecting the fact that some of the design points were picked by a distance-based, rather than D-optimality, criterion. Recall that a-b-c was 6-4-0 in the first example. Setting a-b-c to 10-0-0 in Design-Expert will lead to exactly the same 10-point design as the design in MINITAB.

Table 6.9 displays design criteria for the 10- and 14-point designs for poultry-feed example 2. In this case, APVs, MPVs, and G efficiencies estimated over the design points are much better estimates of the values over the design space than was the case with poultry-feed example 1. Comparing the design criteria for the two 14-point designs (Tables 6.8 and 6.9), we see that the design in Table 6.9 has a lower maximum prediction variance,

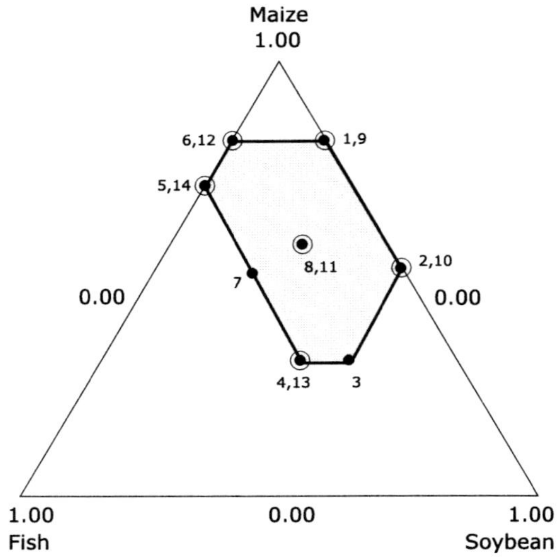

Figure 6.2. *Design points. Poultry-feed example* 2.

Table 6.9. *Prediction-oriented criteria. Poultry-feed example* 2

Criterion	Design Size: 10 pts	14 pts
Over design points		
MPV	0.826	0.554
APV	0.600	0.429
G eff	72.7	77.4
SDC	89.8	88.8
Over candidate points		
MPV	0.826	0.554
APV	0.566	0.372
G eff	72.7	77.4
Over grid of points		
MPV	0.838	0.554
APV	0.532	0.331
G eff	71.6	77.4
Over random points		
MPV	0.838	0.549
APV	0.533	0.328
G eff	71.6	78.0

higher G-efficiency, and a lower scaled D-optimality criterion. On the basis of prediction variances, the results imply that poultry-feed example 2 is the better design. Keep in mind, however, that poultry-feed example 1 has a better sampling of the interior of the design region and will provide better protection against bias in the event that the response surface is of a higher order. The two designs illustrate the variance-bias trade-off that one constantly encounters.

A graphical approach for evaluating mixture designs that has certain advantages over the single-number summaries such as APV and MPV has been described by Vining, Cornell, and Myers [166]. This amounts to plotting h_{00} (Eq. 6.8, page 105) along the Cox-effect directions (cf. page 83). Such plots have been referred to as **prediction variance trace plots**. The Cox-effect directions for the poultry-feed example are illustrated in Fig. 6.3. The dashed lines labeled A, B, and C are the Cox-effect directions for maize, fish, and soybean, respectively. The point where the three Cox-effect directions cross is termed the **reference blend**, or **base point**. (The term "base point" should not to be confused with base points in the XVERT algorithm.) The base point can be chosen anywhere within the design region, although most often it is taken as the overall centroid of the design space. In this example, the overall centroid has been chosen, the composition of which is maize : fish : soybean = 0.5667 : 0.1667 : 0.2667.

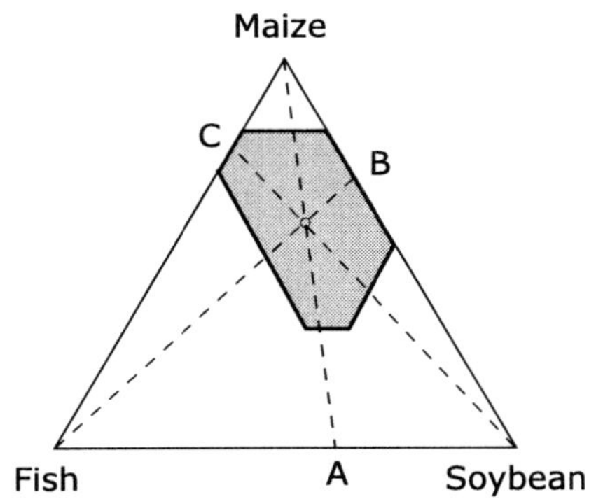

Figure 6.3. *Cox-effect directions. Poultry-feed examples.*

Along the maize Cox-effect direction, the proportions of fish and soybean are in a constant ratio to one another (0.1667/0.2667 = 0.625). Along the fish Cox-effect direction, maize and soybean are in a fixed ratio to one another (0.5667/0.2667 = 2.125). And along the soybean Cox-effect direction, maize and fish are in a constant ratio to one another (0.5667/0.1667 = 3.40).

Fig. 6.4 shows Design-Expert plots of $\sqrt{h_{00}}$ along the three effects directions for the 10- and 14-point designs in Fig. 6.2 and Table 6.9. The principal difference between these plots and those of Vining, Cornell, and Myers [166] is that the ordinate is equal to $\sqrt{h_{00}}$ rather than h_{00}. The value of the ordinate at the intersection of the dashed lines and the three curves is equal to $\sqrt{h_{00}}$ at the base point, which in this case is the overall centroid. The curves to the right of the dashed lines show how $\sqrt{h_{00}}$ changes as one proceeds in a positive direction toward each X_i along the three Cox-effect directions, and

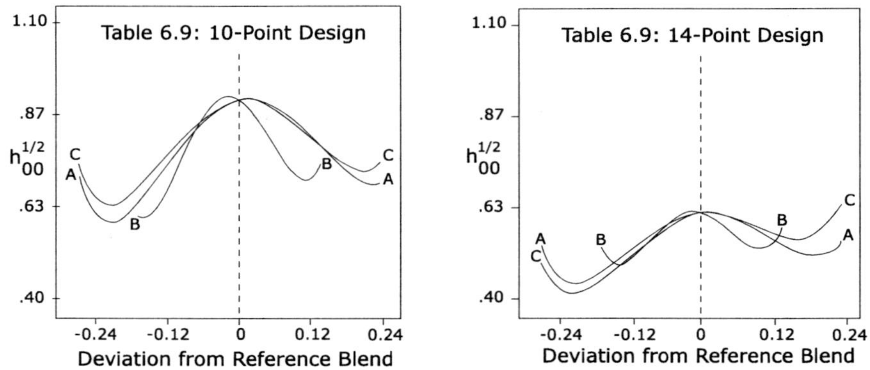

Figure 6.4. *Trace plots of the standard error of prediction. Poultry-feed example* 2.

of course vice versa to the left of the dashed lines. While such plots may not be a perfect solution to understanding how prediction variances (or standard errors of prediction) change throughout a design region, they do have the advantage of affording a two-dimensional plot for multidimensional problems. Contour and response-surface plots are of limited value, as contours or surfaces can be viewed for only three components at a time. Of the software products used for examples in this book, Design-Expert is the only product that outputs trace plots like those in Fig. 6.4. Version 7 has options to plot either h_{00} or $\sqrt{h_{00}}$, and either can be scaled by n (cf. the discussion on page 114).

The most obvious feature in Fig. 6.4 is the large hump in the traces at the center of the 10-point design. Standard errors of prediction have decreased and been evened out rather well in the 14-point design. A less obvious feature of both plots in Fig. 6.4 is that the traces for B (fish) are much shorter than those for A and C (maize and soybean, respectively). This is a reflection of the relative ranges of the components (Fig. 6.3). A subtle feature of the plots is that the traces are longer on the left side of the dashed lines than on the right side. This is because the averaged-extreme-vertices (AEV) centroid is not perfectly centered in the design region. Careful inspection of Fig. 6.3 will reveal that the distances between the centroid and the low ends of the ranges of all three components is slightly greater than the distances between the centroid and the high ends of the ranges.

Another approach to evaluating the variance properties of mixture designs is the **variance dispersion graph** (VDG). VDGs were introduced by Giovannitti-Jensen and Myers [54] in the context of nonmixture response surface designs, such as the central composite design. The method has been adapted to irregularly shaped regions by Piepel et al. [125, 126]. The algorithm for generating the plots is described in detail in the 1992 article [125], while several examples are presented in the 1993 article [126]. Briefly, the method consists of generating a large number of points on the boundary of the design region of interest. The boundary points are then progressively shrunk by discrete increments towards the center of the design to create a series of "shells". This is illustrated in Fig. 6.5 for the poultry-feed design region. The outer shell is the boundary of the design region. Points on the inner shells are calculated using the equation

$$\mathbf{s} = \mathbf{c} + f(\mathbf{x} - \mathbf{c})$$

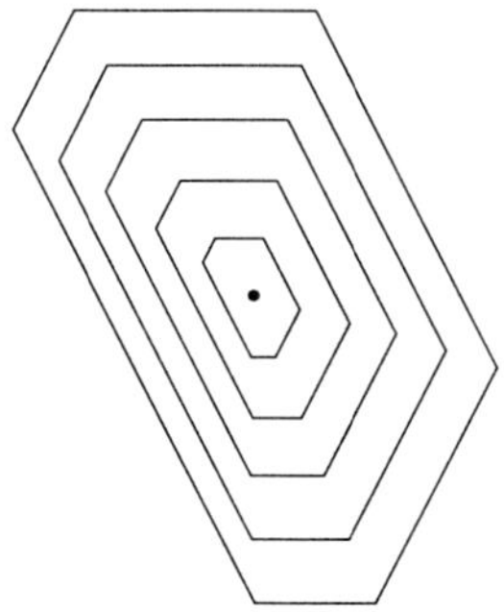

Figure 6.5. *Shrunken regions.*

where **x** is a point on the boundary, **c** is the centroid of the design region, and f is a shrinkage factor. The shrinkage factors in the illustration are 0 (the centroid), 0.2, 0.4, 0.6, 0.8, and 1.0 (the boundary).

To create a VDG, the maximum, minimum, and average prediction variances (apart from σ^2) are evaluated on each shell using Eq. 6.8, page 105. The results are then plotted against the shrinkage factor, f. Figure 6.6 displays the results for the 10- and 14-point poultry-feed designs illustrated in Fig. 6.2. Calculations for the plot were based on 270 points on each of 50 shells.

When comparing competing designs with different numbers of runs, some workers use **scaled prediction variances** for single-number summaries or VDGs [107]. This is done

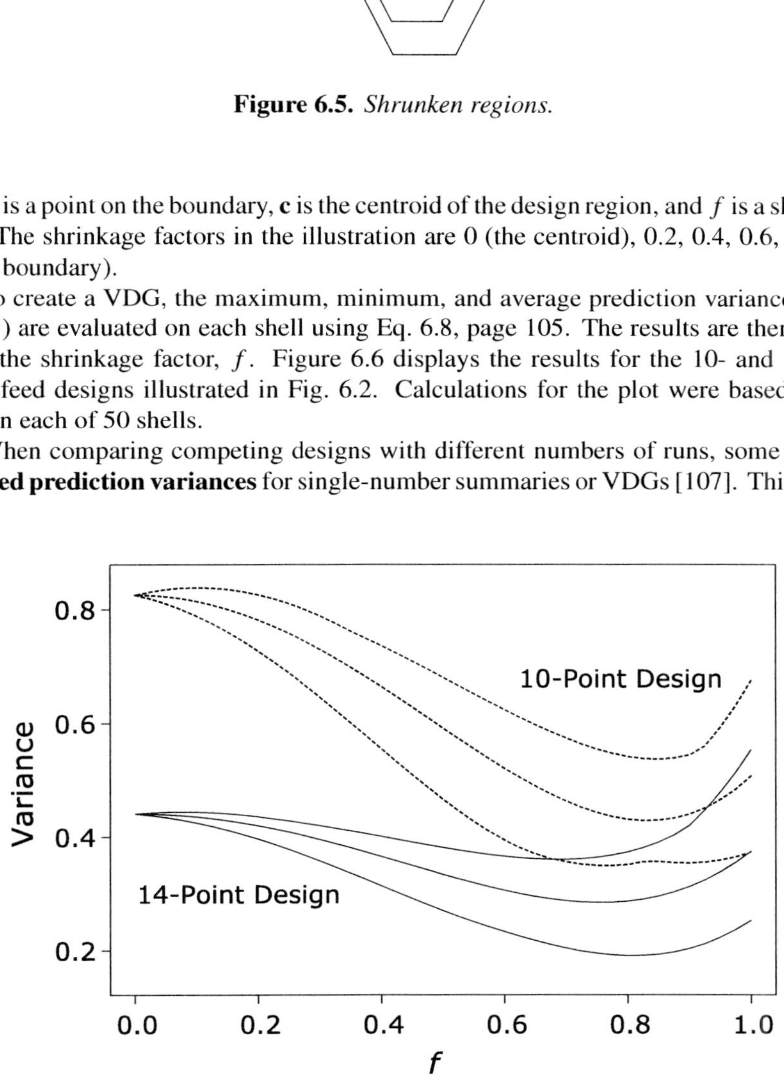

Figure 6.6. *Variance dispersion graph.* 10- *and* 14-*point designs for poultry-feed example* 2.

by modifying Eq. 6.8 as follows:

$$\frac{n \cdot \operatorname{var}(\hat{Y}_i \,|\, \mathbf{x}_0)}{\sigma^2.} = n \cdot \mathbf{x}_0'(\mathbf{X}'\mathbf{X})^{-1}\mathbf{x}_0 \tag{6.11}$$

where n is equal to the number of runs. The rationale for this is that additional runs usually come at a cost. The scaling by n penalizes the design with the more runs.

Prediction variance trace plots only sample the variance along an effect direction, and so one cannot be sure what the variances might be in other regions of the design space. Variance dispersion graphs, on the other hand, give one a more global view of how variances change throughout a design region than do trace plots. Khuri, Harrison, and Cornell [80] claim that even more information about the prediction capability of a model can be obtained by knowing the *distribution* of the prediction variances than by examining variance dispersion graphs. The authors propose plotting quantiles of the scaled prediction variances for a large number of points (10,000 in their example) within a constrained region, and they refer to these plots as SPVQ (scaled prediction variance quantile) plots. The authors demonstrate their method by comparing SPVQ plots for four designs for a fertilizer experiment. An S-PLUS program for constructing SPVQ plots is available at

http://ifasstat.ufl.edu/spvplots/

Leverages also play a part in influencing the sensitivity (or conversely, robustness) of fitted values to outliers. The following discussion is based on an article by Box and Draper [11]. Referring to Eq. 6.2, page 100, the ith fitted value, $\hat{Y}_i$, is given by

$$\hat{Y}_i = \begin{bmatrix} h_{i1} & \dots & h_{iu} & \dots & h_{in} \end{bmatrix} \begin{bmatrix} Y_1 \\ \vdots \\ Y_u \\ \vdots \\ Y_n \end{bmatrix}$$

Thus, a $\hat{Y}_i$ is a linear combination of the Y_i, $i = 1, \dots, n$, and the coefficients in the linear combination are the elements in the ith row of the hat matrix. Suppose that the uth observation, Y_u, has added to it an aberration, c, and so $Y_u + c$ is now an outlier. The *change* in $\hat{Y}_i$ caused by c will be $\hat{\delta}_i = c \cdot h_{iu}$. However, all n fitted values — not just $\hat{Y}_i$ — will be affected by c. As a result, we can write

$$\hat{\boldsymbol{\delta}}_u = \begin{bmatrix} \hat{\delta}_1 \\ \hat{\delta}_2 \\ \vdots \\ \hat{\delta}_n \end{bmatrix} = c \cdot \begin{bmatrix} h_{1u} \\ h_{2u} \\ \vdots \\ h_{nu} \end{bmatrix} = c \cdot \mathbf{h}_u$$

where $\hat{\delta}_1$ is the change in $\hat{Y}_1$ caused by the aberration in Y_u, $\hat{\delta}_2$ is the change in $\hat{Y}_2$ caused by the aberration in $Y_u, \dots,$ and $\hat{\delta}_n$ is the change in $\hat{Y}_n$ caused by the aberration in Y_u. Note that $\hat{\boldsymbol{\delta}}_u$ is boldface to symbolize the vector of elements $\hat{\delta}_1, \hat{\delta}_2, \ldots, \hat{\delta}_n$. The column vector, $\mathbf{h}_u$, is the uth column of the hat matrix, and it is the elements of this vector that determine the relative magnitudes of the $\hat{\delta}_i$ s as a consequence of the aberration, c.

It would be useful to have a single-number summary of the individual $\hat{\delta}_i$s, $i = 1, 2, \ldots, n$. Let us define the overall discrepancy, d_u, caused by the effect of c on the n fitted values as the squared Euclidean length of the vector, $\hat{\boldsymbol{\delta}}_u$. Thus we have $d_u = \hat{\boldsymbol{\delta}}'_u \hat{\boldsymbol{\delta}}_u$, which in scalar notation is

$$d_u = c^2 \cdot \sum_{i=1}^{n} h_{iu}^2 = c^2 \cdot h_{uu} \tag{6.12}$$

In other words, the *overall discrepancy* d_u caused by the aberration c in the uth observation on the n fitted values is proportional to the leverage of the uth observation.

The equality $\sum_{i=1}^{n} h_{iu}^2 = h_{uu}$ in Eq. 6.12 is a consequence of the fact that the hat matrix is an **idempotent matrix**, which has the property $\mathbf{HH} = \mathbf{H}$. An example of such a matrix is

$$\mathbf{H} = \begin{bmatrix} 5/6 & 1/3 & -1/6 \\ 1/3 & 1/3 & 1/3 \\ -1/6 & 1/3 & 5/6 \end{bmatrix} = \frac{1}{6}\begin{bmatrix} 5 & 2 & -1 \\ 2 & 2 & 2 \\ -1 & 2 & 5 \end{bmatrix}$$

for which

$$\begin{aligned}
\mathbf{HH} &= \frac{1}{6}\begin{bmatrix} 5 & 2 & -1 \\ 2 & 2 & 2 \\ -1 & 2 & 5 \end{bmatrix} \frac{1}{6}\begin{bmatrix} 5 & 2 & -1 \\ 2 & 2 & 2 \\ -1 & 2 & 5 \end{bmatrix} \\
&= \frac{1}{36}\begin{bmatrix} 30 & 12 & -6 \\ 12 & 12 & 12 \\ -6 & 12 & 30 \end{bmatrix} \\
&= \frac{1}{6}\begin{bmatrix} 5 & 2 & -1 \\ 2 & 2 & 2 \\ -1 & 2 & 5 \end{bmatrix}
\end{aligned}$$

To illustrate the equality $\sum_{i=1}^{n} h_{iu}^2 = h_{uu}$ in Eq. 6.12, take column 3 as an example. For this column,

$$\begin{aligned}
\sum_{i=1}^{3} h_{i3}^2 &= \left(-\frac{1}{6}\right)^2 + \left(\frac{2}{6}\right)^2 + \left(\frac{5}{6}\right)^2 \\
&= \frac{30}{36} \\
&= \frac{5}{6} \\
&= h_{33}
\end{aligned}$$

Equation 6.12 assumes that it is the uth observation that experiences the aberration, c. Suppose, however, that it is equally likely that the aberration could occur with any of the n observations, Y_i. This would give rise to n mutually exclusive discrepancies d_u, $u = 1, 2, \ldots, n$, the average of which would be

$$\begin{aligned}\bar{d} &= \frac{1}{n}\sum_{u=1}^{n} d_u \\ &= c^2 \cdot \left(\frac{\sum_{u=1}^{n} h_{uu}}{n}\right) \\ &= c^2 \cdot \bar{h}_{uu}\end{aligned}$$

A worthwhile endeavor would be to minimize the maximum d_u, which is equivalent to the G-optimality criterion. Another worthwhile endeavor would be to ensure that the individual d_u s are as close to the average, $\bar{d}$, as possible. This is equivalent to maximizing $\bar{d}/d_u$ (or equivalently, $\bar{h}_{ii}/h_{ii}$), which is the G-efficiency *over the design points*. Stated in other words, the more closely the maximum leverage approaches the average leverage, the more evenly will an aberration in any one observation be spread over the $\hat{Y}_i$ s. One way to reduce the maximum leverage is, of course, to replicate. Furthermore, recalling that $\bar{h}_{ii} = p/n$, this provides further impetus for parsimonious models.

An illustrative example should prove helpful. Consider the hat matrix $\mathbf{H}_1$ (Eq. 6.4, page 101). For simplicity, assume an aberration c = 1.0. If the aberration occurs with observation number one, then we can write

$$\hat{\boldsymbol{\delta}}_1 = \begin{bmatrix} .8333 & .3333 & -.1667 \\ .3333 & .3333 & .3333 \\ -.1667 & .3333 & .8333 \end{bmatrix}\begin{bmatrix} 1 \\ 0 \\ 0 \end{bmatrix} = \begin{bmatrix} .8333 \\ .3333 \\ -.1667 \end{bmatrix}$$

The column vector, $\hat{\boldsymbol{\delta}}_1$, is equal to the first column of the hat matrix. The elements of this vector, $\hat{\delta}_1$, $\hat{\delta}_2$, and $\hat{\delta}_3$, can be plotted as a single point in a coordinate system composed of three orthogonal axes. The squared distance of this point from the origin is equal to $\hat{\boldsymbol{\delta}}_1'\hat{\boldsymbol{\delta}}_1 = 0.8333$, and so $d_1 = 0.8333 = h_{11}$.

If the aberration occurs with observation number two, then we have

$$\hat{\boldsymbol{\delta}}_2 = \begin{bmatrix} .8333 & .3333 & -.1667 \\ .3333 & .3333 & .3333 \\ -.1667 & .3333 & .8333 \end{bmatrix}\begin{bmatrix} 0 \\ 1 \\ 0 \end{bmatrix} = \begin{bmatrix} .3333 \\ .3333 \\ .3333 \end{bmatrix}$$

and with observation number three,

$$\hat{\boldsymbol{\delta}}_3 = \begin{bmatrix} .8333 & .3333 & -.1667 \\ .3333 & .3333 & .3333 \\ -.1667 & .3333 & .8333 \end{bmatrix}\begin{bmatrix} 0 \\ 0 \\ 1 \end{bmatrix} = \begin{bmatrix} -.1667 \\ .3333 \\ .8333 \end{bmatrix}$$

Thus

$$\begin{aligned} d_2 &= \hat{\boldsymbol{\delta}}_2' \hat{\boldsymbol{\delta}}_2 \\ &= 0.3333 \\ &= h_{22} \end{aligned}$$

and

$$\begin{aligned} d_3 &= \hat{\boldsymbol{\delta}}_3' \hat{\boldsymbol{\delta}}_3 \\ &= 0.8333 \\ &= h_{33} \end{aligned}$$

As a result, $\bar{d} = 0.6666$ and $\bar{d}/d_{max} = 0.800$.

The net result of this and the previous chapter is that when designing for irregularly shaped regions using computer-aided design, most of the commonly available software packages will use either the D-optimality criterion or a distance-based criterion. Design evaluation and improvement can be achieved using elements of the G- and V-optimality criteria. As referenced above, Snee's [154] suggestion of a G-efficiency $\geq 50\%$ as well as Belsley, Kuh, and Welsch's [6] suggestion that $h_{ii} < 2p/n$ provide reasonable norms against which to evaluate irregularly shaped designs. The importance of replication cannot be underestimated, as it not only provides degrees of freedom for estimating pure error, but it also plays an important role in minimizing prediction variances and providing robustness to wild observations.

It is worth reiterating that for those software packages that do not output the diagonal elements of the covariance and hat matrices, one should generate a dummy response that is normally distributed with a standard deviation of ~ 1.0. The simulated data can then be fit to models of interest, with a request that leverages also be output. One can then examine the standard errors of the coefficient estimates and the leverages and make decisions as to whether the range of one or more of the components should perhaps be increased or whether there are any high influence points.

Chapter 7

Blocking Mixture Experiments

It is often the case that an experiment is too large to be run on one day, to be run by a single operator, to be analyzed on a single instrument, or to be completed with one batch of raw material. One can envision other scenarios in which all of the runs cannot be performed under homogeneous conditions. Known or suspected discrete changes, such as day-to-day or batch-to-batch variations, are *controllable* nuisance variables — controllable in the sense that one can choose to assign some of the treatment combinations to one day (for example), others to a different day, and remove the effect of day by a method called **blocking**. Preplanned blocking is a design approach that, insofar as possible, minimizes the effect of variability that could arise from step changes.

Preplanned blocking may also be used as a means for sequentially building designs of increasing order. For example, one could design an experiment in two blocks capable of supporting a second-order mixture model, although each block may only support a first-order model. If one has degrees of freedom for lack of fit and pure error built into each block, then one could begin with an experiment that uses only one block. Should a lack-of-fit test suggest the need for model augmentation, then one can run the second block. The focus of this chapter will be on preplanned blocking.

Other situations can benefit from blocking. A common situation would be when there is lack of fit in a fitted model, and there are insufficient design points to support a higher-order model. The difference between this scenario and that in the previous paragraph is that in this case one has not planned in advance on blocking and discovers the need a posteriori. If one desires to fit a higher-order model, then the design needs to be augmented. In all likelihood the additional mixture blends will be prepared on a different day, possibly with a different operator or perhaps using a different batch of raw material. In these situations one would also want to consider blocking the additional runs. This situation is discussed at the end of Section 10.2.

7.1 Symmetrically Shaped Design Regions

The blocking method described in this section requires that each of the q components be able to assume each of the component proportions. For example, if $X_1 = 0.7$ in one formulation, then it is required that there be formulations where $X_2, X_3, \ldots, X_q$ assume the value 0.7. As the design points must be symmetrically disposed throughout the design region, this implies that these designs will "fit" best in design regions that are symmetrical or nearly symmetrical. Simplexes would be a subset of this class. More generally, any constrained region that arises as a result of applying the same constraints to each component will lead to a symmetrically shaped design region.

To illustrate the consequences that can ensue if one neglects to compensate for unwanted discrete changes, consider a hypothetical four-component mixture experiment (Table 7.1). Assume that the data were collected in two different ways. In one case, the data were collected on a single day (day 1), leading to the response labeled Y_1. In the second case, the data were collected on two other days (days 2 and 3), leading to the response Y_2. In this second case, runs 1–5 were carried out on day 2, while runs 6–10 were carried out on day 3. To simulate a day effect (whatever the cause may be), 2.0 has been added to Y_1 (runs 1–5) and subtracted from Y_1 (runs 6–10), giving the results for Y_2. The response Y_1 was simulated using the model

$$Y_1 = 10A + 20B + 30C + 40D$$

with added noise having a standard deviation of 1.0.

Table 7.1. *Hypothetical design. Blocking arrangement* A

Run		Proportion					
No.	Block	A	B	C	D	Y_1	Y_2
1	1	0.10	0.20	0.30	0.40	30.93	32.93
2	1	0.20	0.30	0.40	0.10	25.25	27.25
3	1	0.30	0.40	0.10	0.20	22.30	24.30
4	1	0.10	0.40	0.30	0.20	26.16	28.16
5	1	0.20	0.10	0.40	0.30	27.81	29.81
6	2	0.40	0.10	0.20	0.30	24.01	22.01
7	2	0.25	0.25	0.25	0.25	26.61	24.61
8	2	0.30	0.20	0.10	0.40	26.41	24.41
9	2	0.40	0.30	0.20	0.10	20.29	18.29
10	2	0.25	0.25	0.25	0.25	24.45	22.45

Fitting a Scheffé linear model to Y_1 values leads to

$$\hat{Y}_1 = 9.434A + 21.22B + 30.86C + 40.17D \tag{7.1}$$

which is a reasonable estimate of the simulation model. Summary statistics are $R^2 = 0.9530$, $R^2_{adj} = 0.9294$, $R^2_{pred} = 0.9016$, and $s^2 = 0.612$.[1] There is no indication of lack of fit. Fitting a Scheffé linear model to the Y_2 values leads to

$$\hat{Y}_2 = -0.566A + 26.22B + 35.86C + 40.17D \tag{7.2}$$

[1]The summary statistics R^2, R^2_{adj}, and R^2_{pred} are defined in Section 8.3, beginning on page 172.

which is a disappointing estimate of the simulation model. Summary statistics are $R^2 = 0.8775$, $R^2_{adj} = 0.8163$, $R^2_{pred} = 0.6969$, and $s^2 = 3.282$.

Realizing that the source of the problem is the day effect, one could instead fit the data to a linear Scheffé model augmented by a term to account for the day effect. That is, one could *block on days*. The model is

$$Y_2 = \beta_A A + \beta_B B + \beta_C C + \beta_D D + \gamma z \tag{7.3}$$

where z is a categorical variable that codes the effect of day, and γ is a parameter. The model assumes that the blending properties of the mixture are not affected by day — that is, day does not interact with the $X_i = A, \ldots, D$. The effect of day is simply to offset the responses by a fixed amount.

A common method for coding (used by Design-Expert and MINITAB, for example) is called **effect coding** [91]. (More is said about coding below.) For this example, z would take values of $+1$ for runs 1–5 (which we shall call block 1) and values of -1 for runs 6–10 (block 2).

Fitting the Y_2 data to model 7.3 leads to

$$\hat{Y}_2 = 8.884A + 21.50B + 31.13C + 40.17D + 1.890z \tag{7.4}$$

Summary statistics are $R^2 = 0.9457$, $R^2_{adj} = 0.9132$, $R^2_{pred} = 0.8405$, and $s^2 = 0.723$. While the mixture part of this model is not a perfect representation of the simulation model, it is certainly a vast improvement over Eq. 7.2. The *fitted* values $\hat{Y}_2$ are calculated using all five terms. If one were to use the model for interpretation or to *predict* how the response depends on the blending properties of the components, one would ignore the day effect and use the mixture part of the model only.

Suppose instead of blocking according to Table 7.1, the blocking was carried out according to Table 7.2. The mixture blends and the responses Y_1 are the same in Tables 7.1 and 7.2, but the run order has been changed. As before, 2.0 has been added to Y_1 (runs 1–5) and subtracted from Y_1 (runs 6–10) to simulate a day (block) effect, giving the results for Y_2.

Table 7.2. *Hypothetical design. Blocking arrangement* B

Run No.			Proportion					
New	Original	Block	A	B	C	D	Y_1	Y_2
1	1	1	0.10	0.20	0.30	0.40	30.93	32.93
2	2	1	0.20	0.30	0.40	0.10	25.25	27.25
3	3	1	0.30	0.40	0.10	0.20	22.30	24.30
4	6	1	0.40	0.10	0.20	0.30	24.01	26.01
5	7	1	0.25	0.25	0.25	0.25	26.61	28.61
6	4	2	0.10	0.40	0.30	0.20	26.16	24.16
7	5	2	0.20	0.10	0.40	0.30	27.81	25.81
8	8	2	0.30	0.20	0.10	0.40	26.41	24.41
9	9	2	0.40	0.30	0.20	0.10	20.29	18.29
10	10	2	0.25	0.25	0.25	0.25	24.45	22.45

Fitting model 7.3 to the Y_2 data leads to

$$\hat{Y}_2 = 9.434A + 21.22B + 30.86C + 40.17D + 2.398z \qquad (7.5)$$

The mixture part of this model exactly reproduces model 7.1. The effect of day has been removed from the mixture part of the model. Summary statistics are $R^2 = 0.9727$, $R^2_{adj} = 0.9563$, $R^2_{pred} = 0.8985$, and $s^2 = 0.417$.

To understand how the different models resulting from designs A and B came about, let us examine the correlation matrix of regression coefficients for the two design/model combinations (Table 7.3). In design A, the coefficient for block (γ) is correlated with the linear coefficients of the Scheffé model (with the exception of β_D), or equivalently, the coefficients of the linear mixture model are correlated with the block effect. This means that the coefficients for the mixture part of the model are contaminated by the block effect. In design B, the coefficient for block is uncorrelated with the linear coefficients of the Scheffé model. Block and mixture are *orthogonal* to one another. As a result, the coefficients in the mixture part of the model are unaffected by the block effect. *This is desirable.*

Table 7.3. *Correlation matrices of regression coefficients for blocking arrangements* A *and* B

	Block	*A*	*B*	*C*	*D*
		Blocking arrangement A			
Block	1.000				Symmetric
A	0.6831	1.000			
B	−0.4236	−0.6045	1.000		
C	−0.4236	−0.2790	−0.2114	1.000	
D	0.0000	−0.3479	0.01427	−0.4315	1.000
		Blocking arrangement B			
Block	1.000				Symmetric
A	0.0000	1.000			
B	0.0000	−0.4764	1.000		
C	0.0000	0.01575	−0.4764	1.000	
D	0.0000	−0.4764	0.01575	−0.4764	1.000

Before explaining how the two run orders lead to different results, a word should be said about coding methods for categorical variables in general and blocks in particular. The method used here has been called effect coding [91] or **sum-to-zero coding** [96]. Another method, set-to-zero coding, is explained below. With either coding, the number of categorical variables is always one fewer than the number of blocks. For example, two blocks are coded with a single variable (say z), three blocks are coded by two variables (say z_1 and z_2), four blocks by three variables, and so on. For these three cases, the effect coding would be

Block	z
1	+1
2	−1

Block	z_1	z_2
1	+1	0
2	0	+1
3	−1	−1

Block	z_1	z_2	z_3
1	+1	0	0
2	0	+1	0
3	0	0	+1
4	−1	−1	−1

As an example, when there are three blocks, the average of the block means is

$$\bar{\bar{Y}} = \frac{\bar{Y}_{Block1} + \bar{Y}_{Block2} + \bar{Y}_{Block3}}{3}$$

For each of the three blocks, there is a block *effect*, which is the difference between each block mean and the average of the block means ($\bar{\bar{Y}}$). For Blocks 1 and 2 the block effects are

$$\gamma_1 = \bar{Y}_{Block1} - \bar{\bar{Y}}$$
$$\gamma_2 = \bar{Y}_{Block2} - \bar{\bar{Y}}$$

The terms $\gamma_1 z_1$ and $\gamma_2 z_2$ will appear in the model equation. A term for Block 3, however, will not be present in the model. The block with -1s is called the *omitted level.*

Because the block effects are deviations of the block means from their average, some will be positive and some will be negative. Overall the block effects will *sum to zero.*

$$\gamma_1 + \gamma_2 + \gamma_3 = 0$$

Because of this equality, the implied coefficient for Block 3 is $-(\gamma_1 + \gamma_2)$. In the general case, the implied coefficient for the omitted level will be $-(\gamma_1 + \gamma_2 + \cdots + \gamma_{b-1})$, where b is the number of blocks.

The other common method of coding (used by SAS, for example) has been called **dummy-variable coding** [91] or **set-to-zero coding** [96]. As with effect coding, the number of variables is one less than the number of blocks. The coding for the three examples would be

Block	z
1	1
2	0

Block	z_1	z_2
1	1	0
2	0	1
3	0	0

Block	z_1	z_2	z_3
1	1	0	0
2	0	1	0
3	0	0	1
4	0	0	0

In this case the omitted level is a *comparison* level, and its value is *set to zero* on each dummy variable. With the exception of the comparison level, a dummy variable is coded 1 for all observations in the same block and 0 for all other observations. Again, there will be one less coefficient in the block part of the model than there are blocks. The mixture part of the model describes the blending properties of the components *at the comparison level.*

To illustrate the difference between the two methods of coding, if the data for Y_2 in Table 7.2 are fit to model 7.3 with dummy-variable coding for z, then the resulting model is

$$\hat{Y}_2 = 7.037A + 18.824B + 28.461C + 37.774D + 4.796z \tag{7.6}$$

which does not look anything like model 7.5. In fact, the two models describe exactly the same response surface. If half of the coefficient for z (2.398) is added to each of the linear coefficients in the mixture part of the model, one will reproduce the coefficients in the mixture part of model 7.5. Notice that half of the coefficient for z in this model is exactly equal to the coefficient for z in model 7.5.

Usually the choice of coding method will be driven by the software that one is using. If one is doing one's own programming (as in S-PLUS, GAUSS, or MATLAB), then one has the advantage of being able to choose. Effect coding is the most logical choice when one wants to "regress out" the effect of block. One's interest is usually focused on *removing* the effect of block rather than studying the effect of block.

Dummy-variable coding, on the other hand, has the advantage of referencing all fitted values to a comparison level. This can be a useful coding for categorical process variables when one knows (or is willing to assume) that the process variable does not interact with the X_is. A block is simply one level of a categorical variable that one could call *block*. Examples of other types of categorical variables are *flour type* in the making of bread, *number of passes* through a dispersion mill in the making of photographic dispersions, or *annealing oven* in the heat treatment of alloys. With dummy variable coding, one may be interested in testing the significance of the γ_js to see if the levels of the categorical process variable influence the response relative to a comparison, or control, level.

To simplify the interpretation of a model such as Eq. 7.3, it is desirable to have the coefficients of the mixture part of the model (the β_i) estimated independently of the block parameters (the γ_j). That is, ideally the design should be orthogonally blocked, as in Table 7.2. The conditions for doing this were established by John [75], who modified earlier work by Nigam [114]. Since then, several articles have appeared on this subject in journal articles (see, for example, [15, 48, 90, 138, 139]).

If one associates the letters a, b, c, and d with the proportions 0.1, 0.2, 0.3, and 0.4 in Table 7.2, then, apart from the center points (0.25, 0.25, 0.25, 0.25), the blocks can be represented as in Eq. 7.7. The patterns in Blocks 1 and 2 are **Latin squares**. A Latin square has the property that each symbol appears once and only once in each row and column.

$$
\begin{array}{cc}
\text{Block 1} & \text{Block 2} \\
a\ b\ c\ d & a\ d\ c\ b \\
b\ c\ d\ a & b\ a\ d\ c \\
c\ d\ a\ b & c\ b\ a\ d \\
d\ a\ b\ c & d\ c\ b\ a
\end{array}
\tag{7.7}
$$

The condition for orthogonal blocks when fitting a *linear* Scheffé model is [75]

$$
\sum_{u=1}^{n_w} X_{ui} = k_i \quad \text{for each block and } i = 1, 2, \ldots, q \tag{7.8}
$$

where u is the run number in the wth block, the wth block contains n_w runs, and the k_i are constants. In other words, if one sums the component proportions for the ith component in each block, the sums for each block are equal to one another. This does not mean that the sum for X_1 must equal the sum for X_2 (for example) but only that the sum for X_1 is the same across blocks. As every column in a Latin square contains the same letters, Eq. 7.8 will automatically be satisfied no matter how the Latin square is picked. Further conditions apply when one desires orthogonal blocking for quadratic mixture models (see below).

To illustrate with a numerical example, consider the component proportions in Table 7.1, page 120. The column sums in Block 1 are 0.9, 1.4, 1.5, and 1.2 for A, B, C, and D, respectively. In Block 2 the sums are 1.6, 1.1, 1.0, and 1.3. Eq. 7.8 is not satisfied, and the design does not block orthogonally.

On the other hand, consider the component proportions in Table 7.2, page 121. For each block, $\sum_{u=1}^{5} X_{ui} = 1.25$ for $i = 1, 2, 3, 4$, and so the design blocks orthogonally. Assume that we include the additional mixture $(A, B, C, D) = (0.25, 0.15, 0.40, 0.20)$ in each block. Now $\sum_{u=1}^{6} X_{ui} = 1.50, 1.40, 1.65$, and 1.45 for $i = 1, 2, 3, 4$ for each block. This illustrates the concept that one can augment the runs in each block with additional runs beyond those based on the Latin squares. As long as the additional runs have the same composition in all blocks, Eq. 7.8 will hold and the design will block orthogonally.

For 3×3, 4×4, and 5×5 Latin squares (squares of order 3, 4, and 5), one has a choice of 12, 576, and 161,280 squares, respectively, so there exist many choices [152, 167]. Table 7.4 displays four Latin squares of order 4. One could, for example, place Squares 1 and 2 in Block 1, and Squares 3 and 4 in Block 2. In addition, we could change the letters in one square in each block from $abcd$ to $efgh$ and assign a different set of proportions to e, f, g, and h than to a, b, c, and d. And in addition to this, we could add any composition to Block 1 (such as 0.25, 0.25, 0.25, 0.25) as long as we also add it to Block 2, so that the equality Eq. 7.8 is maintained.

Table 7.4. *Standard Latin squares for $q = 4$*

Square 1	Square 2	Square 3	Square 4
a b c d	*a b c d*	*a b c d*	*a b c d*
b a d c	*b d a c*	*b c d a*	*b a d c*
c d a b	*c a d b*	*c d a b*	*c d b a*
d c b a	*d c b a*	*d a b c*	*d c a b*

The requirements for orthogonal blocking are more stringent when one plans on fitting a *quadratic* Scheffé model. In addition to condition 7.8, the following condition must be satisfied [75]:

$$\sum_{u=1}^{n_w} X_{ui} X_{uj} = k_{ij} \quad \text{for each block and } i, j = 1, 2, \ldots, q \tag{7.9}$$

where u is the run number in the wth block, the wth block contains n_w runs, and where the k_{ij} are any constants. This condition means that each of the crossproducts, $X_i X_j$, must sum to the same constant, k_{ij}, in all blocks. In other words, the crossproduct sum for $X_1 X_2$ must be the same in all blocks but does not have to equal the crossproduct sum for $X_1 X_3$ (say).

To illustrate, and using the shorthand notation 12 for $\sum_{u=1}^{n_w} X_1 X_2$, 13 for $\sum_{u=1}^{n_w} X_1 X_3$, etc., the crossproduct sums for Square 3, Table 7.4, are

$$12 = ab + bc + cd + da = ab + ad + bc + cd$$

$$13 = ac + bd + ca + db = 2(ac + bd)$$

$$14 = ad + ba + cb + dc = ab + ad + bc + cd$$

$$23 = bc + cd + da + ab = ab + ad + bc + cd$$

$$24 = bd + ca + db + ac = 2(ac + bd)$$

$$34 = cd + da + ab + bc = ab + ad + bc + cd$$

This pattern of crossproduct sums can be symbolized by letter codes

12	13	14	23	24	34
C	D	C	C	D	C

where $C = ab + ad + bc + cd$ and $D = 2(ac + bd)$. The letters C and D have no special meaning.

To satisfy Eq. 7.9, one needs to find another Latin square that will lead to the same crossproduct sums 12, 13, ..., 34. The method involves choosing pairs of Latin squares from those produced by permutations of **standard Latin squares** [15, 48, 75, 90, 138, 139]. Standard Latin squares have the first row and the first column in alphabetical order. The number of standard Latin squares of order 3, 4, 5, and 6 is 1, 4, 56, and 9408 [152, 167]. Table 7.4 displays the standard squares for $q = 4$.

A Latin square is **row-reduced** if the elements in the first column are in natural order, such as alphabetical. Standard Latin squares are row reduced. Latin squares of order q are **mates** if they are row-reduced and have identical crossproduct structures [90]. If one permutes the last three columns of each standard square in Table 7.4 (thus keeping the first column in alphabetic order), then one will generate four sets of row-reduced Latin squares. Each set will be of size 6, because the number of permutations of n things taken n at a time is $n!$ (n factorial), and $3! = 6$.

Permuting the last three columns of Square 3, Table 7.4, leads to the six squares below. The first two are mates, the third and fourth are mates, and the fifth and sixth are mates. The squares in Eq. 7.7 are the first two squares.

a b c d	*a d c b*	*a b d c*	*a d b c*	*a c b d*	*a c d b*
b c d a	*b a d c*	*b c a d*	*b a c d*	*b d c a*	*b d a c*
c d a b	*c b a d*	*c d b a*	*c b d a*	*c a d b*	*c a b d*
d a b c	*d c b a*	*d a c b*	*d c a b*	*d b a c*	*d b c a*

Tables 7.15–7.17 in Appendix 7A, page 146, summarize the mates and patterns of crossproduct sums derived from standard Latin squares of order 4. Square 1, Table 7.4, is missing from these tables because this square has no mates. The letters in the columns labeled "square" and "mate" give the order for row one of each Latin square.

To illustrate with a second example, consider the Latin squares in the following pair of mates:

Block 1	Block 2	
a b d c	*a c d b*	
b d c a	*b a c d*	(7.10)
c a b d	*c d b a*	
d c a b	*d b a c*	

The square on the left (Block 1) is a column permutation of Square 2 in Table 7.4. The square on the right is its mate. This pair is the first entry in Table 7.16. Crossproduct sums for both are

$$12 = 14 = 23 = 34 = ab + ac + bd + cd$$

$$13 = 24 = 2(ad + bc)$$

If we assign the letter codes A $= ab + ac + bd + cd$ and B $= 2(ad + bc)$, then the pattern of crossproduct sums is

12	13	14	23	24	34
A	B	A	A	B	A

The crossproduct sums in this example differ from the crossproduct sums for the pair in Eq. 7.7, but the *pattern* of crossproduct sums is the same. For this reason, these two examples appear in the same table (Table 7.16, entries 1 and 2). The letter codes have no meaning whatsoever other than to identify crossproduct sums.

There are two further considerations. First, to support a $q = 4$ Scheffé quadratic model *and* block the runs, one needs at least 11 distinct mixture blends — 10 to support the model and an additional degree of freedom to support blocks (assuming two blocks, one categorical variable). Two blocks of four runs each (as in pairs 7.7 or 7.10) consist of only eight discrete runs, not enough to support the intended model.

One solution to this problem is to combine pairs of blocks. If we combine pairs 7.7 (Pattern 2, Square 3) and 7.10 (Pattern 2, Square 2), then we have the 16-run blocked design

$$
\begin{array}{cc}
\text{Block 1} & \text{Block 2} \\
a\ b\ c\ d & a\ d\ c\ b \\
b\ c\ d\ a & b\ a\ d\ c \\
c\ d\ a\ b & c\ b\ a\ d \\
d\ a\ b\ c & d\ c\ b\ a \\
a\ b\ d\ c & a\ c\ d\ b \\
b\ d\ c\ a & b\ a\ c\ d \\
c\ a\ b\ d & c\ d\ b\ a \\
d\ c\ a\ b & d\ b\ a\ c
\end{array}
\tag{7.11}
$$

Careful examination of this design will reveal that there are no repeat treatment combinations. Designs formed from pairs of mates with the same pattern have no runs that are repeated, either in the same block or in different blocks [48, 138]. A numerical example of this design (Table 7.5) will be discussed below.

The previous example combined pairs of mates with the same pattern but from different standard squares. To illustrate combining pairs of mates with different patterns but from the same standard square, consider combining the second pair in Table 7.16 (Pattern 2, Square 3) with the second pair in Table 7.17 (Pattern 3, Square 3). Although both pairs are derived from Square 3, they have different patterns.

$$
\begin{array}{cc}
\text{Block 1} & \text{Block 2} \\
a\ b\ c\ d & a\ d\ c\ b \\
b\ c\ d\ a & b\ a\ d\ c \\
c\ d\ a\ b & c\ b\ a\ d \\
d\ a\ b\ c & d\ c\ b\ a \\
a\ c\ b\ d & a\ c\ d\ b \\
b\ d\ c\ a & b\ d\ a\ c \\
c\ a\ d\ b & c\ a\ b\ d \\
d\ b\ a\ c & d\ b\ c\ a
\end{array}
\tag{7.12}
$$

Again, there are no repeat treatment combinations.

If pairs of mates are combined from different standard squares *and* with different patterns of crossproduct sums, then there will be runs that are repeated either in the same block, called **repeats**, and/or in different blocks, called **cancelations** [48, 138]. This can be a useful strategy, because repeats provide a means for replication, while cancelations provide a means for reducing the size of the experiment (however, see the caveat below).

To illustrate, consider combining the second pair in Table 7.16 (Pattern 2, Square 3) with the third pair in Table 7.17 (Pattern 3, Square 4). The design is illustrated in Eq. 7.13.

$$\begin{array}{cc} \text{Block 1} & \text{Block 2} \\ a\ b\ c\ d & a\ d\ c\ b \\ b\ c\ d\ a & b\ a\ d\ c \\ c\ d\ a\ b & c\ b\ a\ d \\ d\ a\ b\ c & d\ c\ b\ a \\ a\ b\ c\ d & a\ b\ d\ c \\ b\ a\ d\ c & b\ a\ c\ d \\ c\ d\ b\ a & c\ d\ a\ b \\ d\ c\ a\ b & d\ c\ b\ a \end{array} \tag{7.13}$$

The sequence $c\ d\ a\ b$ appears in the first pair in Block 1 and the second pair in Block 2; the sequence $b\ a\ d\ c$ appears in the first pair in Block 2 and the second pair in Block 1. These are *cancelations* and could be deleted without violating Eq. 7.9. There are also two repeats in this design. The sequence $a\ b\ c\ d$ is repeated in Block 1; the sequence $d\ c\ b\ a$ is repeated in Block 2. Switching the first or the second pair of squares (but not both) reverses the repeats and cancelations.

There is one remaining (but extremely important) consideration when fitting quadratic models to blocked designs based on Latin squares. In designs exemplified by Eqs. 7.11, 7.12, and 7.13, for each blend in each block, it is true that

$$\begin{aligned} X_1X_2 + X_1X_3 + \cdots + X_3X_4 &= ab + ac + ad + bc + bd + cd \\ &= \text{Constant} \end{aligned} \tag{7.14}$$

For example, if $a = 0.1$, $b = 0.2$, $c = 0.3$, and $d = 0.4$, then the Constant is 0.350. Equation 7.14 may also be written

$$X_1X_2 + X_1X_3 + \cdots + X_3X_4 = \text{Constant} \cdot (X_1 + X_2 + X_3 + X_4)$$

because $\sum_{i=1}^{4} X_i = 1.0$. As there exists an exact linear dependency between the crossproduct terms and the linear terms, the $\mathbf{X'X}$ matrix is **singular**, which means that it does not have an inverse. This is critical, because to estimate the least-squares coefficients in a linear regression model requires solution of the equation

$$\mathbf{b} = (\mathbf{X'X})^{-1}\,\mathbf{X'Y} \tag{7.15}$$

where $\mathbf{b}$ is a $p \times 1$ vector of coefficient estimates and $\mathbf{Y}$ is a $n \times 1$ vector of observations. See, for example, Draper and Smith [49], Montgomery, Peck, and Vining [100], or Myers [105]. The inverse of $\mathbf{X'X}$ is, of course, the variance-covariance matrix (apart from σ^2). To achieve nonsingularity while still preserving the orthogonality conditions Eqs. 7.8 and 7.9, a new common blend or blends whose levels cannot be a, b, c, and d in any order must be added to each block [48, 138]. A reasonable blend to add would be (0.25, 0.25, 0.25, 0.25). Canceling runs can reintroduce a singularity, in which case one would also have to add a new common blend or blends to each block (thus eliminating the benefit of cancelations). For example, the cancelations illustrated with the design in Eq. 7.13 result in a singular matrix if one intends to support a quadratic model.

Table 7.5 provides a numerical example of the design in Eq. 7.11, page 127. Mixture components A, B, C, and D correspond to columns 1–4, respectively, in the Latin squares of Eq. 7.11. In this example, $a = 0.1, b = 0.2, c = 0.3$, and $d = 0.4$. To avoid singularity, the common run $(A, B, C, D) = (0.5, 0.0, 0.2, 0.3)$ has been added to each block (the last run in each block). The equivalence of the column sums for the linear terms is a confirmation of condition 7.8, page 124; the equivalence of the column sums for the crossproduct terms

Table 7.5. *Orthogonally blocked mixture design for $q = 4$*

Block	A	B	C	D	AB	AC	AD	BC	BD	CD
1	0.1	0.2	0.3	0.4	0.02	0.03	0.04	0.06	0.08	0.12
1	0.2	0.3	0.4	0.1	0.06	0.08	0.02	0.12	0.03	0.04
1	0.3	0.4	0.1	0.2	0.12	0.03	0.06	0.04	0.08	0.02
1	0.4	0.1	0.2	0.3	0.04	0.08	0.12	0.02	0.03	0.06
1	0.1	0.2	0.4	0.3	0.02	0.04	0.03	0.08	0.06	0.12
1	0.2	0.4	0.3	0.1	0.08	0.06	0.02	0.12	0.04	0.03
1	0.3	0.1	0.2	0.4	0.03	0.06	0.12	0.02	0.04	0.08
1	0.4	0.3	0.1	0.2	0.12	0.04	0.08	0.03	0.06	0.02
1	0.5	0.0	0.2	0.3	0.00	0.10	0.15	0.00	0.00	0.06
Sums:	2.5	2.0	2.2	2.3	0.49	0.52	0.64	0.49	0.42	0.55
2	0.1	0.4	0.3	0.2	0.04	0.03	0.02	0.12	0.08	0.06
2	0.2	0.1	0.4	0.3	0.02	0.08	0.06	0.04	0.03	0.12
2	0.3	0.2	0.1	0.4	0.06	0.03	0.12	0.02	0.08	0.04
2	0.4	0.3	0.2	0.1	0.12	0.08	0.04	0.06	0.03	0.02
2	0.1	0.3	0.4	0.2	0.03	0.04	0.02	0.12	0.06	0.08
2	0.2	0.1	0.3	0.4	0.02	0.06	0.08	0.03	0.04	0.12
2	0.3	0.4	0.2	0.1	0.12	0.06	0.03	0.08	0.04	0.02
2	0.4	0.2	0.1	0.3	0.08	0.04	0.12	0.02	0.06	0.03
2	0.5	0.0	0.2	0.3	0.00	0.10	0.15	0.00	0.00	0.06
Sums:	2.5	2.0	2.2	2.3	0.49	0.52	0.64	0.49	0.42	0.55

Source	*df*
Blocks	1
Model	9
Residuals	7
Lack of Fit	7
Pure Error	0
Corr. total	17

is a confirmation of condition 7.9. The breakdown of the degrees of freedom for this design and a quadratic Scheffé model is summarized below the table. Degrees of freedom for pure error could be increased by replicating any added blend(s) in both blocks.

Prescott et al. [138] have shown that of the 56 standard Latin squares of order 5, no pairs of mates can be generated from 50 of these. The remaining six are summarized in Table 7.6. Permuting the last four columns of each of the six squares in this table will lead to six sets of 24 row-reduced squares (total = 144 squares). Each set of 24 is composed of 12 pairs of mates (total = 72 pairs). These are summarized in Tables 7.18–7.23 in Appendix 7A, page 146, where they are tabulated by crossproduct pattern. The same considerations apply for repeats and cancelations as discussed on page 128.

Table 7.6. *Standard Latin squares for $q = 5$*

Square 1	Square 2	Square 3
a b c d e	*a b c d e*	*a b c d e*
b c e a d	*b c d e a*	*b d e c a*
c e d b a	*c d e a b*	*c e b a d*
d a b e c	*d e a b c*	*d c a e b*
e d a c b	*e a b c d*	*e a d b c*
Square 4	Square 5	Square 6
a b c d e	*a b c d e*	*a b c d e*
b d a e c	*b e d a c*	*b e a c d*
c a e b d	*c d b e a*	*c a d e b*
d e b c a	*d a e c d*	*d c e b a*
e c d a b	*e c a b d*	*e d b a c*

Prescott et al. [138] describe a method for finding the mate of any standard Latin square of order 3–5:

> *The order of the letters in the first row of the mate of a square corresponds to the rows in which the letter* ***a*** *occurs taken in column order.*

To illustrate, consider the first square in Block 1 in Eq. 7.13, page 128. The letter *a* appears in columns 1, 2, 3, and 4 in rows 1, 4, 3, and 2. Therefore, the first row of the mate is *a d c b*, and the full square is the first square in Block 2 of Eq. 7.13. In a similar fashion, consider the second square in Block 1 of Eq. 7.13. The letter *a* appears in columns 1, 2, 3, and 4 in rows 1, 2, 4, and 3, and so the first row of the mate is *a b d c*.

Applying this method to the first square in Table 7.6, the order of the rows in which the letter *a* occurs in column order 1–5 is 1, 4, 5, 2, 3. The first row of the mate is therefore *a d e b c*. This pair is the first entry in Table 7.18, page 147.

Square 3 in Table 7.4, page 125, and Square 2 in Table 7.6 are **standard cyclic Latin squares** of order 4 and 5, respectively. If the letters *a, b, c, …* in a Latin square are replaced by the elements 0, 1, 2, …, then a standard cyclic Latin square is one in which the (i, j)th element is the sum of the ith element in column 1 and the jth element in row 1, reduced modulo q. For example, replacing a, b, c, d, and e in Square 2, Table 7.6 with 0, 1, 2, 3, and 4, respectively, leads to

$$\begin{array}{ccccc} 0 & 1 & 2 & 3 & 4 \\ 1 & 2 & 3 & 4 & 0 \\ 2 & 3 & 4 & 0 & 1 \\ 3 & 4 & 0 & 1 & 2 \\ 4 & 0 & 1 & 2 & 3 \end{array}$$

The (4,5)th element (say) is equal to $3 + 4$ reduced modulo 5, which is 2 (corresponding to the letter c). Modulo division returns the *remainder* of division. Thus, dividing 7 by 5 leads to a remainder of 2.

Permuting the last $q - 1$ columns of standard cyclic Latin squares leads to **cyclic equivalent Latin squares** [90]. For $q > 5$, the method described above for finding mates when $q = 3$–5 applies as well to cyclic equivalent Latin squares of any order [90]. For noncyclic equivalent Latin squares of order higher than five, the reader is referred to Lewis et al. [90].

7.2 Asymmetrically Shaped Design Regions

When the design space is asymmetrical, the Latin-square approach to blocking is not always suitable. For example, if the ranges of components A and B in a mixture are 0.1 and 0.5, respectively, then because of the symmetry of the Latin-square designs, only about 20% of the range of B can be investigated. One would prefer a design that is more space filling but at the same time will block orthogonally.

The method to be described in this section utilizes **projection designs**, originally described in a series of three reports by Hau and Box [65, 66, 67]. As of this writing, the reports are available as PDF files at the Web site of the Center for Quality and Productivity Improvement, University of Wisconsin (http://www.engr.wisc.edu/centers/cqpi). Reference will also be made to more recent work by Prescott [137].

To the author's knowledge, the projection-design approach to blocking is not implemented automatically in any commercially available DOE software. However, the procedures are easy to implement in software that has matrix algebra functionality such as JMP (JMP Scripting Language), MINITAB (Session commands), or S-PLUS (S programming language).

Readers without the requisite software may prefer to skip this section. It is suggested, however, that at least the first few paragraphs be read. A simple method that offers a practical approach to the problem will first be described. Following this, some graphics will be presented that provide a nonmathematical, conceptual description of the projection-design procedure.

When blocking experiments, whether or not they are mixture experiments, the goal is to have the correlations between the block regression coefficient(s) and the response-surface regression coefficients as close to zero as possible. The larger the correlations, the more the regression coefficients for the mixture part of the model will be affected, both in terms of accuracy and precision. (See page 99 for a discussion of the correlation matrix of regression coefficients.)

Many DOE software packages provide a means for rerandomizing experimental runs. Absent a capability (or desire) to run projection designs, a reasonable approach would be to try more than one randomization and check correlation coefficients after each randomization. Assume for example that there are 20 runs, and the experimenter intends to assign 10 runs to each of two blocks. Runs 1–10 might be assigned to Block 1 (say), while runs 11–20 might be assigned to Block 2. After checking correlations between the block regression coefficient and the response-surface model coefficients, one could rerandomize all 20 runs. Again, assign runs 1–10 to Block 1 and runs 11–20 to Block 2. Some of the runs in each block may be the same as in the first randomization, but some will be different. If the correlation coefficients are lower, then the second randomization would be preferred. One could continue with this until satisfied.

Carrying out such a procedure in a spreadsheet setting is admittedly tedious. If one foresees a need for blocking and has software with programming capabilities, then one could write a script, macro, function, or program to carry out repeated randomizations and check correlation coefficients. As an example of the advantage to be gained, consider the 20-run design in Table 7.7. This design is based on the hypothetical constraints

$$\begin{aligned} 0.05 &\leq X_1 \leq 0.55 \\ 0.10 &\leq X_2 \leq 0.50 \\ 0.30 &\leq X_3 \leq 0.50 \end{aligned} \tag{7.16}$$

A D-optimal design to support a quadratic Scheffé model was requested using a popular computing package. IDs 1–6 were assigned to Block 1, IDs 7–12 to Block 2, and IDs 13–20 to Block 3. This is referred to as the "Original" order in Table 7.7. For this order and blocking arrangement, $|r_{max}| = 0.3943$, where $|r_{max}|$ is the maximum absolute correlation between the block coefficients and the model coefficients.

Several thousand rerandomizations in GAUSS led to the range of $|r_{max}|$ values

$$0.0385 \leq |r_{max}| \leq 0.9927$$

The "Revised" order in Table 7.7, one of 20! (20 factorial) possible permutations and not necessarily the optimum permutation, is the run order for which $|r_{max}| = 0.0385$. The rerandomization procedure found a run order that reduced the maximum absolute correlation between the block and model coefficients by approximately one order of magnitude. Block 1 consists of IDs 1, 5, 12, 13, 14, and 19; Block 2 of IDs 7, 9, 11, 15, 18, and 20; and Block 3 of IDs 2, 3, 4, 6, 8, 10, 16, and 17. In both the original and revised orders, correlation coefficients are based on expressing the component proportions in the natural variables.

As $|r_{max}| \to 0$, the diagonal elements of the $(\mathbf{X}'\mathbf{X})^{-1}$ matrix (the c_{ii}) for the model coefficients will approach those in the D-optimal design. For example, Table 7.8 compares the variances (apart from σ^2) of the model coefficients based on the design points in Table 7.7 for three situations: (i) run everything in one block; (ii) run in three blocks in the original order; (iii) run in three blocks in the revised order.

Turning to projection designs, one can understand the general idea of this approach by considering Fig. 7.1. Ignoring the bold diagonal line for the moment, the figure is intended to represent a $k = 2$ central composite design (CCD), where k is the number of factors. Assume that there are also some center points, the exact number not being important for this example.

Table 7.7. *Effect of run order on a blocked $q = 3$ D-optimal design*

	ID			Proportions		
Block	Original	Revised	ID	X_1	X_2	X_3
1	1	1	1	0.1000	0.5000	0.4000
	2	5	2	0.1667	0.3333	0.5000
	3	19	3	0.0500	0.4500	0.5000
	4	14	4	0.1000	0.5000	0.4000
	5	13	5	0.0500	0.4500	0.5000
	6	12	6	0.5500	0.1000	0.3500
2	7	11	7	0.4000	0.1000	0.5000
	8	20	8	0.2000	0.5000	0.3000
	9	9	9	0.3167	0.3833	0.3000
	10	7	10	0.4000	0.1000	0.5000
	11	15	11	0.2125	0.4000	0.3875
	12	18	12	0.3000	0.3000	0.4000
3	13	4	13	0.0500	0.4500	0.5000
	14	3	14	0.3750	0.3250	0.3000
	15	16	15	0.4333	0.2667	0.3000
	16	10	16	0.3875	0.2000	0.4125
	17	8	17	0.2833	0.2167	0.5000
	18	2	18	0.4750	0.1000	0.4250
	19	6	19	0.2000	0.5000	0.3000
	20	17	20	0.5500	0.1000	0.3500

Table 7.8. *Coefficient variances (apart from σ^2) for three run order – blocking arrangements*

	Blocking (# blocks)		
Term	None (1)	Original (3)	Revised (3)
X_1	106.66	119.11	107.42
X_2	123.88	128.10	123.92
X_3	325.13	334.39	326.27
X_1X_2	209.38	248.54	215.05
X_1X_3	2850.42	2935.79	2850.73
X_2X_3	2974.19	3112.77	2977.44

High and low factor levels for Z_1 and Z_2 in the factorial part of the design are coded to ± 1. Coding is accomplished using the expression

$$Z_i = \frac{\xi_i - c_i}{r_i} \tag{7.17}$$

where ξ_i is the actual or natural levels of the factors, c_i is the center point, taken as the average of $[\max(\xi_i) + \min(\xi_i)]$ in the factorial part of the design, and r_i is the half range.

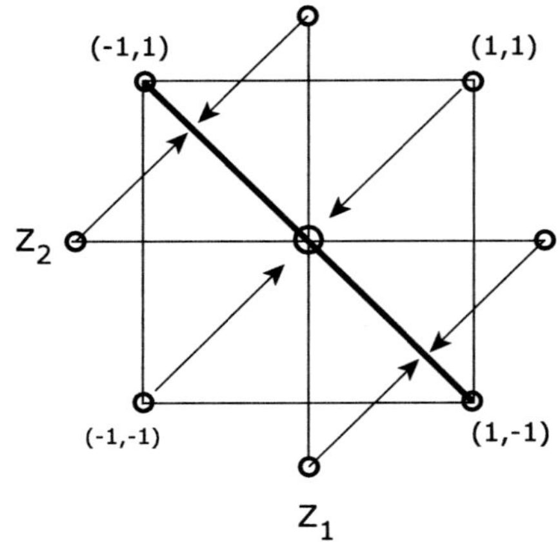

Figure 7.1. *Projection geometry and design for a $k = 2$ central composite design.*

This coding is commonly used in response-surface methodology and causes all the values for Z_i in the factorial part of the design to assume values of -1, $+1$, or 0 (when center points are included) [107]. In this example the axial design points are located at a coded distance of $\sqrt{2}$ from the center of the design. This is therefore a rotatable central composite design [104, 107].

The bold diagonal line in Fig. 7.1 is intended to represent a 2-simplex. The CCD exists in a space called the **unconstrained space**, while the 2-simplex is referred to as the **constrained space**. The arrows in Fig. 7.1 show the perpendicular projection of the design points in unconstrained space onto the constrained space.

Let us use the symbol $\tilde{Z}_i$, $i = 1, 2$, to represent a coordinate system on the 2-simplex to distinguish it from the coordinate system in the Z_i s. One end of the line (the upper left) will be assigned the coordinates $(\tilde{Z}_1, \tilde{Z}_2) = (-1, 1)$, while the other end will be assigned $(\tilde{Z}_1, \tilde{Z}_2) = (1, -1)$. The idea is to determine the coordinates $(\tilde{Z}_1, \tilde{Z}_2)$ on the simplex that result from the perpendicular projection of each point in the CCD onto the simplex. For example, the points $(Z_1, Z_2) = (-1, -1)$ and $(Z_1, Z_2) = (1, 1)$ have coordinates $(\tilde{Z}_1, \tilde{Z}_2) = (0, 0)$ on the simplex. The axial point with coordinates $(Z_1, Z_2) = (\sqrt{2}, 0)$ has coordinates $(\tilde{Z}_1, \tilde{Z}_2) \approx (0.7, -0, 7)$ on the simplex. The points $(Z_1, Z_2) = (1, -1)$, $(-1, 1)$, and $(0, 0)$ have the same values for their coordinates in the $\tilde{Z}_i$ scale.

The overall goal of the projection-design approach is to take a design such as a factorial design or CCD in k-dimensional space, with coordinates expressed in the Z_i, project the design onto a $(k - 1)$-dimensional simplex or constrained region within a simplex, with coordinates expressed in the $\tilde{Z}_i$, and finally to convert the coordinates expressed in the $\tilde{Z}_i$ into coordinates expressed in the X_i, the component proportions.[2] To convert the coordinates expressed in the $\tilde{Z}_i$ into coordinates expressed in the X_i we can use an expression similar to Eq. 7.17, page 133:

$$\tilde{Z}_i = \frac{X_i - c_i}{r_i} \tag{7.18}$$

[2]It will become clear below that certain of the "good" properties of *orthogonally blocked* conventional designs are retained in the projected designs.

Here $\tilde{Z}_i$ is a coded value for X_i, c_i is a **center of interest**, and r_i again represents a half range. Both c_i and r_i are expressed in terms of component proportions. In the context of this example, $(c_1, c_2) = (0.5, 0.5)$ and $(r_1, r_2) = (0.5, 0.5)$.

For the upper left end of the 2-simplex in Fig. 7.1, $(\tilde{Z}_1, \tilde{Z}_2) = (-1, 1)$. Therefore

$$X_1 = c_1 + r_1(-1) = 0.5 + 0.5(-1) = 0$$

$$X_2 = c_2 + r_2(+1) = 0.5 + 0.5(+1) = 1$$

For the lower right end of the 2-simplex in Fig. 7.1, $(\tilde{Z}_1, \tilde{Z}_2) = (1, -1)$. Therefore

$$X_1 = c_1 + r_1(+1) = 0.5 + 0.5(+1) = 1$$

$$X_2 = c_2 + r_2(-1) = 0.5 + 0.5(-1) = 0$$

This procedure can be summarized as the two-step sequence

$$Z_i \xrightarrow{1} \tilde{Z}_i \xrightarrow{2} X_i \tag{7.19}$$

The method for accomplishing step 1 is explained following another example. To accomplish step 2, the constraints on the component proportions must be specified as $c_i \pm r_i$, $i = 1, 2, \ldots, q$. In the context of Fig. 7.1, the mixture constraints are $X_i = 0.5 \pm 0.5$, $i = 1, 2$, or in terms of coded variables, $\tilde{Z}_i = 0 \pm 1$, $i = 1, 2$. Table 7.9 summarizes the settings in the three metrics for the example in Fig. 7.1.

When the constrained space falls within a mixture simplex, the component proportions are subject to the equality constraint

$$\sum_{i=1}^{q} X_i = 1$$

Table 7.9. *Projection of a $k = 2$ central composite design onto a 2-simplex*

	Unconstrained space		Constrained space			
	Z		$\tilde{\mathbf{Z}}$		**X**	
ID	Z_1	Z_2	$\tilde{Z}_1$	$\tilde{Z}_2$	X_1	X_2
1	−1	−1	0	0	0.5	0.5
2	1	−1	1	−1	1	0
3	−1	1	−1	1	0	1
4	1	1	0	0	0.5	0.5
5	−1.4142	0	−0.7071	0.7071	0.1464	0.8536
6	1.4142	0	0.7071	−0.7071	0.8536	0.1464
7	0	−1.4142	0.7071	−0.7071	0.8536	0.1464
8	0	1.4142	−0.7071	0.7071	0.1464	0.8536
9	0	0	0	0	0.5	0.5
10	0	0	0	0	0.5	0.5

(Eq. 2.1, page 9). As the $\tilde{Z}_i$ are simply a rescaling of the X_i, with a little algebra we can reexpress this constraint in terms of the $\tilde{Z}_i$. To do this, we substitute X_i from Eq. 7.18 into the equality constraint, leading to

$$\sum_{i=1}^{q} r_i \tilde{Z}_i + \sum_{i=1}^{q} c_i = 1$$

As $\sum_{i=1}^{q} c_i = 1$, the constraint can be written in terms of the $\tilde{Z}_i$ as

$$\sum_{i=1}^{q} r_i \tilde{Z}_i = 0 \tag{7.20}$$

In the special case where the r_i, $i = 1, 2, \ldots, q$, are equal to one another, the constraint takes the form

$$\sum_{i=1}^{q} \tilde{Z}_i = 0 \tag{7.21}$$

Note that in Table 7.9, $\sum_{i=1}^{2} \tilde{Z}_i = 0$ because $r_1 = r_2$.

As a second example, Fig. 7.2 illustrates the projection of a 2^3 factorial design onto a 3-simplex [137]. In the top diagram, the constrained space has been displaced from the unconstrained space to facilitate illustration. In addition, the Z axes for the factorial design have been slightly rotated clockwise about Z_1. Consequently the Z_3 axis in the right figure is not exactly collinear with the Z_3 axis in the left figure. Again, this is to facilitate illustration.

The 2^3 factorial design is blocked into two blocks of four runs each. Runs 1, 4, 6, and 7 are in Block 1 (filled circles), while runs 2, 3, 5, and 8 are in Block 2 (open circles). In terms of the mixture variables, the constrained region in the $c_i \pm r_i$-notation is $X_i = 0.333 \pm 0.333$, which is a circle, not a simplex (lower figure). This illustrates a limitation to the projection-design approach. Because of the mathematics (below), and for the design to "fit" within the constrained region, the constrained region must be specified in terms of a center of interest and half ranges. For simplex-shaped design regions, this means that as q gets larger and larger, a smaller and smaller fraction of the simplex can be explored. The strong point of projection designs, however, is that they are more space filling than Latin-square designs in *irregularly* shaped design regions.

Design settings in the three metrics for the illustration in Fig. 7.2 are displayed in Table 7.10. The symbol α, a real number that is explained below, should be ignored for the moment. Note that $\alpha \sum_{i=1}^{3} \tilde{Z}_i = 0$ because $r_1 = r_2 = r_3$.

In their first report [65], Hau and Box show that projection of an unconstrained design, $\mathbf{Z}$, onto a constrained space is accomplished by right multiplying the $n \times k$ matrix $\mathbf{Z}$ by a suitable $k \times k$ projection matrix, $\mathbf{P}$.[3]

$$\tilde{\mathbf{Z}}_{n\times k} = \mathbf{Z}_{n\times k}\mathbf{P}_{k\times k} \tag{7.22}$$

The projection matrix is given by

$$\mathbf{P} = \mathbf{I} - \mathbf{r}(\mathbf{r}'\mathbf{r})^{-1}\mathbf{r}' \tag{7.23}$$

[3]For a discussion of projection matrices and their application to the geometry of ordinary least squares, see Rawlings, Pantula, and Dickey [143].

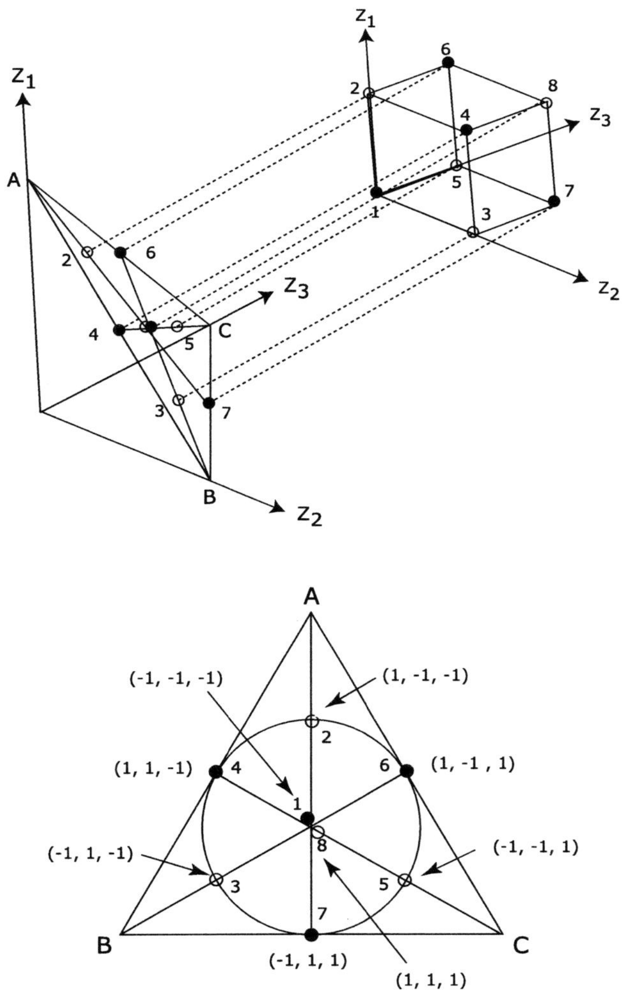

Figure 7.2. *Projection geometry and design for a* 2^3 *factorial design. Adapted from* [137].

where $\mathbf{I}$ is a $k \times k$ identity matrix and $\mathbf{r}$, a $k \times 1$ vector, is given by

$$\mathbf{r} = \begin{bmatrix} r_1 \\ r_2 \\ \vdots \\ r_k \end{bmatrix}$$

The r_i, $i = 1, 2, \ldots, k$, are the half ranges.

Table 7.10. *Projection of a 2^3 factorial design onto a 3-simplex*

	Unconstrained space			Constrained space					
	$\mathbf{Z}$			$\alpha\tilde{\mathbf{Z}}$			$\mathbf{X}$		
ID	Z_1	Z_2	Z_3	$\alpha\tilde{Z}_1$	$\alpha\tilde{Z}_2$	$\alpha\tilde{Z}_3$	X_1	X_2	X_3
1	−1	−1	−1	0	0	0	0.333	0.333	0.333
2	1	−1	−1	1	−0.5	−0.5	0.667	0.166	0.166
3	−1	1	−1	−0.5	1	−0.5	0.166	0.667	0.166
4	1	1	−1	0.5	0.5	−1	0.5	0.5	0
5	−1	−1	1	−0.5	−0.5	1	0.166	0.166	0.667
6	1	−1	1	0.5	−1	0.5	0.5	0	0.5
7	−1	1	1	−1	0.5	0.5	0	0.5	0.5
8	1	1	1	0	0	0	0.333	0.333	0.333

In the example of Fig. 7.1 (page 134) and Table 7.9 (page 135), $\mathbf{r} = [\ 0.5, 0.5\ ]'$, and so

$$\mathbf{P} = \begin{bmatrix} 1 & 0 \\ 0 & 1 \end{bmatrix} - \begin{bmatrix} 0.5 \\ 0.5 \end{bmatrix} \left([\ 0.5 \quad 0.5\ ] \begin{bmatrix} 0.5 \\ 0.5 \end{bmatrix} \right)^{-1} [\ 0.5 \quad 0.5\ ]$$

$$= \begin{bmatrix} 1 & 0 \\ 0 & 1 \end{bmatrix} - \begin{bmatrix} 0.5 & 0.5 \\ 0.5 & 0.5 \end{bmatrix}$$

$$= \begin{bmatrix} 0.5 & -0.5 \\ -0.5 & 0.5 \end{bmatrix}$$

Right multiplying the design matrix $\mathbf{Z}$ (Table 7.9) by the projection matrix $\mathbf{P}$ leads to

$$\tilde{\mathbf{Z}} = \begin{bmatrix} -1 & -1 \\ 1 & -1 \\ -1 & 1 \\ 1 & 1 \\ -1.414 & 0 \\ 1.414 & 0 \\ 0 & -1.414 \\ 0 & 1.414 \\ 0 & 0 \\ 0 & 0 \end{bmatrix} \begin{bmatrix} 0.5 & -0.5 \\ -0.5 & 0.5 \end{bmatrix} = \begin{bmatrix} 0 & 0 \\ 1 & -1 \\ -1 & 1 \\ 0 & 0 \\ -.707 & .707 \\ .707 & -.707 \\ .707 & -.707 \\ -.707 & .707 \\ 0 & 0 \\ 0 & 0 \end{bmatrix}$$

which is the constrained design in Table 7.9.

Turning to Fig. 7.2 and Table 7.10, the **P** matrix is

$$\mathbf{P} = \begin{bmatrix} 1 & 0 & 0 \\ 0 & 1 & 0 \\ 0 & 0 & 1 \end{bmatrix} - \begin{bmatrix} 1/3 \\ 1/3 \\ 1/3 \end{bmatrix} \left(\begin{bmatrix} 1/3 & 1/3 & 1/3 \end{bmatrix} \begin{bmatrix} 1/3 \\ 1/3 \\ 1/3 \end{bmatrix} \right)^{-1} \begin{bmatrix} 1/3 & 1/3 & 1/3 \end{bmatrix}$$

$$= \begin{bmatrix} 1 & 0 & 0 \\ 0 & 1 & 0 \\ 0 & 0 & 1 \end{bmatrix} - \begin{bmatrix} 1/3 & 1/3 & 1/3 \\ 1/3 & 1/3 & 1/3 \\ 1/3 & 1/3 & 1/3 \end{bmatrix}$$

$$= \begin{bmatrix} 2/3 & -1/3 & -1/3 \\ -1/3 & 2/3 & -1/3 \\ -1/3 & -1/3 & 2/3 \end{bmatrix} = \frac{1}{3} \begin{bmatrix} 2 & -1 & -1 \\ -1 & 2 & -1 \\ -1 & -1 & 2 \end{bmatrix}$$

As a result, $\tilde{\mathbf{Z}}$ is given by

$$\tilde{\mathbf{Z}} = \frac{1}{3} \begin{bmatrix} -1 & -1 & -1 \\ 1 & -1 & -1 \\ -1 & 1 & -1 \\ 1 & 1 & -1 \\ -1 & -1 & 1 \\ 1 & -1 & 1 \\ -1 & 1 & 1 \\ 1 & 1 & 1 \end{bmatrix} \begin{bmatrix} 2 & -1 & -1 \\ -1 & 2 & -1 \\ -1 & -1 & 2 \end{bmatrix} = \frac{1}{3} \begin{bmatrix} 0 & 0 & 0 \\ 4 & -2 & -2 \\ -2 & 4 & -2 \\ 2 & 2 & -4 \\ -2 & -2 & 4 \\ 2 & -4 & 2 \\ -4 & 2 & 2 \\ 0 & 0 & 0 \end{bmatrix}$$

The calculated $\tilde{\mathbf{Z}}$ matrix differs from that in Table 7.10 (labeled $\alpha\tilde{\mathbf{Z}}$). The maximum absolute value for $\alpha\tilde{Z}_i$ in the table is 1.0, whereas in the $\tilde{\mathbf{Z}}$ matrix it is 4/3. However, if every element in the $\tilde{\mathbf{Z}}$ matrix is multiplied by $\alpha = 3/4$, then it will be seen that the calculated $\alpha\tilde{\mathbf{Z}}$ matrix matches that in Table 7.10. Generalizing, when the absolute maximum value in a $\tilde{\mathbf{Z}}$ matrix does not equal 1.0, then the matrix is scaled by the reciprocal of the maximum absolute value, symbolized by α. Scaling by α results in at least one of the (scaled) $\tilde{Z}_i$ to range between -1 and $+1$.

Because of this, Eq. 7.18, page 134, should be modified to read

$$\alpha\tilde{Z}_i = \frac{X_i - c_i}{r_i} \tag{7.24}$$

The inclusion of α in Eq. 7.24 has no effect on constraint Eq. 7.20, page 136. One must simply remember that when calculating the X_i from the $\tilde{Z}_i$ (step 2 in Eq. 7.19, page 135), one should use the equation

$$X_i = c_i + \alpha r_i \tilde{Z}_i \tag{7.25}$$

The ID of the design points in Table 7.10 (page 138) corresponds to the labeling of the points in Fig. 7.2 (page 137). The block generator for this design is $B_1 = Z_1Z_2Z_3$, which means that those points for which $Z_1Z_2Z_3 = -1$ (points 1, 4, 6, and 7) are in Block 1, and those for which $Z_1Z_2Z_3 = +1$ (points 2, 3, 5, and 8) are in Block 2. This arrangement leads to *orthogonal blocking* of the two-factor interaction model

$$Y = \alpha_0 + \alpha_1 Z_1 + \alpha_2 Z_2 + \alpha_3 Z_3 + \alpha_{12} Z_1 Z_2 + \alpha_{13} Z_1 Z_3 + \alpha_{23} Z_2 Z_3 + \gamma z$$

The Z_i in Table 7.10 are arranged in what is called **standard order** [13]. A factorial design is in standard order when the first column consists of alternating minus and plus signs, the second column of successive pairs of minus and plus signs, the third column of four minus signs followed by four plus signs, and so on. Let us rearrange the observations in Table 7.10 so that they are sorted instead by block (Table 7.11).

Table 7.11. *Blocked projection design in Table* 7.10

			Z			**X**	
Block	ID	Z_1	Z_2	Z_3	X_1	X_2	X_3
1	1	−1	−1	−1	0.333	0.333	0.333
1	4	1	1	−1	0.5	0.5	0
1	6	1	−1	1	0.5	0	0.5
1	7	−1	1	1	0	0.5	0.5
2	2	1	−1	−1	0.667	0.166	0.166
2	3	−1	1	−1	0.166	0.667	0.166
2	5	−1	−1	1	0.166	0.166	0.667
2	8	1	1	1	0.333	0.333	0.333

In Table 7.11, the column sums of the X_i, $i = 1, 2, 3$, in each block are equal to 1.333 in all cases. Although the three crossproduct terms are not shown, their column sums are equal to 0.361 in each block. Thus both Eqs. 7.8 and 7.9, pages 124 and 125, hold and the design blocks orthogonally. *The orthogonality condition in the Z_i s has been carried over to the X_i s.* More will be said below about the conditions under which one may expect this to occur.

Let us now apply the projection-design approach to an asymmetrical design region. Assume that we would like to project an unconstrained design onto a subspace defined by the constraints

$$\begin{aligned} X_1 &= 0.30 \pm 0.25 \\ X_2 &= 0.30 \pm 0.20 \\ X_3 &= 0.40 \pm 0.10 \end{aligned} \tag{7.26}$$

These constraints are the same as those in Eq. 7.16, page 132, only they are expressed in the form $c_i \pm r_i$. The vector $\mathbf{r}$ is

$$\mathbf{r} = \begin{bmatrix} 0.25 \\ 0.20 \\ 0.10 \end{bmatrix}$$

and so the projection matrix is given by

$$\mathbf{P} = \begin{bmatrix} 1 & 0 & 0 \\ 0 & 1 & 0 \\ 0 & 0 & 1 \end{bmatrix} - \begin{bmatrix} .25 \\ .20 \\ .10 \end{bmatrix} \left([\ .25 \quad .20 \quad .10\ ] \begin{bmatrix} .25 \\ .20 \\ .10 \end{bmatrix} \right)^{-1} [\ .25 \quad .20 \quad .10\ ]$$

$$= \begin{bmatrix} 0.4444 & -0.4444 & -0.2222 \\ -0.4444 & 0.6444 & -0.1777 \\ -0.2222 & -0.1777 & 0.9111 \end{bmatrix}$$

Figure 7.3 illustrates projection of the *same* blocked 2^3 factorial design onto the constrained region defined by Eq. 7.26. The only difference is that two center points have been added to each block. Design settings in unconstrained and constrained space (the Z_i and X_i, respectively) are summarized in Table 7.12. The α value for this design is 0.7627. It is interesting to note that whereas points 1 and 8 projected as centroid points in Fig. 7.1 and Table 7.10, they are no longer centroid points in Fig. 7.3 and Table 7.12.

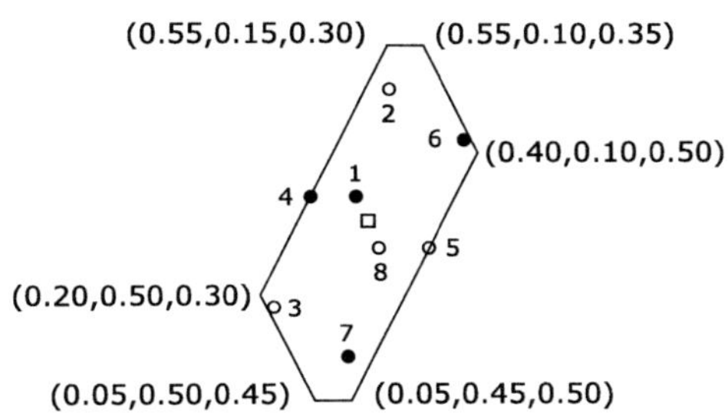

Figure 7.3. *Projection design for a blocked* 2^3 *factorial design. Filled circles, block* 1; *open circles, block* 2; *square, four center points (two from each block).*

The sums of the X_i in each block of Table 7.12 are 1.8, 1.8, and 2.4 for X_1, X_2, and X_3, respectively. Thus, condition 7.8, page 124, for orthogonal blocking of a linear Scheffé model is satisfied. Although the crossproduct terms are not shown, the sums of the $X_i X_j$ in each block are 0.4883, 0.7071, and 0.7117 for $X_1 X_2$, $X_1 X_3$, and $X_2 X_3$, respectively. Condition 7.9, page 125, for orthogonal blocking of a quadratic Scheffé model is also satisfied.

Factorial designs support models with crossproduct (i.e., interaction) terms but do not support models with terms of the form Z_i^2 (full second-order models). Hau and Box showed that projection of a factorial design onto a constrained mixture space leads to designs that still support models with crossproduct terms [66]. Hau and Box refer to such designs as "economical second-order projection designs" because curvature in quadratic *mixture* models is taken care of by the crossproduct terms. In other words, to support a quadratic mixture model it is *not* necessary to project a response-surface design such as a central composite, three-level factorial, or Box–Behnken design.

Table 7.12. *Projected 2^3 factorial design*

		Z			X		
Block	ID	Z_1	Z_2	Z_3	X_1	X_2	X_3
1	1	−1	−1	−1	0.3424	0.2966	0.3610
1	4	1	1	−1	0.3424	0.3576	0.3000
1	6	1	−1	1	0.4271	0.1068	0.4661
1	7	−1	1	1	0.0881	0.4390	0.4729
1		0	0	0	0.3000	0.3000	0.4000
1		0	0	0	0.3000	0.3000	0.4000
2	2	1	−1	−1	0.5119	0.1610	0.3271
2	3	−1	1	−1	0.1729	0.4932	0.3339
2	5	−1	−1	1	0.2576	0.2424	0.5000
2	8	1	1	1	0.2576	0.3034	0.4390
2		0	0	0	0.3000	0.3000	0.4000
2		0	0	0	0.3000	0.3000	0.4000

There are two important considerations when projecting blocked factorial or blocked fractional factorial designs onto simplexes [66]. (1) Two-way interactions must not be confounded with main effects or other two-way interactions, although they may be confounded with three-way interactions. In other words, the design must be of resolution V or higher. (2) No two-way interaction(s) should be lost to blocks. As long as these two conditions are fulfilled, the designs will block orthogonally after projection onto a simplex or onto asymmetrically shaped design regions within a simplex. Replication can be achieved by including the same number of center points in each block.

Some examples of factorial and fractional factorial designs that block orthogonally and, when projected onto mixture space, support quadratic Scheffé models with orthogonal blocking are summarized in Table 7.13. For discussions of generators for fractional factorials and blocked designs see Box and Draper, Chapter 5 [12], Box, Hunter, and Hunter, Chapters 10 and 12 [13], and/or Myers and Montgomery, Chapters 3 and 4 [107].

The last column in Table 7.13 gives the degrees of freedom for lack of fit assuming a quadratic Scheffé model. A more detailed breakdown of the degrees of freedom is given below the table for the even-numbered designs. In calculating degrees of freedom, one should divorce one's thinking from the unconstrained design and focus only on the mixture design. There will be $n - 1$ degrees of freedom for the total corrected sum of squares, $b - 1$ degrees of freedom for blocks, where b is the number of blocks, and $(q - 1) + [q(q - 1)/2]$ degrees of freedom for the quadratic Scheffé models.

Degrees of freedom for pure error can be achieved by adding replicate points of the *same composition* to each block. For example, in the case of design 4, if eight additional points of the same composition were added to the design, two to each block, and if the composition of the added points did not duplicate any composition already in the design, then this would lead to four additional degrees of freedom for lack of fit (one from each block)

Table 7.13. *Blocked factorial and fractional factorial designs*

			Total		Generators		
	q	Design	Runs	Blocks	Factor	Block	df_{LOF}
1	3	2^3	8	2		$B_1 = ABC$	1
2	4	2^4	16	2		$B_1 = ABCD$	5
3	5	2^5	32	2		$B_1 = ABCDE$	16
4	5	2^5	32	4		$B_1 = ABC$ $B_2 = ADE$	14
5	6	2^{6-1}	32	2	$F = ABCD$	$B_1 = ABCE$	10
6	7	2^{7-1}	64	2	$G = ABCDE$	$B_1 = ABCF$	35
7	7	2^{7-1}	64	4	$G = ABCD$	$B_1 = ABCE$ $B_2 = ABDF$	33
8	7	2^{7-1}	64	8	$G = ABCDEF$	$B_1 = ABC$ $B_2 = ADE$ $B_3 = BDF$	29
9	7	2^{7-2}	32	2	$G = ABCDE$	$B_1 = ABCF$	3

	ID			
Source	2	4	6	8
Blocks	1	3	1	7
Model	9	14	27	27
Residuals	5	14	35	29
Lack of Fit	5	14	35	29
Pure Error	0	0	0	0
Corr. total	15	31	63	63

plus four additional degrees of freedom for pure error (one from each block). Formulations of the same composition in *different* blocks are not replicates.

Although there is no defined boundary between what is "economical" and what is not, when $q \geq 6$, full factorials lead to an inordinate number of degrees of freedom for lack of fit. For example, when $q = 6$, if one considered a 2^6 factorial design in eight blocks, then this would lead to $63 - 20 - 7$ (for blocks) $= 36$ degrees of freedom for lack of fit. It would be much more economical to run a half-fraction in two blocks (design 5), in which case there would be $31 - 20 - 1$ (for blocks) $= 10$ degrees of freedom for lack of fit.

When $q \geq 7$, even half-fractions are no longer economical, and one needs to consider quarter-fractions. Comparing design 9 to design 7 or 8, there is a considerable drop in the degrees of freedom for lack of fit — in fact, too much of a drop. In some cases it may prove more advantageous to consider projecting orthogonally blocked CCDs.

Table 7.14. *Blocked central composite designs*

			No. of points				*df*		
ID	q	CCD	fact_1	fact_2	axial	Total	LOF	PE	$\lvert r_{max}\rvert^{\dagger}$
A	3	full	8		6	14	7	0	0.0115
B	3	full	8 0		6 2	16	8	1	$\sim 10^{-16}$
C	3	full	8 2		6 4	20	9	4	$\sim 10^{-15}$
D	3	full	4 4	4 4	6 2	24	9	7	$\sim 10^{-7}$
E	4	full	8	8	8	24	12	0	$\sim 10^{-15}$
F	5	half	16		10	26	10	0	0.0032
G	5	half	16 0		10 6	32	11	5	$\sim 10^{-8}$
H	6	small	16		12	28	6	0	0.0010
I	6	small	16 0		12 4	32	7	3	$\sim 10^{-7}$
J	7	small	22		14	36	7	0	0.0416
K	7	small	22 0		14 8	44	8	7	0.0539

† After projection onto a simplex

One might ask whether *orthogonally blocked* (not necessarily rotatable) CCDs lead to orthogonally blocked mixture designs. The answer is "yes" in some cases and "nearly" in others. Table 7.14 displays some blocking information for CCDs (designs A–E), half-fractions of CCDs (designs F and G), and small composite designs (designs H–K). (See [12] and [107] for discussion of small composite designs and leading references.) The columns labeled "fact_1", "fact_2", and "axial" give the number of design points in the factorial block (or blocks, if the factorial points are divided into two blocks) and the axial block, respectively. If center points are included in any blocks, these are given on the second line of the design. For example, design C has 2 and 4 center points in the factorial and axial blocks, respectively.

Degrees-of-freedom calculations assume that the designs are projected onto an *asymmetrically shaped design region*. This means that replicates in the mixture design will arise only from the replicates in the unconstrained designs in the Zs, and not from projection of factorial points (as when projecting onto a simplex; cf. Tables 7.9 and 7.10). The degrees of freedom for design C (as an example) are calculated as follows: 19 for the total corrected sum of squares, less 1 for blocks (as there are two blocks), less 5 for the quadratic Scheffé model (one fewer than the number of model terms), leaving 13 degrees of freedom for the residuals. The residual degrees of freedom are partitioned into 9 for lack of fit and 4 for pure error (1 degree of freedom for pure error in the factorial block and 3 in the axial block).

Values for $|r_{max}|$ are based on projection *onto a simplex*, but the values are little changed when projection is onto an asymmetrically shaped design region. For example, projecting design D onto the poultry design region (page 106), design E onto the surfactant design region (page 87), or designs F and G onto the alloy design region (page 67) leads to nearly the same $|r_{max}|$ values as listed in Table 7.14.[4]

Considering the possible blocking arrangements of CCDs, the number of center points that might be added to blocks, the constraints on the mixture components, and so forth, the number of possible combinations becomes very large. Table 7.14 is thus only a sampling of the possibilities. The five designs without center points (A, E, F, H, and J) lead to aliasing of one or more of the Z_i^2 terms in a full second-order model in the Zs. However, our concern is only that the designs support two-factor interaction models (terms up to $Z_i Z_j$) without aliasing, and all of the designs in the table will do that. In addition, all but J and K block orthogonally. In the case of J and K, $|r_{max}| = 0.0303$ and 0.0413, respectively, for the blocked two-factor-interaction models in the Zs; in the blocked quadratic *mixture* models, there is a modest increase in the $|r_{max}|$ values to 0.0416 and 0.0539, respectively.

A trend that appears up to (but not including) $q = 7$ is that adjusting the number of runs in blocks so that they are equal to one another leads to orthogonal designs. For example, compare design A with B–D, F with G, and H with I. The trend falls apart at $q = 7$, but designs J and K are the only designs in the table that do not block orthogonally to begin with. Even so, $|r_{max}|$ values for all designs in the table are still so low that for practical purposes they may be considered to block orthogonally in mixture space.

In some cases there is a clear advantage to projecting a blocked CCD (or a fraction thereof) vs. a blocked factorial or blocked fractional factorial design. For example, designs B, C, and D in Table 7.14, when compared to design 1 in Table 7.13, bring in additional degrees of freedom for lack of fit and pure error. Designs G and 4 have the same number of runs (32), but G provides degrees of freedom for pure error. The same can be said about designs I and 5. Although in the case of design K, $|r_{max}| \not\approx 0$, the design provides a more modest number of degrees of freedom for lack of fit than designs 7 or 8, plus the additional benefit of degrees of freedom for pure error.

Summarizing, one has three strategies for orthogonal, or nearly orthogonal, blocking of mixture experiments. The Latin-square approach is probably the best approach for symmetrically shaped design regions. The projection-design approach can be used in either symmetrically or asymmetrically shaped design regions, but it is the best approach for the latter. The rerandomization approach can be used no matter what the shape of the design region. This approach is worth considering when designing mixture-process variable experiments, as the number of runs in such experiments can become quite large. Another advantage of the rerandomization approach is that one can start with a D-optimal design. However, at best one can only hope for nearly orthogonal blocking when using this method.

---

[4]The design regions were approximated using the following center points ($\mathbf{c}$) and half ranges ($\mathbf{r}$). Poultry: $\mathbf{c} = [0.55, 0.15, 0.30]'$, $\mathbf{r} = [0.25, 0.15, 0.30]'$; surfactant: $\mathbf{c} = [0.66, 0.15, 0.16, 0.03]'$, $\mathbf{r} = [0.20, 0.15, 0.16, 0.03]'$; alloy: $\mathbf{c} = [0.06, 0.08, 0.09, 0.23, 0.54]'$, $\mathbf{r} = [0.03, 0.08, 0.09, 0.13, 0.19]'$.

Appendix 7A. Mates for Latin Squares of Order 4 and 5

Table 7.15. *Pattern #1 for* $q = 4$

	First row of		Crossproduct sums					
Square	square	mate	12	13	14	23	24	34
2	*abcd*	*acbd*	A	A	B	B	A	A
3	*abdc*	*adbc*	C	C	D	D	C	C
4	*acdb*	*adcb*	E	E	F	F	E	E

Table 7.16. *Pattern #2 for* $q = 4$

	First row of		Crossproduct sums					
Square	square	mate	12	13	14	23	24	34
2	*abdc*	*acdb*	A	B	A	A	B	A
3	*abcd*	*adcb*	C	D	C	C	D	C
4	*acbd*	*adbc*	E	F	E	E	F	E

Table 7.17. *Pattern #3 for* $q = 4$

	First row of		Crossproduct sums					
Square	square	mate	12	13	14	23	24	34
2	*adbc*	*adcb*	B	A	A	A	A	B
3	*acbd*	*acdb*	D	C	C	C	C	D
4	*abcd*	*abdc*	F	E	E	E	E	F

Table 7.18. *Pattern #1 for* $q = 5$

		First row of		Crossproduct sums									
Square	Pair	square	mate	12	13	14	15	23	24	25	34	35	45
1	a	*abcde*	*adebc*	A	B	A	B	A	B	B	B	A	A
1	b	*acdeb*	*aebcd*	B	A	B	A	B	A	A	A	B	B
2	a	*abced*	*aedbc*	C	D	C	D	C	D	D	D	C	C
2	b	*acedb*	*adbce*	D	C	D	C	D	C	C	C	D	D
3	a	*abdec*	*aecbd*	E	F	E	F	E	F	F	F	E	E
3	b	*acbde*	*adecb*	F	E	F	E	F	E	E	E	F	F
4	a	*abdce*	*acebd*	G	H	G	H	G	H	H	H	G	G
4	b	*adceb*	*aebdc*	H	G	H	G	H	G	G	G	H	H
5	a	*abedc*	*adcbe*	F	E	F	E	F	E	E	E	F	F
5	b	*acbed*	*aedcb*	E	F	E	F	E	F	F	F	E	E
6	a	*abecd*	*acdbe*	A	B	A	B	A	B	B	B	A	A
6	b	*adbec*	*aecdb*	B	A	B	A	B	A	A	A	B	B

Table 7.19. *Pattern #2 for* $q = 5$

		First row of		Crossproduct sums									
Square	Pair	square	mate	12	13	14	15	23	24	25	34	35	45
1	a	*abced*	*adecb*	A	B	B	A	A	B	B	A	B	A
1	b	*acdbe*	*aebdc*	B	A	A	B	B	A	A	B	A	B
2	a	*abcde*	*aedcb*	C	D	D	C	C	D	D	C	D	C
2	b	*acebd*	*adbec*	D	C	C	D	D	C	C	D	C	D
3	a	*abdce*	*aecdb*	E	F	F	E	E	F	F	E	F	E
3	b	*acbed*	*adebc*	F	E	E	F	F	E	E	F	E	F
4	a	*abdec*	*acedb*	G	H	H	G	G	H	H	G	H	G
4	b	*adcbe*	*aebcd*	H	G	G	H	H	G	G	H	G	H
5	a	*abecd*	*adceb*	F	E	E	F	F	E	E	F	E	F
5	b	*acbde*	*aedbc*	E	F	F	E	E	F	F	E	F	E
6	a	*abedc*	*acdeb*	A	B	B	A	A	B	B	A	B	A
6	b	*adbce*	*aecbd*	B	A	A	B	B	A	A	B	A	B

Table 7.20. *Pattern #3 for $q = 5$*

Square	Pair	First row of square	First row of mate	Crossproduct sums 12	13	14	15	23	24	25	34	35	45
1	a	*abdce*	*adbec*	A	A	B	B	B	A	B	B	A	A
1	b	*acedb*	*aecbd*	B	B	A	A	A	B	A	A	B	B
2	a	*abecd*	*aebdc*	C	C	D	D	D	C	D	D	C	C
2	b	*acdeb*	*adcbe*	D	D	C	C	C	D	C	C	D	D
3	a	*abedc*	*aebcd*	E	E	F	F	F	E	F	F	E	E
3	b	*acdbe*	*adceb*	F	F	E	E	E	F	E	E	F	F
4	a	*abcde*	*acbed*	G	G	H	H	H	G	H	H	G	G
4	b	*adecb*	*aedbc*	H	H	G	G	G	H	G	G	H	H
5	a	*abdec*	*adbce*	F	F	E	E	E	F	E	E	F	F
5	b	*acebd*	*aecdb*	E	E	F	F	F	E	F	F	E	E
6	a	*abced*	*acbde*	A	A	B	B	B	A	B	B	A	A
6	b	*adebc*	*aedcb*	B	B	A	A	A	B	A	A	B	B

Table 7.21. *Pattern #4 for $q = 5$*

Square	Pair	First row of square	First row of mate	Crossproduct sums 12	13	14	15	23	24	25	34	35	45
1	a	*abdec*	*adbce*	A	A	B	B	B	B	A	A	B	A
1	b	*acebd*	*aecdb*	B	B	A	A	A	A	B	B	A	B
2	a	*abedc*	*aebcd*	C	C	D	D	D	D	C	C	D	C
2	b	*acdbe*	*adceb*	D	D	C	C	C	C	D	D	C	D
3	a	*abecd*	*aebdc*	E	E	F	F	F	F	E	E	F	E
3	b	*acdeb*	*adcbe*	F	F	E	E	E	E	F	F	E	F
4	a	*abced*	*acbde*	G	G	H	H	H	H	G	G	H	G
4	b	*adebc*	*aedcb*	H	H	G	G	G	G	H	H	G	H
5	a	*abdce*	*adbec*	F	F	E	E	E	E	F	F	E	F
5	b	*acedb*	*aecbd*	E	E	F	F	F	F	E	E	F	E
6	a	*abcde*	*acbed*	A	A	B	B	B	B	A	A	B	A
6	b	*adecb*	*aedbc*	B	B	A	A	A	A	B	B	A	B

Table 7.22. *Pattern #5 for $q = 5$*

		First row of		Crossproduct sums									
Square	Pair	square	mate	12	13	14	15	23	24	25	34	35	45
1	a	*abecd*	*adceb*	A	B	B	A	B	A	B	A	A	B
1	b	*acbde*	*aedbc*	B	A	A	B	A	B	A	B	B	A
2	a	*abdce*	*aecdb*	C	D	D	C	D	C	D	C	C	D
2	b	*acbed*	*adebc*	D	C	C	D	C	D	C	D	D	C
3	a	*abcde*	*aedcb*	E	F	F	E	F	E	F	E	E	F
3	b	*acebd*	*adbec*	F	E	E	F	E	F	E	F	F	E
4	a	*abedc*	*acdeb*	G	H	H	G	H	G	H	G	G	H
4	b	*adbce*	*aecbd*	H	G	G	H	G	H	G	H	H	G
5	a	*abced*	*adecb*	F	E	E	F	E	F	E	F	F	E
5	b	*acdbe*	*aebdc*	E	F	F	E	F	E	F	E	E	F
6	a	*abdec*	*acedb*	A	B	B	A	B	A	B	A	A	B
6	b	*adcbe*	*aebcd*	B	A	A	B	A	B	A	B	B	A

Table 7.23. *Pattern #6 for $q = 5$*

		First row of		Crossproduct sums									
Square	Pair	square	mate	12	13	14	15	23	24	25	34	35	45
1	a	*abedc*	*adcbe*	A	B	A	B	B	B	A	A	A	B
1	b	*acbed*	*aedcb*	B	A	B	A	A	A	B	B	B	A
2	a	*abdec*	*aecbd*	C	D	C	D	D	D	C	C	C	D
2	b	*acbde*	*adecb*	D	C	D	C	C	C	D	D	D	C
3	a	*abced*	*aedbc*	E	F	E	F	F	F	E	E	E	F
3	b	*acedb*	*adbce*	F	E	F	E	E	E	F	F	F	E
4	a	*abecd*	*acdbe*	G	H	G	H	H	H	G	G	G	H
4	b	*adbec*	*aecdb*	H	G	H	G	G	G	H	H	H	G
5	a	*abcde*	*adebc*	F	E	F	E	E	E	F	F	F	E
5	b	*acdeb*	*aebcd*	E	F	E	F	F	F	E	E	E	F
6	a	*abdce*	*acebd*	A	B	A	B	B	B	A	A	A	B
6	b	*adceb*	*aebdc*	B	A	B	A	A	A	B	B	B	A

Part III

Analysis

Chapter 8

Building Models in a Mixture Setting

This chapter and the succeeding two constitute what is often lumped together as "model building". I have chosen to use the word "build" in a slightly more restrictive sense, however, and to apply it to the first stage of a three-stage process. The stages are usually carried out sequentially after the experiment has been carried out and the data collected and can be briefly described as follows.

1. In this chapter we focus on the "building up" process, and thus the chapter title. At this stage we ask questions like, "Is a linear model adequate?" If the answer is "No", then we augment the linear model with the quadratic terms and then ask the question, "Is a quadratic model adequate?" And so on. At each stage, if the answer is "No", we augment the model with an additional, higher-order *group* of terms. The decision process involves hypothesis testing. This implies that the model assumptions (Section 3.1) are not violated, an implication that is really not checked until the second stage. At the conclusion of this stage we have a *tentative model*.

2. The second stage consists of *model evaluation*. In this stage (the subject of Chapter 9), we check for possible violations of the model assumptions, we check for outliers or suspect data points, and we check for influential data points. Plotting of residuals plays a major role. In the absence of prior subject-matter knowledge, we get a first impression of whether a transformation of the response might be needed.

3. The third stage (Chapter 10) consists of *fine-tuning* the model. Based on what is learned in the second stage, one may want to modify the tentative model. In this stage we decide what to do with outliers and/or influential points, whether or not to transform the response, and, if so, specifically what transformation is needed. If the model has been overfit, then we elect regressors for removal from the model. If the model has been underfit and we do not have sufficient data points to support a higher-order model, then we need to consider design augmentation, which means we must consider blocking a posteriori. We may cycle through stages 2 and 3 one or more times before we are satisfied with our model.

This chapter focuses on the application of ordinary least squares (OLS) to the analysis of mixture data. A point mentioned earlier in this book bears repeating. Although we will not go through the development of OLS, we will draw generously on the *results*. The matrix formulation of OLS is covered in several texts on linear regression, and there is no need to repeat what has been well presented elsewhere. See, for example, Draper and Smith [49], Montgomery, Peck, and Vining [100], Myers [105], and Neter et al. [113].

To illustrate some of the principles involved, we will draw on an experiment performed at Stepan Company by Hillshafer, O'Brien, and Williamson [70] to study certain polyurethane-reactive hot-melt adhesive formulations.[1] The analyses using these data, however, are the author's approach to the problem. The experimental setting is illustrated in Fig. 8.1.

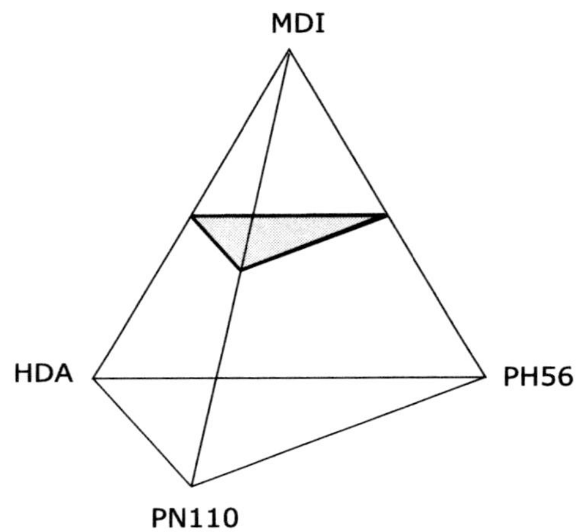

Figure 8.1. *Hot-melt adhesive experimental setting.*

The formulations consisted of four components: two *ortho*-phthalic-based polyols, STEPANOL PN-110 and STEPANOL PH-56, a 4000-molecular weight 1,6-hexanediol adipate (HDA), and 4,4'-diphenylmethane diisocyanate (MDI). The proportion of MDI was maintained at a fixed level in all formulations. This means that the experimental space was restricted to a two-dimensional 3-simplex suspended within the 4-simplex. *We shall henceforth refer to the vertices of the* 3-*simplex as reals* and use the labels HDA, PN-110, and PH-56 for these vertices. We must keep in mind, however, that each of these "reals" has a fixed proportion of MDI (unspecified in the report). This means that every mixture blend will also have the same fixed proportion of MDI and that observed and predicted responses (after a model is fit) are conditional on this fixed proportion of MDI being present. It is a very common experimental situation to have one or more components of a formulation (such as a solvent or solvents) held constant throughout an experiment.

Other than the normal mixture constraints (Eqs. 2.1 and 2.2, page 9), the only bound on a component proportion was a lower bound of 0.5 on HDA, and thus the constrained

[1]The author is indebted to Kip Hillshafer for providing the data.

region is also simplex-shaped. The lower bound on HDA does, however, impart implied upper bounds on PN-110 and PH-56, and so the complete set of constraints is

$$\begin{aligned} 0.50 &\le \text{HDA} \le 1.0 \\ 0.00 &\le \text{PN-110} \le 0.50 \\ 0.00 &\le \text{PH-56} \le 0.50 \end{aligned}$$

Figure 8.2 shows the design in the context of the pseudocomponent simplex. Filled circles represent design points, while a filled circle surrounded by an open circle is a replicated point. Numbers on the outside of the triangle give the component proportions in the reals (as defined above) at the vertices and edges. Numbers in the interior of the triangle label contours of equal $\sqrt{h_{00}}$ (standard errors of prediction, apart from σ). These contours depend, of course, not only on the design but also on the model. For illustrative purposes, a quadratic model has been assumed.

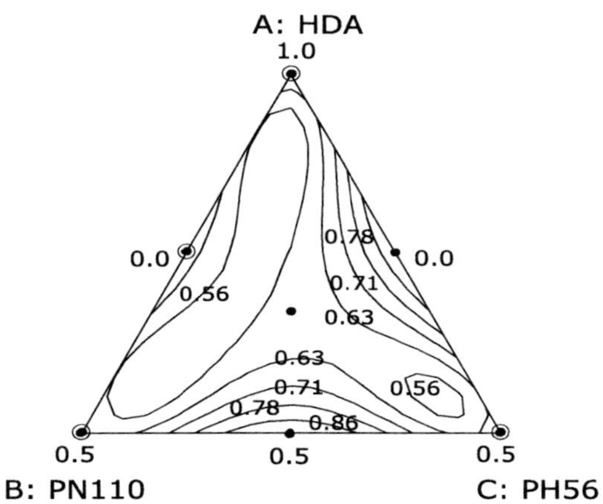

Figure 8.2. *Hot-melt adhesive design. Modified Design-Expert plot.*

The reason that the contours are asymmetrical with respect to the triangle is because the replicate design points are not symmetrically located with respect to the triangle (although they are symmetrically located with respect to the PH-56 axis). Had the replicate at the midpoint of the HDA–PN-110 edge been placed at the overall centroid, then the contours would have been symmetrical with respect to the triangle. However, this would come at a price because $|(\mathbf{X}'\mathbf{X})^{-1}|$ would increase from 169.28 to 230.40 ($\mathbf{X}$ expressed in the pseudocomponent metric). A bit of algebra indicates that this would result in about a 16.7% increase in the volume of the confidence ellipsoid for the parameter estimates (cf. the discussion in Section 5.4.2, page 76). For the quadratic model and the design in Fig. 8.2, it can be seen that precision in prediction will be most precise near the HDA–PN-110 edge and the PH-56 vertex.

Component proportions and two responses are displayed in Table 8.1. Although lacking axial check blends, the design points nevertheless lie on the component axes, and so a simplex-screening plot is worth examining. A plot for the viscosity data, which is

Table 8.1. *Hot-melt adhesive experiment*

	Proportions							
	Component			Pseudocomponent			Response	
ID	HDA	PN-110	PH-56	HDA	PN-110	PH-56	Visc†	GS3‡
1	1.00	0.00	0.00	1.00	0.00	0.00	8.00	94
2	1.00	0.00	0.00	1.00	0.00	0.00	4.80	120
3	0.75	0.25	0.00	0.50	0.50	0.00	19.20	47
4	0.75	0.25	0.00	0.50	0.50	0.00	18.15	44
5	0.75	0.00	0.25	0.50	0.00	0.50	8.60	154
6	0.50	0.50	0.00	0.00	1.00	0.00	51.10	75
7	0.50	0.50	0.00	0.00	1.00	0.00	42.90	47
8	0.50	0.25	0.25	0.00	0.50	0.50	18.28	42
9	0.50	0.00	0.50	0.00	0.00	1.00	7.14	25
10	0.50	0.00	0.50	0.00	0.00	1.00	6.94	29
11	0.67	0.17	0.17	0.33	0.33	0.33	12.48	16

† Viscosity, cP $\times 10^{-3}$, 120°C ‡ Green strength, psi @ 3 min.

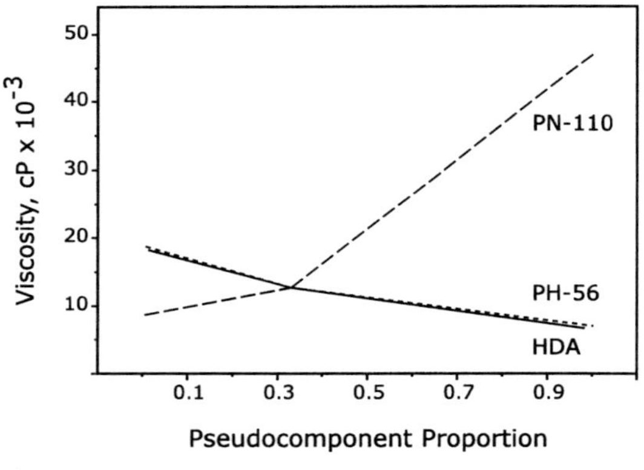

Figure 8.3. *Adhesive viscosity response. Simplex-screening plot.*

the response that will be used to illustrate model fitting, is shown in Fig. 8.3. Responses at replicated points were averaged for this plot. The trace for PN-110 appears definitely nonlinear, providing some justification for the choice of a quadratic model for the contours in Fig. 8.2.

8.1 Partitioning Total Variability. Sequential Sums of Squares

Much space was devoted in Chapter 3 to polynomial models of varying degree — linear, quadratic, cubic, etc. One model that was completely neglected, however, was the very important **null model**. The null model is the model that results if the explanatory variables have no effect on the response. In a mixture setting, this would mean that varying the composition of the mixture does not significantly alter the response. Under these circumstances, then, we would expect there to be no terms in the model higher than linear (no curvature) and that the linear estimators in a Scheffé model would be equal to one another ($\beta_1 = \beta_2 = \cdots = \beta_q$), and symbolized β_0 [92]. If the response is invariant to the levels of the component proportions, then the expectation function is

$$\begin{aligned} E(Y) &= \sum_{i=1}^{q} \beta_i X_i \\ &= \beta_0 \sum_{i=1}^{q} X_1 \\ &= \beta_0 \end{aligned}$$

and the least-squares estimate of β_0 is $\bar{Y}$, the average overall response.

For the viscosity response in the hot-melt adhesive experiment, the response surface for the null model would look like Fig. 8.4. In this figure and the figures to follow, X_1 = HDA, X_2 = PN-110, and X_3 = PH-56. The response surface in this case is a horizontal plane located at the mean viscosity, 17963 cP. Generalizing, *in a mixture setting*, the null model is

$$\hat{Y} = \bar{Y} \tag{8.1}$$

where $\bar{Y}$ is the *average overall response*.

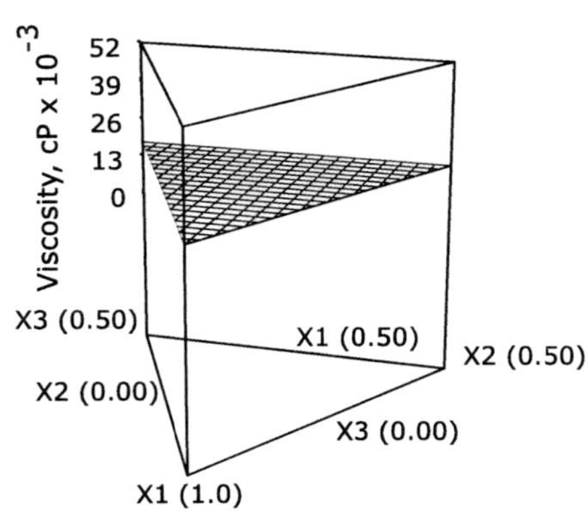

Figure 8.4. *Viscosity response surface. Null model.*

The method of least squares is designed to find parameter estimates for a specified model using the "one-step" estimator Eq. 7.15, page 128, such that the **residual sum of squares** $\sum_{i=1}^{n}(Y_i - \hat{Y}_i)^2$ is minimized. In the special case of the null model, $\hat{Y}_i = \bar{Y}$, and so the sum of squares that is minimized is given by $\sum_{i=1}^{n}(Y_i - \bar{Y})^2$. This is called the **corrected total sum of squares**. It is corrected because variability of the Y_i values is measured about the mean, which is always the case in a Scheffé mixture model. Although in the case of the null model the residual sum of squares and the corrected total sum of squares are equal to one another, in general this is not the case (see below).

On the left in Fig. 8.5, a vertical plane (dashed outline) has been passed through the response surface for the null model, bisecting it along the X_2 (PN-110) axis. The figure on the right is a profile view of the response surface (dashed line) along this axis. The filled circles are observed viscosities along the PN-110 axis. The numbers next to the points are ID numbers in Table 8.1. The solid triangles lying vertically above or below the circles are fitted values based on the null model. In every case, of course, the fitted values will be equal to $\bar{Y}$. The quantity that is minimized by the least-squares procedure, $\sum_{i=1}^{n}(Y_i - \hat{Y}_i)^2$, is the sum of the squared vertical distances between each of the circle symbols and the corresponding triangle symbols, $\sum_{i=1}^{n}(Y_i - \bar{Y})^2$. However, the sum is taken over all $n = 11$ data points, not just the four in this illustration.

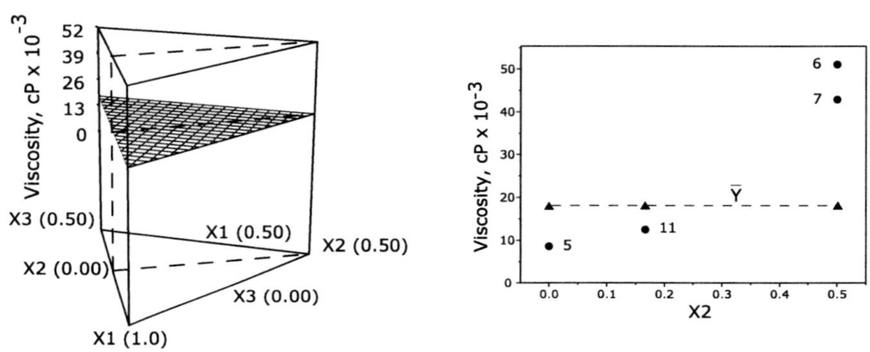

Figure 8.5. *Adhesive viscosity response. PN-*110 *axis. Null model.*

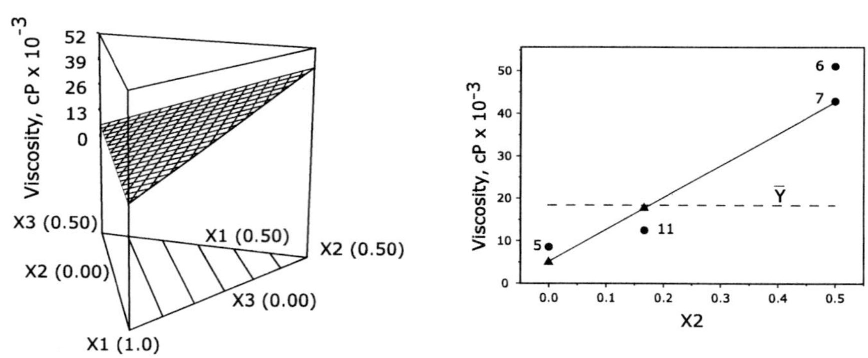

Figure 8.6. *Adhesive viscosity response. PN-*110 *axis. Linear model.*

Consider now Fig. 8.6, which is similar to Fig. 8.5 except that a *linear* (first-order) model has been fit to the data. The response surface is still a flat plane, but the plane is tipped upward in the X_2 direction. The model for the surface is

$$\hat{Y} = 4.18\,(\text{HDA}^*) + 42.4\,(\text{PN-110}^*) + 5.69\,(\text{PH-56}^*) \qquad (8.2)$$

where the superscript asterisk denotes pseudocomponent proportions. The coefficients estimate the response at the vertices of the pseudocomponent simplex.

The profile plot (again along the PN-110 (X_2) component axis) shows that the response surface (solid line) comes much closer to the observed responses than in the case of the null model. The fitted values $\hat{Y}_i$ are again the solid triangles, but only two are visible because the third and fourth are hidden under point 7.

For each point in the figure on the right (as well as for each of the 11 design points), we can partition the distance between an observation (filled circle) and the mean (horizontal dashed line) into two parts as follows:

$$Y_i - \bar{Y} = (Y_i - \hat{Y}_i) + (\hat{Y}_i - \bar{Y}) \tag{8.3}$$

The first term on the right is equal to the distance from an observed value (circle) to its fitted value (triangle); the second term is equal to the distance from a fitted value to the mean. For example, based on model 8.2, $\hat{Y}_5 = 4.934$. Substituting into Eq. 8.3, we have

$$\begin{aligned} 8.60 - 17.96 &= (8.60 - 4.934) + (4.934 - 17.96) \\ -9.36 &= 3.666 - 13.026 \\ -9.36 &= -9.36 \end{aligned}$$

To calculate $\sum_{i=1}^{n}(Y_i - \hat{Y}_i)^2$, the sum of squared residuals, let us square both sides of Eq. 8.3 and sum over all of the data points.

$$\begin{aligned} \sum_{i=1}^{n}(Y_i - \bar{Y})^2 &= \sum_{i=1}^{n}\left[(Y_i - \hat{Y}_i) + (\hat{Y}_i - \bar{Y})\right]^2 \\ &= \sum_{i=1}^{n}(Y_i - \hat{Y}_i)^2 + 2\sum_{i=1}^{n}(Y_i - \hat{Y}_i)(\hat{Y}_i - \bar{Y}) + \sum_{i=1}^{n}(\hat{Y}_i - \bar{Y})^2 \end{aligned}$$

The middle term on the right can be written

$$\begin{aligned} 2\sum_{i=1}^{n}(Y_i - \hat{Y}_i)(\hat{Y}_i - \bar{Y}) &= 2\sum_{i=1}^{n}(Y_i - \hat{Y}_i)\hat{Y}_i - 2\sum_{i=1}^{n}(Y_i - \hat{Y}_i)\bar{Y} \\ &= 2\sum_{i=1}^{n}\hat{Y}_i e_i - 2\bar{Y}\sum_{i=1}^{n}e_i \\ &= 0 \end{aligned} \tag{8.4}$$

Whenever variation is measured about the mean, it is a property of least-squares fitted models that the sum of the residuals weighted by the corresponding fitted values (first summation on the right in Eq. 8.4) as well as the sum of the residuals (second summation on the right) is equal to zero [49, 100, 113]. For this reason, the cross term in Eq. 8.4 vanishes.

We are left, then, with the following important identity for partitioning the total variability:

$$\sum_{i=1}^{n}(Y_i - \bar{Y})^2 = \sum_{i=1}^{n}(Y_i - \hat{Y}_i)^2 + \sum_{i=1}^{n}(\hat{Y}_i - \bar{Y})^2 \tag{8.5}$$

which we shall abbreviate

$$\text{SST} = \text{SSE} + \text{SSR}$$

or equivalently

$$\boxed{\text{SST} = \text{SSR} + \text{SSE}} \tag{8.6}$$

which is the order that shall be used henceforth. In this expression, SST stands for the **corrected total sum of squares**, SSR stands for the **regression** or **model sum of squares**, and SSE stands for the **error** or **residual sum of squares**.

It is important to realize that in any sequential model-building process, the left side of Eq. 8.6 *will remain constant*. The net effect of fitting graduated polynomial functions (such as null, linear, quadratic, cubic, etc.) is to change the allocation of the corrected total sum of squares between the two terms on the right-hand side of the equation. This is in fact the subject of this chapter section — the partitioning of total variability into variability explained by the model (SSR) and random variability (SSE).

On page 158 it was stated that in the case of the null model, the residual sum of squares (SSE) was equal to the total sum of squares (SST), which implies that the regression sum of squares (SSR) must be equal to zero. From Eq. 8.5, we see that the regression sum of squares is given by $\sum_{i=1}^{n}(\hat{Y}_i - \bar{Y})^2$. As $\hat{Y}_i = \bar{Y}$ in the null model, this term obviously must be equal to zero. For the null model and the viscosity data, we have

$$\text{SSR} + \text{SSE} = 0 + 2350.48 \ (= \ 2350.48) \qquad \text{(null model)}$$

In the case of the linear model (Fig. 8.6), the triangles have moved closer to the circles, which means that variability is shifting from the SSE term into the SSR term. For this model, SST is partitioned as follows:

$$\text{SSR} + \text{SSE} = 2141.63 + 208.85 \ (= \ 2350.48) \qquad \text{(linear model)}$$

Fitting the viscosity data to a quadratic model leads to

$$\begin{aligned}\hat{Y} &= 6.44\,(\text{HDA}^*) + 47.0\,(\text{PN-110}^*) + 7.08\,(\text{PH-56}^*)\\ &\quad - 32.9\,(\text{HDA}^*)(\text{PN-110}^*) + 6.13\,(\text{HDA}^*)(\text{PH-56}^*)\\ &\quad - 36.4\,(\text{PN-110}^*))(\text{PH-56}^*)\end{aligned}$$

The response surface and profile plot for this model are illustrated in Fig. 8.7. Both quadratic terms involving PN-110 (X_2) are negative, and this is reflected in the response surface where the X_1–X_2 and X_2–X_3 edges curve *downward* (antagonistic blending). The HDA–PH-56 cross term is slightly positive, and this is reflected in a slight *upward* curvature of the X_1–X_3 edge. This is not at all obvious on the response surface but can be inferred from the curved contour line near the X_1–X_3 edge.

The fitted regression line in the profile plot is a smoothed version of the line for PN-110 in Fig. 8.3, page 156. Compared to the linear model, the regression line has moved even closer to the observed values. This means that part of SSE in the linear model has shifted into SSR in the quadratic model. The partitioning of the total sum of squares is now

$$\text{SSR} + \text{SSE} = 2310.44 + 40.04 \ (= \ 2350.48) \qquad \text{quadratic model}$$

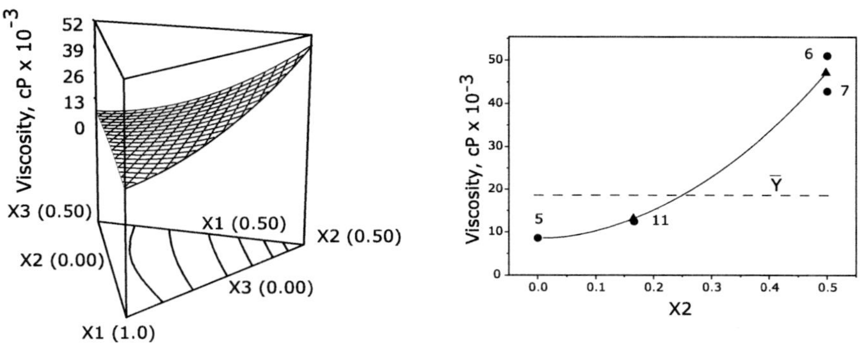

Figure 8.7. *Adhesive viscosity response. PN-110 axis. Quadratic model.*

And finally, if one fits the special cubic model, the breakdown is

$$\text{SSR} + \text{SSE} = 2311.17 + 39.31 \ (= \ 2350.48) \qquad \text{special cubic model}$$

Table 8.2 displays a summary table of the **sequential sums of squares** for the viscosity data along with other pertinent information. The format is much like that of an ANOVA table, although ANOVA tables are more model specific and summarize *partial* sums of squares (Section 8.2, page 165). The meanings of the various columns in this table are explained in succeeding paragraphs.

Table 8.2. *Adhesive viscosity response. Sequential sums of squares*

Terms	Sum of Squares	df	Mean Square	F Value	Prob> F
Linear	2141.63	2	1070.81	41.02	<0.0001
Quadratic	168.81	3	56.27	7.03	0.0305
Special cubic	0.73	1	0.73	0.075	0.7984
Residual	39.31	4	9.83		
Corrected total	2350.48	10	235.05		

The numbers in column 2 for the quadratic and special cubic models appear to be at odds with those in the SSR + SSE breakdowns. The numbers in this column are the *additional* sums of squares for regression as the terms are *sequentially* brought into the model. For the linear model, only the linear terms are in the model, and so SSR = 2141.63; the quadratic and special cubic terms are not in the model, and so SSE = 168.81 + 0.73 + 39.31 = 208.85. For the quadratic model, the linear and quadratic terms are in the model, and so SSR = 2141.63 + 168.81 = 2310.44; the special cubic term is not in the quadratic model, and so SSE = 0.73 + 39.31 = 40.04. For the special cubic model, the linear, quadratic, and special cubic terms are in the model, and so SSR = 2141.63 + 168.81 + 0.73 = 2311.17; SSE = 39.31. In short, a table of sequential sums of squares provides the

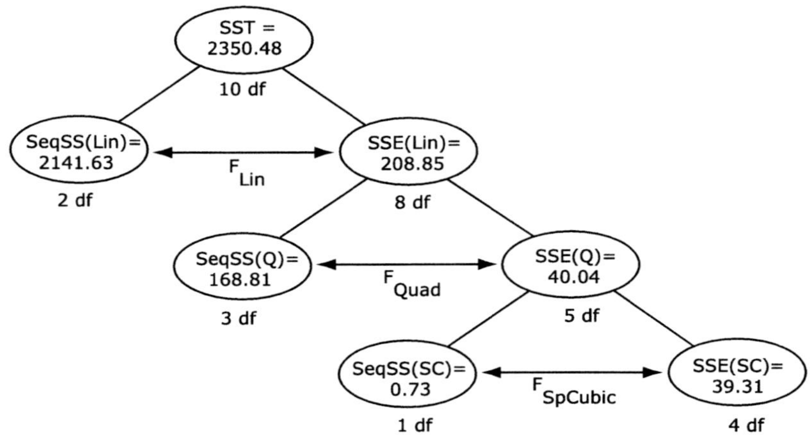

Figure 8.8. *Adhesive viscosity data. Sums-of-squares tree.*

analyst with a summary of the SSR ↔ SSE breakdown as terms of higher order are brought into a model.

The reader may find the illustration in Fig. 8.8 helpful.[2] The cell at the top gives the corrected total sum of squares (2350.48) for the adhesive viscosity data. Beneath this cell, the cells along the left branches give the sequential sums of squares (SeqSSs) for the linear, quadratic, and special cubic *terms*. The residual sums of squares for the linear, quadratic, and special cubic *models* are given along the right branch. Cells at the same horizontal level apply to terms and models of the same order. For example, SeqSS(Q) = 168.81 is the sequential sum of squares for entry of the three quadratic terms into the linear model; SSE(Q) = 40.04 is the residual sum of squares (the unexplained variability) for the full quadratic model. The meaning of the Fs beneath the arrows is explained below.

The SeqSSs for the linear terms (2141.63) is also the SSs for the linear *model*. The SSs for the quadratic model is equal to the SeqSSs for the linear terms (2141.63) plus the SeqSSs for the quadratic terms (168.81). The SSs for the special cubic model is equal to the SeqSSs for the linear terms (2141.63) plus the SeqSSs for the quadratic terms (168.81) plus the SeqSSs for the special cubic terms (0.73).

The degrees-of-freedom column in Table 8.2 gives the number of *additional* degrees of freedom as groups of terms are brought into the model. For example, the null model has one term (the mean), while the linear model has three terms. The difference is 2, and so there are 2 degrees of freedom for the linear terms. Bringing the quadratic terms into the model adds three more terms, and so there are 3 additional degrees of freedom. Finally, bringing the special cubic term into the model adds one additional degree of freedom.

In both Table 8.2 and Fig. 8.8, summing the sums of squares for the linear, quadratic, and special cubic terms *and* the residual for the special cubic model (labeled "Residual" in Table 8.2 and "SSE(SC)" in Fig. 8.8) gives SST, the corrected total sum of squares. Similarly, summing the degrees of freedom for the model terms and the residual gives the

[2]The author is indebted to Stanley Deming for suggesting this diagram.

degrees of freedom for SST. Because the mean has been used to calculate SST, there is one fewer degree of freedom for SST than there are observations. (If one knows the mean, the nth observation can always be calculated from the remaining $n - 1$ observations and the mean.)

Mean squares in column 4 of Table 8.2 are calculated by dividing the sequential sum of squares for a group of terms by its respective degrees of freedom (a kind of averaging process). Under the assumption that the $\epsilon_i \sim \text{NID}(0, \sigma^2)$ (an assumption that has not yet been checked), a statistical theorem informs us that the ratio

$$\begin{aligned} F &= \frac{SS_{terms}/\nu_1}{SS_{residual}/\nu_2} \\ &= \frac{MS_{terms}}{s^2} \end{aligned}$$

follows an F distribution with ν_1 degrees of freedom in the numerator and ν_2 degrees of freedom in the denominator [102]. The F ratio compares the average, or mean, variability explained by the model to the average variability not explained by the model.

To illustrate, for the special cubic term, $MS_{term} = 0.73/1$, and for the special cubic model, the estimate of s^2 is 39.31/4. (Reference to Fig. 8.8 will help to make this clearer.) The F ratio for the special cubic term is then

$$F \text{ (special cubic term)} = \frac{0.73/1}{39.31/4} = 0.074 \tag{8.7}$$

which is equal within rounding error to the value in the table. For the quadratic terms, $MS_{terms} = 168.81/3$ and the estimate of s^2 is $(0.73 + 39.31)/(1 + 4)$. The F ratio is then

$$F \text{ (quadratic terms)} = \frac{168.81/3}{(0.73 + 39.31)/(1 + 4)} = 7.03 \tag{8.8}$$

In a similar fashion, the F ratio for the linear terms is

$$F \text{ (linear terms)} = \frac{2141.63/2}{(168.81 + 0.73 + 39.31)/(3 + 1 + 4)} = 41.02 \tag{8.9}$$

The sums of the SSs in the denominators of each of the three ratios Eqs. 8.7, 8.8, and 8.9 are given in the three cells beneath the SST cell in Fig. 8.8 that lie on the right branch of the sums-of-squares tree. The sums of the degrees of freedom in the denominators of each of the ratios are given beneath the same cells.

Referring to Fig. 8.8 again, to calculate an F-statistic for the quadratic *model* (for example), as opposed to the quadratic terms, one would first sum the SeqSSs for the linear and quadratic terms (2141.63 + 168.61) and divide by the sum of their degrees of freedom (2 + 3). This number, which is the *model* mean square, would then be divided by mean square for the quadratic model (40.04/5).

The F ratios for the linear, quadratic, and special cubic terms are collected in column 5 of Table 8.2. They are test statistics that test the truth of a null hypothesis that can be expressed in words as follows: H_0 : *the response is invariant to the presence or absence of the* ***group*** *of terms that has been added to the model.* Failure to reject this null hypothesis does *not* mean that the least-squares estimates for the terms are necessarily equal to zero. The word *not* in the previous sentence is quadruply emphasized in this paragraph because this is a trap that is easily fallen into. It is true that for special cubic terms, the null hypothesis is (algebraically) $H_0 : \beta_{ijk} = 0$, all $i < j < k$, and for quadratic terms, the null hypothesis is $H_0 : \beta_{ij} = 0$, all $i < j$. However, for linear terms, the null hypothesis is *not* $H_0 : \beta_i = 0$, all i, but rather $H_0 : \beta_1 = \beta_2 = \cdots = \beta_q$. The reader should keep in mind that the linear estimators simply estimate the response at the vertices of the simplex (or pseudocomponent simplex) and *do not estimate the effects*.

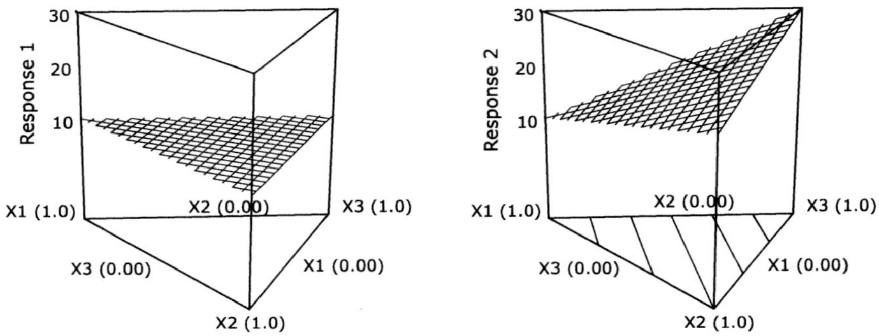

Figure 8.9. *Two linear response surfaces.*

To help grasp this concept, consider the two response surfaces in Fig. 8.9. The surface on the left is modeled by the equation

$$Y = 10X_1 + 10X_2 + 10X_3$$

while the surface on the right is modeled by

$$Y = 10X_1 + 20X_2 + 30X_3$$

If we test the null hypothesis $H_0 : \beta_{X_1} = 10$, we will not reject it in both cases. However, all we have learned is something about the estimated response at $X_1 = 1.0$. We have not learned anything about the effect of X_1.

In this age of computers, it is seldom that we need to consult a table of F values. Instead we usually check what are called p values, which are the numbers in Table 8.2 in the column labeled "Prob> F". When the p value is less than some preselected cutoff (usually 0.10, 0.05, or 0.01), we reject the null hypothesis and conclude that the term or group of terms should be included in the model. The p values in Table 8.2 inform us that the linear and quadratic terms are at or above the 95% level of confidence ($p < 0.05$) but that the cubic term is not ($p \gg 0.05$). (Hereafter when stating a term or set of terms is significant, unless stated otherwise we shall mean at a level of significance ≤ 0.05.) The fact that the group of quadratic terms is significant is not a guarantee that *all* of the quadratic terms are significant. Information about this is found in the ANOVA table, the subject of the next section.

The F ratios in Eqs. 8.7–8.9 can be cast in a general form that we shall find extremely useful when we consider model reduction. Known as the **extra sum-of-squares principle** [49], it is one of the most important formulas for building linear regression models.

$$F = \frac{(\text{SSR}_{fuller\ model} - \text{SSR}_{less\ full\ model})/\Delta df}{s^2} \tag{8.10}$$

Here the nomenclature of Lunneborg has been adopted to describe the two models [91]. For example, if a quadratic model were fit to the viscosity data and one wanted to test the null hypothesis H_0 : all $\beta_{ij} = 0$, the *fuller* model would be the quadratic model and the *less full* model would be the linear model.[3] Δdf is the difference in the model degrees of freedom in the fuller vs. the less full model. The denominator, s^2, is the mean square error (MSE) for the fuller model. Applying Eq. 8.10 to the viscosity data, we have

$$\begin{aligned} F &= \frac{(\text{SSR}_{quadratic\ model} - \text{SSR}_{linear\ model})/\Delta df}{s^2} \\ &= \frac{(2310.44 - 2141.63)/(5-2)}{40.04/5} \\ &= \frac{168.81/3}{8.008} \\ &= 7.03 \end{aligned}$$

which is the same result as Eq. 8.8, page 163. Note that the 6-term quadratic Scheffé model has 5 degrees of freedom, and the 3-term linear Scheffé model has 2 degrees of freedom.

The advantage of Eq. 8.10 is that it is applicable to cases where less than a full group of terms might be considered for removal from a model. For example, if one fit a quadratic Scheffé model to a $q = 5$ mixture problem and wanted to test the null hypothesis that all quadratic terms except those involving X_1 were equal to zero, then Eq. 8.10 would take the form

$$F = \frac{(\text{SSR}_{quadratic\ model} - \text{SSR}_{reduced\ quadratic\ model})/(14-8)}{s^2}$$

The full quadratic model has 15 terms and 14 degrees of freedom. There are a total of 10 quadratic terms. Removing six of these (those not involving X_1) leaves us with a model that has 9 terms and 8 degrees of freedom.

Discussions of the extra sum-of-squares principle can be found in Cornell [29], Draper and Smith [49], Lunneborg [91], Myers [105], Montgomery, Peck, and Vining [100], and Neter et al. [113] as well as in a variety of other texts on linear regression.

8.2 The ANOVA Table. Partial Sums of Squares

It was pointed out in Chapter 1 that before one can really propose a design, one must have some idea of the model that one intends to support. However, things do not always turn out

[3] A variety of other descriptors are used to describe fuller/less full model pairs, such as complete/reduced, full/reduced, and expanded/reduced.

as planned. For example, if experimentation is not expensive, one might design to support a higher-order model, perhaps even a quartic model, because anything of order lower than quartic would also be supported. On the other hand, one might have a preconceived idea that a linear model will be adequate but design for a quadratic model "just in case", and then discover that in fact a higher-order model is required. By examining a table of sequential sums of squares with associated F tests, one can make an *informed choice* as to the probable degree of the model. The sequential sums of squares provide one with an overview of the model landscape. Many DOE packages (not all) output sequential-sums-of-squares tables after a model has been selected, when it is least useful. The ANOVA table, the subject of this section, is a table that is always output after one selects a tentative model and provides detailed, useful information about the model.

The format of an ANOVA table varies slightly from software to software, but the basic structure remains much the same. Table 8.3 is representative and is adapted from Design-Expert output. The table is based on a quadratic Scheffé model that has been fit to the adhesive viscosity data in Table 8.1, page 156.

Table 8.3. *Adhesive viscosity response. Partial sums of squares, quadratic model*

Term(s)	Sum of Squares	df	Mean Square	F Value	Prob> F
Model	2310.44	5	462.09	57.70	0.0002
Linear Mixture	2141.63	2	1070.81	133.71	<0.0001
AB	96.63	1	96.63	12.07	0.0178
AC	2.16	1	2.16	0.27	0.6254
BC	76.14	1	76.14	9.51	0.0274
Residual	40.04	5	8.01		
Lack of Fit	0.73	1	0.73	0.075	0.7984
Pure Error	39.31	4	9.83		
Corrected total	2350.48	10			

Perhaps the first thing to notice is that the table is titled **partial sums of squares** rather than sequential sums of squares (as in Table 8.2, page 161). Summing the sum of squares for Model + Residual *does* equal the Corrected total. However, summing the sum of squares for Linear Mixture + $AB + AC + BC$ gives 2316.56, which does *not* add to the model sum of squares (2310.44) or to anything meaningful. While it is possible to generate an ANOVA table similar in format to Table 8.3 but with sequential sums of squares replacing partial sums of squares, such a table would not be of great use.

The reason for this is the following. In building graduated polynomial models, it makes sense to bring in *groups* of terms in the order linear before quadratic, quadratic before cubic, and cubic before quartic. This is the rationale for sequential sums of squares. However, when we get to the *individual* terms that comprise a group, there are several possible ways to enter the terms. If there are k terms in a group, then there are $k\,!$ possible orders in which the terms could be entered. Sequential sums of squares are dependent on the order in which the variables are entered into the model. If there are k ! terms in a group, then there are k ! sets of sums of squares for the terms in the group, depending on the order in which they are entered into the model.

Consider the quadratic model for the viscosity data. Given that the linear terms are already in the model, the sequential sums of squares for each of the three quadratic terms is tabulated in Table 8.4. The cell entry "90.34|AC" (for example) should be read "90.34 given that AC is already in the model". If the term enters last (fourth column), it is implied that the other two terms are already in the model. Note that the sum of squares in column 4 is equal to the partial sums of squares for the corresponding terms in Table 8.3. In other words, partial sums of squares are the sums of squares that are moved from SSE $\rightarrow$ SSR when a term enters the model last. For example, if AB enters the model last, then SSR increases by 96.63 and SSE decreases by 96.63. Just as important, however, is that partial sums of squares are the sums of squares that are moved from SSR $\rightarrow$ SSE when a term is removed from the model first, *all other terms remaining*. In other words, if AB is removed from the model first, then SSR decreases by 96.63 and SSE increases by 96.63.

Table 8.4. *Effect of order of entry of terms on sequential sums of squares*

	Order of entry		
Term	First	Second	Last
AB	91.35	90.34 $\|AC$ 97.88 $\|BC$	96.63
AC	2.33	1.32 $\|AB$ 3.41 $\|BC$	2.16
BC	68.77	75.30 $\|AB$ 69.85 $\|AC$	76.14

F values in Table 8.3 are calculated by dividing the mean square for a term (quadratic) or group of terms (linear mixture) by s^2, which is the mean square for the residual (8.01). Examination of the p values for the quadratic terms in Table 8.3 indicates that the AC term (HDA* $\times$ PH-56*) is not significant at the $\alpha = 0.05$ level of significance and could be removed from the model. The slight upward curvature on the X_1–X_3 edge in Fig. 8.7, page 161, is thus not significant, and the response could just as well be modeled without this term in the model.

One's interest should focus not only on the sequential construction of models by bringing in groups of terms but also on the possibility of reducing a model to a more parsimonious one by eliminating terms that have little or no explanatory value (Occam's razor). Sequential sums of squares for groups of terms are valuable because they provide guidance about the probable order of the model. For example, the p values in Table 8.2, page 161, suggest that one should begin with the full quadratic model since the special cubic term is not significant. Partial sums of squares are valuable to study after choosing the probable degree of the model. They provide guidance about which terms in a full model may not be needed in the final model. More will be said about removing terms from models in Chapter 10.

The F value for the full quadratic model (57.70) is calculated using the extra sum-of-squares principle (as are all the other F values). The fuller model is the quadratic model, and the less full model is the null model. The regression sum of squares, SSR, is defined as $\sum_{i=1}^{n}(\hat{Y}_i - \bar{Y})^2$. For the null model, $\hat{Y}_i = \bar{Y}$, and so the regression sum of squares is equal to zero. Consequently for the full quadratic model we have

$$\begin{aligned} F &= \frac{(\mathrm{SSR}_{\textit{quadratic model}} - \mathrm{SSR}_{\textit{null model}})/\Delta df}{s^2} \\ &= \frac{(2310.44 - 0)/(5 - 0)}{8.01} \\ &= 57.70 \end{aligned}$$

The hypothesis that is being tested is [92]

$$\begin{aligned} H_0: \quad & \beta_A = \beta_B = \beta_C && \text{(linear terms)} \\ & \beta_{AB} = \beta_{AC} = \beta_{BC} = 0 && \text{(quadratic terms)} \end{aligned}$$

and so there are five numerator degrees of freedom associated with this test. The F value for Linear Mixture in Table 8.3 (133.71) is a test statistic for just the linear terms in the above hypothesis.

The reader will recall that in the design of the hot-melt adhesive experiment (Fig. 8.2, page 155), there were 11 observations, of which eight were replicates (that is, four *pairs* of replicates). Six design points are needed to support the quadratic model, leaving five for the residual (SSE). Of these five, four degrees of freedom are allocated to pure error (corresponding to the four pairs of replicates) and one degree of freedom to lack of fit. The F value for lack of fit in Table 8.3 is calculated by dividing the mean square for lack of fit by the mean square for pure error. If the p value is less than some cut-off value (say 0.05), then we conclude that there is significant lack of fit, and we should consider augmenting the model with higher-order terms. In this particular case the p value is much larger than 0.05, and we therefore conclude that there is no evidence of lack of fit.

The breakdown of the residual sum of squares into lack-of-fit ($\mathrm{SS}_{\mathrm{LOF}}$) and pure error ($\mathrm{SS}_{\mathrm{PE}}$) sums of squares comes about as follows. Table 8.5 tabulates IDs and component proportions for the hot-melt adhesive data. The data are copied from Table 8.1, page 156. The variable $i = 1, 2, \ldots, 7$ identifies each uniquely different design point, while $j = 1, 2$ identifies the replicate number. Reference to Fig. 8.10 should help to clarify these indices. Y_{ij} values are the viscosities, again copied from Table 8.1. $\bar{Y}_i$ values are means of the Y_{ij} s, averaged over replicates.

Assume that a row, $\mathbf{x}_i'$, in the $\mathbf{X}$ matrix, is replicated r_i times. In the hot-melt adhesive experiment, $r_i = 1$ for $i = 3$, 5, and 7, but $r_i = 2$ for $i = 1$, 2, 4, and 6. For the jth replicate of the ith design point, a residual $(Y_{ij} - \hat{Y}_i)$ can be expressed by the following identity:

$$Y_{ij} - \hat{Y}_i = (Y_{ij} - \bar{Y}_i) + (\bar{Y}_i - \hat{Y}_i) \tag{8.11}$$

Table 8.5. *Hot-melt adhesive experiment. Viscosity data*

ID	i	j	Proportions HDA	PN-110	PH-56	Y_{ij}	$\bar{Y}_i$	$\hat{Y}_i$
1	1	1	1.00	0.00	0.00	8.00	6.40	6.44
2	1	2	1.00	0.00	0.00	4.80	6.40	6.44
3	2	1	0.75	0.25	0.00	19.20	18.675	18.52
4	2	2	0.75	0.25	0.00	18.15	18.675	18.52
5	3	1	0.75	0.00	0.25	8.60	8.60	8.29
6	4	1	0.50	0.50	0.00	51.10	47.00	47.04
7	4	2	0.50	0.50	0.00	42.90	47.00	47.04
8	5	1	0.50	0.25	0.25	18.28	18.28	17.97
9	6	1	0.50	0.00	0.50	7.14	7.04	7.08
10	6	2	0.50	0.00	0.50	6.94	7.04	7.08
11	7	1	0.60	0.17	0.17	12.48	12.48	13.18

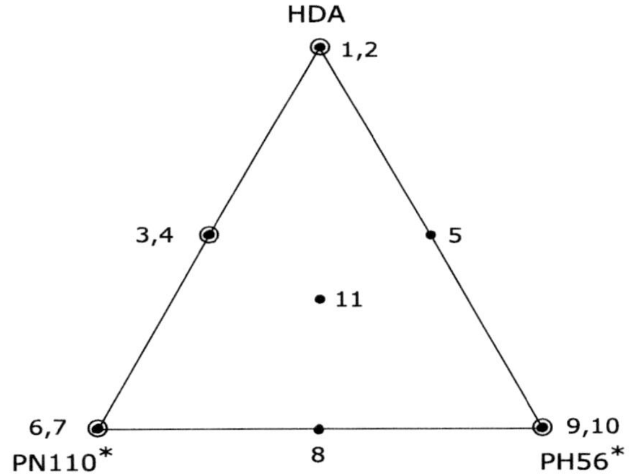

Figure 8.10. *Hot-melt adhesive design. Numbers are IDs in Table* 8.5.

where $\bar{Y}_i$ is the average overall response of the r_i replicates at the ith design point. This expression uses the fact that repeat points Y_{ij} at any $\mathbf{x}_i'$ will have the same mean, $\bar{Y}_i$, and the same predicted value, $\hat{Y}_i$. For example, for the point ID = 1 ($i = j = 1$, $r_i = 2$), we have

$$8.00 - 6.44 = (8.00 - 6.40) + (6.40 - 6.44)$$

$$1.56 = 1.60 - 0.04$$

$$1.56 = 1.56$$

If both sides of Eq. 8.11 are squared, and sums taken over both i and j, we have

$$\sum_{i=1}^{m}\sum_{j=1}^{r_i}(Y_{ij} - \hat{Y}_i)^2 = \sum_{i=1}^{m}\sum_{j=1}^{r_i}(Y_{ij} - \bar{Y}_i)^2 + \sum_{i=1}^{m}\sum_{j=1}^{r_i}(\bar{Y}_i - \hat{Y}_i)^2$$

$$= \sum_{i=1}^{m}\sum_{j=1}^{r_i}(Y_{ij} - \bar{Y}_i)^2 + \sum_{i=1}^{m} r_i(\bar{Y}_i - \hat{Y}_i)^2 \qquad (8.12)$$

As with Eq. 8.4, page 159, the crossproduct term can be shown to vanish. The left side of Eq. 8.12 is equal to SSE, the residual sum of squares. It is exactly equal to the first term on the right in Eq. 8.5, page 159. Note, however, that the summation over i is taken from $i = 1$ to m, not n. Here m is equal to the total number of distinctly different design points (seven in this example). The first term on the right is a measure of the variation of the replicate points, Y_{ij}, about their means, $\bar{Y}_i$. This is pure error. The second term on the right is a weighted sum of the deviations of the replicate means, $\bar{Y}_i$, about the fitted values, $\hat{Y}_i$, the weights being the number of replicates at the ith design point. This is lack of fit.

In simple terms, then, we can abbreviate Eq. 8.12 as follows:

$$\boxed{\text{SSE} = \text{SS}_{\text{PE}} + \text{SS}_{\text{LOF}}} \qquad (8.13)$$

The question arises how to allocate degrees of freedom between lack of fit and pure error when designing an experiment. Assume, by way of example, that we have 10 degrees of freedom over and above those needed to fit a model. How should we allocate these between lack of fit and pure error? Critical values for $F_{0.05;\nu_1,\nu_2}$, where $\nu_1 =$ the numerator degrees of freedom (lack of fit), $\nu_2 =$ the denominator degrees of freedom (pure error), and $\nu_1 + \nu_2 = 10$, are

ν_1	ν_2	$F_{0.05;\nu_1,\nu_2}$
9	1	240.5
8	2	19.37
7	3	8.89
6	4	6.16
5	5	5.05
4	6	4.53
3	7	4.35
2	8	4.46
1	9	5.12

As degrees of freedom are moved from the numerator to the denominator, the critical value drops rapidly and the better the chance one has of detecting real differences. The reason for this can be seen by examining part of an F-table for $p = 0.05$.

	ν_1				
ν_2	1	2	3	4	5
1	161.4	199.5	215.7	224.6	230.2
2	18.51	19.00	19.16	19.25	19.30
3	10.13	9.55	9.28	9.12	9.01
4	7.71	6.94	6.59	6.39	6.26
5	6.61	5.79	5.41	5.19	5.05

As one moves down a column (pure error degrees of freedom), critical values drop rapidly, but as one moves along a row (lack-of-fit degrees of freedom), critical values change slowly. In choosing how to partition degrees of freedom between lack of fit and pure error, it is worth consulting a table of F values.

Typically below an ANOVA table one will find a table of parameter estimates along with other useful information. Table 8.6 is a hybrid of those output by Design-Expert, JMP, and MINITAB. JMP and MINITAB output t and p values. Design-Expert does not output either because p values are in Design-Expert's ANOVA table. Instead, Design-Expert outputs 95% confidence limits on the parameter estimates. Note that the confidence limits for AC include zero, as one would expect based on the relatively large p value.

Table 8.6. *Adhesive viscosity response. Parameter estimates*

Term	Coef	SE Coef	t	p	95% CI Low	95% CI High
A–HDA	6.439	2.00	*	*	1.31	11.57
B–PN-110	47.04	2.00	*	*	41.91	52.17
C–PH-56	7.079	2.00	*	*	1.95	12.21
AB	−32.87	9.46	−3.47	0.0178	−57.20	−8.65
AC	6.128	11.79	0.52	0.6254	−24.18	36.43
BC	−36.35	11.79	−3.08	0.0274	−66.66	−6.05

Let β symbolize a model parameter (linear, quadratic, cubic, etc.), and let b be its least-squares estimate. A $100(1-\alpha)$ percent confidence interval for the model parameter β is

$$b - t_{\alpha/2,n-p} \cdot \text{se}(b) \;\le\; \beta \;\le\; b + t_{\alpha/2,n-p} \cdot \text{se}(b)$$

where $n-p$ is the number of degrees of freedom associated with the residual. In the viscosity experiment, $n = 11$ and $p = 6$, and so one would enter a t table with 5 degrees of freedom. The t value for a two-tailed 95% confidence interval is 2.571, and so the confidence intervals in Table 8.6 were calculated using

$$b - (2.571) \cdot \text{se}(b) \;\le\; \beta \;\le\; b + (2.571) \cdot \text{se}(b)$$

The reader is reminded that the standard errors of the coefficient estimates are given by (Section 5.4.3, page 84)

$$\begin{aligned} \text{se}(b) &= \sqrt{s^2 \cdot c_{ii}} \\ &= \sqrt{(8.01) \cdot c_{ii}} \end{aligned}$$

The value $s^2 = 8.01$ can be found in the table of partial sums of squares (Table 8.3, page 166). The quantity c_{ii} stands for a diagonal element of the variance-covariance matrix, $(\mathbf{X}'\mathbf{X})^{-1}$.

One may wonder why there are no t tests on the linear terms in the model. MINITAB includes an asterisk in the table of parameter estimates (as in Table 8.6). In Design-Expert, there is a padlock icon next to the linear terms, implying that the terms are locked into the model. The reason for this is that a test statistic, such as a t or F value, on a linear term tests $H_0 : \beta_i = 0$. This would make sense if the null model were $\hat{Y} = 0$, but the null model is instead $\hat{Y} = \bar{Y}$ (see the discussion on page 164). Typically the linear terms are *always* retained in Scheffé models. JMP outputs t and p values on linear terms in Scheffé models, but these statistics should be ignored.

8.3 Summary Statistics

Accompanying the ANOVA table and the table of coefficient estimates will be a collection of summary statistics. Typically these will include R^2, R^2_{adj} (adjusted R^2), R^2_{pred} (R^2 for prediction), and PRESS.

8.3.1 The R^2 Statistic

In a multiple regression setting, the **coefficient of determination**, R^2, is defined as

$$R^2 = \frac{\text{SSR}}{\text{SST}} \tag{8.14}$$

Because SST = SSR + SSE this is equivalent to

$$R^2 = \frac{\text{SST} - \text{SSE}}{\text{SST}} \tag{8.15}$$

R^2 can be interpreted as the proportion of the corrected total sum of squares that is accounted for by the model (Eq. 8.14) or as the proportional reduction in the corrected total sum of squares resulting from fitting the model (Eq. 8.15). As SSR $\to 0$, $R^2 \to 0$; as SSE $\to 0$, $R^2 \to 1$. Thus, $0 \le R^2 \le 1.0$.

There is, however, a caveat to the above. JMP reports a statistic called Max R^2. This is the maximum *achievable* R^2, not the maximum R^2 achieved. No model, no matter how complete, can pick up variation due to pure error [49]. The maximum achievable R^2 is therefore given by

$$\text{Max } R^2 = \frac{\text{SST} - \text{SS}_{\text{PE}}}{\text{SST}}$$

When a full quadratic Scheffé model is fit to the viscosity data (Table 8.3, page 166), the R^2 statistic is

$$R^2 = \frac{2350.48 - 40.04}{2350.48} = 0.9830 \tag{8.16}$$

However, the maximum achievable R^2 is

$$\text{Max } R^2 = \frac{2350.48 - 39.31}{2350.48} = 0.9833$$

and so the model has actually explained 100·(0.9830/0.9833) = 99.97% of the variability that can be explained.

Some software products give incorrect results for mixture models. For example, fitting a Scheffé quadratic model to the viscosity data in S-PLUS leads to the following result:

```
Residual standard error:  2.83 on 5 degrees of freedom
Multiple R-Squared:  0.9932
F-statistic:  121.9 on 6 and 5 degrees of freedom, the p value is
0.0000296
```

Clearly, there is a problem. The R^2 value is slightly inflated (0.9932 instead of 0.9830), the F statistic is considerably higher than the model F in Table 8.3 (121.9 instead of 57.70), there are 6 instead of 5 numerator degrees of freedom for the model F-statistic, and the p value is much too low (0.0000296 instead of 0.0002). If one uses the regression platform rather than the mixture platform in MINITAB, a similar result will be obtained, with the exception that R^2 will not be output. Similar results will be obtained in JMP if one does not declare the variables as mixture variables.

The problem is not unusual and arises from the fact that S-PLUS, as well as some other software products, does not recognize no-intercept mixture models as mixture models. Consider the following nonmixture polynomial model involving main effects and interactions:

$$Y = \alpha_0 + \alpha_1 Z_1 + \alpha_2 Z_2 + \alpha_3 Z_3 + \alpha_{12} Z_1 Z_2 + \alpha_{13} Z_1 Z_3 + \alpha_{23} Z_2 Z_3$$

Suppose that we know *in advance*, based on our subject-matter knowledge, that the intercept is equal to some number, say 5.0. In other words, we want to force the intercept through 5.0. We could reparameterize the model as follows:

$$\begin{aligned} Y' &= (Y - 5.0) \\ &= \alpha_1 Z_1 + \alpha_2 Z_2 + \alpha_3 Z_3 + \alpha_{12} Z_1 Z_2 + \alpha_{13} Z_1 Z_3 + \alpha_{23} Z_2 Z_3 \end{aligned} \tag{8.17}$$

Fitting no-intercept models such as Eq. 8.17 is known as **regression through the origin**. Note that the form of the model is the same as a $q = 3$ quadratic Scheffé model.

The null hypothesis for a model such as Eq. 8.17 is

$$\begin{aligned} H_0: \quad & \alpha_i = 0 \quad && \text{all } i \\ & \alpha_{ij} = 0 && \text{all } i, j \end{aligned}$$

In the example being considered here, this is a 6 degree-of-freedom test. When the null hypothesis is satisfied (i.e., the response is invariant to the levels of the factors), then the null model is

$$\hat{Y}' = 0$$

In the case of regression through the origin, then, variation is measured about the origin (zero) instead of about the mean. SSR and SST are redefined as follows:

$$\text{SSR} = \sum_{i=1}^{n} \hat{Y}_i^2 \tag{8.18}$$

$$\text{SST} = \sum_{i=1}^{n} Y_i^2 \tag{8.19}$$

Depending on the software, the redefined SST may be referred to as the **total sum of squares** or the **uncorrected total sum of squares**. For the viscosity data and the quadratic model, the redefined values for SSR and SST are, respectively, 5859.69 and 5899.73. The ratio of these two numbers is 0.9932, which is the number output for R^2 by S-PLUS. To distinguish R^2 values calculated using uncorrected sums of squares from values calculated using corrected sums of squares, the former shall be referred to as $R^2_{(0)}$ values [105].

If one is using software that does not have dedicated mixtures capabilities, there are two workarounds for this problem. The simplest of these is the following. The reader will find it helpful to refer to the discussion of the intercept model 3.30 discussed on page 26. One can reparameterize the quadratic Scheffé model as an intercept model:

$$Y = \beta_0 + \beta_2 X_2 + \beta_3 X_3 + \beta_{12} X_1 X_2 + \beta_{13} X_1 X_3 + \beta_{23} X_2 X_3$$

Here the term in X_1 has been replaced by an intercept, but either of the other linear terms could have been selected as well. In this form, the software thinks that the model is

$$Y = \alpha_0 + \alpha_2 Z_2 + \alpha_3 Z_3 + \alpha_{12} Z_1 Z_2 + \alpha_{13} Z_1 Z_3 + \alpha_{23} Z_2 Z_3$$

and that the explanatory variables are factors rather than components. The null hypothesis that will be tested will be

$$\begin{aligned} H_0: \quad & \alpha_i = 0 \qquad i = 2, 3 \\ & \alpha_{ij} = 0 \qquad \text{all } i, j \end{aligned}$$

When the null hypothesis is satisfied, then the model becomes

$$Y = \alpha_0$$

The least-squares estimate of α_0 is $\bar{Y}$, the mean of the Ys. Consequently, variation will be measured about the mean and the correct statistics will be calculated. Applying this strategy to the viscosity data in S-PLUS leads to the output:

```
Residual standard error:  2.83 on 5 degrees of freedom
Multiple R-Squared:  0.983
F-statistic:  57.7 on 5 and 5 degrees of freedom, the p value is
0.0002021
```

Note that R^2 agrees with the value in Eq. 8.16, page 172, and the F-statistic and degrees of freedom agree with Table 8.3, page 166.

The parameter estimates for the linear terms in the intercept model have different meanings and therefore different values from the same terms in the Scheffé model. However, the model will predict exactly the same response surface as the Scheffé model. The best way to handle the problem is to fit the model twice — once with the intercept and without X_1 (to get the correct statistics) and once without the intercept and with X_1 (to get the parameter estimates for the Scheffé model).

The second workaround is the following. It is easily shown that the uncorrected sums of squares for SSR and SST can be used to give the corrected sums of squares using the relationships

$$\text{SSR(corr)} = \sum_{i=1}^{n}(\hat{Y}_i - \bar{Y})^2 = \sum_{i=1}^{n}\hat{Y}_i^2 - n\bar{Y}^2 \tag{8.20}$$

$$\text{SST(corr)} = \sum_{i=1}^{n}(Y_i - \bar{Y})^2 = \sum_{i=1}^{n}Y_i^2 - n\bar{Y}^2 \tag{8.21}$$

where $n\bar{Y}^2$ is called the **correction for the mean**. One can then use the corrected sums of squares to calculate R^2 and the model F statistic. One will, however, have to consult an F table to estimate the p value.

8.3.2 The Adjusted R^2 Statistic

If variances (mean squares) are used instead of sums of squares, a new statistic can be defined, R^2_{adj}. The adjusted R^2 statistic is defined as

$$R^2_{adj} = \frac{\text{MST} - \text{MSE}}{\text{MST}} \tag{8.22}$$

The **total mean square** is defined as

$$\text{MST} = \frac{\sum_{i=1}^{n}(Y_i - \bar{Y})^2}{n-1}$$

which is simply the variance of the observed Y values about their mean. MSE is mean square error, often abbreviated s^2 or $\hat{\sigma}^2$, and is the residual variance after the model has been fit. Thus the adjusted R^2 statistic can be viewed as the proportional reduction in the variance resulting from fitting a model. Recall that R^2 can be interpreted as the proportional reduction in the sum of squares as a result of fitting a model. For any given model, R^2_{adj} will always be less than R^2.

Figure 8.11 displays a useful feature of R^2_{adj}. In a sequential model-building process, R^2 will continue to increase as more and more variables are put into a model, regardless of whether the variables have any explanatory value. While big gains in R^2 are usually indicative of a "better" model, there often comes a point where the statistic loses its usefulness for discriminating between models. On the other hand, the adjusted R^2 statistic will often — although not always — go through a maximum as one builds a model (or reduces a model by removing unnecessary terms). The model that has the maximum value of R^2_{adj} is often taken by model builders as the "best" model. This will also be the point where the difference between R^2 and R^2_{adj} is a minimum.

This effect is illustrated in Fig. 8.11 using the viscosity data from the hot-melt adhesive experiment. The abscissa in this plot shows the total number of terms in the model. The three linear terms are present in all models. When the number of terms is 4, there is one quadratic term in the model; when 5, there are two quadratic terms in the model; when 6, all the quadratic terms are in the model; and when 7, the model is the special cubic model. The 4- and 5-term models selected for this plot are those that have the maximum R^2 for that size model. For the 4-term case, AB is in the model; for the 5-term case, AB and BC are

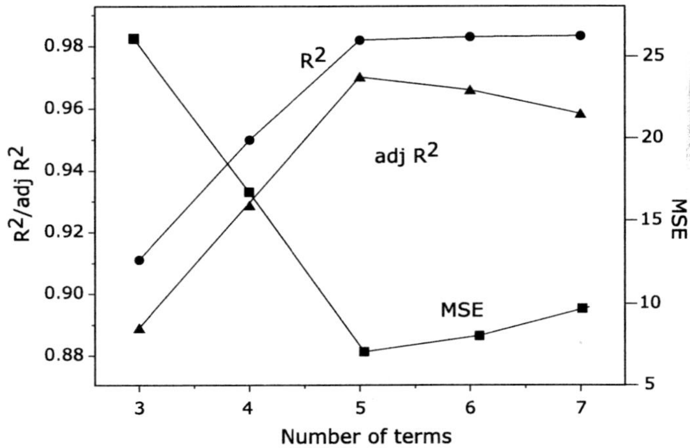

Figure 8.11. *Adhesive viscosity response. Some summary statistics.*

in the model. R^2_{adj} peaks when $p = 5$, and this is the point where the difference between R^2 and R^2_{adj} is minimal ($R^2 = 0.9830$, $R^2_{adj} = 0.9659$, $\Delta = 0.0171$).

The point where R^2_{adj} peaks coincides with the point where MSE is a minimum. This is not an accident. Equation 8.22 can be written in the form

$$R^2_{adj} = 1 - \frac{\text{MSE}}{\text{MST}}$$

Since in any sequential model building process, SST (and therefore MST) is a constant, R^2_{adj} will be at a maximum when MSE is at a minimum. Many model builders like to focus on MSE, and choose a model with the minimum MSE. This criterion for choosing a model leads to the same choice as the R^2_{adj} criterion.

This behavior occurs because when one adds or removes a term from a model, one moves part of the total sum of squares from SSE $\rightarrow$ SSR or SSR $\rightarrow$ SSE, respectively. In addition, however, one also moves a degree of freedom from SSE $\rightarrow$ SSR or SSR $\rightarrow$ SSE. When the new value for SSE is divided by the new value for its degrees of freedom, it is sometimes the case that the resulting ratio (MSE/MST) has *decreased* (and as a consequence, R^2_{adj} has increased).

8.3.3 PRESS and R^2 for Prediction

Consider the following thought experiment. Remove observation 1 from the viscosity data and refit a model of some specified degree (say quadratic). Use the model to predict $\hat{Y}_1$, but instead call it $\hat{Y}_{1,-1}$ to indicate that it is the predicted value for point 1 based on point 1 being removed from the regression. This is a true predicted value, as opposed to $\hat{Y}_1$, which is actually a *fitted* value (although we casually refer to the $\hat{Y}$s as predicted values).

The difference between the observed value Y_1 and the predicted value $\hat{Y}_{1,-1}$ is called a **PRESS residual** (PRESS to be defined in the next paragraph) or sometimes a deleted residual. A PRESS residual is symbolized $e_{i,-i}$, indicating that it is the residual for the ith observation based on the ith point being removed from the regression. Thus for point 1, the PRESS residual is $e_{1,-1}$. Now replace point 1 and remove point 2 and repeat the procedure. This will lead to $e_{2,-2}$. And so on until all n observations have been successively removed from the regression.

The **prediction error sum of squares**, or PRESS, is defined as

$$\text{PRESS} = \sum_{i=1}^{n}(Y_i - \hat{Y}_{i,-i})^2$$
$$= \sum_{i=1}^{n} e_{i,-i}^2 \tag{8.23}$$

It would appear from the thought experiment that calculating PRESS would be a computer-intensive project. It turns out that the PRESS residuals are easily calculated from the ordinary residuals using the expression [100, 105]

$$e_{i,-i} = \frac{e_i}{1 - h_{ii}} \tag{8.24}$$

where h_{ii} is the leverage of the ith data point. In view of the relationship between the two types of residuals, PRESS is actually evaluated using the expression

$$\text{PRESS} = \sum_{i=1}^{n}\left(\frac{e_i}{1 - h_{ii}}\right)^2 \tag{8.25}$$

The expression for R^2 in Eq. 8.15, page 172, can be written in the form

$$R^2 = 1 - \frac{\text{SSE}}{\text{SST}}$$

Replacing SSE with PRESS leads to the expression for R^2 for prediction, R^2_{pred}.

$$R^2_{pred} = 1 - \frac{\text{PRESS}}{\text{SST}} \tag{8.26}$$

From these two expressions it is easy to see that as PRESS approaches SSE, R^2_{pred} will approach R^2. One must keep in mind that PRESS can exceed SST, and thus R^2_{pred} can take on negative values. The disparity between PRESS and SSE provides some idea of how well a model performs in *prediction*. In some respects it is easier to compare R^2_{pred} to R^2 than it is to compare PRESS to SSE because R^2_{pred} is scale free.

A data point with a high leverage value will lead to a sizable difference between the ordinary residual and the PRESS residual, implying that the particular data point has a large influence on the regression. Although information about influence is already present in the leverage value, the effect on a residual can be quite striking. Table 8.7 compares the two types of residuals for the hot-melt adhesive data. The results are based on fitting a full six-term quadratic model to both the viscosity and 3-minute green strength (GS3) data. The ID column identifies the mixture in Table 8.1, page 156. For points 5 and 8, the PRESS residuals are more than seven times as large as the ordinary residuals. These are the unreplicated edge centroids in Fig. 8.2, page 155.

Table 8.7. *Hot-melt adhesive responses. Ordinary and PRESS residuals*

		Viscosity residuals		GS3 residuals	
ID	Leverage	Ordinary	PRESS	Ordinary	PRESS
1	0.4980	1.5613	3.1100	−15.559	−30.992
2	0.4980	−1.6387	−3.2640	10.441	20.797
3	0.4673	0.6796	1.2760	11.737	22.035
4	0.4673	−0.3704	−0.6953	8.737	16.402
5	0.8694	0.3092	2.3681	20.469	156.757
6	0.4980	4.0613	8.0897	11.441	22.790
7	0.4980	−4.1387	−8.2437	−16.559	−32.983
8	0.8694	0.3093	2.3677	20.473	156.729
9	0.4980	0.0613	0.1222	−4.559	−9.082
10	0.4980	−0.1387	−0.2762	−0.559	−1.114
11	0.3388	−0.6959	−1.0524	−46.062	−69.660

For the viscosity data and a quadratic Scheffé model, PRESS $= 168.3$, SSE $= 40.04$, $R^2_{pred} = 0.9284$, and $R^2 = 0.9830$. The difference between R^2 and R^2_{pred} would not be considered large, and we conclude that the quadratic model performs well in prediction and is reasonably robust to point deletion.

The story is somewhat different for the 3-minute green strength data. Fitting a quadratic model to this data leads to PRESS $= 57828$, SSE $= 3951.2$, $R^2_{pred} = -2.083$, and $R^2 = 0.7894$. Diagnostics to be introduced in Chapter 9 (specifically Cook's D and DFBETAS) point to IDs #5 and #8 as the culprits — a conclusion that can be gleaned by examining the PRESS residuals as well. Ignoring ID #5 leads to a huge change in the coefficient estimate of AC; ignoring ID #8 leads to a huge change in the coefficient estimate for BC. Thus, the fitted model is heavily dependent upon the presence of these two data points in the data set.

Another point worth bringing up is that lack-of-fit tests for the GS3 data (Table 8.9, page 182) suggest lack of fit for the linear and quadratic models. Lack of fit cannot be tested in the special cubic model because there are no degrees of freedom for lack of fit. Also the PRESS statistic and R^2_{pred} cannot be evaluated for the special cubic model because three of the data points have leverages of 1.0 (cf. Eq. 8.24). The reader should be able to figure out which three data points have $h_{ii} = 1.0$ based upon leverage property 6.1, page 103.

The PRESS statistic can be viewed as a form of **cross validation**. Cross validation is accomplished by splitting a data set into two parts, one part being the estimation data set and the other the prediction data set. With small- to medium-sized data sets, however, splitting into two halves is really not practical because it can significantly reduce the precision with which regression coefficients are estimated. Snee [155] recommends that the total number of observations (estimation + prediction data sets) be at least equal to $2p+25$ if the two sets are to be of approximately the same size. In the case of the viscosity data fit to a quadratic model, this recommendation would lead to an experiment size of 37 observations. The PRESS statistic provides a reasonable alternative that is readily available in most software packages. Although the prediction data "set" is only of size 1, there are a total of n sets. See Montgomery, Peck, and Vining for an excellent discussion of data splitting, as well as a description of the DUPLEX algorithm, which has been used for splitting data for cross validation purposes [100].

Case Study

Let us apply some of the procedures discussed in this chapter to the hot-melt adhesive GS3 data (Table 8.1). This particular example was chosen to illustrate that things do not always turn out as planned. Despite computers, there are times when judgment is required.

Table 8.8. *Adhesive GS3 response. Sequential model sums of squares*

Terms	Sum of Squares	df	Mean Square	F Value	Prob> F
Mean	43659.00	1	43659.00		
Linear	6600.15	2	3300.08	2.17	0.1765
Quadratic	8206.70	3	2735.57	3.46	0.1074
Special cubic	3208.65	1	3208.65	17.29	0.0142
Residual	742.50	4	185.62		
Total	62417.00	11	5674.27		

Table 8.8 displays the sequential sums of squares for this response. The table has been patterned after Design-Expert's output. At first glance the results appear a bit puzzling. The first thing to notice is that instead of the corrected total sum of squares in the last line (as in Table 8.2, page 161, for the viscosity data), the *uncorrected* total sum of squares (Eq. 8.19, page 173) is reported. The corrected total sum of squares has been reexpressed as an uncorrected total and a correction for the mean, which is the entry in the first line of the table. This reexpression has been done using Eq. 8.21 on page 175. Subtracting the correction for the mean (first line) from the uncorrected total (last line) leads to the corrected total (SST).

As we start to build our model, the first group of terms that comes into the model contains the linear terms. The F value for these terms tests $H_0 : \beta_1 = \beta_2 = \beta_3$. Judging from the p value (0.1765), it appears that the linear terms are not significantly different from one another, and we may elect not to reject the null hypothesis. If we do, then our model is simply $\hat{Y} = \bar{Y}$. The residual sum of squares (SSE) for this model will be equal to the total uncorrected sum of squares less the correction for the mean — in other words, SST, the total corrected sum of squares. Recalling that SST = SSR + SSE, for the null model we have SST = 0 + 18758.00.

The story might end there, except for the fact that entry of the special cubic term into the model appears to be significant, despite the fact that the linear and quadratic terms are apparently not significant. Normally we would expect to see significance of lower-order terms first and eventually reach a point where higher-order terms are not significant.

If we decide to do a bit of data snooping, we might have a look at the ANOVA tables for the three models (page 182). We notice that the p values for lack of fit are 0.0107 and 0.0142 for the linear and quadratic models, respectively — both well below a cutoff of (say) 0.05. On the one hand, then, the sequential sums of squares are telling us that the linear and quadratic terms are not needed, but on the other hand the lack-of-fit tests tell us that both models are underfit. This is a situation that one occasionally meets, and it illustrates that judgment is sometimes necessary.

From the three ANOVA tables, we notice that a significant amount of SSE has moved into SSR in the series linear → quadratic → special cubic. For the linear model, SSR + SSE = 6600.15 + 12157.85 (= 18758.0), for the quadratic model, SSR + SSE = 14806.85 + 3951.15 (= 18758.0), and for the special cubic model, SSR + SSE = 18015.50 + 742.50 (= 18758.0). R^2 has risen from 0.3519 for the linear model to 0.7894 for the quadratic model and to 0.9604 for the special cubic model. At the same time, MSE has decreased from 1519.73 for the linear model to 790.23 for the quadratic model and to 185.62 for the special cubic model. The steady decrease in MSE indicates that we have not gone through a minimum, as was the case with the viscosity data (Fig. 8.11). However, we are unable to test for lack of fit in the cubic model, because we have used all of the degrees of freedom available for the model.

The reason for this unusual behavior is that regressors in mixture models are not orthogonal to one another but are nearly always correlated. The significance of a term or group of terms in a mixture model depends very much on what else is in the model. For example, on entry into the model, the linear terms are not significant ($p = 0.1765$). Once the quadratic terms enter the model, the linear terms become more significant ($p = 0.0858$). On entry of the special cubic term, the linear terms become even more significant ($p = 0.0102$).

The same situation applies to the quadratic terms. On entry into the model the p value is 0.1074, but when the special cubic term enters the model, the p value for the quadratic terms drops to 0.0133 (calculated using other software).

Does this mean that sequential sums of squares are suspect? The answer is "No". Sequential sums of squares are meaningful for entry into or exit from a model given that *the term or group of terms is the last to enter or the first to exit*. As higher-order terms are brought into a model, p values derived from sequential sums of squares for the lower-order terms no longer apply.

These considerations apply as well to the reverse process, model reduction, a topic that is taken up in Chapter 10. Removing a term or group of terms from a model can cause the p values of other terms to change, sometimes significantly. Terms that were otherwise not significant may become significant, and vice versa. This behavior is exacerbated by collinearity, the topic of Chapter 14.

Another factor that comes into play in the GS3 analysis, and that probably serves to distinguish the GS3 analysis from the viscosity analysis, is the signal-to-noise ratio of the two responses. For fitted models, Design-Expert reports **adequate precision**, defined as [12, 107]

$$\text{Adequate precision} = \frac{\text{range}(\hat{Y}_i)}{\sqrt{s^2(p/n)}} \tag{8.27}$$

The reader should recognize the denominator as the square root of the average prediction variance. To ensure that the model will yield satisfactory predictions, Design-Expert recommends that adequate precision be ≥ 4. The tabular summary below shows that for all three models, signal to noise is considerably lower in the GS3 data than in the viscosity data. This means that test statistics such as t and F tests will not be as powerful in the GS3 case.

	Adequate precision	
Model	Viscosity	GS3
Linear	14.33	3.23
Quadratic	19.42	5.39
Special cubic	16.23	12.70

Table 8.9. *Adhesive GS3 response. ANOVA tables for linear (top), quadratic (middle), and special cubic (bottom) models*

Term(s)	Sum of Squares	df	Mean Square	F Value	Prob> F
Model	6600.15	2	3300.08	2.17	0.1765
Linear Mixture	6600.15	2	3300.08	2.17	0.1765
Residual	12157.85	8	1519.73		
Lack of Fit	11415.35	4	2853.84	15.37	0.0107
Pure Error	742.50	4	185.62		
Corrected total	18758.00	10			

Term(s)	Sum of Squares	df	Mean Square	F Value	Prob> F
Model	14806.85	5	2961.37	3.75	0.0867
Linear Mixture	6600.15	2	3300.08	4.18	0.0858
AB	3764.46	1	3764.46	4.76	0.0808
AC	3772.74	1	3772.74	4.77	0.0806
BC	577.66	1	577.66	0.731	0.4316
Residual	3951.15	5	790.23		
Lack of fit	3208.65	1	3208.65	17.29	0.0142
Pure Error	742.50	4	185.62		
Corrected total	18758.00	10			

Term(s)	Sum of Squares	df	Mean Square	F Value	Prob> F
Model	18015.50	6	3002.58	16.18	0.0089
Linear Mixture	6600.15	2	3300.08	17.78	0.0102
AB	1976.33	1	1976.33	10.65	0.0310
AC	6055.20	1	6055.20	32.62	0.0046
BC	3.20	1	3.20	0.017	0.9019
ABC	3208.65	1	3208.65	17.29	0.0142
Pure Error	742.50	4	185.62		
Corrected total	18758.00	10			

Chapter 9

Model Evaluation

In the last chapter much space was devoted to summary statistics such as R^2, R^2_{adj}, PRESS, and R^2_{pred}. These statistics do not require the assumption that the $\epsilon_i \sim \text{NID}(0, \sigma^2)$, nor do they measure any departure from this assumption. Also in the last chapter, F tests were used in the sequential and partial sums of squares tables. These F tests *do* require that the $\epsilon_i \sim \text{NID}(0, \sigma^2)$. By engaging in hypothesis tests, we in effect "jumped the gun" and assumed that the residuals were normally distributed.

Diagnostics to be discussed in this chapter are based heavily on the study of residuals, because it is the residuals (e_i) that assume the role of surrogates for the conceptual errors (ϵ_i). Much use will be made of plots, which usually can quickly highlight a trend or deviant points. We shall use these diagnostics to help us uncover

- departures from the model assumptions,
- outliers or suspect data points,
- high influence data points,
- unplanned systematic variability.

There is a plethora of diagnostic tests available for testing linear regression models. The discussion will be limited to those that are available in commercial DOE packages.

9.1 Scaling Residuals

The question arises what type of residual — ordinary or standardized — would be most useful for diagnostic purposes. To answer this question, we need to take a close look at the variances of the residuals, because it is the square roots of the variances (the standard errors) that are used to standardize the residuals. One should keep in mind the distinction between the e_i (the ordinary residuals), which are measurable quantities, and the ϵ_i, which are the conceptual errors. It is possible to derive a relationship between the two that will provide a means for standardizing the residuals.

Recall from Eq. 6.2, page 100, that $\hat{\mathbf{Y}} = \mathbf{HY}$, where $\mathbf{Y}$ is a $n \times 1$ vector of observed values of Y (the response), and $\hat{\mathbf{Y}}$ is a $n \times 1$ vector of fitted values of Y. This means that we can write the vector of ordinary residuals as

$$\begin{aligned} \mathbf{e} &= \mathbf{Y} - \hat{\mathbf{Y}} \\ &= \mathbf{Y} - \mathbf{HY} \\ &= (\mathbf{I} - \mathbf{H})\mathbf{Y} \end{aligned}$$

where $\mathbf{I}$ is a $n \times n$ identity matrix with diagonal elements equal to one and off-diagonal elements equal to zero. This result says that the residuals are linear combinations of the observed responses.

In place of $\mathbf{Y}$, we can substitute the expression for the general linear model (Eq. 3.11, page 18):

$$\begin{aligned} \mathbf{e} &= (\mathbf{I} - \mathbf{H})(\mathbf{X}\boldsymbol{\beta} + \boldsymbol{\epsilon}) \\ &= \mathbf{X}\boldsymbol{\beta} - \mathbf{HX}\boldsymbol{\beta} + (\mathbf{I} - \mathbf{H})\boldsymbol{\epsilon} \\ &= \mathbf{X}\boldsymbol{\beta} - \mathbf{X}(\mathbf{X}'\mathbf{X})^{-1}\mathbf{X}'\mathbf{X}\boldsymbol{\beta} + (\mathbf{I} - \mathbf{H})\boldsymbol{\epsilon} \\ &= \mathbf{X}\boldsymbol{\beta} - \mathbf{X}\boldsymbol{\beta} + (\mathbf{I} - \mathbf{H})\boldsymbol{\epsilon} \\ &= (\mathbf{I} - \mathbf{H})\boldsymbol{\epsilon} \end{aligned} \tag{9.1}$$

The third and fourth lines of this development make use of the definition of the hat matrix (Eq. 6.3, page 100) plus the fact that $(\mathbf{X}'\mathbf{X})^{-1}(\mathbf{X}'\mathbf{X}) = \mathbf{I}$. Thus the residuals are the same linear transformation of the conceptual errors as they are of the observed responses.

The relationship $\mathbf{e} = (\mathbf{I} - \mathbf{H})\boldsymbol{\epsilon}$ provides a means to get at the variances of the ordinary residuals. The quantity on the right side of this expression is known as a **linear estimator**. Abbreviating the right side as $\mathbf{A}\boldsymbol{\epsilon}$, the variance-covariance matrix of a linear estimator such as $\mathbf{A}\boldsymbol{\epsilon}$ is given by [100, 105]

$$\text{var}(\mathbf{A}\boldsymbol{\epsilon}) = \mathbf{A}\text{var}(\boldsymbol{\epsilon})\mathbf{A}' \tag{9.2}$$

(A scalar analogy to this would be $\text{var}(c\,y) = c^2\text{var}(y)$, where c is a constant and y is a random variable.) Applying Eq. 9.2 to the expression $\mathbf{e} = (\mathbf{I} - \mathbf{H})\boldsymbol{\epsilon}$ leads to

$$\begin{aligned} \text{var}(\mathbf{e}) &= \text{var}\left[(\mathbf{I} - \mathbf{H})\boldsymbol{\epsilon}\right] \\ &= (\mathbf{I} - \mathbf{H})\text{var}(\boldsymbol{\epsilon})(\mathbf{I} - \mathbf{H})' \\ &= \sigma^2(\mathbf{I} - \mathbf{H})(\mathbf{I} - \mathbf{H})' \\ &= \sigma^2(\mathbf{I} - \mathbf{H}) \end{aligned} \tag{9.3}$$

The third and fourth lines of this derivation make use of the second Gauss–Markov condition ($\text{var}(\epsilon_i) = \sigma^2$ for all i, page 17) and the fact that $\mathbf{I} - \mathbf{H}$ is symmetric and idempotent.

The matrices $\mathbf{I}$ and $\mathbf{H}$ in Eq. 9.3 are of dimension $n \times n$. The hat matrix $\mathbf{H}$ has h_{ii} for the diagonal elements and h_{ij} for the off-diagonal elements. This means that the variance-covariance matrix $\mathbf{I} - \mathbf{H}$ will have $1 - h_{ii}$ for the diagonal elements and $-h_{ij}$ for the off-diagonal elements. As a result, the variances and covariances of the residuals are given in scalar notation by

$$\text{var}(e_i) = \sigma^2 (1 - h_{ii}) \tag{9.4}$$

$$\text{cov}(e_i, e_j) = -\sigma^2 (h_{ij}) \tag{9.5}$$

The properties of the e_i are then quite different from the properties of the ϵ_i. Equation 9.5 informs us that the residuals are correlated, in contrast to the conceptual errors, which are assumed to be uncorrelated (third Gauss–Markov condition, page 17). Furthermore, it should be clear from Eq. 9.4 that *the variance of e_i depends on where the point $\mathbf{x}_i$ lies in* $\mathbf{X}$ *space*. A large h_{ii} will lead to a small var(e_i) and consequently a small residual. Recall that a leverage of 1.0 will result in the fitted value for a data point to equal the observed value. This will make the fit look good and possibly hide a potential problem.

A correct standardization of the e_i takes location into consideration and leads to an expression for the **standardized residuals**:

$$\frac{e_i}{\text{s.e.}(e_i)} = \frac{e_i}{\sigma\sqrt{1 - h_{ii}}} \tag{9.6}$$

A sample-based estimate, known as a **studentized residual**, is given by the expression

$$r_i = \frac{e_i}{s\sqrt{1 - h_{ii}}} \tag{9.7}$$

The advantages of using the r_i rather than the e_i is that the r_i have zero mean and unit variance, they are scale free, and they remove the effect of location.

Other adjectives that have been applied to this residual are standardized, internally studentized, and standardized PRESS. The reason for the adjective *internally studentized* will become clearer below when we discuss externally studentized residuals. The reason for the adjective *standardized PRESS* is because standardization of the PRESS residuals leads to r_i. This can be seen as follows (cf. Section 8.3.3 and Eq. 8.24):

$$\begin{aligned} \text{var}(e_{i,-1}) &= \text{var}\left[\frac{e_i}{1 - h_{ii}}\right] \\ &= \frac{1}{(1 - h_{ii})^2}\left[\sigma^2(1 - h_{ii})\right] \\ &= \frac{\sigma^2}{1 - h_{ii}} \end{aligned}$$

The second line of this development makes use of the fact that var($c\,y$) $= c^2$var(y), where c is a constant and y is a random variable. A standardized PRESS residual is then given by

$$\begin{aligned} \frac{e_{i,-i}}{\text{s.e.}(e_{i,-i})} &= \frac{e_i/(1 - h_{ii})}{\sqrt{\sigma^2/(1 - h_{ii})}} \\ &= \frac{e_i}{\sigma\sqrt{1 - h_{ii}}} \end{aligned}$$

which is the same result as Eq. 9.6.

Another method for scaling residuals that is often used takes the *deletion* approach. In this case, the estimate of σ in Eq. 9.6 is s_{-i} rather than s. The term s_{-i} is the residual standard deviation based upon the ith data point being excluded from the regression. The resulting statistic, often called ***R*-student**, is given by

$$t_i = \frac{e_i}{s_{-i}\sqrt{1 - h_{ii}}} \tag{9.8}$$

In addition to the name R-student, a host of other names have been applied to this statistic. MINITAB refers to these as deleted residuals, while Design-Expert Version 6 calls t_i the outlier t statistic. Other names include deletion, externally studentized, cross-validatory, jack-knife, and studentized deleted residuals. Because of the fuzzy nomenclature for r_i and t_i, it is *very important* when reading a journal article or a book to check to be sure you understand the author's terminology. The same applies to software, of course. We shall adopt the ***R*-student** terminology.

Like PRESS residuals, it would appear that n regressions need to be run to obtain values for s_{-i}, $i = 1, 2, \ldots, n$. But like the PRESS residuals, the values for s_{-i} can be calculated from the results of a single regression on the full data set. It can be shown [2, 22] that the following relationship exists between SSE_{-i}, the residual sum of squares with the ith observation deleted, and SSE, the residual sum of squares using all observations:

$$\text{SSE}_{-i} = \text{SSE} - \frac{e_i^2}{1 - h_{ii}}$$

Thus, a relatively large residual combined with high leverage will lead to a large "correction factor" for SSE. However, it is not the sum of squares that we are interested in but *mean* square error (s_{-i}^2), so degrees of freedom need to be taken into account. For SSE we can substitute $(n - p)s^2$, and for SSE_{-i}, $(n - p - 1)s_{-i}^2$, there being one less degree of freedom for s_{-i} than for s. With these substitutions and a bit of rearrangement, we have

$$s_{-i} = \sqrt{\frac{(n - p)s^2 - e_i^2/(1 - h_{ii})}{n - p - 1}} \tag{9.9}$$

The second term in the numerator will tend to be large if there is a large residual, large h_{ii}, or a combination of the two. This will tend to make s_{-i} small, which in turn will tend to lead to $t_i > r_i$, or possibly even $t_i \gg r_i$. Thus the magnitudes of e_i and h_{ii} influence t_i in two ways: once through the terms in Eq. 9.8 apart from s_{-i} and again through the expression for s_{-i} in Eq. 9.9. It should also be clear that there *will* be cases where r_i and t_i will not differ by much.

9.2 Plotting Residuals

Plots of residuals play an important role in model adequacy checking. Much space will be devoted here to *useful* residual plots, but let us begin with residual plots that are not particularly useful in a mixture setting. In a nonmixture setting, it is common practice to plot the **residuals vs. the factor levels**. Such plots can be very useful for uncovering model underspecification. Curvature suggests the need for higher-order terms in the model. In a mixture setting, on the other hand, the "factors" (which are now the X_i and any higher-order terms) are always correlated with one another. As the proportion of a component changes throughout the design region, the proportions of the other $q-1$ components undergo offsetting changes. It is extremely difficult to mentally deconvolute plots of e_i vs. X_i, and one's time is better spent examining other residual plots. Furthermore, if the analyst has included sufficient points for lack of fit and pure error, he or she has an alternative method for judging whether or not a model is underspecified (a formal lack-of-fit test).

9.2.1 Checking Assumptions

Diagnostic checks for the model assumptions are divided into two parts: (i) checks for the homogeneous variance assumption (Gauss–Markov) and (ii) checks for the normality assumption. The diagnostic normally used to uncover a violation of the homogeneous variance assumption is a plot of the **residuals vs. the fitted values.**[1] Usually the ordinary residuals or the studentized residuals are used for this plot.

Figure 9.1 displays a Design-Expert plot of studentized residuals vs. fitted values for a situation where the homogeneous variance assumption is *not* violated. In this situation, the variance of the residuals appears to be independent of the size of the fitted value, and as a result the residuals fall randomly within a horizontal band. This is the pattern that one would like to see.

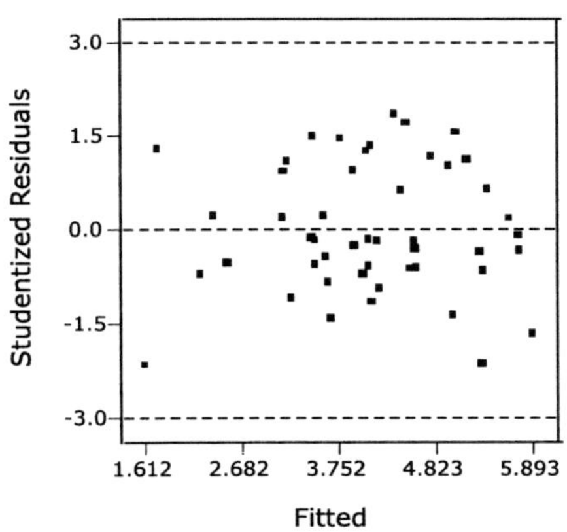

Figure 9.1. *Homogeneous variance assumption not violated.*

It is not at all uncommon, however, for the size of an error to depend on the size of a measurement. This is apt to be the case when the ratio of the maximum to the minimum response is an order of magnitude or more. The behavior is characterized by a funnel-like pattern opening to the right, as suggested by Fig. 9.2.

Problems such as this are usually handled by a power transformation of the response, one of the subjects discussed in Chapter 10.

[1]Plotting residuals vs. the observed Y_is is not done because they are correlated ($r_{eY} = \sqrt{1 - R^2}$). The residuals are uncorrelated with the $\hat{Y}_i$s. See Draper and Smith for proofs [49].

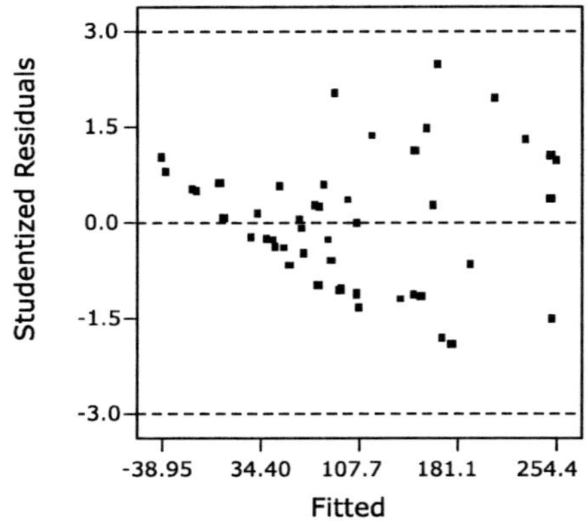

Figure 9.2. *Homogeneous variance assumption violated.*

The usual approach for checking the normality assumption is a **normal probability plot**. These plots can be a bit confusing because different software packages have different methods of choosing and labeling axes.

Table 9.1 tabulates studentized residuals (r_i) along with some normal order statistics resulting from fitting a quadratic Scheffé model to the adhesive viscosity data. The r_i values in column 2 have been sorted from smallest to largest and are indexed by i (column 3). The fourth column, labeled "P_i", is the cumulative probability (area) under the normal curve for 11 *normally distributed* data points, the number of observations in the adhesive data set. If the unit area under a normal curve is divided into n equal areas (where $n = 11$ in this example), then we might expect that the ith observation would fall near the middle of the ith section.

Table 9.1. *Studentized residuals, cumulative probability, and expected values of studentized residuals. Adhesive viscosity data*

ID[†]	r_i	i	P_i	$E(r_i)$
7	−2.064	1	0.0556	−1.593
2	−0.8172	2	0.1444	−1.061
11	−0.3024	3	0.2333	−0.7279
4	−0.1793	4	0.3222	−0.4615
10	−0.06915	5	0.4111	−0.2247
9	0.03059	6	0.5000	0.0000
5	0.3024	7	0.5889	0.2247
8	0.3024	8	0.6778	0.4615
3	0.3291	9	0.7667	0.7279
1	0.7786	10	0.8556	1.061
6	2.025	11	0.9444	1.593

[†] cf. Table 8.1, page 156

There are different formulas for estimating this, among them $P_i = (i - 0.375)/(n + 0.25)$, $P_i = (i - 0.5)/n$, and $P_i = (i - 1/3)/(n + 1/3)$, where i is the index in column three [49]. Using the first expression, the expected value of the ith-order statistic (column 5 in the table) is given by

$$E(r_i) = \phi^{-1}\left(\frac{i - 0.375}{n + 0.25}\right) \tag{9.10}$$

where ϕ^{-1} is the inverse cumulative distribution function of the normal distribution [2]. Values for $E(r_i)$ are the distances in units of standard deviation that the ith residual lies with respect to the center of the normal curve. These are called **z-scores** or **normal scores**.

Dotplots (MINITAB) of the studentized residuals and their expected values are displayed in Fig. 9.3. Normal probability plots of the studentized residuals are plots of $E(r_i)$ against r_i, but the choice of the x- and y-axes is software dependent. We shall adopt the convention that the y-axis is $E(r_i)$ and the x-axis is r_i. An additional confusing issue is that the labeling of the axis selected for $E(r_i)$ is also software dependent. Some products label this axis "Normal score" and show z-scores, while others label it "Percent" or "Normal % Probability" and show $100 \times P_i$ values.

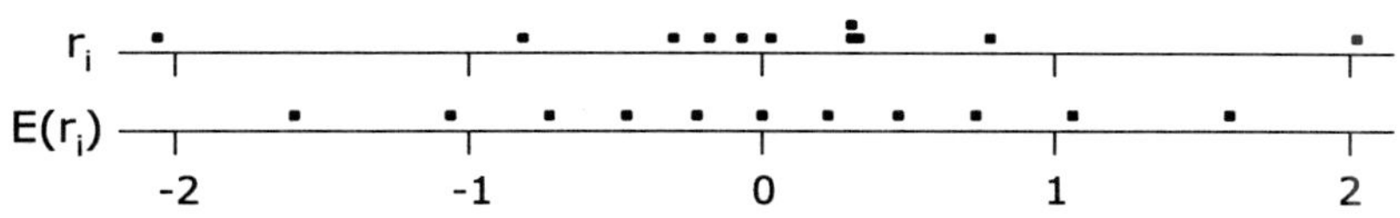

Figure 9.3. *Dotplots of data from Table* 9.1.

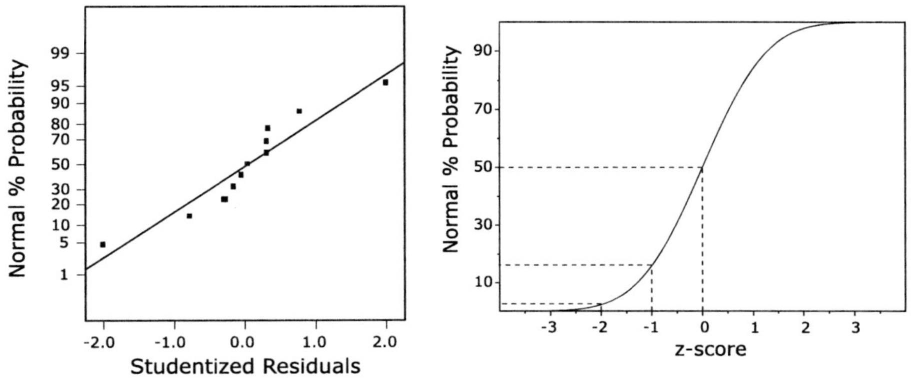

Figure 9.4. *Normal probability plot of studentized residuals. Adhesive viscosity data.*

A case in point of the latter is Design-Expert. The left figure in Fig. 9.4 displays a Design-Expert normal probability plot of the studentized residuals in Table 9.1. The y-axis is actually linear in z-scores but is labeled in units of cumulative probability, which makes the scale nonlinear. The relationship between the two scales is shown on the right. With reference to the dashed lines, the percentage area under a normal probability curve between $z = -2$ and $z = -1$ is $15.87 - 2.275 = 12.595\%$, while the area under the curve between

$z = -1$ and $z = 0$ is $50.00 - 15.87 = 34.13\%$. Thus, when labeling a normal-score axis in units of "Normal % Probability", percentages will be compressed in the middle of the scale.

The studentized residuals assume an S shape in the normal probability plot on the left in Fig. 9.4. This is one of four prototypical deviations from normality that one might observe. These are shown in the MINITAB plots of Fig. 9.5. Proceeding clockwise from upper left, the shapes can be described as (i) concave down, (ii) concave up, (iii) S-shaped, and (iv) reverse S-shaped.

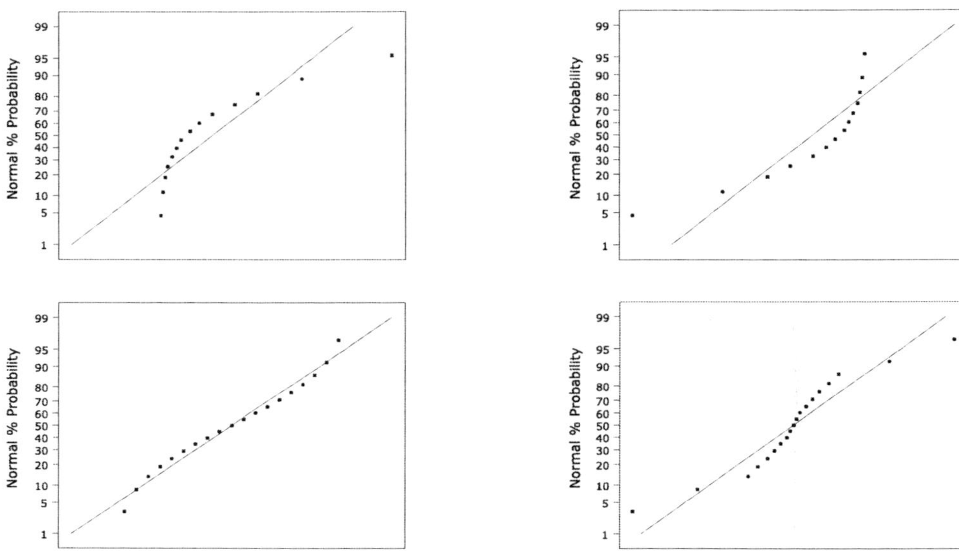

Figure 9.5. *Prototypical normal probability plots.*

The following associations can be made:

Concave down → skewed right Concave up → skewed left

Reverse S-shaped → light tailed S-Shaped → heavy tailed

The shapes can be inferred by looking at the plotted points at the extremes of the normal probability plots and asking the question, "In what direction should I move the points to bring them back to the straight line?" Points can be moved left or right, but not up or down, because in these examples the vertical axis has been selected for the expected values (the $E(r_i)$). For example, in the S-shaped curve in the lower right of Fig. 9.5, points at the lower end need to be moved to the right because the r_i values are too negative. Those at the upper end are too positive and need to be moved to the left. In both cases, this brings the points closer to the center of the normal curve (located at zero on the abscissa), and so the distribution is heavy in the tails.

The question arises, When does a normal probability plot imply nonnormality? Daniel [38] gives 40 plots of 16 independent random normal deviates drawn from a published table. A study of these leaves one with the impression that rather large departures from linearity are not at all uncommon. In the realm of formal statistical inference, one might consider a test such as the Shapiro–Wilk statistic. Unfortunately, formal statistical inference is undermined by the following problem. Equation 9.1, page 184, can be written in scalar form as follows:

$$e_i = \epsilon_i - \sum_{j=1}^{n} h_{ij}\epsilon_j \tag{9.11}$$

The implication of 9.11 (as well as Eq. 9.1) is that a specific residual is a linear combination of the conceptual errors. For small samples, the h_{ij} can be relatively large, and the second term on the right may dominate. Even if the conceptual errors are not normally distributed, the residuals will have a distribution that is closer to normal because of the central limit theorem (see, for example, Montgomery [102] or Moore and McCabe [103]). This phenomenon, termed **supernormality**, is discussed in many regression texts (see, for examples, [2, 22, 100, 105, 144]). As the sample size gets larger, the h_{ij} will become smaller, and eventually a point will be reached where the first term on the right dominates. Under these conditions, the e_i will have approximately the same distribution as the ϵ_i, and formal statistical inference becomes more reasonable.

A form of Monte Carlo testing in which an envelope is constructed for the normal probability plot by simulation has been suggested by Atkinson [2]. The procedure can be implemented in MINITAB using a MINITAB macro supplied on the diskette accompanying the regression text by Ryan [144].

The first step in Atkinson's procedure is to generate 19 sets, each of size n, of simulated Y values. Each sample is generated from the standard normal distribution so that $Y \sim \text{NID}(0, \sigma^2)$. It is not important what the value of σ^2 is because the residuals will be studentized (see discussion of Eq. 9.7, page 185). Each of the 19 sets is regressed on the X variables, leading to 19 sets of residuals. Each set of n residuals is studentized and ordered from smallest to largest. The envelope boundaries are given by

$$r_{l(i)} = \min_m r_{m(i)}$$

$$r_{u(i)} = \max_m r_{m(i)}$$

where $m = 1, 2, \ldots, 19$ and $r_{l(i)}$ and $r_{u(i)}$ are, respectively, the lower and upper boundaries for the ith observation. The simulation is repeated 19 times to give estimates of the 5th and 95th percentiles of the distribution of the ordered residuals. Ryan [144] has an excellent discussion of this procedure, some of the problems associated with it, and an outlier-resistant modification published by Flack and Flores [52]. Ryan also supplies a MINITAB macro to implement the Flack and Flores procedure.

The plot in Fig. 9.6 shows an Atkinson simulation envelope for the adhesive viscosity data fitted to a quadratic Scheffé model. The smaller filled points are the envelope boundaries, while the open circles are the studentized residuals, r_i. Careful examination will reveal that there are only 10, rather than 11, open circles. The reason for this is that in MINITAB, if two or more observations are equal, then they are all given the same normal

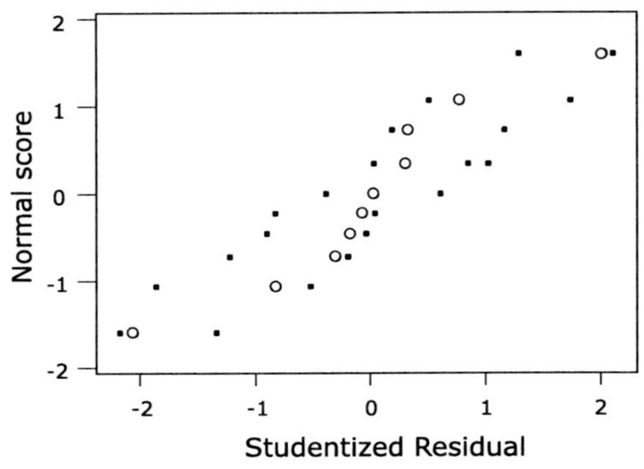

Figure 9.6. *Adhesive viscosity response. Simulation envelope.*

score, and this is based on the average of their ranks. Observations for which ID = 5 and 8 in Table 9.1 have the same value for r_i. Rerunning the MINITAB macro more than once will give a different envelope each time, and so one should run it several times to get a feeling for the results. The plot in this figure is representative. One would conclude from the plot that the residuals, although heavy in the tails, are (marginally) normally distributed.

9.2.2 Outlier Detection

Referring once more to the normal probability plots for the adhesive viscosity data (Figs. 9.4 and 9.6, pages 189 and 192), there are two residuals (one at each end of the plot) whose absolute values are rather large. The point at the lower left is ID #7, and the point at the upper right is ID #6 (cf. Table 8.1, page 156). Identifying potential outliers from normal probability plots is similar to the method used to identify significant effects in unreplicated factorials [13, 38].

Another graphical approach to outlier detection is to use the same kind of plot as used to test the homogeneous variance assumption (Figs. 9.1 and 9.2, page 188). However, it is often more useful to view an **index plot**. The *index* on the x-axis could be the run order or the standard order. Such a plot immediately identifies the aberrant point, should there be one.

Index plots of r_i (Eq. 9.7) and t_i (Eq. 9.8) for the adhesive viscosity residuals are displayed in Fig. 9.7. The increase in the t_i values over the r_i values for points 6 and 7 is apparent (the y-axes are scaled equivalently). No reference lines other than $r_i = 0$ have been added to the plot for the studentized residuals. The r_i do not exactly follow the t-distribution [105], and so formal hypothesis testing is only marginally useful. Nonetheless, an r_i value that lies 3–4 standard deviations from the mean should certainly raise a flag. t_i values, on the other hand, *do* follow the t-distribution [105], and so if one insists on formal hypothesis testing rather than simply taking a diagnostic viewpoint, cut-off points are more meaningful with plots of t_i than with plots of r_i.

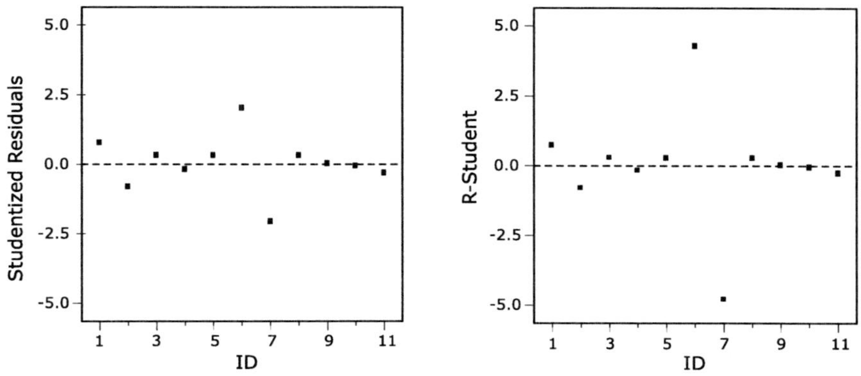

Figure 9.7. *Viscosity response. Index plots of studentized residuals and R-student.*

To test a single R-student statistic, one would enter a t table at the chosen significance level with $n - p - 1$ degrees of freedom. As the R-student statistics is a deletion residual, there is one less degree of freedom for s^2 than there is when all observations are included. In the adhesive viscosity data, $n - p - 1 = 11 - 6 - 1 = 4$. The two-tailed critical value for $t_{.05,4}$ is 2.776. On this basis, one might conclude that points 6 and 7 are outliers. Unfortunately, this conclusion would be incorrect.

The reason for this is that when one examines a plot such as the one for R-student in Fig. 9.7, one is making n inferences, informally or formally, and under these circumstances a critical value based on a single t test does not apply. In the adhesive experiment, if we conduct 11 t tests, each with a **comparisonwise error rate** of 5%, the overall or **experimentwise error rate** will be much larger [96, 103]. One approach to handling this problem is to use the **Bonferroni method**. If one is conducting n two-tailed t tests and wishes to control the experimentwise error rate so that it is *no greater* than $\alpha/2$, then one would use the $(\alpha/2n) \times 100\%$ point of the t-distribution based on $n - p - 1$ residual degrees of freedom. Myers [105] includes tables of Bonferroni critical values for $\alpha = 0.05$ and $\alpha = 0.01$. See Ott, Schilling, and Neubauer [116] for an interesting discussion of comparisonwise vs. experimentwise error rates.

For the adhesive data, the Bonferroni critical point for $n = 11$ and $p = 6$ is 5.75, and so points 6 and 7 would not be classified as outliers. The R-student values for points 6 and 7 are 4.28 and -4.80, respectively.

9.3 Measuring Influence

Clearly, we want the fitted model to reflect most of the data and not be highly influenced by one or two errant observations. When we think about influence, we ask ourselves whether there might be observations that when deleted would produce a substantial change in one or more of the coefficients and/or one or more of the fitted responses. For example, if deletion of an observation changed the sign of a parameter estimate, then inferences concerning that parameter would be questionable. Most influence diagnostics are based in one way or another on the *algebra of deletion*. "The Algebra of Deletion" is, in fact, a chapter title in Atkinson [2], and the monograph by Cook and Weisberg [22] is devoted almost entirely to this topic. Diagnostics that have evolved from this algebra provide us with measures of the extent to which parameter estimates and fitted values are influenced by individual observations.

Points that are extreme in the y-direction are generally referred to as *outliers*, while those that are extreme in the x-direction are referred to as *high-leverage points*. Data points with inflated t_i or $e_{i,-i}$ values exemplify the former; these tend to be influential regardless of their leverage. Generalities about leverage are hard to make, however. A high-leverage data point may or may not be influential. The difference is illustrated in Fig. 9.8 for a single regressor. In both cases the lone data point is a high-leverage point. In the example on the left, the point follows the general trend of the data, and is not influential. Deletion of point A would have little effect on the slope and intercept. In the example on the right, deletion of point A would have a huge effect on the slope and intercept. In this case, point A is not only a high-leverage point but also a high influence point. We shall see examples in the adhesive viscosity data of low-leverage/high-influence as well as high-leverage/low-influence data points.

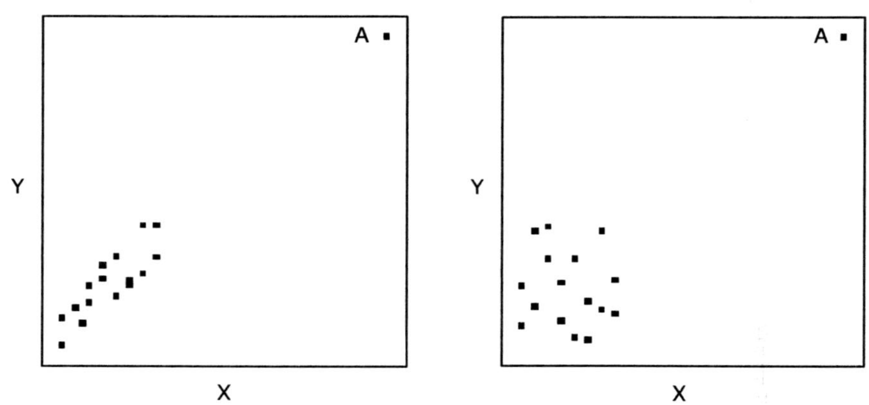

Figure 9.8. *Low- vs. high-influence data point.*

There are many diagnostics for detecting influential data points, but we shall limit ourselves to three measures that are widely used in practice — **Cook's Distance,** ***DFFITS***, and ***DFBETAS***.

9.3.1 Cook's Distance

Cook's Distance can be formulated in three different ways, two of which provide insight into its meaning and the third of which sheds light on how extreme data points affect its calculation. Recall the ellipsoidal confidence regions displayed in Fig. 5.3, page 77. Figure 9.9 shows 50%, 95%, and 99% confidence ellipses for the parameter estimates in the model

$$\hat{Y} = 10b_1 + 20b_2$$

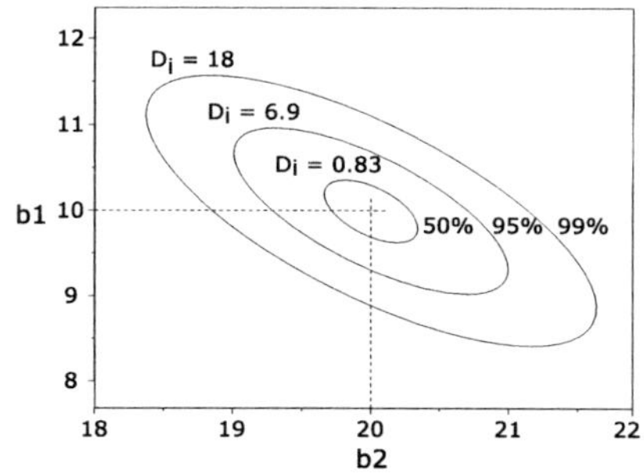

Figure 9.9. *Joint confidence regions for b_1 and b_2.*

assuming $s = 0.25$ and that the design is design A in Table 5.5, page 77. The 95% confidence ellipse in Fig. 9.9 is the same as the ellipse labeled "A" in Fig. 5.3. The "$D_i =$" labels on the ellipses in Fig. 9.9 are explained below.

It is a result of mathematical statistics that a joint confidence region (such as the ellipses in Fig. 9.9) at level $100(1-\alpha)\%$ for a parameter vector $\mathbf{b}$ is given by those sets of parameters $\mathbf{b}^*$ for which

$$\frac{(\mathbf{b}^* - \mathbf{b})'\mathbf{X}'\mathbf{X}(\mathbf{b}^* - \mathbf{b})}{ps^2} \leq F_{\alpha;p,n-p} \tag{9.12}$$

where $F_{\alpha;p,n-p}$ is the value of the F-distribution with p numerator and $n-p$ denominator degrees of freedom that leaves probability α in the upper tail. See, for example, [2, 22, 100, 105, 113].

This expression is not as difficult to use as one might think. The quantity on the left is a scalar, which means that both the numerator and denominator are scalars. The 2×1 parameter vector $\mathbf{b}$ for this small example is

$$\mathbf{b} = \begin{bmatrix} b_1 \\ b_2 \end{bmatrix} = \begin{bmatrix} 10 \\ 20 \end{bmatrix}$$

while the parameter vector $\mathbf{b}^*$ is

$$\mathbf{b}^* = \begin{bmatrix} \text{any value} \\ \text{any value} \end{bmatrix}$$

In general terms, then, the dimensions of the difference vector, $(\mathbf{b}^* - \mathbf{b})$, will be $p \times 1$. Thus the dimensions of the three terms in the numerator of Eq. 9.12 are $1 \times p$, $p \times p$, and $p \times 1$, the product of which is 1×1 (a scalar).

The ellipses in Fig. 9.9 were generated by doing a two-dimensional grid search using different values for b_1^* and b_2^*, which are the elements of $\mathbf{b}^*$. For each $\mathbf{b}^*$, the left side of

Eq. 9.12 was evaluated and the result compared to $F_{\alpha,2,4}$, since for this particular example $p = 2$ and $n - p = 6 - 2 = 4$. The F values are $F_{0.01;2,4} = 18.00$, $F_{0.05;2,4} = 6.94$, and $F_{0.50;2,4} = 0.8284$.

Cook [20] proposed that a natural extension of Eq. 9.12 would be to replace $\mathbf{b}^*$ with $\mathbf{b}_{-i}$, where $\mathbf{b}_{-i}$ is the least-squares estimate of $\boldsymbol{\beta}$ with the ith data point deleted (and, of course, $\mathbf{b}$ is the least-squares estimate of $\boldsymbol{\beta}$ using all the observations). For our two-component example, $\mathbf{b}_{-i}$ would be

$$\mathbf{b}_{-i} = \begin{bmatrix} b_1 \\ b_2 \end{bmatrix}_{-i} \tag{9.13}$$

where i takes on values from 1 to n. The suggested measure, known as **Cook's** $\boldsymbol{D}$, was defined to be

$$D_i = \frac{(\mathbf{b}_{-i} - \mathbf{b})'\mathbf{X}'\mathbf{X}(\mathbf{b}_{-i} - \mathbf{b})}{ps^2} \tag{9.14}$$

D_i is a scaled, squared distance measure and can be thought of as representing the distance between the vectors $\mathbf{b}$ and $\mathbf{b}_{-i}$.

This means that instead of labeling the three ellipses in Fig. 9.9 50%, 95%, and 99%, we can instead label them, respectively, $D_i = 0.83$, 6.9, and 18. If deletion of a data point perturbs (i.e., influences) the 2×1 parameter vector such that $D_i = 0.83$, for example, then the perturbed parameter vector lies somewhere on the 50% confidence ellipse. The locus of this ellipse encompasses the universe of all 2×1 parameter vectors for which $D_i = 0.83$. The larger D_i, the further the ellipse will be from $\mathbf{b}$, the parameter vector calculated using all of the data points.

In the original article on Cook's D [20], Cook suggested that for an *uncomplicated analysis* (author's italics), one would like $\mathbf{b}_{-i}$ to stay "well within" a 10% confidence region. Cook and Weisberg [24] suggest it is generally "useful" to study cases that have $D_i > 0.5$, and it is always "important" to study cases that have $D_i > 1$. Montgomery, Peck, and Vining [100] note that $F_{0.50;\nu_1,\nu_2} \simeq 1$ for many values of ν_1 and ν_2 and suggest that points for which $D_i > 1$ be considered influential. As there is no formal significance test associated with Cook's D [24], in the end the analyst should simply view D_i as a diagnostic that helps to "flag" potentially influential data points.

In the small example that we are using here, if deletion of the ith data point caused Eq. 9.14 to evaluate to 0.5, then $\mathbf{b}_{-i}$ is located on a 36% confidence ellipse (because $F_{0.64;2,4} = 0.5$ and $100(1 - 0.64)\% = 36\%$); if D_i evaluated to 1.0, then $\mathbf{b}_{-i}$ is located even further from $\mathbf{b}$, on a 56% confidence ellipse (because $F_{0.44;2,4} = 1.0$ and $100(1 - 0.44)\% = 56\%$). (F-tables do not, of course, have tables for upper-tail probabilities of 0.64 or 0.44, and a computer program is required to interpolate. The regression package *Arc*, discussed in Chapter 1, can do such calculations not only for the F-distribution but also for the normal, t-, and χ^2-distributions.)

Two points need to be made about Eq. 9.14. First, it is important to note the subscript on D_i. This means that there are $i = 1, 2, \ldots n$ values for Cook's D, one for each observation. Second, Cook's D is a *composite measure* of the effect of deleting the ith data point on the least-squares *vector* of coefficient estimates.

A second expression for Cook's D can be arrived at by a rearrangement of the terms in the numerator of Eq. 9.14:

$$D_i = \frac{[\mathbf{X}(\mathbf{b}_{-i} - \mathbf{b})]' \, [\mathbf{X}(\mathbf{b}_{-i} - \mathbf{b})]}{ps^2} = \frac{(\hat{\mathbf{Y}}_{-i} - \hat{\mathbf{Y}})' \, (\hat{\mathbf{Y}}_{-i} - \hat{\mathbf{Y}})}{ps^2} \tag{9.15}$$

This development makes use of the property $B'A' = (AB)'$ of transposed matrices and of the general linear model (Eq. 3.11, page 18). In this expression, $\hat{\mathbf{Y}}$ is a $n \times 1$ vector of fitted values using all the observations, while $\hat{\mathbf{Y}}_{-i}$ is a $n \times 1$ vector of fitted values (one of which is predicted) when the model is fit without the ith data point. The dimensions of the first parenthetical term in the numerator are then $1 \times n$, of the second parenthetical term, $n \times 1$, and of the product of the two, 1×1 (a scalar).

In this interpretation, Cook's D is a *composite measure* of the change in the vector of n fitted values when observation i is not used in estimating $\mathbf{b}$. Apart from ps^2, D_i is the squared Euclidean distance that the vector of fitted values moves when the ith observation is deleted.

It is possible to derive yet another expression for Cook's D (see, for example, Myers [105]). The result is

$$D_i = \left(\frac{r_i^2}{p}\right)\left(\frac{h_{ii}}{1 - h_{ii}}\right) \tag{9.16}$$

This is the expression that is usually used to compute D_i. Thus Cook's D is a function of the studentized residual for the ith data point *and* the leverage for the same point. A large studentized residual, high leverage, or a combination of the two will lead to inflated D_i values. If one has carefully designed an experiment, then hopefully one has taken care of any leverage problems at the design stage, and so a high value for D_i would put the onus on the studentized residual. Typically one would examine an index plot or table of Cook's D together with an index plot or table of leverages.

Table 9.2 summarizes a variety of diagnostic data, some of which have appeared in earlier tables, for the adhesive viscosity response. Values for observations 6 and 7 stand out for several of the diagnostics, particularly t_i, $e_{i,-i}$, D_i, and *DFFITS* (a diagnostic that remains to be discussed). Points 6 and 7 illustrate relatively low-leverage/high-influence data points. Points 5 and 8, on the other hand, are examples of relatively high-leverage/low-influence data points. Various options for handling points 6 and 7 will comprise part of the discussion in the next chapter.

Atkinson [2] defines an analog of D_i, symbolized $D_{\mathbf{I}}$, for *multiple deletion* of observations. The subscript $\mathbf{I}$ denotes a $m \times 1$ vector of indices identifying the points to be deleted. $D_{\mathbf{I}}$ is given by

$$D_{\mathbf{I}} = \frac{(\mathbf{b}_{-\mathbf{I}} - \mathbf{b})'\mathbf{X}'\mathbf{X}(\mathbf{b}_{-\mathbf{I}} - \mathbf{b})}{ps^2}$$

Deletion of points 6 and 7 from the adhesive viscosity data and refitting a quadratic model leads to $D_{\mathbf{I}} = 3.719$, which means that $b_{-\mathbf{I}}$ is located on a 91% confidence ellipse (because $F_{0.09;6,5} = 3.719$ and $100(1 - 0.09)\% = 91\%$). There are other consequences to deleting points 6 and 7, and these will be discussed in the next chapter.

Table 9.2. *Residuals, leverages, Cook's D, and DFFITS. Adhesive viscosity data. Quadratic Scheffé model, real metric*

ID	e_i	r_i	t_i	$e_{i,-i}$	h_{ii}	D_i	$DFFITS_i$
1	1.561	0.779	0.743	3.110	0.498	0.100	0.740
2	−1.639	−0.817	−0.785	−3.264	0.498	0.110	−0.782
3	0.680	0.329	0.298	1.276	0.467	0.016	0.279
4	−0.370	−0.179	−0.161	−0.695	0.467	0.005	−0.151
5	0.309	0.302	0.273	2.368	0.869	0.102	0.704
6	4.061	2.025	4.276	8.090	0.498	0.678	4.258
7	−4.139	−2.064	−4.799	−8.244	0.498	0.704	−4.779
8	0.309	0.302	0.273	2.368	0.869	0.101	0.704
9	0.061	0.031	0.027	0.122	0.498	0.000	0.027
10	−0.139	−0.069	−0.062	−0.276	0.498	0.001	−0.062
11	−0.696	−0.302	−0.273	−1.052	0.339	0.008	−0.195

9.3.2 DFFITS

We can write the $n \times 1$ vector $\hat{\mathbf{Y}}_{-i}$ in Eq. 9.15 in general terms as follows:

$$\hat{\mathbf{Y}}_{-i} = \begin{bmatrix} \hat{Y}_1 \\ \hat{Y}_2 \\ \vdots \\ \hat{Y}_i \\ \vdots \\ \hat{Y}_n \end{bmatrix}_{-i} \equiv \begin{bmatrix} \hat{Y}_{1,-i} \\ \hat{Y}_{2,-i} \\ \vdots \\ \hat{Y}_{i,-i} \\ \vdots \\ \hat{Y}_{n,-i} \end{bmatrix}$$

The focus of *DFFITS* is on a *single element* of $\hat{\mathbf{Y}}_{-i}$, specifically the element $\hat{Y}_{i,-i}$. This diagnostic was proposed by Belsley, Kuh, and Welsch [6]. Like Cook's *D*, *DFFITS* can be expressed in more than one way. One gains some intuitive insight when it is expressed as follows:

$$DFFITS_i = \frac{\hat{Y}_i - \hat{Y}_{i,-i}}{s_{-i}\sqrt{h_{ii}}} \tag{9.17}$$

The prefix "DF" stands for the *difference* between the fitted value $\hat{Y}_i$ for the ith case when all observations are used in fitting the model and the predicted value $\hat{Y}_{i,-i}$ for the ith case when the ith observation is omitted in fitting the model. Note that if $\hat{Y}_{i,-i} > \hat{Y}_i$, then $DFFITS_i$ will be negative, and if $\hat{Y}_{i,-i} < \hat{Y}_i$, $DFFITS_i$ will be positive. In other words, the direction of the change in $\hat{Y}_i$ will be of opposite sign to $DFFITS_i$.

The denominator provides a standardization, since $\text{var}(\hat{Y}) = \sigma^2 h_{ii}$. Note, however, that the ith observation is not used in the computation of s, and so s_{-i} replaces s in the denominator. Thus the value of $DFFITS_i$ is equal to the number of estimated standard errors of $\hat{Y}_i$ that the fitted value changes when the ith point is removed from the data set. Belsley, Kuh, and Welsch [6] suggest that observations that have $|DFFITS_i| > \sim 2\sqrt{p/n}$ warrant examination. For the adhesive viscosity response (quadratic model), this would be ~ 1.5.

To illustrate the calculation of *DFFITS* using Eq. 9.17, consider omitting observation 7 from the adhesive viscosity data. The actual values for $\hat{Y}_7$ and $\hat{Y}_{7,-7}$ are 47.0386 and 51.1437, respectively. In addition, $s^2_{-7} = 1.4815$ and $h_{77} = 0.49796$ (values in Table 9.2 are rounded values). Therefore

$$\begin{aligned} DFFITS_7 &= \frac{47.0386 - 51.1437}{\sqrt{(1.4815)(0.49796)}} \\ &= -4.7794 \end{aligned}$$

It can be shown that *DFFITS* values can also be computed from the results of a single regression (see Montgomery, Peck, and Vining [100] or Myers [105] for derivations). The result is

$$DFFITS_i = t_i \left(\frac{h_{ii}}{1 - h_{ii}} \right)^{1/2} \tag{9.18}$$

If $h_{ii} > 0.5$, then $DFFITS_i > t_i$, while if $h_{ii} < 0.5$, then $DFFITS_i < t_i$. Thus, $DFFITS_i$ can be viewed as essentially the R-student statistic inflated or deflated according to the leverage value. To illustrate again using observation 7

$$\begin{aligned} DFFITS_7 &= -4.7989 \left(\frac{0.49796}{1 - 0.49796} \right)^{1/2} \\ &= -4.7794 \end{aligned}$$

9.3.3 DFBETAS

Let us generalize the 2×1 vector $\mathbf{b}_{-i}$ in Eq. 9.13, page 196, and write it as

$$\mathbf{b}_{-i} = \begin{bmatrix} b_1 \\ b_2 \\ \vdots \\ b_j \\ \vdots \\ b_p \end{bmatrix}_{-i} \equiv \begin{bmatrix} b_{1,-i} \\ b_{2,-i} \\ \vdots \\ b_{j,-i} \\ \vdots \\ b_{p,-i} \end{bmatrix}$$

Each $b_{j,-i}$ ($j = 1, 2, \ldots, p$) is a least-squares estimate of b_j given that the ith observation ($i = 1, 2, \ldots, n$) is deleted. For each $b_{j,-i}$ there is a $DFBETAS_{j,i}$, and consequently there are $p \times n$ *DFBETAS* values. Like *DFFITS*, *DFBETAS* was proposed by Belsley, Kuh, and Welsch [6] and is defined in a fashion that is quite similar to *DFFITS*.

$$DFBETAS_{j,i} = \frac{b_j - b_{j,-i}}{s_{-i}\sqrt{c_{jj}}} \tag{9.19}$$

Here c_{jj} is the jth diagonal element of the $(\mathbf{X}'\mathbf{X})^{-1}$ matrix, which is equal to the variance of b_j apart from σ^2. Thus $DFBETAS_{j,i}$ is equal to the number of estimated standard errors of b_j that the coefficient estimate changes when the ith point is removed from the data set. Because of the way *DFBETAS* is defined, the change will be in a direction that is *opposite* to the sign of $DFBETAS_{j,i}$. Belsley, Kuh, and Welsch [6] suggest that if $|DFBETAS_{j,i}| > \sim 2/\sqrt{n}$, then the ith observation deserves examination. For the adhesive viscosity data, $2/\sqrt{n} \simeq 0.6$.

DFBETAS values are actually calculated from an expression that requires only a single regression on all the observations (see Montgomery, Peck, and Vining [100] or Myers [105]). Unless the reader is comfortable with the matrix formulation of OLS, he or she may prefer to think of it in terms of Eq. 9.19. The expression that is used is

$$DFBETAS_{j,i} = \left(\frac{r_{j,i}}{\sqrt{\mathbf{r}_j'\mathbf{r}_j}}\right)\left(\frac{t_i}{\sqrt{1-h_{ii}}}\right) \tag{9.20}$$

The term $r_{j,i}$ in the numerator of the first fraction is the (j, i)th element of the $p \times n$ matrix

$$\mathbf{R} = (\mathbf{X}'\mathbf{X})^{-1}\mathbf{X}'$$

The term $\mathbf{r}_j'$ in the denominator of the first fraction denotes the jth row of $\mathbf{R}$. See Myers [105] for an explanation of the meaning of these terms. Equation 9.20 indicates that a large residual in combination with a relatively high leverage will lead to large values of this statistic.

Table 9.3 displays $DFBETAS_{j,i}$ values based on the reals for the adhesive viscosity data. It should not take the reader long to search through the 66 $DFBETAS_{j,i}$ values and find those whose absolute value is > 1 (say). Consider, however, a situation in which one carries out a six-component mixture experiment. A $q = 6$ quadratic Scheffé model has 21 terms. If we support the model with a 31-observation data set — five additional runs for lack of fit, and five for pure error — then a table of $DFBETAS_{j,i}$ will have 651 entries. Now the search process takes on a greater degree of difficulty.

Table 9.3. *DFBETAS. Adhesive viscosity data. Quadratic Scheffé model, real metric*

ID	A = HDA	B = PN-110	C = PH-56	AB	AC	BC
1	0.740	0.208	0.163	−0.301	−0.233	0.018
2	−0.782	−0.220	−0.172	0.318	0.246	−0.020
3	0.005	−0.209	0.028	0.224	−0.029	−0.029
4	−0.003	0.113	−0.015	−0.121	0.016	0.016
5	0.017	0.071	−0.591	−0.074	0.619	−0.107
6	−0.017	2.463	−0.103	−1.731	0.106	−1.342
7	0.020	−2.764	0.116	1.943	−0.119	1.506
8	0.017	0.071	0.104	−0.074	−0.107	0.619
9	0.000	0.000	0.013	0.000	−0.009	−0.009
10	0.000	0.001	−0.029	−0.001	0.019	0.019
11	0.017	0.071	0.104	−0.074	−0.107	−0.107

Table 9.4. *Binary representation of DFBETAS. Adhesive viscosity data*

ID	A = HDA	B = PN-110	C = PH-56	AB	AC	BC	Total
1	0	0	0	0	0	0	0
2	0	0	0	0	0	0	0
3	0	0	0	0	0	0	0
4	0	0	0	0	0	0	0
5	0	0	0	0	0	0	0
6	0	1	0	1	0	1	3
7	0	1	0	1	0	1	3
8	0	0	0	0	0	0	0
9	0	0	0	0	0	0	0
10	0	0	0	0	0	0	0
11	0	0	0	0	0	0	0
Total	0	2	0	2	0	2	

What is needed is a simpler way of tabulating the data and/or a graphical approach to presentation of the data. One approach to the former would be to check the absolute value of each element in the table against some cut-off value using a relational operator. If the entry is greater than the cut-off value, then the result evaluates to 1; otherwise it evaluates to 0. Table 9.4 was constructed in just this way using a cut-off value of 1. One can quickly make inferences from the column and row sums. The column sums indicate that all terms containing B are influenced by case deletion, and the row sums indicate that cases 6 and 7 are the influential cases. A graphical method of presenting the data is illustrated in Fig. 9.10. The numbers in the plot are the IDs in Table 8.1, page 156, of the influential observations.

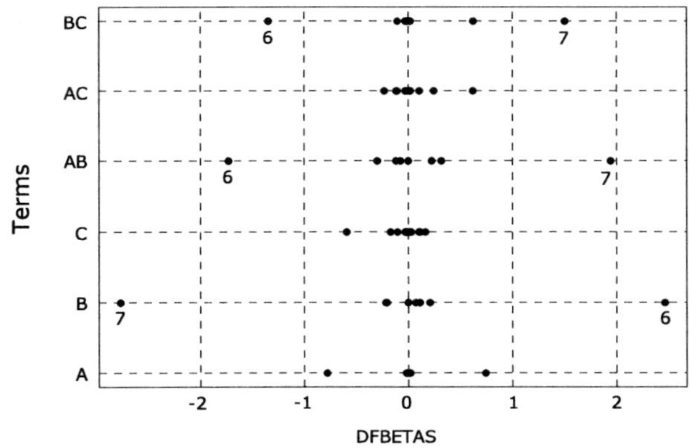

Figure 9.10. *DFBETAS$_{j,i}$. Adhesive viscosity data.*

At the time of this writing influence diagnostics available in MINITAB Release 13 are ordinary residuals, internally and externally studentized residuals, leverages, Cook's D, and $DFFITS$. Others can be programmed using MINITAB's macro language. JMP Release 5 can output ordinary residuals, internally studentized residuals, leverages, and Cook's D. Others can be programmed using the JMP Scripting Language (cf. Freund, Littell, and Creighton [53]). Design-Expert Version 7 outputs ordinary residuals, internally and externally studentized residuals, leverages, Cook's D, $DFFITS$, and $DFBETAS$.

Case Study

Let us return to the analysis of the adhesive 3-minute green strength data (Table 8.1, page 156). Some discussion can be found on page 178, but now that we have some additional diagnostics, it is interesting to look a bit further into the analysis.

Table 9.5. *Residuals, leverages, Cook's D, and DFFITS. Adhesive 3-minute green strength data. Quadratic Scheffé model, real metric*

ID	e_i	r_i	t_i	$e_{i,-i}$	h_{ii}	D_i	$DFFITS_i$
1	−15.559	−0.781	−0.746	−30.992	0.498	0.101	−0.743
2	10.441	0.524	0.482	20.797	0.498	0.045	0.480
3	11.737	0.572	0.529	22.034	0.467	0.048	0.496
4	8.737	0.426	0.388	16.402	0.467	0.026	0.363
5	20.469	2.015	4.158	156.757	0.869	4.506	10.728
6	11.441	0.574	0.532	22.790	0.498	0.054	0.530
7	−16.559	−0.831	−0.801	−32.983	0.498	0.114	−0.798
8	20.473	2.015	4.158	156.729	0.869	4.504	10.726
9	−4.559	−0.229	−0.206	−9.081	0.498	0.009	−0.205
10	−0.559	−0.028	−0.025	−1.114	0.498	0.000	−0.025
11	−46.062	−2.015	−4.158	−69.660	0.339	0.347	−2.976

Table 9.5 summarizes several diagnostics for the green strength data. The data are for a quadratic model expressed in the reals despite the fact that there is evidence that the quadratic model exhibits lack of fit. If because of this lack of fit one fits the GS3 data to a special cubic model, one finds three observations that have a leverage of 1.0. This is not a good situation for two reasons: (1) Any diagnostic that has the term $(1 - h_{ii})$ will be indeterminate. This would include PRESS and R^2_{pred} as well as r_i, t_i, and $e_{i,-i}$. (2) Any point that has a leverage of 1.0 exerts the ultimate leverage, forcing the response surface to pass through that point. The residual will be zero. This is similar to, although not as extreme as, fitting a straight line to two points. It works, but is it a good idea?

Recall (page 178) that $R^2_{pred} = -2.083$ for a quadratic model fitted to this response although $R^2 = 0.7894$. This is because the PRESS statistic is quite large (57,828) when compared with SSE (3951.2). The largest ordinary residual is that for case 11 (−46.062). The PRESS residual $e_{i,-i}$ for this observation is among the largest, being surpassed (in absolute magnitude) only by observations 5 and 8, the ordinary residuals of which are *half* the size of that for observation 11. Thus, in the case of observations 5 and 8, we have a situation where a fairly large ordinary residual combined with a high-leverage value leads to the highest values for Cook's D and $DFFITS$. On the other hand, the largest residual in absolute magnitude (case 11) has the smallest leverage value because it is the centroid of the design region. As a result its influence in terms of Cook's D and $DFFITS$ is less than that of cases 5 and 8.

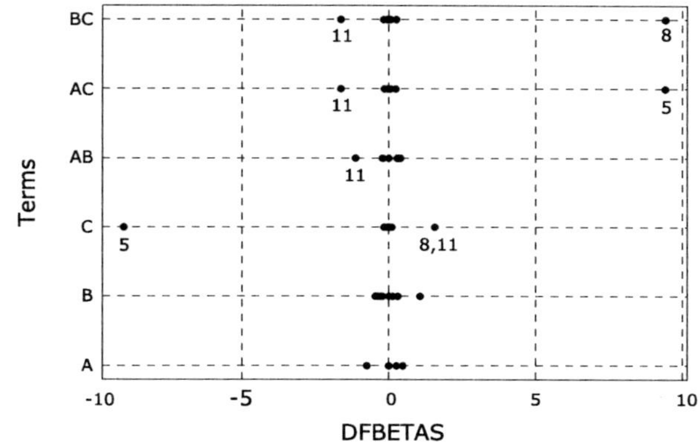

Figure 9.11. *$DFBETAS_{j,i}$ for adhesive green strength data.*

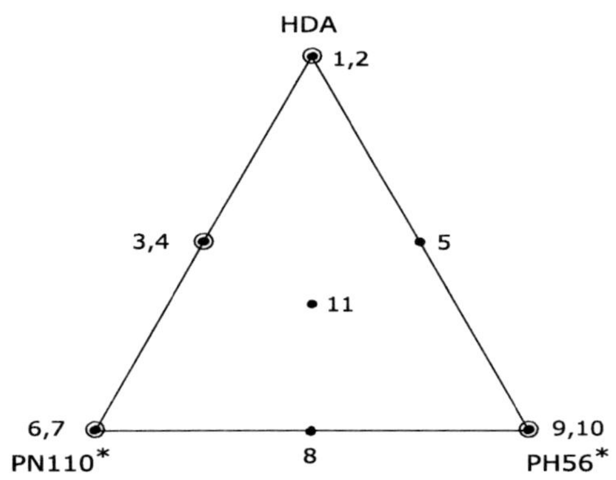

Figure 9.12. *Hot-melt adhesive design. Numbers are IDs in Table* 9.5 *and Fig.* 9.11.

The $DFBETAS$ plot in Fig. 9.11 demonstrates what a huge impact observations 5 and 8 have on terms that contain C. For example, if observation 8 is deleted from the data set, the coefficient for BC changes from -400 to -2540; if observation 5 is deleted, the coefficient for C changes from -562 to 505 and that for AC from 1024 to -1116. These numbers are based on the model being expressed in the reals. In addition, deletion of 5 causes the leverages of points 8 and 11 to become 1; deletion of 8 causes the leverages of points 5 and 11 to become 1.

These results are not meant to imply that one should entertain deleting observation 5 or 8. Neither observation was flagged as an outlier using the R-student test with a Bonferroni critical value (page 193). This discussion is simply meant to underscore how sensitive the model is to cases 5 and 8. Observation 5 is located at the midpoint of the HDA*–PH-56* edge (Fig. 9.12) while observation 8 is located at the midpoint of the PN-110*–PH-56* edge (asterisks denoting pseudocomponents). Both are unreplicated. Replication of *either* would have reduced h_{ii}, $i = 5$ or 8, from 0.869 to 0.465; replication of *both* would have reduced h_{ii}, $i = 5$ and 8, to 0.462. While this may not seem overly impressive, consider the fact that $h_{ii}/(1 - h_{ii})$ would change from $6.63 \rightarrow 0.869$ (replicate either) $\rightarrow 0.859$ (replicate both) — and this could be known *before* doing any experiments.

One can see in the *DFBETAS* plot that although point 11 is influential for all the quadratic terms, the effect is not as large as in the case of observations 5 and 8. The reason for this is that the leverage for point 11 is so much lower than that for 5 and 8 that even though 11 has the largest ordinary residual, its influence turns out to be much less.

Chapter 10
Model Revision

During model evaluation, one or more situations may have arisen: the potential impact of outliers or influential observations may need to be investigated; hypothesis tests may indicate that some of the model terms are not statistically significant suggesting the adequacy of a more parsimonious model; a lack-of-fit test may indicate the need for design/model augmentation; there may be evidence for nonnormality and/or nonconstant error variance suggesting the need for a transformation. In short, there are several things that may need to be considered.

Application of the methods to be described in this chapter requires *judgment* on the part of the analyst. Model building is in some respects a bit of an art form, which is what makes it so interesting. It is not unusual for experienced analysts to take different approaches to building and refining a model. The best way to develop your own skills is to do it over and over again.

10.1 Remedial Measures for Outliers

To this point, the word "outlier" has implicitly been taken to mean what we might call a **residual outlier**. In the context of the discussion in this section, and in the more general context of robust regression (to be discussed), the word "outlier" is sometimes broadened to include ***X* outliers** and ***Y* outliers** [100, 144]. X, Y, and residual outliers are not, however, mutually exclusive classes. A data point could be both a Y outlier and a residual outlier, or it could be one and not the other. A simple linear regression setting, as in Fig. 10.1, can be used to illustrate the differences. We can then consider graphical extensions to the multiple regression case.

Assume that we have 12 observations and that 11 of these constitute the cluster that follows an approximate linear trend in the lower left of Fig. 10.1. Let us further assume that the 12th observation is, *in turn*, point A, B, C, D, or E, so that we have five sets of 12 observations. Point A would not be considered a X or Y outlier because, relative to the cluster of 11 points, neither of its coordinates is extreme. Points B, C, and D would be classified as Y outliers because, relative to the cluster of 11 points, they are remote in the Y-direction. Points C, D, and E would be classified as X outliers because they are remote

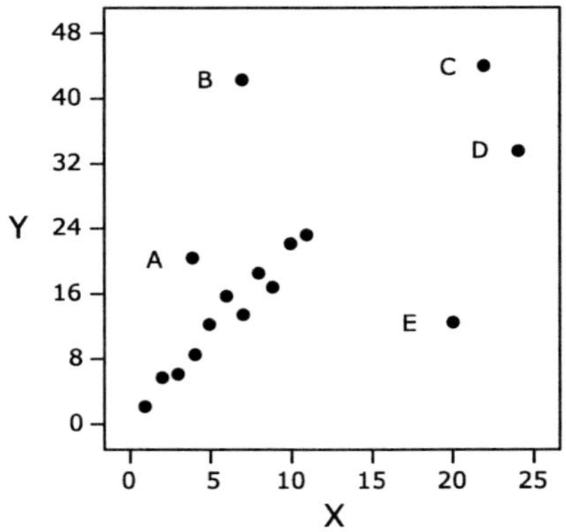

Point	Outlier type X	Y	Residual
A			√
B		√	√
C	√	√	
D	√	√	√
E	√		√

Figure 10.1. *X, Y, and residual outliers.*

in the X-direction — again relative to the cluster of 11 points. Points C and D are both X and Y outliers.

Consider now fitting the linear (first-degree) regression model $Y = \alpha_0 + \alpha_1 X$ to each of the five sets of 12 observations. (The symbol "α" is used here to represent a coefficient in a nonmixture model.) In each case the average leverage will be equal to $p/n = 2/12 = 0.167$. The data that were used by the author to create the plot in Fig. 10.1 lead to the leverages (h_{ii}) in Table 10.1, column 2, for points A–E. Thus the leverages for points A and B are less than the average leverage, while the leverages of points C–E exceed $2p/n$.

Table 10.1. *Residuals, leverages, Cook's D, and DFFITS. Simple regression example*

ID	h_{ii}	e_i	r_i	t_i	$e_{i,-i}$	D_i	$DFFITS_i$
A	0.113	10.2	2.93	7.37	11.5	0.546	2.63
B	0.0909	24.9	3.12	17.8	27.4	0.486	5.62
C	0.707	−0.370	−0.483	−0.464	−1.26	0.282	−0.721
D	0.752	−3.98	−2.77	−5.42	−16.0	11.6	−9.45
E	−0.652	10.1	−3.06	−11.6	−28.9	8.78	−15.9

Based on the ordinary residuals (e_i), points A, B, and E are clearly residual outliers, while D is marginally a residual outlier. When leverage enters the picture (as with r_i, t_i, $e_{i,-1}$, D_i, and $DFFITS_i$), point D would definitely be classified as a residual outlier. Because point C follows the general trend of the other 11 observations, this point would not be a residual outlier even when leverage is taken into account.

Echoing the statement made earlier, one can see from the table accompanying Fig. 10.1 that a data point could be both a Y outlier and a residual outlier, or it could be one and not the other. One can also see that a data point could be both a X outlier and a residual outlier, or it could be one and not the other. In addition, a point that is both an X and Y outlier may or may not be a residual outlier.

It is good to keep in mind that the influence of an outlier on a regression model is dependent on more than one factor. An observation may have a large ordinary (absolute) residual, $|e_i|$, but this is not a guarantee that it will exert a significant effect on the regression model. Its effect will depend in part on its leverage. Should the leverage be relatively small, its influence may be negligible. On the other hand, a point with a relatively small $|e_i|$ *could* have a disproportionate influence on the regression model if it happens to have a relatively large leverage.

Turning to a multiple regression setting, one approach to identifying Y outliers would be to make a boxplot of the Y values. Figure 10.2 displays box plots for the adhesive viscosity and 3-minute green strength data. The Y axes cover the full range of the responses, with the boxes representing the interquartile ranges (IQR) and the horizontal line within the boxes representing the medians. In the viscosity plot, the two horizontal lines that are not connected to the IQR are points 6 (topmost) and 7. The reason the points are not connected to the IQR is because they lie more than 1.5 $\times$ the interquartile range above the third quartile and are therefore considered Y outliers [103]. The same would be true for any point that might lie more than 1.5 $\times$ the IQR below the first quartile.

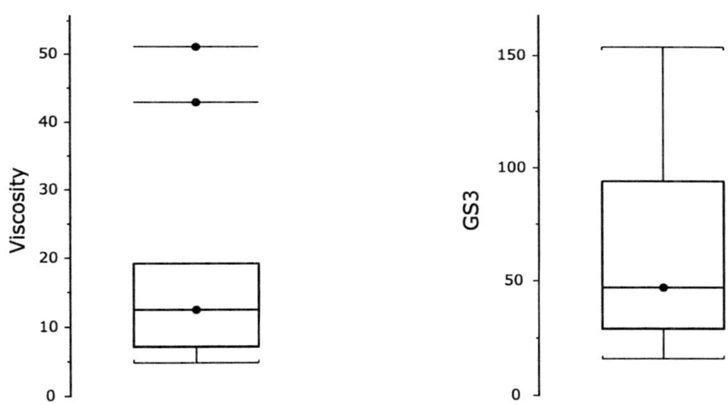

Figure 10.2. *Box plots. Adhesive viscosity and GS3 data.*

In the case of the GS3 box plot, the top and bottom horizontal lines correspond to observations 5 and 11, respectively. Judging from the results in Table 9.5, page 202, both of these points appear to be residual outliers. However, point 8 also appears to be a residual outlier, yet its Y value lies within the IQR between the median and the first quartile. If we use the 1.5 $\times$ IQR rule as a guide, then none of these points is classified as a Y outlier.

To identify potential X outliers, an index plot of leverages is useful (Fig. 10.3). The average leverage is $p/n = 6/11 = 0.545$, identified in the figure by the upper dashed line. Twice the average leverage equals 1.09, but since leverages cannot exceed 1.0, no leverage value can exceed this. Nonetheless, observations 5 and 8 clearly stand out as outlying in the X direction, even if they do not strictly qualify as X outliers in the sense of the $2p/n$ rule (page 106).

Figure 9.7, page 193, displays an index plot of the R-student statistic for the viscosity response. The $|t_i|$ values for points 6 and 7 do not exceed the Bonferroni critical value (5.75 for $n = 11$ and $p = 6$ [105]), so on this basis neither would be classified as a residual

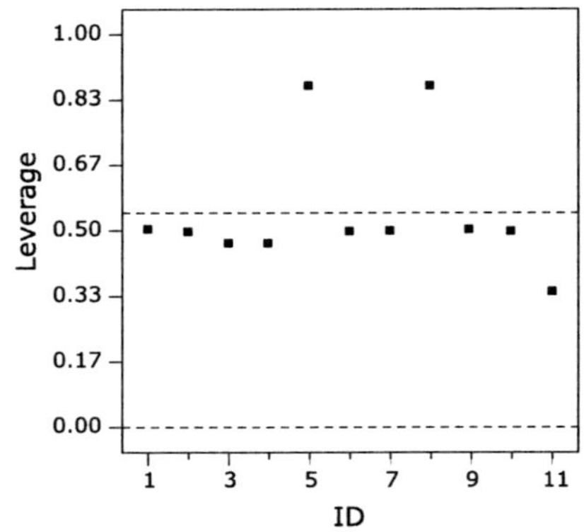

Figure 10.3. *Leverages in the adhesive experiment.*

outlier. However, points 6 and 7 are clearly *outlying*, and for our purposes we shall consider them residual outliers as they have by far the largest $|e_i|$ of the 11 data points (Table 9.2, page 198). Their leverages are only moderate (Fig. 10.3).

An index plot of the R-student statistic for the adhesive GS3 data is shown in Fig. 10.4. $|t_i|$ values for points 5, 8, and 11 do not exceed the Bonferroni critical value (5.75), but like points 6 and 7 in the viscosity response, they appear to be clearly outlying. Points 5 and 8 are residual outliers because of the combined effect of their relatively large h_{ii} (Fig. 10.3) and $|e_i|$ (Table 9.5, page 202). Point 11, on the other hand, is a residual outlier solely because of its very large value for $|e_i|$. Point 11 actually has the lowest leverage of the 11 data points, but because the value of $|e_i|$ is so large, it has a large effect on diagnostics such as t_i, $e_{i,-1}$, D_i, and $DFFITS_i$.

Putting the foregoing into a table leads to the tabular summary beneath Fig. 10.4. Points 6 and 7 in the viscosity data are multiple regression analogs of point B in Fig. 10.1. Points 5 and 8 in the GS3 data are multiple-regression analogs of point E in the same figure. Point 11 in the GS3 data is a multiple-regression analog of point A.

One's first reaction to an extreme observation, such as a Y value or residual, is that there is something wrong with the datum. In fact, the datum may be perfectly good and simply signal that the response is extreme in certain regions of the design space. Whether the response is desirable or undesirable, something has been learned — if the response is desirable, explore the region more thoroughly; if it is undesirable, avoid formulations that lead to the undesirable response.

If one were to prepare a checklist of what to do with apparent outliers, the first item on the list would be to check data entry. Assuming the data were correctly entered into a laboratory notebook, could an error have been made when transferring the data into a software program? This is easy to check and to correct.

At the other extreme in terms of effort would be repeating the measurement. This might include remaking the formulation, as there is no guarantee that the source of an extreme value is necessarily instrument related. If the repeat measurement or experiment supports the original datum, then one has no justification for ignoring the point when fitting the model. This might indicate an inadequacy in the model, and one may need to consider

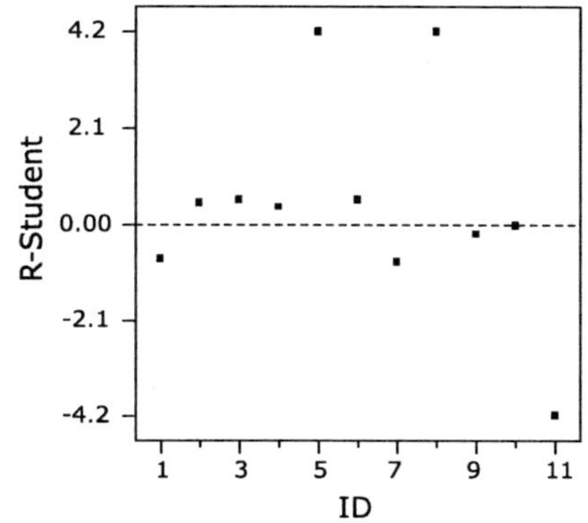

Figure 10.4. *R-student statistic. Adhesive GS3 response.*

		Outlier type		
Response	Point	*X*	*Y*	Residual
Viscosity	5	√		
	6		√	√
	7		√	√
	8	√		
GS3	5	√		√
	8	√		√
	11			√

a different model or possibly a transformation. If the repeat measurement or experiment is *not* in agreement with the original datum, then one cannot be sure which value is erroneous without an additional replicate to help resolve the discrepancy. If it is finally concluded that the original datum was in error, then one has a valid reason for ignoring the aberrant data point.[1]

In between these extremes we have an existing data set with one or more troubling observations that we can neither ignore nor correct. We must live with the data and try to make some sense out of it. One way to determine how a suspected outlier may impact a model is simply to model the data with and without the observation and confront the difference. This has been called **parallel modeling** by Lunneborg [91]. We already suspect from the tabulated diagnostics (pages 198 and 202) and the plots of *DFBETAS* (pages 201 and 203) that points 6 and 7 may have a significant effect on the model for the adhesive viscosity data and that points 5, 8, and 11 may impact the model for the 3-minute green strength data. Table 10.2 summarizes the actual parameter estimates (in the reals, rounded to whole numbers) for the two responses and for various single-point deletions. Note that the changes in the parameter estimates on point deletion are in the directions predicted by the values for *DFBETAS*, i.e., of opposite sign to *DFBETAS*.

[1]Some regulatory agencies require that *every* data point be reported whether or not a point has been deemed an outlier. Depending on one's circumstances, it would be prudent to check into this.

Table 10.2. *Effect of point deletion on the parameter estimates for the quadratic Scheffé model. Adhesive viscosity and GS3 data*

	Viscosity			3-min Green Strength			
Term	All	−6	−7	All	−5	−8	−11
A	6	6	6	110	107	107	107
B	153	130	177	428	323	323	323
C	−5	−3	−6	−562	505	−749	−749
AB	−131	−100	−163	−821	−616	−616	−616
AC	25	22	27	1023	−1116	1392	1392
BC	−145	−115	−176	−401	−32	−2540	−32
R^2	0.983	0.994	0.996	0.789	0.923	0.959	0.954
R^2_{pred}	0.928	.	.	−2.083	.	.	.

In the case of the viscosity response, the changes in the parameter estimates on deletion of point 6 or 7 are relatively small. Contour plots and trace plots undergo only small changes if either point is deleted. In other words, the model is robust to point deletion, a fact that is reflected in the R^2_{pred} value for the model (0.928) based on using all the data points. (R^2_{pred} values are missing for the models with deleted points because there are points with leverages equal to 1.0.)

Such is not the case with the 3-minute green strength data which has $R^2_{pred} = -2.083$. Point deletion leads to large changes in the parameter estimates for the quadratic terms, suggesting that it would be wise to have a look at some graphics based on the different models. Particularly dramatic changes are apparent if one views the response traces. This provides an opportunity to introduce an effects direction known as the **Piepel-effect direction** (described in more detail in Section 11.3).

Figure 10.5 shows the adhesive pseudocomponent simplex ($A'B'C'$) embedded within a HDA–PN-110–PH-56 simplex (ABC). Recall that the lower bound on component A

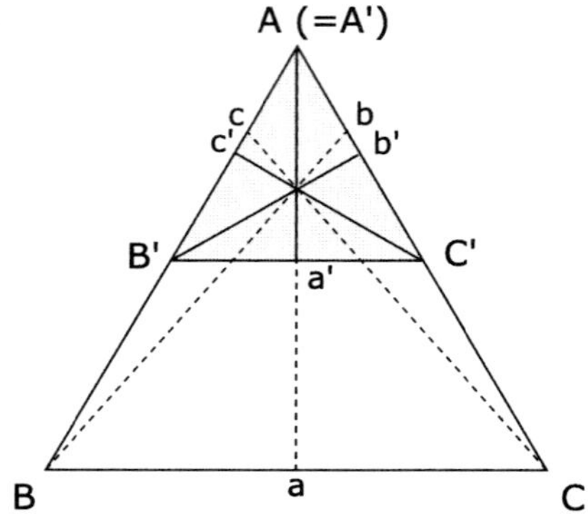

Figure 10.5. *Cox- vs. Piepel-effect directions.*

(HDA) was 0.5. *With respect to a chosen base point* (the centroid of the pseudocomponent simplex in this example), the three Cox-effect directions are aA, bB, and cC. The three Piepel-effect directions are $a'A'$, $b'B'$, and $c'C'$. $a'A'$ overlays aA but is shorter. In other words, the Piepel-effect directions are the loci of points drawn from points (a', b', and c') on the boundaries of the pseudocomponent simplex through a reference, or base, point within the pseudocomponent simplex and terminating at the pseudocomponent vertices (A', B', and C'). In this particular example, the reference point is the centroid of the pseudocomponent simplex, but that does not have to be the case.

Trace plots in Fig. 10.6 show the change in the GS3 response along the Piepel-effect directions. The point where the three curves cross (the point where "Deviation from Base Point" is equal to zero) corresponds to the composition of the base point, which in this case is the overall centroid of the pseudocomponent simplex. When all 11 data points are used to fit the model, the curves for A' and C' (pseudo HDA and PS-56, respectively) are concave down. When point 5 is not used to fit the model, the curves for A' and C' are concave up, indicating a dramatic change in the shape of the response surface. When all points are used, the curve for B' (pseudo PN-110) is concave up but trends downward from low to high B'. When point 5 is not used, the B' trace is approximately linear and trends *upward* from low to high B'.

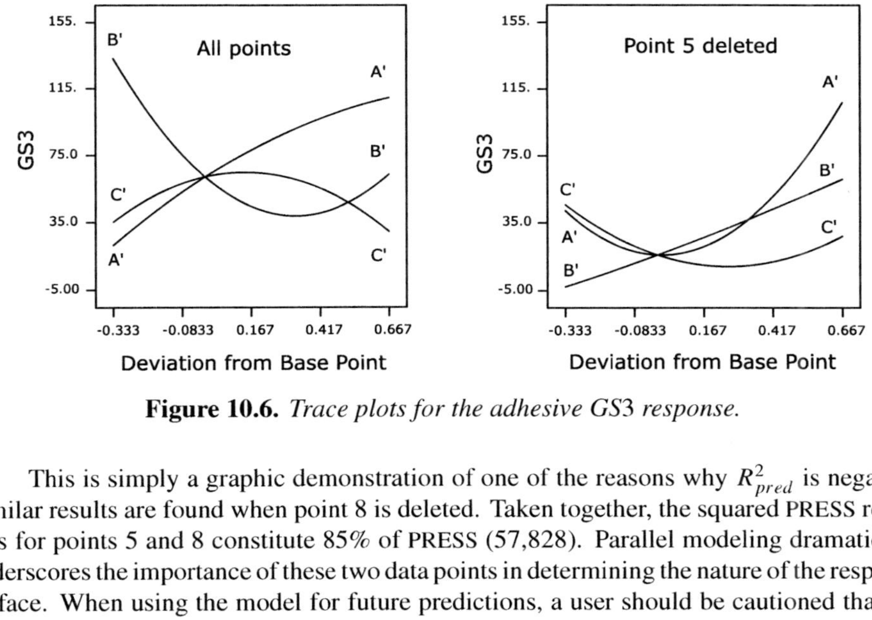

Figure 10.6. *Trace plots for the adhesive GS3 response.*

This is simply a graphic demonstration of one of the reasons why R^2_{pred} is negative. Similar results are found when point 8 is deleted. Taken together, the squared PRESS residuals for points 5 and 8 constitute 85% of PRESS (57,828). Parallel modeling dramatically underscores the importance of these two data points in determining the nature of the response surface. When using the model for future predictions, a user should be cautioned that the model hinges heavily on the responses for points 5 and 8. In terms of the pseudocomponent simplex illustrated in Fig. 8.10, page 169, these are the edge centroids of the HDA–PH56* and PN110*–PH56* edges, respectively.

There is another way to think about parallel modeling. One could view it as a case of weighted least squares where the weight given to an aberrant data point is zero. In OLS, of course, each data point is given unit weight. Thus we might ask if there is a compromise position, where instead of giving an aberrant point zero or unit weight, we adopt a middle ground. What we are looking for, then, is a method for fitting models that diminishes the weight of data points that have large residuals.

In OLS, the estimators are found by finding the regression model coefficients that satisfy

$$\underset{\boldsymbol{\beta}}{\text{Minimize}} \sum_{i=1}^{n} e_i^2$$

In weighted least squares, estimators are found that satisfy

$$\underset{\boldsymbol{\beta}}{\text{Minimize}} \sum_{i=1}^{n} w_i e_i^2$$

where w_i is the weight assigned to the ith data point. Many **robust regression** procedures make use of the second equation, the weights being determined by an iterative procedure known as **iteratively reweighted least squares** (IRLS). In outline form, these procedures consist of the following steps:

1. Carry out an (unweighted) OLS regression using the least-squares estimator of the coefficients (Eq. 7.15, page 128).
2. Use the residuals from the regression model to form weights.
3. Use these weights to carry out a weighted least-squares regression. Rather than use the least-squares estimator of the coefficients, the coefficients are estimated using

$$\mathbf{b} = (\mathbf{X}'\mathbf{W}\mathbf{X})^{-1}\mathbf{X}'\mathbf{W}\mathbf{Y} \tag{10.1}$$

   where $\mathbf{b}$ is a $p \times 1$ vector of coefficient estimates, $\mathbf{W}$ is a $n \times n$ diagonal matrix of weights, and $\mathbf{Y}$ is a $n \times 1$ vector of observations.
4. Repeat steps 2 and 3 until some convergence criterion is satisfied. Typically the convergence criterion takes the form of the maximum percentage change in a coefficient estimate.

Note that step 2 of the procedure makes use of *residual outliers*. The feature that distinguishes many robust regression procedures from one another is the method by which this step — forming weights — is carried out. The procedures can be classified according to their **influence function**, as it is the influence function that determines the weight, or influence, that the ith data point has on the fitted model. The influence function can be symbolized $\psi(e_i^*)$, where e_i^* is the ith scaled residual and the method of scaling is explained beginning on page 214.

A variety of influence functions have been proposed in the literature. One popular influence function, the **Huber influence function** [73], is given by

$$\psi(e_i^*) = \begin{cases} -r, & e_i^* < -r \\ e_i^*, & -r \le e_i^* \le r \\ r, & e_i^* > r \end{cases} \tag{10.2}$$

where r is a *tuning constant*. A value for r that is often used is 1.345. If for seven data points, $e_i^* = -3, -2, -1, 0, 1, 2$, and 3, and $r = 1.345$, then the Huber influence function would be calculated as follows (first two columns):

	Huber		Ramsay	
e_i^*	$\psi(e_i^*)$	$w_i = \psi(e_i^*)/e_i^*$	$\psi(e_i^*)$	$w_i = \psi(e_i^*)/e_i^*$
−3	−1.345	0.448	−1.220	0.407
−2	−1.345	0.673	−1.098	0.549
−1	−1	1.0	−0.741	0.741
0	0	1.0	0	1.0
1	1	1.0	0.741	0.741
2	1.345	0.673	1.098	0.549
3	1.345	0.448	1.220	0.407

This means that whatever the value of r, the maximum absolute value of $\psi(e_i^*)$ will be equal to r. The *weight* assigned to the ith data point is given by

$$w_i = \frac{\psi(e_i^*)}{e_i^*} \tag{10.3}$$

which, for this small example, leads to the weights in the third column of the table. When $e_i^* = 0$, which would be the case when $h_{ii} = 1.0$, the weight is taken to be 1.0.

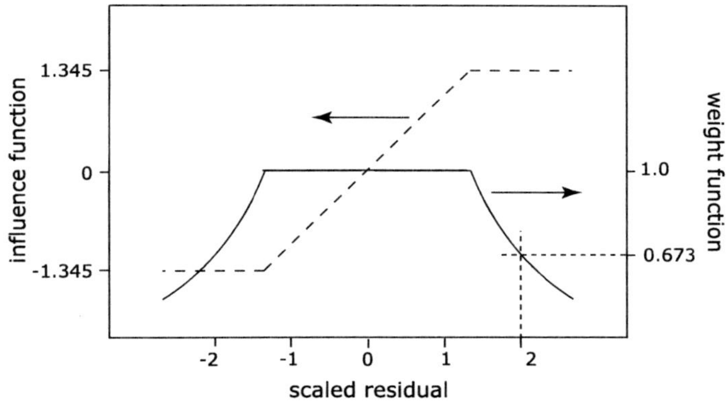

Figure 10.7. *Huber influence and weight functions.*

Figure 10.7 illustrates the Huber influence and weight functions for the case where $r = 1.345$. The influence function (long dashed line) has a slope of one between $\psi(e_i^*) = -1.345$ and $\psi(e_i^*) = 1.345$ and a slope of zero elsewhere. Between $\psi(e_i^*) = -1.345$ and $\psi(e_i^*) = 1.345$, $w_i = 1$ (solid line), but outside of these extremes the weights fall off asymptotically toward zero for large $|e_i^*|$. The intersecting short dashed lines illustrates the weight function when $e_i^* = 2$.

As a second example, the Ramsay influence function [142] is given by

$$\psi(e_i^*) = e_i^* \exp(-r|e_i^*|) \tag{10.4}$$

where r is a tuning constant, often taken to be 0.3 for this function. As with the Huber criterion, the Ramsay weight function is calculated using Eq. 10.3. Like the Huber weight function, the Ramsay weight function asymptotically approaches zero for large $|e_i^*|$. Figure 10.8 illustrates the Ramsay influence and weight functions for the case where $r = 0.3$, while the tabulation on page 213 illustrates its calculation for the series $e_i^* = -3, -2, \ldots, 2, 3$. As with the Huber function, when $e_i^* = 0$ the weight is taken to be 1.0.

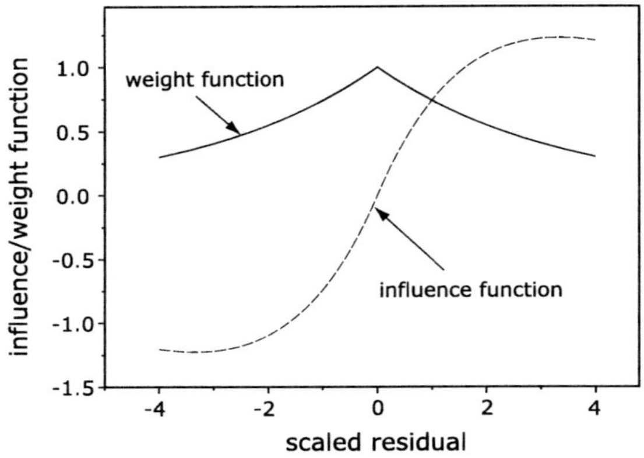

Figure 10.8. *Ramsay influence and weight functions.*

Several other influence functions have been proposed in the literature, and for some of these the weight functions become equal to (rather than asymptotically approach) zero at sufficiently large $|e_i^*|$. For overviews, see Montgomery, Peck, and Vining [100], Neter et al. [113], Ryan [144], and Wilcox [172].

To this point, nothing has been said about the scaled residuals, the e_i^*, that are used by the influence functions. When the distribution of the residuals is not normal, then s (i.e., $\sqrt{\text{MSE}}$) is not a robust estimator of σ. A frequently used robust estimate of scale is the **median absolute deviation**, defined as

$$\text{MAD} = \text{median}\,\{|\,e_i - \text{median}(e_i)|\,\}$$

When sampling from a normal distribution, however, MAD estimates $z_{0.75}\sigma$ rather than σ, where $z_{0.75}$ is the 0.75 lower-tail probability of the standard normal distribution [100, 113, 172]. Typically, then, MAD is rescaled so that it estimates σ when sampling from a normal distribution.

$$\text{MADN} = \frac{\text{MAD}}{z_{0.75}} = \frac{\text{MAD}}{0.6745}$$

If MADN is used as the robust estimate of scale, then the scaled residuals e_i^* are given by

$$e_i^* = \frac{e_i}{\text{MADN}} \tag{10.5}$$

The following table illustrates the calculation of the e_i^* from the ordinary residuals resulting from a least-squares fit of a quadratic Scheffé model to the adhesive GS3 data. Column 2 contains the ranked residuals, e_i, the median of which is 8.737. Column 4 lists the ranked values of $|e_i - \text{median}(e_i)|$, the median of these values being 11.732 (= MAD). The value of MADN for the first iteration is then 11.732/0.6745 = 17.394. The e_i^* values in column 6 are obtained by dividing the e_i values in column 2 by 17.394. Note that the ranking of the e_i^* is the same as the ranking of the e_i. This procedure is repeated for each iteration. As convergence is approached, the value of MADN also converges.

ID	e_i	ID	$\lvert e_i - \text{median}(e_i)\rvert$	ID	e_i^*
11	−46.062	4	0.000	11	−2.648
7	−16.559	2	1.704	7	−0.9520
1	−15.559	6	2.704	1	−0.8945
9	−4.559	3	3.000	9	−0.2621
10	−0.559	10	9.296	10	−0.03214
4	8.737	8	11.732	4	0.5023
2	10.441	5	11.736	2	0.6003
6	11.441	9	13.296	6	0.6578
3	11.737	1	24.296	3	0.6748
5	20.469	7	25.296	5	1.177
8	20.473	11	54.799	8	1.177

Table 10.3 displays the weights from IRLS regressions of the adhesive GS3 data using the Huber and Ramsay influence functions with tuning constants of 1.345 and 0.3, respectively. The meanings of $k = 0$ and $k = 0.5$ are explained below. In all cases, iterations were carried out until the maximum percentage change in a coefficient estimate was 0.1%. This required 14, 24, 8, and 6 iterations for columns 2–5, respectively.

Table 10.3. *Data point weights. Robust regression of adhesive GS3 data*

	Weights			
	$k = 0$		$k = 0.5$	
ID	Huber	Ramsay	Huber	Ramsay
1	0.4220	0.1065	1.0000	0.6522
2	1.0000	0.7174	1.0000	0.7255
3	1.0000	0.8599	1.0000	0.8676
4	1.0000	0.8643	1.0000	0.9435
5	1.0000	0.9962	0.3571	0.2967
6	1.0000	0.7665	1.0000	0.7078
7	0.3818	0.0818	1.0000	0.6311
8	1.0000	0.9962	0.3571	0.2967
9	1.0000	0.8199	1.0000	0.9200
10	1.0000	0.8211	1.0000	0.9688
11	0.1241	0.0013	0.3571	0.2967

Ignoring for the moment the (unexplained) distinction between $k = 0$ and $k = 0.5$, within a single value for k one can clearly see the difference between the Huber and Ramsay functions. In the case of the Huber function, several values have $w_i = 1.0$ and receive the same weight as in OLS regression. The *only* way that a data point can have a weight of 1.0 with the Ramsay function is if $e_i^* = 0$ (cf. Eq. 10.4, page 213). This would be the case if there were data points with leverages of 1.0. The weights in Table 10.3 are based on the quadratic Scheffé model, and for this model there are no data points with $h_{ii} = 1.0$.

The columns labeled $k = 0$ are the weights resulting from IRLS regression using the methods described to this point. We note that neither point 5 nor point 8 is significantly downweighted, despite the fact that these points have the largest PRESS residuals, Cook's distance, and $DFFITS_i$ (Table 9.5, page 202). We *know*, however, that points 5 and 8 are very influential in the fitted OLS model for the GS3 data, but we also know that one reason for this is the relatively high leverages for these two data points. Any robust procedure that does not downweight the importance of these points is of questionable use.

The reason that these points have not been downweighted is that the procedures described to this point reduce the influence of observations that are outlying with respect to the iteratively reweighted e_i only and are *not* sensitive to observations that are outlying with respect to their X values (i.e., their leverages). As a result, there is the likelihood that a high-leverage observation might receive a weight of 1.0 (or a relatively high weight) simply because the ordinary residual is small. Estimators described to this point are called ***M* estimators**, and M estimators are known to have a "breakdown point" of $1/n$.

In the context of robust regression, the **breakdown point** is the smallest fraction of anomalous data that can distort the estimator so badly that it is of no use to the model builder. The concept of breakdown point can be illustrated by considering two measures of location, the sample mean and median. The breakdown point of the mean is $1/n$, because if any single point goes to $\pm\infty$, then the mean goes to $\pm\infty$. On the other hand, the sample median has a breakdown point of nearly 1/2 because it is possible to set nearly one half of the observations to $\pm\infty$ without affecting the median.

One way to combat this problem would be to do robust regression as described above but with the ordinary residuals replaced by studentized or perhaps even PRESS residuals. We might redefine the scaled residuals e_i^* as

$$e_i^* = \frac{e_i/(1-h_{ii})^k}{\text{MADN}} \tag{10.6}$$

When $k = 0$ the scaled residuals are defined as in Eq. 10.5, page 214 (i.e., ordinary residuals scaled by MADN). When $k = 0.5$, the scaled residuals are the studentized residuals scaled by MADN rather than by s. And when $k = 1.0$, the scaled residuals become the PRESS residuals scaled by MADN. For a given value of k, as h_{ii} becomes larger and larger, the numerator $(e_i/(1-h_{ii})^k)$ and e_i^* become larger and larger. The net effect will be that, for a given e_i, a point with a relatively large h_{ii} will tend to be downweighted more than one whose h_{ii} is relatively small. One can view the exponent k as a secondary tuning constant that "dials in" the effect of leverage. The larger the value of k, the greater the impact of h_{ii} on e_i^*.

In the absence of high-leverage points (X outliers), residual outliers can still arise simply because one or more e_i is large. In these circumstances M estimation may work well, in which case one would set $k = 0$. Robust regression procedures that take allowance of leverage are known as **bounded influence estimators** or ***GM* estimators** (for generalized M estimators). In making a choice between M and GM estimators, one should be guided by the importance of leverage as evidenced by diagnostics such as PRESS, the R-student statistic, Cook's D, $DFFITS$, and $DFBETAS$. The weights in Table 10.3 that fall under the heading $k = 0.5$ use scaled residuals calculated using Eq. 10.6 with $k = 0.5$. The results satisfy our intuition that points 5, 8, and 11 should be downweighted.

Table 10.4. *Adhesive GS3 response. OLS and IRLS parameter estimates using the Huber and Ramsay influence functions*

		IRLS($k = 0.5$)	
Term	OLS	Huber	Ramsay
A	109.56	107.96	108.86
B	63.56	61.96	63.00
C	29.56	27.96	27.90
AB	−205.18	−173.23	−176.09
AC	255.89	257.98	256.44
BC	−100.13	−98.04	−99.87
$\sum_{i=1}^{n} \lvert e_i \rvert$	166.595	157.264	156.625

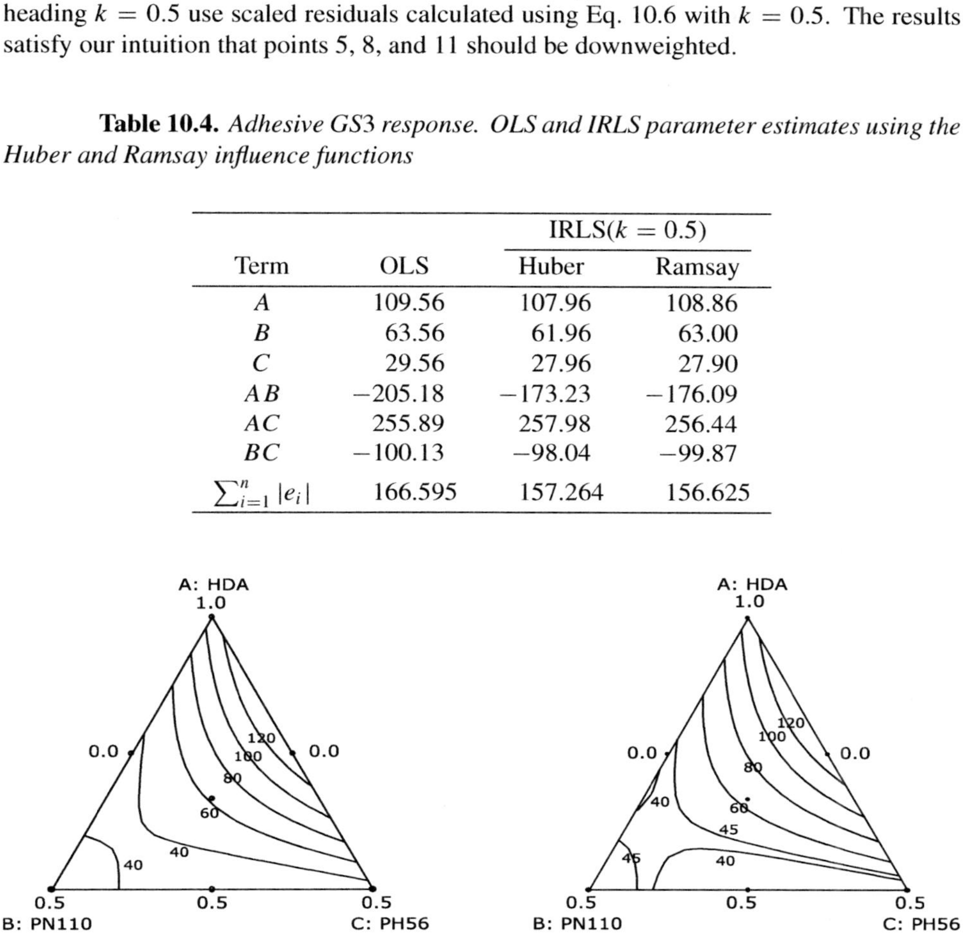

Figure 10.9. *Adhesive GS3 response. Contour plots for OLS (left) and IRLS (right) models.*

Parameter estimates (in the pseudocomponent metric) for the OLS and IRLS ($k = 0.5$) regression of the GS3 data are displayed in Table 10.4. One can infer from the relatively small changes in the parameter estimates that trace plots should be similar for the three models. Although the trace plots are not reproduced here, this turns out to be true. In their place, contour plots are displayed in Fig. 10.9 for the OLS and IRLS models.

The plots clearly show that the general features of the response surface are maintained despite the fact that the weights for points 5, 8, and 11 are slightly less than 0.3. This result makes one somewhat more comfortable about using the OLS response surface for prediction.

The sums of the absolute residuals for the OLS model and for the IRLS models at the final iteration are summarized in the last line of the table. The smaller values for the iteratively reweighted least-squares models suggest a better overall quality of fit than that provided by OLS.

Whether to use a Huber function, a Ramsay function, or some other function for IRLS estimation is a judgment that must be left to the analyst. The effectiveness of different influence functions and tuning constants is best determined by trying them on the data and then examining the weights, the residuals, and the models. In the example illustrated here, despite the differences in the weights (Table 10.3), there is little to choose between the models (Table 10.4) based on the Huber and Ramsay functions. In other cases there may be significant differences. In such cases one may need to assess the predictive capabilities of the models in further studies.

Many of the popular computing packages do not have robust regression capabilities. An exception to this is S-PLUS, which has several robust procedures. The S-PLUS function `rreg` performs M estimation using any one of eleven different weight functions, including Huber but excluding Ramsay. The functions `lmRobMM`, `lmsreg`, and `ltsreg` perform MM-regression, least median of squares regression, and least trimmed squares regression, respectively. These are high-breakdown estimators that generally require at least twice as many observations as regressor variables. See Montgomery, Peck, and Vining [100] and Ryan [144] for discussion of these algorithms. Wilcox provides many S-PLUS functions for robust estimation, and their use is explained in his text [172]. At the time of this writing, these were available at

http://www.academicpress.com/updates/ireht.htm

It should be noted that Wilcox's function `bmreg` for bounded-influence M-regression is written for models that contain an intercept. To use this function for no-intercept mixture models, one of the linear terms must be dropped and an intercept introduced in its place, as discussed on page 26.

10.2 Variable Selection

The word *overfitting* is used to describe the practice of fitting a model that contains terms that have little or no explanatory value. *Underfitting* refers to the practice of fitting a model in which terms that *do* have explanatory value have been removed or are missing from the model. Overfitted models are sometimes referred to as overspecified and underfitted models as underspecified. Overfitting should not be confused with overparameterization. In an overparameterized model, there are more terms than can uniquely be estimated. In an overfitted model, all terms can be uniquely estimated, but some are unnecessary. In the interest of parsimony (and for other reasons cited in the following paragraph), we seek a model that strikes a balance between over- and underfitting.

The following is stated without proof. (See Myers [105] for mathematical details.)

- Overfitted models result in inflated variances in both the coefficient estimates and the fitted values (the $\hat{Y}$s).
- Underfitted models lead to bias in the coefficient estimates, in prediction, and in the estimate of error variance (s^2).

Thus one desires a model that is a compromise between inflated variances and bias in the coefficients and in prediction. Decisions about whether or not to remove a term from a model are usually made by examining the p values for the model terms. Terms that have p values greater than some preselected significance level (say 0.05 or 0.10) are considered unnecessary and can be removed from the model.

Table 10.5. *H&M surfactant design*

ID	A (Nonionic A)	B (Nonionic B)	C (Anionic)	D (Zwitter)	Lather Units
1	1.000	0.0000	0.0000	0.0000	7.17
2	0.5000	0.0000	0.5000	0.0000	2.68
3	0.5000	0.5000	0.0000	0.0000	3.08
4	0.9500	0.0000	0.0000	0.0500	6.99
5	0.5000	0.0000	0.4500	0.0500	2.92
6	0.5000	0.4500	0.0000	0.0500	2.89
7	0.7500	0.0000	0.2500	0.0000	4.83
8	0.7500	0.2500	0.0000	0.0000	3.85
9	0.5000	0.2500	0.2500	0.0000	3.13
10	0.7250	0.0000	0.2250	0.0500	4.43
11	0.7250	0.2250	0.0000	0.0500	3.60
12	0.6500	0.1500	0.1500	0.0500	3.75
13	0.8292	0.0792	0.0792	0.0125	5.39
14	0.5792	0.0792	0.3292	0.0125	2.64
15	0.5792	0.3292	0.0792	0.0125	3.56
16	0.8042	0.0792	0.0792	0.0375	5.23
17	0.5792	0.0792	0.3042	0.0375	3.22
18	0.5792	0.3042	0.0792	0.0375	3.52
19	0.6583	0.1583	0.1583	0.0250	4.31
20	0.6500	0.1500	0.1500	0.0500	3.26

Table 10.5 displays the surfactant design of Heinsman and Montgomery [68] plus their response called "lather units". Fitting the response data to the 10-term quadratic Scheffé model expressed in the pseudocomponent metric leads to p values for the quadratic terms displayed in Table 10.6, column 2. The p values for the four linear terms are absent from the table for reasons related to the discussion on page 164. *Testing the significance of linear parameter estimates in Scheffé models is of little interest and should be avoided.*

On the other hand, it does make sense to test the significance of higher-order terms, as these terms carry information about nonlinear blending. Examples of these types of tests are

$$\begin{aligned} \mathrm{H}_0 &: \quad b_{ij} = 0 \\ \mathrm{H}_0 &: \quad b_{ijk} = 0 \\ \mathrm{H}_0 &: \quad b_{ij(i-j)} = 0 \end{aligned}$$

where b_{ij}, b_{ijk}, and $b_{ij(i-j)}$ are coefficient estimates for β_{ij}, β_{ijk}, and $\beta_{ij(i-j)}$.

Table 10.6. *Surfactant example. Analyses for lather units, pseudocomponent metric*

	Total Number of Terms in the Model							
	10		9		8		5	
Term	p	se[†]	p	se	p	se	p	se
AB	0.002	1.29	0.001	1.23	0.001	1.25	0.001	1.18
AC	.269	1.29	.245	1.23				
AD	.206	96.9	.134	86.3	.210	85.3		
BC	.812	1.48						
BD	.215	96.4	.144	86.2	.226	85.1		
CD	.202	96.4	.133	86.2	.209	85.1		
SSR	31.540		31.532		31.357		31.081	
SSE	1.2725		1.2802		1.4554		1.7316	
s^2	.1273		.1164		.1213		.1154	
R^2	.9612		.9610		.9556		.9472	
R^2_{adj}	.9263		.9326		.9298		.9332	
R^2_{pred}	.8112		.8660		.8671		.9170	

[†] Standard errors of coefficient estimates

The p values output by software (as exemplified in Table 10.6) test the significance of *single-term deletions*. When a variable is deleted, the sum of squares for the deleted variable is pooled into residual error. Because of this, and because regressor variables in Scheffé models are always correlated with one another, the removal of a term from the model will result in changes in the p values of the remaining terms. A remaining term may become less significant or more significant. For example, the largest p value for the 10-term model in Table 10.6 is that for BC (0.812). Removal of this term from the model causes the p values for terms remaining in the 9-term model to change.

For deletions of more than one term, one could use a manual approach, such as removal of terms one by one, or an automated approach using an algorithm, such as backward elimination (explained below). Let us use the manual procedure and continue with the example in Table 10.6. In the 9-term model, the least-significant term is AC ($p = 0.245$), and so we remove this term from the model, leading to the 8-term model. We are now confronted with a dilemma.

To this point, the terms that have been removed from the model all had comparatively small standard errors. The standard errors of all terms containing D, on the other hand, are inflated because of the small range of D (compare the Design Study at the end of Chapter 5, and particularly Table 5.11, page 92). Is the reason that all quadratic terms in D appear not to be significant because they really are not significant, or is it because the standard errors of these terms are so inflated that the terms simply *appear* not to be significant? If the latter and we delete terms, then we are in danger of making a Type II error (Chapter 6, page 98). Unfortunately, there is no simple solution to this dilemma. It is a problem that one often encounters in a mixture setting.

One thing we might do is to see what happens to the model summary statistics (summarized in the bottom panel of Table 10.6) if we proceed on the assumption that the terms in D are indeed not significant. We could continue deleting terms one by one, checking p values as we go along, or take a chance and consider a *multiple-term deletion*. By "take a chance" we mean that we have no idea what is going to happen to the p values of terms containing D when one or more are removed from the model. Are they going to remain insignificant, or will one of them all of a sudden become significant?

To test the **composite null hypothesis** $H_0 : b_{AD} = b_{BD} = b_{CD} = 0$, we use the extra sum-of-squares principle. The fuller model is the 8-term model, and the less full is the 5-term model containing the four linear terms plus the term in AB. The calculations are

$$\begin{aligned} F &= \frac{(\text{SSR}_{fuller\ model} - \text{SSR}_{less\ full\ model})/\Delta df}{s^2} \\ &= \frac{(31.357 - 31.081)/3}{0.1213} \\ &= 0.7584 \end{aligned}$$

Because the tabled value for $F_{0.05;3,12}$ is 3.49 and $0.7584 < 3.49$, we do not reject the null hypothesis $H_0 : b_{AD} = b_{BD} = b_{CD} = 0$. (The p value for this test is 0.5387.) Because of the inflated variances for the quadratic crossproduct terms containing D, we can never be sure about their significance. However, some reassurance for their removal is provided by the model summary statistics. The model with five terms has the highest R^2_{adj} and R^2_{pred} values of the four models. Note that R^2 steadily decreases as terms are removed from the model, as expected.

In going from the 10-term model to the 5-term model, the error sum of squares (SSE) increases from 1.2725 to 1.7316. At the same time, however, the error degrees of freedom have increased from 10 to 15 (because the model degrees of freedom have decreased by 5). The result is that *mean* square error (s^2) is less for the 5-term model than for the 10-term model.

The final model in terms of pseudocomponents is

$$\widehat{\text{lather units}} = 7.26A^* + 3.20B^* + 2.64C^* + 2.09D^* - 4.92A^*B^*$$

while in terms of reals the model is

$$\widehat{\text{lather units}} = 7.26A + 8.98B - 1.98C - 3.09D - 19.7AB$$

The forms of the reduced models and the summary statistics (cf. Table 10.6, bottom panel) are the same whether one carries out the removal process starting with a full model in the pseudos or in the reals.

As a second example, let us fit a model to the 3-minute green strength response from the Stepan hot-melt adhesive experiment (Table 8.1, page 156). Table 10.7 displays the sequential model sums of squares. The p values imply that the special cubic term is significant, but possibly all the quadratic terms may not be because of the marginal significance ($p = 0.1074$) of this group. (It is good to keep in mind, however, that the sums of squares in this table are *sequential* sums of squares. The significance of the quadratic terms assumes that the linear terms are the only other terms in the model.)

Table 10.7. *Adhesive GS3 response. Sequential model sums of squares*

Terms	Sum of Squares	df	Mean Square	F Value	Prob> F
Linear	6600.2	2	3300.1	2.1715	0.1765
Quadratic	8206.7	3	2735.6	3.4617	0.1074
Special cubic	3208.6	1	3208.6	17.286	0.0142
Residual	742.5	4	185.62		
Corrected total	18758.	10	1875.8		

Table 10.8. *Adhesive example. Effect of hierarchy on the GS3 analysis*

	p Values			
	Pseudos		Reals	
Term	7-Term Model	6-Term Model	7-Term Model	6-Term Model
AB	0.031	0.015	0.031	0.073
AC	.005	.001	.005	.066
BC	.902		.017	
ABC	.014	.004	.014	.305
SSE	742.5	745.7	742.5	3593.0
s	13.62	12.21	13.62	26.81
R^2	.9604	.9602	.9604	.8085
R^2_{adj}	.9010	.9205	.9010	.6169

The p values for individual quadratic and special cubic terms are presented in Table 10.8. Values for the full special cubic model are given in column 2 for the model expressed in the pseudos and in column 4 for the model expressed in the reals. If one were model fitting in terms of the pseudos, then one would conclude that the term in BC was unnecessary ($p = 0.902$). Removal of this term leads to a slight increase in the residual sum of squares (from 742.5 to 745.7), but because of the additional degree of freedom for error, the value of s decreases slightly while R^2_{adj} increases a bit.

The situation is quite different in the reals. If one assumes that a deleted term in the pseudos should also be deleted in the reals, then one should remove the term in BC. Yet this term is highly significant in the reals, having a p value of 0.017. Removal of this term causes the residual sum of squares (SSE) to soar and R^2 and R^2_{adj} to plummet.

This result — scale-dependent model summary statistics — is related to a principle originally put forth by Peixoto [117] and discussed further by Nelder [110, 111]. A polynomial model such as

$$y = \beta_0 + \beta_1 x + \beta_2 x^2 + \beta_3 x^3 + \epsilon$$

is said to be **hierarchically well-formulated** if all terms "contained" by each term in the model are also in the model. For example, x^3 contains x^2 and x, and both x^2 and x are in the model; x^2 contains x, and x is in the model. The model

$$y = \beta_0 + \beta_1 x + \beta_3 x^3 + \epsilon$$

is not well-formulated because x^2 is contained in x^3, but x^2 is missing from the model.

A linear transformation of the x variables of a not-well-formulated polynomial regression model may lead to scale-dependent model summary statistics. In going from the reals to the pseudos (or vice versa), the X_i undergo linear transformations. To see this, recall that pseudocomponent proportions are calculated from the proportions in the reals using the expression (Eq. 4.21, page 58)

$$X_i^* = \frac{X_i - L_i}{1 - L}$$

This can be cast in the form

$$X_i^* = a_i + bX_i \tag{10.7}$$

where $a_i = -L_i/(1 - L)$, $b = 1/(1 - L)$, and X_i^* is the proportion of the ith pseudocomponent.

Application of this principle to a special cubic model for three components, A, B, and C, implies that if the special cubic term ABC is in the model, then to be hierarchically well-formulated, the model should contain AB, AC, BC, A, B, and C. One can see why this is so by carrying out a bit of algebra. Let us express the pseudocomponent proportions using Eq. 10.7 but retain the A-B-C notation. The special cubic term in the pseudos is then

$$A^* B^* C^* = (a_A + bA)(a_B + bB)(a_C + bC)$$

One can see by inspection that when this expression is multiplied out, there will be terms in A, B, C, AB, AC, BC, and ABC. (There will also be a constant term, $a_A a_B a_C$. Any constant term can be written as a constant times 1, and since $\sum_i^q X_i = 1$, a constant term can be reexpressed as a sum of linear terms.)

In light of this, algebraic back transformation of the 6-*term* model for GS3 in the pseudos (with B^*C^* missing, as in Table 10.8) will introduce BC and lead to a 7-*term* model in the reals. That is, the transformed model is not of the same form as the untransformed model. The transformed model is

$$\widehat{\text{GS3}} = 107A + 321B - 751C - 613AB + 1395AC + 7610BC - 15219ABC$$

This model has SSE = 745.7 — the same value as the 6-term model in the pseudos (cf. Table 10.8). On the other hand, least-squares fitting of the response values to the 7-term special cubic model in the reals *directly* leads to

$$\widehat{\text{GS3}} = 107A + 323B - 749C - 616AB + 1392AC + 7491AC - 15047ABC$$

This model has SSE = 742.5 (Table 10.8), slightly less than the algebraically back-transformed model. This is not unexpected because OLS will find parameter estimates that minimize the residual sum of squares. The differences in the coefficients in the two models are small but real and are not due to rounding error.

Peixoto's hierarchy principle was developed in the context of polynomial models. Scheffé canonical polynomials that contain terms such as $AB(A - B)$ and $AB(A - B)^2$ do not really fall into this definition. Because of this, linear transformation of the pseudos to the reals can lead to unexpected results in the model form in the reals. For example, if a Scheffé model expressed in pseudocomponent proportions has a cubic term such as $A^*B^*(A^* - B^*)$, then for the model to be well-formulated it must not only contain A^*B^* but also all quadratic crossproduct terms in A^* and B^* — that is A^*C^*, A^*D^*, etc. and B^*C^*, B^*D^*, etc.

One can draw on Eq. 10.7 and write a cubic term as

$$A^*B^*(A^* - B^*) = (a_A + bA)(a_B + bB)[(a_A - a_B) + b(A - B)]$$

Expanding the terms on the right side is tedious, and we will not carry it out here. In so doing, however, one will end up with an expression that contains terms in A^2 and B^2. Squared terms can be reexpressed as

$$\begin{aligned} A^2 &= A \cdot (1 - B - C - D - \ldots) \\ B^2 &= B \cdot (1 - A - C - D - \ldots) \end{aligned}$$

and as a result, all quadratic crossproduct terms in A and B must be present.

In a term like $A^*B^*(A^* - B^*)$, all the quadratic crossproduct terms in A and B other than AB are "masked". It is not at all obvious by inspection that BC, for example, is contained in $A^*B^*(A^* - B^*)$. When talking about Scheffé canonical polynomials, it is probably better to drop the potentially confusing description "hierarchically well-formulated" and refer to the models as "scale-independent" (or "scale-dependent", as the case may be). A scale-dependent model, then, would be one that exhibits different summary statistics in the pseudos and reals, assuming that the same terms are in both models.

In Scheffé models, scale-dependency issues do not arise until one fits models of order greater than two. Although the Scheffé quartic model (Eq. 3.32, page 28) is scale-independent, certain subsets of the model are not. One in particular that is not is the special quartic model (Eq. 3.35, page 29). Some software products such as Design-Expert and MINITAB warn the user when a model is scale-dependent. If there is any doubt, one should fit the model in both metrics and check the summary statistics. If one's intention is to work in both metrics, then one must be careful not to remove terms from a model that are required to maintain scale independency.

The question naturally arises, "Which metric should I choose?" The answer to this depends in part on the reason for fitting a model. If the primary reason is to obtain a better understanding of the system — how the real-world components interact with one another and control the response(s) — then the answer would be to fit a model that has all terms significant in the reals and bring subject-matter knowledge to bear on coefficient interpretation. Pseudocomponents are mathematical artifacts and frequently do not even lie within the design region. It is usually not at all obvious what the physical meaning is (for example) of pseudocomponent A blending nonlinearly with pseudocomponent B. Thus model interpretation would favor fitting a model in the reals.

A word of caution is in order here, however. We have not discussed problems that arise in the presence of severe collinearity, nor have we even defined collinearity. That discussion is relegated to Chapter 14. If one knows that a data set is **ill conditioned** — which is the condition that leads to collinearity — then one must be cautious about interpreting coefficient estimates. In these situations, the signs and magnitudes of coefficient estimates may have no physical significance, yet the model may perform well in prediction!

If the primary purpose of the model is for prediction and/or optimization, then one would want to look carefully at the summary statistics for the models in the two metrics. In this case, one may not be so concerned about understanding the system but perhaps more concerned about getting a product "out the door" within a time frame (a not-unusual situation in an industrial setting). If the model has a higher R^2_{pred} and lower s in one scale than in another, then one would opt for the better fitting model.

When there are a large number of terms in a model, the one-by-one manual approach to removing unwanted terms from a model can become tedious. In such circumstances it may be convenient to adopt an automated approach called **backward elimination**. Backward elimination is nothing more than an automation of the manual procedure described previously. In adopting such an approach, one must be cognizant of the risks involved. Because of automation of the variable selection procedure, one will have no knowledge about changes in the p values of terms as terms are removed during the elimination process. Drawing from examples already discussed in this section, one may end up removing terms that are not significant simply because their standard errors are inflated. In addition, one could very well end up with a scale-dependent model.

Despite these risks, backward elimination is often employed by model builders. The algorithm begins with all the candidate regressors in the model. At each step, the one variable whose deletion will cause the smallest increase in the residual sum of squares (SSE), and therefore maintain the largest possible sum of squares for regression (SSR), is removed from the model. This is equivalent to removing the variable with the largest p value. The algorithm continues until all terms in the model have p values less than some prespecified significance level, α_{out}.

Another algorithm, called **forward selection**, is the opposite of backward elimination. Forward selection begins with no regressors in the model and inserts variables one at a time until a suitable model is obtained. At each step, the variable selected for entry is the one that will cause the largest decrease in SSE and largest increase in SSR. This is equivalent to selecting the variable that will lead to the smallest p value. The algorithm stops when the next variable to enter has a p value greater than some prespecified significance level, α_{in}.

In either forward selection or backward elimination, the addition or deletion of a variable can have an effect on the significance of other variables that are in the model. This is particularly true in the case of mixture models, where the regressors are always correlated to some degree. Because of this, a variable added early in forward selection could become unimportant after other variables are added. Conversely, a variable dropped in backward elimination could become significant after other variables are dropped from the model.

To combat this problem, an algorithm called **stepwise selection** or **stepwise regression** can be used. The procedure is a modification of the forward selection process. After the addition of each variable, the algorithm checks the importance of any previously included variables. If any are no longer significant, the algorithm switches to backward elimination and drops variables one at a time until all are significant. Forward selection then resumes. In stepwise selection, significance levels to enter (α_{in}) and leave (α_{out}) must both be specified, with $\alpha_{in} \leq \alpha_{out}$. The reason for this is because if $\alpha_{in} > \alpha_{out}$, the rule for entering is less stringent than the rule for remaining. Thus a variable could enter and immediately be removed, which could end up in an endless loop.

Backward elimination, forward selection, and stepwise selection will not necessarily lead to the same model. It is difficult to give a specific recommendation. With today's powerful computers, it is a simple matter to run all three algorithms as well as take the manual approach (if the model is not too large) and compare the results. The author's personal experience has been that forward selection is the least satisfactory of the three. In the final analysis one must pay attention to summary statistics and, as far as possible, bring subject-matter knowledge or intuition to bear.

Taken together, backward elimination, forward selection, and stepwise selection comprise what are sometimes called **stepwise regression procedures**. Thus the adjective *stepwise* has a twofold meaning, referring not only to the set of three algorithms but also to one specific member of the set. All three procedures are implemented in most popular computing packages.

The expression "variable selection" ordinarily implies the selection of variables that have explanatory value from a longer list of variables, some of which may have no explanatory value. The longer list would consist of the terms in the tentative model. A more liberal interpretation of "variable selection" would include the selection of variables that may be *missing* from the tentative model. When the p value for a lack-of-fit test is small (indicating significance), one would conclude that the tentative model does not adequately describe the data. When this situation arises, one must consider augmenting the model and possibly the design.

In cases where the design must be augmented, it is probable that the additional runs are going to have to be prepared and analyzed on different days than the original set of runs. In addition, instruments may have to be reset, different operators may be involved, a new batch of a mixture component may be needed, and so forth. All of these factors can lead to unplanned systematic variation. Unless one considers blocking a posteriori, this variation will end up as bias in the estimate of σ^2, and this in turn will desensitize hypothesis testing. A better approach is to introduce a blocking variable, which will have the effect of converting the unplanned systematic variation into a factor of the design. The augmented model will then be of the form

$$\widehat{Y} = \text{Original (tentative) model} + \text{additional terms} + \widehat{\gamma} z \tag{10.8}$$

In these situations, one does not have the luxury of using the preplanned blocking procedures described in Chapter 7. One is forced to live with nonzero correlation coefficients between $\widehat{\gamma}$, the coefficient estimate for block, and the coefficient estimates for the mixture model. However, with many commercially available computing packages, one does have the benefit of built-in procedures for D-optimal design augmentation with blocking. One can also use the DMAX procedure in MIXSOFT as long as one recognizes that a block is simply one level of a categorical variable called *block* (cf. page 124). One would select `Mixture-process variable` for the model type, and `Mixture model + no-intercept linear polynomial in the PVs` for the model form.

There are a few things that one should keep in mind when augmenting a design to support a higher-order polynomial. The minimum number of additional runs required will be equal to the number of additional terms in the model plus one additional run for blocks.

Beyond these, more runs could, of course, be added to provide extra degrees of freedom for lack of fit. If replication is desired, then some of the additional runs must be replicated. Formulations in different blocks that have the same composition are *not* replicates because they are in different blocks.

10.3 Partial Quadratic Mixture Models

Much of the discussion in Section 10.2 centered around tests of hypotheses such as H_0 : $b_{ij} = 0$. When a full quadratic mixture model is fit to data and this null hypothesis is not rejected, then the term $b_{ij}X_iX_j$ is removed from the model. We say that the resulting model is a **reduced model**. Another way of looking at this is to say that when this null hypothesis is not rejected, we *restrict* the value of the parameter β_{ij} to zero, in which case we could call the reduced model a **restricted model**. In a quadratic model, however, restrictions of the form $b_{ij} = 0$ are not necessarily the only restrictions of interest. For example, suppose one would like to test the null hypothesis $H_0 : b_{ij} = b_{ik}$. That is, one suspects that component i exhibits the same nonlinear blending behavior towards components j and k. Or perhaps one would like to test the null hypothesis $H_0 : b_{ij} = b_{ik} + b_{jk}$, i.e., a parameter estimate is equal to the sum of two other parameter estimates.

One approach to fitting models with restrictions on parameters is to use **restricted least squares** (see, for example, Cornell [29, Appendix 6B]). This procedure requires software capable of carrying out matrix manipulations as well as an understanding of restricted least squares. An alternative approach, requiring only OLS, is to fit **partial quadratic mixture models**, a model form introduced by Piepel, Szychowski, and Loeppky [135]. The reader's attention is called to the word *quadratic*. This section is about quadratic mixture models only — not linear models, not cubic models, etc. — but only quadratic models.

It will be helpful in this section to adopt several abbreviations used by Piepel et al. in their paper, and so these are listed here. Their meanings will become clear as they arise in the discussion.

CSQ model - complete (full) quadratic Scheffé model

RSQ model - restricted Scheffé quadratic model

RRSQ model - restricted and reparameterized Scheffé quadratic model

PQM model - partial quadratic mixture model

The relationship between these various models is illustrated below. Again, this will be explained in the discussion to follow.

$$\text{CSQ model} \quad \overset{\text{restrict}}{\longrightarrow} \quad \text{RSQ model}$$

$$\text{RSQ model} \quad \overset{\text{reparameterize}}{\longrightarrow} \quad \text{RRSQ model}$$

$$\text{RRSQ model} \quad = \quad \text{PQM model}$$

Finally, following Piepel et al., we shall adopt the symbol δ to represent coefficients in PQM models, retaining the symbol β for Scheffé models.

Suppose that one had fit the complete Scheffé quadratic (CSQ) model

$$E(Y) = \beta_1 X_1 + \beta_2 X_2 + \beta_3 X_3 + \beta_{12} X_1 X_2 + \beta_{13} X_1 X_3 + \beta_{23} X_2 X_3 \tag{10.9}$$

to a set of data, and one suspected that $b_{12} = b_{13}$. Perhaps components X_2 and X_3 have similar chemical structures, and it is reasonable to posit that they may exhibit the same nonlinear blending behavior toward component X_1. If this were the case, then one could rewrite the CSQ model as the restricted Scheffé quadratic (RSQ) model

$$E(Y) = \beta_1 X_1 + \beta_2 X_2 + \beta_3 X_3 - \delta_{11} X_1 X_2 - \delta_{11} X_1 X_3 + \beta_{23} X_2 X_3 \tag{10.10}$$

where at this point $-\delta_{11}$ is *nothing more than a symbol* to represent β_{12} and β_{13}. This is a RSQ model because of the restriction $\beta_{12} = \beta_{13}$ $(= -\delta_{11})$. Although there are six terms in the model, because of the restriction there are only five unique terms, and therefore there are 4, not 5, model degrees of freedom. (Recall the discussion on page 162 regarding model degrees of freedom in Scheffé models.)

The terms in δ_{11} can be reparameterized as follows:

$$\begin{aligned} -\delta_{11} X_1 X_2 - \delta_{11} X_1 X_3 &= -\delta_{11} X_1 (X_2 + X_3) \\ &= -\delta_{11} X_1 (1 - X_1) \\ &= -\delta_{11} X_1 + \delta_{11} X_1^2 \end{aligned}$$

We can therefore rewrite the RSQ model 10.10 as the restricted and reparameterized Scheffé quadratic (RRSQ) model 10.11.

$$\begin{aligned} E(Y) &= (\beta_1 - \delta_{11}) X_1 + \beta_2 X_2 + \beta_3 X_3 + \delta_{11} X_1^2 + \beta_{23} X_2 X_3 \\ &= \sum_{i=1}^{3} \delta_i X_i + \delta_{11} X_1^2 + \delta_{23} X_2 X_3 \end{aligned} \tag{10.11}$$

where $\delta_1 = \beta_1 - \delta_{11}$, $\delta_i = \beta_i$ for $i = 2, 3$, and $\delta_{23} = \beta_{23}$. Partial quadratic mixture (PQM) models are RRSQ models, and so model 10.11 is a PQM model. There are no restrictions on the δs in the PQM model. The model may be fit by ordinary, rather than restricted, least squares. Like model 10.10, model 10.11 will have four degrees of freedom. To test the null hypothesis $H_0 : b_{12} = b_{13}$, one would use the extra sum-of-squares principle. The fuller model would be model 10.9 and the less full would be model 10.11.

The RSQ model 10.10 and the PQM model 10.11 will give *identical* $E(Y)$ values for *any* given mixture and given set of parameter values $\delta_i (i = 1, 2, 3)$, δ_{11}, and δ_{23}. On the basis of prediction, then, there is little to choose between the two model forms. This equivalence does not extend to the interpretation of the two models. The RSQ model 10.10 implies equal nonlinear blending behavior of component X_1 with components X_2 and X_3. The PQM model 10.11 implies that component 1 exhibits a quadratic curvature effect on the response independent of any interaction with components 2 or 3. One would use subject-matter knowledge or intuition to choose between the two interpretations and the two model forms. An example of choosing one interpretation over another is provided by the example at the end of this section.

The different interpretations of models 10.10 and 10.11 also extend to the meanings of the coefficients for the linear term in X_1 in the two models. In the RSQ model 10.10, as

$X_1 \to 1.0$, $E(Y) \to \beta_1$. In the PQM model 10.11, as $X_1 \to 1.0$, $E(Y) \to \delta_1 + \delta_{11}$. The $E(Y)$ at $X_1 = 1.0$ is the same in either case, because $\delta_1 + \delta_{11} = (\beta_1 - \delta_{11}) + \delta_{11} = \beta_1$. In the PQM model, the $E(Y)$ at the X_1 vertex is made up of two parts: the expected response in the absence of a quadratic effect of X_1 (δ_1) and the additional response due to the quadratic effect of X_1 (δ_{11}). A three-dimensional graphic of this situation is shown in the Case Study at the end of this chapter (Fig. 10.24, page 253).

The discussion to this point has implied model development by applying restrictions to the parameters in a CSQ model. However, one can take a different viewpoint and look at PQM models as Scheffé linear models augmented with squared and/or quadratic crossproduct terms. The *general* form of a PQM model as presented by Piepel et al. is

$$E(Y) = \sum_{i=1}^{q} \delta_i X_i + \sum_{i=1}^{q-1} \sum_{j=i+1}^{q} \delta_{ij} X_i X_j + \sum_{i=1}^{q} \delta_{ii} X_i^2 \tag{10.12}$$

where at least $q + 1$ of appropriately chosen δ_{ij} and δ_{ii} are zero.

The model is composed of two parts: the model *equation* 10.12 and the model *assumptions* stated beneath the model equation.

Consider the requirement that at least $q + 1$ squared and/or crossproduct terms be zero, ignoring for the moment "appropriately chosen". A CSQ model has $q + q(q-1)/2$ terms (Section 3.3, page 23), while a *restricted* Scheffé quadratic (RSQ) model has *at most* $(q-1) + q(q-1)/2$ unique terms, the difference arising from at least one restriction on the CSQ crossproduct terms. The PQM model Eq. 10.12 *appears* to contain $2q + q(q-1)/2$ terms. For the PQM model to be equivalent to the RSQ model, then, requires the elimination of at least $q + 1$ terms from Eq. 10.12. Algebraically,

$$\left[2q + \frac{q(q-1)}{2}\right] - \left[(q-1) + \frac{q(q-1)}{2}\right] = q + 1$$

As for "appropriately chosen", consider the equality

$$\begin{aligned} X_k^2 &= X_k \cdot (1 - X_1 - \cdots - X_{k-1} - X_{k+1} - \cdots X_q) \\ &= X_k - \sum_{\substack{i=1 \\ i \neq k}}^{q} X_i X_k \end{aligned} \tag{10.13}$$

Including X_k^2 as a regressor along with X_k and all quadratic crossproduct terms in X_k will lead to an exact dependency among the regressors and an overparameterized model. Thus, "appropriately chosen" means that if $\delta_{kk} \neq 0$ for some k, then at least one δ_{ik} $(i \neq k)$ must equal zero.

An approach suggested by Piepel, Szychowski, and Loeppky [135] for developing these models consists of the following four steps:

1. Fit a Scheffé linear model to serve as a basepoint.

2. Use a variable selection procedure such as stepwise regression to augment the linear model with squared and/or crossproduct terms. The candidate set of regressors would

consist of all squared and quadratic crossproduct terms or a subset of these. If any squared terms are selected, go to step 4; otherwise finish with step 3.

3. Use the resulting reduced quadratic Scheffé model for model interpretation and prediction.

4. Rewrite the PQM model containing one or more squared terms in its equivalent RSQ form. The latter will contain only quadratic crossproduct terms. Use subject-matter knowledge to choose between the PQM and RSQ models and their interpretations.

To illustrate this procedure, assume one followed these steps for a three-component mixture setting and discovered that the fitted model was of the form

$$E(Y) = \sum_{i=1}^{3} \delta_i X_i + \delta_{11} X_1^2 + \delta_{22} X_2^2 \tag{10.14}$$

This is a PQM model, but it is also a RRSQ model. Following the advice in step 4, one would reparameterize the RRSQ model as an RSQ model and compare the latter with the PQM model. The terms in X_i^2 may be reparameterized as follows:

$$\begin{aligned}
\delta_{11} X_1^2 &= \delta_{11} X_1 (1 - X_2 - X_3) \\
&= \delta_{11} X_1 - \delta_{11} X_1 X_2 - \delta_{11} X_1 X_3 \\
\delta_{22} X_2^2 &= \delta_{22} X_2 (1 - X_1 - X_3) \\
&= \delta_{22} X_2 - \delta_{22} X_1 X_2 - \delta_{22} X_2 X_3
\end{aligned}$$

Substituting these expressions into Eq. 10.14 leads to the RSQ model 10.15:

$$\begin{aligned}
E(Y) &= (\delta_1 + \delta_{11}) X_1 + (\delta_2 + \delta_{22}) X_2 + \delta_3 X_3 \\
&\quad - (\delta_{11} + \delta_{22}) X_1 X_2 - \delta_{11} X_1 X_3 - \delta_{22} X_2 X_3 \\
&= \sum_{i=1}^{3} \beta_i X_i + \sum_{i=1}^{2} \sum_{j=2}^{3} \beta_{ij} X_i X_j
\end{aligned} \tag{10.15}$$

where $\beta_i = \delta_i + \delta_{ii}$ for $i = 1, 2$, $\beta_3 = \delta_3$, $\beta_{12} = -\delta_{11} - \delta_{22}$, $\beta_{13} = -\delta_{11}$, and $\beta_{23} = -\delta_{22}$. Model 10.15 appears to be a CSQ model, but it is really a RSQ model because of the restriction $\beta_{12} = \beta_{13} + \beta_{23}$. Comparing models 10.14 and 10.15, one might decide on the basis of subject-matter knowledge that there is no physical reason whatsoever why one should expect β_{12} to be equal to the sum of β_{13} and β_{23}. One would therefore conclude on the basis of the PQM model that components 1 and 2 exhibit quadratic curvature effects independent of any nonlinear blending with each other or with component 3.

PQM models 10.11 and 10.14 are equivalent, respectively, to the restrictions

$$\begin{aligned}
\beta_{12} &= \beta_{13} \\
\beta_{12} &= \beta_{13} + \beta_{23}
\end{aligned}$$

on the parameters in the CSQ model. Fitting reduced Scheffé quadratic models limits the user to restrictions of the form $\beta_{ij} = 0$. Allowing crossproduct terms *as well as* squared terms to augment linear terms opens up a broader class of RSQ models. Furthermore, fitting PQM models does not require restricted least squares, which would be the case if one fit data directly to CSQ models subject to parameter restrictions.

With a bit of algebra, one can show that the general form of the PQM model 10.12, page 229, can be rewritten as an equivalent RSQ model. To do this, the last term in model 10.12 is reparameterized as

$$\begin{aligned}\sum_{i=1}^{q} \delta_{ii} X_i^2 &= \sum_{i=1}^{q} \delta_{ii} X_i \left(1 - \sum_{\substack{j=1 \\ j \neq i}}^{q} X_j\right) \\ &= \sum_{i=1}^{q} \delta_{ii} X_i - \sum_{i=1}^{q} \sum_{\substack{j=1 \\ j \neq i}}^{q} \delta_{ii} X_i X_j\end{aligned}$$

Substitution of the reparameterized squared term into model 10.12 leads to

$$E(Y) = \sum_{i=1}^{q} \beta_i X_i + \sum_{i=1}^{q-1} \sum_{j=i+1}^{q} \beta_{ij} X_i X_j \tag{10.16}$$

where $\beta_i = \delta_i + \delta_{ii}$ and $\beta_{ij} = \delta_{ij} - \delta_{ii} - \delta_{jj}$.

Returning to the first example in this section, where $\beta_{12} = \beta_{13}$, if

$$\begin{aligned}\beta_{12} &= \delta_{12} - \delta_{11} - \delta_{22} \\ \beta_{13} &= \delta_{13} - \delta_{11} - \delta_{33}\end{aligned}$$

then for β_{12} to equal β_{13} requires that

$$\delta_{12} - \delta_{11} - \delta_{22} = \delta_{13} - \delta_{11} - \delta_{33}$$

This will be true if

$$\delta_{12} = \delta_{22} = \delta_{13} = \delta_{33} = 0$$

which can be seen to be the case by consulting model 10.11, page 228.

The second example, where $\beta_{12} = \beta_{13} + \beta_{23}$, requires that

$$\delta_{12} - \delta_{11} - \delta_{22} = (\delta_{13} - \delta_{11} - \delta_{33}) + (\delta_{23} - \delta_{22} - \delta_{33})$$

This will be true if

$$\delta_{12} = \delta_{13} = \delta_{23} = \delta_{33} = 0$$

which can be seen to be the case by consulting model 10.14, page 230.

An example where a PQM model is much preferred over a reduced Scheffé model should prove helpful. Portland cement concrete is a mixture of water, portland cement, fine aggregate, and coarse aggregate. Additional components may be added to the basic mixture to enhance certain performance criteria, such as compressive strength, elastic modulus, or

Table 10.9. *Concrete mixture experiment*

Water	Cement	Silica	HRWRA	Coarse	Fine	RCT
0.1850	0.1474	0.0130	0.0046	0.4000	0.2500	1278
0.1600	0.1500	0.0130	0.0074	0.4098	0.2598	862
0.1850	0.1300	0.0130	0.0046	0.4174	0.2500	1162
0.1600	0.1300	0.0270	0.0046	0.4284	0.2500	387
0.1600	0.1300	0.0130	0.0046	0.4424	0.2500	776
0.1850	0.1300	0.0130	0.0074	0.4073	0.2573	1027
0.1712	0.1500	0.0130	0.0046	0.4112	0.2500	744
0.1850	0.1300	0.0270	0.0074	0.4003	0.2503	492
0.1720	0.1395	0.0130	0.0060	0.4097	0.2598	842
0.1850	0.1300	0.0130	0.0046	0.4000	0.2674	903
0.1600	0.1500	0.0130	0.0060	0.4210	0.2500	583
0.1600	0.1300	0.0130	0.0046	0.4000	0.2924	684
0.1600	0.1300	0.0270	0.0046	0.4000	0.2784	292
0.1600	0.1400	0.0130	0.0046	0.4324	0.2500	604
0.1720	0.1395	0.0130	0.0060	0.4097	0.2598	847
0.1600	0.1400	0.0130	0.0046	0.4000	0.2824	720
0.1656	0.1343	0.0165	0.0067	0.4233	0.2536	554
0.1712	0.1500	0.0130	0.0046	0.4000	0.2612	792
0.1600	0.1300	0.0270	0.0046	0.4284	0.2500	348
0.1850	0.1300	0.0130	0.0046	0.4174	0.2500	968
0.1600	0.1300	0.0130	0.0046	0.4212	0.2712	700
0.1600	0.1300	0.0270	0.0074	0.4128	0.2628	316
0.1767	0.1417	0.0270	0.0046	0.4000	0.2500	390
0.1725	0.1300	0.0270	0.0074	0.4131	0.2500	302
0.1725	0.1300	0.0130	0.0074	0.4000	0.2771	682
0.1600	0.1500	0.0130	0.0060	0.4000	0.2710	505
0.1600	0.1400	0.0270	0.0074	0.4000	0.2656	245
0.1600	0.1400	0.0270	0.0074	0.4156	0.2500	310
0.1720	0.1395	0.0130	0.0060	0.4097	0.2598	636
0.1725	0.1300	0.0270	0.0074	0.4000	0.2631	356
0.1850	0.1300	0.0130	0.0046	0.4000	0.2674	820
0.1600	0.1500	0.0130	0.0074	0.4098	0.2598	553
0.1600	0.1400	0.0200	0.0060	0.4120	0.2620	340
0.1720	0.1395	0.0130	0.0060	0.4097	0.2598	640
0.1656	0.1500	0.0270	0.0074	0.4000	0.2500	239
0.1767	0.1417	0.0270	0.0046	0.4000	0.2500	332

chloride ion permeability. Typically high-performance concrete mixtures contain at least six components.

A six-component mixture experiment was carried out at the Federal Highway Administration (FHWA) and the National Institutes of Standards and Technology [140]. The concrete mixtures consisted of six components: water, cement, microsilica, high-range water-reducing admixture (HRWRA), coarse aggregate, and fine aggregate. The compositions of the 36 blends are summarized in Table 10.9 in units of volume fraction.[2] The goal

[2]The author is indebted to the senior author, Marcia Simon, for providing additional data beyond that published in [140].

of the experiment was to find the optimum proportions for a concrete mix meeting specified conditions for fresh-concrete slump, compressive strength after 1 day and after 28 days, a target 42-day rapid-chloride test (ASTM C1202), and minimum cost (dollars per cubic meter). The results of the rapid-chloride test (RCT) are included in Table 10.9.

The lower and upper bounds on the volume fractions in Table 10.9 are

$$
\begin{array}{ccccc}
0.160 & \le & \text{Water} & \le & 0.185 \\
0.130 & \le & \text{Cement} & \le & 0.150 \\
0.013 & \le & \text{Silica} & \le & 0.027 \\
0.0046 & \le & \text{HRWRA} & \le & 0.0074 \\
0.400 & \le & \text{Coarse} & \le & 0.4424 \\
0.250 & \le & \text{Fine} & \le & 0.2924
\end{array}
$$

From the discussion in Section 10.2, we know that components with relatively small ranges will have inflated standard errors. Inspection of the lower and upper bounds on the volume fractions suggests that terms containing HRWRA should have inflated standard errors. This is indeed the case. If a CSQ model is fit to ln(RCT) (cf. the following paragraph), the standard errors of terms that contain HRWRA range between 82.2–88.7. All other terms have standard errors in the range 0.13–5.93, at least one order of magnitude less. The numbers are based on the proportions being expressed as pseudocomponent proportions.

The CSQ model for the RCT response contains 21 terms, of which 15 are quadratic crossproduct terms. The Box–Cox procedure (explained in Section 10.4) suggests a log transformation. Summary statistics for the CSQ model fit to ln(RCT) are $R^2 = 0.956$, $R^2_{adj} = 0.898$, $R^2_{pred} = 0.804$, and $s = 0.148$. Despite these reasonably good summary statistics, none of the 15 quadratic terms have p values less than 0.05! The smallest value is for the water–fine aggregate nonlinear blending term, for which $p = 0.1178$. The reason for this situation is due in part to the small range of HRWRA. Because all terms in HRWRA have inflated standard errors, it is perhaps not surprising that at least five of the 15 crossproduct terms (i.e., those involving HRWRA) are *apparently* not statistically significant.

We can apply backward elimination to the CSQ model, but caution is in order. If terms in HRWRA are eliminated early in the process, then it would be wise not to use this procedure because of the likelihood of making a Type II error. Surprisingly, it turns out that *no terms* in HRWRA are removed from the model and that only those terms with the smaller standard errors are eliminated — a counterintuitive outcome. The reason for this may be that all quadratic crossproduct terms other than those containing HRWRA are, in truth, not statistically significant. Coefficient estimates and p values for the reduced Scheffé model in the pseudos are summarized in columns 2 and 3 of Table 10.10.

As the number of components in a mixture goes up, the likelihood of a component exhibiting the same nonlinear blending behavior with all other components diminishes. The similarity of the nonlinear blending coefficients for all terms containing D suggests that a better interpretation might be that component D is exhibiting a quadratic curvature effect. When the regression analysis was repeated using stepwise regression (with $\alpha_{in} = \alpha_{out} = 0.10$), including as candidate regressors not only all quadratic crossproduct terms but also a term in D^2, the results in Table 10.10 for the PQM model were obtained. The quadratic crossproduct term BC is now in the model, but all of the crossproduct terms involving D have been replaced by a single term in D^2. This model has the best summary statistics

Table 10.10. *Concrete mixture experiment. Scheffé vs. PQM models for ln(RCT)*

Reduced Scheffé Model			PQM Model		
Component	Parameter Estimate	p Value	Component	Parameter Estimate	p Value
A-Water	7.149	n.a.	A	7.197	n.a.
B-Cement	6.493	n.a.	B	6.577	n.a.
C-Silica	4.109	n.a.	C	4.265	n.a.
D-HRWRA	149.7	n.a.	D	−3.481	n.a.
E-Coarse	6.634	n.a.	E	6.612	n.a.
F-Fine	6.484	n.a.	F	6.436	n.a.
AD	−151.7	0.033	BC	−2.031	0.030
BD	−155.6	0.031	D^2	138.3	0.023
CD	−158.3	0.030			
DE	−150.9	0.033			
DF	−154.9	0.031			
R^2	0.923		R^2	0.931	
R^2_{adj}	0.892		R^2_{adj}	0.913	
R^2_{pred}	0.825		R^2_{pred}	0.887	
s	0.152		s	0.136	

of all three models — the CSQ model, the reduced CSQ model, and the PQM model. In addition, the model provides a much more reasonable interpretation of the results than does the reduced CSQ model. Although the analysis was carried out using pseudocomponents, the form of the final model is the same if the analysis is carried out in the reals.

To perform a statistical test of significance of the 13 degree-of-freedom composite null hypothesis on the full CSQ model:

$$\begin{aligned} \mathrm{H}_0: \quad & \beta_{AB} = \beta_{AC} = \beta_{AE} = \beta_{AF} = \beta_{BE} = \beta_{BF} = \beta_{CE} = \beta_{CF} = \beta_{EF} = 0 \\ & \beta_{AD} = \beta_{BD} = \beta_{CD} = \beta_{DE} = \beta_{DF} \end{aligned}$$

one would proceed as follows. The full 21-term CSQ model has SSR = 7.132 with 20 degrees of freedom. The 8-term PQM model has SSR = 6.941 with 7 degrees of freedom. Using the extra sum-of-squares principle, the calculations would be

$$F = \frac{(7.132 - 6.941)/(20 - 7)}{0.02174} = 0.6758$$

The denominator in this expression is the MSE for the fuller (CSQ) model and is based on 15 degrees of freedom. The tabled value for $F_{0.05;13,15} = 2.45$, and since $0.6758 \ll 2.45$, the composite null hypothesis is not rejected. (The p value is 0.7580.)

The paper by Piepel, Szychowski, and Loeppky [135] contains a comprehensive discussion of augmenting linear mixture models with squared and/or crossproduct terms.

In addition, application of PQM models to examples from the field of glass technology are presented.[3]

10.4 Transformation of the Response

In this section we focus on transformations in *linear* regression models. This means that transformations of certain nonlinear regression models to linear models are excluded.

There are three basic reasons for undertaking a transformation in linear regression models. The two cited most often are to remedy inhomogeneity of variance and nonnormality of errors. Less often cited is what has been termed **simplicity of structure** or **additivity of structure** [2]. Polynomial models are additive in the sense that they are a sum of terms, but the word "additivity" is used here in a more restrictive sense. Additivity (or simplicity) of structure implies that a first-order model, with few if any second-order terms, is adequate. The interpretation of models with terms higher than first order is not as straightforward as the interpretation of first-order models. In higher-order terms the effect of one component depends on the level of one or more other components. Additivity of structure in a transformed model is not necessarily always "better" than nonadditivity in an untransformed model. One needs to weigh the advantage of a simple, perhaps first-order, model form against the disadvantage of modeling in a scale that may not be easily interpretable. Examples of simplicity of structure will be provided with some of the examples later in this chapter.

One can envisage transforming the left side of the model (the response), the right side (the regressors), or both sides. Analytical procedures exist for all three [144], but we shall focus on the first. With regard to the right side, certain nonpolynomial model forms *are* known to be effective in some mixtures settings, and the reader is referred to Cornell [29] for details of these models. The models include those with inverse terms of the form X_i^{-1}, those with terms of the form $\ln(X_i)$ and $(\ln(X_i) - \ln(X_j))^2$ (used in log contrast models), and those with terms that are homogeneous of degree one (for modeling additive blending). A term is homogeneous of degree one if multiplying each of the variables that comprise a term by a constant, t, is equivalent to multiplying the term by t. This means that

$$f(tx_1, tx_2, \ldots, tx_k) = tf(x_1, x_2, \ldots, x_k)$$

Examples of terms that are homogeneous of degree one are $(X_iX_j)/(X_i + X_j)$ and $\sqrt{X_iX_j}$. For example,

$$\begin{aligned} \frac{(tX_i)(tX_j)}{tX_i + tX_j} &= \frac{t^2X_iX_j}{t(X_i + X_j)} \\ &= t \cdot \frac{X_iX_j}{X_i + X_j} \end{aligned}$$

It is not uncommon for errors in measurement (of a response, for example) to be proportional to the magnitude of the measured value. This is often the case when the ratio of the maximum to the minimum response is large, say one or more orders of magnitude.

[3]PQM models can be fit by stepwise regression in Design-Expert. To add squared terms to the candidate list of regressors requires using the `Add Term` dialog in the model-specification screen.

Such situations lead to inhomogeneity of variance and nonnormality of the errors. Under these conditions, a transformation of the response may be effective. There is no guarantee that *both* inhomogeneity of variance and nonnormal errors will be corrected by the same transformation. It has been empirically observed, however, that in a surprisingly large number of cases, variance stabilization and approximate normality can be accomplished with the same transformation.

Because of the supernormality property of residuals (cf. page 191), a normal plot is often not the best diagnostic for identifying the need for a transformation. Usually a plot of the studentized residuals vs. the fitted values is the primary diagnostic. In such a plot, the "funnel effect", as illustrated in Fig. 9.2, page 188, is indicative of a dependence of σ on $E(Y)$.

When the "funnel effect" is observed in a residuals plot, it suggests that the standard deviation (σ) is proportional to ($\propto$) some (unknown) power (α) of $E(Y)$. This can be represented algebraically as

$$\sigma \propto \eta^{\alpha}$$

where, for notational simplicity, the symbol $E(Y)$ has been replaced by the Greek letter η. In these circumstances, the transformation $Y' = Y^{(1-\alpha)} = Y^{\lambda}$ will tend to stabilize the variance [12, 13, 49]. Such transformations are called **power transformations**. Some common power transformations are summarized in Table 10.11. In most cases these transformations require nonnegative responses, which means that when some are negative then a fixed positive quantity must be added to all responses.

Table 10.11. *Some common power transformations*

Relationship	α	$\lambda = (1-\alpha)$	Transformation
$\sigma \propto \eta^{0}$	0	1	none
$\sigma \propto \eta^{0.5}$	0.5	0.5	$\sqrt{Y}$
$\sigma \propto \eta^{1.0}$	1.0	0	$\ln(Y)$
$\sigma \propto \eta^{1.5}$	1.5	−0.5	$1/\sqrt{Y}$
$\sigma \propto \eta^{2.0}$	2.0	−1	$1/Y$

Table 10.11 prompts two questions. First, when $\alpha = 1.0$, $\lambda = 0$ and $Y^0 = 1$. It does not make sense to have all of the transformed response values equal to one. Why then is the logarithmic transformation associated with $\lambda = 0$? Second, how does one determine the value of α or, equivalently, λ?

To take care of the discontinuity problem when $\lambda = 0$, Box and Cox [10] considered the parametric family of power transformations

$$Y^{(\lambda)} = \begin{cases} \dfrac{Y^{\lambda} - 1}{\lambda} & \lambda \neq 0 \\ \ln(Y) & \lambda = 0 \end{cases} \qquad (10.17)$$

where $Y^{(\lambda)}$ means that Y is raised to a power that is a function of λ. The reason this solves the discontinuity problem is because, according to l'Hôpital's rule,

$$\lim_{x \to c} \frac{f(x)}{g(x)} = \lim_{x \to c} \frac{f'(x)}{g'(x)}$$

which means that

$$\lim_{\lambda \to 0} \left(\frac{Y^\lambda - 1}{\lambda} \right) = \lim_{\lambda \to 0} \left(\frac{Y^\lambda \ln(Y)}{1} \right)$$
$$= \ln(Y)$$

To solve the second problem — choosing a value for λ — Box and Cox defined a *normalized* power transformation, a slight modification of the power transformation in Eq. 10.17.

$$Y^{(\lambda)} = \begin{cases} \dfrac{Y^\lambda - 1}{\lambda \dot{Y}^{\lambda - 1}} & \lambda \neq 0 \\ \dot{Y} \ln(Y) & \lambda = 0 \end{cases} \tag{10.18}$$

In these equations, $\dot{Y}$ is the geometric mean of the observations, given by

$$\dot{Y} = \left(\prod_{i=1}^{n} Y_i \right)^{1/n}$$
$$\equiv \ln^{-1} \left(\frac{1}{n} \sum_{i=1}^{n} \ln(Y_i) \right)$$

The reason for this is that as λ changes, one cannot compare model summary statistics because the scale for $Y^{(\lambda)}$ in Eq. 10.17 changes. The divisor $\dot{Y}^{\lambda-1}$ (when $\lambda \neq 0$) and the multiplier $\dot{Y}$ (when $\lambda = 0$) are factors that put everything on the same scale so that summary statistics *can* be compared between different values of λ.

The **Box–Cox procedure** uses the method of maximum likelihood to find a value for λ that, *insofar as possible*, satisfies the aims of homogeneity of variance and normality [13]. The procedure involves fitting the *same* model to $Y^{(\lambda)}$, as expressed in Eqs. 10.18, for a grid of λ values. The maximum-likelihood estimate of λ is that value of λ that minimizes the error sum of squares of the fitted model. Typically what is done is to plot SSE or ln(SSE) vs. λ, and to read the value of λ that minimizes SSE from the graph. For the mathematical details underlying this procedure, the reader is referred to books dedicated to regression analysis, such as Draper and Smith [49], Montgomery, Peck, and Vining [100], and Myers [105].

Figure 10.10 displays a Design-Expert Box–Cox plot. The legend to the left of the figure indicates that the minimum in ln(SSE) occurs when $\lambda = -0.21$, identified in the plot by the long vertical dashed line. The points where the solid horizontal line cuts the U-shaped curve (identified by the short vertical dashed lines) delineate the upper and lower 95% confidence intervals for λ. In this example, the interval is $-1.35 \leq \lambda \leq 0.81$. The interval does not include the value $\lambda = 1$ (marked by the solid vertical line in the figure), and so we conclude that a transformation may be helpful. Although the minimum occurs

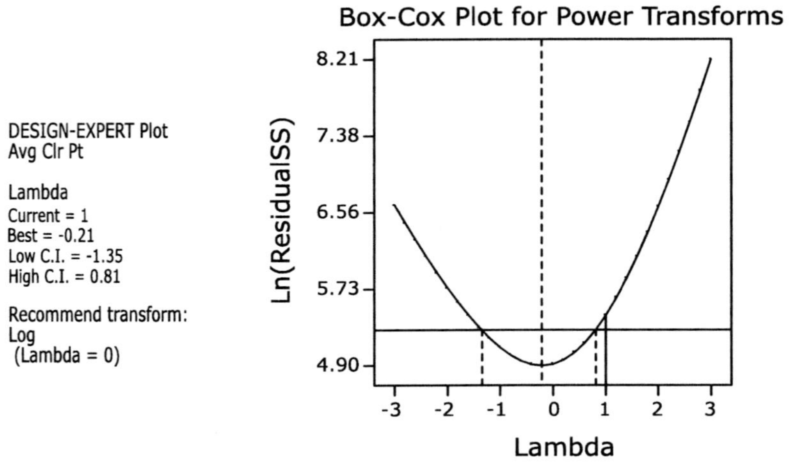

Figure 10.10. *Representative Box–Cox plot.*

at $\lambda = -0.21$, rather than using $Y^{-0.21}$ as the response, for simplicity one usually chooses the nearest half-fraction, such as one of the values in Table 10.11. Thus, in this particular example, Design-Expert recommends a log transformation ($\lambda = 0$), although one cannot rule out a square-root, inverse square-root, or inverse transformation, as these all fall within the 95% confidence region.

An example may be helpful. The data in Table 10.12 are for a light-duty liquid detergent (LDLD) experiment [109]. The authors were interested in formulating an LDLD consisting of the four components water, ethanol, urea, and sodium xylene sulfonate (SXS). The only constraints on the component proportions were lower bounds on water and ethanol of 0.92 and 0.02, respectively. As this is a "lower-bounds-only" problem, the experimental region is shaped like a simplex. The lower bounds on water and ethanol lead, however, to implied constraints, and so the complete set of constraints is

$$\begin{array}{ccccc} 0.92 & \leq & \text{water} & \leq & 0.98 \\ 0.02 & \leq & \text{ethanol} & \leq & 0.08 \\ 0 & \leq & \text{urea} & \leq & 0.06 \\ 0 & \leq & \text{SXS} & \leq & 0.06 \end{array}$$

Responses of interest were the viscosity and what is called the "clear point". The latter is the temperature at which the solution becomes clear after a previous freezing. Aims were a viscosity in the range of 180–200 cP and a clear point of 5°C–10°C.

The Box–Cox plot displayed in Fig. 10.10 is actually a plot for the average clear-point data fitted to the 10-term quadratic Scheffé model. Before rushing into a transformation, however, it is always wise to first have a look at some diagnostic plots. The ratio of the maximum to minimum response in this case is only 7.3, and so it is a bit surprising that a transformation is called for, although one cannot rule out a transformation on this basis alone.

Table 10.12. *LDLD experiment*

ID	Percentage				Avg Clr Pt	Viscosity
	Water	Ethanol	Urea	SXS	°C	cP
1	98	2	0	0	36.5	362.5
2	96	4	0	0	8	199
3	96	2	2	0	5.3	431
4	96	2	0	2	7	341
5	94	6	0	0	7.3	129
6	94	4	2	0	5	265
7	94	4	0	2	11.3	200
8	94	2	4	0	14.2	497
9	94	2	2	2	8.6	492.5
10	94	2	0	4	11.9	332
11	92	8	0	0	11.5	78
12	92	6	2	0	8.3	119
13	92	6	0	2	12.7	98
14	92	4	4	0	8.6	232
15	92	4	2	2	10	235
16	92	4	0	4	14.2	166
17	92	2	6	0	23.7	1630
18	92	2	4	2	10.8	427
19	92	2	2	4	13.6	358
20	92	2	0	6	12.3	365

The plot on the left in Fig. 10.11 suggests that ID $= 1$ is an outlier. This is because the Bonferroni critical value for controlling the experimentwise error rate at ≤ 0.025 (two-tailed), based on a sample size of $n = 20$ and a model with $p = 10$ terms, is equal to 4.15 [105]. In addition, Cook's D for the same point is very much out of line with the other observations. Additional evidence that this point may be an outlier is provided by ignoring this point and generating a second Box–Cox plot for the remaining 19 observations. When one does this, it is found that the resulting 95% confidence interval on λ is $-0.58 \leq \lambda \leq 1.71$. As the interval includes 1.0, no transformation is recommended. Although this does not prove that observation 1 is an outlier, it does suggest that the log transformation recommended for the 20-observation data set is being driven by observation 1. A preferable situation is for *all* of the data points to suggest that a transformation is needed.

A Box–Cox plot for the viscosity data fitted to a quadratic Scheffé model leads to a minimum SSE at $\lambda = -0.8$, with a 95% confidence interval of $-1.51 \leq \lambda \leq -0.2$. Thus, $\lambda = 1$ is not in the interval, and the recommended transformation would be the inverse of the response ($\lambda = -1$).

As with the clear-point data, let us have a look at index plots of the R-student statistic and Cook's D (Fig. 10.12). ID $= 17$ now appears to be an outlier and highly influential. The R-student statistic is even more extreme than was the case with the clear-point data.

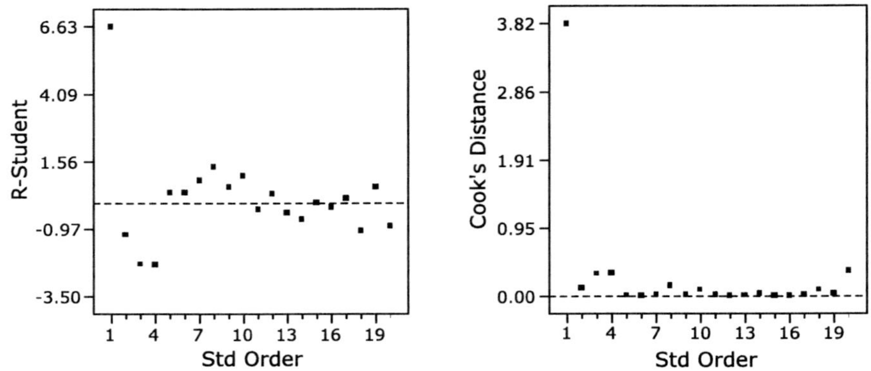

Figure 10.11. *R-student and Cook's D. Quadratic Scheffé model for average clear point.*

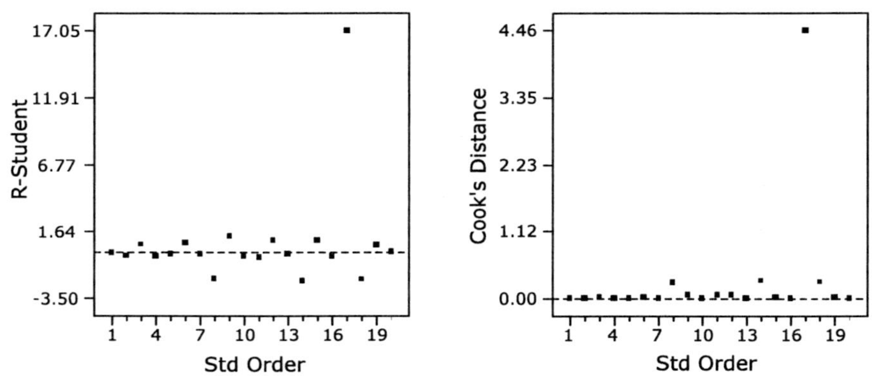

Figure 10.12. *R-student and Cook's D. Quadratic Scheffé model for LDLD viscosity data.*

When observation 17 is set aside and the regression rerun, the outlier problem disappears. The plot of the studentized residuals vs. the fitted values assumes a classic funnel shape (Fig. 10.13), indicating inhomogeneity of variance. The minimum in the Box–Cox plot occurs at $\lambda = 0.11$, with a 95% confidence interval of $-0.87 \leq \lambda \leq 0.94$. This suggests the feasibility of a log transformation, although a square-root or inverse square-root transformation might also be considered.

When the 95% confidence interval on λ includes a choice of transformations, a **lambda plot** is a helpful aid for understanding how a choice of transformation can sometimes lead to simplicity of structure (page 235) [9, 60]. A lambda plot is a plot of the t values for the parameter estimates (y-axis) vs. the λ values (x-axis). Figure 10.14 shows a lambda plot for the LDLD viscosity data fitted to a quadratic Scheffé model.

The curved lines in Fig. 10.14 trace the t values for the six quadratic terms in the quadratic Scheffé model as the value of λ changes. The two horizontal dashed lines at $t = \pm 2.26$ delineate the critical values for t at the 95% level based on $n - p = 19 - 10 = 9$

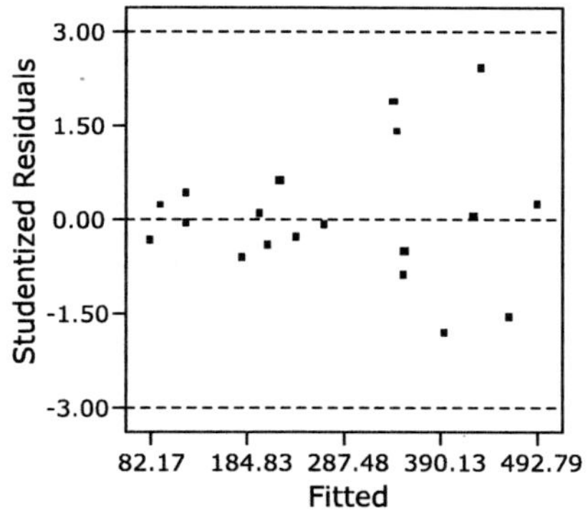

Figure 10.13. *LDLD viscosity response. Homogeneous variance assumption violated.*

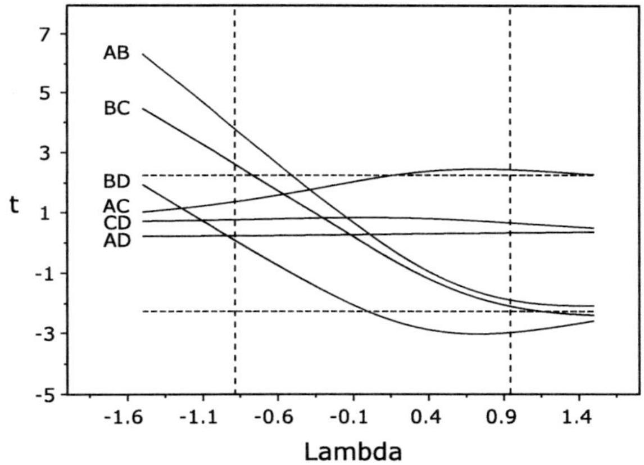

Figure 10.14. *LDLD viscosity response. Lambda plot.*

degrees of freedom. When a trace falls either above the upper or below the lower critical value, then the p value for that term in the quadratic Scheffé model is ≤ 0.05. If a trace falls between the critical values, then that term is not significant. The two vertical dashed lines delineate the 95% confidence interval for λ based on the Box–Cox plot.

For simplicity of structure, one would look for λ values within the 95% confidence region for λ where all the traces (or as many as possible) fall between the critical values. In Fig. 10.14, this happens to be the case for the approximate interval $-0.5 \leq \lambda \leq -0.1$ but not for $\lambda = 0$, the maximum likelihood estimate of λ based on the Box–Cox plot. One might therefore consider the possibility of choosing a λ value in the more restricted interval between -0.5 and -0.1.

One should keep in mind that the t values in the lambda plot pertain to the full 10-term quadratic model. There is always the possibility that on variable selection, a term that was not significant in the full model may become significant in the reduced model.

Columns 2 and 3 in Table 10.13 tabulate the number of significant quadratic terms in full and reduced models for several λ values. The number of terms in the full model can be inferred from the lambda plot. The number of terms in the reduced models is obtained by fitting reduced models to the transformed response using backward elimination. When $\lambda = -0.2$ or -0.333, the reduced models have *simplicity of structure* in the sense used by Atkinson [2], which in this case means a first-degree model.

Table 10.13. *LDLD viscosity response. Effect of* λ *on model forms*

λ	Quadratic terms with $\lvert t\rvert \geq 2.26$ Full	Reduced	R^2	$r^2_{Y\hat{Y}_{raw}}$
1	2	3	0.9625	0.9625
0.5	2	3	.9783	.9700
0.0	1	2	.9832	.9694
−0.1	0	2	.9836	.9691
−0.2	0	0	.9721	.9520
−0.333	0	0	.9732	.9505
−0.5	0	1	.9789	.9460

The question arises, "How well do the models based on the transformed responses fit the data?" R^2 values in Table 10.13 apply to the fitted model *in the transformed scale* (except for $\lambda = 1$, which corresponds to the untransformed response). One might consider calculating fitted values in the original scale, $\hat{Y}_{raw}$, from fitted values in the transformed scale, $\hat{Y}'$, and using the results to calculate an R^2 value in the original scale. Unfortunately, this strategy will not work. This is because only in the scale used to fit the model will the equality SST = SSR + SSE hold. For example, when $Y' = Y^{-1/3}$, the back-calculated values in the original scale are SST = 316354, SSR = 314258, and SSE = 15809, and SST < SSR + SSE.

In OLS, the coefficient of determination, R^2, is equal to the squared correlation coefficient between Y and $\hat{Y}$, $r^2_{Y\hat{Y}}$, where both Y and $\hat{Y}$ are in the same metric. A procedure suggested by Ryan [144] for comparing the observed vs. the fitted values in the original scale, when the fitted values in the original scale are calculated from those in the transformed scale, is to calculate the squared correlation coefficient between Y and $\hat{Y}_{raw}$, $r^2_{Y\hat{Y}_{raw}}$. $\hat{Y}_{raw}$ denotes the fitted values transformed back to the original scale. Column 5 in Table 10.13 tabulates $r^2_{Y\hat{Y}_{raw}}$ values for each λ value.

This is not a perfect solution, however. Assume, for example, that there are two values of $\hat{Y}'$ that are the same except for sign — say ± 2 — and that the λ value is 0.5. Values for Y are calculated as $Y = (Y')^2$, which will be 4 in both cases. Ryan recommends that a simple (but not very satisfying) way to avoid this problem is not to use data points in the calculation of $r^2_{Y\hat{Y}_{raw}}$ for which Y' values are negative [144].

In light of Fig. 10.13 (page 241) and Table 10.13, one might choose either $\lambda = 0$ (close to the maximum-likelihood value for λ) or $\lambda = -0.333$ (for simple structure).

The two models (in terms of pseudocomponents) are

$$\widehat{\ln Y} = 5.85A + 4.30B + 6.09C + 5.88D + 0.965AC - 0.883BD$$

$$\widehat{Y^{-1/3}} = 0.136A + 0.237B + 0.121C + 0.146D$$

where A = water, B = ethanol, C = urea, and D = SXS. In either case, the residuals display homogeneity of variance (Fig. 10.15).

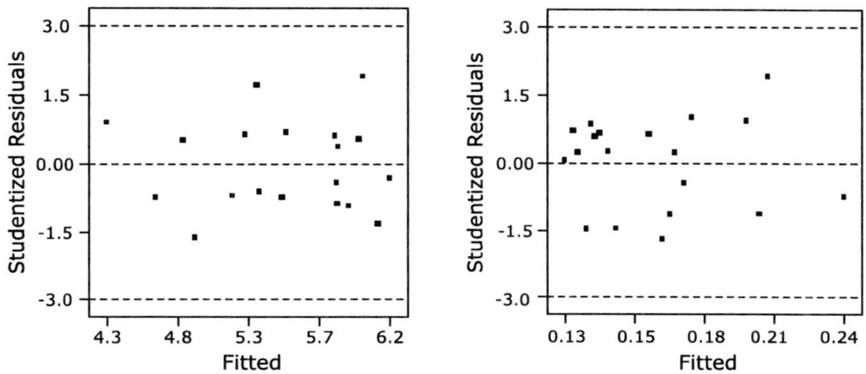

Figure 10.15. *Studentized residuals for* ln(y) *(left) and* $y^{-1/3}$ *(right). Reduced models, LDLD experiment.*

Design-Expert and JMP can determine the likelihood estimate of λ for regression applications. On request, JMP will insert into the data table $Y^{(\lambda)}$ based on Eqs. 10.18, page 237. Once a value of λ is selected, one is of course free to fit the model using either $Y^{(\lambda)}$ (Y raised to a power that is a function of λ) or Y^{λ} (Y raised to the λ power), the latter being simpler. In either case, if the selected value for λ is zero, one should use $\ln(Y)$ or $\log(Y)$ as the response.

MINITAB does not have the Box–Cox procedure for regression applications built into the software (although it does provide a Box–Cox power-transformation procedure for control chart data). At the time of this writing a MINITAB macro called `BCtrans` was available at

http://www.minitab.com/support/macros/index.asp

The MINITAB macro is written for models that include an intercept. To use it for a mixture model, one of the linear terms should be dropped. The macro will automatically add an intercept.

To the author's knowledge, lambda plots are currently not offered by any of the popular computing packages. The lambda plots included in this chapter were constructed by generating the data in GAUSS, importing the results into S-PLUS, and using S-PLUS's graphics capabilities to make the plot.

Responses can sometimes occur as proportions. For example, the yield in a chemical reaction, although usually expressed as "percent yield", could also be expressed as a proportion. Yields are bounded by 0% and 100%, but any response that is bounded by a lower and upper limit can be reexpressed as a proportion (P) using the expression

$$P = \frac{Y - \text{lower limit}}{\text{upper limit} - \text{lower limit}}$$

Another example where responses occur in proportions is *grouped binomial data*. A binomial response is a **dichotomous variable**, meaning that it can fall into only one of two categories. Responses that can have only two experimental outcomes are sometimes called **quantal responses**. Examples of binomial responses are "success" or "failure" and "dead" or "alive". A trial or experiment with only two possible outcomes is sometimes called a **Bernoulli trial**. When testing chemically induced tumors in rats, for example, often the same treatment is applied to a group of rats. An individual rat is an example of a Bernoulli trial, as the response for the rat will be either "tumor" or "no tumor". In terms of the *group* of rats, the response is the proportion of rats in the group that develop a tumor (or the proportion of rats in the group that do not develop a tumor). By grouping the binomial data into batches, we change the response from a dichotomous variable to a proportion. Thus responses that are proportions can be simply bounded in origin (as in the chemical yield example) or binomial in origin (as in the rat example).

Both of these examples could be cast into a mixture setting. For instance, the yield in a chemical reaction could be measured as a function of a mixture of two catalysts. The response would be the yield (expressed as a proportion), which would be measured as a function of blends of the two catalysts. In a similar manner, suppose one suspected that two chemicals could have a synergistic effect on chemically induced tumors. Blends of the two chemicals could then be used to study the probability of occurrence of tumors in rats.

Responses that are proportions tend to exhibit S-shaped response curves such as those exhibited in Fig. 10.16. The illustration on the left in Fig. 10.16 is for the case where there is a single explanatory variable, X, which need not be a mixture variable. The curve on the right is for a three-component mixture setting. The S-shaped character of the raw

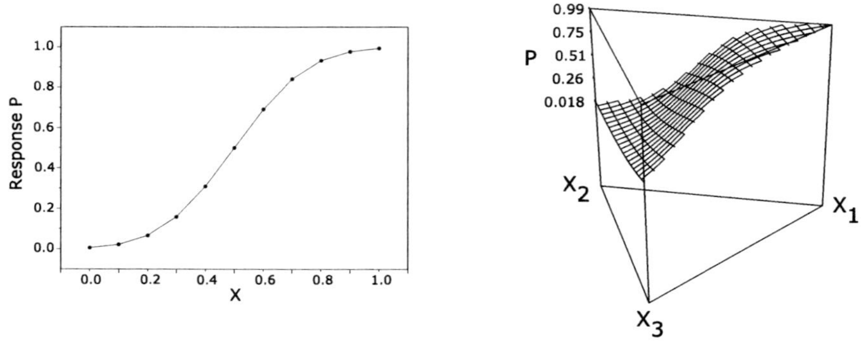

Figure 10.16. *Proportions as responses.*

data is evident especially in the tails, as in both illustrations the proportions asymptotically approach 0 on the left and 1 on the right.

Two popular transformations for proportional-type responses are the **logit** and the **arcsine square-root** transformations. Letting P_i be the ith observed response, with $0 \leq P_i \leq 1$, then the logit transformation is

$$\text{logit}(P) = \ln\left(\frac{P_i}{1 - P_i}\right) \tag{10.19}$$

and the arcsine square-root transformation is

$$\arcsin(P) = \sin^{-1}\sqrt{P_i} \tag{10.20}$$

where $\arcsin(P)$ is expressed in radians.[4] These two transformations have the property that they will stretch the upper and lower tails in S-shaped curves, making the relationship more linear over a broader domain of the explanatory variable(s). When these transformations are applied to the data in the left illustration in Fig. 10.16, the result is the "stretched" curves in Fig. 10.17. In neither case has exact linearity been achieved. This is because the curve in Fig. 10.16 (left) was generated using the cumulative normal distribution function. Applying the inverse of this function, called the **probit** ($\boldsymbol{\phi}^{-1}$ in Eq. 9.10, page 189), will in fact lead to exact linearity with X.

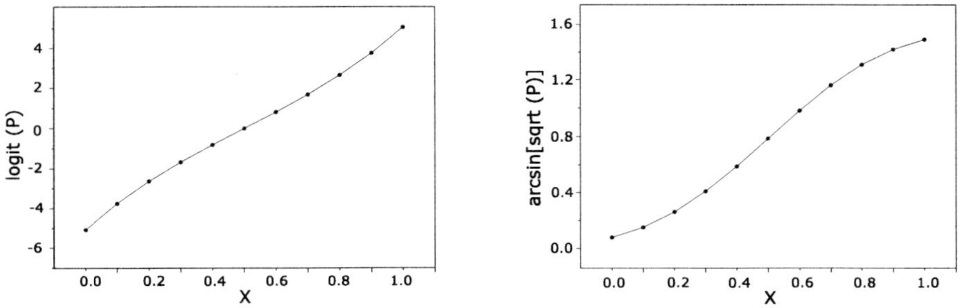

Figure 10.17. *Logit (left) and arcsine square-root (right) transformations of the curve in Fig.* 10.16 *(left).*

When the data are binomial in origin, then the ratio $P_i/(1 - P_i)$ is called the **odds** or **odds ratio**. The odds ratio has no fixed maximum value, but like the probability, it has a minimum value of zero. The **log odds**, or logit, becomes increasingly negative as the odds decrease from 1 to 0 and becomes increasingly positive as the odds increase from 1 to infinity. The arcsine square-root, on the other hand, varies between 0 and $\pi/2$ radians as P varies from 0 to 1.

[4] When $P = 0$ the proportion is sometimes computed as $1/4n$ and when $P = 1$, as $(n - 1/4)/n$ before applying the transformation. This has the effect of improving the homogeneity of variances in the angle metric.

The relationship between the logit and *logistic regression* can be seen as follows. Let $\xi(\mathbf{x}_i)$ be shorthand representation for a linear (as opposed to nonlinear) model for the ith observation. For example, if the linear model were a $q = 2$ quadratic Scheffé model, then

$$\xi(\mathbf{x}_i) = \beta_1 X_{1(i)} + \beta_2 X_{2(i)} + \beta_{12} X_{1(i)} X_{2(i)}$$

where $i = 1, 2, \ldots, n$ and n is the number of observations. The following algebra shows the relationship between the logit normal model (Eq. 10.21) and the **logistic regression model** (Eq. 10.22), sometimes called the logistic growth model:

$$\ln\left(\frac{P_i}{1 - P_i}\right) = \xi(\mathbf{x}_i) \tag{10.21}$$

$$\frac{P_i}{1 - P_i} = \exp^{\xi(\mathbf{x}_i)}$$

$$P_i = \exp^{\xi(\mathbf{x}_i)} - \exp^{\xi(\mathbf{x}_i)} P_i$$

$$P_i + \exp^{\xi(\mathbf{x}_i)} P_i = \exp^{\xi(\mathbf{x}_i)}$$

$$P_i = \frac{\exp^{\xi(\mathbf{x}_i)}}{1 + \exp^{\xi(\mathbf{x}_i)}}$$

$$= \frac{1}{1 + \exp^{-\xi(\mathbf{x}_i)}} \tag{10.22}$$

The logistic regression model (Eq. 10.22) is a nonlinear model, and the method used to fit data to this model is maximum likelihood. This topic is beyond the scope of this book, and the interested reader is referred to specialized texts on the subject, such as Hosmer and Lemeshow [72] or Myers, Montgomery, and Vining [108]. The response surface in Fig. 10.16 (right) was generated using Eq. 10.22 with

$$\xi(\mathbf{x}_i) = 5X_1 + 4X_2 - X_3$$

Figure 10.18 shows how logit(P) varies with mixture composition. The surface is now linear because it is described by Eq. 10.21.

When the proportions are binomial in origin, it can be shown [49] that the variance of P_i is given by

$$\text{var}(P_i) = \frac{1}{m_i}\pi_i(1 - \pi_i)$$

where m_i is the number of Bernoulli trials for the ith proportion and $\pi_i = E(P_i)$. The maximum value of var(P_i) will occur when $\pi_i = 0.5$. This means that if one were to fit a linear model to P_i as the response, then nonconstant error variance should be evidenced by midscale "swelling" in a plot of the residuals vs. the fitted values. To the author's knowledge, this has never been observed in a mixture setting, probably because there are so few examples of binomial responses in the published mixture literature. Nonetheless, if

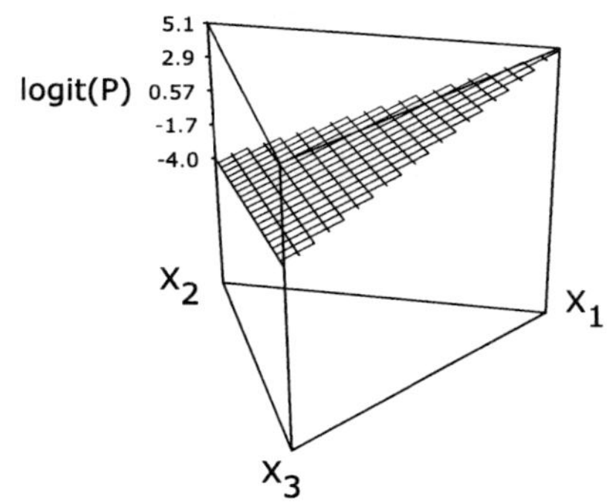

Figure 10.18. *Logit transformation of data used for Fig.* 10.16 *(right).*

this were the case, then one approach would be to use weighted least squares when fitting the logit model, the weights being given by (see Myers [105])

$$\frac{1}{m_i \hat{P}_i(1 - \hat{P}_i)}$$

An alternative approach is to use the arcsine square-root transformation, as it is claimed that this transformation will stabilize the variance if the data are binomial in origin [23, 49, 143]. When the number of Bernoulli trials, m_i, is not constant throughout the data, the transformation should be replaced by $2m^{1/2} \sin^{-1} \sqrt{P_i}$ [49].

What is often done in practice is to simply fit the logit normal model after checking that the residuals are approximately normally distributed and the variance is reasonably constant. A small example from the literature may be helpful. Chen, Li, and Jackson [17] studied the effects of dietary fat, carbohydrate, and fiber on the proportion of rats out of groups of 30 exhibiting drug-induced mammary gland tumors under isocaloric consumption. The drug used was 7,12-dimethylbenz(a)anthracene (DMBA), a common substance in these types of studies. Data are summarized in Table 10.14.

Table 10.14. *DMBA-induced mammary gland tumors experiment*

	Proportion			P_i	$\widehat{P}_i$ based on		
ID	Fat	Carb	Fiber	(observed)	Normal	Logit	Logistic
1	0.175	0.775	0.050	0.567	0.582	0.593	0.585
2	0.153	0.820	0.027	0.500	0.496	0.483	0.488
3	0.133	0.863	0.004	0.567	0.545	0.542	0.549
4	0.491	0.470	0.039	0.800	0.740	0.738	0.733
5	0.440	0.538	0.022	0.700	0.740	0.745	0.744
6	0.390	0.607	0.003	0.767	0.815	0.814	0.813
7	0.701	0.267	0.032	0.600	0.652	0.660	0.645
8	0.638	0.343	0.019	0.767	0.724	0.740	0.729
9	0.576	0.421	0.003	0.867	0.841	0.841	0.833

Letting $\xi(\mathbf{x}_i)$ represent the terms in a $q = 3$ quadratic Scheffé model for the ith observation,

$$\xi(\mathbf{x}_i) = \beta_1 X_{1(i)} + \beta_2 X_{2(i)} + \beta_3 X_{3(i)} \\ + \beta_{12} X_{1(i)} X_{2(i)} + \beta_{13} X_{1(i)} X_{3(i)} + \beta_{23} X_{2(i)} X_{3(i)} \quad (10.23)$$

the authors fit the P_i data to model 10.22 using maximum likelihood, and the logit data to model 10.21 using OLS. The coefficients β_i and β_{ij} differed slightly in the two models, but the differences were negligible.

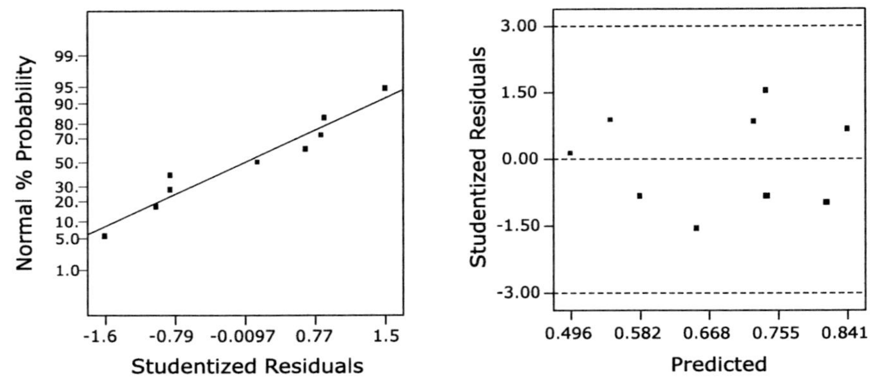

Figure 10.19. *Diagnostic plots for a quadratic Scheffé model fit to P_i data for DMBA-induced tumors.*

When one is dealing with the midrange of proportions (as in the DMBA-induced tumors example), or when proportions are distributed over a relatively narrow range, the data are likely to be normally distributed with constant variance. In these circumstances, not only is a transformation unnecessary, but a transformation will have little effect. Figure 10.19 displays a normal probability plot of the studentized residuals and a plot of the studentized residuals vs. the fitted values, $\hat{P}_i$, when the Scheffé polynomial Eq. 10.23 is fit *directly* to the observed P_i data in Table 10.14 (no transformation). Although there are too few points to accurately assess normality, the residuals do appear to be reasonably well behaved. In this example, nothing is gained by fitting the logit normal or logistic regression models in preference to a normal error regression model. Predicted probabilities based on the three regression models

$$P_i = \xi(\mathbf{x}_i) \qquad \text{normal error}$$

$$\ln\left(\frac{P_i}{1 - P_i}\right) = \xi(\mathbf{x}_i) \qquad \text{logit normal}$$

$$P_i = \frac{\exp^{\xi(\mathbf{x}_i)}}{1 + \exp^{\xi(\mathbf{x}_i)}} \qquad \text{logistic}$$

are summarized in the table. This data set will be revisited in much more depth in Chapter 14.

In another example, Shelton [151] studied the efficacy of blends of two herbicides on the control of the weed species yellow foxtail. In addition to the proportions of the two herbicides, the *amount* of the herbicide blend applied to a weed plot was also a factor. Each treatment was defined therefore by a herbicide blend and an amount. Each of 18 different treatments was applied to plots of yellow foxtail. There were three blocks of 18 plots, for a total of 54 plots. This is a **mixture-amount experiment**, a topic discussed in Chapter 13.

The response in this experiment was the percent weed control in a plot, and this varied between 0% and 100%. In this case it was important to transform the data, because if one fits a mixture-amount model to the untransformed P_i data, some of the fitted values are negative and some exceed 1.0. The authors chose to fit the logit of P_i to a mixture-amount model.

As a final example, Claringbold [18] studied the percentage response of groups of 12 mice to the joint intravaginal administration of estrone, estradiol, and estriol. The quantal response was cornification of the vaginal epithelium. In addition to the proportions of the three hormones, there were also three dosages of the estrogen blends, and so this is also a mixture-amount experiment. The data have been analyzed by Piepel and Cornell [130], who chose to fit the arcsine square-root transformed response to a mixture-amount model.

Case Study

For the case study we return to the adhesive viscosity data (Table 8.1, page 156). Using some of the methods described in this chapter, we shall see how one might approach simplification of the quadratic Scheffé model. As several models will be fit to the data, the reader may find it helpful to occasionally refer to Table 10.15, page 256, where some summary statistics are collected for each model. For notational simplicity, let A = HDA*, B = PN-110*, C = PH-56*, and the superscript asterisk designate pseudocomponents.

The complete quadratic Scheffé model for the adhesive viscosity data is (Table 8.6, page 171)

$$\widehat{\text{visc}} = 6.439A + 47.04B + 7.079C - 32.87AB + 6.128AC - 36.35BC \qquad (10.24)$$

This is model a in Table 10.15. Summary statistics were $R^2 = 0.9830$, $R^2_{adj} = 0.9659$, and $R^2_{pred} = 0.9284$. Referring to Table 8.3, page 166, the p value for the AC nonlinear blending term in this model is equal to 0.6254, indicating that the term is not significant. Removing this term and refitting the simplified model leads to

$$\widehat{\text{visc}} = 6.766A + 47.01B + 7.406C - 33.06AB - 36.12BC \qquad (10.25)$$

This is model b in Table 10.15. Summary statistics were $R^2 = 0.9820$, $R^2_{adj} = 0.9701$, and $R^2_{pred} = 0.9328$.

A Box–Cox plot for the *unreduced* quadratic model is displayed in Fig. 10.20. The minimum in the curve occurs at $\lambda = 0.44$, with a 95% confidence interval of $-0.39 \leq \lambda \leq 1.12$. As $\lambda = 1$ falls within this interval, a transformation is not suggested. However, the interval is relatively wide, and it would be interesting to examine a lambda plot.

The lambda plot in Fig. 10.21 indicates that as one moves toward negative values of λ, the significance of the two quadratic terms AB and BC diminishes. Below $\lambda \approx 0.7$, both terms are no longer significant, and simple structure results. Because $\lambda = 0$ is in the

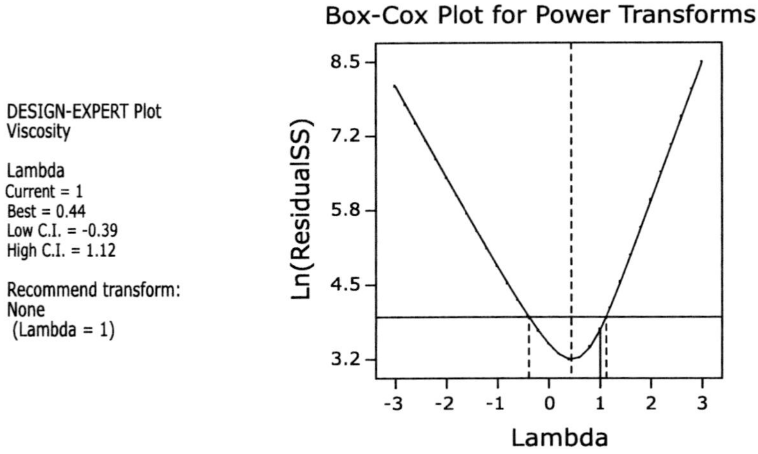

Figure 10.20. *Adhesive viscosity response. Box–Cox plot.*

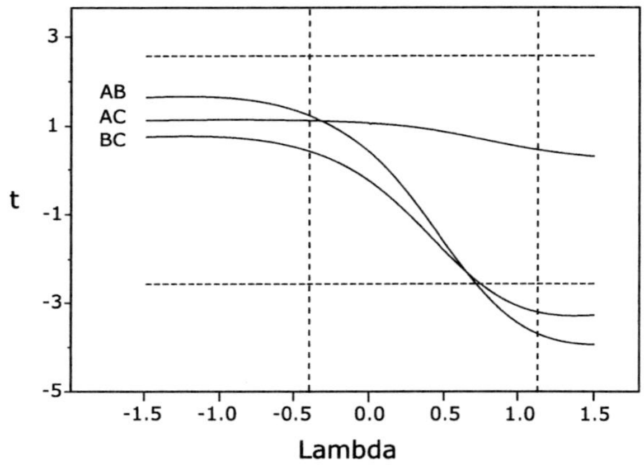

Figure 10.21. *Adhesive viscosity response. Lambda plot.*

interval, it would be of interest to investigate what might happen if a linear Scheffé model were fit to the natural log of the viscosity data.

The linear model fit to the log viscosity data (model c, Table 10.15) is

$$\widehat{\ln(\text{visc})} = 1.893A + 3.858B + 1.990C \tag{10.26}$$

In the log scale, the values for R^2, R^2_{adj}, and R^2_{pred} are, respectively, 0.9633, 0.9541, 0.9175, and so the model fits the data well. The coefficients for A and C are quite similar, and there is substantial overlap of their 95% confidence intervals:

$$1.664 \leq \beta_A \leq 2.122$$

$$1.754 \leq \beta_C \leq 2.227$$

In the contour plot for model 10.26 (Fig. 10.22), the contour lines are almost parallel to the AC subsimplex. This means that once a level of B is selected, one can vary the relative amounts of A and C with virtually no effect on the response. This suggests that the effects of A and C are similar to one another, an observation that is reinforced by the trace plot in Fig. 10.22. The effects directions in the trace plot are the Piepel-effect directions (cf. Fig. 10.5, page 210).

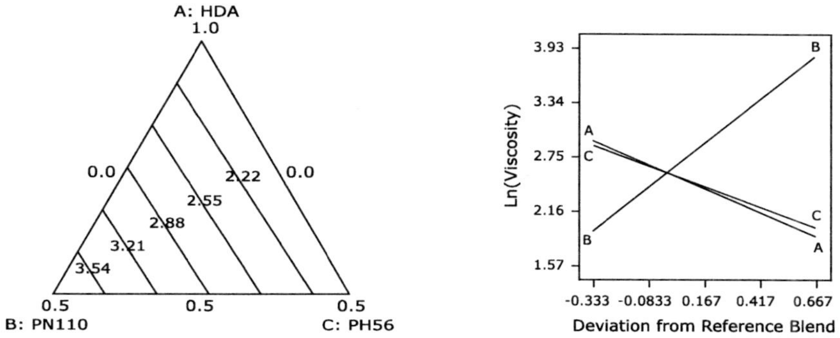

Figure 10.22. *Contour and trace plots for the adhesive ln(viscosity) model.*

In light of this, one could consider further simplification of the model by combining the proportions of A and C. A formal statistical test of this procedure is made using the extra sum-of-squares principle (Eq. 8.10, page 165). The fuller model is the $q = 3$ linear Scheffé model, and the less full or **restricted model** is the $q = 2$ linear model. The calculations are

$$\begin{aligned} F &= \frac{(\mathrm{SSR}_{\textit{fuller model}} - \mathrm{SSR}_{\textit{restricted model}})/\Delta df}{s^2} \\ &= \frac{(5.5995 - 5.5883)/1}{0.026684} \\ &= 0.4197 \end{aligned}$$

The tabled F value for $F_{0.05;1,8}$ is 5.32, and since $0.4197 \ll 5.32$, we do not reject the null hypothesis $\mathrm{H}_0 : \beta_A = \beta_C$, inferring the adequacy of the restricted model. (In terms of p values, $p = 0.5352$, and $0.5352 \gg 0.0500$.) The simplified model is

$$\ln(\widehat{\text{visc}}) = 1.940(A + C) + 3.854B \tag{10.27}$$

which is model d in Table 10.15. *In the log scale*, the values for R^2, R^2_{adj}, and R^2_{pred} are, respectively, 0.9614, 0.9571, and 0.9413.

In the case of model 10.27, the largest $|t_i|$ is that for observation 2, for which $t_2 = -4.78$. The next largest $|t_i|$ is that for observation 5, for which $t_5 = 1.59$. The Bonferroni critical value for the R-student statistic ($\alpha = 0.05$, $n = 11$ observations, and $p = 2$ parameters) is 3.90 [105]. As $|t_2| > 3.90$, observation 2 appears to be an outlier. To reduce the influence of observation 2, a robust regression was carried out using the Huber influence function with $r = 1.345$ (Eq. 10.2, page 212), $k = 0$ (Eq. 10.6, page 216), and a tolerance of $1e - 3$. Four iterations led to the model

$$\ln(\widehat{\text{visc}}) = 1.981(A + C) + 3.840B \tag{10.28}$$

which is model e in Table 10.15. The weights of observations 2 and 5 were reduced to 0.312 and 0.744, respectively, the remaining nine observations having weights of 1.0.

Model 10.28 is expressed in terms of pseudocomponent proportions. The equivalent model in the reals is (cf. Section 4.5)

$$\ln(\widehat{\text{visc}}) = 1.981(\text{HDA} + \text{PH-56}) + 5.699(\text{PN-110}) \tag{10.29}$$

As this is a "two-component" system, we need only specify the level of one component to define a mixture. It is convenient, then, to reparameterize Eq. 10.29 as an intercept mixture model (model 10.30).

$$\ln(\widehat{\text{visc}}) = 1.981 + 3.718(\text{PN-110}). \tag{10.30}$$

The response surface is plotted in Fig. 10.23. Because the original constraints on PN-110 were $0 \le \text{PN-110} \le 0.5$, the scale of the x-axis runs from 0.0 to 0.5. Considering that we started with a model (model a) with $r^2_{Y\hat{Y}raw} = 0.9830$ and ended with a model (model e) with $r^2_{Y\hat{Y}raw} = 0.9815$, very little information has been lost in the process of model simplification.

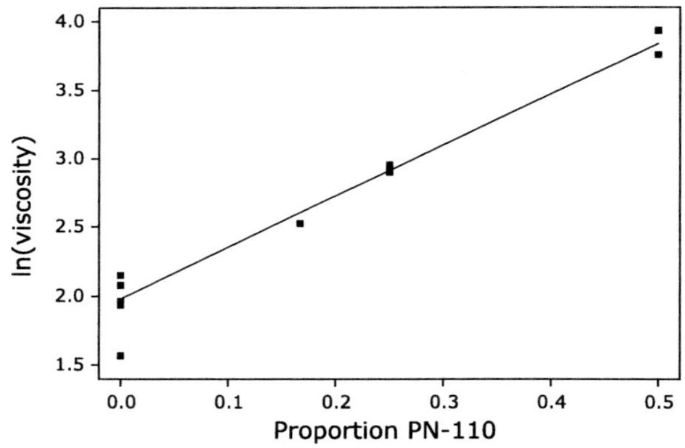

Figure 10.23. *Adhesive ln(viscosity) response surface.*

A somewhat different approach will now be used to arrive at a model and a plot for viscosity rather than ln(viscosity). Returning to the reduced quadratic Scheffé model (Eq. 10.25, page 249), we note that the coefficient estimates for AB and BC are rather similar to one another. This suggests that either (1) pseudocomponent B exhibits nearly the same nonlinear blending behavior with pseudocomponents A and C, or (2) pseudocomponent B exhibits a quadratic curvature effect independent of any nonlinear blending with components

A and C. The same possibilities arise when the model is expressed in terms of the reals (Eq. 10.31).

$$\widehat{\text{visc}} = 6.766(\text{HDA}) + 153.4(\text{PN-110}) + 8.046(\text{PH-56}) - 132.2(\text{HDA})(\text{PN-110}) - 144.5(\text{PN-110})(\text{PH-56}) \quad (10.31)$$

While possibility (1) cannot be ruled out, it seems somewhat fortuitous that PN-110 should exhibit the *same* nonlinear blending behavior towards components of different chemical types (cf. page 154). To explore the second possibility, the regressor B^2 was included with the set of three nonlinear blending terms AB, AC, and BC, and a quadratic model fit using stepwise regression. The model that resulted from this approach was the PQM model

$$\widehat{\text{visc}} = 6.874A + 12.769B + 7.291C + 34.248B^2 \quad (10.32)$$

which is model f in Table 10.15. Summary statistics were $R^2 = 0.9819$, $R^2_{adj} = 0.9741$, and $R^2_{pred} = 0.9340$.

Figure 10.24 displays a three-dimensional response surface for model 10.32. This surface is virtually identical to the response surface for model 10.25, page 249. In terms of fitting or prediction, then, it makes little difference which model is used. It is the *interpretation* of the two surfaces that differs. In the case of model 10.25, the estimated response at the B vertex is equal to the linear coefficient for component B (47.04). In the case of model 10.32, the estimated response at the B vertex is divided into two pieces: that due to the linear coefficient for B (12.769) plus that due to the quadratic coefficient for B^2 (34.248). The dashed lines in Fig. 10.24 outline the linear surface defined by the first three terms in Eq. 10.32.

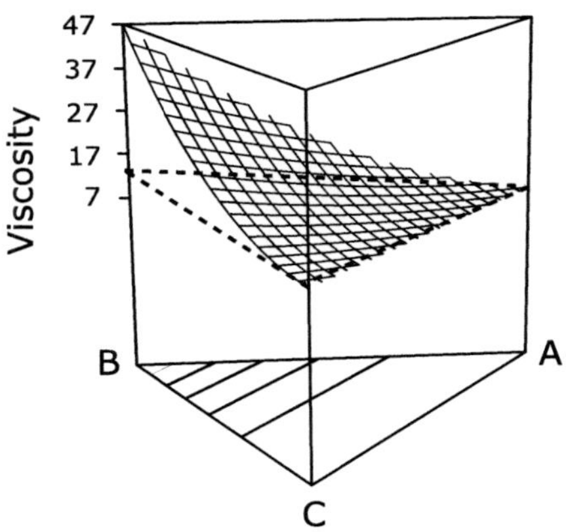

Figure 10.24. *Adhesive viscosity response surface.*

The interpretation of model 10.32 is then that the curvature in the response surface is due to the quadratic effect of pseudocomponent B (or equivalently, the component PN-110) in and of itself and has nothing to do with its interaction with pseudocomponents A or C. Whether the curvature is due to the quadratic effect of pseudocomponent B or approximately equivalent, nonlinear blending of B with A and C is a decision to be made by the experimenter based on his or her subject-matter knowledge. Again, although models 10.25 and 10.32 have different interpretations, they lead to virtually identical predictions.

Note in model 10.32 that the estimated coefficients for A and C are nearly equal to one another. The same can be said about the corresponding coefficients when the model is expressed in the reals:

$$\widehat{\text{visc}} = 6.874(\text{HDA}) + 28.66(\text{PN-110}) + 7.707(\text{PH-56}) + 137.0(\text{PN-110})^2 \qquad (10.33)$$

The contour plot in Fig. 10.25 shows that the contours are again nearly parallel to the AC subsimplex. In this case, however, and in contrast to Fig. 10.22 (where the response is the natural log of the viscosity rather than the viscosity), the contour *spacing* changes although the *increments* between each contour remain the same. This is simply a reflection of the quadratic effect of B. The trace plots in this figure show that the effects of A and C virtually overlay each other. The reason they are curved and not linear is because in the Piepel-effect directions, the traces cross the contour levels *that are determined by B alone.*

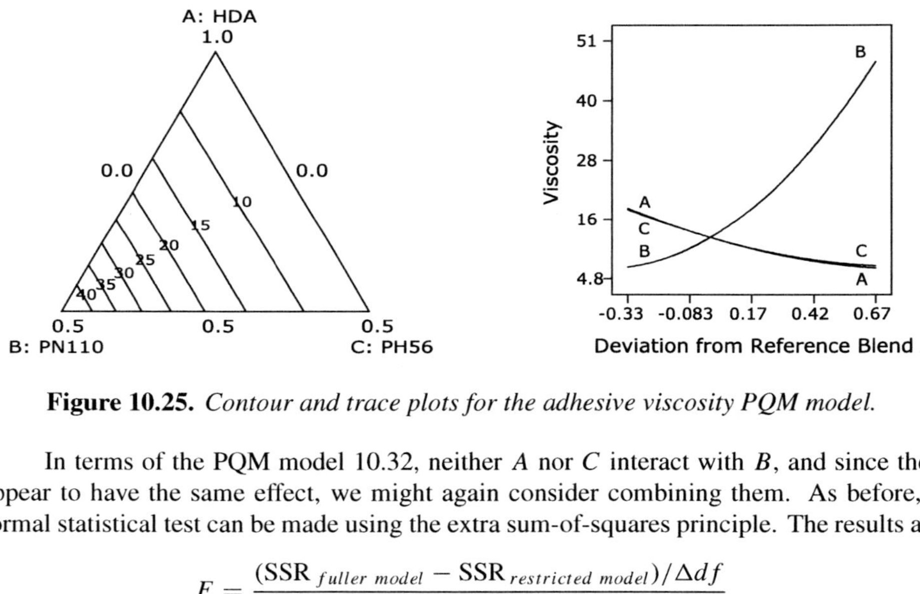

Figure 10.25. *Contour and trace plots for the adhesive viscosity PQM model.*

In terms of the PQM model 10.32, neither A nor C interact with B, and since they appear to have the same effect, we might again consider combining them. As before, a formal statistical test can be made using the extra sum-of-squares principle. The results are

$$\begin{aligned} F &= \frac{(\text{SSR}_{\textit{fuller model}} - \text{SSR}_{\textit{restricted model}})/\Delta df}{s^2} \\ &= \frac{(2307.93 - 2307.73)/1}{6.0786} \\ &= 0.0329 \end{aligned}$$

The tabled F value for $F_{0.05;1,7}$ is 5.59, and since $0.0329 \ll 5.59$, we do not reject the null hypothesis $\text{H}_0 : \beta_A = \beta_C$. The restricted PQM model is

$$\widehat{\text{visc}} = 7.084(A + C) + 12.66B + 34.36B^2 \qquad (10.34)$$

which is model g in Table 10.15. Fitting statistics for this restricted model are $R^2 = 0.9818$, $R^2_{adj} = 0.9773$, and $R^2_{pred} = 0.9372$.

When fitting the full quadratic Scheffé model to the viscosity data, $|t_i|$ values for observations 6 and 7 were the highest of the 11 observations (4.276 and 4.799, respectively) but did not exceed the Bonferroni critical value at the $\alpha = 0.05$ level (5.75). In the case of model 10.34, observations 6 and 7 again have the highest $|t_i|$ values (4.93 and 5.11, respectively), but in this case their values *do* exceed the Bonferroni critical value (4.10). Also, in contrast to the case for the full quadratic model, observations 6 and 7 now have the highest leverage of the 11 observations. To dampen the influence of these observations on the fitted model, a robust regression was carried out using the regressors $(A + C)$, B, and B^2. The Huber influence function was used with $r = 1.345$, $k = 0.5$, and a tolerance of $1e - 3$. The resulting model, which is model h in Table 10.15, is

$$\widehat{\text{visc}} = 7.292(A + C) + 9.902B + 39.68B^2 \tag{10.35}$$

The weights for observations 2, 6, and 7 were 0.599, 0.771, and 0.177, respectively; all other observations had weights of 1.0.

Reexpressing model 10.35 in terms of the reals leads to

$$\widehat{\text{visc}} = 7.292(\text{HDA} + \text{PH-56}) + 12.512(\text{PN-110}) + 158.72(\text{PN-110})^2 \tag{10.36}$$

To get rid of the term in (HDA + PH-56), this model can be reparameterized to the intercept model

$$\widehat{\text{visc}} = 7.292 + 5.220(\text{PN-110}) + 158.72(\text{PN-110})^2 \tag{10.37}$$

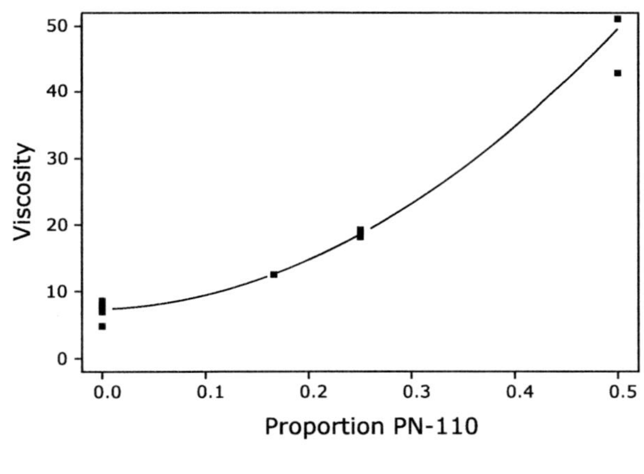

Figure 10.26. *Adhesive viscosity response surface.*

Figure 10.26 is a plot of the response surface corresponding to model 10.37. The scale of the y-axis in this plot is, of course, different from the scale of the y-axis in Fig. 10.23, page 252. Allowing for the scale difference, the two plots lead to nearly equivalent predicted viscosities. There are small differences because of the robust regression procedures that were applied.

The reader should not get the idea that mixture models in general are as amenable to simplification as in this case study. In this example, a fortuitous confluence of circumstances led to a situation where model simplification was rather straightforward. It is hoped, however, that the methodology illustrated here will pique the reader's curiosity to try different model forms when fitting models to mixture data.

Table 10.15. *Adhesive viscosity response. Model summary statistics for several models*

Model	Response	Model	$r^2_{Y\hat{Y}raw}$	$\sum \lvert e_i \rvert$
a	visc	$q = 3$ quadratic	0.9830	13.96
b	visc	reduced $q = 3$ quadratic	.9820	15.03
c	ln(visc)	$q = 3$ linear	.9811	16.53
d	ln(visc)	$q = 2$ linear	.9815	15.19
e	ln(visc)	robust (Huber) $q = 2$ linear	.9815	15.31
f	visc	$q = 3$ PQM	.9819	15.08
g	visc	$q = 2$ PQM	.9818	14.63
h	Visc	robust (Huber) $q = 2$ PQM	.9813	14.59

Chapter 11

Effects

This chapter is devoted to properties of response surfaces associated with *first-degree polynomial models*. We are therefore dealing with surfaces that are planes ($q = 3$) or hyperplanes ($q > 3$). In the discussion surrounding Fig. 8.9, page 164 (Chapter 8), it was emphasized that parameter estimates in linear Scheffé models do not estimate the effects of components but rather estimate the response at the component or pseudocomponent vertices. The *effect* of a component (ΔY) is the change in response for some stated change in composition. It depends on both the gradient ($\Delta Y/\Delta X$) and on the change in the proportion of the component (ΔX) in a specified effect direction. There are three effect directions in a mixture setting — orthogonal, Cox, and Piepel — and these are explained in the sections that follow.

A planar surface for a $q = 3$ mixture setting is displayed in Fig. 11.1, the generating first-degree polynomial being given in the figure caption. This surface will be used to illustrate the various meanings of an "effect". Notice that the contour lines are straight and equally spaced. Also notice that the surface slopes downward from right to left and that the responses at the X_1, X_2, and X_3 vertices are 1, −2, and 3, respectively. Points $\mathbf{x}_M$ and $\mathbf{x}_N$ are explained in Section 11.1.

11.1 Orthogonal Effects

A useful effects direction, particularly in simplex-shaped design regions, is the **orthogonal-effect direction**. The orthogonal-effect direction for a component X_i is the direction that is orthogonal (or perpendicular) to the subsimplex that does not contain X_i. For example, for X_2 in Fig. 11.1, the subsimplex not containing X_2 is the side or edge of the triangle connecting the X_1 and X_3 vertices. The orthogonal-effect direction for X_2 would be any vector that is perpendicular to the X_1–X_3 subsimplex. The arrows in Fig. 11.1 illustrate three orthogonal-effect vectors for component X_2.

Because it is a linear surface, the response at the tail of the long arrow (where $X_2 = 0$ and $X_1 = X_3$) is equal to the average of the responses at the X_1 and X_3 vertices, which is $(1 + 3)/2 = 2$. As one proceeds from the tail to the head of the long arrow, one is moving along an orthogonal-effect vector that is on the component axis. One moves from a point where $X_2 = 0$ to the point where $X_2 = 1$. In traversing this path, the response

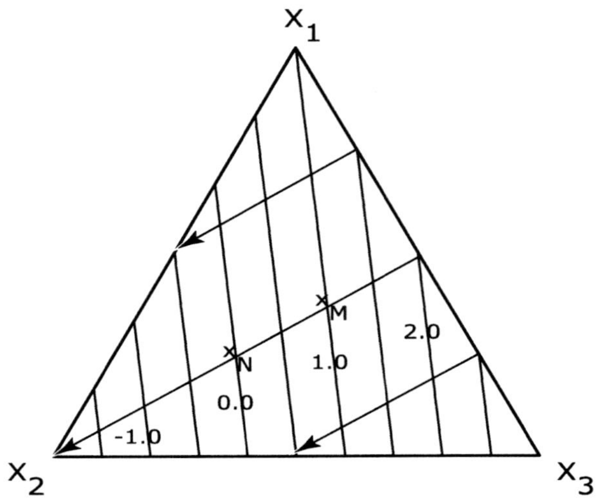

Figure 11.1. *Surface:* $X_1 - 2X_2 + 3X_3$. *Arrows: orthogonal-effect directions.*

changes from +2 to -2, and so the change in the response per unit change in X_2 (i.e., the slope, or *gradient*) is equal to -4. The reason that the slope is negative is not because β_2 is negative. Exactly the same gradient would result if the generating polynomial had been $101X_1 + 98X_2 + 103X_3$.

The reason that the slope is negative is because of the *relationship* between the coefficients. The method that was used in the previous paragraph to calculate the gradient suggests a general method for calculating gradients in the orthogonal-effect directions for any q-simplex:

$$G_i = b_i - \frac{1}{q-1}\sum_{j \neq i} b_j \tag{11.1}$$

The first term in this expression is the least-squares estimate for the response at the ith vertex. The second term estimates the response at the AEV (averaged-extreme-vertex) centroid of the subsimplex that does not contain X_i. The gradient is therefore a *linear combination of parameter estimates*. Applying this formula to the example in Fig. 11.1 leads to the following gradients (column 3) in the orthogonal-effect directions:

i	b_i	G_i	TE_i
1	1	0.5	0.5
2	-2	-4.0	-4.0
3	3	3.5	3.5

The meaning of TE_i is explained below.

Ignoring multicomponent constraints for the moment, the experimental region in a mixture setting is defined by restrictions of the form

$$L_i \leq X_i \leq U_i \qquad i = 1, 2, 3$$

where L_i is the lower bound on the ith component and U_i is the upper bound. The maximum possible range for X_i is then $U_i - L_i$, which shall be symbolized R_i. The **total effect** of a component in the orthogonal-effect direction is then defined to be [118]

$$TE_i = R_i G_i = R_i \left(b_i - \frac{1}{q-1} \sum_{j \neq i} b_j \right) \tag{11.2}$$

Let $\mathbf{x}_L$ and $\mathbf{x}_H$ be 3×1 vectors representing compositions *on an effect vector* where a component is at its lowest (L_i) and highest (U_i) values, respectively. In the case of orthogonal effects (as opposed to Cox and Piepel effects), a unique $\mathbf{x}_L$, $\mathbf{x}_H$ pair is not explicitly defined. All that matters is that the pair lie on the same effect vector. For component X_2 in Fig. 11.1, the only effect vector on which the total orthogonal effect makes any physical sense is the X_2 component axis. On every other effect vector $\mathbf{x}_H$ would lie outside the simplex and the total effect, despite being defined, could not be observed experimentally.

Figure 11.2 illustrates the situation in the context of a constrained region lying within a simplex. The constrained region is the parallelogram. The upper and lower bounds on component X_1 are identified by the horizontal lines cutting through the simplex. Two $\mathbf{x}_L$, $\mathbf{x}_H$ pairs are shown lying at the ends of two orthogonal-effect vectors. For these two vectors, as well as for *any* orthogonal-effect vector, either $\mathbf{x}_L$ or $\mathbf{x}_H$ will lie outside of the constrained region. In this example, then, the total orthogonal effect, despite being defined, is experimentally unobservable.

Assume that one's interest instead was on the effect of a component over a narrower range $\Delta_i < R_i$, say between compositions $\mathbf{x}_M$ and $\mathbf{x}_N$, both of which must lie on the same orthogonal-effect vector. At point $\mathbf{x}_M$ in Fig. 11.1, $(X_1, X_2, X_3) = (0.375, 0.25, 0.375)$, and at point $\mathbf{x}_N$, $(X_1, X_2, X_3) = (0.25, 0.5, 0.25)$. In this case, the change in the proportion of X_2 is not $R_2 = 1.0$ but only $\Delta_2 = 0.25$. The **partial effect** of a component in the

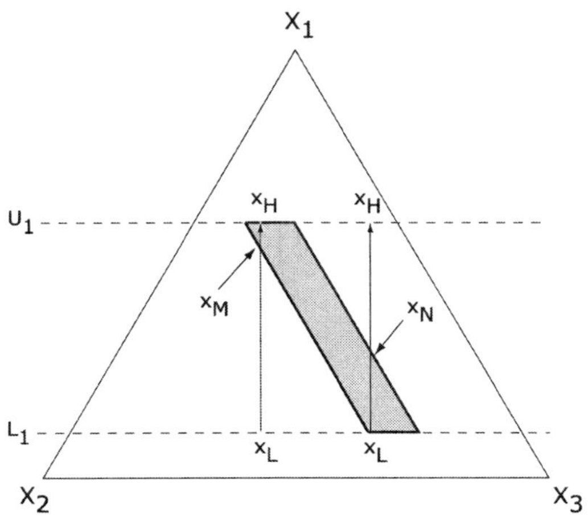

Figure 11.2. *Orthogonal effects in a constrained region.*

orthogonal-effect direction is given by a slightly modified version of Eq. 11.2 [118]:

$$\begin{aligned} PE_i &= \Delta_i G_i \\ &= \Delta_i \left(b_i - \frac{1}{q-1} \sum_{j \neq i} b_j \right) \end{aligned} \tag{11.3}$$

As $\Delta_2 = 0.25$ between points $\mathbf{x}_M$ and $\mathbf{x}_N$, the partial effect of component 2 is then equal to $0.25 \times (-4) = -1.0$. Note that two components might have the same partial orthogonal effect but for entirely different reasons. Δ_i for one component might be large but the gradient small, or Δ_i might be small and the gradient large.

Another way to calculate this is to write

$$PE_i = \hat{Y}(\mathbf{x}_N) - \hat{Y}(\mathbf{x}_M)$$

where both $\mathbf{x}_M$ and $\mathbf{x}_N$ must lie on the same effect vector for the ith component. For component X_2 in Fig. 11.1, $\mathbf{x}_M = (0.375, 0.25, 0.375)$ and $\mathbf{x}_N = (0.25, 0.5, 0.25)$. The partial effect of component 2 is then

$$\begin{aligned} PE_2 &= [1(.25) - 2(.5) + 3(.25)] - [1(.375) - 2(.25) + 3(.375)] \\ &= [0] - [1.0] \\ &= -1.0 \end{aligned}$$

Note that in Fig. 11.1 the contour lines for $\hat{Y} = 0$ and $\hat{Y} = 1.0$ cross the long arrow at points $\mathbf{x}_N$ and $\mathbf{x}_M$, respectively.

In the context of the constrained region in Fig. 11.2, let $\mathbf{x}_N \equiv \mathbf{x}_H$ on the left vector (arrow) and $\mathbf{x}_M \equiv \mathbf{x}_L$ on the right vector (arrow). Having specified a $\mathbf{x}_M$, $\mathbf{x}_N$ pair on each vector, the partial orthogonal effect of X_1 on the left vector is less than the partial orthogonal effect of X_1 on the right vector because Δ_1 on the left is less than Δ_1 on the right. As the end points $\mathbf{x}_M$ and $\mathbf{x}_N$ on both effect vectors lie on the boundary of the constrained region, these effects are also **constraint-region-bounded** effects [123].

Points $\mathbf{x}_M$ and $\mathbf{x}_N$ in Fig. 11.1 do not lie on a boundary, and so the partial orthogonal effect of X_2 defined by this pair is not a constraint-region-bounded effect. The three arrows in this figure represent constraint-region-bounded effects, but two of these are partial orthogonal effects and one is a total orthogonal effect.

Consider now the constrained region for the Stepan hot-melt adhesive experiment (page 154). Figure 11.3 is the same as Fig. 10.5, page 210, except that the A,B,C nomenclature in Fig. 10.5 has been replaced by the X_1,X_2,X_3 nomenclature in Fig. 11.3. In addition, points b'' and c'' are labeled for reference in Section 11.2.

The constraints on the component proportions were

$$\begin{array}{ccccc} 0.50 & \leq & X_1 & \leq & 1.0 \\ 0.00 & \leq & X_2 & \leq & 0.50 \\ 0.00 & \leq & X_3 & \leq & 0.50 \end{array}$$

where $X_1 =$ HDA, $X_2 =$ PN-110, and $X_3 =$ PH-56. The total orthogonal effect for X_1 is equal to the difference in response between formulations X_1' and a'. This is a total effect

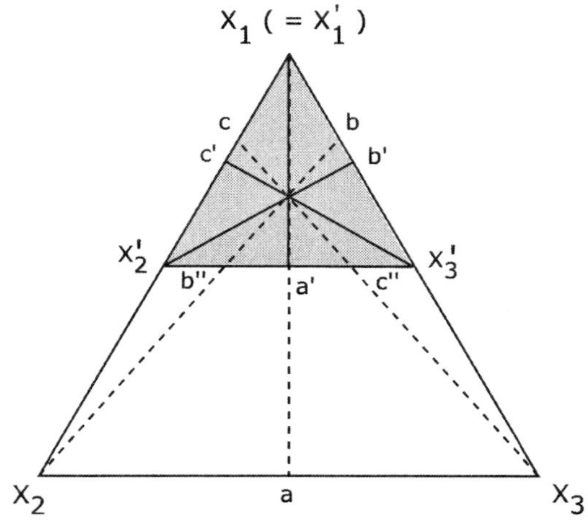

Figure 11.3. *Effects directions in the hot-melt adhesive experiment.*

because the point X_1' is a point on U_1 and a' is a point on L_1. For X_2, the total orthogonal effect is equal to the difference in response between formulations X_2' and b' because X_2' lies on U_2 and b' lies on L_2. For X_3, the total orthogonal effect is equal to the difference in response between formulations X_3' and c' because X_3' lies on U_3 and c' lies on L_3. If a response surface is generated over the simplex using the same first-degree polynomial as in Fig. 11.1, the gradients in the orthogonal-effect directions will be the same in the two figures. As the ranges R_i, $i = 1, 2, 3$, in Fig. 11.3 are half the corresponding ranges in Fig. 11.1, the total orthogonal effects in Fig. 11.3 are equal to half the total orthogonal effects in Fig. 11.1. The total orthogonal effects in this case are $TE_1 = 0.25$, $TE_2 = -2$, and $TE_3 = 1.75$.

Another difficulty with the orthogonal-effect directions can be illustrated with Fig. 11.1. As one proceeds along the long arrow, the relative proportions of X_1 to X_3 remain constant at 1:1. However, as one proceeds along the upper arrow in this figure, the relative proportions of X_1 to X_3 change from 3:1 to 1:0; along the lower arrow the relative proportions change from 1:3 to 0:1. Thus, not only is the proportion of X_2 changing, but so too is the ratio of X_1 to X_3. This is an unappealing situation, and it would be useful to be able to define an effect direction in which the relative proportions of all components except the one whose effect is being measured remain constant. This goal motivates the discussion in the next section.

11.2 Cox Effects

Cox-effect directions were introduced in the discussions surrounding Figs. 5.5 (page 83) and 6.3 (page 111). Figure 11.4 is the same as Fig. 11.1 except that the arrows in the orthogonal-effect direction have been replaced by arrows in Cox-effect directions.

Assume that the formulation $(X_1, X_2, X_3) = (0.2, 0.6, 0.2)$ is of interest (solid circle in Fig. 11.4). Perhaps it is the composition at which a product is currently being manufactured. Whatever the reason, we designate this point as the **base point** with reference

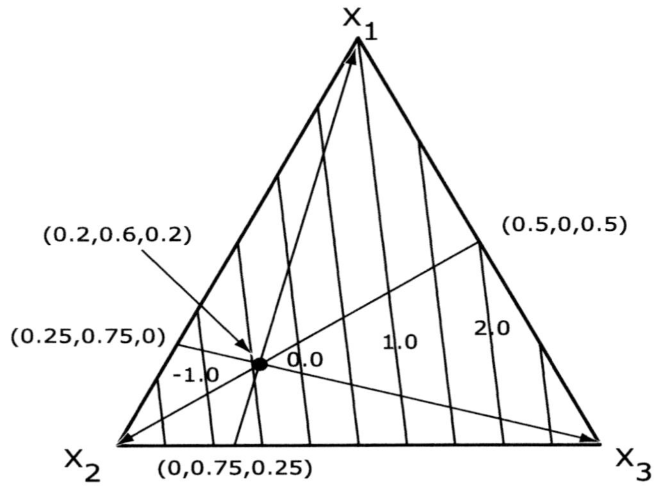

Figure 11.4. *Surface:* $X_1 - 2X_2 + 3X_3$. *Arrows: Cox-effect directions.*

to which Cox effects will be estimated. The base point can be chosen anywhere within the design region. We then pass arrows through this point such that the tips extend to the vertices and the tails terminate at the boundary of the simplex. These are the **Cox-effect directions**, introduced by Cox in 1971 [35]. Along the Cox-effect direction for component X_i, the relative proportions of the X_j, $j \neq i$, remain constant.

Of particular interest are the proportions of the X_j at the end points, because these are needed to estimate the gradients. Designating the composition at a *base* point (not end point) as $\mathbf{s} = (s_1, s_2, \ldots, s_q)$, then at the *end* point for the ith Cox-effect direction, the proportion of the jth component will be

$$X_j = \frac{s_j}{1 - s_i}$$

For example, at the end point for the X_1 Cox-effect direction, where $X_1 = 0$, the proportion of X_2 will be equal to $0.6/(1-0.2) = 0.75$, the proportion of X_3 will be $0.2/(1-0.2) = 0.25$, and the relative proportions $X_2 : X_3 = 3 : 1$, the same as at the base point.

The gradient for the ith component in the Cox-effect direction will be equal to the difference in the response at $X_i = 1$ and $X_i = 0$. The points where $X_i = 1$ and $X_i = 0$ must both lie on the Cox-effect vector for the ith component. The gradient is still the change in response per unit change in the proportion of the ith component, but the direction in which this is measured is defined differently. Consequently one can write an expression for the gradient in the Cox-effect directions that is somewhat similar to Eq. 11.1 for the gradient in the orthogonal-effect directions.

$$G_i = b_i - \sum_{j \neq i}^{q} b_j \frac{s_j}{1 - s_i} \tag{11.4}$$

The first term is the response at the vertex of the ith component, while the second term gives the response at the end point of the ith Cox-effect direction. Equation 11.4 is sometimes reexpressed in the form

$$G_i = \frac{1}{1-s_i}\left(b_i - \sum_{j=1}^{q} b_j s_j\right) \tag{11.5}$$

The two expressions are equivalent to one another, Eq. 11.5 being simply an algebraic rearrangement of Eq. 11.4. Note that the summation in Eq. 11.4 is taken over all j except i, whereas in Eq. 11.5 the summation is taken over all j, including i.

Using either Eq. 11.4 or 11.5, the three Cox-effect gradients for the base point $\mathbf{s} = (0.2, 0.6, 0.2)$ in Fig. 11.4 are

i	b_i	G_i
1	1	1.75
2	−2	−4.00
3	3	4.25

For example, for component X_1 and Eq. 11.4, the calculations are

$$G_1 = 1 - (-2)(0.75) - (3)0.25) = 1.75$$

Using Eq. 11.5, the calculations would be

$$G_1 = \frac{1}{(1-0.2)}[1 - (1)(0.2) - (-2)(0.6) - (3)(0.2)] = 1.75$$

Aside from the possible differences in effect directions for orthogonal vs. Cox effects, there is another important difference. Referring to Eq. 11.1, page 258, if two components have the same coefficient estimates, then they must have the same gradient in the orthogonal-effect direction. In the Cox-effect directions, however, the fact that two components may have the same coefficient estimates is not a sufficient (nor even necessary) condition for the two components to have the same gradient. This is because of the presence of s_i and s_j in Eq. 11.5. In the event that two components have the same coefficient estimates *and* the same proportions at the chosen base point, then they will have the same gradients in the Cox-effect directions.

As with orthogonal effects, one can talk about total, partial, and constraint-region-bounded Cox effects. The expression for a total Cox effect, for example, would be

$$\begin{aligned} TE_i &= R_i G_i \\ &= \frac{R_i}{1-s_i}\left(b_i - \sum_{j=1}^{q} b_j s_j\right) \end{aligned} \tag{11.6}$$

with Δ_i replacing R_i in the case of partial Cox effects. Once a base point is specified, there is one unique effect vector for each component. Both $\mathbf{x}_L$ and $\mathbf{x}_H$ for the ith component must lie on the ith effect vector, and therefore both $\mathbf{x}_L$ and $\mathbf{x}_H$ are explicitly defined. Despite being defined, however, the total effect may or may not be an observable.

In Fig. 11.3, page 261, the base point for the hot-melt adhesive experiment is taken as the centroid of the pseudocomponent simplex, with composition $\mathbf{s} = (2/3, 1/6, 1/6)$. With this point as the reference blend, the three Cox-effect directions are aX_1, bX_2, and cX_3. The compositions at points a, b, and c are easy to work out because the proportion of one component must always be zero, and the relative proportions of the other two components must be in the same ratio as their proportions at **s**. The end point compositions for a, b, and c are, respectively, $(0, 0.5, 0.5)$, $(0.8, 0, 0.2)$, and $(0.8, 0.2, 0)$. Knowing the compositions of the end points, one can calculate the three gradients using Eq. 11.4 or 11.5. The results are

i	b_i	G_i
1	1	0.5
2	−2	−3.4
3	3	2.6

These results assume that the response surface is defined by the same polynomial used in Figs. 11.1 and 11.4. Note that these gradients differ from those for the base point $\mathbf{s} = (0.2, 0.6, 0.2)$.

In Fig. 11.3 the effect of component 1 in the Cox-effect direction (which is the same as the orthogonal-effect direction) is equal to the difference in response between the points X_1 ($\equiv X_1'$) and a'. This is a total Cox effect and a constraint-region-bounded effect. The effect of component 2 in the Cox-effect direction (which is not the same as the orthogonal-effect direction) is equal to the difference in response between the points b'' and b. This is a partial Cox effect and a constraint-region-bounded effect. The effect of component 3 in the Cox-effect direction is equal to the difference in response between the points c'' and c, and this too is a partial Cox effect and a constraint-region-bounded effect.

Let us work out the constraint-region-bounded Cox effect of component 2. At the point b'', the proportion of X_1 is 0.5. We know that the ratio of X_1 to X_3 at b'' must be 4:1 (the ratio at **s**), and so the composition at b'' must be $(0.5, 0.375, 0.125)$. The constraint-region-bounded effect of X_2 is then

$$\begin{aligned} CE_2 &= [1(.5) - 2(.375) + 3(.125)] - [1(.8) - 2(0) + 3(.2)] \\ &= 0.125 - 1.4 \\ &= -1.275 \end{aligned}$$

where CE_2 symbolizes the constraint-region-bounded effect of X_2. Alternatively, one could calculate the effect as the product $\Delta_2 G_2 = (0.375)(-3.4) = -1.275$. The reader is encouraged to work out the effects of X_1 and X_3. (Answers: 0.25 and 0.975, respectively.)

Figure 11.5 displays the same parallelogram-shaped constrained region as in Fig. 11.2, page 259. With the overall centroid of the constrained region taken as the base point, Cox-effect directions have been drawn through this point. Although the total Cox effects for components X_1 and X_3 are not experimentally realizable, the situation is improved over that in the orthogonal-effect directions. Further improvement is possible by considering the Piepel-effect directions.

11.3 Piepel Effects

In an effort to incorporate more information about the size of the constrained region into an effects measure, Piepel proposed a third effects direction. Called the **Piepel-effect direction**, it is similar to a Cox-effect direction except that arrows drawn through a base

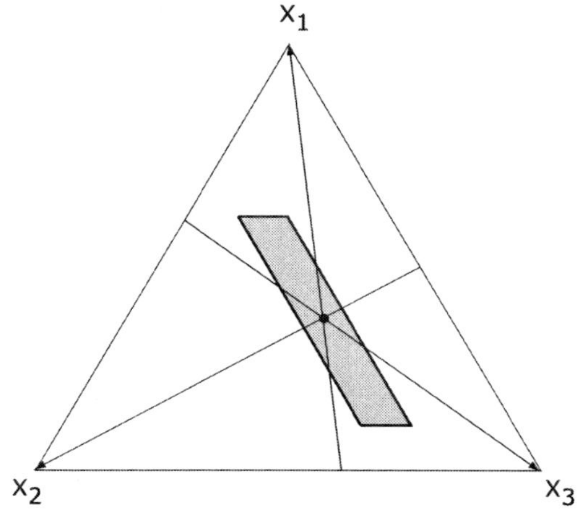

Figure 11.5. *Total Cox effects for X_1 and X_3 unrealizable.*

point extend from the boundaries of the pseudocomponent simplex to the vertices of the pseudocomponent simplex. For example, consider the constrained region for the hot-melt adhesive experiment (Fig. 11.3, page 261). With the centroid of the constrained region taken as the AEV centroid, the Piepel-effect directions for components X_1, X_2, and X_3 are $a'X_1'$, $b'X_2'$, and $c'X_3'$, respectively. In this particular example these happen to be the same as the orthogonal-effect directions.

In general this is not the case, however. Consider the parallelogram-shaped region in Fig. 11.6. The pseudocomponent simplex is represented by the dashed triangle. Taking the AEV centroid of the constrained region as the base point, three effect vectors have

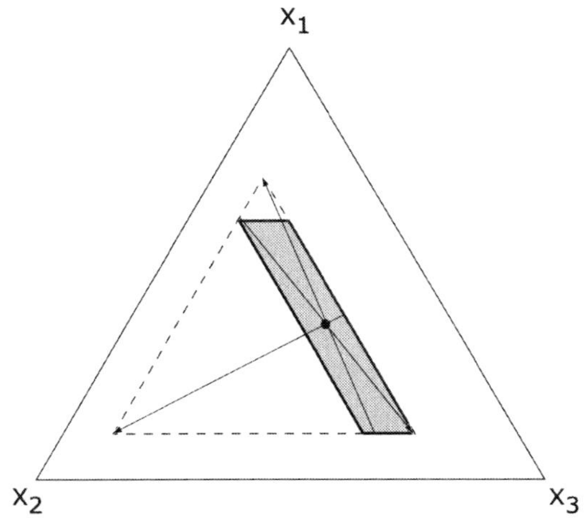

Figure 11.6. *Piepel-effect directions.*

been passed through the base point in the Piepel-effect directions. One end of each vector terminates at a pseudocomponent vertex, and the other end terminates at the boundary of the pseudocomponent simplex. Along each of the vectors the range of the components is $R_i = U_i - L_i$, and so the total Piepel effects are experimentally realizable.

To calculate gradients in the Piepel-effect directions, one can simply replace Eqs. 11.4 and 11.5 with Eqs. 11.7 and 11.8. The only difference between the two sets of equations is that in Eqs. 11.7 and 11.8, the b_i s and the s_i s have been primed to indicate coefficient estimates and base-point proportions in the pseudocomponent metric. The gradients, however, are now expressed in terms of *unit changes in the pseudocomponent proportions*, not the component proportions.

$$G'_i = b'_i - \sum_{\substack{j=1 \\ j \neq i}}^{q} b'_j \frac{s'_j}{1 - s'_i} \tag{11.7}$$

$$G'_i = \frac{1}{1 - s'_i} \left(b'_i - \sum_{j=1}^{q} b'_j s'_j \right) \tag{11.8}$$

To calculate effects, ranges (R'_i) and differences (Δ'_i) should also be expressed in units of pseudocomponent proportions. For example, a total Piepel effect would be given by

$$TE_i = \frac{R'_i}{1 - s'_i} \left(b'_i - \sum_{j=1}^{q} b'_j s'_j \right) \tag{11.9}$$

To reexpress the gradient in terms of unit changes in the *component* proportions, one can simply divide G'_i by $1 - L$, where L is defined in Eq. 4.20, page 58:

$$L = \sum_{i=1}^{q} L_i \qquad i = 1, 2, \ldots, q$$

The quantity $1 - L$ is equal to the height of the pseudocomponent simplex relative to a height of one for the simplex in the reals. Thus, the gradients in the Piepel-effect directions in terms of unit changes in the reals is

$$G_i = \frac{G'_i}{1 - L} = \frac{1}{(1 - s'_i)(1 - L)} \left(b'_i - \sum_{j=1}^{q} b'_j s'_j \right) \tag{11.10}$$

It is possible, as well as more convenient, to reexpress Eq. 11.10 in terms of coefficient estimates taken from the model fitted in the reals. An expression that is equivalent to Eq. 11.10 is [118]

$$G_i = \frac{1}{1 - s'_i} \left(b_i - \sum_{j=1}^{q} b_j s'_j \right) \tag{11.11}$$

where the b_is are no longer primed since they are based on the model fitted in the reals. The s_i, however, are still primed to indicate pseudocomponent proportions at the base point.

When using Eq. 11.11 to calculate effects, ranges (R_i) and differences (Δ_i) should now be expressed in units of component proportions. For example, a total Piepel effect would be given by

$$TE_i = \frac{R_i}{1 - s_i'}\left(b_i - \sum_{j=1}^{q} b_j s_j'\right) \tag{11.12}$$

with $R_i = U_i - L_i$. As $R_i G_i = R_i' G_i'$, Eq. 11.12 gives the same result as Eq. 11.9. However Eq. 11.12 is somewhat more convenient to use.

A problem with the concept of Piepel-effect directions is that while it addresses the problem of keeping the effect vectors within the design region, it defeats the principal advantage of the Cox-effect directions. That advantage is that along the Cox-effect direction for component X_i, the relative proportions of the X_j, $j \neq i$, remain constant. Very few experimenters are interested in asking the question, "What will happen to my response if I enrich or deplete my reference mixture with pseudocomponent X_i^*?" Furthermore, particularly with multicomponent mixtures, it is difficult to mentally picture what direction the Piepel-effect directions are going in.

11.4 Calculating/Displaying Effects

Determining when an effects direction encounters a boundary of a constrained region is not a trivial matter, particularly when there are several components, when lower or upper bounds are not simple proportions, or when there are multicomponent constraints. Fortunately in these situations one can rely on software for help. Nonetheless it is important that the practitioner understand what the software is calculating or displaying in a graph.

MIXSOFT has two routines related to effects. The first calculates orthogonal, Cox, and Piepel constraint-region-bounded end points and effects as well as effect standard deviations. Gradients can be calculated by specifying the lower and upper bounds as zero and one, respectively. Although MIXSOFT currently does not have graphics capabilities, the second routine outputs a file that can be used with software that does have graphics capabilities, such as JMP, MINITAB, or S-PLUS, to obtain up to 10 trace plots corresponding to different responses or models.

To illustrate, the parallelogram-shaped region in Figs. 11.2, 11.5, and 11.6 is approximated by the constraints

$$\begin{aligned} 0.1 &\leq X_1 \leq 0.6 \\ 0.2 &\leq X_2 \leq 0.3 \\ 0.1 &\leq X_3 \leq 0.7 \end{aligned}$$

The AEV centroid has composition $(X_1, X_2, X_3) = (0.350, 0.250, 0.400)$ in the reals or $(0.4167, 0.0833, 0.5000)$ in the pseudos. Taking this as the base point and assuming the linear model $\hat{Y} = X_1 - 2X_2 + 3X_3$, the gradients, total effects, and constraint-region-bounded effects are summarized in Table 11.1. The column headed "Percent of total" is equal to $100 \times$ the constraint-region-bounded effect divided by the total effect. The percentages for X_1 and X_3 increase in the sequence orthogonal $\rightarrow$ Cox $\rightarrow$ Piepel because an increasing amount of information about the shape of the design region is incorporated into the estimates. The small differences in the three constraint-region-bounded

Table 11.1. *Comparison of gradients vs. effects*

Component	Gradient† O	Gradient† C	Gradient† P	Total effect O	Total effect C	Total effect P
X_1	0.500	−0.077	−1.286	0.250	−0.038	−0.643
X_2	−4.000	−4.067	−4.091	−0.400	−0.407	−0.409
X_3	3.500	3.250	2.500	2.100	1.950	1.500
Component	**Constraint-region-bounded effect O**	**Constraint-region-bounded effect C**	**Constraint-region-bounded effect P**	**Percent of total O**	**Percent of total C**	**Percent of total P**
X_1	0.100	−0.020	−0.643	40	52	100
X_2	−0.400	−0.407	−0.409	100	100	100
X_3	0.700	0.780	1.500	33.3	40	100

† O = orthogonal, C = Cox, P = Piepel

effects for X_2 arise because the three effects directions are not exactly collinear with one another.

The information in Table 11.1 for the Cox- and Piepel-effect directions is displayed graphically in Fig. 11.7. These plots are called **effect plots** or **response-trace plots**. Such plots are output by (for example) Design-Expert, MINITAB, and JMP, although JMP separates the individual lines into separate plots using JMP's Prediction Profiler.

The abscissa in each plot is in units of pseudocomponent proportions. The abscissa scale is expanded in the Piepel-effect directions relative to the Cox-effect directions because more information about the shape of the design region has been incorporated into the estimates. The slopes are equal to the gradients, G_i'. The constraint-region-bounded effects are equal to the difference in the response values at the end points of each line.

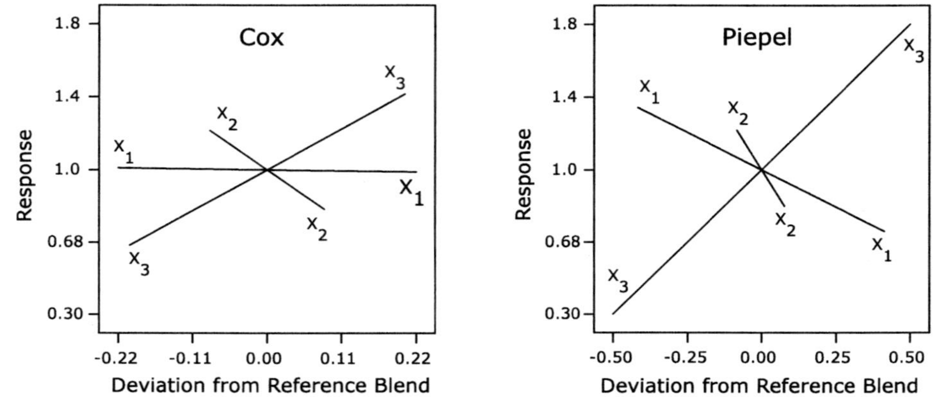

Figure 11.7. *Response trace plots.*

11.5 Inferences

Two hypotheses of particular interest in a mixture setting are

$$H_0 : E_i = 0 \tag{11.13}$$

where E_i is usually taken to be a constraint-region-bounded effect and could be a orthogonal, Cox, or Piepel effect, and

$$H_0 : G_i = G_j \tag{11.14}$$

A horizontal or near-horizontal trace for the ith component in a trace plot would suggest the truth of the first hypothesis. In many cases this is a reasonable assumption, but one can sometimes be fooled. If an effect is estimated with high precision, then a near-horizontal trace *could* be statistically significant. Conversely, the effect of a component with a trace that is clearly not horizontal might not be statistically significant if the variance of the effect is large. Thus, it is desirable whenever possible to do t tests on effects.

If the effect of the ith component is not statistically significant, then one can consider fitting a reduced first-order model to the data. By "reduced" we mean that the term for the ith component can be dropped from the model, the proportions of the remaining components renormalized so they add to one, and then a first-degree model in $q - 1$ variables fitted to the data. A question that may arise is whether or not the ith component can be dropped from the formulation in future makes. The answer would depend on the range of the ith component that was explored. If the range included formulations with 0% of the ith component and changing its level does not have a statistically significant effect on the response, then presumably it could be dropped from the formulation. On the other hand, if the range did *not* include formulations with 0% of the ith component, then concluding that the component can be removed from the formulation is equivalent to making an extrapolation and should be used with caution. It would be wise to prepare a test formulation omitting the component in question to confirm the prediction.

If two traces in a trace plot (say for the ith and jth components) have nearly the same slope, one might conjecture that the components have the same effect. As we are comparing slopes, what we really mean is that the two components may have the same gradient. If for components i and j the first hypothesis is rejected but the second (Eq. 11.14) is not rejected, then when fitting models one could consider combining the proportions of the two components into a single component. This would make particularly good sense if the two components had similar chemical structures or similar functions in the formulation. In future makes, one might consider increasing the proportion of the less expensive component and reducing the proportion of the more expensive component, possibly even eliminating it.

If we are to perform a t test on an effect or gradient, then we need to estimate its standard error. Recalling that $\text{var}(c\,y) = c^2\text{var}(y)$, where c is a constant and y is a random variable, we can write the variance of a total effect as

$$\text{var}(TE_i) = \text{var}(R_i G_i) = R_i^2 \cdot \text{var}(G_i)$$

The t-statistic would then be

$$t = \frac{R_i G_i}{\sqrt{R_i^2 \cdot \text{var}(G_i)}} = \frac{G_i}{\text{se}(G_i)}$$

where "se" stands for standard error. Thus a t test on an effect is equivalent to a t test on a gradient, and so we need a method for calculating the standard error of a gradient.

Without loss of generality, let us use gradients in the Cox-effect directions to illustrate the calculations. Consider Fig. 11.4, page 262, where the compositions at the tails of the arrows are labeled. Define $\boldsymbol{\delta}_i$ as the *difference* in composition between the tips and tails of the arrows (i.e., the two end points). Then for this example

$$\begin{aligned}\boldsymbol{\delta}_1' &= (\ 1.0 \quad -0.75 \quad -0.25\ )\\ \boldsymbol{\delta}_2' &= (\ -0.50 \quad 1.0 \quad -0.50\ )\\ \boldsymbol{\delta}_3' &= (\ -0.25 \quad -0.75 \quad 1.0\ )\end{aligned}$$

where the $\boldsymbol{\delta}_i$s are primed because vectors are commonly expressed as column vectors. The $\boldsymbol{\delta}_i'$s, $i = 1, 2, 3$, can be assembled into a 3×3 matrix, Δ:

$$\Delta_{3\times 3} = \begin{pmatrix} 1.0 & -0.75 & -0.25 \\ -0.50 & 1.0 & -0.50 \\ -0.25 & -0.75 & 1.0 \end{pmatrix}$$

The three gradients are then

$$\mathbf{G} = \Delta\mathbf{b} = \begin{pmatrix} 1.0 & -0.75 & -0.25 \\ -0.50 & 1.0 & -0.50 \\ -0.25 & -0.75 & 1.0 \end{pmatrix}\begin{pmatrix} 1 \\ -2 \\ 3 \end{pmatrix} = \begin{pmatrix} 1.75 \\ -4.00 \\ 4.25 \end{pmatrix}$$

Generalizing from 3 to q, the q gradients in a mixture setting can be expressed in the form

$$\mathbf{G}_{q\times 1} = \Delta_{q\times q}\mathbf{b}_{q\times 1} \tag{11.15}$$

where $\mathbf{G}$ is a $q \times 1$ vector of gradients, and $\mathbf{b}$ is a $q \times 1$ vector of parameter estimates. Equation 11.15 says that the gradients are linear combinations of the parameter estimates. The equation is nothing more than a matrix formulation for the expressions previously derived for calculating gradients in the orthogonal-, Cox-, or Piepel-effect directions (Eqs. 11.1, 11.4, and 11.7, respectively).

The quantity on the right side of Eq. 11.15 is a **linear estimator** (cf. page 184), and its variance is given by

$$\begin{aligned}\text{var}(\mathbf{G}) &= \Delta\,\text{var}(\mathbf{b})\Delta' \\ &= \Delta(\mathbf{X}'\mathbf{X})^{-1}\Delta' s^2 \end{aligned} \tag{11.16}$$

Keep in mind that although $\mathbf{G}$ is a $q \times 1$ vector, var($\mathbf{G}$) is a $q \times q$ matrix with diagonal elements equal to the variances and off-diagonal elements equal to the covariances.

If one wanted to test the hypothesis $H_0 : G_i = \boldsymbol{\delta}_i'\mathbf{b} = 0$, where $\boldsymbol{\delta}_i'$ is the ith row of the matrix Δ, then a t test for the hypothesis can be made using the formula

$$t = \frac{\boldsymbol{\delta}_i'\mathbf{b}}{\sqrt{d_{ii}s^2}} \tag{11.17}$$

where d_{ii} is the ith diagonal element of the matrix $\Delta(\mathbf{X}'\mathbf{X})^{-1}\Delta'$. The degrees of freedom for the t test are the degrees of freedom associated with the estimate of s^2.

Suppose that one wanted to test $H_0 : G_i = G_j$ (Eq. 11.14). This is equivalent to testing $H_0 : G_i - G_j = 0$, and so we need an estimate of the standard error of $G_i - G_j$. As G_i and G_j are random variables, it is true that

$$\mathrm{var}(G_i - G_j) = \mathrm{var}(G_i) + \mathrm{var}(G_j) - 2 \cdot \mathrm{cov}(G_i, G_j)$$

where "cov" stands for covariance. (See, for example, Montgomery [102, page 26].) Consequently, a t test for this hypothesis can be made using the formula

$$t = \frac{(\boldsymbol{\delta}_i' - \boldsymbol{\delta}_j')\mathbf{b}}{\sqrt{(d_{ii} + d_{jj} - 2d_{ij})s^2}} \tag{11.18}$$

where d_{ii} and d_{jj} are the ith and jth diagonal elements of the matrix $\Delta(\mathbf{X}'\mathbf{X})^{-1}\Delta'$, and d_{ij} is the ijth off-diagonal element of the same matrix.

At the time of this writing, the hypothesis $H_0 : G_i = 0$ (or the equivalent hypothesis $H_0 : E_i = 0$) could be tested in MIXSOFT and Design-Expert Version 7. Currently tests of the form $H_0 : G_i = G_j$ have not been programmed into popular computing packages, and so one must resort to writing a macro, function, or script.

Clearly, an effect in a mixture setting is a fuzzier concept than an effect in a nonmixture setting. With three types of effect to choose from, where does one begin? Orthogonal effects are the easiest to calculate and conceptualize in a simplex-shaped design region. However, it is the author's feeling that in general Cox-effect directions are probably the most useful from the standpoint of practical formulation development. Often one is interested in finding answers to questions such as, "If I enrich my test mixture with component X_i, will my response improve?" One is seldom interested in what will happen if a test mixture is enriched with *pseudo*component X_i.

In making decisions about whether a component does or does not have an effect, it is probably a good idea to test this using orthogonal, Cox, and Piepel effects. Generally the three effects directions will not be collinear with one another, but they point in the same general direction. If a Cox effect were not statistically significant, then one might pause before concluding that the component has "no effect". It would be wise to check this in all three effect directions.

Case Study

The Hald cement data are a famous data set that has been analyzed in a variety of books and papers to illustrate collinearity and variable selection in linear regression analysis. The data set is called the "Hald cement data" because it was discussed in a text authored by Hald [61]. However, the source of the data, a paper by Woods, Steinour, and Starke [173], preceded Hald's book by 20 years.

An interesting aspect of the various textbook analyses is that the regressor variables were always treated as independent variables. It was not until Piepel and Redgate's paper [134] appeared in 1998 that it was pointed out that the data actually constituted a mixture experiment. The reasons for the confusion are discussed in detail by Piepel and Redgate, and the reader is referred to this article for an interesting discussion of the history of the data set, plus a variety of detailed analyses.

As the purpose here is simply to illustrate effects plots and inferences about effects, we shall begin with the data in Table 11.2, based on data from the paper by Piepel and Redgate. The table expresses the cement formulations in terms of normalized versions of analyzed oxide compositions. The response, H_{180}, is the heat of hardening after the cement cured for 180 days, expressed in calories/gram.

Table 11.2. *Hald cement data*

ID	SiO_2	Al_2O_3	Fe_2O_3	MgO	CaO	H_{180}
1	0.2743	0.03762	0.01980	0.02475	0.6436	78.5
2	0.2600	0.03500	0.05100	0.02300	0.6310	74.3
3	0.2181	0.05677	0.02789	0.04980	0.6474	104.3
4	0.2465	0.05812	0.02806	0.02405	0.6433	87.6
5	0.2500	0.03900	0.02100	0.02400	0.6660	95.9
6	0.2226	0.06188	0.02794	0.02395	0.6637	109.2
7	0.2098	0.04618	0.05723	0.02108	0.6657	102.7
8	0.2357	0.04814	0.07222	0.02207	0.6219	72.5
9	0.2220	0.04642	0.06155	0.02321	0.6468	93.1
10	0.2129	0.08756	0.01194	0.02488	0.6627	115.9
11	0.2252	0.05005	0.07508	0.02202	0.6276	83.8
12	0.2132	0.06106	0.02903	0.02603	0.6707	113.3
13	0.2183	0.05583	0.02692	0.02393	0.6750	109.4

Fitting a linear Scheffé model to the H_{180} data leads to reasonably encouraging summary statistics: $R^2 = 0.986$, $R^2_{adj} = 0.978$, and $R^2_{pred} = 0.843$. Examination of various regression diagnostics indicates that all observations but one have Cook's D values less than one. Cook's D for formulation #3, however, is 14.1, and this observation is therefore extremely influential. The source of the problem lies in the proportion of MgO in this formulation, which is approximately twice as large as the proportions in the other 12 formulations.

Withholding observation #3 from the analysis and refitting the model leads to about the same values for R^2 and R^2_{adj} (0.987 and 0.979, respectively) but a significantly improved value for R^2_{pred} (0.970). Expressed in the reals, the model is

$$\widehat{H}_{180} = -456.4SiO_2 + 24.29Al_2O_3 - 237.2Fe_2O_3 + 755.9MgO + 293.6CaO$$

A trace plot for the model is shown in Fig. 11.8. The abscissa in this figure is expressed in units of component proportions, while the reference blend is taken as the average of the 12 blends used in the analysis. Its composition is

$$\begin{aligned} \mathbf{s} &= (SiO_2, Al_2O_3, Fe_2O_3, MgO, CaO) \\ &= (0.2325, 0.0522, 0.0402, 0.0236, 0.6515) \end{aligned} \tag{11.19}$$

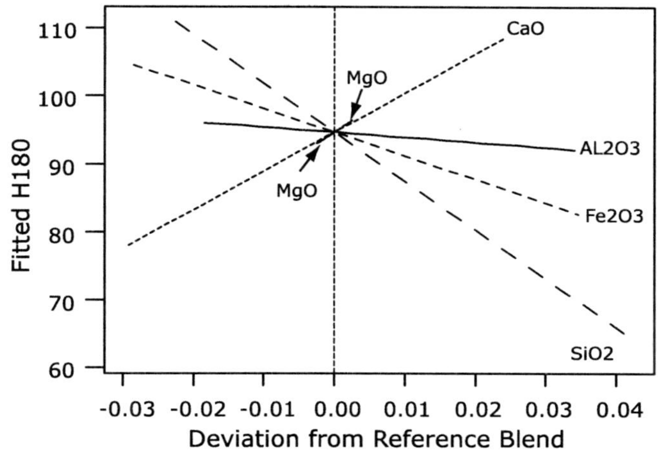

Figure 11.8. *Hald cement experiment.* 5 *components,* 12 *observations.*

One would infer from examining this figure that the effect of Al_2O_3 is possibly negligible, while the gradients for MgO and CaO are probably significant and approximately equal to one another. MIXSOFT or Design-Expert could be used to check the first inference, but we shall show how to check both inferences.

The Δ matrix, which is the matrix of differences between the effect-directions end points, is

$$\Delta = \begin{pmatrix} 1.000 & -0.0681 & -0.0523 & -0.0307 & -0.849 \\ -0.245 & 1.0000 & -0.0424 & -0.0249 & -0.687 \\ -0.242 & -0.0544 & 1.0000 & -0.0246 & -0.679 \\ -0.238 & -0.0535 & -0.0411 & 1.0000 & -0.667 \\ -0.667 & -0.1500 & -0.115 & -0.0677 & 1.000 \end{pmatrix}$$

The off-diagonal elements of this matrix are calculated from the elements of **s** (Eq. 11.19). For example, $\delta_{15} = -0.6515/(1 - 0.2325)$, while $\delta_{51} = -0.2325/(1 - 0.6515)$.

Substituting Δ and **b** into Eq. 11.15, page 270, leads to

$$\mathbf{G} = \Delta\mathbf{b} = \Delta \cdot \begin{pmatrix} -456.4 \\ 24.29 \\ -237.2 \\ 755.9 \\ 293.6 \end{pmatrix} = \begin{pmatrix} -718.1 \\ -74.4 \\ -346.0 \\ 677.1 \\ 570.5 \end{pmatrix} \tag{11.20}$$

The elements of **G** are the slopes of the traces in Fig. 11.8 in the order SiO_2, Al_2O_3, Fe_2O_3, MgO, and CaO.

The variance-covariance matrix of **G** is found by solving Eq. 11.16, page 270. The result is

$$\text{var}(\mathbf{G}) = \begin{pmatrix} 3211.79 & 3169.11 & -418.697 & -29311.7 & -171.839 \\ 3169.11 & 6231.09 & 148.091 & -32329.1 & -596.241 \\ -418.697 & 148.091 & 1527.37 & 16668.7 & -1652.71 \\ -29311.7 & -32329.1 & 16668.7 & 721512. & -45906.2 \\ -171.839 & -596.241 & -1652.71 & -45906.2 & 5200.53 \end{pmatrix} \quad (11.21)$$

The value of s^2 in Eq. 11.16, which is 5.0458, has been incorporated into the elements of Eq. 11.21.

The significance of each slope in Fig. 11.8 is checked by dividing the elements of **G** in Eq. 11.20 by the square roots of the diagonal elements in Eq. 11.21, which are the standard errors of the gradients. The results are

Oxide	t
SiO_2	−12.7
Al_2O_3	−0.9
Fe_2O_3	−8.8
MgO	0.8
CaO	7.9

One does not even need to check a t-table to conclude that the effects of Al_2O_3 and MgO are not statistically significant. Despite the fact that MgO and CaO have similar slopes, their statistical significance is entirely different. This is a case where visual inspection of the trace plot has led to an erroneous inference. The reason, of course, is because the variance of the gradient for MgO is more than two orders of magnitude larger than the variance of any of the other components. The inflated variance for MgO is a consequence of the small range for this component.

A t-statistic for the hypothesis $H_0 : G_{MgO} = G_{CaO}$ is made using Eq. 11.18, page 271. The result is

$$\begin{aligned} t &= \frac{G_{MgO} - G_{CaO}}{\left[\text{var}(G_{MgO}) + \text{var}(G_{CaO}) - 2 \cdot \text{cov}(G_{MgO}, G_{CaO})\right]^{0.5}} \\ &= \frac{677.1 - 570.5}{[721512 + 5200 - 2(-45906)]^{0.5}} \\ &= 0.1 \end{aligned}$$

Thus while the gradients for MgO and CaO are statistically indistinguishable from one another, that for the former is not statistically significant, while that for the latter is significant.

Based on these conclusions, a reduced first-order model can be fit to the data. One can renormalize the data in Table 11.2 so that the sum of the component proportions for SiO_2, Al_2O_3, and Fe_2O_3 add to one and fit a model in three components. The result is

$$\widehat{H}_{180} = -383.3SiO_2 - 214.3Fe_2O_3 + 284.1CaO$$

with $R^2 = 0.985$, $R^2_{adj} = 0.981$, and $R^2_{pred} = 0.973$. Compared to the model fitted to five components, nothing has been lost by fitting the more parsimonious model.

For an additional example illustrating how to reduce the number of components in mixture models, the reader is referred to another article by Piepel and Redgate [133]. Their data consisted of composition and response data for 58 simulated high-level nuclear waste glasses, each formulation containing nine oxides. Using methods described in this chapter, the authors were able to show that one component had no effect and three components exhibited statistically indistinguishable gradients. In addition to this, three other components also exhibited statistically indistinguishable gradients, although the gradients in this set differed from those in the other set. The net effect was that the nine-component problem could be reduced to a four-component problem (two sets of three combined components plus two individual components).

Chapter 12

Optimization

Among the reasons for fitting models to data, certainly an important one is that a model can lead to a better understanding of the system. Do the components blend linearly or nonlinearly? Are one or more of the components inactive? Do some of the components exhibit the same effect? To which components is the response most (least) sensitive?

A second important reason for fitting models to data is to find combinations of the input variables — component proportions in a mixture setting — where one can come as close as possible to a goal. A goal could be a minimum, maximum, range, or target value. In an industrial setting it is usually the case that there is not one but several responses of importance. This has led to a variety of approaches to **multiple-response optimization**. Of particular interest in recent years — an outgrowth of Taguchi's robust parameter design (RPD) — is the **dual-response problem**, in which the goal is to simultaneously optimize the mean and minimize the variance. (See Myers [106], Myers and Montgomery [107], and Kacker [77] for an overview of RPD and Köksoy and Doganaksoy [81] for a review of the dual-response problem.) The overall goal then is one of product or process improvement or optimization.

Optimization methods have evolved through a method known as **steepest ascent** (or descent, as the case may be). This is a single-response, gradient-based approach requiring a continuously differentiable first-order model. There is assumed to be a region of operability where it is theoretically possible to experiment and measure the response. It is often the case, particularly with process variables, that one does not always explore the region of operability in a single designed experiment. In such cases a smaller region, called the region of interest, is defined and a design capable of supporting a first-order model is placed within this region. Steepest ascent provides an efficient means, through the sequential acquisition of data, for moving the region of interest through the region of operability to a point where there is evidence of curvature. At this point one augments the design so that it will support a second-order model and applies a procedure, known as canonical analysis, that facilitates interpretation of the response surface as a maximum, minimum, or saddle point. An alternative procedure, ridge analysis (not to be confused with ridge regression), is sometimes applied to the second-order model. This is also a gradient-based technique requiring a second-order model that has continuous first derivatives. These procedures are covered in detail in texts on response-surface methodology such as Box and Draper [12], Khuri and Cornell [79], and Myers and Montgomery [107].

There are cases in a mixture setting where the sequential acquisition of data may not be practical because of time or possibly economic constraints. For example, color photographic dispersions coated on paper base are subjected to a variety of relatively long-term tests, such as four to six weeks incubation in a wet or dry oven and two to four weeks exposure to high-intensity daylight radiation. The pressures of industry are such that time often does not permit the possibly slow sequential acquisition of data. In these circumstances the region of interest must be enlarged so that it covers as closely as possible the region of operability, and a designed experiment must be planned that can be carried out as a single event. In a mixture setting, this would be done by adjusting the constraints on the component proportions so that the ranges of the components are the maximum feasible ranges.

When there are several response variables, other techniques must be applied as well. One approach is to specify the multiple-response problem as a constrained optimization problem. One of the responses is selected as the primary response, and this is maximized or minimized subject to constraints on the other responses. The method of maximization or minimization may involve gradient-based methods, such as the generalized reduced gradient (GRG) procedure [42], or derivative-free methods, such as the Nelder–Mead simplex procedure [112]. The software package EXCEL employs a version of the GRG called the Solver. At the time of this writing a step-by-step tutorial for using the Solver can be found at the Web site

http://www.solver.com/tutorial.htm

Del Castillo and Montgomery [42] illustrate the GRG method for a multiple-response three-component mixture experiment.

An alternative to the constrained optimization approach that has become very popular is to combine the responses into a single composite response and then to optimize the composite response. The method of combining the responses uses the **desirability function**, originally proposed by Harrington [64]. The method was later modified by Derringer and Suich [44], and it is their modification that is used by popular computing packages such as Design-Expert, JMP, and MINITAB. Details are provided in Section 12.2.

Gradient-based optimization methods require that the objective function (that which is to be optimized) have continuous first derivatives. The nature of the desirability function is such that there will be a point or points where first derivatives do not exist. (This will become clearer in Section 12.2, where the desirability function is explained.) In these cases direct-search methods, such as the previously mentioned Nelder–Mead simplex procedure, must be used. Del Castillo, Montgomery, and McCarville [43] describe modified desirability functions that are everywhere differentiable so that gradient-based optimization methods can be used. The method consists of approximating the nondifferentiable points in the desirability functions with local fourth-order polynomials. The authors also show how to enter the requisite information into an EXCEL spreadsheet so that the EXCEL Solver can be used for optimization.

The emphasis in this chapter will be on describing the optimization procedures available in computing packages such as Design-Expert, JMP, and MINITAB. Section 12.1 describes a graphical approach to optimization. Section 12.2 describes numerical optimization using the desirability function. Section 12.3 discusses the inclusion of mixing measurement error into the desirability function.

For a recent overview of various optimization procedures in statistics, see Carlyle, Montgomery, and Runger [14].

12.1 Graphical Optimization

For single-response optimization, the most straightforward graphical approach is simply to examine trilinear contour plots or three-dimensional response surfaces. This obviously has its limitations, because one can look at only three components at a time. However, judging from *A Catalog of Mixture Experiment Examples* by Piepel and Cornell [131], the largest proportion of reported mixture experiments have been those for three components.

Beyond four components this approach is impractical. The number of plots one must examine is $\binom{q}{3}$ (q things taken three at a time). For $q = 4$, 5, and 6 one would need to examine 4, 10 , and 20 plots, respectively, to fully assess the situation. When $q \geq 4$ it can sometimes prove helpful to first examine a trace plot of the response. If a component appears to have little effect, then this would be a first choice to leave out of a ternary plot.

When there is more than one response, the graphical approach would entail looking at overlaid contour plots. Of course the problems discussed in the previous paragraph remain. Both Design-Expert and MINITAB have the capability of displaying overlaid contour plots with areas of undesirable responses grayed out.

Multiple-response graphical optimization will be illustrated using an example from the chromatographic literature [171]. The experiment involved reverse-phase high-performance liquid chromatography (RP-HPLC) of saccharin, caffeine, and benzoic acid. Eluents were mixtures of water, acetonitrile (CH_3CN), and methanol (CH_3OH). Constraints on the component proportions were

$$\begin{aligned} 0.4 &\leq H_20 \leq 0.8 \\ 0.2 &\leq CH_3CN \leq 0.6 \\ 0.0 &\leq CH_3OH \leq 0.2 \end{aligned}$$

Data are summarized in Table 12.1. R_{12} and R_{23} are measures of the separation, or resolution, between the saccharin and caffeine peaks and between the caffeine and benzoic acid peaks, respectively, while t is the time to elute 99.8% of the last peak. Larger values of R_{12} and/or R_{23} imply better separations.

The following models were fit to each of the three responses (expressed in terms of the natural variables, or reals):

$$\begin{aligned} \hat{R}_{12} &= 2.50(H_2O) - 0.0321(CH_3CN) + 1.86(CH_3OH) \\ \hat{R}_{23} &= 9.30(H_2O) + 3.93(CH_3CN) - 2.99(CH_3OH) - 21.6(H_20)(CH_3CN) \\ \hat{t} &= 20.3(H_2O) + 13.1(CH_3CN) - 7.80(CH_3OH) - 55.1(H_20)(CH_3CN) \end{aligned}$$

Design-Expert contour plots based on each of these models are displayed in Fig. 12.1. The triangles in these figures are *pseudocomponent* simplexes, but the labeling is in units of the reals. For example, the bottom edge of the triangle corresponds to 40% water, the right edge to 20% acetonitrile, and the left edge to 0% methanol. Proceeding counterclockwise from the water vertex, the vertices correspond to 80% water, 60% acetonitrile, and 40% methanol.

Table 12.1. *RP-HPLC separation of saccharin, caffeine, and benzoic acid*

ID	Volume Fraction Water	Acetonitrile	Methanol	R_{12}	R_{23}	t, min
1	0.80	0.20	0.00	1.96	4.80	10.13
2	0.60	0.20	0.20	2.00	3.19	6.50
3	0.40	0.40	0.20	1.19	1.24	3.06
4	0.40	0.60	0.00	0.94	0.88	2.63
5	0.60	0.40	0.00	1.36	1.82	4.07
6	0.467	0.466	0.066	1.47	1.27	3.08
7	0.50	0.50	0.00	1.33	1.40	3.00
8	0.50	0.40	0.10	1.29	1.70	3.66
9	0.40	0.50	0.10	1.23	1.02	2.85

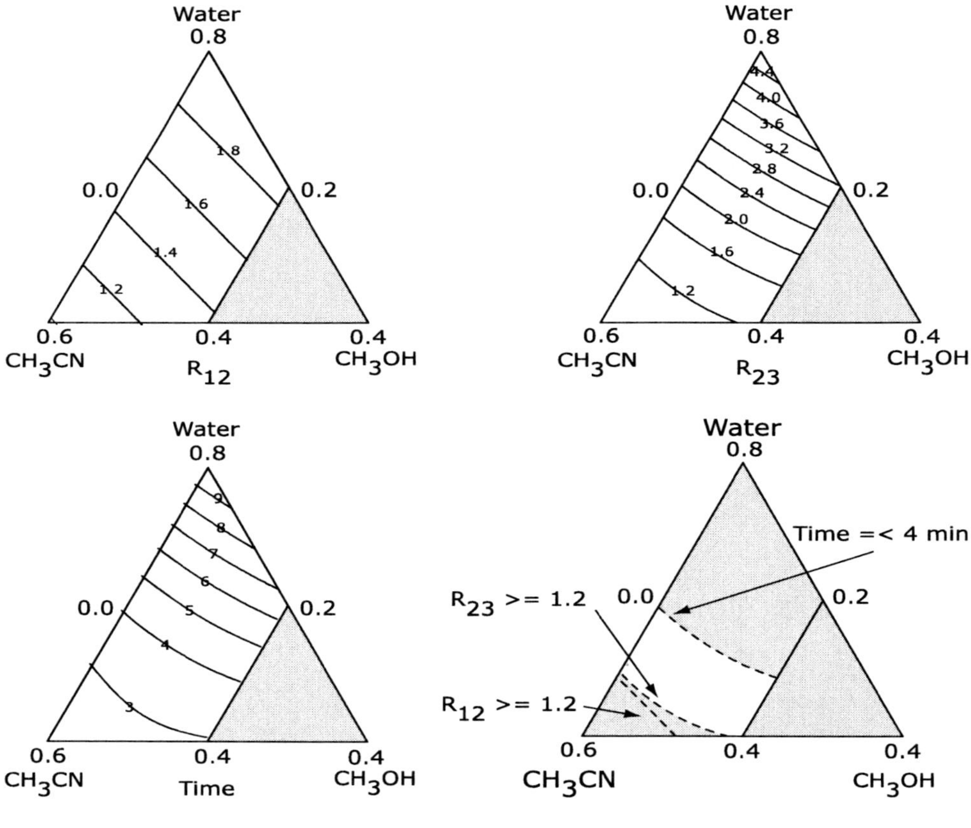

Figure 12.1. *Chromatography experiment. Contour and overlay plots.*

Because the upper bound on methanol is 0.2 (corresponding to 20%), the lower right part of each triangle is grayed out.

Both R_{12} and R_{23} increase (better separation) as the proportion of water increases. Note that the contour interval for R_{12} is 0.2, whereas that for R_{23} is 0.4. In addition, the contour lines are closer together in the plot for R_{23} than in the R_{12} plot, indicating that R_{23} is much more sensitive to the amount of water than is R_{12}. The effect of water on t is to increase the time necessary to elute the last component. Thus there is a trade-off, better separations being achieved at the expense of longer elution times.

In both Design-Expert or MINITAB one can specify an acceptable range for each response. This can be either one-sided (open-ended) or two-sided. Assume that the following one-sided ranges are requested:

$$R_{12} > 1.2 \qquad R_{23} > 1.2 \qquad t < 4 \text{ min}$$

That is, we desire that the resolutions be greater than 1.2 and that we achieve this in less than 4 minutes.

The lower right triangle in Fig. 12.1 displays what amounts to overlaid contour plots. Unacceptable regions have been grayed out, leaving a window of acceptability. Boundaries for each of the three responses are indicated, that for R_{12} falling within the grayed out region. The region of acceptability is determined by R_{23} and time.

It is good to keep in mind that the models used to construct individual and overlaid contour plots are based on $\hat{Y}$ values and that $\hat{Y}$ values have variances. Thus there is always some uncertainty associated with the predicted response(s) as well as the location of the optimal region. It is always good practice to follow up an optimization exercise with confirmatory experiments.

12.2 Numerical Optimization

When $q \geq 4$, a numerical procedure using the desirability function provides a more efficient approach to optimization. This approach involves the conversion of each response $\hat{Y}_i$ into a desirability d_i that varies over the interval $0 \leq d_i \leq 1$. If $d_i = 1$, then this implies that the response is desirable, satisfactory, or acceptable. Conversely, if $d_i = 0$, then the response is considered undesirable, unsatisfactory, or unacceptable. The bounds and their interpretation are obviously subjective.

If one is measuring m responses in a mixture experiment, each mixture blend will have desirabilities $d_1, d_2, \ldots, d_m$ associated with the corresponding responses. The overall desirability, D, is given by the geometric mean of the individual desirabilities.

$$D = (d_1 d_2 \cdots d_m)^{1/m} \tag{12.1}$$

The rationale for using a geometric rather than an arithmetic mean is that if any individual desirability d_i is equal to zero, then the overall desirability will also be equal to zero. The maximum desirability is found either by a direct search or gradient-based procedure, as explained on page 278.

A method for converting $\hat{Y}_i$ values to individual desirabilities is illustrated in Fig. 12.2 [44]. As indicated in the figure caption, each of the four figures represents a different goal. Note that in each figure the $\hat{Y}$ value is on the x-axis and the individual desirability is on the

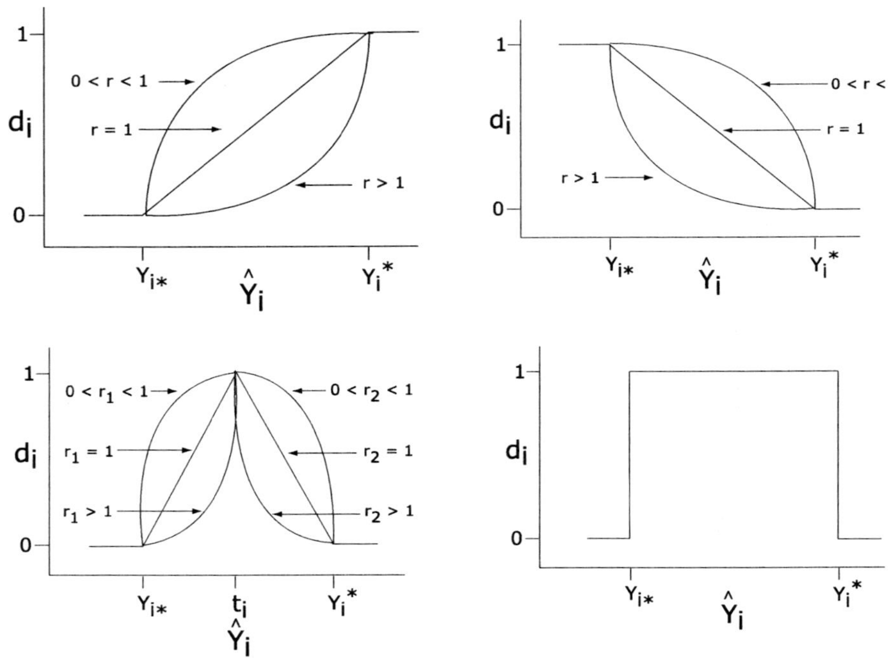

Figure 12.2. *Desirability function. Goal is maximum (upper left); minimum (upper right); target (lower left); range (lower right).*

y-axis. For purposes of explanation, let us focus on the upper left figure (goal is maximum) and confine ourselves to the straight (diagonal) line in the figure (labeled $r = 1$), ignoring for the moment the curved lines.

In this figure the symbol Y_{i*} represents the minimum acceptable value of $\hat{Y}_i$ and is assigned a desirability of zero. The symbol Y_i^* represents the point on the $\hat{Y}_i$ scale beyond which higher $\hat{Y}_i$ values have little or no additional merit and is assigned a desirability of 1.0. Between Y_{i*} and Y_i^* the desirability ranges between zero and one. We thus have three regions: $\hat{Y}_i < Y_{i*}$; $Y_{i*} \leq \hat{Y}_i \leq Y_i^*$; and $\hat{Y}_i > Y_i^*$. Corresponding to these three regions, the desirability can be expressed analytically as follows:

$$d_i = \begin{cases} 0 & \hat{Y}_i < Y_{i*} \\ \left(\dfrac{\hat{Y}_i - Y_{i*}}{Y_i^* - Y_{i*}} \right)^r & Y_{i*} \leq \hat{Y}_i \leq Y_i^* \\ 1 & \hat{Y}_i > Y_i^* \end{cases} \tag{12.2}$$

When $r = 1$ the relationship between d_i and $\hat{Y}_i$ is linear. Because desirability values between zero and one are fractional, raising a desirability to a power $r > 1$ will cause the desirability to become smaller, leading to a family of curves that are concave up. Conversely, raising a desirability to a power $0 < r < 1$ will cause the desirability to become larger, leading to a family of curves that are concave down. An example of each is illustrated in the figure. Choices for Y_{i*}, Y_i^*, and r are all subjective.

When the goal is a minimum (upper right figure in Fig. 12.2) or a target (lower left figure in Fig. 12.2), the expressions for desirability are simply adjusted to reflect the different goals (Eq. 12.3 when goal is minimum, Eq. 12.4 when goal is target). The fourth possibility (goal is a range, lower right figure) is included as an additional possibility in the Design-Expert software package.

$$d_i = \begin{cases} 1 & \hat{Y}_i < Y_{i*} \\ \left(\dfrac{Y_i^* - \hat{Y}_i}{Y_i^* - Y_{i*}}\right)^r & Y_{i*} \leq \hat{Y}_i \leq Y_i^* \\ 0 & \hat{Y}_i > Y_i^* \end{cases} \tag{12.3}$$

$$d_i = \begin{cases} 0 & \hat{Y}_i < Y_{i*} \\ \left(\dfrac{\hat{Y}_i - Y_{i*}}{t_i - Y_{i*}}\right)^{r_1} & Y_{i*} \leq \hat{Y}_i \leq t_i \\ \left(\dfrac{\hat{Y}_i - Y_i^*}{t_i - Y_i^*}\right)^{r_2} & t_i < \hat{Y}_i \leq Y_i^* \\ 0 & \hat{Y}_i > Y_i^* \end{cases} \tag{12.4}$$

The exponent r is referred to as the *weight*. A weight greater than 1 places more emphasis on the goal; a response value must be close to the goal to have a high desirability. Conversely, a weight value less than 1 places less emphasis on the target; a response value that is remote from the goal may have a high desirability.

A further refinement, called *importance*, is available in many software packages. The equation for calculating the overall desirability is modified as follows:

$$D = (d_1^{t_1} d_2^{t_2} \cdots d_m^{t_m})^{1/(t_1+t_2+\cdots+t_m)}$$

where t_1 is the relative importance of response 1, t_2 is the relative importance of response 2, and so forth. t_i values can take on integer values in the range 1–5 in Design-Expert, assume any values in JMP, and range from 0.1 to 10 in MINITAB. If $t_1 = t_2 = \cdots = t_m$, then the overall desirability reduces to that given by Eq. 12.1, page 281.

Table 12.2 displays data for a four-component coating experiment [69]. The experiment involved a search for a combination of prime pigment (titanium dioxide), vehicle (latex emulsion), and two extender pigments that would maximize hiding power and at the same time minimize scrub loss (mg/100 cycles, based on ASTM methods D2805-70 and D2846-74). The bounds for the component proportions were

$$\begin{array}{ccccc} 0.05 & \leq & TiO_2 & \leq & 0.45 \\ 0.20 & \leq & \text{Vehicle} & \leq & 0.60 \\ 0.30 & \leq & \text{Extender}A & \leq & 0.70 \\ 0.05 & \leq & \text{Extender}B & \leq & 0.45 \end{array}$$

Table 12.2. *Coating experiment*

	Component				Hiding	Scrub
ID	TiO_2	Vehicle	Extender A	Extender B	Power	Loss
1	0.050	0.200	0.300	0.450	7.8953	533.67
2	0.450	0.200	0.300	0.050	32.862	749.00
3	0.050	0.600	0.300	0.050	3.7210	39.50
4	0.050	0.200	0.700	0.050	9.2751	203.25
5	0.250	0.200	0.300	0.250	20.132	555.25
6	0.050	0.400	0.300	0.250	4.7137	51.75
7	0.050	0.200	0.500	0.250	8.3829	342.75
8	0.250	0.400	0.300	0.050	16.245	84.75
9	0.250	0.200	0.500	0.050	22.639	360.75
10	0.050	0.400	0.500	0.050	5.4645	48.00
11	0.050	0.333	0.433	0.183	5.8882	76.00
12	0.183	0.200	0.433	0.183	17.256	386.25
13	0.183	0.333	0.300	0.183	12.351	136.00
14	0.183	0.333	0.433	0.050	14.499	75.50
15	0.100	0.250	0.350	0.300	10.548	325.75
16	0.300	0.250	0.350	0.100	22.096	359.00
17	0.100	0.450	0.350	0.100	6.2888	40.75
18	0.100	0.250	0.550	0.100	10.629	136.67
19	0.150	0.300	0.400	0.150	11.777	114.00

Although not obvious, this is a lower-bounds-only problem, the upper bounds being implied by the lower bounds. (Cf. Chapter 4, page 38, for a discussion of implied constraints.) The design region is therefore a simplex. IDs 1–4 are vertices of the simplex, 5–10 are edge centroids, 11–14 are two-dimensional constraint-plane centroids, 15–18 are axial check blends, and 19 is the overall centroid.

Reduced quadratic models were fit to each response. In terms of the reals, the coefficient estimates were

	Coefficient Estimates	
Term	Hiding power	Scrub loss
TiO_2	67.35	3959
Vehicle	7.55	918
Extdr A	11.14	429
Extdr B	11.05	2019
$TiO_2 \times$ Vehicle	−63.86	−8235
$TiO_2 \times$ Extdr A	36.61	−3277
$TiO_2 \times$ Extdr B	·	−2397
Vehicle × Extdr A	−30.97	−2381
Vehicle × Extdr B	−37.65	−6209
Extdr A × Extdr B	·	·

Values for Y_{i*} and Y_i^* were set to the minimum and maximum observed responses in both cases. Goals were to maximize hiding power and minimize scrub loss, both the weight and importance being set equal to 1.

Different software products have different methods of displaying the results of numerical optimization. Figure 12.3 displays one of three methods used by Design-Expert (top) and the method used by JMP (bottom). The method used by MINITAB is somewhat similar to that used by JMP. The Design-Expert graphs are called *ramps*, while the JMP display is called the *Profiler*.

The location of the filled circles in the Design-Expert ramps graphically illustrates the results of the numerical optimization. Not only can one specify a goal for a response, but one can also specify goals for the components. These can be set to a minimum, maximum, target, full range, narrower range, or "is equal to". A minimum, for example, would be useful if one component were very expensive. In this example the goal for each component was set to its full range. The filled circles for hiding power and scrub loss are not at the top of their respective ramps, indicating that an overall desirability of 1 was not achieved. The overall desirability found by Design-Expert was 0.656, and this is printed beneath the ramps. In addition to the ramps display, Design-Expert also presents the results as a histogram and as a textual report.

In the JMP Profiler, the first four plots in the top row are trace plots in the Cox-effect directions for hiding power; the first four plots in the second row are trace plots in the Cox-effect directions for scrub loss; the four plots in the bottom row are trace plots in the Cox-effect directions for the overall desirability. The four vertical dashed lines identify the composition at the point where the desirability is maximized. Note that the vertical dashed lines intersect the desirability traces at their maxima. The composition at the maximum desirability is indicated by the *horizontal* numbers beneath each abscissa. The *vertical* numbers on each abscissa label the lower and upper bounds of each component, identified by tick marks; the center tick marks the midrange. The numerical value beside the word "Desirability" on the ordinate is the overall desirability (0.643 in this example). Note that the composition and desirability found by JMP differ somewhat from those found by Design-Expert.

The reader may find it helpful to view Fig. 12.4. These are edited MINITAB trace plots in the Cox-effect directions for the base point (TiO_2, Vehicle, Extdr A, Extdr B) = (0.268, 0.357, 0.325, 0.050), which is the composition found by JMP for maximum desirability (Fig. 12.3). The curved lines for hiding power in Fig. 12.3 are the same as those in Fig. 12.4 (left); the curved lines for scrub loss in Fig. 12.3 are the same as those in Fig. 12.4 (right). The only difference is that the lines have been separated in Fig. 12.3.

Referring again to Fig. 12.3, the rightmost box in the top row of the Profiler is a plot of hiding power (ordinate) vs. desirability (abscissa), and thus the d_i- and $\hat{Y}_i$-axes are switched relative to those in Fig. 12.2, page 282. Similarly, the rightmost box in the second row of the Profiler is a plot of scrub loss vs. desirability. Each of these plots has three small boxes. The goal (maximum, minimum, or target) for a response is set by positioning these boxes either with the mouse or in a dialog box. Values for r are set interactively by adjusting the curve shapes with the mouse. In this example, hiding power has been set to "goal is maximum" and scrub loss to "goal is minimum".

The mouse can be used to move any of the vertical dashed lines in Fig. 12.3. In so doing, one will move along a Cox-effect direction, and as a result the proportions of the

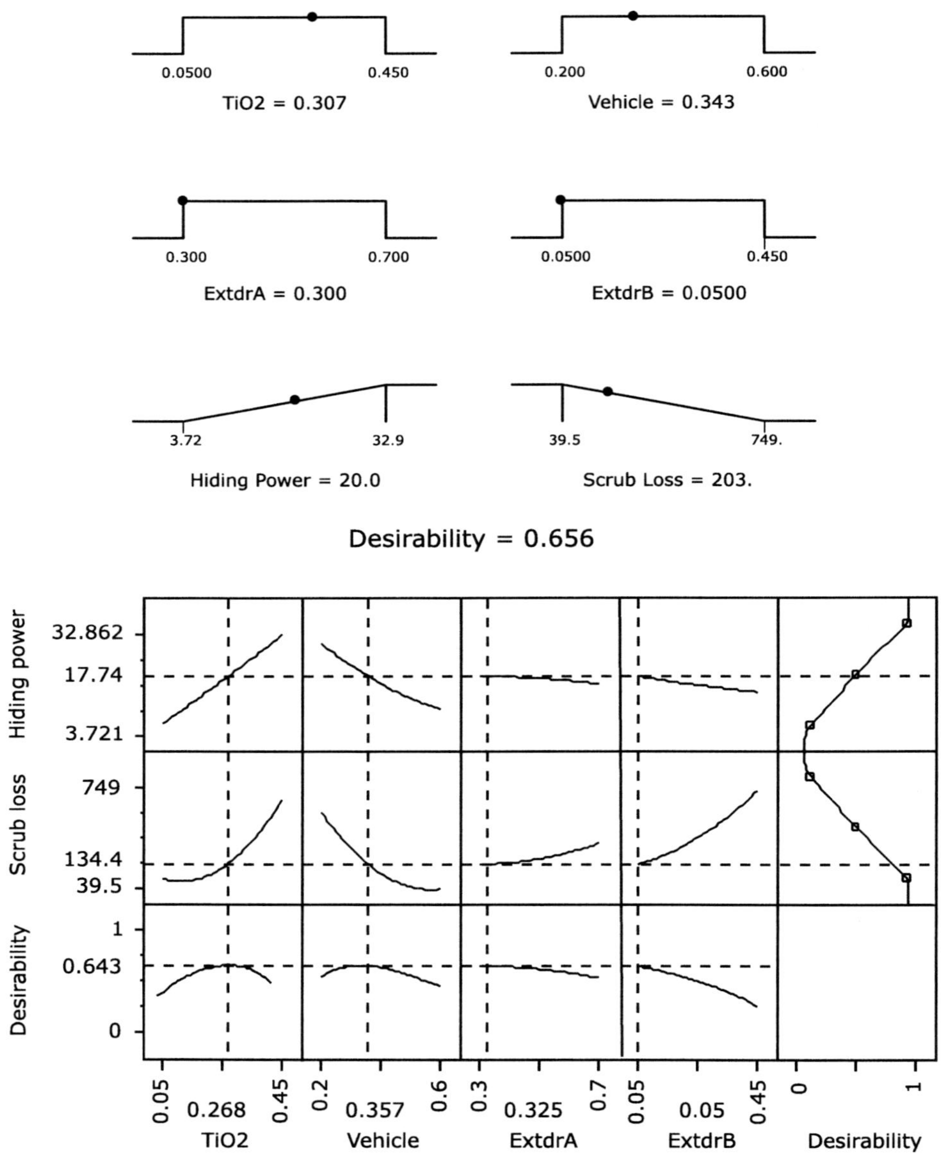

Figure 12.3. *Design-Expert's ramps (top) and JMP's Profiler (bottom) for the coating experiment.*

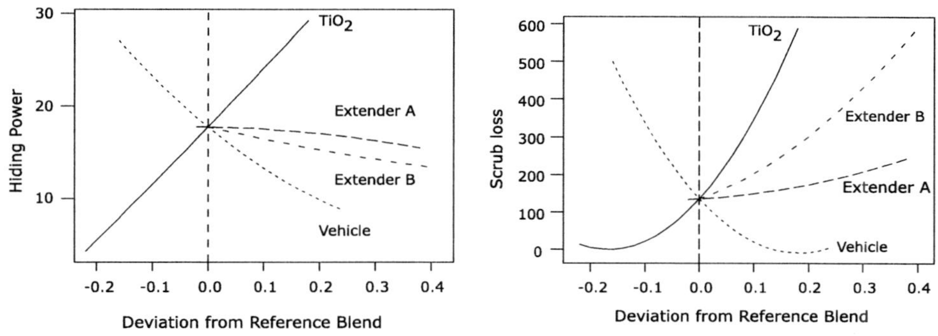

Figure 12.4. *Response trace plots for hiding power (left) and scrub loss (right). Cox-effect directions. Base point: (TiO_2, Vehicle, Extdr A, Extdr B) =* $(0.268, 0.357, 0.325, 0.050)$.

other components will change but remain in constant ratio to one another. Because the desirability has been maximized, changing the composition by moving a vertical dashed line will decrease the desirability.

A cautionary word is in order: The abscissa in the Profiler in Fig. 12.3 would lead one to believe that at the base point (TiO_2, Vehicle, Extdr A, Extdr B) = (0.268, 0.357, 0.325, 0.050), each component can be varied over its full range. The full range of each component is 0.4, as can be deduced from the constraints given on page 283 or the labeling of the abscissa in the Profiler. However, the constraint-region-bounded ranges, *given this base point*, are

$$\begin{array}{rcccl} 0.05 & \leq & TiO_2 & \leq & 0.268 \\ 0.20 & \leq & \text{Vehicle} & \leq & 0.357 \\ 0.30 & \leq & \text{Extender}A & \leq & 0.325 \\ 0.05 & \leq & \text{Extender}B & \leq & 0.123 \end{array}$$

These were determined using MIXSOFT's `EFFECTS` routine and cannot be arrived at by a simple hand calculation. Note that at the JMP optimum the component proportions for TiO_2, Vehicle, and Extdr A are at the *upper* end of their respective constraint-region-bounded ranges, while the proportion for Extdr B is at the *lower* end of its range. Recall that when changing the proportion of a component in its Cox-effect direction, the proportions of the $q-1$ other components remain in constant ratio to one another. For example, if we increased the proportion of TiO_2 from 0.268 to 0.450, then the proportions of Vehicle, Extdr A, and Extdr B would drop to 0.268, 0.244, and 0.038, respectively, in which case the lower bound on Extdr B has been violated.

Figure 12.5 may help to visualize the situation. This is a Design-Expert contour plot of hiding power for the case where the proportion of TiO_2 is 0.268. The box is similar to the "flags" that may be added at any point in a Design-Expert contour plot. Entries 2–5 in the box are explained below, but all of the entries apply to the filled circle in the contour plot. Note that the filled circle is located at the lower bound of Extdr B, and so any increase in TiO_2, Vehicle, or Extdr A along its Cox-effect direction *must* result in offsetting decreases in the remaining $q - 1$ components, including Extdr B.

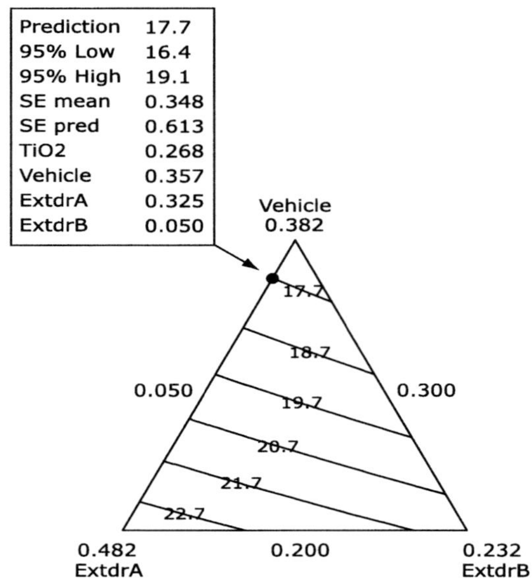

Figure 12.5. *Contour plot for hiding power when* $TiO_2 = 0.268$.

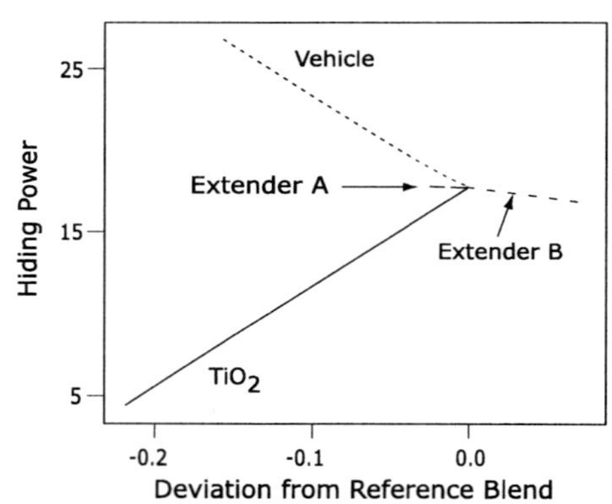

Figure 12.6. *Constraint-region bounded trace plot for hiding power. Cox-effect directions. Base point: (*TiO_2*, Vehicle, Extdr A, Extdr B)* $= (0.268, 0.357, 0.325, 0.050)$.

In light of all this, Fig. 12.6 is a corrected trace plot in the Cox-effect directions for the base point corresponding to the JMP optimum. It was constructed in MINITAB using data output from the MIXSOFT routine EFFPLT. The abscissa is labeled in terms of reals.

Referring back to Fig. 12.3, page 286, we see that Design-Expert and JMP have arrived at somewhat different compositions and predicted values at the optimum. In the general case, it is highly likely that the mixture blend at the optimum will not be a blend that was part of the experiment. The question arises whether one should put a confidence interval or a prediction interval around a predicted response.

A confidence interval is an interval estimate of the *fitted mean* response ($\mu(\mathbf{x}_0)$) based on the model equation for particular values of the regressor variables $\mathbf{x}_0$ and is retrospective. It is not a suitable interval for a *predicted future* observation ($Y(\mathbf{x}_0)$), which is prospective. Prediction implies that the experimenter is going to perform a confirmatory experiment at the particular values of the regressor variables. A future estimate of $Y(\mathbf{x}_0)$, namely $\hat{Y}(\mathbf{x}_0)$, will have variance s^2, which will be in addition to the variance associated with estimating $\mu(\mathbf{x}_0)$.

A $100(1-\alpha)\%$ *confidence* interval on $\mu(\mathbf{x}_0)$ is given by

$$\hat{Y}(\mathbf{x}_0) - t_{\alpha/2,n-p}\sqrt{s^2\, h_{00}} \le \mu(\mathbf{x}_0) \le \hat{Y}(\mathbf{x}_0) + t_{\alpha/2,n-p}\sqrt{s^2\, h_{00}}$$

while a $100(1-\alpha)\%$ *prediction* interval on $Y(\mathbf{x}_0)$ is given by

$$\hat{Y}(\mathbf{x}_0) - t_{\alpha/2,n-p}\sqrt{s^2\,(1+h_{00})} \le Y(\mathbf{x}_0) \le \hat{Y}(\mathbf{x}_0) + t_{\alpha/2,n-p}\sqrt{s^2\,(1+h_{00})}$$

where

$$h_{00} = \mathbf{x}_0'(\mathbf{X}'\mathbf{X})^{-1}\mathbf{x}_0 \tag{12.5}$$

The quantity $\mathbf{x}_0'(\mathbf{X}'\mathbf{X})^{-1}\mathbf{x}_0$ is like a leverage value, except that in the case of a leverage value $\mathbf{x}_0$ is replaced by $\mathbf{x}_i$, $i = 1, 2, \ldots, n$. (Cf. Eq. 6.8, page 105, and the discussion leading up to this equation beginning on page 100.) Apart from σ^2, $\mathbf{x}_0'(\mathbf{X}'\mathbf{X})^{-1}\mathbf{x}_0$ is the variance of prediction at the particular values of the regressor variables, $\mathbf{x}_0$.

The quantities $\sqrt{s^2 h_{00}}$ and $\sqrt{s^2(1+h_{00})}$ in the expressions for the confidence and prediction intervals are called the *standard error of the mean* and the *standard error of prediction*, respectively. These are entries 4 and 5 in the box in Fig. 12.5 and were calculated using $h_{00} = 0.4752$ and $s2 = 0.2550$. The value for h_{00} was calculated using GAUSS.

$$\begin{aligned}
\text{SE mean} &= \sqrt{(0.2550)(0.4752)} \\
&= 0.3481 \\
\text{SE pred} &= \sqrt{(0.2550)(1+0.4752)} \\
&= 0.6133
\end{aligned}$$

The coating experiment had 19 observations (Table 12.2, page 284), and the model for hiding power had 8 terms (page 284). Thus the residual has $19 - 8 = 11$ degrees of freedom. The tabled value for $t_{.025,11}$ is 2.201, and so the estimated 95% confidence and prediction intervals are

$$\begin{aligned}
17.71 - (2.201)(0.3481) \le \mu(\mathbf{x}_0) &\le 17.71 + (2.201)(0.3481) \\
16.97 \le \mu(\mathbf{x}_0) &\le 18.50 \\
17.73 - (2.201)(0.6133) \le Y(\mathbf{x}_0) &\le 17.73 + (2.201)(0.6133) \\
16.38 \le Y(\mathbf{x}_0) &\le 19.08
\end{aligned}$$

The values for "95% Low" and "95% High" in the box in Fig. 12.5 refer to the prediction intervals.

In addition to uncertainty in prediction of a response at an optimum, there is also uncertainty about the *location* of the optimum. In a mixture setting, this translates to uncertainty in the composition of the mixture blend at the optimum. It is possible to calculate a confidence region on the location of an optimum for a single response, but this is beyond the scope of this book and is not an option in popular software packages. The reader is referred to Myers and Montgomery [107], Sections 6.5.2 and 6.5.3, for a discussion of the procedure and an illustrative example.

12.3 Propagation of Error

Assume that one is going to prepare a three-component mixture blend and that the aim composition in terms of the reals is $(A, B, C) = (0.50, 0.25, 0.25)$. Assume further that 400 gm of the blend will be prepared, in which case the aim composition in terms of the actuals is $(A, B, C) = (200, 100, 100)$. In the course of preparing the mixture, there will almost certainly be mixing measurement error. Instead of the actual weights (or volumes) being (200, 100, 100), they may turn out to be (199.1, 99.2, 100.5). In terms of the reals, the true composition will be $(A, B, C) = (0.499, 0.249, 0.252)$ rather than (0.50, 0.25, 0.25). Because of **propagation of** (mixing measurement) **error**, this will have an effect on the response. If the particular mixture blend is in a region where the response is very sensitive to composition, then this could have a large effect on the response. On the other hand, if the blend is in a region where the response is insensitive (robust) towards composition, then the propagated error will be much less.

In the previous paragraph, the true composition in terms of the reals was calculated using the expression

$$X_i = \frac{A_i + e_i}{\sum_{i=1}^{q}(A_i + e_i)} \quad \text{for } i = 1, \ldots, q \tag{12.6}$$

where X_i is the true composition in terms of the reals, A_i is the aim amount in terms of the actuals, and e_i is the measurement error made in the aim amount of component i. This type of error has been referred to as **absolute measurement error** and is to be contrasted with **relative measurement error** that leads to actual component amounts of the form $A_i(1 + e_i)$ [164]. In this case the true composition in terms of the reals would be calculated using the expression

$$X_i = \frac{A_i(1 + e_i)}{\sum_{i=1}^{q} A_i(1 + e_i)} \quad \text{for } i = 1, \ldots, q \tag{12.7}$$

Using fitted models, Design-Expert will calculate propagation of error (POE) resulting from absolute measurement errors in component amounts and/or in the levels of process variables. The method of calculation is beyond the scope of this book, but interested readers may download a PDF file containing analysis details at http://www.statease.com. In order to output POEs, the user must supply an estimate of the standard deviation of the e_i, $i = 1, \ldots, q$.

Propagation of error can be illustrated by doing a simulation experiment. For this purpose we shall draw on the polyurethane reactive (PUR) hot-melt adhesive experiment performed by Hillshafer, O'Brien, and Williamson [70] (cf. page 154ff). According to Hillshafer et al., 400 gram samples of each PUR formulation were prepared. Table 12.3

Table 12.3. *One-minute green strength data for the hot-melt adhesive experiment*

	Component			
ID	HDA	PN110	PH56	GS1[†]
1	400	0	0	48.0
2	400	0	0	58.0
3	300	100	0	24.0
4	300	100	0	18.0
5	300	0	100	36.0
6	200	200	0	49.0
7	200	200	0	46.0
8	200	100	100	32.0
9	200	0	200	11.0
10	200	0	200	16.0
11	266.66	66.67	66.67	21.0

[†] Green strength, psi @ 1 min.

summarizes the aim compositions in terms of the actuals. The compositions and their standard orders are identical to those in Table 8.1, page 156, where the compositions are expressed in terms of the reals and the pseudos. The response GS1 stands for 1-minute green strength, measured as the required force per square inch necessary for separation of two glued 1-inch cubes of wood. The blocks were pulled 1 minute after the time that the adhesive was applied.

Fitted models in terms of the reals (Eq. 12.8) and the actuals (Eq. 12.9) were

$$\widehat{\text{GS1}} = 53.4(\text{HDA}) + 285(\text{PN110}) - 25.8(\text{PH56}) - 485(\text{HDA})(\text{PN110}) \tag{12.8}$$

$$\widehat{\text{GS1}} = 0.133(\text{HDA}) + 0.711(\text{PN110}) - 0.0645(\text{PH56}) - 0.00303(\text{HDA})(\text{PN110}) \tag{12.9}$$

with $R^2 = 0.9566$, $R^2_{adj} = 0.9381$, $R^2_{pred} = 0.8760$, and $s^2 = 15.526$. Diagnostics such as a normal probability plot, a plot of residuals vs. predicted, the R-student, and Cook's D statistics were all satisfactory. Figure 12.7 displays contour and three-dimensional surface plots based on the reduced quadratic model(s). The illustrated simplex is the pseudocomponent simplex, but the labeling is in terms of the actuals.

The saddle point located near the overall centroid of the pseudocomponent simplex is clearly the region where the response would be least sensitive to mixing measurement error. Moving away from this region towards the HDA vertex or the PN110 pseudocomponent vertex causes the response to rise; moving toward the PH56 pseudocomponent vertex causes the response to decrease. In each case, the response will become more sensitive to mixing measurement error than in the region of the saddle point. This is also evident from the Design-Expert contour plot of POE in Fig. 12.8. In the vicinity of the saddle point (and near the overall centroid), POE is a minimum, increasing as one moves away from this region. The plot is based on assumed standard deviations of 4.0 for HDA, PN110, and PH56, respectively.[1] Thus there exists a trade-off between maximizing the response (at least 25 psi according to Hillshafer et al.) and minimizing POE (smaller is better).

[1] Hillshafer et al. do not report s_i values. The value 4.0 for s_i, $i = 1, 2, 3$, is an assumption made by the author for purposes of illustration.

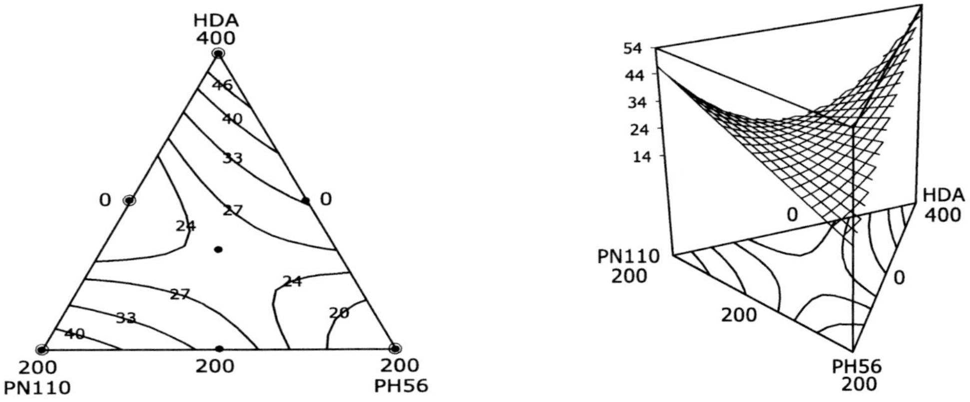

Figure 12.7. *Contour and 3D surface plot for 1-minute green strength. Reduced quadratic model.*

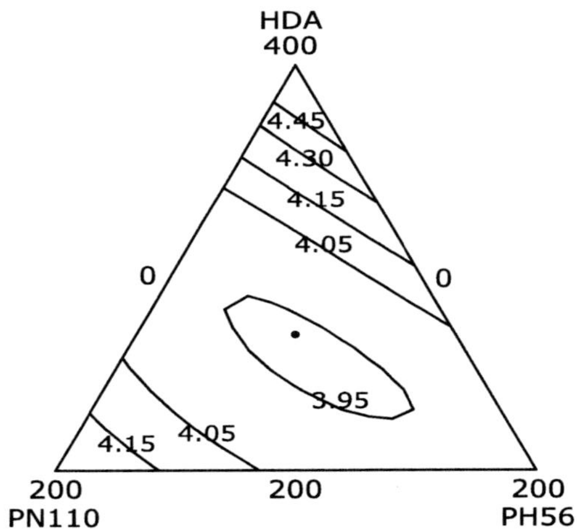

Figure 12.8. *Propagation of error: 1-minute green strength. Reduced quadratic model.*

For simulation purposes, three mixture blends were selected, all of which lie along the Piepel-effect direction for HDA when the base point is the overall centroid. In terms of the actuals, the blends were (HDA, PN110, PH56) = (380, 10, 10)[2], (266.66, 66.67, 66.67) (ID 11, the overall centroid), and (200, 100, 100) (ID 8, the midpoint of the PN110–PH56 edge). In the simulation experiment, each blend was assumed to be made 10 times. A random normal generator was used to generate the e_i, $i = 1, 2, 3$, with a mean of 0 and a standard deviation of 4. For each blend, the errors were added to the aim amounts in terms

[2]The composition (400, 0, 0) (IDs 1 and 2) was not selected as an example blend for the simulation for the following reason. When a component is absent from the desired blend, errors should really not be added to zero because there would be no error in the measured amount. However, POE is related to the *slope* of the response in certain directions, and as a result errors must be added to zero to pick up variation around the point. This will lead to positive and negative amounts of components whose aim level is zero. This physically impossible outcome is a disturbing way to illustrate simulation.

of the actuals. Results are summarized in columns 2–4 of Table 12.4 at the end of this chapter (page 296). Proportions were then calculated using Eq. 12.6, page 290 (columns 5–7 in the table). Finally, $\widehat{\text{GS1}}$ was estimated for each mixture using model 12.8.

The standard deviations of the actuals and the $\widehat{\text{GS1}}$ are summarized beneath each set of 10 "makes" in Table 12.4. The POE for the ith data point is defined as the square root of the sum of the variance of the simulated response *arising from mixing measurement error* plus s^2 from the analysis of variance.

$$\text{POE}_i = \sqrt{\text{var}(\hat{Y}_i) + s^2} \tag{12.10}$$

For the specific example used here, this can be reexpressed as

$$\text{POE}_i = \sqrt{\text{var}(\widehat{\text{GS1}}_i) + 15.526}$$

Keep in mind that $\text{var}(\widehat{\text{GS1}})$ is the variance arising from mixing measurement error, *not* prediction variance based on the model.

For the three design points in Table 12.4 the calculated POEs are

$$\begin{aligned}\text{POE}_{\text{ID}\,.} &= \sqrt{2.2884^2 + 15.526}\\ &= 4.56\end{aligned}$$

$$\begin{aligned}\text{POE}_{\text{ID 11}} &= \sqrt{0.1134^2 + 15.526}\\ &= 3.94\end{aligned}$$

$$\begin{aligned}\text{POE}_{\text{ID 8}} &= \sqrt{0.8209^2 + 15.526}\\ &= 4.02\end{aligned}$$

Values reported by Design-Expert are 4.51, 3.94, and 4.02. The small difference between the simulated value for the first point and the value calculated analytically by Design-Expert arises because of the small sample size for the simulation. If instead of 10 samples, several thousand had been simulated, then the simulated result would asymptotically approach the analytical value.

To illustrate the effect of incorporating POE into an optimization procedure, consider the following three scenarios:

1. Maximize GS1, ignore POE.
2. Maximize GS1 (importance = 1), minimize POE (importance = 1).
3. Maximize GS1 (importance = 1), minimize POE (importance = 5).

Scenario 3 is included simply to illustrate the effect that differences in relative importance can have on the location of an optimum.

For each scenario, let

$$\begin{aligned}\text{GS1}_* &= 25\\ \text{GS1}^* &= 58\\ \text{POE(GS1)}_* &= 3.9403\\ \text{POE(GS1)}^* &= 4.7561\end{aligned}$$

The minimum value for GS1 (25 psi) is the minimum desired value reported by Hillshafer et al.; the maximum value is the maximum observed value (Table 12.3, page 291). The minimum and maximum values for POE are the minimum and maximum values calculated by Design-Expert assuming the response surface is defined by model 12.8, page 291. The compositions corresponding to the minimum and maximum POEs were (HDA, PN110, PH56) $= (255.8, 65.3, 78.9)$ and $(400, 0, 0)$, respectively. Note that the composition at the minimum POE is very close to the composition for the overall centroid (Table 12.3, page 291, ID 11).

Numerical optimization for these three scenarios leads to the following results:

1.
$$(\text{HDA}, \text{PN110}, \text{PH56}) = (400.00, 0, 0)$$
$$\widehat{\text{GS1}} = 53.39$$
$$\widehat{\text{POE}} = 4.7561$$
$$D = \left[\left(\frac{53.39 - 25}{58 - 25}\right)\left(\frac{4.7561 - 4.7561}{4.7561 - 3.9403}\right)^0\right]$$
$$= ((0.8603)(1.0000)) = 0.8603$$

2.
$$(\text{HDA}, \text{PN110}, \text{PH56}) = (200.00, 196.14, 3.86)$$
$$\widehat{\text{GS1}} = 46.96$$
$$\widehat{\text{POE}} = 4.2785$$
$$D = \left[\left(\frac{46.96 - 25}{58 - 25}\right)\left(\frac{4.7561 - 4.2785}{4.7561 - 3.9403}\right)\right]^{1/2}$$
$$= ((0.6655)(0.5854))^{1/2} = 0.6242$$

3.
$$(\text{HDA}, \text{PN110}, \text{PH56}) = (200.00, 127.77, 72.23)$$
$$\widehat{\text{GS1}} = 35.40$$
$$\widehat{\text{POE}} = 4.0759$$
$$D = \left[\left(\frac{35.40 - 25}{58 - 25}\right)\left(\frac{4.7561 - 4.0759}{4.7561 - 3.9403}\right)^5\right]^{1/6}$$
$$= \left((0.3152)(0.8338)^5\right)^{1/6} = 0.7090$$

Figure 12.9 displays three-dimensional surface plots for desirability for each scenario. The plateau in each surface (more obvious in scenarios 2 and 3 than in scenario 1) identifies regions where the overall desirability is equal to zero. The filled circle in each contour plot shows the location of the optimum. The highest point on each desirability surface lies directly above each filled circle.

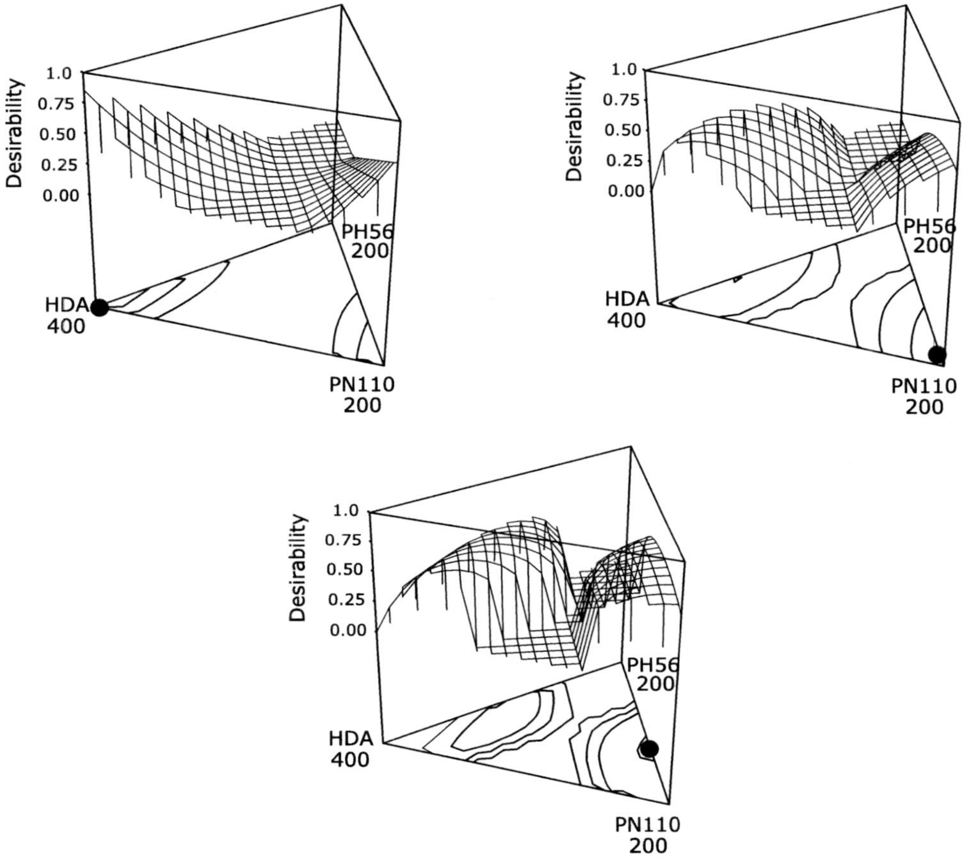

Figure 12.9. *3D surface plots of desirability. Upper left, scenario 1; upper right, scenario 2; bottom, scenario 3.*

As the importance of POE relative to GS1 rises from 0 → 1 → 5, the predicted values for $\widehat{GS1}$ at the point of maximum desirability decrease (become less desirable) from 53.39 → 46.96 → 35.40. At the same time, the predicted values for $\widehat{POE}$ decrease (become more desirable) from 4.7561 → 4.2785 → 4.0759. Note in Fig. 12.9 the dramatic shift in the location of the point of maximum desirability when minimization of POE is assumed to be as important as maximization of GS1.

For a different approach to the inclusion of robustness into optimization procedures in mixture design problems, see de Boer, Smilde, and Doornbos [39, 40, 41], who define a **robustness coefficient** that is based on proportional measurement error (Eq. 12.7, page 290) rather than absolute measurement error (Eq. 12.6, page 290). Details are not provided here because at the present time their method has not been incorporated into software with experimental design functionality. Their methods can be implemented using software with matrix algebra capabilities.

Table 12.4. *Simulation: 1-minute green strength*

ID	Actuals HDA	Actuals PN110	Actuals PH56	Reals HDA	Reals PN110	Reals PH56	$\widehat{GS1}$
.	380.81	7.5324	0.9464	0.978	0.019	0.002	48.483
	378.92	14.045	6.5201	0.949	0.035	0.016	44.037
	380.70	9.9510	5.5747	0.961	0.025	0.014	46.369
	389.48	7.6070	8.3673	0.961	0.019	0.021	47.345
	389.19	4.7402	7.3085	0.970	0.012	0.018	49.117
	377.90	16.996	14.905	0.922	0.041	0.036	41.532
	382.95	8.9400	5.9189	0.963	0.022	0.015	46.906
	385.59	11.995	12.713	0.940	0.029	0.031	44.358
	383.97	7.8054	15.519	0.943	0.019	0.038	46.034
	383.27	10.779	15.380	0.936	0.026	0.038	44.538
s	3.95	3.58	5.02	.	.	.	2.2884
11	267.17	61.124	63.729	0.682	0.156	0.163	24.976
	265.18	69.051	73.227	0.651	0.169	0.178	24.793
	268.49	71.474	63.954	0.665	0.177	0.158	24.658
	269.36	65.867	63.804	0.675	0.165	0.160	24.794
	261.33	61.557	73.182	0.660	0.155	0.185	24.905
	263.16	68.657	69.026	0.657	0.171	0.172	24.760
	272.71	74.868	62.515	0.665	0.183	0.152	24.586
	261.09	66.457	71.761	0.654	0.166	0.180	24.805
	271.59	65.283	69.762	0.668	0.161	0.172	24.864
	266.91	64.198	64.931	0.674	0.162	0.164	24.845
s	4.02	4.283	4.25	.	.	.	0.1134
8	203.73	98.394	100.49	0.506	0.244	0.250	30.087
	200.56	102.62	99.915	0.498	0.255	0.248	31.122
	203.50	100.73	108.53	0.493	0.244	0.263	30.576
	206.53	98.012	100.39	0.510	0.242	0.248	29.782
	198.08	100.33	98.880	0.499	0.253	0.249	30.939
	210.47	100.08	102.46	0.510	0.242	0.248	29.815
	202.07	93.280	105.13	0.505	0.233	0.263	29.395
	196.97	97.372	101.30	0.498	0.246	0.256	30.528
	201.57	101.79	92.148	0.510	0.257	0.233	30.762
	199.92	108.29	99.744	0.490	0.265	0.244	32.244
s	4.00	3.92	4.24	.	.	.	0.8209

Part IV

Special Topics

Chapter 13

Including Process Variables

It is generally true — and particularly so in an industrial setting — that the performance of a formulation will depend not only on the mixture blend but also on the settings of certain **process variables**. Examples of process variables are mixing time and temperature, cooking time and temperature (as in food preparations), mold temperature and pressure (as in plastics formulations), melt temperature and hold time (as in photographic coatings and ink jet formulations), coating thickness (as in ablative, ink jet, powder, protective, or UV curable coatings), amount (as in drug dosage or fertilizer and pesticide applications), and so forth.

As was the case in the mixture setting, models for mixture-amount (MA) and mixture-process variable (MPV) experiments will be developed first, and then designs to support the models will be considered.

13.1 Models

Historically models and designs that include "amount" were developed separately from models and designs that include other process variables. Nonetheless, "amount" can be considered as simply a special case of a process variable. The methodology for model development was put forth by Piepel and Cornell [128] in the context of MA experiments, but it is applicable to MPV experiments in general.

It is convenient to illustrate the development by example and then to generalize the result. Without loss of generality, then, let us hypothesize that a linear Scheffé model will satisfactorily describe the response in a three-component mixture setting. The model is

$$E(Y)_{\mathrm{M}} = \beta_1 X_1 + \beta_2 X_2 + \beta_3 X_3 \tag{13.1}$$

where the subscript M simply denotes the model as the mixture part of the MPV model. Let us also assume that we would like to include two process variables, Z_1 and Z_2, in both the model and design but that a linear Scheffé model will continue to adequately model the response at the levels of Z_j, $j = 1, 2$. However, it is suspected that as the levels of Z_1 and Z_2 are changed, the values of the linear coefficients β_i, $i = 1, 2, 3$, may change. Notationally, we could represent the coefficients in Eq. 13.1 as $\beta_i(Z_j)$, where the parenthetical Z_j means that the β_i are *functions* of the levels of the process variables. The functional dependence could take any of several forms, but one of the simplest would be a polynomial form such as

$$\beta_i(Z_j) = \beta_i^0 + \beta_i^1 Z_1 + \beta_i^2 Z_2 + \beta_i^{12} Z_1 Z_2 \qquad i = 1, 2, 3 \tag{13.2}$$

In this notation, the subscript i identifies the mixture component, X_i, while the superscript identifies the process-variable term(s). A zero superscript means that there is no process variable in the term. In this particular example we are restricting the process-variable polynomial to a two-factor interaction model, but higher-order terms could be included as well.

Substituting Eq. 13.2 for $i = 1, 2, 3$ into 13.1 leads to the MPV model

$$\begin{aligned} E(Y) = \quad & (\beta_1^0 + \beta_1^1 Z_1 + \beta_1^2 Z_2 + \beta_1^{12} Z_1 Z_2) X_1 \\ & + (\beta_2^0 + \beta_2^1 Z_1 + \beta_2^2 Z_2 + \beta_2^{12} Z_1 Z_2) X_2 \\ & + (\beta_3^0 + \beta_3^1 Z_1 + \beta_3^2 Z_2 + \beta_3^{12} Z_1 Z_2) X_3 \end{aligned} \tag{13.3}$$

Although this is a specific example, we can use it to make several generalizations about MPV models.

First, if we decide in advance that a two-factor interaction model in the process variables will adequately describe the dependence of the response on the process variables, then the process-variable model can be written

$$E(Y)_{\text{PV}} = \alpha_0 + \alpha_1 Z_1 + \alpha_2 Z_2 + \alpha_{12} Z_1 Z_2 \tag{13.4}$$

where the subscript PV on the left-hand side denotes the model as the process-variable part of the MPV model. Equation 13.3 can be derived simply by multiplying Eq. 13.1 by Eq. 13.4 and then cleaning up the notation. For example, multiplying the first term in Eq. 13.1 by Eq. 13.4 leads to

$$\begin{aligned} (\alpha_0 + \alpha_1 Z_1 + \alpha_2 Z_2 + \alpha_{12} Z_1 Z_2)\beta_1 X_1 &= (\alpha_0\beta_1 + \alpha_1\beta_1 Z_1 + \alpha_2\beta_1 Z_2 + \alpha_{12}\beta_1 Z_1 Z_2) X_1 \\ &\equiv (\beta_1^0 + \beta_1^1 Z_1 + \beta_1^2 Z_2 + \beta_1^{12} Z_1 Z_2) X_1 \end{aligned}$$

Note that the subscripts on the αs in Eq. 13.4 become the superscripts on the βs in Eq. 13.3. Generalizing, one can derive a MPV model by multiplying a proposed mixture model, $E(Y)_{\text{M}}$, by a proposed process-variable model, $E(Y)_{\text{PV}}$, and then simplifying the notation. In short

$$\text{MPV model} = \text{Mixture model} \times \text{Process-variable model}$$

Second, one could in effect "transpose" model 13.3 and write it in the equivalent form

$$\begin{aligned} E(Y) = \quad & \beta_1^0 X_1 + \beta_2^0 X_2 + \beta_3^0 X_3 \\ & + (\beta_1^1 X_1 + \beta_2^1 X_2 + \beta_3^1 X_3) Z_1 \\ & + (\beta_1^2 X_1 + \beta_2^2 X_2 + \beta_3^2 X_3) Z_2 \\ & + (\beta_1^{12} X_1 + \beta_2^{12} X_2 + \beta_3^{12} X_3) Z_1 Z_2 \end{aligned} \tag{13.5}$$

Each line in Eq. 13.5 displays the parametric dependence of one term in the process-variable model (including the intercept) on the mixture blends. On the other hand, each line in Eq. 13.3 displays the parametric dependence of one term in the mixture model on the levels of the process variables. It will be convenient to refer to MPV model forms exemplified by Eqs. 13.3 and 13.5 as types X and Z, respectively, the letter designating the variate (mixture or process variable) that is outside of the parentheses. Keep in mind that these are *not* different parameterizations but simply two different ways of writing the same model.

Third, one could write models 13.3 and 13.5 in multiplied-out form:

$$\begin{aligned} E(Y) = {} & \beta_1^0 X_1 + \beta_2^0 X_2 + \beta_3^0 X_3 + \beta_1^1 X_1 Z_1 + \beta_2^1 X_2 Z_1 + \beta_3^1 X_3 Z_1 + \beta_1^2 X_1 Z_2 \\ & + \beta_2^2 X_2 Z_2 + \beta_3^2 X_3 Z_2 + \beta_1^{12} X_1 Z_1 Z_2 + \beta_2^{12} X_2 Z_1 Z_2 + \beta_3^{12} X_3 Z_1 Z_2 \end{aligned} \tag{13.6}$$

This is a confusing way to write the model and becomes even more confusing as the size of the mixture and/or process-variable part of the model increases. For example, if one combined a six-term quadratic Scheffé model with a 10-term quadratic model in three process variables, then the multiplied-out form would have a string of 60 terms. Of course writing the model in either of the forms of Eqs. 13.3 or 13.5 does not reduce the number of terms, but it does present the model in a form that is structurally interpretable. Furthermore, once one understands the model structure, it is easy to write out a MPV model without going through the multiplication step.

Fourth, with reference to the model example Eq. 13.5, assume that

$$\begin{aligned} \beta_1^1 &= \beta_2^1 = \beta_3^1 \\ \beta_1^2 &= \beta_2^2 = \beta_3^2 \\ \text{and } \beta_1^{12} &= \beta_2^{12} = \beta_3^{12} \end{aligned} \tag{13.7}$$

Each of the three sets of assumptions has a single superscript — 1 in the first set, 2 in the second set, and 12 in the third set. For the first set, and with reference to Eq. 13.5, one could write

$$\begin{aligned} (\beta_1^1 X_1 + \beta_2^1 X_2 + \beta_3^1 X_3) Z_1 &= \beta_{1,2,3}(X_1 + X_2 + X_3) Z_1 \\ &= \beta_0^1 Z_1 \end{aligned}$$

where the zero subscript means that there is no mixture component in the term, just as a zero superscript means that there is no process variable in a term. Another way of looking at this assumption is to think of the process variable term interacting identically with each mixture variable, which will have the effect of simply moving the response surface up or down but not changing its shape. The same considerations apply to the second and third assumptions in Eqs. 13.7.

If the above assumptions hold, then the MPV model would assume the form

$$E(Y) = \beta_1^0 X_1 + \beta_2^0 X_2 + \beta_3^0 X_3 + \beta_0^1 Z_1 + \beta_0^2 Z_2 + \beta_0^{12} Z_1 Z_2 \tag{13.8}$$

Note that every term in the model has either a zero subscript or a zero superscript. The model is considered an *additive* model because the levels of the process variables do not influence the blending properties of the mixture components as embodied in the β_i^0, $i = 1, 2, 3$. Models 13.3, 13.5, and 13.6 are *interaction* models because the blending properties of the mixture are influenced by the levels of the process variables (Eq. 13.3), or equivalently, the effect of the process variables is influenced by the composition of the mixture (Eq. 13.5). In a model such as 13.8, the effect of the process variables is simply to offset the mixture response surface but not to change its shape. An additive model such as Eq. 13.8 has fewer terms than an interaction model such as Eq. 13.5 and therefore requires fewer design points to support the model. It is not uncommon for additive MPV models to be proposed, but in so doing one is assuming a priori that equalities such as those in Eq. 13.7 hold.

Because MPV models can have a large number of terms — thus requiring many design points to support the models — there is interest in finding ways to minimize the size of the model. Consider crossing the mixture model

$$E(Y)_{\text{M}} = \beta_1 X_1 + \beta_2 X_2 + \beta_3 X_3 + \beta_{12} X_1 X_2 + \beta_{13} X_1 X_3 + \beta_{23} X_2 X_3 \qquad (13.9)$$

with the process-variable model

$$E(Y)_{\text{PV}} = \alpha_0 + \alpha_1 Z_1 + \alpha_2 Z_2 + \alpha_{12} Z_1 Z_2 + \alpha_{11} Z_1^2 + \alpha_{22} Z_2^2 \qquad (13.10)$$

The resulting 36-term MPV model will contain terms such as $X_1 X_2 Z_1$ and $X_1 Z_1 Z_2$, which are of order 3, as well as terms like $X_1 X_2 Z_1 Z_2$ and $X_1 X_2 Z_1^2$, which are of order 4. One might postulate instead a smaller MPV model with terms up to order 2 only, with the realization that if lack of fit is indicated, the model and possibly the design will need augmentation. A model with terms up to order 2 in this example would be

$$\begin{aligned} E(Y) = \quad & \beta_1^0 X_1 + \beta_2^0 X_2 + \beta_3^0 X_3 + \beta_{12}^0 X_1 X_2 + \beta_{13}^0 X_1 X_3 + \beta_{23}^0 X_2 X_3 \\ & +(\beta_1^1 X_1 + \beta_2^1 X_2 + \beta_3^1 X_3) Z_1 \\ & +(\beta_1^2 X_1 + \beta_2^2 X_2 + \beta_3^2 X_3) Z_2 \\ & +\beta_0^{12} Z_1 Z_2 + \beta_0^{11} Z_1^2 + \beta_0^{22} Z_2^2 \end{aligned} \qquad (13.11)$$

The model includes the quadratic Scheffé mixture model (line 1), two-way interactions between the linear terms in the mixture variables and the main effects in the process variables (lines 2 and 3), the two-factor interaction term between the process variables (line 4), and the pure quadratic terms in the process variables (line 4). The terms in line 4 — all with zero subscripts — do not affect the blending properties of the mixture and are thus additive in the sense described above. Admittedly this is not a small model, but it does have 21 terms less than the full quadratic $\times$ quadratic model.

Models such as Eq. 13.11, in which additive terms in the process variables are combined with interaction terms between mixture and process variables, shall be referred to as composite MPV models. Composite models of order 2 have been proposed by Kowalski, Cornell, and Vining [83] to combat the problem of large models and designs in mixture-process variable settings. The authors recommend several efficient designs that support these models for cases where the mixture space is simplex-shaped. One can envisage models such as Eq. 13.11 as members of a ladder of models. When resources are limited, the lowest rung of the ladder could be an additive model such as Eq. 13.8 but with first-order terms only in the process variables. The second rung might be a model containing terms up to order 2 such as Eq. 13.11. Higher rungs would contain terms of higher order still. When lack of fit is indicated, one must weigh cost and time constraints against moving up a rung of the ladder.

If resources were *not* limited, then one could of course begin with a design to support a model with higher-order terms and then use variable selection and/or the extra sum-of-squares principle to reduce the model to a more parsimonious one. More will be said about this in Section 13.4.

13.2 Designs

Candidate designs to support MPV models are constructed by performing what is called a **Cartesian join** ($\otimes$) of a mixture design with a process-variable design. In a Cartesian join, a mixture design is assigned to every treatment combination of a process-variable design, or equivalently, a process-variable design is assigned to every mixture blend of a mixture design. This is symbolized

$$\text{MPV design} = \text{Mixture design} \otimes \text{Process-variable design}$$

To illustrate, consider the two designs in Fig. 13.1. The design on the left is a simplex centroid design, and it will support the quadratic Scheffé model Eq. 13.9, page 302, as well as the special cubic model. The design on the right is a face-centered central composite design (or a 3-level factorial) and will support the quadratic model Eq. 13.10, page 302. A Cartesian join of the two designs is illustrated in Fig. 13.2. In this illustration the mixture design has been assigned to every treatment combination of the process-variable design. The crossed design could have been illustrated just as well by one large triangle with rectangles located at the seven designs points of the simplex centroid design. In either case, there are $9 \times 7 = 63$ design points in the candidate design.

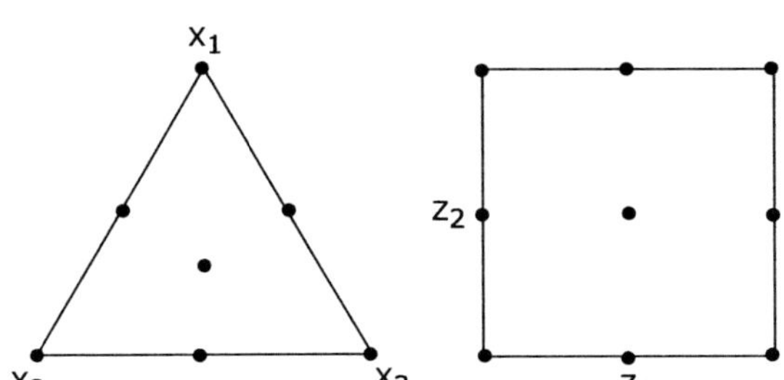

Figure 13.1. *Mixture and process-variable designs.*

The adjective "candidate" is used because in most cases this will lead to designs with many more points than necessary. For example, the 63-point design in Fig. 13.2 will support a quadratic $\times$ quadratic MPV model which has $6 \times 6 = 36$ terms. This leaves 27 degrees of freedom for lack of fit, many more than necessary.

As cost usually enters into the picture, rather than designing to support a full 36-term quadratic $\times$ quadratic MPV model at the outset, let us instead consider designing to support a somewhat smaller model. If we choose a model with terms up to order 2, then this would

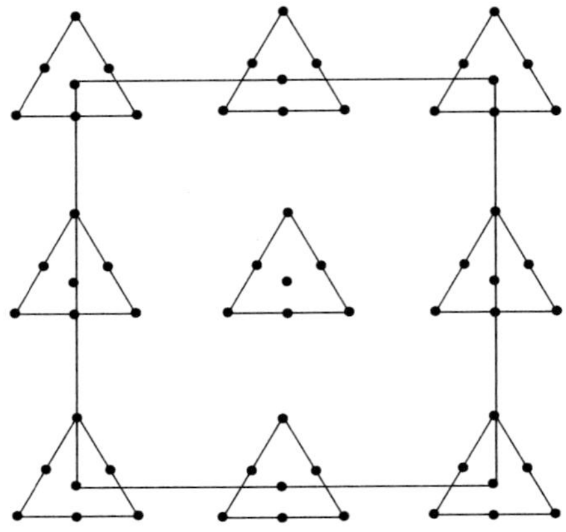

Figure 13.2. *63-point mixture-process variable design.*

be the 15-term model on page 302 (model 13.11). This is a specific example of the models proposed by Kowalski, Cornell, and Vining (KCV).

The two designs in Fig. 13.3 are both 23-point designs, and both will support the 15-term MPV model Eq. 13.11. The design on the left was suggested by Kowalski, Cornell, and Vining, while the design on the right is a D-optimal design created using Design-Expert (the arrows are explained below). The KCV design exemplifies the type of designs proposed by these authors for a variety of mixture-process variable settings. For the process-variable part of the design, a face-centered central composite design is proposed (Fig. 13.1, right, illustrates the design for two factors). The vertices of the simplex centroid design (Fig. 13.1, left, illustrates the design for three mixture components) are then run at half of the factorial points (A and C) and the edge centroids run at the other half (B and D). Because they alternate, if the design is collapsed across the levels of either process variable, then one

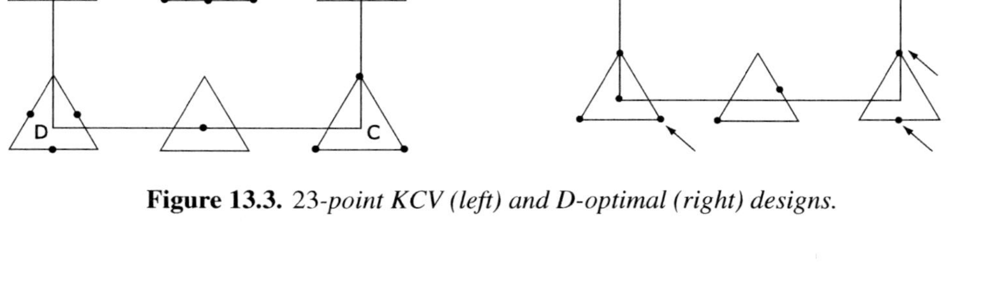

Figure 13.3. *23-point KCV (left) and D-optimal (right) designs.*

gets a full simplex centroid design at the low and high levels of the remaining process variables. An attractive feature of the design is its symmetry. The highest leverage points in the design are the 12 points at the vertices of the rectangle, and so these are good candidates for replication.

In the general case, KCV designs are constructed by appropriate combination of a 2-level fraction of a simplex centroid design (Section 4.3.2, page 50) with a face-centered central composite design in the process variables. (An example of a 2-level fraction of a simplex centroid design is illustrated in Fig. 4.11, page 52, for the {4,2} design.) "Appropriate combination" means the following:

1. The vertices of the $\{q, 2\}$ simplex centroid design are run at alternating vertices of the central composite design.

2. The edge centroids of the $\{q, 2\}$ simplex centroid design are run at the remaining vertices of the central composite design.

3. The overall centroid of the $\{q, 2\}$ simplex centroid design is run at the axial points of the central composite design.

4. Either the full $\{q, 2\}$ simplex centroid design, or only the overall centroid, is run at the center of the central composite design.

The design on the right in Fig. 13.3 is a D-optimal design. The design was selected by Design-Expert using the 63 design points in Fig. 13.2 as candidates to support model 13.11, page 302. The points with arrows are the five highest leverage points and would be good candidates for replication.

There are pros and cons to the D-optimal design. The scaled D-optimality criterion (page 80) for the KCV design is 12.09, while that for the D-optimal design is 10.11. This means that the confidence ellipsoid for the parameter estimates using the D-optimal design will be somewhat smaller than that for the KCV design. One could say that the efficiency of the KCV design relative to the D-optimal design is equal to $100 \times (10.11/12.09) \approx 83.6\%$.

On the other hand, the dissymmetry of the D-optimal design leads to contours of the standard error of prediction that are quite asymmetrical. The plots in Fig. 13.4 are Design-Expert contour plots of the standard errors of prediction apart from σ (cf. Eq. 6.8, page 105). The contours are based on the assumption that the model to be fit is the 15-term model 13.11, page 302. One can choose whether to display such plots in the context of either the process-variable design or the mixture design. In either case one has to choose the levels of the mixture variables (when viewing the process design) or the process variables (when viewing the mixture design). The plots in Fig. 13.4 are for the case where $(X_1, X_2, X_3) = (0..333, 0.333, 0.333)$. The two "flags" in each design are explained below.

When viewed in the context of the process design, there are more design points in the center of the KCV design than in the D-optimal design, but there are more design points on the boundary of the D-optimal design than the KCV design (cf. Fig. 13.3). This is reflected in the standard errors of prediction, the KCV design having lower standard errors of prediction near the center but higher standard errors of prediction near the boundary. The contour intervals in Fig. 13.4 are the same for both designs, suggesting that the standard errors of prediction are more "even" throughout the design region in the case of the D-optimal design.

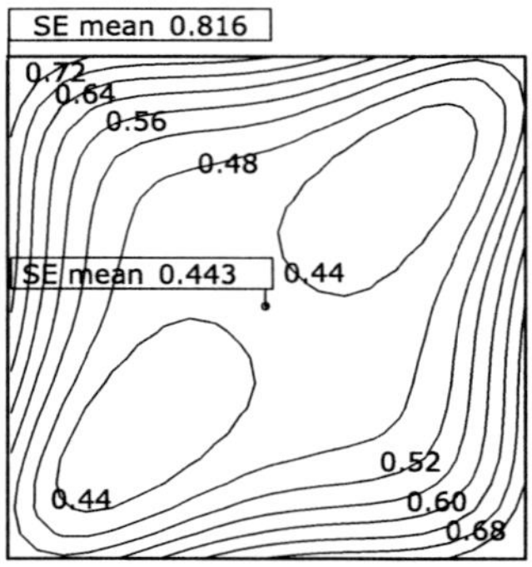

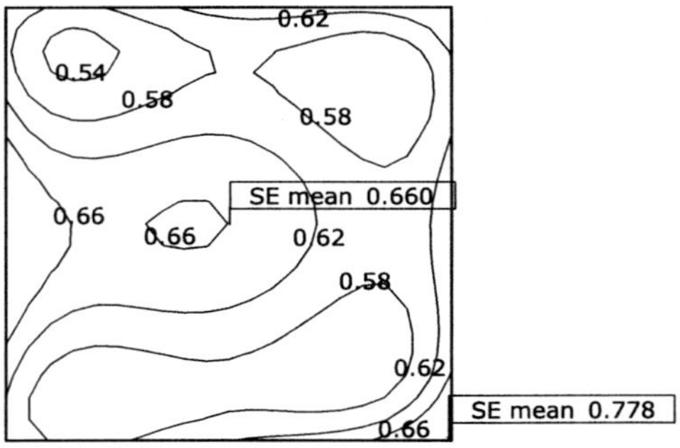

Figure 13.4. *Design-Expert standard error plots. KCV design (left) vs. D-optimal design (right).* $(X_1, X_2, X_3) = (0.333, 0.333, 0.333)$.

A more global view of precision in prediction can be gained by viewing variance dispersion graphs, introduced in Section 6.2, page 100. Goldfarb et al. [56] describe an approach for mixture-process variable experiments in which shrinkage factors are applied to *both* the mixture ($f_{mixture}$) *and* the process-variable ($f_{process}$) parts of a design. The plots illustrated in Fig. 13.5 were created using this method.

To create these plots, 60 equally spaced points were first generated on the boundary of the 3-simplex (20 per edge) and 80 equally spaced points on the boundary of the process design (20 per edge). Shrinkage values of 1.0, 0.9, 0.8, ..., 0.0 were then applied to both design spaces, leading to $11 \times 11 = 121$ possible combinations of $f_{mixture}$ and $f_{process}$. When $f_{mixture} = f_{process} = 0$ there will be only a single point of interest, the overall centroid of the mixture process-variable design. When $f_{mixture} = 0$ but $f_{process} \neq 0$, there will be 80 points of interest per shrinkage value for $f_{process}$. When $f_{process} = 0$ but $f_{mixture} \neq 0$, there will be 60 points of interest per shrinkage value for $f_{mixture}$. And when *both* $f_{mixture}$ *and* $f_{process}$ are nonzero, there will be $60 \times 80 = 4800$ points of interest per combination of shrinkage values. For each combination of shrinkage values, the maximum, minimum, and average prediction variances (apart from σ^2) were calculated (although only the maximum and average variances are discussed here). With reference to Eq. 6.8, page 105, the maximum prediction variance is max $(\mathbf{x}_0'(\mathbf{X}'\mathbf{X})^{-1}\mathbf{x}_0)$ and the average prediction variance is avg $(\mathbf{x}_0'(\mathbf{X}'\mathbf{X})^{-1}\mathbf{x}_0)$. $\mathbf{x}_0'$ is a 15×1 vector whose elements are the values of the regressor variables in model Eq. 13.11, page 302, for each point of interest. $\mathbf{X}$ is the model matrix (cf. page 78) for the KCV design or the D-optimal design.

The top two plots in Fig. 13.5 display contours of maximum prediction variance, while the bottom two plots display contours of average variance. The left two plots are for the KCV design, and the right two plots are for the D-optimal design. Contour intervals are the same for the two top plots and the same for the two bottom plots, but the interval in the bottom plots is half that in the top plots. The only exception to this is the dashed contours, which were included to fill out the region near $f_{mixture} = f_{process} = 0$ in the Dopt-max plot. The KCV design appears to have generally lower variances yet wider "swings" in the prediction variances, whereas the variances in the D-optimal design, although generally larger, are more "evened out". This is particularly apparent in the plots of average prediction

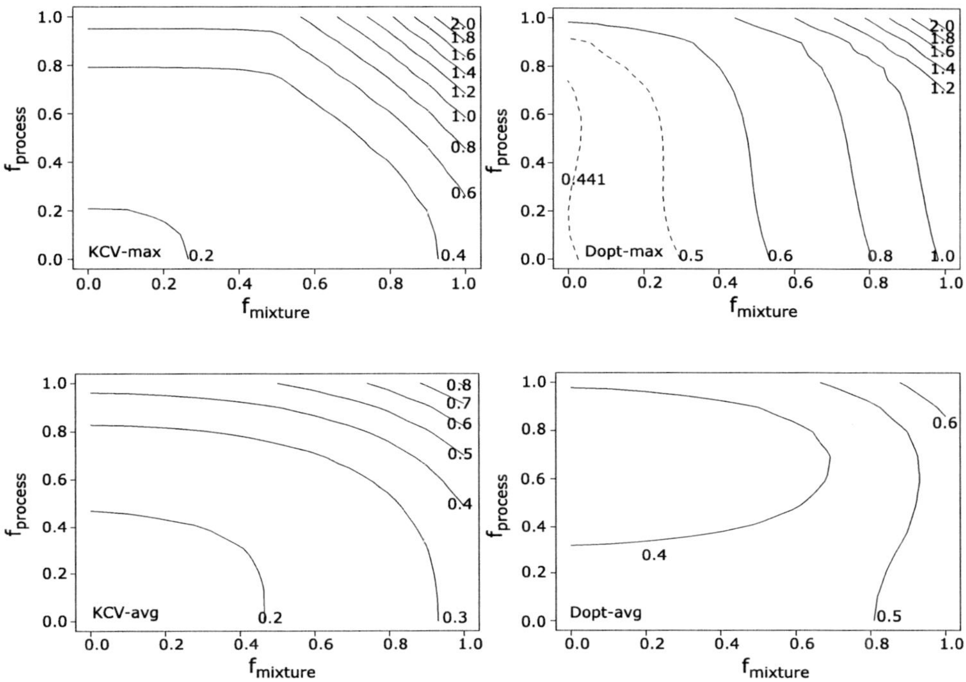

Figure 13.5. *Variance dispersion contour plots. Top left: 23-point KCV, maximum prediction variance; top right: 23-point D-optimal, maximum prediction variance; bottom left: 23-point KCV, average prediction variance; bottom right: 23-point D-optimal, average prediction variance.*

variance. Because both the KCV and D-optimal designs have the same number of design points, it is fair to make comparisons such as these. When comparing designs with different numbers of design points, one might prefer to use scaled prediction variances as discussed on page 114.

A question that naturally arises is how the plots in Fig. 13.4 are related to the plots in Fig. 13.5. Keep in mind that contours in Fig. 13.4 are contours of standard errors of prediction (apart from σ), whereas contours in Fig. 13.5 are contours of prediction variance (apart from σ^2). Also keep in mind that the plots in Fig. 13.4 are for the case where $(X_1, X_2, X_3) = (0.333, 0.333, 0.333)$, which means that $f_{mixture} = 0$. The flag at the center of the left figure in Fig. 13.4 identifies the single point in Fig. 13.5 located at $f_{mixture} = f_{process} = 0.0$ for the KCV design. Similarly, the flag at the center of the right figure in Fig. 13.4 identifies the single point in Fig. 13.5 located at $f_{mixture} = f_{process} = 0.0$ for the D-optimal design. The flag at the boundary of the left figure in Fig. 13.4 identifies the maximum standard error of prediction on the boundary when $f_{mixture} = 0$, and similarly for the flag at the boundary of the right figure. These points are equivalent to the points in the top two contour plots in Fig. 13.5 where $f_{mixture} = 0.0$ and $f_{process} = 1.0$.

When there are lower and upper bounds on the component proportions, more often than not the mixture design region will be asymmetrical. In these circumstances the D-optimal design approach becomes an attractive approach to MPV design. In cases where there are three or more process variables, one does have an additional recourse. By using fractional design plans in the process variables, savings in the number of experiments can be achieved at the sacrifice of estimating certain (usually higher-order) interaction effects among the process variables. This approach has been discussed by Cornell and Gorman [32] and Cornell [29].

13.3 Collecting Data

Milliken and Johnson devote a section in their book, *Analysis of Messy Data*, to an interesting discussion of the structure of an experimental design (cf. [96, Section 4.2]). These authors point out that an experimental design consists of two parts, the **treatment structure** and the **design structure**. Often the treatment structure by itself is referred to as the "experimental design". The treatment structure consists of the set of treatments or treatment combinations that an experimenter has selected to study. For example, in a mixture setting, a simplex lattice design is an example of a treatment structure, while in a process-variable setting, a factorial design is an example of a treatment structure. The design structure, on the other hand, has to do with how the experiment is carried out. For the designs discussed up to this point, the tacit assumption has been made that the design structure is a completely randomized design structure.

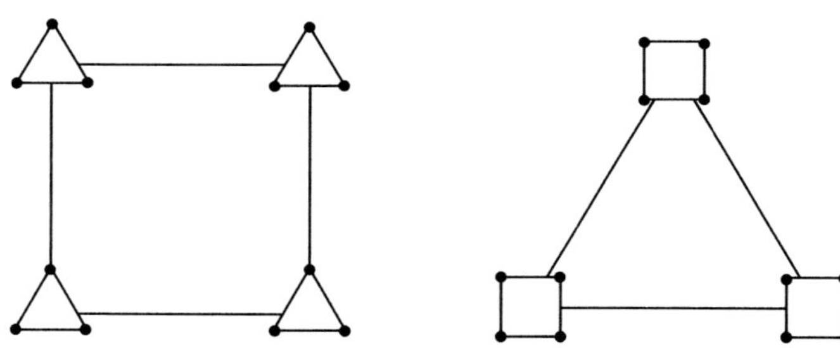

Figure 13.6. *A* 12*-run mixture-process variable design.*

Consider the 12-run crossed design in Fig. 13.6. This can be viewed as a two-way treatment structure in which the levels of the simplex design have been crossed with the levels of a factorial design. The treatment combinations can be displayed in two apparently equivalent fashions. Depending on the design structure, one can envisage four different ways of carrying out the experiment.

1. If the mixture blends are easy to prepare and the treatment combinations of the process variables are easy to set, then one of the 12 design points is selected at random. The mixture is prepared, the levels of the process variables set, and a response measurement made. A second design point is then selected at random, and the procedure repeated. This is continued until the responses for the 12 design points have been measured. This is a completely randomized design structure.

2. If the treatment combinations of the process variables are hard to set, an experimenter may prefer to randomly select a combination of levels of the process variables and then prepare and run all of the mixture blends at this combination. Next, a second combination of levels of the process variables would be randomly chosen, and then all of the mixture blends prepared again and run at the second combination. The procedure is repeated until all possible treatment combinations have been run. This approach is suggested by the illustration on the left in Fig. 13.6. This is known as a *split-plot design* and has been referred to as embedding the mixture blends within each process variable combination [27]. In the context of a split-plot design, the four settings of the two process variables (i.e., low–low, low–high, high–low, and high–high) are called whole-plot experimental units. The mixture blends in the simplex design are called subplot experimental units.

3. If it is easier to reset the process variables than it is to prepare the mixture blends, then an experimenter may opt to prepare just three mixture blends. One of the blends would be randomly selected and divided into four samples. A response measurement would be made for each sample at each of the four treatment combinations of the process variable. This approach is suggested by the right-hand illustration in Fig. 13.6. This is also a split-plot design and has been referred to as embedding the process variable combinations within each mixture blend [27]. In this case the whole-plot and subplot experimental units are the reverse of case (2).

4. If it is difficult to reset the process variables *and* to prepare the mixture blends, then an experimenter may prepare three mixture blends and set the process variables to a randomly chosen treatment combination. One-fourth of each mixture blend would then be tested in random order at the chosen combination of levels of the process variables. Following this, a second combination of the process variables would be randomly chosen and one-fourth of each mixture blend tested in random order at these settings. The process would continue until all the measurements were complete. This is known as a *strip-plot design* [96].

It is often the case that the only practical way to carry out a MPV experiment is in a split-plot mode. When this is the case, there are certain consequences that the reader should be aware of. First, owing to the **restricted randomization** of the process variables (Case 2) or the mixture blends (Case 3), two sources of error exist. One source of error exists among the whole-plot experimental units and a second source among the subplot experimental units. The situation is somewhat more complicated than this, however, because to have a measure of the whole-plot error requires replication. To keep the analysis relatively simple (or perhaps less complex), what is often done is simply to replicate the entire design, leading to what is called a **balanced design**. With reference to Fig. 13.6, there would be 24 design points, two at each of the treatment combinations. Other replication methods have been suggested by Kowalski, Cornell, and Vining [84]. The gain in efficiency by restricting randomization is consequently offset to some degree by the need for additional experiments.

A split-plot approach to analyzing mixture experiments containing process variables has been described by Cornell [27] and by Kowalski, Cornell, and Vining [84]. See also Cornell, Sections 7.3 and 7.4 [29]. The method requires software capable of doing matrix calculations and is somewhat complex. Readers using point-and-click software may wonder what the consequences are if a MPV experiment carried out in a split-plot mode is analyzed as though it had a completely randomized design structure. To the author's knowledge there has been no definitive study of this, but the following three examples provide some clues.

1. Cornell provides a hypothetical example in which a {3,1} simplex lattice design is crossed with a 2^2 factorial design [27]. The result is the left illustration in Fig. 13.6, i.e., the whole-plot experimental units are the combined levels of the process variables. The entire design was replicated so that there are 24 responses. Using a split-plot analysis, Cornell found that the terms X_1Z_2, X_2Z_2, X_3Z_2, and $X_2Z_1Z_2$ were significantly different from zero at the $\alpha = 0.05$ level. (Remember that the linear terms in the mixture variables are not tested against zero.) Assuming a completely randomized design structure, the same terms (and no more) are found significant.

2. Cornell provides another example in which blends of three plasticizers are tested at four combinations of two process variables (extrusion rate and drying temperature) in a factorial treatment structure [27]. The response is scaled vinyl thickness. The crossed design is similar to that in Fig. 13.6, except that there were lower and upper bounds on the proportions of each plasticizer. The mixture design region is shaped like a parallelogram, with a design point at each of the four vertices of the parallelogram and an additional point at the overall centroid. The design was replicated, and so there is a total of 2 replicates $\times$ 5 mixture blends $\times$ 4 process-variable treatment combinations $=$ 40 design points. The candidate model was model 13.11, page 302, without the terms in X_1X_3 and X_2X_3. Using a split-plot approach, significant terms at the $\alpha = 0.05$ level were X_1Z_2, X_2Z_2, X_3Z_2, and $X_2Z_1Z_2$. Assuming a completely randomized design structure, the same terms (and no more) are found significant.

3. A third example is provided in Section 7.4 of Cornell [29]. The mixture components and process variables were the same as in example 2. In this case, however, the mixture design was a {3,2} simplex lattice design. The plasticizers constituted 40.7% of all formulations, the remainder consisting of components such as stabilizers, lubricants, etc. The design was replicated, and so there were 2 replicates $\times$ 6 mixture blends $\times$ 4 process-variable treatment combinations $=$ 48 design points. The candidate model was the quadratic Scheffé model crossed with a two-factor interaction model in the process variables, for a total of $6 \times 4 = 24$ terms. Using a split-plot approach, significant terms at the $\alpha = 0.05$ level were X_1X_2, X_1X_3, $X_2X_3Z_2$, $X_1Z_1Z_2$, $X_2Z_1Z_2$, and $X_1X_2Z_1Z_2$. Assuming a completely randomized design structure, all but $X_2X_3Z_2$ are found significant. This term has a p value of 0.0685 and is the next most significant term.

In the final analysis the experimenter is going to have to make his or her own decision whether to analyze a MPV experiment conducted in split-plot mode as a completely randomized design. There is no doubt that the more accurate analysis is the split-plot analysis. Weighed against this, however, is the increase in experiment size coupled with the complexity of the analysis.

13.4 Analysis

Fitting MPV models to data is more complex, but at the same time more interesting, than fitting mixture models to data. The steps remain much the same: sequential sums of squares to assist in making an informed choice of model order, and partial sums of squares for model refinement. Different software products handle these in different ways, and so it is worth looking at the output of two (Design-Expert and MINITAB) that have dedicated MPV functionality.

Consider the case where a $q = 3$ quadratic Scheffé model is crossed with a two-factor interaction model in three process variables. The model will have $6 \times 7 = 42$ terms. MINITAB sequentially builds the MPV model line by line as follows:

$$
\begin{aligned}
& (\text{linear terms} + \text{quadratic terms} + \cdots) \\
+ \quad & (\text{linear terms} + \text{quadratic terms} + \cdots) \cdot Z_1 \\
+ \quad & (\text{linear terms} + \text{quadratic terms} + \cdots) \cdot Z_2 \\
+ \quad & (\text{linear terms} + \text{quadratic terms} + \cdots) \cdot Z_1 Z_2 \\
+ \quad & (\text{linear terms} + \text{quadratic terms} + \cdots) \cdot Z_3 \\
+ \quad & (\text{linear terms} + \text{quadratic terms} + \cdots) \cdot Z_1 Z_3 \\
+ \quad & (\text{linear terms} + \text{quadratic terms} + \cdots) \cdot Z_2 Z_3 \\
+ \quad & \text{Residual}
\end{aligned}
$$

This is a type Z model form (cf. page 301), each line displaying the parametric dependence of a process-variable term (including the intercept, line 1) on the mixture-variable terms. In this example the parenthetical terms would be truncated after the quadratic terms, and each of the seven lines would be split into two lines, one for the linear mixture terms and the other for the quadratic mixture terms. If the mixture model were a special cubic model, each parenthetical mixture model would consist of three groups of terms, and the MPV model would be built by sequentially adding seven sets of three terms. In the example considered here there would be a total of 14 lines, each line showing the sequential sum of squares plus an F ratio and p value for the additional terms. In MINITAB the denominator for the F ratio is taken as the residual in line 8 rather than as a pooled residual that would include the sum of squares for those terms not included in the particular model.

In the approach discussed above, a fourth-order term such as $X_1 X_2 Z_1 Z_2$ will enter the model before a second-order term such as $X_1 Z_3$. A different approach, which is the approach used by Design-Expert, is to first sort the rows in a type Z model form by the degree of the process-variable terms, so that the model would be expressed as follows:

$$
\begin{aligned}
& (\text{linear terms} + \text{quadratic terms} + \cdots) \\
+ \quad & (\text{linear terms} + \text{quadratic terms} + \cdots) \cdot Z_1 \\
+ \quad & (\text{linear terms} + \text{quadratic terms} + \cdots) \cdot Z_2 \\
+ \quad & (\text{linear terms} + \text{quadratic terms} + \cdots) \cdot Z_3 \\
+ \quad & (\text{linear terms} + \text{quadratic terms} + \cdots) \cdot Z_1 Z_2 \\
+ \quad & (\text{linear terms} + \text{quadratic terms} + \cdots) \cdot Z_1 Z_3 \\
+ \quad & (\text{linear terms} + \text{quadratic terms} + \cdots) \cdot Z_2 Z_3 \\
+ \quad & \text{Residual}
\end{aligned}
$$

Terms are sequentially brought into the model according to the order of the process-variable terms. Thus row 1 would be brought in first, followed by rows 2–4 as a group, followed by rows 5–7 as a group, for a total of three sequential sums of squares and p values. However, in the example used here this would be broken into two sets of three steps. The first set would contain only the parenthetical linear terms, while the second set would *include* the quadratic terms. If the mixture model contained special cubic terms, then there would be a third set that would include these terms, and so on. The sequential sums of squares are therefore calculated assuming different model forms for the mixture part of the model.

Design-Expert also sequentially builds the model using the type X model form (cf. page 301). The model can be expressed as follows:

$$\begin{aligned} & (\text{intercept} + \text{linear terms} + \text{2FI terms} + \cdots) \cdot X_1 \\ + \quad & (\text{intercept} + \text{linear terms} + \text{2FI terms} + \cdots) \cdot X_2 \\ + \quad & (\text{intercept} + \text{linear terms} + \text{2FI terms} + \cdots) \cdot X_3 \\ + \quad & (\text{intercept} + \text{linear terms} + \text{2FI terms} + \cdots) \cdot X_1X_2 \\ + \quad & (\text{intercept} + \text{linear terms} + \text{2FI terms} + \cdots) \cdot X_1X_3 \\ + \quad & (\text{intercept} + \text{linear terms} + \text{2FI terms} + \cdots) \cdot X_2X_3 \\ + \quad & \text{Residual} \end{aligned}$$

In the example used here, the parenthetical process-variable terms would be truncated after the two-factor interaction ("2FI") terms, for a total of seven terms within each set of parentheses. In this case terms are brought into the model according to the order of the mixture-variable terms. This means that rows 1–3 are brought in as a group, followed by rows 4–6 as a group. As in the type Z case, this is broken into sets according to the order of the parenthetical process-variable model. In the example used here, there would be three sets of two. The first set would contain only the parenthetical intercepts, which means that the sequential sum of squares would be calculated first for the linear mixture model and then for the additional terms in the quadratic mixture model. The second set of two would *include* the linear terms in the process variables, while the third set of two would include the two-factor interaction terms.

As a result, Design-Expert Version 6 displays a summary table that looks like Table 13.1. Apart from the first row and the first column, each cell entry contains two p values, both of which are based on sequential sums of squares. Using square brackets to distinguish mixture-variable terms from process-variable terms, consider the two p values for the [Quadratic] $\times$ 2FI cell. The [Quadratic] by 2FI model could arise sequentially by (1) augmentation of a [Quadratic] $\times$ Linear model with the two-factor interaction terms or (2) augmentation of a [Linear] $\times$ 2FI model with the quadratic mixture terms. The p values in the cells are based on F ratios calculated using a residual sum of squares that pools the sum of squares for higher-order terms that are not in the model, plus any residual sum of squares left over after fitting the highest-order model. Design-Expert recommends that the analyst choose a candidate model where the additional terms are significant for *both* mixture (augmentation by row) *and* process (augmentation by column). An example of using this type of table will be given in the Case Study at the end of the chapter.

Turning to partial sums of squares, in the course of variable selection one may be confronted with model terms of the form

$$(\beta_1^k X_1 + \cdots + \beta_q^k X_q + \text{possibly higher-order terms})Z_k \tag{13.12}$$

Table 13.1. *Design-Expert Fit Summary table*

	Process				
Mixture	Mean	Linear	2FI	Quadratic	Cubic
Mean					
Linear					
Quadratic					
Sp. Cubic					
Cubic					

Let Z_k be a generic notation for *any* process variable term, whether it be linear, two-factor interaction, quadratic, etc. In addition, let β_j be a generic representation for the coefficients in "possibly higher-order terms". Thus, β_j represents the coefficient for any quadratic, special cubic, cubic, etc. parenthetical term. When there are terms such as in Eq. 13.12 in the model, there are at least two types of hypotheses that may be of interest. The first is

$$\mathrm{H}_0: \quad \beta_i^k = 0 \tag{13.13}$$

and the second is

$$\begin{aligned} \mathrm{H}_0: \quad & \beta_1^k = \beta_2^k = \cdots = \beta_q^k && \text{(Linear Terms)} \\ & \beta_j^k = 0 && \text{(Higher-order Terms)} \end{aligned} \tag{13.14}$$

With regard to the first hypothesis, while it does not make sense to test H_0: $\beta_i^0 = 0$ (where β_i^0 is a linear coefficient in a Scheffé mixture model), it *does* make sense to test H_0: $\beta_i^k = 0$, $k \neq 0$, in a MPV model. Consider the simple MPV model

$$\begin{aligned} E(Y) = \quad & \beta_1^0 X_1 + \beta_2^0 X_2 + \beta_{12}^0 X_1 X_2 \\ & + (\beta_1^1 X_1 + \beta_2^1 X_2 + \beta_{12}^1 X_1 X_2) Z_1 \end{aligned}$$

When Z_1 is coded as ± 1, the terms β_1^1 and β_2^1 are simply quantities that are added to $(Z_1 = +1)$ or subtracted from $(Z_1 = -1)$ the respective β_i^0. If it were true that

$$\beta_1^1 = \beta_2^1 = 0$$

then the resulting MPV model can be written

$$E(Y) = \beta_1^0 X_1 + \beta_2^0 X_2 + (\beta_{12}^0 + \beta_{12}^1 Z_1) X_1 X_2$$

This model differs from the quadratic Scheffé model only in the magnitude of the nonlinear blending term. The interpretation would be that the process variable does have an effect on the nonlinear blending of components 1 and 2.

Removing linear terms of the form $\beta_i^k X_i$ from MPV models can lead to scale-dependency issues (cf. page 224). When there is a string of terms such as in Eq. 13.12, the

MPV model will usually be scale-dependent if any one of the q linear terms is removed from the string. An exception to this is when $q = 2$. This means that the summary statistics in a model fitted in the pseudos will differ from the summary statistics for the same model fitted in the reals. This applies not only to interaction models but to composite models as well. Unless summary statistics in both metrics have been checked, one should exercise caution interpreting a model in the pseudos. An easy way to get into this problem is when using automated model building algorithms such as backward elimination.

With respect to the second hypothesis, if the effect of Z_k were simply additive, then the string of terms in Eq. 13.12 could be replaced by a single term, $\beta_0^k Z_k$, resulting in considerable simplification of the model. However, the single term $\beta_0^k Z_k$ is not a subset of the terms in Eq. 13.12. One has two recourses in this situation. One is to fit both models and use the extra sum-of-squares principle to make a decision. The fuller model would be the model containing the terms in Eq. 13.12, and the less full model would replace all terms in Eq. 13.12 with the single term, $\beta_0^k Z_k$.

An alternative approach has been suggested by Gorman and Cornell [58]. The reader is reminded of the discussion of intercept mixture models in Chapter 3. One can *replace* any linear term in a mixture model by an intercept term. Component 1 is a logical choice, and so the linear terms in Eq. 13.12 can be reparameterized as

$$(\delta_0^k + \delta_2^k X_2 + \cdots + \delta_q^k X_q + \text{possibly higher-order terms}) Z_k \tag{13.15}$$

where δ_0^k replaces $\beta_1^k X_1$ and $\delta_i^k = (\beta_i^k - \beta_1^k)$ for $i = 2, \ldots, q$. One can now use variable selection techniques to test whether all terms other than δ_0^k are equal to zero. If so, then the string of terms in Eq. 13.12 reduces to $\delta_0^k Z_k \equiv \beta_0^k Z_k$.

13.5 Related Applications

It has been implied in the previous sections that the process variables were *controllable* process variables. An important area of industrial experimentation is that of robust design. Robust design takes into consideration *uncontrollable* process variables, often referred to as **noise variables**. Although uncontrollable in normal operation, for purposes of experimentation it is assumed that the levels of noise variables can be controlled. Designs and models that include noise variables have the same structure as designs and models that include controllable process variables, although the analyses are somewhat more complex.

Including noise variables in an MPV experiment leads to models that contain interactions between mixture variables, controllable process variables, and noise variables. Such models are not useful for optimization per se because the noise variables are uncontrollable in normal operation. Using expectation and variance operators, the fitted model can be used to derive one model for the process mean and another for the process variance. These can then be jointly optimized using response surface methods. For an excellent overview of the methodology, see Myers and Montgomery [107]. A recent application to a soap formulation experiment is described in Goldfarb, Borror, and Montgomery [55].

Recently there has been increasing interest in **mixture-of-mixtures** problems. Interest in this subject goes back to the seminal papers by Lambrakis [86, 89], in which a theory was developed for mixtures of "major" components, where each major component is itself a mixture of minor components. When the major components constitute categories, then this has been called the **categorized-components problem** [31, 34]. For example, in formulating a medication two major categories might be *drug* and *excipient*,[1] the drug category consisting of two or more drugs, and the excipient category of two or more excipients. When the major components do not constitute categories, the more general term *mixture-of-mixtures* (MoM) has been suggested [124].

Historically there have been two types of MoM experiments, referred to by Piepel as types A and B [124]. In a type A MoM experiment, the proportions of the major components are fixed, and only the proportions of the minor components are varied. This approach is described in articles by Lambrakis [86, 89], Cornell and Good [31], Piepel [124] and Dingstad, Egelandsdal, and Næs [45]. A type B MoM is one in which the proportions of *both* the major and the minor components are varied. These types of experiments are of interest when one would like to understand the effect of intercategory blending on intracategory blending, or vice versa. An example of this experimental setting can be found in Cornell and Ramsey [34]. See also Cornell, Sections 4.13.2 and 4.13.3 [29].

Designs and models for mixture-amount and mixture-process variable experiments, mixture experiments that include noise variables, and mixture-of-mixtures experiments, although differing in detail, do have many things in common, and they can all be classified under the general heading of **crossed experiments**.

Case Study

Chardon et al. [16] were interested in developing a finishing product for polyester-cotton cloth used as bed linen in the hotel trade. The finishing product formula was to be based on the use of three fabric softeners and a catalyzed resin. The design for the experiment is displayed in Fig. 13.7.

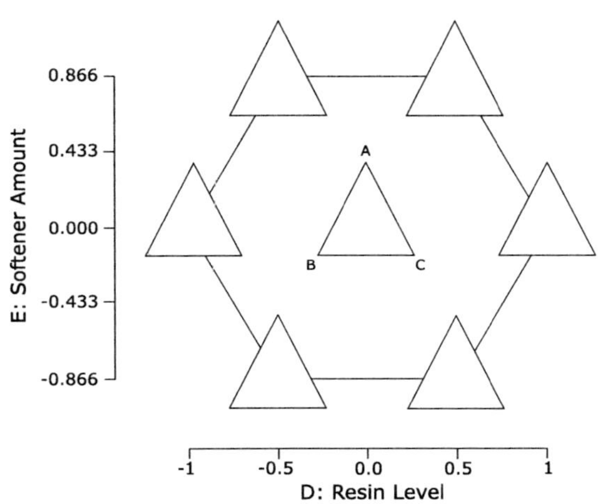

Figure 13.7. *Design for the textile finishing-product experiment.*

[1] An excipient is a pharmacologically inert adhesive substance used to bind the contents of a tablet or pill.

The mixture design for the softeners (labeled A, B, and C) consisted of the seven blends of a simplex centroid design. One process variable was resin level (D), and the other was total amount of softener(s) (E). The experiment can be viewed as a combined MA-MPV experiment. The low and high levels of resin were 30 and 150 g/l, respectively, while for the softener they were 45 and 105 g/l.

The process-variable part of the design is a hexagon design, a special case of an **equiradial design** [46, 47, 107]. The design has the desirable property of being rotatable, which means that contours of $\sqrt{h_{00}}$ are spherical about the center of the design. This property is retained in the MA-MPV design as well.

Crossing the seven mixture blends with the seven treatment combinations of the hexagon design leads to a total of 49 experiments. A table summarizing the design settings and the responses is displayed near the end of this section (Table 13.3, page 321). Coded values are used for the process-variable settings. A response of interest was the hydrophilicity of the polyester-cotton cloth, evaluated by the absorption time for a drop of water. It is not clear from the article whether the authors' "hydrophilicity property value" is the same thing as the absorption time. In any case, the authors sought a finishing product formula that would minimize their hydrophilicity property value.

The hydrophilicity property values in Table 13.3 range over more than two orders of magnitude. In such a situation it is almost certain that a transformation will be required. This was confirmed by fitting various MPV models to the untransformed hydrophilicity property values. Plots of the scaled residuals vs. the fitted values exhibited the classic funnel pattern, and the Box–Cox procedure (page 237) recommended a log transformation. As a result, all analyses were carried out using ln(hydrophilicity) as the response.

Table 13.2. *Finishing product experiment. Fit Summary table for mixture × process variable models*

	Process			
Mixture	Mean	Linear	2FI	Quadratic
Mean		0.0582	0.6015	0.2496
Linear	<0.0001	<0.0001	0.5087	0.0182
		<0.0001	<0.0001	<0.0001
Quadratic	0.1321	<0.0001	0.7608	0.0070
		0.0162	0.0800	0.0111
Sp. Cubic	0.6148	0.0002	0.8838	0.0898
		0.8631	0.9572	0.8496

Table 13.2 is a Design-Expert table of p values based on the sequential building of various crossed models. The top number in each cell with two entries is for augmentation by column (process), while the bottom number is for augmentation by row (mixture). The highest-order polynomial where the additional terms are significant for both process and mixture is the [Quadratic] × Quadratic model. Using square brackets to designate

the mixture part of the model, the value 0.0070 is the p value for augmentation of the [Quadratic] × 2FI crossed model

$$\begin{aligned} E(Y) = \; & (\beta_A^0 A + \beta_B^0 B + \beta_C^0 C + \beta_{AB}^0 AB + \beta_{AC}^0 AC + \beta_{BC}^0 BC) \\ & + (\beta_A^D A + \beta_B^D B + \beta_C^D C + \beta_{AB}^D AB + \beta_{AC}^D AC + \beta_{BC}^D BC)D \\ & + (\beta_A^E A + \beta_B^E B + \beta_C^E C + \beta_{AB}^E AB + \beta_{AC}^E AC + \beta_{BC}^E BC)E \\ & + (\beta_A^{DE} A + \beta_B^{DE} B + \beta_C^{DE} C + \beta_{AB}^{DE} AB + \beta_{AC}^{DE} AC + \beta_{BC}^{DE} BC)DE \end{aligned}$$

with all terms that are quadratic in the process variables:

$$\begin{aligned} & (\beta_A^{D^2} A + \beta_B^{D^2} B + \beta_C^{D^2} C + \beta_{AB}^{D^2} AB + \beta_{AC}^{D^2} AC + \beta_{BC}^{D^2} BC)D^2 \\ & + (\beta_A^{E^2} A + \beta_B^{E^2} B + \beta_C^{E^2} C + \beta_{AB}^{E^2} AB + \beta_{AC}^{E^2} AC + \beta_{BC}^{E^2} BC)E^2 \end{aligned}$$

As before, subscripts identify the mixture component(s) and superscripts identify the process variable(s). When augmenting by process variable, it is helpful to have in mind a type Z model form (page 301) for the MPV model.

The value 0.0111 is the p value for augmentation of the [Linear] × Quadratic crossed model

$$\begin{aligned} E(Y) = \; & (\beta_A^0 + \beta_A^D D + \beta_A^E E + \beta_A^{DE} DE + \beta_A^{D^2} D^2 + \beta_A^{E^2} E^2)A \\ & + (\beta_B^0 + \beta_B^D D + \beta_B^E E + \beta_B^{DE} DE + \beta_B^{D^2} D^2 + \beta_B^{E^2} E^2)B \\ & + (\beta_C^0 + \beta_C^D D + \beta_C^E E + \beta_C^{DE} DE + \beta_C^{D^2} D^2 + \beta_C^{E^2} E^2)C \end{aligned}$$

with all terms that are quadratic in the mixture variables:

$$\begin{aligned} & (\beta_{AB}^0 + \beta_{AB}^D D + \beta_{AB}^E E + \beta_{AB}^{DE} DE + \beta_{AB}^{D^2} D^2 + \beta_{AB}^{E^2} E^2)AB \\ & + (\beta_{AC}^0 + \beta_{AC}^D D + \beta_{AC}^E E + \beta_{AC}^{DE} DE + \beta_{AC}^{D^2} D^2 + \beta_{AC}^{E^2} E^2)AC \\ & + (\beta_{BC}^0 + \beta_{BC}^D D + \beta_{BC}^E E + \beta_{BC}^{DE} DE + \beta_{BC}^{D^2} D^2 + \beta_{BC}^{E^2} E^2)BC \end{aligned}$$

When augmenting by mixture variable, it is helpful to have in mind a type X model form for the MPV model. In either case the final model is the same. It has simply been built in two different ways. Because both p values in the [Quadratic] × Quadratic cell are significant, and because this is the highest-order polynomial where this is true, the [Quadratic] × Quadratic model should be a good candidate with which to begin the model-building process.

The [Quadratic] × Quadratic model has $6 \times 6 = 36$ model terms. Table 13.4, page 323, summarizes the OLS parameter estimates and p values for the 36-term model. The table is divided into three sections corresponding to three different sorting methods. The data in columns 1–3 are *exactly* the same as the data in columns 4–6 and *exactly* the same as the data in columns 7–9. The reason for sorting the results by three different methods is because different views of the same model can sometimes provide insights into what the model is trying to tell us. The model is clearly overspecified, as 23 of the 36 terms have $p > 0.05$, and of these 22 have $p > 0.10$.

In columns 1–3 the terms are sorted by mixture-variable order, corresponding to a type X model structure. Starting with the highest-order terms in the mixture variables (bottom of the table), one would look to see if perhaps all terms involving either AB, AC, or BC might not be statistically significant. (A p value of 0.0111 is not a guarantee that *all* terms in AB, AC, and BC will necessarily be significant.) This is not the case, and so this does not help very much.

In columns 4–6 the terms are sorted by process-variable order, corresponding to a type Z model structure. Again, one would start with the highest-order terms in the process variables (bottom of the table) to see if any generalizations can be made. (As before, a p value of 0.0070 is not a guarantee that *all* terms in D^2 and E^2 will necessarily be significant.) The results are again mixed, and no sweeping simplification appears obvious.

In columns 7–9 the terms are sorted by total order, from first through fourth. Here there is a pattern: all terms of order 4 are not statistically significant at the 0.05 level of significance. As a first step, then, one might try removing these nine terms from the model. Summary statistics for the full (36-term) model are $R^2 = 0.9829$, $R^2_{adj} = 0.9368$, and $R^2_{pred} = 0.3679$; for the reduced 27-term model the results are $R^2 = 0.9749$, $R^2_{adj} = 0.9453$, and $R^2_{pred} = 0.7947$.

In the reduced model there are still several terms with p values $\gg 0.05$, most of which are third-order terms. If backward elimination is applied to this model, eight additional terms are removed (seven of which are third order) leading to a 19-term model with $R^2 = 0.9693$, $R^2_{adj} = 0.9509$, and $R^2_{pred} = 0.9068$.

In the 19-term model the coefficient estimates for AE^2, BE^2, and CE^2 have 95% confidence intervals that overlap substantially, suggesting that they might be replaced by the single term E^2.

			95% CI	
Coefficient	Estimate	p	Low	High
AE^2	−0.6343	0.0047	−1.0583	−0.2104
BE^2	−0.8095	0.0005	−1.2334	−0.3855
CE^2	−0.9573	0.0005	−1.4607	−0.4539

If these three terms are replaced by E^2, the resulting 17-term model has $R^2 = 0.9684$, $R^2_{adj} = 0.9526$, and $R^2_{pred} = 0.9197$. In going from the 36-term → 27-term → 19-term → 17-term model, the difference between R^2 and R^2_{adj} has decreased from 0.0461 → 0.0296 → 0.0184 → 0.0158. Because R^2_{adj} has steadily increased, mean square error has steadily decreased.

Using the extra sum-of-squares principle, the F ratio for the 19 degree-of-freedom composite hypothesis corresponding to all of the assumptions made in the 17-term model is 0.579 with 19 numerator and 13 denominator degrees of freedom. This corresponds to a p value of 0.8647. Of the 19 degrees of freedom, 17 are for the hypothesis that 17 parameter estimates are equal to zero, and two are for the hypothesis that three parameter estimates are equal to one another.

The final model with coefficient estimates to two significant digits expressed in type X format is

$$
\begin{aligned}
\hat{Y} = & \quad (2.7 + 0.77D - 0.92E - 0.79E^2)A \\
+ & \quad (5.4 + 0.72E + 0.69DE - 0.79E^2)B \\
+ & \quad (5.7 + 0.41D + 0.30E - 0.64D^2 - 0.79E^2)C \\
+ & \quad (-1.7 + 1.8E)AB \\
+ & \quad (1.8 + 2.7E)AC \\
+ & \quad (-1.1 + 2.0D)BC
\end{aligned}
\tag{13.16}
$$

while in type Z format the model is

$$
\begin{aligned}
\hat{Y} = & \quad 2.7A + 5.4B + 5.7C - 1.7AB + 1.8AC - 1.1BC \\
+ & \quad (0.77A + 0.41C + 2.0BC)D \\
+ & \quad (-0.92A + 0.72B + 0.30C + 1.8AB + 2.7AC)E \\
+ & \quad (0.69B)DE \\
+ & \quad (-0.64C)D^2 \\
& \quad -0.79E^2
\end{aligned}
\tag{13.17}
$$

Although the model could be more compactly written on perhaps two lines, it is written in these forms to display the model structure. The type X model structure Eq. 13.16 emphasizes the dependence of the coefficients in the mixture model on the process variables; the type Z model structure Eq. 13.17 emphasizes the dependence of the coefficients in the process-variable model on the mixture blends.

In the type X model structure we see that the effect of the process variables on the linear terms of the mixture model is more complex than their effect on the quadratic terms of the mixture model. The former involves terms of order 2 in the process variables, whereas the latter involves only terms of order 1 in the process variables. This observation is obscured in the type Z model formulation.

In the type Z model structure we see that the effect of the mixture variables on the intercept and linear terms of the process-variable model involves terms of order 2 in the mixture variables. The effect of the mixture variables on the two-factor interaction and pure quadratic terms in the process variables involves only terms of order 1 in the mixture variables. This observation is not obvious in the type X formulation.

Even so, it is difficult to tell from either model formulation exactly what is going on. One can, however, gain much insight by looking at contour plots. Two approaches to presenting the plots are displayed in Fig. 13.8, page 324. In the top figure one can visualize how the response changes with mixture composition for each of the seven treatment combinations of the process variables. This figure emphasizes the type X model formulation. In the bottom figure one can visualize how the response changes with treatment combination of the process variables at each of the seven formulations of the mixture design. This figure emphasizes the type Z model formulation. The corners of the squares should be ignored in this figure because the design region is actually hexagonal. As in the top figure, resin amount is on the x-axis of the squares and softener amount on the y-axis. In both figures the contour interval is 0.5 and the numbers are equal to the predicted values for ln(hydrophilicity).

One obvious difference between the two figures is that there is much less variability between contour plots in the upper figure than in the lower figure. In the upper figure, all the response surfaces slope downhill towards the A vertex, somewhat more steeply at the high level of softener than at the low level. For any of the seven treatment combinations of the process variables, the lowest predicted hydrophilicity (the goal of the experiment) would occur at or near the vertex for surfactant A. Considering all the treatment combinations, the combination of factors that would produce the lowest predicted hydrophilicity would be softener A at a high softener amount and a low resin level. The same conclusion was reached using the numerical optimizer in Design-Expert.

Suppose, however, that softener B was significantly less costly than softener A. If this were the case, then the contour plot at the B vertex in the lower figure implies that one would be better off operating at a low softener amount and a high resin level. If softener C were the least costly of the three, then the contour plot implies that one should operate at a low softener amount and low resin level.

In conclusion, when fitting large MPV models it can be helpful to first sort the terms in the complete model into structurally interpretable forms, as in Table 13.4. In most cases the complete model will be overspecified, but one can nevertheless learn something about what may or may not be important by studying p values. It is also recommended that any reduced models be written out in a structurally interpretable form, as in Eqs. 13.16 and 13.17. Contour plots and/or trace plots (when there are a large number of mixture components or factors) can be of great help in understanding these relatively complicated models as well as in communicating conclusions and recommendations to colleagues and management. Readers who anticipate analyzing MPV experiments should also consider recommendations in Cornell [28]. See also Cornell, Section 7.12 [29].

Table 13.3. *Finishing product experiment*

	Mixture			Process		
ID	*A*	*B*	*C*	*D*	*E*	Hydrophilicity
1	1	0	0	−1	0	10
2	0	1	0	−1	0	185
3	0	0	1	−1	0	90
4	0.5	0.5	0	−1	0	25
5	0.5	0	0.5	−1	0	40
6	0	0.5	0.5	−1	0	70
7	0.333	0.333	0.333	−1	0	30
8	1	0	0	−0.5	0.866	2
9	0	1	0	−0.5	0.866	160
10	0	0	1	−0.5	0.866	175
11	0.5	0.5	0	−0.5	0.866	25
12	0.5	0	0.5	−0.5	0.866	65
13	0	0.5	0.5	−0.5	0.866	75
14	0.333	0.333	0.333	−0.5	0.866	45
15	1	0	0	−0.5	−0.866	10
16	0	1	0	−0.5	−0.866	75
17	0	0	1	−0.5	−0.866	110
18	0.5	0.5	0	−0.5	−0.866	15
19	0.5	0	0.5	−0.5	−0.866	30
20	0	0.5	0.5	−0.5	−0.866	60
21	0.333	0.333	0.333	−0.5	−0.866	25
22	1	0	0	0	0	12
23	0	1	0	0	0	210
24	0	0	1	0	0	300
25	0.5	0.5	0	0	0	40
26	0.5	0	0.5	0	0	90
27	0	0.5	0.5	0	0	240
28	0.333	0.333	0.333	0	0	120
29	1	0	0	0.5	0.866	7
30	0	1	0	0.5	0.866	300
31	0	0	1	0.5	0.866	180
32	0.5	0.5	0	0.5	0.866	45
33	0.5	0	0.5	0.5	0.866	105
34	0	0.5	0.5	0.5	0.866	270
35	0.333	0.333	0.333	0.5	0.866	90

cont'd

Table 13.3. *cont'd*

	Mixture			Process		
ID	*A*	*B*	*C*	*D*	*E*	Hydrophilicity
36	1	0	0	0.5	−0.866	35
37	0	1	0	0.5	−0.866	45
38	0	0	1	0.5	−0.866	85
39	0.5	0.5	0	0.5	−0.866	20
40	0.5	0	0.5	0.5	−0.866	55
41	0	0.5	0.5	0.5	−0.866	75
42	0.333	0.333	0.333	0.5	−0.866	35
43	1	0	0	1	0	30
44	0	1	0	1	0	300
45	0	0	1	1	0	300
46	0.5	0.5	0	1	0	40
47	0.5	0	0.5	1	0	190
48	0	0.5	0.5	1	0	300
49	0.333	0.333	0.333	1	0	85

Table 13.4. *Finishing product experiment. Parameter estimates and p values*

By Process variable			By Mixture variable			By Order		
Term	Coeff	p	Term	Coeff	p	Term	Coeff	p
A	2.470		A	2.470		A	2.470	
AD	0.791	0.0003	B	5.332		B	5.332	
AE	−0.924	0.0001	C	5.689		C	5.689	
ADE	0.001	0.9969	AB	−0.614	0.6417	AB	−0.613	0.6417
AD^2	0.404	0.2607	AC	1.917	0.1606	AC	1.917	0.1606
AE^2	−0.579	0.1153	BC	0.116	0.9298	BC	0.116	0.9298
B	5.332		AD	0.791	0.0003	AD	0.791	0.0003
BD	0.188	0.2660	BD	0.188	0.2660	BD	0.188	0.2660
BE	0.772	0.0004	CD	0.370	0.0395	CD	0.370	0.0395
BDE	0.659	0.0625	ABD	−0.864	0.2662	AE	−0.924	0.0001
BD^2	0.152	0.6661	ACD	0.362	0.6341	BE	0.772	0.0004
BE^2	−0.843	0.0289	BCD	1.710	0.0387	CE	0.356	0.0464
C	5.689		AE	−0.924	0.0001	ABD	−0.864	0.2662
CD	0.370	0.0395	BE	0.772	0.0004	ACD	0.362	0.6341
CE	0.356	0.0464	CE	0.356	0.0464	BCD	1.710	0.0387
CDE	0.166	0.6157	ABE	1.741	0.0358	ABE	1.741	0.0358
CD^2	−0.565	0.1234	ACE	2.686	0.0032	ACE	2.686	0.0032
CE^2	−0.880	0.0235	BCE	−0.608	0.4280	BCE	−0.608	0.4280
AB	−0.614	0.6417	ADE	0.001	0.9969	ADE	0.002	0.9969
ABD	−0.864	0.2662	BDE	0.659	0.0625	BDE	0.659	0.0625
ABE	1.741	0.0358	CDE	0.166	0.6157	CDE	0.166	0.6157
$ABDE$	−0.648	0.6702	$ABDE$	−0.648	0.6702	AD^2	0.404	0.2607
ABD^2	−2.635	0.1187	$ACDE$	−0.648	0.6703	BD^2	0.152	0.6661
ABE^2	0.072	0.9645	$BCDE$	0.772	0.6126	CD^2	−0.565	0.1234
AC	1.917	0.1606	AD^2	0.404	0.2607	AE^2	−0.579	0.1153
ACD	0.362	0.6341	BD^2	0.152	0.6661	BE^2	−0.843	0.0289
ACE	2.686	0.0032	CD^2	−0.565	0.1234	CE^2	−0.880	0.0235
$ACDE$	−0.648	0.6703	ABD^2	−2.635	0.1187	$ABDE$	−0.648	0.6702
ACD^2	−0.388	0.8093	ACD^2	−0.388	0.8093	$ACDE$	−0.648	0.6703
ACE^2	0.227	0.8877	BCD^2	−1.775	0.2809	$BCDE$	0.772	0.6126
BC	0.116	0.9298	AE^2	−0.579	0.1153	ABD^2	−2.635	0.1187
BCD	1.710	0.0387	BE^2	−0.843	0.0289	ACD^2	−0.388	0.8093
BCE	−0.608	0.4280	CE^2	−0.880	0.0235	BCD^2	−1.775	0.2809
$BCDE$	0.772	0.6126	ABE^2	0.072	0.9645	ABE^2	0.072	0.9645
BCD^2	−1.775	0.2809	ACE^2	0.227	0.8877	ACE^2	0.227	0.8877
BCE^2	−1.059	0.5138	BCE^2	−1.059	0.5138	BCE^2	−1.059	0.5138

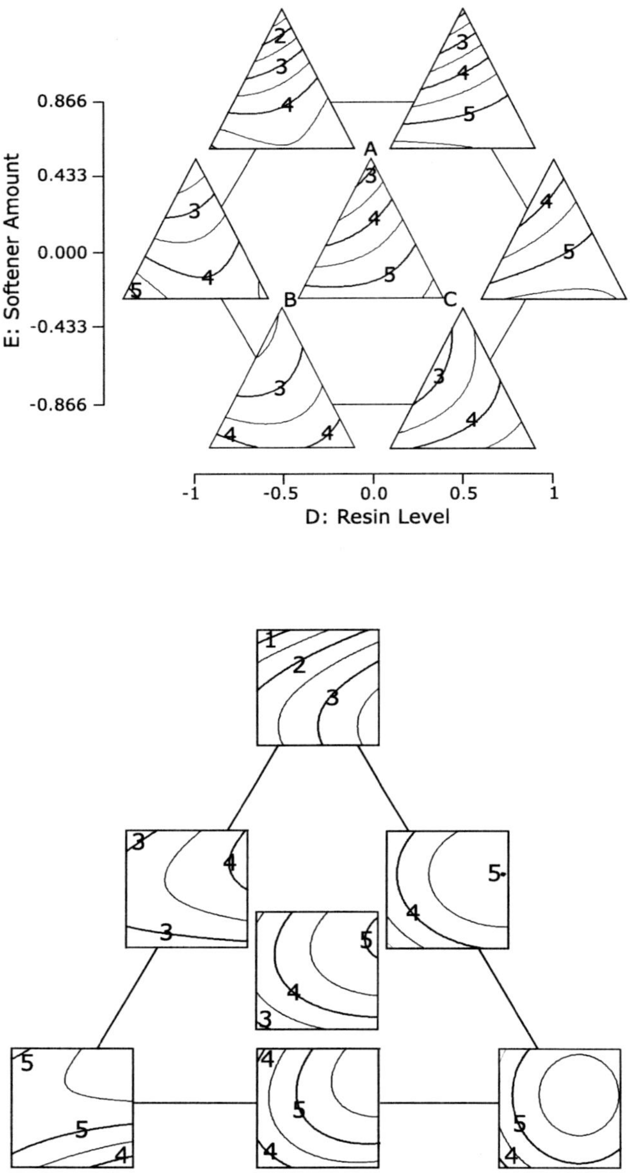

Figure 13.8. *Ln(hydrophilicity) contours.*

Chapter 14

Collinearity

Several references have been made to collinearity in previous chapters, the inferences having been that it is something to be avoided. It is now time to define exactly what collinearity is and to illustrate what its impact can be when fitting models to data. This chapter is divided into four sections. In the first section collinearity will be defined and its potential impact illustrated by revisiting an earlier data set. The second section will discuss various warnings and diagnostics that one can use to signal the presence of collinearity and to identify the underlying source(s) of the problem. The third section will provide some suggestions for combatting the problem. The chapter closes with a case study of a data set that was discussed earlier in this book.

14.1 Definition and Impact

Two situations where collinearity often arises are (1) when the *proportions* of one or more components are significantly larger (e.g., by an order of magnitude) than the proportions of one or more other components, and (2) when the *range* of one or more components is significantly larger than the range of one or more other components. The data in Table 14.1 illustrate the first situation. The column labeled A in the upper table is a set of random numbers that lie between 0 and 0.05. The column labeled B is another set of random numbers that lie between 0.9 and 1.0. The range of B is roughly 1.8 times the range of A, so there is not a large disparity in their ranges. On the other hand, the mean level of B is roughly 45 times as large as the mean level of A. Because $A + B < 1$, A and B could be thought of as random component proportions in a mixture experiment that has more than two components.

Multiplying the relatively small numbers in column A by numbers that are fairly close to 1 (column B) leads to a column of numbers (AB) that have very much the same character as column A. This is easily seen in Fig. 14.1, where AB is plotted against A. It is also apparent from inspection of the correlation matrix (the lower tabulation in Table 14.1) that A and AB are highly correlated. As regressors in a linear model, A and AB would be carriers of virtually the same information.

Table 14.1. *Hypothetical component proportions and correlation matrix*

A	B	AB
0.02816	0.9476	0.02668
0.01228	0.9749	0.01198
0.02218	0.9769	0.02167
0.04684	0.9114	0.04269
0.01293	0.9223	0.01192
0.01833	0.9741	0.01786
0.04751	0.9010	0.04281
0.006246	0.9622	0.006011
0.01005	0.9552	0.009601
0.005919	0.9662	0.005719

	A	B	AB
A	1.0000		Symmetric
B	−0.7288	1.0000	
AB	0.9992	−0.7051	1.0000

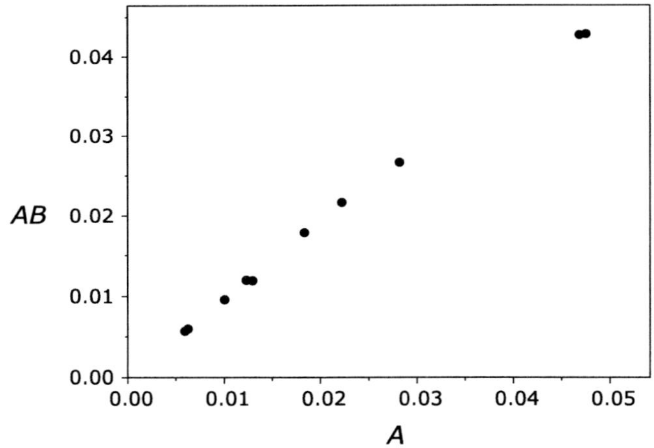

Figure 14.1. *Scatter plot of* AB *vs.* A, *Table* 14.1.

Collinearity is a condition among a set of regressor variables where one or more *near* linear dependencies exist. A **near linear dependency** among regressor variables exists whenever

$$\sum_{i=1}^{p} c_i \mathbf{x}_i \approx \mathbf{0} \qquad (14.1)$$

where p is the number of terms in the model, the c_i, $i = 1, 2, \ldots, p$, are coefficients that are not all zero, and $\mathbf{0}$ is a $n \times 1$ vector of zeros. The c_i are not arbitrary constants but are arrived at by a procedure to be fully explained in Section 14.2.[1] The symbol $\mathbf{x}_i$ (which, it should be noted, is boldface) is a column of the $n \times p$ **X** matrix and represents what can be called a *generalized* regressor — a linear, quadratic, or any higher-order term as well as an intercept should one be present. It is important to note that the "$\approx$" sign is not an "$=$" sign. Had there been an "$=$" sign, then the **X** matrix would be overparameterized because at least one column of the **X** matrix would be an *exact*, rather than a near, linear combination of one or more of the other columns.

Although Table 14.1 is not an **X** matrix, a near linear dependency does exist between columns A and AB. Letting **a** and **ab** represent 10×1 vectors with elements equal to those in columns A and AB, respectively, we can write

$$0.9234\,\mathbf{a} - \mathbf{ab} \approx \mathbf{0} \tag{14.2}$$

which is of the form of Eq. 14.1. The relationship was found by least-squares fitting of the no-intercept model $AB = \beta_A A$.

In designed experiments, and especially in designed mixture experiments, collinearities are more likely to occur when one is fitting models of order higher than one. This is because higher-order models contain crossproduct terms, such as AB in this simple illustration. (Collinearity also often creeps into observational data where the levels of the explanatory variables are observed rather than assigned.)

Table 14.2. *DMBA-induced mammary gland tumors experiment (cont'd)*

	Proportion						
ID	Fat(A)	Carb(B)	Fiber(C)	AB	AC	BC	P_i
1	0.175	0.775	0.050	0.1356	0.0087	0.0388	0.567
2	0.153	0.820	0.027	0.1255	0.0041	0.0221	0.500
3	0.133	0.863	0.004	0.1148	0.0005	0.0035	0.567
4	0.491	0.470	0.039	0.2308	0.0191	0.0183	0.800
5	0.440	0.538	0.022	0.2367	0.0097	0.0118	0.700
6	0.390	0.607	0.003	0.2367	0.0012	0.0018	0.767
7	0.701	0.267	0.032	0.1872	0.0224	0.0085	0.600
8	0.638	0.343	0.019	0.2188	0.0121	0.0065	0.767
9	0.576	0.421	0.003	0.2425	0.0017	0.0013	0.867

To illustrate the impact that collinearity can have on model fitting and interpretation, we return to the DMBA (7,12-dimethylbenz(a)anthracene)-induced tumor example (page 247). Chen, Li, and Jackson [17] studied the effect of dietary fat, carbohydrate, and fiber on the proportion (P_i) of rats out of groups of 30 exhibiting DMBA-induced mammary gland tumors under isocaloric consumption. The data in the first five columns of Table 10.14 on page 247 are repeated in Table 14.2 (columns 1–4 and 8, respectively) for ease of reference. Figure 14.2 displays the design region in the context of the pseudocomponent simplex, although numeric labels are in terms of the reals. The nine design points are indicated by filled circles.

[1]Briefly, the procedure calculates variance decomposition proportions followed by least-squares model fitting.

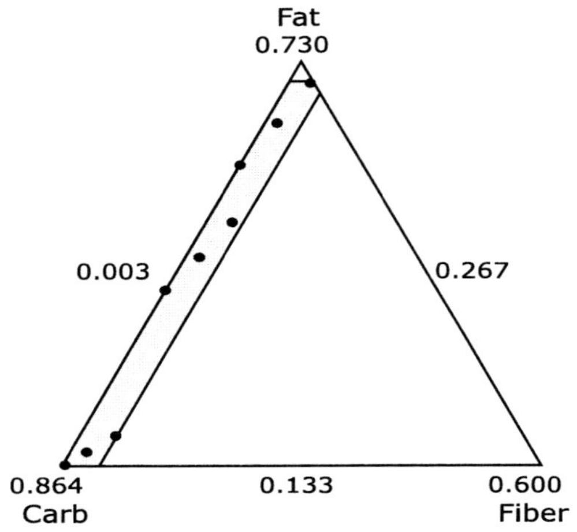

Figure 14.2. *DMBA-induced mammary gland tumors. Pseudocomponent simplex.*

The constraints on the component proportions for this experiment were

$$\begin{array}{rcccl} 0.133 & \leq & \text{Fat} & \leq & 0.701 \\ 0.267 & \leq & \text{Carb} & \leq & 0.863 \\ 0.003 & \leq & \text{Fiber} & \leq & 0.050 \end{array}$$

The mean levels of fat, carbohydrate, and fiber in the experiment are 0.411, 0.567, and 0.022, respectively. Additionally, the relative ranges for the three components are 0.953, 1.000, and 0.079, respectively. On both accounts then — disparities in proportions as well as disparities in ranges — this data set should be a good candidate for collinearity.

As discussed previously, when proportions (as a response) are distributed over a narrow range as in this example, a transformation such as a logit transformation is not only unnecessary but will have little effect. Although the authors fitted the logit normal and logistic regression models to the data, for the sake of simplicity let us proceed by fitting a normal regression model to the P_i data. However, following Chen, Li, and Jackson let us choose a Scheffé quadratic model. The columns labeled AB, AC, and BC in Table 14.2 are the Fat $\times$ Carb, Fat $\times$ Fiber, and Carb $\times$ Fiber terms, respectively.

In this example, it can be shown using the methods to be described in Section 14.2 that a near linear dependency exists between all terms that contain C (i.e., fiber), the component with the smallest range and the smallest proportions. The following approximate relationship holds:

$$\mathbf{c} \approx 1.0219\,\mathbf{ac} + 1.0508\,\mathbf{bc} \tag{14.3}$$

where **c**, **ac**, and **bc** are 9×1 vectors with elements equal to those in the columns labeled C, AC, and BC in Table 14.2. The relationship was found by least-squares fitting of the no-intercept model $C = \beta_{AC} AC + \beta_{BC} BC$. In terms of Eq. 14.1,

$$\mathbf{c} - 1.0219\,\mathbf{ac} - 1.0508\,\mathbf{bc} \approx \mathbf{0} \tag{14.4}$$

Table 14.3. *Collinearity in the DMBA-induced tumor data, quadratic Scheffé model*

ID	C	$1.0219(AC) + 1.0508(BC)$	Δ
1	0.0500	0.0497	0.0003
2	0.0270	0.0275	−0.0005
3	0.0040	0.0042	−0.0002
4	0.0390	0.0388	0.0002
5	0.0220	0.0223	−0.0003
6	0.0030	0.0031	−0.0001
7	0.0320	0.0319	0.0001
8	0.0190	0.0192	−0.0002
9	0.0030	0.0031	−0.0001

The multiple correlation coefficient between the left and right sides of Eq. 14.3 is 0.9999. To see that this approximation holds, Table 14.3 summarizes the left and right sides of Eq. 14.3, and the difference between the two.

Uncovering near dependencies such as that exemplified by Eq. 14.3 is not a trivial matter. As mentioned above, methods will be described in the next section. For an in-depth presentation of the subject, the reader is referred to the text *Conditioning Diagnostics* by Belsley [5]. The title of Belsley's book derives from the fact that data sets with near dependencies are called **ill-conditioned** data sets. While uncovering near dependencies is not trivial, fortunately several "red flags" are available to warn the experimenter when collinearity lurks within a data set. These will be discussed later in this chapter.

To this point we have still not said anything about the impact of ill conditioning on fitted regression models. To do this, we shall carry out a simulation experiment using the DMBA design (Table 14.2). The procedure will be to use a simulation model to generate several data sets and then to fit models to the data and see how well we recover the simulation model.

For our simulation model we shall use a Scheffé quadratic model in the reals fitted to the experimental P_i values in Table 14.2. Table 14.4 summarizes parameter estimates (b), standard errors (SE), t values, and p values for this model. Summary statistics are $R^2 = 0.8955$, $R^2_{adj} = 0.7213$, $R^2_{pred} = -0.3458$, and $s = 0.0671$.[2] The pattern of p values is similar to that found by Chen, Li, and Jackson [17] when the response was logit(P_i). These workers concluded that the best model was a reduced quadratic model in which the only retained quadratic terms was that for AB (Fat $\times$ Carb). We shall proceed using the full 6-term quadratic model and untransformed P_i values.

Table 14.5 summarizes six sets of simulated P_i values using the model coefficients in Table 14.4 in combination with a standard deviation of 0.0671. Experimental P_i values are included in the table for ease of reference. Correlation coefficients between the experimental

[2]If the model is refit to the data with the exclusion of ID 7, summary statistics are substantially improved: $R^2 = 0.9797$, $R^2_{adj} = 0.9291$, $R^2_{pred} = 0.4189$, and $s = 0.0351$. Nonetheless, we shall continue with this example using all of the data.

Table 14.4. *Parameter estimates in the reals for a quadratic Scheffé model fitted to the DMBA-induced tumor P_i data*

Term	b	SE	t	p
A	0.4121	0.319	1.291	0.2873
B	0.3324	0.126	2.638	0.0778
C	123.4	111.2	1.110	0.3481
AB	1.996	0.768	2.599	0.0805
AC	−130.8	114.5	−1.143	0.3360
BC	−130.1	118.0	−1.103	0.3504

Table 14.5. *Experimental and simulated P_i data. DMBA-induced tumors example*

Exptl	ID, Simulated P_i					
P_i	1	2	3	4	5	6
0.567	0.4840	0.5968	0.5546	0.6317	0.6284	0.5207
0.500	0.5598	0.6331	0.6463	0.5203	0.5017	0.3481
0.567	0.4417	0.5808	0.5726	0.5322	0.6298	0.5881
0.800	0.7623	0.6683	0.7218	0.7072	0.7669	0.7380
0.700	0.8515	0.7978	0.7746	0.8555	0.7480	0.5920
0.767	0.6805	0.7826	0.8572	0.7488	0.7724	0.8341
0.600	0.7345	0.6533	0.6833	0.6317	0.6221	0.6161
0.767	0.8131	0.6941	0.6322	0.6693	0.8144	0.6461
0.867	0.7768	0.7900	0.8738	0.9193	0.8483	0.7720

P_i values and each of the six sets of simulated values are 0.7258, 0.7385, 0.7362, 0.8133, 0.9637, and 0.8597 for simulations 1–6, respectively.

Table 14.6 summarizes least-squares parameter estimates for the quadratic Scheffé model fitted to the simulated data in Table 14.5. Parameter estimates based on the experimental P_i values are included in Table 14.6 for ease of reference. The most striking result is the instability of the coefficient estimates for terms that contain C vs. the relative stability of the estimates for A, B, and AB. Estimates for the coefficients for C, AC, and BC vary more than two orders of magnitude on either side of zero. R^2 values range from 0.8038 (simulation 3) to 0.9866 (simulation 1), so all models fit reasonably well.

One might justifiably ask (for example), "Are the Fat × Fiber (AC) and the Carb × Fiber (BC) terms synergistic or antagonistic?" Or perhaps, "Are the absolute values of the Fat × Fiber and the Carb × Fiber terms large or small?" This typifies one of the dilemmas that one is often confronted with when fitting models to ill-conditioned data sets. In a "one-off" situation, where a model is fitted to a single set of experimental data, one may be unaware that some coefficients are unstable — and consequently suspect — unless one has previously checked some collinearity diagnostics (to be discussed shortly).

Table 14.6. *Parameter estimates using the simulated data in Table* 14.5

	Estimated coefficients					
ID	b_A	b_B	b_C	b_{AB}	b_{AC}	b_{BC}
Exptl	0.4121	0.3324	123.4	1.996	−130.8	−130.1
1	0.3789	0.1259	−216.2	1.915	224.6	229.7
2	0.4561	0.3983	−71.50	1.467	69.39	76.70
3	0.3535	0.3998	1.984	1.938	−5.821	−2.776
4	0.4416	0.2330	53.80	2.184	−64.10	−53.07
5	0.5404	0.4199	123.3	1.590	−131.4	−129.8
6	0.2387	0.4862	374.6	2.031	−383.7	−401.1

Term	$\sqrt{c_{ii}}$
A	4.759
B	1.878
C	1656.
AB	11.45
AC	1706.
BC	1757.

There is another unfortunate consequence of experimental designs that produce excessive collinearity. The reader is reminded of discussions in Sections 5.4.3 (page 84) and 6.1 (page 95), where it was pointed out that the diagonal elements of the $(\mathbf{X}'\mathbf{X})^{-1}$ matrix, symbolized c_{ii}, are equal to the variances of the coefficient estimates (apart from σ^2). The value for s, the estimate of σ, will vary somewhat from simulation to simulation — in this example between 0.0281 for simulation 1 and 0.0856 for simulation 4. On the other hand, $\sqrt{c_{ii}}$ will be the same from simulation to simulation and will be proportional to the standard errors of the coefficient estimates for each simulation.

$\sqrt{c_{ii}}$ values are tabulated beneath Table 14.6. Terms that contain C have standard errors that are from 2–3 orders of magnitude larger than terms that do not contain C. The impact of such large standard errors will be to desensitize t tests of coefficient estimates. The result will be an increased risk of making a Type II error — that is, not claiming that a parameter estimate is significant when in fact it is significant.

Within rounding error, the standard errors of the parameters estimates in Table 14.4, page 330, are equal to 0.0671 times the $\sqrt{c_{ii}}$ values in the above table. The question arises, then, Is the reason that the estimates for AC and BC in Table 14.4 do not appear to be significant because they really are not significant, or simply because the standard errors are inflated? Without prior subject-matter knowledge, it is difficult to know.

Recapping, parameter estimates of terms involved in collinearity can

- be sensitive to small changes in the responses,
- have inflated standard errors,
- have incorrect signs,
- be of the wrong magnitude.

14.2 Warnings and Diagnostics

The discovery of damaging collinearity *after* a design has been executed is an unfortunate circumstance. It is better to discover it before the design is executed so that other strategies (Section 14.3) can be considered. Fortunately several warning signs and formal diagnostics are available to alert the experimenter to the presence of collinearity in a proposed experimental design. Not all diagnostics are available in every DOE package, but generally enough are available to raise a warning flag.

Drawing from the discussion earlier in this chapter, one approach would be to examine the standard errors of the coefficient estimates that arise from the proposed experimental design. Ideally one would like a table of c_{ii} or $\sqrt{c_{ii}}$ values, but in lieu of this one could adopt the procedure described in Section 6.1, page 95. Briefly, this involves generating a normally distributed dummy response with a mean of any value but with a standard deviation of ~ 1.0. One then fits a model to the data and examines the standard errors of the coefficient estimates. The results should be reasonable approximations of the $\sqrt{c_{ii}}$ values. Terms whose standard errors appear out of line should send up a warning flag. This should not be considered a formal diagnostic but simply a warning that something may be amiss with the proposed design. A related but more refined approach is the calculation of **variance inflation factors** (VIFs).

If the **X** matrix is first standardized by scaling each column to unit length — that is, each column element is divided by the square root of the sum of squares of that column — then VIFs are equal to the diagonal elements of $(\mathbf{X}_n'\mathbf{X}_n)^{-1}$, $\mathbf{X}_n$ being the column-normalized **X** matrix.[3] VIFs for the DMBA-induced tumor data are

Term	VIF
A	42.77
B	11.51
C	18264.
AB	46.42
AC	3518.
BC	8031.

Like standard errors of parameter estimates, VIFs do not identify the number of collinearities, but they do help to identify terms that are involved in any near dependencies.

VIFs can be calculated by an alternative method that provides a different insight into their meaning. The VIF for the ith regressor is given by

$$\mathrm{VIF}_i = \frac{1}{1 - R^2_{(0)}} \tag{14.5}$$

[3]When the model contains an intercept, there has been controversy in the literature over whether the **X** matrix should first be centered and then scaled. Centering makes the intercept orthogonal to the other regressors, thereby eliminating any collinearities involving the constant term. In the case of slack-variable mixture models (page 25), VIFs will be lower if the **X** matrix is centered and scaled than if the matrix is simply scaled. Because the intercept in slack-variable mixture models is an important part of the model, it is the author's opinion that the **X** matrix for such models should not be centered. With no-intercept mixture models, such as Scheffé models, centering and scaling leads to $\mathbf{X}'\mathbf{X}$ matrices that are singular.

where $R^2_{(0)}$ is the coefficient of determination when $\mathbf{X}_i$ is regressed on the remaining regressors *without the inclusion of an intercept*. In this case the usual R^2 has been redefined as

$$R^2_{(0)} = \frac{\sum_{i=1}^{n} \hat{X}_i^2}{\sum_{i=1}^{n} X_i^2}$$

That is, variation is measured about zero rather than about the mean (as explained on page 173). A large VIF_i implies that much of the variability in a regressor $\mathbf{X}_i$ can be explained by the other regressors.

Another signal for collinearity is the presence of large off-diagonal elements in either the **correlation matrix of regressors** (commonly referred to as simply the correlation matrix) or the **correlation matrix of regression coefficients**. Referring back to the lower table in Table 14.1, page 326, the large correlation coefficient between A and AB indicates a collinearity between the two. Off-diagonal elements in a correlation matrix whose absolute values are large are a sufficient, but not a necessary, condition for collinearity. This is because simple correlations signal only *pairwise* collinearities, as exemplified by Eq. 14.2, page 327, but not linear combinations between more than two regressors, as exemplified by Eq. 14.4, page 328.

Table 14.7 displays the correlation matrix of regressors for the DMBA-induced tumor data. There is little hint that there is a collinearity between C, AC, and BC, and in fact the strongest pairwise correlation is between A and B ($r_{AB} = -0.9970$). The negative correlation indicates that when the proportion of A increases, then the proportion of B will usually decrease, and vice versa. This is because C is such a small proportion of the total mixture. However, despite the value for r_{AB}, the situation is dominated by the collinearity between terms involving C. This will become clearer later in this chapter when variance-decomposition proportions are introduced.

Table 14.7. *Correlation matrix of regressors. DMBA-induced tumor data*

	A	B	C	AB	AC	BC
A	1.0000					Symmetric
B	−0.9970	1.0000				
C	−0.0710	−0.0070	1.0000			
AB	0.7520	−0.7341	−0.2536	1.0000		
AC	0.5914	−0.6447	0.6628	0.2139	1.0000	
BC	−0.4696	0.4020	0.8794	−0.4677	0.2266	1.0000

Table 14.8 displays the correlation matrix of regression coefficients for the DMBA-induced tumor data. The elements of this matrix can be calculated from the elements of the $(\mathbf{X}'\mathbf{X})^{-1}$ matrix using Eq. 6.1, page 98.[4] Here we see that the coefficient estimates for terms that contain C are highly correlated with one another. In addition, when the signs of the correlation coefficients are taken into consideration, we see that C will vary in a direction

[4]In matrix algebra terms, the correlation matrix of regression coefficients is equal to $\mathbf{C}^{-1}(\mathbf{X}'\mathbf{X})^{-1}\mathbf{C}^{-1}$, where $\mathbf{C}^2$ is a $p \times p$ diagonal matrix whose diagonal elements are equal to those of $(\mathbf{X}'\mathbf{X})^{-1}$.

Table 14.8. *Correlation matrix of regression coefficients. DMBA-induced tumor data*

	A	B	C	AB	AC	BC
A	1.0000					Symmetric
B	0.3368	1.0000				
C	0.0649	0.3687	1.0000			
AB	−0.8990	−0.6146	−0.0350	1.0000		
AC	−0.1013	−0.3612	−0.9987	0.0559	1.0000	
BC	−0.0528	−0.3803	−0.9997	0.0296	0.9974	1.0000

that is opposite to AC and BC, while AC and BC will tend to covary. That such is the case can be seen by inspection of the coefficient estimates for C, AC, and BC in Table 14.6, page 331.

The above warnings (standard errors of parameter estimates and correlation coefficients) and diagnostic (VIFs) all suffer from the absence of meaningful boundaries between values that can be considered high and those that can be considered low. Many texts and technical articles suggest a cutoff of ~ 10 for VIFs, but this recommendation is practically always made in the context of nonmixture experiments. Higher cutoffs have been suggested for mixture settings [57, 162].

In addition, the number of near dependencies present in an **X** matrix is not always clear from an examination of VIFs. In the DMBA example it is quite clear from examining the VIFs and the correlation matrix of regression coefficients that there is one near dependency. However, consider a situation where four regressors, say X_1–X_4, have very large VIFs. One possibility would be a single near dependency involving X_1, X_2, X_3, and X_4. Another possibility (among several) would be two near dependencies, one involving X_1, X_2, and X_3, and the second, X_2, X_3, and X_4.

Condition indices are collinearity diagnostics that have meaningful cut-off values and that can identify the number of near dependencies [5]. Because of their importance as a diagnostic, much space will be devoted in this section to the topic. Data matrices that differ only by the scale assigned to the columns will lead to different condition indices, and so it is necessary to standardize matrices in a manner that makes interpretation of the indices meaningful. In a mixture setting, for example, expressing composition in terms of reals or actuals would make a difference in the calculated value(s) of this diagnostic. The standardization method most commonly used is **unit-length scaling**, in which the columns of the **X** matrix are normalized to 1.0.

As before, let us represent the column-normalized **X** matrix by $\mathbf{X}_n$. Using eigenanalysis, a crossproduct matrix such as $\mathbf{X}_n'\mathbf{X}_n$ may be decomposed as

$$\mathbf{X}_n'\mathbf{X}_n = \mathbf{V}\boldsymbol{\Lambda}\mathbf{V}' \tag{14.6}$$

where $\mathbf{V}$ is a $p \times p$ orthogonal matrix whose columns are the eigenvectors of $\mathbf{X}_n'\mathbf{X}_n$ and $\boldsymbol{\Lambda}$ is a $p \times p$ diagonal matrix whose diagonal elements are the eigenvalues of $\mathbf{X}_n'\mathbf{X}_n$. Representing the eigenvalues of $\mathbf{X}_n'\mathbf{X}_n$ from largest to smallest as $\lambda_1, \lambda_2, \ldots, \lambda_p$, the condition indices of the $\mathbf{X}_n'\mathbf{X}_n$ matrix are given by

$$\eta_i = \frac{\lambda_{max}}{\lambda_i} \qquad i = 1, 2, \ldots, p \tag{14.7}$$

where $\lambda_{max} \equiv \lambda_1$. When $\lambda_i = \lambda_p$, then the **condition number** κ is defined as

$$\kappa = \frac{\lambda_{max}}{\lambda_{min}} \tag{14.8}$$

since $\lambda_{min} \equiv \lambda_p$. The condition number is therefore a measure of the spread of the condition indices.

Condition indices are useful because whenever near linear dependencies exist in a data matrix, one or more of the eigenvalues will be "small". A small eigenvalue will result in large condition indices, and so the question becomes what is "large". A cutoff that has been demonstrated to work in practice is $\sim$ 1000 [5].[5] The number of condition indices $\geq$ 1000 identifies the number of near dependencies among the columns of the data matrix. The magnitudes of the high condition indices provide a measure of their relative severity. Some software packages report only the largest condition index (i.e., the condition number, κ) which is helpful but does not provide any information about the number of near dependencies.

An equivalent but slightly different approach to calculating condition indices is to use the **singular-value decomposition** of $\mathbf{X}_n$ rather than the eigendecomposition of $\mathbf{X}_n'\mathbf{X}_n$. Belsley [5] offers several reasons for preferring the singular-value decomposition, one of which is that the algorithms that exist for computing this decomposition are numerically more stable than those for computing the eigensystem of $\mathbf{X}_n'\mathbf{X}_n$, particularly when $\mathbf{X}$ is ill conditioned. The singular-value decomposition of $\mathbf{X}_n$ can be written

$$\mathbf{X}_n = \mathbf{UDV}' \tag{14.9}$$

where $\mathbf{U}'\mathbf{U} = \mathbf{V}'\mathbf{V} = \mathbf{I}$ and $\mathbf{D}$ is a $p \times p$ diagonal matrix with elements $\mu_1, \mu_2, \ldots, \mu_p$ called the *singular values* of $\mathbf{X}_n$. The $p \times p$ matrix $\mathbf{V}$ in Eq. 14.9 is the same as the matrix $\mathbf{V}$ in Eq. 14.6, i.e., the matrix of eigenvectors of $\mathbf{X}_n'\mathbf{X}_n$. The dimensions of $\mathbf{U}$ are $n \times p$.

An important relationship exists between the matrices $\mathbf{D}$ in Eq. 14.9 and $\mathbf{\Lambda}$ in Eq. 14.6, as the following algebra shows:

$$\begin{aligned} \mathbf{X}_n'\mathbf{X}_n &= (\mathbf{UDV}')'(\mathbf{UDV}') \\ &= (\mathbf{VDU}')(\mathbf{UDV}') \\ &= \mathbf{VD}^2\mathbf{V}' \end{aligned} \tag{14.10}$$

Comparing Eq. 14.10 with 14.6, we see that the singular values are equal to the square roots of the eigenvalues.

Condition indices and condition numbers are sometimes based on singular values rather than eigenvalues. To identify the difference, a tilde will be used for those based on

[5]This is based on Belsley's extensive simulation studies in *Conditioning Diagnostics*, Chapter 4 ("The Experimental Experience"). Belsley's cutoff is based on singular values rather than eigenvalues, explained later in this section.

singular values. Thus, the condition indices and condition number using singular values are defined as follows:

$$\tilde{\eta}_i = \frac{\mu_{max}}{\mu_i} \qquad i = 1, 2, \ldots, p \tag{14.11}$$

$$\tilde{\kappa} = \frac{\mu_{max}}{\mu_{min}} \tag{14.12}$$

Belsley suggests using $\tilde{\eta} \geq\sim 30$ to identify the number of near dependencies. Since 30^2 is of the same order of magnitude as 1000, this is equivalent to using $\eta \geq\sim 1000$ for the same purpose. Because of the difference between κ and $\tilde{\kappa}$, it is important to check which of the two diagnostics a text book, article, or DOE software product may be reporting.

The eigenanalysis of $\mathbf{X}_n'\mathbf{X}_n$ (or the singular value decomposition of $\mathbf{X}_n$) provides a means for decomposing VIFs into a sum of terms, each associated with one of the p eigenvalues of $\mathbf{X}_n'\mathbf{X}_n$ (or singular values of $\mathbf{X}_n$). Taking the inverse of both sides of Eq. 14.6, page 334, leads to

$$\begin{aligned}(\mathbf{X}_n'\mathbf{X}_n)^{-1} &= (\mathbf{V}\mathbf{\Lambda}\mathbf{V}')^{-1} \\ &= (\mathbf{V}')^{-1}\mathbf{\Lambda}^{-1}\mathbf{V}^{-1} \\ &= \mathbf{V}\mathbf{\Lambda}^{-1}\mathbf{V}' \end{aligned} \tag{14.13}$$

The first step uses the property that the inverse of a product is the product of the inverses taken in reverse order; the second step makes use of the fact that because $\mathbf{V}$ is an orthogonal matrix, $\mathbf{V}^{-1} = \mathbf{V}'$ [149]. The diagonal elements on the left (which are the VIFs) must be equal to the diagonal elements on the right. Equating the jth VIF on the left with the jth diagonal element on the right leads to

$$\begin{aligned}\text{VIF}_j &= \frac{v_{j1}^2}{\lambda_1} + \frac{v_{j2}^2}{\lambda_2} + \cdots + \frac{v_{jp}^2}{\lambda_p} \\ &= \sum_{i=1}^{p} \frac{v_{ji}^2}{\lambda_i} \qquad j = 1, 2, \ldots, p \end{aligned} \tag{14.14}$$

where v_{ji} is the jith element of $\mathbf{V}$.[6] Because $\mathbf{V}$ is a square matrix, both j (the coefficient index) and i (the eigenvalue index) run from 1 to p. Belsley defines **variance-decomposition proportions** as

$$\pi_{ij} = \frac{v_{ji}^2/\lambda_i}{\text{VIF}_j} \tag{14.15}$$

where i is still the eigenvalue index and j, the coefficient index. Variance-decomposition proportions are then summarized in a $\mathbf{\Pi}$ matrix. The $\mathbf{\Pi}$ matrix for the DMBA-induced tumor data is displayed in Table 14.9. Note that within rounding error, the proportions in each column sum to one.

[6]By convention, matrix elements are doubly subscripted, the first subscript usually identifying the row and the second the column.

Table 14.9. *Scaled condition indices and variance-decomposition proportions. DMBA-induced tumor data*

Condition	Variance-Decomposition Proportions					
Index, $\tilde{\eta}$	A	B	C	AB	AC	BC
1	0.001	0.003	0.000	0.001	0.000	0.000
3	0.010	0.004	0.000	0.004	0.000	0.000
3	0.000	0.082	0.000	0.005	0.000	0.000
9	0.038	0.548	0.000	0.045	0.001	0.000
21	0.946	0.225	0.000	0.944	0.001	0.000
378	0.004	0.138	1.000	0.001	0.997	0.999

The condition indices in Table 14.9 are $\tilde{\eta}$ values, and therefore values $\geq\sim 30$ identify near dependencies in the data. Belsley makes several practical suggestions for displaying these tables. One is to "keep the numbers simple". There is no need to display any fractional part of the indices, and three digits to the right of the decimal point are adequate for the variance proportions. Another is to "look at the high scaled condition indices first", specifically those that exceed the threshold of ~ 30.

Following the second bit of advice, one should focus on the condition index 378. A regressor is considered involved in at least one near dependency if its variance proportions *summed over the set of high condition indices* ($\tilde{\eta} \geq\sim 30$) exceeds a threshold of ~ 0.5. In this particular example, the "set" is a set of one. The variance proportions associated with $\tilde{\eta} = 378$ indicate that C, AC, and BC are the regressors involved in the dominant near dependency. To understand regressor involvement, Belsley recommends carrying out **auxiliary regressions**. Regressing C, for example, on AC and BC leads to the model in Eq. 14.3, page 328.

The variance proportions for A and AB associated with the condition index $\tilde{\eta} = 21$ suggest a further problem. If one regresses A on the other five variates, one finds $R^2_{(0)} = 0.9766$, corresponding to a VIF of 42.73. If one regresses A on AB only, one finds $R^2_{(0)} = 0.9055$, corresponding to a (partial) VIF of 10.48. Thus, the dependency between A and AB accounts for only about 25% of the VIF for A, the remainder being accounted for by other dependencies. A and AB are not as highly dependent as one might think. It is better to follow Belsley's advice and focus on condition numbers that exceed the threshold of ~ 30.

The steps involved can be summarized as follows: (1) Normalize the $\mathbf{X}$ matrix; (2) Calculate the condition number (κ or $\tilde{\kappa}$) and/or the condition indices (η or $\tilde{\eta}$); (3) Determine the number of near dependencies; (4) Determine regressor involvement; (5) Carry out auxiliary regressions. Depending on the level of detail to which one cares to take the procedure, it could be truncated at any point after step 2.

The reader may justifiably feel that this is a great deal of work to go to when the answer seemed clear from a simple inspection of the standard errors of the parameter estimates or the VIFs in the DMBA experiment. This is not always the case, however. Consider the hypothetical 16-point design in Fig. 14.3. Constraints on the component proportions are given beneath the figure.

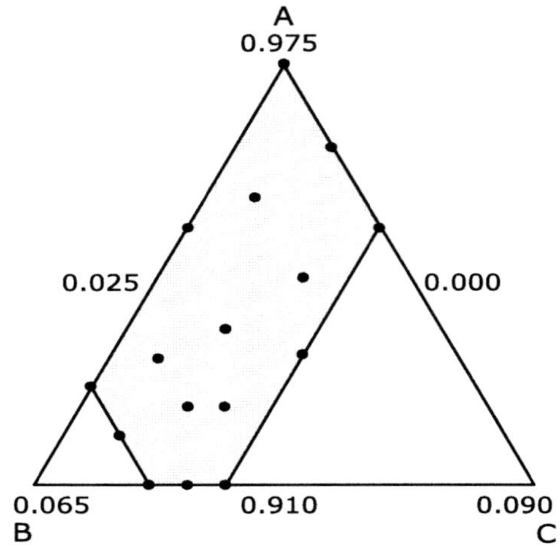

Figure 14.3. *Hypothetical design. Pseudocomponent simplex.*

$$\begin{array}{ccccc} 0.910 & \le & A & \le & 0.975 \\ 0.000 & \le & B & \le & 0.050 \\ 0.025 & \le & C & \le & 0.050 \end{array}$$

The constraints indicate a disparity in the component proportions but not in the ranges. Although the design is shown in the context of the pseudocomponent simplex, numeric labels are in terms of the reals. Component proportions are summarized in Table 14.10. IDs 1–5 are vertices, 6–10 are edge centroids, 11–15 are axial check blends, and 16 is the overall centroid.

Standard errors of parameter estimates (apart from σ) and VIFs for a quadratic model are summarized beneath Table 14.10, page 339. Clearly, the data are ill conditioned, an observation that is supported by the condition number, $\tilde{\kappa} = 1879$ (Table 14.11, page 340). Two condition indices clearly exceed a threshold of ~ 30, while another is marginal ($\tilde{\eta} = 29$). Summing the variance proportions of the two largest $\tilde{\eta}$ values leads to the conclusion that all the regressors appear to be involved in near dependencies in one way or another.

Table 14.11 displays a problem that can arise in interpreting variance decomposition proportions. Dependencies that differ by more than half an order of magnitude (say an order of magnitude) are called **dominant dependencies**. In Table 14.11, for example, $\log(414/29) = 1.15$ and $\log(1879/29) = 1.8$. Dominant dependencies can mask the simultaneous involvement of regressors in other near dependencies. The possibility exists that a regressor having most of its variance determined by a near dependency with a high condition index is also involved in one or more near dependencies with lower condition indices. This situation can make it very difficult to select regressors and regressands for auxiliary regressions.[7] Belsley proposes that a relatively problem-free progression of condition indices would be 1, 3, 10, 30, 100, 300, ..., called the **progression of 10/30** [5].

[7] Dependencies of roughly equal magnitude, called **competing dependencies**, can also cause problems. Variance proportions associated with competing dependencies can become confounded [5].

Table 14.10. *Component proportions. Hypothetical 3-component experiment*

ID	A	B	C
1	0.9750	0.0000	0.0250
2	0.9100	0.0500	0.0400
3	0.9250	0.0500	0.0250
4	0.9500	0.0000	0.0500
5	0.9100	0.0400	0.0500
6	0.9500	0.0250	0.0250
7	0.9625	0.0000	0.0375
8	0.9175	0.0500	0.0325
9	0.9100	0.0450	0.0450
10	0.9300	0.0200	0.0500
11	0.9545	0.0140	0.0315
12	0.9220	0.0390	0.0390
13	0.9295	0.0390	0.0315
14	0.9420	0.0140	0.0440
15	0.9220	0.0340	0.0440
16	0.9340	0.0280	0.0380

Term	$\sqrt{c_{ii}}$	VIF
A	6	442
B	960	16146
C	3639	321579
AB	1019	15420
AC	3932	325527
BC	4109	430

The progression corresponds to intervals in $\log \tilde{\eta}$ of ~ 0.5, or half an order of magnitude in terms of $\tilde{\eta}$.[8]

Belsley recommends a relatively simple procedure for untangling situations like this. First, guess a set of regressors equal in number to the number of near dependencies, r, that are likely to be involved in the near dependencies. Second, find the condition number of the $\mathbf{X}$ matrix composed of the remaining $p - r$ regressors, where p is equal to the number of rows in the $\mathbf{\Pi}$ matrix. If the condition number is of the same order of magnitude as the $(p - r)$th in the full $p \times p$ $\mathbf{\Pi}$ matrix, then a subset has been found that possess no near dependencies. This set can be used as the set of regressors for the auxiliary regressions.

In the present example, let us guess that $r = 3$ and that the subset possessing no near dependencies is composed of the $p - r = 3$ variables A, B, and AB. The condition number, $\tilde{\kappa}$, for this subset is 227, and so this is not a satisfactory subset. As a second guess, let us try the set AB, AC, and BC. The condition indices and variance-decomposition proportions for this subset are given in Table 14.12. This subset fulfills the requirement and may be

[8] Equivalent to intervals of one order of magnitude if the condition indices are based on eigenvalues.

Table 14.11. *Scaled condition indices and variance-decomposition proportions. Hypothetical 3-component example*

Condition	Variance-Decomposition Proportions					
Index, $\tilde{\eta}$	A	B	C	AB	AC	BC
1	0.000	0.000	0.000	0.000	0.000	0.000
3	0.001	0.000	0.000	0.000	0.000	0.001
10	0.016	0.000	0.000	0.000	0.000	0.014
29	0.103	0.001	0.000	0.001	0.000	0.129
414	0.003	0.998	0.000	0.998	0.000	0.037
1879	0.877	0.001	1.000	0.001	1.000	0.820

Table 14.12. **Π** *Matrix for a subset of terms in Table* 14.11

$\tilde{\eta}$	AB	AC	BC
1	0.006	0.034	0.007
4	0.027	0.943	0.044
10	0.967	0.023	0.950

Table 14.13. *Auxiliary regressions of A, B, and C on AB, AC, and BC. Hypothetical 3-component experiment*

	Coefficient Estimates†			
Regressand	AB	AC	BC	$R^2_{(0)}$
A	25.05	24.88	−578.6	0.9818
	[0.0004]	[0.0000]	[0.0010]	
B	1.0525	−0.0079	1.0230	0.9999
	[0.0000]	[0.0505]	[0.0035]	
C	−0.0368	1.0445	1.8695	1.0000
	[0.0006]	[0.0000]	[0.0000]	

† Numbers in square brackets are p values

used as the set of regressors. Regressing the remaining variates (A, B, and C) on this subset leads to the results in Table 14.13.

In terms of Eq. 14.1, page 326, the near dependencies are

$$\mathbf{a} - 25.05\mathbf{ab} - 24.88\mathbf{ac} + 578.6\mathbf{bc} \approx \mathbf{0} \tag{14.16}$$

$$\mathbf{b} - 1.0525\mathbf{ab} + 0.0079\mathbf{ac} - 1.0230\mathbf{bc} \approx \mathbf{0} \tag{14.17}$$

$$\mathbf{c} + 0.0368\mathbf{ab} - 1.0445\mathbf{ac} - 1.8695\mathbf{bc} \approx \mathbf{0} \tag{14.18}$$

In this notation the boldface terms are 16×1 vectors.

14.3 Dealing with Collinearity

A close look at Fig. 14.3, page 338, reveals that the hypothetical 16-point design comprises a significant proportion of the content of the pseudocomponent simplex. Had the design space been displayed in the context of the full simplex (i.e., where the vertices represent the *real* components), then because of the small ranges of the components the design space would occupy only a small portion of the simplex. If collinearity is a problem related to small ranges, then this suggests that if instead of expressing the component proportions in terms of the reals they were instead expressed in terms of the pseudos, then the ill conditioning of the **X** matrix should be reduced. This strategy would not be effective, however, for the DMBA-induced tumor design (Fig. 14.2, page 328) because the range of fiber is small even in the context of the pseudocomponent simplex. What follows therefore does not apply to that experiment.

Table 14.14 displays variance inflation factors for a quadratic model expressed in terms of the reals and the pseudos for the hypothetical design in Table 14.10. By coding in terms of pseudocomponents the condition number of the **X** matrix is reduced from 1879 in the reals to 27 in the pseudos.

Table 14.14. *Variance inflation factors. Reals vs. pseudos. Hypothetical* 3-*component experiment, Table* 14.10

	VIF	
Term	Reals	Pseudos
A	442	3
B	16146	10
C	321579	103
AB	15420	4
AC	325527	36
BC	430	63

A simulation experiment similar to that in the case of the DMBA-induced tumor experiment was carried out. Specifically, the DMBA model in Table 14.4, page 330, was used to generate response data with a standard deviation of ~ 0.0671 using the hypothetical design in Table 14.10, page 339, rather than the DMBA design. Table 14.15 displays estimated coefficients in the pseudos for quadratic models fitted to six sets of simulated data.

Comparing these results with those in Table 14.6, page 331, for the DMBA experiment, there is considerably less variation in the coefficient estimates, although there is still some switching of signs in the A^*B^* and A^*C^* quadratic terms. The coefficient estimates for C^* are significantly more stable than in the DMBA simulation. It appears that reparameterization has taken us "from darkness into light" [5].

The linear regression model in matrix algebra terms is

$$\mathbf{Y} = \mathbf{X}\boldsymbol{\beta} + \boldsymbol{\epsilon}$$

(cf. Eq. 3.11, page 18). A reparameterized version of this model can be written in the form

$$\begin{aligned}\mathbf{Y} &= (\mathbf{X}\mathbf{G}^{-1})(\mathbf{G}\boldsymbol{\beta}) + \boldsymbol{\epsilon} \\ &= \mathbf{X}^*\boldsymbol{\delta} + \boldsymbol{\epsilon}\end{aligned}$$

Table 14.15. *Coefficient estimates, pseudocomponent metric. Hypothetical 3-component experiment*

ID	b_{A^*}	b_{B^*}	b_{C^*}	$b_{A^*B^*}$	$b_{A^*C^*}$	$b_{B^*C^*}$
1	0.2616	0.4465	1.6975	0.0882	−2.2678	−2.0369
2	0.2482	0.3257	0.7262	0.4287	−0.6669	−0.3831
3	0.2732	0.5295	0.4604	−0.0542	0.2476	−0.7333
4	0.2456	0.4484	0.9419	0.1570	−0.9094	−0.9679
5	0.1951	0.4051	1.3146	0.0505	−0.7556	−1.5996
6	0.4328	0.1612	1.3545	0.3049	−1.9521	−0.7448

where $\mathbf{G}$ is a $p \times p$ nonsingular matrix, $\boldsymbol{\beta}$ is a $p \times 1$ vector of parameters in the reals, and $\boldsymbol{\delta}$ is a $p \times 1$ vector of parameters in the transformed metric. As $\mathbf{G}$ is nonsingular, we may write

$$\boldsymbol{\beta} = \mathbf{G}^{-1}\boldsymbol{\delta} \tag{14.19}$$

To convert a $p \times 1$ vector of parameter estimates expressed in the pseudos to a vector of estimates expressed in the reals, we need only multiply $\boldsymbol{\delta}$ by $\mathbf{G}^{-1}$. As $\mathbf{X}^* = \mathbf{X}\mathbf{G}^{-1}$, $\mathbf{G}^{-1}$ is easily derived.

$$\begin{aligned} \mathbf{X}^* &= \mathbf{X}\mathbf{G}^{-1} \\ \mathbf{X}'\mathbf{X}^* &= \mathbf{X}'\mathbf{X}\mathbf{G}^{-1} \\ (\mathbf{X}'\mathbf{X})^{-1}\mathbf{X}'\mathbf{X}^* &= \mathbf{G}^{-1} \end{aligned} \tag{14.20}$$

Applying Eq. 14.19 to the parameter estimates expressed in terms of the pseudos (Table 14.15) leads to the parameter estimates in terms of the reals (Table 14.16). These estimates are, in fact, exactly the same estimates that one would obtain if the Y data were fitted directly to the model in the reals. With the exception of the estimates for A, the parameter estimates in the reals are not at all stable. The transformation from the pseudos to the reals has simply reintroduced the ill conditioning.

A question that one might ask is, "Would variable selection be safer in the pseudos than in the reals?" In other words, might one consider testing the significance of the quadratic

Table 14.16. *Coefficient estimates, real metric. Hypothetical 3-component experiment*

ID	b_A	b_B	b_C	b_{AB}	b_{AC}	b_{BC}
1	0.9170	−16.5954	498.0321	20.8683	−536.7514	−482.0978
2	0.4195	−92.3920	147.4778	101.4563	−157.8574	−90.6809
3	0.0694	21.4877	−48.9196	−12.8258	58.6085	−173.5534
4	0.4622	−29.8794	201.6615	37.1505	−215.2407	−229.0970
5	0.1669	−2.4929	175.6537	11.9621	−178.8294	−378.5957
6	1.1179	−75.8747	424.1975	72.1644	−462.0338	−176.2946

terms in terms of the pseudos, fitting a reduced model, and then transforming back to the reals? Unfortunately, in quadratic Scheffé models t tests of the quadratic terms are invariant to linear transformations (of which the pseudocomponent transformation is a special case), a result that has been analytically derived by Cornell and Gorman [33]. Thus inferences about the significance of quadratic terms are going to be the same in the two metrics, and one will still be exposed to the risk of making a Type II error.

Historically, one of the reasons why the pseudocomponent transformation was introduced was to combat computer round-off error. Inversion of $\mathbf{X}'\mathbf{X}$ matrices when $\mathbf{X}$ is ill conditioned can be subject to round-off error. With the advent of modern computers and the increased accuracy of matrix inversion routines, this is seldom a problem. If one's interest in model fitting is model interpretation, most practitioners would undoubtedly have more interest in the natural variables (the reals) than in linear transformations of the natural variables (such as the pseudos). At some point, then, a practitioner will want to convert coefficient estimates in the pseudos into estimates in the reals. The consequences are well summarized by Belsley's observation, "It is true that one can often find a reparameterization ... that takes us from darkness into light, but should we desire to return, the lights must again be dimmed." Thus if one is interested in interpreting a model in the natural variables, a linear transformation of the variates is a nonsolution to the problem.

Another approach for dealing with the collinearity problem that is often suggested is to remove some of the collinear regressors from the model.[9] If in the DMBA study, for example, one had prior knowledge that the Fat × Fiber and Carbohydrate × Fiber terms were negligible, then this would be a legitimate approach. Without prior knowledge, however, which term(s) should be eliminated? Presumably collinear variates are present because the practitioner had reason for including them in the first place.

The simplest way (not necessarily the best way) to handle the collinearity problem is to avoid hypothesis testing and live with the full model. In severely ill-conditioned data sets it is not unusual to find that *all* parameter estimates lack significance, yet R^2 may exceed 0.9. The result will be an inability to safely interpret the meanings of the coefficients, but prediction does not necessarily suffer. The model may still predict well, provided one restricts prediction to regions where the near dependencies hold. This will generally be the case in the design region if the design points adequately cover the region. Extrapolation should always be avoided but particularly so when collinearity is present.

An approach to combating collinearity that can be very successful, although it is not a panacea, is respecification of the regression model, preferably in a form that is *physically interpretable*. For example, in the DMBA-induced tumor experiment the range of fiber is so small relative to the other two components that one could do the following thought experiment: remove fiber from the mixture and reintroduce it as a process variable. The proportions for fat and carbohydrate must, of course, be renormalized so that they sum to 1.0.

Table 14.17 displays the restructured data. Fat and carbohydrate are mixture components A and B, their proportions summing to one; fiber is process variable Z. A 6-term composite MPV model containing terms to order 2 is

$$P_i = \beta_A^0 A + \beta_B^0 B + \beta_{AB}^0 AB + \beta_A^Z AZ + \beta_B^Z BZ + \beta_0^{Z^2} Z^2 \qquad (14.21)$$

[9] Keep in mind possible hierarchy considerations (Chapter 10).

Table 14.17. *Respecification of the DMBA regressors*

	Fat	Carb	Fiber	
ID	A	B	Z	P_i
1	0.1842	0.8158	0.050	0.567
2	0.1572	0.8428	0.027	0.500
3	0.1335	0.8665	0.004	0.567
4	0.5109	0.4891	0.039	0.800
5	0.4499	0.5501	0.022	0.700
6	0.3912	0.6088	0.003	0.767
7	0.7242	0.2758	0.032	0.600
8	0.6504	0.3496	0.019	0.767
9	0.5777	0.4223	0.003	0.867

The condition number $\tilde{\kappa}$ of the uncoded **X** matrix is 20, significantly less than the value 378 for the quadratic Scheffé model.

Fitting the data in Table 14.17 to model 14.21 leads, after variable selection, to the reduced model

$$\hat{P}_i = 0.193(\text{Fat}) + 0.250(\text{Carb}) + 2.270(\text{Fat})(\text{Carb}) \qquad (14.22)$$

with no detectable involvement of fiber. This is not to say that fiber may not be involved in DMBA-induced tumor development, but the data are inadequate to detect any statistically significant involvement. Summary statistics for the full quadratic Scheffé model, the full MPV model Eq. 14.21, and the reduced MPV model Eq. 14.22 are

Model	R^2	R^2_{adj}	R^2_{pred}	s
quadratic Scheffé	0.8955	0.7213	−0.3458	0.0671
Model 14.21	.9020	.7388	−.2841	.0650
Model 14.22	.7842	.7123	.3277	.0682

The reader is reminded that including nonexplanatory variables in a model will always increase R^2, which is probably why the R^2 values for the quadratic and full MPV models are higher than that for the reduced MPV model Eq. 14.22. Reducing the full MPV model to the simpler model Eq. 14.22 significantly improves R^2_{pred} with little effect on R^2_{adj} and s.

Figure 14.4 (left) displays a trace plot for model 14.22. The plot is invariant to the level of fiber because fiber does not appear in the model. According to this picture, the tumor rate maximizes at roughly a 50:50 blend of fat and carbohydrate, falling off with either a high-fat or a high-carbohydrate diet.

Inferring from model 14.22 that fiber does not interact with fat or carbohydrate, one has some justification for removing the Fat × Fiber and Carb × Fiber terms from the full quadratic Scheffé model, leading to the reduced Scheffé model

$$\hat{P}_i = 0.267(\text{Fat}) + 0.298(\text{Carb}) - 0.762(\text{Fiber}) + 2.185(\text{Fat})(\text{Carb}) \qquad (14.23)$$

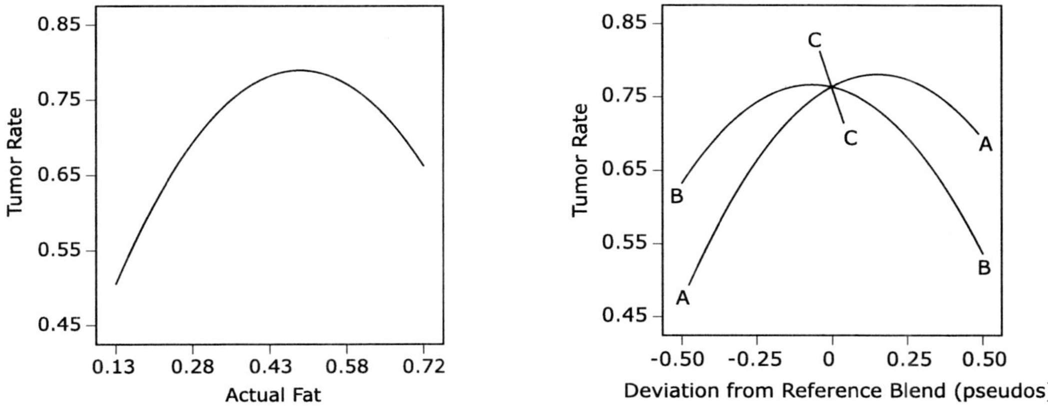

Figure 14.4. *Trace plots. DMBA experiment. Reduced mixture-process variable model (left) and reduced quadratic Scheffé model (right).*

with $R^2 = 0.8410$, $R^2_{adj} = 0.7456$, $R^2_{pred} = 0.4025$, and $s = 0.0641$. Comparing the coefficient for Fiber in the reduced model to that in the full model (Table 14.4, page 330), we see roughly a two-order of magnitude change. This is because by removing the terms in AC and BC the near dependencies have been eliminated, and the condition number $\tilde{\kappa}$ of the reduced **X** matrix is now only 15.

A trace plot in the Cox-effect directions for model 14.23 is shown in Fig. 14.4 (right). With the exception of the trace for fiber, the trace plots on the left and the right are telling about the same story, i.e., the predicted tumor rate maximizes in the vicinity of a 50:50 mixture of fat (A) and carbohydrate (B). The problem is that there appears to be a very significant effect of fiber. This is one of those cases where visual inferences made from a trace plot can be misleading. When methods described in Section 11.5, page 269, are used to test the significance of the fiber effect, it is found that the effect of fiber is not statistically significant. Thus we are left with the result that perhaps high fiber reduces the tumor rate, but the data are inadequate to support this on a statistical basis.

The Design-Expert three-dimensional response surface in Fig. 14.5 graphically conveys these results. The top of the bow-shaped surface is located at roughly a 50:50 blend of fat and carbohydrate. The downward tip in the direction of the fiber vertex indicates a reduction in the tumor rate at high fiber, but this is the effect that is not statistically significant.

The hypothetical 3-component design in Fig. 14.3 (page 338) and Table 14.10 (page 339) presents a more difficult challenge. In this case all three components have very small ranges. When one or more components have very small ranges, it is sometimes the case that **ratio designs and models** can prove useful. Ratios are certainly physically interpretable variates.

In a q-component mixture, compositions can be represented by $q - 1$ ratios. The mixture blends in Table 14.10, page 339, for example, can be described by two ratios R_1 and R_2, where these remain to be defined. Ratios are independent variables, and so a quadratic model in R_1 and R_2 would be of the form

$$Y = \alpha_0 + \alpha_1 R_1 + \alpha_2 R_2 + \alpha_{11} R_1^2 + \alpha_{22} R_2^2 + \alpha_{12} R_1 R_2 \tag{14.24}$$

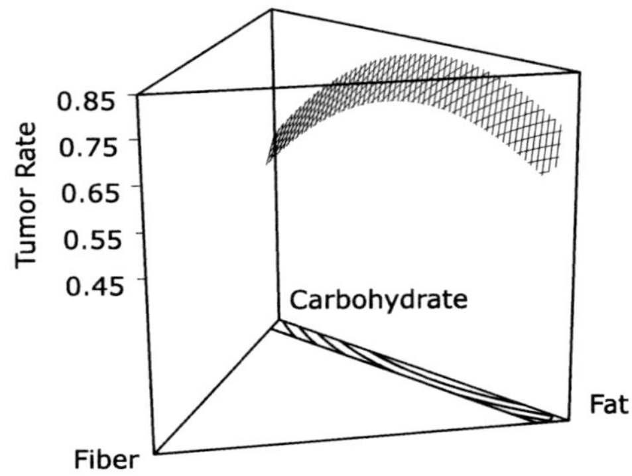

Figure 14.5. *Response surface. DMBA experiment. Reduced quadratic Scheffé model.*

One might consider two types of ratios. An example of *simple* ratios would be

$$\frac{X_1}{X_2}, \frac{X_2}{X_3}, \ldots, \frac{X_{q-1}}{X_q}$$

or perhaps

$$\frac{X_1}{X_q}, \frac{X_2}{X_q}, \ldots, \frac{X_{q-1}}{X_q}$$

Hierarchical ratios are of the form

$$\frac{X_1}{X_2}, \frac{X_3}{X_1 + X_2}, \ldots, \frac{X_q}{(X_1 + X_2 + \cdots + X_{q-1})}$$

Table 14.18 summarizes a variety of ways to restructure the data in Table 14.10 in the form of ratios. B does not appear by itself in the denominator of any of the ratios because its minimum level is zero.[10] In the best case (ID 9), $\tilde{\kappa} = 105$, indicating that there is still a problem with collinearity. Auxiliary regressions for this parameterization indicate that the intercept is involved in a near dependency with the other five regressors ($R^2_{(0)} = 0.9983$).

Although we have reduced the ill conditioning from $\tilde{\kappa} = 1879$ in the quadratic Scheffé model to 105 in the ratio model 9, we have not completely eliminated the problem. If adequate precision (cf. page 181) is high, ill conditioning remaining in the ratio model may not present a problem. That is, as s gets smaller and smaller a point will be reached where the impact of ill conditioning will disappear. Although a cut-off value for adequate precision cannot be given, one way to explore this is to run simulations using the fitted model and the experimental standard deviation. If large variations are not observed in the estimated coefficients, then one has a reasonably well-behaved model.

[10]The suggestion has been made that one could add a small positive quantity to the proportions of any component whose minimum level is zero [153].

Table 14.18. *Condition indices ($\tilde{\eta}_1, \tilde{\eta}_2, \ldots$) for several quadratic ratio models. Hypothetical 3-component experiment*

ID	R_1	R_2	$\tilde{\eta}$
1	B/A	B/C	160, 60, ...
2	B/A	C/A	127, 32, ...
3	B/C	C/A	126, 38, ...
4	B/A	A/C	119, 27, ...
5	B/C	A/C	116, 27, ...
6	B/A	$C/(A+B)$	125, 32, ...
7	C/A	$B/(A+C)$	122, 32, ...
8	A/C	$B/(A+C)$	119, 27, ...
9	B/C	$A/(B+C)$	105, 19, ...

Increasing the range of one or more of the components can also help the situation. For example, perhaps one could operate with the constraints

$$\begin{aligned} 0.900 &\le A \le 1.000 \\ 0.000 &\le B \le 0.050 \\ 0.000 &\le C \le 0.050 \end{aligned}$$

rather than those on page 338. These constraints lead to a 13-point design composed of 4 vertices, 4 edge centroids, 4 axial check blends, and the overall centroid. In this case components B and C have zero levels, and so neither can appear in the denominator of a ratio. The choice of regressors would be limited to ratio sets 2, 6, or 7 in Table 14.18. Assuming the quadratic model 14.24, the condition number $\tilde{\kappa}$ for each of these is 18. Compared to the condition number for the quadratic Scheffé model, $\tilde{\kappa} = 371$, this represents a significant improvement.

Other methods for dealing with collinearity are *biased regression methods*, but these fall beyond the scope of this text. Examples are ridge regression and principal components regression. For details, the reader is referred to texts such as Draper and Smith [49], Montgomery, Peck, and Vining [100], Myers [105], Myers and Montgomery [107], or Rawlings, Pantula, and Dickey [143]. See Jackson [74] for an insightful review of principal components regression. St. John illustrates ridge regression in mixture settings [162].

Case Study

The previously discussed surfactant design presents another example that is amenable to restructuring. The disparity in the ranges of the components (page 87), $\sqrt{c_{ii}}$ values (page 96), the correlation matrix of regression coefficients (page 99), and the analysis for lather units (pages 220) all point to a near dependency among terms containing D (the zwitterionic surfactant) in the quadratic Scheffé model. In the variable selection procedure (page 220), concern was raised whether it was correct to test the composite hypothesis $H_0 : b_{AD} = b_{BD} = b_{CD} = 0$ because of the inflated standard errors of terms containing D.

The quadratic Scheffé model for this design has a condition number $\tilde{\kappa}$ of 478. If the model is restructured as a mixture-process variable model with D being the process variable Z, then a 10-term composite MPV model with terms up to order 2 is

$$\text{lather units} = \beta_A^0 A + \beta_B^0 B + +\beta_C^0 C + \beta_{AB}^0 AB + +\beta_{AC}^0 AC + \beta_{BC}^0 BC$$
$$+ \beta_A^Z AZ + \beta_B^Z BZ + +\beta_C^Z CZ + \beta_0^{Z^2} Z^2 \tag{14.25}$$

where A is the nonionic surfactant A, B is the nonionic surfactant B, and C is the anionic surfactant. The proportions of A, B, and C are, of course, renormalized so that they sum to one. This model has a condition number $\tilde{\kappa}$ of only 29.

Applying variable selection methods with $\alpha_{out} = 0.10$ leads to the reduced model

$$\widehat{\text{lather units}} = 7.16A + 8.97B - 1.91C - 19.1AB - 125Z^2 \tag{14.26}$$

The p values for all terms except Z^2 are < 0.002; that for Z^2 is 0.0813, and so this term is of marginal significance. As terms are removed from model 14.25, the condition number $\tilde{\kappa}$ slowly decreases from 29 (model 14.25) to 19 (model 14.26). Significance testing of crossproduct terms involving Z in Eq. 14.25 is safer than significance testing of crossproduct terms involving D in the Scheffé model because of the vastly improved conditioning in the MPV model Eq. 14.25.

When variable selection procedures were applied to the quadratic Scheffé model on page 220, the final model in terms of the reals was

$$\widehat{\text{lather units}} = 7.26A + 8.98B - 1.98C - 3.09D - 19.7AB$$

There is no hint of a quadratic effect of D in this model, which if present would be evidenced by crossproduct terms in D.

It is interesting to go back and reexamine the variable selection procedure starting with the quadratic Scheffé model (Table 10.6, page 220). The first two terms to be removed from the model are the AC and BC terms, leaving the eight-term model

$$\widehat{\text{lather units}} = 7.17A + 10.0B - 2.21C - 429D - 21.2AB$$
$$+ 451AD + 434BD + 452CD \tag{14.27}$$

The striking similarity of the crossproduct terms involving D suggests either equal nonlinear blending behavior of the zwitterionic surfactant with the nonionic and anionic surfactants *or* that the zwitterionic surfactant is exhibiting a quadratic curvature effect independent of any interaction with the other surfactants. The latter interpretation suggests a partial quadratic mixture model, an observation that would have been missed had the automated backward elimination procedure been used. Refitting of the data to a PQM model leads to

$$\widehat{\text{lather units}} = 7.22A + 9.71B - 2.12C + 19.13D - 21.3AB - 441D^2 \tag{14.28}$$

The condition number $\tilde{\kappa}$ of the PQM model 14.28 is 23, while that for the reduced Scheffé model 14.27 is 397. The large estimates for terms containing D in Eq. 14.27 is a reflection of the ill conditioning. Note how the coefficient estimate for D has diminished by over an order of magnitude in Eq. 14.28 when compared with Eq. 14.27.[11]

[11] Part of this can be attributed to the different meaning of the coefficient for D in the PQM model. See the Case Study in Chapter 10 and particularly Fig. 10.24, page 253.

The large coefficient for D^2 in Eq. 14.28 is a reflection of a small dependency ($D^2 \approx 0.0462D$, $R^2_{(0)} = 0.9724$). Because of this, the standard error for D^2 is inflated ($\sqrt{c_{ii}} = 976$), as is that for D ($\sqrt{c_{ii}} = 50$). As a result, the p value for the D^2 term in model 14.28 is 0.1932, and so it would be removed from the model. (The p value for the D term is 0.2687, but we do not remove linear terms from no-intercept mixture models.)

In summary, after removing AC and BC from the full quadratic Scheffé model, we are left with a model that (i) is still ill conditioned but (ii) has crossproduct terms involving D that are surprisingly similar. If we reparameterize the partially reduced Scheffé model 14.27 to the PQM model 14.28, we break up the near dependency between D, AD, BD, and CD. Even though the D^2 term turns out not to be significant, we can relate its presence in the PQM model to the presence of the Z^2 term in the MPV model. The two models are basically telling us the same story.

Bibliography

[1] J. Aitchison. *The Statistical Analysis of Compositional Data*. Chapman and Hall, New York, 1986.

[2] A. C. Atkinson. *Plots, Transformations, and Regression*. Clarendon Press, Oxford, UK, 1985.

[3] A. C. Atkinson and A. N. Donev. *Optimum Experimental Designs*. Oxford University Press, New York, 1996.

[4] R. J. Belloto, A. M. Dean, M. A. Moustafa, A. M. Molokhia, M. W. Gouda, and T. D. Sokoloski. Statistical techniques applied to solubility predictions and pharmaceutical formulations: An approach to problem solving using mixture response surface methodology. *Int. J. Pharm.*, 23:195–207, 1985.

[5] D. A. Belsley. *Conditioning Diagnostics. Collinearity and Weak Data in Regression*. John Wiley & Sons, New York, NY, 1991.

[6] D. A. Belsley, E. Kuh, and R. E. Welsch. *Regression Diagnostics: Identifying Influential Data and Sources of Collinearity*. John Wiley & Sons, New York, NY, 1980.

[7] J. J. Borkowski. A comparison of prediction variance criteria for response surface designs. *J. Qual. Technol.*, 35:70–77, 2003.

[8] G. E. P. Box. Choice of response surface design and alphabetic optimality. *Util. Math.*, 21B:11–55, 1982.

[9] G. E. P. Box. Signal-to-noise ratios, performance criteria, and transformations. *Technometrics*, 30:1–17, 1988.

[10] G. E. P. Box and D. R. Cox. An analysis of transformations. *J. Roy. Statist. Soc. B*, 26:211–243, 1964.

[11] G. E. P. Box and N. R. Draper. Robust designs. *Biometrika*, 62:347–352, 1975.

[12] G. E. P. Box and N. R. Draper. *Empirical Model-Building and Response Surfaces*. John Wiley & Sons, New York, NY, 1987.

[13] G. E. P. Box, W. G. Hunter, and J. S. Hunter. *Statistics for Experimenters*. John Wiley & Sons, New York, NY, 1978.

[14] W. M. Carlyle, D. C. Montgomery, and G. C. Runger. Optimization problems and methods in quality control and improvement. *J. Qual. Technol.*, 32:1–31, 2000.

[15] L. Y. Chan and M. K. Sandhu. Optimal orthogonal block designs for a quadratic mixture model for three components. *J. Appl. Statist.*, 26:19–34, 1999.

[16] J. Chardon, J. Nony, M. Sergent, D. Mathieu, and R. Phan-Tan-Luu. Experimental research methodology applied to the development of a formulation for use with textiles. *Chemometrics and Intelligent Laboratory Systems*, 6:313–321, 1989.

[17] J. J. Chen, L. Li, and C. D. Jackson. Analysis of quantal response data from mixture experiments. *Environmetrics*, 7:503–512, 1996.

[18] P. J. Claringbold. Use of the simplex design in the study of joint action of related hormones. *Biometrics*, 11:174–185, 1955.

[19] G. W. Cobb. *Introduction to Design and Analysis of Experiments*. Springer-Verlag, New York, NY, 1998.

[20] R. D. Cook. Detection of influential observations in linear regression. *Technometrics*, 19:15–18, 1977.

[21] R. D. Cook and C. J. Nachtsheim. A comparison of algorithms for constructing exact D-optimal designs. *Technometrics*, 24:315–324, 1980.

[22] R. D. Cook and S. Weisberg. *Residuals and Influence in Regression*. Chapman and Hall, New York, NY, 1982.

[23] R. D. Cook and S. Weisberg. *An Introduction to Regression Graphics*. John Wiley & Sons, New York, NY, 1994.

[24] R. D. Cook and S. Weisberg. *Applied Regression Including Computing and Graphics*. John Wiley & Sons, New York, NY, 1999.

[25] J. A. Cornell. Process variables in the mixture problem for categorized components. *J. Amer. Statist. Assoc.*, 66:42–48, 1971.

[26] J. A. Cornell. A comparison between two ten-point designs for studying three-component mixture systems. *J. Qual. Technol.*, 18:1–15, 1986.

[27] J. A. Cornell. Analyzing data from mixture experiments containing process variables: A split-plot approach. *J. Qual. Technol.*, 20:2–23, 1988.

[28] J. A. Cornell. Fitting models to data from mixture experiments containing other factors. *J. Qual. Technol.*, 27:13–33, 1995.

[29] J. A. Cornell. *Experiments with Mixtures*. John Wiley & Sons, New York, NY, 3rd edition, 2002.

[30] J. A. Cornell and J. C. Deng. Combining process variables and ingredient components in mixing experiments. *J. Food Sci.*, 47:836–843,848, 1982.

[31] J. A. Cornell and I. J. Good. The mixture problem for categorized components. *J. Amer. Statist. Assoc.*, 65:339–355, 1970.

[32] J. A. Cornell and J. W. Gorman. Fractional design plans for process variables in mixture experiments. *J. Qual. Technol.*, 16:20–38, 1984.

[33] J. A. Cornell and J. W. Gorman. Two new mixture models for living with collinearity but removing its influence. *J. Qual. Technol.*, 35:78–88, 2003.

[34] J. A. Cornell and P. J. Ramsey. A generalized mixture model for categorized-components problems with an application to a photoresist-coating experiment. *Technometrics*, 40:48–61, 1998.

[35] D. R. Cox. A note on polynomial response functions for mixtures. *Biometrika*, 58:155–159, 1971.

[36] R. B. Crosier. Mixture experiments, geometry and pseudocomponents. *Technometrics*, 26:209–216, 1984.

[37] R. B. Crosier. The geometry of constrained mixture experiments. *Technometrics*, 28:95–102, 1986.

[38] C. Daniel. *Applications of Statistics to Industrial Experimentation.* John Wiley & Sons, New York, NY, 1976.

[39] J. H. de Boer, A. K. Smilde, and D. A. Doornbos. Introduction of a robustness coefficient in optimization procedures: Implementation in mixture design problems. Part I: Theory. *Chemometrics and Intelligent Laboratory Systems*, 7:223–236, 1990.

[40] J. H. de Boer, A. K. Smilde, and D. A. Doornbos. Introduction of a robustness coefficient in optimization procedures: Implementation in mixture design problems. Part II: Some practical considerations. *Chemometrics and Intelligent Laboratory Systems*, 10:325–336, 1991.

[41] J. H. de Boer, A. K. Smilde, and D. A. Doornbos. Introduction of a robustness coefficient in optimization procedures: Implementation in mixture design problems. Part III: Validation and comparison with competing criteria. *Chemometrics and Intelligent Laboratory Systems*, 15:13–28, 1991.

[42] E. Del Castillo and D. C. Montgomery. A nonlinear programming solution to the dual response problem. *J. Qual. Technol.*, 25:199–204, 1993.

[43] E. Del Castillo, D. C. Montgomery, and D. R. McCarville. Modified desirability functions for multiple response optimization. *J. Qual. Technol.*, 28:337–345, 1996.

[44] G. Derringer and R. Suich. Simultaneous optimization of several response functions. *J. Qual. Technol.*, 12:214–219, 1980.

[45] G. Dingstad, B. Egelandsdal, and T. Næs. Modeling methods for crossed mixture experiments — a case study from sausage production. *Chemometrics and Intelligent Laboratory Systems*, 66:175–190, 2003.

[46] D. H. Doehlert. Uniform shell designs. *Appl. Stat.*, 19:231–239, 1970.

[47] D. H. Doehlert and V. L. Klee. Experimental designs through level reduction of the d-dimensional cuboctahedron. *Discrete Math.*, 2:309–334, 1972.

[48] N. R. Draper, P. Prescott, S. M. Lewis, A. M. Dean, P. W. M. John, and M. G. Tuck. Mixture designs for four components in orthogonal blocks. *Technometrics*, 35:268–276, 1993.

[49] N. R. Draper and H. Smith. *Applied Regression Analysis*. John Wiley & Sons, New York, NY, 3rd edition, 1998.

[50] O. Dykstra. The augmentation of experimental data to maximize $|\mathbf{X}'\mathbf{X}|$. *Technometrics*, 13:682–688, 1971.

[51] V. V. Fedorov. *Theory of Optimal Experiments*. Academic Press, New York, NY, 1972.

[52] V. F. Flack and R. A. Flores. Using simulated envelopes in the evaluation of normal probability plots of regression residuals. *Technometrics*, 31:219–225, 1989.

[53] R. J. Freund, R. C. Littell, and L. Creighton. *Regression Using JMP*. SAS Publishing, Cary, NC, 2003.

[54] A. Giovannitti-Jensen and R. H. Myers. Graphical assessment of the prediction capability of response surface designs. *Technometrics*, 31:159–171, 1989.

[55] H. B. Goldfarb, C. M. Borror, and D. C. Montgomery. Mixture-process variable experiments with noise variables. *J. Qual. Technol.*, 35:393–405, 2003.

[56] H. B. Goldfarb, C. M. Borror, D. C. Montgomery, and C. M. Anderson-Cook. Three-dimensional variance dispersion graphs for mixture-process experiments. *J. Qual. Technol.*, 36:109–124, 2004.

[57] J. W. Gorman. Fitting equations to mixture data with restraints on composition. *J. Qual. Technol.*, 2:186–194, 1970.

[58] J. W. Gorman and J. A. Cornell. A note on model reduction for experiments with both mixture components and process variables. *Technometrics*, 24:243–247, 1982.

[59] J. W. Gorman and J. E. Hinman. Simplex lattice designs for multicomponents systems. *Technometrics*, 4:463–487, 1962.

[60] P. D. Haaland. *Experimental Design in Biotechnology*. Marcel Dekker, Inc., New York, NY, 1989.

[61] A. Hald. *Statistical Theory with Engineering Applications*. John Wiley & Sons, New York, 1952.

[62] R. H. Hardin and N. J. A. Sloane. *Gosset: A General Purpose Program for Constructing Experimental Designs*. Mathematical Sciences Research Center, AT&T Bell Laboratories, Murray Hill, NJ, 1992.

[63] L. B. Hare. Designs for mixture experiments involving process variables. *Technometrics*, 21:159–173, 1979.

[64] E. C. Harrington Jr. The desirability function. *Ind. Qual. Control*, 2:494–498, 1965.

[65] I. Hau and G. E. P. Box. *Constrained Experimental Designs, Part* I: *Construction of Projection Designs*. Technical Report 53, Center for Quality and Productivity Improvement, University of Wisconsin, Madison, WI, 1990.

[66] I. Hau and G. E. P. Box. *Constrained Experimental Designs, Part* II: *Analysis of Projection Designs*. Technical Report 54, Center for Quality and Productivity Improvement, University of Wisconsin, Madison, WI, 1990.

[67] I. Hau and G. E. P. Box. *Constrained Experimental Designs, Part* III: *Properties of Projection Designs*. Technical Report 55, Center for Quality and Productivity Improvement, University of Wisconsin, Madison, WI, 1990.

[68] J. A. Heinsman and D. C. Montgomery. Optimization of a household product formulation using a mixture experiment. *Quality Engineering*, 7:583–600, 1995.

[69] K. K. Hesler and J. R. Lofstrom. Application of simplex lattice design experimentation to coatings research. *J. Coatings Technol.*, 53:33–40, 1981.

[70] D. K. Hillshafer, M. E. O'Brien, and E. H. Williamson. Aromatic Polyester Polyols Provide a Novel Approach to Reactive Hot Melt Adhesives. Presented at Adhesive and Sealant Council Meeting, Pittsburgh, PA, October 2002.

[71] D. C. Hoaglin and R. E. Welsch. The hat matrix in regression and ANOVA. *Amer. Statist.*, 32:17–22, 1978.

[72] D. W. Hosmer and S. Lemeshow. *Applied Logistic Regression*. John Wiley & Sons, New York, NY, 2000.

[73] P. J. Huber. Robust regression: Asymptotics, conjectures, and Monte Carlo. *Ann. Statist.*, 1:799–821, 1973.

[74] J. E. Jackson. *A User's Guide to Principal Components*. John Wiley & Sons, New York, NY, 1991.

[75] P. W. M. John. *Experiments with Mixtures Involving Process Variables*. Technical Report 8, Center for Statistical Sciences, University of Texas, Austin, TX, 1984.

[76] M. E. Johnson and C. J. Nachtsheim. Some guidelines for constructing exact D-optimal designs on convex mixture spaces. *Technometrics*, 25:271–277, 1983.

[77] R. N. Kacker. Off-line quality control, parameter design, and the Taguchi method. *J. Qual. Technol.*, 17:176–188, 1985.

[78] R. W. Kennard and L. Stone. Computer aided design of experiments. *Technometrics*, 11:137–148, 1969.

[79] A. I. Khuri and J. A. Cornell. *Response Surfaces, Designs and Analyses*. Marcel Dekker, Inc., New York, NY, 2nd edition, 1996.

[80] A. I. Khuri, J. M. Harrison, and J. A. Cornell. Using quantile plots of the prediction variance for comparing designs for a constrained mixture region: An application involving a fertilizer experiment. *Appl. Statist.*, 48:521–532, 1999.

[81] O. Köksoy and N. Doganaksoy. Joint optimization of mean and standard deviation using response surface methods. *J. Qual. Technol.*, 35:239–252, 2003.

[82] G. F. Koons. Effect of sinter composition on emissions: A multi-component, highly-constrained mixture experiment. *J. Qual. Technol.*, 21:261–267, 1989.

[83] S. Kowalski, J. A. Cornell, and C. G. Vining. A new model and class of designs for mixture experiments with process variables. *Commun. Statist.—Theory Meth.*, 29:2255–2280, 2000.

[84] S. Kowalski, J. A. Cornell, and C. G. Vining. Split-plot designs and estimation methods for mixture experiments with process variables. *Technometrics*, 44:72–79, 2002.

[85] I. S. Kurotori. Experiments with mixtures of components having lower bounds. *Ind. Qual. Control*, 22:592–596, 1966.

[86] D. P. Lambrakis. Experiments with mixtures: A generalization of the simplex-lattice design. *J. Roy. Statist. Soc. B*, 30:123–136, 1968.

[87] D. P. Lambrakis. Experiments with p-component mixtures. *J. Roy. Statist. Soc. B*, 30:137–144, 1968.

[88] D. P. Lambrakis. Experiments with mixtures: An alternative to the simplex-lattice design. *J. Roy. Statist. Soc. B*, 31:234–245, 1969.

[89] D. P. Lambrakis. Experiments with mixtures: Estimated regression function of the multiple-lattice design. *J. Roy. Statist. Soc. B*, 31:276–284, 1969.

[90] S. M. Lewis, A. M. Dean, N. R. Draper, and P. Prescott. Mixture designs for q components in orthogonal blocks. *J. Roy. Statist. Soc. B*, 56:457–467, 1994.

[91] C. E. Lunneborg. *Modeling Experimental and Observational Data*. Duxbury Press, Belmont, CA, 1994.

[92] D. W. Marquardt and R. D. Snee. Test statistics for mixture models. *Technometrics*, 16:533–537, 1974.

[93] R. A. McLean and V. L. Anderson. Extreme vertices designs of mixture experiments. *Technometrics*, 8:447–456, 1966.

[94] R. Mead. *The Design of Experiments.* Cambridge University Press, New York, NY, 1988.

[95] R. K. Meyer and C. J. Nachtsheim. The coordinate-exchange algorithm for constructing exact optimal experimental designs. *Technometrics*, 37:60–69, 1995.

[96] G. A. Milliken and D. E. Johnson. *Analysis of Messy Data. Volume* 1: *Designed Experiments.* Van Nostrand Reinhold Company, New York, NY, 1984.

[97] T. J. Mitchell. An algorithm for the construction of D-optimal experimental designs. *Technometrics*, 16:203–210, 1974.

[98] T. J. Mitchell. Computer construction of D-optimal first-order designs. *Technometrics*, 16:211–220, 1974.

[99] T. J. Mitchell and F. L. Miller Jr. Use of "Design Repair" to Construct Designs for Special Linear Models. Technical report, *Math. Div. Ann. Progr. Rept.* (ORNL-4661), Oak Ridge National Laboratory, Oak Ridge, TN, 1970.

[100] D. C. Montgomery, E. A. Peck, and G. G. Vining. *Introduction to Linear Regression Analysis.* John Wiley & Sons, New York, NY, 3rd edition, 2001.

[101] D. C. Montgomery and S. R. Voth. Multicollinearity and leverage in mixture experiments. *J. Qual. Technol.*, 26:96–108, 1994.

[102] D.C. Montgomery. *Design and Analysis of Experiments.* John Wiley & Sons, New York, NY, 4th edition, 1997.

[103] D. S. Moore and G. P. McCabe. *Introduction to the Practice of Statistics.* W. H. Freeman and Company, New York, NY, 1998.

[104] R. H. Myers. *Response Surface Methodology.* Edwards Brothers (distributors), Ann Arbor, MI, 1976.

[105] R. H. Myers. *Classical and Modern Regression with Applications.* PWS-KENT Publishing Company, Boston, MA, 2nd edition, 1990.

[106] R. H. Myers. Response surface methodology — current status and future directions. *J. Qual. Technol.*, 31:30–44, 1999.

[107] R. H. Myers and D. C. Montgomery. *Response Surface Methodology. Process and Product Optimization Using Designed Experiments.* John Wiley & Sons, New York, NY, 2nd edition, 2002.

[108] R. H. Myers, D. C. Montgomery, and G. G. Vining. *Generalized Linear Models.* John Wiley & Sons, New York, NY, 2002.

[109] J. P. Narcy and J. Renaud. Use of simplex experimental design in detergent formulation. *J. Am. Oil Chem. Soc.*, 49:598–608, 1972.

[110] J. A. Nelder. The selection of terms in response-surface models — how strong is the weak-heredity principle? *Amer. Statist.*, 52:315–318, 1998.

[111] J. A. Nelder. Functional marginality and response-surface fitting. *J. App. Statist.*, 27:109–112, 2000.

[112] J. A. Nelder and R. Mead. A simplex method for function minimization. *Comput. J.*, 7:308–313, 1965.

[113] J. Neter, M.H. Kutner, C.J. Nachtsheim, and W. Wasserman. *Applied Linear Statistical Models*. Irwin, Chicago, IL, 4th edition, 1996.

[114] A. K. Nigam. Block designs for mixture experiments. *Ann. Math. Statist.*, 41:1861–1869, 1970.

[115] A. K. Nigam, S. C. Gupta, and S. Gupta. A new algorithm for extreme vertices designs for linear mixture models. *Technometrics*, 25:367–371, 1983.

[116] E. R. Ott, E. G. Schilling, and D. V. Neubauer. *Process Quality Control*. McGraw-Hill, New York, NY, 3rd edition, 2000.

[117] J. L. Peixoto. A property of well-formulated polynomial regression models. *Amer. Statist.*, 44:26–30, 1990.

[118] G. F. Piepel. Measuring component effects in constrained mixture experiments. *Technometrics*, 24:29–39, 1982.

[119] G. F. Piepel. Defining consistent constraint regions in mixture experiments. *Technometrics*, 25:97–101, 1983.

[120] G. F. Piepel. Programs for generating extreme vertices and centroids of linearly constrained experimental regions. *J. Qual. Technol.*, 20:125–139, 1988.

[121] G. F. Piepel. Screening designs for constrained mixture experiments derived from classical screening designs. *J. Qual. Technol.*, 22:23–33, 1990.

[122] G. F. Piepel. Screening designs for constrained mixture experiments derived from classical screening designs — an addendum. *J. Qual. Technol.*, 23:96–101, 1991.

[123] G. F. Piepel. MIXSOFT Version 2.0, Rev. 1 and MIXSOFT User's Guide Version 2.06. MIXSOFT–Mixture Experiment Software, Richland, WA, 1992.

[124] G. F. Piepel. Modeling method for mixture-of-mixtures experiments applied to a tablet formulation problem. *Pharmaceutical Development and Technology*, 4:593–606, 1999.

[125] G. F. Piepel and C. M. Anderson. Variance dispersion graphs for designs on polyhedral regions. In 1992 *Proceedings of the Section on Physical and Engineering Sciences*, pages 111–117, Alexandria, VA, 1992. American Statistical Association.

[126] G. F. Piepel, C. M. Anderson, and P. E. Redgate. Response surface designs for irregularly-shaped regions, Parts 1, 2, and 3. In 1993 *Proceedings of the Section on Physical and Engineering Sciences*, pages 205–227, Alexandria, VA, 1993. American Statistical Association.

[127] G. F. Piepel, S. K. Cooley, and B. Jones. Construction of a 21-Component Layered Mixture Experiment Design. Presented at the 46th Annual Fall Technical Conference, Valley Forge, PA, October 2002.

[128] G. F. Piepel and J. A. Cornell. Models for mixture experiments when the response depends on the total amount. *Technometrics*, 27:219–227, 1985.

[129] G. F. Piepel and J. A. Cornell. Designs for mixture-amount experiments. *J. Qual. Technol.*, 19:11–28, 1987.

[130] G. F. Piepel and J. A. Cornell. Mixture experiment approaches: Examples, discussion, and recommendations. *J. Qual. Technol.*, 26:177–196, 1994.

[131] G. F. Piepel and J. A. Cornell. *A Catalog of Mixture Experiment Examples*. Technical report, Batelle, Pacific Northwest Laboratories, Richland, WA, 2004.

[132] G. F. Piepel, R. D. Hicks, J. M. Szychowski, and J. L. Loepky. Methods for assessing curvature and interaction in mixture experiments. *Technometrics*, 44:161–172, 2002.

[133] G. F. Piepel and T. Redgate. Mixture experiment techniques for reducing the number of components applied for modeling waste glass sodium release. *J. Am. Ceram. Soc.*, 80:3038–3044, 1997.

[134] G. F. Piepel and T. Redgate. A mixture experiment analysis of the Hald cement data. *Amer. Statist.*, 52:23–30, 1998.

[135] G. F. Piepel, J. M. Szychowski, and J. L. Loeppky. Augmenting Scheffé linear mixture models with squared and/or crossproduct terms. *J. Qual. Technol.*, 34:297–314, 2002.

[136] R. L. Plackett and J. P. Burman. The design of optimum multifactorial experiments. *Biometrika*, 33:305–325, 1946.

[137] P. Prescott. Projection designs for mixture experiments in orthogonal blocks. *Commun. Statist.—Theory Meth.*, 29:2229–2253, 2000.

[138] P. Prescott, N. R. Draper, A. M. Dean, and S. M. Lewis. Mixture designs for five components in orthogonal blocks. *J. Appl. Statist.*, 20:105–117, 1993.

[139] P. Prescott, N. R. Draper, S. M. Lewis, and A. M. Dean. Further properties of mixture designs for five components in orthogonal blocks. *J. Appl. Statist.*, 24:147–156, 1997.

[140] Proceedings of the PCI/FHWA International Symposium on High-Performance Concrete. *Concrete Mixture Optimization Using Statistical Mixture Design Methods*, New Orleans, LA, October 20–22, 1997.

[141] F. Pukelsheim. *Optimal Design of Experiments*. John Wiley & Sons, New York, NY, 1993.

[142] J. O. Ramsay. A comparative study of several robust estimates of slope, intercept, and scale in linear regression. *J. Amer. Statist. Assoc.*, 72:608–615, 1977.

[143] J. O. Rawlings, S. G. Pantula, and D. A. Dickey. *Applied Regression Analysis: A Research Tool*. Springer-Verlag, New York, NY, 2nd edition, 1998.

[144] T. P. Ryan. *Modern Regression Methods*. John Wiley & Sons, New York, NY, 1997.

[145] S. K. Saxena and A. K. Nigam. Restricted exploration of mixtures by symmetric-simplex design. *Technometrics*, 19:47–52, 1977.

[146] H. Scheffé. Experiments with mixtures. *J. Roy. Statist. Soc. B*, 20:344–360, 1958.

[147] H. Scheffé. The simplex-centroid design for experiments with mixtures. *J. Roy. Statist. Soc. B*, 25:235–263, 1963.

[148] S. R. Searle. *Linear Models*. John Wiley & Sons, New York, NY, 1971.

[149] S. R. Searle. *Matrix Algebra Useful for Statistics*. John Wiley & Sons, New York, NY, 1982.

[150] S. R. Searle. *Linear Models for Unbalanced Data*. John Wiley & Sons, New York, NY, 1987.

[151] J. T. Shelton. A Response Surface Methodology for Detecting Synergistic Herbicide Mixtures. Presented at the Joint Statistical Meetings, Anaheim, CA. ICI Americas, Richmond, CA, 1990.

[152] N. J. A. Sloane. On-Line Encyclopedia of Integer Sequences. Available from http://www.research.att.com/~njas/sequences/index.html. AT&T Research.

[153] R. D. Snee. Techniques for the analysis of mixture data. *Technometrics*, 15:517–528, 1973.

[154] R. D. Snee. Experimental designs for quadratic models in constrained mixture spaces. *Technometrics*, 17:149–159, 1975.

[155] R. D. Snee. Validation of regression models: Methods and examples. *Technometrics*, 19:415–428, 1977.

[156] R. D. Snee. Experimental designs for mixture systems with multicomponent constraints. *Commun. Statist.—Theory. Meth.*, A8:303–326, 1979.

[157] R. D. Snee. Computer-aided design of experiments — some practical experiences. *J. Qual. Technol.*, 17:222–236, 1985.

[158] R. D. Snee and D. W. Marquardt. Extreme vertices designs for linear mixture models. *Technometrics*, 16:399–408, 1974.

[159] R. D. Snee and D. W. Marquardt. Screening concepts and designs for experiments with mixtures. *Technometrics*, 18:19–29, 1976.

[160] R. D. Snee and A. A. Rayner. Assessing the accuracy of mixture model regression calculations. *J. Qual. Technol.*, 14:67–79, 1982.

[161] H. Soo, E. H. Sander, and D. W. Kess. Definition of a prediction model for determination of the effect of processing and compositional parameters on the textural characteristics of fabricated shrimp. *J. Food Sci.*, 43:1165–1171, 1978.

[162] R. C. St. John. Experiments with mixtures, ill-conditioning and ridge regression. *J. Qual. Technol.*, 16:81–96, 1984.

[163] Stat-Ease, Inc. *Design-Expert* 6 *User's Guide*, Minneapolis, MN, 2000.

[164] S. H. Steiner and M. Hamada. Making mixtures robust to noise and mixing measurement error. *J. Qual. Technol.*, 29:441–450, 1997.

[165] D. J. Van Schalkwyk. *On the Design of Mixture Experiments*. Ph.D. thesis, University of London, 1971.

[166] G. G. Vining, J. A. Cornell, and R. H. Myers. A graphical approach for evaluating mixture designs. *Appl. Statist.*, 42:127–138, 1993.

[167] E. W. Weisstein. Eric Weisstein's World of Mathematics. Available from http://mathworld.wolfram.com. Wolfram Research.

[168] W. J. Welch. Computer-aided design of experiments for response estimation. *Technometrics*, 26:217–224, 1984.

[169] W. J. Welch. ACED: Algorithms for the construction of experimental designs. *Amer. Statist.*, 39:146, 1985.

[170] W. J. Welch. *ACED: Algorithms for the Construction of Experimental Designs and ACED User's Guide Version* 1.6.1, University of Waterloo, Waterloo, ON N2L 3G1, Canada, 1985.

[171] J. W. Weyland, H. Rolink, and D. A. Doornbos. Reverse-phase high-performance liquid chromatographic separation of saccharin, caffeine and benzoic acid using nonlinear programming. *J. Chromatogr.*, 247:221–229, 1982.

[172] R. R. Wilcox. *Introduction to Robust Estimation and Hypothesis Testing*. Academic Press, New York, NY, 1997.

[173] H. Woods, H. H. Steinour, and H. R. Starke. Effect of composition of Portland cement on heat evolved during hardening. *Ind. Eng. Chem.*, 24:1207–1214, 1932.

[174] H. P. Wynn. The sequential generation of D-optimum experimental designs. *Ann. Math. Statist.*, 41:1655–1664, 1970.

[175] H. P. Wynn. Results in the theory and construction of D-optimum experimental designs. *J. Roy. Statist. Soc. B*, 34:133–147, 1972.

Index

Mathematical
Surveys
and
Monographs

Volume 105

Locally Solid Riesz Spaces with Applications to Economics

Second Edition

Charalambos D. Aliprantis
Purdue University

Owen Burkinshaw
Indiana University-Purdue University Indianapolis

American Mathematical Society

2000 *Mathematics Subject Classification.* Primary 46A40, 46B40, 47B60, 47B65, 91B50; Secondary 28A33.

Selected excerpts in this Second Edition are reprinted with the permissions of Cambridge University Press, the *Canadian Mathematical Bulletin*, Elsevier Science/Academic Press, and the *Illinois Journal of Mathematics*.

For additional information and updates on this book, visit
www.ams.org/bookpages/surv-105

Library of Congress Cataloging-in-Publication Data

Aliprantis, Charalambos D.
Locally solid Riesz spaces with applications to economics / Charalambos D. Aliprantis, Owen Burkinshaw.—2nd ed.
p. cm. — (Mathematical surveys and monographs, ISSN 0076-5376 ; v. 105)
Rev. ed. of: Locally solid Riesz spaces. 1978.
Includes bibliographical references and index.
ISBN 0-8218-3408-8 (alk. paper)
1. Riesz spaces. 2. Economics, Mathematical. I. Burkinshaw, Owen. II. Aliprantis, Charalambos D. III. Locally solid Riesz spaces. IV. Title. V. Mathematical surveys and monographs ; no. 105.

QA322.A39 2003
515′.73—dc22 2003057948

∞ The paper used in this book is acid-free and falls within the guidelines
established to ensure permanence and durability.
Visit the AMS home page at http://www.ams.org/

10 9 8 7 6 5 4 3 2 1 08 07 06 05 04 03

To the memory of my parents: *Διονύσιο και Σωφρονία*

Charalambos Dionysios Aliprantis

To my high school mathematics and science teachers,
Elizabeth De Vary and Donald Roush.
Great teachers make a difference in the lives of their students.

Owen Burkinshaw

Contents

Preface of the 1st Edition

In 1928, at the International Mathematical Congress in Bologna, Italy, F. Riesz in a short address [**115**] triggered the investigation of what is today called the theory of Riesz spaces. Soon after, in the mid-thirties, H. Freudenthal [**60**] and L. V. Kantorovich [**69, 70**] independently set up the axiomatic foundation and derived a number of properties dealing with the lattice structure of Riesz spaces. From then on the growth of the subject was rapid. In the forties and early fifties the Japanese school led by H. Nakano, T. Ogasawara, and K. Yosida, and the Russian school, led by L. V. Kantorovich, A. I. Judin, and B. Z. Vulikh, made fundamental contributions.

At the same time a number of books started to appear on the field. Most of the early work on Riesz spaces, as well as most of the books, was devoted to the so-called "algebraic part" of the theory and little attention was given to the "analytical part." The bulk of the existing work on the analytical part is in the area of normed Riesz spaces, and in particular in the theory of Banach lattices; we note the work of W. A. J. Luxemburg and A. C. Zaanen appearing in a series of articles [**89**] and the recent book of H. H. Schaefer [**123**]. The general theory of topological Riesz spaces seems somehow to have been neglected. The recent book by D. H. Fremlin [**58**] is partially devoted to this subject; and there are, of course, a number of books on linear ordered spaces (see [**65, 108, 127, 130**]). But there have been no books completely devoted to topological Riesz spaces.

The purpose of this book is to fill this gap and present a unified approach to the theory of locally solid Riesz spaces. It has been our desire to emphasize the relationship between the order and topological structures.

The present work is written for graduate students and for persons involved in research in functional analysis. The book is self-contained and should be accessible to any student with a standard course in functional analysis. The material has been arranged into seven chapters. As a supplement, at the end of each chapter, we have included a number of exercises of varying degrees of difficulty. We have also listed some open problems to give insight into possible areas of research. Throughout the book the basic concepts are illustrated by numerous examples and counterexamples, and special notes trace the history of the subject. However, it is possible that credit was not given to all contributors to this field.

In Chapter 1 a summary of the basic theory of the lattice structure of Riesz spaces is given, with special emphasis on results that are not easily accessible and to proofs that are not well known. The fundamental properties of locally solid topologies are presented in Chapter 2, while Chapter 3 studies the interaction between order convergence and topological convergence (Lebesgue properties) with applications to the representation theorems of abstract L_p-spaces.

Chapter 4 deals with Fatou topologies and shows the remarkable relationship between order completeness and topological completeness, while Chapter 5 investigates the important properties of metrizable locally solid Riesz spaces.

In Chapter 6 locally solid topologies that are also locally convex are considered. Weakly compact sets are studied via the concept of order-equicontinuity. The interplay between the order structure and the topological concept of compactness is demonstrated extensively, and characterizations of semireflexive Riesz spaces are presented. Finally, Chapter 7 discusses both the remarkable lattice and topological properties of laterally complete Riesz spaces.

This book has been influenced by the discussions the first author had with W. A. J. Luxemburg and the second with S. Kaplan, to whom we express our appreciation for introducing us to the subject.

Finally, we would like to thank our wives, Bernadette and Betty, for their help and understanding during the preparation of this book.

Indianapolis
June 1977

C. D. ALIPRANTIS
O. BURKINSHAW

Preface of the 2nd Edition

This monograph is the 2nd edition of our book *Locally Solid Riesz Spaces* that was published by Academic Press in 1978. As mentioned in the introduction of that book, the monograph presented the theory of locally solid topologies on Riesz spaces in a systematic and thorough manner. However, since the appearance of *Locally Solid Riesz Spaces* in 1978 many developments and changes have occurred in the field of Riesz spaces the most important of which was the application of locally solid Riesz spaces to economics. As it turned out, the lattice and topological structures of ordered spaces were the main ingredients needed for an economic framework on which a fruitful and credible economic analysis can be founded.

The 2nd edition is a complete revision of the 1st one. In order for the book to be as self-contained as possible we have added lots of extra material. Also, to make the book more reader-friendly the old material has been rearranged in many places. The chapter structure of the 2nd edition follows the structure of the 1st edition. Specifically, the first seven chapters of this book are along the spirit of the 1st edition. However, we have added two more chapters; chapters 8 and 9. The new Chapter 8 presents a brief but rather comprehensive treatment of General Economic Equilibrium in the framework of locally solid Riesz spaces. Due to this chapter, we changed the title of the book to *Locally Solid Riesz Spaces with Applications to Economics*.

Chapter 9 presents complete solutions to all exercises in the book. As with the 1st edition, at the end of each chapter there is a collection of exercises of varying degree of difficulty. The purpose of the exercises is to present some supplementary material and to challenge the reader's understanding of the subject matter in the chapter. The reader will discover by himself that the material presented in the solutions is very complementary and supplementary to the material in the text.

We take this opportunity to thank our late colleague and collaborator Yuri Abramovich for reading the manuscript and making numerous suggestions and corrections. The constructive suggestions made by two anonymous referees that resulted in improving the exposition of the material in this monograph are greatly appreciated. Special thanks are due to our student Daniela Puzzello for reading carefully the manuscript and correcting numerous mistakes—mathematical and non-mathematical. Finally, we express our sincerest thanks to Professor Sergei I. Gelfand, the AMS Director of Acquisitions, for his constant support and help during the preparation of this monograph.

It is our hope that this edition of *Locally Solid Riesz Spaces* will not only be beneficial to mathematicians and economists but also to other scientists in many disciplines pure and applied as well.

West Lafayette and Indianapolis
May 2003

C. D. ALIPRANTIS
O. BURKINSHAW

List of Special Symbols

Numbers that follow each entry are the page numbers on which the symbol first occurs.

CHAPTER 1

The Lattice Structure of Riesz spaces

In the first chapter of this book we shall deal with the fundamental properties of Riesz spaces and we hope that the reader is somewhat familiar with the basic theory of ordered vector spaces and Riesz spaces. Most of the elementary properties of Riesz spaces can be found in the excellent book "Riesz Spaces," Vol. I, by W. A. J. Luxemburg and A. C. Zaanen [**90**]. However, for the sake of completeness, we shall state (and provide detailed proofs of) the basic results regarding the theory of abstract Riesz spaces. Special emphasis will be given to proofs of results that are either not easily accessible or according to our judgment greatly simplified proofs.

Throughout this book the symbol $\mathbb{R}$ will denote the set of real numbers, $\mathbb{N}$ will denote the set of natural numbers, $\mathbb{Q}$ will denote the set of rational numbers, and $\mathbb{Z}$ will denote the set of integers.

1.1. Elementary Properties of Riesz spaces

An **ordered vector space** is a real vector space E equipped with an order relation $\geq$ (i.e., with a transitive, reflexive, and antisymmetric relation $\geq$) compatible with the algebraic structure of E in the sense that it possesses the following two properties:

(i) If $u, v \in E$ satisfy $u \geq v$, then $u + w \geq v + w$ for all $w \in E$.
(ii) If $u, v \in E$ satisfy $u \geq v$, then $\lambda u \geq \lambda v$ for each $\lambda \geq 0$.

Occasionally the notation $v \leq u$ will be used instead of $u \geq v$. The relation $u > v$ (or $v < u$) will mean $u \geq v$ and $u \neq v$. The zero element of a vector space will be denoted by 0. The elements u of E with $u \geq 0$ are called **positive elements** or **positive vectors**. The set $E^+ = \{u \in E\colon\ u \geq 0\}$ is called the **positive cone** (or simply the **cone**) of E. The positive cone satisfies the following three properties:

(a) $E^+ + E^+ \subseteq E^+$, where $E^+ + E^+ = \{u + v\colon\ u, v \in E^+\}$.
(b) $\lambda E^+ \subseteq E^+$ for each $0 \leq \lambda \in \mathbb{R}$, where $\lambda E^+ = \{\lambda u\colon\ u \in E^+\}$.
(c) $E^+ \cap (-E^+) = \{0\}$, where $-E^+ = \{-u\colon\ u \in E^+\}$.

Any subset C of a real vector space E satisfying properties (a), (b), and (c) is called a **cone**[1] of E. If C is a cone of E, note that the relation $u \geq v$ whenever $u - v \in C$ makes E an ordered vector space whose positive cone is precisely C.

A nonempty subset A of an ordered vector space E is said to have a **supremum** (or a **least upper bound**) if there exists some $u \in E$ satisfying $u \geq a$ for all $a \in A$ and such that whenever $v \geq a$ holds for some $v \in E$ and all $a \in A$, then $v \geq u$. Clearly, a subset A can have at most one supremum, denoted $\sup A$. A similar

[1] Some authors call a subset that satisfies properties (a), (b), and (c) a **convex cone** or even a **pointed convex cone**.

definition can be given for the **infimum** (or the **greatest lower bound**) of a nonempty subset A of E, denoted by $\inf A$.

DEFINITION 1.1. *A* ***Riesz space*** *(or a* ***vector lattice****) is an ordered vector space L with the additional property that the supremum of every nonempty finite subset of L exists.*[2]

Following the classical lattice notation, we denote the supremum of the set $\{u,v\}$ by $u \vee v$. Similarly, $u \wedge v$ denotes the infimum of the set $\{u,v\}$. That is, we let

$$u \vee v = \sup\{u,v\} \quad \text{and} \quad u \wedge v = \inf\{u,v\}.$$

If $\mathcal{C} = \{u_1, \dots, u_n\}$ is a finite collection of elements in a Riesz space, then we shall denote $\sup \mathcal{C}$ and $\inf \mathcal{C}$ by $\bigvee_{i=1}^n u_i$ and $\bigwedge_{i=1}^n u_i$, respectively. That is, we shall write

$$\bigvee_{i=1}^n u_i = \sup\{u_1, \dots, u_n\} \quad \text{and} \quad \bigwedge_{i=1}^n u_i = \inf\{u_1, \dots, u_n\}.$$

Likewise, if a subset $\mathcal{C}$ in a Riesz space has a supremum or an infimum we shall denote $\sup \mathcal{C}$ and $\inf \mathcal{C}$ by $\bigvee_{u \in \mathcal{C}} u$ and $\bigwedge_{u \in \mathcal{C}} u$, respectively. That is, occasionally we shall write

$$\sup \mathcal{C} = \bigvee_{u \in \mathcal{C}} u \quad \text{and} \quad \inf \mathcal{C} = \bigwedge_{u \in \mathcal{C}} u.$$

It should be clear that an ordered vector space L is a Riesz space if and only if for every pair of vectors $u, v \in L$ the supremum $u \vee v$ exists in L. We also have the following elementary but very useful result.

LEMMA 1.2. *An ordered vector space L is a Riesz space if and only if for every pair of vectors $u, v \in L$ their infimum $u \wedge v$ exists in L. Moreover, if u and v are elements in a Riesz space, then*

$$u \vee v = -\big[(-u) \wedge (-v)\big] \quad \textit{and} \quad u \wedge v = -\big[(-u) \vee (-v)\big].$$

PROOF. Assume first that every pair in an ordered vector space L has an infimum, and let $u, v \in L$. Put $w = (-u) \wedge (-v)$. We shall prove that $u \vee v = -w$. To see this, note first that we have $w \leq -u$ and $w \leq -v$ or $u \leq -w$ and $v \leq -w$. That is, $-w$ is an upper bound of the set $\{u,v\}$. To see that $-w$ is the least upper bound of the set $\{u,v\}$, suppose that a vector t satisfies $u \leq t$ and $v \leq t$. This implies $-t \leq -u$ and $-t \leq -v$, and therefore $-t \leq (-u) \wedge (-v) = w$, i.e., $t \geq -w$. Hence $u \vee v = -w$ holds. The converse can be proven in a similar fashion. ∎

For an element u of a Riesz space L the **positive part** of u is defined by $u^+ = u \vee 0$, the **negative part** by $u^- = (-u) \vee 0$, and the **absolute value** by $|u| = u \vee (-u)$.

The following theorem contains a list of several important lattice identities that hold true for elements in Riesz spaces. These identities will be used freely throughout this monograph.

---

[2] The concept of a Riesz space is due, independently, to F. Riesz [**115, 116**], H. Freudenthal [**60**], and L. V. Kantorovich [**69, 70, 71**].

THEOREM 1.3 (Lattice Identities). *For any elements u, v, and w in a Riesz space we have the following:*

(1) $u+v\vee w=(u+v)\vee(u+w)$ *and* $u+v\wedge w=(u+v)\wedge(u+w)$
(2) $u-v\wedge w=(u-v)\vee(u-w)$ *and* $u-v\vee w=(u-v)\wedge(u-w)$
(3) $u\vee v=(u-v)^+ +v=(v-u)^+ +u$
(4) $\lambda(u\vee v)=(\lambda u)\vee(\lambda v)$ *and* $\lambda(u\wedge v)=(\lambda u)\wedge(\lambda v)$ *for* $\lambda\geq 0$
(5) $|\lambda u|=|\lambda|\,|u|$ *for all* $\lambda\in\mathbb{R}$
(6) $u\vee v=\frac{1}{2}\big(u+v+|u-v|\big)$ *and* $u\wedge v=\frac{1}{2}\big(u+v-|u-v|\big)$
(7) $u+v=u\vee v+u\wedge v$
(8) $u=u^+-u^-$ *and* $u^+\wedge u^-=0$
(9) $|u|=u^++u^-$ *(hence* $|u|=0$ *if and only if* $u=0$*)*
(10) $|u-v|=u\vee v-u\wedge v$
(11) $|u+v|\vee|u-v|=|u|+|v|$
(12) $|u|\vee|v|=\frac{1}{2}\big(|u+v|+|u-v|\big)$ *and* $|u|\wedge|v|=\frac{1}{2}\big||u+v|-|u-v|\big|$

PROOF. (1) The first identity should be interpreted as follows:

- *If two elements v and w in an ordered vector space have a supremum $v\vee w$, then for every element u the set $\{u+v,u+w\}$ also has a supremum and $(u+v)\vee(u+w)=u+(v\vee w)$.*

To see this, let $t=v\vee w$ and fix u. Clearly, $u+v\leq u+t$ and $u+w\leq u+t$. Now assume that $u+v\leq s$ and $u+w\leq s$. It follows that $v\leq s-u$ and $w\leq s-u$, and so $t=v\vee w\leq s-u$ or $u+t\leq s$. This shows that $u+t$ is the supremum of the set $\{u+v,u+w\}$, and this proves the validity of the first identity. The validity of the second identity can be proven in a similar manner.

(2) It follows from part (1) and Lemma 1.2.

(3) To establish the first identity in this part note that

$$(u-v)^++v=(u-v)\vee 0+v=\big[(u-v)+v\big]\vee(0+v)=u\vee v\,.$$

The second identity can be proven in a similar manner.

(4) Assume that $\lambda>0$. From $u\leq u\vee v$ and $v\leq u\vee v$, we get $\lambda u\leq\lambda(u\vee v)$ and $\lambda v\leq\lambda(u\vee v)$. Now assume that $\lambda u\leq w$ and $\lambda v\leq w$. It follows that $u\leq\frac{1}{\lambda}w$ and $v\leq\frac{1}{\lambda}w$, and so $u\vee v\leq\frac{1}{\lambda}w$. Consequently $\lambda(u\vee v)\leq w$, and from this we see that $(\lambda u)\vee(\lambda v)=\lambda(u\vee v)$. The other identity can be established with a similar argument.

(5) If $\lambda\geq 0$, then $|\lambda u|=(\lambda u)\vee(-\lambda u)=\lambda\big[u\vee(-u)\big]=|\lambda||u|$ and if $\lambda<0$, then $|\lambda u|=(\lambda u)\vee(-\lambda u)=\big[(-\lambda)(-u)\big]\vee(-\lambda u)=(-\lambda)\big[(-u)\vee u\big]=|\lambda||u|$.

(6) For the first identity in this part note that

$$\begin{aligned}\tfrac{1}{2}\big(u+v+|u-v|\big) &= \tfrac{1}{2}\big[u+v+(u-v)\vee(v-u)\big]\\ &= \tfrac{1}{2}\big[(2u)\vee(2v)\big]=u\vee v\,.\end{aligned}$$

(7) Adding the identities in (6) yields the desired identity.

(8) Letting $v=0$ in (7) yields

$$u=u\vee 0+u\wedge 0=u\vee 0-(-u)\vee 0=u^+-u^-\,.$$

For the second identity in this part, note that

$$\begin{aligned} u^+ \wedge u^- &= (u^+ - u^-) \wedge 0 + u^- = u \wedge 0 + u^- \\ &= -\big[(-u) \vee 0\big] + u^- = -u^- + u^- = 0\,. \end{aligned}$$

(9) Observe that

$$|u| = u \vee (-u) = (2u) \vee 0 - u = 2u^+ - (u^+ - u^-) = u^+ + u^-\,.$$

(10) Using the definition of the absolute value, (1) and (7), we get

$$\begin{aligned} |u - v| &= (u - v) \vee (v - u) = (2u) \vee (2v) - (u + v) \\ &= 2(u \vee v) - (u \vee v + u \wedge v) \\ &= u \vee v - u \wedge v\,. \end{aligned}$$

(11) Notice that

$$\begin{aligned} |u + v| \vee |u - v| &= \big[(u + v) \vee (-u - v)\big] \vee \big[(u - v) \vee (v - u)\big] \\ &= \big[(u + v) \vee (u - v)\big] \vee \big[(-u - v) \vee (v - u)\big] \\ &= \big[u + v \vee (-v)\big] \vee \big[-u + (-v) \vee v\big] \\ &= \big[u + |v|\big] \vee \big[-u + |v|\big] \\ &= \big[u \vee (-u)\big] + |v| = |u| + |v|\,. \end{aligned}$$

(12) For the first identity in this part note that

$$\begin{aligned} \tfrac{1}{2}\big[|u + v| + |u - v|\big] &= \tfrac{1}{2}\big[(u + v) \vee (-u - v) + |u - v|\big] \\ &= \tfrac{1}{2}\big[\big(u + v + |u - v|\big) \vee \big(-u - v + |u - v|\big)\big] \\ &= \tfrac{1}{2}\Big(\big[2(u \vee v)\big] \vee \big[2\{(-u) \vee (-v)\}\big]\Big) \\ &= u \vee v \vee (-u) \vee (-v) \\ &= \big[u \vee (-u)\big] \vee \big[v \vee (-v)\big] \\ &= |u| \vee |v|\,. \end{aligned}$$

Using (10), (7), and (11) and the preceding identity, we also get

$$\begin{aligned} &\big||u + v| - |u - v|\big| \\ &\quad = |u + v| \vee |u - v| - |u + v| \wedge |u - v| \\ &\quad = |u + v| \vee |u - v| - \big(|u + v| + |u - v| - |u + v| \vee |u - v|\big) \\ &\quad = 2\big(|u + v| \vee |u - v|\big) - \big(|u + v| + |u - v|\big) \\ &\quad = 2\big(|u| + |v|\big) - 2\big(|u| \vee |v|\big) = 2\big(|u| \wedge |v|\big)\,, \end{aligned}$$

and the proof is finished. ■

From property (3) of the preceding theorem it should be clear that:

- *An ordered vector space L is a Riesz space if and only if u^+ exists for each $u \in L$.*

(In conjunction with this claim see also Exercise 1 at the end of the chapter.)

Several lattice identities that are included in Theorem 1.3 can be stated in a more general setting. We leave the proof of the next result to the reader.

LEMMA 1.4. *If a nonempty subset A of a Riesz space L has a supremum (resp. an infimum), then for each $u \in L$ the set $u+A$ has a supremum (resp. an infimum) and the following lattice identities hold:*

$$u + \sup A = \sup(u + A) \quad \textit{and} \quad u + \inf A = \inf(u + A)\,.$$

Similarly, the interpretation of the following lattice identities should be analogous:

$$u - \sup A = \inf(u - A) \quad \textit{and} \quad u - \inf A = \sup(u - A).$$

$$\lambda \sup A = \sup(\lambda A) \quad \textit{and} \quad \lambda \inf A = \inf(\lambda A) \ \textit{for all}\ \lambda \geq 0\,.$$

Riesz spaces considered as lattices satisfy the distributive laws. In fact they satisfy the "infinite distributive laws" as the next lemma shows.

LEMMA 1.5 (The Infinite Distributive Law). *If* $\sup A$ *exists for a nonempty subset A of a Riesz space L, then* $\sup\{u \wedge a\colon\ a \in A\}$ *exists for each $u \in L$ and*

$$u \wedge \sup A = \sup\big\{u \wedge a\colon\ a \in A\big\}\,.$$

Likewise, if $\inf A$ *exists for a nonempty subset A of some Riesz space L, then* $\inf\{u \vee a\colon\ a \in A\}$ *exists for each $u \in L$ and*

$$u \vee \inf A = \inf\big\{u \vee a\colon\ a \in A\big\}\,.$$

In particular, if $u, u_1, u_2, \ldots, u_n$ are elements in a Riesz space, then

$$u \wedge \Big[\bigvee_{i=1}^{n} u_i\Big] = \bigvee_{i=1}^{n} (u \wedge u_i) \quad \textit{and} \quad u \vee \Big[\bigwedge_{i=1}^{n} u_i\Big] = \bigwedge_{i=1}^{n} (u \vee u_i)\,.$$

PROOF. We prove the first formula only. Let $s = \sup A$ and fix $u \in L$. Clearly, $u \wedge a \leq u \wedge s$ for all $a \in A$. Now assume that some $t \in L$ is an upper bound of the set $\{u \wedge a\colon\ a \in A\}$, i.e., $u \wedge a \leq t$ holds for all $a \in A$. Using the lattice identity $x \vee y + x \wedge y = x + y$, we see that $u + a - u \vee a = u \wedge a \leq t$ for all $a \in A$. This implies $a \leq t + u \vee a - u \leq t + u \vee s - u$ for all $a \in A$, and so $s \leq t + u \vee s - u$. Consequently, $u \wedge s = u + s - u \vee s \leq t$, and from this we get $u \wedge s = \sup\big\{u \wedge a\colon\ a \in A\big\}$. ■

As an immediate application of the distributive laws we shall present a classical identity in Riesz spaces known as *Birkhoff's identity* and is due to G. Birkhoff [**39**, Theorem 8, p. 296].

COROLLARY 1.6 (Birkhoff's Identity). *If u, v, and w are arbitrary elements in a Riesz space, then*

$$|u \vee w - v \vee w| + |u \wedge w - v \wedge w| = |u - v|\,.$$

PROOF. Using the lattice identities (10) and (7) of Theorem 1.3 in conjunction with the distributive laws, we see that

$$\begin{aligned}
&|u \vee w - v \vee w| + |u \wedge w - v \wedge w| \\
&\quad = \big[(u \vee w) \vee (v \vee w) - (u \vee w) \wedge (v \vee w)\big] + \cdots \\
&\qquad \cdots + \big[(u \wedge w) \vee (v \wedge w) - (u \wedge w) \wedge (v \wedge w)\big] \\
&\quad = \big[w \vee (u \vee v) - w \vee (u \wedge v)\big] + \big[w \wedge (u \vee v) - w \wedge (u \wedge v)\big] \\
&\quad = \big[w \vee (u \vee v) + w \wedge (u \vee v)\big] - \big[w \vee (u \wedge v) + w \wedge (u \wedge v)\big] \\
&\quad = \big[w + u \vee v\big] - \big[w + u \wedge v\big] = u \vee v - u \wedge v = |u - v|\,,
\end{aligned}$$

as desired. ■

We are now ready to state and prove a number of important lattice inequalities in Riesz spaces.

THEOREM 1.7 (Lattice Inequalities). *The following lattice inequalities hold in Riesz spaces.*

(1) *If two elements u and v in a Riesz space satisfy $u \le v$, then*

$$u^+ \le v^+ \quad \textit{and} \quad v^- \le u^- .$$

(2) (The Triangle Inequality) *If u and v are any two elements in a Riesz space, then*

$$\big||u| - |v|\big| \le |u+v| \le |u| + |v| .$$

(3) (Birkhoff's Inequalities [**39**, p. 296]) *If u, v, and w are elements in a Riesz space, then*

$$|u \vee w - v \vee w| \le |u-v| \quad \textit{and} \quad |u \wedge w - v \wedge w| \le |u-v| .$$

(4) *If $u, u_1, u_2, \ldots, u_n$ are positive elements in a Riesz space, then*

$$u \wedge (u_1 + u_2 + \cdots + u_n) \le u \wedge u_1 + u \wedge u_2 + \cdots + u \wedge u_n .$$

If $u \wedge u_i \wedge u_j = 0$ holds for all $i \ne j$, then equality holds.

(5) *If $u_1, u_2, \ldots, u_n$ are elements in a Riesz space, then*

$$n\big(u_1^+ \wedge \cdots \wedge u_n^+\big) = n(u_1 \wedge \cdots \wedge u_n)^+ \le (u_1 + \cdots + u_n)^+ .$$

PROOF. (1) Let $u \le v$. Then $u \le v \le v \vee 0 = v^+$ and $0 \le v^+$. This implies $u^+ = u \vee 0 \le v^+$. The other inequality follows from the previous one by observing that $-v \le -u$.

(2) Clearly, $u + v \le |u| + |v|$ and $-(u+v) = -u - v \le |u| + |v|$. This implies $|u+v| = (u+v) \vee [-(u+v)] \le |u| + |v|$. For the other inequality note that we have $|u| = \big|(u+v) - v\big| \le |u+v| + |v|$ and so $|u| - |v| \le |u+v|$. Similarly,

$$-\big(|u| - |v|\big) = |v| - |u| \le |u+v| ,$$

and consequently $\big||u| - |v|\big| = (|u| - |v|) \vee (|v| - |u|) \le |u+v|$ is also true.

(3) They follow immediately from Birkhoff's identity; see Corollary 1.6.

(4) Assume that u and u_1, u_2 are all positive vectors. For simplicity, we let $v = u \wedge (u_1 + u_2)$. Then $v \le u_1 + u_2$ and so $v - u_1 \le u_2$. Also we have $v - u_1 \le v \le u$. Consequently $v - u_1 \le u \wedge u_2$. This implies $v - u \wedge u_2 \le u_1$ and so from the inequality $v - u \wedge u_2 \le v \le u$, we infer that $v - u \wedge u_2 \le u \wedge u_1$ or $v \le u \wedge u_1 + u \wedge u_2$.

If $u \wedge u_1 \wedge u_2 = (u \wedge u_1) \wedge (u \wedge u_2) = 0$, then from

$$\begin{aligned} u \wedge (u_1 + u_2) &\le u \wedge u_1 + u \wedge u_2 = (u \wedge u_1) \vee (u \wedge u_2) \\ &= u \wedge (u_1 \vee u_2) \le u \wedge (u_1 + u_2), \end{aligned}$$

it follows that $u \wedge (u_1 + u_2) = u \wedge u_1 + u \wedge u_2$. The general case can be proven by induction.

(5) From the inequality $n(u_1 \wedge \cdots \wedge u_n) \le u_1 + \cdots + u_n$, it easily follows that $n(u_1 \wedge \cdots \wedge u_n)^+ \le (u_1 + \cdots + u_n)^+$. Now note that

$$\begin{aligned} n\big(u_1^+ \wedge u_2^+ \wedge \cdots \wedge u_n^+\big) &= n\big[(u_1 \vee 0) \wedge (u_2 \vee 0) \wedge \cdots \wedge (u_n \vee 0)\big] \\ &= n\big[(u_1 \wedge u_2 \wedge \cdots \wedge u_n) \vee 0\big] \\ &= n(u_1 \wedge u_2 \wedge \cdots \wedge u_n)^+ , \end{aligned}$$

and the proof is finished. ■

We continue by introducing the disjointness (or orthogonality) notion in Riesz spaces.

DEFINITION 1.8. *Two elements u and v in a Riesz space are called* ***disjoint*** *or* ***orthogonal****, in symbols $u \perp v$, if $|u| \wedge |v| = 0$.*

Two nonempty subsets A and B in a Riesz space are said to be **disjoint** or **orthogonal**, in symbols $A \perp B$, if $a \perp b$ for all $a \in A$ and $b \in B$.

The symbol $u \perp D$ means that $u \perp v$ for each $v \in D$. The basic disjointness properties are included in the next lemma.

LEMMA 1.9 (Disjointness Properties). *We have the following:*

(1) *If $u \perp v$ and $u \perp w$ hold in a Riesz space, then $u \perp (\lambda v + \mu w)$ for all $\lambda, \mu \in \mathbb{R}$.*
(2) *Two vectors u and v in an arbitrary Riesz space are disjoint if and only if $|u+v| = |u-v|$.*
(3) *If $u \perp v$ holds in a Riesz space, then*
$$|u+v| = |u-v| = |u| + |v| = \big||u| - |v|\big| = |u| \vee |v|\,.$$
(4) *Every subset of a Riesz space consisting of pairwise disjoint nonzero elements is linearly independent.*

PROOF. (1) Assume that $u \perp v$ and $u \perp w$ and let $\lambda, \mu \in \mathbb{R}$. Then we have

$$\begin{aligned}
0 &\le |u| \wedge |\lambda v + \mu w| \le |u| \wedge \big(|\lambda v| + |\mu w|\big) \\
&= |u| \wedge \big(|\lambda||v| + |\mu||w|\big) \le |u| \wedge \big(|\lambda||v|\big) + |u| \wedge \big(|\mu||w|\big) \\
&\le (1+|\lambda|)|u| \wedge (1+|\lambda|)|v| + (1+|\mu|)|u| \wedge (1+|\mu|)|w| \\
&= (1+|\lambda|)\big[|u| \wedge |v|\big] + (1+|\mu|)\big[|u| \wedge |w|\big] \\
&= (1+|\lambda|)0 + (1+|\mu|)0 = 0\,,
\end{aligned}$$

and so $|u| \wedge |\lambda v + \mu w| = 0$, i.e., $u \perp (\lambda v + \mu w)$.

(2) The second identity in part (12) of Theorem 1.3 states that for any two vectors in a Riesz space we have: $|u| \wedge |v| = \frac{1}{2}\big||u+v| - |u-v|\big|$. From this we infer immediately that $u \perp v$ (i.e., $|u| \wedge |v| = 0$) if and only if $|u+v| = |u-v|$.

(3) Assume that $u \perp v$ holds in a Riesz space. By part (2) we already know that $|u+v| = |u-v|$. The same conclusion applied to the pair of vectors $|u|$ and $|v|$ yields
$$\big||u| - |v|\big| = \big||u| + |v|\big| = |u| + |v| = |u| \vee |v|\,.$$
Now notice that a quick glance at the identity in part (11) of Theorem 1.3 yields $|u+v| = |u-v| = |u+v| \vee |u-v| = |u| + |v|$, and the conclusion follows.

(4) Assume that $u_1, \ldots, u_n$ are pairwise disjoint nonzero elements in a Riesz space and that $\alpha_1 u_1 + \cdots + \alpha_n u_n = 0$. Then by parts (1) and (3) we have

$$0 = |\alpha_1 u_1 + \cdots + \alpha_n u_n| = |\alpha_1 u_1| + \cdots + |\alpha_n u_n| = |\alpha_1||u_1| + \cdots + |\alpha_n||u_n|\,,$$

and from this it follows that $|\alpha_i||u_i| = 0$ for each i. Since $|u_i| > 0$ for each i, we get $|\alpha_i| = 0$ or $\alpha_i = 0$ for each $i = 1, \ldots, n$. Consequently the vectors $u_1, \ldots, u_n$ are linearly independent. ■

The next property of Riesz spaces—known as the **Riesz Decomposition Property**—plays a crucial role in the study of Riesz spaces.

THEOREM 1.10 (The Riesz Decomposition Property). *Assume that in a Riesz space L we have the inequality*

$$|u| \le |v_1 + v_2 + \cdots + v_n|.$$

Then there exist elements $u_1, \ldots, u_n \in L$ satisfying $|u_i| \le |v_i|$ for each $i = 1, \ldots, n$, and $u = u_1 + u_2 + \cdots + u_n$.

In addition, if the vector u is positive, then the vectors u_i can be chosen to be positive too.

PROOF. By using induction it suffices to establish the result for $n = 2$. Thus assume $|u| \le |v_1 + v_2|$. Next put $u_1 = \big[u \vee (-|v_1|)\big] \wedge |v_1|$. From the inequalities $-|v_1| \le u \vee \big(-|v_1|\big)$ and $-|v_1| \le |v_1|$, it follows that $-|v_1| \le u_1$ or $-u_1 \le |v_1|$. On the other hand, from $u_1 \le |v_1|$, we obtain $|u_1| = (-u_1) \vee u_1 \le |v_1|$ (and clearly $0 \le u_1 \le u$ holds if u is also positive). Now if $u_2 = u - u_1$, then

$$u_2 = u - \big[u \vee (-|v_1|)\big] \wedge |v_1| = \big[0 \wedge \big(u + |v_1|\big)\big] \vee \big(u - |v_1|\big).$$

However, $|u| \le |v_1| + |v_2|$ implies $-|v_1| - |v_2| \le u \le |v_1| + |v_2|$, from which we get

$$-|v_2| = \big(-|v_2|\big) \wedge 0 \le \big(u + |v_1|\big) \wedge 0 \le u_2 \le 0 \vee \big(u - |v_1|\big) \le |v_2|.$$

This implies $|u_2| \le |v_2|$, and the proof is finished. ■

The **Cartesian product** $L = \Pi L_\alpha$ of a family of Riesz spaces $\{L_\alpha\}_{\alpha\in A}$, ordered componentwise, that is, ordered by the relation $\{u_\alpha\}_{\alpha\in A} \ge \{v_\alpha\}_{\alpha\in A}$ whenever $u_\alpha \ge v_\alpha$ for all $\alpha \in A$, becomes a Riesz space. The proof is straightforward; note that $\{u_\alpha\}_{\alpha\in A} \vee \{v_\alpha\}_{\alpha\in A} = \{u_\alpha \vee v_\alpha\}_{\alpha\in A}$ holds in L and that a similar formula holds true for the infimum.

A subset S in a Riesz space L is said to be **solid** if it follows from $|u| \le |v|$ in L and $v \in S$ that $u \in S$. Every subset A of L is included in a smallest (with respect to the inclusion) solid set, called the **solid hull** of A and denoted $\mathrm{Sol}\,(A)$. Clearly,

$$\mathrm{Sol}\,(A) = \big\{v \in L\colon\ \exists\, u \in A \text{ such that } |v| \le |u|\,\big\}.$$

Every solid set S is a balanced set, that is, if $u \in S$, then $\lambda u \in S$ for each $\lambda \in \mathbb{R}$ with $|\lambda| \le 1$.

Recall that a subset A in a vector space is said to be **convex** if for all $u, v \in A$ and all $0 \le \lambda \le 1$ we have $\lambda u + (1-\lambda)v \in A$. The **convex hull** $\mathrm{co}\, A$ of a subset A in a vector space E is the smallest convex set that includes A and consists precisely of all elements of E of the form $\sum_{i=1}^n \lambda_i u_i$, where $u_i \in A$ and $\lambda_i \ge 0$ for each $i = 1, \ldots, n$ and $\sum_{i=1}^n \lambda_i = 1$.[3] The next result is due to I. Namioka [**104**, Lemma 7.6, p. 37].

THEOREM 1.11 (Namioka). *In a Riesz space the convex hull of a solid set is a solid set.*

PROOF. Let S be a solid set in a Riesz space L, $u \in \mathrm{co}\, S$ and let $v \in L$ satisfy $|v| \le |u|$. Then there exist vectors $u_1, \ldots, u_n \in S$ and positive scalars $\lambda_1, \ldots, \lambda_n$ with $\sum_{i=1}^n \lambda_i = 1$ such that $u = \sum_{i=1}^n \lambda_i u_i$. By Theorem 1.10 there exist $v_1, \ldots, v_n$ such that $|v_i| \le |\lambda_i u_i| = \lambda_i |u_i|$ for each i and $v = \sum_{i=1}^n v_i$. Put $w_i = \frac{1}{\lambda_i} v_i$ and note that $w_i \in S$ for each $i = 1, \ldots, n$. It follows that $v = \sum_{i=1}^n v_i = \sum_{i=1}^n \lambda_i w_i \in \mathrm{co}\, S$ and hence $\mathrm{co}\, S$ is a solid set. ■

---

[3] Recall that any vector of the form $\sum_{i=1}^n \lambda_i u_i$, where $\lambda_i \ge 0$ for each $i = 1, \ldots, n$ and $\sum_{i=1}^n \lambda_i = 1$, is called a **convex combination** of the vectors $u_1, \ldots, u_n$.

Recall that a binary relation $\succeq$ of a set A **directs** the set A if it satisfies the following three conditions:

(1) $\alpha \succeq \alpha$ for all $\alpha \in A$.
(2) If $\alpha, \beta, \gamma \in A$ satisfy $\alpha \succeq \beta$ and $\beta \succeq \gamma$, then $\alpha \succeq \gamma$.
(3) For any pair $\alpha, \beta \in A$ there exists $\gamma \in A$ with $\gamma \succeq \alpha$ and $\gamma \succeq \beta$.

The set A [more precisely the pair $(A, \succeq)$] is called a **directed set**.

A **net** of a set X is a mapping $u\colon A \to X$ from a directed set A to X. As usual, $u(\alpha)$ will be denoted by u_α and the net $u\colon A \to X$ by $\{u_\alpha\}$, omitting the directed set A if its presence is not necessary. The set A is called the **index set** of the net $\{u_\alpha\}$ and its members are referred to as **indices**.

A **subnet** of a net $\{u_\alpha\}_{\alpha \in A}$ is any net of the form $\{u_{\lambda_\kappa}\}_{\kappa \in K}$, where K is a directed set and $\lambda\colon K \to A$ is a mapping such that for any index $\alpha \in A$ there exists some index $\kappa_\alpha \in K$ such that $\kappa \succeq \kappa_\alpha$ implies $\lambda_\kappa \succeq \alpha$.

According to the definition of a directed set any partially ordered set with a largest element can be viewed as a directed set. However, such a net can be thought of as a trivial net. The non-trivial nets are of course the ones where property (3) can be stated as follows: *For any pair of indices α, β there exists some index $\gamma \in A$ different than α and β satisfying $\gamma \succeq \alpha$ and $\gamma \succeq \beta$.*

A net $\{u_\alpha\}$ in a Riesz space L is called **increasing** (in symbols $u_\alpha \uparrow$) if $u_\beta \geq u_\alpha$ whenever the indices α and β satisfy $\beta \succeq \alpha$. The symbol $u_\alpha \downarrow$ denotes a **decreasing** net and its definition is analogous. The notation $u_\alpha \uparrow u$ means that the net $\{u_\alpha\}$ is increasing and the supremum of the set $\{u_\alpha\colon\ \alpha \in A\}$ exists and equals u. The meaning of $u_\alpha \downarrow u$ is defined similarly. The notations

$$u_\alpha \uparrow \leq u \quad \text{and} \quad u \leq u_\alpha \downarrow$$

mean $u_\alpha \uparrow$ and $u_\alpha \leq u$ for each α (resp. $u_\alpha \downarrow$ and $u_\alpha \geq u$ for each α). Also for two nets $\{u_\alpha\}$ and $\{v_\alpha\}$ with the same index set the symbol $v_\alpha \leq u_\alpha \downarrow u$ means that $v_\alpha \leq u_\alpha$ for each α and $u_\alpha \downarrow u$.

We also say that a nonempty subset D of an ordered vector space is **directed upward**, in symbols $D \uparrow$, if for each pair $u, v \in D$ there exists some $w \in D$ with $w \geq u$ and $w \geq v$. The symbolism $D \uparrow u$ means that $D \uparrow$ and that $u = \sup D$. The definitions of a **downward directed set** D, in symbols $D \downarrow$, and $D \downarrow u$ are defined analogously.

There is a natural association between increasing nets and upward directed sets. If $u_\alpha \uparrow$, then the set $D = \{u_\alpha\colon\ \alpha \in A\}$ is an upward directed set—and $u_\alpha \uparrow u$ holds if and only if $D \uparrow u$. On the other hand, if $D \uparrow$ is an upward directed set, then by letting $u_\alpha = \alpha$ for each $\alpha \in D$ and direct D by letting $\alpha \succeq \beta$ whenever $\alpha \geq \beta$, then the net $\{u_\alpha\}_{\alpha \in D}$ is increasing—and $u_\alpha \uparrow u$ holds if and only if $D \uparrow u$. The reader also should notice that there is a natural correspondence between decreasing nets and downward directed sets.

And now we come to the notion of order convergence in a Riesz space.

DEFINITION 1.12 (Order Convergence). *A net $\{u_\alpha\}_{\alpha \in A}$ in a Riesz space L is **order convergent** to an element $u \in L$, in symbols $u_\alpha \xrightarrow{o} u$, whenever there exists another net $\{v_\alpha\}_{\alpha \in A}$ of L (with the same index set) satisfying $|u_\alpha - u| \leq v_\alpha \downarrow 0$. The element u is called the **order limit** of the net $\{u_\alpha\}$.*

The order limits are uniquely determined.

LEMMA 1.13. *In a Riesz space a net can have at most one order limit.*

PROOF. Assume that $u_\alpha \xrightarrow{o} u$ and $u_\alpha \xrightarrow{o} w$ hold in a Riesz space. Let $\{v_\alpha\}$ and $\{x_\alpha\}$ be two nets satisfying $|u_\alpha - u| \le v_\alpha \downarrow 0$ and $|u_\alpha - w| \le x_\alpha \downarrow 0$. But then for each α we have

$$|u - w| \le |u - u_\alpha| + |u_\alpha - w| \le v_\alpha + x_\alpha .$$

Since $v_\alpha + x_\alpha \downarrow 0$, it follows that $|u - w| = 0$, so that $u = w$. ■

The basic properties of order convergence of nets are summarized in the next theorem whose easy proof is left for the reader; see also [**90**, p. 78].

THEOREM 1.14. *Assume that the nets $\{u_\alpha\}_{\alpha\in A}$ and $\{v_\beta\}_{\beta\in B}$ in a Riesz space satisfy $u_\alpha \xrightarrow{o} u$ and $v_\beta \xrightarrow{o} v$. Then we have:*

(i) $|u_\alpha| \xrightarrow{o} |u|$, $u_\alpha^+ \xrightarrow{o} u^+$, *and* $u_\alpha^- \xrightarrow{o} u^-$.
(ii) $\lambda u_\alpha + \mu v_\beta \xrightarrow[(\alpha,\beta)]{o} \lambda u + \mu v$ *for all* $\lambda, \mu \in \mathbb{R}$.
(iii) $u_\alpha \vee v_\beta \xrightarrow[(\alpha,\beta)]{o} u \vee v$ *and* $u_\alpha \wedge v_\beta \xrightarrow[(\alpha,\beta)]{o} u \wedge v$.
(iv) *If* $u_\alpha \uparrow u$ *or* $u_\alpha \downarrow u$, *then* $u_\alpha \xrightarrow{o} u$.
(v) *If* $u_\alpha \le w$ *for all* $\alpha \succeq \alpha_1$, *then* $u \le w$.
(vi) *If* $u_\alpha \uparrow$ *and* $u_\alpha \xrightarrow{o} u$, *then* $u_\alpha \uparrow u$.

Note. All nets having a double index (α, β) are considered indexed by the product $A \times B$, where as usual the direction of the product $A \times B$ is the componentwise one.

A subset S of a Riesz space L is said to be **order closed** if it follows from $\{u_\alpha\} \subseteq S$ and $u_\alpha \xrightarrow{o} u$ in L that $u \in S$, that is, S is order closed if it contains its order limits. Similarly, a subset S is called **σ-order closed** if it contains its sequential order limits; see [**90**, pp. 78–91] for more about order convergence and order closed sets.

Recall that a subset S of a Riesz space is called **order closed** (resp. σ-order closed) if it follows from $u_\alpha \xrightarrow{o} u$ and $\{u_\alpha\} \subseteq S$ (resp. $u_n \xrightarrow{o} u$ and $\{u_n\} \subseteq S$) that $u \in S$. Clearly, every order closed set is σ-order closed. The order closed solid sets have a nice and useful characterization.

LEMMA 1.15. *A solid subset S of a Riesz space is order closed if and only if $\{u_\alpha\} \subseteq S$ and $0 \le u_\alpha \uparrow u$ imply $u \in S$.*

Similarly, a solid subset S of a Riesz space is σ-order closed if and only if u belongs to S whenever $0 \le u_n \uparrow u$ holds for some sequence $\{u_n\} \subseteq S$.

PROOF. To see this, assume that a solid set S has the stated property. Also, let a net $\{u_\alpha\} \subseteq S$ satisfy $u_\alpha \xrightarrow{o} u$. Pick a net $\{v_\alpha\}$ having the same index set satisfying $v_\alpha \downarrow 0$ and $|u_\alpha - u| \le v_\alpha$ for each α. Now notice that $0 \le \big(|u| - v_\alpha\big)^+ \uparrow |u|$ and $\big(|u| - v_\alpha\big)^+ \le |u_\alpha|$ holds for each α. This implies $\big\{\big(|u| - v_\alpha\big)^+\big\} \subseteq S$, and from this we get $u \in S$. ■

1.2. Ideals, Bands, and Riesz Subspaces

In this section, we shall present several properties of some special vector subspaces of a Riesz space known as ideals and bands. We start our discussion with their definition.

DEFINITION 1.16. *A solid vector subspace of a Riesz space is called an **ideal**. A σ-order closed ideal is called a **σ-ideal** and an order closed ideal is called a **band**.*

Note that if A and B are ideals, then their **algebraic sum**

$$A + B = \{a + b\colon\ a \in A \text{ and } b \in B\}$$

is also an ideal. (For this conclusion we must use the Riesz decomposition property.) A glance at Lemma 1.15 also establishes the following.

LEMMA 1.17. *An ideal A is a band if and only if $0 \le u_\alpha \uparrow u$ in L with $\{u_\alpha\} \subseteq A$ imply $u \in A$. Similarly, an ideal A is a σ-ideal if and only if $0 \le u_n \uparrow u$ and $\{u_n\} \subseteq A$ imply $u \in A$.*

Every subset D of a Riesz space L is included in a smallest ideal L_D, called the **ideal generated by** D. The ideal L_D is the intersection of all ideals that include D. It is easy to see that

$$L_D = \Big\{ u \in L\colon\ \exists u_1, \ldots, u_n \in D \text{ and } \lambda \ge 0 \text{ with } |u| \le \lambda \sum_{i=1}^{n} |u_i| \Big\}.$$

A **principal ideal** of a Riesz space L is any ideal generated by a singleton $\{u\}$. This ideal will be denoted by L_u. Clearly,

$$L_u = \big\{ v \in L\colon\ \exists \lambda \ge 0 \text{ such that } |v| \le \lambda |u| \big\}.$$

An **order unit** (or a **strong unit**) is an element $e > 0$ satisfying $L_e = L$, that is, if for every $u \in L$ there exists some $\lambda > 0$ with $|u| \le \lambda e$ (or, equivalently, if for each $u \in L$ there exists some $\lambda > 0$ with $u \le \lambda e$).

We are now ready to introduce the notion of a Riesz subspace.

DEFINITION 1.18. *A vector subspace K of a Riesz space L is said to be a* ***Riesz subspace*** *(or a* ***vector sublattice****) if for every pair of elements $u, v \in K$ the supremum of u and v taken in L belongs to K.*

From part (3) of Theorem 1.3 it follows that a vector subspace K of a Riesz space is a Riesz subspace if and only if for every $u \in K$ we have $u^+ \in K$. Consequently from the inequality $0 \le u^+ \le |u|$, it follows that every ideal is a Riesz subspace. A Riesz subspace need not be an ideal, as can be easily demonstrated by examples. For instance, the vector subspace $V = \big\{(\lambda, \lambda)\colon\ \lambda \in \mathbb{R}\big\}$ is a Riesz subspace in $\mathbb{R}^2$ but is not an ideal. However, it should be noted that a Riesz subspace K is an ideal if and only if $0 \le u \le v$ and $v \in K$ imply $u \in K$.

Given a Riesz subspace K of a Riesz space L we say that the embedding of K into L **preserves arbitrary suprema and infima** if for every subset of K whose supremum (or infimum) exists in K, the supremum (or infimum) of the same subset exists in L and is the same as that in K. This and related properties are included in the next definition.

DEFINITION 1.19. *A Riesz subspace K of a Riesz space L is said to be:*

(1) ***regular****, if the embedding of K into L preserves arbitrary suprema and infima,*
(2) ***σ-regular****, if the embedding of K into L preserves countable suprema and infima, and*
(3) ***majorizing****, if for each $u \in L$ there exists some $v \in K$ such that $u \le v$.*

The proof of the next theorem is straightforward and is left for the reader.

THEOREM 1.20. *For a Riesz subspace K of a Riesz space L the following statements are equivalent.*

(i) *K is a regular Riesz subspace of L.*
(ii) *If $\{u_\alpha\} \subseteq K$ satisfies $u_\alpha \downarrow 0$ in K, then $u_\alpha \downarrow 0$ also holds in L.*
(iii) *If $\{u_\alpha\} \subseteq K$ satisfies $u_\alpha \xrightarrow{o} u$ in K, then $u_\alpha \xrightarrow{o} u$ also holds in L.*

The reader should be able to formulate by himself the σ-analogue of Theorem 1.20. The proof of the following result is also straightforward.

LEMMA 1.21. *Every ideal is a regular Riesz subspace.*

A density property dealing with Riesz subspaces is described next.

DEFINITION 1.22. *A Riesz subspace K of a Riesz space L is said to be:*

(1) ***Order dense*** *in L, if for every $0 < u \in L$ there exists some $v \in K$ such that $0 < v \leq u$.*
(2) ***Super order dense*** *in L, if for every $0 < u \in L$ there exists a sequence $\{u_n\} \subseteq K$ with $0 \leq u_n \uparrow u$ in L.*

Clearly every super order dense Riesz subspace is order dense. Order dense Riesz subspaces are also regular Riesz subspaces.

THEOREM 1.23. *Every order dense Riesz subspace of a Riesz space is a regular Riesz subspace.*

PROOF. Assume that K is order dense in L and a net $\{u_\alpha\}$ satisfies $u_\alpha \downarrow 0$ in K. If $u_\alpha \downarrow 0$ does not hold in L, then there exists some $0 < u \in L$ with $0 < u \leq u_\alpha$ for all α. But then since K is order dense in L, there exists $v \in K$ such that $0 < v \leq u$. So, $0 < v \leq u_\alpha$ holds in K for all α, contradicting $u_\alpha \downarrow 0$ in K. Theorem 1.20 now shows that K is a regular Riesz subspace. ■

For an interesting and extensive study of regular Riesz subspaces see the paper by W. Wnuk [**129**].

The **disjoint complement** A^{d} of a nonempty subset A of a Riesz space L is defined by

$$A^{\mathrm{d}} = \{u \in L\colon\ u \perp v \text{ for all } v \in A\}.$$

It can be seen easily that A^{d} is an ideal of L, and by using the infinite distributive laws, it can be shown that A^{d} is actually a band of L. The disjoint complement $(A^{\mathrm{d}})^{\mathrm{d}}$ will be denoted by A^{dd}. Note that $A \cap A^{\mathrm{d}} = \{0\}$ and that $A \subseteq A^{\mathrm{dd}}$. Also, if $A \subseteq B$, then $B^{\mathrm{d}} \subseteq A^{\mathrm{d}}$.

THEOREM 1.24. *Every ideal A of a Riesz space L is order dense in A^{dd}. In particular, an ideal A is order dense in L if and only if $A^{\mathrm{d}} = \{0\}$.*

PROOF. Assume that A is not order dense in A^{dd}. This means that there exists some $0 < u \in A^{\mathrm{dd}}$ with no element $v \in A$ satisfying $0 < v \leq u$. In particular (since A is an ideal), we must have $|v| \wedge u = 0$ for each $v \in A$. Thus $u \in A^{\mathrm{d}}$ and hence $u \in A^{\mathrm{d}} \cap A^{\mathrm{dd}} = \{0\}$, a contradiction. This contradiction establishes that A is order dense in A^{dd}.

Now if $A^{\mathrm{d}} = \{0\}$, then $A^{\mathrm{dd}} = L$ and so A is order dense in L. On the other hand, if A is order dense in L and $0 < u \in A^{\mathrm{d}}$, then pick some $v \in A$ with $0 < v \leq u$ and note that $v \in A \cap A^{\mathrm{d}} = \{0\}$, which is impossible. Thus, if A is order dense in L, then $A^{\mathrm{d}} = \{0\}$. ■

The next theorem shows that for every ideal A the ideal $A \oplus A^{\mathrm{d}}$ is an order dense ideal.

THEOREM 1.25. *If A is an ideal of a Riesz space L, then the ideal $A \oplus A^{\mathrm{d}}$ is order dense in L.*

PROOF. If $u \in (A \oplus A^{\mathrm{d}})^{\mathrm{d}}$, then $u \in A^{\mathrm{d}} \cap A^{\mathrm{dd}} = \{0\}$. Therefore, $(A \oplus A^{\mathrm{d}})^{\mathrm{d}} = \{0\}$, and the conclusion follows from Theorem 1.24. ■

We introduce next the class of Archimedean Riesz spaces.

DEFINITION 1.26. *A Riesz space L is said to be **Archimedean** if $u, v \in L^+$ and $nu \leq v$ for each $n \in \mathbb{N}$ imply $u = 0$.*

The best known non-Archimedean Riesz space is the **lexicographic plane** *Lex*, that is, $Lex = \mathbb{R}^2$ ordered as follows: $(x_1, y_1) \geq (x_2, y_2)$ if either $x_1 > x_2$ or else $x_1 = x_2$ and $y_1 \geq y_2$. Observe that the lexicographic plane is totally ordered, that is, every two elements are comparable. In fact the cone of the lexicographic plane is

$$Lex_+ = \big\{(0, y)\colon\ y \geq 0\big\} \bigcup \big\{(x, y) \in \mathbb{R}^2\colon\ x > 0\big\}.$$

It should be clear that Riesz subspaces of Archimedean Riesz spaces are Archimedean and that any Cartesian product of Archimedean Riesz spaces is also Archimedean. In Archimedean Riesz spaces the order denseness has a nice characterization.

THEOREM 1.27. *For a Riesz subspace K of an Archimedean Riesz space L the following statements are equivalent:*

1. *K is order dense in L.*
2. *For each $u \in L^+$ we have $u = \sup\big\{v \in K\colon\ 0 \leq v \leq u\big\}$, or equivalently, for each $u \in L^+$ there exists a net $\{v_\alpha\} \subseteq K$ such that $0 \leq v_\alpha \uparrow u$ in L.*

PROOF. (1) $\Longrightarrow$ (2) Let $0 < u \in L$ and assume by way of contradiction that $u = \sup\{v \in K\colon\ 0 \leq v \leq u\}$ does not hold. This means that there exists $0 < w \in L$ such that $v \in K$ and $0 \leq v \leq u$ imply $v \leq u - w$. Pick some $x \in K$ such that $0 < x \leq w$ and observe that $x \leq w + (u - w) = u$. Therefore $x \leq u - w$. This in turn implies $2x = x + x \leq (u - w) + w = u$ and so $2x \leq u - w$. By an inductive argument, we get $0 < nx \leq u$ for $n = 1, 2, \ldots$, contradicting the Archimedean property of L.

(2) $\Longrightarrow$ (1) Obvious. ■

To see that (1) does not imply (2) in the non-Archimedean case, let $L = Lex$, the lexicographic plane and let K be the "y-axis," i.e., $K = \big\{(0, y)\colon\ y \in \mathbb{R}\big\}$. Note that K is an order dense band of L and that (2) does not hold. Indeed, if we consider the positive element $u = (1, 0)$, then $\{v \in K\colon\ 0 \leq v \leq u\} = \big\{(0, y)\colon\ y \geq 0\big\} = K^+$ and $\sup K^+$ does not exist in the (non-Archimedean) Riesz space L.

By combining Theorems 1.24 and 1.27, we see at once that every band A in an Archimedean Riesz space satisfies $A = A^{\mathrm{dd}}$. Actually, as we shall see next, this condition characterizes the Archimedean property.

THEOREM 1.28. *For a Riesz space L the following are equivalent:*

(i) *L is Archimedean.*

(ii) *If $\{\lambda_\alpha\}$ is a bounded net in $\mathbb{R}$ with $\lambda_\alpha \to 0$, then for each $u \in L$ we have $\lambda_\alpha u \xrightarrow{o} 0$.*

(iii) *If $\lambda_n \to 0$ holds in $\mathbb{R}$, then $\lambda_n u \xrightarrow{o} 0$ in L for each $u \in L$.*

(iv) *For each band B in L we have $B = B^{\mathrm{dd}}$.*

(v) *If S is a nonempty subset of L^+ and*

$$T = \{v \in L\colon\ 0 \le v \le u \text{ for all } u \in S\},$$

then $\inf\{u - v\colon\ u \in S \text{ and } v \in T\} = 0$.

PROOF. (i) $\Longrightarrow$ (ii) Assume that a bounded net $\{\lambda_\alpha\}$ in $\mathbb{R}$ satisfies $\lambda_\alpha \to 0$ and let $u \in L$. For each index α let $\mu_\alpha = \sup_{\beta \succeq \alpha} |\lambda_\alpha|$. Clearly $\mu_\alpha \downarrow 0$ and we claim that $\mu_\alpha |u| \downarrow 0$. Indeed, if $\mu_\alpha|u| \downarrow 0$ is not true, then there exists some $v \in L^+$ such that $0 < v \le \mu_\alpha |u|$ holds for each α. Now note that for each $n \in \mathbb{N}$ there exists some index α_n such that $\mu_\alpha \le \frac{1}{n}$ for all $\alpha \succeq \alpha_n$. This implies $v \le \frac{1}{n}|u|$ or $nv \le |u|$ for each $n \in \mathbb{N}$, contrary to the Archimedean property of L. Hence $\mu_\alpha |u| \downarrow 0$. To verify the validity of (ii) now note that $|\lambda_\alpha u| = |\lambda_\alpha||u| \le \mu_\alpha |u|$ holds for each α.

(ii) $\Longrightarrow$ (iii) Obvious.

(iii) $\Longrightarrow$ (iv) Let B be a band of L and let $0 < u \in B^{\mathrm{dd}}$. We claim that $\sup\{v \in B\colon\ 0 \le v \le u\} = u$. Indeed, if this is not the case, then as in the proof of Theorem 1.27 there exists some $x \in B$ such that $0 < nx \le u$ or $0 < x \le \frac{1}{n}u$ for each $n \in \mathbb{N}$. Now our assumption yields $\frac{1}{n}u \downarrow 0$ and so $x = 0$, which is a contradiction. Hence $u = \sup\{v \in B\colon\ 0 \le v \le u\}$ in L and since B is a band, we obtain $u \in B$. Thus, $B^{\mathrm{dd}} \subseteq B$ and since $B \subseteq B^{\mathrm{dd}}$ is always true, we get $B = B^{\mathrm{dd}}$.

(iv) $\Longrightarrow$ (v) We first claim that if J is an ideal of L, then the band generated by J coincides with J^{dd}. To see this, notice that the band B_J generated by J is precisely the set

$$B_J = \{u \in L\colon\ \text{There exists a net } \{u_\alpha\} \subseteq J \text{ such that } 0 \le u_\alpha \uparrow |u|\}.$$

Since J^{dd} is a band satisfying $J \subseteq J^{\mathrm{dd}}$, it follows that $B_J \subseteq J^{\mathrm{dd}}$. On the other hand, if B is any band satisfying $J \subseteq B$, then it follows from our condition that $J^{\mathrm{dd}} \subseteq B^{\mathrm{dd}} = B$. This implies $B_J = J^{\mathrm{dd}}$.

Now let S be a nonempty subset of L^+, and let

$$T = \{v \in L\colon\ 0 \le v \le u \text{ for all } u \in S\}.$$

To show that $\inf\{u - v\colon\ u \in S \text{ and } v \in T\} = 0$, assume that some $w \in L^+$ satisfies $0 \le w \le u - v$ for all $u \in S$ and all $v \in T$. This implies $w + v \le u$ for all $u \in S$ and all $v \in T$, and so $w + v \in T$ for all $v \in T$. It follows that $w \le u - (w + v)$ or $2w \le u - v$ for all $u \in S$ and all $v \in T$. By an inductive argument, we get

$$0 \le nw \le u - v \le u$$

for all $u \in S$ and all $v \in T$. Now notice that the ideal $L_w \oplus L_w^{\mathrm{d}}$ of L satisfies $(L_w \oplus L_w^{\mathrm{d}})^{\mathrm{dd}} = L$, and consequently by the preceding conclusion any $u \in S$ belongs in the band generated by $L_w \oplus L_w^{\mathrm{d}}$. So, we have

$$u = \sup\{x \in L_w \oplus L_w^{\mathrm{d}} : 0 \le x \le u\}. \tag{$\star$}$$

Now assume that $x = x_1 + x_2$ in $L_w \oplus L_w^{\mathrm{d}}$ satisfies $0 \le x \le u$. Since $x_1 \in L_w$ there exists some $n \in \mathbb{N}$ satisfying $x_1 \le nw$ and therefore we have $x_1 + w \le (n+1)w \le u$

and $x_2 \perp (x_1 + w)$. This implies $x + w = (x_1 + w) + x_2 = (x_1 + w) \vee x_2 \le u$ or $x \le u - w$. Consequently, it follows from $(\star)$ that $u \le u - w$ or $w \le 0$. Hence, $w = 0$ and from this we infer that $\inf\{u - v\colon\ u \in S \text{ and } v \in T\} = 0$.

(v) $\Longrightarrow$ (i) Assume that $0 \le nu \le v$ holds for all n and consider the set $S = \{v - nu\colon\ n \in \mathbb{N}\}$. Consequently, if $T = \{x \in L^+\colon\ x \le v - nu\}$, then we have $\inf\{v - nu - x\colon\ n \in \mathbb{N} \text{ and } x \in T\} = 0$. Now notice that if $x \in T$, then by rewriting $x \le v - (n+1)u$ as $0 \le u \le v - nu - x$, we get $u = 0$. This shows that L is an Archimedean Riesz space. ■

We continue by recalling the notion of an operator.

DEFINITION 1.29. *A linear mapping $T\colon V \to W$ between two vector spaces is called a (linear)* ***operator***.[4]

An operator $T\colon E \to F$ between two ordered vector spaces is said to be a **positive operator**, in symbols $T \ge 0$ or $0 \le T$, if T carries positive elements to positive elements, i.e., $u \ge 0$ in E implies $T(u) \ge 0$ in F. A positive operator $T\colon E \to F$ is called **strictly positive**, in symbols $T \gg 0$ or $0 \ll T$, if $u > 0$ implies $T(u) > 0$.

DEFINITION 1.30. *An operator $\pi\colon L \to M$ between two Riesz spaces is called:*

(1) *a* ***Riesz homomorphism*** *(or a* ***lattice homomorphism****), whenever $u \wedge v = 0$ in L implies $\pi(u) \wedge \pi(v) = 0$ in M.*
(2) *a* ***Riesz σ-homomorphism*** *(or a* ***lattice σ-homomorphism****), if π is a Riesz homomorphism and $u_n \xrightarrow{o} 0$ in L implies $\pi(u_n) \xrightarrow{o} 0$ in M.*
(3) *a* ***normal Riesz homomorphism*** *(or a* ***normal lattice homomorphism****), if π is a Riesz homomorphism and $u_\alpha \xrightarrow{o} 0$ in L implies $\pi(u_\alpha) \xrightarrow{o} 0$ in M.*
(4) *a* ***Riesz isomorphism*** *(or a* ***lattice isomorphism****), if π is a one-to-one (into) Riesz homomorphism.*

Riesz homomorphisms preserve the lattice structure as the following theorem shows whose proof is left to the reader.

THEOREM 1.31. *For an operator $\pi\colon L \to M$ between two Riesz spaces the following statements are equivalent:*

(i) *π is a Riesz homomorphism.*
(ii) *$\pi(u \wedge v) = \pi(u) \wedge \pi(v)$ for all $u, v \in L$.*
(iii) *$\pi(u \vee v) = \pi(u) \vee \pi(v)$ for all $u, v \in L$.*
(iv) *$\pi(u \vee v) = \pi(u) \vee \pi(v)$ whenever $u \wedge v = 0$ holds in L.*
(v) *$\pi(u^+) = [\pi(u)]^+$ for all $u \in L$.*
(vi) *$\pi(|u|) = |\pi(u)|$ for all $u \in L$.*

From the identity $\pi(|u|) = |\pi(u)|$, we see that every Riesz homomorphism satisfies $\pi(u) \in M^+$ if $u \in L^+$, that is, every Riesz homomorphism is a positive operator.

LEMMA 1.32. *An onto Riesz homomorphism $\pi\colon L \to M$ is normal if and only if its* ***kernel*** *$K_\pi = \{u \in L\colon\ \pi(u) = 0\}$ is a band of L.*

Likewise an onto Riesz homomorphism $\pi\colon L \to M$ between two Riesz spaces is a Riesz σ-homomorphism if and only if its kernel is a σ-ideal.

[4] We emphasize here that in this book the word "operator" will be synonymous with the words "linear operator."

PROOF. Let $\pi\colon L \to M$ be an onto Riesz homomorphism. It is easy to check that the kernel K_π of π is an ideal of L.

Assume first that π is a normal Riesz homomorphism. Suppose also that a net $\{u_\alpha\} \subseteq K_\pi$ satisfies $0 \le u_\alpha \uparrow u$ in L. Since π is a normal Riesz homomorphism, it follows that $\pi(u) = \sup_\alpha \pi(u_\alpha) = \sup_\alpha 0 = 0$, and so $u \in K_\pi$. This shows that K_π is a band.

For the converse, assume that K_π is a band and let $u_\alpha \downarrow 0$ in L. Also suppose that $0 \le v \le \pi(u_\alpha)$ holds in M for some v and all α. Since π is onto there exists some $w \in L$ with $\pi(w) = v$. For each α let $x_\alpha = (w^+ - u_\alpha)^+$ and note that

$$\pi(x_\alpha) = \big[\pi(w^+) - \pi(u_\alpha)\big]^+ = \big[v - \pi(u_\alpha)\big]^+ = 0\,,$$

i.e., $x_\alpha \in K_\pi$ for each α. Now observe that $0 \le x_\alpha \uparrow w^+$, and from this—using that K_π is a band—we obtain $w^+ \in K_\pi$. So, $v = \pi(w) = \big[\pi(w)\big]^+ = \pi(w^+) = 0$. This implies $\pi(u_\alpha) \downarrow 0$ in M. ■

Two Riesz spaces L and M are called **Riesz isomorphic** (or **lattice isomorphic**) if there exists a Riesz isomorphism from L onto M.

THEOREM 1.33. *Let $\pi\colon L \to M$ be an onto Riesz homomorphism between two Riesz spaces. Then π carries solid subsets of L to solid subsets of M.*

PROOF. Assume that S is a solid subset of the Riesz space L and that for some $v \in S$ and $u \in L$ we have $|\pi(u)| \le |\pi(v)|$ in M. Put $w = \big[(-|v|) \vee u\big] \wedge |v| \in L$ and note that since $|w| \le |v|$ holds we have $w \in S$. Also,

$$\pi(w) = \big[(-|\pi(v)|) \vee \pi(u)\big] \wedge |\pi(v)| = \pi(u)\,.$$

Thus $\pi(u) \in \pi(S)$, and so $\pi(S)$ is a solid subset of M. ■

We continue our discussion with a brief description of quotient Riesz spaces. Recall that if Y is a vector subspace of a vector space X, then the **quotient vector space** X/Y is the vector space consisting of all equivalence classes of X relative to Y. That is, if $x \in X$, then the **equivalence class** of x is defined by $\dot{x} = \{z \in X\colon\ x - z \in Y\} = x + Y$. As usual, the algebraic operations in X/Y are defined by

$$\dot{x} + \dot{z} = (x+z)^{\boldsymbol{\cdot}} \quad \text{and} \quad \lambda\dot{x} = (\lambda x)^{\boldsymbol{\cdot}}.$$

The **canonical projection** $Q\colon X \to X/Y$ from X onto X/Y, defined by $Q(x) = \dot{x}$, is an onto linear operator.

Now assume that L is a Riesz space and A is an ideal of L. Then we claim that the canonical image of the positive cone L^+, i.e., the set

$$\begin{aligned} Q(L^+) &= \big\{\dot{u} \in L/A\colon\ u \in L^+\big\} \\ &= \big\{\dot{u} \in L/A\colon\ \exists\, v \in L^+ \text{ such that } \dot{u} = \dot{v}\big\}\,, \end{aligned}$$

is a cone of the quotient vector space L/A. Indeed, note first that if $\dot{u}, \dot{v} \in Q(L^+)$ and $\lambda \ge 0$, then $\dot{u} + \dot{v} \in Q(L^+)$ and $\lambda\dot{u} \in Q(L^+)$ are obviously true. To see that $Q(L^+)$ is a pointed convex cone, assume that $\dot{u} \in Q(L^+) \cap [-Q(L^+)]$. So, there exist $v, w \in L^+$ such that $\dot{u} = \dot{v} = -\dot{w}$. This implies that $(v+w)^{\boldsymbol{\cdot}} = 0$ or $u + w \in A$. Now from $0 \le v \le v + w$ and the fact that A is an ideal, we see that $v \in A$ and so $\dot{u} = \dot{v} = 0$. Therefore, $Q(L^+)$ is a cone of L/A.

In the sequel (and everywhere in this monograph) the quotient vector space L/A will be considered ordered by means of the cone $Q(L^+)$. Therefore, L/A is an ordered vector space. Notice that the vector ordering $\geq$ induced by the cone $Q(L^+)$ on L/A is characterized as follows:

$$\dot{u} \geq \dot{v} \iff \text{there exist } u_1 \in \dot{u} \text{ and } v_1 \in \dot{v} \text{ such that } u_1 \geq v_1 .$$

To see this, assume that $\dot{u} \geq \dot{v}$. This means that $(u - v)^{\cdot} \in Q(L^+)$, and so there exists some $w \in L^+$ such that $u - v - w = a \in A$. Letting $u_1 = u - a$ and $v_1 = v$, we see that $u_1 \in \dot{u}$, $v_1 \in \dot{v}$ and $u_1 \geq v_1$. Conversely, if $u_1 \in \dot{u}$ and $v_1 \in \dot{v}$ satisfy $u_1 \geq v_1$, then we have $\dot{u} - \dot{v} = \dot{u}_1 - \dot{v}_1 = (u_1 - v_1)^{\cdot} \in Q(L^+)$ so that $\dot{u} \geq \dot{v}$.

We are now ready to establish that the quotient ordered vector space L/A is itself a Riesz space.

THEOREM 1.34. *If A is an ideal of a Riesz space L, then the quotient ordered vector space L/A is a Riesz space. Moreover, the canonical projection $Q\colon L \to L/A$ is an onto Riesz homomorphism.*

PROOF. Let $u \in L$. From $u \leq u^+$ and $0 \leq u^+$, we see that $\dot{u} \leq (u^+)^{\cdot}$ and $0 \leq (u^+)^{\cdot}$ hold in L/A. On the other hand, assume $\dot{u} \leq \dot{v}$ and $0 \leq \dot{v}$ in L/A. Pick $u_1 \in \dot{u}$, $v_1 \in \dot{v}$, and $v_2 \in \dot{v}$ with $u_1 \leq v_1$ and $0 \leq v_2$. Then the relations

$$\begin{aligned} u &= u_1 + (u - u_1) \leq v_1 \vee v_2 + (u - u_1)^+ \\ &= v_2 + (v_1 - v_2)^+ + (u - u_1)^+ \in v_2 + A \end{aligned}$$

and

$$v_2 + (v_1 - v_2)^+ + (u - u_1)^+ \geq 0$$

imply $u^+ \leq v_2 + (v_1 - v_2)^+ + (u - u_1)^+$, which in turn guarantees that $(u^+)^{\cdot} \leq \dot{v}_2 = \dot{v}$. Therefore, $(\dot{u})^+ = (u^+)^{\cdot}$, or $[Q(u)]^+ = Q(u^+)$, holds in L/A. This establishes that L/A is a Riesz space and that the canonical projection $Q\colon L \to L/A$ is a Riesz homomorphism. ■

It should be noted that according to Lemma 1.32 the canonical projection $Q\colon L \to L/A$ is a normal Riesz homomorphism (resp. a Riesz σ-homomorphism) if and only if the ideal A is a band of L (resp. a σ-ideal of L).

We shall close the section with a discussion regarding normal sequences of sets. Sequences of normal sets will appear frequently in our discussion in this book.

DEFINITION 1.35. *A sequence $\{S_n\}$ of subsets of a vector space is said to **normal** if*

$$S_{n+1} + S_{n+1} \subseteq S_n$$

*holds for each n. The **null set** of a normal sequence $\{S_n\}$ is the set*

$$N = \bigcap_{n=1}^{\infty} S_n .$$

The reader should pause for a few minutes and verify the following properties that are included in the next lemma.

LEMMA 1.36. *For a normal sequence $\{S_n\}$ of subsets of a vector space we have the following properties.*

1. *The null set of $\{S_n\}$ is closed under addition.*
2. *If each S_n is circled, then the null set of $\{S_n\}$ is a vector subspace.*
3. *If each S_n is a solid subset of a Riesz space, then the null set of $\{S_n\}$ is an ideal.*
4. *If each S_n is a solid and order closed subset of a Riesz space, then the null set of $\{S_n\}$ is a band.*

(5) *If each S_n is a solid and σ-order closed subset of a Riesz space, then the null set of $\{S_n\}$ is a σ-ideal.*

Normal sequences of sets also exhibit the following summability property.

LEMMA 1.37. *If $\{S_n\}$ is a normal sequence of circled subsets of a vector space and $1 \le n_1 < n_2 < \cdots < n_k$ are natural numbers, then*

$$S_{n_k} + S_{n_{k-1}} + \cdots + S_{n_2} \subseteq S_{n_1}.$$

PROOF. Notice that

$$\begin{aligned} S_{n_k} + S_{n_{k-1}} + \cdots + S_{n_2} &\subseteq S_{n_k} + S_{n_k} + S_{n_{k-1}} + S_{n_{k-2}} + \cdots + S_{n_2} \\ &\subseteq S_{n_{k-1}} + S_{n_{k-1}} + S_{n_{k-2}} + \cdots + S_{n_2} \\ &\subseteq S_{n_{k-2}} + S_{n_{k-2}} + \cdots + S_{n_2} \\ &\subseteq S_{n_2} + S_{n_2} \subseteq S_{n_1}. \end{aligned}$$

as claimed. ■

1.3. Order Completeness and Projection Properties

Let L be an arbitrary Riesz space. For any two elements $u, v \in L$ with $u \le v$, the **order interval** (or simply the **interval**) $[u, v]$ is the subset of L defined by $[u, v] = \{w \in L\colon\ u \le w \le v\}$. A subset A of L is said to be (order) **bounded from above** if there exists some $u \in L$ satisfying $a \le u$ for all $a \in A$. The bounded from below subsets are defined analogously. A subset A of L is called **order bounded** if it is both bounded from above and from below, or equivalently if A is included in an order interval.

DEFINITION 1.38. *A Riesz space L is said to be:*

(1) ***Dedekind** (or **order**) **complete**, if every nonempty subset of L that is bounded from above has a supremum.*
(2) ***Dedekind** (or **order**) **σ-complete**, if every countable subset of L that is bounded from above has a supremum.*
(3) ***Almost Dedekind** (or **almost order**) **σ-complete**, if L is Riesz isomorphic to a super order dense Riesz subspace of some Dedekind σ-complete Riesz space.*

Clearly, Dedekind completeness implies Dedekind σ-completeness and Dedekind σ-completeness implies almost Dedekind σ-completeness. Note also that every Dedekind σ-complete Riesz space L is Archimedean. (Indeed, if $0 \le nu \le v$ holds for all n, then from the existence of the supremum $w = \sup\{nu\colon\ n = 1, 2, \ldots\}$ in L, it follows from the relation $nu = (n+1)u - u \le w - u$ that $w \le w - u$. Therefore, $u \le 0$ and hence $u = 0$.) Thus all the above classes of Riesz spaces are Archimedean.

LEMMA 1.39. *A Riesz space L is Dedekind complete if and only if for every net $\{u_\alpha\}$ in L satisfying $0 \le u_\alpha \uparrow \le v$ in L we have $u_\alpha \uparrow u$ for some u.*

Similarly, a Riesz space L is Dedekind σ-complete if and only if for every sequence $\{u_n\}$ of L satisfying $0 \le u_n \uparrow \le v$ we have $u_n \uparrow u$ for some u.

PROOF. The following claims can be easily verified.

(a) If D is a nonempty subset in an ordered vector space, then D and the collection of all finite suprema of D (which is an upward directed set) have the same upper bounds.

(b) If D is an upward directed subset in an ordered vector space and $u \in D$ is a fixed element, then $\sup D$ exists if and only if the upward directed subset $D_1 = \{v - u\colon\ v \in D \text{ and } v \geq u\}$ of L^+ has a supremum—in which case we have $\sup D_1 = \sup D - u$.

From these properties the desired conclusions should be immediate. ■

The next simple result will play a fundamental role in the discussion in this monograph.

THEOREM 1.40. *Let K be an order dense Riesz subspace of an Archimedean Riesz space L. If K is Dedekind complete in its own right, then K is an ideal of L.*

PROOF. Assume that $0 \leq u \leq v$ with $v \in K$ and $u \in L$. Since L is Archimedean and K is order dense in L, there exists a net $\{u_\alpha\}$ of K with $0 \leq u_\alpha \uparrow u$ in L; see Theorem 1.27. Now since K is Dedekind complete, $0 \leq u_\alpha \uparrow w$ holds in K for some $w \in K$. But, by Theorem 1.23, K is a regular Riesz subspace of L and thus $u_\alpha \uparrow w$ holds also in L. It follows that $u = w \in K$, so that K is an ideal of L. ■

For a proof of the next basic theorem (due to H. Nakano and A. I. Judin) concerning the existence of the Dedekind completion of an Archimedean Riesz space see [**90**, p. 191]. (See also Theorem 1.49 below.)

THEOREM 1.41 (Nakano–Judin). *Let L be an Archimedean Riesz space. Then there exists a Dedekind complete Riesz space L^δ (unique up to Riesz isomorphism) such that L^δ contains a majorizing order dense Riesz subspace that is Riesz isomorphic to L, which we identify with L. That is, L is identified with a Riesz subspace of L^δ and for each $u \in L^\delta$ we have*

$$u = \sup\{v \in L\colon\ v \leq u\} = \inf\{w \in L\colon\ w \geq u\}.$$

The unique Dedekind complete Riesz space L^δ has a special name.

DEFINITION 1.42 (Dedekind Completion). *If L is an Archimedean Riesz space, then the unique Dedekind complete Riesz space L^δ that satisfies the properties of Theorem 1.41 is called the **Dedekind completion** of L.*

The **Dedekind σ-completion** L^σ of an Archimedean Riesz space L is defined now as follows: L^σ is the intersection of the collection of all Riesz subspaces between L and L^δ that are Dedekind σ-complete in their own right—this collection is nonempty since L^δ is one of them.

Clearly, L^σ is a Dedekind σ-complete Riesz space lying between L and L^δ, i.e., $L \subseteq L^\sigma \subseteq L^\delta$; see also Exercise 32 at the end of the chapter. Categorically, the Dedekind completion and Dedekind σ-completion of an Archimedean Riesz space L are characterized by the following uniqueness property: There exists a Riesz isomorphism π from L into L^δ (resp. L^σ) such that given any normal Riesz homomorphism ϕ from L into any Dedekind complete (resp. σ-complete) Riesz space M, there exists a unique normal Riesz homomorphism ψ from L^δ (resp. L^σ) into M satisfying $\phi = \psi \circ \pi$. For details see [**27**].

Every almost Dedekind σ-complete Riesz space L is super order dense in its Dedekind σ-completion L^σ. In particular, note that if the Riesz space L is almost Dedekind σ-complete, then for every $u \in L^\sigma$ there exist two sequences $\{u_n\}$ and

$\{v_n\}$ in L with $u_n \uparrow u$ and $v_n \downarrow u$ in L^σ; see also Exercise 29 at the end of this chapter. For more about the Dedekind σ-completion see [**112**].

DEFINITION 1.43. *A Riesz space L is said to have the* ***countable sup property*** *if for every subset S of L whose supremum exists in L, there exists an at most countable subset of S having the same supremum as S in L.*

A Dedekind complete Riesz space with the countable sup property is called a ***super Dedekind complete Riesz space****.*

Note that a Riesz space L has the countable sup property if and only if for every net $\{u_\alpha\}$ of L with $u_\alpha \downarrow 0$ there exists an increasing sequence of indices $\{\alpha_n\}$ such that $u_{\alpha_n} \downarrow 0$.

The proof of the following result is straightforward.

LEMMA 1.44. *Every Archimedean Riesz space with the countable sup property is super order dense in L^δ.*

In particular, every Archimedean Riesz space with the countable sup property is almost Dedekind σ-complete and its Dedekind completion is super Dedekind complete.

A useful criterion for the countable sup property, due to W. A. J. Luxemburg and A. C. Zaanen [**89**, Note X, Theorem 31.11, p. 493], is included in the next theorem.

THEOREM 1.45 (Luxemburg–Zaanen). *Let $T\colon L \to M$ be a strictly positive operator between two Riesz spaces. If M is Archimedean with the countable sup property, then L is also Archimedean with the countable sup property.*

PROOF. If $0 \le nv \le u$ for each n, then $0 \le nT(v) \le T(u)$ holds in M for each n. Since M is Archimedean, we get $T(v) = 0$. Now the strict positivity of T implies $v = 0$ and consequently L is an Archimedean Riesz space.

Since M^δ is super Dedekind complete, replacing M by M^δ if necessary, we can assume without loss of generality that M is super Dedekind complete. To see that L has the countable sup property let $u_\alpha \downarrow 0$ in L. Then $T(u_\alpha) \downarrow u^* \ge 0$ holds in M for some u^*. Pick a sequence $\{u_{\alpha_n}\} \subseteq \{u_\alpha\}$ with $u_{\alpha_n} \downarrow$ in L and with $T(u_{\alpha_n}) \downarrow u^*$ in M. To complete the proof it is enough to show that $u_{\alpha_n} \downarrow 0$ holds in L. To this end, assume by way of contradiction that there exists some $u \in L$ satisfying $0 < u \le u_{\alpha_n}$ for each n and some $u \in L^+$. Since $u_\alpha \downarrow 0$, there exists some index β with $u \wedge u_\beta < u$, and so

$$u_{\alpha_n} - u_\beta \wedge u_{\alpha_n} = (u_{\alpha_n} - u_\beta)^+ \ge (u - u_\beta)^+ = u - u \wedge u_\beta = w > 0\,.$$

Thus $u_{\alpha_n} \ge w + u_\beta \wedge u_{\alpha_n}$, and hence

$$T(u_{\alpha_n}) \ge T(w) + T(u_\beta \wedge u_{\alpha_n}) \ge T(w) + u^*$$

for all n. This implies $u^* \ge T(w)+u^*$ or $T(w) \le 0$. However, the latter is impossible since $w > 0$ and the strict positivity of T yields $T(w) > 0$. This contradiction completes the proof of the theorem. ∎

Recall that a **band** is an order closed ideal. Every nonempty subset D of a Riesz space is included in a smallest band, called the **band generated** by D and denoted B_D. A **principal band** is a band that is generated by a singleton; the principal band generated by u will be denoted by B_u.

It should be easy to check that the bands generated by a set D and an element u are given by:

$$\begin{aligned} B_D &= \{v \in L\colon\ \exists\,\{v_\alpha\} \subseteq L_D \text{ such that } 0 \le v_\alpha \uparrow |v|\,\}, \text{ and} \\ B_u &= \{v \in L\colon\ |v| \wedge n|u| \uparrow |v|\,\}, \end{aligned}$$

where L_D denotes the ideal generated by D in L.

A band B of a Riesz space L is said to be a **projection band** if $B \oplus B^{\mathrm{d}} = L$. Every projection band B of a Riesz space L gives rise to a natural projection P_B on L with range B defined by

$$P_B(u) = u_1\,,$$

where $u = u_1 + u_2 \in B \oplus B^{\mathrm{d}} = L$ with $u_1 \in B$ and $u_2 \in B^{\mathrm{d}}$. Any such projection will be referred to as a **band projection** or as an **order projection**. It is not difficult to see that every band projection P_B is a normal Riesz homomorphism. To establish this claim, note first that if $u = u_1 + u_2 \in B \oplus B^{\mathrm{d}} = L$, then

$$|u| = |u_1 + u_2| = |u_1| + |u_2| \in B \oplus B^{\mathrm{d}}$$

and so $P_B(|u|) = |u_1| = \big|P_B(u)\big|$. This shows that P_B is a Riesz homomorphism and, in particular, P_B is also a positive operator. Moreover, it is not difficult to see that for each $u \in L^+$ we have $0 \le P_B(u) \le u$, and from this it easily follows that P_B is also order continuous.

An element u in a Riesz space is called a **projection element** if the principal band B_u that it generates is a projection band. The projection determined by a principal projection band B_u will be denoted by P_u, instead of the most cumbersome symbolism P_{B_u}.

An element $e > 0$ in a Riesz space L is called a **weak order unit** (or simply a **weak unit**) if $B_e = L$. From Theorem 1.24 and condition (iv) of Theorem 1.28 it should be clear that an element $e > 0$ in an Archimedean Riesz space is a weak unit if and only if $u \perp e$ implies $u = 0$.

The next theorem characterizes the bands that are projection bands.

THEOREM 1.46. *For a Riesz space L we have the following.*

(1) *A band B of L is a projection band if and only if for each $u \in L^+$ the supremum* $\sup\{v \in B\colon\ 0 \le v \le u\}$ *exists in L. Moreover, if B is a projection band, then for each $u \in L^+$ we have*

$$P_B(u) = \sup\{v \in B\colon\ 0 \le v \le u\}\,.$$

(2) *A principal band B_u of L is a projection band if and only if for each $v \in L^+$ the supremum* $\sup\{v \wedge n|u|\colon\ n \in \mathbb{N}\}$ *exists in L. Moreover, if B_u is a projection band, then for each $v \in L^+$ we have*

$$P_u(v) = \sup\{v \wedge n|u|\colon\ n \in \mathbb{N}\}\,.$$

PROOF. (1) Assume first that B is a projection band of a Riesz space L and let $u \in L^+$. If a vector $v \in B$ satisfies $0 \le v \le u$, then the positivity of P_B implies $0 \le v = P_B(v) \le P_B(u) \in B$. This immediately guarantees that the supremum of the set $\{v \in B\colon\ 0 \le v \le u\}$ exists in the Riesz space L and that we have $\sup\{v \in B\colon\ 0 \le v \le u\} = \max\{v \in B\colon\ 0 \le v \le u\} = P_B(u)$.

For the converse assume that for every positive vector u in the Riesz space L the supremum $\sup\{v \in B\colon\ 0 \le v \le u\}$ exists in L. Now fix some $u \in L^+$ and let $u_1 = \sup\{v \in B\colon\ 0 \le v \le u\} \le u$. Since the set $\{v \in B\colon\ 0 \le v \le u\}$ is a directed upward subset of the band B, it follows that $u_1 \in B$. Put $u_2 = u - u_1 \geq 0$. To establish that B is a projection band, it suffices to show that $u_2 \in B^{\mathrm{d}}$. To see this, let $0 \leq w \in B$ and put $w_1 = w \wedge u_2 \geq 0$. Then $u_1 + w_1 \in B$ and clearly $u_1 + w_1 \leq u_1 + u_2 = u$ implies that $u_1 + w_1 \leq u_1$. Hence, $w_1 \leq 0$ and thus $w_1 = w \wedge u_2 = 0$, so that $u_2 \in B^{\mathrm{d}}$.

(2) Fix $v \in L^+$ and let $w = \sup\{x \in B_u \colon\ 0 \leq x \leq v\}$. Clearly, for each $n \in \mathbb{N}$ we have $v \wedge n|u| \in B_u$ and $v \wedge n|u| \leq v$, and so $v \wedge n|u| \leq w$ holds for each n. That is, w is an upper bound of the sequence $\{v \wedge n|u|\}$. We claim that w is the least upper bound of $\{v \wedge n|u|\}$, i.e., we claim that $w = \sup\{v \wedge n|u| \colon\ n \in \mathbb{N}\}$.

To see this, assume that some $y \in B_u$ satisfies $v \wedge n|u| \leq y$ for all $n \in \mathbb{N}$. Now let $x \in B_u$ with $0 \leq x \leq v$. It follows that there exists a net $\{x_\alpha\} \subseteq L_u$ such that $0 \leq x_\alpha \uparrow x$. Clearly for each index α there exists some $n \in \mathbb{N}$ such that $x_\alpha \leq n|u|$, and so $x_\alpha \leq v \wedge n|u| \leq y$. This implies $x \leq y$ and so $w \leq y$. Therefore, w is the least upper bound of the sequence $\{v \wedge n|u|\}$. ■

We continue with the introduction of several projection properties in Riesz spaces.

DEFINITION 1.47. *A Riesz space L is said to have:*

(1) *the* ***projection property****, if every band of L is a projection band,*
(2) *the* ***principal projection property****, if every principal band of L is a projection band, and*
(3) ***sufficiently many projections****, if every nonzero band of L includes a nonzero projection band.*

In view of Theorem 1.46 we can see easily that Dedekind completeness implies the projection property and Dedekind σ-completeness the principal projection property.

Clearly, the projection property implies the principal projection property, which in turn implies the property of having sufficiently many projections. We shall see later that no reverse implication holds. The above classes of Riesz spaces are all Archimedean as the following result of W. A. J. Luxemburg and A. C. Zaanen [**90**, p. 174] shows.

THEOREM 1.48 (Luxemburg–Zaanen). *Every Riesz space with sufficiently many projections is Archimedean.*

PROOF. Let L be a Riesz space with sufficiently many projections and assume that for two elements $u, v \in L^+$ we have $0 \leq nu \leq v$ for each $n = 1, 2, \ldots$. Now assume by way of contradiction that $u > 0$. So the principal band B_u is nonzero. According to our hypothesis there exists a nonzero projection band with $B \subseteq B_u$. In particular, if we let $S = \{w \in B \colon\ 0 \leq w \leq v\}$, then the supremum of the set S exists in L and belongs to B. Let $s = \sup S \in B$.

We claim that $B^+ = S$. To see this, let $0 < x \in B$. Since $x \in B_u$, it follows that $x \wedge nu \uparrow x$ in L. Moreover, from $x \wedge nu \leq nu \leq v$ for each n, we infer that $x \leq v$, that is, $x \in S$. In particular, we have $0 < s \in B$. Since $0 < 2s \in B$, we get $2s \leq s$ or $s \leq 0$, which is impossible. This contradiction shows that $u = 0$ and so L is an Archimedean Riesz space. ■

We continue our discussion with a characterization of the Dedekind completion of certain Archimedean Riesz spaces. For every Riesz subspace L of a Riesz space M there exists (at least) one ideal J of M that is maximal with respect to the property

$L \cap J = \{0\}$. In other words, there exists an ideal J of M with $J \cap L = \{0\}$ and such that whenever another ideal I satisfies $J \subseteq I$ and $L \cap I = \{0\}$, then $I = J$. To see this, consider the collection $\mathcal{C}$ of all ideals J of M with $L \cap J = \{0\}$ and note that $\mathcal{C}$ is nonempty (since $\{0\} \in \mathcal{C}$) and partially ordered by the inclusion relation. Now use Zorn's lemma to obtain a maximal element $J^* \in \mathcal{C}$, and note that J^* is maximal with respect to the property $L \cap J^* = \{0\}$.

Next we assume that L is an Archimedean Riesz subspace of a Riesz space M and that J is a maximal ideal in M with respect to the property $L \cap J = \{0\}$. Then we claim that the canonical projection $Q\colon M \to M/J$ restricted to L, i.e., the operator $Q\colon L \to M/J$, is a Riesz isomorphism, and if L is identified with $Q(L)$, the ideal I generated by L in M/J is an Archimedean Riesz space that includes L as an order dense Riesz subspace. Indeed, to see first that $Q\colon L \to M/J$ is one-to-one, note that if some $u \in L$ satisfies $\dot{u} = 0$ in M/J, then $u \in L \cap J = \{0\}$, so that $u = 0$. Now to see that L is order dense in M/J (and hence in I) let $0 < \dot{u} \in M/J$. This implies $u \notin J$ and since J is a maximal ideal in M with respect to the property $L \cap J = \{0\}$ and the ideal $J + L_u$ properly includes J, we infer that $(J + L_u) \cap L \neq \{0\}$. Pick $0 < w \in (J + L_u) \cap L$, write $w = w_1 + w_2$ with $w_1 \in J$ and $|w_2| \le \lambda |u|$ for some $\lambda > 0$, and note that $0 < \dot{w} = \dot{w}_2 \le \lambda \dot{u}$. Consequently, the element $\frac{1}{\lambda} w \in L$ satisfies $0 < \frac{1}{\lambda} \dot{w} \le \dot{u}$ and therefore L is order dense in M/J. Finally, to see that I is Archimedean let $0 \le n\dot{x} \le \dot{y}$ for all n with $y \in L$. If $\dot{x} > 0$ holds, then pick some $z \in L$ with $0 < \dot{z} \le \dot{x}$ and note that $0 < n\dot{z} \le \dot{y}$ holds for all n, contrary to the Archimedean property of L.

If now M is a Dedekind complete Riesz space, then it turns out that the ideal I is the Dedekind completion of L. This result is due to W. Maxey [**95**, Theorem 1.3, p. 18].

THEOREM 1.49 (Maxey). *Let L be a Riesz subspace of a Dedekind complete Riesz space M and let J be an ideal of M that is maximal with respect to the property $L \cap J = \{0\}$. Then the ideal I generated by L in M/J (where L is identified with its canonical image in M/J) is the Dedekind completion of the Archimedean Riesz space L.*

PROOF. According to the preceding discussion, we have to show only that I is Dedekind complete. Since L is order dense and majorizes the Archimedean Riesz space I, it is enough to show that for every nonempty subset S of L^+ the infimum of the set $\dot{S}$ exists in I. To this end, let $S \subseteq L^+$ be nonempty. Since M is Dedekind complete the infimum $u = \inf S$ exists in M, and we claim that $\dot{u} = \inf \dot{S}$ in I.

To see this, let $T = \{t \in L^+\colon\ t \le s \text{ for all } s \in S\}$. It should be clear that $t \le u$ holds in M for all $t \in T$. Now, according to condition (v) of Theorem 1.28, we have $\inf(S - T) = 0$. The latter, in view of the order denseness of L in I, implies

$$\inf(\dot{S} - \dot{T}) = \inf\{\dot{s} - \dot{t}\colon\ s \in S \text{ and } t \in T\} = 0$$

in I. Now assume that some $\dot{v} \in I$ satisfies $\dot{u} \le \dot{v} \le \dot{s}$ for all $s \in S$. Then, from $0 \le \dot{v} - \dot{u} \le \dot{s} - \dot{t}$ for all $s \in S$ and all $t \in T$ and $\inf(\dot{S} - \dot{T}) = 0$, it follows that $\dot{v} - \dot{u} = 0$ or $\dot{v} = \dot{u}$. This shows that $\inf \dot{S} = \dot{u}$ holds in I, so that I is Dedekind complete. ■

We shall close the section with a characterization of H. Nakano [**99**] of the $C(X)$ Riesz spaces that are Dedekind complete. (Recall that a topological space is called **extremally disconnected** if the closure of every open set is open.)

THEOREM 1.50 (Nakano). *For a topological space X we have:*

(a) *If X is extremally disconnected, then the Riesz space $C(X)$ is Dedekind complete.*

(b) *If $C(X)$ is Dedekind complete and X is also completely regular,[5] then X is extremally disconnected.*

PROOF. (a) Assume that X is an extremally disconnected topological space and let $\mathcal{F}$ be a nonempty subset of $C(X)^+$. We must establish that $\inf \mathcal{F}$ exists in $C(X)$.

To see this, for each $r > 0$ define the open set

$$V_r = \{x \in X\colon \text{ there exists } f \in \mathcal{F} \text{ such that } f(x) < r\}.$$

It is easy to see that $0 < r_1 < r_2$ implies $V_{r_1} \subseteq V_{r_2}$ and so $\overline{V}_{r_1} \subseteq \overline{V}_{r_2}$. Since the topological space X is extremally disconnected, the closure $\overline{V}_r$ is also an open set and clearly $X = \bigcup_{r>0} \overline{V}_r$. Now define the function $g\colon X \to [0,\infty)$ via the formula

$$g(x) = \inf\{r > 0\colon\ x \in \overline{V}_r\}.$$

We claim that $g \in C(X)$ and that $g = \inf \mathcal{F}$ holds in $C(X)$.

To see that g is continuous, we start by defining for each $a \in \mathbb{R}$ the sets

$$U_a = \{x \in X\colon\ g(x) < a\} \quad \text{and} \quad W_a = \{x \in X\colon\ g(x) \le a\}.$$

A direct verification shows that

$$U_a = \begin{cases} \bigcup_{0<r<a} \overline{V}_r & \text{if } a > 0 \\ \varnothing & \text{if } a \le 0 \end{cases} \quad \text{and} \quad W_a = \begin{cases} \bigcap_{r>a} \overline{V}_r & \text{if } a \ge 0 \\ \varnothing & \text{if } a < 0\,. \end{cases}$$

It follows that the U_a are open sets and the W_a are closed sets. Now if (a,b) is an arbitrary open interval in $\mathbb{R}$, then $f^{-1}\big((a,b)\big) = U_b \setminus W_a$, an open set. This shows that g is continuous, i.e., $g \in C(X)$.

To see that g is a lower bound for the set $\mathcal{F}$, note that if $f \in \mathcal{F}$, then for each $x \in X$ and each $\epsilon > 0$ we have $f(x) < f(x) + \epsilon$, and so $g(x) \le f(x) + \epsilon$ for all $f \in \mathcal{F}$, all $x \in X$, and all $\epsilon > 0$. This implies $g(x) \le f(x)$ for all $f \in \mathcal{F}$ and all $x \in X$, i.e., g is a lower bound for $\mathcal{F}$ in $C(X)$.

To establish that $g = \inf \mathcal{F}$ holds in $C(X)$, assume that some function $h \in C(X)$ satisfies $h \le f$ for all $f \in \mathcal{F}$. Suppose that $0 < s < r$. Then for each $x \in V_s$ there exists some $f \in \mathcal{F}$ satisfying $h(x) \le f(x) < s$, and so $h(x) < s$. Since h is a continuous function, we see that $h(x) \le s$ for all $x \in \overline{V}_s$. This implies $\overline{V}_s \subseteq \{x \in X\colon\ h(x) < r\}$. Therefore,

$$U_r = \{x \in X\colon\ g(x) < r\} = \bigcup_{0<s<r} \overline{V}_s \subseteq \{x \in X\colon\ h(x) < r\}.$$

Now from $g(x) < g(x) + \epsilon$, we get $h(x) \le g(x) + \epsilon$ for all $\epsilon > 0$, and so $h(x) \le g(x)$ for all $x \in X$. The latter shows that $g = \inf \mathcal{F}$ holds in $C(X)$.

(b) Assume that $C(X)$ is Dedekind complete and that X is completely regular. Let V be a nonempty open subset of X. If $\overline{V} = X$, then $\overline{V}$ is an open set. So, we can assume that $\overline{V} \ne X$. For each $x \notin \overline{V}$ pick a continuous function $f_x\colon X \to [0,1]$ such that $f_x(x) = 0$ and $f_x(y) = 1$ for all $y \in \overline{V}$. Let $f = \inf_{x \in X} f_x$ in $C(X)$.

---

[5] Recall that a topological space Ω is said to be **completely regular** if for each nonempty closed set C and each $x \notin C$ there exists a continuous function $f\colon \Omega \to [0,1]$ such that $f(x) = 1$ and $f(y) = 0$ for all $y \in C$.

Clearly, $f(y) = 0$ holds for all $y \notin \overline{V}$. Now let $z \in V$. Since V is an open set, there exists a continuous function $h_z\colon X \to [0,1]$ such that $h_z(z) = 1$ and $h_z(y) = 0$ for all $y \notin V$. It follows that $h_z \leq f_x$ for all $x \in X$ and all $z \in V$ and so $h_z \leq f$ for all $z \in V$. In particular, from $1 = h_z(z) \leq f(z) \leq 1$ we get $f(z) = 1$ for all $z \in V$, and so by the continuity of f, we see that $f(z) = 1$ for all $z \in \overline{V}$. But then $\chi_{\overline{V}} = f \in C(X)$, and this shows that $\overline{V}$ is a clopen subset of X. This complete the proof. ■

Recall that a **Stonean space** is a Hausdorff compact extremally disconnected topological space. Since every Stonean space is clearly completely regular, the following result is an immediate consequence of the preceding theorem.

COROLLARY 1.51 (Nakano). *A Hausdorff compact topological space X is Stonean if and only if the Riesz space $C(X)$ is Dedekind complete.*

1.4. The Freudenthal Spectral Theorem

In this section we shall discuss some special elements associated with a positive vector u that will be referred to as components of u. Our main objective is to demonstrate that the components can be viewed as the "building blocks" of the elements in an Archimedean Riesz space with the principal projection property. We start our discussion with the notion of a component.

DEFINITION 1.52. *A **component** of a positive vector u in a Riesz space L is any positive element $v \in L$ satisfying $v \wedge (u - v) = 0$.*

The collection of all components of a positive vector u will be denoted by $\mathcal{C}_u$, that is,

$$\mathcal{C}_u = \big\{ v \in L^+ \colon\ v \wedge (u - v) = 0 \big\}.$$

The set $\mathcal{C}_u$ inherits the order of L and so $\mathcal{C}_u$ is a partially ordered set having smallest element the zero vector 0 and largest element the vector u. The basic properties of the partially ordered set $\mathcal{C}_u$ are described in the next result.[6]

THEOREM 1.53. *For a positive element u in a Riesz space L we have:*

(1) $0, u \in \mathcal{C}_u$ *and* $\mathcal{C}_u \subseteq [0, u]$.
(2) *If $v, w \in \mathcal{C}_u$ and $v \leq w$ holds, then $w - v \in \mathcal{C}_u$.*
(3) *If $v_1, v_2, w_1, w_2 \in \mathcal{C}_u$ satisfy the inequalities $v_1 \leq v_2 \leq w_1 \leq w_2$, then $v_2 - v_1 \perp w_2 - w_1$.*
(4) *If $v, w \in \mathcal{C}_u$, then $v \vee w$ and $v \wedge w$ both belong to $\mathcal{C}_u$ (and so $\mathcal{C}_u$ is a Boolean algebra with smallest element 0 and largest element u).*
(5) *If L is Dedekind complete, then for every nonempty subset C of $\mathcal{C}_u$ the elements $\sup C$ and $\inf C$ both belong to $\mathcal{C}_u$ (and so in this case $\mathcal{C}_u$ is an order complete Boolean algebra).*
(6) *The set of components $\mathcal{C}_u$ of u is precisely the set of all extreme points of the convex set $[0, u]$.*[7]

---

[6] Recall that a **Boolean algebra** $\mathcal{B}$ is a distributive complemented lattice with smallest and largest elements. That is, $\mathcal{B}$ is a partially ordered set that is a distributive lattice with a smallest element 0 and a largest element e such that for each $a \in \mathcal{B}$ there exists a (necessarily unique) element $a' \in \mathcal{B}$ (called the *complement* of a) satisfying $a \wedge a' = 0$ and $a \vee a' = e$. A Boolean algebra $\mathcal{B}$ is **Dedekind complete** if every nonempty subset of $\mathcal{B}$ has a supremum.

[7] As usual, a vector x in a convex set C is said to be an **extreme point** of C if it follows from $x = \lambda y + (1 - \lambda)z$ with $y, z \in C$ and $0 < \lambda < 1$ that $y = z = x$.

PROOF. (1) Obvious.

(2) It follows immediately from the inequalities

$$\begin{aligned} 0 &\le (w-v)\wedge\big[u-(w-v)\big] = (w-v)\wedge\big[(u-w)+v\big] \\ &\le (w-v)\wedge(u-w)+(w-v)\wedge v \\ &\le w\wedge(u-w)+(u-v)\wedge v = 0+0=0\,. \end{aligned}$$

(3) Note that $0\le(v_2-v_1)\wedge(w_2-w_1)\le w_1\wedge(u-w_1)=0$.

(4) Let $v,w\in\mathcal{C}_u$. Then, using the distributive laws, we see that

$$\begin{aligned} (v\vee w)\wedge(u-v\vee w) &= (v\vee w)\wedge\big[(u-v)\wedge(u-w)\big] \\ &= \big[v\wedge(u-v)\wedge(u-w)\big]\vee\big[w\wedge(u-v)\wedge(u-w)\big] \\ &= 0\vee 0 = 0\,, \end{aligned}$$

and

$$\begin{aligned} (v\wedge w)\wedge(u-v\wedge w) &= (v\wedge w)\wedge\big[(u-v)\vee(u-w)\big] \\ &= \big[v\wedge w\wedge(u-v)\big]\vee\big[v\wedge w\wedge(u-w)\big] \\ &= 0\vee 0=0\,. \end{aligned}$$

(5) Now assume that L is Dedekind complete and let C be a nonempty set of components of u. Put $s=\sup C$ and $t=\inf C$. Then, using the infinite distributive laws, we get

$$\begin{aligned} 0\le s\wedge(u-s) &= \Big[\bigvee_{c\in C} c\Big]\wedge(u-s) \\ &= \bigvee_{c\in C}\big[c\wedge(u-s)\big] \\ &\le \bigvee_{c\in C}\big[c\wedge(u-c)\big] = 0\,. \end{aligned}$$

Similarly, we have

$$\begin{aligned} 0\le t\wedge(u-t) &= t\wedge\Big[\bigvee_{c\in C}(u-c)\Big] \\ &= \bigvee_{c\in C}\big[t\wedge(u-c)\big] \\ &\le \bigvee_{c\in C}\big[c\wedge(u-c)\big] = 0\,. \end{aligned}$$

(6) Assume first that an element $v\in[0,u]$ is an extreme point of $[0,u]$ and let $w=v\wedge(u-v)\ge 0$. We must show that $w=0$. Clearly, $0\le v-w\le v$ and $0\le v+w\le u$, and from the convex combination $v=\frac{1}{2}(v-w)+\frac{1}{2}(v+w)$ we get $v-w=v$. So $w=0$, as desired.

For the converse, assume that $v\in\mathcal{C}_u$ and let $v=\lambda x+(1-\lambda)y$, where $x,y\in[0,u]$ and $0<\lambda<1$. From $v\wedge(u-v)=0$, it follows that $x\wedge(u-v)=0$, and so from part (4) of Theorem 1.7 we get

$$x=x\wedge u=x\wedge v+x\wedge(u-v)=x\wedge v\le v\,.$$

Similarly, $y\le v$. Now if either $x<v$ or $y<v$ is true, then

$$v=\lambda x+(1-\lambda)y<\lambda v+(1-\lambda)v=v$$

is also true, which is impossible. Consequently $x = y = v$ holds, and therefore v is an extreme point of the order interval $[0, u]$. This completes the proof of the theorem. ∎

In terms of components the projection property has been characterized by K. K. Kutty and J. Quinn [**80**, Theorem 2.6, p. 309] as follows.

THEOREM 1.54 (Kutty–Quinn). *For an Archimedean Riesz space L the following statements are equivalent.*

(1) *L has the projection property.*
(2) *The components of the elements of L^+ in L^δ lie in L. That is, if $u \in L^+$ and $v \in L^\delta$ satisfy $v \wedge (u - v) = 0$, then $v \in L$.*

PROOF. (1) $\Longrightarrow$ (2) Assume that $v \in L^\delta$ satisfies $v \wedge (u - v) = 0$ for some $u \in L$. Let B_v denote the band generated by v in L^δ. Then $B = B_v \cap L$ is a band in L, and hence a projection band of L. Let $u = u_1 + u_2$, where $u_1 \in B \subseteq B_v$ and $u_2 \in B^{\mathrm{d}}$. Since L is order dense in L^δ, we get $u_2 \wedge v = 0$, and from this it easily follows that $v = u_1 \in L$.

(2) $\Longrightarrow$ (1) Let B be an arbitrary band in L and fix some $u \in L^+$. Put $u^* = \sup\{v \in B \colon\ 0 \leq v \leq u\}$ in L^δ. If $u^* \wedge (u - u^*) > 0$, then there exists some $w \in L$ with $0 < w \leq u^* \wedge (u - u^*)$. In particular, it follows that $w \in B$ (since $0 \leq w \leq u^*$ and B is a band) and $w + u^* \leq u$. But then

$$w + u^* = \sup\{w + v \colon\ v \in B \text{ and } 0 \leq v \leq u\} = u^*\,,$$

a contradiction. This shows that $u^* \wedge (u - u^*) = 0$, and thus from our hypothesis $u^* \in L$. Theorem 1.46 now shows that B is a projection band of L. ∎

A sequence $\{x_n\}$ in a Riesz space L is called u-**uniformly Cauchy**, where $u \in L^+$, if there exists a sequence of positive real numbers $\{\epsilon_n\}$ with $\epsilon_n \to 0$ satisfying $|x_{n+p} - x_n| \leq \epsilon_n u$ for all $n, p = 1, 2, \ldots$. A sequence $\{x_n\}$ in L is said to be u-**uniformly convergent** to some $x \in L$ if there exists a sequence $\{\epsilon_n\}$ of positive real numbers with $\epsilon_n \to 0$ and such that $|x_n - x| \leq \epsilon_n u$ holds for each n; the element x is called a **uniform limit** of the sequence $\{x_n\}$. A sequence $\{x_n\}$ is said to be **uniformly convergent**, in symbols $x_n \xrightarrow{u} x$, if there exists some $u \in L^+$ such that $\{x_n\}$ is u-**uniformly convergent**. The reader should verify that if $x_n \xrightarrow{u} x$ and $y_n \xrightarrow{u} y$, then for all $\lambda, \mu \in \mathbb{R}$ we have $\lambda x_n + \mu y_n \xrightarrow{u} \lambda x + \mu y$.

In Archimedean Riesz spaces the uniformly convergent sequences are also order convergent—and consequently their uniform limits are uniquely determined. Indeed, if in an Archimedean Riesz space we have $|x_n - x| \leq \epsilon_n u$ for all n and some sequence $\{\epsilon_n\} \subseteq \mathbb{R}^+$ with $\epsilon_n \to 0$, then put $\delta_n = \sup\{\epsilon_k \colon\ k \geq n\}$ and note that $\delta_n \downarrow 0$ in $\mathbb{R}$. This implies $\delta_n u \downarrow 0$ and the conclusion follows by observing that $|x_n - x| \leq \delta_n u$ holds for all n.

A Riesz space is called **uniformly complete** if every u-uniformly Cauchy sequence has a u-uniform limit.

LEMMA 1.55. *A Riesz space L is uniformly complete if and only if every increasing u-uniformly Cauchy sequence in L^+ has a u-uniform limit.*

PROOF. Assume that in a Riesz space every increasing u-uniformly sequence in L^+ is u-uniformly convergent. Now let $\{x_n\}$ be a u-uniform Cauchy sequence. By passing to a subsequence, we can assume without loss of generality that $|x_{n+1} - x_n| \le \frac{1}{2^n}u$ holds for each n. Now consider the sequences $\{v_n\}$ and $\{w_n\}$ of L^+ defined by

$$v_n = \sum_{i=1}^{n}(x_{i+1} - x_i)^+ \quad \text{and} \quad w_n = \sum_{i=1}^{n}(x_{i+1} - x_i)^- .$$

An easy verification shows that both sequences $\{v_n\}$ and $\{w_n\}$ are u-uniformly Cauchy and increasing. Therefore, according to our hypothesis, both are uniformly convergent, say $v_n \xrightarrow{u} v$ and $w_n \xrightarrow{u} w$. Now notice that for each $n \in \mathbb{N}$ we have $v_n - w_n = \sum_{i=1}^{n}(x_{i+1} - x_i) = x_{n+1} - x_1$, and from this we get $x_n \xrightarrow{u} v - w + x_1$. Thus, L is uniformly complete. ■

LEMMA 1.56. *Every Dedekind σ-complete Riesz space is uniformly complete.*

PROOF. Assume that an increasing sequence $\{u_n\}$ of positive vectors in a Dedekind σ-complete Riesz space L is u-uniformly Cauchy. Pick a sequence $\{\epsilon_n\}$ of positive real numbers with $\epsilon_n \to 0$ and such that $|u_{n+p} - u_n| \le \epsilon_n u$ holds for all $n, p \in \mathbb{N}$. It follows that $0 \le u_n \le u_1 + \epsilon_1 u$ for all n, and since L is Dedekind σ-complete, we infer that $0 \le u_n \uparrow v$ holds in L. This implies $|u_{n+p} - u_n| \xrightarrow[p\to\infty]{o} |v - u_n|$, and so $|v - u_n| \le \epsilon_n u$ holds for all n. Hence, the sequence $\{u_n\}$ is u-uniformly convergent to v. Now a glance at Lemma 1.55 guarantees that L is uniformly complete. ■

Now let L be a Riesz space and let $0 < u \in L$. A u-**step function** is any element $s \in L$ of the form

$$s = \sum_{i=1}^{n} \alpha_i x_i , \tag{$\star$}$$

where the elements $x_1, \ldots, x_n$ are pairwise disjoint components of u such that $u = x_1 + \cdots + x_n$. The expression of s in $(\star)$ is called a **representation** of s as a u-step function.

Clearly, every component of u is a u-step function. Also, it should be clear that every u-step function belongs to the ideal generated by u. Now if s and t are u-step functions with representations

$$s = \sum_{i=1}^{n} \alpha_i x_i \quad \text{and} \quad t = \sum_{j=1}^{m} \beta_j y_j ,$$

then, using that $x_i = x_i \wedge u = x_i \wedge (y_1 + \cdots + y_m) = x_i \wedge y_1 + \cdots + x_i \wedge y_m$ and similarly $y_j = \sum_{i=1}^{n} x_i \wedge y_j$, we can write

$$s = \sum_{i=1}^{n}\sum_{j=1}^{m} \alpha_i (x_i \wedge y_j) \quad \text{and} \quad t = \sum_{i=1}^{n}\sum_{j=1}^{m} \beta_j (x_i \wedge y_j) .$$

These are also representations of s and t as u-step functions. From these formulas we see that:

$$\begin{aligned}
s+t &= \sum_{i=1}^{n}\sum_{j=1}^{m}(\alpha_i+\beta_j)x_i\wedge y_j\\
\lambda s &= \sum_{i=1}^{n}\lambda\alpha_i x_i\\
|s| &= \sum_{i=1}^{n}|\alpha_i|x_i\\
s\vee t &= \sum_{i=1}^{n}\sum_{j=1}^{m}\max\{\alpha_i,\beta_j\}(x_i\wedge y_j)\\
s\wedge t &= \sum_{i=1}^{n}\sum_{j=1}^{m}\min\{\alpha_i,\beta_j\}(x_i\wedge y_j)\,.
\end{aligned}$$

The preceding identities show that the collection S_u of all u-step functions is a Riesz subspace of L. That is, we have the following result.

LEMMA 1.57. *The collection S_u of all u-step functions is a Riesz subspace of L_u (and hence of L). Moreover, for each $s \in S_u$ there exist uniquely determined distinct scalars $\alpha_1,\ldots,\alpha_n$ and pairwise disjoint nonzero components $x_1,\ldots,x_n$ of u with $u = x_1+\cdots+x_n$ such that*

$$s=\sum_{i=1}^{n}\alpha_i x_i\,.$$

PROOF. The first part follows immediately from the identities preceding the statement of the lemma. Only the uniqueness of the representation needs proof. To this end, assume that there exist distinct scalars $\beta_1,\ldots,\beta_m$ and pairwise disjoint nonzero components $y_1,\ldots,y_m$ of the element u with $u=y_1+\cdots+y_m$ such that $s=\sum_{j=1}^{m}\beta_j y_j$. It follows that $x_i=\sum_{j=1}^{m}x_i\wedge y_j$ and $y_j=\sum_{i=1}^{n}y_j\wedge x_i$, and from this we get $\sum_{i=1}^{n}\sum_{j=1}^{m}(\alpha_i-\beta_j)(x_i\wedge y_j)=0$. Taking absolute values yields $\sum_{i=1}^{n}\sum_{j=1}^{m}|\alpha_i-\beta_j|(x_i\wedge y_j)=0$ and hence $|\alpha_i-\beta_j|(x_i\wedge y_j)=0$ for all i and j. Since for each i there exists some j with $x_i\wedge y_j\neq 0$, it follows that for each i there exists exactly one j with $\alpha_i=\beta_j$ and for this j we have $x_i=y_j$. Since a similar claim is true for each j, it follows that the x_i coincide with the y_j and the α_i with the β_j in a one-to-one fashion. This establishes the uniqueness of the representation with the above properties. ■

The unique representation of a step function s as described in Lemma 1.57 is called the **standard representation** of s. Clearly, the standard representation of the step function $s=0$ is $0=0u$.

When L has the principal projection property, it turns out that the Riesz subspace S_u of all u-step functions is "uniformly dense" in the principal ideal L_u. This is a basic result in the theory of Riesz spaces known as the Freudenthal Spectral Theorem. It was proven by H. Freudenthal in [**60**].

THEOREM 1.58 (The Freudenthal Spectral Theorem). *If L is a Riesz space with the principal projection property and $u \in L^+$, then for each positive element $x \in L_u$ there exists a sequence $\{s_n\}$ of u-step functions such that:*

(1) $0 \le x - s_n \le \frac{1}{n}u$ *for each n, and*
(2) $0 \le s_n \uparrow x$.

In particular, S_u is u-uniformly dense in L_u.

PROOF. Without loss of generality we can assume that $0 < x \le u$ holds. For each $0 \le \alpha \le 1$ denote by P_α the band projection of the Riesz space L onto the principal band generated by $(x - \alpha u)^+$. The family $\{P_\alpha\colon\ 0 \le \alpha \le 1\}$ of band projections satisfies the following properties.

(i) $P_0 = P_x$ and $P_1 = 0$.
(ii) If $0 \le \alpha \le \beta \le 1$ holds, then
  (a) $0 \le P_\beta \le P_\alpha \le I$, where I is the identity operator on L, and
  (b) $\alpha(P_\alpha - P_\beta)(u) \le (P_\alpha - P_\beta)(x) \le \beta(P_\alpha - P_\beta)(u)$.
(iii) If $0 \le \alpha \le \beta \le \alpha_1 \le \beta_1 \le 1$ holds, then $(P_\alpha - P_\beta)(u)$ and $(P_{\alpha_1} - P_{\beta_1})(u)$ are two disjoint components of u.

Let us verify these properties. Property (i) should be obvious. For (ii) and (iii) assume that $0 \le \alpha \le \beta \le \alpha_1 \le \beta_1 \le 1$. The verification of the inequalities $0 \le P_\beta \le P_\alpha \le I$ should be straightforward. For the other inequalities note first that

$$\begin{aligned}(P_\alpha - P_\beta)(x) - \alpha(P_\alpha - P_\beta)(u) &= P_\alpha(x - \alpha u) - P_\beta(x - \alpha u)\\ &\ge (x - \alpha u)^+ - P_\beta(x - \alpha u)^+ \ge 0\,,\end{aligned}$$

and so $\alpha(P_\alpha - P_\beta)(u) \le (P_\alpha - P_\beta)(x)$. Also, from

$$\begin{aligned}(P_\alpha - P_\beta)(x) - \beta(P_\alpha - P_\beta)(u) &= P_\alpha(x - \beta u) - P_\beta(x - \beta u)\\ &= P_\alpha(x - \beta u) - (x - \beta u)^+\\ &\le P_\alpha(x - \beta u)^+ - (x - \beta u)^+ \le 0\,,\end{aligned}$$

we see that $(P_\alpha - P_\beta)(x) \le \beta(P_\alpha - P_\beta)(u)$. This establishes (ii). The claim in (iii) follows immediately from part (3) of Theorem 1.53.

Now let $\epsilon > 0$. Since the collection S_u of all u-step functions is a Riesz subspace, it suffices to prove the existence of some u-step function s such that $0 \le x - s \le \epsilon u$. To this end, suppose that $0 = \alpha_0 < \alpha_1 < \cdots < \alpha_n = 1$ is a partition of $[0, 1]$ with mesh less than ϵ. Put $x_i = (P_{\alpha_{i-1}} - P_{\alpha_i})(u)$ for each $i = 1, \ldots, n$, and note that $x_1 + x_2 + \cdots + x_n \le u$ and that $x_i \wedge x_j = 0$ holds for $i \ne j$. On the other hand, using (ii) above it is easy to see that the u-step function $s = \sum_{i=1}^n \alpha_{i-1}x_i$ satisfies

$$\begin{aligned}s = \sum_{i=1}^n \alpha_{i-1}x_i &\le \sum_{i=1}^n (P_{\alpha_{i-1}} - P_{\alpha_i})(x)\\ &= P_0(x) - P_1(x) = P_0(x) = x\\ &\le \sum_{i=1}^n \alpha_i x_i \le \sum_{i=1}^n (\alpha_{i-1} + \epsilon)x_i \le s + \epsilon u\,,\end{aligned}$$

and so $0 \le x - s \le \epsilon u$ holds, as desired. ■

We are now in a position to state and prove another important characterization of Dedekind completeness.

THEOREM 1.59. *For a Riesz space L the following are equivalent.*

(a) *L is Dedekind complete.*

(b) *L has the projection property and is uniformly complete.*

PROOF. (a) $\Longrightarrow$ (b) If a Riesz space is Dedekind complete, then it has the projection property and (by Lemma 1.56) is uniformly complete.

(b) $\Longrightarrow$ (a) Let L be a uniformly complete Riesz space with the projection property. Then L is Archimedean and so its Dedekind completion L^δ exists. We must show that $L = L^\delta$.

To this end, let $0 \leq u^* \in L^\delta$. Pick some $u \in L^+$ such that $0 \leq u^* \leq u$. According to the Freudenthal Spectral Theorem 1.58 there exists a sequence $\{s_n\}$ of u-step functions in L^δ that converges u-uniformly to u^* in L^δ. Since L has the projection property, it follows from Theorem 1.54 that the sequence $\{s_n\}$ is a sequence of u-step functions in L. Now an easy argument shows that $\{s_n\}$ is a u-uniformly sequence in L, and so by the uniform completeness of L the sequence $\{s_n\}$ is u-uniformly convergent to some $v \in L$. Now notice that $\{s_n\}$ is also u-uniformly convergent to v in L^δ, and from the uniqueness of the uniform limits in Archimedean Riesz spaces, we see that $u^* = v \in L$. Therefore, $L = L^\delta$, and thus L is Dedekind complete. ■

1.5. The Main Inclusion Theorem

The following theorem, known as the main inclusion theorem, shows the various interrelationships between the different classes of Archimedean Riesz spaces.

THEOREM 1.60 (The Main Inclusion Theorem). *Consider the following properties that a Riesz space may possess: super Dedekind completeness (SDC), Dedekind completeness (DC), Dedekind σ-completeness ($D\sigma C$), almost Dedekind σ-completeness ($AD\sigma C$), the projection property (PP), the principal projection property (PPP), the property of having sufficiently many projections (SMP), and the property of being Archimedean (Arch). Then the following (and only these) implications hold:*

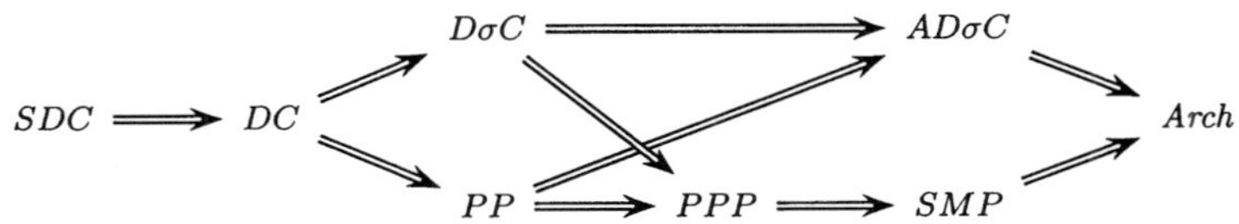

The only implication that has not been already mentioned is the implication $PP \Longrightarrow AD\sigma C$. To see this, note first that by Theorem 1.54 the components of the elements of L^+ in L^δ lie in L, and then use Freudenthal's Spectral Theorem 1.58 to conclude that L is super order dense in L^δ.

Next we exhibit examples to show that no other implication holds. The details in the examples are left as exercises to the reader.

EXAMPLE 1.61. (1) *A Dedekind complete Riesz space that is not super Dedekind complete.* Take L to be the Dedekind complete Riesz space consisting of all real-valued functions defined on some uncountable set X, that is, $L = \mathbb{R}^X$. The vector space L becomes a Dedekind complete Riesz space if ordered by the pointwise ordering, that is, $u \geq v$ in L whenever $u(x) \geq v(x)$ for all $x \in X$. Note that the collection S of all characteristic functions of the singletons has as supremum the constant function one. However, observe that no countable subset of S has as supremum the function one.

(2) *A Dedekind σ-complete Riesz space that is not Dedekind complete and without the projection property.* Consider the vector space L consisting of all real-valued Lebesgue measurable functions defined on $[0,1]$; L becomes a Riesz space if ordered by the pointwise ordering. Observe that $u_\alpha \xrightarrow{o} u$ in L implies $u_\alpha(x) \to u(x)$ for each $x \in [0,1]$. Thus, if E is a non-Lebesgue measurable subset of $[0,1]$ and $\{\chi_\alpha\}$ is the net of all characteristic functions of the finite subsets of E, then $0 \leq \chi_\alpha \uparrow \leq \mathbf{1}$ holds, where $\mathbf{1}(x) = 1$ for all $x \in [0,1]$, but the net $\{\chi_\alpha\}$ does not have a supremum since $\chi_\alpha(x) \uparrow \chi_{_E}(x)$ holds for all $x \in [0,1]$. This shows that L is not Dedekind complete.

To see that the Riesz space L does not have the projection property note that the band $B = \{u \in L\colon\ u(x) = 0 \text{ for all } x \in E\}$ is not a projection band. For the Dedekind σ-completeness let $0 \leq u_n \uparrow \leq u$ in L, and note that the function $v\colon [0,1] \to \mathbb{R}$ defined by $v(x) = \sup\{u_n(x)\colon\ n \in \mathbb{N}\} < \infty$ for each $x \in [0,1]$ satisfies $v \in L$ and $u_n \uparrow v$ in L. This example also shows that PPP (or SMP) does not imply PP.

(3) *A Riesz space with the projection property that is not Dedekind complete.* Consider an infinite set X, and let L be the Riesz space of all real-valued functions defined on X whose range is finite, with the pointwise ordering. Then L is not a Dedekind complete Riesz space, but it satisfies the projection property.

(4) *An almost Dedekind σ-complete Riesz space with sufficiently many projections but without the principal projection property.* Let L be the Riesz space of all convergent real sequences with the pointwise ordering, i.e., $L = c$. The Riesz space L is almost Dedekind σ-complete since it is super order dense in ℓ_∞, the super Dedekind complete Riesz space of all order bounded real sequences. Note also that L has sufficiently many projections but L lacks the principal projection property, since if $u = (1, 0, \frac{1}{2}, 0, \frac{1}{3}, 0, \ldots)$ then the band generated by u is $B_u = \{(v_1, v_2, \ldots) \in L\colon\ v_n = 0 \text{ if } n \text{ is even}\}$, which is not a projection band since $\mathbf{1} = (1,1,1,\ldots) \in L$ has no decomposition with respect to this band.

(5) *An almost Dedekind σ-complete Riesz space with the principal projection property that is not Dedekind σ-complete and that lacks the projection property.* Let L be the Riesz space of all ultimately constant sequences with the pointwise ordering. The Riesz space L is super order dense in ℓ_∞ and satisfies the principal projection property, but is not Dedekind σ-complete and fails to satisfy the projection property.

(6) *A Riesz space with the principal projection property that is not almost Dedekind σ-complete and that lacks the projection property.* Let L denote the vector space of all functions $f\colon \mathbb{R} \to \mathbb{R}$ with the property that there exists a two-tailed sequence $\{x_n\}_{n=-\infty}^{\infty}$ of real numbers with

$$\cdots < x_{-1} < x_0 < x_1 < \cdots$$

such that $x_n \to \infty$ as $n \to \infty$ and $x_{-n} \to -\infty$ as $n \to \infty$ and such that f is constant on the open intervals (x_k, x_{k+1}) for all integers $k \in \mathbb{Z}$. Note that L is a Riesz space under the pointwise ordering, and that order convergence in L implies pointwise convergence. The Riesz space L does not have the projection property, for if

$$B = \{u \in L\colon\ u(x) = 0 \text{ for all irrational numbers } x\},$$

then B is not a projection band. It is clear though that the Riesz space L does have the principal projection property, for if $v \in L$, then the principal band B_v is $B_v = \{u \in L\colon\ u(x) = 0 \text{ whenever } v(x) = 0\}$. By the restrictions on functions in L, the band B_v will always be a projection band. Note, however, that L is not almost Dedekind σ-complete.

(7) *An almost Dedekind σ-complete Riesz space without sufficiently many projections.* Let $L = C[0,1]$, the Riesz space of continuous real-valued functions on $[0,1]$ with the pointwise ordering. Note that L does not have sufficiently many projections, but it does satisfy the countable sup property, since $\phi\colon C[0,1] \to \mathbb{R}$, defined by $\phi(u) = \int_0^1 u(x)\,dx$, is a strictly positive operator; see Theorem 1.45. Therefore, L is super order dense in its Dedekind completion (see Lemma 1.44) so that L is almost Dedekind σ-complete. ■

Note. The main inclusion theorem was stated and proven without the class of almost Dedekind σ-complete Riesz spaces by W. A. J. Luxemburg and A. C. Zaanen [**90**, pp. 137, 174]. The class of almost Dedekind σ-complete Riesz spaces was introduced by C. D. Aliprantis and E. Langford [**26**] and independently by J. Quinn [**112**].

For a proof of the next result see Exercise 8 at the end of the chapter.

THEOREM 1.62. *Let J be an ideal in a Riesz space L. If L satisfies any of the properties of the main inclusion theorem, then J considered as a Riesz space in its own right also satisfies the same property.*

1.6. Order Bounded Operators

In this section we shall discuss the basic properties of order bounded operators. We start with their definition.

DEFINITION 1.63. *An operator $T\colon L \to M$ between two Riesz spaces is said to be* ***order bounded*** *if T carries order bounded subsets of L to order bounded subsets of M, that is, if $T(A)$ is an order bounded subset of M whenever A is an order bounded subset of L.*

The collection of all order bounded operators from L to M will be denoted by $\mathcal{L}_b(L,M)$. We shall write $\mathcal{L}_b(L)$ instead of $\mathcal{L}_b(L,L)$.

It should be clear that $\mathcal{L}_b(L,M)$ under the usual pointwise algebraic operations is a real vector space. Recall that an operator $T\colon L \to M$ is called **positive** (denoted symbolically by $T \geq 0$ or $0 \leq T$) if $T(u) \geq 0$ holds in M whenever $u \geq 0$ holds in L. Clearly, every positive operator is order bounded and likewise is every operator that can be written as a difference of two positive operators.[8] Observe that the relation $T_1 \geq T_2$ whenever $T_1 - T_2 \geq 0$ defines an order relation on $\mathcal{L}_b(L,M)$. It is easy to see that:

- *If L and M are Riesz spaces, then the real vector space $\mathcal{L}_b(L,M)$ is an ordered vector space whose positive cone consists of all positive operators.*

[8] An operator between ordered vector spaces that can be written as a difference of two positive operators is called a **regular operator**.

We start our discussion with a basic result concerning extensions of positive operators.

LEMMA 1.64. *Suppose that L and M are two Riesz spaces with M Archimedean. Assume also that $T\colon L^+ \to M^+$ is an **additive mapping**, that is, $T(u+v) = T(u) + T(v)$ holds for all $u, v \in L^+$. Then T has a unique extension to a positive operator from L to M. Moreover, the extension (denoted by T again) is given by*

$$T(u) = T(u^+) - T(u^-)$$

for all $u \in L$.

PROOF. Let $T\colon L^+ \to M^+$ be an additive mapping. Consider the mapping $S\colon L \to M$ defined by

$$S(u) = T(u^+) - T(u^-).$$

Clearly, $S(u) = T(u)$ for each $u \in L^+$. So, the mapping S extends T to all of L. Since $u = u^+ - u^-$ for each $u \in L$, it follows that S is the only possible linear extension of T to all of L. Therefore, in order to complete the proof, we must show that S is linear. That is, we must prove that S is additive and homogeneous.

For the additivity of S start by observing that if any vector $u \in L$ can be written as a difference of two positive vectors, say $u = v_1 - v_2$ with $v_1, v_2 \in L^+$, then $S(u) = T(v_1) - T(v_2)$ holds. To see this, fix any $u \in L$ and assume that $u = u^+ - u^- = v_1 - v_2$, where $v_1, v_2 \in L^+$. Then $u^+ + v_2 = v_1 + u^-$, and so the additivity of T on L^+ yields

$$T(u^+) + T(v_2) = T(u^+ + v_2) = T(v_1 + u^-) = T(v_1) + T(u^-)$$

or $S(u) = T(u^+) - T(u^-) = T(v_1) - T(v_2)$. From this property, we can easily establish that S is additive. Indeed, if $u, w \in L$, then note that

$$\begin{aligned} S(u+w) &= S\big(u^+ + w^+ - (u^- + w^-)\big) \\ &= T(u^+ + w^+) - T(u^- + w^-) \\ &= T(u^+) + T(w^+) - T(u^-) - T(w^-) \\ &= \big[T(u^+) - T(u^-)\big] + \big[T(w^+) - T(w^-)\big] \\ &= S(u) + S(w). \end{aligned}$$

In particular, the additivity of S implies that $S(ru) = rS(u)$ holds for all $u \in L$ and all rational numbers r.

It remains to show that S is homogeneous. For this, we need to prove first that S is monotone. That is, $u \geq v$ in L implies $S(u) \geq S(v)$ in M. Indeed, if $u \geq v$, then $u - v \in L^+$, and so by the additivity of S we get

$$S(u) = S\big((u-v) + v\big) = S(u-v) + S(v) = T(u-v) + S(v) \geq S(v).$$

Now fix $u \in L^+$ and let $\lambda \in \mathbb{R}$. Pick two sequences of rational numbers $\{r_n\}$ and $\{t_n\}$ such that $r_n \uparrow \lambda$ and $t_n \downarrow \lambda$. From $r_n u \leq \lambda u \leq t_n u$ and the monotonicity of S, we obtain

$$r_n S(u) = S(r_n u) \leq S(\lambda u) \leq S(t_n u) = t_n S(u)$$

for each n. Using that M is Archimedean, we easily get $\lambda S(u) = S(\lambda u)$; see Theorems 1.14(vi) and 1.28. Finally, if $\lambda \in \mathbb{R}$ and $u \in L$, then

$$\begin{aligned} S(\lambda u) &= S(\lambda u^+ + (-\lambda)u^-) = S\big(\lambda u^+\big) + S\big((-\lambda)u^-\big) \\ &= \lambda S(u^+) - \lambda S(u^-) = \lambda\big[T(u^+) - T(u^-)\big] = \lambda S(u). \end{aligned}$$

So, S is also homogeneous, and the proof is finished. ■

The preceding lemma is not true if M is not Archimedean.

EXAMPLE 1.65. Let $\phi\colon \mathbb{R} \to \mathbb{R}$ be an additive function that is not linear, i.e., not of the form $\phi(x) = cx$, and let M be the lexicographic plane. Consider the mapping $T\colon \mathbb{R}^+ \to M^+$ defined by $T(x) = \big(x, \phi(x)\big)$ for each x in $\mathbb{R}^+$. Note that T is additive and that if T could be extended to an operator from $\mathbb{R}$ to M, then ϕ itself should be linear, which is impossible. ■

The rest of the section is devoted to several applications of Lemma 1.64. The first one deals with extension properties of operators.

THEOREM 1.66. *Let L and M be two Riesz spaces with M Dedekind complete, let A be an ideal of L, and let $T\colon A \to M$ be an operator that has an extension to an order bounded operator from L to M. Define the mapping $T_A\colon L^+ \to M^+$ via the formula*

$$T_A(u) = \sup\big\{T(v)\colon\ v \in A \ \text{and}\ 0 \le v \le u\big\}, \quad u \in L^+.$$

Then we have the following:

1. *The mapping $T_A\colon L^+ \to M^+$ is additive—and so it extends uniquely to a positive operator from L to M.*
2. *If $u \in A^{\mathrm{d}}$, then $T_A(u) = 0$.*
3. *If $S \in \mathcal{L}_{\mathrm{b}}(L, M)$ is a positive operator that dominates T on A (that is, $T(u) \le S(u)$ holds for all $u \in A^+$), then $T_A(u) \le S_A(u)$ also holds for each $u \in L^+$.*
4. *If the operator $T\colon A \to M$ is positive, then T_A is a positive linear extension of T and is the* ***minimal extension*** *of T in the sense that if $S\colon L \to M$ is any positive operator that extends $T\colon A \to M$, then $T_A \le S$.*

PROOF. (1) Notice that if $R \in \mathcal{L}_{\mathrm{b}}(L, M)$ is an extension of the operator $T\colon A \to M$, then $\big\{T(v)\colon\ v \in A \text{ and } 0 \le v \le u\big\} \subseteq R\big([0, u]\big)$, and so the set $\big\{T(v)\colon\ v \in A \text{ and } 0 \le v \le u\big\}$ is order bounded in M. This implies that T_A is well defined by the indicated formula and that T_A maps indeed L^+ to M^+.

Now to see that T_A is additive, let $u, w \in L^+$. If two arbitrary elements $v_1, v_2 \in A$ satisfy $0 \le v_1 \le u$ and $0 \le v_2 \le w$, then we have $0 \le v_1 + v_2 \le u + w$ and $T(v_1) + T(v_2) = T(v_1 + v_2) \le T_A(u + w)$. This implies $T_A(u) + T_A(w) \le T_A(u + w)$.

To establish the reverse inequality, assume that some vector $v \in A$ satisfies $0 \le v \le u + w$. By the Riesz decomposition property, we can write $v = v_1 + v_2$, where $0 \le v_1 \le u$ and $0 \le v_2 \le w$. Since A is an ideal, it follows that $v_1, v_2 \in A$, and so

$$T(v) = T(v_1 + v_2) = T(v_1) + T(v_2) \le T_A(u) + T_A(w)\,.$$

This implies $T_A(u+w) \le T_A(u) + T_A(w)$. Consequently, $T_A(u+w) = T_A(u) + T_A(w)$ so that T_A is additive.

(2) Let $0 \le u \in A^{\mathrm{d}}$. Then we have $\{v \in A\colon\ 0 \le v \le u\} = \{0\}$, and from this we see that $T_A(u) = 0$.

(3) Assume that a positive operator $S \in \mathcal{L}_{\mathrm{b}}(L, M)$ satisfies $T(v) \le S(v)$ for all $v \in A^+$. If $0 \le v \le u$ with $v \in A$, then $T(v) \le S(v) \le S_A(u)$. This implies $T_A(u) \le S_A(u)$ for all $u \in L^+$.

(4) Suppose that $T\colon A \to M$ is a positive operator. It should be clear that $T_A(u) = T(u)$ for all $u \in A^+$, and so T_A is a positive linear extension of T. The

fact that T_A is the minimal extension of T follows immediately from the preceding part. ■

The following classical theorem, due to F. Riesz [**116**] and L. V. Kantorovich [**70**], shows that if M is Dedekind complete, then the ordered vector space $\mathcal{L}_b(L, M)$ is a Dedekind complete Riesz space.

THEOREM 1.67 (Riesz–Kantorovich). *If L and M are two Riesz spaces with M Dedekind complete, then the ordered vector space $\mathcal{L}_b(L, M)$ is a Dedekind complete Riesz space. Its lattice operations for all $T \in \mathcal{L}_b(L, M)$ and every $u \in L^+$ are given by*

(1) $T^+(u) = \sup\{T(v)\colon\ 0 \le v \le u\}$,
(2) $T^-(u) = \sup\{-T(v)\colon\ 0 \le v \le u\}$, *and*
(3) $|T|(u) = \sup\{|T(v)|\colon\ |v| \le u\}$.

In particular, for each pair $S, T \in \mathcal{L}_b(L, M)$ and all $u \in L^+$ we have:

$$[S \vee T](u) = \sup\{S(v) + T(w)\colon\ v, w \in L^+ \text{ and } v + w = u\}$$
$$[S \wedge T](u) = \inf\{S(v) + T(w)\colon\ v, w \in L^+ \text{ and } v + w = u\}.$$

Moreover, we have the following properties.

(a) $|T(u)| \le |T|(|u|)$ *for every $T \in \mathcal{L}_b(L, M)$ and each $u \in L$.*
(b) *An increasing net $\{T_\alpha\} \subseteq \mathcal{L}_b(L, M)$ satisfies $T_\alpha \uparrow T$ in $\mathcal{L}_b(L, M)$ if and only if $T_\alpha(u) \uparrow T(u)$ holds in M for all $u \in L^+$.*
(c) *$T_\alpha \xrightarrow{o} T$ in $\mathcal{L}_b(L, M)$ implies $T_\alpha(u) \xrightarrow{o} T(u)$ in M for all $u \in L$.*

PROOF. Let $T \in \mathcal{L}_b(L, M)$. If we put $A = L$, then the operator T_A defined in Theorem 1.66 is clearly the supremum of T and 0 in $\mathcal{L}_b(L, M)$. That is, $T^+ = T_A$, and the desired formula follows. This shows that $\mathcal{L}_b(L, M)$ is a Riesz space. The formulas in (2) and (3) now easily follow from the lattice identities $T^- = (-T)^+$ and $|T| = T^+ + T^-$.

To see that $\mathcal{L}_b(L, M)$ is Dedekind complete let $0 \le T_\alpha \uparrow \le S$ in $\mathcal{L}_b(L, M)$. For each $u \in L^+$ put $T(u) = \sup_\alpha T(u_\alpha)$ in M, and note that $T\colon L^+ \to M^+$ is additive and hence extendable, according to Lemma 1.64, to all of L. Clearly, $T_\alpha \uparrow T$ holds in $\mathcal{L}_b(L, M)$ and $T_\alpha(u) \uparrow T(u)$ in M for each $u \in L^+$. From this conclusion the validity of the claim regarding the order limits follows easily. The formulas for the suprema follow from the lattice identities $S \vee T = (S - T)^+ + T$ and $S \wedge T = -[(-S) \vee (-T)]$. ■

The operator $|T| = T \vee (-T)$ is known as the **modulus** of the operator T. Formulas (1), (2), and (3) as stated in Theorem 1.67 are referred to as the **Riesz–Kantorovich formulas**.

When L has the principal projection property, Y. A. Abramovich has expressed in [**1**] the lattice operations of $\mathcal{L}_b(L, M)$ in terms of components as follows.

THEOREM 1.68 (Abramovich). *If L has the principal projection property and M is a Dedekind complete Riesz space, then for each pair $S, T \in \mathcal{L}_b(L, M)$ and each $u \in L^+$ we have:*

$$[S \vee T](u) = \sup\{S(v) + T(w)\colon\ v \wedge w = 0 \text{ and } v + w = u\},$$
$$[S \wedge T](u) = \inf\{S(v) + T(w)\colon\ v \wedge w = 0 \text{ and } v + w = u\}.$$

PROOF. Notice that the first formula follows from the second by using the identity $S \vee T = -\big[(-S) \wedge (-T)\big]$. Also, if the second formula is true for the special case $S \wedge T = 0$, then it is true in general. This claim follows easily from the identity $(S - S \wedge T) \wedge (T - S \wedge T) = 0$. To complete the proof, assume that $S \wedge T = 0$ in $\mathcal{L}_b(L, M)$. Fix $u \in L^+$ and put

$$x = \inf\big\{S(y) + T(u-y)\colon\ y \wedge (u-y) = 0\big\}.$$

We must show that $x = 0$.

To this end, fix any $0 \le v \in L^+$ satisfying $0 \le v \le u$. Let P denote the band projection of L onto the band generated by $(2v-u)^+$ and put $w = P(u)$. From the inequality $u \le 2v + (u-2v)^+$ and $(u-2v)^+ \wedge (2v-u)^+ = 0$, it follows that $P(u) \le 2P(v) + P\big((u-2v)^+\big) = 2P(v) \le 2v$. Therefore,

$$w \le 2v\,. \tag{$\star$}$$

Also, from $(2v-u)^+ \le (2u-u)^+ = u$ we see that

$$2v - u \le (2v-u)^+ = P\big((2v-u)^+\big) \le P(u) = w\,,$$

and consequently

$$u - w \le 2(u-v)\,. \tag{$\star\star$}$$

Now combining ($\star$) and ($\star\star$), we get

$$0 \le x \le S(w) + T(u-w) \le 2\big[S(v) + T(u-v)\big]\,, \tag{$\star\star\star$}$$

for all elements $v \in L^+$ with $0 \le v \le u$. Taking into consideration that (according to Theorem 1.67) we have

$$\inf\big\{S(v) + T(u-v)\colon\ 0 \le v \le u\big\} = 0\,,$$

it follows from ($\star\star\star$) that $x = 0$, and the proof is finished. ■

A mapping $p\colon X \to Y$ from a vector space to an ordered vector space is called **sublinear** if $p(u+v) \le p(u) + p(v)$ and $p(\lambda u) = \lambda p(u)$ hold for all $u, v \in X$ and all $\lambda \ge 0$. The next result is a generalization (due to L. V. Kantorovich) of the classical Hahn–Banach theorem and its proof is similar.

THEOREM 1.69 (Hahn–Banach–Kantorovich). *Let $p\colon L \to M$ be a sublinear mapping between two Riesz spaces with M Dedekind complete. Assume that X is a vector subspace of L and that $T\colon X \to M$ is an operator satisfying $T(u) \le p(u)$ for each $u \in X$. Then T can be extended to an operator $T^*\colon L \to M$ satisfying $T^*(u) \le p(u)$ for all $u \in L$.*[9]

We continue with the "dual" theorem of Theorem 1.67. It is essentially due to I. Namioka [**104**, Theorem 7.8, p. 38] and W. A. J. Luxemburg and A. C. Zaanen [**89**, Note VI, Theorem 19.6, p. 662]; see also [**5**, p. 669].

---

[9] This result is also true if p is merely a **convex function**, that is, if the mapping $p\colon L \to M$ satisfies $p(\lambda u + (1-\lambda)v) \le \lambda p(u) + (1-\lambda)p(v)$ for all $u, v \in L$ and all $0 \le \lambda \le 1$; see [**8**, Theorem 5.40, p. 184].

THEOREM 1.70 (Namioka–Luxemburg–Zaanen). *Let L and M be two Riesz spaces with M Dedekind complete. Then for all operators $0 \le T \in \mathcal{L}_b(L,M)$ and all $u, v \in L$ the following formulas hold:*

$$\begin{aligned} T(u^+) &= \max\bigl\{S(u)\colon\ 0 \le S \le T\bigr\},\\ T(u^-) &= \max\bigl\{-S(u)\colon\ 0 \le S \le T\bigr\},\\ T(|u|) &= \max\bigl\{|S(u)|\colon\ |S| \le T\bigr\},\\ T(u \vee v) &= \max\bigl\{S(u)+R(v)\colon\ S,R \ge 0 \text{ and } S+R=T\bigr\},\\ T(u \wedge v) &= \min\bigl\{S(u)+R(v)\colon\ S,R \ge 0 \text{ and } S+R=T\bigr\}. \end{aligned}$$

PROOF. Notice that all formulas follows from the first one. Therefore, we need only to verify the first formula. To this end, let $T\colon L \to M$ be a positive operator and fix $u \in L$.

Define $p\colon L \to M$ by $p(v) = T(v^+)$. Clearly p is a sublinear mapping satisfying $T(v) \le p(v)$ for all $v \in L$. Consider next the vector subspace $X = \{\lambda u\colon\ \lambda \in \mathbb{R}\}$, and let $S\colon X \to M$ be the operator defined by $S(\lambda u) = \lambda p(u) = \lambda T(u^+)$. Clearly, $S(\lambda u) \le p(\lambda u)$ holds for each $\lambda \in \mathbb{R}$. By Theorem 1.69 there exists an extension $S^*\colon L \to M$ of S such that $S^*(v) \le p(v)$ for all $v \in L$. In particular, it follows that $0 \le S^* \le T$. Now since $S^*(u) = S(u) = T(u^+)$ and

$$\sup\bigl\{S(u)\colon\ S \in \mathcal{L}_b(L,M) \text{ and } 0 \le S \le T\bigr\} \le T(u^+),$$

we easily obtained the desired identity. ■

The important notion of order continuity is defined as follows.

DEFINITION 1.71. *An operator $T\colon L \to M$ between two Riesz spaces is said to be **order continuous**, if $u_\alpha \xrightarrow{o} 0$ in L implies $T(u_\alpha) \xrightarrow{o} 0$ in M.*

*Similarly, an operator $T\colon L \to M$ between two Riesz spaces is called **σ-order continuous**, if $u_n \xrightarrow{o} 0$ in L implies $T(u_n) \xrightarrow{o} 0$ in M.*

Notice that a positive operator $T\colon L \to M$ is order continuous if and only if $u_\alpha \downarrow 0$ in L implies $T(u_\alpha) \downarrow 0$ in M. A similar statement holds true for σ-order continuous positive operators.

LEMMA 1.72. *Every order continuous operator is order bounded.*

PROOF. Let $T\colon L \to M$ be an order continuous operator between two Riesz spaces and let $0 \le u \in L$. Then the set

$$D = [0,u] \cup \{\lambda u\colon\ -1 < \lambda < 0\}$$

satisfies $D \downarrow -u$ in L. If we write D as a net $\{w_\alpha\}_{\alpha \in A}$, then from $w_\alpha \downarrow -u$ and the order continuity of T there is a net $\{v_\alpha\}_{\alpha \in A}$ in M satisfying $\bigl|T(w_\alpha) + T(u)\bigr| \le v_\alpha$ for each $\alpha \in A$. In particular, if $\alpha_0 \in A$ satisfies $w_{\alpha_0} = u$, then $v_\alpha \le v_{\alpha_0}$ holds true for all $\alpha \in A$, and from this we get $\bigl|T(w_\alpha)\bigr| \le \bigl|T(u)\bigr| + v_{\alpha_0}$ for all α. This implies $\bigl|T(x)\bigr| \le \bigl|T(u)\bigr| + v_{\alpha_0}$ for all $x \in [0,u]$. Hence, T is order bounded. ■

It should be noted that a σ-order continuous operator need not be order bounded; see Exercise 30 at the end of the chapter. The collection of all order continuous operators in $\mathcal{L}_b(L,M)$ will be denoted by $\mathcal{L}_n(L,M)$, and the collection of all σ-order continuous operators in $\mathcal{L}_b(L,M)$ by $\mathcal{L}_c(L,M)$. Both collections $\mathcal{L}_n(L,M)$ and $\mathcal{L}_c(L,M)$ are vector subspaces of $\mathcal{L}_b(L,M)$. T. Ogasawara [**105**] has shown that they are also bands.

THEOREM 1.73 (Ogasawara). *Let L and M be two Riesz spaces with M Dedekind complete. Then $\mathcal{L}_n(L,M)$ and $\mathcal{L}_c(L,M)$ are bands in $\mathcal{L}_b(L,M)$.*

PROOF. We show the result for $\mathcal{L}_n(L,M)$; the case $\mathcal{L}_c(L,M)$ is similar. Note first that $0 \le S \le T \in \mathcal{L}_n(L,M)$ and $u_\alpha \downarrow 0$ implies $S(u_\alpha) \downarrow 0$ in M, which means that $S \in \mathcal{L}_n(L,M)$. Assume now that $R \in \mathcal{L}_n(L,M)$ and that $u_\alpha \uparrow u$ holds in the Riesz space L. We shall show that $R^+(u_\alpha) \uparrow R^+(u)$, and this will suffice to establish that $\mathcal{L}_n(L,M)$ is an ideal. To this end, let $R^+(u_\alpha) \uparrow h \le R^+(u)$ in M.

If $0 \le v \le u_\alpha$ holds for some α, then $R(v) \le R^+(v) \le R^+(u_\alpha) \le h$. Now if $0 \le w \le u$, then $u_\alpha \wedge w \uparrow w$ holds in L, and therefore [since $R \in \mathcal{L}_n(L,M)$] we have $R(u_\alpha \wedge w) \xrightarrow{o} R(w)$. But then from the above remark, we have $R(u_\alpha \wedge w) \le h$ for all α and so by Theorem 1.14(v) we obtain $R(w) \le h$. Consequently, by Theorem 1.67, we get $R^+(u) \le h$, that is, $h = R^+(u)$.

To see that $\mathcal{L}_n(L,M)$ is a band of $\mathcal{L}_b(L,M)$, let $0 \le T_\alpha \uparrow T$ in $\mathcal{L}_b(L,M)$ with $\{T_\alpha\} \subseteq \mathcal{L}_n(L,M)$ and $0 \le x_\lambda \uparrow x$ in L. Also, assume that $T(x_\lambda) \uparrow f \le T(x)$ in M. Note that $T_\alpha(x_\lambda) \le f$ for each α and λ, and that for each fixed α we have $T_\alpha(x_\lambda) \uparrow_\lambda T_\alpha(x)$. Now by Theorem 1.14(v) we obtain $T_\alpha(x) \le f$ for all α and consequently $T(x) = \sup_\alpha T_\alpha(x) \le f$. Hence $f = T(x)$, so that $T \in \mathcal{L}_n(L,M)$. ■

As one more application of Lemma 1.64 we shall obtain an important embedding theorem for Archimedean Riesz spaces. But first we need some preliminary discussion. We start with a definition.

DEFINITION 1.74. *A* ***complete disjoint system*** *$\{e_i\colon\ i \in I\}$ in a Riesz space L is a maximal pairwise disjoint collection of nonzero positive elements; that is, $e_i > 0$ for each i, $e_i \wedge e_j = 0$ if $i \ne j$, and if $u \wedge e_i = 0$ for all $i \in I$, then $u = 0$.*

A basic property of the complete disjoint systems is described in the next result.

THEOREM 1.75. *Let $\{e_i\colon\ i \in I\}$ be a complete disjoint system in a Riesz space L consisting of projection elements. Then for every $u \in L^+$ we have*

$$u = \sup\{P_{e_i}(u)\colon\ i \in I\}.$$

PROOF. Let $u \in L^+$ and assume that some $w \in L^+$ satisfies $P_{e_i}(u) \le u - w$ for all $i \in I$. It follows that $0 \le w \le u - P_{e_i}(u) \in B_{e_i}^{\mathrm{d}}$ holds for each $i \in I$, that is, $w \wedge e_i = 0$ for all $i \in I$. Hence, $w = 0$ and so $u = \sup\{P_{e_i}(u)\colon\ i \in I\}$ holds. ■

Assume that $\{e_\alpha\colon\ \alpha \in A\}$ is a complete disjoint system of projection elements of an Archimedean Riesz space L. Let B_α denote the band generated by e_α in L, and let K be the Cartesian product ΠB_α. It can be readily seen that the mapping $\pi\colon L^+ \to K^+$ defined by $\pi(u) = (P_{e_\alpha}(u))_{\alpha \in A}$ is additive. Hence, by Lemma 1.64, the mapping $\pi\colon L \to K$ defined by $\pi(u) = \pi(u^+) - \pi(u^-) = (P_{e_\alpha}(u))_{\alpha \in A}$ is the unique linear positive extension of π. Now by using Theorem 1.75, we conclude that π is a Riesz isomorphism (into). Note, moreover, that $\pi(L)$ is order dense in K and that $(e_\alpha)_{\alpha \in A}$ is a weak unit of K. By applying the above technique to L^δ (and taking into consideration that each Dedekind complete Riesz space admits a complete disjoint system), we see that every Archimedean Riesz space can be embedded as an order dense Riesz subspace in a Dedekind complete Riesz space with a weak order unit. We have thus proven the following theorem.

THEOREM 1.76. *If $\{e_\alpha\colon\ \alpha \in A\}$ is a complete disjoint system of projection elements in an Archimedean Riesz space L, then the operator $\pi\colon L \to \Pi B_\alpha$, defined by $\pi(u) = (P_{e_\alpha}(u))_{\alpha\in A}$, is a Riesz isomorphism (into), and $\pi(L)$ is an order dense Riesz subspace of ΠB_α.*

In particular, every Archimedean Riesz space can be embedded as an order dense Riesz subspace in a Dedekind complete Riesz space with a weak order unit.

The next application of Lemma 1.64 deals with the characterization of discrete Archimedean Riesz spaces. A **discrete element** of a Riesz space L is an element $u > 0$ whose generating ideal L_u coincides with the vector subspace generated by u in L, that is, $L_u = \{\lambda u\colon\ \lambda \in \mathbb{R}\}$. A discrete element $u > 0$ is also called an **extremal element** and its half-ray $\{\lambda u\colon\ \lambda \geq 0\}$ is referred to as an **extremal half-ray** of the positive cone L^+. Clearly, if u is a discrete element, then any other vector v satisfying $0 < v \leq u$ is also a discrete element. In Archimedean Riesz spaces the discrete elements have an additional interesting property.

THEOREM 1.77. *If u is a discrete element in an Archimedean Riesz space L, then its principal ideal L_u is a projection band. That is, every discrete element is a projection element.*

PROOF. Let $u > 0$ be a discrete element. To see that L_u is a band let $0 \leq v_\alpha \uparrow v$ in L with $\{v_\alpha\} \subseteq L_u$. For each index α pick some $\lambda_\alpha \in \mathbb{R}$ such that $v_\alpha = \lambda_\alpha u$ and note that $0 \leq \lambda_\alpha \uparrow$ holds in $\mathbb{R}$. Since L is Archimedean and $0 \leq \lambda_\alpha u \leq v$ holds for each α, there exists some $0 \leq \lambda \in \mathbb{R}$ such that $\lambda_\alpha \uparrow \lambda$. From Theorem 1.28, it follows that $v_\alpha = \lambda_\alpha u \uparrow \lambda u$, and so $v = \lambda u \in L_u$. This shows that L_u is a band.

To show that L_u is a projection band, let $w \in L^+$. Since we have $w \wedge nu \in L_u$ for each n, it follows that there exists some $\mu_n \in \mathbb{R}$ such that $w \wedge nu = \mu_n u$. Clearly, $0 \leq \mu_n \uparrow \mu$ for some $\mu \in \mathbb{R}$. It is now easy to see that $\mu u = \sup\{w \wedge nu\colon\ n \in \mathbb{N}\}$, so that by Theorem 1.46 the band L_u is a projection band. ■

A **discrete Riesz space** is a Riesz space L having a complete disjoint system consisting of discrete elements. Note that a Riesz space L is discrete if and only if for every $0 < u \in L$ there exists a discrete element $v \in L$ satisfying $0 < v \leq u$. Indeed, if L is a discrete Riesz space, then there exists a complete disjoint system $\{e_i\colon\ i \in I\}$ of discrete elements; thus if $u > 0$, then $u \wedge e_i > 0$ must hold for at least one i and so $u \wedge e_i$ is a discrete element satisfying $0 < u \wedge e_i \leq u$. For the converse, assume that the condition is satisfied. Using Zorn's lemma we can choose a maximal family $\{e_i\colon\ i \in I\}$ of pairwise disjoint discrete elements. If $u \geq 0$ satisfies $u \wedge e_i = 0$ for each i, then $u = 0$, since otherwise by the condition, there should exist a discrete element v satisfying $0 < v \leq u$, and thus $v \wedge e_i = 0$ for all i, contrary to the choice of the family $\{e_i\colon\ i \in I\}$.

The discrete Archimedean Riesz spaces are characterized as follows.

THEOREM 1.78. *For an Archimedean Riesz space L the following statements are equivalent.*

(i) *L is a discrete Riesz space.*
(ii) *L is Riesz isomorphic to an order dense Riesz subspace of some Riesz space of the form $\mathbb{R}^X$.*

PROOF. (i) $\Longrightarrow$ (ii) Let $\{e_i\colon\ i \in I\}$ be a complete disjoint system of discrete elements. From Theorems 1.77 and 1.75 it follows that $u = \sup\{P_{e_i}(u)\colon\ i \in I\}$ for each $u \in L^+$. Also, if $u \in L^+$, then for each i there exists a unique $\lambda_i^u \in \mathbb{R}$ with

$P_{e_i}(u) = \lambda_i^u e_i$. Consider now the mapping $\pi\colon L^+ \to (\mathbb{R}^I)^+$ given by $\pi(u) = \{\lambda_i^u\}_{i\in I}$ for $u \in L^+$ and note that π is additive. So, by Lemma 1.64, we can extend π from L to $\mathbb{R}^I$ by letting $\pi(u) = \pi(u^+) - \pi(u^-)$ for $u \in L$. Clearly π is a Riesz isomorphism (into) and $\pi(L)$ is an order dense Riesz subspace of $\mathbb{R}^I$.

(ii) $\Longrightarrow$ (i) Straightforward. ∎

It should be noted that not every Riesz subspace of an Archimedean discrete Riesz space is discrete. For instance, consider the discrete Riesz space $\mathbb{R}^{[0,1]}$ and its Riesz subspace $C[0,1]$, the Riesz space of all continuous real-valued functions on $[0,1]$. Note that $C[0,1]$ is not discrete—actually $C[0,1]$ has no discrete elements at all.

1.7. The Order Dual of a Riesz space

This section discusses a few basic properties of the order dual of a Riesz space. We start by introducing the order dual of a Riesz space.

DEFINITION 1.79. *For any Riesz space L the Dedekind complete Riesz space $\mathcal{L}_{\mathrm{b}}(L,\mathbb{R})$ is called the* ***order dual*** *of L and is denoted by $L^\sim$, i.e., $L^\sim = \mathcal{L}_{\mathrm{b}}(L,\mathbb{R})$.*

Similarly, if L is a Riesz space, then its ***order continuous dual*** *is defined by $L_{\mathrm{n}}^\sim = \mathcal{L}_{\mathrm{n}}(L,\mathbb{R})$, and its σ-**order continuous dual** is defined by $L_{\mathrm{c}}^\sim = \mathcal{L}_{\mathrm{c}}(L,\mathbb{R})$.*[10]

The **double order dual** of a Riesz space L is the order dual of $L^\sim$ and is denoted $L^{\sim\sim}$, i.e., $L^{\sim\sim} = (L^\sim)^\sim$.

For a Riesz space L and any $\phi \in L^\sim$ the **null ideal** of ϕ is defined by

$$N_\phi = \left\{u \in L\colon\ |\phi|(|u|) = 0\right\}.$$

Note that N_ϕ is actually an ideal of L, and that if ϕ is order continuous (i.e., $\phi \in L_{\mathrm{n}}^\sim$), then N_ϕ is a band of L. The **carrier** C_ϕ of any $\phi \in L^\sim$ is simply the disjoint complement of N_ϕ, i.e., $C_\phi = N_\phi^{\mathrm{d}}$. Clearly, for each $\phi \in L^\sim$ we have $N_\phi \cap C_\phi = \{0\}$.

Since the linear functional $|\phi|$ is always strictly positive on C_ϕ, we have immediately (from Theorem 1.45) the following useful result.

LEMMA 1.80. *If L is a Riesz space, then for every $\phi \in L^\sim$ its carrier C_ϕ has the countable sup property.*

We continue with a basic property of order continuous linear functionals.

THEOREM 1.81. *Let L be an Archimedean Riesz space, $\phi \in L_{\mathrm{n}}^\sim$ and $\psi \in L^\sim$. Then the following statements are equivalent.*

1. *$\phi \perp \psi$ in $L^\sim$, that is, $|\phi| \wedge |\psi| = 0$.*
2. *$C_\psi \subseteq N_\phi$.*
3. *$C_\phi \perp C_\psi$, that is, $|u| \wedge |v| = 0$ for all $u \in C_\phi$ and $v \in C_\psi$.*

PROOF. (1) $\Longrightarrow$ (2) Assume that $|\phi| \wedge |\psi| = 0$ holds in the order dual $L^\sim$. Let $0 < u \in C_\psi = N_\psi^{\mathrm{d}}$ and fix $\epsilon > 0$. Pick a sequence $\{v_n\} \subseteq L^+$ satisfying $0 \le v_n \le u$ and $|\psi|(v_n) + |\phi|(u - v_n) \le 2^{-n}\epsilon$ for all n; see Theorem 1.67. Put $w_n = \inf\{v_i\colon\ 1 \le i \le n\}$ and note that $w_n \downarrow 0$ in L. Indeed, if $0 \le w \le w_n$

[10] In the terminology of W. A. J. Luxemburg and A. C. Zaanen an order continuous linear functional is called a **normal integral** and a σ-order continuous linear functional an **integral**; see [**89**, Notes VI and VIII]. The order continuous terminology is due to H. Nakano [**100**].

holds for all n, then $0 \le |\psi|(w) \le |\psi|(v_n) \le 2^{-n}\epsilon$ for all n, so that $|\psi|(w) = 0$ and hence $w \in N_\psi \cap C_\psi = \{0\}$. Therefore $w = 0$, and consequently $w_n \downarrow 0$ in L. Since $|\phi| \in L_{\rm n}^\sim$, it follows that $|\phi|(u - w_n) \uparrow |\phi|(u)$, and from

$$0 \le |\phi|(u - w_n) = |\phi|\Big(\bigvee_{i=1}^{n}(u - v_i)\Big) \le \sum_{i=1}^{n} |\phi|(u - v_i) \le \epsilon\,,$$

we get $0 \le |\phi|(u) \le \epsilon$ for all $\epsilon > 0$. Thus $|\phi|(u) = 0$ and so $C_\psi \subseteq N_\phi$.

(2) $\Longrightarrow$ (3) Let $u \in C_\phi$ and $v \in C_\psi$. Then we have $|u| \wedge |v| \in C_\psi \subseteq N_\phi$, and thus $|u| \wedge |v| \in C_\phi \cap N_\phi = \{0\}$. Hence, $|u| \wedge |v| = 0$ so that $C_\phi \perp C_\psi$.

(3) $\Longrightarrow$ (1) Note that $C_\phi \perp C_\psi$ implies $C_\psi \subseteq C_\phi^{\rm d} = N_\phi^{\rm dd} = N_\phi$, where the last equality holds by virtue of the Archimedean property of L and the fact that $\phi \in L_{\rm n}^\sim$. If $0 \le u = u_1 + u_2$, where $u_1 \in N_\psi$ and $u_2 \in C_\psi$, then

$$0 \le [|\phi| \wedge |\psi|](u) = [|\phi| \wedge |\psi|](u_1) + [|\phi| \wedge |\psi|](u_2) \le |\psi|(u_1) + |\phi|(u_2) = 0\,,$$

and thus $|\phi| \wedge |\psi| = 0$ on the order dense ideal $N_\psi \oplus C_\psi$. Now use Theorem 1.27 and that (by Theorem 1.73) $|\phi| \wedge |\psi| \in L_{\rm n}^\sim$ to obtain $|\phi| \wedge |\psi| = 0$ in $L^\sim$. ■

An immediate consequence of the preceding result is the following important theorem that was proven first by H. Nakano [**100**, Theorem 20.1, p. 74] for Dedekind σ-complete Riesz spaces and it was generalized by W. A. J. Luxemburg and A. C. Zaanen [**89**, Note IX, Theorem 31.2(ii), p. 373].

THEOREM 1.82 (Nakano–Luxemburg–Zaanen). *If L is an Archimedean Riesz space and $\phi, \psi \in L_{\rm n}^\sim$, then the following statements are equivalent:*

(1) $\phi \perp \psi$
(2) $C_\phi \subseteq N_\psi$
(3) $C_\psi \subseteq N_\phi$
(4) $C_\phi \perp C_\psi$

The definitions of the null ideal and carrier can be defined for any order bounded operator. If L and M are Riesz spaces with M Dedekind complete, then for each operator $T \in \mathcal{L}_{\rm b}(L, M)$ its **null ideal** is defined by

$$N_T = \{u \in L\colon\ |T|(|u|) = 0\}\,,$$

and its **carrier** by $C_T = N_T^{\rm d}$. It should be clear that a positive operator is strictly positive if and only if $N_T = \{0\}$.

A generalization of Theorem 1.82 to $\mathcal{L}_{\rm n}(L, M)$ is not possible as the following example shows. Let $L = M = L_1[0, 1]$ (= the Riesz space of the equivalence classes of all Lebesgue integrable functions on $[0, 1]$), and note that both L and M are super Dedekind complete Riesz spaces.[11] Now let S and T be the positive operators from L to M defined by

$$S(u) = u \quad\text{and}\quad T(u) = \Big[\int_0^1 u(x)\,dx\Big]\mathbf{1}\,,$$

where $\mathbf{1}$ is the constant function one on $[0, 1]$, i.e., $\mathbf{1}(x) = 1$ for each $x \in [0, 1]$. Note that $S, T \in \mathcal{L}_{\rm n}(L, M)$ and $N_T = N_S = \{0\}$. Thus, $C_T = C_S = L$. It can be shown, however, that $S \wedge T = 0$ in $\mathcal{L}_{\rm n}(L, M)$.

---

[11] The Lebesgue integral $\phi\colon L_1[0, 1] \to \mathbb{R}$, defined by $\phi(u) = \int_0^1 u(x)\,dx$, is a strictly positive linear functional and so by Theorem 1.45 the Riesz space $L_1[0, 1]$ has the countable sup property.

Every element u in a Riesz space L gives rise to an order continuous linear functional $\widetilde{u}\colon L^\sim \to \mathbb{R}$ defined via the formula

$$\widetilde{u}(f) = f(u)\,, \quad f \in L^\sim\,.$$

To see that $\widetilde{u}$ is order bounded notice that $\widetilde{u} = \widetilde{u^+} - \widetilde{u^-}$. To see that $\widetilde{u}$ is order continuous for each $u \geq 0$ notice that if $f_\alpha \downarrow 0$ holds in $L^\sim$, then $\widetilde{u}(f_\alpha) = f_\alpha(u) \downarrow 0$.

If $\mathcal{I}$ is an arbitrary ideal of $L^\sim$, then each $\widetilde{u}$ can be considered restricted to $\mathcal{I}$, and so there exists a natural mapping $u \mapsto \widetilde{u}$ from L to $\mathcal{I}^\sim_{\rm n}$, which is referred to as the **natural embedding** of L into $\mathcal{I}^\sim_{\rm n}$. The properties of this natural embedding $u \mapsto \widetilde{u}$ are included in the next basic theorem of H. Nakano [**100**, Theorem 22.6, p. 83].

THEOREM 1.83 (Nakano). *Let L be a Riesz space, let $\mathcal{I}$ be an ideal of $L^\sim$, and let $u \mapsto \widetilde{u}$ be the natural embedding of L to $\mathcal{I}^\sim_{\rm n}$. Then we have:*

(1) *The natural embedding $u \mapsto \widetilde{u}$ is a Riesz homomorphism. In particular, its range is a Riesz subspace of $\mathcal{I}^\sim_{\rm n}$.*
(2) *The natural embedding $u \mapsto \widetilde{u}$ is a Riesz isomorphism if and only if $\mathcal{I}$ separates the points of L.*[12]
(3) *If L is Archimedean and $\mathcal{I} \subseteq L^\sim_{\rm n}$, then the natural embedding $u \mapsto \widetilde{u}$ is a normal Riesz homomorphism whose range is order dense in $\mathcal{I}^\sim_{\rm n}$.*

PROOF. (1) Let $0 \leq \phi \in \mathcal{I}$ and $u \in L$. From Theorems 1.70 and 1.67 and the fact that $\mathcal{I}$ is an ideal in $L^\sim$, we get

$$\begin{aligned}
\widetilde{u^+}(\phi) &= \phi(u^+) \\
&= \max\big\{\psi(u)\colon\ \psi \in L^\sim \text{ and } 0 \leq \psi \leq \phi\big\} \\
&= \max\big\{\widetilde{u}(\psi)\colon\ \psi \in L^\sim \text{ and } 0 \leq \psi \leq \phi\big\} \\
&= \max\big\{\widetilde{u}(\psi)\colon\ \psi \in \mathcal{I} \text{ and } 0 \leq \psi \leq \phi\big\} \\
&= (\widetilde{u})^+(\phi)\,,
\end{aligned}$$

so that $\widetilde{u^+} = (\widetilde{u})^+$ holds in $\mathcal{I}^\sim_{\rm n}$. This shows that $u \mapsto \widetilde{u}$ is a Riesz homomorphism.

(2) Assume first that $u \mapsto \widetilde{u}$ is a Riesz isomorphism and let $u \neq 0$. It follows that $\widetilde{u} \neq 0$, and consequently there exists some $\phi \in \mathcal{I}$ such that $\phi(u) = \widetilde{u}(\phi) \neq 0$. This shows that $\mathcal{I}$ separates the points of L.

For the converse, assume that $\mathcal{I}$ separates the points of L, and let $\widetilde{u} = 0$. This implies $\phi(u) = \widetilde{u}(\phi) = 0$ for each $\phi \in \mathcal{I}$, and from this we infer that $u = 0$. Therefore, $u \mapsto \widetilde{u}$ is a Riesz isomorphism.

(3) To see that $u \mapsto \widetilde{u}$ is a normal Riesz homomorphism let $u_\alpha \downarrow 0$ in L and note that since $\widetilde{u}_\alpha(\phi) = \phi(u_\alpha) \downarrow 0$ for all $0 \leq \phi \in \mathcal{I}$, it follows from Theorem 1.67 that $\widetilde{u}_\alpha \downarrow 0$ in $\mathcal{I}^\sim_{\rm n}$.

So, it remains to show that the range of L is order dense in $\mathcal{I}^\sim_{\rm n}$. To this end, let $0 < \phi \in \mathcal{I}^\sim_{\rm n}$. Without loss of generality, we can assume that there exists some $u \in L$ with $\phi \leq \widetilde{u}$. To see this, note first that $C_\phi \neq \{0\}$. Indeed, if $C_\phi = N^{\rm d}_\phi = \{0\}$, then $N_\phi = N^{\rm dd}_\phi = L$, and so $\phi = 0$, a contradiction. Now pick $0 < f \in C_\phi$, and then choose $0 < u \in C_f$; again if $C_f = \{0\}$, then $f = 0$, which is impossible. If

[12] Recall that a collection $\mathcal{F}$ of real-valued functions on a set X is said to **separate the points of** X if for every pair of distinct points $x_1, x_2 \in X$ there exists some function $f \in \mathcal{F}$ such that $f(x_1) \neq f(x_2)$.

$\widetilde{u} \wedge \phi = 0$, then by Theorem 1.82 we have $\widetilde{u}(C_\phi) = \{0\}$ so that $\widetilde{u}(f) = f(u) = 0$, contrary to the choices of u and f. Replacing now ϕ by $\phi \wedge \widetilde{u}$, we can establish our claim.

Now choose $0 < \lambda < 1$ with $\psi = (\phi - \lambda\widetilde{u})^+ > 0$. Notice that if $(\phi - \lambda\widetilde{u})^+ = 0$ for each $0 < \lambda < 1$, then $0 < \phi \leq \lambda\widetilde{u}$ holds true for each $0 < \lambda < 1$, contradicting the Archimedean property of $L^\sim$. Next select $0 < g \in C_\psi$ and then some $0 < v \in C_g$. Put $w = v \wedge \lambda u$ and note that $0 \leq \widetilde{w} \leq \widetilde{v}$ and $\widetilde{w} \leq \lambda\widetilde{u}$. We shall finish the proof by establishing that $0 < \widetilde{w} \leq \phi$.

To see that $\widetilde{w} > 0$, assume by way of contradiction that $\widetilde{w} = 0$. Then $\widetilde{v} \wedge \widetilde{u} = 0$ and so (since $0 \leq \psi \leq \widetilde{u}$) we have $\widetilde{v} \wedge \psi = 0$. But then Theorem 1.82 implies $\widetilde{v}(C_\psi) = \{0\}$ so that $\widetilde{v}(g) = g(v) = 0$, contrary to $0 < v \in C_g$.

To see that $\widetilde{w} \leq \phi$ assume that $\widetilde{w} \not\leq \phi$. So, $\phi_1 = (\widetilde{w} - \phi)^+ > 0$. Choose some $0 < h \in C_\phi$, and note [since $0 \leq \phi_1 \leq (\lambda\widetilde{u} - \phi)^+ = (\phi - \lambda\widetilde{u})^-$] that $\phi_1 \perp \psi$. So, by Theorem 1.82 (again!), $C_{\phi_1} \perp C_\psi$. Hence $h \perp g$ and so by using Theorem 1.82 once more, we get $h(C_g) = \{0\}$. In particular, we have $h(v) = 0$. Now note that

$$0 < \phi_1(h) = (\widetilde{w} - \phi)^+(h) \leq \widetilde{w}(h) \leq \widetilde{v}(h) = h(v) = 0\,,$$

which is impossible. This contradiction completes the proof. ■

If L is an Archimedean Riesz space, M a Dedekind complete Riesz space, and $0 \leq T \in \mathcal{L}_{\mathrm{n}}(L, M)$, we define $T^\delta \colon L^\delta \to M$ by

$$T^\delta(u^*) = \sup\bigl\{T(v)\colon\ v \in L \text{ and } v \leq u^*\bigr\} \tag{$\star$}$$

for every $u^* \in L^\delta$, the Dedekind completion of L. The order continuity of T easily implies that

$$T^\delta(u^*) = \inf\bigl\{T(w)\colon\ w \in L \text{ and } u^* \leq w\bigr\}\,.$$

These show that $T^\delta \colon L^\delta \to M$ is a positive operator satisfying $T^\delta(u) = T(u)$ for all $u \in L$. Also, if $u^*_\alpha \downarrow 0$ in L^δ, then there exists a net $\{v_\lambda\}$ in L with $v_\lambda \downarrow 0$ in L and such that for each λ there exists an index α with $v_\lambda \geq u^*_\alpha$; so by the order continuity of T, we get $T^\delta(u^*_\alpha) \downarrow 0$ in M, that is, $T^\delta \in \mathcal{L}_{\mathrm{n}}(L^\delta, M)$. Therefore, we have defined a mapping $T \mapsto T^\delta$, from $[\mathcal{L}_{\mathrm{n}}(L, M)]^+$ to $[\mathcal{L}_{\mathrm{n}}(L^\delta, M)]^+$, which is onto and one-to-one. Furthermore, $T \mapsto T^\delta$ is additive on $[\mathcal{L}_{\mathrm{n}}(L, M)]^+$, and hence by Lemma 1.64, it extends to a linear operator from $\mathcal{L}_{\mathrm{n}}(L, M)$ onto $\mathcal{L}_{\mathrm{n}}(L^\delta, M)$, via the usual formula $T^\delta = (T^+)^\delta - (T^-)^\delta$.

Now if $S \wedge T = 0$ in $\mathcal{L}_{\mathrm{n}}(L, M)$, then for every $0 \leq u \in L$ we have

$$\begin{aligned} 0 &\leq [S^\delta \wedge T^\delta](u) \\ &= \inf\bigl\{S^\delta(v^*) + T^\delta(w^*)\colon\ 0 \leq v^*, w^* \in L^\delta \text{ and } v^* + w^* = u\bigr\} \\ &\leq \inf\bigl\{S(v) + T(w)\colon\ v, w \in L^+ \text{ and } v + w = u\bigr\} \\ &= [S \wedge T](u) = 0\,. \end{aligned}$$

So, $[S^\delta \wedge T^\delta](u) = 0$ for all $u \in L$ and hence [since L is order dense in L^δ and $S^\delta \wedge T^\delta \in \mathcal{L}_{\mathrm{n}}(L^\delta, M)$], we have $[S^\delta \wedge T^\delta](u^*) = 0$ for all $u^* \in L^\delta$, that is, $S^\delta \wedge T^\delta = 0$ in $\mathcal{L}_{\mathrm{n}}(L^\delta, M)$. We have therefore established the following result.

THEOREM 1.84. *If L is an Archimedean Riesz space and M is a Dedekind complete Riesz space, then the mapping $T \mapsto T^\delta$ (defined by ($\star$) above) is a Riesz isomorphism between $\mathcal{L}_{\mathrm{n}}(L, M)$ and $\mathcal{L}_{\mathrm{n}}(L^\delta, M)$.*

In particular, for every Archimedean Riesz space L the Riesz spaces $L^\sim_{\mathrm{n}}$ and $(L^\delta)^\sim_{\mathrm{n}}$ are Riesz isomorphic via $\phi \mapsto \phi^\delta$.

The final result of this section reveals the relationship between the Riesz homomorphisms of a Riesz space and the discrete elements of its order dual.

THEOREM 1.85. *For a Riesz space L and some $0 < \phi \in L^\sim$ the following statements are equivalent.*

(i) *ϕ is a Riesz homomorphism (from L to $\mathbb{R}$).*
(ii) *ϕ is a discrete element of $L^\sim$, i.e., $L^\sim_\phi = \{\lambda\phi\colon\ \lambda \in \mathbb{R}\}$.*

PROOF. (i) $\Longrightarrow$ (ii) Assume that ϕ is a Riesz homomorphism. Let $\psi \in L^\sim$ satisfy $0 \le \psi \le \phi$. If $u \in \operatorname{Ker}\phi$, then we have

$$\big|\psi(u)\big| \le \psi(|u|) \le \phi(|u|) = \big|\phi(u)\big| = 0\,,$$

and hence $\psi(u) = 0$, so that $\operatorname{Ker}\phi \subseteq \operatorname{Ker}\psi$. The last inclusion is, however, equivalent to the property that $\psi = \lambda\phi$ for some $\lambda \in \mathbb{R}$.

(ii) $\Longrightarrow$ (i) Suppose that ϕ is a discrete element. Note first that since $|\psi| \le \phi$ implies $\psi = \lambda\phi$ for some $|\lambda| \le 1$, we have

$$\big|\psi(u)\big| = \big|\lambda\phi(u)\big| = |\lambda|\,\big|\phi(u)\big| \le \big|\phi(u)\big|$$

for all $u \in L$. It now follows from Theorem 1.70 that for each $u \in L$ we have

$$\phi(|u|) = \max\big\{\big|\psi(u)\big|\colon\ \psi \in L^\sim \text{ and } |\psi| \le \phi\big\} \le \big|\phi(u)\big|\,,$$

and hence (since $|\phi(u)| \le \phi(|u|)$ is always true) we get $\phi(|u|) = |\phi(u)|$ for all $u \in L$. Thus ϕ is a Riesz homomorphism, and the proof is finished. ■

Exercises

1. Show that an ordered vector space E is a Riesz space if and only if there exists some element $a \in E$ such that $u \vee a$ exists in E for each $u \in E$.
2. For a nonempty subset A of a Riesz space establish the following.
   (a) If A has a supremum, then the set $-A = \{-a\colon\ a \in A\}$ has an infimum and $\inf(-A) = -\sup A$.
   (b) If the set A has an infimum, then the set $-A$ has a supremum and $\sup(-A) = -\inf A$.
3. Establish the following identities for elements u and v in a Riesz space.
   (a) $u = (u - v)^+ + u \wedge v$.
   (b) $|u + v| \wedge |u - v| = \big||u| - |v|\big|$.
4. Give an example of a convex subset of an Archimedean Riesz space whose solid hull is not convex.
5. Prove Theorem 1.14.
6. Let B be the band generated by an ideal J in an arbitrary Riesz space L and let $U = \big\{u \in L^+\colon\ \exists\ \{u_\alpha\} \subseteq J \text{ with } u_\alpha \uparrow u \text{ in } L\big\}$. Establish that $B = \{u - v\colon\ u, v \in U\}$ and that $B^+ = U$. State and prove a similar statement for the σ-ideal generated by J in L.
7. Prove Theorem 1.31.
8. Prove Theorem 1.62.
9. Consider an uncountable set X and let L be the Riesz space consisting of all real-valued functions defined on X such that there exists a real number $f(\infty)$ (depending on f) for which the set $\{x \in X\colon\ f(x) \ne f(\infty)\}$ is finite. Determine L^σ and L^δ. Is L almost Dedekind σ-complete?

10. Let L be a majorizing Riesz subspace of a Dedekind complete Riesz space M. Show that if an ideal J in M is maximal with respect to the property $L \cap J = \{0\}$, then M/J is the Dedekind completion of the Riesz space L.

11. Show that the Riesz space of Example 1.61(6) is not almost Dedekind σ-complete.

12. An element $u > 0$ in a Riesz space L is called an **atom** whenever $v \wedge w = 0$, $v \leq u$, and $w \leq u$ imply either $v = 0$ or $w = 0$.

(i) Show that every discrete element of a Riesz space is an atom and give an example of an atom that is not a discrete element.

(ii) If L is Archimedean, show that an element is an atom if and only if it is a discrete element.

(iii) Show that $C[0,1]$ does not have any discrete elements.

13. (Judin) If every disjoint subset of an Archimedean Riesz space L is finite, show that L is Riesz isomorphic to some $\mathbb{R}^n$. [*Hint*: Use the previous exercise to show that L is a discrete Riesz space and then use Theorem 1.78.]

14. Show that an Archimedean Riesz space that is **totally ordered** (i.e., every two elements are comparable) is either trivial or else is Riesz isomorphic to $\mathbb{R}$.

15. Show that for an Archimedean Riesz space L the following statements are equivalent.

(a) L has the countable sup property.

(b) For each nonempty bounded above subset A of L there exists an at most countable subset B of A having the same upper bounds as A.

(c) Every nonempty bounded above disjoint subset of L^+ is at most countable.

16. Let $T\colon L \to M$ be a one-to-one operator from a Riesz space L onto the Riesz space M. Show that T is a Riesz isomorphism if and only if T and T^{-1} are both positive operators.

17. Show that for an operator $T\colon L \to L$ on a Riesz space the following statements are equivalent.

(a) T is a band projection.

(b) T is a projection satisfying $0 \leq T \leq I$, where $I\colon L \to L$ denotes the identity operator on L.

(c) The ranges of T and $I - T$ are disjoint, i.e., $T(u) \perp [v - T(v)]$ holds for all $u, v \in L$.

18. If A and B are projection bands in a Riesz space L, then show that:

(a) The vector subspaces A^{d}, $A \cap B$, and $A + B$ are projection bands and their order projections satisfy:

(i) $P_{A^{\mathrm{d}}} = I - P_A$.

(ii) $P_{A\cap B} = P_A P_B = P_B P_A$.

(iii) $P_{A+B} = P_A + P_B - P_A P_B$.

(b) For all $u, v \in L^+$ we have:

$P_{A\cap B}(u) = P_A(u) \wedge P_B(u)$ and $P_A(u) \wedge P_B(v) = [P_A P_B](u \wedge v)$.

19. Let L and M be two Riesz spaces with M Dedekind complete and let A be an ideal in L. Establish the following.

(a) The mapping $\mathcal{P}\colon \mathcal{L}_b^+(L,M) \to \mathcal{L}_b^+(L,M)$, defined by $\mathcal{P}(T) = T_A$ (=the minimal extension of T to L when T is considered restricted to A), is additive—and so it extends uniquely to a positive operator from $\mathcal{L}_b(L,M)$ to $\mathcal{L}_b(L,M)$.

(b) The positive operator $\mathcal{P}$ is a band projection on $\mathcal{L}_b(L,M)$.

(c) For each $T \in \mathcal{L}_b^+(L,M)$ the positive operator T_A is a component of T in $\mathcal{L}_b(L,M)$.

20. For an order bounded operator $T\colon L \to M$ between two Riesz spaces with M Dedekind complete show that the following statements are equivalent.

(a) T is order continuous.

(b) If $u_\alpha \downarrow 0$ holds in L, then $T(u_\alpha) \xrightarrow{o} 0$ holds in M.

(c) If $u_\alpha \downarrow 0$ holds in L, then $\inf_\alpha |T(u_\alpha)| = 0$ holds in M.

(d) T^+ and T^- are both order continuous.

(e) $|T|$ is order continuous.

21. (Aliprantis [**5**]) Let L be a Riesz space with $L^\sim \neq \{0\}$ and let M be any Riesz space. Consider $\mathcal{L}_b(L,M)$ partially ordered by $S \leq T$ whenever $S(u) \leq T(u)$ for each $u \in L^+$. Pick $0 < \phi \in L^\sim$ and define $u \mapsto T_u$ from M to $\mathcal{L}_b(L,M)$ by $T_u(v) = \phi(v)u$ for all $v \in L$ and $u \in M$.

(i) Show that $T\colon M \to \mathcal{L}_b(L,M)$ is a positive, one-to-one linear mapping.

(ii) If M is Dedekind complete show that $T\colon M \to \mathcal{L}_b(L,M)$ is a normal Riesz isomorphism (into); and thus if $L^\sim \neq \{0\}$ we can consider M as a regular Riesz subspace of $\mathcal{L}_b(L,M)$.

(iii) By using the embedding $T\colon M \to \mathcal{L}_b(L,M)$ show that if $\mathcal{L}_b(L,M)$ is a Dedekind complete Riesz space, then M is also Dedekind complete (the converse of Theorem 1.67).

22. For a Riesz space L and a super Dedekind complete Riesz space M establish the following.

(a) If L has an order unit, then $\mathcal{L}_b(L,M)$ is super Dedekind complete.

(b) If L is Archimedean with a weak order unit, then $\mathcal{L}_n(L,M)$ is super Dedekind complete.

23. Let L be a Dedekind σ-complete Riesz space, M a super Dedekind complete Riesz space and $T \in \mathcal{L}_c(L,M)$.

(i) Show that $C_T = N_T^d$ is super Dedekind complete and that T restricted to C_T is an order continuous operator.

(ii) Show that C_T is a projection band of L.

(iii) Show that $T \in \mathcal{L}_n(L,M)$ if and only if N_T is a band of L.

24. Let $L = C(X_\infty)$, where X_∞ is the one-point compactification of an uncountable set X, considered with the discrete topology. If $\{x_1, x_2, \ldots\}$ is a countable subset of X and $\phi\colon L \to \mathbb{R}$ is defined by

$$\phi(u) = u(\infty) + \sum_{n=1}^{\infty} 2^{-n} u(x_n)\,, \quad u \in L\,,$$

show that $\phi \in L_c^\sim$, that N_ϕ is a band in L, and that $\phi \notin L_n^\sim$. Why does the last claim not contradict part (iii) of the previous exercise?

25. Let $L = C[0,1]$. Show that $L_n^\sim = L_c^\sim = \{0\}$. Determine also $L^\sim$. [*Hint*: For $L^\sim$ use the Riesz representation theorem.]

26. (Luxemburg–Zaanen [**89**, Note VI]; Kaplan [**73**, Note I]) Assume that $L = C(X)$, where X is an arbitrary topological space, and let a sequence $\{u_n\}$ in L satisfy $u_n(x) \downarrow 0$ for every $x \in X$. Show that $\phi(u_n) \downarrow 0$ holds for every $0 \le \phi \in L^\sim$. [*Hint*: If $\phi(u_n) \ge \epsilon > 0$ holds for all n, then note that the vector $w = \sum_{n=1}^\infty \bigl[u_n - \frac{\epsilon u_1}{2\phi(u_1)}\bigr]^+ \in C(X)$ satisfies $\phi(w) = \infty$.]

27. Suppose that L is a Riesz space such that its order continuous dual $L_n^\sim$ separates the points of L. Establish the following.
(a) The Riesz space L is Archimedean.
(b) If $\psi \in (L_n^\sim)^{\rm d}$, then its null ideal N_ψ is order dense in L (or equivalently $C_\psi = \{0\}$).

28. Give an example of a Riesz homomorphism π from a Dedekind complete Riesz space L onto itself, which is not a discrete element in $\mathcal{L}_{\rm b}(L)$ and deduce from this that Theorem 1.85 cannot be generalized. [*Hint*: Consider $\pi\colon \mathbb{R}^2 \to \mathbb{R}^2$ defined by $\pi(u,v) = (u, 2v)$.]

29. For a Riesz space L show that the following statements are equivalent.
(i) L is almost Dedekind σ-complete.
(ii) Whenever $0 \le u_n \downarrow$ holds in L, there exists a sequence $\{v_n\}$ in L^+ satisfying $0 \le v_n \uparrow$, $v_k \le u_n$ for all $n, k \in \mathbb{N}$ and such that $u_n - v_n \downarrow 0$.

30. (Lozanovsky [**83**]) Consider the standard "**Fourier coefficients**" operator $T\colon L_1[0,1] \to c_0$ defined by

$$T(u) = \Bigl(\int_0^1 u(x)\sin x\,dx, \int_0^1 u(x)\sin 2x\,dx, \ldots\Bigr).$$

Establish the following properties.
(a) T is a continuous operator.
(b) T is not order bounded.
(c) T is σ-order continuous but fails to be order continuous.

31. For a Hausdorff compact topological space X establish the following.
(a) Every norm dense Riesz subspace of $C(X)$ is super order dense.
(b) A nonzero mapping $\pi\colon C(X) \to \mathbb{R}$ is a Riesz homomorphism if and only if there exist a real number $\alpha > 0$ and some $\xi \in X$ (both uniquely determined) such that $\pi(u) = \alpha u(\xi)$ for all $u \in C(X)$.

32. Establish that the Dedekind σ-completion of an Archimedean Riesz space is indeed a σ-Dedekind complete Riesz space by proving the following result regarding an Archimedean Riesz space L: *If $\{L_i\}_{i\in I}$ is a family of Riesz subspaces between L and L^δ such that each L_i is Dedekind σ-complete in its own right, then the intersection of the family $\bigcap_{i\in I} L_i$ is a Riesz subspace between L and L^δ that is also a Dedekind σ-complete Riesz space in its own right.*

CHAPTER 2

Locally Solid Topologies

This chapter presents the basic theory of locally solid Riesz spaces. As we shall see, this concept is a natural topological condition relating to the vector ordering. This chapter will illustrate the interplay between the order and the topological structures of Riesz spaces. We begin by characterizing a locally solid topology in terms of the uniform continuity of the lattice operations, and then study topological convergence, bands, ideals, and solid sets. We then focus our attention on locally convex-solid Riesz spaces and draw a number of interesting results about such spaces and their duals. Finally, we investigate the order properties of a Hausdorff locally solid Riesz space that are inherited by its topological completion.

2.1. Linear Topologies on Vector Spaces

In this section we shall discuss briefly (mainly without proofs) the basic properties of the linear topologies needed for this book. For a complete study of the theory of topological vector spaces, we refer the reader to [**55, 64, 75, 78, 118, 122**]. All vector spaces will be assumed to be real vector spaces.

By a **linear topology** τ on a vector space E we mean a topology τ on E that makes the addition and the scalar multiplication continuous, that is, the topology τ makes the two functions

$$\begin{array}{lll}(x,y)\mapsto x+y & \text{from} & E\times E \ \text{ to } \ E\\ (\lambda,x)\mapsto \lambda x & \text{from} & \mathbb{R}\times E \ \text{ to } \ E\end{array}$$

continuous. A **topological vector space** (E,τ) is a vector space E equipped with a linear topology τ.

Every linear topology τ on a vector space E has a base $\mathcal{N}$ for the neighborhoods of zero satisfying the following properties:

(a) Each $V\in\mathcal{N}$ is a **balanced set**, that is, $\lambda u\in V$ holds for all $u\in V$ and every $\lambda\in\mathbb{R}$ with $|\lambda|\le 1$.

(b) Each $V\in\mathcal{N}$ is an **absorbing set**, that is, for every $u\in E$, there exists some $\lambda>0$ such that $\lambda u\in V$.

(c) For each $V\in\mathcal{N}$ there exists some $W\in\mathcal{N}$ with $W+W\subseteq V$.

Conversely, if a nonempty collection $\mathcal{B}$ of subsets of a vector space E is a **filter base**[1] satisfying properties (a), (b), and (c), then there exists a unique linear topology τ on E whose family of neighborhoods at zero have $\mathcal{B}$ as a base.

Property (c) guarantees that normal sequences of neighborhoods are abundant. Since for a linear topology on a vector space and any neighborhood V of zero there

[1] This means that every $V\in\mathcal{B}$ is nonempty and for every pair $V,W\in\mathcal{B}$ there exists some $U\in\mathcal{B}$ with $U\subseteq V\cap W$.

exists [according to property (c)] another neighborhood W of zero with $W+W \subseteq V$, an easy inductive argument guarantees that:

- *For any neighborhood V of zero there exists a normal sequence $\{V_n\}$ of neighborhoods with $V_1 = V$.*

A topological vector space (E, τ) is called **metrizable** if there exists a metric generating the topology τ. The metrizable topological vector spaces are characterized as follows.

THEOREM 2.1. *A Hausdorff topological vector space (E, τ) is metrizable if and only if τ has a countable neighborhood base at zero.*

The locally convex topologies are defined as follows.

DEFINITION 2.2. *A **locally convex topology** on a vector space is a linear topology that has a base at zero consisting of convex sets.*[2]

A **seminorm** ρ on a vector space E is a function $\rho\colon E \to \mathbb{R}$ satisfying the following three properties.

(1) $\rho(u) \geq 0$ for all $u \in E$.
(2) $\rho(u+v) \leq \rho(u) + \rho(v)$ for all $u, v \in E$.
(3) $\rho(\lambda u) = |\lambda|\rho(u)$ for all $\lambda \in \mathbb{R}$ and $u \in E$.

A **norm** on a vector space is a seminorm ρ satisfying $\rho(u) = 0$ if and only if $u = 0$.

If ρ is a seminorm on an arbitrary vector space E, then for any $\epsilon > 0$ the set $\{u \in E\colon\ \rho(u) \leq \epsilon\}$, the **$\epsilon$-closed ball** of ρ at zero, is a convex subset of E.

Every family $\{\rho_\alpha\}_{\alpha \in A}$ of seminorms on a vector space E defines a locally convex topology on E by taking as a base for the neighborhoods at zero the finite intersections of the closed balls of the seminorms ρ_α. In other words, a subset V of E is a neighborhood of zero for this topology if there exist some $\epsilon > 0$ and a finite number of indices $\alpha_1, \ldots, \alpha_n$ such that

$$\bigl\{u \in E\colon\ \rho_{\alpha_m}(u) \leq \epsilon \text{ for each } m = 1, 2, \ldots, n\bigr\} \subseteq V\,.$$

This linear topology is called the **locally convex topology generated by the family of seminorms** $\{\rho_\alpha\}_{\alpha \in A}$.

It turns out that every locally convex topology is generated by a family of seminorms.

THEOREM 2.3. *A linear topology τ on a vector space E is locally convex if and only if there is a family $\{\rho_\alpha\}_{\alpha \in A}$ of seminorms on E that generates τ.*

A subset A in a topological vector space (E, τ) is called **topologically bounded** (or simply **τ-bounded**) if for every τ-neighborhood V of zero there exists some $\lambda > 0$ such that $\lambda A \subseteq V$. Note that a subset A in a locally convex space (E, τ) whose topology τ is generated by a family of seminorms $\{\rho_\alpha\}_{\alpha \in A}$ is τ-bounded if and only if $\rho_\alpha(A)$ is a bounded subset of $\mathbb{R}$ for each α.

The **topological dual** E' of a topological vector space (E, τ) is the collection of all τ-continuous linear functionals on E. With the usual algebraic operations, E' is a vector space. Actually, E' is a vector subspace of the **algebraic dual** E^*, the vector space of all linear functionals on E.

The **weak topology** $\sigma(E, E')$ of a locally convex space (E, τ) is the locally convex topology generated on the vector space E by the family of seminorms

---

[2] As usual, a subset A of E is called **convex** if $\lambda u + (1-\lambda)v \in A$ holds for all $u, v \in A$ and all $0 \leq \lambda \leq 1$.

$\{\rho_f\colon\ f \in E'\}$, where $\rho_f(u) = |f(u)|$ for $u \in E$. The weak topology $\sigma(E, E')$ is the smallest locally convex topology on E for which every $f \in E'$ is continuous. The topological dual of $\big(E, \sigma(E, E')\big)$ is precisely E'. The weak topology $\sigma(E, E')$ is also called the **topology of pointwise convergence** on E since a net $\{u_\alpha\}$ in E satisfies $u_\alpha \xrightarrow{\sigma(E,E')} u$ if and only if $f(u_\alpha) \to f(u)$ holds in $\mathbb{R}$ for each $f \in E'$.

Similarly, the weak topology $\sigma(E', E)$ is the locally convex topology on the vector space E' generated by the family of seminorms $\{\rho_u\colon\ u \in E\}$; where we let $\rho_u(f) = \big|f(u)\big|$ for $f \in E'$. The topology $\sigma(E', E)$ is also called the **topology of pointwise convergence** on E' since a net $\{f_\alpha\}$ in E' satisfies $f_\alpha \xrightarrow{\sigma(E',E)} f$ in E' if and only if $f_\alpha(u) \to f(u)$ holds in $\mathbb{R}$ for each $u \in E$.

The **strong topology** $\beta(E', E)$ on E' is the locally convex topology generated by the family of seminorms $\{\rho_A\colon\ A \in \mathcal{B}\}$, where

$$\rho_A(f) = \sup\big\{|f(u)|\colon\ u \in A\big\}$$

and $\mathcal{B}$ is the collection of all $\sigma(E, E')$-bounded subsets of E. The strong topology $\beta(E', E)$ is the topology of uniform convergence on the $\sigma(E, E')$-bounded subsets of E. The topology $\beta(E, E')$ is defined similarly. The **Mackey topology** $\tau(E, E')$ on E is the locally convex topology of uniform convergence on the balanced, convex and $\sigma(E', E)$-compact subsets of E'. If (E, τ) is Hausdorff, then we have $\tau \subseteq \tau(E, E')$. For a locally convex topology τ^* on E, in order for the topological dual of (E, τ^*) to coincide with E' it is necessary and sufficient for τ^* to satisfy

$$\sigma(E, E') \subseteq \tau^* \subseteq \tau(E, E')\,.$$

The Mackey topology $\tau(E', E)$ on E' is defined similarly.

THEOREM 2.4. *For a Hausdorff locally convex space (E, τ) and a subset A of E we have the following.*

(i) *The set A is τ-bounded if and only if A is $\sigma(E, E')$-bounded.*

(ii) *If in addition A is a convex set, then A is τ-closed if and only if A is $\sigma(E, E')$-closed.*

A net $\{u_\alpha\}$ in a topological vector space (E, τ) is called a τ-**Cauchy net** if for every τ-neighborhood V of zero there exists an index α_0 such that $u_\alpha - u_\beta \in V$ for all $\alpha, \beta \succeq \alpha_0$. The following characterization of Cauchy nets will be very useful in our discussion in this monograph.

LEMMA 2.5. *A net $\{x_\alpha\}$ in a topological vector space (E, τ) is a τ-Cauchy net if and only if it satisfies the condition:*

- For each index α and each τ-neighborhood V of zero there exists another index $\beta \succeq \alpha$ such that for all $\gamma \succeq \beta$ we have $x_\gamma - x_\beta \in V$.

In particular, if a net $\{u_\alpha\}$ in a topological vector space (E, τ) is not a τ-Cauchy net, then there exist a τ-neighborhood U of zero and an increasing sequence $\{\alpha_n\}$ of indices such that $u_{\alpha_{n+1}} - u_{\alpha_n} \notin U$ holds for each n.

PROOF. Let $\{x_\alpha\}$ be a net in a topological vector space (E, τ). If $\{x_\alpha\}$ is a τ-Cauchy net, then it should be clear that it satisfies the stated condition. For the converse, assume that $\{x_\alpha\}$ satisfies the condition and let V be a τ-neighborhood of zero. Choose a balanced τ-neighborhood W of zero such that $W + W \subseteq V$. Fix some index α and then choose another index $\beta_0 \succeq \alpha$ such that $\beta \succeq \beta_0$ implies $x_\beta - x_{\beta_0} \in W$. So if $\beta, \gamma \succeq \beta_0$, then

$$x_\beta - x_\gamma = (x_\beta - x_{\beta_0}) + (x_{\beta_0} - x_\gamma) \in W + W \subseteq V\,.$$

This shows that $\{x_\alpha\}$ is a τ-Cauchy net. ■

A subset A of (E,τ) is called **τ-complete** if every τ-Cauchy net of the set A is τ-convergent to some element in A. In particular, (E,τ) is called **topologically complete** (or briefly **τ-complete**) if each τ-Cauchy net of E is τ-convergent in E.

A function $f\colon A \to (E_1,\tau_1)$, from a subset A of a topological vector space (E,τ) to another topological vector space (E_1,τ_1), is called **uniformly continuous** if for every τ_1-neighborhood V of zero of E_1 there exists a τ-neighborhood W of zero of E such that $f(u)-f(v)\in V$ whenever $u,v\in A$ satisfy $u-v\in W$. Concerning uniformly continuous functions, we have the following useful extension theorem.

THEOREM 2.6. *Let $f\colon A\subseteq(E,\tau)\to(E_1,\tau_1)$ be a uniformly continuous function. If (E_1,τ_1) is a Hausdorff τ_1-complete topological vector space, then there exists a unique uniformly continuous extension of f to the τ-closure $\overline{A}$ of A.*

PROOF. Let $u\in\overline{A}$. Pick a net $\{u_\alpha\}\subseteq A$ such that $u_\alpha \xrightarrow{\tau} u$. From the uniform continuity of f it follows that the net $\{f(u_\alpha)\}$ is τ_1-Cauchy. Therefore, there exists some $v\in E_1$ such that $f(u_\alpha)\xrightarrow{\tau_1} v$. Now assume that another net $\{x_\lambda\}\subseteq A$ satisfies $x_\lambda\xrightarrow{\tau} u$. As above, let $f(x_\lambda)\xrightarrow{\tau_1} x$ in E_1. Using the uniform continuity of f, we get $x=v$. That is, for each $u\in\overline{A}$ there exists a unique $\overline{f}(u)\in E_1$ such that whenever a net $\{u_\alpha\}\subseteq A$ satisfies $u_\alpha\xrightarrow{\tau} u$, then $f(u_\alpha)\xrightarrow{\tau_1}\overline{f}(u)$.

Clearly, the function $\overline{f}\colon\overline{A}\to E_1$ is an extension of f. We claim that $\overline{f}$ is uniformly continuous. To see this, let V be a τ_1-closed neighborhood of zero in E_1 and then select a τ-open neighborhood W of zero in E such that $u,v\in W$ with $u-v\in W$ imply $f(u)-f(v)\in V$. Now let $x,y\in\overline{A}$ satisfy $x-y\in W$. Pick two nets $\{x_\alpha\}$ and $\{y_\lambda\}$ of A with $x_\alpha\xrightarrow{\tau} x$ and $y_\lambda\xrightarrow{\tau} y$. Clearly, $x_\alpha-y_\lambda\xrightarrow[(\alpha,\lambda)]{\tau} x-y\in W$ in E and therefore there exist indices α_0 and λ_0 such that $\alpha\succeq\alpha_0$ and $\lambda\succeq\lambda_0$ imply $x_\alpha-y_\lambda\in W$. It follows that $f(x_\alpha)-f(y_\lambda)\in V$ for all $\alpha\succeq\alpha_0$ and $\lambda\succeq\lambda_0$. Since $f(x_\alpha)-f(y_\lambda)\xrightarrow[(\alpha,\lambda)]{\tau_1}\overline{f}(x)-\overline{f}(y)$ holds in E_1 and V is τ_1-closed, we easily obtain that $\overline{f}(x)-\overline{f}(y)\in V$. This shows that $\overline{f}$ is a uniformly continuous function—and the only continuous extension of f to all of $\overline{A}$. ■

A subset A of a topological vector space (E,τ) is called **totally bounded** if for each τ-neighborhood V of zero there exists a finite subset $\{x_1,\dots,x_n\}$ of A such that $A\subseteq\bigcup_{i=1}^n(x_i+V)$. It should be clear that subsets of totally bounded sets are also totally bounded. The following result dealing with totally bounded sets is well-known and it will be needed; for a proof see, for instance [**118**, p. 60].

THEOREM 2.7. *Let (E,τ) be a Hausdorff topological vector space and let A be a subset of E. Then A is τ-compact if and only if A is totally bounded and τ-complete.*

Two topological vector spaces (E,τ) and (E_1,τ_1) are called **isomorphic** if there exists a one-to-one and onto operator $T\colon E\to E_1$ that is at the same time a homeomorphism, in other words, if there exists a one-to-one mapping from E onto E_1 preserving both structures, the algebraic and the topological.

For the next theorem dealing with the topological completion of a Hausdorff topological vector space see in particular [**78**, p. 33].

THEOREM 2.8. *For every Hausdorff topological vector space* (E, τ) *there exists a unique (up to isomorphism) Hausdorff topological vector space* $(\widehat{E}, \widehat{\tau})$ *having the following properties.*

1. *The topological vector space* $(\widehat{E}, \widehat{\tau})$ *is* $\widehat{\tau}$*-complete.*
2. *There exists a vector subspace* E_1 *of* $\widehat{E}$ *that is algebraically isomorphic to* E *(thus identifying* E *with* E_1 *we consider* E *as a vector subspace of* $\widehat{E}$*).*
3. *The topology* $\widehat{\tau}$ *induces* τ *on* E.
4. *The vector subspace* E *is* $\widehat{\tau}$*-dense in* $\widehat{E}$.

In particular, whenever $\mathcal{N}$ *is a base of zero for* τ*, then the collection* $\{\overline{V}\colon V \in \mathcal{N}\}$ *of the* $\widehat{\tau}$*-closures in* $\widehat{E}$ *of the neighborhoods* $V \in \mathcal{N}$ *is a base at zero for* $\widehat{\tau}$.

The unique Hausdorff topological vector space $(\widehat{E}, \widehat{\tau})$ has a special name.

DEFINITION 2.9. *If* (E, τ) *is a Hausdorff topological vector space, then the unique (up to isomorphism) Hausdorff topological vector space* $(\widehat{E}, \widehat{\tau})$ *described in Theorem* 2.8 *is called the* ***topological completion*** *of* (E, τ).

Recall that for every mapping f from a topological space (X, τ) to a set Y, the finest topology τ_f on Y for which the function f is continuous is called the **quotient topology** determined by f. A moment's thought reveals that we have $\tau_f = \{A \subseteq Y\colon f^{-1}(A) \in \tau\}$.

If now (E, τ) is a topological vector space, F is a vector space, and $f\colon E \to F$ is an onto linear mapping, then the quotient topology τ_f on F is a linear topology. The details are included in the next theorem.

THEOREM 2.10. *Let* $f\colon (E, \tau) \to F$ *be an operator from a topological vector space* (E, τ) *onto a vector space* F*. Then the following statements hold:*

(i) *The quotient topology* τ_f *on* F *is a linear topology and the operator* $f\colon (E, \tau) \to (F, \tau_f)$ *is an open mapping.*
(ii) *For every base* $\mathcal{N}$ *at zero in* E*, the collection* $\{f(V)\colon V \in \mathcal{N}\}$ *is a base for the* τ_f*-neighborhoods at zero in* F.

As a special case of the preceding theorem consider a topological vector space (E, τ) and a vector subspace A of E. Then the canonical projection $Q\colon E \to E/A$ defined by $Q(u) = \dot{u} = u + A$ for $u \in E$, is an onto operator, and hence its quotient topology τ_Q (which we shall denote by τ/A) is a linear topology on the quotient space E/A. We summarize this discussion in the next theorem.

THEOREM 2.11. *For every vector subspace* A *of a topological vector space* (E, τ) *the quotient topology* τ/A *on* E/A *satisfies the following properties.*

(a) *The canonical projection* $Q\colon (E, \tau) \to (E/A, \tau/A)$ *is continuous and open.*
(b) *For every base* $\mathcal{N}$ *for the* τ*-neighborhoods at zero of* E*, the collection* $\{Q(V)\colon V \in \mathcal{N}\}$ *is a base for the* τ/A*-neighborhoods at zero in* E/A.
(c) $(E/A, \tau/A)$ *is a Hausdorff space if and only if* A *is* τ*-closed.*

Finally, we shall close the section by mentioning a few more basic results that will be needed for our discussion. But first, let us take another look at the weak topologies. For a nonempty set X the real vector space $\mathbb{R}^X$ equipped with its product topology becomes a locally convex vector space (actually a Hausdorff locally convex-solid Riesz space). The product topology on $\mathbb{R}^X$ is called the **topology of pointwise convergence**, and is generated by the family of Riesz seminorms

$\{\rho_x\colon\ x \in X\}$, where for each $f \in \mathbb{R}^X$ we have

$$\rho_x(f) = |f(x)|\,.$$

For any nonempty set A of real-valued functions defined on X (i.e., $A \subseteq \mathbb{R}^X$), the **weak topology** $\sigma(A, X)$ on A (also called the **topology of pointwise convergence on** A) is the Hausdorff topology induced on A by the product topology of $\mathbb{R}^X$. Note that indeed a net $\{f_\alpha\} \subseteq A$ satisfies $f_\alpha \xrightarrow{\sigma(A,X)} f$ in A if and only if $f_\alpha(x) \to f(x)$ holds in $\mathbb{R}$ for each $x \in X$. If A is a vector subspace of $\mathbb{R}^X$, then $\sigma(A, X)$ is a Hausdorff locally convex topology on A.

If E is a vector space, then its algebraic dual E^* (the vector space of all linear functionals on E) is clearly a vector subspace of $\mathbb{R}^E$. Obviously, E^* is a closed vector subspace of $\mathbb{R}^E$. Moreover, for every linear topology τ on E we have the following vector subspace inclusions: $E' \subseteq E^* \subseteq \mathbb{R}^E$. This implies that the weak topology $\sigma(E', E)$ is nothing else than the restriction of the product topology of $\mathbb{R}^E$ to E'. Similarly, considering E as a vector subspace of $\mathbb{R}^{E'}$, we see that the weak topology $\sigma(E, E')$ on E is the restriction of the product topology of $\mathbb{R}^{E'}$ to E.

For a subset A of a topological vector space (E, τ) the **polar** of A is defined by

$$A^{\circ} = \{\phi \in E'\colon\ |\phi(u)| \le 1 \text{ for all } u \in A\}\,.$$

It should be obvious that A° is a $\sigma(E', E)$-closed set. If V is also a τ-neighborhood of zero, then V° is pointwise closed in $\mathbb{R}^E$. Indeed, if $\{\phi_\alpha\} \subseteq V^{\circ}$ satisfies $\phi_\alpha \to \phi$ in $\mathbb{R}^E$, then $\phi \in E^*$ and $|\phi(u)| \le 1$ holds for every $u \in V$ and hence $\phi \in E'$, since ϕ is bounded on V. On the other hand, as it is readily seen, V° is topologically bounded in $\mathbb{R}^E$ and thus V° is a compact subset of $\mathbb{R}^E$ [and hence of E' for $\sigma(E', E)$] by Tychonoff's classical compactness theorem. We have, thus, proven the famous Alaoglu–Bourbaki theorem.

THEOREM 2.12 (Alaoglu–Bourbaki). *For every τ-neighborhood V of zero of an arbitrary topological vector space (E, τ), its polar V° is a $\sigma(E', E)$-compact subset of E'.*

The polar of a subset B in the topological dual E' of a topological vector space (E, τ) is defined by

$$B^{\circ} = \{u \in E\colon\ |\phi(u)| \le 1 \text{ for all } \phi \in B\}\,.$$

Clearly, B° is $\sigma(E, E')$-closed. The **bipolar** of a subset A of E is the set $(A^{\circ})^{\circ}$, the polar of A°, and is denoted by $A^{\circ\circ}$.

The following basic theorems can be found in most books on topological vector spaces. See, for instance [**8, 55, 64, 78, 118, 122**].

THEOREM 2.13 (The Bipolar Theorem). *For each subset A of a topological vector space (E, τ) the bipolar $A^{\circ\circ}$ is the balanced, convex, $\sigma(E, E')$-closed hull of A. In particular, if (E, τ) is a Hausdorff locally convex space and A is a balanced, convex, and τ-closed subset of E, then $A = A^{\circ\circ}$.*

THEOREM 2.14 (Banach–Steinhaus). *Let E be an arbitrary Banach space. Then a subset A of E' is norm bounded if and only if A is $\sigma(E', E)$-bounded.*[3]

Recall that a subset A of a topological space (X, τ) is said to be **relatively τ-compact** if the τ-closure of A is τ-compact.

---

[3] This theorem is also known as the **Principle of Uniform Boundedness**.

THEOREM 2.15 (Eberlein–Šmulian). *A subset A of a Banach space E is relatively $\sigma(E, E')$-compact if and only if every sequence of A has a subsequence that is $\sigma(E, E')$-convergent to an element of E.*

2.2. The Basic Properties of Locally Solid Topologies

We start with the definition of locally solid Riesz spaces. They were introduced for the first time by G. T. Roberts in [**117**] with a different name. (Keep in mind that a subset S of a Riesz space L is said to be solid if $|u| \leq |v|$ and $v \in S$ imply $u \in S$.)

DEFINITION 2.16 (Roberts). *A (not necessarily Hausdorff) linear topology τ on a Riesz space L is said to be* ***locally solid*** *if τ has a base at zero consisting of solid sets.*

A ***locally solid Riesz space*** *(L, τ) is a Riesz space L equipped with a locally solid topology τ.*

The next result, due to G. T. Roberts [**117**] and I. Namioka [**104**, Theorem 8.1, p. 40], characterizes the linear topologies that are locally solid.

THEOREM 2.17 (Roberts–Namioka). *For a linear topology τ on a Riesz space L the following statements are equivalent.*

1. *(L, τ) is a locally solid Riesz space.*
2. *The mapping $(u, v) \mapsto u \vee v$, from $(L, \tau) \times (L, \tau)$ to (L, τ), is uniformly continuous.*
3. *The mapping $(u, v) \mapsto u \wedge v$, from $(L, \tau) \times (L, \tau)$ to (L, τ), is uniformly continuous.*
4. *The mapping $u \mapsto |u|$, from (L, τ) to (L, τ), is uniformly continuous.*
5. *The mapping $u \mapsto u^-$, from (L, τ) to (L, τ), is uniformly continuous.*
6. *The mapping $u \mapsto u^+$, from (L, τ) to (L, τ), is uniformly continuous.*

PROOF. (1) $\Longrightarrow$ (2) Use the inequality

$$\big|u \vee v - f \vee g\big| \leq |u - f| + |v - g| .$$

(2) $\Longrightarrow$ (3) It follows from the identity $u \wedge v = -\big[(-u) \vee (-v)\big]$.

(3) $\Longrightarrow$ (4) Use the identity $|u| = -\big[u \wedge (-u)\big]$.

(4) $\Longrightarrow$ (5) It follows from the identity $u^- = \frac{1}{2}\big(|u| - u\big)$.

(5) $\Longrightarrow$ (6) It follows from the identity $u^+ = (-u)^-$.

(6) $\Longrightarrow$ (1) Let U be an arbitrary τ-neighborhood of zero. Choose a balanced τ-neighborhood V of zero such that $V + V \subseteq U$. By the uniform continuity of the function $u \mapsto u^+$, there exists a balanced τ-neighborhood W of zero such that $u - v \in W$ implies $u^+ - v^+ \in V$. In particular, note that if $u \in W$, then $u^+, u^- \in V$.

Next pick a balanced τ-neighborhood W_1 of zero with $W_1 + W_1 \subseteq W$, and then pick another balanced τ-neighborhood W_2 of zero such that $u - v \in W_2$ guarantees that $u^+ - v^+ \in W_1$. We claim that the solid hull of W_2 satisfies $\mathrm{Sol}\,(W_2) \subseteq U$. Indeed, if $|u| \leq |v|$ holds with $v \in W_2$, then $v^+, v^- \in W_1$ and so $|v| = v^+ + v^- \in W$. Therefore, from $|v| = u^+ - \big(u^+ - |v|\big) \in W$, we get $u^+ = (u^+)^+ - \big(u^+ - |v|\big)^+ \in V$. Similarly, $u^- \in V$ and thus $u = u^+ - u^- \in V + V \subseteq U$. Consequently, τ is a locally solid topology, and the proof is finished. ■

It may be asked at this point whether the continuity of $u \mapsto u^+$ at zero is enough to ensure that the linear topology τ is locally solid. The following example shows that the answer is negative.

EXAMPLE 2.18. Let L be the Riesz space consisting of all continuous piecewise linear functions on $[0,1]$, ordered with the pointwise ordering. Define a norm ρ on L by

$$\rho(u) = |u(0)| + \int_0^1 |u'(x)|\,dx\,,$$

where u' is the derivative of u (note that u' exists for all but a finite number of points and that u' is a step function). It is left to the reader to verify that ρ is a norm on L satisfying $\rho(u) = \rho(|u|)$ for all $u \in L$. Note also that $0 \le u \le v$ in L does not imply $\rho(u) \le \rho(v)$, that is, ρ is not a Riesz norm.[4]

Let τ be the Hausdorff linear topology generated by ρ. Since for each $u \in L$ we have $|u'(x)| = \big|[u^+(x)]'\big| + \big|[u^-(x)]'\big|$ for all but a finite number of points $x \in [0,1]$, it follows that $\lim \rho(u_n) = 0$ implies $\lim \rho(u_n^+) = 0$, that is, the function $u \mapsto u^+$ from (L,τ) to (L,τ) is continuous at zero. However, it is easy to see that τ is not a locally solid topology. ■

We continue with a number of properties of locally solid Riesz spaces.

THEOREM 2.19. *If (L,τ) is a locally solid Riesz space, then:*

(i) *The order bounded subsets of L are τ-bounded.*
(ii) *The τ-closure of any solid subset of L is also a solid subset.*
(iii) *The τ-closure of a Riesz subspace of L is a Riesz subspace (and hence the τ-closure of an ideal is likewise an ideal).*

PROOF. (i) Let $[u,v]$ be an order interval of the Riesz space L and let V be a solid τ-neighborhood of zero. Pick $\lambda > 0$ such that $\lambda(|u|+|v|) \in V$. Now since $u \le w \le v$ implies $|w| \le |u| + |v|$ and V is a solid set, we easily get $\lambda[u,v] \subseteq V$, so that order interval $[u,v]$ is τ-bounded.

(ii) Assume that S is a solid subset of L and $|u| \le |v|$ with $v \in \overline{S}$. Pick a net $\{v_\alpha\} \subseteq S$ with $v_\alpha \xrightarrow{\tau} v$ and define $u_\alpha = \big(u \wedge |v_\alpha|\big) \vee \big(-|v_\alpha|\big)$ for each α. Clearly, $|u_\alpha| \le |v_\alpha|$ holds for all α. Thus $\{u_\alpha\} \subseteq S$ and by Theorem 2.17 we have $u_\alpha \xrightarrow{\tau} u$. Hence $u \in \overline{S}$ and so $\overline{S}$ is solid.

(iii) Let K be a Riesz subspace of L. Clearly, $\overline{K}$ is a vector subspace of L. Now if $u \in \overline{K}$, then there exists a net $\{u_\alpha\}$ of K with $u_\alpha \xrightarrow{\tau} u$. But then, by Theorem 2.17, the net $\{u_\alpha^+\} \subseteq K$ satisfies $u_\alpha^+ \xrightarrow{\tau} u^+$, and so $u^+ \in \overline{K}$. This establishes that $\overline{K}$ is a Riesz subspace of L. The parenthetical part easily follows from part (ii). ■

Assume now that $\{(L_\alpha,\tau_\alpha)\colon\ \alpha \in A\}$ is a family of locally solid Riesz spaces. Let $L = \Pi L_\alpha$ be the Cartesian product of the family of Riesz spaces $\{L_\alpha\colon\ \alpha \in A\}$. Then L ordered componentwise is a Riesz space, and if it is equipped with the product topology $\tau = \Pi\tau_\alpha$, it is easy to see that it becomes a locally solid Riesz space. Hence we have the following result.

THEOREM 2.20. *Let $(L_\alpha,\tau_\alpha)_{\alpha\in A}$ be a family of locally solid Riesz spaces and let $L = \Pi L_\alpha$ and $\tau = \Pi\tau_\alpha$. Then (L,τ) is a locally solid Riesz space.*

---

[4] As we shall see later in Section 2.3, a norm ρ on a Riesz space L will be called a **Riesz norm** (or a **lattice norm**) if $|u| \le |v|$ implies $\rho(u) \le \rho(v)$.

More properties for Hausdorff locally solid Riesz spaces are included in the next theorem.

THEOREM 2.21. *For a Hausdorff locally solid Riesz space* (L, τ) *we have the following:*

(a) *The Riesz space L is Archimedean.*
(b) *The cone L^+ of L is τ-closed.*
(c) *If $u_\alpha \xrightarrow{\tau} u$ and $u_\alpha \uparrow$ in L, then $u_\alpha \uparrow u$ holds in L. Similarly, $u_\alpha \xrightarrow{\tau} u$ and $u_\alpha \downarrow$ imply $u_\alpha \downarrow u$.*
(d) *Every band of L is τ-closed.*
(e) *If two arbitrary nets $\{u_\alpha\}$ and $\{v_\alpha\}$ (with the same index set) satisfy $u_\alpha \downarrow \leq v_\alpha \downarrow$ and $v_\alpha - u_\alpha \xrightarrow{\tau} 0$, then $u_\alpha \downarrow u$ if and only if $v_\alpha \downarrow u$.*

PROOF. (a) Assume $0 \leq nu \leq v$ holds for all n and some $u, v \in L^+$. Then $0 \leq u \leq \frac{1}{n} v \xrightarrow{\tau} 0$, and hence $u = 0$, since τ is a Hausdorff locally solid topology. Hence L is Archimedean.

(b) By Theorem 2.17 we know that the function $u \mapsto u^-$, from (L, τ) to (L, τ), is continuous. The desired conclusion now follows immediately from the identity $L^+ = \{u \in L\colon\ u^- = 0\}$.

(c) Fix an index α. Then for $\beta \succeq \alpha$ we have

$$0 \leq u \vee u_\alpha - u \leq u \vee u_\beta - u \leq |u_\beta - u| \xrightarrow[\beta \succeq \alpha]{\tau} 0$$

and hence $u \vee u_\alpha - u = 0$, that is, $u_\alpha \leq u$ holds for all α. Now let $u_\alpha \leq v$ for all α and some $v \in L$. Then $0 \leq v - u_\alpha \xrightarrow{\tau} v - u$, and so (since L^+ is τ-closed) we get $v - u \in L^+$, i.e., $u \leq v$. This shows that $u_\alpha \uparrow u$ in L. For the other case apply the preceding arguments to the net $\{-u_\alpha\}$.

(d) Observe that for every nonempty subset D of L its disjoint complement $D^{\mathrm{d}} = \{u \in L\colon\ |u| \wedge |v| = 0 \text{ for all } v \in D\}$ is τ-closed. Indeed, if $u \in \overline{D^{\mathrm{d}}}$ then $u_\alpha \xrightarrow{\tau} u$ holds for some net $\{u_\alpha\} \subseteq D^{\mathrm{d}}$, and hence $0 = |v| \wedge |u_\alpha| \xrightarrow{\tau} |v| \wedge |u|$ holds for every $v \in D$. Since τ is a Hausdorff topology, $|v| \wedge |u| = 0$ for all $v \in D$, that is, $u \in D^{\mathrm{d}}$ and therefore D^{d} is τ-closed.

To obtain the desired conclusion, observe that since L is Archimedean every band B satisfies $B = B^{\mathrm{dd}}$ (see Theorem 1.28) and thus B is τ-closed.

(e) If $u_\alpha \downarrow u$ and $v_\alpha \geq v \geq u$ for all α, then $v_\alpha - u_\alpha \geq (v - u_\alpha)^+ \geq 0$ for all α and hence $(v - u_\alpha)^+ \xrightarrow{\tau} 0$. But $(v - u_\alpha)^+ \uparrow (v - u)^+ = v - u$ holds and hence by part (c) we have $v - u = 0$, i.e., $u = v$. Thus $v_\alpha \downarrow u$.

Now if $v_\alpha \downarrow u$ holds in L, then the inequalities $0 \leq u - u \wedge u_\alpha \leq v_\alpha - u_\alpha$ implies that $u - u \wedge u_\alpha \xrightarrow{\tau} 0$. Thus, since $u - u \wedge u_\alpha \uparrow$, we get by part (c) that $0 \leq u - u \wedge u_\alpha \uparrow 0$, and therefore $u - u \wedge u_\alpha = 0$ for all α, or $u \leq u_\alpha$ for each α. From this it is easy to see now that $u_\alpha \downarrow u$ holds in L. ■

H. Nakano has shown in [**100**, Theorem 31.1, p. 134] and [**103**, Theorem 31.1, p. 134] that the norm dual of normed Riesz space is an ideal in its order dual. This was generalized for locally solid Riesz spaces by G. T. Roberts [**117**, Theorem 2, p. 536]. (Keep in mind that the **topological dual** E' of a topological vector space (E, τ) is the vector space consisting of all τ-continuous linear functionals on E.)

THEOREM 2.22 (Nakano–Roberts). *Let (L, τ) be a locally solid Riesz space. Then the topological dual L' of (L, τ) is an ideal of the order dual $L^\sim$ (and hence it is a Dedekind complete Riesz space in its own right).*

PROOF. Observe first that L' is a vector subspace of $L^\sim$. To see this, assume by way of contradiction that for some $\phi \in L'$ we have $\phi \notin L^\sim$. Then there exists some $u \in L^+$ and a sequence $\{u_n\} \subseteq [0,u]$ satisfying $|\phi(u_n)| \geq n$ for each n. It now follows from the relations $0 \leq \frac{1}{n}u_n \leq \frac{1}{n}u \xrightarrow{\mathcal{T}} 0$ that $\frac{1}{n}u_n \xrightarrow{\mathcal{T}} 0$ and so (since $\phi \in L'$) we get $\lim \phi(\frac{1}{n}u_n) = 0$, contrary to $|\phi(\frac{1}{n}u_n)| \geq 1$ for each n.

To see that L' is an ideal of $L^\sim$, assume $|\phi| \leq |\psi|$ with $\psi \in L'$ and $\phi \in L^\sim$. Let $u_\alpha \xrightarrow{\mathcal{T}} 0$ and $\epsilon > 0$. By Theorem 1.67 there exists a net $\{v_\alpha\}$ satisfying $|v_\alpha| \leq |u_\alpha|$ and $|\psi|(|u_\alpha|) \leq |\psi(v_\alpha)| + \epsilon$ for all α. Clearly, $v_\alpha \xrightarrow{\mathcal{T}} 0$, and since

$$|\phi(u_\alpha)| \leq |\phi|(|u_\alpha|) \leq |\psi|(|u_\alpha|) \leq |\psi(v_\alpha)| + \epsilon,$$

we get $0 \leq \limsup |\phi(u_\alpha)| \leq \epsilon$ for all $\epsilon > 0$. Thus, $\limsup |\phi(u_\alpha)| = 0$, and so $\lim \phi(u_\alpha) = 0$. Hence $\phi \in L'$, and the proof is finished. ■

The following example shows that L' in general is not a band of $L^\sim$ and need not be even a σ-ideal of $L^\sim$.

EXAMPLE 2.23. Let L denote the Riesz space consisting of all real sequences that are eventually zero, with the pointwise ordering. Now consider the linear functional $\phi\colon L \to \mathbb{R}$ defined for each $u \in L$ by

$$\phi(u) = \sum_{n=1}^{\infty} nu(n).$$

Clearly, ϕ is a positive linear functional and so $\phi \in L^\sim$. Now let τ be the Hausdorff locally solid topology generated by the (Riesz) seminorm

$$\rho(u) = \sup\{|u(n)|\colon\ n = 1, 2, 3, \dots\}.$$

Next consider the sequence $\{u_n\}$ in L defined by $u_n(k) = 0$ whenever $k \neq n$ and $u_n(n) = n^{-\frac{1}{2}}$, and note that $\rho(u_n) \to 0$. But $\phi(u_n) = n^{\frac{1}{2}} \to \infty$ and so $\phi \notin L'$. On the other hand, if we consider the sequence $\{\phi_n\} \subseteq L^\sim$, defined by $\phi_n(u) = \sum_{k=1}^{n} ku(k)$ for each $u \in L$, then it follows from $|\phi_n(u)| \leq (\sum_{k=1}^{n} k)\rho(u)$ that $\{\phi_n\} \subseteq L'$. Moreover, we have $0 \leq \phi_n \uparrow \phi$ in $L^\sim$. This shows that L' is not a σ-ideal of $L^\sim$. ■

The next theorem informs us that quotients of locally solid Riesz spaces equipped with their quotient topologies are locally solid Riesz spaces.

THEOREM 2.24. *Let (L,τ) be a locally solid Riesz space and let Q be a Riesz homomorphism from L onto a Riesz space M. Then M equipped with the quotient topology τ_Q is a locally solid Riesz space. In particular, if A is an ideal of L, then $(L/A, \tau/A)$ is a locally solid Riesz space.*

PROOF. Let $\mathcal{N}$ be a base at zero for τ consisting of solid sets. By Theorem 2.10 we know that $\{Q(V)\colon\ V \in \mathcal{N}\}$ is a base at zero for τ_Q. But by Theorem 1.33 each $Q(V)$ is a solid set, and hence the linear topology τ_Q is locally solid. The last part follows by observing that the canonical projection of L onto L/A is a Riesz homomorphism. (In this case we denote the quotient topology by τ/A.) ■

2.3. Locally Convex-solid Topologies

A linear topology τ on a Riesz space L that is at the same time locally solid and locally convex will be called a **locally convex-solid topology**. A **locally convex-solid Riesz space** (L, τ) is a Riesz space L equipped with a locally convex-solid topology τ. According to Theorems 1.11 and 2.19(ii) every locally convex-solid Riesz space (L, τ) has a base at zero consisting of absorbing, τ-closed, convex, and solid sets.

A seminorm ρ on a Riesz space L is said to be a **Riesz** (or a **lattice**) **seminorm** if $|u| \le |v|$ in L implies $\rho(u) \le \rho(v)$, or equivalently, whenever $\rho(u) = \rho(|u|)$ holds for all $u \in L$ and $0 \le u \le v$ in L implies $\rho(u) \le \rho(v)$. A **Riesz** (or a **lattice**) **norm** is a Riesz seminorm that is also a norm. A **seminormed** (resp. **normed**) **Riesz space** L_ρ is a Riesz space L equipped with a Riesz seminorm (resp. norm) ρ. A complete normed Riesz space is called a **Banach lattice**.

Observe that a seminorm ρ on a Riesz space L is a Riesz seminorm if and only if its open unit ball $B_\rho = \{u \in L\colon\ \rho(u) < 1\}$ is a solid subset of L. Indeed, note first that if ρ is a Riesz seminorm, then clearly B_ρ is a solid subset of L. On the other hand, if B_ρ is a solid set and $|u| \le |v|$ holds in L, then $\big|[\rho(v)+\epsilon]^{-1}u\big| \le \big|[\rho(v)+\epsilon]^{-1}v\big|$ holds for all $\epsilon > 0$, and since $[\rho(v) + \epsilon]^{-1}v \in B_\rho$ we get $[\rho(v) + \epsilon]^{-1}u \in B_\rho$. Thus, $\rho(u) < \rho(v) + \epsilon$ holds for all $\epsilon > 0$, and therefore $\rho(u) \le \rho(v)$, i.e., ρ is a Riesz seminorm. Similarly, a seminorm ρ is a Riesz seminorm if and only if its closed unit ball $U_\rho = \{u \in L\colon\ \rho(u) \le 1\}$ is a solid set.

It should be clear that any collection of Riesz seminorms generates (in the usual manner) a locally-convex solid topology. The point is that every locally convex-solid topology on a Riesz space L is generated by a family of Riesz seminorms.

THEOREM 2.25. *Let L be a Riesz space and τ a linear topology on L. Then the following statements are equivalent.*

(i) *τ is a locally convex-solid topology.*
(ii) *There exists a family $\{\rho_\alpha\}_{\alpha \in A}$ of Riesz seminorms that generates the topology τ.*

PROOF. (i) $\Longrightarrow$ (ii) Let $\mathcal{N}$ be a base of zero for τ consisting of τ-closed, convex, and solid sets. For each $V \in \mathcal{N}$, let ρ_V denote the **Minkowski seminorm** generated by V, that is,

$$\rho_V(u) = \inf\{\lambda > 0\colon\ u \in \lambda V\}.$$

Then ρ_V is a τ-continuous seminorm whose closed unit ball coincides with the solid set V. By the above discussion each ρ_V is a Riesz seminorm. The desired conclusion now follows immediately by observing that the family $\{\rho_V\colon\ V \in \mathcal{N}\}$ generates the topology τ.

(ii) $\Longrightarrow$ (i) Obvious. ■

Assume now that L is a Riesz space and ρ is a Riesz seminorm on L. If A is an ideal of L, we denote by $\dot{u}$ the equivalence class of L/A whose representative is u, i.e., $\dot{u} = Q(u) = u + A$, where Q is the canonical projection of L onto L/A (we also write $\dot{V} = Q(V)$ for $V \subseteq L$).

It is well known that the function $\dot{\rho}$ defined on L/A by

$$\dot{\rho}(\dot{u}) = \inf\{\rho(v)\colon\ v \in L \text{ and } v \in \dot{u}\},\quad \dot{u} \in L/A,$$

is a seminorm (called the **quotient seminorm** of ρ), which is actually a Riesz seminorm. This follows immediately from Theorem 1.33 by noting that the open unit ball of $\dot{\rho}$ is solid since it is the image of B_ρ under the canonical projection of L onto L/A. The following result should now be clear.

THEOREM 2.26. *Let (L, τ) be a locally convex-solid Riesz space whose topology τ is generated by a family of Riesz seminorms $\{\rho_i\}_{i \in I}$, and let A be an ideal of L. Then the quotient topology τ/A on L/A is a locally convex-solid topology generated by the family of Riesz seminorms $\{\dot{\rho}_i\}_{i \in I}$.*

The discussion next will give the reader an idea as to how far the locally solid topologies are from the locally convex-solid ones.

DEFINITION 2.27. *A **Riesz pseudonorm** ρ is a real-valued function defined on a Riesz space L (i.e., $\rho\colon L \to \mathbb{R}$) satisfying the following properties.*

(1) *$\rho(u) \geq 0$ for each $u \in L$.*
(2) *$\rho(u + v) \leq \rho(u) + \rho(v)$ for all $u, v \in L$.*
(3) *$\rho(\lambda u) \to 0$ as $\lambda \to 0$ for each $u \in L$.*
(4) *$\rho(u) \leq \rho(v)$ whenever $|u| \leq |v|$ holds in L.*

Clearly, every Riesz seminorm is a Riesz pseudonorm. The converse is, of course, false. For instance, if ρ is a nonzero Riesz pseudonorm (in particular, if ρ is a nonzero Riesz seminorm) on a Riesz space L, then the function $\rho_1\colon L \to \mathbb{R}$ defined by $\rho_1(u) = \frac{\rho(u)}{1+\rho(u)}$ is a nonzero Riesz pseudonorm that fails to be a Riesz seminorm. From property (4) of Definition 2.27, it should be immediate that if ρ is a Riesz pseudonorm on a Riesz space L, then

$$\rho(u) = \rho(|u|)$$

holds for all $u \in L$.

The next result, due to D. H. Fremlin [**58**, 22C Proposition, p. 39], asserts that every locally solid topology is generated by a family of Riesz pseudonorms.

THEOREM 2.28 (Fremlin). *A linear topology on a Riesz space is locally solid if and only if it is generated by a family of Riesz pseudonorms.*

Moreover, if a family $\{\rho_i\}_{i \in I}$ of Riesz pseudonorms generates a locally solid topology τ on a Riesz space L, then $u_\alpha \xrightarrow{\tau} u$ in L if and only if $\rho_i(u_\alpha - u) \xrightarrow[\alpha]{} 0$ for each $i \in I$.

PROOF. If a linear topology τ is generated by a family of Riesz pseudonorms, then τ is clearly locally solid. For the converse, assume that τ is a locally solid topology on a Riesz space L and let V be an arbitrary solid τ-neighborhood of zero. To finish the proof, it suffices to show that there is a τ-continuous Riesz pseudonorm $\rho\colon L \to [0, \infty)$ and some $\epsilon > 0$ such that $B_\rho(\epsilon) = \{u \in L\colon\ \rho(u) < \epsilon\} \subseteq V$.

To establish this, start by choosing a sequence $\{V_n\}$ of solid τ-neighborhoods of zero such that $V_{n+1} + V_{n+1} + V_{n+1} \subseteq V_n$ holds for each n and with $V_1 \subseteq V$. Now define the function $d\colon L \to [0, \infty)$ via the formula

$$d(u) = \begin{cases} 1 & \text{if } u \notin V_1, \\ 2^{-n} & \text{if } u \in V_n \setminus V_{n+1} \\ 0 & \text{if } u \in \bigcap_{n=1}^{\infty} V_n\,. \end{cases}$$

The solidness of the neighborhoods V_n imply immediately that whenever $|v| \leq |u|$ holds true, then $d(v) \leq d(u)$.

Now by means of d define the function $\rho\colon L \to [0,\infty)$ via the formula:

$$\rho(u) = \inf\Big\{\sum_{i=1}^{n} d(u_i)\colon\ u_1,\ldots,u_n \in L \text{ and } \sum_{i=1}^{n} u_i = u\Big\}.$$

We claim that ρ is a Riesz pseudonorm with the desired properties. The verification of the properties will be done by steps.

(1) $\rho(u) \geq 0$ *for each* $u \in L$ *and* $\rho(0) = 0$.

This follows immediately from the definition of ρ.

(2) $|v| \leq |u|$ *implies* $\rho(v) \leq \rho(u)$.

To see this, assume that $|v| \leq |u|$ and let $u = \sum_{i=1}^{n} u_i$. By the Riesz Decomposition Property (see Theorem 1.10) there exist elements $v_1,\ldots,v_n$ satisfying $v = \sum_{i=1}^{n} v_i$ and $|v_i| \leq |u_i|$ for each i. This implies

$$\rho(v) \leq \sum_{i=1}^{n} d(v_i) \leq \sum_{i=1}^{n} d(u_i),$$

and from this we easily get $\rho(v) \leq \rho(u)$.

(3) $\rho(u+v) \leq \rho(u) + \rho(v)$ *for all* $u, v \in L$.

This is straightforward.

(4) ρ *is a* τ*-continuous Riesz pseudonorm.*

From the inequality $\rho(u) \leq d(u)$ for each $u \in L$ and the fact that $d(u) \leq 2^{-n}$ for each $u \in V_n$, it follows that ρ is τ-continuous at zero. In particular, $\lim_{\lambda\to 0} \rho(\lambda u) = 0$ holds for each $u \in L$. Now a glance at the triangle inequality guarantees that ρ is indeed a τ-continuous Riesz pseudonorm.

(5) $B_\rho\big(\frac{1}{2}\big) = \big\{u \in L\colon\ \rho(u) < \frac{1}{2}\big\} \subseteq V_1 \subseteq V$.

The proof of (5) will be based upon the following property:

$$\sum_{i=1}^{n} d(u_i) < \tfrac{1}{2^m} \implies \sum_{i=1}^{n} u_i \in V_m\,. \tag{$\star$}$$

To verify ($\star$), we use induction on n. For $n = 1$ we have $d(u_1) < 2^{-m}$, and so $u_1 \in V_{m+1} \subseteq V_m$ is trivially true. For the induction step, assume that if $\{u_i\colon\ i \in I\}$ is a collection of at most n elements satisfying $\sum_{i\in I} d(u_i) < 2^{-m}$ for some $m \in \mathbb{N}$, then $\sum_{i\in I} u_i \in V_m$.

Suppose that $\sum_{i=1}^{n+1} d(u_i) < \frac{1}{2^m}$ for some $m \in \mathbb{N}$. Clearly, we have $d(u_i) \leq \frac{1}{2^{m+1}}$, and so $u_i \in V_{m+1}$ for each $1 \leq i \leq n+1$. We now distinguish the following two cases.

CASE 1: $\sum_{i=1}^{n+1} d(u_i) < \frac{1}{2^{m+1}}$

Then $\sum_{i=1}^{n} d(u_i) < \frac{1}{2^{m+1}}$, and so by the induction hypothesis we have $\sum_{i=1}^{n} u_i \in V_{m+1}$. This implies

$$\sum_{i=1}^{n+1} u_i = \sum_{i=1}^{n} u_i + u_{n+1} \in V_{m+1} + V_{m+1} \subseteq V_m\,.$$

CASE 2: $\sum_{i=1}^{n+1} d(u_i) \geq \frac{1}{2^{m+1}}$

Let $1 \le k \le m+1$ be the largest k such that $\sum_{i=k}^{n+1} d(u_i) \ge \frac{1}{2^{m+1}}$. In view of $d(u_{m+1}) < \frac{1}{2^{m+1}}$, we see that $k < m+1$.

Assume first that $k > 1$. From $\sum_{i=1}^{n+1} d(u_i) < \frac{1}{2^m}$, we get $\sum_{i=1}^{k-1} d(u_i) < \frac{1}{2^{m+1}}$. Consequently, our induction hypothesis yields $\sum_{i=1}^{k-1} u_i \in V_{m+1}$. Also, by the choice of k we have $\sum_{i=k+1}^{n+1} d(u_i) < \frac{1}{2^{m+1}}$, and thus by our induction hypothesis also we have $\sum_{i=k+1}^{n+1} u_i \in V_{m+1}$. Therefore, in this case we obtain

$$\sum_{i=1}^{n+1} u_i = \sum_{i=1}^{k-1} u_i + u_k + \sum_{i=k+1}^{n+1} u_i \in V_{m+1} + V_{m+1} + V_{m+1} \subseteq V_m\,.$$

If $k = 1$, then we have $\sum_{i=2}^{n+1} d(u_i) < \frac{1}{2^{m+1}}$, and so $\sum_{i=2}^{n+1} \in V_{m+1}$. This implies $\sum_{i=1}^{n+1} u_i = u_1 + \sum_{i=2}^{n+1} u_i \in V_{m+1} + V_{m+1} \subseteq V_m$. This completes the induction and the proof of ($\star$).

The validity of (5) now should be immediate and this completes the proof of the theorem. ■

Recall that a locally convex space (E, τ) is said to be a **barrelled space** whenever every absorbing, balanced, convex, and τ-closed subset of E is already a τ-neighborhood of zero. Barrelled Riesz spaces have the property that their topological duals are bands in their order duals.

THEOREM 2.29. *If (L, τ) is a barrelled locally convex-solid Riesz space, then L' is a band in $L^\sim$.*

PROOF. Assume that $0 \le \phi_\alpha \uparrow \phi$ holds in $L^\sim$ with $\{\phi_\alpha\} \subseteq L'$. Then the set $V = \{u \in L\colon\ |\phi_\alpha(u)| \le 1 \text{ for all } \alpha\}$ is an absorbing, balanced, convex and τ-closed subset of L, as can be easily verified. Thus V is a τ-neighborhood of zero. But $|\phi_\alpha(u)| \le 1$ for each $u \in V$ and all α implies that $|\phi(u)| \le 1$ for each $u \in V$, since $\phi_\alpha \uparrow \phi$; see Theorem 1.67. So, $\phi \in L'$, since ϕ is bounded on a τ-neighborhood of zero. This shows that L' is a band of $L^\sim$. ■

The band generated by the topological dual L' in $L^\sim$ has been characterized by W. A. J. Luxemburg [**88**, Theorem 5.3, p. 127] as follows.

THEOREM 2.30 (Luxemburg). *If (L, τ) is a locally convex-solid Riesz space, then the band generated by L' in $L^\sim$ consists of all order bounded linear functionals that are τ-continuous on the order intervals of L.*

PROOF. We consider the following vector subspace of $L^\sim$:

$$\mathcal{B} = \{\phi \in L^\sim\colon\ \phi \text{ is } \tau\text{-continuous on the order intervals of } L\}\,.$$

We shall establish that the band generated by L' coincides with $\mathcal{B}$. This will be done in three steps.

STEP 1: *The vector subspace $\mathcal{B}$ is a band in $L^\sim$ and $L' \subseteq \mathcal{B}$.*

Clearly, $\mathcal{B}$ is a vector subspace of $L^\sim$ satisfying $L' \subseteq \mathcal{B}$. To see that $\mathcal{B}$ is an ideal, suppose $|\phi| \le |\psi|$ holds in $L^\sim$ with $\psi \in \mathcal{B}$. Now assume that an order bounded net $\{u_\alpha\}$ in L satisfies $u_\alpha \xrightarrow{\tau} 0$, and let $\epsilon > 0$. For each α pick some v_α with $|v_\alpha| \le |u_\alpha|$ such that

$$|\phi(u_\alpha)| \le |\phi|(|u_\alpha|) \le |\psi(v_\alpha)| + \epsilon\,. \qquad (\star)$$

Since $|v_\alpha| \le |u_\alpha|$ for each α and $u_\alpha \xrightarrow{\tau} 0$, it follows from the local solidness of τ that $v_\alpha \xrightarrow{\tau} 0$, and so $\psi(v_\alpha) \to 0$. Now a glance at $(\star)$ shows that $\limsup_\alpha |\phi(u_\alpha)| \le \epsilon$ for all $\epsilon > 0$. This implies $\phi(u_\alpha) \to 0$. That is, $\phi \in \mathcal{B}$ so that $\mathcal{B}$ is an ideal. It is straightforward to verify that $\mathcal{B}$ is also order closed in $L^\sim$. Consequently, $\mathcal{B}$ is a band.

STEP 2: *Let $\phi \in (L')^{\mathrm{d}}$. If some $u > 0$ satisfies $\phi(u) > 0$, then for each $0 < \epsilon < \phi(u)$ the convex set $C_\epsilon = \{v \in [0,u]\colon\ \phi(v) \le \epsilon\}$ is τ-dense in the order interval $[0,u]$.*

If C_ϵ is not τ-dense in $[0,u]$, then it is easy to see that u does not belong to the τ-closure $\overline{C_\epsilon}$ of C_ϵ. But then, by the separating hyperplane theorem, there exists some $\psi \in L'$ such that $\psi(u) > 1$ and $\psi(v) \le 1$ for each $v \in C_\epsilon$. Replacing ψ by ψ^+, we can assume that $\psi \ge 0$. From $\psi \wedge \phi = 0$, it follows that there exists some sequence $\{v_n\} \subseteq [0,u]$ such that $\phi(v_n) + \psi(u - v_n) \to 0$. In particular, we have $v_n \in C_\epsilon$ for all $n \ge n_0$, and so

$$\psi(u) = \psi(v_n) + \psi(u - v_n) \le 1 + \psi(u - v_n) \to 1\,,$$

contrary to $\psi(u) > 1$. This shows that C_ϵ is τ-dense in $[0,u]$.

STEP 3: *If $\phi \in \mathcal{B} \cap (L')^{\mathrm{d}}$, then $\phi = 0$ (and hence $\mathcal{B} = (L')^{\mathrm{dd}}$, the band generated by L' in $L^\sim$).*

Let $0 \le \phi \in \mathcal{B} \cap (L')^{\mathrm{d}}$ and assume by way of contradiction that there exists some $u > 0$ with $\phi(u) = 1$. Let $\epsilon = \frac{1}{2}$ in STEP 2, and then select a net $\{u_\alpha\} \subseteq C_\epsilon$ such that $u_\alpha \xrightarrow{\tau} u$. From $\phi \in \mathcal{B}$, we get $\phi(u_\alpha) \to \phi(u) = 1$. However, this contradicts $\phi(u_\alpha) \le \frac{1}{2}$ for each α. ■

The following example illustrates the preceding result.

EXAMPLE 2.31. Let L be the Riesz space of all eventually zero sequences, and let τ be the locally convex-solid topology generated on L by the Riesz norm

$$\|(x_1, x_2, \ldots)\|_\infty = \max\{|x_n|\colon\ n = 1, 2, \ldots\}\,.$$

It is easy to see that $L^\sim = \mathbb{R}^{\mathbb{N}}$, the Riesz space of all sequences. The topological dual L' coincides with ℓ_1. The band generated by ℓ_1 in $\mathbb{R}^{\mathbb{N}}$ is precisely equal to $\mathbb{R}^{\mathbb{N}}$. Now it can be verified directly that every sequence in $\mathbb{R}^{\mathbb{N}}$ defines a linear functional on L that is indeed τ-continuous on the order intervals of L. ■

For any Riesz space L observe that for each $\phi \in L^\sim$ the function $\rho_\phi\colon L \to \mathbb{R}$, defined by $\rho_\phi(u) = |\phi|(|u|)$, is a Riesz seminorm on L. Thus any nonempty subset A of $L^\sim$ defines a (not necessarily Hausdorff) locally convex-solid topology on L via the family of Riesz seminorms $\{\rho_\phi\colon\ \phi \in A\}$. This topology is called the absolute weak topology generated by A on L.

DEFINITION 2.32. *Let L be a Riesz space and let A be a nonempty subset of the order dual $L^\sim$. The **absolute weak topology** $|\sigma|(L,A)$ generated by A on L is the locally convex-solid topology generated by the collection of Riesz seminorms $\{\rho_\phi\colon\ \phi \in A\}$, where $\rho_\phi(u) = |\phi|(|u|)$.*

Observe that $|\sigma|(L,A) = |\sigma|(L, L^\sim_A)$, where $L^\sim_A$ is the ideal generated by A in $L^\sim$. Indeed, note first that $|\sigma|(L,A) \subseteq |\sigma|(L, L^\sim_A)$ is trivially true and then that for every $\psi \in L^\sim_A$ there exist $\{\phi_1, \ldots, \phi_n\} \subseteq A$ and some $\lambda > 0$ such that $|\psi| \le \lambda \sum_{i=1}^n |\phi_i|$; hence $\rho_\psi \le \lambda \sum_{i=1}^n \rho_{\phi_i}$ holds and consequently ρ_ψ is a $|\sigma|(L,A)$-continuous Riesz seminorm. This yields $|\sigma|(L, L^\sim_A) \subseteq |\sigma|(L,A)$ and so $|\sigma|(L,A) =$

$|\sigma|(L, L_A^\sim)$. Note also that the inequalities $|\psi(u)| \le |\psi|(|u|) \le \lambda\sum_{i=1}^n \rho_{\phi_i}(u)$ imply that every $\psi \in L_A^\sim$ is a $|\sigma|(L,A)$-continuous linear functional on L.

The next theorem tells us that $L_A^\sim$ consists precisely of all $|\sigma|(L,A)$-continuous linear functionals on L. This result appeared in various forms in the literature. Its formulation below is due to S. Kaplan [**73**, Note II, Proposition 10.1, p. 349].

THEOREM 2.33 (Kaplan). *Let L be a Riesz space, A a nonempty subset of $L^\sim$, and $L_A^\sim$ the ideal generated by A in $L^\sim$. Then we have*

$$\big(L, |\sigma|(L,A)\big)' = L_A^\sim .$$

PROOF. As noted above every $\phi \in L_A^\sim$ is $|\sigma|(L,A)$-continuous and so we need only to show that every $\phi \in \big(L, |\sigma|(L,A)\big)'$ belongs to $L_A^\sim$. To this end, suppose that some linear functional $\phi\colon L \to \mathbb{R}$ is $\big(L, |\sigma|(L,A)\big)$-continuous. Then there exist $\psi \in L_A^\sim$ and $\lambda > 0$ such that $|\phi(u)| \le \lambda\rho_\psi(u)$ for all $u \in L$. Consequently, $0 \le |\phi|(u) \le \lambda\rho_\psi(u) = \lambda|\psi|(u)$ holds by Theorem 1.67(a) for all $u \in L^+$, and hence $|\phi| \le \lambda|\psi|$. This shows that $\phi \in L_A^\sim$ and the proof is finished. ■

DEFINITION 2.34. *Now let (L,τ) be a locally convex-solid space. We define the* ***absolute weak topology*** *on L to be the topology $|\sigma|(L,L')$. That is, $|\sigma|(L,L')$ is the locally convex-solid topology on L generated by the family of Riesz seminorms $\{\rho_\phi\colon\ \phi \in L'\}$.*

We have the following relationships between the various topologies.

THEOREM 2.35. *If (L,τ) is a locally convex-solid Riesz space, then*

$$\sigma(L,L') \subseteq |\sigma|(L,L') \subseteq \tau .$$

In particular, $|\sigma|(L,L')$ is Hausdorff if and only if τ is likewise Hausdorff.

PROOF. Assume $u_\alpha \xrightarrow{|\sigma|(L,L')} 0$. Then for each $\phi \in L'$ it follows from the inequality $|\phi(u_\alpha)| \le |\phi|(|u_\alpha|)$ that $\lim \phi(u_\alpha) = 0$. Therefore, $u_\alpha \xrightarrow{\sigma(L,L')} 0$ and so $\sigma(L,L') \subseteq |\sigma|(L,L')$.

On the other hand, if $u_\alpha \xrightarrow{\tau} 0$ holds, then $|u_\alpha| \xrightarrow{\tau} 0$ also holds since τ is locally solid. But then $\lim \rho_\phi(u_\alpha) = \lim |\phi|(|u_\alpha|) = 0$ for each $\phi \in L'$ and so $u_\alpha \xrightarrow{|\sigma|(L,L')} 0$. This implies $|\sigma|(L,L') \subseteq \tau$, and the proof is finished. ■

Some conditions for the absolute weak topology $|\sigma|(L,L')$ to coincide with the weak topology $\sigma(L,L')$ are included in the next result.

THEOREM 2.36. *For a Hausdorff locally convex-solid Riesz space (L,τ) the following statements are equivalent.*

(i) $\sigma(L,L') = |\sigma|(L,L')$.
(ii) *The weak topology $\sigma(L,L')$ is locally solid.*
(iii) *The mapping $u \mapsto |u|$ is weakly continuous at zero.*
(iv) *Every order interval of L' lies in a finite dimensional vector space.*
(v) *Every principal ideal of L' is of finite dimension.*

PROOF. The implications (i) $\Longrightarrow$ (ii) $\Longrightarrow$ (iii) are obvious.

(iii) $\Longrightarrow$ (iv) It is enough to show that the order interval $[0,\phi]$ spans a finite dimensional vector space for each $0 \le \phi \in L'$. To this end, assume that $0 \le \phi \in L'$. Then the set $V = \{u \in L\colon\ \phi(|u|) \le 1\}$ is a $\sigma(L,L')$-neighborhood of zero, since the weak continuity of $u \mapsto |u|$ at zero implies the continuity at zero of the function

$u \mapsto \phi(|u|)$ from $(L, \sigma(L, L'))$ to $\mathbb{R}$. Thus there exist $\epsilon > 0$ and $\phi_1, \ldots, \phi_n \in L'$ such that

$$\{u \in L: |\phi_i(u)| \le \epsilon \text{ for } i = 1, \ldots, n\} \subseteq V.$$

Now assume that $\psi \in L'$ satisfies $0 \le \psi \le \phi$. If $v \in \operatorname{Ker} \phi_i$ for each $i = 1, \ldots, n$ [that is, $\phi_i(v) = 0$ for each $i = 1, \ldots, n$], then $\phi_i(kv) = 0$ for $i = 1, \ldots, n$ and all k, and so $kv \in V$ for all k. Thus $k\phi(|v|) \le 1$ for all k and so $\phi(|v|) = 0$. It follows from $0 \le |\psi(v)| \le \psi(|v|) \le \phi(|v|) = 0$ that $v \in \operatorname{Ker} \psi$. Consequently, we have

$$\bigcap_{i=1}^{n} \operatorname{Ker} \phi_i \subseteq \operatorname{Ker} \psi.$$

Now observe that the last condition is equivalent to the property that ψ lies in the linear span of $\{\phi_1, \ldots, \phi_n\}$; see, for example, [**64**, p. 186]. Hence $[0, \phi]$ spans a finite dimensional vector subspace in L'.

(iv) $\Longrightarrow$ (v) Note that for any $\phi \in L'$ the ideal generated by ϕ in L' coincides with the vector subspace generated by the order interval $[0, |\phi|]$ in L', and the desired conclusion follows.

(v) $\Longrightarrow$ (i) Observe first that if $0 < \phi \in L'$ is a discrete element of L', then $\phi(|u|) = |\phi(u)|$ for all $u \in L$ (since ϕ is also a discrete element in $L^\sim$), and hence a Riesz homomorphism by Theorem 1.85.

Now let $0 < \phi \in L'$ and let L'_ϕ be the principal ideal generated by ϕ in L'. By our hypothesis, L'_ϕ is of finite dimension and hence L'_ϕ is Riesz isomorphic to some $\mathbb{R}^n$; see Exercise 13 of Chapter 1. Thus, there exists a Hamel basis $\phi_1, \ldots, \phi_n$ of L'_ϕ consisting of positive discrete elements of L'. In particular, we have $\phi = \sum_{i=1}^n \lambda_i \phi_i$ for some appropriate real numbers $\lambda_1, \ldots, \lambda_n$. Thus if $u_\alpha \xrightarrow{\sigma(L,L')} 0$ holds, then from

$$\phi(|u_\alpha|) = \sum_{i=1}^{n} \lambda_i \phi_i(|u_\alpha|) = \sum_{i=1}^{n} \lambda_i |\phi_i(u_\alpha)|$$

we easily get $|u_\alpha| \xrightarrow{|\sigma|(L,L')} 0$. This implies $\sigma(L, L') = |\sigma|(L, L')$, and the proof is finished. ∎

A glance at the preceding proof also yields the following.

COROLLARY 2.37. *If in a Hausdorff locally convex-solid Riesz space (L, τ) the weak topology $\sigma(L, L')$ is locally solid, then L' is a discrete Riesz space.*

We are now in the position to characterize the finite dimensional normed Riesz spaces in terms of the weak topologies. The next result is due to A. L. Peressini [**107**, Theorem 2.4, p. 208]; see also [**108**, 2.14 Proposition, p. 134].

THEOREM 2.38 (Peressini). *For a normed Riesz space L the following statements are equivalent.*

(1) *L is of finite dimension, that is, L is Riesz isomorphic to some $\mathbb{R}^n$ with its natural ordering.*
(2) *The weak topology $\sigma(L, L')$ is a locally solid topology.*

PROOF. (1) $\Longrightarrow$ (2) Obvious.

(2) $\Longrightarrow$ (1) It is enough to show that L' is of finite dimension. To this end, assume that L' contains a sequence $\{\phi_n\}$ of linearly independent linear functionals. Without loss of generality, we can assume that each ϕ_n has norm one. Since L' is norm complete the series $\sum_{n=1}^{\infty} 2^{-n} |\phi_n|$ is norm convergent in L'. Put

$\phi = \sum_{n=1}^{\infty} 2^{-n}|\phi_n|$ and note that $[0,\phi]$ spans an infinite dimensional vector space, contrary to Theorem 2.36. This contradiction establishes that L' is finite dimensional and completes the proof. ■

The following example shows that the implication (2) $\Longrightarrow$ (1) in Theorem 2.38 is no longer true if the hypothesis of L being a normed Riesz space is replaced by local convexity.

EXAMPLE 2.39. Let X be an infinite set, and let $L = \mathbb{R}^X$ and τ be the Hausdorff locally convex-solid topology of pointwise convergence. In other words, τ is generated by the family of Riesz seminorms $\{\rho_x\colon\ x \in X\}$, where $\rho_x(u) = \big|u(x)\big|$ for each $u \in L$.

If $x_1, \dots, x_n$ is a finite number of points of X and $\lambda_1, \dots, \lambda_n$ are real numbers, then the linear functional $\phi\colon L \to \mathbb{R}$, defined by

$$\phi(f) = \sum_{i=1}^{n} \lambda_i f(x_i)\,,$$

is τ-continuous. We claim that every $\phi \in L'$ is of the above form.

To see this, note that if $\phi \in L'$, then there exist $\epsilon > 0$ and a finite number of points $x_1, \dots, x_n$ in X such that $\big|\phi(f)\big| \le 1$ holds whenever $f \in L$ satisfies $\big|f(x_i)\big| \le \epsilon$ for each $i = 1, \dots, n$. Put $\lambda_i = \phi\big(\chi_{\{x_i\}}\big)$ and note that for each positive k the function $g_k = k\big[f - \sum_{i=1}^{n} f(x_i)\chi_{\{x_i\}}\big]$ satisfies $g_k(x_i) = 0$ for each $i = 1, \dots, n$. Thus $k\big|\phi(f) - \sum_{i=1}^{n} \lambda_i f(x_i)\big| \le 1$ holds for each $k \in \mathbb{N}$, and so $\phi(f) = \sum_{i=1}^{n} \lambda_i f(x_i)$ for all $f \in L$. It follows now easily that $\sigma(L, L') = |\sigma|(L, L') = \tau$. Note, however, that L is infinite dimensional. ■

2.4. Topological Completion of a Locally Solid Riesz Space

As we saw in Theorem 2.8 every Hausdorff topological vector space (E,τ) has a unique topological completion $(\widehat{E},\widehat{\tau})$ in the sense that $(\widehat{E},\widehat{\tau})$ is $\widehat{\tau}$-complete, containing (E,τ) as a vector and topological $\widehat{\tau}$-dense subspace. In the next theorem we shall show that if in addition E is a Riesz space and τ is a locally solid topology, then $(\widehat{E},\widehat{\tau})$ has a natural Riesz space structure that induces the original Riesz space structure on E, i.e., E is a Riesz subspace of $\widehat{E}$. The details are included in the next theorem whose proof is taken from [**3**, Theorem 2.1, p. 109].

THEOREM 2.40. *Let (L,τ) be a Hausdorff locally solid Riesz space and let $(\widehat{L},\widehat{\tau})$ be its topological completion. Then the $\widehat{\tau}$-closure of L^+ is a cone of $\widehat{L}$ and $(\widehat{L},\widehat{\tau})$ equipped with this cone is a Hausdorff locally solid Riesz space containing L as a Riesz subspace.*

Moreover, the $\widehat{\tau}$-closure of a solid subset of L is a solid subset of $\widehat{L}$. In particular, if $\mathcal{N}$ is a base at zero for (L,τ) consisting of solid sets, then $\{\overline{V}\colon\ V \in \mathcal{N}\}$ is also a base at zero for $(\widehat{L},\widehat{\tau})$ consisting of solid sets.

PROOF. We show first that $\widehat{L}^+ = \overline{L^+}$ (the $\widehat{\tau}$-closure of L^+ in $\widehat{L}$) is a cone of $\widehat{L}$. To see this, note that the inclusions $\widehat{L}^+ + \widehat{L}^+ \subseteq \widehat{L}^+$ and $\lambda\widehat{L}^+ \subseteq \widehat{L}^+$ for $\lambda \ge 0$ follow immediately from the corresponding inclusions of L^+, and the continuity of the addition and scalar multiplication in $\widehat{L}$. To establish that $\widehat{L}^+ \cap (-\widehat{L}^+) = \{0\}$, let $\widehat{u} \in \widehat{L}^+ \cap (-\widehat{L}^+)$, and then choose two nets $\{u_\alpha\}$ and $\{w_\beta\}$ of L^+ with $u_\alpha \xrightarrow{\widehat{\tau}} \widehat{u}$

and $w_\beta \xrightarrow{\widehat{\tau}} -\widehat{u}$. But then from the property $0 \le u_\alpha \le u_\alpha + w_\beta \xrightarrow[(\alpha,\beta)]{\tau} 0$ we obtain $u_\alpha \xrightarrow{\tau} 0$, and thus $\widehat{u} = 0$. Hence $\widehat{L}^+$ is a cone in $\widehat{L}$.

We shall show next that $\widehat{L}$ is a Riesz space under this cone. The mapping $u \mapsto u^+$, from (L, τ) to $(\widehat{L}, \widehat{\tau})$, is uniformly continuous, and since $(\widehat{L}, \widehat{\tau})$ is $\widehat{\tau}$-complete, it can be extended uniquely to a uniformly continuous mapping p from L to $\widehat{L}$; see Theorem 2.6. We shall show next that $p(\widehat{u}) = \widehat{u} \vee 0$ holds in $\widehat{L}$ for each $\widehat{u} \in \widehat{L}$. Indeed, if $\widehat{u} \in \widehat{L}$, then pick a net $\{u_\alpha\} \subseteq L$ with $u_\alpha \xrightarrow{\widehat{\tau}} \widehat{u}$ and observe that $p(u_\alpha) - u_\alpha = u_\alpha^+ - u_\alpha \in L^+$ implies $p(\widehat{u}) - \widehat{u} \in \widehat{L}^+$, i.e., $p(\widehat{u}) \ge \widehat{u}$ holds in $\widehat{L}$. Clearly $p(\widehat{u}) \ge 0$ also holds in $\widehat{L}$. On the other hand, if $\widehat{u} \le \widehat{v}$ and $0 \le \widehat{v}$ hold in $\widehat{L}$, then pick two nets $\{w_\alpha\}$ and $\{v_\alpha\}$ in L^+ (with the same index set) such that $w_\alpha \xrightarrow{\widehat{\tau}} \widehat{v} - \widehat{u}$ and $v_\alpha \xrightarrow{\widehat{\tau}} \widehat{v}$. Note that $f_\alpha = v_\alpha - w_\alpha \xrightarrow{\widehat{\tau}} \widehat{u}$ and $v_\alpha - p(f_\alpha) = v_\alpha - f_\alpha^+ \ge 0$. Hence by taking the $\widehat{\tau}$-limits, we get $\widehat{v} - p(\widehat{u}) \in \widehat{L}^+$ or $\widehat{v} \ge p(\widehat{u})$ in $\widehat{L}$. So $p(\widehat{u}) = \widehat{u}^+$ holds in $\widehat{L}$. This shows that $\widehat{L}$ is a Riesz space containing L as a Riesz subspace. Now, according to Theorem 2.17, the uniform continuity of $\widehat{u} \mapsto \widehat{u}^+$ guarantees that $(\widehat{L}, \widehat{\tau})$ is a locally solid Riesz space.

For the last assertion, let $\overline{S}$ denote the $\widehat{\tau}$-closure of a solid subset S of L. Assume that $|\widehat{u}| \le |\widehat{v}|$ with $\widehat{v} \in \overline{S}$. Pick a net $\{v_\alpha\}$ of S with $v_\alpha \xrightarrow{\widehat{\tau}} \widehat{v}$ and another net $\{u_\beta\}$ in L with $u_\beta \xrightarrow{\widehat{\tau}} \widehat{u}$. For each α and β put $w_{\alpha,\beta} = \big[(-|v_\alpha|) \vee u_\beta\big] \wedge |v_\alpha|$, and observe that $|w_{\alpha,\beta}| \le |v_\alpha|$ holds for all α and β. Thus $\{w_{\alpha,\beta}\} \subseteq S$. But $w_{\alpha,\beta} \xrightarrow[(\alpha,\beta)]{\widehat{\tau}} \widehat{u}$, so that $\widehat{u} \in \overline{S}$ and hence $\overline{S}$ is a solid subset of $\widehat{L}$. ■

The next two theorems deal with the lattice relation of L with respect to $\widehat{L}$. The first one was proven by I. Kawai [**74**, Theorem 4.1, p. 296] for the locally convex case and by C. D. Aliprantis [**3**, Theorem 2.3, p. 110] in general.

THEOREM 2.41 (Kawai–Aliprantis). *For a Hausdorff locally solid Riesz space (L, τ) the following statements are equivalent.*

(i) *L is a regular Riesz subspace of $\widehat{L}$.*
(ii) *Every τ-Cauchy net $\{u_\alpha\} \subseteq L^+$ with $u_\alpha \xrightarrow{o} 0$ in L satisfies $u_\alpha \xrightarrow{\tau} 0$.*

PROOF. (i) $\Longrightarrow$ (ii) Let $\{u_\alpha\} \subseteq L^+$ be a τ-Cauchy net such that $u_\alpha \xrightarrow{o} 0$ holds in L. Pick a net $\{v_\alpha\} \subseteq L^+$ such that $0 \le u_\alpha \le v_\alpha \downarrow 0$. By hypothesis $v_\alpha \downarrow 0$ also holds in $\widehat{L}$. Since $\{u_\alpha\}$ is a τ-Cauchy net of L, there exists some $\widehat{u} \in \widehat{L}$ such that $u_\alpha \xrightarrow{\widehat{\tau}} \widehat{u}$. But for each fixed β we have $0 \le u_\alpha \le v_\alpha \le v_\beta$ for all $\alpha \ge \beta$, and thus $0 \le \widehat{u} \le v_\beta$ holds for all β. Hence $\widehat{u} = 0$, and consequently $u_\alpha \xrightarrow{\tau} 0$.

(ii) $\Longrightarrow$ (i) Let $u_\alpha \downarrow 0$ in L. Assume that $0 \le \widehat{v} \le u_\alpha$ holds in $\widehat{L}$ for each α. Pick a net $\{v_\lambda\}$ in L^+ with $v_\lambda \xrightarrow{\widehat{\tau}} \widehat{v}$ and note that

$$|v_\lambda \wedge u_\alpha - \widehat{v}| = |v_\lambda \wedge u_\alpha - \widehat{v} \wedge u_\alpha| \le |v_\lambda - \widehat{v}|$$

for all α, λ. This shows that $v_\lambda \wedge u_\alpha \xrightarrow[(\lambda,\alpha)]{\widehat{\tau}} \widehat{v}$. But we also have the relation $0 \le v_\lambda \wedge u_\alpha \le u_\alpha \downarrow_\alpha 0$ in L so that $v_\lambda \wedge u_\alpha \xrightarrow[(\lambda,\alpha)]{o} 0$ in L. By our hypothesis we get $v_\lambda \wedge u_\alpha \xrightarrow[(\lambda,\alpha)]{\widehat{\tau}} 0$. This implies $\widehat{v} = 0$ and hence $u_\alpha \downarrow 0$ holds in $\widehat{L}$. Therefore L is a regular Riesz subspace of $\widehat{L}$. ■

Recall that a subset A of a topological vector space (E,τ) is said to be τ-**complete** whenever every τ-Cauchy net of A is τ-convergent to some element (necessarily unique if τ is a Hausdorff topology) in A.

The next result was proven by M. Duhoux [**54**] and independently by C. D. Aliprantis [**3**].

THEOREM 2.42 (Aliprantis–Duhoux). *For a Hausdorff locally solid Riesz space (L,τ) the following statements are equivalent.*

(a) *L is an ideal of $\widehat{L}$ (and hence an order dense ideal).*
(b) *Every order interval of L is τ-complete.*

In particular, if L is an ideal of $\widehat{L}$, then for every $0 \le \widehat{u} \in \widehat{L}$ there exists a net $\{u_\alpha\} \subseteq L^+$ with $u_\alpha \uparrow \widehat{u}$ and $u_\alpha \xrightarrow{\widehat{\tau}} \widehat{u}$.

PROOF. (a) $\Longrightarrow$ (b) If $0 \le \widehat{u} \in \widehat{L}$, pick a net $\{u_\alpha\} \subseteq L^+$ with $u_\alpha \xrightarrow{\widehat{\tau}} \widehat{u}$ and note that since L is an ideal in $\widehat{L}$ we have $\{u_\alpha \wedge \widehat{u}\} \subseteq L^+$. Thus by taking the finite suprema of the net $\{u_\alpha \wedge \widehat{u}\}$, we get a net $\{v_\lambda\} \subseteq L^+$ such that $v_\lambda \uparrow \widehat{u}$ and $v_\lambda \xrightarrow{\widehat{\tau}} \widehat{u}$, and so in particular L is order dense in $\widehat{L}$.

Now since $[u,v] = u + [0, v-u]$, it is enough to establish that for each $u \in L^+$ the interval $[0,u]$ is τ-complete. To this end, let $\{u_\alpha\}$ be a τ-Cauchy net in $[0,u]$. Then $u_\alpha \xrightarrow{\widehat{\tau}} \widehat{u}$ holds in $\widehat{L}$. But then $0 \le \widehat{u} \le u$ also holds, and since L is an ideal of $\widehat{L}$, we have $\widehat{u} \in L$. Therefore, $[0,u]$ is τ-complete.

(b) $\Longrightarrow$ (a) Assume that $0 \le \widehat{u} \le u \in L$. Pick a net $\{u_\alpha\} \subseteq L^+$ with $u_\alpha \xrightarrow{\widehat{\tau}} \widehat{u}$. We can suppose that $0 \le u_\alpha \le u$ holds for each α; otherwise we can replace (if necessary) the net $\{u_\alpha\}$ by $\{u_\alpha \wedge u\}$. It follows now that $\widehat{u} \in L$, and from this (since L is a Riesz subspace of $\widehat{L}$), we get that L is an ideal of $\widehat{L}$. ■

Given a topological vector space (E,τ) it is natural to ask what conditions will guarantee that every τ-Cauchy net of E converges with respect to τ to some element of E. That is: *When is (E,τ) topologically complete?* Clearly for a Hausdorff topological vector space (E,τ) this question can be phrased as follows: *When does $E = \widehat{E}$ hold?*

The following discussion will answer this question for the Hausdorff locally solid Riesz spaces. But first we give a definition.

DEFINITION 2.43. *A locally solid Riesz space (L,τ) is said to satisfy the* ***monotone completeness property*** *(denoted by* MCP*), if every monotone τ-Cauchy net of L is τ-convergent in L.*

Note that a locally solid Riesz space (L,τ) satisfies MCP if and only if every increasing τ-Cauchy net of L^+ is τ-convergent in L. It should be also clear that every topologically complete locally solid Riesz space satisfies MCP. However, there exist Hausdorff locally solid Riesz spaces satisfying MCP or having their order intervals topologically complete without being topologically complete. Some examples follow.

EXAMPLE 2.44. Let X be an infinite set and let $L = \ell_\infty(X)$, i.e., let L be the Riesz space consisting of all bounded real-valued functions defined on X with the pointwise ordering. Let τ be the Hausdorff locally solid topology of pointwise convergence. Note that the order intervals of L are τ-complete, but (L,τ) is not τ-complete. Observe also that $\widehat{L} = \mathbb{R}^{[0,1]}$. ■

The next example is due to D. H. Fremlin (personal communication).

EXAMPLE 2.45 (Fremlin). Let $L = C[0,1]$ and let $\mathcal{C}$ denote the collection of all at most countable and compact subsets of $[0,1]$. For each $K \in \mathcal{C}$ define the Riesz seminorm $\rho_K : L \to \mathbb{R}$ by

$$\rho_K(u) = \sup\{|u(x)|: \ x \in K\}.$$

Define also $\rho\colon L \to \mathbb{R}$ by $\rho(u) = \int_0^1 |u(x)|\,dx$. Now let τ be the (nonmetrizable) Hausdorff locally solid topology generated by the family of Riesz seminorms $\{\rho_K: \ K \in \mathcal{C}\} \cup \{\rho\}$.

In particular, observe that $u_\alpha \xrightarrow{\tau} u$ implies $u_\alpha(x) \to u(x)$ for each x in $[0,1]$. The locally solid Riesz space (L,τ) has the following properties.

(1) *(L,τ) is not τ-complete.*

To see this, we proceed as follows. For each $\epsilon > 0$ and every $K \in \mathcal{C}$, pick $u_{\epsilon,K} \in C[0,1]$ such that $0 \le u_{\epsilon,K} \le \mathbf{1}$, $\int_0^1 u_{\epsilon,K}\,dx \ge 1-\epsilon$ and $u_{\epsilon,K}(x) = 0$ for all $x \in K$. (Since the Lebesgue measure of K is zero and K is compact, functions with the above properties exist.) Next consider the net $\{u_{\epsilon,K}: \ (\epsilon,K) \in (0,\infty)\times\mathcal{C}\}$, directed as usual by $(\epsilon_1,K_1) \succeq (\epsilon_2,K_2)$ whenever $\epsilon_1 \le \epsilon_2$ and $K_1 \supseteq K_2$. It can easily be seen that $\{u_{\epsilon,K}\}$ is a τ-Cauchy net. However, if $u_{\epsilon,K} \xrightarrow{\tau} u$ holds in L, then since $u_{\epsilon,K}(x) \to u(x)$ must hold for each $x \in [0,1]$, we must have $u = 0$, contradicting the fact $\rho(u_{\epsilon,K}) \xrightarrow[(\epsilon,K)]{} 1$. This shows that (L,τ) is not τ-complete.

We shall identify next the topological completion of (L,τ). Let τ_1 be the Hausdorff locally solid topology generated by the family of Riesz seminorms $\{\rho_K: \ K \in \mathcal{C}\}$ defined above. Note that $(C[0,1],\tau_1)$ is τ_1-complete. Let τ_2 be the Hausdorff locally solid topology generated by the integral norm ρ on $L_1[0,1]$. Obviously $(L_1[0,1],\tau_2)$ is τ_2-complete. Let $M = C[0,1]\times L_1[0,1]$ and $\tau_3 = \tau_1\times\tau_2$ (the product topology). We shall show that (M,τ_3) is isomorphic to $(\widehat{L},\widehat{\tau})$. To see this, consider the mapping $f \mapsto \phi(f) = (f,\dot f)$ from L to M, where $\dot f$ denotes the equivalence class determined by f in $L_1[0,1]$. Note that $f \mapsto \phi(f)$ is a Riesz isomorphism from L onto $\phi(L)$, which is in addition a homeomorphism from (L,τ) onto $(\phi(L),\tau_3)$. Thus, it remains to be shown that $\phi(L)$ is τ_3-dense in M. To this end, let $(g,\dot f) \in M$, $\epsilon > 0$, and $K \in \mathcal{C}$. Select a function $u \in C[0,1]$ with $u(x) = g(x)$ for each $x \in K$ and $\int_0^1 |f(x)-u(x)|\,dx < \epsilon$, and note that $(u,\dot u) \in \phi(L)$ satisfies $\rho_K(u-g) = 0$ and $\rho(\dot f - \dot u) < \epsilon$. Since $\{\rho_K: \ K \in \mathcal{C}\}$ is closed under finite maxima, the latter shows that $\phi(L)$ is τ_3-dense in M.

(2) *L is not a σ-regular subspace of $\widehat{L}$.*

To establish this, pick a sequence $\{u_n\}$ in L with $u_n \downarrow 0$ and with the set $\{x \in [0,1]: \ \inf\{u_n(x)\} > 0\}$ having positive Lebesgue measure. Then $u_n \downarrow 0$ does not hold in $\widehat{L}$.

(3) *If $u_\alpha \downarrow 0$ in L and $\{u_\alpha\}$ is a τ-Cauchy net, then $u_\alpha \xrightarrow{\tau} 0$.*

Indeed, if $u_\alpha \downarrow 0$ and $\{u_\alpha\}$ is a τ-Cauchy net, then $\{u_\alpha\}$ converges uniformly on every countable compact subset of $[0,1]$, say to some function u. But then $u_\alpha(x) \downarrow u(x)$ holds for every $x \in [0,1]$, and since u is continuous on every countable compact subset of $[0,1]$, we have $u \in C[0,1]$. Hence $u = 0$ (since $u_\alpha \downarrow u$ also holds in L). Now from Dini's classical theorem $\{u_\alpha\}$ converges uniformly to zero on $[0,1]$, and from this it follows that $u_\alpha \xrightarrow{\mathcal{T}} 0$.

(4) (L, τ) *satisfies the monotone completeness property* (MCP).

Indeed, if $0 \leq u_\alpha \uparrow$ and $\{u_\alpha\}$ is a τ-Cauchy net, then $u_\alpha(x) \uparrow u(x)$ holds for each $x \in [0,1]$. Since $\{u_\alpha\}$ converges uniformly on every countable compact subset of $[0,1]$, it follows that $u \in C[0,1]$. From Dini's classical theorem, it follows that $\{u_\alpha\}$ converges uniformly to u on $[0,1]$, and from this it follows that $u_\alpha \xrightarrow{\mathcal{T}} u$. ■

We present next a general characterization for the topological completeness of Hausdorff locally solid Riesz spaces. The characterization was obtained by I. Amemiya [**31**] for Dedekind σ-complete spaces and it was generalized by the authors [**16**, Theorem 5.3, p. 32].

THEOREM 2.46 (Amemiya–Aliprantis–Burkinshaw). *For a Hausdorff locally solid Riesz space (L, τ) the following statements are equivalent.*

(i) *(L, τ) is τ-complete, i.e., $L = \widehat{L}$.*
(ii) *(L, τ) satisfies* MCP *and every order interval of L is τ-complete.*

PROOF. (i) $\Longrightarrow$ (ii) Obvious.

(ii) $\Longrightarrow$ (i) Since the order intervals of L are τ-complete, L is an ideal of $\widehat{L}$ by Theorem 2.42. Let $0 \leq \widehat{u} \in \widehat{L}$. Then, by Theorem 2.42 again, there exists a net $\{u_\alpha\} \subseteq L^+$ such that $0 \leq u_\alpha \uparrow \widehat{u}$ and $u_\alpha \xrightarrow{\widehat{\mathcal{T}}} \widehat{u}$. But then $\{u_\alpha\}$ is a τ-Cauchy net of L, and so by our hypothesis $u_\alpha \xrightarrow{\mathcal{T}} u$ holds in L for some u. It follows that $\widehat{u} = u \in L$, so that $L = \widehat{L}$. This proves that (L, τ) is τ-complete. ■

The next theorem gives a condition for the Dedekind and Dedekind σ-completeness to be inherited by $\widehat{L}$. It is essentially due to W. A. J. Luxemburg [**87**, Note XVI, Theorem 66.5, p. 665].

THEOREM 2.47 (Luxemburg). *For a Hausdorff locally solid Riesz space (L, τ) with τ-complete order intervals we have the following.*

(i) *If L is Dedekind σ-complete, then so is $\widehat{L}$.*
(ii) *If L is Dedekind complete, then so is $\widehat{L}$.*

PROOF. Note first that, according to Theorem 2.42, L is an ideal of $\widehat{L}$.

(i) Assume $0 \leq \widehat{u}_n \uparrow \widehat{u}$ holds in $\widehat{L}$. Choose a net $\{w_\alpha\} \subseteq L^+$ with $0 \leq w_\alpha \uparrow \widehat{u}$ and $w_\alpha \xrightarrow{\widehat{\mathcal{T}}} \widehat{u}$. For each α the sequence $\{w_\alpha \wedge \widehat{u}_n\}$ is a sequence in L^+ satisfying $0 \leq w_\alpha \wedge \widehat{u}_n \uparrow_n \leq w_\alpha$. Put $v_\alpha = \sup_n w_\alpha \wedge \widehat{u}_n$ in L. Now since for each α, β and n we have $|w_\alpha \wedge \widehat{u}_n - w_\beta \wedge \widehat{u}_n| \leq |w_\alpha - w_\beta|$, it follows that $|v_\alpha - v_\beta| \leq |w_\alpha - w_\beta|$, which shows that $\{v_\alpha\}$ is a τ-Cauchy net of L. Let $0 \leq v_\alpha \xrightarrow{\widehat{\mathcal{T}}} \widehat{v} \in \widehat{L}$, and therefore by Theorem 2.21(c), we have $v_\alpha \uparrow \widehat{v}$ in $\widehat{L}$. Thus $w_\alpha \wedge \widehat{u}_n \leq v_\alpha \leq \widehat{v}$ holds for each α and n, and so $\widehat{u}_n \leq \widehat{v}$ for all n. Now assume that $\widehat{u}_n \leq \widehat{w}$ holds in $\widehat{L}$ for $n = 1, 2, \ldots$. Then we have $w_\alpha \wedge \widehat{u}_n \leq \widehat{w}$ for all α and all n and so $v_\alpha \leq \widehat{w}$ for all α. Thus $\widehat{v} \leq \widehat{w}$ and hence $\widehat{u}_n \uparrow \widehat{v}$ holds in $\widehat{L}$, so that $\widehat{L}$ is Dedekind σ-complete.

(ii) Repeat the arguments of the preceding proof. ■

The next result also was established for normed Riesz spaces by W. A. J. Luxemburg [**87**, Note XVI, Theorem 66.3, p. 664] and was generalized by C. D. Aliprantis [**3**, Lemma 7.1, p. 122].

THEOREM 2.48 (Luxemburg–Aliprantis). *Let (L, τ) be a Hausdorff locally solid Riesz space and let B be a projection band of L. Then the $\widehat{\tau}$-closure $\overline{B}$ of B in $\widehat{L}$ is a projection band of $\widehat{L}$. Moreover, we have $\overline{B^{\mathrm{d}}} = (\overline{B})^{\mathrm{d}}$, where B^{d} is the disjoint complement of B in L and $(\overline{B})^{\mathrm{d}}$ is the disjoint complement of $\overline{B}$ in $\widehat{L}$.*

PROOF. Combining the last part of Theorem 2.40 with Theorem 2.19(ii), we get immediately that both $\overline{B}$ and $\overline{B^{\mathrm{d}}}$ are ideals in $\widehat{L}$. Now let $\widehat{u} \in \widehat{L}$. Then there exists a net $\{u_\alpha\}$ in L with $u_\alpha \xrightarrow{\widehat{\tau}} \widehat{u}$. Write $u_\alpha = v_\alpha + w_\alpha$ with $\{u_\alpha\} \subseteq B$ and $\{w_\alpha\} \subseteq B^{\mathrm{d}}$; this is possible since $B \oplus B^{\mathrm{d}} = L$. Since now $|u_\alpha - u_\beta| = |v_\alpha - v_\beta| + |w_\alpha - w_\beta|$, it follows that both nets $\{v_\alpha\}$ and $\{w_\alpha\}$ are τ-Cauchy nets in L. So, there exist $\widehat{v} \in \overline{B}$ and $\widehat{w} \in \overline{B^{\mathrm{d}}}$ such that $v_\alpha \xrightarrow{\widehat{\tau}} \widehat{v}$ and $w_\alpha \xrightarrow{\widehat{\tau}} \widehat{w}$. Thus $\widehat{u} = \widehat{v} + \widehat{w} \in \overline{B} + \overline{B^{\mathrm{d}}}$, that is $\widehat{L} = \overline{B} + \overline{B^{\mathrm{d}}}$.

To complete the proof, we must show that $\overline{B} \cap \overline{B^{\mathrm{d}}} = \{0\}$. To establish this, let $0 \le \widehat{u} \in \overline{B} \cap \overline{B^{\mathrm{d}}}$ and then pick two nets $\{u_\alpha\} \subseteq B$ and $\{v_\alpha\} \subseteq B^{\mathrm{d}}$ with $u_\alpha \xrightarrow{\widehat{\tau}} \widehat{u}$ and $v_\alpha \xrightarrow{\widehat{\tau}} \widehat{u}$. But then it follows that $0 = |u_\alpha| \wedge |v_\alpha| \xrightarrow{\widehat{\tau}} \widehat{u}$, and so $\widehat{u} = 0$. The identity $\overline{B^{\mathrm{d}}} = (\overline{B})^{\mathrm{d}}$ should now be immediate. ■

The next result is due to C. D. Aliprantis [**3**, Theorem 7.4, p. 123].

THEOREM 2.49 (Aliprantis). *For a Hausdorff locally solid Riesz space (L, τ) such that L is order dense in $\widehat{L}$ we have the following.*

(i) *If L has sufficiently many projections, then $\widehat{L}$ likewise has sufficiently many projections.*
(ii) *If L has the projection property, then $\widehat{L}$ is Dedekind complete.*

PROOF. (i) Let B be a nonzero band in $\widehat{L}$. Then the order denseness of L in $\widehat{L}$ implies that $B \cap L$ is a nonzero band in L. Pick a nonzero projection band A in L with $A \subseteq B \cap L \subseteq B$. By Theorem 2.48 we know that $\overline{A}$ is a projection band of $\widehat{L}$ and by Theorem 2.21(d) we have $\overline{A} \subseteq B$.

(ii) Let B be a band in $\widehat{L}$. Then $B \cap L$ is a projection band in L and so Theorem 2.48 implies that $\overline{B \cap L}$ $(\subseteq B)$ is a projection band in $\widehat{L}$. Now if $0 \le \widehat{u} \in B$, then $0 \le u_\alpha \uparrow \widehat{u}$ holds in $\widehat{L}$ for some net $\{u_\alpha\} \subseteq B \cap L$. Hence $\widehat{u} \in \overline{B \cap L}$, and consequently $B = \overline{B \cap L}$. This shows that $\widehat{L}$ has the projection property. Now since the Riesz space $\widehat{L}$ is uniformly complete, $\widehat{L}$ must be Dedekind complete according to Theorem 1.59. ■

The order denseness of L in $\widehat{L}$ does not guarantee that the principal projection property is inherited by $\widehat{L}$. An example, due to W. A. J. Luxemburg [**87**, Note XVI, Example 65.6, p. 663], clarifies the situation.

EXAMPLE 2.50 (Luxemburg). Let X be an infinite set and let L be the Riesz space of all functions $f\colon X \to \mathbb{R}$ for which there exists a constant $f(\infty)$ such that $\{x \in X\colon\ f(x) \neq f(\infty)\}$ is a finite subset of X, with the pointwise ordering. Let τ be the Hausdorff locally solid topology generated by the sup norm. The topological completion $\widehat{L}$ of (L,τ) is the Riesz space consisting of all functions $f\colon X \to \mathbb{R}$ for which there exists a constant $f(\infty)$ such that for every $\epsilon > 0$ the set $\{x \in X\colon\ |f(x) - f(\infty)| > \epsilon\}$ is finite. Alternatively, $\widehat{L} = C(X_\infty)$, where X_∞ is the one-point compactification of the set X considered with the discrete topology. Clearly $\widehat{\tau}$ is the topology generated by the sup norm on $\widehat{L}$.

Note that L has the principal projection property, that L is order dense in $\widehat{L}$, but $\widehat{L}$ does not have the principal projection property. However, it does have sufficiently many projections; compare this conclusion with Theorem 2.49. ■

Exercises

1. (Namioka [**104**]) A subset A of an ordered vector space is said to be **full** if for each $u, v \in A$ with $u \leq v$ we have $[u,v] \subseteq A$. A linear topology τ on an ordered vector space is said to be **locally full** if it has a base for the τ-neighborhoods at zero consisting of full neighborhoods.
   (a) Give an example of a full set in a Riesz space that is not solid.
   (b) Give an example of a solid set in a Riesz space that is not full.
   (c) Show that a linear topology τ on a Riesz space is locally solid if and only if it is locally full and for every τ-neighborhood V of zero there exists another τ-neighborhood U of zero such that $u \in U$ implies $u^+ \in V$.

2. Let (L,τ) be a locally solid Riesz space. Denote by $\mathcal{N}$ the collection of all τ-neighborhoods of zero and define the **kernel** of the topology τ by $K_\tau = \bigcap_{V\in\mathcal{N}} V$. Show that K_τ is a τ-closed ideal of L. Show also that τ is a Hausdorff topology if and only if $K_\tau = \{0\}$.

3. Let $\{u_\alpha\}$ be a net in a Hausdorff locally solid Riesz space (L,τ) such that $u_\alpha \uparrow$. Show that if $\{u_\alpha\}$ has an accumulation point u, then $u_\alpha \uparrow u$ in L.

4. Let (L,τ) be a Hausdorff locally solid Riesz space, and let $A \subseteq L$. If $u \in \overline{A}$, show that $u = \inf\{u \vee v\colon\ v \in A\} = \sup\{u \wedge v\colon\ v \in A\}$.

5. Let (L,τ) be a Hausdorff locally solid Riesz space, and let L be order dense in a Riesz space M. If τ can be extended to a locally solid topology τ^* on M, show that τ^* is necessarily a Hausdorff topology.

6. If W denotes the collection of all weak order units in a locally solid Riesz space (L,τ), show that either $W = \emptyset$ or else W is τ-dense in L^+.

7. Let (L,τ) be a Hausdorff locally solid Riesz space. Show that every τ-dense ideal of L is order dense.

8. If (L, τ) is an Archimedean locally solid Riesz space, show that there exists a locally solid topology τ^* on L^δ that induces τ on L.

9. Let (E, τ) be a Hausdorff locally convex space and let $\{p_\alpha\}_{\alpha \in A}$ be a family of τ-continuous seminorms that generate τ. For each α let $\widehat{p}_\alpha \colon \widehat{E} \to \mathbb{R}$ be the unique uniformly $\widehat{\tau}$-continuous extension of the seminorm $p_\alpha \colon E \to \mathbb{R}$. Establish that the family $\{\widehat{p}_\alpha\}_{\alpha \in A}$ of $\widehat{\tau}$-continuous seminorms on $\widehat{E}$ generates the locally convex topology $\widehat{\tau}$.

If (E, τ) is a Hausdorff locally convex-solid space and $\{p_\alpha\}_{\alpha \in A}$ is a family of τ-continuous Riesz seminorms that generates τ, then show that $\{\widehat{p}_\alpha\}_{\alpha \in A}$ is a family of $\widehat{\tau}$-continuous Riesz seminorms on $\widehat{E}$ that generates the locally convex-solid topology $\widehat{\tau}$.

10. Let (L, τ) be a Hausdorff locally solid Riesz space, and let L' and $(\widehat{L})'$ denote the topological duals of (L, τ) and $(\widehat{L}, \widehat{\tau})$, respectively. For every $\phi \in L'$ let $\widehat{\phi}$ be the unique continuous extension of ϕ to $\widehat{L}$ (see Theorem 2.6). Show that $\phi \mapsto \widehat{\phi}$ is a Riesz isomorphism from L' onto $(\widehat{L})'$.

11. Show that a Hausdorff locally solid Riesz space (L, τ) is sequentially τ-complete if and only if every order interval of L is sequentially τ-complete and each increasing τ-Cauchy sequence of L^+ is τ-convergent.

12. Let L_ρ be a seminormed Riesz space and let $\{u_n\} \subseteq L^+$ be a disjoint sequence. Show that $\lim_{n\to\infty} \rho(u - u \wedge u_n) = \rho(u)$ for every $u \in L^+$.

13. (Luxemburg) Show that every seminorm ρ on a finite dimensional Archimedean Riesz space L that satisfies $\rho(u) = \rho(|u|)$ for all $u \in L$ is a Riesz seminorm. (Compare this with Example 2.18.)

14. Let L be the Riesz space consisting of all real sequences with only a finite number of nonzero terms, equipped with the sup norm. Show that $L' = \ell_1$, and give an example of some $\phi \in L_n^\sim$ with $\phi \notin L'$.

15. Show that if the topological dual L' of a Hausdorff locally convex-solid Riesz space (L, τ) admits an order unit, then the topology $|\sigma|(L, L')$ is generated by a Riesz norm and that $\tau = |\sigma|(L, L')$.

16. Consider the Riesz space $L = C[0,1]$ equipped with the Hausdorff locally solid topology τ generated by the family of Riesz seminorms $\{\rho_x \colon x \in [0,1]\} \cup \{\rho\}$, where $\rho_x(u) = |u(x)|$ and $\rho(u) = \int_0^1 |u(x)|\,dx$ for all $u \in C[0,1]$ and $x \in [0,1]$. Determine $(\widehat{L}, \widehat{\tau})$. [*Hint*: Show that $\widehat{L} = \mathbb{R}^{[0,1]} \times L_1[0,1]$.]

17. Let (L, τ) be a locally solid Riesz space. If L is separable, i.e., if it contains a countable τ-dense subset, show that L' is super Dedekind complete.

18. Let L be an Archimedean Riesz space. Consider $L_n^\sim$ and $(L^\delta)_n^\sim$ identified as in Theorem 1.84. If $\{\phi_n\} \subseteq (L_n^\sim)^+$ show that $\{\phi_n\}$ converges to ϕ for $\sigma(L_n^\sim, L)$ if and only if $\{\phi_n\}$ converges to ϕ for $\sigma(L_n^\sim, L^\delta)$. [*Hint*: Approximate an element of L^δ from above and below in the order sense.]

19. Let (L, τ) be a locally solid Riesz space. If L^+ has a nonempty interior, show that $L' = L^\sim$.

20. Prove the following Krein–Milman type of result for locally convex-solid Riesz spaces. *If τ is a (not necessarily Hausdorff) locally convex-solid topology on a Riesz space L with the principal projection property, then for each $u \in L^+$ the convex hull of all extreme points of $[0, u]$ is τ-dense in the order interval $[0, u]$.* [Hint: Use Theorems 1.53 and 1.68 in conjunction with the separation theorem.]

CHAPTER 3

Lebesgue Topologies

In this chapter we shall consider locally solid topologies satisfying an additional continuity property that will be referred to as the Lebesgue property. A Lebesgue topology τ will be defined to be a locally solid topology satisfying $u_\alpha \xrightarrow{\tau} 0$ whenever $u_\alpha \downarrow 0$; and a pre-Lebesgue topology will be a locally solid topology τ such that $\{u_n\}$ is a τ-Cauchy sequence whenever $0 \leq u_n \uparrow \leq u$ holds.

These additional properties on the topology will impose a number of interesting conditions on the structure of the Riesz space. We shall see, for instance, that the Lebesgue property is equivalent to the property that every topologically closed ideal is a band and that the pre-Lebesgue property becomes equivalent to the property that every disjoint order bounded sequence of the Riesz space converges topologically to zero. On the other hand, a number of results will be obtained relating the above properties to the Dedekind and topological completeness.

3.1. Examples and Properties of Lebesgue Topologies

The relation so far between order and topology has been through the locally solid topologies. In this section we shall describe a number of properties connecting order and topological continuity. These properties are described in the following definition.

DEFINITION 3.1. *Let (L, τ) be a locally solid Riesz space. Then we say that:*

(1) *(L, τ) satisfies the **σ-Lebesgue property** (or that τ is a **σ-Lebesgue topology**), if $u_n \downarrow 0$ in L implies $u_n \xrightarrow{\tau} 0$.*

(2) *(L, τ) satisfies the **Lebesgue property** (or that τ is a **Lebesgue topology**), if $u_\alpha \downarrow 0$ in L implies $u_\alpha \xrightarrow{\tau} 0$.*

(3) *(L, τ) satisfies the **pre-Lebesgue property** (or that τ is a **pre-Lebesgue topology**), if $0 \leq u_n \uparrow \leq u$ in L implies that $\{u_n\}$ is a τ-Cauchy sequence.*

It should be clear from the preceding definition that a locally solid Riesz space (L, τ) satisfies the σ-Lebesgue property if and only if $u_n \xrightarrow{\tau} 0$ whenever $u_n \xrightarrow{o} 0$, (L, τ) satisfies the Lebesgue property if and only if $u_\alpha \xrightarrow{\tau} 0$ whenever $u_\alpha \xrightarrow{o} 0$, and (L, τ) satisfies the pre-Lebesgue property if and only if $\{u_n\}$ is a τ-Cauchy sequence whenever $0 \leq u_n \downarrow$ holds.

Historical note. The σ-Lebesgue property seems to have appeared first in the work of L. V. Kantorovich [**71**, p. 153]. For normed Riesz spaces H. Nakano was the first to make a systematic study of these properties. In his terminology a Riesz norm with the σ-Lebesgue property is referred to as a continuous norm, a Riesz norm with the Lebesgue property as a universally continuous norm, and a Riesz norm with the pre-Lebesgue property as a strictly monotone norm; see [**100**, pp. 126–134]. In the terminology of W. A. J. Luxemburg and A. C. Zaanen the Lebesgue and pre-Lebesgue properties are referred to as the (A,i), (A,ii), and (A,iii) properties, respectively [**89**, Note X, p. 501]. The above terminology is due to D. H. Fremlin [**58**]. The names "σ-Lebesgue topology" and "Lebesgue topology" are justified by the well-known Lebesgue dominated convergence theorem from the theory of integration. For a justification of the pre-Lebesgue name see Theorem 3.26.

In modern terminology, a Banach lattice with the Lebesgue property is called a Banach lattice with order continuous norm—and a Riesz norm that generates a Lebesgue topology is called an *order continuous norm*. For a complete discussion of Banach lattices with order continuous norms, see the monograph by W. Wnuk [**128**].

We next illustrate the above properties with examples.

EXAMPLE 3.2. Take $L = C[0,1]$, the Riesz space of all continuous real-valued functions on the interval $[0,1]$ with the pointwise ordering. Pick $1 \leq p < \infty$ and let τ be the Hausdorff locally solid topology generated by the Riesz norm

$$\rho(u) = \left[\int_0^1 |u(x)|^p \, dx\right]^{\frac{1}{p}}.$$

Then τ is a pre-Lebesgue topology on L. Indeed, if $0 \leq u_n \uparrow \leq u$ holds in L, then $u_n(x) \uparrow v(x) < \infty$ holds for each $x \in [0,1]$. Clearly, $v \in L_p[0,1]$ and so

$$\rho(u_m - u_n) \leq \rho(u_m - v) + \rho(u_n - v) \xrightarrow[n,m\to\infty]{} 0.$$

This shows that $\{u_n\}$ is a τ-Cauchy sequence. Note that the topological completion $(\widehat{L}, \widehat{\tau})$ of (L, τ) is $L_p[0,1]$ with $\widehat{\tau}$ being the topology generated by the L_p-norm. It is easy to see that τ is not a σ-Lebesgue topology. Actually, a stronger result (essentially due to W. A. J. Luxemburg and A. C. Zaanen [**89**, Note VII, Example 24.5(ii), p. 674]) holds true:

- *There is no Hausdorff locally solid topology on $C[0,1]$ that satisfies the σ-Lebesgue property.*

To see this, assume by way of contradiction that there exists a Hausdorff locally solid topology τ on $C[0,1]$ with the σ-Lebesgue property. Let $\{r_1, r_2, \ldots\}$ be an enumeration of the rational numbers in $[0,1]$. Now for each pair m and n of natural numbers pick a function $u_{m,n} \in C[0,1]$ such that:

(1) $0 \leq u_{m,n}(x) \leq 1$ for all $x \in [0,1]$.
(2) $u_{m,n}(r_n) = 1$.
(3) $u_{m,n}(x) = 0$ for all $x \in [0,1]$ with $|x - r_n| > \frac{1}{m2^n}$.

Put $v_{m,n} = \sup\{u_{m,i} \colon i = 1, \ldots, n\}$. Then $v_{m,n} \uparrow_n$ for each fixed m and since $v_{m,n}(r_n) = 1$ holds for each n, we see that $v_{m,n} \uparrow_n \mathbf{1}$ holds in L for each fixed m, where $\mathbf{1}(x) = 1$ for each $x \in [0,1]$.

Now let V be a solid τ-neighborhood of zero. Choose a sequence $\{V_n\}$ of solid τ-neighborhoods of zero such that $V_{n+1} + V_{n+1} \subseteq V_n$ holds for each n and with $V_1 = V$. Since $v_{m,n} \uparrow_n \mathbf{1}$ and τ has the σ-Lebesgue property, it follows that $v_{m,n} \xrightarrow[n\to\infty]{\tau} \mathbf{1}$ for each fixed m. Therefore for each $m \in \mathbb{N}$ there exists some $n_m \in \mathbb{N}$ with $\mathbf{1} - v_{m,n_m} \in V_{m+1}$. Observe now that, according to (3), the set $\{x \in [0,1] \colon v_{m,n}(x) > 0\}$ is at most of Lebesgue measure $\frac{1}{m}$ for every n, and in particular for $n = n_m$. Now for each n put $w_n = \inf\{v_{m,n_m} \colon m = 1, 2, \ldots, n\}$. Then $w_n \downarrow$ and the set $\{x \in [0,1] \colon w_n(x) > 0\}$ is at most of Lebesgue measure $\frac{1}{n}$.

Hence $w_n \downarrow 0$ holds in L and so $w_n \xrightarrow{\tau} 0$. Now notice that

$$\begin{aligned} 0 \le \mathbf{1} - w_n &= \bigvee_{m=1}^{n} (\mathbf{1} - v_{m,n_m}) \\ &\le \sum_{m=1}^{n} (\mathbf{1} - v_{m,n_m}) \in V_2 + \cdots + V_{n+1} \subseteq V_1 = V\,. \end{aligned}$$

This implies $\mathbf{1} \in V$ for every solid closed τ-neighborhood V of zero, and consequently $\mathbf{1} = 0$, which is a contradiction. This establishes that the Riesz space $C[0,1]$ does not admit any Hausdorff σ-Lebesgue locally solid topology. ■

EXAMPLE 3.3. Again take L to be $C[0,1]$ and let τ be the Hausdorff locally solid topology generated by the Riesz norm

$$\rho(u) = \max\bigl\{|u(x)| \colon\ x \in [0,1]\bigr\}\,.$$

It can easily be seen that (L, τ) is τ-complete, i.e., L is a Banach lattice without the pre-Lebesgue property. ■

EXAMPLE 3.4. Once more let $L = C[0,1]$, and let τ be the Hausdorff locally solid topology of pointwise convergence, i.e., the topology τ is generated by the family of Riesz seminorms $\{\rho_x\}_{x\in[0,1]}$, where

$$\rho_x(u) = |u(x)|\,,\quad u \in L\,.$$

In this case, (L, τ) has the pre-Lebesgue property, it fails to satisfy the Lebesgue Property, and is not τ-complete. Note that $\widehat{L} = \mathbb{R}^{[0,1]}$ and that $\widehat{\tau}$ is the topology of pointwise convergence in $\mathbb{R}^{[0,1]}$. ■

EXAMPLE 3.5. Let $L = \ell_\infty(X)$, where X is an infinite set, i.e., let L be the Riesz space of all bounded real-valued functions defined on X, with the pointwise ordering. Again let τ be the Hausdorff locally solid topology of pointwise convergence. Then τ is a Lebesgue topology. Note also that $\widehat{L} = \mathbb{R}^X$ with $\widehat{\tau}$ being the topology of pointwise convergence. ■

EXAMPLE 3.6 (Aliprantis [**2**]). Fix $0 < p < \infty$ and let L be the Riesz space consisting of all Lebesgue measurable functions $f \colon [0,1] \to \mathbb{R}$ such that $\int_0^1 |f(x)|^p\,dx < \infty$, with the pointwise ordering. Note that the elements of L are functions and not equivalence classes, that is, functions differing at one point are already considered to be distinct.

Now for each $n \in \mathbb{N}$, each $\epsilon > 0$, and any finite subset F of $[0,1]$, define the set

$$W_{F,n,\epsilon} = \Bigl\{ f \in L \colon\ \int_0^1 |f(x)|^p\,dx < \tfrac{1}{n} \text{ and } |f(x)| < \epsilon \text{ for all } x \in F \Bigr\}\,.$$

As F runs over the finite subsets of $[0,1]$, n over the natural numbers, and ϵ over $(0,\infty)$, we obtain a family of sets $\{W_{F,n,\epsilon}\}$ that is a filter base for a linear topology τ on L. Clearly each $W_{F,n,\epsilon}$ is a solid set. It should also be clear that τ is a Hausdorff topology. Thus, (L, τ) is a Hausdorff locally solid Riesz space. Clearly $u_\alpha \xrightarrow{\tau} 0$ if and only if $u_\alpha(x) \to 0$ for each $x \in [0,1]$ and $\int_0^1 |u_\alpha(x)|^p\,dx \to 0$. Observe also that for $p \ge 1$ the topology τ is a locally convex-solid topology. The locally solid Riesz space (L, τ) has the following interesting properties.

(1) *L is Dedekind σ-complete but not Dedekind complete.*

Indeed, if $0 \le u_n \uparrow \le u$ in L, then $u_n(x) \uparrow v(x)$ holds for each $x \in [0,1]$. This implies that $v \in L$ and that $u_n \uparrow v$ in L so that the Riesz space L is Dedekind

σ-complete. To see that L is not Dedekind complete notice first that $u_\alpha \uparrow u$ in L is equivalent to $u_\alpha(x) \uparrow u(x)$ for each $x \in [0,1]$. Now consider a Lebesgue nonmeasurable subset E of $[0,1]$, and define the net $\{u_\alpha\}$ by letting $u_\alpha = \chi_\alpha$, where α is a finite subset of E. Clearly, $\{u_\alpha\} \subseteq L$, $0 \le u_\alpha \uparrow \le \mathbf{1}$, where $\mathbf{1}(x) = 1$ for all $x \in [0,1]$, and $u_\alpha(x) \uparrow \chi_E(x)$ for all x. It follows that $\sup\{u_\alpha\}$ does not exist in L, and so L is not Dedekind complete.

(2) *(L,τ) satisfies the σ-Lebesgue and pre-Lebesgue properties but not the Lebesgue property.*

If $u_n \downarrow 0$ in L, then $u_n(x) \downarrow 0$ for each $x \in [0,1]$, and the σ-Lebesgue property follows from the Lebesgue dominated convergence theorem. For the pre-Lebesgue property note that if $0 \le u_n \uparrow \le u$, then $u_n \uparrow v$ must hold in L for some v and so $u_n \xrightarrow{\tau} v$, from which it follows that $\{u_n\}$ is a τ-Cauchy sequence. To see that (L,τ) does not satisfy the Lebesgue property consider the net $\{\chi_\alpha\}$, where α runs over all finite subsets of $[0,1]$ directed by letting $\alpha \succeq \beta$ if $\alpha \supseteq \beta$. Note that $0 \le \chi_\alpha \uparrow \mathbf{1}$ in L but $\chi_\alpha \xrightarrow{\tau} \mathbf{1}$ does not hold since $\int_0^1 \big|\mathbf{1}(x) - \chi_\alpha(x)\big|^p\,dx = 1$ for all α.

(3) *L contains a τ-closed σ-ideal that is not a band.*

Let $J = \big\{u \in L\colon\ u = 0 \text{ a.e.}\big\}$. Obviously J is a σ-ideal of L. To see that J is τ-closed let $u_\alpha \xrightarrow{\tau} u$ in L with $\{u_\alpha\} \subseteq J$. This implies

$$\int_0^1 |u(x)|^p\,dx = \int_0^1 |u(x) - u_\alpha(x)|^p\,dx \to 0\,,$$

from which it follows that $\int_0^1 \big|u(x)\big|^p\,dx = 0$, so that $u \in J$. To see that J is not a band of L use the net $\{\chi_\alpha\}$ in (2) above. (Compare this conclusion with Theorems 3.7 and 3.9.)

(4) *(L,τ) is sequentially τ-complete but not τ-complete.*

To see that (L,τ) is not τ-complete use the net $\{\chi_\alpha\}$ of (2). The sequential τ-completeness follows from the Lebesgue dominated convergence theorem.

Next we shall determine the topological completion $(\widehat{L},\widehat{\tau})$ of (L,τ). Consider the Riesz space $K = \mathbb{R}^{[0,1]} \times L_p[0,1]$, where $\mathbb{R}^{[0,1]}$ is equipped with the topology of pointwise convergence and $L_p[0,1]$ is equipped with the topology generated by the "L_p-norm." Note that K with the product topology, is a topologically complete Hausdorff locally solid Riesz space. Define now the mapping $\pi\colon L \to K$ via the formula $\pi(u) = \big(u, \dot{u}\big)$, where $\dot{u}$ denotes the equivalence class determined by u in $L_p[0,1]$. It is not difficult to see that π is a Riesz isomorphism (into) whose image $\pi(L)$ is τ-dense in K. Hence, $\widehat{L} = K$. Observe also that since, for $0 < p < 1$ the topological completion $(\widehat{L},\widehat{\tau})$ is not locally convex-solid, neither is (L,τ). For the computation of the topological dual L' of (L,τ) see Example 5.25. ■

We continue with a characterization, due to T. Andô (see [**87**, Note XIV, Theorem 47.3, p. 244]), of the Lebesgue and σ-Lebesgue topologies.

THEOREM 3.7 (Andô). *For a locally solid Riesz space (L,τ) the following statements are equivalent.*

(i) *(L,τ) satisfies the Lebesgue property.*
(ii) *Every τ-closed ideal of L is a band.*

PROOF. (i) $\Longrightarrow$ (ii) Let J be a τ-closed ideal of L, and let $0 \le u_\alpha \uparrow u$ in L with $\{u_\alpha\} \subseteq J$. Then $u_\alpha \xrightarrow{\tau} u$ and so $u \in J$. Therefore J is a band of L.

(ii) $\Longrightarrow$ (i) Assume that $0 \leq u_\alpha \uparrow u$ holds in L and that U is a solid τ-neighborhood of zero. Pick another solid τ-neighborhood V of zero such that $V + V \subseteq U$ and then pick some $0 < \delta < 1$ with $(1-\delta)u \in V$. Now for each α let $v_\alpha = (u_\alpha - \delta u)^+$ and let J be the ideal generated in L by the net $\{v_\alpha\}$. Since $0 \leq v_\alpha \uparrow (1-\delta)u$ holds in L, it follows that $u \in B_J$, the band generated by J in L. By our hypothesis the closure $\overline{J}$ of J [which is an ideal by Theorem 2.19(iii)] is a band of L and hence $B_J \subseteq \overline{J}$. In particular, it follows that $u \in \overline{J}$ and thus $(u+V) \cap J \neq \emptyset$. Pick some $0 \leq w \in J$ such that $u - w \in V$. Now since $w \in J$, there exists some positive integer n and some index β with $0 \leq w \leq nv_\beta$. But then $(u_\beta - \delta u)^- \wedge v_\beta = 0$ implies $(u_\beta - \delta u)^- \wedge w = 0$, so that $(u_\beta - \delta u)^- + w \wedge u = (u_\beta - \delta u)^- \vee (w \wedge u) \leq u$ and hence $0 \leq (u_\beta - \delta u)^- \leq u - u \wedge w \leq |u - w| \in V$. Thus $(u_\beta - \delta u)^- \in V$. Now note that for each $\alpha \succeq \beta$ we have

$$\begin{aligned} 0 \leq u - u_\alpha &\leq u - u_\beta = (1-\delta)u + (\delta u - u_\beta) \\ &\leq (1-\delta)u + (u_\beta - \delta u)^- \in V + V \subseteq U . \end{aligned}$$

Thus, $u - u_\alpha \in U$ for all $\alpha \succeq \beta$. That is, $u_\alpha \xrightarrow{\tau} u$ and the proof is finished. ■

For Archimedean Riesz space we can add one more characterization of the Lebesgue property.

THEOREM 3.8. *Let (L, τ) be an Archimedean locally solid Riesz space (in particular a Hausdorff locally solid Riesz space). Then the following statements are equivalent.*

(i) *(L, τ) satisfies the Lebesgue property.*
(ii) *Every order dense ideal of L is τ-dense.*
(iii) *Every τ-closed ideal of L is a band.*

PROOF. The implication (i) $\Longrightarrow$ (ii) follows immediately from Theorem 1.27 and (iii) $\Longrightarrow$ (i) was proven in Theorem 3.7. Thus, to complete the logical cycle we have to show that (ii) $\Longrightarrow$ (iii). To this end, let J be a τ-closed ideal of L. By Theorem 1.25 we know that $J \oplus J^{\mathrm{d}}$ is order dense in L, and hence by our hypothesis $J \oplus J^{\mathrm{d}}$ is τ-dense in L. In particular, if $u \in J^{\mathrm{dd}}$, then there exists a net $\{u_\alpha + v_\alpha\}$, where $u_\alpha \in J$ and $v_\alpha \in J^{\mathrm{d}}$, with $u_\alpha + v_\alpha \xrightarrow{\tau} u$. Now from $|u_\alpha + v_\alpha - u| = |u_\alpha - u| + |v_\alpha|$, we easily get $u_\alpha \xrightarrow{\tau} u$. Therefore $u \in \overline{J} = J$. This implies $J = J^{\mathrm{dd}}$, so that J is a band of L, and the proof is finished. ■

The sequential analogue of Theorem 3.7 is as follows—its proof is similar and is omitted.

THEOREM 3.9. *A locally solid Riesz space (L, τ) satisfies the σ-Lebesgue property if and only if every τ-closed ideal of L is a σ-ideal.*

The next result deals with quotient locally solid Riesz spaces and is taken from [**81**].

THEOREM 3.10 (Aliprantis–Langford). *For a locally solid Riesz space (L, τ) and an ideal A of L we have the following.*

(a) *If (L, τ) has the pre-Lebesgue property, then the quotient locally solid Riesz space $(L/A, \tau/A)$ also has the pre-Lebesgue property.*
(b) *If L is Archimedean and (L, τ) has the Lebesgue property, then $(L/A, \tau/A)$ likewise has the Lebesgue property.*

PROOF. (a) Suppose that $0 \le \dot{u}_n \downarrow$ in L/A. Put $v_n = \inf\{u_k^+ \colon\ 1 \le k \le n\}$ for all n and note that $\dot{v}_n = \dot{u}_n$ for each n and that $0 \le v_n \downarrow$ holds in L. It now follows from the pre-Lebesgue property of (L, τ) that $\{v_n\}$ is a τ-Cauchy sequence. From this it easily follows that $\{\dot{u}_n\}$ is a τ/A-Cauchy sequence of L/A. Therefore, $(L/A, \tau/A)$ has the pre-Lebesgue property.

(b) Assume that $\dot{u}_\alpha \downarrow 0$ holds in L/A. Consider the net $\{v_\lambda\}$ consisting of all finite infima of the net $\{u_\alpha^+\}$. Then it should be clear that $0 \le v_\lambda \downarrow$ holds in L and $\dot{v}_\lambda \downarrow 0$ in L/A. In particular, note that if $v \in L$ satisfies $0 \le v \le v_\lambda$ for all λ, then $\dot{v} = 0$ and so $v \in A$. On the other hand, note that since L is Archimedean, there exists an increasing net $\{w_\beta\}$ in L^+ satisfying $0 \le w_\beta \uparrow \le v_\lambda$ for each λ and with $v_\lambda - w_\beta \downarrow_{(\lambda,\beta)} 0$ in L; see statement (v) of Theorem 1.28. By the Lebesgue property of (L, τ), we now infer that $v_\lambda - w_\beta \xrightarrow[(\lambda,\beta)]{\tau} 0$. Since $\dot{v}_\lambda = \dot{v}_\lambda - \dot{w}_\beta$, we get $\dot{v}_\lambda \xrightarrow{\tau/A} 0$. Now it follows from the construction of $\{v_\lambda\}$ that $\dot{u}_\alpha \xrightarrow{\tau/A} 0$, and the proof is finished. ■

If $\{(L_\alpha, \tau_\alpha)\}_{\alpha \in A}$ is a family of locally solid Riesz spaces such that each topology τ_α is a Lebesgue topology, then it is easy to see that the product topology $\tau = \Pi\tau_\alpha$ on $L = \Pi L_\alpha$ is likewise a Lebesgue topology. A similar conclusion holds true if each τ_α satisfies the σ-Lebesgue or the pre-Lebesgue property. Thus we have the following theorem.

THEOREM 3.11. *If every member of an arbitrary family of locally solid Riesz spaces $\{(L_\alpha, \tau_\alpha)\}_{\alpha \in A}$ satisfies the Lebesgue (resp. the σ-Lebesgue or the pre-Lebesgue) property, then the product Riesz space $(\Pi L_\alpha, \Pi\tau_\alpha)$ also satisfies the Lebesgue (resp. the σ-Lebesgue or the pre-Lebesgue) property.*

3.2. Locally Convex-solid Lebesgue Topologies

Recall that for any given Hausdorff locally convex space (E, τ), the **strong topology** $\beta(E', E)$ on its topological dual E' is the locally convex topology of uniform convergence on the τ-bounded subsets of E, where, as usual, a subset A of E is τ-bounded if for every τ-neighborhood V of zero there exists some $\lambda > 0$ such that $\lambda A \subseteq V$. The strong topology $\beta(E', E)$ is the Hausdorff locally convex topology on E' generated by the family of seminorms $\{\rho_B \colon\ B \in \mathcal{B}\}$, where $\mathcal{B}$ is the collection of all τ-bounded subsets of E and

$$\rho_B(\phi) = \sup\{|\phi(u)| \colon\ u \in B\}, \phi \in E',$$

for each $B \in \mathcal{B}$. The **bidual** E'' of E is the topological dual of the Hausdorff locally convex space $(E', \beta(E', E))$.

If now (L, τ) is a Hausdorff locally convex-solid Riesz space, then in order to generate the topology $\beta(L', L)$ it suffices to restrict ourselves to the collection $\mathcal{B}_s$ of all τ-bounded solid subsets of L. In this case each ρ_A becomes a Riesz seminorm. To see this, assume that $A \in \mathcal{B}_s$ and $|\psi| \le |\phi|$ holds in L'. Then, by part (c) of Theorem 1.67 and the solidness of A, for every $v \in A$ we have

$$|\psi(v)| \le |\psi|(|v|) \le |\phi|(|v|) = \sup\{|\phi(u)| \colon\ |u| \le |v|\} \le \rho_A(\phi),$$

and hence $\rho_A(\psi) \le \rho_A(\phi)$, so that ρ_A is a Riesz seminorm. Thus $\beta(L', L)$ is a Hausdorff locally convex-solid topology on L', and hence L'' is an ideal of $(L')^\sim$.

There exists a natural embedding $u \mapsto \widetilde{u}$ of L into its bidual L'' [that is, the topological dual of $\big(L', \beta(L', L)\big)$] defined by

$$\widetilde{u}(\phi) = \phi(u)\,, \quad \phi \in L'\,.$$

By the Hahn–Banach theorem it is easy to see that $u \mapsto \widetilde{u}$ is an algebraic isomorphism (into) that is also a Riesz isomorphism. To see this, assume that $0 \le \phi \in L'$ and $u \in L$, and then use Theorems 1.70 and 1.67 to obtain

$$\begin{aligned} \widetilde{u^+}(\phi) &= \phi(u^+) = \sup\{\psi(u)\colon\ \psi \in L' \text{ and } 0 \le \psi \le \phi\} \\ &= \sup\{\widetilde{u}(\psi)\colon\ \psi \in L' \text{ and } 0 \le \psi \le \phi\} = (\widetilde{u})^+(\phi)\,. \end{aligned}$$

Therefore $\widetilde{u^+} = (\widetilde{u})^+$; see also Theorem 1.83.

In the sequel we shall consider L embedded in L'' according to the above Riesz isomorphism $u \mapsto \widetilde{u}$ and will make no distinction between an element of L and its image in L''. Note that under this identification we have the inclusion $L \subseteq (L')^\sim_{\rm n}$. Indeed, if $\phi_\alpha \downarrow 0$ holds in L', then $u(\phi_\alpha) = \phi_\alpha(u) \downarrow 0$ holds for each $u \in L^+$ by Theorem 1.67 and the desired conclusion follows.

The Lebesgue property in Hausdorff locally convex-solid Riesz spaces is characterized as follows.

THEOREM 3.12. *For a Hausdorff locally convex-solid Riesz space (L, τ) the following statements are equivalent.*

1. *(L, τ) satisfies the Lebesgue property.*
2. *L' is an ideal of $L^\sim_{\rm n}$, i.e., $L' \subseteq L^\sim_{\rm n}$.*
3. *Every band of L' is $\sigma(L', L)$-closed.*
4. *L is a regular Riesz subspace of L''.*
5. *L is an order dense Riesz subspace of $(L')^\sim_{\rm n}$.*
6. *The null ideal N_ϕ is a band of L for each $\phi \in L'$.*
7. *Every order dense ideal of L is τ-dense.*
8. *Every τ-closed ideal of L is a band.*

PROOF. (1) $\Longrightarrow$ (2) Obvious.

(2) $\Longrightarrow$ (3) Let B be a band of L', and let $\{\phi_\alpha\} \subseteq B$ be such that $\phi_\alpha \xrightarrow{\sigma(L',L)} \phi$ holds in L'. Since L' is Dedekind complete (see Theorem 2.22), we can write $\phi = \phi_1 + \phi_2$ with $\phi_1 \in B$ and $\phi_2 \in B^{\rm d}$. It follows that the net $\{\psi_\alpha\}$, where $\psi_\alpha = \phi_\alpha - \phi_1 \in B$, satisfies $\psi_\alpha \xrightarrow{\sigma(L',L)} \phi_2$. Now if $\phi_2 \ne 0$, then either $\phi_2^- \ne 0$ or $\phi_2^+ \ne 0$; we shall assume the latter. Since $\phi_2^+ \in L^\sim_{\rm n}$, we have $C_{\phi_2^+} \ne \{0\}$. Choose $0 < u \in C_{\phi_2^+}$ and note that Theorem 1.82 implies $\phi_2^-(u) = 0$ and $\psi_\alpha(u) = 0$ for all α. Consequently, it follows from

$$\lim_\alpha \psi_\alpha(u) = \phi_2(u) = \phi_2^+(u) - \phi_2^-(u)$$

that $\phi_2^+(u) = 0$, contrary to the choice of u. So, $\phi_2 = 0$ and $\phi = \phi_1 \in B$. Therefore the band B is $\sigma(L', L)$-closed.

(3) $\Longrightarrow$ (4) Let $u_\alpha \downarrow 0$ in L and assume that $0 < \phi \le u_\alpha$ holds in L'' for all α and some ϕ. Fix an index α and let $u = u_\alpha$; thus $0 < \phi \le u$ holds in L''. Next choose some $0 < \delta < 1$ such that $\psi = (\phi - \delta u)^+ > 0$ and then select $0 < f \in C_\psi$. Observe that $f \notin B_f^{\rm d}$ (where B_f is the principal band generated by f in L') and that $B_f^{\rm d}$ is $\sigma(L', L)$-closed by our hypothesis. It now follows from the separation theorem that there exists $v \in L$ such that $f(v) = 1$ and $g(v) = 0$ for all $g \in B_f^{\rm d}$;

we can assume $v > 0$. Put $w = v \wedge \delta u$ and note that $w \neq 0$. [Indeed, if $w = 0$, then $v \wedge u = 0$ and so $v \wedge \psi = 0$, since $0 < \psi \leq u$; but then Theorem 1.82 implies $f(v) = v(f) = 0$, a contradiction.]

Now let $\psi_1 = (w - \phi)^+$ and note that the inequalities

$$0 \leq (w - \phi)^+ \leq (\delta u - \phi)^+ = (\phi - \delta u)^-$$

imply $(w - \phi)^+ \wedge \psi = 0$. So, for every $h \in C_{\psi_1}$ we have $h \perp f$ and thus (by the choice of v above) $h(v) = 0$. Consequently, if $0 \leq h \in C_{\psi_1}$, then we have $0 \leq \psi_1(h) \leq w(h) \leq v(h) = h(v) = 0$ and so $\psi_1(h) = 0$. But then $C_{\psi_1} = \{0\}$, and therefore $\psi_1 = 0$. Hence $0 < w \leq \phi$ holds and thus $0 < w \leq u_\alpha$ for all α, a contradiction. Thus $u_\alpha \downarrow 0$ in L'', so that L is a regular Riesz subspace of L''.

(4) $\Longrightarrow$ (5) Assume $u_\alpha \downarrow 0$ in L. Then by our hypothesis we have $u_\alpha \downarrow 0$ in L'' and so $u_\alpha(\phi) = \phi(u_\alpha) \downarrow 0$ holds for every $0 \leq \phi \in L'$. It now follows that $L' \subseteq L_n^\sim$ and the desired conclusion follows from Theorem 1.83.

(5) $\Longrightarrow$ (6) Let $0 \leq \phi \in L'$ and let $0 \leq u_\alpha \uparrow u$ in L with $\{u_\alpha\} \subseteq N_\phi$. Then $0 \leq u_\alpha \uparrow u$ holds in L'', and so $0 = \phi(u_\alpha) = u_\alpha(\phi) \uparrow u(\phi) = \phi(u)$. Therefore $\phi(u) = 0$ and so $u \in N_\phi$. Thus N_ϕ is a band of L.

(6) $\Longrightarrow$ (7) Let A be an order dense ideal of L and let $\overline{A}$ be its τ-closure. If $\overline{A} \neq L$, then there exists $0 < u \in L$ with $u \notin \overline{A}$. Now by the separation theorem there exists $\phi \in L'$ with $\phi(u) \neq 0$ and $\phi(v) = 0$ for all $v \in A$. Since A is an ideal it follows that $A \subseteq N_\phi$, and so by the order denseness of A in L and the fact that (by our assumption) N_ϕ is a band, we infer that $N_\phi = L$. This implies $\phi = 0$, which is a contradiction. Therefore, $\overline{A} = L$ for each order dense ideal A of L.

(7) $\Longrightarrow$ (8) and (8) $\Longrightarrow$ (1) follow immediately from Theorem 3.8, and the proof is finished. ■

The next result is an easy consequence of the preceding theorem and Theorem 1.40.

THEOREM 3.13. *For a Dedekind complete Hausdorff locally convex-solid Riesz space (L, τ) the following statements are equivalent.*

(i) *(L, τ) satisfies the Lebesgue property.*
(ii) *L is an order dense ideal of $(L')_n^\sim$.*

Historical Note. Most of the equivalent statements of Theorem 3.12 have been obtained by many authors. They have appeared (with a different terminology) in the paper by I. Kawai [**74**, Theorem 5.1, p. 299] and at the same time in the works of H. H. Schaefer [**120**]. Statement (3) of Theorem 3.12 is due to W. A. J. Luxemburg and A. C. Zaanen [**89**, Note XI, Theorem 36.2, p. 517]. Other places where similar results can be found include [**108**, pp. 90–100] and [**130**, Chapter 13].

As an application of Theorem 3.12 we shall characterize the perfect Riesz spaces. To do this, we need their definition and some preliminary discussion.

For a Riesz space L let us denote the Dedekind complete Riesz space $(L_n^\sim)_n^\sim$ by L_{nn}. By Theorem 1.83 we know that there exists a natural Riesz homomorphism $u \mapsto \widetilde{u}$ from L into L_{nn} defined by $\widetilde{u}(f) = f(u)$ for each $f \in L_n^\sim$.

DEFINITION 3.14. *A Riesz space L is said to be a* ***perfect Riesz space*** *if $u \mapsto \widetilde{u}$ is an onto Riesz isomorphism (from L onto L_{nn}).*

The following result should be immediate from the preceding definition.

COROLLARY 3.15. *Every perfect Riesz space is Dedekind complete.*

We continue with the definition of the Levi properties.

DEFINITION 3.16. *A locally solid Riesz space (L,τ) is said to satisfy the* ***Levi property*** *(or that τ is a* ***Levi topology****) if every increasing τ-bounded net of L^+ has a supremum in L.*[1]

By replacing the word "net" with "sequence" in the preceding definition we obtain the **σ-Levi property** (or a **σ-Levi topology**). Clearly, every locally solid Riesz space with the Levi (resp. σ-Levi) property must be Dedekind complete (resp. Dedekind σ-complete). Also a σ-Levi topology on a Riesz space must be Hausdorff; see Exercise 1 at the end of this chapter.

The next three results deal with perfect Riesz spaces and they are due to H. Nakano [**100**]; see also [**89**, Note VIII]. The first one characterizes the perfect Riesz spaces.

THEOREM 3.17 (Nakano). *A Riesz space L is perfect if and only if the absolute weak topology $|\sigma|(L,L_n^\sim)$ is a Levi topology.*

PROOF. Assume first that L is a perfect Riesz space. Also, suppose that a $|\sigma|(L,L_n^\sim)$-bounded net $\{u_\alpha\}_{\alpha\in A}$ in the Riesz space L satisfies $0\le u_\alpha\uparrow$. Then for each $0\le\phi\in L_n^\sim$ we have $F(\phi)=\sup\{\phi(u_\alpha)\colon\ \alpha\in A\}<\infty$, and as is easily checked, the mapping $F\colon (L_n^\sim)^+\to\mathbb{R}^+$ is additive. Therefore, by Lemma 1.64, the mapping F is extendable to a (unique) positive linear functional on all of $L_n^\sim$, that is, $0\le F\in (L_n^\sim)^\sim$. Now observe that if each u_α is identified with its image in L_{nn}, then $0\le u_\alpha\uparrow F$ holds in $(L_n^\sim)^\sim$, and so $F\in L_{nn}$ [since L_{nn} is a band of $(L_n^\sim)^\sim$]. Thus $F=u$ must be the case for some $u\in L$, since L is perfect. But then $u_\alpha\uparrow u$ holds in L. This shows that $|\sigma|(L,L_n^\sim)$ is a Levi topology.

For the converse, assume that $|\sigma|(L,L_n^\sim)$ is a Levi topology. According to Theorem 2.33, we know that $|\sigma|(L,L_n^\sim)$ is a Hausdorff locally convex-solid topology with the Lebesgue property whose topological dual is $L_n^\sim$. It is easy to see now that $u\mapsto\widehat{u}$ is a Riesz isomorphism from L into L_{nn} and so L (identified with its image) is an order dense ideal of L_{nn} by Theorem 3.13. We need only to show that $L=L_{nn}$. To this end, let $0\le F\in L_{nn}$. Pick a net $\{u_\alpha\}$ in L^+ with $0\le u_\alpha\uparrow F$ in L_{nn}. Since $|\phi|(u_\alpha)=u_\alpha(|\phi|)\le F(|\phi|)$ holds for each α and all $\phi\in L_n^\sim$, it follows that $\{u_\alpha\}$ is $|\sigma|(L,L_n^\sim)$-bounded. Therefore, by our hypothesis, $0\le u_\alpha\uparrow u$ holds in L for some u. But then since L is order dense in L_{nn}, $u_\alpha\uparrow u$ also holds in L_{nn}. This implies $F=u\in L$, that is, $L=L_{nn}$. Consequently, L is a perfect Riesz space and the proof is finished. ■

THEOREM 3.18 (Nakano). *The order dual of any Riesz space is a perfect Riesz space.*

PROOF. Let L be an arbitrary Riesz space. According to Theorem 3.17 we must establish that $|\sigma|\big(L^\sim,(L^\sim)_n^\sim\big)$ is a Levi topology. To this end, note that if $0\le\phi_\alpha\uparrow$ holds in $L^\sim$ with the net $\{\phi_\alpha\}$ being $|\sigma|\big(L^\sim,(L^\sim)_n^\sim\big)$-bounded, then $\phi(u)=\sup\{\phi_\alpha(u)\}<\infty$ for all $u\in L^+$ and moreover the mapping $\phi\colon L^+\to\mathbb{R}^+$

[1] Levi topologies were first considered by H. Nakano [**101**]. The term "Levi topology" is due to D. H. Fremlin [**58**, 23I Definition, p. 49], although Fremlin's definition is slightly different than the one given here. The terminology is justified from Beppo Levi's classical theorem from the theory of integration.

is additive. Thus, by Lemma 1.64, it can be extended uniquely to a positive linear functional on L, that is, $\phi \in L^\sim$. It follows now easily that $0 \le \phi_\alpha \uparrow \phi$ holds in $L^\sim$ and so $|\sigma|\big(L^\sim, (L^\sim)^\sim_{\rm n}\big)$ is a Levi topology. ■

THEOREM 3.19 (Nakano). *Every band of a perfect Riesz space is a perfect Riesz space in its own right.*

PROOF. Let B be a band of a perfect Riesz space L. It suffices to show that $|\sigma|(B, B^\sim_{\rm n})$ is a Levi topology. To this end, note first that every $\phi \in L^\sim_{\rm n}$ restricted to B is an order continuous linear functional. Now assume that a net $\{u_\alpha\} \subseteq B$ is $|\sigma|(B, B^\sim_{\rm n})$-bounded and satisfies $0 \le u_\alpha \uparrow$. Then it follows from the above observation that $\{u_\alpha\}$ is also a $|\sigma|(L, L^\sim_{\rm n})$-bounded net. Since L is a perfect Riesz space there exists some $u \in L$ such that $u_\alpha \uparrow u$. But then, since B is a band, we have $u \in B$ and $u_\alpha \uparrow u$ in B. This shows that B is a perfect Riesz space. ■

The sequential analogue of Theorem 3.12 is as follows.

THEOREM 3.20. *For a Hausdorff locally convex-solid Riesz space (L, τ) the following statements are equivalent.*

(1) *(L, τ) satisfies the σ-Lebesgue property.*
(2) *The topological dual of L consists of σ-order continuous linear functionals, i.e., $L' \subseteq L^\sim_{\rm c}$.*
(3) *L is a σ-regular Riesz subspace of $(L')^\sim_{\rm n}$.*
(4) *The null ideal N_ϕ is a σ-ideal of L for each $\phi \in L'$.*
(5) *Every τ-closed ideal of L is a σ-ideal.*

PROOF. (1) $\Longrightarrow$ (2) Obvious.

(2) $\Longrightarrow$ (3) Assume that $u_n \downarrow 0$ in L. Clearly, $\phi(u_n) \downarrow 0$ for each $0 \le \phi \in L'$, and so $u_\alpha \downarrow 0$ holds also in $(L')^\sim_{\rm n}$. That is, L is a σ-regular Riesz subspace of $(L')^\sim_{\rm n}$.

(3) $\Longrightarrow$ (4) Assume that $0 \le \phi \in L'$ and let a sequence $\{u_n\} \subseteq N_\phi$ satisfy $0 \le u_n \uparrow u$ in L. Since $u_n \uparrow u$ holds in $(L')^\sim_{\rm n}$ by our hypothesis, it follows that $0 = \phi(u_\alpha) \uparrow \phi(u)$, and so $\phi(u) = 0$. Hence $u \in N_\phi$ and therefore N_ϕ is a σ-ideal.

(4) $\Longrightarrow$ (5) Let J be a τ-closed ideal of L, and let $\{u_n\} \subseteq J$ satisfy $0 \le u_n \uparrow u$ in L. If $u \notin J$, then by the separation theorem there exists some $\phi \in L'$ such that $\phi(v) = 0$ for all $v \in J$ and with $\phi(u) \ne 0$. But then, since J is an ideal of L, it follows from part (3) of Theorem 1.67 that $J \subseteq N_\phi$ and so (since N_ϕ is a σ-ideal) $u \in N_\phi$. Thus $\phi(u) = 0$, a contradiction. So, J is a σ-ideal.

(5) $\Longrightarrow$ (1) This follows immediately from Theorem 3.9. ■

The final result in this section deals with a useful continuity property of locally convex-solid Riesz spaces.

THEOREM 3.21. *Let (L, τ) be a Hausdorff locally convex-solid Riesz space and let $u_\alpha \downarrow 0$ in L. Then $u_\alpha \xrightarrow{\tau} 0$ holds in L if and only if $u_\alpha \xrightarrow{\sigma(L,L')} 0$.*

PROOF. It should be clear that $u_\alpha \xrightarrow{\tau} 0$ implies $u_\alpha \xrightarrow{\sigma(L,L')} 0$. For the converse, assume that $u_\alpha \xrightarrow{\sigma(L,L')} 0$.

Note first that 0 belongs to the $\sigma(L, L')$-closure of the convex hull $\mathrm{co}\,\{u_\alpha\}$ of the net $\{u_\alpha\}$. But by Theorem 2.4(ii) the $\sigma(L, L')$-closure of $\mathrm{co}\,\{u_\alpha\}$ coincides with the τ-closure of $\mathrm{co}\,\{u_\alpha\}$.

Now let V be a solid τ-neighborhood of zero. By the preceding remark, we have $V \cap \operatorname{co}\{u_\alpha\} \neq \emptyset$, and so there exist a finite number of indices $\alpha_1, \ldots, \alpha_n$ and positive constants $\lambda_1, \ldots, \lambda_n$ with $\sum_{i=1}^n \lambda_i = 1$ such that $\sum_{i=1}^n \lambda_i u_{\alpha_i} \in V$. Now fix any index β such that $\beta \succeq \alpha_i$ for each $i = 1, \ldots, n$ and note that for each $\alpha \succeq \beta$ we have

$$0 \leq u_\alpha \leq u_\beta = \sum_{i=1}^n \lambda_i u_\beta \leq \sum_{i=1}^n \lambda_i u_{\alpha_i} \in V\,.$$

This, coupled with the solidness of V, implies that $u_\alpha \in V$ for all $\alpha \succeq \beta$. That is, $u_\alpha \xrightarrow{\tau} 0$, and the proof is finished. ■

3.3. Lebesgue Properties and L_p-Spaces

Recall that a sequence $\{u_n\}$ in a Riesz space is said to be **disjoint** whenever $|u_n| \wedge |u_m| = 0$ holds for all $n \neq m$. We shall say that a sequence $\{u_n\}$ in a Riesz space is k-**disjoint**, where $k \in \mathbb{N}$, if for every subset I of $\mathbb{N}$ with at least k elements we have $\bigwedge_{i \in I} |u_i| = 0$.

The following important result characterizes the pre-Lebesgue property in terms of disjoint and k-disjoint sequences. It was proven for Banach lattices by P. Meyer-Nieberg [**97**] (see also [**123**, Lemma 5.13, p. 92]) and independently by D. H. Fremlin [**58**, 24H Proposition, p. 56] for locally solid Riesz spaces.

THEOREM 3.22 (Fremlin–Meyer-Nieberg). *For a locally solid Riesz space (L, τ) the following statements are equivalent.*

(1) *(L, τ) satisfies the pre-Lebesgue property.*
(2) *If $0 \leq u_\alpha \uparrow \leq u$ holds in L, then $\{u_\alpha\}$ is a τ-Cauchy net of L.*
(3) *Every order bounded disjoint sequence of L is τ-convergent to zero.*
(4) *Every order bounded k-disjoint sequence of L is τ-convergent to zero.*

PROOF. (1) $\Longrightarrow$ (2) Let $0 \leq u_\alpha \uparrow \leq u$ in L. If the net $\{u_\alpha\}$ is not a τ-Cauchy net, then according to Lemma 2.5 there exists a τ-neighborhood V of zero and an increasing sequence $\{\alpha_n\}$ of indices such that $u_{\alpha_{n+1}} - \alpha_{\alpha_n} \notin V$ holds for each n. But then we have $0 \leq u_{\alpha_n} \uparrow \leq u$ and $\{u_{\alpha_n}\}$ is not a τ-Cauchy sequence, contrary to our hypothesis.

(2) $\Longrightarrow$ (3) Let $\{u_n\}$ be an order bounded disjoint sequence of L. If for each $n \in \mathbb{N}$ we let $w_n = \sum_{i=1}^n |u_i| = \sup\{|u_i|\colon\ i = 1, \ldots, n\}$, then $\{w_n\}$ is an increasing order bounded sequence of L^+ and hence $\{w_n\}$ is a τ-Cauchy sequence of L. It follows now easily that $|u_n| \xrightarrow{\tau} 0$ and from this (since τ is a locally solid topology) we infer that $u_n \xrightarrow{\tau} 0$.

(3) $\Longrightarrow$ (4) We shall use induction on k. Since the only 1-disjoint sequence is the zero sequence, the claim is true for $k = 1$. Now for the induction step, assume that every order bounded k-disjoint sequence of the Riesz space L is τ-convergent to zero.

Let $\{u_n\}$ be a $(k+1)$-disjoint sequence of $[0, u]$. Put $v_1 = 0$ and for $n \geq 2$ let $v_n = \left(u_n - n\sum_{i=1}^{n-1} u_i - \frac{1}{n}u\right)^+$. Clearly $0 \leq v_n \leq u$ for all n and we claim that $\{v_n\}$

is a disjoint sequence. For if $m > n > 1$, then the inequalities

$$\begin{aligned} 0 &\le \tfrac{1}{m}v_m = \Big[\tfrac{1}{m}u_m - \sum_{i=1}^{m-1} u_i - \tfrac{1}{m^2}u\Big]^+ \le \Big[\tfrac{1}{m}u - \sum_{i=1}^{m-1} u_i\Big]^+ \\ &\le \Big[\tfrac{1}{m}u - \sum_{i=1}^{n} u_i\Big]^+ = \Big[\tfrac{1}{m}u - \sum_{i=1}^{n-1} u_i - u_n\Big]^+ \\ &\le \Big[\tfrac{1}{n}u + n\sum_{i=1}^{n-1} u_i - u_n\Big]^+ = \Big[u_n - n\sum_{i=1}^{n-1} u_i - \tfrac{1}{n}u\Big]^- \end{aligned}$$

easily imply that $v_m \wedge v_n = 0$. Hence, by our assumption $v_n \xrightarrow{\mathcal{T}} 0$.

Now put $w_1 = 0$ and $w_n = u_n \wedge \big(n\sum_{i=1}^{n-1} u_i\big)$ for $n \ge 2$. We claim that $\{w_n\}$ is also a k-disjoint sequence of $[0, u]$. Indeed, for any k natural numbers $1 \le n_1 < n_2 < \cdots < n_k$ we have trivially $\inf\{w_{n_i}\colon\ i = 1, \ldots, k\} = 0$ if $n_1 = 1$, and if $n_1 > 1$, then

$$\begin{aligned} 0 &\le \bigwedge_{i=1}^{k} w_{n_i} \le u_{n_1} \wedge u_{n_2} \wedge \cdots \wedge u_{n_k} \wedge \Big(n_1 \sum_{i=1}^{n_1-1} u_i\Big) \\ &\le n_1 \sum_{i=1}^{n_1-1} (u_{n_1} \wedge u_{n_2} \wedge \cdots \wedge u_{n_k} \wedge u_i) = 0\,, \end{aligned}$$

where every term in the last sum is zero by the $(k+1)$-disjointness of $\{u_n\}$. Thus by the induction hypothesis $w_n \xrightarrow{\mathcal{T}} 0$.

Now for $n \ge 2$ the inequalities

$$\begin{aligned} 0 \le u_n &= v_n + u_n \wedge \Big(n\sum_{i=1}^{n-1} u_i + \tfrac{1}{n}u\Big) \\ &\le v_n + u_n \wedge \Big(n\sum_{i=1}^{n-1} u_i\Big) + u_n \wedge \tfrac{1}{n}u \\ &\le v_n + w_n + \tfrac{1}{n}u \end{aligned}$$

imply $u_n \xrightarrow{\mathcal{T}} 0$. The induction is now complete and the validity of (4) has been established.

(4) $\Longrightarrow$ (1) Assume that $0 \le u_n \uparrow \le u$ holds in L. If $\{u_n\}$ is not a τ-Cauchy sequence, then (by passing to a subsequence if necessary) we can assume without loss of generality that there exists a solid τ-neighborhood V of zero such that $u_{n+1} - u_n \notin V$ for all n. Next choose a solid τ-neighborhood W of zero with $W + W \subseteq V$ and then pick some $k \in \mathbb{N}$ such that $\frac{1}{k}u \in W$.

Now for each $n \in \mathbb{N}$ let $w_n = \big(u_{n+1} - u_n - \frac{1}{k}u\big)^+$. We claim that the sequence $\{w_n\}$ is a k-disjoint order bounded sequence of L^+. Indeed, observe first that $0 \le w_n \le u$ holds for all n and then (in view of part (5) of Theorem 1.7) if

$1 \le n_1 < n_2 < \cdots < n_k$ are k arbitrary natural numbers, then we have

$$\begin{aligned} 0 &\le w_{n_1} \wedge \cdots \wedge w_{n_k} \le k\big(w_{n_1} \wedge \cdots \wedge w_{n_k}\big) \\ &\le \Big[\sum_{i=1}^{k} \big(u_{n_{i+1}} - u_{n_i} - \tfrac{1}{k}u\big)\Big]^+ = \big(u_{n_{k+1}} - u_{n_1} - u\big)^+ \\ &\le \big(u_{n_{k+1}} - u\big)^+ = 0\,. \end{aligned}$$

This implies that the order bounded sequence $\{w_n\}$ is k-disjoint.

Now from our hypothesis we get $w_n \xrightarrow{\tau} 0$ and so there exists some n_0 such that $w_n \in W$ holds for all $n \ge n_0$. But then for all $n \ge n_0$ we have

$$u_{n+1} - u_n = w_n + (u_{n+1} - u_n) \wedge \tfrac{1}{k}u \in W + W \subseteq V\,,$$

a contradiction. This contradiction establishes the validity of (1) and completes the proof of the theorem. ∎

The next two results deal with interrelationships between the Lebesgue properties.

THEOREM 3.23. *In Archimedean locally solid Riesz spaces the Lebesgue property implies the pre-Lebesgue property.*

PROOF. Let (L, τ) be an Archimedean locally solid Riesz space with the Lebesgue property and let $0 \le u_n \downarrow$ in L. Since L is Archimedean there exists a net $\{v_\alpha\}$ of L such that $u_n - v_\alpha \downarrow_{(\alpha,n)} 0$ holds in L; see Theorem 1.28. But then, since τ is a Lebesgue topology, it follows that $u_n - v_\alpha \xrightarrow[(\alpha,n)]{\tau} 0$, and from this we easily infer that $\{u_n\}$ is a τ-Cauchy sequence. ∎

Observe that Example 3.2 shows that the pre-Lebesgue property does not imply the Lebesgue property even in the Archimedean case.

THEOREM 3.24. *For a topologically complete Hausdorff locally solid Riesz space (L, τ) the following statements are equivalent.*

(i) *(L, τ) satisfies the Lebesgue property.*

(ii) *(L, τ) satisfies the pre-Lebesgue property.*

Moreover, in this case, L is Dedekind complete.

PROOF. (i) $\Longrightarrow$ (ii) It follows immediately from Theorem 3.23 by observing that L is Archimedean.

(ii) $\Longrightarrow$ (i) Let $u_\alpha \downarrow 0$ in L. Then, according to Theorem 3.22, the net $\{u_\alpha\}$ is τ-Cauchy and hence (since L is τ-complete) $u_\alpha \xrightarrow{\tau} u$ holds in L for some u. Now by part (c) of Theorem 2.21 we have $u_\alpha \downarrow u$ and so $u = 0$. Thus, $u_\alpha \xrightarrow{\tau} 0$ and so τ is a Lebesgue topology.

To see that in this case L is Dedekind complete, let $0 \le u_\alpha \uparrow \le u$ in L. Then again $\{u_\alpha\}$ is a τ-Cauchy net and hence τ-convergent in L, say $u_\alpha \xrightarrow{\tau} v$. This implies $u_\alpha \uparrow v$ in L, and so the Riesz space L is Dedekind complete. ∎

The next example shows that the σ-Lebesgue property does not imply the pre-Lebesgue property.

EXAMPLE 3.25. Let X be an uncountable set and let L be the Riesz space consisting of all functions $f\colon X \to \mathbb{R}$ for which there exists a real number $f(\infty)$ such that the set $\{x \in X\colon\ |f(x) - f(\infty)| > \epsilon\}$ is finite for every $\epsilon > 0$. In other words, $L = C(X_\infty)$, where X_∞ is the one-point compactification of the set X equipped with the discrete topology; see also Example 2.50.

Let τ be the Hausdorff locally convex-solid topology generated by the Riesz norm $\rho(u) = \sup\{|u(x)|\colon\ x \in X\}$, $u \in L$. Then τ is a σ-Lebesgue topology. To see this let $u_n \downarrow 0$ in L. Observe that $u_n(x) \downarrow 0$ for each $x \in X$ and since X is uncountable, there exists some $y \in X$ with $u_n(y) = u_n(\infty)$ for all n. Thus $u_n(x) \downarrow 0$ holds for every $x \in X_\infty$. It follows now from Dini's classical theorem that $u_n \xrightarrow{\tau} 0$.

To establish that (L, τ) does not satisfy the pre-Lebesgue property note that $0 \leq \chi_\alpha \uparrow \mathbf{1}$ holds in L; where α runs over the finite subsets of X and $\mathbf{1}(x) = 1$ for all $x \in X$. However, $\{\chi_\alpha\}$ is not a τ-Cauchy net. ■

The next theorem justifies the pre-Lebesgue name.

THEOREM 3.26. *For a Hausdorff locally solid Riesz space (L, τ) the following statements are equivalent.*

(i) *(L, τ) satisfies the pre-Lebesgue property.*
(ii) *$(\widehat{L}, \widehat{\tau})$ satisfies the pre-Lebesgue property.*
(iii) *$(\widehat{L}, \widehat{\tau})$ satisfies the Lebesgue property.*

PROOF. (i) $\Longrightarrow$ (ii) Assume $0 \leq \widehat{u}_n \downarrow$ holds in $\widehat{L}$ and that U is an arbitrary solid $\widehat{\tau}$-neighborhood of zero. Pick a solid $\widehat{\tau}$-neighborhood V of zero such that $V + V + V \subseteq U$. Next choose a sequence $\{V_n\}$ of solid $\widehat{\tau}$-neighborhoods of zero satisfying $V_1 = V$ and $V_{n+1} + V_{n+1} \subseteq V_n$ for all n.

For each $n \in \mathbb{N}$ choose $v_n \in L^+$ such that $\widehat{u}_n - v_n \in V_{n+1}$ and then let $u_n = \inf\{v_i\colon\ i = 1, \ldots, n\}$ for all n. It should be clear that $0 \leq u_n \downarrow$ holds in L and so since (L, τ) satisfies the pre-Lebesgue property, $\{u_n\}$ is a τ-Cauchy sequence of L. In particular, we have $u_n - u_m \in V$ for all $n, m \geq n_0$. But since

$$\widehat{u}_n - u_n = \widehat{u}_n - \bigwedge_{i=1}^{n} v_i = \bigvee_{i=1}^{n} (\widehat{u}_n - v_i) \leq \bigvee_{i=1}^{n} (\widehat{u}_i - v_i) \leq \sum_{i=1}^{n} |\widehat{u}_i - v_i|$$

and $u_n - \widehat{u}_n \leq \sum_{i=1}^{n} |\widehat{u}_i - v_i|$, it follows that

$$\big|\widehat{u}_n - u_n\big| \leq \sum_{i=1}^{n} |\widehat{u}_i - v_i| \in V_2 + \cdots + V_{n+1} \subseteq V\,,$$

so that $|\widehat{u}_n - u_n| \in V$ holds for each n. Finally for $n, m \geq n_0$ we have

$$|\widehat{u}_n - \widehat{u}_m| \leq |\widehat{u}_n - u_n| + |u_n - u_m| + |u_m - \widehat{u}_m| \in V + V + V \subseteq U\,,$$

and hence $\widehat{u}_n - \widehat{u}_m \in U$ for all $n, m \geq n_0$. Thus $\{\widehat{u}_n\}$ is a $\widehat{\tau}$-Cauchy sequence.

(ii) $\Longrightarrow$ (iii) It follows from Theorem 3.24.

(iii) $\Longrightarrow$ (i) It follows from Theorem 3.23. ■

Combining Theorems 3.26 and 3.24 we also obtain the following result.

THEOREM 3.27. *If a Hausdorff locally solid Riesz space (L, τ) satisfies either the pre-Lebesgue or the Lebesgue property, then $(\widehat{L}, \widehat{\tau})$ satisfies the Lebesgue property and $\widehat{L}$ is Dedekind complete.*

Note. The equivalence of statements (i) and (ii) of Theorem 3.26 as well as Theorem 3.27 were proven first by W. A. J. Luxemburg for normed Riesz spaces [**87**, Note XVI, Theorems 64.1 and 64.3, p. 658] and by I. Kawai in [**74**]. They were generalized in [**3**, p. 112].

Let us call two locally solid Riesz spaces (L,τ) and (L_1,τ_1) **isomorphic** if there exists a Riesz isomorphism from L onto L_1 that is also a homeomorphism; in other words if there exists a mapping from L onto L_1 that preserves all structures, the algebraic, the lattice, and the topological. Let us also say that a locally solid Riesz space (L,τ) is **embeddable** into another locally solid Riesz space (L_1,τ_1) if there exists a Riesz subspace M of L_1 such that (L,τ) is isomorphic to (M,τ_1).

The best-known Banach lattice (i.e., a complete normed Riesz space) without the pre-Lebesgue property is ℓ_∞, that is, ℓ_∞ is the Riesz space consisting of all bounded real sequences with the pointwise ordering and equipped with the sup norm. To see this, consider the sequence $\{u_n\}$ of ℓ_∞ defined by $u_n(k)=0$ if $1\le k\le n$ and $u_n(k)=1$ for $n>k$. Clearly, $u_n\downarrow 0$ in ℓ_∞ and

$$\rho_\infty(u_{n+p}-u_n)=\sup\{|u_{n+p}(m)-u_n(m)|\colon\ m\in\mathbb{N}\}=1$$

holds for all $n,p\in\mathbb{N}$.

Surprisingly, the pre-Lebesgue property in Dedekind σ-complete locally solid Riesz spaces depends upon the embeddability of ℓ_∞ into those spaces. The details follow.

THEOREM 3.28. *For a Dedekind σ-complete locally solid Riesz space (L,τ) the following statements are equivalent.*

(i) *(L,τ) satisfies the pre-Lebesgue property.*
(ii) *ℓ_∞ is not embeddable in (L,τ).*

PROOF. (i) $\Longrightarrow$ (ii) Assume that ℓ_∞ can be embedded in (L,τ). Consider a sequence $\{u_n\}$ in ℓ_∞ with $0\le u_n\downarrow$ in ℓ_∞ that is not a norm Cauchy sequence. But then $\{u_n\}$ (considered as a sequence of L under the embedding) is not a τ-Cauchy sequence either. This shows that (L,τ) does not satisfy the pre-Lebesgue property, a contradiction. This contradiction establishes that ℓ_∞ is not embeddable in (L,τ).

(ii) $\Longrightarrow$ (i) Assume now that τ is not a pre-Lebesgue topology. According to Theorem 3.22 there exist a disjoint sequence $\{u_n\}$ in L^+ and some $u\in L^+$ satisfying $0\le u_n\le u$ for all n and with $\{u_n\}$ not τ-convergent to zero.

By passing to a subsequence if necessary, we can assume that there exists a solid τ-neighborhood V of zero such that $u_n\notin V$ for all n. Next observe that for every $a=(a_1,a_2,\ldots)\in\ell_\infty$ the sequence $\{v_n\}$ of L defined by $v_n=\sum_{i=1}^n a_iu_i$ is order convergent in L (here we use that L is Dedekind σ-complete) and as usual we denote its order limit by $\sum_{n=1}^\infty a_nu_n$. We have thus defined a mapping $\pi\colon \ell_\infty\to L$ via the formula $\pi(a)=\sum_{n=1}^\infty a_nu_n$ for $a=(a_1,a_2,\ldots)\in\ell_\infty$. It can be readily seen that π is an into Riesz isomorphism, which is also continuous, because

$$|\pi(a)|=\Big|\sum_{n=1}^\infty a_nu_n\Big|=\sum_{n=1}^\infty |a_n|u_n=\sup_{n\in\mathbb{N}}|a_n|u_n\le\rho_\infty(a)u\,.$$

On the other hand, to show that π^{-1} is continuous consider a net $\{a_\lambda\}$ of ℓ_∞ satisfying $\pi(a_\lambda)\xrightarrow{\tau}0$ and assume (by way of contradiction) that $\{a_\lambda\}$ does not converge to zero in ℓ_∞. By passing to a subnet if necessary, we can assume that there exists some $\epsilon>0$ such that $\rho_\infty(a_\lambda)\ge 2\epsilon$ for all λ. Now choose an index λ_0 such that $\pi(a_\lambda)\in\epsilon V$ for all $\lambda\succeq\lambda_0$ and then fix any index $\lambda\succeq\lambda_0$. If we let $a_\lambda=(a_1^\lambda,a_2^\lambda,\ldots)$, then choose some k such that $|a_k^\lambda|>\epsilon$ and then note that

$$|\pi(a_\lambda)|=\Big|\sum_{n=1}^\infty a_n^\lambda u_n\Big|=\sum_{n=1}^\infty |a_n^\lambda|u_n\ge|a_k^\lambda|u_k\ge\epsilon u_k\ge 0\,.$$

Thus $\epsilon u_k \in \epsilon V$ and so $u_k \in V$, contrary to the choice of the neighborhood V. Hence the net $\{a_\lambda\}$ converges to zero in ℓ_∞, and therefore π embeds ℓ_∞ in (L, τ). The last conclusion, however, contradicts our hypothesis and the proof is finished. ■

Since by Theorem 3.24 the Lebesgue and pre-Lebesgue properties are equivalent in topologically complete Hausdorff locally solid Riesz spaces, the following result should be immediate.

THEOREM 3.29. *For a topologically complete Dedekind σ-complete Hausdorff locally solid Riesz space (L, τ) the following statements are equivalent.*

(i) *(L, τ) satisfies the Lebesgue property.*
(ii) *ℓ_∞ is not embeddable in (L, τ).*

As an application of Theorem 3.28 we also have the following interesting result.

THEOREM 3.30. *Let (L, τ) be a Dedekind σ-complete locally solid Riesz space that is also separable (in the sense that it contains a countable τ-dense subset). Then (L, τ) satisfies the pre-Lebesgue property.*

PROOF. According to Theorem 3.28 we need only to show that ℓ_∞ is not embeddable in L. So, assume by way of contradiction that ℓ_∞ is embeddable in L. Let A denote the points of ℓ_∞ whose coordinates are either zero or one. Clearly, A is an uncountable set and $\rho_\infty(x-y) = 1$ holds for every pair $x, y \in A$ with $x \neq y$.

Now choose a solid τ-neighborhood V of L such that $V \cap \ell_\infty \subseteq B$, where $B = \{u \in \ell_\infty \colon \ \rho_\infty(u) < 1\}$ is the open unit ball of ℓ_∞. Next select another solid τ-neighborhood W of zero with $W + W \subseteq V$. Clearly, $x - y \notin V$ for every pair $x, y \in A$ with $x \neq y$. On the other hand, if C is a countable τ-dense subset of L, then for each $x \in A$ choose some $u_x \in C$ with $x - u_x \in W$ and note that if $x \neq y$, then $u_x \neq u_y$. Indeed, if $x, y \in A$ satisfy $x \neq y$ and $u_x = u_y$, then it follows from

$$\begin{aligned} |x-y| &\leq |x-u_x| + |u_x - u_y| + |y - u_y| \\ &= |x - u_x| + |y - u_y| \in W + W \subseteq V \end{aligned}$$

that $x - y \in V$, a contradiction. But then it readily follows that C must be an uncountable set, which is impossible. Hence ℓ_∞ is not embeddable in (L, τ), and the proof is finished. ■

Note. For Banach lattices Theorem 3.29 can be found in [**82**, Proposition 2.1, p. 121], [**97**], and [**123**, Theorem 5.14, p. 94]. Theorem 3.30 for normed Riesz spaces has appeared in [**87**, Note XIV, Theorem 43.1, p. 229]; see also [**123**, p. 95].

The rest of the section is devoted to the study of abstract L_p-spaces. Recall that a Riesz seminorm ρ on a Riesz space L is called **p-additive**, where $1 \leq p < \infty$, if $[\rho(u+v)]^p = [\rho(u)]^p + [\rho(v)]^p$ holds for every $u, v \in L^+$ with $u \wedge v = 0$. A Banach lattice L_ρ is called an **abstract L_p-space** if the Riesz norm ρ is p-additive for some $1 \leq p < \infty$. For the rest of this chapter instead of the letter ρ, we shall use the conventional symbol $\|\cdot\|$ to indicate the norm.

The next theorem informs us that normed Riesz spaces with p-additive norms always satisfy the pre-Lebesgue property.

THEOREM 3.31. *Every normed Riesz space with p-additive norm satisfies the pre-Lebesgue property. In particular, any abstract L_p-space is Dedekind complete and satisfies the Lebesgue property.*

PROOF. Let L be a normed Riesz space with p-additive norm, and let $\{u_n\}$ be a disjoint sequence satisfying $0 \leq u_n \leq u$ for all n. From the p-additivity of the norm, it follows that

$$\sum_{n=1}^{k} \|u_n\|^p = \Big\| \sum_{n=1}^{k} u_n \Big\|^p = \Big\| \bigvee_{n=1}^{k} u_n \Big\|^p \leq \|u\|^p$$

for all k and hence the series $\sum_{n=1}^{\infty} \|u_n\|^p$ converges. In particular, $\|u_n\| \to 0$. Now a glance at Theorem 3.22 guarantees that the normed Riesz space L satisfies the pre-Lebesgue property. The last part follows immediately from Theorem 3.24. ■

THEOREM 3.32. *The norm completion of a normed Riesz space with a p-additive norm is a Dedekind complete abstract L_p-space with the Lebesgue property.*

PROOF. Let L be a normed Riesz space with norm completion $\widehat{L}$, and let $\widehat{u}, \widehat{v} \in \widehat{L}$ satisfy $\widehat{u} \wedge \widehat{v} = 0$. Pick two sequences $\{u_n\}$ and $\{v_n\}$ in L^+ such that $\|u_n - \widehat{u}\| \to 0$ and $\|v_n - \widehat{v}\| \to 0$ and observe that $\|u_n \wedge v_n\| \to \|\widehat{u} \wedge \widehat{v}\| = 0$. Next note that $(u_n - u_n \wedge v_n) \wedge (v_n - u_n \wedge v_n) = 0$ holds for each n and so by the p-additivity of the norm we have

$$\big\|(u_n - u_n \wedge v_n) + (v_n - u_n \wedge v_n)\big\|^p = \|u_n - u_n \wedge v_n\|^p + \|v_n - u_n \wedge v_n\|^p$$

for each n. Letting $n \to \infty$, we get

$$\|\widehat{u} + \widehat{v}\|^p = \|\widehat{u}\|^p + \|\widehat{v}\|^p .$$

This shows that $\widehat{L}$ is an abstract L_p-space and the rest of the proof follows immediately from Theorem 3.31. ■

Examples of normed Riesz spaces with p-additive norm are provided by the classical Banach lattices $L_p(X, \Sigma, \mu)$, where $1 \leq p < \infty$ and (X, Σ, μ) is an arbitrary measure space and the norm on $L_p(X, \Sigma, \mu)$ is defined by

$$\|f\|_p = \Big[\int_X |f|^p \, d\mu \Big]^{\frac{1}{p}} .$$

It turns out that these are the only abstract L_p-spaces. More precisely, we shall show next that every abstract L_p-space is lattice isometric to some $L_p(X, \Sigma, \mu)$. (As usual, two normed Riesz spaces are called **lattice isometric** if there exists a norm-preserving Riesz isomorphism from one onto the other.)

Recall that for any non empty subset A of $\mathbb{R}^+$ we put

$$\sum_{a \in A} a = \sup \Big\{ \sum_{a \in F} a \colon \ F \subseteq A \text{ and } F \text{ is finite} \Big\} .$$

Clearly, $\sum_{a \in A} a \leq \infty$.

The following famous result asserts that every abstract L_p-space has a representation of the form $L_p(X, \Sigma, \mu)$. For abstract L_1-spaces it was proven by S. Kakutani [**67**, Theorem 7, p. 532], while H. F. Bohnenblust [**40**] and H. Nakano [**102**, p. 121] established the general result under some additional conditions.

THEOREM 3.33 (Kakutani–Bohnenblust–Nakano). *Every abstract L_p-space is lattice isometric to a concrete Banach lattice $L_p(X, \Sigma, \mu)$.*

PROOF. Let L be an abstract L_p-space. Then, by Theorem 3.31, the Banach lattice L is Dedekind complete and satisfies the Lebesgue property. Assume first that L contains a weak order unit $e > 0$. Let $\mathcal{C}_e$ denote the Boolean algebra of all components of the vector e, that is $\mathcal{C}_e = \{u \in L\colon\ u \wedge (e-u) = 0\}$, and observe that $\mathcal{C}_e$ is Dedekind complete; see Theorem 1.53. Now let S_e be the Riesz subspace of L consisting of all e-step functions, i.e., an element $f \in L$ belongs to S_e if it can be written in the form $f = \sum_{i=1}^{n} \alpha_i u_i$, where $u_1, u_2, \ldots, u_n \in \mathcal{C}_e$ are pairwise disjoint. By the Freudenthal Spectral Theorem 1.58 we know that S_e is order dense in L_e (and hence it is order dense and norm dense in L).

By the well-known Stone's representation theorem (see, for instance, [**90**, Theorem 6.6, p. 27]) there exists a set X and a Boolean algebra $\mathcal{A}$ of subsets of X such that $\mathcal{C}_e$ is isomorphic to $\mathcal{A}$, say via $u \mapsto A_u$. Now by defining $\mu(A_u) = \|u\|^p$ we can easily see that μ defines a σ-additive measure on $\mathcal{A}$, in the sense that if $A_u = \bigcup_{n=1}^{\infty} A_{u_n}$ and the A_{u_n} are pairwise disjoint, then $\mu(A_u) = \sum_{n=1}^{\infty} \mu(A_{u_n})$ holds. This follows easily from the Lebesgue property. Next, by standard arguments, one can generate the exterior measure μ^* of μ and then let Σ be the σ-algebra of all measurable subsets of X. Now, consider the Banach lattice $L_p(X, \Sigma, \mu)$ and define $\pi\colon S_e \to L_p(X, \Sigma, \mu)$ via the formula $\pi(f) = \sum_{i=1}^{n} \alpha_i \chi_{A_{u_i}}$ for all $f = \sum_{i=1}^{n} \alpha_i u_i \in S_e$. Note that π is a well-defined norm preserving Riesz isomorphism (into) whose image is order dense (and hence norm dense) in $L_p(X, \Sigma, \mu)$. It now easily follows from the last observation that π extends (in the obvious manner) to a norm-preserving Riesz isomorphism from L onto $L_p(X, \Sigma, \mu)$ and so the result is established for this case.

Now if L does not contain a weak order unit, consider a complete disjoint system $\{e_i\colon\ i \in I\}$ of non-zero positive elements and note that by what was shown before the band $B_i = B_{e_i}$ is lattice isometric to some $L_p(X_i, \Sigma_i, \mu_i)$, say via $\pi_i\colon B_i \to L_p(X_i, \Sigma_i, \mu_i)$. Without loss of generality we can assume that $X_i \cap X_j = \emptyset$ if $i \neq j$. Put $X = \bigcup_{i \in I} X_i$,

$$\Sigma = \big\{A \subseteq X\colon\ A \cap X_i \in \Sigma_i \text{ for all } i \in I\big\},$$

$\mu(A) = \sum_{i \in I} \mu_i(A \cap X_i)$ for each $A \in \Sigma$ and note that (X, Σ, μ) is a measure space. Now by Theorem 1.75 for each $u \in L^+$ we have $u = \bigvee_{i \in I} u_i$, where $u_i = P_{e_i}(u) \in B_i$, and consequently by the p-additivity of $\|\cdot\|$ we obtain $\|u\|^p = \sum_{i \in I} \|u_i\|^p$. On the other hand, if we define $\pi\colon L^+ \to L_p^+(X, \Sigma, \mu)$ by $\pi(u) = \sum_{i \in I} \pi_i(u_i)$, then π is one-to-one, onto, additive and norm preserving. We leave the details for the reader to complete the proof by verifying (using Lemma 1.64) that π extends to a norm-preserving Riesz isomorphism from L onto $L_p(X, \Sigma, \mu)$. ■

By Theorem 3.32 and the above result the following theorem should be clear now.

THEOREM 3.34. *For every normed Riesz space L whose norm is p-additive there exists a norm-preserving Riesz isomorphism from L onto a norm dense Riesz subspace of some $L_p(X, \Sigma, \mu)$.*

If L is a normed Riesz space, then the strong topology $\beta(L', L)$ on L' is generated by the Riesz norm

$$\begin{aligned} \|\phi\| &= \sup\big\{|\phi(u)|\colon\ u \in L \text{ and } \|u\| \le 1\big\} \\ &= \sup\big\{|\phi|(u)\colon\ u \in L^+ \text{ and } \|u\| = 1\big\} = \big\||\phi|\big\|\,. \end{aligned}$$

Observe that L' is norm complete (i.e., L' is a Banach lattice) and that the natural embedding of L into L'' is norm preserving.

DEFINITION 3.35. *A normed Riesz space L is called an M-**space** (and the norm of L is called an M-**norm**) if:*

$$u \wedge v = 0 \quad \textit{implies} \quad \|u+v\| = \max\{\|u\|, \|v\|\}.$$

*A Banach lattice that is at the same time an M-space is called an **abstract** M-**space** or simply an AM-**space**.*

A typical AM-space with an order unit is the classical Riesz space $C(X)$, where X is a compact Hausdorff topological space, equipped with the sup norm $\|f\|_\infty = \sup\{|f(x)|: x \in X\}$; note that the closed unit ball is simply the order interval $[-\mathbf{1}, \mathbf{1}]$ ($\mathbf{1}(x) = 1$ for all $x \in X$) and that every Riesz subspace of $C(X)$ (with the induced norm) is also an M-space. Another class of M-spaces is provided by the Riesz subspaces of the AM-spaces of the form

$$c_0(A) = \{f \in \mathbb{R}^A \mid \text{The set } \{x \in A: |f(x)| > \epsilon\} \text{ is finite for each } \epsilon > 0\},$$

with the sup norm. Observe that each $c_0(A)$ satisfies the Lebesgue property and that it has a strong unit if and only if the set A is finite (see also Exercise 13 at the end of this chapter.)

LEMMA 3.36. *Let L be a normed Riesz space, let $\phi, \psi \in L'$ satisfy $\phi \wedge \psi = 0$ and let $\epsilon > 0$. Then there exist $u, v \in L^+$ with $u \wedge v = 0$ such that $\|u\| \le 1$, $\|v\| \le 1$, and*

$$\|\phi\| \le \phi(u) + \epsilon \quad \textit{and} \quad \|\psi\| \le \psi(v) + \epsilon.$$

PROOF. Assume that $\phi, \psi \in L'$ satisfy $\phi \wedge \psi = 0$, and let $\epsilon > 0$. Choose x, y in L^+ with $\|x\| = 1$, $\|y\| = 1$ and such that $\|\phi\| \le \phi(x) + \frac{\epsilon}{3}$ and $\|\psi\| \le \psi(y) + \frac{\epsilon}{3}$. Since $[\phi \wedge \psi](x) = 0$, there exist $x_1, x_2 \in L^+$ with $x_1 + x_2 = x$ satisfying $\phi(x_1) + \psi(x_2) < \frac{\epsilon}{3}$. Similarly, there exist $y_1, y_2 \in L^+$ with $y_1 + y_2 = y$ and such that $\phi(y_1) + \psi(y_2) < \frac{\epsilon}{3}$. Put $u = x_2 - y_1 \wedge x_2$, $v = y_1 - y_1 \wedge x_2$, and note that $u, v \in L^+$, $\|u\| \le 1$, $\|v\| \le 1$, $u \wedge v = 0$, and that

$$\begin{aligned} \phi(u) &= \phi(x_2) - \phi(y_1 \wedge x_2) \ge \phi(x_2) - \tfrac{\epsilon}{3} \\ &= \phi(x) - \phi(x_1) - \tfrac{\epsilon}{3} \ge \left[\|\phi\| - \tfrac{\epsilon}{3}\right] - \tfrac{\epsilon}{3} - \tfrac{\epsilon}{3} = \|\phi\| - \epsilon, \end{aligned}$$

and similarly $\psi(v) \ge \|\psi\| - \epsilon$. ■

We now define the class of L-spaces.

DEFINITION 3.37. *A normed Riesz space L is called an L-**space** (and the norm of L is called an L-**norm**) if:*

$$u \wedge v = 0 \quad \textit{implies} \quad \|u+v\| = \|u\| + \|v\|.$$

*A Banach lattice that is at the same time an L-space is called an **abstract** L-**space** or simply an AL-**space**.*

In other words, an AL-space is an abstract L_1-space. The L- and M-spaces are in a dual relationship to each other.

THEOREM 3.38. *For a normed Riesz space L we have the following.*

(i) *L is an L-space if and only if L' is an M-space.*
(ii) *L is an M-space if and only if L' is an L-space.*

PROOF. We shall show first that the norm dual of an L-space is an M-space. To this end, let L be an L-space and assume that $\phi, \psi \in L'$ satisfy $\phi \wedge \psi = 0$. Since the inequality $\|\phi + \psi\| \geq \max\{\|\phi\|, \|\psi\|\}$ is obvious, it suffices to show that $\|\phi + \psi\| \leq \max\{\|\phi\|, \|\psi\|\}$. To establish this inequality, let $\epsilon > 0$. Choose some $w \in L^+$ with $\|w\| \leq 1$ such that $\|\phi+\psi\| \leq [\phi+\psi](w)+\epsilon$. Now since $[\phi\wedge\psi](w) = 0$, there exist $u, v \in L^+$ with $u + v = w$ such that $\phi(u) + \psi(v) < \epsilon$. But then we have

$$\begin{aligned}\|\phi + \psi\| &= \phi(w) + \psi(w) + \epsilon = \phi(v) + \psi(u) + \phi(u) + \psi(v) + \epsilon \\ &\leq \phi(v) + \psi(u) + 2\epsilon \leq \phi(v - v \wedge u) + \psi(u - v \wedge u) + 3\epsilon \\ &\leq \max\{\|\phi\|, \|\psi\|\}\big(\|v - v \wedge u\| + \|u - v \wedge u\|\big) + 3\epsilon \\ &\leq \max\{\|\phi\|, \|\psi\|\} + 3\epsilon\,,\end{aligned}$$

where the last inequality holds by virtue of

$$\|v - v \wedge u\| + \|u - v \wedge u\| = \|u + v - 2(v \wedge u)\| \leq \|w\| \leq 1\,,$$

due to $(v - v \wedge u) \wedge (u - v \wedge u) = 0$ and the fact that the norm of L is an L-norm. The desired conclusion now should be immediate.

Next we show that the norm dual of an M-space is an L-space. To see this, let L be an M-space and assume that $\phi, \psi \in L'$ satisfy $\phi \wedge \psi = 0$. By virtue of the triangle inequality, it is enough to establish that we have $\|\phi\| + \|\psi\| \leq \|\phi + \psi\|$. To this end, let $\epsilon > 0$. By Lemma 3.36 there exist $u, v \in L^+$ with $u \wedge v = 0$, $\|u\| \leq 1$, $\|v\| \leq 1$, and such that $\|\phi\| \leq \phi(u) + \epsilon$ and $\|\psi\| \leq \psi(v) + \epsilon$. Now, since the normed Riesz space L is an M-space, we have $\|u + v\| = \max\{\|u\|, \|v\|\} \leq 1$ and so

$$\|\phi\| + \|\psi\| \leq \phi(u + v) + \psi(u + v) + 2\epsilon \leq \|\phi + \psi\| + 2\epsilon\,,$$

and the desired conclusion follows.

To complete the proof note that if the topological dual L' of a normed Riesz space L is an L-space, then its double norm dual L'' is an M-space and hence the Riesz subspace L of L'' is also an M-space. A similar observation holds true if L' is an M-space, and the proof is finished. ∎

COROLLARY 3.39. *If L is an L-space, then for every $u, v \in L^+$ we have*

$$\|u + v\| = \|u\| + \|v\|\,,$$

That is, the norm is an additive function on the positive cone L^+.

PROOF. Assume that L is an L-space. By Theorem 3.38 the double norm dual L'' of L is an AL-space, which according to Theorem 3.33 is lattice isometric to some $L_1(X, \Sigma, \mu)$-space. Since the norm is additive on the positive cone of $L_1(X, \Sigma, \mu)$ and L can be considered as a Riesz subspace of L'', we easily infer that the norm of the Riesz space L is additive on the positive cone L^+. ∎

An M-space L is said to be an **AM-space with unit** if in addition to being an M-space, L also has an order unit. If L is an AM-space with a unit e, then the formula

$$\|u\|_\infty = \inf\{\lambda > 0\colon\ |u| \leq \lambda e\} = \min\{\lambda \geq 0\colon\ |u| \leq \lambda e\} \qquad (\star)$$

defines a Riesz norm on L that is equivalent to $\|\cdot\|$.[2] Under this new norm, L remains an M-space and its closed unit ball coincides with the order interval $[-e, e]$;

---

[2] This conclusion is not that trivial. We need to show first that the $(L, \|\cdot\|_\infty)$ is a Banach lattice and then invoke Theorem 5.21.

see Exercise 17 at the end of this section. Unless otherwise stated, we will always assume that an M-space with unit is equipped with the norm given by ($\star$) for some order unit e.

The next result presents concrete representations of AM-spaces with order units. It is due to S. Kakutani [**68**] and H. F. Bohnenblust and S. Kakutani [**41**] and independently to Mark and Selim Krein [**79**].

THEOREM 3.40 (Kakutani–Bohnenblust–M. and S. Krein). *A Banach lattice L is an AM-space with unit e if and only if L is lattice isometric to some $C(X)$-space for a unique (up to homeomorphism) compact Hausdorff space X in such a way that the unit e is identified with the constant function $\mathbf{1}$ on X.*

Furthermore, a normed Riesz space L is an M-space if and only if L is lattice isometric to a Riesz subspace of a $C(X)$-space.

PROOF. If $L = C(X)$ for some compact topological space X, then it should be clear that L equipped with the sup norm $\|f\|_\infty = \sup\{|f(x)|\colon\ x \in X\}$ is an AM-space with unit the constant function $\mathbf{1}$.

For the converse assume that L is an AM-space with unit e. Consider the set $Y = \{\phi \in L'_+\colon\ \|\phi\| = 1\}$ and note that

$$Y = \{\phi \in L'_+\colon\ \phi(e) = \|\phi\| = 1\}.$$

Clearly, Y is nonempty, convex, and $\sigma(L', L)$-compact. We shall verify next that the extreme points[3] of Y are precisely the Riesz homomorphisms from L to $\mathbb{R}$ that have the value 1 on e. Indeed, if $\phi \in Y$ is an extreme point of Y and $\psi \in L'$ satisfies $0 < \psi < \phi$, then

$$\phi = \|\psi\|\left(\tfrac{\psi}{\|\psi\|}\right) + \|\phi - \psi\|\left(\tfrac{\phi-\psi}{\|\phi-\psi\|}\right).$$

By Theorem 3.38 the Banach space L' is an AL-space and so, by Corollary 3.39, we have $\|\psi\| + \|\phi - \psi\| = \|\phi\| = 1$ and thus $\phi = \frac{\psi}{\|\psi\|}$. This implies $\psi = \|\psi\|\phi$, and so by Theorem 1.85 the linear functional ϕ is a Riesz homomorphism. On the other hand, if $\phi \in Y$ is a Riesz homomorphism and $\phi = \lambda\phi_1 + (1-\lambda)\phi_2$ with $0 < \lambda < 1$ and $\phi_1, \phi_2 \in Y$, then by Theorem 1.85 we have $\phi = \alpha\lambda\phi_1 = \beta(1-\lambda)\phi_2$ for some $\alpha, \beta \in \mathbb{R}^+$. It follows that $\alpha\lambda = \beta(1-\lambda) = 1$ and so $\phi = \phi_1 = \phi_2$, so that ϕ is an extreme point of Y.

Now let X denote the collection of all extreme points of Y equipped with the $\sigma(L', L)$-topology. Note that (by virtue of the Krein–Milman theorem) X is nonempty and (as easily checked) compact. Next define the mapping $\pi\colon L \to C(X)$ by $\pi(u)(\phi) = \phi(u)$ for all $\phi \in X$ and all $u \in L$. It should be clear that π is a norm-preserving Riesz isomorphism (the preservation of the norm follows from the well-known Krein–Milman theorem; see for example [**21**, Theorem 9.14, p. 137]). Now since $\pi(e)(\phi) = 1$ for all $\phi \in X$, and since the closed Riesz subspace $\pi(L)$ of $C(X)$ separates the points of X, it follows from the lattice version of the classical Stone–Weierstrass theorem (see [**24**, Theorem 11.3, p. 88]) that π is onto, and the proof of the representation is finished.

Now if L is an M-space, then L' is an AL-space, and therefore the formula $e(f) = \|f^+\| - \|f^-\|$ for $f \in L'$ defines a continuous linear functional on L'; see Lemma 1.64. Since $|\phi| \le \|\phi\|e$ for all $\phi \in L''$, it follows that e is an order unit of the AM-space L''. Also, the closed unit ball of L'' is precisely $[-e, e]$. By the above,

[3] Recall that $\phi \in Y$ is an **extreme point** of Y if $\phi = \lambda\phi_1 + (1-\lambda)\phi_2$ with $0 < \lambda < 1$ and $\phi_1, \phi_2 \in Y$ imply $\phi_1 = \phi_2 = \phi$.

L'' is lattice isometric to some $C(X)$-space. Identifying L with a Riesz subspace of L'', we see that L is lattice isometric to a Riesz subspace of $C(X)$. ■

We have seen so far that the M-spaces and the normed Riesz spaces with p-additive norms have the property that the expression $\|u+v\|$ is a function of $\|u\|$ and $\|v\|$ if $u \wedge v = 0$. We shall show next that those are the only Riesz norms which have this characteristic property. To do this we need a lemma due to H. F. Bohnenblust [**40**].

LEMMA 3.41 (Bohnenblust). *Assume that $f\colon \mathbb{R}^+ \times \mathbb{R}^+ \to \mathbb{R}^+$ is a continuous function. Then f is of the form $f(\xi,\eta) = (\xi^p + \eta^p)^{\frac{1}{p}}$ for some $0 < p < \infty$ or $f(\xi,\eta) = \max\{\xi,\eta\}$ if and only if the function f satisfies the following five properties.*

(1) $f(\tau\xi,\tau\eta) = \tau f(\xi,\eta)$ *for all* $\tau,\xi,\eta \in \mathbb{R}^+$.
(2) $f(\xi,\eta) \le f(\xi_1,\eta_1)$ *if* $0 \le \xi \le \xi_1$ *and* $0 \le \eta \le \eta_1$.
(3) $f(\xi,\eta) = f(\eta,\xi)$ *for all* $\xi,\eta \in \mathbb{R}^+$.
(4) $f(0,1) = 1$.
(5) $f\big(\xi, f(\eta,\tau)\big) = f\big(f(\xi,\eta),\tau\big)$ *for all* $\xi,\eta,\tau \in \mathbb{R}^+$.

PROOF. It should be clear that if f is of the described form, then f satisfies the listed properties. Next, we shall establish the converse. So, assume that a continuous function $f\colon \mathbb{R}^+ \times \mathbb{R}^+ \to \mathbb{R}^+$ satisfies the five properties.

Define the sequence $\{a_n\}$ of real numbers recursively by

$$a_1 = 1 \quad \text{and} \quad a_n = f(1, a_{n-1}) \text{ for } n = 2, 3, \ldots .$$

The sequence $\{a_n\}$ has the following two properties.

(i) $a_{n+m} = f(a_n, a_m)$ for all $n, m \in \mathbb{N}$.

To see this, we use induction on the natural number m. For $m = 1$ we have $a_{n+1} = f(1,a_n) = f(a_n,1)$ and if $a_{n+m} = f(a_n,a_m)$, then

$$\begin{aligned} a_{n+m+1} &= f(1,a_{n+m}) = f\big(1, f(a_m,a_n)\big) = f\big(f(1,a_m),a_n\big) \\ &= f(a_{m+1},a_n) = f(a_n,a_{m+1}) . \end{aligned}$$

(ii) $a_{nm} = a_n a_m$ for all $n, m \in \mathbb{N}$.

Again we use induction on m. For $m = 1$ the result is trivial and if $a_{nm} = a_n a_m$ holds, then

$$a_n a_{m+1} = a_n f(1,a_m) = f(a_n, a_n a_m) = f(a_n, a_{nm}) = a_{n+nm} = a_{n(m+1)} .$$

Now since $\{a_n\}$ is a nondecreasing sequence two cases arise.

Case 1. $a_2 = 1$, *that is,* $f(1,1) = 1$.

In this case, notice that for $\xi \ge \eta > 0$ we have

$$\xi = \xi f(1,0) \le \xi f\big(1,\tfrac{\eta}{\xi}\big) = f(\xi,\eta) \le \xi f(1,1) = \xi ,$$

and hence $f(\xi,\eta) = \max\{\xi,\eta\}$.

Case 2. $a_2 > 1$ (*and hence* $a_n > 1$ *for each* n).

Let $m, n > 1$. Then for every $k \in \mathbb{N}$ let h be the unique natural number satisfying $m^h \le n^k \le m^{h+1}$, and note that since $(a_m)^h = a_{m^h}$, and similarly for

$h+1$, we get $h\ln a_m \le k\ln a_n \le (h+1)\ln a_m$. Now $h\ln m \le k\ln n \le (h+1)\ln m$ implies that

$$\frac{\ln a_m}{\ln m} - \frac{\ln a_m}{k\ln n} \le \frac{(h+1)\ln a_m}{k\ln n} - \frac{\ln a_m}{k\ln n} = \frac{h\ln a_m}{k\ln n} \le \frac{k\ln a_n}{k\ln n} = \frac{\ln a_n}{\ln n}.$$

Similarly, we get $\frac{\ln a_m}{\ln m} + \frac{\ln a_m}{k\ln n} \ge \frac{\ln a_n}{\ln n}$, and consequently

$$\frac{\ln a_m}{\ln m} - \frac{\ln a_m}{k\ln n} \le \frac{\ln a_n}{\ln n} \le \frac{\ln a_m}{\ln m} + \frac{\ln a_m}{k\ln n}$$

for all k. This implies $\frac{\ln a_m}{\ln m} = \frac{\ln a_n}{\ln n}$. That is, the value of $\frac{\ln a_n}{\ln n}$ is independent of n. Put $p = \frac{\ln 2}{\ln a_2}$ and observe that $a_n = n^{\frac{1}{p}}$. But then it easily follows that $f(1, r^{\frac{1}{p}}) = (1+r)^{\frac{1}{p}}$ holds for all positive rational numbers r and from this and the continuity of f we get that $f(\xi,\eta) = \big(\xi^p + \eta^p\big)^{\frac{1}{p}}$. ■

The next result also is due to H. F. Bohnenblust [**40**].

THEOREM 3.42 (Bohnenblust). *For a normed Riesz space L of dimension at least three, the following statements are equivalent.*

(i) *The norm of L is either an M-norm or a p-additive norm.*
(ii) *There exists a function $f\colon \mathbb{R}^+ \times \mathbb{R}^+ \to \mathbb{R}^+$ satisfying*

$$\|u+v\| = f\big(\|u\|, \|v\|\big)$$

for every pair $u, v \in L^+$ with $u \wedge v = 0$.

PROOF. (i) $\Longrightarrow$ (ii) If the norm of the Riesz space L is an M-norm let $f(\xi,\eta) = \max\{\xi,\eta\}$, and if the norm of L happens to be p-additive let $f(\xi,\eta) = \big(\xi^p+\eta^p\big)^{\frac{1}{p}}$.

(ii) $\Longrightarrow$ (i) Assume that the norm satisfies (ii). Since L has dimension at least three, there exist three pairwise disjoint elements u, v, and w in L^+ each of norm one. But then $f(\xi,\eta) = \|\xi u + \eta v\|$ holds for all $\xi,\eta \in \mathbb{R}^+$. It follows from the last formula that f is continuous satisfying the five conditions of Lemma 3.41; where the fifth property holds true by virtue of

$$\|\xi u + \eta v + \tau w\| = f\big(f(\xi,\eta),\tau\big) = f\big(\xi, f(\eta,\tau)\big).$$

So, by Lemma 3.41, we have either $f(\xi,\eta) = \big(\xi^p+\eta^p\big)^{\frac{1}{p}}$ for some $0 < p < \infty$ or $f(\xi,\eta) = \max\{\xi,\eta\}$. The result now follows by observing also that in the first case $p \ge 1$ must hold true due to the convexity of the norm. ■

Exercises

1. Show that every locally solid Riesz space (L,τ) with the σ-Levi property is a Hausdorff topological space. [*Hint*: Consider the intersection K_τ of all τ-neighborhoods of zero and note that if $0 \le u \in K_\tau$, then $\{nu\}$ is an increasing τ-bounded sequence.]
2. Show that a Hausdorff locally convex-solid topology τ on a Riesz space L is Lebesgue if and only if $|\sigma|(L,L')$ is Lebesgue. [*Hint*: Use Theorem 3.21; for another proof see Theorem 4.9.]

3. Show that a Hausdorff locally convex-solid Riesz space (L, τ) satisfies the Lebesgue property if and only if $C_\phi \neq \{0\}$ for each $0 < \phi \in L'$. [*Hint*: Show that every order dense ideal is τ-dense.]

4. Let (L, τ) and (M, τ^*) are two Hausdorff locally convex-solid Riesz spaces such that τ is a σ-Lebesgue topology and let $T\colon \bigl(L, \sigma(L, L')\bigr) \to \bigl(M, \sigma(M, M')\bigr)$ be a continuous positive operator. Show that T is σ-order continuous.

5. Let (L, τ) be a Hausdorff locally solid Riesz space with the pre-Lebesgue property. Show that if L is infinite dimensional, then the positive cone L^+ does not have interior points. [*Hint*: If $0 < u$ is an interior point of L^+, then $[-u, u]$ is a τ-neighborhood of zero and u is an order unit.]

6. By a **topological completion** of a locally solid Riesz space (L, τ) we mean any topologically complete locally solid Riesz space (L^*, τ^*) (not necessarily Hausdorff) including (L, τ) as a topologically τ^*-dense Riesz subspace (that is, L is a τ^*-dense Riesz subspace of L^* and τ^* induces τ on L).

(i) Give an example of a non-Hausdorff topological completion of some Hausdorff locally solid Riesz space.

(ii) If (L^*, τ^*) is a topological completion of a locally solid Riesz space (L, τ), show that (L, τ) satisfies the pre-Lebesgue property if and only if (L^*, τ^*) does.

7. Show by means of a counterexample that the hypothesis of Dedekind σ-completeness in Theorem 3.30 cannot be dropped.

8. Let $L = \ell_\infty$ and let A be the ideal of L consisting of all null sequences, that is, $A = c_0$. Show that $A^\sim_{\rm n} = \ell_1$ and $A_{\rm nn} = \ell_\infty$, and conclude from this example that an ideal of a perfect Riesz space need not be perfect.

9. Show that a Riesz space L is perfect if and only if it admits a Lebesgue locally convex-solid topology with the Levi property.

10. Generalize Theorem 3.18. That is, show that if L is any Riesz space and M any perfect Riesz space, then $\mathcal{L}_{\rm b}(L, M)$ is a perfect Riesz space.

11. Show that a normed Riesz space L is an M-space if and only if for all $u, v \in L^+$ we have $\|u \vee v\| = \max\bigl\{\|u\|, \|v\|\bigr\}$.

12. Let L be a Riesz space, A an ideal of L, and ρ a p-additive Riesz seminorm on L. Show that the quotient Riesz seminorm $\dot\rho$ is p-additive on L/A. Prove a similar statement if ρ is an M-seminorm.

13. (Bohnenblust) Show that an abstract M-space L_ρ satisfies the Lebesgue property if and only if L_ρ is lattice isometric to some $c_0(X)$, equipped with the sup norm. [*Hint*: Show first that L is a discrete Riesz space and then use Theorem 1.78.]

14. Show that every abstract L_p-space is a perfect Riesz space.

15. Let L be an almost Dedekind σ-complete Riesz space. Show that every σ-Lebesgue topology on L is necessarily a pre-Lebesgue topology.

16. Let M be a regular Riesz subspace of a Riesz space L. If τ is a Hausdorff locally convex-solid Lebesgue topology on L, show that M is order dense in L if and only if M is τ-dense in L.

17. Assume that L is an AM-space with an order unit e. Show that the formula

$$\|u\|_\infty = \inf\bigl\{\lambda > 0\colon\ |u| \le \lambda e\bigr\} = \min\bigl\{\lambda \ge 0\colon\ |u| \le \lambda e\bigr\}$$

defines a Riesz norm on L that is equivalent to $\|\cdot\|$.

CHAPTER 4

Fatou Topologies

As we have seen in the previous chapter the full richness of locally solid Riesz spaces can be realized when there is a further interaction between the order structure and the topological structure.

In this chapter we shall see that when the topology has a base at zero consisting of solid and order closed neighborhoods we are led to some of the deepest results of the theory of locally solid Riesz spaces. Two of the major results of this chapter are the striking theorem that all Hausdorff Lebesgue topologies on a Riesz space induce the same topology on the order bounded sets, and Nakano's theorem relating Dedekind completeness and topological completeness.

4.1. Basic Properties of the Fatou Topologies

We start with the introduction of the Fatou topologies.

DEFINITION 4.1. *A locally solid Riesz space (L, τ) is said to satisfy the* ***Fatou property*** *(or that τ is a* ***Fatou topology****) if τ has a base at zero consisting of solid and order closed sets.*

Similarly, (L, τ) satisfies the ***σ-Fatou property*** *(or that τ is a* ***σ-Fatou topology****) if τ has a base at zero consisting of solid and σ-order closed sets.*

A solid and order closed neighborhood of zero will be called a **Fatou neighborhood**. A **σ-Fatou neighborhood** is a solid and σ-order closed neighborhood of zero. Clearly, the Fatou property implies the σ-Fatou property.

LEMMA 4.2. *Every Lebesgue topology is a Fatou topology.*

PROOF. If a locally solid Riesz space (L, τ) satisfies the Lebesgue property, then every solid and τ-closed subset of L is order closed. ■

A Riesz pseudonorm ρ on a Riesz space L is said to be a **Fatou pseudonorm** (or that ρ satisfies the **Fatou property**) if $0 \leq u_\alpha \uparrow u$ in L implies $\rho(u_\alpha) \uparrow \rho(u)$. Similarly, a Riesz pseudonorm ρ on a Riesz space L is said to be a **σ-Fatou pseudonorm** (or that ρ satisfies the **σ-Fatou property**) whenever $0 \leq u_n \uparrow u$ in L implies $\rho(u_n) \uparrow \rho(u)$.

EXAMPLE 4.3. It should be clear that any topology generated by a Fatou (resp. σ-Fatou) pseudonorm is a Fatou (resp. σ-Fatou) topology. The sup norm on $C[0, 1]$ (and on any $C(X)$-space with X Hausdorff and compact) is a Fatou Riesz norm, and so the Hausdorff locally solid topology τ generated by the sup norm on $C[0, 1]$ is a Fatou topology. However, note that τ is not a Lebesgue topology. ■

EXAMPLE 4.4. Take $L = C[0, 1] \times \mathbb{R}^{[0,1]}$ and let τ be the product topology generated by the sup norm on $C[0, 1]$, and the topology of pointwise convergence on $\mathbb{R}^{[0,1]}$. Observe that (L,τ) satisfies the Fatou property, but it does not satisfy the Lebesgue property. ■

The next three results will explain why the Fatou properties are so named.

THEOREM 4.5. *For a Riesz seminorm ρ on a Riesz space L the following statements are equivalent.*

(i) *The Riesz seminorm ρ satisfies the σ-Fatou property, that is, $0 \leq u_n \uparrow u$ in L implies $\rho(u_n) \uparrow \rho(u)$.*
(ii) *If $u_n \xrightarrow{o} u$ holds in L, then $\rho(u) \leq \liminf \rho(u_n)$.*
(iii) *The closed unit ball $U_\rho = \{u \in L\colon\ \rho(u) \leq 1\}$ is σ-order closed.*

PROOF. (i) $\Longrightarrow$ (ii) Assume that $u_n \xrightarrow{o} u$ holds in L. Pick a sequence $\{v_n\} \subseteq L$ with $|u_n - u| \leq v_n \downarrow 0$ in L. Now use the inequalities $\big(|u| - v_n\big)^+ \leq |u_n|$ and $0 \leq \big(|u| - v_n\big)^+ \uparrow |u|$ to obtain that $\rho(u) \leq \liminf \rho(u_n)$.

(ii) $\Longrightarrow$ (iii) Assume that a sequence $\{u_n\} \subseteq U_\rho$ satisfies $u_n \xrightarrow{o} u$. Note that $\rho(u_n) \leq 1$ for all n and $\rho(u) \leq \liminf \rho(u_n)$ imply $\rho(u) \leq 1$ and so $u \in U_\rho$. This shows that U_ρ is a σ-order closed set.

(iii) $\Longrightarrow$ (i) Let $0 \leq u_n \uparrow u$ in L. Put $a = \lim \rho(u_n)$. We need only to show that $\rho(u) \leq a$. Observe that given any $\epsilon > 0$ the relations $(a+\epsilon)^{-1}u_n \xrightarrow{o} (a+\epsilon)^{-1}u$ and $(a+\epsilon)^{-1}u_n \in U_\rho$ for all n imply easily that $(a+\epsilon)^{-1}u \in U_\rho$, and hence $\rho(u) \leq a+\epsilon$ holds for all $\epsilon > 0$. Thus, $\rho(u) \leq a$, and the proof is finished. ■

The analogue of Theorem 4.5 for Fatou seminorms is as follows. Its proof is similar and is omitted.

THEOREM 4.6. *For a Riesz seminorm ρ on a Riesz space L the following statements are equivalent.*

(i) *The Riesz seminorm ρ satisfies the Fatou property, that is, $0 \leq u_\alpha \uparrow u$ in L implies $\rho(u_\alpha) \uparrow \rho(u)$.*
(ii) *If $u_\alpha \xrightarrow{o} u$ holds in L, then $\rho(u) \leq \liminf \rho(u_\alpha)$.*
(iii) *The closed unit ball $U_\rho = \{u \in L\colon\ \rho(u) \leq 1\}$ is order closed.*

The proof of the next result (that is due to D. H. Fremlin) can be found in [**58**, 23B Theorem, p. 44].

THEOREM 4.7 (Fremlin). *For a locally solid Riesz space (L,τ) the following statements are equivalent.*

(1) *τ is a Fatou topology.*
(2) *There exists a family $\{\rho_\alpha\}$ of Fatou pseudonorms that generates the topology τ.*

Historical Note. The term "Fatou seminorm" is due to W. A. J. Luxemburg and A. C. Zaanen [**89**, Notes II and III] and it was obviously inspired from the well-known Fatou's lemma from the theory of integration, where $L = L_1(X,\Sigma,\mu)$ and $\rho(u) = \int |u|\,d\mu$. The term "Fatou topology" was subsequently introduced by D. H. Fremlin in [**57**]. Fatou and σ-Fatou norms were first studied by H. Nakano [**103**, § 30] under a different terminology.

We have seen so far that the Lebesgue property implies the Fatou property. The next theorem characterizes the Lebesgue property in terms of the Fatou property.

THEOREM 4.8. *For an Archimedean locally solid Riesz space* (L, τ) *the following statements are equivalent.*

(i) (L, τ) *satisfies the Lebesgue property.*

(ii) (L, τ) *satisfies both the Fatou and the pre-Lebesgue properties.*

PROOF. (i) $\Longrightarrow$ (ii) It follows immediately from Theorem 3.23 and Lemma 4.2.

(ii) $\Longrightarrow$ (i) Assume that $u_\alpha \downarrow 0$, and let V be an order closed τ-neighborhood of zero. By Theorem 3.22 the net $\{u_\alpha\}$ is τ-Cauchy, and so there exists an index α_0 such that $u_\alpha - u_\beta \in V$ for all $\alpha, \beta \succeq \alpha_0$. Now notice that for each fixed $\alpha \succeq \alpha_0$ we have $0 \le u_\alpha - u_\beta \uparrow_{\beta \succeq \alpha_0} u_\alpha$ and so, since V is order closed, we see that $u_\alpha \in V$ for all $\alpha \succeq \alpha_0$. Hence $u_\alpha \xrightarrow{\tau} 0$ so that (L, τ) satisfies the Lebesgue property. ■

As an application of the preceding result we have the following.

THEOREM 4.9. *For a Hausdorff locally convex-solid Riesz space* (L, τ) *the following statements are equivalent.*

(i) $|\sigma|(L, L')$ *is a Fatou topology.*

(ii) $|\sigma|(L, L')$ *is a Lebesgue topology.*

(iii) $L' \subseteq L_n^\sim$.

(iv) τ *is a Lebesgue topology.*

PROOF. To see that (i) $\Longrightarrow$ (ii) observe that $\big(L, |\sigma|(L, L')\big)$ satisfies the pre-Lebesgue property. Indeed, if $0 \le u_n \uparrow \le u$ holds in L, then $\{\phi(u_n)\}$ is a Cauchy sequence for every $0 \le \phi \in L'$, and so $\{u_n\}$ is a $|\sigma|(L, L')$-Cauchy sequence. The result now follows from Theorem 4.8. For the other implications use Theorem 3.12 in conjunction with the inclusions $\sigma(L, L') \subseteq |\sigma|(L, L') \subseteq \tau$. ■

THEOREM 4.10. *Let* (L, τ) *be a locally solid Riesz space with the Fatou property, and let* M *be a regular Riesz subspace of* L *(in particular, let* M *be an order dense Riesz subspace or an ideal of* L*). Then* τ *induces a Fatou topology on* M.

PROOF. If V is a Fatou τ-neighborhood of zero, then $V \cap M$ is a solid and order closed subset of M since M is a regular Riesz subspace of L. ■

The next example shows that the hypothesis of M being a regular Riesz subspace in the preceding theorem cannot be dropped.

EXAMPLE 4.11. Let $L = \mathbb{R}^{[0,1]}$ and let τ be the Hausdorff Lebesgue topology of pointwise convergence. Take $M = C[0,1]$ and note that the Riesz subspace M is not regular. Now observe that τ restricted to M is not a Fatou topology. ■

We continue with a useful extension theorem. It was proved first by I. Kawai [**74**, Theorem 1.5, p. 288] for locally convex-solid topologies with the Lebesgue property and was generalized by the authors [**16**, Theorem 2.1, p. 26].

THEOREM 4.12 (Kawai–Aliprantis–Burkinshaw). *Let* L *be an Archimedean Riesz space. If* τ *is a Fatou topology on* L*, then there exists a unique Fatou topology* τ^δ *on the Dedekind completion* L^δ *of* L *that induces* τ *on* L.

If in addition τ *is Hausdorff, then so is* τ^δ *and if* τ *is a Lebesgue topology, then likewise is* τ^δ.

PROOF. Let $\mathcal{N}$ be a base at zero for τ consisting of solid and order closed sets. For each $V \in \mathcal{N}$ we define the subset V^δ of L^δ by

$$V^\delta = \big\{u^* \in L^\delta \colon \exists \{u_\alpha\} \subseteq V \text{ such that } 0 \le u_\alpha \uparrow |u^*| \text{ in } L^\delta\big\}.$$

Observe that since V is solid and order closed and L is order dense in L^δ, we have

$$V^\delta = \{u^* \in L^\delta\colon\ v \in L \text{ and } 0 \le v \le |u^*| \text{ implies } v \in V\}.$$

Note that each V^δ is a solid, absorbing, and order closed subset of L^δ. To see that V^δ is an order closed set, assume that a net $\{u^*_\alpha\} \subseteq V^\delta$ satisfies $0 \le u^*_\alpha \uparrow u^*$ in L^δ. Since L is order dense in L^δ, for each u^*_α there exists a net of $V \cap L^+$ that increases to u^*_α. The net consisting of the finite suprema of these nets is a net of $V \cap L^+$ that increases to u^*, so that $u^* \in V^\delta$. Observe also that $(V \cap W)^\delta \subseteq V^\delta \cap W^\delta$ and that $W + W \subseteq V$ in L implies $W^\delta + W^\delta \subseteq V^\delta$ in L^δ. Moreover note that $V^\delta \cap L = V$. Thus, the collection $\mathcal{N}^\delta = \{V^\delta\colon\ V \in \mathcal{N}\}$ is a base at zero for a Fatou topology τ^δ on L^δ that induces τ on L.

Now let τ^* be a Fatou topology on L^δ that induces τ on L. If W is a Fatou τ^*-neighborhood of zero, then $V = W \cap L$ is a Fatou τ-neighborhood of zero for L, and hence $V^\delta \subseteq W$, that is, W is a τ^δ-neighborhood of zero. If $V \in \mathcal{N}$, then $U \cap L \subseteq V$ for some Fatou τ^*-neighborhood of zero U. Note that if $u^* \in U$, then $0 \le u_\alpha \uparrow |u^*|$ for some net $\{u_\alpha\} \subseteq L$. So, $\{u_\alpha\} \subseteq U \cap L$ and hence $u^* \in V^\delta$, i.e., $U \subseteq V^\delta$, which shows that V^δ is a τ^*-neighborhood of zero. Thus, $\tau^* = \tau^\delta$ and hence τ^δ is uniquely determined.

If τ is Hausdorff, note that the order denseness of L in L^δ implies that τ^δ is also a Hausdorff topology. If now (L, τ) satisfies the Lebesgue property, then it also satisfies the Fatou property, and hence τ can be extended to τ^δ as above on L^δ. It easily follows that τ^δ is likewise a Lebesgue topology. ■

Following the arguments of the preceding proof we obtain the following sequential analogue of Theorem 4.12.

THEOREM 4.13. *Let L be an almost Dedekind σ-complete Riesz space. If τ is a σ-Fatou topology on L, then there exists a unique σ-Fatou topology τ^σ on L^σ (the Dedekind σ-completion of L) that induces τ on L.*

If in addition τ is a Hausdorff topology, then so is τ^σ and if (L, τ) satisfies the σ-Lebesgue property, then likewise does (L^σ, τ^σ).

4.2. The Structure of the Fatou Topologies

We start our discussion with one of the basic properties needed for the discussion in this section.

LEMMA 4.14. *Let L be a Dedekind σ-complete Riesz space, let $\{S_n\}$ be a normal sequence of solid and σ-order closed subsets of L, and let $\{u_n\}$ be an order bounded sequence of L satisfying*

$$u_{n+1} - u_n \in S_{n+2}$$

for all n. Then the elements $w = \liminf u_n$ and $u = \limsup u_n$ satisfy

$$w - u_n \in S_n \quad \text{and} \quad u - u_n \in S_n$$

for each $n = 1, 2, \ldots$.

PROOF. We shall show that $w - u_n \in S_n$ for each n. The proof for the lim sup is similar. Start by observing that

$$\begin{aligned} 0 \;\le\; u_n - \bigwedge_{m=n}^{n+p} u_m &= \bigvee_{m=n}^{n+p} (u_n - u_m) \le \bigvee_{m=n}^{n+p} |u_n - u_m| \\ &\le \sum_{m=n}^{n+p} |u_{m+1} - u_m| \in S_{n+2} + \cdots + S_{n+p+2} \subseteq S_{n+1} \end{aligned}$$

implies $u_n - \bigwedge_{m=n}^{n+p} u_m \in S_{n+1}$ for all $n, p = 1, 2, \ldots$. Therefore, if we define $w_n = \bigwedge_{m=n}^{\infty} u_m$, then it follows from $0 \le u_n - \bigwedge_{m=n}^{n+p} u_m \uparrow_p u_n - w_n$ and the σ-order closedness of S_{n+1} that $u_n - w_n \in S_{n+1}$ for all n.

On the other hand, if $w = \liminf u_n = \bigvee_{n=1}^{\infty} w_n$, then it follows from

$$\begin{aligned} |w_{n+p} - u_n| &\le |w_{n+p} - u_{n+p}| + |u_{n+p} - u_n| \\ &\le |w_{n+p} - u_{n+p}| + \sum_{m=n}^{n+p-1} |u_{m+1} - u_m| \in S_{n+p+1} + S_{n+1} \subseteq S_n, \end{aligned}$$

and

$$|w_{n+p} - u_n| \xrightarrow[p\to\infty]{o} |w - u_n|,$$

in conjunction with the σ-order closedness of S_n that $|w - u_n| \in S_n$ for all n. Since each S_n is also solid, we get $w - u_n \in S_n$ for all n, and the proof is finished. ■

We now continue with the introduction of the carrier of a locally solid topology. The notion of the carrier appeared implicitly in the work of I. Kawai [**74**] and later in the book of D. H. Fremlin [**58**, p. 50].

DEFINITION 4.15. *Let (L, τ) be a locally solid Riesz space, and let $\mathcal{N}$ denote the collection of all normal sequences consisting of solid τ-neighborhoods of zero. We define the* ***carrier*** *C_τ of the topology τ by*

$$C_\tau = \bigcup \{ N^{\mathrm{d}} \colon \; \exists \, \{V_n\} \in \mathcal{N} \text{ such that } N = \textstyle\bigcap_{n=1}^{\infty} V_n \} .$$

The disjoint complements N^{d} appearing in the definition of the carrier will be referred to as the ***components*** *of C_τ.*

An immediate consequence of the preceding definition is the following.

COROLLARY 4.16. *Every locally solid topology induces a Hausdorff locally solid topology on its carrier.*

PROOF. Let τ be a locally solid topology on a Riesz space L and let $u \neq 0$ belong to C_τ. Pick a normal sequence $\{V_n\}$ of solid τ-neighborhoods of zero with $u \in N^{\mathrm{d}}$, where $N = \bigcap_{n=1}^{\infty} V_n$. If $u \in V_n$ holds for each n, then $u \in N \cap N^{\mathrm{d}} = \{0\}$, which is impossible. Hence $u \notin V_n$ must hold for some n, and this shows that τ restricted to C_τ is a Hausdorff topology. ■

The basic properties of C_τ are included in the next result.

THEOREM 4.17. *For a locally solid Riesz space (L, τ) we have the following.*

(a) *The carrier C_τ is a σ-ideal.*
(b) *If τ is a Hausdorff Fatou topology, then C_τ is order dense in L (and so if L has the countable sup property, then $C_\tau = L$).*
(c) *If τ is a Lebesgue topology, then C_τ has the countable sup property.*

PROOF. (a) Assume that $\{u_n\}$ is a sequence in C_τ. For each k pick a normal sequence $\{V_n^k\colon\ n = 1, 2, \ldots\}$ of solid τ-neighborhoods of zero such that for each k we have $u_k \in N_k^{\rm d}$, where $N_k = \bigcap_{n=1}^{\infty} V_n^k$. Now for each $n \in \mathbb{N}$ let $W_n = \bigcap_{i=1}^{n} V_n^i$. Put $U_1 = W_1$ and then use an inductive argument to construct a sequence $\{U_n\}$ of solid τ-neighborhoods of zero such that $U_{n+1} + U_{n+1} \subseteq U_n \cap W_n$. In particular, note that $\{U_n\}$ is a normal sequence of solid τ-neighborhoods of zero whose null ideal satisfies

$$N = \bigcap_{n=1}^{\infty} U_n \subseteq \bigcap_{n=1}^{\infty} W_n \subseteq \bigcap_{n=1}^{\infty} V_n^k = N_k$$

for all k. Thus $N_k^{\rm d} \subseteq N^{\rm d}$ for each k, and so $\{u_n\} \subseteq N^{\rm d}$. Now if $0 \le u_n \uparrow u$ holds, then $u \in N^{\rm d} \subseteq C_\tau$, so that C_τ is a σ-ideal.

(b) Let $0 < u \in L$. Since τ is a Hausdorff Fatou topology there exists a Fatou τ-neighborhood V of zero with $u \notin V$. Next choose a normal sequence $\{V_n\}$ of Fatou τ-neighborhoods of zero with $V_1 = V$. Note that $N = \bigcap_{n=1}^{\infty} V_n$ is a band of L and so $N = N^{\rm dd}$, since L is Archimedean. But then since $u \notin N = N^{\rm dd}$, there exists some $0 < v \in N^{\rm d}$ such that $w = u \wedge v > 0$. Therefore, $0 < w \le u$ and $w \in C_\tau$ hold in L and so C_τ is order dense in L. The parenthetical part follows immediately from part (a).

(c) Assume that $u_\alpha \downarrow 0$ in C_τ (and so $u_\alpha \downarrow 0$ also holds in L since C_τ is an ideal of L). We can assume that $0 \le u_\alpha \le u$ holds in C_τ for each α and some $u \in C_\tau$. Pick a normal sequence $\{V_n\}$ of Fatou τ-neighborhoods of zero with $u \in N^{\rm d}$, where $N = \bigcap_{n=1}^{\infty} V_n$. Thus, $\{u_\alpha\} \subseteq N^{\rm d}$. Since τ is a Lebesgue topology we have $u_\alpha \xrightarrow{\tau} 0$. Now for each n pick an index α_n such that $u_{\alpha_n} \in V_n$. We can assume that the sequence of indices $\{\alpha_n\}$ is increasing and so $u_{\alpha_n}\!\downarrow$ in C_τ. If $0 \le w \le u_{\alpha_n}$ holds for all n, then $w \in N \cap N^{\rm d} = \{0\}$. This shows that $u_{\alpha_n} \downarrow 0$. Hence C_τ has the countable sup property. ■

We continue with a metrizability property of the Fatou topologies.

THEOREM 4.18. *Let L be a Riesz space equipped with a Hausdorff Fatou (resp. Lebesgue) topology τ. If L has the countable sup property and admits an at most countable complete disjoint system (in particular, if L has a weak order unit), then L also admits a metrizable Fatou (resp. Lebesgue) topology τ' coarser than τ, that is $\tau' \subseteq \tau$.*

PROOF. By part (b) of Theorem 4.17 we have $C_\tau = L$. Let $\{u_n\}$ be a countable complete disjoint system of the Riesz space L. As in the proof of part (a) of Theorem 4.17 there exists a normal sequence $\{V_n\}$ of Fatou τ-neighborhoods of zero such that $\{u_n\} \subseteq N^{\rm d}$, where $N = \bigcap_{n=1}^{\infty} V_n$. Since $\{u_n\}$ is a complete disjoint system, it follows that $L = N^{\rm d}$. Clearly, the sequence $\{V_n\}$ defines a metrizable Fatou topology τ' on L (which must necessarily be Lebesgue, if τ is Lebesgue) satisfying $\tau' \subseteq \tau$. ■

As a first application of Lemma 4.14 we shall present a condition under which topological convergence implies order convergence.

THEOREM 4.19. *Let (L, τ) be a Hausdorff locally solid Riesz space with the Fatou property. Assume that L has the countable sup property and that $\{u_\alpha\}$ is an order bounded net of L.*

If $u_\alpha \xrightarrow{\tau} u$ holds in L, then there exists an increasing sequence of indices $\{\alpha_n\}$ such that $u_{\alpha_n} \xrightarrow{o} u$ in L.

PROOF. Assume that a net $\{u_\alpha\}$ in L satisfies $u_\alpha \xrightarrow{\tau} u$ and $|u_\alpha| \leq v$ for all α. By Theorem 4.12 the topology τ extends to a Fatou topology τ^δ on the Dedekind completion L^δ of L [the Riesz space L is Archimedean by Theorem 2.21(a)]. Obviously L^δ has the countable sup property and so by part (b) of Theorem 4.17 we have $C_{\tau^\delta} = L^\delta$.

Next choose a normal sequence $\{V_n\}$ of Fatou τ^δ-neighborhoods of zero such that $v \in N^{\mathrm{d}}$, where $N = \bigcap_{n=1}^\infty V_n$. Since $\{u_\alpha\}$ is a τ^δ-Cauchy net of L^δ, there is an increasing sequence $\{\alpha_n\}$ of indices such that $u_{\alpha_{n+1}} - u_{\alpha_n} \in V_{n+2}$ and $u_{\alpha_n} - u \in V_n$ for all n. Put $u^* = \limsup u_{\alpha_n}$ and $w^* = \liminf u_{\alpha_n}$ in L^δ. By Lemma 4.14 we get $u_{\alpha_n} - u^* \in V_n$ for all n and thus $u^* - u \in V_n$ for each n. So, we have $u^* - u \in N \cap N^{\mathrm{d}}$, and therefore $u^* = u$. Similarly, $w^* = u$. But then $u_{\alpha_n} \xrightarrow{o} u$ holds in L^δ, and consequently $u_{\alpha_n} \xrightarrow{o} u$ holds also in L (since L has the countable sup property), and the proof is finished. ■

The next result, due to I. Amemiya [**31**, Theorem (B,6), p. 22] and D. H. Fremlin [**58**, 23L Proposition, p. 50], asserts that the solid and order closed sets of a Hausdorff locally solid Riesz space with the Fatou property are topologically closed.

THEOREM 4.20 (Amemiya–Fremlin). *Let (L, τ) be a Hausdorff locally solid Riesz space with the Fatou property. Then every solid and order closed subset of L is τ-closed.*

PROOF. Assume first that L is also Dedekind complete. Let $S \subseteq L$ be a solid and order closed subset and let $u \in \overline{S}$, the τ-closure of S. Assume also that an element $0 \leq v \in C_\tau$ satisfies $0 \leq v \leq |u|$. Since $\overline{S}$ is a solid set (see Theorem 2.19) we have $v \in \overline{S}$.

Pick a normal sequence $\{V_n\}$ of Fatou τ-neighborhoods of zero with $v \in N^{\mathrm{d}}$, where $N = \bigcap_{n=1}^\infty V_n$. Then for each n there exists some $v_n \in S$ with $v_n - v \in V_{n+3}$. Without loss of generality we can assume that $0 \leq v_n \leq v$ holds for all n (and thus $\{v_n\} \subseteq N^{\mathrm{d}}$). Note next that $v_{n+1} - v_n \in V_{n+2}$, and so by Lemma 4.14 we get that the element $w = \liminf v_n$ satisfies $w - v_n \in V_n$ for all n. Consequently, $w - v \in V_n$ for each $n \in \mathbb{N}$. But then $w - v \in N \cap N^{\mathrm{d}}$, and therefore $v = w \in S$. Now by Theorem 4.17(b) there exists a net $\{v_\alpha\} \subseteq C_\tau$ with $0 \leq v_\alpha \uparrow |u|$. By the above we have $\{v_\alpha\} \subseteq S$ and so, by the order closedness of S, we get $|u| \in S$. Since S is also solid, the latter implies $u \in S$. Therefore S is τ-closed.

Now consider the case where L is not Dedekind complete. Use Theorem 4.12 to extend τ to τ^δ on L^δ, and then let

$$S^\delta = \left\{ u^* \in L^\delta \colon \exists \{u_\alpha\} \subseteq S \text{ with } 0 \leq u_\alpha \uparrow |u^*| \right\}.$$

Then S^δ is solid and order closed in L^δ, and so S^δ is τ^δ-closed. This implies that $S = S^\delta \cap L$ is τ-closed in L, and the proof is finished. ■

The following result presents a striking relationship between Fatou and Lebesgue topologies. It was proven for Dedekind complete Riesz spaces by I. Amemiya and T. Mori [**32**, p. 135] and by I. Amemiya [**31**, Theorems (B,4) and (B,5), pp. 21–22] by employing a Baire category-type argument; the general case is due to the authors [**18**].

THEOREM 4.21 (Amemiya–Mori). *Assume that a Riesz space L is equipped with a Hausdorff Fatou topology τ_1 and a Lebesgue topology τ_2. Then the topology induced by τ_1 on any order bounded subset of L is finer that the topology induced by τ_2 on the same set.*

PROOF. According to Theorem 4.12, we can assume without loss of generality that L is Dedekind complete. Let $0 < u \in L$. It is enough to show that the identity mapping from $[0, u]$ to itself is τ_1-τ_2 continuous at zero. So, let a net $\{u_\alpha\} \subseteq [0, u]$ satisfy $u_\alpha \xrightarrow{\tau_1} 0$ and assume that $\{u_\alpha\}$ is not τ_2-convergent to zero. Then there exists a solid τ_2-neighborhood of zero U such that given any α we can find some $\beta \succeq \alpha$ satisfying $u_\beta \notin U$.

Choose another solid τ_2-neighborhood V of zero such that $V+V \subseteq U$. Now since (by part (b) of Theorem 4.17) the carrier C_{τ_1} of τ_1 is order dense in L and since τ_2 is a Lebesgue topology, there is some $w \in C_{\tau_1}$ with $0 < w \le u$ and $u - w \in V$. Pick a normal sequence $\{V_n\}$ of Fatou τ_1-neighborhoods of zero with $w \in N^{\mathrm{d}}$, where $N = \bigcap_{n=1}^{\infty} V_n$. Since $w \wedge u_\alpha \xrightarrow{\tau_1} 0$, there exists an increasing sequence $\{\alpha_n\}$ of indices such that $u_{\alpha_n} \notin U$ and $w \wedge u_{\alpha_n} \in V_{n+1}$ for all n. But for all n and p we have

$$0 \le \bigvee_{m=n}^{n+p} w \wedge u_{\alpha_m} \le \sum_{m=n}^{n+p} w \wedge u_{\alpha_m} \in V_{n+1} + \cdots + V_{n+p+1} \subseteq V_n\,,$$

and so $\bigvee_{m=n}^{\infty} w \wedge u_{\alpha_m} \in V_n$ for all n. This implies $w_1 = \limsup w \wedge u_{\alpha_n} \in V_n$ for each n, and consequently $w_1 \in N \cap N^{\mathrm{d}}$, i.e., $w_1 = 0$. This shows that $w \wedge u_{\alpha_n} \xrightarrow{o} 0$ holds in L and so $w \wedge u_{\alpha_n} \xrightarrow{\tau_2} 0$ must also hold since τ_2 is a Lebesgue topology. But then we have

$$\begin{aligned} 0 \le u_{\alpha_n} &= (u_{\alpha_n} - w)^+ + w \wedge u_{\alpha_n} \\ &\le (u - w) + w \wedge u_{\alpha_n} \in V + V \subseteq U \end{aligned}$$

for all sufficiently large n, contrary to the choice of the sequence $\{u_{\alpha_n}\}$. This shows that $u_\alpha \xrightarrow{\tau_2} 0$ and the proof is finished. ∎

The next result is an obvious consequence of the preceding theorem and describes a basic property of the Hausdorff Lebesgue topologies.

THEOREM 4.22 (Amemiya–Mori). *All Hausdorff Lebesgue topologies on a Riesz space L induce the same topology on the order bounded subsets of L.*

Another consequence of Theorem 4.21 is the following continuity property.

THEOREM 4.23. *Let $T\colon (L, \tau) \to (M, \tau_1)$ be an order continuous Riesz homomorphism. Assume that τ is a Hausdorff Fatou topology and that τ_1 is a Lebesgue topology. Then T restricted to any order bounded subset of L is continuous.*

PROOF. Let $\mathcal{N}$ be a base at zero for τ_1. Then $\{T^{-1}(V)\colon\ V \in \mathcal{N}\}$ is a base for a locally solid topology τ^* on L. Since T is order continuous and τ_1 is a Lebesgue topology, it follows that τ^* is also a Lebesgue topology. Thus the operator $T\colon (L, \tau^*) \to (M, \tau_1)$ is continuous and by Theorem 4.21 we have $\tau^* \subseteq \tau$ on every order bounded subset of L. The conclusion now should be immediate. ∎

The fact that all Hausdorff Lebesgue topologies agree on the order bounded subsets is by no means sufficient to guarantee that a Riesz space can admit at most one Hausdorff Lebesgue topology. The following examples illustrate this possibility as well as Theorem 4.21.

EXAMPLE 4.24. Let $L = L_\infty[0,1]$ be the super Dedekind complete Riesz space consisting of all equivalence classes of the (almost everywhere) bounded Lebesgue measurable functions on $[0,1]$. Let τ_1 and τ_2 be the Hausdorff Lebesgue topologies on L generated respectively by the Riesz norms

$$\rho_1(u) = \int_0^1 |u(x)|\,dx \quad \text{and} \quad \rho_2(u) = \Big[\int_0^1 |u(x)|^2\,dx\Big]^{\frac{1}{2}}.$$

Note that $\tau_1 \neq \tau_2$. Actually the Hausdorff Lebesgue topologies generated by the Riesz norms

$$\rho_p(u) = \Big[\int_0^1 |u(x)|^p\,dx\Big]^{\frac{1}{p}}$$

for $1 \le p < \infty$ are all distinct. ■

EXAMPLE 4.25. Fix an infinite set X and put $L = \ell_\infty(X)$. Let τ_1 be the Hausdorff Fatou (but not Lebesgue) topology generated by the sup norm and let τ_2 be the Hausdorff Lebesgue topology of the pointwise convergence. Then $\tau_2 \subseteq \tau_1$, but $\tau_1 \neq \tau_2$. Compare this conclusion with Theorem 4.21. ■

The next result presents a relationship between the countable sup property and the Hausdorff Lebesgue topologies.

THEOREM 4.26. *For a Hausdorff locally solid Riesz space (L, τ) with the Lebesgue property the following statements are equivalent.*

(i) *L has the countable sup property.*
(ii) *$C_\tau = L$, where C_τ is the carrier of τ.*
(iii) *τ induces a metrizable topology on each order interval of L.*

PROOF. (i) $\Longleftrightarrow$ (ii) It follows immediately from parts (b) and (c) of Theorem 4.17.

(ii) $\Longrightarrow$ (iii) Let $u \in L^+$. Then $[0,u] \subseteq N^{\mathrm{d}}$ must hold for some normal sequence $\{V_n\}$ of Fatou τ-neighborhoods of zero, where $N = \bigcap_{n=1}^\infty V_n$. Now τ restricted to N^{d} satisfies the Lebesgue property. Also $\{V_n\}$ defines a metrizable Lebesgue topology on N^{d}. But then, according to Theorem 4.22, both topologies agree on $[0,u]$ and consequently τ induces a metrizable topology on $[0,u]$.

(iii) $\Longrightarrow$ (i) Assume that $u_\alpha \downarrow 0$ in L. Without loss of generality we can assume that $0 \le u_\alpha \le u$ holds in L for all α and some $u \in L$. Let $\{V_n\}$ be a normal sequence of Fatou τ-neighborhoods of zero such that $\{V_n \cap [0,u]\}$ is a base at zero for the topology induced on $[0,u]$ by τ. For each n pick an index α_n with $u_{\alpha_n} \in V_n \cap [0,u]$ (this is possible since the topology τ is Lebesgue). We can assume that $u_{\alpha_n} \downarrow$ in L. If $0 \le v \le u_{\alpha_n}$ holds for all n, then $v \in V_n \cap [0,u]$ holds for all n. Hence $v = 0$ and so $u_{\alpha_n} \downarrow 0$ in L. This shows that L has the countable sup property. ■

We close this section with a comparison property of the Fatou topologies.

THEOREM 4.27. *Assume that an Archimedean Riesz space L is equipped with a Fatou topology τ and a locally solid topology τ'. Then the following statements are equivalent.*

(i) *τ is coarser than τ', that is, $\tau \subseteq \tau'$.*
(ii) *τ induces a coarser topology than τ' on its carrier C_τ.*

(iii) *τ induces a coarser topology than τ' on each component N^{d} of its carrier C_τ.*

PROOF. The implications (i) $\Longrightarrow$ (ii) $\Longrightarrow$ (iii) are obvious.

(iii) $\Longrightarrow$ (i) Let V be a Fatou τ-neighborhood of zero. Pick a normal sequence $\{V_n\}$ of Fatou τ-neighborhoods of zero with $V_1 + V_1 \subseteq V$ and note that $N \oplus N^{\mathrm{d}}$ is order dense in L, where $N = \bigcap_{n=1}^{\infty} V_n$. By hypothesis there exists a solid τ'-neighborhood U of zero such that $U \cap N^{\mathrm{d}} \subseteq V_1 \cap N^{\mathrm{d}}$.

Now let $u \in U$. Since L is Archimedean and $N \oplus N^{\mathrm{d}}$ is order dense in L, there exists a net $\{u_\alpha + v_\alpha\} \subseteq N \oplus N^{\mathrm{d}}$ such that $0 \le u_\alpha + v_\alpha \uparrow |u|$ in L. But then $\{u_\alpha\} \subseteq N \subseteq V_1$ and since $\{v_\alpha\} \subseteq N^{\mathrm{d}}$, $0 \le v_\alpha \le |u|$ for each α, and U is solid, we get $\{v_\alpha\} \subseteq U \cap N^{\mathrm{d}} \subseteq V_1 \cap N^{\mathrm{d}} \subseteq V_1$. Thus $\{u_\alpha + v_\alpha\} \subseteq V_1 + V_1 \subseteq V$ and hence $|u| \in V$, since V is order closed. Therefore, $u \in V$ and so $U \subseteq V$. This shows that $\tau \subseteq \tau'$. ■

4.3. Topological Completeness and Fatou Topologies

We start this section with one of the deepest results of the theory of locally solid Riesz spaces. It is due to H. Nakano [**101**, Theorem 3.3, p. 93]. Nakano's original proof, however, had a number of gaps that were filled later by H. H. Schaefer [**121**, Theorem 1, p. 304]. The proof below is due to the authors [**16**, Theorem 3.2, p. 28]; see also [**17**].

THEOREM 4.28 (Nakano). *If (L, τ) is a Dedekind complete locally solid Riesz space with the Fatou property, then the order intervals of L are τ-complete.*

PROOF. Assume first that τ is also a Hausdorff topology. By Theorem 2.42 we have only to show that L is an ideal of $\widehat{L}$. We shall establish this by showing that L is order dense in J, the ideal generated by L in $\widehat{L}$. It will then follow from Theorem 1.40 that $L = J$.

To this end, let $0 < \widehat{u} \le u \in L$. Pick a normal sequence of Fatou τ-neighborhoods $\{V_n\}$ of zero of L with $\widehat{u} \notin \overline{V}_1$ (the $\widehat{\tau}$-closure of V_1 in $\widehat{L}$). Pick a net $\{v_\alpha\}$ of L^+ satisfying $0 \le v_\alpha \le u$ for all α and $v_\alpha \xrightarrow{\widehat{\tau}} \widehat{u}$. Next, choose a sequence $\{\alpha_n\}$ of indices such that the elements $u_n = v_{\alpha_n}$ satisfy

$$\widehat{u} - u_n \in \overline{V}_n \quad \text{and} \quad u_{n+p} - u_n \in V_{n+2} \text{ for all } n, p.$$

It follows from Lemma 4.14 that $w = \limsup u_n$ satisfies $w - u_n \in V_n$ for all n and thus $\widehat{u} - w \in \overline{V}_n$ for each n. Now let

$$S = \{v \in L^+ \colon\ v - w \in N = \textstyle\bigcap_{n=1}^{\infty} V_n\}.$$

Obviously S is nonempty. Put $s = \inf S$ in L and note that since N is a band of L, we have $s - w \in N$ and thus $s > 0$ (otherwise, $s = 0$ implies $w \in N$ and consequently $\widehat{u} \in \overline{V}_1$, a contradiction). Now let W_1 be a Fatou τ-neighborhood of zero of L with $W_1 \subseteq V_1$. Pick a normal sequence $\{W_n\}$ of Fatou τ-neighborhoods of zero of L with $W_n \subseteq V_n$ for $n = 1, 2, \ldots$. Applying the same arguments as above, we obtain an element $a \in L^+$ with $\widehat{u} - a \in \overline{W}_n$ for all n. Next use Theorem 4.20 in conjunction with

$$a - w = (a - \widehat{u}) + (\widehat{u} - w) \in \overline{W}_{n-1} + \overline{V}_{n-1} \subseteq \overline{V}_n,$$

to get $a - w \in V_n$ for each n and consequently $a - w \in N$. Thus $s \le a$ and so $0 \le (s - \widehat{u})^+ \le (a - \widehat{u})^+ \in \overline{W}_1$. Thus $(s - \widehat{u})^+ \in \overline{W}_1$ for all Fatou τ-neighborhoods of zero W_1 with $W_1 \subseteq V_1$. Therefore $(s - \widehat{u})^+ = 0$ and hence $0 < s \le \widehat{u}$. This shows that L is order dense in J.

Now assume that τ is not a Hausdorff topology. Consider the kernel of the topology $K_\tau = \bigcap\{V\colon\ V \text{ is a } \tau\text{-neighborhood of zero}\}$ and note that K_τ is a band of L. Thus $L = K_\tau \oplus K_\tau^{\mathrm{d}}$ and by Theorem 4.10 the topology τ induces a Hausdorff Fatou topology on K_τ^{d}. Since K_τ^{d} is Dedekind complete, by the above proof, the order intervals of K_τ^{d} are τ-complete. Finally, since every net of K_τ is τ-convergent to zero and the order intervals of K_τ^{d} are τ-complete, we infer that the order intervals of L are τ-complete. ■

For some interesting applications of Theorem 4.28 see [**63**]. A sequential analogue of Theorem 4.28 is presented next.

THEOREM 4.29. *Let (L, τ) be a Dedekind complete Hausdorff locally solid Riesz space with the σ-Fatou property. Then the order intervals of the Riesz space L are sequentially τ-complete.*

PROOF. Put $L_\sigma = \big\{\widehat{u} \in \widehat{L}\colon\ \exists\{u_n\} \subseteq L \text{ with } u_n \xrightarrow{\widehat{\tau}} \widehat{u}\big\}$. Clearly, L_σ is a Riesz subspace of $\widehat{L}$ including L as a Riesz subspace and L is sequentially τ-complete if and only if $L = L_\sigma$. Now assume that $0 < \widehat{u} \le u \in L$ with $\widehat{u} \in L_\sigma$. Pick a sequence $\{u_n\} \subseteq L^+$ with $0 \le u_n \le u$ for all n and $u_n \xrightarrow{\widehat{\tau}} \widehat{u}$. Let V be a σ-Fatou neighborhood of zero of L with $\widehat{u} \notin \overline{V}$. Choose a normal sequence $\{V_n\}$ of σ-Fatou τ-neighborhoods of zero of L with $V_1 = V$. By passing to a subsequence (if necessary) we can suppose $u_{n+1} - u_n \in V_{n+2}$ for all n. By Lemma 4.14 the element $w = \liminf u_n \in L$ satisfies $w - u_n \in V_n \subseteq V$ for each n. Note that $w > 0$, for otherwise $w = 0$ implies $\widehat{u} \in \overline{V}$, a contradiction.

Now let W be any σ-Fatou τ-neighborhood of zero of L, and then choose a normal sequence $\{W_n\}$ of σ-Fatou τ-neighborhoods of zero with $W_1 = W$. Next pick a subsequence $\{v_n\}$ of $\{u_n\}$ with $v_{n+1} - v_n$ in W_{n+2} for all n. Let $v = \liminf v_n \in L$ and note that (by Lemma 4.14 again) we have $0 < w \le v$ and $v - v_n \in W_n$ for all n. But then $0 \le (w - \widehat{u})^+ \le (v - \widehat{u})^+ \in \overline{W}$ implies $(w - \widehat{u})^+ = 0$, i.e., $0 < w \le \widehat{u}$. Hence L is order dense in L_σ and so by Theorem 1.40 the Riesz space L is an ideal in L_σ. This shows that the order intervals of L are sequentially τ-complete. ■

Following closely the preceding proof, we readily see that a more general result holds. Namely:

- *If (L, τ) is a Hausdorff Dedekind σ-complete Riesz space with the σ-Fatou property, then for every $0 \le \widehat{u} \in L_\sigma$ there exists a net $\{u_\alpha\} \subseteq L$ satisfying $0 \le u_\alpha \uparrow \widehat{u}$ and $u_\alpha \xrightarrow{\widehat{\tau}} \widehat{u}$.*

The next example shows that the Dedekind completeness in Theorem 4.28 cannot be weakened to the projection property.

EXAMPLE 4.30. Let L be the Riesz space of all real-valued functions defined on an infinite set X whose range is finite, with the pointwise ordering. It can be seen easily that L is not Dedekind complete but it has the projection property. Let τ be the Hausdorff Lebesgue topology of pointwise convergence. Clearly, the order intervals of L are not τ-complete. Moreover, $\widehat{L} = \mathbb{R}^X$ and $L^\delta = \ell_\infty(X)$. ■

Additional important properties of the Fatou topologies are included in the next theorem, which is due to I. Kawai [**74**, p. 298] and D. H. Fremlin [**57**, Theorem 1, p. 343].

THEOREM 4.31 (Kawai–Fremlin). *For a Hausdorff locally solid Riesz space* (L, τ) *with the Fatou property, we have the following:*

(i) *L is order dense in its topological completion $\widehat{L}$.*
(ii) *The topological completion $(\widehat{L}, \widehat{\tau})$ of (L, τ) also satisfies the Fatou property.*

PROOF. The proof below is due to the authors [**16**, Theorem 4.1]. Assume first that L is also Dedekind complete. Then, according to Theorem 4.28, the Riesz space L is an ideal of $\widehat{L}$ and it follows from Theorem 2.42 that L is order dense in $\widehat{L}$. Now let $\mathcal{N}$ be a base of Fatou τ-neighborhoods at zero and for each $V \in \mathcal{N}$ let $\overline{V}$ denote its $\widehat{\tau}$-closure in $\widehat{L}$. We know that $\overline{V}$ is a solid subset of $\widehat{L}$. To see that $\overline{V}$ is also order closed let $0 \leq \widehat{u}_\alpha \uparrow \widehat{u}$ in $\widehat{L}$ with $\{\widehat{u}_\alpha\} \subseteq \overline{V}$. According to Theorem 2.42 there exists a net $\{v_\lambda\}$ in L with $0 \leq v_\lambda \uparrow \widehat{u}$ and $v_\lambda \xrightarrow{\widehat{\tau}} \widehat{u}$. We shall show that $\{v_\lambda\} \subseteq V$ and this will imply that $\widehat{u} \in \overline{V}$. To see this, notice that for each α there exists a net of L^+ that increases to $\widehat{u}_\alpha$. Since $\widehat{u}_\alpha \in \overline{V}$ any such net must lie in V. Taking the finite suprema of all those nets, we get a net $\{w_\delta\} \subseteq V$ with $0 \leq w_\delta \uparrow \widehat{u}$. But then for each fixed λ we have $0 \leq w_\delta \wedge v_\lambda \uparrow_\delta v_\lambda$ in L and since V is order closed, we get $v_\lambda \in V$ for each λ. Now since the collection $\{\overline{V}\colon\ V \in \mathcal{N}\}$ is a base at zero for $\widehat{\tau}$ consisting of solid and order closed sets, it follows that $\widehat{\tau}$ is a Fatou topology.

If L is not Dedekind complete, extend τ to τ^δ on L^δ according to Theorem 4.12. Then as above (L^δ, τ^δ) satisfies the Fatou property and L^δ is order dense in $\widehat{L^\delta}$. Now observe that the $\widehat{\tau^\delta}$-closure of L in $\widehat{L^\delta}$ is the topological completion of (L, τ). Hence L is order dense in $\widehat{L}$ and from Theorem 4.10 it follows that $(\widehat{L}, \widehat{\tau})$ satisfies the Fatou property.

Finally, observe that if $\mathcal{N}$ is a base of Fatou τ-neighborhoods of zero, then the collection $\{\widehat{V}\colon\ V \in \mathcal{N}\}$, where

$$\widehat{V} = \big\{\widehat{u} \in \widehat{L}\colon\ v \in L \text{ and } 0 \leq v \leq |\widehat{u}| \text{ implies } v \in V\big\},$$

is a base of Fatou neighborhoods at zero for $\widehat{\tau}$. ■

As an application of Theorem 4.31 we present the following result.

THEOREM 4.32. *Let (L, τ) be a Hausdorff locally convex-solid Riesz space whose topology is generated by a family of Fatou seminorms $\{\rho_\alpha\}$. For each α let $\widehat{\rho}_\alpha$ denote the unique continuous extension of ρ_α to $\widehat{L}$. Then we have:*

(i) *Each $\widehat{\rho}_\alpha$ satisfies $\widehat{\rho}_\alpha(\widehat{u}) = \sup\{\rho_\alpha(u)\colon\ u \in L$ and $0 \leq u \leq |\widehat{u}|\}$ for each $\widehat{u} \in \widehat{L}$.*
(ii) *Each $\widehat{\rho}_\alpha$ is a Fatou seminorm of $\widehat{L}$.*
(iii) *The family $\{\widehat{\rho}_\alpha\}$ generates the Fatou topology $\widehat{\tau}$.*

PROOF. (i) For each τ-continuous Fatou seminorm ρ on L define $\rho^*\colon \widehat{L} \to \mathbb{R}$ by

$$\rho^*(\widehat{u}) = \sup\big\{\rho(u)\colon\ u \in L \text{ and } 0 \leq u \leq |\widehat{u}|\big\},$$

and note that since $\widehat{\rho}$ is a Riesz seminorm on $\widehat{L}$ we have $\rho^*(\widehat{u}) \leq \widehat{\rho}(\widehat{u})$ for all $\widehat{u} \in \widehat{L}$. We shall show that $\rho^* = \widehat{\rho}$ on $\widehat{L}$, and this suffices to establish (i).

To do this, we need to establish the triangle inequality for ρ^*. So, let $\widehat{u}, \widehat{v} \in \widehat{L}$. By Theorem 4.31 there exist two nets $\{u_\beta\}$ and $\{v_\gamma\}$ of L with $0 \le u_\beta \uparrow |\widehat{u}|$ and $0 \le v_\gamma \uparrow |\widehat{v}|$ in $\widehat{L}$. Now if a vector $w \in L$ satisfies $0 \le w \le |\widehat{u}| + |\widehat{v}|$, then $0 \le (u_\beta + v_\gamma) \wedge w \uparrow_{(\beta,\gamma)} w$ holds in L and so since ρ is a Fatou seminorm of L, we get $\rho\big((u_\beta + v_\gamma) \wedge w\big) \uparrow_{(\beta,\gamma)} \rho(w)$. But then, from

$$\rho\big((u_\beta + v_\gamma) \wedge w\big) \le \rho(u_\beta + v_\gamma) \le \rho(u_\beta) + \rho(v_\gamma) \le \rho^*(\widehat{u}) + \rho^*(\widehat{v}),$$

we get $\rho(w) \le \rho^*(\widehat{u}) + \rho^*(\widehat{v})$ so that $\rho^*(|\widehat{u}| + |\widehat{v}|) \le \rho^*(\widehat{u}) + \rho^*(\widehat{v})$. Now the desired inequality follows from $\rho^*(\widehat{u} + \widehat{v}) \le \rho^*(|\widehat{u}| + |\widehat{v}|)$.

Now let $\widehat{u} \in \widehat{L}$. Choose a net $\{u_\lambda\} \subseteq L$ with $u_\lambda \xrightarrow{\widehat{\tau}} \widehat{u}$, and note that the inequalities

$$\begin{aligned} \rho^*(\widehat{u}) &\ge \rho^*(u_\lambda) - \rho^*(u_\lambda - \widehat{u}) \\ &\ge \rho(u_\lambda) - \widehat{\rho}(u_\lambda - \widehat{u}) \\ &= \widehat{\rho}(u_\lambda) - \widehat{\rho}(u_\lambda - \widehat{u}) \end{aligned}$$

imply $\rho^*(\widehat{u}) \ge \widehat{\rho}(\widehat{u})$. Hence $\rho^* = \widehat{\rho}$ on $\widehat{L}$.

(ii) Let ρ be a τ-continuous Fatou seminorm on L, and let $0 \le \widehat{u}_\lambda \uparrow \widehat{u}$ in $\widehat{L}$. Put $a = \lim \widehat{\rho}(\widehat{u}_\lambda)$. We must show that $\widehat{\rho}(\widehat{u}) \le a$. For each $\widehat{u}_\lambda$ there exists a net of L^+ that increases to $\widehat{u}_\lambda$. By taking the finite suprema of these nets we get a net $\{v_\nu\}$ of L^+ satisfying $0 \le v_\nu \uparrow \widehat{u}$. In particular, it follows that $\lim \rho(v_\nu) = a$. Now if $w \in L$ satisfies $0 \le w \le \widehat{u}$, then from

$$0 \le v_\nu \wedge w \uparrow w \text{ in } L \quad \text{and} \quad \rho(v_\nu \wedge w) \le \rho(v_\nu) \le a,$$

and the fact that ρ is a Fatou seminorm of L, we see that $\rho(w) \le a$. Thus $\widehat{\rho}(\widehat{u}) \le a$, as desired.

(iii) Obvious. ■

THEOREM 4.33. *For a Riesz space L with the projection property we have the following.*

(i) *If τ is a Hausdorff Fatou topology on L, then $\widehat{L}$ is Dedekind complete.*
(ii) *If L admits a Hausdorff locally solid topology τ for which the order intervals of L are τ-complete, then both L and $\widehat{L}$ are Dedekind complete.*

PROOF. (i) Apply Theorem 2.49.

(ii) By Theorem 2.49 we know that $\widehat{L}$ is Dedekind complete and so is L, since L is an ideal of $\widehat{L}$; see Theorem 2.42. ■

COROLLARY 4.34. *For a Hausdorff locally solid Riesz space (L, τ) with the Fatou property we have the following.*

(1) *If L has the projection property, then:*
  (a) *The Dedekind completion L^δ of L can be taken to be $\mathcal{I}_L$, the ideal generated by L in $\widehat{L}$, and*
  (b) *The topology τ^δ is the restriction of $\widehat{\tau}$ to $\mathcal{I}_L$ and we have the order dense Riesz subspace inclusions: $L \subseteq L^\delta \subseteq \widehat{L}$.*
(2) *If L is Dedekind complete, then $\widehat{L}$ is also Dedekind complete.*

Recall that a locally solid Riesz space (L, τ) satisfies the Levi property if every increasing τ-bounded net of L^+ has a supremum in L; see Definition 3.16. Note that if (L, τ) satisfies the Levi property, then L is Dedekind complete and τ is a Hausdorff topology (see Exercise 1 of Chapter 3).

DEFINITION 4.35. *A locally solid Riesz space (L, τ) is said to be a* ***Nakano space*** *if (L, τ) satisfies both the Levi property and the Fatou property.*

The next three theorems give conditions for the topological completeness of locally solid Riesz spaces. The first one is due to the authors [**16**, Theorem 5.4, p. 32].

THEOREM 4.36 (Aliprantis–Burkinshaw). *Let (L, τ) be a locally solid Riesz spaces with the Levi property. If the order intervals of L are τ-complete, then (L, τ) is τ-complete.*

PROOF. Let $0 \le \widehat{u} \in \widehat{L}$. By Theorem 2.42 there exists a net $\{u_\alpha\}$ of L^+ with $0 \le u_\alpha \uparrow \widehat{u}$. Since $\{u_\alpha\}$ is τ-bounded, $0 \le u_\alpha \uparrow u$ must hold in L for some $u \in L$, and so since L is an ideal of $\widehat{L}$ we have $u_\alpha \uparrow u$ in $\widehat{L}$. Hence $\widehat{u} = u \in L$ and so $\widehat{L} = L$, i.e., (L, τ) is τ-complete. ■

The next result is due to H. Nakano [**101**, Theorem 4.2, p. 96]. Its proof below is due to the authors [**16**, Corollary 5.5, p. 32].

THEOREM 4.37 (Nakano). *Every Nakano space (L, τ) is τ-complete.*

PROOF. Apply Theorems 4.28 and 4.36. ■

Recall also that a locally solid Riesz space (L, τ) satisfies the **Monotone Completeness Property** (MCP) if every monotone τ-Cauchy net of L is τ-convergent; see Definition 2.43.

THEOREM 4.38. *Every Hausdorff locally solid Dedekind complete Riesz space with the Fatou property that satisfies* MCP *is topologically complete.*

PROOF. Let (L, τ) be a Hausdorff locally solid Riesz space with the Fatou property that also satisfies MCP. By Theorem 4.28 the Riesz space L is an ideal of $\widehat{L}$ and the desired conclusion follows from Theorem 2.46. ■

COROLLARY 4.39. *Every Hausdorff locally solid Riesz space with the Lebesgue property that satisfies* MCP *is Dedekind complete and topologically complete.*

PROOF. Let (L, τ) be a Hausdorff locally solid Riesz space with the Lebesgue property that also satisfies MCP. We need to show only that L is Dedekind complete.

To this end, let $0 \le u_\alpha \uparrow \le u$ in L. By Theorem 3.22 the net $\{u_\alpha\}$ of L is τ-Cauchy and therefore by our hypothesis $u_\alpha \xrightarrow{\tau} v$ holds in L. It follows now from Theorem 2.21 that $0 \le u_\alpha \uparrow v$ in L and the proof is finished. ■

We continue with a result regarding sequential topological completeness.

THEOREM 4.40. *A Dedekind σ-complete Hausdorff locally solid Riesz space (L, τ) with the σ-Fatou property and the Monotone Completeness Property is sequentially τ-complete.*

PROOF. Let $L_\sigma = \{\widehat{u} \in \widehat{L}\colon\ \exists\{u_n\} \subseteq L \text{ such that } u_n \xrightarrow{\widehat{\tau}} \widehat{u}\}$. By the remark following the proof of Theorem 4.29 we know that for every $0 \le \widehat{u} \in L_\sigma$ there exists a net $\{v_\alpha\}$ in L^+ with $0 \le v_\alpha \uparrow \widehat{u}$ and $v_\alpha \xrightarrow{\widehat{\tau}} \widehat{u}$. In particular, note that $\{v_\alpha\}$ is a τ-Cauchy net of L. Now since $\{v_\alpha\}$ is monotone and L satisfies MCP, it follows that $v_\alpha \xrightarrow{\tau} v$ holds for some $v \in L$. Thus, $\widehat{u} = v \in L$ and hence $L = L_\sigma$. This shows that L is sequentially τ-complete. ∎

We close this section with an example, due to D. H. Fremlin, that shows that a Hausdorff locally solid Riesz space with the Fatou property and with MCP need not be topologically complete.

EXAMPLE 4.41 (Fremlin). Let K_1 be the set of all real-valued functions f defined on $[0,1]$ such that $\lim_{t\downarrow x} f(t)$ exists in $\mathbb{R}$ for every $x \in [0,1)$. Alternatively, f belongs to K_1 whenever $\lim f(t_n)$ exists in $\mathbb{R}$ for every strictly decreasing sequence $\{t_n\} \subseteq [0,1]$, i.e., $t_{n+1} < t_n$ for all n. With the usual operations of addition and scalar multiplication and with the pointwise ordering K_1 is an order dense Riesz subspace of $\mathbb{R}^{[0,1]}$ (for this conclusion use that $\chi_{\{t\}} \in K_1$ for every $t \in [0,1]$). Note that for $f \in K_1$ and $\epsilon > 0$ the set

$$\mathcal{O}_\epsilon = \{t \in [0,1]\colon \text{ The oscillation of } f \text{ at } t \text{ is greater than } \epsilon\}$$

does not contain strictly decreasing sequences and thus $\mathcal{O}_\epsilon$ must be countable. Therefore, the set of discontinuities of f is at most countable.

Now let S_1 be the set of all sequences $s = (t_0, t_1, \ldots)$ such that $0 \le t_n \le 1$ for all $n = 0, 1, \ldots$ and with $t_n \downarrow t_0$. The collection S_1 contains all constant sequences. For every sequence $s = (t_0, t_1, \ldots) \in S_1$ define the Fatou seminorm ρ_s on K_1 by $\rho_s(f) = \sup\{|f(t_k)|\colon\ k \ge 0\}$ for $f \in K_1$ [the supremum is finite since $\lim f(t_n)$ exists in $\mathbb{R}$]. Clearly the family of Fatou seminorms $\{\rho_s\colon\ s \in S_1\}$ defines a Hausdorff Fatou topology τ_1 on K_1 and, as can be easily seen, (K_1, τ_1) is τ_1-complete.

Similarly, let K_2 be the order dense Riesz subspace of $\mathbb{R}^{[0,1]}$ consisting of all functions g for which $\lim_{t\uparrow x} g(t)$ exists for every $x \in (0,1]$. Let S_2 be the collection of all sequences $s = (t_0, t_1, \ldots)$ of $[0,1]$ satisfying $t_n \uparrow t_0$ and define $\rho_s(g) = \sup\{|g(t_k)|\colon\ k = 0, 1, \ldots\}$ for $g \in K_2$. Let τ_2 be the Hausdorff Fatou topology on K_2 generated by the family of Riesz seminorms $\{\rho_s\colon\ s \in S_2\}$. Note that (K_2, τ_2) is τ_2-complete.

Now put $K = K_1 \times K_2$ and $\tau_3 = \tau_1 \times \tau_2$. Then (K, τ_3) is a Hausdorff locally solid Riesz space with the Fatou property that is also τ_3-complete. Next define

$$L = \{(u,v) \in K\colon\ u(t) = v(t) \text{ if } u \text{ and } v \text{ are both continuous at } t\}.$$

Then L is a Riesz subspace of K (this requires the fact that the functions in both K_1 and K_2 are continuous except on countable sets). Also, since L contains $(\chi_{\{t\}}, 0)$ and $(0, \chi_{\{t\}})$ for every $t \in [0,1]$, it follows that L is order dense in K and hence the topology τ induced by τ_3 on L is a Hausdorff Fatou topology. We leave it to the reader as an exercise to verify the following.

(1) (L, τ) satisfies the Monotone Completeness Property.
(2) (L, τ) is not τ-complete.
(3) L is τ_3-dense in K, that is, $(\widehat{L}, \widehat{\tau}) = (K, \tau_3)$.

Note also that the set

$$S = \{(u,0) \in L\colon\ u(1) = 0 \text{ and } |u(t)| \le 1 \text{ for } t \in [0,1]\}$$

is a solid and order closed subset of L, which is also τ_3-closed in $K = \widehat{L}$. However, S is not order closed in $\widehat{L}$. Thus, the closure in $\widehat{L}$ of a solid and order closed subset of a Hausdorff locally solid Riesz space (L, τ) with the Fatou property need not be an order closed subset of $\widehat{L}$. ■

4.4. Quotient Riesz Spaces and Fatou Properties

In this section we shall present conditions under which the quotient Riesz spaces of a locally solid Riesz space (L, τ) with the Fatou property (or the σ-Fatou property) inherit the same property. To indicate that general results cannot be obtained, we shall start with two examples. In the first example we shall present a normed Riesz space L_ρ with the Fatou property that is almost Dedekind σ-complete with the principal projection property and a band A of L such that the quotient Riesz space L/A (with the quotient topology) fails to satisfy the σ-Fatou property.

EXAMPLE 4.42. Let L be the Riesz space consisting of all real sequences that are ultimately constant with the pointwise ordering. Note that L is almost Dedekind σ-complete with the principal projection property but without the projection property and without being Dedekind σ-complete. Let τ be the Hausdorff Fatou topology on L generated by the Fatou norm

$$\rho(u) = \sup\{|u_n|\colon\ n = 2, 4, 6, \ldots\} + \sup\{2^{-n}|u_n|\colon\ n = 1, 3, 5, \ldots\}$$

for all $u = (u_1, u_2, \ldots) \in L$. Now let A be the band

$$A = \{u \in L\colon\ u_n = 0 \text{ for } n = 1, 3, 5, \ldots\}.$$

Observe that the quotient Riesz norm $\dot{\rho}$ that generates the Hausdorff locally solid topology τ/A on L/A is given by

$$\dot{\rho}(\dot{u}) = |u(\infty)| + \sup\{2^{-n}|u_n|\colon\ n = 1, 3, 5, \ldots\},$$

where $u(\infty)$ denotes the ultimate constant value of u.

Now by letting $e_n = (1, 1, \ldots, 1, 0, 0, \ldots)$, where the 1 occupy the first n positions, we see that $e_n \uparrow e = (1, 1, 1, \ldots)$ in L and since A is a band of L, it follows that $0 \le \dot{e}_n \uparrow \dot{e}$ in L/A.

Next, note that $\dot{\rho}(\dot{e}_{n+p} - \dot{e}_n) \le \frac{1}{2^n}$ implies that $\{\dot{e}_n\}$ is a $\dot{\rho}$-Cauchy sequence and consequently for a σ-order closed τ/A-neighborhood V of zero the property $0 \le \dot{e}_{n+p} - \dot{e}_n \uparrow_p \dot{e} - \dot{e}_n$ in L/A implies $\dot{e} - \dot{e}_n \in V$ for all sufficiently large n. Thus, if L/A satisfies the σ-Fatou property, then $\dot{e}_n \xrightarrow{\tau/A} \dot{e}$, i.e., $\dot{\rho}(\dot{e}_n - \dot{e}) \to 0$, and in particular, $\dot{\rho}(\dot{e}_n) \to \dot{\rho}(\dot{e})$. However, the last conclusion is impossible since $\dot{\rho}(\dot{e}_n) = \frac{1}{2}$ for all n and $\dot{\rho}(\dot{e}) = \frac{3}{2}$, so that $(L/A, \tau/A)$ does not satisfy the σ-Fatou property. ■

As a second example we present a Dedekind σ-complete locally solid Riesz space (L, τ) with the Fatou property and a band A of L such that the quotient locally solid Riesz space $(L/A, \tau/A)$ fails to satisfy the Fatou property.

EXAMPLE 4.43. Let L be the Riesz space of all functions $u\colon [0, 1] \to \mathbb{R}$ such that $u(x) \ne u(0)$ holds for at most countably many $x \in [0, 1]$. Note that L is a Dedekind σ-complete Riesz space. Let τ be the locally solid topology generated by the Riesz seminorm

$$\rho(u) = \sup\{|u(x)|\colon\ \tfrac{1}{2} \le x \le 1\}.$$

Now consider the band

$$A = \left\{u \in L\colon\; u(x) = 0 \text{ for } x \in \left[0, \tfrac{1}{2}\right)\right\}.$$

Even though $0 \leq u_\alpha \uparrow u$ in L does not necessarily imply $u_\alpha(x) \uparrow u(x)$ for all $x \in [0,1]$, it can be seen that ρ is a Fatou seminorm and so τ is a Fatou topology. But $(L/A, \tau/A)$ fails to have the Fatou property as we now show. Let α run over the family of finite subsets of $(0,1]$, and let $u_\alpha = \chi_\alpha$, the characteristic function of α. Clearly $0 \leq u_\alpha \uparrow \mathbf{1}$, where $\mathbf{1}(x) = 1$ for all x, holds in L and so $0 \leq \dot{u}_\alpha \uparrow \dot{\mathbf{1}}$ holds in L/A, since A is a band. Observe now that $\dot{\rho}(\dot{u}_\alpha) = 0$ and $\dot{\rho}(\dot{\mathbf{1}}) = 1$. This implies that $(L/A, \tau/A)$ does not satisfy the Fatou property. ■

We continue with a few affirmative results.

THEOREM 4.44. *Let (L, τ) be a Dedekind σ-complete Riesz space satisfying the σ-Fatou property and let A be a σ-ideal of L. Then the quotient Riesz space $(L/A, \tau/A)$ satisfies the σ-Fatou property.*

PROOF. It is enough to establish that the canonical image of any σ-Fatou τ-neighborhood of L is a σ-Fatou τ/A-neighborhood of L/A. To this end, let V be a σ-Fatou τ-neighborhood of L. Note first that $\dot{V} = Q(V)$ is a solid subset of L/A; see Theorem 1.33. On the other hand, observe that if $0 \leq \dot{u}_n \uparrow \dot{u}$ holds in L/A with $\{\dot{u}_n\} \subseteq \dot{V}$, then we can assume without loss of generality that $0 \leq u_n \leq u$ and that $\{u_n\} \subseteq V$ both hold in L, and so by the σ-order closedness of V, we get $w = \liminf u_n \in V$. But since A is a σ-ideal the canonical projection of L onto L/A is a Riesz σ-homomorphism and thus

$$\dot{u} = \sup_n \dot{u}_n = \liminf \dot{u}_n = \dot{w} \in \dot{V}.$$

This shows that $\dot{V}$ is σ-order closed and the proof is finished. ■

Now assume that A is a projection band of a Riesz space L, i.e., $L = A \oplus A^{\mathrm{d}}$. If q denotes the restriction to A^{d} of the canonical projection $Q\colon L \to L/A$, then $q\colon A^{\mathrm{d}} \to L/A$ is an onto Riesz isomorphism. If in addition τ is a locally solid topology on L, then q is also continuous if A^{d} is equipped with the induced topology τ and L/A with τ/A. Now consider the scheme of operators

$$(L, \tau) \xrightarrow{\;Q\;} (L/A, \tau/A) \xrightarrow{\;q^{-1}\;} (A^{\mathrm{d}}, \tau).$$

It follows that $P = q^{-1} \circ Q$, where P denotes the projection of L onto A^{d}. The basic properties of the quotient topologies imply that q^{-1} is also continuous, and hence $(L/A, \tau/A)$ and (A^{d}, τ) are Riesz homeomorphic. In particular, it follows that if τ is a Fatou (or σ-Fatou) topology, then so is its restriction to A^{d}. Therefore, we have proven the following result.

THEOREM 4.45. *Let (L, τ) be a locally solid Riesz space with the Fatou (resp. σ-Fatou property), and let A be a projection band of L. Then $(L/A, \tau/A)$ satisfies the Fatou (resp. σ-Fatou) property.*

There is one more positive result concerning quotient spaces and the σ-Fatou property that is stated next.

THEOREM 4.46. *Let (L, τ) be an almost Dedekind σ-complete Riesz space with the σ-Fatou property and let A be a σ-ideal of L. Suppose that the ideal A^* generated by A in the Dedekind σ-completion L^σ of L is likewise a σ-ideal of L^σ. Then $(L/A, \tau/A)$ satisfies the σ-Fatou property.*

PROOF. By Theorem 4.13 we can extend τ to a σ-Fatou topology on L^σ; where if V is a solid σ-Fatou neighborhood of zero for τ, then

$$V^* = \{u \in L^\sigma\colon\ \exists\{u_n\} \subseteq V \text{ with } 0 \le u_n \uparrow |u|\,\}$$

is a σ-Fatou τ^σ-neighborhood of L^σ and $V = V^* \cap L$.

Now let V be a σ- Fatou τ-neighborhood of zero of L; we need to show that $Q(V) = \dot{V}$ is a σ-Fatou τ/A-neighborhood of L/A. To this end, let $0 \le \dot{u}_n \uparrow \dot{u}$ in L/A with $\{\dot{u}_n\} \subseteq \dot{V}$; without loss of generality we can assume that $0 \le u_n \le u$ for all n and $\{u_n\} \subseteq V$. Then $v = \liminf u_n \in V^*$ and we claim $u - v \in A^*$. Indeed, since L is super order dense in L^σ, there exists a sequence $\{x_k\} \subseteq L^+$ with $x_k \uparrow u - v$. We shall show that $\{x_k\} \subseteq A$ and this will imply that $u - v \in A^*$ since A^* is a σ-ideal. For each fixed n let $y_m = \bigwedge_{i=n}^{n+m} u_i$ and note that for each k the relation

$$0 \le x_k \le u - v \le u - \bigwedge_{i=n}^{\infty} u_i = u - \bigwedge_{m=1}^{\infty} y_m = \bigvee_{m=1}^{\infty} (u - y_m)$$

implies $(u - y_m) \wedge x_k \uparrow_m x_k$ in L and so $(\dot{u} - \dot{y}_m) \wedge \dot{x}_k \uparrow \dot{x}_k$ in L/A. But then $(\dot{u} - \dot{u}_n) \wedge \dot{x}_k = (\dot{u} - \dot{y}_m) \wedge \dot{x}_k$ for each m and so $(\dot{u} - \dot{u}_n) \wedge \dot{x}_k = \dot{x}_k$ for all n. It follows now from $\dot{u_n} \uparrow \dot{u}$ that $\dot{x}_k = 0$, and so $x_k \in A$ holds for all k.

Since A^* is the ideal generated by A in L^σ and $u - v \in A^*$, there exists $w \in A$ with $u - v \le w$. Thus $(u - w)^+ \le v \in V^*$ and so $(u - w)^+ \in V$. But then we have $\dot{u} = (\dot{u} - \dot{w})^+ \in \dot{V}$. Hence $\dot{V}$ is σ-order closed, and therefore $(L/A, \tau/A)$ has the σ-Fatou property. ■

The next result presents a condition for A^* to be a σ-ideal in L^σ.

THEOREM 4.47. *Let L be an almost Dedekind σ-complete Riesz space, and let A be a σ-ideal in L with the property that every band B contained in A^{dd} is a projection band.*[1] *Then the ideal A^* generated by A in L^σ is a σ-ideal.*

PROOF. Suppose that $0 \le u_n^* \uparrow u^*$ with $\{u_n^*\} \subseteq A^*$. Since each u_n^* is the upward limit of a sequence of positive elements of L, using a diagonal process, we can find a sequence $\{u_n\}$ of elements of A with $0 \le u_n \uparrow u^*$. There exists some $v \in L$ with $u^* \le v$; let u be the projection of v onto the band A^{dd}. Since $\{u_n\} \subseteq A \subseteq A^{\mathrm{dd}}$, certainly $u_n \le u$ for all n and thus $u^* \le u \in A^{\mathrm{dd}}$. Let B_{u^*} be the band generated in L^σ by u^* and let $B = B_{u^*} \cap A^{\mathrm{dd}}$. Then B is a band of L and since $B \subseteq A^{\mathrm{dd}}$, the band B is a projection band. Let $x = P_B(u)$.

Clearly, we have $u_n \in B$ for every n so that $P_B(u_n) = u_n$, and consequently, $u_n = P_B(u_n) \le P_B(u) = x \in B$, implying that $u^* \le x$. Now note that $x \in B \subseteq B_{u^*}$ implies $x = \sup\{x \wedge nu^*\colon\ n = 1, 2, \ldots\}$ in L^σ. Let $t_n = x \wedge nu_n \in A$. Since $u_n \uparrow u^*$, it follows that $0 \le t_n \uparrow x$ in L. But $\{t_n\} \subseteq A$ and A is a σ-ideal so that $x \in A$. Since $u^* \le x$, necessarily $u^* \in A^*$, and the proof is finished. ■

As an application of the last two results we have the following theorem.

THEOREM 4.48. *Let (L, τ) be a locally solid Riesz space with the σ-Fatou property and let A be a σ-ideal of L. If L has the projection property then $(L/A, \tau/A)$ satisfies the σ-Fatou property.*

---

[1] It should not be that difficult to verify that this condition is equivalent to the assumption that A^{dd} is a projection band of L and that A^{dd}, considered as a Riesz space in its own right, has the projection property.

PROOF. Apply Theorems 4.46 and 4.47 by noting that L is almost Dedekind σ-complete; see Theorem 1.60. ■

Note. The results in this section were first proven by E. Langford and C. D. Aliprantis for seminormed Riesz spaces in [**81**, Section 5, pp. 205–210]; see also [**4**].

Exercises

1. Prove Theorem 4.13.
2. Let (L,τ) be a Hausdorff almost Dedekind σ-complete Riesz space with the σ-Fatou property. Show that τ can be extended to a σ-Fatou Hausdorff topology on L^δ.
3. Show that the following stronger version of Theorem 4.23 holds true: *If $T\colon (L,\tau) \to (M,\tau')$ is a positive order continuous operator and if τ is a Hausdorff Fatou topology and τ' a Lebesgue topology, then T restricted to the order bounded subsets of L is continuous.* [Hint: If V is a solid τ'-neighborhood of zero let $V^* = \{u \in L\colon\ T(|u|) \in V\}$.]
4. Let (L,τ) be a locally solid Riesz space and let $\{N_\alpha\colon\ \alpha \in A\}$ denote the family of all null ideals of the normal sequences of solid τ-neighborhoods of zero. Show that C_τ is order dense in L if and only if $\bigcap_{\alpha\in A} N_\alpha^{\mathrm{dd}} = \{0\}$.
5. Let (L,τ) be a Hausdorff locally solid Riesz space with the Fatou property and consider the Fatou extension τ^δ of τ to L^δ (see Theorem 4.12). Show that $C_{\tau^\delta} \cap L = C_\tau$.
6. If τ is a Fatou topology on a Riesz space L with the projection property, show that C_τ is τ-dense in L.
7. Consider the Hausdorff locally solid Riesz space (L,τ) of Example 3.6. Determine the carrier C_τ and verify that C_τ is order dense but not τ-dense in L. Is τ a Fatou topology?
8. For a Hausdorff locally solid Riesz space (L,τ) with the Fatou property show that $C_\phi \subseteq C_\tau$ holds for every $\phi \in L^\sim$.
9. Let L be a Riesz space with the projection property. If τ is a locally solid topology on L with the σ-Fatou property, then show that:
   (i) The kernel K_τ of τ (i.e., the intersection of all τ-neighborhoods of zero) is a τ-closed σ-ideal of L.
   (ii) The quotient space $(L/K_\tau, \tau/K_\tau)$ is a Hausdorff locally solid Riesz space with the σ-Fatou property.
10. If τ is a Fatou topology on a Riesz space L with the projection property, show that $(L/K_\tau, \tau/K_\tau)$ is a Hausdorff locally solid Riesz space with the Fatou property.
11. Consider the Hausdorff locally solid Riesz space (L,τ) of Example 4.41. Verify the three claims about (L,τ).

12. If ρ is a Fatou norm on a Riesz space L, that is, if $0 \le u_\alpha \uparrow u$ in L implies $\rho(u_\alpha) \uparrow \rho(u)$, then show that the norm extension $\widehat{\rho}$ of ρ to the norm completion of $\widehat{L}_\rho$ is a Fatou norm and that

$$\widehat{\rho}(\widehat{u}) = \sup\{\rho(u)\colon\ u \in L \text{ and } 0 \le u \le |\widehat{u}|\,\}$$

holds for all $\widehat{u} \in \widehat{L}_\rho$.

13. Let L be an Archimedean Riesz space and ρ a Fatou seminorm on L. Define the ideal $I_\rho = \{u \in L\colon\ \rho(u) = 0\}$ and show that the quotient Riesz norm $\dot{\rho}$ on L/I_ρ is a Fatou norm.

14. (Aliprantis [**4**]) Let L be an almost Dedekind σ-complete Riesz space and let ρ be a σ-Fatou seminorm on L, i.e., $0 \le u_n \uparrow u$ implies $\rho(u_n) \uparrow \rho(u)$. If $I_\rho = \{u \in L\colon\ \rho(u) = 0\}$, show that the quotient Riesz norm $\dot{\rho}$ on L/I_ρ is a σ-Fatou norm.

15. (Aliprantis [**4**]) Let X be an uncountable set, X_∞ the one-point compactification of X considered with the discrete topology, and $\{x_1, x_2, \ldots\}$ a countable subset of X. Define the Riesz seminorm ρ on $L = C(X_\infty)$ by

$$\rho(u) = |u(\infty)| + \sup\big\{|u(x_i)|\colon\ i = 1, 2, \ldots\big\}$$

and show that:

(i) ρ satisfies the σ-Lebesgue property but ρ is not a Fatou seminorm.

(ii) $I_\rho = \big\{u \in L\colon\ \rho(u) = 0\big\} = \big\{u \in L\colon\ u(x_n) = 0\ \forall\ n \in \mathbb{N}\big\}$, and infer from this that I_ρ is a band of L.

(iii) The quotient norm $\dot{\rho}$ on L/I_ρ is not a σ-Fatou seminorm. Why does this not contradict the previous exercise?

16. If $\rho\colon L \to \mathbb{R}$ is a Riesz pseudonorm on a Riesz space, then its **null ideal** is defined by

$$I_\rho = \big\{u \in L\colon\ \rho(u) = 0\big\}\,.$$

Establish the following.

(a) I_ρ is indeed an ideal of L.

(b) If $\{\rho_n\}$ is a sequence of Riesz pseudonorms on a Riesz space L, then the function $\rho\colon L \to \mathbb{R}$, defined by

$$\rho(u) = \sum_{n=1}^{\infty} \tfrac{1}{2^n} \cdot \tfrac{\rho_n(u)}{1+\rho_n(u)}$$

for each $u \in L$, is a Riesz pseudonorm (which is in addition τ-continuous if each ρ_n is τ-continuous for some locally solid topology τ on L) such that $I_\rho = \bigcap_{n=1}^{\infty} I_{\rho_n}$.

(c) If τ is a locally solid topology on a Riesz space L and $\mathcal{F}$ denotes the collection of all τ-continuous Riesz pseudonorms on the Riesz space L, then $C_\tau = \bigcup_{\rho\in\mathcal{F}} I_\rho^{\mathrm{d}}$.

17. Show that a Fatou topology τ on a Riesz space L is a Hausdorff topology if and only if its carrier C_τ is order dense in L.

CHAPTER 5

Metrizability

In this chapter we shall deal with metrizable locally solid topologies, that is, with topologies that in addition to being locally solid are also generated by a metric. This is equivalent, of course, for Hausdorff linear topologies to the existence of a countable base at zero consisting of solid sets (Theorem 2.1). In particular, the class of metrizable locally solid Riesz spaces includes the class of normed Riesz spaces.

The metrizability of the topology has a number of interesting consequences. Among the remarkable results concerning a metrizable locally solid Riesz space (L,τ), we shall find that the upper and lower elements of L are enough to describe the structure of $\widehat{L}$; the order denseness of L in $\widehat{L}$ becomes equivalent to a simple continuity property ($u_n \downarrow 0$ and $\{u_n\}$ τ-Cauchy implies $u_n \xrightarrow{\tau} 0$) and the Lebesgue property becomes equivalent to the pre-Lebesgue and σ-Lebesgue properties taken together. On the other hand, we shall see that a Riesz space can admit at most one topologically complete metrizable locally solid topology.

5.1. Upper and Lower Elements

We start with the definition of the upper and lower elements of a Hausdorff locally solid Riesz space.

DEFINITION 5.1. *Let (L,τ) be a Hausdorff locally solid Riesz space. We shall call an element $0 \leq \widehat{u} \in \widehat{L}$ an* ***upper element*** *(resp. a* ***lower element****) of L if there exists a net $\{u_\alpha\} \subseteq L^+$ with $u_\alpha \uparrow \widehat{u}$ and $u_\alpha \xrightarrow{\widehat{\tau}} \widehat{u}$ (resp. $u_\alpha \downarrow \widehat{u}$ and $u_\alpha \xrightarrow{\widehat{\tau}} \widehat{u}$).*

Note that by Theorem 2.21 the conditions $u_\alpha \uparrow$ and $u_\alpha \xrightarrow{\widehat{\tau}} \widehat{u}$ are actually enough to ensure that $u_\alpha \uparrow \widehat{u}$ in $\widehat{L}$. It should be also clear that finite sums, finite infima, and finite suprema of upper elements (resp. lower elements) are upper elements (resp. lower elements).

The next theorem tells us that for metrizable locally solid Riesz spaces the upper and lower elements can be described in terms of sequences.

THEOREM 5.2. *For a metrizable locally solid Riesz space (L,τ) an element $0 \leq \widehat{u} \in \widehat{L}$ is an upper element (resp. a lower element) of L if and only if there exists a sequence $\{u_n\} \subseteq L^+$ satisfying $u_n \uparrow \widehat{u}$ (resp. $u_n \downarrow \widehat{u}$) and $u_n \xrightarrow{\widehat{\tau}} \widehat{u}$.*

PROOF. Let $\{V_n\}$ be a normal countable base at zero for τ consisting of solid sets and let $0 \leq \widehat{u} \in \widehat{L}$ be an upper element of L. Pick a net $\{u_\alpha\} \subseteq L^+$ with $u_\alpha \uparrow \widehat{u}$ and $u_\alpha \xrightarrow{\widehat{\tau}} \widehat{u}$. For each positive integer n pick an index α_n with $\widehat{u} - u_{\alpha_n} \in \overline{V}_n$. We can assume that the sequence of indices $\{\alpha_n\}$ is increasing, i.e., $\alpha_{n+1} \succeq \alpha_n$ holds for each n. It now easily follows that $u_{\alpha_n} \xrightarrow{\widehat{\tau}} \widehat{u}$ and $0 \leq u_{\alpha_n} \uparrow \widehat{u}$ in $\widehat{L}$, and the desired conclusion follows. ■

The next theorem gives some information about the structure of the topological completion of a metrizable locally solid Riesz space.

THEOREM 5.3. *Let (L, τ) be a metrizable locally solid Riesz space. Then every element of $\widehat{L}$ can be written as a difference of two upper elements of L.*

PROOF. It is enough to establish the result for positive elements of $\widehat{L}$. So, let $0 \leq \widehat{u} \in \widehat{L}$. Pick a normal countable base $\{V_n\}$ at zero for τ consisting of solid sets. Choose a sequence $\{u_n\} \subseteq L^+$ with $u_n \xrightarrow{\widehat{\tau}} \widehat{u}$. By passing to a subsequence (if necessary), we can assume without loss of generality that $u_{n+1} - u_n \in V_n$ holds for all n. Now put $v_1 = w_1 = 0$ and for each $n \geq 2$ let

$$v_n = u_1^+ + \sum_{k=1}^{n-1}(u_{k+1} - u_k)^+ \quad \text{and} \quad w_n = u_1^- + \sum_{k=1}^{n-1}(u_{k+1} - u_k)^- .$$

It is easy to check that both $\{v_n\}$ and $\{w_n\}$ are increasing τ-Cauchy sequences of L^+. Hence there exist $\widehat{v}, \widehat{w} \in \widehat{L}$ such that $v_n \xrightarrow{\widehat{\tau}} \widehat{v}$ and $w_n \xrightarrow{\widehat{\tau}} \widehat{w}$. In particular, note that both $\widehat{v}$ and $\widehat{w}$ are upper elements of L.

Now observe that $u_n = v_n - w_n$ holds for all $n \geq 2$ and so by taking the $\widehat{\tau}$-limits we get $\widehat{u} = \widehat{v} - \widehat{w}$, and the proof is finished. ■

We continue with two important approximation properties concerning metrizable locally solid Riesz spaces. They are essentially due to W. A. J. Luxemburg [**87**, Note XVI, Theorems 60.3 and 60.6, p. 648].

THEOREM 5.4. *Let (L, τ) be a metrizable locally solid Riesz space. Assume that $0 \leq \widehat{f} \in \widehat{L}$ and V is a τ-neighborhood of zero for (L, τ). Then there exists an upper element $\widehat{u}$ of L with $\widehat{f} \leq \widehat{u}$ and $\widehat{u} - \widehat{f} \in \overline{V}$.*

In particular, there exists a sequence $\{\widehat{u}_n\}$ of upper elements of L with $\widehat{u}_n \downarrow \widehat{f}$ and $\widehat{u}_n \xrightarrow{\widehat{\tau}} \widehat{f}$ in $\widehat{L}$.

PROOF. Pick a normal countable base $\{V_n\}$ at zero consisting of solid sets. Also choose a sequence $\{f_n\} \subseteq L^+$ with $f_n \xrightarrow{\widehat{\tau}} \widehat{f}$. Now fix a neighborhood V_k and then select a subsequence $\{g_n\}$ of $\{f_n\}$ such that $g_{n+p} - g_n \in V_{k+n+2}$ holds for all n and p. Next, for each n let $u_n = g_{n+1} + \sum_{i=1}^{n} |g_{i+1} - g_i|$ and note that since

$$u_{n+1} - u_n = g_{n+2} - g_{n+1} + |g_{n+2} - g_{n+1}| \geq 0\,,$$

we have $0 \leq u_n \uparrow$. On the other hand, the relation

$$0 \leq u_{n+p} - u_n = g_{n+p+1} - g_{n+1} + \sum_{i=n+1}^{n+p} |g_{i+1} - g_i| \in V_{k+n+3} + V_{k+n+2} \subseteq V_n$$

for all n and p implies that $\{u_n\}$ is a τ-Cauchy sequence. Hence $u_n \xrightarrow{\widehat{\tau}} \widehat{u}$ holds in $\widehat{L}$ for some $\widehat{u}$, and therefore $u_n \uparrow \widehat{u}$ also holds in $\widehat{L}$. Thus $\widehat{u}$ is an upper element of L.

Now since $0 \leq u_n - g_{n+1} = \sum_{i=1}^{n} |g_{i+1} - g_i| \in V_k$ holds for each n, it follows that $0 \leq \widehat{u} - \widehat{f} \in \overline{V}_k$, and the validity of the first claim has been established. The last part follows from the first by noting that the infimum of any finite collection of upper elements is again an upper element. ■

THEOREM 5.5. *Let (L,τ) be a metrizable locally solid Riesz space. Assume that $0 \leq \widehat{f} \in \widehat{L}$ and let V be a τ-neighborhood of zero for (L,τ). Then there exists a lower element $\widehat{u}$ of L with $\widehat{u} \leq \widehat{f}$ and $\widehat{f} - \widehat{u} \in \overline{V}$.*

In particular, there exists a sequence $\{\widehat{u}_n\}$ of lower elements of L with $\widehat{u}_n \uparrow \widehat{f}$ and $\widehat{u}_n \xrightarrow{\widehat{\tau}} \widehat{f}$ in $\widehat{L}$.

PROOF. Assume first that $\widehat{f} \leq u$ also holds for some $u \in L$. Pick an upper element $\widehat{w}$ of L with $u - \widehat{f} \leq \widehat{w}$ and $\widehat{w} - (u - \widehat{f}) = \widehat{f} - (u - \widehat{w})$ in $\overline{V}$. Put $\widehat{u} = (u - \widehat{w})^+$ and note that $\widehat{u}$ is a lower element of L satisfying $\widehat{u} \leq \widehat{f}$ and $\widehat{f} - \widehat{u} \in \overline{V}$.

Now pick a solid τ-neighborhood U of zero with $U + U \subseteq V$ and a sequence $\{f_n\} \subseteq L^+$ with $f_n \xrightarrow{\widehat{\tau}} \widehat{f}$. Fix some $k \in \mathbb{N}$ such that $f_k - \widehat{f} \in \overline{U}$ and note that $f_k \wedge \widehat{f} - \widehat{f} \in \overline{U}$. But then, by what was shown above, there exists a lower element $\widehat{u}$ of L with $\widehat{u} \leq f_k \wedge \widehat{f}$ and such that $f_k \wedge \widehat{f} - \widehat{u} \in \overline{U}$. Consequently, $\widehat{f} - \widehat{u} = (\widehat{f} - \widehat{f} \wedge f_k) + (\widehat{f} \wedge f_k - \widehat{u}) \in \overline{U} + \overline{U} \subseteq \overline{V}$ and so $\widehat{f} - \widehat{u} \in \overline{V}$, and the proof is finished. ■

Recall that a Riesz subspace K of a Riesz space L is said to be **majorizing** L (or that it is **full** in L) if for every $u \in L$ there exists some $v \in K$ with $u \leq v$. Let us also call a subset A of L **majorizing** L (or that it is **full** in L) if the ideal generated by A in L is a majorizing Riesz subspace.

The next result is due to A. L. Peressini [**107**, Proposition 1.3, p. 204].

THEOREM 5.6 (Peressini). *For a Hausdorff locally convex-solid Riesz space (L,τ) the following statements are equivalent.*

(i) *The absolute weak topology $|\sigma|(L,L')$ is metrizable.*
(ii) *L' contains a countable majorizing subset.*

PROOF. (i) $\Longrightarrow$ (ii) There exists a sequence $\{\phi_n\} \subseteq L'$ such that the family of Riesz seminorms $\{\rho_{\phi_n}\}$ generates $|\sigma|(L,L')$, where $\rho_{\phi_n}(u) = |\phi_n|(|u|)$. For each n let $\psi_n = \sup\{|\phi_k|\colon 1 \leq k \leq n\}$. Then $0 \leq \psi_n \uparrow$ holds in L' and $\{\rho_{\psi_n}\}$ generates the topology $|\sigma|(L,L')$. Now since each $\phi \in L'$ is bounded on the closed unit ball of some ρ_{ψ_n}, there exists some $\lambda > 0$ (depending on ϕ) such that $|\phi| \leq \lambda\psi_n$. This shows that sequence $\{\psi_n\}$ is majorizing in L'.

(ii) $\Longrightarrow$ (i) Observe that the inclusions $\sigma(L,L') \subseteq |\sigma|(L,L') \subseteq \tau$ imply that $|\sigma|(L,L')$ is a Hausdorff topology. On the other hand, if $\{\phi_n\}$ is a countable majorizing subset of L', then the sequence $\{\rho_{\phi_n}\}$ generates $|\sigma|(L,L')$ and hence $|\sigma|(L,L')$ is metrizable. ■

For a Riesz space L let $\mathcal{S}_L$ denote the collection of all Riesz seminorms on L. Clearly the relation $\rho_1 \leq \rho_2$ if $\rho_1(u) \leq \rho_2(u)$ for all $u \in L^+$ partially orders $\mathcal{S}_L$. It should be also clear that for every pair $\rho_1, \rho_2 \in \mathcal{S}_L$ the function $\rho'\colon L \to \mathbb{R}$, defined by

$$\rho'(u) = \max\{\rho_1(u), \rho_2(u)\},$$

is a Riesz seminorm on L that, as easily seen, is the least upper bound of ρ_1 and ρ_2 in $\mathcal{S}_L$. Also if $\rho_1, \rho_2 \in \mathcal{S}_L$, then the function $\rho''\colon L \to \mathbb{R}$ defined by

$$\rho''(u) = \inf\{\rho_1(u_1) + \rho_2(u_2)\colon\ u_1, u_2 \in L^+ \text{ and } u_1 + u_2 = |u|\},$$

is a Riesz seminorm that is the greatest lower bound of ρ_1 and ρ_2 in $\mathcal{S}_L$.

To see this, note first that we have $\rho''(u) = \rho''(|u|)$ for all $u \in L$. Now if $0 \le v \le u = u_1 + u_2$ with $u_1, u_2 \in L^+$, write $v = v_1 + v_2$ with $0 \le v_1 \le u_1$, $0 \le v_2 \le u_2$ and note that

$$\rho''(v) \le \rho_1(v_1) + \rho_2(v_2) \le \rho_1(u_1) + \rho_2(u_2),$$

so that $\rho''(v) \le \rho''(u)$. To establish the triangle inequality, assume that $u = u_1 + u_2$ and $v = v_1 + v_2$ with the u_i and v_i in L^+, and then note that

$$\begin{aligned}\rho''(u+v) &\le \rho_1(u_1+u_2) + \rho_2(v_1+v_2)\\ &\le \big[\rho_1(u_1) + \rho_2(u_2)\big] + \big[\rho_1(v_1) + \rho_2(v_2)\big],\end{aligned}$$

from which it follows that $\rho''(u+v) \le \rho''(u) + \rho''(v)$. Thus, for any vectors $u, v \in L$ we have

$$\rho''(u+v) = \rho''\big(|u+v|\big) \le \rho''\big(|u| + |v|\big) \le \rho''(u) + \rho''(v).$$

It should be also clear that ρ'' is the infimum of ρ_1 and ρ_2 in $\mathcal{S}_L$. We have thus established the following result.

THEOREM 5.7. *If $\mathcal{S}_L$ is the collection of all Riesz seminorms on a Riesz space L ordered by $\rho_1 \le \rho_2$ whenever $\rho_1(u) \le \rho_2(u)$ for all $u \in L^+$, then $\mathcal{S}_L$ is a lattice. Moreover, the lattice operations in $\mathcal{S}_L$ are given by the formulas*

$$\begin{aligned}[\rho_1 \vee \rho_2](u) &= \max\{\rho_1(u), \rho_2(u)\}, \text{ and}\\ [\rho_1 \wedge \rho_2](u) &= \inf\{\rho_1(u_1) + \rho_2(u_2):\ u_1, u_2 \in L^+ \text{ and } u_1 + u_2 = |u|\,\}\end{aligned}$$

for all $\rho_1, \rho_2 \in \mathcal{S}_L$ and all $u \in L$.

It should be noted that $\mathcal{S}_L$ is in general a non-distributive lattice; see Exercise 10 at the end of the chapter.

Recall that for a Riesz space L the band consisting of all σ-order continuous linear functionals in $L^\sim$ is denoted $L^\sim_{\rm c}$ and the band consisting of all order continuous linear functionals in $L^\sim$ is denoted $L^\sim_{\rm n}$.

DEFINITION 5.8. *For a locally solid Riesz space (L, τ) we put*

$$L'_{\rm c} = L' \cap L^\sim_{\rm c} \quad \text{and} \quad L'_{\rm n} = L' \cap L^\sim_{\rm n},$$

and call them the ***topological σ-order continuous dual*** *and the* ***topological order continuous dual****, respectively, of (L, τ).*

The next result is due to W. A. J. Luxemburg and A. C. Zaanen [**89**, Note XII, Theorem 37.1, p. 520].

THEOREM 5.9 (Luxemburg–Zaanen). *If (L, τ) is a metrizable locally convex-solid Riesz space, then $L'_{\rm c}$ is an order dense ideal of $L^\sim_{\rm c}$ and $L'_{\rm n}$ is an order dense ideal of $L^\sim_{\rm n}$.*

PROOF. We shall prove that $L'_{\rm c}$ is an order dense ideal of $L^\sim_{\rm c}$, the other case can be established in a similar manner. We need to show that for every $0 < \phi \in L^\sim_{\rm c}$ there exists some $\psi \in L'_{\rm c}$ with $0 < \psi \le \phi$.

To this end, let $\{\rho_n\}$ be a sequence of Riesz seminorms generating τ such that $\rho_n \le \rho_{n+1}$ for all n and fix $0 < \phi \in L^\sim_{\rm c}$. Choose some $u \in L^+$ such that $0 < \phi(u) < 1$ and then select $\delta > 0$ so that $\phi(u) > \delta > 0$. Observe that $\rho(v) = \phi(|v|)$ is a Riesz seminorm of L. For each n let $\rho^*_n = \rho_n \wedge \rho$, and we claim that $\rho^*_n(u) > 0$ for at least one n.

For suppose if possible that we have $\rho^*_n(u) = 0$ for all $n \in \mathbb{N}$. Let $\epsilon_1 = \phi(u) - \delta > 0$, and then choose a vector $0 \le w_1 \le u$ such that $\rho_1(w_1) + \phi(u - w_1) < \epsilon_1$. Then

$\phi(w_1) = \phi(u) - \phi(u - w_1) > \phi(u) - \epsilon_1 = \delta$ and $\rho_1(w_1) < 1$. Assume that the element w_n has been chosen to satisfy $0 \le w_n \le u$, $w_n \le w_{n-1}$ (where we let $w_0 = u$), $\phi(w_n) > \delta$, and $\rho_n(w_n) < \frac{1}{n}$. Put $\epsilon_{n+1} = \phi(w_n) - \delta > 0$ and observe that since $\rho^*_{n+1}(w_n) = 0$, there is $0 \le w_{n+1} \le w_n$ such that $\rho_{n+1}(w_{n+1}) + \phi(w_n - w_{n+1}) \le \frac{\epsilon_{n+1}}{n+1}$. Thus $\rho_{n+1}(w_{n+1}) < \frac{1}{n+1}$ and

$$\phi(w_{n+1}) = \phi(w_n) - \phi(w_n - w_{n+1}) > \phi(w_n) - \epsilon_{n+1} = \delta\,.$$

Therefore, we can construct by induction a sequence $0 \le w_n \downarrow$ such that $\phi(w_n) > \delta > 0$ and $\rho_n(w_n) < \frac{1}{n}$ for all n. It follows that $w_n \downarrow 0$. Indeed, if $0 \le w \le w_n$ for all n, then for each fixed k, we have $\rho_k(w) \le \rho_{k+n}(w_{n+k}) \le \frac{1}{n+k}$ for all n, so that $\rho_k(w) = 0$ for all k and hence $w = 0$. But now $\phi(w_n) > \delta$ for each n contradicts $\phi \in L^\sim_c$, and our claim has been established.

Next pick n so that $\rho^*_n(u) > 0$ and let X be the vector subspace generated by u, i.e., $X = \{\lambda u\colon\ \lambda \in \mathbb{R}\}$. Define the linear mapping $\psi\colon X \to \mathbb{R}$ by $\psi(\lambda u) = \lambda\rho^*_n(u)$. By Theorem 1.69 we know that ψ has a linear extension to L (denoted by ψ again) satisfying $\psi \le \rho^*_n$. In particular, it follows that $\psi \in L'$ and $\psi \le \phi$. But then $\psi^+ \le \phi$ also holds and hence $\psi^+ \in L'_c$. The desired conclusion now follows by observing that $\psi^+(u) \ge \psi(u) = \rho^*_n(u) > 0$ implies $\psi^+ > 0$. ■

The next results describe some properties of the metrizable σ-Fatou topologies.

THEOREM 5.10. *Let (L, τ) be a metrizable locally solid Riesz space with the σ-Fatou property, and let $\{u_n\}$ be a sequence of L satisfying $0 \le u_n \le u$ for all n and $u_n \xrightarrow{\tau} u$. Then there exists a subsequence $\{v_n\}$ of $\{u_n\}$ and a net $\{w_\alpha\}$ of L^+ such that:*

(i) *For each index α there exists some $n_\alpha \in \mathbb{N}$ (depending on α) satisfying $w_\alpha \le v_n$ for all $n \ge n_\alpha$, and*
(ii) *$0 \le w_\alpha \uparrow u$ holds in L.*

PROOF. Fix a countable normal base $\{V_n\}$ at zero for the topology τ consisting of σ-Fatou τ-neighborhoods. Also, choose a subsequence $\{v_n\}$ of $\{u_n\}$ satisfying $u - v_n \in V_n$ for all n.

Now let $A = \{a \in L^+\colon\ \exists n$ such that $a \le v_k$ for all $k \ge n\}$. Observe that $a, b \in A$ imply $a \vee b \in A$. In particular, A is directed upward. Thus by defining $w_\alpha = \alpha$ for each $\alpha \in A$, we get an increasing net $\{w_\alpha\}$ of L^+ that is bounded from above by u and satisfies (i) by construction. So, it remains to be shown that $0 \le w_\alpha \uparrow u$ holds in L. To establish this let w be an upper bound of the net $\{w_\alpha\}$. By replacing w by $w \wedge u$ (if necessary), we can assume that $w_\alpha \le w \le u$ holds for all α. Now for $m \in \mathbb{N}$ put $S = \{s \in L^+\colon\ s \le v_k$ for all $k \ge m\}$, and note that since L is Archimedean, it follows from Theorem 1.28 that

$$\inf\{v_k - s\colon\ k \ge m \text{ and } s \in S\} = 0\,.$$

But then the inequality $0 \le (v_k - w)^+ \le v_k - s$ for $k \ge m$ and $s \in S$ implies

$$\Big[\bigwedge_{k=m}^{\infty} v_k \vee w\Big] - w = \bigwedge_{k=m}^{\infty} (v_k - w)^+ = 0\,.$$

and so $w = \bigwedge_{k=m}^{\infty} v_k \vee w$. Now observe that

$$0 \le u - w = \bigvee_{k=m}^{\infty} (u - v_k \vee w) = \bigvee_{k=1}^{\infty} x_k\,,$$

where $x_k = \bigvee_{i=m}^{m+k}(u - v_i \vee w)$. On the other hand, $|u-v_i \vee w| \le u-v_i \in V_i$ for all i implies

$$0 \le x_k \le \sum_{i=m}^{m+k} |u - v_i| \in V_m + \cdots + V_{m+k} \subseteq V_{m-1}$$

for all $m \ge 2$. Now from $0 \le x_k \uparrow u - w$ and the σ-order closedness of V_{m-1} we get $u - w \in V_{m-1}$ for all $m \ge 2$. Thus, $u - w = 0$ and hence $w = u$. This shows that $0 \le w_\alpha \uparrow u$ holds in L, and the proof is finished. ■

For metrizable locally solid Riesz spaces Theorem 4.20 can be strengthened as follows.

THEOREM 5.11. *Let (L, τ) be a metrizable locally solid Riesz space with the σ-Fatou property. Then every solid and order closed subset of L is τ-closed.*

PROOF. Let S be a solid and order closed subset of L and let $u \in \overline{S}$. Pick a sequence $\{u_n\} \subseteq L^+ \cap S$ with $u_n \xrightarrow{\tau} |u|$. Replacing $\{u_n\}$ by $\{u_n \wedge |u|\}$ (if necessary), we can assume that $0 \le u_n \le |u|$ holds for all n, and so according to Theorem 5.10 there exists a net $\{w_\alpha\}$ of S with $w_\alpha \uparrow |u|$ in L. But then $|u| \in S$ and so $u \in S$. This shows that S is τ-closed. ■

THEOREM 5.12. *Let (L, τ) be a metrizable locally solid Riesz space with the σ-Fatou property. Then the topology induced by τ on any order bounded subset of L is finer than the topology induced by any Lebesgue topology of L on the same set.*

PROOF. Let τ^* be a Lebesgue topology on L, and let $\{u_n\} \subseteq [0, u]$ be such that $u_n \xrightarrow{\tau} 0$. It is enough to show that $u_n \xrightarrow{\tau^*} 0$. Assume by way of contradiction that $u_n \xrightarrow{\tau^*} 0$ does not hold. Then there exists a τ^*-neighborhood V of zero and a subsequence $\{v_n\}$ of $\{u_n\}$ with $v_n \notin V$ for each n. Put $x_n = u - v_n$ and note that $0 \le x_n \le u$ holds for each n and $x_n \xrightarrow{\tau} u$. By Theorem 5.10 there exists a subsequence $\{x_{k_n}\}$ of $\{x_n\}$ and a net $\{w_\alpha\}$ of L with $0 \le w_\alpha \uparrow u$ and such that for each α there exists some $k \in \mathbb{N}$ satisfying $w_\alpha \le x_{k_n}$ for all $n \ge k$. Now since τ^* is a Lebesgue topology, $w_\alpha \xrightarrow{\tau^*} u$ holds and therefore $x_{k_n} = u - v_{k_n} \xrightarrow{\tau^*} u$ also holds true. But then $v_{k_n} \xrightarrow{\tau^*} 0$ holds, in contradiction to $v_{k_n} \notin V$ for all n, and the proof is finished. ■

Recall that a seminorm ρ on a Riesz space L satisfies the **σ-Fatou property** if $0 \le u_n \uparrow u$ implies $\rho(u_n) \uparrow \rho(u)$. The next result is taken from [**4**].

THEOREM 5.13 (Aliprantis). *If a Riesz norm ρ on a Riesz space L satisfies the σ-Fatou property, then the continuous extension of ρ to its norm completion $\widehat{L}_\rho$ also satisfies the σ-Fatou property.*

PROOF. We denote the continuous extension of ρ to $\widehat{L}_\rho$ by ρ again. Assume $0 \le \widehat{u}_n \uparrow \widehat{u}$ in $\widehat{L}_\rho$. Since $\lim \rho(\widehat{u}_n) \le \rho(\widehat{u})$ we need only to establish the inequality $\rho(\widehat{u}) \le \lim \rho(\widehat{u}_n)$.

For each n pick an upper element $\widehat{v}_n$ of L satisfying $0 \le \widehat{u} - \widehat{u}_n \le \widehat{v}_n$ and $\rho(\widehat{u} - \widehat{u}_n - \widehat{v}_n) < \frac{1}{n}$. We can assume that $\widehat{v}_n \downarrow$ and so by Theorem 2.21(e) we have $\widehat{v}_n \downarrow 0$. Now fix $\epsilon > 0$ and then choose some $u \in L^+$ with $\rho(\widehat{u} - u) < \epsilon$. For each n pick $u_n \in L^+$ such that $\rho(\widehat{u}_n - u_n) < \epsilon$. In particular, it follows that $\rho(u_n) \le \rho(\widehat{u}_n) + \epsilon \le \lim \rho(\widehat{u}_n) + \epsilon$ and that $\rho(\widehat{u}) - \epsilon < \rho(u)$.

Next, put $\widehat{w}_n = |u - \widehat{u}| + |u_n - \widehat{u}_n|$ and note that $\rho(\widehat{w}_n) < 2\epsilon$ and

$$u - u_n \le |u - u_n| \le |u - \widehat{u}| + |u_n - \widehat{u}_n| + |\widehat{u} - \widehat{u}_n| \le \widehat{w}_n + \widehat{v}_n\,.$$

For each n pick some vector $y_n \in L^+$ with $0 \le y_n \le \widehat{v}_n$ and $\rho(\widehat{v}_n - y_n) < 2^{-n}\epsilon$. Let $x_n = \bigwedge_{i=1}^n y_i$ for all n. Then $0 \le x_n \le \widehat{v}_n$, $\rho(\widehat{v}_n - x_n) < \epsilon$, and $x_n \downarrow 0$ in L. Consequently, we have $u - u_n \le \widehat{w}_n + \widehat{v}_n = \widehat{w}_n + \widehat{v}_n - x_n + x_n$ and therefore $(u - x_n)^+ \le \widehat{w}_n + \widehat{v}_n - x_n + u_n$. This implies

$$\rho\big((u - x_n)^+\big) \le \rho(\widehat{w}_n) + \rho(\widehat{v}_n - x_n) + \rho(u_n) < 4\epsilon + \lim \rho(\widehat{u}_n)\,.$$

Since $(u - x_n)^+ \uparrow u$ in L and ρ is a σ-Fatou norm, it follows from the last relation that $\rho(u) \le 4\epsilon + \lim \rho(\widehat{u}_n)$, and hence $\rho(\widehat{u}) - \epsilon < 4\epsilon + \lim \rho(\widehat{u}_n)$. Therefore, $\rho(\widehat{u}) \le 5\epsilon + \lim \rho(\widehat{u}_n)$ for each $\epsilon > 0$ so that $\rho(\widehat{u}) \le \lim \rho(\widehat{u}_n)$, as desired. ■

5.2. Fréchet Topologies

In this section we shall deal with metrizable locally solid topologies that are in addition topologically complete. We start with some characterizations of this property.

THEOREM 5.14. *For a metrizable locally solid Riesz space (L, τ) the following statements are equivalent.*

(i) *(L, τ) is τ-complete, that is, $L = \widehat{L}$.*
(ii) *Every increasing τ-Cauchy net of L^+ is τ-convergent, that is, (L, τ) satisfies* MCP.
(iii) *Every increasing τ-Cauchy sequence of L^+ is τ-convergent.*
(iv) *Every upper element of L lies in L.*

PROOF. The implications (i) $\Longrightarrow$ (ii) $\Longrightarrow$ (iii) $\Longrightarrow$ (vi) should be obvious. To see that (iv) $\Longrightarrow$ (i) holds true use Theorem 5.3. ■

Note that Example 2.45 shows that statements (i) and (ii) of Theorem 5.14 are no longer equivalent if the metrizability assumption is dropped; although (i) $\Longrightarrow$ (ii) is obvious.

In the next theorem, we shall characterize the norm completeness of normed Riesz spaces. Recall that for a positive sequence $\{u_n\}$ we denote by $\sum_{n=1}^\infty u_n$ the supremum (if it exists) of the sequence of partial sums $\{\sum_{k=1}^n u_k\}$, that is, we put $\sum_{n=1}^\infty u_n = \sup\{\sum_{k=1}^n u_k\colon\ n \in \mathbb{N}\}$.

The next result is due to W. A. J. Luxemburg and A. C. Zaanen [**89**, Note I, Theorem 4.2, p. 141] and I. Halperin and W. A. J. Luxemburg [**89**, Note VIII, Theorem 26.2].

THEOREM 5.15 (Halperin–Luxemburg–Zaanen). *For a normed Riesz space L_ρ the following statements are equivalent.*

(i) *L_ρ is a Banach lattice, that is, L_ρ is norm complete.*
(ii) *If $\{u_n\} \subseteq L^+$ satisfies $\sum_{n=1}^\infty \rho(u_n) < \infty$, then $\sum_{n=1}^\infty u_n$ exists.*
(iii) *If $\{u_n\} \subseteq L^+$ satisfies $\sum_{n=1}^\infty \rho(u_n) < \infty$, then $\sum_{n=1}^\infty u_n$ exists and $\rho(\sum_{n=1}^\infty u_n) \le \sum_{n=1}^\infty \rho(u_n)$.*[1]

PROOF. (i) $\Longrightarrow$ (ii) Assume that a sequence $\{u_n\} \subseteq L^+$ satisfies $\sum_{n=1}^\infty \rho(u_n) < \infty$. Put $w_n = \sum_{k=1}^n u_k$ for each n, and note that $\{w_n\}$ is a ρ-Cauchy sequence. Hence since L_ρ is ρ-complete, there exists some $u \in L_\rho$ with $\lim \rho(w_n - u) = 0$, and therefore $u = \sum_{n=1}^\infty u_n$ holds in L according to Theorem 2.21(c).

[1] This property is known as the **Riesz–Fischer property**.

(ii) $\Longrightarrow$ (iii) Assume by way of contradiction that some sequence $\{u_n\}$ in L^+ satisfies $\sum_{n=1}^{\infty}\rho(u_n)<\infty$ and $\rho\big(\sum_{n=1}^{\infty}u_n\big)>\sum_{n=1}^{\infty}\rho(u_n)$. Then there exists some $\epsilon>0$ such that $\rho\big(\sum_{n=1}^{\infty}u_n\big)>\epsilon+\sum_{n=1}^{\infty}\rho(u_n)$. Observe that by the triangle inequality we have

$$\rho\Big(\sum_{n=m}^{\infty}u_n\Big)>\epsilon+\sum_{n=m}^{\infty}\rho(u_n)$$

for all m. Next choose a strictly increasing sequence of positive integers $\{m_k\}$ such that $\sum_{n=m_k}^{\infty}\rho(u_n)<\frac{1}{k^3}$ for all k. Let $w_k=k\sum_{n=m_k}^{m_{k+1}}u_n$ and note that $\sum_{k=1}^{\infty}\rho(w_k)\le\sum_{k=1}^{\infty}\frac{1}{k^2}<\infty$; and so, by our hypothesis, $w=\sum_{k=1}^{\infty}w_k$ exists. But then $w\ge k\sum_{n=m_k}^{\infty}u_n$ for each k implies $\rho(w)\ge k\rho\big(\sum_{n=m_k}^{\infty}u_n\big)\ge k\epsilon$ for all k, which is impossible. Hence (iii) is true.

(iii) $\Longrightarrow$ (i) Assume that $\{u_n\}$ is a ρ-Cauchy sequence of L^+ satisfying $0\le u_n\uparrow$ in L. Pick a subsequence $\{v_n\}$ of $\{u_n\}$ with $\rho(v_{n+1}-v_n)<\frac{1}{2^n}$ for all n and note that $\sum_{n=1}^{\infty}\rho(v_{n+1}-v_n)<\infty$. Consequently, $\sum_{n=k}^{\infty}(v_{n+1}-v_n)$ exists in the Riesz space L for all $k\in\mathbb{N}$. Put $v=\sum_{n=1}^{\infty}(v_{n+1}-v_n)$, and note that since

$$\sum_{n=k}^{\infty}(v_{n+1}-v_n)=\sum_{n=1}^{\infty}(v_{n+1}-v_n)-\sum_{n=1}^{k-1}(v_{n+1}-v_n)=v+v_1-v_k$$

holds for all $k\ge 2$, we obtain

$$\rho(v+v_1-v_k)=\rho\Big(\sum_{n=k}^{\infty}(v_{n+1}-v_n)\Big)\le\sum_{n=k}^{\infty}\rho(v_{n+1}-v_n)\,,$$

from which it follows that $\{v_n\}$ is norm convergent to $v+v_1$. This implies that $\{u_n\}$ is norm convergent in L. Now use Theorem 5.14 to complete the proof of the theorem. ∎

We continue with the introduction of Fréchet spaces.

DEFINITION 5.16. *A metrizable locally solid topology τ on a Riesz space L is said to be a* ***Fréchet topology*** *if (L,τ) is τ-complete. Moreover,*

(1) *A* ***Fréchet space*** *(L,τ) is a Riesz space L equipped with a Fréchet topology τ, and*
(2) *A* ***Fréchet lattice*** *(L,τ) is a Fréchet space that is in addition locally convex.*

The Hausdorff quotient spaces of Fréchet spaces are also Fréchet spaces.

THEOREM 5.17. *If (L,τ) is a Fréchet space and A is a τ-closed ideal, then $(L/A,\tau/A)$ is likewise a Fréchet space.*

PROOF. Observe that τ/A is a Hausdorff metrizable locally solid topology. In particular, note that if $\{V_n\}$ is a normal base at zero for (L,τ) consisting of solid sets, then $\{\dot{V}_n\}$ is a normal base at zero for $(L/A,\tau/A)$ consisting of solid sets.

We need to show that $(L/A,\tau/A)$ is τ/A-complete. To this end, let $\{\dot{u}_n\}$ be a τ/A-Cauchy sequence in L/A. Pick a subsequence $\{\dot{v}_n\}$ of $\{\dot{u}_n\}$ with $\dot{v}_{n+1}-\dot{v}_n\in\dot{V}_n$ for all n. For each n choose $x_n\in V_n$ with $\dot{x}_n=\dot{v}_{n+1}-\dot{v}_n$ and then let $w_1=v_1$ and $w_n=v_1+x_1+\cdots+x_{n-1}$ for $n\ge 2$. Clearly, $\{w_n\}$ is a τ-Cauchy sequence of L. Hence there exists some $v\in L$ with $w_n\xrightarrow{\tau}v$ in L, and so $\dot{w}_n=\dot{v}_n\xrightarrow{\tau/A}\dot{v}$ also holds in L/A. This shows that $(L/A,\tau/A)$ is τ/A-complete, and the proof is finished. ∎

The following result should be obvious.

THEOREM 5.18. *For an at most countable family $\{(L_n, \tau_n)\}$ of Fréchet spaces the Cartesian product $(\Pi L_n, \Pi\tau_n)$ is a Fréchet space.*

Concerning Fréchet spaces the following remarkable theorem holds.

THEOREM 5.19. *Assume that (L, τ_1) is a Fréchet space and that (M, τ_2) is a locally solid Riesz space. Then every positive operator $T\colon (L, \tau_1) \to (M, \tau_2)$ is continuous.*

PROOF. Let $\{V_n\}$ be a normal base at zero for τ_1 consisting of solid neighborhoods. Assume that $T\colon (L, \tau_1) \to (M, \tau_2)$ is a positive operator that fails to be continuous. Then there exists a sequence $\{u_n\} \subseteq L^+$ with $u_n \xrightarrow{\tau_1} 0$ and a solid τ_2-neighborhood W of zero such that $T(u_n) \notin W$ for each n. By passing to a subsequence (if necessary) we can assume that $u_n \in \frac{1}{n}V_n$ (or $nu_n \in V_n$) for each n.

Next put $w_n = \sum_{k=1}^{n} ku_k$ for each n and note that $\{w_n\}$ is a τ_1-Cauchy sequence satisfying $0 \leq w_n \uparrow$. Now since (L, τ_1) is a Fréchet space, there exists some $w \in L$ such that $w_n \xrightarrow{\tau_1} w$ and so by Theorem 2.21 we get $0 \leq w_n \uparrow w$ in L. In particular, it follows that $0 \leq nu_n \leq w$ for all n, and therefore (since T is positive) $0 \leq T(u_n) \leq \frac{1}{n}T(w)$ holds for all n. But then, it follows that $T(u_n) \xrightarrow{\tau_2} 0$, contradicting the choice of the sequence $\{u_n\}$. This contradiction completes the proof. ■

THEOREM 5.20. *Let (L, τ) be a Fréchet space. Then any other locally solid topology τ' on L is coarser than τ, i.e., $\tau' \subseteq \tau$.*

PROOF. Consider the identity operator $I\colon (L, \tau) \to (L, \tau')$ and apply Theorem 5.19. ■

THEOREM 5.21. *A Riesz space L admits at most one Fréchet topology.*

The next result is an immediate consequence of the preceding result and is due to C. Goffman [**62**, Theorem 2, p. 537].

COROLLARY 5.22 (Goffman). *All Riesz norms on a Riesz space L that make L a Banach lattice are equivalent. That is, a Riesz space admits at most one (up to an equivalence) Riesz norm under which it is a Banach lattice.*

We already know that $L' \subseteq L^\sim$. If, in addition, (L, τ) is a Fréchet space, then by Theorem 5.19 every $\phi \in L^\sim$ is continuous and hence $L' = L^\sim$. We thus have the following result.

THEOREM 5.23. *For every Fréchet space (L, τ) we have $L' = L^\sim$. In particular, if L_ρ is a Banach lattice, then $L'_\rho = L^\sim_\rho$.*

As an application of the preceding theorem we shall exhibit an example of a Dedekind complete Riesz space without nonzero order bounded linear functionals. Recall that $L_p[0,1]$, where $0 < p < \infty$, denotes the Riesz space consisting of all equivalence classes of real-valued Lebesgue measurable functions on $[0,1]$ such that $\int_0^1 |f(x)|^p\,dx < \infty$.

For each n let

$$V_n = \Big\{ f \in L_p[0,1] \colon \int_0^1 |f(x)|^p\,dx \leq \tfrac{1}{n} \Big\}.$$

Then each V_n is an absorbing and solid set, $\bigcap_{n=1}^\infty V_n = \{0\}$, and for each n there exists some m such that $V_m + V_m \subseteq V_n$. Thus $\{V_n\}$ defines a metrizable locally solid topology τ on $L = L_p[0,1]$. Using Theorem 5.14 or standard arguments from the theory of integration one can show that (L,τ) is τ-complete. So, according to Theorem 5.21, the topology τ is the only Fréchet topology that $L_p[0,1]$ can carry.

If $1 \le p < \infty$, then (L,τ) is a Fréchet lattice (actually a Banach lattice). For $0 < p < 1$ it turns out that (L,τ) is a Fréchet space but not a Fréchet lattice as the next result, due to M. M. Day [**48**], will show.

THEOREM 5.24 (Day). *If $L = L_p[0,1]$ for some $0 < p < 1$, then $L^\sim = \{0\}$. In particular, L does not admit any Hausdorff locally convex-solid topology.*

PROOF. Fix $0 < p < 1$. Since $t^p \le t+1$ holds for all $t \ge 0$, it easily follows that $L_1[0,1]$ is an ideal of $L_p[0,1]$. Also, since the constant function one is a weak order unit, it follows from the Lebesgue dominated convergence theorem that $L_1[0,1]$ is τ-dense in $L_p[0,1]$. Now by Theorem 5.23 we have $L' = L^\sim$, and so every positive linear functional on L is τ-continuous. Thus, in order to show that $L^\sim = \{0\}$, it suffices to establish that every $0 \le \phi \in L'$ vanishes on $L_1[0,1]$.

To this end let $0 \le \phi \in L'$. Choose some $n \in \mathbb{N}$ such that if $f \in L$ satisfies $\int_0^1 |f(x)|^p\,dx \le \frac{1}{n}$, then $|\phi(f)| \le 1$, and therefore for every function $f \in L$ we have $|\phi(f)| \le n^{\frac{1}{p}}\big[\int_0^1 |f(x)|^p\,dx\big]^{\frac{1}{p}}$. In particular, if $f \in L_1[0,1]$ satisfies $\int_0^1 |f(x)|\,dx \le 1$, then the last inequality implies

$$|\phi(f)| \le n^{\frac{1}{p}}\Big[\int_0^1 |f(x)|\,dx + 1\Big]^{\frac{1}{p}} \le (2n)^{\frac{1}{p}}\,.$$

Hence ϕ restricted to $L_1[0,1]$ is norm continuous and thus by the Riesz representation theorem there exists some $0 \le g \in L_\infty[0,1]$ such that

$$\phi(f) = \int_0^1 g(x)f(x)\,dx$$

holds for all $f \in L_1[0,1]$. We claim that $g = 0$.

If $g \ne 0$, then there is a Lebesgue measurable subset E of $[0,1]$ with $\lambda(E) > 0$ (where λ denotes the Lebesgue measure on $[0,1]$) such that $g(x) \ge \epsilon$ holds for some $\epsilon > 0$ and all $x \in E$. Now for each positive integer k pick a measurable subset E_k of E with $0 < \lambda(E_k) \le \frac{1}{k}$, and note that

$$\begin{aligned} \epsilon\lambda(E_k) &\le \int_0^1 g(x)\chi_{E_k}(x)\,dx = \phi(\chi_{E_k}) \\ &\le n^{\frac{1}{p}}\Big[\int_0^1 \chi_{E_k}(x)\,dx\Big]^{\frac{1}{p}} = n^{\frac{1}{p}}\big[\lambda(E_k)\big]^{\frac{1}{p}}\,. \end{aligned}$$

Hence $k^{\frac{1}{p}-1} \le \big[\lambda(E_k)\big]^{1-\frac{1}{p}} \le \epsilon^{-1}n^{\frac{1}{p}}$ for all k, a contradiction. Thus $g = 0$ and so $\phi = 0$; so that $L^\sim = \{0\}$. The last assertion follows immediately from the Hahn–Banach extension theorem. ■

As an application of the above results we shall determine the topological dual of the locally solid Riesz space studied in Example 3.6.

APPLICATION 5.25. Let (L,τ) be the Hausdorff locally solid Riesz space described in Example 3.6. We shall show that:

(1) If $0 < p < 1$, then for every $\phi \in L'$ there exist a finite number of points $x_1,\dots,x_n$ in $[0,1]$ and real constants $\alpha_1,\dots,\alpha_n$ (all depending on ϕ) such

that

$$\phi(u) = \sum_{i=1}^{n} \alpha_i u(x_i)$$

holds for all $u \in L$.

(2) If $1 \le p < \infty$, then for each $\phi \in L'$ there exist a finite number of points $x_1, \ldots, x_n$ in $[0,1]$, real constants $\alpha_1, \ldots, \alpha_n$, and some $v \in L_q[0,1]$, where $\frac{1}{p} + \frac{1}{q} = 1$, (all depending on ϕ) such that

$$\phi(u) = \sum_{i=1}^{n} \alpha_i u(x_i) + \int_0^1 u(x)v(x)\,dx$$

holds for all $u \in L$.

To establish the above claims let $\phi \in L'$, and note that by the continuity of ϕ there exist a finite number of points $x_1, \ldots, x_n$ in $[0,1]$ and some $\epsilon > 0$ such that $|\phi(u)| \le 1$ holds whenever $u \in L$ satisfies $|u(x_i)| \le \epsilon$ for all $i = 1, \ldots, n$ and $\int_0^1 |u(x)|^p\,dx \le \epsilon$. For each i put $\alpha_i = \phi(\chi_{\{x_i\}})$. Now if $u, v \in L$ satisfy $u(x) = v(x)$ for almost all x, let

$$w = \Big[u - \sum_{i=1}^{n} u(x_i)\chi_{\{x_i\}}\Big] - \Big[v - \sum_{i=1}^{n} v(x_i)\chi_{\{x_i\}}\Big] \in L\,,$$

and note that the function $g_k = kw$ satisfies $g_k(x_i) = 0$ for $i = 1, \ldots, n$ and $\int_0^1 |g_k(x)|^p\,dx = 0$ for each $k \in \mathbb{N}$. In particular, we get $|\phi(g_k)| = k|\phi(w)| \le 1$ for each $k \in \mathbb{N}$ and so

$$\phi(u) - \sum_{i=1}^{n} \alpha_i u(x_i) = \phi(v) - \sum_{i=1}^{n} \alpha_i v(x_i)\,.$$

This shows that the linear functional $\psi\colon L_p[0,1] \to \mathbb{R}$, defined for each $u \in L$ by

$$\psi([u]) = \phi(u) - \sum_{i=1}^{n} \alpha_i u(x_i)\,,$$

where $[u]$ is the equivalence class determined by $u \in L$ in $L_p[0,1]$, is well-defined.

Next observe that ψ is a continuous linear functional on $L_p[0,1]$. Indeed, assume that $[u] \in L_p[0,1]$ (where $u \in L$) satisfies $\int_0^1 |u(x)|^p\,dx \le \epsilon$ and consider $v \in L$ defined by $v(x_i) = 0$ for each $i = 1, \ldots, n$ and $v(x) = u(x)$ otherwise. Then $[v] = [u]$ and

$$\big|\psi([u])\big| = \big|\psi([v])\big| = \big|\phi(v)\big| \le 1\,,$$

and hence ψ is bounded on a neighborhood of zero; therefore ψ is continuous.

If now $0 < p < 1$ holds, then by Theorem 5.24 we get $\psi = 0$ and the validity of (1) follows. If $1 \le p < \infty$ holds, then by the Riesz representation theorem we know that $\psi([u]) = \int_0^1 u(x)v(x)\,dx$ holds for some $[v] \in L_q[0,1]$, where $\frac{1}{p} + \frac{1}{q} = 1$, and hence the validity of (2) also follows. ■

5.3. The Pseudo Lebesgue Properties

We start this section by introducing another continuity property, that is weaker than the σ-Lebesgue property. It was introduced as property (A,0) by W. A. J. Luxemburg [**87**, Note XVI, p. 650].

DEFINITION 5.26 (Luxemburg). *Let (L,τ) be a locally solid Riesz space. Then we say that:*

(i) *(L,τ) satisfies the* ***pseudo σ-Lebesgue property*** *if for each τ-Cauchy sequence $\{u_n\}$ that satisfies $u_n \downarrow 0$ we have $u_n \xrightarrow{\tau} 0$.*

(ii) *(L,τ) satisfies the* ***pseudo Lebesgue property*** *if for each τ-Cauchy net $\{u_\alpha\}$ that satisfies $u_\alpha \downarrow 0$ we have $u_\alpha \xrightarrow{\tau} 0$.*

It should be clear that the pseudo Lebesgue property implies the pseudo σ-Lebesgue property. The converse is, however, false as Example 3.6 shows. Every topologically complete Hausdorff locally solid Riesz space (L,τ) satisfies the pseudo Lebesgue property. Indeed, if $u_\alpha \downarrow 0$ and $\{u_\alpha\}$ is τ-Cauchy, then $u_\alpha \xrightarrow{\tau} u$ holds for some $u \in L$ and so by Theorem 2.21 we have $u_\alpha \downarrow u$; thus $u = 0$ and hence $u_\alpha \xrightarrow{\tau} 0$. Similarly, every sequentially τ-complete Hausdorff locally solid Riesz space satisfies the pseudo σ-Lebesgue property.

Also if (L,τ) is a Hausdorff locally solid Riesz space such that L is a regular (resp. σ-regular) Riesz subspace of $\widehat{L}$, then (L,τ) satisfies the pseudo Lebesgue (resp. pseudo σ-Lebesgue) property. Indeed, if L is a regular Riesz subspace of $\widehat{L}$ and $\{u_\alpha\}$ is a τ-Cauchy net of L with $u_\alpha \downarrow 0$ in L, then $u_\alpha \xrightarrow{\widehat{\tau}} \widehat{u}$ holds in $\widehat{L}$ for some $\widehat{u}$. But then $u_\alpha \downarrow \widehat{u}$ holds in $\widehat{L}$. So, $\widehat{u} = 0$ and the claim is established.

For metrizable locally solid Riesz spaces the pseudo Lebesgue and pseudo σ-Lebesgue properties are equivalent.

THEOREM 5.27. *For a metrizable locally solid Riesz space (L,τ) the following statements are equivalent.*

(i) *(L,τ) satisfies the pseudo Lebesgue property.*

(ii) *(L,τ) satisfies the pseudo σ-Lebesgue property.*

PROOF. (i) $\Longrightarrow$ (ii) Obvious.

(ii) $\Longrightarrow$ (i) Assume that a τ-Cauchy net $\{u_\alpha\}$ satisfies $u_\alpha \downarrow 0$ in L. Then $u_\alpha \xrightarrow{\widehat{\tau}} \widehat{u}$ holds in $\widehat{L}$ for some $\widehat{u} \geq 0$; in particular, we have $u_\alpha \downarrow \widehat{u}$ in $\widehat{L}$. If $0 < \widehat{u}$ holds in $\widehat{L}$, then by Theorem 5.5 there exists a strictly positive lower element $\widehat{w}$ of L such that $0 < \widehat{w} \leq \widehat{u}$. Let $\{w_n\} \subseteq L^+$ be such that $w_n \downarrow \widehat{w}$ and $w_n \xrightarrow{\widehat{\tau}} \widehat{w}$ in $\widehat{L}$. By virtue of $u_\alpha \downarrow 0$ in L we get $w_n \downarrow 0$ in L. Now since $\{w_n\}$ is τ-Cauchy and (L,τ) satisfies the pseudo σ-Lebesgue property, we get $w_n \xrightarrow{\tau} 0$. Hence $\widehat{w} = 0$, contradicting $\widehat{w} > 0$. Therefore, $\widehat{u} = 0$ and so $u_\alpha \xrightarrow{\tau} 0$. ■

THEOREM 5.28. *For a locally solid Riesz space (L,τ) we have:*

(i) *If (L,τ) satisfies the Fatou property, then (L,τ) also satisfies the pseudo Lebesgue property.*

(ii) *If (L,τ) satisfies the σ-Fatou property, then (L,τ) also satisfies the pseudo σ-Lebesgue property.*

PROOF. (i) Assume that a τ-Cauchy net $\{u_\alpha\}$ satisfies $u_\alpha \downarrow 0$ in L. Let V be a Fatou τ-neighborhood of zero. Then there exists an index α_1 such that $u_\alpha - u_\beta \in V$ for all $\alpha, \beta \succeq \alpha_1$. But for each fixed $\alpha \succeq \alpha_1$ we have $u_\alpha - u_\beta \xrightarrow[\beta \succeq \alpha]{o} u_\alpha$ and hence $u_\alpha \in V$ for all $\alpha \succeq \alpha_1$. Therefore, $u_\alpha \xrightarrow{\tau} 0$.

(ii) Repeat the arguments of the previous part. ■

The following theorem gives a few more characterizations of the pseudo σ-Lebesgue property in metrizable locally solid Riesz spaces. The characterizations were proven by W. A. J. Luxemburg for the normed case [**87**, Note XVI, Theorem 61.5, p. 652] and they were generalized by C. D. Aliprantis [**3**, Theorem 6.1, p. 119].

THEOREM 5.29 (Luxemburg–Aliprantis). *For a metrizable locally solid Riesz space (L,τ) the following statements are equivalent.*

(i) *(L,τ) satisfies the pseudo σ-Lebesgue property.*
(ii) *L is order dense in $\widehat{L}$.*
(iii) *L is a regular Riesz subspace of $\widehat{L}$.*
(iv) *L is a σ-regular Riesz subspace of $\widehat{L}$.*

PROOF. (i) $\Longrightarrow$ (ii) Let $0<\widehat{u}\in\widehat{L}$. By Theorem 5.5 there exists a lower element $\widehat{w}$ of L such that $0<\widehat{w}\leq\widehat{u}$. Next pick a sequence $\{w_n\}\subseteq L^+$ with $w_n\downarrow\widehat{w}$ and $w_n\xrightarrow{\widehat{\tau}}\widehat{w}$; in particular, note that $\{w_n\}$ is a τ-Cauchy sequence of L. We now claim that there exists an element $0<w\in L$ such that $0<w\leq\widehat{w}$. If not, then $w_n\downarrow 0$ holds in L and so by the pseudo σ-Lebesgue property of (L,τ) we get $w_n\xrightarrow{\tau}0$. But then $\widehat{w}=0$, a contradiction. Thus L is order dense in $\widehat{L}$.

(ii) $\Longrightarrow$ (iii) It follows immediately from Theorem 1.23.

(iii) $\Longrightarrow$ (iv) Obvious.

(iv) $\Longrightarrow$ (i) Assume $u_n\downarrow 0$ in L and $\{u_n\}$ is τ-Cauchy. Then $u_n\xrightarrow{\widehat{\tau}}\widehat{u}$ holds in $\widehat{L}$ and so $u_n\downarrow\widehat{u}$ also holds in $\widehat{L}$. But then since L is a σ-regular Riesz subspace of $\widehat{L}$, we get $\widehat{u}=0$ and so $u_n\xrightarrow{\tau}0$. ■

It may be asked at this point whether statements (ii) and (iii) of the preceding theorem are equivalent without assuming metrizability. Clearly, (ii) $\Longrightarrow$ (iii) but (iii) no longer implies (ii), as the following example of D. H. Fremlin (oral communication) shows.

EXAMPLE 5.30 (Fremlin). Let $L=C[0,1]$ and let τ be the Hausdorff locally solid topology generated by the Riesz seminorms

$$\rho_{K_1,K_2}(u)=\sup\{|u(x)|:\ x\in K_1\cup K_2\}\,,\quad u\in L\,,$$

where K_1 is a compact subset of rational numbers of $[0,1]$ and K_2 is a compact subset of irrational numbers of $[0,1]$.

Observe that τ is nonmetrizable. By the classical Tietze extension theorem it is easy to see that $\widehat{L}$ can be identified with the Riesz space of all real-valued functions on $[0,1]$ whose restrictions to the rational and irrational numbers of $[0,1]$ are both continuous.

Note that $0<\widehat{u}=\chi_E\in\widehat{L}$, where $E=\mathbb{Q}\cap[0,1]$ and $\mathbb{Q}$ denotes the set of all rational numbers of $\mathbb{R}$, but there is no nonzero function of $C[0,1]$ that is below $\widehat{u}$. That is, L is not order dense in $\widehat{L}$.

On the other hand, if $u_\alpha\downarrow 0$ in L and $0<\widehat{u}\leq u_\alpha$ holds in $\widehat{L}$ for all α, then there exists a real number x_0 such that $\widehat{u}(x_0)>0$. If x_0 is rational, then there exist some $\epsilon>0$ and a neighborhood V of x_0 such that $\widehat{u}(x)\geq\epsilon$ for all rational numbers $x\in V$. But then $u_\alpha(x)\geq\epsilon>0$ also holds for all $x\in V$ and all α. It now easily follows that $u_\alpha\downarrow 0$ does not hold in L, a contradiction. The same argument can be applied if x_0 is an irrational number. Hence $u_\alpha\downarrow 0$ holds in $\widehat{L}$.

The above arguments show that L is a regular Riesz subspace of $\widehat{L}$, but L is not order dense in $\widehat{L}$. ■

THEOREM 5.31. *Let (L,τ) be a metrizable locally solid Riesz space with the pseudo σ-Lebesgue property. If L is Dedekind σ-complete, then L is an ideal of $\widehat{L}$ (and hence the order intervals of L are τ-complete). In particular, in this case, L is super order dense in $\widehat{L}$.*

PROOF. Note first that if $\widehat{u}$ is a lower element of L, then $u_n \downarrow \widehat{u}$ and $u_n \xrightarrow{\widehat{\tau}} \widehat{u}$ hold in $\widehat{L}$ for some sequence $\{u_n\} \subseteq L^+$. Now since the Riesz space L is Dedekind σ-complete, $u_n \downarrow u$ must hold in L for some $u \in L^+$. From Theorem 5.29 it follows that $u_n \downarrow u$ must also hold in $\widehat{L}$ and hence $\widehat{u} = u \in L$. Therefore, every lower element of L lies in L.

Now let $0 \leq \widehat{u} \in \widehat{L}$. By Theorem 5.5 there exists a sequence $\{u_n\}$ in L^+ with $u_n \uparrow \widehat{u}$ and $u_n \xrightarrow{\widehat{\tau}} \widehat{u}$. Now if $\widehat{u} \leq u$ holds for some $u \in L^+$, then $u_n \uparrow w$ holds in L for some $w \in L^+$ and so by Theorem 5.29 we get $\widehat{u} = w \in L$. Thus L is an ideal of $\widehat{L}$. The last claim should be obvious from the preceding arguments. ■

The next result is essentially due to W. A. J. Luxemburg [**87**, Note XVI, Theorem 66.5, p. 665].

THEOREM 5.32 (Luxemburg). *For a metrizable locally solid Riesz space (L,τ) with the pseudo σ-Lebesgue property we have the following:*

(i) *If L is Dedekind σ-complete, then so is $\widehat{L}$.*
(ii) *If L is Dedekind complete, then so is $\widehat{L}$.*
(iii) *If L is super Dedekind complete, then so is $\widehat{L}$.*

PROOF. Statements (i) and (ii) follow from Theorems 5.31 and 2.47. To show that (iii) is true, note first that $\widehat{L}$ is Dedekind complete and that L is a super order dense ideal of $\widehat{L}$ by Theorem 5.31. Now assume $0 \leq \widehat{u}_\alpha \uparrow \widehat{u}$ holds in $\widehat{L}$. Pick a sequence $\{u_n\} \subseteq L^+$ with $0 \leq u_n \uparrow \widehat{u}$ and note that since for each k we have $0 \leq u_k \wedge \widehat{u}_\alpha \uparrow_\alpha u_k$ in the super Dedekind complete Riesz space L, there exists a sequence of indices $\{\alpha_{k,n}\colon\ n = 1,2,\ldots\}$ with

$$\sup\{u_k \wedge \widehat{u}_{\alpha_{k,n}}\colon\ n = 1,2,\ldots\} = u_k$$

in L. It now easily follows that $\sup\{\widehat{u}_{\alpha_{k,n}}\colon\ k,n = 1,2,\ldots\} = \widehat{u}$ holds in $\widehat{L}$. This shows that $\widehat{L}$ is super Dedekind complete. ■

The next result characterizes the Lebesgue property in terms of the pseudo σ-Lebesgue property in metrizable locally solid Riesz spaces. For normed Riesz spaces it is due to W. A. J. Luxemburg and A. C. Zaanen [**89**, Note X, Theorem 33.8, p. 505]; it was generalized in [**3**].

THEOREM 5.33 (Luxemburg–Zaanen). *For a metrizable locally solid Riesz space (L,τ) the following statements are equivalent.*

(1) *(L,τ) satisfies the Lebesgue property.*
(2) *(L,τ) satisfies the pseudo σ-Lebesgue property and the pre-Lebesgue property.*
(3) *(L,τ) satisfies the σ-Lebesgue property and the pre-Lebesgue property.*

In particular, in this case L has the countable sup property.

PROOF. (1) $\Longrightarrow$ (2) Obvious.

(2) $\Longrightarrow$ (3) Assume $u_n \downarrow 0$ in L. Since (L, τ) has the pre-Lebesgue property, $\{u_n\}$ is a τ-Cauchy sequence; and so $u_n \xrightarrow{\tau} 0$. That is, (L, τ) satisfies the σ-Lebesgue property.

(3) $\Longrightarrow$ (1) Assume $u_\alpha \downarrow 0$ in L. Then since (L, τ) satisfies the pre-Lebesgue property, $\{u_\alpha\}$ is a τ-Cauchy net; see Theorem 3.22. So, $u_\alpha \downarrow \widehat{u}$ and $u_\alpha \xrightarrow{\widehat{\tau}} \widehat{u}$ holds in $\widehat{L}$ for some $\widehat{u}$. But then, by Theorem 5.29, we have $u_\alpha \downarrow 0$ in $\widehat{L}$. This implies $\widehat{u} = 0$ and so $u_\alpha \xrightarrow{\tau} 0$. ■

The next result is essentially due to H. Nakano [**102**, p. 321].

THEOREM 5.34 (Nakano). *For a metrizable locally solid Riesz space (L, τ) the following statements are equivalent.*

(i) *(L, τ) satisfies the Lebesgue property and L is super Dedekind complete.*
(ii) *(L, τ) satisfies the σ-Lebesgue property and L is σ-Dedekind complete.*

PROOF. (i) $\Longrightarrow$ (ii) Obvious.

(ii) $\Longrightarrow$ (i) Note that (L, τ) must satisfy the pre-Lebesgue property and so by Theorem 5.33 it follows that (L, τ) satisfies the Lebesgue property. To see that L is super Dedekind complete assume $0 \leq u_\alpha \uparrow \leq u$ in L. From Theorem 3.22 it follows that $\{u_\alpha\}$ is a τ-Cauchy net and therefore $u_\alpha \xrightarrow{\widehat{\tau}} \widehat{u}$ must hold in $\widehat{L}$, and clearly $u_\alpha \uparrow \widehat{u}$ in $\widehat{L}$. Since (L, τ) is metrizable, there exists a subsequence $\{u_{\alpha_n}\} \subseteq \{u_\alpha\}$ satisfying $u_{\alpha_n} \xrightarrow{\widehat{\tau}} \widehat{u}$ and $0 \leq u_{\alpha_n} \uparrow \widehat{u}$ in $\widehat{L}$. But then since L is Dedekind σ-complete, there exists some $v \in L^+$ such that $0 \leq u_{\alpha_n} \uparrow v \leq u$ in L, and so by Theorem 5.29 we have $\widehat{u} = v \in L$. This shows that L is super Dedekind complete, and the proof is finished. ■

Note that Example 3.6 shows that Theorems 5.33 and 5.34 are false for non-metrizable Hausdorff locally solid Riesz spaces. By comparing Theorems 3.27 and 5.34 the following result should be clear now.

THEOREM 5.35. *If a metrizable locally solid Riesz space (L, τ) satisfies the pre-Lebesgue property, then $(\widehat{L}, \widehat{\tau})$ satisfies the Lebesgue property and $\widehat{L}$ is super Dedekind complete.*

As an illustration of the preceding result and Theorem 3.27 consider the Riesz space $L = C[0,1]$. If τ is the metrizable locally solid topology generated by the Riesz norm $\rho(u) = \int_0^1 |u(x)|\,dx$, then $\widehat{L} = L_1[0,1]$, a super Dedekind complete Riesz space. On the other hand, if τ^* is the Hausdorff locally solid topology of pointwise convergence on L, then $\widehat{L} = \mathbb{R}^{[0,1]}$, a Dedekind complete Riesz space that is not super Dedekind complete.

The next theorem shows that for metrizable locally solid Riesz spaces the σ-Lebesgue property is preserved under the process of topological completion.

THEOREM 5.36. *If a metrizable locally solid Riesz space (L, τ) satisfies the σ-Lebesgue property, then so does its topological completion $(\widehat{L}, \widehat{\tau})$.*

PROOF. Let $\{V_n\}$ be a normal base at zero for τ consisting of solid sets. Assume that $\widehat{f}_n \downarrow 0$ holds in $\widehat{L}$. We must show that $\widehat{f}_n \xrightarrow{\widehat{\tau}} 0$ in $\widehat{L}$.

For each n let $\widehat{u}_n$ be an upper element of L such that $\widehat{f}_n \leq \widehat{u}_n$ and $\widehat{u}_n - \widehat{f}_n \in \overline{V}_n$; see Theorem 5.4. Put $\widehat{w}_n = \bigwedge_{i=1}^n \widehat{u}_i$ for all n and note that each $\widehat{w}_n$ is an upper element of L, $\widehat{w}_n \downarrow$ and $\widehat{w}_n - \widehat{f}_n \in \overline{V}_n$. Since $\widehat{f}_n \downarrow 0$, it follows from Theorem 2.21(e) that $\widehat{w}_n \downarrow 0$ also holds in $\widehat{L}$. This shows that we can assume without loss in generality that $\{\widehat{f}_n\}$ is a sequence of upper elements of L.

Now let V_k be a neighborhood from $\{V_n\}$. For each n choose an element $g_n \in L^+$ such that $\widehat{f}_n - g_n \in \overline{V}_{k+n+2}$ and $0 \le g_n \le \widehat{f}_n$. Let $v_n = \bigwedge_{i=1}^n g_i$ and note that $v_n \downarrow 0$ holds in L. It follows from the σ-Lebesgue property of L that $v_n \in V_{k+1}$ for all $n \ge n_0$. Now since

$$\begin{aligned} 0 &\le \widehat{f}_n - v_n = \bigvee_{i=1}^n (\widehat{f}_n - g_i) \le \bigvee_{i=1}^n (\widehat{f}_i - g_i) \\ &\le \sum_{i=1}^n (\widehat{f}_i - g_i) \in \overline{V}_{k+3} + \cdots + \overline{V}_{k+n+2} \subseteq \overline{V}_{k+1}\,, \end{aligned}$$

we get $\widehat{f}_n - v_n \in \overline{V}_{k+1}$ for each $n \in \mathbb{N}$. Thus

$$\widehat{f}_n = v_n + (\widehat{f}_n - v_n) \in \overline{V}_{k+1} + \overline{V}_{k+1} \subseteq \overline{V}_k$$

for all $n \ge n_0$. Therefore $\widehat{f}_n \xrightarrow{\widehat{\tau}} 0$, and the proof is finished. ∎

The last two results of this section are due to E. Langford and C. D. Aliprantis [**81**, pp. 210–211]. They deal both with the pseudo σ-Lebesgue property in quotient locally solid Riesz spaces. The first one follows immediately from the discussion preceding Theorem 4.45 and its proof is omitted.

THEOREM 5.37 (Langford–Aliprantis). *Let (L, τ) be a locally solid Riesz space with the pseudo σ-Lebesgue property, and let A be a projection band of L. Then the quotient locally solid Riesz space $(L/A, \tau/A)$ satisfies the pseudo σ-Lebesgue property.*

In Theorem 4.44 we saw that the Dedekind σ-completeness of L was sufficient to guarantee that the σ-Fatou property was inherited by $(L/A, \tau/A)$. This works also in the pseudo σ-Lebesgue case. Interestingly enough, we had to assume that A was a σ-ideal in Theorem 4.44, but it suffices to have a simple ideal here, provided that the topology is metrizable.

THEOREM 5.38 (Langford–Aliprantis). *Suppose that (L, τ) is a Dedekind σ-complete metrizable locally solid Riesz space with the pseudo σ-Lebesgue property, and that A is an ideal of L. Then $(L/A, \tau/A)$ satisfies the pseudo σ-Lebesgue property.*

PROOF. Suppose $\dot{u}_n \downarrow 0$ in L/A and that $\{\dot{u}_n\}$ is a τ/A-Cauchy sequence. Pick a normal base $\{V_n\}$ at zero for τ and note that $\{\dot{V}_n\}$ is a normal base at zero in L/A for τ/A. By passing to a subsequence (if necessary) we can assume $\dot{u}_n - \dot{u}_{n+1} \in \dot{V}_{n+2}$ for all n.

Now for each n choose $0 \le x_n \in V_{n+2}$ such that $\dot{x}_n = \dot{u}_n - \dot{u}_{n+1}$ and put $v_1 = u_1^+$ and $v_n = u_1^+ - (x_1 + \cdots + x_{n-1})$ for $n \ge 2$. Observe that $v_n \downarrow$ in L and that $\dot{v}_n = \dot{u}_n$ for all n. Taking positive parts we get $0 \le v_n^+ \downarrow$ and so by the Dedekind σ-completeness of L there exists some $v \in L^+$ such that $v_n^+ \downarrow v \ge 0$ in L. Now $0 \le v \le v_n^+$ for all n, so that $0 \le \dot{v} \le \dot{v}_n^+ = \dot{u}_n$ holds for all n. It follows that $\dot{v} = 0$ and thus if $w_n = v_n^+ - v$, then $w_n \downarrow 0$ in L and $\dot{w}_n = \dot{u}_n$ for all n. But $\{w_n\}$ is a τ-Cauchy sequence, since if n and k are arbitrary, then

$$\begin{aligned} 0 &\le w_n - w_{n+k} = v_n^+ - v_{n+k}^+ \le |v_n - v_{n+k}| \\ &= x_n + \cdots + x_{n+k-1} \in V_{n+2} + \cdots + V_{n+k+1} \subseteq V_n\,. \end{aligned}$$

By the pseudo σ-Lebesgue property of (L,τ) we have $w_n \xrightarrow{\tau} 0$ and consequently $\dot{u}_n = \dot{w}_n \xrightarrow{\tau/A} 0$. ■

5.4. The B-Property

In this section we shall introduce a property that is stronger than the pre-Lebesgue property and independent from the Lebesgue property. This property was introduced as property (B,iii) by W. A. J. Luxemburg and A. C. Zaanen [**89**, Note XI, p. 507]

DEFINITION 5.39. *A locally solid Riesz space (L,τ) is said to satisfy the B-**property** if it follows from $0 \le u_n \uparrow$ in L and $\{u_n\}$ τ-bounded that $\{u_n\}$ is a τ-Cauchy sequence.*

It should be clear that the B-property implies the pre-Lebesgue property. The Lebesgue property is, however, independent from the B-property, as the following examples show.

EXAMPLE 5.40. Let X be an infinite set, and let L be the Riesz space consisting of all real-valued functions defined on X that vanish outside finite sets, with the pointwise ordering. Let τ be the Hausdorff locally solid topology on L generated by the sup norm.

Note that (L,τ) satisfies the Lebesgue property, but it does not satisfy the B-property. For, if $\{x_1, x_2, \ldots\}$ is a countable subset of X, then the sequence $u_n = \chi_{\{x_1,\ldots,x_n\}}$, $n = 1, 2, \ldots$, satisfies $0 \le u_n \uparrow$ and is τ-bounded. However, $\{u_n\}$ is not a τ-Cauchy sequence. ■

EXAMPLE 5.41. Consider the Riesz space $L = C[0,1]$ equipped with the Hausdorff locally solid topology generated by the Riesz norm $\rho(u) = \int_0^1 |u(x)|\,dx$. Clearly, L satisfies the B-property, but it does not satisfy the σ-Lebesgue property; see Example 3.2. ■

We continue with a simple characterization of the B-property, the easy proof of which is left to the reader.

THEOREM 5.42. *For a locally solid Riesz space (L,τ) the following statements are equivalent.*

(i) *(L,τ) satisfies the B-property.*
(ii) *If a net $\{u_\alpha\}$ is τ-bounded and satisfies $0 \le u_\alpha \uparrow$, then $\{u_\alpha\}$ is τ-Cauchy.*

It should be noted that if a Hausdorff locally solid Riesz space satisfies the B-property, then it also satisfies the following "weak Lebesgue property:" *If a net $\{u_\alpha\} \subseteq L^+$ satisfies $0 \le u_\alpha \uparrow \widehat{u}$ in $\widehat{L}$, then $u_\alpha \xrightarrow{\widehat{\tau}} \widehat{u}$.*

The next result is due to C. D. Aliprantis [**3**, Theorem 5.3, p. 118].

THEOREM 5.43 (Aliprantis). *For a metrizable locally solid Riesz space (L,τ) the following statements are equivalent.*

(i) *(L,τ) satisfies the B-property.*
(ii) *$(\widehat{L},\widehat{\tau})$ satisfies the B-property.*

In particular, if (L,τ) satisfies the B-property, then $(\widehat{L},\widehat{\tau})$ is super Dedekind complete with the Lebesgue property.

PROOF. (i) $\Longrightarrow$ (ii) Pick a normal base $\{V_n\}$ at zero for τ consisting of solid sets. Assume that $0 \le \widehat{f}_n \uparrow$ is a $\widehat{\tau}$-bounded sequence in $\widehat{L}$. Pick a V_k from $\{V_n\}$.

For each n choose some $g_n \in L^+$ such that $\widehat{f}_n - g_n \in \overline{V}_{n+1}$. Next, define $u_n = 0$ if $n < k+3$ and $u_n = \bigvee_{i=k+3}^n g_i$ if $n \geq k+3$. Clearly, $0 \leq u_n \uparrow$ in L. For each $n \geq k+3$ we have

$$u_n - \widehat{f}_n = \bigvee_{i=k+3}^{n} (g_i - \widehat{f}_n) \leq \bigvee_{i=k+3}^{n} (g_i - \widehat{f}_i) \leq \sum_{i=k+3}^{n} \big|g_i - \widehat{f}_i\big| ,$$

and

$$\widehat{f}_n - u_n \leq \widehat{f}_n - g_n \leq \sum_{i=k+3}^{n} \big|g_i - \widehat{f}_i\big| .$$

Consequently, we have

$$\big|u_n - \widehat{f}_n\big| \leq \sum_{i=k+3}^{n} \big|g_i - \widehat{f}_i\big| \in \overline{V}_{k+4} + \cdots + \overline{V}_{n+1} \subseteq \overline{V}_{k+2} .$$

Therefore, $u_n - \widehat{f}_n \in \overline{V}_{k+2}$ for all $n \geq k+3$.

Now for each n let $\widehat{w}_n = \sum_{i=1}^n |\widehat{f}_i - g_i|$ and observe that $\{\widehat{w}_n\}$ is a $\widehat{\tau}$-Cauchy increasing sequence of $\widehat{L}$. Therefore, $\widehat{w}_n \xrightarrow{\widehat{\tau}} \widehat{w}$ holds in $\widehat{L}$, and thus $\widehat{w}_n \uparrow \widehat{w}$ also holds. In particular, we get $|\widehat{f}_n - u_n| \leq \widehat{w}$ for all n and thus $\{\widehat{f}_n - u_n\}$ is $\widehat{\tau}$-bounded. Consequently, from $u_n = \widehat{f}_n - (\widehat{f}_n - u_n)$ we see that $\{u_n\}$ is τ-bounded and so by our hypothesis $\{u_n\}$ is a τ-Cauchy sequence. In particular, we have $|u_n - u_m| \in V_{k+2}$ for all $n, m \geq n_0 \geq k+3$. Now if $n, m \geq n_0$, then

$$\big|\widehat{f}_n - \widehat{f}_m\big| \leq \big|\widehat{f}_n - u_n\big| + |u_n - u_m| + \big|\widehat{f}_m - u_m\big| \in \overline{V}_{k+2} + \overline{V}_{k+2} + \overline{V}_{k+2} \subseteq \overline{V}_k .$$

Hence, $\widehat{f}_n - \widehat{f}_m \in \overline{V}_k$ for all $n, m \geq n_0$. That is, $\{\widehat{f}_n\}$ is a $\widehat{\tau}$-Cauchy sequence, as desired.

(ii) $\Longrightarrow$ (i) Obvious.

The last part follows immediately from Theorem 5.35. ■

5.5. Comparing Locally Solid Topologies

The results in this section are due to the authors [**19**] and deal with comparison properties of locally solid topologies on a Riesz space. The main tool that will be used for the comparisons is the notion of the carrier of a locally solid topology. In Section 4.2 we saw that if a Hausdorff locally solid topology has the Fatou property, then its carrier is order dense; see Theorem 4.17. We start by naming the locally solid topologies with order dense carriers.

DEFINITION 5.44. *A locally solid topology τ on a Riesz space L is said to be **entire** if its carrier C_τ is order dense in L.*

Two immediate consequences of the above definition are the following.

(1) Every entire topology is Hausdorff, and
(2) Only Archimedean Riesz spaces can admit entire topologies.

To see (1), assume that τ is an entire topology on a Riesz space L and let $u \neq 0$. There exists some $v \in C_\tau$ such that $0 < v \leq |u|$. Pick a normal sequence $\{V_n\}$ of τ-neighborhoods of zero with $v \in N^{\mathrm{d}}$, where $N = \bigcap_{n=1}^\infty V_n$. If $v \in V_n$ for each n, then $v \in N \cap N^{\mathrm{d}} = \{0\}$, which is impossible. Hence, $v \notin V_n$ for some n, and this shows that τ must be a Hausdorff topology.

Invoking Theorem 4.17 we see that the following topologies are entire:

- Metrizable locally solid topologies
- Hausdorff Fatou topologies
- Hausdorff Lebesgue topologies

Entire topologies when restricted to ideals or order dense Riesz subspaces remain entire.

LEMMA 5.45. *Let τ be an entire topology on a Riesz space L. If K is either an ideal or an order dense Riesz subspace of L, then τ restricted to K is entire.*

PROOF. Let τ_1 denote the restriction of τ to K and let $0 < u \in K$. Since τ is entire, there exists some $v \in C_\tau$ with $0 < v \leq u$. Notice that in either case, we can assume that $v \in K$. Indeed, if K is an ideal, then $v \in K$, and if K is order dense in L, then pick some $w \in K$ with $0 < w \leq v$ and then replace v by $w \in C_\tau$.

Now choose a normal sequence $\{V_n\}$ of solid τ-neighborhoods of zero with $v \in N^{\mathrm{d}}$, where $N = \bigcap_{n=1}^{\infty} V_n$. Then $\{V_n \cap K\}$ is a normal sequence of solid τ_1-neighborhoods of K and from

$$N_1 = \bigcap_{n=1}^{\infty} (V_n \cap K) \subseteq \bigcap_{n=1}^{\infty} V_n = N$$

we get $N^{\mathrm{d}} \subseteq N_1^{\mathrm{d}}$ in L. This implies that v belongs to the disjoint complement of N_1 in K. Thus, $v \in C_{\tau_1}$ and consequently the topology τ_1 induces an entire topology on K. ■

Another class of entire topologies is provided by the next result. Keep in mind that if X is a topological space, then $C_c(X)$ denotes the Riesz space of all continuous real-valued functions on X with compact support.[2]

THEOREM 5.46. *If X is a separable metrizable topological space, then the carrier of every Hausdorff locally solid topology τ on $C(X)$ coincides with $C(X)$, i.e., $C_\tau = C(X)$; and hence τ is entire.*

PROOF. Let d be a metric generating the topology of X and let $\{x_1, x_2, \ldots\}$ be a countable dense subset of X. For each pair $k, n \in \mathbb{N}$ pick a function $f_{k,n} \in C(X)$ such that:

(1) $0 \leq f_{k,n}(x) \leq \frac{1}{n}$ for all $x \in X$.
(2) $f_{k,n}(x) = 0$ if $d(x_k, x) \geq \frac{1}{n}$.
(3) $f_{k,n}(x_k) = \frac{1}{n}$.

Let $\{u_1, u_2, \ldots\}$ be an enumeration of the double sequence $\{f_{k,n}\}$.

Now let τ be a Hausdorff locally solid topology on $C(X)$. Using induction, it is easy to see that we can pick a normal sequence $\{V_n\}$ of solid τ-neighborhoods of zero such that $u_n \notin V_n$ holds for each n. We claim that $N = \bigcap_{n=1}^{\infty} V_n = \{0\}$. To see this, let $0 \leq f \in N$. If $f \neq 0$, then $f(x_i) > 0$ must hold for some i, and from this we see that for some k we have $0 < u_k \leq f$. So, $u_k \in N \subseteq V_k$, contrary to $u_k \notin V_k$. Thus, $N = \{0\}$ and hence $N^{\mathrm{d}} = C(X)$. In particular, $C_\tau = C(X)$. ■

[2]The **support** of any function $f \in C(X)$ is the closure of the set $\{x \in X\colon\ f(x) \neq 0\}$.

The next result is an immediate consequence of Theorem 5.46; see also Exercise 20 at the end of the chapter.

COROLLARY 5.47. *If X is a metrizable topological space, then every Hausdorff locally solid topology on $C_c(X)$ is entire.*

The next result characterizes the entire topologies in terms of Fatou topologies on certain Riesz spaces.

THEOREM 5.48. *If a Riesz space L admits a Hausdorff Lebesgue topology, then a locally solid topology τ on L is entire if and only if L admits a Hausdorff Fatou topology coarser than τ.*

PROOF. Assume that τ_1 is a Hausdorff Fatou topology on L that is coarser than τ, i.e., $\tau_1 \subseteq \tau$. Clearly $C_{\tau_1} \subseteq C_\tau$ and since (by Theorem 4.17) the carrier C_{τ_1} is order dense in L, it follows that C_τ is also order dense in L.

For the converse, assume that L admits a Hausdorff Lebesgue topology τ^* and let τ be an entire topology on L. Choose a base $\mathcal{N}$ at zero for τ consisting of solid τ-neighborhoods. For each $V \in \mathcal{N}$, let V^* denote the τ^*-closure of V. Since τ^* is a Lebesgue topology, it follows that V^* is a solid and order closed set. Consequently, the filter base $\{V^*\colon\ V \in \mathcal{N}\}$ defines a Fatou topology which is coarser than τ. To complete the proof we shall show that this topology is Hausdorff.

Assume by way of contradiction that there exists some vector $u \in L$ satisfying $0 < u \in V^*$ for each $V \in \mathcal{N}$. Since C_τ and C_{τ^*} are order dense ideals, so is their intersection $C_\tau \cap C_{\tau^*}$. Therefore, there exists some $v \in C_\tau \cap C_{\tau^*}$ satisfying $0 < v \leq u$. Pick a normal sequence $\{V_n\}$ of solid τ-neighborhoods of zero with $v \in N^{\mathrm{d}}$, where $N = \bigcap_{n=1}^{\infty} V_n$. Similarly, pick another normal sequence $\{W_n\}$ of solid τ^*-neighborhoods of zero with $v \in M^{\mathrm{d}}$, where $M = \bigcap_{n=1}^{\infty} W_n$. Clearly, $v \in V_n^*$ for all n. Now for each n choose some $v_n \in V_n$ with $0 \leq v_n \leq v$ and $v - v_n \in W_n$. Then $\{v_n\} \subseteq M^{\mathrm{d}}$ and $v_n \xrightarrow{\tau_1} v$ in M^{d}, where τ_1 is the metrizable Lebesgue topology on M^{d} generated by $\{W_n \cap M^{\mathrm{d}}\}$. Now, according to Theorem 5.10, there exists a subsequence $\{w_n\}$ of $\{v_n\}$ and a net $\{u_\alpha\}$ satisfying $0 \leq u_\alpha \uparrow v$ in M^{d}, and such that given any index α, there exists some n_α with $u_\alpha \leq w_n$ for all $n \geq n_\alpha$. It follows that $\{u_\alpha\} \subseteq N$. But since $\{u_\alpha\} \subseteq N^{\mathrm{d}}$, we get $u_\alpha = 0$ for all α. Hence $v = 0$, a contradiction. This contradiction shows that the locally solid topology generated by the filter base $\{V^*\colon\ V \in \mathcal{N}\}$ is Hausdorff. ■

LEMMA 5.49. *If τ is a Hausdorff σ-Fatou topology on a Riesz space L, then $C_\phi \subseteq C_\tau$ holds for each $\phi \in L^\sim$.*

PROOF. Let $\phi \in L^\sim$ and pick some $0 < u \in C_\phi$. Since $C_\tau \oplus C_\tau^{\mathrm{d}}$ is order dense in L and C_ϕ has (by Lemma 1.80) the countable sup property, there exists a sequence $\{u_n + v_n\} \subseteq C_\tau \oplus C_\tau^{\mathrm{d}}$ with $0 \leq u_n + v_n \uparrow u$ in L. We shall show that $v_n = 0$ for all n. It will then follow from the σ-order closedness of C_τ that $u \in C_\tau$.

Assume by way of contradiction that $v_k \neq 0$ for some k (note that $v_k \in C_\phi \cap C_\tau^{\mathrm{d}}$). Choose a normal sequence $\{V_n\}$ of σ-order closed solid τ-neighborhoods of zero with $v_k \notin V_1$. Since $N \oplus N^{\mathrm{d}}$ is order dense in L, where $N = \bigcap_{n=1}^{\infty} V_n$, there exists a sequence $\{x_n + y_n\} \subseteq N \oplus N^{\mathrm{d}}$ with $0 \leq x_n + y_n \uparrow v_k$. It follows that $y_n = 0$ for all n, and so $v_k \in V_1$, which is a contradiction, and the proof is finished. ■

THEOREM 5.50. *If $L_n^\sim$ separates the points of a Riesz space L, then every Hausdorff σ-Fatou topology on L is entire.*

PROOF. Let τ be a Hausdorff σ-Fatou topology on L, and assume $0 < u \in L$. Choose $0 < \phi \in L_n^\sim$ with $\phi(u) > 0$, and then pick a net $\{u_\alpha + v_\alpha\} \subseteq N_\phi \oplus C_\phi$ with $0 \le u_\alpha + v_\alpha \uparrow u$. Since N_ϕ is a band of L, there exists an index α with $0 < v_\alpha \le u$. By Lemma 5.49, we have $v_\alpha \in C_\phi \subseteq C_\tau$, so that C_τ is order dense in L. ■

Recall that for a locally solid Riesz space (L, τ) we let $L'_c = L' \cap L_c^\sim$ and $L'_n = L' \cap L_n^\sim$. According to Theorem 5.9 when τ is a metrizable locally convex-solid topology the ideal L'_c is order dense in $L_c^\sim$ and L'_n is order dense in $L_n^\sim$. The next theorem gives some conditions sufficient for L'_n to be order dense in $L_n^\sim$.

THEOREM 5.51. *Let (L, τ) be a Hausdorff locally convex-solid Riesz space. If τ is either entire or a σ-Fatou topology, then L'_n is order dense in $L_n^\sim$.*

PROOF. Let $0 < \phi \in L_n^\sim$. We claim that in either case there exists some $0 < u \in C_\tau$ with $\phi(u) > 0$. Indeed, this is clearly true if τ is entire. If τ is a σ-Fatou topology, then choose some $0 < u \in C_\phi$, and note that according to Lemma 5.49 we have $u \in C_\tau$.

Next pick a normal sequence $\{V_n\}$ of solid and convex τ-neighborhoods of zero with $u \in N^d$, where $N = \bigcap_{n=1}^\infty V_n$. Clearly, $\{V_n \cap N^d\}$ defines a metrizable locally convex-solid topology τ_m on N^d, and the restriction of ϕ to N^d is order continuous. By Theorem 5.9 there exists a τ_m-continuous linear functional ϕ_1 on N^d with $0 < \phi_1 \le \phi$ on N^d. Now since τ_m is coarser than the restriction of τ on N^d, it follows that ϕ_1 is τ-continuous on N^d, and hence by the Hahn–Banach theorem ϕ_1 has a τ-continuous linear extension to L, which we denote by ϕ_1 again. Put $\psi = |\phi_1| \wedge \phi$ in $L^\sim$, and note that $\psi \in L'_n$ and that $0 < \psi \le \phi$, and the proof is complete. ■

COROLLARY 5.52. *Let (L, τ) be a Hausdorff locally convex-solid Riesz space. If L'_n separates the points of L, then τ is entire.*

PROOF. Note that $|\sigma|(L, L'_n)$ is a Hausdorff Lebesgue topology satisfying $|\sigma|(L, L'_n) \subseteq \tau$. The conclusion now follows from $C_{|\sigma|} \subseteq C_\tau$ and the order denseness of $C_{|\sigma|}$. ■

The Amemiya–Mori Theorem 4.21 presented an important comparison between Fatou and Lebesgue topologies. The next result of the authors [**19**, Theorem 4.2, p. 190] generalizes this theorem.

THEOREM 5.53 (Aliprantis–Burkinshaw). *If a Hausdorff locally solid topology τ on a Riesz space L is either entire or a σ-Fatou topology, then τ induces a finer topology on the order bounded subsets of L than any Lebesgue topology on L.*

PROOF. Let τ_1 be a (not necessarily Hausdorff) Lebesgue topology on L, and let $0 \le x \in C_{\tau_1}$. Note that τ_1 restricted to C_{τ_1} is a Hausdorff Lebesgue topology, and moreover, from Theorem 4.17 it follows that C_{τ_1} has the countable sup property. Thus, if τ is a Hausdorff σ-Fatou topology, then τ restricted to C_{τ_1} is a Fatou topology. On the other hand, if τ is entire, then by Lemma 5.45 the topology τ restricted to C_{τ_1} is also entire and thus by Theorem 5.48 there exists a Hausdorff Fatou topology on C_{τ_1} that is coarser than the restriction of τ on C_{τ_1}. Therefore, in either case, there exists a Hausdorff Fatou topology τ^* on C_{τ_1} which is coarser than τ on C_{τ_1}. But then it follows from Theorem 4.21 that τ^* (and hence) τ restricted to $[0, x]$ is finer than τ_1 on $[0, x]$. Thus, the desired conclusion is true for the order bounded subsets of C_{τ_1}.

Now assume that an order bounded net $\{u_\alpha\}$ satisfies $0 \leq u_\alpha \leq u$ for all α and $u_\alpha \xrightarrow{\tau} 0$. To finish the proof we must show that $u_\alpha \xrightarrow{\tau_1} 0$. To this end, let V be an arbitrary solid τ_1-neighborhood of zero. Choose another solid τ_1-neighborhood W of zero with $W + W + W \subseteq V$. Now select a normal sequence $\{W_n\}$ of solid τ_1-neighborhoods of zero with $W_1 = W$ and put $N = \bigcap_{n=1}^\infty W_n$. Since $N \oplus N^{\rm d}$ is order dense in L and τ_1 is a Lebesgue topology, there exists some $0 \leq v + w \in N \oplus N^{\rm d}$ with $0 \leq v + w \leq u$ and $u - (v + w) \in W$. Clearly $w \in C_{\tau_1}$ and $(u - w)^+ = (u - v - w) + v \in W + W$. Now, according to the first part of the proof, there exists some index β such that $u_\alpha \wedge w \in W$ for all $\alpha \succeq \beta$. But then we have

$$0 \leq u_\alpha = (u_\alpha - w)^+ + u_\alpha \wedge w \leq (u - w)^+ + u_\alpha \wedge w \in W + W + W \subseteq V$$

for all $\alpha \succeq \beta$, and so $u_\alpha \in V$ for all $\alpha \succeq \beta$. Hence $u_\alpha \xrightarrow{\tau_1} 0$. ■

COROLLARY 5.54. *If τ is a metrizable locally solid topology on a Riesz space L, then τ induces a finer topology on the order bounded subsets of L than any Lebesgue topology on L.*

The next result is a converse of Theorem 5.53 and it demonstrates that Theorem 5.53 is the best possible generalization of Theorem 4.21.

THEOREM 5.55. *Let τ^* be a Hausdorff Lebesgue topology on a Riesz space L. If τ is a locally solid topology on L that induces a finer topology than τ^* on the order bounded subsets of L, then τ must be entire.*

PROOF. We claim that $C_{\tau^*} \subseteq C_\tau$. To see this, let $0 \leq u \in C_{\tau^*}$. Choose a normal sequence $\{V_n\}$ of solid τ^*-neighborhoods of zero with $u \in N^{\rm d}$, where $N = \bigcap_{n=1}^\infty V_n$. Since $\tau^* \subseteq \tau$ holds on $[0, u]$, there exists a normal sequence $\{W_n\}$ of solid τ-neighborhoods of zero with $W_n \cap [0, u] \subseteq V_n \cap [0, u]$ for all n. Put $M = \bigcap_{n=1}^\infty W_n$, and note that $u \in M^{\rm d}$. Indeed, since $M \oplus M^{\rm d}$ is order dense in L, there exists a net $\{w_\alpha + v_\alpha\} \subseteq M \oplus M^{\rm d}$ with $0 \leq w_\alpha + v_\alpha \uparrow u$ in L. Since

$$w_\alpha \in M \cap [0, u] \subseteq N \cap [0, u] = \{0\},$$

we get $w_\alpha = 0$ for all α. Therefore, $0 \leq v_\alpha \uparrow u$ in L, and so $u \in M^{\rm d}$. Thus, $u \in C_\tau$, so that $C_{\tau^*} \subseteq C_\tau$. To complete the proof notice that the carrier C_{τ^*} is order dense in the Riesz space L. ■

From what we have seen in this section most "reasonable" locally solid topologies on a Riesz space seem to be entire. So, one can view the nonentire topologies as "pathological" cases. One such example is the Hausdorff locally convex-solid topology $\tau = |\sigma|\big(L_\infty[0,1], L_1^{\rm d}[0,1]\big)$ on $L_\infty[0,1]$. It can be shown that for this topology τ we have $C_\tau = \{0\}$. For details see Exercise 19 at the end of this chapter and Exercise 20 of Chapter 6.

Exercises

1. Show that if (L, τ) is a metrizable locally solid Riesz space with the pre-Lebesgue property, then every order bounded disjoint subset of L is at most countable. [*Hint*: If $\{V_n\}$ is a base at zero for τ and A is an order bounded disjoint subset of L, then $\{u \in A\colon\ u \notin V_n\}$ is finite for each n.]

2. Show that if (L, τ) satisfies the hypotheses of the preceding exercise, then L has the countable sup property.

3. Show that a Hausdorff locally solid Riesz space (L, τ) satisfies the Lebesgue property if and only if it satisfies the pre-Lebesgue property and pseudo Lebesgue properties together.

4. Let L be a Riesz space and τ a metrizable linear topology on L. Establish that if the solid hull of every τ-bounded subset of L is τ-bounded, then τ is a locally solid topology. [*Hint*: Pick a base $\{V_n\}$ at zero for τ, and then show that given any τ-neighborhood V of zero there exists n such that $\mathrm{Sol}\,(\frac{1}{n}V_n) \subseteq V$.]

5. Let (L, τ) be a metrizable locally solid Riesz space satisfying the σ-Lebesgue property, and let A be an ideal of L. Show that A is τ-closed if and only if A is a σ-ideal.

6. Let (L, τ) be a metrizable locally solid Riesz space and let (M, τ^*) be a Hausdorff locally solid Riesz space satisfying the σ-Lebesgue property. Assume that $T\colon (L, \tau) \to (M, \tau^*)$ is a continuous operator which is also positive and order continuous. Show that the unique continuous extension $\widehat{T}\colon (\widehat{L}, \widehat{\tau}) \to (\widehat{M}, \widehat{\tau}^*)$ is also a positive order continuous operator.

7. Show that every separable Fréchet space contains a weak order unit.

8. Let (L, τ) be a Fréchet space. Show that if $u_n \xrightarrow{\tau} u$ holds in L, then there exists a subsequence $\{v_n\}$ of $\{u_n\}$ and some $v \in L^+$ such that $|v_n - u| \leq \frac{1}{n}v$ holds for all n.

9. A proper ideal A of a Riesz space L is called **maximal** if it follows from $A \subseteq B$, $B \neq A$, and B an ideal of L that $B = L$. Show that every maximal ideal in a Fréchet space (L, τ) is τ-closed.

10. Consider the super Dedekind complete Riesz space $L = L_\infty[0, 1]$ and the three Riesz norms defined on L by:

$$\begin{aligned} \rho_1(u) &= \int_0^{\frac{1}{2}} |u(x)|\,dx + \Big[\int_{\frac{1}{2}}^1 u^2(x)\,dx\Big]^{\frac{1}{2}} \\ \rho_2(u) &= \Big[\int_0^{\frac{1}{2}} u^2(x)\,dx\Big]^{\frac{1}{2}} + \int_{\frac{1}{2}}^1 |u(x)|\,dx \\ \rho_3(u) &= \operatorname{ess\,sup} |u|\,. \end{aligned}$$

Establish the following properties:

(i) $[\rho_1 \wedge \rho_2](u) = \int_0^1 |u(x)|\,dx$ for all $u \in L$.

(ii) $[(\rho_1 \wedge \rho_2) \vee \rho_3](\mathbf{1}) = 1$, where $\mathbf{1}(x) = 1$ for each $x \in [0, 1]$.

(iii) $(\rho_1 \vee \rho_3) \wedge (\rho_2 \vee \rho_3)(\mathbf{1}) \geq \frac{2(1+\sqrt{2})}{3+\sqrt{2}} > 1$, and conclude from this that $\mathcal{S}_L$ is not a distributive lattice.

11. Let X be a nonempty set and let L be the Riesz space consisting of all real-valued functions on X whose range is finite.

(a) Show that L satisfies the projection property.

(b) If ρ denotes the sup norm on L, then show that L_ρ satisfies the Fatou property.

(c) Determine the norm completion of L_ρ. (Compare your answer with Theorems 5.29 and 2.49.)

12. Show that the topological completion of the Hausdorff locally solid Riesz space of Example 5.30 consists of all real-valued functions defined on $[0, 1]$ whose restrictions to the rationals and irrationals of $[0, 1]$ are both continuous. [*Hint*: Use Tietze's extension theorem.]

13. Generalize Theorem 5.24 as follows: *If (X, Σ, μ) is a nonatomic finite measure space, then for each $0 < p < 1$ the Riesz space $L = L_p(X, \Sigma, \mu)$ has trivial order dual, i.e., $L^\sim = \{0\}$.*

14. Consider the Riesz space $L = C[0,1]$ equipped with the Riesz norm $\rho(u) = \int_0^1 |u(x)|\,dx$. Show that L_ρ satisfies the pre-Lebesgue property and that it fails to satisfy the pseudo σ-Lebesgue property.

15. Consider the Riesz spaces $L = L_p[0,1]$, where $0 < p < 1$, and $M = C[0,1]$. Show that $\mathcal{L}_b(L,M) = \{0\}$, a super Dedekind complete Riesz space. Why does this not contradict Exercise 21(ii) of Chapter 1?

16. Let (L,τ) be a metrizable locally convex-solid Riesz space. Show that the bidual L'' is a σ-ideal of $(L')^{\sim}$. [*Hint*: Use the fact that topological bidual L'' is sequentially $\sigma(L'',L')$-complete; see [**78**, p. 396].]

17. Let ρ be a σ-Fatou norm on a Riesz space L. Show that the continuous extension $\widehat{\rho}$ of ρ to $\widehat{L}_\rho$ is given by

$$\widehat{\rho}(\widehat{u}) = \inf\{\lim_{n\to\infty} \rho(u_n)\colon \{u_n\} \subseteq L^+,\ u_n \uparrow \text{ and } u_n \wedge |\widehat{u}| \uparrow |\widehat{u}| \text{ in } \widehat{L}_\rho\}$$

for all $\widehat{u} \in \widehat{L}_\rho$.

18. Show that a Hausdorff topologically complete locally convex-solid Riesz space (L,τ) satisfies the B-property if and only if L is a band in $(L')^{\sim}_{\mathrm{n}}$.

19. Let L be a Riesz space and let B the disjoint complement of $L^{\sim}_{\mathrm{n}}$ in $L^{\sim}_{\mathrm{c}}$, i.e., B is the band of $L^{\sim}$ such that $L^{\sim}_{\mathrm{c}} = L^{\sim}_{\mathrm{n}} \oplus B$. Show that if B separates the points of L, then $L^{\sim}_{\mathrm{n}} = \{0\}$.

20. Prove Corollary 5.47.

21. Assume that X is a metrizable topological space and let a function $g \in C(X)$ have a separable support. If B denotes the band generated by g in $C(X)$, then show that every Hausdorff locally solid topology on B is entire.

CHAPTER 6

Weak Compactness in Riesz Spaces

A number of properties of locally convex-solid Riesz spaces have been presented in the previous chapters as they were needed. In this chapter a systematic study of these spaces will be made.

Some of the basic properties of locally convex-solid Riesz spaces are derived first, and then the compact subsets of $L_c^\sim$ and $L_n^\sim$ are characterized (in various topologies) in terms of the concept of order-equicontinuity. Results concerning the weak convergence of sequences in the topological dual will be obtained and conditions for weak sequential compactness will be given. The solid τ-compact subsets as well as the weakly compact solid subsets will be characterized, and some of the most fascinating results will deal with compact sets. For instance, we shall establish that if a subset A of a topologically complete Hausdorff locally convex-solid Riesz space (L,τ) has the property that every disjoint sequence in its solid hull is τ-convergent to zero, then A must be $\sigma(L,L')$-relatively compact.

Finally, these results will be used to give necessary and sufficient conditions, in terms of disjoint sequences, for a locally convex-solid Riesz space to be semireflexive.

6.1. Topologies on the Duals of a Riesz Space

We shall discuss here a few basic results concerning several natural locally convex-solid topologies on the duals of a Riesz space. We have already mentioned a number of them in Section 2.3. Most of the results in this section have appeared in many forms in the works of several authors; see for instance [**89**], [**102**], [**107**], and [**130**].

We start our discussion with a simple but very important compactness property of the order intervals of the topological dual of a Hausdorff locally convex-solid Riesz space.

THEOREM 6.1. *For every Hausdorff locally convex-solid Riesz space (L,τ) the order intervals of L' are $\sigma(L',L)$-compact.*

PROOF. Let $[0,\phi]$ be an order interval of $L' \subseteq \mathbb{R}^L$. We must show that $[0,\phi]$ is bounded and closed in $\mathbb{R}^L$. It should be clear that $[0,\phi]$ is bounded in $\mathbb{R}^L$. Now if $\phi_\alpha \to \psi$ holds in $\mathbb{R}^L$ with $\{\phi_\alpha\} \subseteq [0,\phi]$, then from the pointwise convergence in $\mathbb{R}^L$ it follows that $0 \le \psi \le \phi$ and so $\psi \in L'$, that is, $[0,\phi]$ is closed in $\mathbb{R}^L$ and consequently the order interval $[0,\phi]$ is $\sigma(L',L)$-compact. ∎

For a locally solid Riesz space (L,τ), the polar of any solid subset of L is a solid subset of L'. Indeed, by Theorem 1.67 we have

$$|\phi|\big(|u|\big) = \sup\big\{|\phi(v)|\colon\ |v| \le |u|\big\}$$

for all $u \in L$ and $\phi \in L'$, and so if S is a solid subset of L, then

$$\begin{aligned} S^{\text{o}} &= \{\phi \in L' \colon |\phi(u)| \leq 1 \text{ for } u \in S\} \\ &= \{\phi \in L' \colon |\phi|(|u|) \leq 1 \text{ for } u \in S\}, \end{aligned}$$

so that S^{o} is a solid subset of L'.

Likewise, if S is a solid subset of L', then S^{o} is a solid subset of L; to see this we must use Theorem 1.70. Finally, if A is an ideal of L, then A^{o} is a band of L' and moreover

$$\begin{aligned} A^{\text{o}} &= \{\phi \in L' \colon \phi(u) = 0 \text{ for } u \in A\} \\ &= \{\phi \in L' \colon |\phi|(|u|) = 0 \text{ for } u \in A\}. \end{aligned}$$

THEOREM 6.2. *Let (L, τ) be a locally solid Riesz space, and let two bands A and B satisfy $L = A \oplus B$. Then we have:*

(1) *$L' = A^{\text{o}} \oplus B^{\text{o}}$.*
(2) *The projection P of L' onto A^{o} satisfies $P\phi(u) = \phi(u)$ for all $u \in B$ and all $\phi \in L'$.*

PROOF. By the above discussion A^{o} and B^{o} are both bands of L' satisfying $A^{\text{o}} \cap B^{\text{o}} = \{0\}$, as easily checked. Let Q be the projection of L onto B. For $\phi \in L'$, let $\phi_1 = \phi \circ Q$ and $\phi_2 = \phi - \phi_1$. Then $\phi_1(u) = 0$ for all $u \in A$ and $\phi_2(u) = 0$ for all $u \in B$ and thus $\phi_1 \in A^{\text{o}}$ and $\phi_2 \in B^{\text{o}}$. Hence $\phi = \phi_1 + \phi_2 \in A^{\text{o}} \oplus B^{\text{o}}$, and therefore $L' = A^{\text{o}} \oplus B^{\text{o}}$. For the proof of part (2) note that $\phi_1 = P\phi$ and that $\phi_2(u) = 0$ for all $u \in B$. ■

Now let L be a Riesz space, X a subset of L, and A an ideal of $L^\sim$.

DEFINITION 6.3. *The **absolute weak topology** $|\sigma|(A, X)$ on A is the locally convex-solid topology generated by the family of Riesz seminorms $\{\rho_u \colon u \in X\}$, where $\rho_u(\phi) = |\phi|(|u|)$ for all $\phi \in A$.*

It should be clear that $|\sigma|(A, X) = |\sigma|(A, L_X)$, where L_X is the ideal generated by X in L.

THEOREM 6.4. *For an arbitrary Riesz space L and any ideal A of $L^\sim$ the absolute weak topology $|\sigma|(A, L)$ is a Hausdorff Lebesgue topology on A satisfying $\sigma(A, L) \subseteq |\sigma|(A, L)$.*

PROOF. To see that $|\sigma|(A, L)$ is Hausdorff and finer than $\sigma(A, L)$ use the inequality $|\phi(u)| \leq |\phi|(|u|)$ and the fact that $\sigma(A, L)$ is a Hausdorff topology. Now if $\phi_\alpha \downarrow 0$ in A, then $\phi_\alpha \downarrow 0$ in $L^\sim$ and so $\phi_\alpha(u) \downarrow 0$ for every $u \in L^+$. Thus $\{\phi_\alpha\}$ is $|\sigma|(A, L)$-convergent to zero and so $|\sigma|(A, L)$ is also a Lebesgue topology. ■

The basic properties of the absolute weak topology $|\sigma|(L^\sim, L)$ are included in the next theorem.

THEOREM 6.5. *For every Riesz space L the absolute weak topology $|\sigma|(L^\sim, L)$ is a Hausdorff, Lebesgue, and Levi topology. Moreover, the order dual $L^\sim$ is $|\sigma|(L^\sim, L)$-complete.*

PROOF. To see that $|\sigma|(L^\sim, L)$ is a Hausdorff Lebesgue topology use Theorem 6.4. For the Levi property consider $0 \leq \phi_\alpha \uparrow$ in $L^\sim$ with the net $\{\phi_\alpha\}$ being $|\sigma|(L^\sim, L)$-bounded. Put $\phi(u) = \sup\{\phi_\alpha(u)\}$ for every $u \in L^+$, and note that $\phi \colon L^+ \to \mathbb{R}^+$ is an additive mapping. By Lemma 1.64, ϕ can be extended to all of L by $\phi(u) = \phi(u^+) - \phi(u^-)$. It is now readily seen that $\phi_\alpha \uparrow \phi$ holds in $L^\sim$ and so $|\sigma|(L^\sim, L)$ is a Levi topology. The topological completeness of $|\sigma|(L^\sim, L)$ now follows from Nakano's Theorem 4.37. ∎

THEOREM 6.6. *Let (L, τ) be a locally convex-solid Riesz space. Then the $|\sigma|(L', L)$-closure of an ideal A of L' coincides with the band generated by A in L'.*

PROOF. By Theorem 1.24 the ideal A is order dense in A^{dd} and consequently, since $|\sigma|(L', L)$ is a Lebesgue topology, it follows from Theorem 1.27 that the ideal A is $|\sigma|(L', L)$-dense in A^{dd}. The result now follows by observing that A^{dd} is $|\sigma|(L', L)$-closed; see part (d) of Theorem 2.21. ∎

THEOREM 6.7. *Let (L, τ) be a locally convex-solid Riesz space. Then the topological completion of $\big(L', |\sigma|(L', L)\big)$ is the band generated by L' in $L^\sim$. In particular, L' is $|\sigma|(L', L)$-complete if and only if L' is a band of $L^\sim$.*

PROOF. Note that $|\sigma|(L^\sim, L)$ induces the topology $|\sigma|(L', L)$ on L' and so by Theorem 6.5 the topological completion of $\big(L', |\sigma|(L', L)\big)$ must coincide with the $|\sigma|(L^\sim, L)$-closure of L'. Now since $|\sigma|(L^\sim, L)$ is a Lebesgue topology, this closure is a band of $L^\sim$, which is precisely the band generated by L' in $L^\sim$. ∎

COROLLARY 6.8. *If L is a Riesz space and A is an ideal of $L^\sim$, then the topological completion of $\big(A, |\sigma|(A, L)\big)$ is the band generated by A in $L^\sim$.*

PROOF. Note that if $\tau = |\sigma|(L, A)$, then the topological dual of (L, τ) is exactly A. So, by Theorem 6.7, the topological completion of $\big(A, |\sigma|(A, L)\big)$ is the band generated by A in $L^\sim$. ∎

Recall that for a Riesz space L and any ideal $\mathcal{I}$ of $L^\sim$, there exists a natural embedding $u \mapsto \widetilde{u}$ of L into $\mathcal{I}_n^\sim$ defined by $\widetilde{u}(\phi) = \phi(u)$ for all $\phi \in \mathcal{I}$. By Theorem 1.83 this embedding is a Riesz homomorphism that is one-to-one if and only if $\mathcal{I}^\circ = \{0\}$. So, if $\mathcal{I}^\circ = \{0\}$, then we can identify L with a Riesz subspace of $\mathcal{I}_n^\sim$, and as usual we shall make no distinction between an element u in L and its image $\widetilde{u}$ in $\mathcal{I}_n^\sim$.

Now if (L, τ) is a Hausdorff locally convex-solid Riesz space, then by the Hahn–Banach theorem we have $(L')^\circ = \{0\}$, and thus as above, L is a Riesz subspace of $(L')_n^\sim$, which is not necessarily an order dense Riesz subspace. (However, by Theorem 3.12 the Riesz space L is an order dense Riesz subspace of $(L')_n^\sim$ if and only if τ is a Lebesgue topology.)

The next result informs us that the Riesz space L is $|\sigma|\big((L')_n^\sim, L'\big)$-dense in the band $(L')_n^\sim$.

THEOREM 6.9. *If (L, τ) is a Hausdorff locally convex-solid Riesz space, then L is $|\sigma|\big((L')_n^\sim, L'\big)$-dense in $(L')_n^\sim$.*

PROOF. Put $M = (L')_n^\sim$. Then L' (embedded in $M_n^\sim$ according to Theorem 1.83) is an order dense Riesz subspace of $M_n^\sim$ and so by Theorem 1.40 the Riesz space L' is an ideal of $M_n^\sim$ (and hence of $M^\sim$). Now by Theorem 2.33 the topological dual of $\big(M, |\sigma|(M, L')\big)$ coincides with the ideal generated by L' in $M^\sim$, which is exactly L'. Now let $\overline{L}$ denote the $|\sigma|(M, L')$-closure of L in M. If $\overline{L} \neq M$, then there exists some $0 < v \in M$ with $v \notin \overline{L}$, and so by the classical separation theorem there exists a $|\sigma|(M, L')$-continuous linear functional ϕ on M, that is, some

$\phi \in L'$, satisfying $\phi(u) = 0$ for all $u \in L$ and $v(\phi) > 0$. But $\phi(u) = 0$ for all $u \in L$ implies $\phi = 0$, a contradiction. Therefore, L is $|\sigma|\big((L')^\sim_{\mathrm{n}}, L'\big)$-dense in $(L')^\sim_{\mathrm{n}}$. ■

Now if (L, τ) is a Hausdorff locally convex-solid Riesz space, then the topology $|\sigma|\big((L')^\sim_{\mathrm{n}}, L'\big)$ induces the topology $|\sigma|(L, L')$ on L; this follows immediately from the fact that $u_\alpha \to 0$ holds in L for the topology $|\sigma|\big((L')^\sim_{\mathrm{n}}, L'\big)$ if and only if $\lim |\phi|(|u_\alpha|) = 0$ for every $\phi \in L'$, which is equivalent to $u_\alpha \to 0$ in L for $|\sigma|(L, L')$. Note next that, by Corollary 6.8, we know that $(L')^\sim_{\mathrm{n}}$ is $|\sigma|\big((L')^\sim_{\mathrm{n}}, L'\big)$-complete and hence by combining this with the last theorem, we obtain the following result.

THEOREM 6.10. *For every Hausdorff locally convex-solid Riesz space (L, τ) the topological completion of $\big(L, |\sigma|(L, L')\big)$ is $(L')^\sim_{\mathrm{n}}$ equipped with the absolute weak topology $|\sigma|\big((L')^\sim_{\mathrm{n}}, L'\big)$.*

We saw in Section 3.2 that for a Hausdorff locally convex-solid Riesz space (L, τ) the strong topology $\beta(L', L)$ is a locally convex-solid topology on the topological dual L' that is generated by the family of Riesz seminorms $\{\rho_A\colon\ A \in \mathcal{B}_s\}$, where $\rho_A(\phi) = \sup\{|\phi(u)|\colon\ u \in A\}$, and $\mathcal{B}_s$ is the collection of all τ-bounded solid subsets of L. Since for every $u \in L$ the set $A = \big[-|u|, |u|\big]$ belongs to $\mathcal{B}_s$ and $|\phi|\big(|u|\big) = \rho_A(\phi)$ for all $\phi \in L'$, it follows easily that $\beta(L', L)$ is finer than $|\sigma|(L', L)$. Thus, the following inclusions hold:

$$\sigma(L', L) \subseteq |\sigma|(L', L) \subseteq \beta(L', L)\,.$$

It turns out that for each $A \in \mathcal{B}_s$ the Riesz seminorm ρ_A has the Fatou property.

LEMMA 6.11. *If (L, τ) is a Hausdorff locally convex-solid Riesz space and A is a nonempty τ-bounded solid subset of L, then the Riesz seminorm ρ_A is a Fatou seminorm.*

PROOF. Let A be a nonempty τ-bounded solid subset of L and assume that $0 \leq \phi_\alpha \uparrow \phi$ in L'. If $\rho_A(\phi_\alpha) \uparrow \rho_A(\phi)$ does not hold, then there exists some $\epsilon > 0$ such that $\rho_A(\phi_\alpha) \leq \rho_A(\phi) - \epsilon$ for all α. Now pick $0 \leq u \in A$ with $\rho_A(\phi) - \epsilon < \phi(u)$, and note that the net $\{\phi_\alpha(u)\}$ does not converge to $\phi(u)$, contradicting $0 \leq \phi_\alpha \uparrow \phi$ in L'. Thus $\rho_A(\phi_\alpha) \uparrow \rho_A(\phi)$ holds and the proof is finished. ■

COROLLARY 6.12. *For every Hausdorff locally convex-solid Riesz space (L, τ) the strong topology $\beta(L', L)$ is a Fatou topology.*

Another useful property of the order intervals of the topological dual of a Hausdorff locally convex-solid Riesz space is described in the next result.

THEOREM 6.13. *If (L, τ) is a Hausdorff locally convex-solid Riesz space, then the order intervals of L' are $\beta(L', L)$-complete. If, in addition, L' is a band in $L^\sim$, then L' is $\beta(L', L)$-complete.*

PROOF. By Corollary 6.12 the topology $\beta(L', L)$ is a Fatou topology. The first part follows now immediately from Theorem 4.28 by observing that L' is Dedekind complete.

Now if L' is a band of $L^\sim$, then $\beta(L', L)$ is a Levi topology. To see this, let $0 \leq \phi_\alpha \uparrow$ in L' with $\{\phi_\alpha\}$ being $\beta(L', L)$-bounded. Then there exists some $\phi \in L^\sim$ such that $\phi_\alpha \uparrow \phi$ and hence $\phi \in L'$, since L' is a band of $L^\sim$, so that $\beta(L', L)$ is a Levi topology. By Theorem 4.37 it follows that L' is $\beta(L', L)$-complete. ■

Recall that a Riesz space L is called **uniformly complete** if for every sequence $\{u_n\}$ of L that satisfies $|u_{n+p} - u_n| \leq \epsilon_n v$ for all n and p, some $v \in L^+$, and some real sequence $0 \le \epsilon_n \to 0$, there exists some $u \in L^+$ satisfying $|u_n - u| \le \epsilon_n v$ for all n.

THEOREM 6.14. *Let (L, τ) be an Archimedean locally convex-solid Riesz space and let $A \subseteq L'$. If L is uniformly complete, then the following statements are equivalent.*

(i) *A is $|\sigma|(L', L)$-bounded.*
(ii) *A is $\sigma(L', L)$-bounded.*

PROOF. (i) $\Longrightarrow$ (ii) Use the inclusion $\sigma(L', L) \subseteq |\sigma|(L', L)$.

(ii) $\Longrightarrow$ (i) Let A be $\sigma(L', L)$-bounded and fix $u \in L^+$. Define the Riesz norm ρ on the principal ideal L_u generated by u in L by

$$\rho(v) = \inf\{\lambda > 0\colon\ |v| \le \lambda u\}, \quad v \in L_u,$$

and note that $[-u, u]$ is the closed unit ball of ρ in L_u. Now since L is uniformly complete it easily follows that L_u is ρ-complete.

On the other hand, if $\phi \in A$, then $\sup\{|\phi(v)|\colon\ v \in [-u, u]\} < \infty$ and so ϕ restricted to L_u is bounded on the closed unit ball $[-u, u]$ of L_u, and therefore is ρ-continuous. Next note that the restriction of the elements of A to L_u are $\sigma(L'_u, L_u)$-bounded (where L'_u is the norm dual of L_u). Hence by the Banach–Steinhaus Theorem 2.14, the set A restricted to L_u is norm bounded. That is, there exists a number $c > 0$ such that $|\phi(v)| \le c$ for all $v \in [-u, u]$ and all $\phi \in A$. But then, by Theorem 1.67(3), we have $|\phi|(u) = \sup\{|\phi(v)|\colon\ v \in [-u, u]\} \le c$ for all $\phi \in A$. This shows that A is $|\sigma|(L', L)$-bounded, and the proof is finished. ■

THEOREM 6.15. *Let (L, τ) be a Hausdorff locally convex-solid Riesz space. If L' is a σ-ideal of $L^\sim$, then for a subset A of L' the following statements are equivalent.*

(i) *The set A is $|\sigma|(L', L)$-bounded.*
(ii) *The set A is $\beta(L', L)$-bounded.*

PROOF. (i) $\Longrightarrow$ (ii) If the set A is not $\beta(L', L)$-bounded, then neither is its solid hull $S = \mathrm{Sol}\,(A)$. This means that there exists a solid $\sigma(L, L')$-bounded subset B of L such that

$$\sup\{|\phi(u)|\colon\ \phi \in S \text{ and } u \in B\} = \infty.$$

For each n pick some $0 \le u_n \in B$ and $0 \le \phi_n \in S$ with $\phi_n(u_n) \ge n^3$. Now for each n let $\psi_n = \sum_{k=1}^{n} \frac{1}{k^2}\phi_k$ and note that since A (and hence S) is $|\sigma|(L', L)$-bounded, the sequence $\{\psi_n\}$ is $|\sigma|(L', L)$-bounded. Consequently, $0 \le \psi_n \uparrow \psi$ holds in $L^\sim$ for some $\psi \in L^\sim$, and therefore since L' is a σ-ideal of $L^\sim$ we have $\psi \in L'$. Now observe that since the sequence $\{u_n\}$ is $|\sigma|(L, L')$-bounded, the sequence $\{\frac{1}{n}u_n\}$ necessarily must be $|\sigma|(L, L')$-convergent to zero in L, which is in contradiction with

$$\psi\big(\tfrac{1}{n}u_n\big) \ge \tfrac{1}{n^2}\phi_n\big(\tfrac{1}{n}u_n\big) = \tfrac{1}{n^3}\phi_n(u_n) \ge 1$$

for each n. Hence the set A is $\beta(L', L)$-bounded.

(ii) $\Longrightarrow$ (i) It follows from the inclusion $|\sigma|(L', L) \subseteq \beta(L', L)$. ■

Recall that the **bidual** L'' of a Hausdorff locally convex-solid Riesz space (L, τ) is the topological dual of $\big(L', \beta(L', L)\big)$. Since $\beta(L', L)$ is a locally solid topology, it follows that L'' is an ideal of $(L')^\sim$. Moreover, L is a Riesz subspace of L'' under its natural embedding; see Theorem 1.83.

THEOREM 6.16. *Let (L, τ) be a Hausdorff locally convex-solid Riesz space, and let $0 \le u_\alpha \uparrow$ be a τ-bounded net of L. Then $0 \le u_\alpha \uparrow F$ holds in L'' for some $F \in L''$.*

PROOF. For every $0 \le \phi \in L'$ define $F(\phi) = \sup\{\phi(u_\alpha)\}$ and note that $F(\phi)$ is a real number since $\{u_\alpha\}$ is τ-bounded. Obviously, $F\colon L'_+ \to \mathbb{R}^+$ defines an additive function. By Lemma 1.64, F extends to a positive linear functional from L' to $\mathbb{R}$. Note that $0 \le u_\alpha \uparrow F$ holds in $(L')^\sim$. To complete the proof we shall show that $F \in L''$.

To this end, let A be the solid hull of $\{u_\alpha\}$ in L and note that A is τ-bounded. Now for $0 \le \phi \in L'$ we have

$$0 \le \phi(u_\alpha) \le \sup\{\phi(u)\colon\ u \in A\} = \rho_A(\phi)$$

for all α and so $F(\phi) \le \rho_A(\phi)$. But then $|F(\phi)| \le F(|\phi|) \le \rho_A(\phi)$ holds for all $\phi \in L'$ and so F is $\beta(L', L)$-continuous, because the Riesz seminorm ρ_A is $\beta(L', L)$-continuous. ■

The next result describes an important property of Hausdorff locally convex-solid Riesz spaces with the pre-Lebesgue property. It is due to I. Kawai [**74**, Theorem 5.3, p. 301].

THEOREM 6.17 (Kawai). *If a Hausdorff locally convex-solid Riesz space (L, τ) satisfies the pre-Lebesgue property, then τ and $|\sigma|(L, L')$ induce the same topology on the order bounded subsets of L.*

PROOF. Let A be an order bounded subset of L. By Theorem 3.27 the topological completion $(\widehat{L}, \widehat{\tau})$ of (L, τ) satisfies the Lebesgue property and hence by Theorem 4.9 the Hausdorff locally solid topology $|\sigma|\big(\widehat{L}, \widehat{L}'\big)$ likewise does. But A is also an order bounded subset of $\widehat{L}$ and hence by Theorem 4.22 the topologies $\widehat{\tau}$ and $|\sigma|\big(\widehat{L}, \widehat{L}'\big)$ both induce the same topology on A. Now since each $\phi \in L'$ has a unique continuous linear extension $\widehat{\phi}$ to all of $\widehat{L}$ and $\phi \mapsto \widehat{\phi}$ is a Riesz isomorphism between L' and $\widehat{L}'$ (see Exercise 10 of Chapter 2), we see that $|\sigma|(L, L')$ and $|\sigma|\big(\widehat{L}, \widehat{L}'\big)$ both induce the same topology on A and our conclusion follows. ■

If a Hausdorff locally convex-solid Riesz space (L, τ) satisfies the Lebesgue property, then L is order dense in $(L')^\sim_{\mathrm{n}}$ (see Theorem 3.12) and hence the ideal generated by L in $(L')^\sim_{\mathrm{n}}$ (which is the same as the ideal generated by L in L'') is the Dedekind completion of L. However, if τ is not a Lebesgue topology, then L is no longer order dense in $(L')^\sim_{\mathrm{n}}$. In this case the Dedekind completion of L can be obtained from L'' by using Theorem 1.49 as follows.

THEOREM 6.18. *Assume that (L, τ) is a Hausdorff locally convex-solid Riesz space and let A be an ideal of L'' that is maximal with respect to the property $L \cap A = \{0\}$. Then the Dedekind completion of L is the ideal generated by L in L''/A (where L is identified with its canonical image in L''/A).*

6.2. Weak Compactness in the Order Duals

The objective of this section is to characterize the weakly compact subsets of $L^\sim_{\mathrm{n}}$, $L^\sim_{\mathrm{c}}$, and $L^\sim$ in terms of the order structure of L. We start our discussion with some basic properties of $L^\sim$.

THEOREM 6.19. *Let L be a Riesz space. Then for a subset A of $L^\sim$ the following statements are equivalent:*

(i) *The set A is $|\sigma|(L^\sim, L)$-bounded.*
(ii) *The solid hull of A is relatively $\sigma(L^\sim, L)$-compact.*

PROOF. (i) $\Longrightarrow$ (ii) Since $|\sigma|(L^\sim, L)$ is a locally solid topology, the solid hull S of A is also $|\sigma|(L^\sim, L)$-bounded and hence the set S is $\sigma(L^\sim, L)$-bounded. Thus, in order to establish that S is relatively $\sigma(L^\sim, L)$-compact, it suffices to show that the $\sigma(L^*, L)$-closure of S in L^* lies in $L^\sim$ (since a $\sigma(L^*, L)$-closed and bounded subset of the algebraic dual L^* is $\sigma(L^*, L)$-compact).

To this end, let ϕ be in the $\sigma(L^*, L)$-closure of S, fix some $u \in L^+$ and put $\delta = \sup\{|\psi|(u)\colon\ \psi \in S\} < \infty$. Now for each $v \in [0, u]$ choose some $\psi \in S$ with $|\phi(v) - \psi(v)| < 1$ and observe that we have $|\phi(v)| \le |\phi(v) - \psi(v)| + |\psi(v)| \le 1 + \delta$. Thus ϕ carries order bounded subsets of L to bounded subsets of $\mathbb{R}$, and hence $\phi \in L^\sim$.

(ii) $\Longrightarrow$ (i) Let S be the solid hull of A. By our assumption, S is relatively $\sigma(L^\sim, L)$-compact and so for each $u \in L^+$ there exists some $\delta > 0$ such that $|\psi(u)| \le \delta$ for all $\psi \in S$. But if $\phi \in A$, then $|\phi| \in S$ and thus $0 \le |\phi|(u) \le \delta$ for $\phi \in A$. So, A is $|\sigma|(L^\sim, L)$-bounded. ■

THEOREM 6.20. *If L is an Archimedean uniformly complete Riesz space, then $L^\sim$ is sequentially $\sigma(L^\sim, L)$-complete.*

PROOF. Let $\{\phi_n\}$ be a $\sigma(L^\sim, L)$-Cauchy sequence of $L^\sim$. Then $\{\phi_n\}$ is $\sigma(L^\sim, L)$-bounded and consequently by Theorem 6.14 the sequence $\{\phi_n\}$ is $|\sigma|(L^\sim, L)$-bounded. But then, by Theorem 6.19, the sequence $\{\phi_n\}$ is relatively $\sigma(L^\sim, L)$-compact and consequently $\{\phi_n\}$ has a $\sigma(L^\sim, L)$-accumulation point $\phi \in L^\sim$. Since the sequence $\{\phi_n\}$ is $\sigma(L^\sim, L)$-Cauchy, we see that the sequence $\{\phi_n\}$ is $\sigma(L^\sim, L)$-convergent to ϕ. ■

Let L be a Riesz space. Recall that for a nonempty subset A of $L^\sim$ we define $\rho_A\colon L \to \mathbb{R}$ for each $u \in L$ by

$$\rho_A(u) = \sup\{|\phi(u)|\colon\ \phi \in A\} \le \infty\,.$$

Note that we have the following properties.

(1) If $A \subseteq B$, then $\rho_A(u) \le \rho_B(u)$ for all $u \in L$.
(2) If A is $\sigma(L^\sim, L)$-bounded, then ρ_A is a seminorm on L.
(3) If A is solid and $\sigma(L^\sim, L)$-bounded, then ρ_A is a Riesz seminorm.

For the above properties see the discussion at the beginning of Section 3.2. As usual, a sequence $\{u_n\}$ in L is called ρ_A-**Cauchy** if for every $\epsilon > 0$ there exists some k such that $\rho_A(u_n - u_m) < \epsilon$ for all $n, m \ge k$.

The $\sigma(L^\sim_{\rm c}, L)$-compact subsets of $L^\sim_{\rm c}$ will be characterized first for Archimedean Riesz spaces. The next theorem is, in fact, a very weak result, as Theorem 6.26 and Example 6.27 will later show. As we shall see, if in addition L is assumed to be almost Dedekind σ-complete, then much stronger results hold true.

THEOREM 6.21. *Let L be an Archimedean Riesz space. For a nonempty subset A of $L_c^\sim$ the following statements are equivalent.*

(i) *The solid hull S of A is relatively $\sigma(L_c^\sim, L)$-compact.*
(ii) *If $u_n \downarrow 0$ holds in L, then $\lim \rho_A(u_n) = 0$.*
(iii) *If $u_n \downarrow 0$ holds in L, then $\lim \rho_S(u_n) = 0$.*

PROOF. (i) $\Longrightarrow$ (ii) Suppose that (ii) is not true. Then there exist some $\epsilon > 0$ and a sequence $\{u_n\}$ of L satisfying $u_n \downarrow 0$ and $\rho_A(u_n) > 2\epsilon$ for all n. For each n pick some $\phi_n \in A$ such that $|\phi_n(u_n)| > 2\epsilon$. Let $0 \leq \phi \in L_c^\sim$ be a $\sigma(L_c^\sim, L)$-accumulation point of the sequence $\{|\phi_n|\}$. Given n, choose some $m \geq n$ such that $(|\phi_m| - \phi)(u_n) < \epsilon$ and note that since

$$\begin{aligned} 2\epsilon &< \big|\phi_m(u_m)\big| \leq |\phi_m|(u_m) \leq |\phi_m|(u_n) \\ &= \big(|\phi_m| - \phi\big)(u_n) + \phi(u_n) \leq \epsilon + \phi(u_n)\,, \end{aligned}$$

we have $\phi(u_n) > \epsilon$ for all n. But this contradicts $u_n \downarrow 0$ and $\phi \in L_c^\sim$.

(ii) $\Longrightarrow$ (iii) If (iii) is false, there exist sequences $\{u_n\}$ of L with $u_n \downarrow 0$ and $\{\phi_n\} \subseteq A$, and some $\epsilon > 0$ such that $|\phi_n|(u_n) > 2\epsilon$ holds for each n. By Theorem 1.67(3) for each n there exists some w_n in L satisfying $|w_n| \leq u_n$ and $|\phi_n(w_n)| > 2\epsilon$. So, either $|\phi_n(w_n^+)| > \epsilon$ or $|\phi_n(w_n^-)| > \epsilon$. Let v_n equal to w_n^+ or w_n^- so that $|\phi_n(v_n)| > \epsilon$. Now since $v_1 \vee u_n \downarrow v_1$ and $\phi_1 \in L_c^\sim$, there exists some k_1 such that $\big|\phi_1(v_1 \vee u_{k_1})\big| > \epsilon$. Similarly, since $v_{k_1} \vee u_n \downarrow v_{k_1}$ there exists $k_2 > k_1$ such that $\big|\phi_{k_1}(v_{k_1} \vee u_{k_2})\big| > \epsilon$. Proceeding inductively, we obtain a sequence $k_1 < k_2 < k_3 < \cdots$ satisfying $\big|\phi_{k_n}(v_{k_n} \vee u_{k_{n+1}})\big| > \epsilon$ for all n. Put $x_n = v_{k_n} \vee u_{k_{n+1}}$, and note that $x_n \downarrow 0$; this follows from $0 \leq x_{n+1} \leq u_{k_{n+1}} \leq x_n \leq u_n \downarrow 0$. But then $\rho_A(x_n) \geq \big|\phi_{k_n}(x_n)\big| > \epsilon$ holds for all n, contradicting (ii).

(iii) $\Longrightarrow$ (i) Note first that $S = \operatorname{Sol}(A)$ is $|\sigma|(L_c^\sim, L)$-bounded. Indeed, for $u \in L^+$ we have $\frac{1}{n}u \downarrow 0$ since L is Archimedean; so there exists k such that $\rho_S\big(\frac{1}{n}u\big) \leq 1$ for all $n \geq k$, and thus $|\phi|(u) \leq k$ for all $\phi \in S$.

By Theorem 6.19 the set S is relatively $\sigma(L^\sim, L)$-compact and to complete the proof we shall show that the $\sigma(L^\sim, L)$-closure of S in $L^\sim$ is included in $L_c^\sim$. To this end, let ψ be in the $\sigma(L^\sim, L)$-closure of S and fix some $\epsilon > 0$. Now given $u \in L$ pick some $\phi \in S$ such that $\big|(\phi - \psi)(u)\big| < \epsilon$, and note that from

$$\big|\psi(u)\big| \leq \big|(\phi - \psi)(u)\big| + \big|\phi(u)\big| \leq \epsilon + \rho_S(u)\,,$$

it follows that $\big|\psi(u)\big| \leq \rho_S(u)$ for each $u \in L$. Therefore, if $u_n \downarrow 0$ holds in L, then $\lim \psi(u_n) = \lim \rho_S(u_n) = 0$ so that $\psi \in L_c^\sim$. ■

To study the deeper properties of the compact subsets of $L_n^\sim$ and $L_c^\sim$ we need to introduce the concept of order-equicontinuity. This fundamental notion appeared first in the works of H. Nakano under the term "universal-equicontinuity."

DEFINITION 6.22 (Nakano). *Let L be a Riesz space. A nonempty subset A of $L^\sim$ is called* ***order-equicontinuous*** *on L, if $0 \leq u_n \uparrow \leq u$ in L implies that $\{u_n\}$ is a ρ_A-Cauchy sequence.*

It should be clear that a nonempty $\sigma(L^\sim, L)$-bounded subset A of $L^\sim$ is order-equicontinuous if and only if $0 \leq u_\alpha \uparrow \leq u$ in L implies that $\{u_\alpha\}$ is a ρ_A-Cauchy net; see Theorem 3.22.

Three basic properties of the order-equicontinuous sets are included in the next result.

THEOREM 6.23. *For a Riesz space L and an order-equicontinuous subset A of $L^{\sim}$ we have the following.*

(i) *Any nonempty subset of A is order-equicontinuous on L.*
(ii) *The set A is $|\sigma|(L^{\sim}, L)$-bounded.*
(iii) *The convex solid hull of A is order-equicontinuous (and hence the solid hull of A is also order-equicontinuous).*

PROOF. (i) Obvious.

(ii) If A is not $|\sigma|(L^{\sim}, L)$-bounded, then there exist $u \in L^+$ and a sequence $\{\phi_n\} \subseteq A$ such that $|\phi_n|(u) > 2^{n+1}$ for all n. Now for each n there exists, by part (3) of Theorem 1.67, some $0 \leq v_n \leq u$ such that $\big|\phi_n(v_n)\big| > 2^n$. Put $w_n = \sum_{i=1}^{n} 2^{-i} v_i$ and note that for each $n \geq 2$ we have

$$0 \leq w_n \uparrow \leq u \quad \text{and} \quad \rho_A(w_n - w_{n-1}) \geq \big|\phi_n(2^{-n} v_n)\big| > 1\,.$$

This shows that $\{w_n\}$ is not a ρ_A-Cauchy sequence, a contradiction.

(iii) Let S and C be the solid and convex-solid hulls of A, respectively. Since $S \subseteq C$, we have $\rho_S(u) \leq \rho_C(u)$ for all $u \in L$. On the other hand, if $\phi \in C$, then there exist positive constants $\lambda_1, \ldots, \lambda_n$ with $\sum_{i=1}^{n} \lambda_i = 1$ and $\{\phi_1, \ldots, \phi_n\} \subseteq S$ such that $\phi = \sum_{i=1}^{n} \lambda_i \phi_i$ (see Theorem 1.11) and so

$$\big|\phi(u)\big| = \Big|\sum_{i=1}^{n} \lambda_i \phi_i(u)\Big| \leq \sum_{i=1}^{n} \lambda_i \big|\phi_i(u)\big| \leq \sum_{i=1}^{n} \lambda_i \rho_S(u) = \rho_S(u)$$

holds for all $u \in L$, from which it follows $\rho_C(u) \leq \rho_S(u)$. Hence, $\rho_S = \rho_C$, and so it is enough to show that S is order-equicontinuous.

Assume by way of contradiction that S is not an order-equicontinuous set. Then there exist a sequence $\{u_n\} \subseteq L$ and some $\epsilon > 0$ satisfying $0 \leq u_n \uparrow \leq u$ in L and $\rho_S(u_{n+1} - u_n) > 2\epsilon$. For each n choose some $\phi_n \in A$ with $|\phi_n|(u_{n+1} - u_n) > 2\epsilon$, and then use part (3) of Theorem 1.67 to select some vector $v_n \in L$ satisfying $0 \leq v_n \leq u_{n+1} - u_n$ and $\big|\phi_n(v_n)\big| > \epsilon$ for each n. But then $0 \leq w_n = \sum_{i=1}^{n} v_i \uparrow \leq u$ and $\rho_A(w_n - w_{n+1}) \geq \big|\phi_n(v_n)\big| > \epsilon$ holds for all n, contrary to the fact that $\{w_n\}$ must be a ρ_A-Cauchy sequence. ■

The following result should be now obvious.

LEMMA 6.24. *Let L be a Riesz space and let A be a nonempty solid $\sigma(L^{\sim}, L)$-bounded subset of $L^{\sim}$. Then A is order-equicontinuous if and only if the Riesz seminorm ρ_A generates a pre-Lebesgue topology on L.*

The next result, due to O. Burkinshaw [**44**, Theorem 3.2, p. 192] and D. H. Fremlin [**58**, Theorem 81H, p. 223], presents several characterizations of the order-equicontinuity and demonstrates the power of this property. It will later become evident that the concept of order-equicontinuity is the single thread that ties together the various results on weak compactness.

THEOREM 6.25 (Burkinshaw–Fremlin). *Let L be a Riesz space. Then for a nonempty subset A of $L^\sim$ the following statements are equivalent.*

(i) *The set A is order-equicontinuous on L.*

(ii) *For each $u \in L^+$ and each $\epsilon > 0$ there exist some $\delta > 0$ and some finite subset $\{\phi_1, \ldots, \phi_n\}$ of A such that:*

- $v \in [0, u]$ and $|\phi_i|(v) < \delta$ for all i imply $|\phi|(v) < \epsilon$ for all $\phi \in A$.

(iii) *For each $u \in L^+$ and each $\epsilon > 0$ there exists some $0 \leq \psi \in L^\sim_A$, the ideal generated by A in $L^\sim$, such that if $\phi \in A$, then*

$$\big(|\phi| - \psi\big)^+(u) < \epsilon\,.$$

(iv) *For each $u \in L^+$ and each $\epsilon > 0$ there exists some $0 \leq \psi \in L^\sim$ such that for each $\phi \in A$ we have*

$$\big(|\phi| - \psi\big)^+(u) < \epsilon\,.$$

(v) *The set A is $|\sigma|(L^\sim, L)$-bounded and every disjoint sequence in the solid hull of A is $|\sigma|(L^\sim, L)$-convergent to zero.*

(vi) *The set A is $|\sigma|(L^\sim, L)$-bounded and $\lim \rho_A(u_n) = 0$ holds whenever $\{u_n\}$ is an order bounded disjoint sequence of L.*

PROOF. (i) $\Longrightarrow$ (ii) By Theorem 6.23, we can assume that the set A is solid. If (ii) is not valid, then there exist $u \in L^+$, $\epsilon > 0$, and sequences $\{u_n\} \subseteq [0, u]$ and $\{\phi_n\} \subseteq A \cap L^\sim_+$ such that for each n we have:

$$\phi_i(u_n) < \tfrac{1}{2^n} \ \text{ for all } \ i = 1, \ldots, n \quad \text{and} \quad \phi_n(u_n) > 3\epsilon\,.$$

For each fixed $n \in \mathbb{N}$ consider the sequence $\{v_{k,n}\colon k = 1, 2, \ldots\}$ defined by $v_{k,n} = \bigvee_{i=n}^{n+k} u_i$ and note that $0 \leq v_{k,n} \uparrow_k \leq u$. So, by the order-equicontinuity of A there exists some $r_n \in \mathbb{N}$ such that $\phi(v_{k,n} - v_{r_n,n}) < \frac{\epsilon}{2^n}$ for every $0 \leq \phi \in A$ and every $k \geq r_n$. Now for each $n \in \mathbb{N}$ let $w_n = v_{r_n,n}$ and observe that the inequality $(u_m - w_n)^+ \leq (v_{m,n} - w_n)^+$ for $m \geq n$ implies $\phi\big((u_m - w_n)^+\big) \leq \frac{\epsilon}{2^n}$ for every $0 \leq \phi \in A$. For each n put $z_n = \bigwedge_{i=1}^n w_i$. From $(u_n - z_n)^+ \leq \sum_{i=1}^n (u_n - w_i)^+$, we get

$$\phi_n\big((u_n - z_n)^+\big) \leq \sum_{i=1}^n \phi_n\big((u_n - w_i)^+\big) \leq \sum_{i=1}^n \tfrac{\epsilon}{2^i} \leq \epsilon\,.$$

So, $\phi_n(z_n) \geq \phi_n(u_n) - \phi_n\big((u_n - z_n)^+\big) \geq 3\epsilon - \epsilon = 2\epsilon$ for all n. Now from

$$\phi_n(z_{n+1}) \leq \phi_n(w_{n+1}) \leq \sum_{i=n+1}^{\infty} \phi_n(u_i) \leq \sum_{i=n+1}^{\infty} \tfrac{1}{2^i} = \tfrac{1}{2^n}\,,$$

we obtain that $\phi_n(z_n - z_{n+1}) \geq 2\epsilon - \frac{1}{2^n}$ holds for all n, and in particular $\rho_A(z_n - z_{n+1}) \geq \phi_n(z_n - z_{n+1}) > \epsilon$ for all sufficiently large n. But then we have $0 \leq z_1 - z_n \uparrow \leq z_1$ and $\{z_1 - z_n\}$ is not a ρ_A-Cauchy sequence, contrary to the order-equicontinuity of A.

(ii) $\Longrightarrow$ (iii) Let $u \in L^+$ and fix $0 < \epsilon < 1$. Choose $0 < \delta < 1$ and $\{\phi_1, \ldots, \phi_n\} \subseteq A$ that satisfy (ii). Subsequently pick some $0 < \lambda < 1$ satisfying $0 < \lambda\epsilon < \delta$ and $|\phi_i|(\lambda u) < \delta$ for all $i = 1, \ldots, n$. Clearly, $|\phi|(\lambda u) < \epsilon$ for all $\phi \in A$. Put $\psi = \frac{1}{\lambda^2}\big(|\phi_1| + \cdots + |\phi_n|\big)$ and note that $0 \leq \psi \in L^\sim_A$.

Now note that if $\phi \in A$ and $0 \leq v \leq u$ satisfy $\big(|\phi| - \psi\big)(v) \geq 0$, then we have $|\phi|(v) \geq \psi(v) \geq \frac{1}{\lambda^2}|\phi_i|(v)$ and therefore $|\phi_i|(v) \leq \lambda^2|\phi|(v) < \lambda\epsilon < \delta$ for

$i = 1, \ldots, n$. So, $|\phi|(v) < \epsilon$ and hence $\big(|\phi| - \psi\big)(v) \le |\phi|(v) < \epsilon$. By part (1) of Theorem 1.67, we obtain $\big(|\phi| - \psi\big)^+(u) \le \epsilon$, as desired.

(iii) $\Longrightarrow$ (iv) Obvious.

(iv) $\Longrightarrow$ (v) Let $\{\phi_n\}$ be a disjoint sequence in the solid hull of A. For each n pick some $\psi_n \in A$ such that $|\phi_n| \le |\psi_n|$. Now let $u \in L^+$ and fix $\epsilon > 0$. Choose $0 \le \psi \in L^\sim$ such that $\big(|\phi| - \psi\big)^+(u) < \epsilon$ for all $\phi \in A$, and note that $\big(|\phi_n| - \psi\big)^+(u) \le \big(|\psi_n| - \psi\big)^+(u) < \epsilon$ for all n. Since $\big\{|\phi_n| \wedge \psi\big\}$ is a disjoint sequence lying in $[0, \psi]$, we see that $\sum_{k=1}^{n}(|\phi_k| \wedge \psi)(u) \le \psi(u)$ for all n and $u \in L^+$, and so $\lim(|\phi_n| \wedge \psi)(u) = 0$. Pick some k such that $(|\phi_n| \wedge \psi)(u) < \epsilon$ for all $n \ge k$, and note that

$$0 \le |\phi_n|(u) = \big(|\phi_n| - \psi\big)^+(u) + (|\phi_n| \wedge \psi)(u) < 2\epsilon$$

for all $n \ge k$. This shows that $\{\phi_n\}$ is $|\sigma|(L^\sim, L)$-convergent to zero.

The $|\sigma|(L^\sim, L)$-boundedness of the set A follows immediately from the identity $|\phi|(u) = \big(|\phi| - \psi\big)^+(u) + (|\phi| \wedge \psi)(u)$.

(v) $\Longrightarrow$ (vi) If (vi) is false, then there exist a disjoint sequence $\{u_n\}$ in some $[0, u]$, some sequence $\{\phi_n\} \subseteq A$ and some $\epsilon > 0$ satisfying $|\phi_n|(u_n) > \epsilon$ for each n. Consider L embedded in $(L^\sim)_{\mathrm{n}}^\sim$ (see part (1) of Theorem 1.83) and for each n denote by P_n the projection of $L^\sim$ onto the carrier band C_{u_n}; recall that $L^\sim = N_{u_n} \oplus C_{u_n}$, where as usual $N_{u_n} = \big\{\phi \in L^\sim\colon\ |\phi|(u_n) = 0\big\}$ and $C_{u_n} = N_{u_n}^{\mathrm{d}}$. For each n put also $\psi_n = P_n\big(|\phi_n|\big) \ge 0$. Since $u_n \wedge u_m = 0$ for $n \ne m$, it follows from Theorems 1.83(1) and 1.81 that $C_{u_n} \perp C_{u_m}$ for $n \ne m$ and so $\{\psi_n\}$ is a disjoint sequence lying in the solid hull of A. From our hypothesis, it follows that $\psi_n \xrightarrow{|\sigma|(L^\sim, L)} 0$. Now notice that for all n we have

$$\psi_n(u) \ge \psi_n(u_n) = \big[P_n\big(|\phi_n|\big)\big](u_n) = |\phi_n|(u_n) > \epsilon\,,$$

contrary to $\psi_n \xrightarrow{|\sigma|(L^\sim, L)} 0$, and the desired conclusion follows.

(vi) $\Longrightarrow$ (i) Let $S = \mathrm{Sol}\,(A)$, the solid hull of A. Since A is $|\sigma|(L^\sim, L)$-bounded, ρ_S is a Riesz seminorm on L. Now let $\{u_n\}$ be an order bounded disjoint sequence and let $\epsilon > 0$. From

$$\rho_S(u_n) = \sup\big\{|\phi|(|u_n|)\colon\ \phi \in A\big\} = \sup\big\{|\phi(v)|\colon\ \phi \in A \text{ and } |v| \le |u_n|\big\},$$

we see that for each n there exist some $\phi_n \in A$ and an element v_n satisfying $|v_n| \le |u_n|$ and $\rho_S(u_n) - \epsilon \le |\phi_n(v_n)| \le \rho_A(v_n)$. Since $\{v_n\}$ is an order bounded disjoint sequence of L, it follows from our assumption that $\lim \rho_A(v_n) = 0$, and so $\limsup \rho_S(u_n) \le \epsilon$ for all $\epsilon > 0$, so that $\lim \rho_S(u_n) = 0$. Thus, by Theorem 3.22, ρ_S generates a pre-Lebesgue topology. So, S is order-equicontinuous and hence A is likewise order-equicontinuous on L. ■

For an almost Dedekind σ-complete Riesz space L the order-equicontinuous subsets of $L_{\mathrm{c}}^\sim$ are characterized as follows.

THEOREM 6.26. *Let L be an almost Dedekind σ-complete Riesz space. Then for a nonempty subset A of $L_{\mathrm{c}}^\sim$ the following statements are equivalent.*

(i) *The set A is order-equicontinuous on L.*
(ii) *The solid hull S of A is relatively $\sigma(L_{\mathrm{c}}^\sim, L)$-compact.*
(iii) *If $u_n \downarrow 0$ holds in L, then $\lim \rho_A(u_n) = 0$.*
(iv) *If $u_n \downarrow 0$ holds in L, then $\lim \rho_S(u_n) = 0$.*

PROOF. The equivalence of (ii), (iii), and (iv) has already been established in Theorem 6.21.

(i) $\Longrightarrow$ (iii) Let $u_n \downarrow 0$ in L and fix $\epsilon > 0$. For each n choose some $\phi_n \in A$ with $\rho_A(u_n) \leq \bigl|\phi_n(u_n)\bigr| + \epsilon$. Now pick k so that $\rho_A(u_n - u_m) < \epsilon$ for $n, m \geq k$, and then for any $n > k$ choose $m_n > n$ such that $|\phi_n(u_{m_n})| < \epsilon$ (this is possible since $\phi_n \in L_c^\sim$). Now notice that for $n \geq k$ we have

$$\rho_A(u_n) \leq \bigl|\phi_n(u_n)\bigr| + \epsilon \leq \bigl|\phi_n(u_n - u_{m_n})\bigr| + \bigl|\phi_n(u_{m_n})\bigr| + \epsilon \leq 3\epsilon\,.$$

Therefore, $\lim \rho_A(u_n) = 0$.

(iv) $\Longrightarrow$ (i) Assume that $0 \leq u_n \uparrow \leq u$ holds in L. Since L is almost Dedekind σ-complete, there exists a sequence $\{v_n\} \subseteq L^+$ satisfying $v_n \downarrow$ and $u_m \leq v_n$ for all n and m and such that $v_n - u_n \downarrow 0$; see Definition 1.38(3) and related discussion. Now since ρ_S is a Riesz seminorm and $0 \leq u_{n+p} - u_n \leq v_n - u_n$ holds, we get

$$\rho_A(u_{n+p} - u_n) \leq \rho_S(u_{n+p} - u_n) \leq \rho_S(v_n - u_n) \downarrow 0\,,$$

from which it follows that $\{u_n\}$ is a ρ_A-Cauchy sequence. ■

The following example shows that the assumption of almost Dedekind σ-completeness in the preceding theorem cannot be weakened to the Archimedean property. (Compare this example with Theorem 6.21.)

EXAMPLE 6.27. Consider a normed Riesz space L_ρ with the σ-Lebesgue property but without the pre-Lebesgue property—the normed Riesz space of Example 3.25 is such a space. Note that under these circumstances L is not almost Dedekind σ-complete and that $L' \subseteq L_c^\sim$; see Exercise 15 of Chapter 3.

Now if U_ρ is the closed unit ball of L_ρ, then since U_ρ is solid it follows that its polar $A = U_\rho^\circ$ is a solid subset of L' and by Theorem 2.12 the set A is also $\sigma(L', L)$-compact. Thus, A is a solid $\sigma(L_c^\sim, L)$-compact subset of $L_c^\sim$. On the other hand, since $\rho = \rho_A$ holds, the set A cannot be order-equicontinuous on L, since otherwise L_ρ would satisfy the pre-Lebesgue property. Therefore, there exist solid $\sigma(L_c^\sim, L)$-compact sets that are not order-equicontinuous. ■

The next important theorem shows that the concepts of order-equicontinuity and relative $\sigma(L_c^\sim, L)$-compactness in $L_c^\sim$ coincide for Dedekind σ-complete Riesz spaces.

THEOREM 6.28. *Let L be a Dedekind σ-complete Riesz space. Then for a nonempty subset A of $L_c^\sim$ the following statements are equivalent.*

(i) *The set A is relatively $\sigma(L_c^\sim, L)$-compact.*
(ii) *The set A is order-equicontinuous on L.*

PROOF. (i) $\Longrightarrow$ (ii) If A is not order-equicontinuous, then there are $\epsilon > 0$, $0 \leq u_n \uparrow\leq u$ in L, and $\{\phi_n\} \subseteq A$ such that $\bigl|\phi_n(u_{n+1} - u_n)\bigr| > 4\epsilon$ for all n. Put $w_n = u_{n+1} - u_n$, and observe that $w_n \xrightarrow{o} 0$. Indeed, since the Riesz space L is Dedekind σ-complete we have $0 \leq u_n \uparrow v$ in L and so $0 \leq w_n \leq v - u_n \downarrow 0$. Let ϕ be a $\sigma(L_c^\sim, L)$-accumulation point of the sequence $\{\phi_n\}$.

Next, we shall construct by induction a strictly increasing sequence $\{k_n\}$ of natural numbers such that for each $n \geq 2$ we have $\bigl|(\phi_i - \phi)(w_{k_n})\bigr| < 2^{-n}\epsilon$ and $\bigl|(\phi_{k_n} - \phi)(w_i)\bigr| < 2^{-n}\epsilon$ for $i = 1, \ldots, k_{n-1}$. To see this, start with $k_1 = 1$ and assume that k_n has been chosen. By virtue of $w_n \xrightarrow{o} 0$ and $\phi_i - \phi \in L_c^\sim$ there exists

$m > k_n$ with $|(\phi_i - \phi)(w_k)| < 2^{-(n+1)}\epsilon$ for all $i = 1, \ldots, k_n$ and all $k \geq m$. Now since ϕ is a $\sigma(L_c^\sim, L)$-accumulation point of $\{\phi_n\}$, there exists some $k_{n+1} > m > k_n$ so that $|(\phi_{k_{n+1}} - \phi)(w_i)| < 2^{-(n+1)}\epsilon$ for $i = 1, \ldots, k_n$. Thus, the positive integer k_{n+1} can be chosen to satisfy the above properties and the induction is complete.

Now for each n let $\psi_n = \phi_{k_n}$ and $z_n = \sum_{i=1}^n w_{k_i}$. The Dedekind σ-completeness implies $0 \leq z_n \uparrow z$ in L. Now notice that for $m > n \geq 2$ we have

$$\begin{aligned}
|(\psi_n - \phi)(z_{n-1})| &\leq \sum_{i=1}^{n-1} |(\psi_n - \phi)(w_{k_i})| < \sum_{i=1}^{n-1} 2^{-i}\epsilon < \epsilon, \text{ and} \\
|(\psi_n - \phi)(z_m - z_n)| &\leq \sum_{i=n+1}^{m} |(\psi_n - \phi)(w_{k_i})| < \sum_{i=n+1}^{m} 2^{-i}\epsilon < \epsilon.
\end{aligned}$$

Therefore, if $m > n \geq 2$, then

$$\begin{aligned}
|(\psi_n - \phi)(z_m)| &\geq |\psi_n(w_{k_n})| - |\phi(w_{k_n})| - |(\psi_n - \phi)(z_{n-1})| - |(\psi_n - \phi)(z_m - z_n)| \\
&> 4\epsilon - |\phi(w_{k_n})| - \epsilon - \epsilon = 2\epsilon - |\phi(w_{k_n})|.
\end{aligned}$$

Taking into account that $\psi_n - \phi \in L_c^\sim$ and letting $m \to \infty$, it follows from the preceding inequalities that $|(\psi_n - \phi)(z)| \geq 2\epsilon - |\phi(w_{k_n})|$ for all n. Using that $w_{k_n} \xrightarrow{o} 0$ and $\phi \in L_c^\sim$, we infer that there exists some $k \in \mathbb{N}$ such that

$$|(\psi_n - \phi)(z)| > \epsilon \text{ for all } n \geq k. \tag{$\star$}$$

Now let $\psi \in L_c^\sim$ be a $\sigma(L_c^\sim, L)$-accumulation point of the sequence $\{\psi_n\}$. Then since

$$\begin{aligned}
|(\psi - \phi)(w_{k_i})| &\leq |(\psi - \psi_n)(w_{k_i})| + |(\psi_n - \phi)(w_{k_i})| \\
&\leq |(\psi - \psi_n)(w_{k_i})| + 2^{-n}\epsilon
\end{aligned}$$

holds true for all n sufficiently large, we get $\psi(w_{k_i}) = \phi(w_{k_i})$ for all i and so since $\psi, \phi \in L_c^\sim$, we get $\psi(z) = \phi(z)$. However, the last conclusion contradicts ($\star$), and the desired conclusion follows.

(ii) $\Longrightarrow$ (i) It follows immediately from Theorem 6.26. ■

Since a set is order-equicontinuous if and only if its solid-convex hull is order-equicontinuous, we have the following consequence of the preceding result.

COROLLARY 6.29. *Let L be a Dedekind σ-complete Riesz space and let A be a nonempty subset of $L_c^\sim$. Then the set A is relatively $\sigma(L_c^\sim, L)$-compact if and only if its solid-convex hull is relatively $\sigma(L_c^\sim, L)$-compact.*

We now turn our attention to the solid $\sigma(L_n^\sim, L)$-compact subsets of $L_n^\sim$. The next theorem, due to H. Nakano [**103**, § 28], tells us that the solid $\sigma(L_n^\sim, L)$-compact subsets of $L_n^\sim$ are always order-equicontinuous on L and hence they satisfy the equivalent conditions of Theorem 6.25.

THEOREM 6.30 (Nakano). *Let L be an Archimedean Riesz space. Then for a nonempty subset A of $L_n^\sim$ the following statements are equivalent.*

(i) *The set A is order-equicontinuous on L.*
(ii) *The solid hull S of A is relatively $\sigma(L_n^\sim, L)$-compact.*
(iii) *If $u_\alpha \downarrow 0$ holds in L, then $\lim \rho_A(u_\alpha) = 0$.*

PROOF. (i) $\Longrightarrow$ (ii) Note first that since $L_n^\sim$ is the topological dual of the locally convex-solid Riesz space $(L, |\sigma|(L, L_n^\sim))$, it follows from Theorem 6.23 that the solid hull $S = \mathrm{Sol}\,(A)$ of A is $|\sigma|(L_n^\sim, L)$-bounded and consequently $|\sigma|(L^\sim, L)$-bounded.

So, by Theorem 6.19, S is relatively $\sigma(L^\sim, L)$-compact. To finish the proof we shall show that the $\sigma(L^\sim, L)$-closure of S lies in $L_n^\sim$.

To this end, let ϕ be in the $\sigma(L^\sim, L)$-closure of S and let $u_\alpha \downarrow 0$ in L. Since S is also order-equicontinuous, it follows that $\{u_\alpha\}$ is a ρ_S-Cauchy net, and so given $\epsilon > 0$ there exists an index α_0 such that $\rho_S(u_\alpha - u_\beta) < \epsilon$ for $\alpha, \beta \succeq \alpha_0$. But for each fixed $\alpha \succeq \alpha_0$ there exists some $\psi \in S$ such that $\left|(\psi - \phi)(u_\alpha)\right| < \epsilon$, and so if $\beta \succeq \alpha$, then

$$\begin{aligned} \left|\phi(u_\alpha)\right| &\le \left|(\psi-\phi)(u_\alpha)\right| + \left|\psi(u_\alpha)\right| < \epsilon + \left|\psi(u_\alpha - u_\beta)\right| + \left|\psi(u_\beta)\right| \\ &\le \epsilon + \rho_S(u_\alpha - u_\beta) + \left|\psi(u_\beta)\right| \le 2\epsilon + \left|\psi(u_\beta)\right|. \end{aligned}$$

Now from $\lim_{\beta\succeq\alpha} \psi(u_\beta) = 0$, it follows that $\left|\phi(u_\alpha)\right| \le 2\epsilon$ for all $\alpha \succeq \alpha_0$. Therefore, $\phi \in L_n^\sim$, as desired.

(ii) $\Longrightarrow$ (iii) Note that ρ_S is a Riesz seminorm on L. Now let $u_\alpha \downarrow 0$ in L and fix $\epsilon > 0$. For each index α choose some $0 \le \phi_\alpha \in S$ such that $\rho_S(u_\alpha) \le \phi_\alpha(u_\alpha) + \epsilon$. Next pick some $\sigma(L_n^\sim, L)$-accumulation point $0 \le \phi \in L_n^\sim$ of $\{\phi_\alpha\}$, and then select an index α_1 so that $\phi(u_\alpha) < \epsilon$ for all $\alpha \succeq \alpha_1$. Now choose $\alpha_2 \succeq \alpha_1$ such that $\left|(\phi_{\alpha_2} - \phi)(u_{\alpha_1})\right| < \epsilon$, and note that for $\alpha \succeq \alpha_2$ we have

$$\begin{aligned} \rho_A(u_\alpha) &\le \rho_S(u_\alpha) \le \rho_S(u_{\alpha_2}) \le \phi_{\alpha_2}(u_{\alpha_2}) + \epsilon \le \phi_{\alpha_2}(u_{\alpha_1}) + \epsilon \\ &= (\phi_{\alpha_2} - \phi)(u_{\alpha_1}) + \phi(u_{\alpha_1}) + \epsilon < 3\epsilon. \end{aligned}$$

Thus $\lim \rho_A(u_\alpha) = 0$.

(iii) $\Longrightarrow$ (i) Use the arguments in the proof of Theorem 3.23. ■

COROLLARY 6.31 (Nakano). *Let L be a Dedekind σ-complete Riesz space. Then a subset of $L_n^\sim$ is relatively $\sigma(L_n^\sim, L)$-compact if and only if its convex-solid hull is relatively $\sigma(L_n^\sim, L)$-compact.*

By Theorem 1.83 for every locally convex-solid Riesz space (L, τ) there exists a Riesz homomorphism from L into $(L')_n^\sim$. If in addition τ is a Hausdorff topology, then this embedding is a Riesz isomorphism (into) and in this case as usual we consider L as a Riesz subspace of $(L')_n^\sim$. Now since L'' is an ideal of $(L')^\sim$, the ideal I generated by the image of L in $(L')_n^\sim$ is the same as the ideal generated by L in L''. Applying once more Theorem 1.83 to the pair L' and I, we obtain that the embedding of L' into $I_n^\sim$ is a Riesz isomorphism. Moreover, using Theorem 1.40, we see that L' is an order dense ideal of $I_n^\sim$.

The next theorem gives a condition for L' to coincide with $I_n^\sim$.

THEOREM 6.32. *Let (L, τ) be a locally convex-solid Riesz space and let I be the ideal generated by L in L''. If L' is a band of $L^\sim$, then $L' = I_n^\sim$.*

PROOF. As observed above, L' is always an order dense ideal of $I_n^\sim$. Now let $0 \le \phi \in I_n^\sim$. Pick a net $\{\phi_\alpha\} \subseteq L'$ with $0 \le \phi_\alpha \uparrow \phi$ in $I_n^\sim$ and note that since $\{\phi_\alpha\}$ is $|\sigma|(I_n^\sim, I)$-bounded, the net $\{\phi_\alpha\}$ is also $|\sigma|(L', L)$-bounded. Therefore, by Theorem 6.5, we have $\phi_\alpha \uparrow \psi$ in L' for some $0 \le \psi \in L'$. Since L' is order dense in $I_n^\sim$, it follows that $\phi = \psi \in L'$, so that $L' = I_n^\sim$. ■

The solid compact subsets of $L_c^\sim$ were characterized in terms of the σ-Lebesgue property, while those in $L_n^\sim$ were characterized by means of the Lebesgue property. Order-equicontinuity has been the central property relating weak compactness in the spaces $L_c^\sim$ and $L_n^\sim$. The next theorems will show that there is actually a deeper relationship.

THEOREM 6.33 (Burkinshaw [**44**]). *Let (L, τ) be an Archimedean locally convex-solid Riesz space such that L' is a band in $L^{\sim}$ and let I be the ideal generated by L in L''. Then for a subset A of L' the following statements are equivalent.*

(i) *The set A is order-equicontinuous on L.*
(ii) *The set A is relatively $\sigma(L', I)$-compact.*

PROOF. (i) $\Longrightarrow$ (ii) Let $0 \leq v \in I$ and fix $\epsilon > 0$. Pick $u \in L$ with $v \leq u$, and then use Theorem 6.25 to select some $\psi \in L'$ such that $(|\phi| - \psi)^+(u) < \epsilon$ holds for all $\phi \in A$. Hence $(|\phi| - \psi)^+(v) < \epsilon$ for all $\phi \in A$ and thus A is order-equicontinuous on I (by Theorem 6.25 again). Now by Theorem 6.32 we have $L' = I_{\mathrm{n}}^{\sim}$, and therefore A is relatively $\sigma(I_{\mathrm{n}}^{\sim}, I)$-compact; see Theorem 6.30. This implies that A is relatively $\sigma(L', I)$-compact.

(ii) $\Longrightarrow$ (i) Since $L \subseteq I$ and $L' = I_{\mathrm{n}}^{\sim}$, it follows from Theorem 6.30 that A is order-equicontinuous on I and consequently the set A is also order-equicontinuous on L. ■

By observing that the topological dual of $(L, |\sigma|(L, L^{\sim}))$ is $L^{\sim}$, we have the following important special case of Theorem 6.33.

THEOREM 6.34 (Burkinshaw [**44**]). *Let L be an Archimedean Riesz space, and let I be the ideal generated by L in $(L^{\sim})_{\mathrm{n}}^{\sim}$. Then for a nonempty subset A of $L^{\sim}$ the following statements are equivalent.*

(i) *A is order-equicontinuous on L.*
(ii) *A is relatively $\sigma(L^{\sim}, I)$-compact.*

Now consider an arbitrary Hausdorff compact topological space X, and let $L = C(X)$ be the Riesz space of all real-valued continuous functions on X equipped with the sup norm. If $\mathbf{1}$ denotes the constant function one on X, then $\mathbf{1}$ is an order unit of L and, in addition, $\mathbf{1}$ is an order unit in the norm bidual L''. Indeed, since $\|\phi\| = \phi(\mathbf{1})$ holds for all $0 \leq \phi \in L'$, it follows that for each $0 \leq f \in L''$ we have $0 \leq f(\phi) \leq \|f\| \cdot \|\phi\| = \|f\|\phi(\mathbf{1})$ for all $0 \leq \phi \in L'$, so that $0 \leq f \leq \|f\|\mathbf{1}$ holds in L'', and therefore $\mathbf{1}$ is an order unit of L''. In particular, the ideal generated by L in L'' coincides with L'' and thus by virtue of $L' = L^{\sim}$, it follows from Theorem 6.33 that a subset of L' is relatively $\sigma(L', L'')$-compact if and only if it is order-equicontinuous on L. Finally, by noting that a subset of $C(X)$ is order bounded if and only if it is norm bounded, we get the following classical result of A. Grothendieck on weak compactness in the norm dual of $C(X)$; see [**55**, p. 389] or [**123**, Proposition 9.8, p. 126].

THEOREM 6.35 (Grothendieck). *Consider $L = C(X)$ with the sup norm, where X is a Hausdorff compact topological space. Then for a norm bounded subset A of L' the following statements are equivalent.*

(i) *The set A is relatively $\sigma(L', L'')$-compact.*
(ii) *For every positive norm bounded disjoint sequence $\{f_n\}$ of $C(X)$ we have $\lim \phi(f_n) = 0$ uniformly for $\phi \in A$.*

If L is Dedekind σ-complete, then the order-equicontinuous subsets of $L_{\mathrm{c}}^{\sim}$ and $L_{\mathrm{n}}^{\sim}$ are actually relatively compact in a number of topologies. By combining Theorems 6.28 and 6.34 we have the following.

THEOREM 6.36. *Let L be a Dedekind σ-complete Riesz space and let I be the ideal generated by L in $(L^\sim)^\sim_{\rm n}$. For a nonempty subset A of $L^\sim_{\rm c}$ the following statements are equivalent.*

(i) *A is order-equicontinuous on L.*
(ii) *A is relatively $\sigma(L^\sim_{\rm c}, L)$-compact.*
(iii) *A is relatively $\sigma(L^\sim_{\rm c}, I)$-compact.*

Now if we identify $L^\sim_{\rm n}$ and $(L^\delta)^\sim_{\rm n}$ according to Theorem 1.84 and take into consideration Theorems 6.30 and 6.34, we obtain the following result concerning Archimedean Riesz spaces.

THEOREM 6.37. *Let L be an Archimedean Riesz space, and let I be the ideal generated by L in $(L^\sim)^\sim_{\rm n}$. For a nonempty subset A of $L^\sim_{\rm n}$ the following statements are equivalent.*

(i) *The set A is order-equicontinuous on L.*
(ii) *The solid hull S of A is relatively $\sigma(L^\sim_{\rm n}, L)$-compact.*
(iii) *The set A is relatively $\sigma(L^\sim_{\rm n}, L^\delta)$-compact.*
(iv) *The set A is relatively $\sigma(L^\sim_{\rm n}, I)$-compact.*

We close the section by mentioning that weakly compact subsets of $L^\sim_{\rm n}$ have appeared in the works of S. Kaplan and D. H. Fremlin. An extensive study of weakly compact sets in Riesz spaces can be found in [**58**]. The Dedekind σ-completeness assumption can be weakened in many of the results of this section; for such generalizations see the paper by P. G. Dodds [**53**].

6.3. Weak Sequential Convergence

We start this section with a theorem relating weak sequential convergence with the concept of order-equicontinuity. It is a result of basic importance and will be used to characterize weak sequential compactness. To prove this theorem we need the following simple lemma; keep in mind that in an arbitrary Riesz space we write $\sum_{i=1}^\infty u_i = \sup_n \sum_{i=1}^n u_i$ if the supremum exists.

LEMMA 6.38. *Let L be a Dedekind σ-complete Riesz space and assume that a sequence $\{u_n\} \subseteq L^+$ satisfies $\sum_{n=1}^\infty u_n \le u$. Then for every $0 \le \phi \in L^\sim$ and each $\epsilon > 0$ there exists a subsequence $\{v_n\}$ of $\{u_n\}$ (depending on ϕ and ϵ) such that the vector $v = \sum_{n=1}^\infty v_n$ satisfies $\phi(v) < \epsilon$.*

PROOF. Let $0 \le \phi \in L^\sim$ and $\epsilon > 0$. Choose a countable partition $\{\mathbb{N}_n\}$ of $\mathbb{N}$ such that each N_n is infinite and for each n let $w_n = \sum_{i \in \mathbb{N}_n} u_i$. Since $\mathbb{N}_i \cap \mathbb{N}_j = \emptyset$ holds for $i \ne j$, it follows that $0 \le \sum_{i=1}^n w_i \le \sum_{i=1}^\infty u_i \le u$ for all n. Thus,

$$\sum_{i=1}^\infty \phi(w_i) \le \phi(u) < \infty\,,$$

and hence $\phi(w_i) < \epsilon$ for some i. Now write $\mathbb{N}_i = \{k_1, k_2, \ldots\}$ with $k_1 < k_2 < \cdots$ and let $v_n = u_{k_n}$ for each n. ■

The proof of the next result that is due to O. Burkinshaw [**44**] is based on some techniques introduced by P. G. Dodds [**53**].

THEOREM 6.39 (Burkinshaw). *If L is a Dedekind σ-complete Riesz space, then every $\sigma(L^\sim, L)$-convergent sequence of $L^\sim$ is order-equicontinuous on L.*

PROOF. Without loss of generality we can only consider sequences that are $\sigma(L^\sim, L)$-convergent to zero. If the result is false, then there exist some $\epsilon > 0$, a sequence $\{u_n\}$ of L satisfying $0 \le u_n \uparrow u$ in L, and some $\sigma(L^\sim, L)$-convergent to zero sequence $\{\phi_n\} \subseteq L^\sim$ satisfying $\big|\phi_n(u_{n+1} - u_n)\big| > 3\epsilon$ for all n. For each n put $w_n = u_{n+1} - u_n \ge 0$, and note that $\sum_{n=1}^{\infty} w_n = u - u_1 \le u$ and that $\big|\phi_n(w_n)\big| > 3\epsilon$.

Next, we shall use induction to construct subsequences $\{v_n^k\}$ $(k = 0, 1, 2, \dots)$ of $\{w_n\}$ and natural numbers $n_0 < n_1 < n_2 < \cdots$ such that for each $k \ge 1$ we have:

(1) $\{v_n^k\}$ is a subsequence of $\{v_n^{k-1}\}$.
(2) The vector w_{n_k} is a member of the subsequence $\{v_n^{k-1}\}$.
(3) $\sum_{i=0}^{k-1} \big|\phi_{n_k}(w_{n_i})\big| < \epsilon$, and
(4) $|\phi_{n_k}|(v_k) < \epsilon$, where $v_k = \sum_{n=1}^{\infty} v_n^k$.

We start the induction process by letting $v_n^0 = w_n$ for all n and $n_0 = 1$. From $\phi_n(w_{n_0}) \to 0$, it follows that there is some $n_1 > n_0$ such that $|\phi_{n_1}(w_{n_0})| < \epsilon$. By Lemma 6.38 there exists a subsequence $\{v_n^1\}$ of $\{w_n\}$ with $v_1 = \sum_{n=1}^{\infty} v_n^1$ satisfying $|\phi_{n_1}|(v_1) < \epsilon$. Now for the inductive step, assume that $\{v_n^k\}$ and n_k have been selected satisfying properties (1) through (4). Since $\lim \phi_n(w_i) = 0$ for $1 \le i \le n_k$ there is some $n_{k+1} > n_k$ large enough so that $\sum_{i=0}^{k} |\phi_{n_{k+1}}(w_{n_i})| < \epsilon$ and $w_{n_{k+1}}$ is a member of the subsequence $\{v_n^k\}$. Now apply Lemma 6.38 to the sequence $\{v_n^k\}$ to choose a subsequence $\{v_n^{k+1}\}$ such that the element $v_{k+1} = \sum_{n=1}^{\infty} v_n^{k+1}$ satisfies $|\phi_{n_{k+1}}|(v_{k+1}) < \epsilon$, and the inductive construction is complete.

Next let $w = \sum_{i=0}^{\infty} w_{n_i}$. Then

$$0 \le w = \sum_{i=0}^{k} w_{n_i} + \sum_{i=k+1}^{\infty} w_{n_i} \le \sum_{i=0}^{k} w_{n_i} + v_k$$

holds for all $k \ge 1$, and hence

$$|\phi_{n_k}|\Big(w - \sum_{i=0}^{k} w_{n_i}\Big) \le |\phi_{n_k}|(v_k) < \epsilon$$

for all $k \ge 1$. Now notice that for all $k \ge 1$ we have

$$\begin{aligned} \big|\phi_{n_k}(w)\big| &= \Big|\phi_{n_k}\Big(w_{n_k} + \Big[w - \sum_{i=0}^{k} w_{n_i}\Big] + \sum_{i=0}^{k-1} w_{n_i}\Big)\Big| \\ &\ge \big|\phi_{n_k}(w_{n_k})\big| - |\phi_{n_k}|\Big(w - \sum_{i=0}^{k} w_{n_i}\Big) - \sum_{i=0}^{k-1} \big|\phi_{n_k}(w_{n_i})\big| \\ &> 3\epsilon - \epsilon - \epsilon = \epsilon\,, \end{aligned}$$

contrary to $\lim_k \phi_{n_k}(w) = 0$. The proof is finished. ■

As an application of Theorem 6.39 we have the following result of H. H. Schaefer [**123**, Theorem 10.5, p. 133].

THEOREM 6.40 (Schaefer). *Let (L, τ) be a Dedekind σ-complete locally convex-solid Riesz space and let I be the ideal generated by L in L''. Then every $\sigma(L', L)$-convergent to zero sequence of L' is also $\sigma(L', I)$-convergent to zero.*

PROOF. Assume that $\{\phi_n\} \subseteq L'$ satisfies $\phi_n \xrightarrow{\sigma(L',L)} 0$. By Theorem 6.39 the sequence $\{\phi_n\}$ is order-equicontinuous on L. Now notice that I also coincides with the ideal generated by L in $(L^\sim)^\sim_{\rm n}$. So, by Theorem 6.34 the sequence $\{\phi_n\}$ is relatively $\sigma(L^\sim, I)$-compact and hence $\phi_n \xrightarrow{\sigma(L',I)} 0$. ■

COROLLARY 6.41 (Schaefer). *If (L, τ) is a Dedekind σ-complete locally convex-solid Riesz space, then every band projection on L' is sequentially $\sigma(L', L)$-continuous. That is, if P is the projection of L' onto a band B and a sequence $\{\phi_n\}$ in L' satisfies $\phi_n \xrightarrow{\sigma(L',L)} 0$, then $P(\phi_n) \xrightarrow{\sigma(L',L)} 0$.*

PROOF. Assume $\phi_n \xrightarrow{\sigma(L',L)} 0$. Let I denote the ideal generated by L in L''. Since $L' = B \oplus B^{\rm d}$ by applying Theorem 6.2 to $(L', |\sigma|(L', I))$, we obtain that $I = B^{\rm o} \oplus (B^{\rm d})^{\rm o}$. Now for $u \in L$ write $u = F + G$ with $F \in B^{\rm o}$ and $G \in (B^{\rm d})^{\rm o}$, and note that

$$\begin{aligned} P\phi_n(u) &= F(P\phi_n) + G(P\phi_n) = G(P\phi_n) \\ &= G(\phi_n) - G(\phi_n - P\phi_n) = G(\phi_n) . \end{aligned}$$

By Theorem 6.40 we have $\lim G(\phi_n) = 0$ and thus $\lim P\phi_n(u) = 0$ for each $u \in L$. Hence $P\phi_n \xrightarrow{\sigma(L',L)} 0$. ■

Using the preceding result we get the following version of Theorem 6.20.

THEOREM 6.42. *If L is a Dedekind σ-complete Riesz space, then every band of $L^\sim$ is sequentially $\sigma(L^\sim, L)$-complete. In particular, $L^\sim_{\rm c}$ and $L^\sim_{\rm n}$ are both sequentially $\sigma(L^\sim, L)$-complete.*

The next result, due to O. Burkinshaw [**44**, Proposition 4.6, p. 199], shows the close relationship between order-equicontinuity and sequential compactness.

THEOREM 6.43 (Burkinshaw). *If L is Dedekind σ-complete, then for a non-empty subset A of $L^\sim$ the following statements are equivalent.*

(i) *The set A is order-equicontinuous on L.*
(ii) *For each element $u \in L^+$ and each sequence $\{\phi_n\}$ of A there exist some $\phi \in L^\sim$ and a subsequence $\{\phi_{k_n}\}$ of $\{\phi_n\}$ satisfying $\lim \phi_{k_n}(v) = \phi(v)$ for all $v \in [0, u]$.*

PROOF. (i) $\Longrightarrow$ (ii) Let I be the ideal generated by L in $(L^\sim)^\sim_{\rm n}$. For $u \in L^+$ let I_u be the ideal generated by u in I. Then we have $L^\sim = N_u \oplus C_u$, where $N_u = \{\phi \in L^\sim\colon\ |\phi|(u) = 0\}$ and $C_u = N_u^{\rm d}$. Now let $\{\phi_n\} \subseteq A$. For each n write $\phi_n = f_n + g_n$ with $f_n \in N_u$ and $g_n \in C_u$ and note that $\phi_n(v) = g_n(v)$ for all $v \in [0, u]$.

Since the solid hull of A is order-equicontinuous on L, it follows that $\{g_n\}$ is also order-equicontinuous on L, and thus by Theorem 6.34 the sequence $\{g_n\}$ is relatively $\sigma(L^\sim, I)$-compact. Now observe that C_u is $|\sigma|(L^\sim, I)$-closed and consequently by part (ii) of Theorem 2.4 is also $\sigma(L^\sim, I)$-closed (since the topological dual of $(L^\sim, |\sigma|(L^\sim, I))$ is exactly I).

It now follows that the set $\{g_1, g_2, \ldots\}$ is relatively $\sigma(C_u, I)$-compact and hence it is also a relatively $\sigma(C_u, I_u)$-compact set. Let $G = \overline{\{g_n\}}$, the $\sigma(C_u, I)$-closure of the set $\{g_1, g_2, \ldots\}$, and note that G is a $\sigma(C_u, I_u)$-compact subset of C_u.

Next, equip C_u with the Riesz norm $\rho_u(\phi) = |\phi|(u)$ $(\phi \in C_u)$ and notice that (according to Theorem 2.33) the norm dual of (C_u, ρ_u) is precisely I_u. This means that I_u also can be identified with the norm dual of the norm completion

$(\widehat{C}_u, \widehat{\rho}_u)$ (which is a Banach lattice) of (C_u, ρ_u). In particular, G is also a weakly compact subset of the Banach lattice $\widehat{C}_u$, and so according to the Eberlein–Šmulian Theorem 2.15 there exists a subsequence $\{g_{k_n}\}$ of $\{g_n\}$ and some $\phi \in G \subseteq C_u$ such that $g_{k_n} \xrightarrow{\sigma(C_u, I_u)} \phi$. This implies that for each vector $v \in [0, u]$ we have $\lim \phi_{k_n}(v) = \lim g_{k_n}(v) = \phi(v)$.

(ii) $\Longrightarrow$ (i) If A is not order-equicontinuous on L, then there exist some $\epsilon > 0$, some $\phi \in L^\sim$, a sequence $\{u_n\}$ in L with $0 \leq u_n \uparrow \leq u$, and a sequence $\{\phi_n\} \subseteq A$ satisfying $\big|\phi_n(u_{n+1} - u_n)\big| > \epsilon$ for each n and $\lim \phi_n(v) = \phi(v)$ for all $v \in [0, u]$. Let L_u be the principal ideal generated by u in L. For each n let ψ_n denote the restriction of ϕ_n to L_u. Then $\{\psi_n\} \subseteq L_u^\sim$ and $\{\psi_n\}$ is $\sigma(L_u^\sim, L_u)$-convergent to ϕ. Thus, by Theorem 6.39, the sequence $\{\psi_n\}$ is order-equicontinuous on L_u, which is in contradiction to

$$\big|\psi_n(u_{n+1} - u_n)\big| = \big|\phi_n(u_{n+1} - u_n)\big| > \epsilon$$

for all n. The proof of the theorem is finished. ■

We continue with a useful variation of Theorem 4.18. Recall that for each $\phi \in L^\sim$ the Riesz seminorm ρ_ϕ on L is defined by $\rho_\phi(u) = |\phi|\big(|u|\big)$ for all $u \in L$.

LEMMA 6.44. *Let (L, τ) be a Hausdorff locally convex-solid Riesz space with the Lebesgue property. If L has the countable sup property, then for every sequence $\{u_n\}$ of L there exists a sequence $\{\phi_n\}$ of L' such that $\{\rho_{\phi_n}\}$ defines a metrizable Lebesgue topology on the band generated by $\{u_n\}$ in L.*

PROOF. Let B be the band generated by a sequence $\{u_n\}$ of L. Since $|\sigma|(L, L')$ is a Lebesgue topology, it follows from Theorem 4.26 that the carrier $C_{|\sigma|}$ equals L. Thus, as in the proof of Theorem 4.17(a), there exists a normal sequence $\{V_n\}$ of $|\sigma|(L, L')$-neighborhoods of zero such that $\{u_n\} \subseteq N^{\mathrm{d}}$, where $N = \bigcap_{n=1}^\infty V_n$. For each n choose $\phi_n \in L'$ such that the open unit ball of ρ_{ϕ_n} is contained in V_n. It now easily follows from $B \subseteq N^{\mathrm{d}}$ that the sequence $\{\rho_{\phi_n}\}$ defines a metrizable Lebesgue topology on B. ■

We close the section with a characterization of the weak sequential completeness. It is due to O. Burkinshaw and P. G. Dodds [**46**, Corollary 4.2] and P. Meyer-Nieberg [**97**, Theorem 1.7, p. 308].

THEOREM 6.45 (Burkinshaw–Dodds–Meyer-Nieberg). *Let (L, τ) be a Hausdorff locally convex-solid Riesz space with the Lebesgue property. If L has the countable sup property, then the following statements are equivalent.*

(i) *L is sequentially $\sigma(L, L')$-complete.*
(ii) *Every monotone $\sigma(L, L')$-Cauchy sequence of L^+ is necessarily $\sigma(L, L')$-convergent.*
(iii) *(L, τ) satisfies the σ-Levi property, that is, whenever $0 \leq u_n \uparrow$ holds in L and $\{u_n\}$ is τ-bounded, it follows that $\sup\{u_n\}$ exists in L.*

PROOF. The implications (i) $\Longrightarrow$ (ii) $\Longrightarrow$ (iii) are obvious.

(iii) $\Longrightarrow$ (i) Let $\{u_n\}$ be a $\sigma(L, L')$-Cauchy sequence of L. By Theorem 3.12 we know that L is order dense in $(L')_{\mathrm{n}}^\sim$. It follows that $\{u_n\}$ is a $\sigma\big((L')_{\mathrm{n}}^\sim, L'\big)$-Cauchy sequence and thus by Theorem 6.42 there exists some $f \in (L')_{\mathrm{n}}^\sim$ such that $\{u_n\}$ is $\sigma\big((L')_{\mathrm{n}}^\sim, L'\big)$-convergent to f. To complete the proof we shall show that f belongs to L.

To this end, let A denote the ideal generated by $\{u_n\}$ in L and let B denote the band generated by A in $(L')^\sim_{\rm n}$. By Theorem 3.12 we know that the band B is $\sigma\big((L')^\sim_{\rm n}, L'\big)$-closed [recall that $(L')^\sim_{\rm n}$ is the topological dual of $\big(L', \sigma\big(L', (L')^\sim_{\rm n}\big)\big)$] and thus $f \in B$. Pick a net $\{w_\alpha\} \subseteq L$ with $0 \le w_\alpha \uparrow f^+$ in $(L')^\sim_{\rm n}$. Now by Lemma 6.44 there exists a sequence $\{\phi_n\} \subseteq L'$ such that $\{\rho_{\phi_n}\}$ generates a metrizable topology on A. But $\{\rho_{\phi_n}\}$ generates (always) a Lebesgue topology on B [where $\rho_{\phi_n}(g) = |g|\big(|\phi_n|\big)$ for all $g \in (L')^\sim_{\rm n}$] that is also metrizable due to the order denseness of L in $(L')^\sim_{\rm n}$, and so there exists a sequence $\{w_{\alpha_n}\} \subseteq \{w_\alpha\}$ such that $0 \le w_{\alpha_n} \uparrow f^+$ in $(L')^\sim_{\rm n}$. But then since clearly $\{w_{\alpha_n}\}$ is $\sigma(L, L')$-bounded, $\{w_{\alpha_n}\}$ is τ-bounded and hence by our hypothesis $w_{\alpha_n} \uparrow w$ holds in L. Taking into consideration that L is order dense in $(L')^\sim_{\rm n}$ once more, we get $f^+ = w \in L$. Similarly, $f^- \in L$. Therefore, $f = f^+ - f^- \in L$, and the proof is finished. ■

6.4. Compact Solid Sets

The weakly compact (respectively, τ-compact) solid subsets of a locally convex-solid Riesz space (L, τ) will be characterized in terms of disjoint sequences in this section. We start with a basic result concerning weakly compact intervals.

THEOREM 6.46. *Let (L, τ) be a Hausdorff locally convex-solid Riesz space. Then for a vector $u \in L^+$ the following statements are equivalent.*

(i) *The order interval $[0, u]$ is $\sigma(L, L')$-compact.*
(ii) *The order interval $[-u, u]$ is a solid subset of $(L')^\sim_{\rm n}$.*
(iii) *The order interval $[-u, u]$ is a solid subset of L''.*
(iv) *The principal ideal L_u generated by u in the Riesz space L is Dedekind complete and τ-induces a Lebesgue topology on L_u.*
(v) *Every disjoint sequence of the order interval $[0, u]$ is τ-convergent to zero and $[0, u]$ is τ-complete.*
(vi) *Every disjoint sequence of the order interval $[0, u]$ is τ-convergent to zero and every increasing τ-Cauchy net of $[0, u]$ is τ-convergent.*

PROOF. (i) $\Longrightarrow$ (ii) It follows from Theorem 6.9 that the order interval $[0, u]$ of L is $|\sigma|\big((L')^\sim_{\rm n}, L'\big)$-dense in the order interval

$$\langle 0, u\rangle = \big\{v \in (L')^\sim_{\rm n}:\ 0 \le v \le u\big\}.$$

Indeed, if $v \in \langle 0, u\rangle$ and $\{v_\alpha\} \subseteq [0, u]$ is $|\sigma|\big((L')^\sim_{\rm n}, L'\big)$-convergent to v, then the net $\{|v_\alpha| \wedge u\}$ of $[0, u]$ is $|\sigma|\big((L')^\sim_{\rm n}, L'\big)$-convergent to v. Now, since $[0, u]$ is $\sigma(L, L')$-compact (and hence $\sigma\big((L')^\sim_{\rm n}, L'\big)$-compact), it follows that $[0, u] = \langle 0, u\rangle$. This implies that the order interval $[-u, u]$ is a solid subset of $(L')^\sim_{\rm n}$.

(ii) $\Longrightarrow$ (iii) This follows from the fact that L'' is an ideal of $(L')^\sim$.

(iii) $\Longrightarrow$ (iv) Clearly, the principal ideal L_u generated by u in L coincides with the principal ideal generated by u in L'' and hence L_u is Dedekind complete. Now if $\{u_\alpha\} \subseteq L_u$ satisfies $u_\alpha \downarrow 0$ in L, then it also satisfies $u_\alpha \downarrow 0$ in L'' and consequently $\lim \phi(u_\alpha) = 0$ for each $\phi \in L'$. By Theorem 3.21 we have $u_\alpha \xrightarrow{\tau} 0$, and so τ induces a Lebesgue topology on L_u.

(iv) $\Longrightarrow$ (v) Since L_u is Dedekind complete and τ induces a Lebesgue topology on L_u, it follows from Theorem 4.28 that each order interval of L_u is τ-complete.

Since the topology induced by τ on L_u is a Lebesgue topology (and hence a pre-Lebesgue topology by Theorem 3.23), we easily infer that each disjoint sequence of $[0,u]$ is τ-convergent to zero.

(v) $\Longrightarrow$ (vi) Obvious.

(vi) $\Longrightarrow$ (i) By Theorem 3.22 the topology induced by τ on L_u is a pre-Lebesgue topology. Therefore, each increasing net of $[0,u]$ is τ-Cauchy and thus by hypothesis is τ-convergent. Now, it follows from Theorem 2.21 that L_u is Dedekind complete and that τ induces a Lebesgue topology on L_u.

Furthermore, since $|\sigma|(L,L') \subseteq \tau$, it follows that $|\sigma|(L,L')$ also induces a Lebesgue topology on L_u and consequently by Theorem 4.28 the order interval $[0,u]$ is $|\sigma|(L,L')$-complete. But, by Theorem 6.9, the order interval $[0,u]$ is $|\sigma|\big((L')_n^\sim, L'\big)$-dense in $\langle 0,u\rangle$ and thus $[0,u] = \langle 0,u\rangle$.

Finally, by applying Theorem 6.1 to L' equipped with the topology $|\sigma|\big(L', (L')_n^\sim\big)$, we obtain that each order interval of $(L')_n^\sim$ is $\sigma\big((L')_n^\sim, L'\big)$-compact, and so $[0,u]$ is itself $\sigma\big((L')_n^\sim, L'\big)$-compact. Hence $[0,u]$ is $\sigma(L,L')$-compact, and the proof is finished. ■

The previous characterizations of the weakly compact order intervals give an interesting insight into the convex and solid weakly compact subsets of locally convex-solid Riesz spaces.

THEOREM 6.47. *Let (L,τ) be a Hausdorff locally convex-solid Riesz space. Then for a convex and solid subset A of L the following statements are equivalent.*

(i) *The set A is $\sigma(L,L')$-compact.*
(ii) *Every disjoint sequence of A is $\sigma(L,L')$-convergent to zero and each increasing net in A^+ is τ-convergent to an element in A.*

PROOF. (i) $\Longrightarrow$ (ii) Note that the set A considered as a subset of $(L')_n^\sim$ is $\sigma\big((L')_n^\sim, L'\big)$-compact. Hence, by Theorems 6.30 and 6.25, each disjoint sequence of A is $\sigma\big((L')_n^\sim, L'\big)$-convergent to zero and thus it is also $\sigma(L,L')$-convergent to zero. Finally, A is $\sigma\big((L')_n^\sim, L'\big)$-bounded and therefore if $\{u_\alpha\} \subseteq A$ satisfies $0 \le u_\alpha \uparrow$ in L, then by Theorem 6.16 the supremum $u = \sup\{u_\alpha\}$ exists in $(L')_n^\sim$. Also $\phi(u_\alpha) \uparrow u(\phi)$ for each $0 \le \phi \in L'$, and so $u \in A$ since A is $\sigma\big((L')_n^\sim, L'\big)$-closed. Now Theorem 3.21 implies $u_\alpha \xrightarrow{\tau} u$.

(ii) $\Longrightarrow$ (i) Observe first that the set A is $|\sigma|(L,L')$-bounded, for suppose that $\sup\{\phi(u)\colon\ u \in A\} = \infty$ for some $0 \le \phi \in L'$. Choose $0 \le u_n \in A$ such that $\phi(u_n) > n2^n$ for each n. But then since A is solid and convex, we have $\sum_{i=1}^n 2^{-i}u_i \in A$ for each n and hence $\sum_{i=1}^n 2^{-i}u_i \uparrow u$ for some $u \in A$. This implies $\phi(u) \ge \phi(2^{-n}u_n) > n$ for all n, a contradiction.

Now consider A as a subset of $(L')_n^\sim$. Then, according to condition (v) of Theorem 6.25, the set A is order-equicontinuous on L' and thus (by Theorem 6.30) the set A is relatively $\sigma\big((L')_n^\sim, L'\big)$-compact. To complete the proof we shall show that A is $\sigma\big((L')_n^\sim, L'\big)$-closed.

Since for every $u \in A$ the order interval $\big[0,|u|\big]$ satisfies statement (vi) of Theorem 6.46, it follows that A is a solid subset of $(L')_n^\sim$. Also it follows from our hypothesis that A is an order closed subset of $(L')_n^\sim$, and hence since the topology $|\sigma|\big((L')_n^\sim, L'\big)$ is a Hausdorff Fatou topology, A is $|\sigma|\big((L')_n^\sim, L'\big)$-closed by Theorem 4.20. Now by Theorem 2.33 the topological dual of $(L')_n^\sim$ with the topology $|\sigma|\big((L')^\sim_{\rm n}, L'\big)$ is L', and since A is convex, it follows from Theorem 2.4(ii) that the set A is necessarily $\sigma\big((L')^\sim_{\rm n}, L'\big)$-closed, and the proof is finished. ■

LEMMA 6.48. *Let (L,τ) be a Hausdorff locally convex-solid Riesz space and let B be the band generated by a sequence $\{\phi_n\}$ in L'. If A is a relatively $\sigma(L,L')$-compact subset of L, then for every sequence $\{u_n\}\subseteq A$ there exists a subsequence $\{w_n\}$ of $\{u_n\}$ and some $u\in L$ such that $w_n \xrightarrow{\sigma(L,B)} u$.*

PROOF. Let A be a relatively $\sigma(L,L')$-compact subset of L and consider A as a subset in $(L')^\sim_{\rm n}$. Then the set A is necessarily relatively $\sigma\big((L')^\sim_{\rm n}, L'\big)$-compact and so, by Theorem 6.30 and Corollary 6.31, the set A is order-equicontinuous on L' (since L' is Dedekind complete). Now if $\{u_n\}\subseteq A$, then by Theorem 6.43 there exists a subsequence $\{v_n^1\}$ of $\{u_n\}$ such that the limit $\lim \phi(v_n^1)$ exists for all $\phi\in\big[0,|\phi_1|\big]$. Next if the sequence $\{v_n^i\}$ has been chosen, then use Theorem 6.43 again to obtain a subsequence $\{v_n^{i+1}\}$ of $\{v_n^i\}$ such that $\lim_n \phi(v_n^{i+1})$ exists for all $\phi\in\big[0,|\phi_{i+1}|\big]$. Thus, by induction, we can select the sequences $\{v_n^i\}$ $(i=1,2,\dots)$ with the above properties. Put $w_n=v_n^n$ (the "diagonal subsequence"), and note that $\{w_n\}$ is indeed a subsequence of $\{u_n\}$ such that $\lim\phi(w_n)$ exists in $\mathbb{R}$ for all ϕ in the ideal I generated by $\{\phi_n\}$ in L'.

Now let u be a $\sigma(L,L')$-accumulation point of the sequence $\{w_n\}$. Clearly, $\lim\phi(w_n)=\phi(u)$ for all $\phi\in I$. Also since I is order dense in the band B generated by $\{\phi_n\}$ in L', for any $0\le\psi\in B$ there exists a net $\{\psi_\alpha\}\subseteq I$ with $0\le\psi_\alpha\uparrow\psi$ in L'. Now since $\{w_n\}$ is order-equicontinuous on L', it follows from Theorem 6.30 [applied to L' and $(L')^\sim_{\rm n}$] that the net $\{\psi_\alpha\}$ converges uniformly on $\{w_n\}$, and therefore $\lim\psi(w_n)=\psi(u)$, and the proof is finished. ■

Recall that a subset A in a topological vector space (E,τ) is called **relatively sequentially τ-compact** whenever each sequence of A has a subsequence that is τ-convergent to some element of E.

THEOREM 6.49. *Let (L,τ) be a Hausdorff locally convex-solid Riesz space. If L has the countable sup property, then each solid relatively $\sigma(L,L')$-compact subset of L is relatively sequentially $\sigma(L,L')$-compact.*

PROOF. Let S be a solid relatively $\sigma(L,L')$-compact set and let $\{u_n\}\subseteq S$. Since each order interval of S is $\sigma(L,L')$-compact, it follows from Theorem 6.46 that S is a solid subset of $(L')^\sim_{\rm n}$. Hence $|\sigma|(L,L')$ induces a Lebesgue topology on the ideal A generated by $\{u_n\}$ in L. Also, by Lemma 6.44, there exists a positive sequence $\{\phi_n\}\subseteq L'$ such that $\{\rho_{\phi_n}\}$ defines a metrizable Lebesgue topology on A.

Let B be the band generated by $\{\phi_n\}$ in L'. Then $L'=B\oplus B^{\rm d}$ and by Lemma 6.48 there exists a subsequence $\{w_n\}$ of $\{u_n\}$ such that $\lim\phi(w_n)$ exists in $\mathbb{R}$ for each $\phi\in B$. Also, according to Theorem 6.2, we have $(L')^\sim_{\rm n}=B^{\rm o}\oplus(B^{\rm d})^{\rm o}$, and we claim that $\{w_n\}\subseteq(B^{\rm d})^{\rm o}$. To see this, fix $n\in\mathbb{N}$ and write $|w_n|=v_1+v_2$ with $v_1\in B^{\rm o}$ and $v_2\in(B^{\rm d})^{\rm o}$. Since S is a solid subset of $(L')^\sim_{\rm n}$, it follows that $v_1\in L$, and since $v_1\in B^{\rm o}$, we have $\rho_{\phi_m}(v_1)=\phi_m(v_1)=0$ for all $m\in\mathbb{N}$. Consequently, $v_1=0$ and so $\{w_n\}\subseteq(B^{\rm d})^{\rm o}$. Now assume that $u\in L$ is an arbitrary $\sigma(L,L')$-accumulation point of $\{w_n\}$. Then $\lim\phi(w_n)=\phi(u)$ holds for each $\phi\in B$ and also $\psi(u)=\psi(w_n)=0$ for all $\psi\in B^{\rm d}$. Therefore, $\{w_n\}$ is $\sigma(L,L')$-convergent to u, and the proof is complete. ■

To continue our discussion we shall need a result from the theory of locally convex topological spaces. For a proof of the following result see [**78**, Theorem 24.7, p. 313].

LEMMA 6.50. *Let (E, τ) be a metrizable locally convex topological vector space and let C be a relatively $\sigma(E, E')$-compact subset of E. If u belongs to the $\sigma(E, E')$-closure of C, then there exists a sequence $\{u_n\} \subseteq C$ such that $u_n \xrightarrow{\sigma(E,E')} u$.*

The next theorem presents a condition under which the notions of weak compactness and weak sequential compactness coincide.

THEOREM 6.51. *Let (L, τ) be a Hausdorff locally convex-solid Riesz space and suppose that L' has a countable complete disjoint system. Then for a subset A of L the following statements are equivalent.*

(i) *The set A is relatively $\sigma(L, L')$-compact.*
(ii) *The set A is relatively sequentially $\sigma(L, L')$-compact.*

PROOF. Fix a complete disjoint sequence $\{\phi_n\} \subseteq L'$ and let I be the ideal generated in L' by $\{\phi_n\}$.

(i) $\Longrightarrow$ (ii) Note that the band generated by $\{\phi_n\}$ in L' is all of L'. Now if $\{u_n\} \subseteq A$, then by Lemma 6.48 there exists a subsequence $\{w_n\}$ of $\{u_n\}$ that is $\sigma(L, L')$-convergent to some $u \in L$.

(ii) $\Longrightarrow$ (i) Consider $A \subseteq (L')^{\sim}_{\mathrm{n}}$. Since each sequence of A contains a $\sigma(L, L')$-convergent subsequence, it follows from Theorem 6.43 that A is order-equicontinuous on L'. Thus, by Theorem 6.30, the set A is relatively $\sigma\big((L')^{\sim}_{\mathrm{n}}, L'\big)$-compact. Note that since $\sigma\big((L')^{\sim}_{\mathrm{n}}, I\big)$ is a coarser topology than $\sigma\big((L')^{\sim}_{\mathrm{n}}, L'\big)$, it follows that the set A is relatively $\sigma\big((L')^{\sim}_{\mathrm{n}}, I\big)$-compact. But, by Theorem 5.6, the Riesz space $(L')^{\sim}_{\mathrm{n}}$ is metrizable under the topology $|\sigma|\big((L')^{\sim}_{\mathrm{n}}, I\big)$. Now by applying Theorem 2.33 to $(L')^{\sim}_{\mathrm{n}}$ with $|\sigma|\big((L')^{\sim}_{\mathrm{n}}, I\big)$, we see that the topological dual of $(L')^{\sim}_{\mathrm{n}}$ with the topology $|\sigma|\big((L')^{\sim}_{\mathrm{n}}, I\big)$ is exactly I. To complete the proof, we shall show that the $\sigma\big((L')^{\sim}_{\mathrm{n}}, I\big)$-closure of A is contained in L.

To this end, let $v \in (L')^{\sim}_{\mathrm{n}}$ be in the $\sigma\big((L')^{\sim}_{\mathrm{n}}, I\big)$-closure of A. By Lemma 6.50 [applied to the metrizable topology $|\sigma|\big((L')^{\sim}_{\mathrm{n}}, I\big)$] there exists a sequence $\{u_n\} \subseteq A$ such that $\lim \phi(u_n) = \phi(v)$ for all $\phi \in I$. But since A is relatively sequentially $\sigma(L, L')$-compact, there exists some vector $u \in L$ and a subsequence $\{v_n\}$ of the sequence $\{u_n\}$ such that $\lim \psi(v_n) = \psi(u)$ for all $\psi \in L'$. Clearly, $\phi(u) = \phi(v)$ for all $\phi \in I$ and since I is order dense in L', it follows that $\phi(u) = \phi(v)$ for all $\phi \in L'$, i.e., $v = u \in L$, as desired. ■

In order to investigate additional properties of solid compact sets we need the following extension of Theorem 6.25 that is due to O. Burkinshaw and P. G. Dodds [**45**, Proposition 2.2].

THEOREM 6.52 (Burkinshaw–Dodds). *Let (L, τ) be a Hausdorff locally convex-solid Riesz space with the monotone completeness property. Let A be a nonempty solid $\sigma(L, L')$-bounded subset of L and B a nonempty solid $\sigma(L', L)$-bounded subset of L' such that*

$$\sup\big\{|\phi|(|u|)\colon\ \phi \in B \text{ and } u \in A\big\} < \infty\,.$$

Then the following statements are equivalent.

(1) *For each disjoint sequence $\{\phi_n\} \subseteq B$ we have $\lim \rho_A(\phi_n) = 0$.*
(2) *For each disjoint sequence $\{u_n\} \subseteq A$ we have $\lim \rho_B(u_n) = 0$.*

(3) *For each* $\epsilon > 0$ *there exist some* $\psi \geq 0$ *in the ideal generated by* B *in* L' *and some* $v \geq 0$ *in the ideal generated by* A *in* L *such that:*
   (a) $\rho_A\big((|\phi| - \psi)^+\big) < \epsilon$ *for all* $\phi \in B$.
   (b) $\rho_B\big((|u| - v)^+\big) < \epsilon$ *for all* $u \in A$.

(4) *For each* $\epsilon > 0$ *there exist* $0 \leq \psi \in L'$ *and* $v \in L^+$ *such that:*
   (a) $\rho_A\big((|\phi| - \psi)^+\big) < \epsilon$ *for all* $\phi \in B$.
   (b) $\rho_B\big((|u| - v)^+\big) < \epsilon$ *for all* $u \in A$.

PROOF. (1) $\Longrightarrow$ (2) If (2) were not true, then there exist $\epsilon > 0$, a positive sequence $\{\phi_n\} \subseteq B$, and a positive disjoint sequence $\{u_n\} \subseteq A$ such that $\phi_n(u_n) > \epsilon$ for all n. Denote by P_n the projection of L' onto the carrier band C_{u_n} (recall that $L' = N_{u_n} \oplus C_{u_n}$, where as usual $N_{u_n} = \big\{\phi \in L'\colon\ |\phi|(u_n) = 0\big\}$ and $C_{u_n} = N_{u_n}^{\mathrm{d}}$). Put $\psi_n = P_n(\phi_n)$ for each n and observe that $\{\psi_n\} \subseteq B$, since the set B is solid. But then, according to Theorem 1.82, the sequence $\{\psi_n\}$ is disjoint and satisfies $\rho_A(\psi_n) \geq P_n(\phi_n)(u_n) = \phi_n(u_n) > \epsilon$ for all n, contrary to our hypothesis.

(2) $\Longrightarrow$ (3) Put $\delta = \sup\big\{|\phi|(|u|)\colon\ \phi \in B \text{ and } u \in A\big\}$. Assume by way of contradiction that there exists some $\epsilon > 0$ such that for every $v \geq 0$ in the ideal generated by A there exists some $u \in A$ (depending on v) satisfying $\rho_B\big((|u| - v)^+\big) > 3\epsilon$. It follows that there exists a positive sequence $\{u_n\} \subseteq A$ such that

$$\rho_B\Big(\Big[u_{n+1} - 4^n \sum_{i=1}^{n} u_i\Big]^+\Big) > 3\epsilon$$

for all n. Since τ is a Hausdorff locally convex topology and $\{u_n\} \subseteq A$ is τ-bounded, it follows that $\sum_{i=1}^{n} 2^{-i}u_i \uparrow$ and that $\big\{\sum_{i=1}^{n} 2^{-i}u_i\big\}$ is a τ-Cauchy sequence. Hence $u = \sum_{i=1}^{\infty} 2^{-i}u_i = \sup_n \sum_{i=1}^{n} 2^{-i}u_i$ exists in L. For each n let $v_n = \big(u_{n+1} - 4^n \sum_{i=1}^{n} u_i - 2^{-n}u\big)^+$. Clearly, $\{v_n\} \subseteq A$ and we claim that $\{v_n\}$ is a disjoint sequence. Indeed, for $m > n \geq 1$ we have

$$\begin{aligned} 0 \ &\leq\ 4^{-m}v_m = \Big[4^{-m}u_{m+1} - \sum_{i=1}^{m} u_i - 2^{-3m}u\Big]^+ \leq \big[4^{-m}u_{m+1} - u_{n+1}\big]^+ \\ &\leq\ \big[2^{-n}2^{-(m+1)}u_{m+1} - u_{n+1}\big]^+ \leq \Big[2^{-n}u + 4^n \sum_{i=1}^{n} u_i - u_{n+1}\Big]^+ \ \perp\ v_n\,, \end{aligned}$$

and so $v_m \wedge v_n = 0$. Now from $v_n + 2^{-n}u \geq \big[u_{n+1} - 4^n \sum_{i=1}^{n} u_i\big]^+$, we obtain $\rho_B(v_n) + \rho_B(2^{-n}u) \geq \rho_B\big(\big[u_{n+1} - 4^n \sum_{i=1}^{n} u_i\big]^+\big)$ and hence $\rho_B(v_n) \geq 3\epsilon - 2^{-n}\delta$. This implies $\rho_B(v_n) > \epsilon$ for all sufficiently large n, in contradiction to (2).

Thus, we have shown that given $\epsilon > 0$ there exists some $v \geq 0$ in the ideal generated by A in L such that $\rho_B\big((|u| - v)^+\big) < \epsilon$ for all $u \in A$. It is now a simple consequence of Theorem 1.10 and the latter statement that $\lim \rho_B(v_n) = 0$ for each order bounded disjoint sequence in the ideal generated by A in L. But then by Theorem 6.25 there exists some $\psi \geq 0$ in the ideal generated by B in L' such that $\big(|\phi| - \psi\big)^+(v) < \epsilon$ for all $\phi \in B$. So, for all $u \in A$ and all $\phi \in B$ we have

$$\begin{aligned} \big(|\phi| - \psi\big)^+(|u|) \ &=\ \big(|\phi| - \psi\big)^+\big((|u| - v)^+\big) + \big(|\phi| - \psi\big)^+(|u| \wedge v) \\ &\leq\ \rho_B\big((|u| - v)^+\big) + \big(|\phi| - \psi\big)^+(v) < 2\epsilon\,. \end{aligned}$$

Thus, $\rho_A\big((|\phi| - \psi)^+\big) \leq 2\epsilon$ for all $\phi \in B$.

(3) $\Longrightarrow$ (4) Obvious.

(4) $\Longrightarrow$ (1) Let $\epsilon > 0$ and let $\{\phi_n\} \subseteq B$ be a positive disjoint sequence. By our hypothesis there exist $v \in L^+$ and $0 \le \psi \in L'$ such that $\rho_A\big((|\phi| - \psi)^+\big) < \epsilon$ and $\rho_B\big((|u| - v)^+\big) < \epsilon$ for every $\phi \in B$ and $u \in A$.

Clearly, $\{\phi_n \wedge \psi\}$ is an order bounded disjoint sequence. From this it easily follows that $\lim[\phi_n \wedge \psi](v) = 0$. Now using the inequalities

$$\begin{aligned}[\phi_n \wedge \psi](|u|) &\le [\phi_n \wedge \psi]\big((|u| - v)^+\big) + [\phi_n \wedge \psi](v)\\ &\le \rho_B\big((|u| - v)^+\big) + [\phi_n \wedge \psi](v),\end{aligned}$$

we infer that $\rho_A(\phi_n \wedge \psi) \le \epsilon + [\phi_n \wedge \psi](v)$. Since ρ_A is a Riesz seminorm, we get $\rho_A(\phi_n) \le \rho_A\big((\phi_n - \psi)^+\big) + \rho_A(\phi_n \wedge \psi) \le 2\epsilon + [\phi_n \wedge \psi](v)$, from which the validity of (1) follows, and the proof is finished. ■

It should be clear that if two nonempty solid sets $A \subseteq L$ and $B \subseteq L'$ satisfy any one of the equivalent conditions of Theorem 6.52, then their convex hulls also satisfy the equivalent statements of Theorem 6.52.

The next result presents a sufficient condition for a solid set to be weakly compact. It is due to O. Burkinshaw and P. G. Dodds [**45**, Proposition 3.2] and P. Meyer-Nieberg [**97**, Theorem 2.6, p. 313].

THEOREM 6.53 (Burkinshaw–Dodds–Meyer-Nieberg). *Let (L, τ) be a Hausdorff locally convex-solid Riesz space with the monotone completeness property and let $S \subseteq L$ be solid and τ-bounded. If every disjoint sequence of S is τ-convergent to zero, then the set S is relatively $\sigma(L, L')$-compact.*

PROOF. Let $\overline{S}$ be the τ-closure of S. Then every disjoint sequence of $\overline{S}$ is also τ-convergent to zero. To see this, let $\{u_n\} \subseteq \overline{S}$ be a positive disjoint sequence and let V be a solid τ-neighborhood of zero. For each n choose some $0 \le v_n \in S$ such that $u_n - v_n \in V$. Then $u_n \wedge v_n \xrightarrow{\tau} 0$, since $\{u_n \wedge v_n\}$ is a disjoint sequence of S. Choose $m \in \mathbb{N}$ so that $u_n \wedge v_n \in V$ holds for all $n > m$ and note that $0 \le u_n = (u_n - v_n)^+ + u_n \wedge v_n \in V + V$ for $n > m$. This implies that $u_n \xrightarrow{\tau} 0$.

Thus, without loss of generality, we can assume that S is τ-closed and by the remark following Theorem 6.52 we can also assume that S is convex. By Theorem 6.47 it is enough to show that if $\{v_\alpha\} \subseteq S$ satisfies $0 \le v_\alpha \uparrow$, then $v_\alpha \xrightarrow{\tau} u$ for some $u \in S$. To this end, let $\epsilon > 0$ and let V be a closed convex and solid τ-neighborhood of zero. By Theorem 6.52 there exists a positive element v in the ideal generated by S in L such that $|\phi|\big(|u| - |u| \wedge v\big) < \epsilon$ for all $u \in S$ and all $\phi \in V^\circ$. By Theorem 6.46 we know that S is a solid subset of $(L')^\sim_n$ and thus the interval $[0, u] \subseteq L$ is $\sigma(L, L')$-compact. Hence the net $\{v_\alpha \wedge v\}$ is τ-Cauchy by Theorem 6.46 again. This implies the existence of an index α_1 such that $|\phi|\big(|v_\alpha \wedge v - v_\beta \wedge v|\big) < \epsilon$ for all $\phi \in V^\circ$ and all $\alpha, \beta \succeq \alpha_1$, and so

$$|\phi|\big(|v_\alpha - v_\beta|\big) \le |\phi|\big(v_\alpha - v_\alpha \wedge v\big) + |\phi|\big(|v_\alpha \wedge v - v_\beta \wedge v|\big) + |\phi|\big(v_\beta - v_\beta \wedge v\big) \le 3\epsilon$$

for all $\alpha, \beta \succeq \alpha_1$ and all $\phi \in V^\circ$. Therefore, $\{v_\alpha\}$ is a τ-Cauchy net and so $v_\alpha \xrightarrow{\tau} u$ holds in L. Since S is τ-closed we get $u \in S$, and the proof is finished. ■

A sufficient condition for an arbitrary solid subset of L' to be $\sigma(L', L'')$-compact in terms of (L, τ) is presented next. It is due to O. Burkinshaw and P. G. Dodds [**45**, Proposition 3.3].

THEOREM 6.54 (Burkinshaw–Dodds). *Let the Hausdorff locally convex-solid Riesz space (L,τ) be τ-complete and assume that L' is $\beta(L',L)$-complete. Also suppose that a subset S of L' is solid and $\beta(L',L)$-bounded. If $\lim \rho_S(u_n) = 0$ whenever $\{u_n\}$ is a τ-bounded disjoint sequence of L, then the set S is relatively $\sigma(L',L'')$-compact.*

PROOF. By Theorem 6.52 for every τ-bounded subset B of L we have $\lim \rho_B(\phi_n) = 0$ whenever $\{\phi_n\}$ is a disjoint sequence of S. But this merely says that each disjoint sequence of S is $\beta(L',L)$-convergent to zero and the result follows from Theorem 6.53 by observing that the topological dual of $\big(L', \beta(L',L)\big)$ is precisely L''. ■

Recall that an AL-space is a Banach lattice L_ρ whose norm is additive on L_ρ^+, that is, it satisfies $\rho(u+v) = \rho(u) + \rho(v)$ for all $u, v \in L_\rho^+$. In the next theorem we shall show that for AL-spaces the converse of Theorem 6.53 holds true and that for a weakly compact set its closed solid hull is also weakly compact.

THEOREM 6.55. *For a nonempty subset A of an AL-space L_ρ the following statements are equivalent.*

(i) *A is relatively $\sigma(L_\rho, L_\rho')$-compact.*
(ii) *A is norm bounded and each disjoint sequence in the solid hull of A is norm convergent to zero.*

PROOF. First define $\phi(u) = \rho(u^+) - \rho(u^-)$ for all $u \in L_\rho$ and note that $0 \le \phi \in L_\rho'$; see Lemma 1.64. From this it follows easily that the norm topology coincides with $|\sigma|(L_\rho, L_\rho')$. Therefore, $|\sigma|(L_\rho, L_\rho')$ is a complete topology and consequently $L_\rho = (L_\rho')_{\rm n}^\sim$ by Theorem 6.10. The equivalence of (i) and (ii) now follows from Theorem 6.30 and Corollary 6.31 applied to the pair of Riesz spaces L_ρ' and $(L_\rho')_{\rm n}^\sim$. ■

Recall that a **discrete element** of a Riesz space L is any element $u > 0$ such that $L_u = \{\lambda u\colon\ \lambda \in \mathbb{R}\}$. According to Exercise 12 of Chapter 1, an element $u > 0$ in an Archimedean Riesz space is a discrete element if and only if:

$$w, v \in [0,u] \ \text{ and } \ w \perp v \implies v = 0 \ \text{ or } \ w = 0\,.$$

If $u > 0$ and $v > 0$ are discrete elements, then it is easy to check that either $u \wedge v = 0$ or $u = \lambda v$ for some $\lambda > 0$. Also, it should be kept in mind that if $0 < u \in L$ is a discrete element, then from part (3) of Theorem 1.67 it follows that $\big|\phi(u)\big| = |\phi|(u)$ for all $\phi \in L^\sim$.

It turns out that the ideal generated by the collection of all discrete elements coincides with the vector subspace generated by the collection of all discrete elements. To see this, let $|u| \le \sum_{i=1}^n \lambda_i v_i$, where $\lambda_i > 0$ and v_i is a discrete element for each i. Then, by the Riesz Decomposition Property, we can write $u = \sum_{i=1}^n u_i$ with $|u_i| \le \lambda_i v_i$ for each i, and so using that each v_i is a discrete element, we have $u_i = \mu_i v_i$ for each i so that $u = \sum_{i=1}^n \mu_i v_i$.

The band generated by the discrete elements of a Riesz space L will be denoted by $\mathfrak{D}_L$ or simply by $\mathfrak{D}$ if there is no ambiguity about the Riesz space L. Note that an Archimedean Riesz space L is discrete if and only if $\mathfrak{D}_L = L$.

THEOREM 6.56. *For a Hausdorff locally convex-solid Riesz space* (L,τ) *and a vector* $u \in L^+$ *the following statements are equivalent.*

(i) *The order interval* $[0,u]$ *is* $\sigma(L,L')$*-compact and* $u \in \mathfrak{D}$.
(ii) *Each* $\sigma(L,L')$*-Cauchy net of* $[0,u]$ *is* τ*-convergent.*
(iii) *Each* $\sigma(L,L')$*-convergent sequence of* $[0,u]$ *is* τ*-convergent and* $[0,u]$ *is* τ*-complete.*
(iv) *The order interval* $[0,u]$ *is* τ*-compact.*
(v) *The order interval* $[0,u]$ *is* $|\sigma|(L,L')$*-compact.*

PROOF. (i) $\Longrightarrow$ (ii) Assume by way of contradiction that (ii) is false. Then there exists a net $\{v_\alpha\} \subseteq [0,u]$ and a τ-closed convex and solid τ-neighborhood V of zero satisfying $v_\alpha \xrightarrow{\sigma(L,L')} v$ and $v_\alpha - v \notin V$ for all α. By the Bipolar Theorem 2.13, for each index α there exists some $\phi_\alpha \in V^\circ$ such that $|\phi_\alpha(v_\alpha - v)| > 1$. Let ϕ be an arbitrary $\sigma(L',L)$-accumulation point of the net $\{\phi_\alpha\}$.

Pick an index α_1 such that $|\phi(v_\alpha - v)| < \frac{1}{2}$ for all $\alpha \succeq \alpha_1$, and note that for $\alpha \succeq \alpha_1$ we have $\big|(\phi_\alpha - \phi)(v_\alpha - v)\big| \geq 1 - \frac{1}{2} = \frac{1}{2}$. Consequently, from $|v - v_\alpha| \leq u$ it follows that for all $\alpha \succeq \alpha_1$ we have

$$|\phi_\alpha - \phi|(u) \geq |\phi_\alpha - \phi|\big(|v_\alpha - v|\big) \geq \tfrac{1}{2}\,. \tag{$\star$}$$

Since, by Theorem 1.24, the ideal generated by the discrete elements of L is order dense in $\mathfrak{D}$, there exists a net $\{u_\lambda\}$ such that $0 \leq u_\lambda \uparrow u$, where each u_λ is a linear combination of discrete elements. By part (iv) of Theorem 6.46, it follows that $u_\lambda \xrightarrow{\tau} u$ and therefore there exists an index λ_0 satisfying $|\phi_\alpha - \phi|(u - u_{\lambda_0}) < \frac{1}{4}$ for all α. So, $|\phi_\alpha - \phi|(u_{\lambda_0}) \geq \frac{1}{2} - \frac{1}{4} = \frac{1}{4}$ for all $\alpha \succeq \alpha_1$. Let $u_{\lambda_0} = \sum_{i=1}^n c_i x_i$, where each x_i is a discrete element, and note that for each $\alpha \succeq \alpha_1$ we have

$$\sum_{i=1}^n \big|(\phi_\alpha - \phi)(c_i x_i)\big| = \sum_{i=1}^n |c_i| \cdot |\phi_\alpha - \phi|(x_i) \geq |\phi_\alpha - \phi|(u_{\lambda_0}) \geq \tfrac{1}{4}\,.$$

However, this contradicts the fact that ϕ is a $\sigma(L',L)$-accumulation point of the net $\{\phi_\alpha\}$, and the validity of (ii) has been established.

(ii) $\Longrightarrow$ (iii) Obvious.

(iii) $\Longrightarrow$ (iv) Each disjoint sequence of $[0,u]$ is τ-convergent to zero and so by Theorem 6.46 the interval $[0,u]$ is $\sigma(L,L')$-compact. We will show that $[0,u]$ is totally bounded for τ and the conclusion will follow from Theorem 2.7.

If $[0,u]$ is not a τ-totally bounded set, there exist a closed convex solid τ-neighborhood V of zero and a sequence $\{u_n\} \subseteq [0,u]$ such that $u_n - u_m \notin V$ for $n \neq m$. Now fix $\epsilon > 0$. Then, according to Theorem 6.52, there is some $0 \leq \psi \in L'$ such that $\big(|\phi| - \psi\big)^+(u) < \epsilon$ holds for every $\phi \in V^\circ$. Let B be the band generated by $\psi \in L'$ and write $L' = B \oplus B^{\mathrm{d}}$. By Theorem 6.46 the ideal L_u generated by u in L is Dedekind complete and τ induces a Lebesgue topology on L_u. So, the ideal $J = \big\{v \in L_u\colon\ |\phi|(|v|) = 0 \text{ for all } \phi \in B^{\mathrm{d}}\big\}$ is a band of L_u (for if $\{v_\alpha\} \subseteq J$ satisfies $0 \leq v_\alpha \uparrow v$ in L_u, then $v_\alpha \xrightarrow{\tau} v$ and so $v \in J$). Let P denote the projection of L_u onto J, so that $[0, P(u)] \subseteq [0,u]$. By Lemma 6.48 there exists a subsequence $\{P(v_n)\}$ of the sequence $\{P(u_n)\}$ such that the sequence $\{\phi(Pv_n)\}$ converges for all $\phi \in B$. But $\phi(Pv_n) = 0$ for all $\phi \in B^{\mathrm{d}}$, so that $\{P(v_n)\}$ is $\sigma(L,L')$-convergent.

By our hypothesis $\{P(v_n)\}$ is also τ-convergent. This implies

$$\begin{aligned} |\phi|\big(|u_n - u_m|\big) &= \big(|\phi| - \psi\big)^+\big(|u_n - u_m|\big) + \big[|\phi| \wedge \psi\big]\big(|u_n - u_m|\big) \\ &\leq 2\epsilon + \psi\big(|u_n - u_m|\big) = 2\epsilon + \psi\big(|P(u_n) - P(u_m)|\big) \\ &\leq 3\epsilon \end{aligned}$$

for all $\phi \in V^\circ$ and some m, n sufficiently large. Therefore $u_n - u_m \in V$ for some $n \neq m$, a contradiction. Thus $[0, u]$ is τ-totally bounded and the desired conclusion follows.

(iv) $\Longrightarrow$ (v) It follows immediately from $|\sigma|(L, L') \subseteq \tau$.

(v) $\Longrightarrow$ (i) Suppose that $0 < u \notin \mathfrak{D}$. Then, by Theorem 1.25, there exists $0 < v \leq u$ with $v \in \mathfrak{D}^{\mathrm{d}}$. Thus, the interval $[0, v]$ contains no discrete elements and is $|\sigma|(L, L')$-compact. So replacing u by v if necessary, we can assume that $[0, u]$ contains no discrete elements. Clearly $[0, u]$ is $\sigma(L, L')$-compact, and so by Theorem 6.46 the ideal L_u generated by u is Dedekind complete and each $\phi \in L'$ is order continuous on L_u. Choose $0 \leq \phi \in L'$ such that $\phi(u) > 0$.

Next we claim that given $0 \leq \delta \leq \phi(u)$ there exists a component v of u (i.e., $v \wedge (u - v) = 0$) with $\phi(v) = \delta$. To see this, let C be the collection of all components w of u with $\phi(w) \leq \delta$. Since $0 \in C$ we have $C \neq \emptyset$ and clearly C is partially ordered by $\leq$. Also if K is a chain of C, then $w = \sup K$ is a component of u and since ϕ is order continuous on L_u, it follows that $\phi(w) = \sup\{\phi(v)\colon\ v \in K\} \leq \delta$, i.e., $w \in C$. It now follows from Zorn's lemma that C has a maximal element, say v. If $\phi(v) < \delta$, then choose $\epsilon > 0$ such that $\phi(v) + \epsilon < \delta$ and note that there exists a strictly positive component w of $u - v$ (and hence of u) with $\phi(w) \leq \epsilon$. Indeed, since the vector $u - v$ is not a discrete vector, there exists a component $0 < v_1 \leq u - v$ with $\phi(v_1) \leq \frac{1}{2}\phi(u - v)$, and so repeating the process the existence of w can be established. But then the component $v + w$ of u satisfies $v < v + w$ and $\phi(v + w) = \phi(v) + \phi(w) \leq \phi(v) + \epsilon < \delta$, contradicting the maximality property of v. Hence $\phi(v) = \delta$.

Now by using induction on k, for each k we can choose a finite collection of elements $\{z_i^k\colon\ i = 1, \ldots, 2^k\}$ satisfying the following properties:

(a) $z_i^k \wedge z_j^k = 0$ if $i \neq j$.
(b) $\sum_{i=1}^{2^k} z_i^k = u$.
(c) $\phi(z_i^k) = 2^{-k}\phi(u)$ for $i = 1, 2, \ldots, 2^k$, and
(d) $z_i^k = z_{2i-1}^{k+1} + z_{2i}^{k+1}$.

For each $k \in \mathbb{N}$ let $x_k = \sum_{j=1}^{2^k}(-1)^j z_j^k \in [-u, u] = -u + 2[0, u]$ and note that for $n \neq m$ we have $\phi\big(|x_n - x_m|\big) = \frac{1}{2}\phi(u)$, in contradiction with the $|\sigma|(L, L')$-compactness of $[0, u]$. Thus $u \in \mathfrak{D}$ and the proof is finished. ■

The next result, due to I. Kawai [**74**, Theorem 5.4, p. 302], is an immediate consequence of Theorem 6.56.

COROLLARY 6.57 (Kawai). *For a Hausdorff locally convex-solid Riesz space (L, τ) the following statements are equivalent.*

(i) *L is discrete, Dedekind complete, and τ is a Lebesgue topology.*
(ii) *Each order interval of L is $|\sigma|(L, L')$-compact.*
(iii) *Each order interval of L is τ-compact.*

(iv) *Each order bounded $\sigma(L, L')$-convergent sequence is τ-convergent and each order bounded increasing τ-Cauchy net is τ-convergent.*
(v) *Each order bounded $\sigma(L, L')$-Cauchy net of L is τ-convergent.*

If a Hausdorff locally convex-solid Riesz space (L, τ) satisfies statement (i) of Corollary 6.57, then it follows from Theorems 1.40 and 1.78 that L can be identified with an order dense ideal of some $\mathbb{R}^X$ Riesz space.

The next result is due to P. G. Dodds [**53**, Theorem 4.7, p. 400].

THEOREM 6.58 (Dodds). *Let (L, τ) be a Hausdorff locally convex-solid Riesz space with the Lebesgue property. For a sequence $\{u_n\}$ of $\mathfrak{D}$ the following statements are equivalent.*

(i) *$\{u_n\}$ is $\sigma(L, L')$-convergent to zero.*
(ii) *$\{u_n\}$ is $|\sigma|(L, L')$-convergent to zero.*

PROOF. (i) $\Longrightarrow$ (ii) Let A be the set of all discrete elements of L, and let Λ be the collection of all finite subsets of A. For each $\lambda \in \Lambda$ let $P_\lambda = \sum_{a \in \lambda} P_a$, where P_a is the projection of L onto the band generated by a; see Theorem 1.77. Note that $u = \sup\{P_\lambda(u)\colon\ \lambda \in \Lambda\}$ for each element $0 \le u \in \mathfrak{D}$. Let Q_λ denote the projection of L' onto the band $[(P_\lambda(L))^\circ]^{\mathrm{d}}$, and let Q be the projection of L' onto the band $(\mathfrak{D}^\circ)^{\mathrm{d}}$. It follows from Theorem 6.39 [applied to the pair of Riesz spaces L' and $(L')^\sim_{\mathrm{n}}$] that the sequence $\{u_n\}$ is order-equicontinuous on L'. Since $Q_\lambda(\phi) \uparrow Q(\phi)$ for each $0 \le \phi \in L'$, it follows from Theorem 6.30 that for each $\epsilon > 0$ there exists λ_1 such that $[Q(\phi) - Q_{\lambda_1}(\phi)](|u_n|) < \epsilon$ for all n. But then since $|\phi(u)| = |\phi|(u)$ for each discrete element u and each $\phi \in L'$, we have

$$[Q_a(\phi)](|u_n|) = \phi(P_a(|u_n|)) = |\phi(P_a(u_n))| = |[Q_a(\phi)](u_n)|$$

for all n, all positive discrete elements a, and all $0 \le \phi \in L'$. Thus

$$\begin{aligned} \phi(|u_n|) &= [Q(\phi)](|u_n|) = [Q(\phi) - Q_{\lambda_1}(\phi)](|u_n|) + [Q_{\lambda_1}(\phi)](|u_n|) \\ &\le \epsilon + \sum_{a \in \lambda_1} [Q_a(\phi)](|u_n|) = \epsilon + \sum_{a \in \lambda_1} |[Q_a(\phi)](u_n)| \end{aligned}$$

for all $0 \le \phi \in L'$. Now since $\{u_n\}$ is $\sigma(L, L')$-convergent to zero and λ_1 is a finite set, there exists some m so that for any fixed $0 \le \phi \in L'$ we have $\phi(|u_n|) < 2\epsilon$ for $n > m$, that is, $\{u_n\}$ is $|\sigma|(L, L')$-convergent to zero.

(ii) $\Longrightarrow$ (i) Obvious, from the inclusion $\sigma(L, L') \subseteq |\sigma|(L, L')$. ■

The characterizations of the τ-compact order intervals presented in Theorem 6.56 will now be used to study the τ-compact solid subsets of a locally convex-solid Riesz space (L, τ). A surprisingly simple relationship between a solid τ-compact subset and its disjoint sequences is included in the next result.

THEOREM 6.59. *If (L, τ) is a τ-complete Hausdorff locally convex-solid Riesz space, then for a τ-bounded solid subset S of L the following statements are equivalent.*

(i) *The solid set S is relatively τ-compact.*
(ii) *Each disjoint sequence of S is τ-convergent to zero and every element $0 < u \in S$ belongs to $\mathfrak{D}$.*
(iii) *Each $\sigma(L, L')$-convergent sequence of S is τ-convergent and S is relatively $\sigma(L, L')$-compact.*

PROOF. (i) $\Longrightarrow$ (ii) Let $\{u_n\} \subseteq S$ be a disjoint sequence, and let V be a solid τ-neighborhood of zero. Since $\{u_n\}$ lies in the τ-totally bounded set S, it follows that $\{u_n\}$ is a τ-totally bounded sequence and so there exists some m such that $\{u_n\} \subseteq \bigcup_{i=1}^m (u_i + V)$. If $n > m$, then $u_n - u_i \in V$ for some $1 \le i \le m$. Since the sequence $\{u_n\}$ is disjoint, it follows that $|u_n| \le |u_n| + |u_i| = |u_n - u_i| \in V$ for $n > m$, so that $u_n \in V$ for $n > m$, and therefore $u_n \xrightarrow{\tau} 0$. Finally, since each order interval of S is τ-compact it follows from Theorem 6.56 that every element $0 < u \in S$ belongs to $\mathfrak{D}$.

(ii) $\Longrightarrow$ (iii) Note first that by Theorem 6.53 the set S is relatively $\sigma(L, L')$-compact. Now, as in the proof of Theorem 6.53, we can show that every disjoint sequence in the closed convex hull of the set S is also τ-convergent to zero. Thus we can assume without loss of generality that S is convex and τ-closed.

Assume that a sequence $\{u_n\} \subseteq S$ satisfies $u_n \xrightarrow{\sigma(L,L')} u$. Fix a closed convex solid τ-neighborhood V of zero and let $\epsilon > 0$. By Theorem 6.52 there exists $0 \le \psi \in L'$ such that $\big(|\phi| - \psi\big)^+\big(|u|\big) < \epsilon$ for all $\phi \in V^\circ$ and all $u \in S$. Since by Theorem 6.46 each interval of S is $\sigma(L, L')$-compact, it follows (again from Theorem 6.46) that τ induces a Lebesgue topology on the ideal A generated by S in L. But then $|\sigma|(A, L')$ is also a Lebesgue topology and hence by Theorem 6.58, applied to $\big(A, |\sigma|(A, L')\big)$, the sequence $\{|u_n - u|\}$ is $|\sigma|(A, L')$-convergent to zero. Pick some m such that $\psi\big(|u_n - u|\big) < \epsilon$ for $n \ge m$. Then

$$|\phi|\big(|u_n - u|\big) \le \big(|\phi| - \psi\big)^+\big(|u_n - u|\big) + \psi\big(|u_n - u|\big) < 3\epsilon$$

for all $\phi \in V^\circ$ and all $n \ge m$. From the Bipolar Theorem 2.13, we get $u_n - u \in V$ for $n \ge m$ and hence $u_n \xrightarrow{\tau} u$.

(iii) $\Longrightarrow$ (i) We will show that S is τ-totally bounded. If not, there exists a closed convex solid τ-neighborhood V of zero and a sequence $\{u_n\} \subseteq S$ such that $u_n - u_m \notin V + V + V$ for $n \ne m$. Then, by Theorem 6.52, there exists some $v \ge 0$ in the ideal generated by S such that $|\phi|\big((|u| - v)^+\big) < 1$ for all $\phi \in V^\circ$ and all $u \in S$. Therefore, $\big(|u| - v\big)^+ \in V$ for all $u \in S$.

Write $v = \sum_{i=1}^r \lambda_i v_i$ with $0 \le v_i \in S$ and $\lambda_i > 0$ for each i. From Theorems 6.56 and 6.46 we see that each order interval $[0, \lambda_i v_i]$ is τ-compact and therefore, since a finite sum of totally bounded sets is totally bounded, the order interval $[0, v] = \sum_{i=1}^r [0, \lambda_i v_i]$ is τ-totally bounded. Hence the sequence $\{u_n \wedge v\} \subseteq [0, v]$ is τ-totally bounded and so there exists some $m \in \mathbb{N}$ such that $\{u_n \wedge v\} \subseteq \bigcup_{i=1}^m (u_i \wedge v + V)$. But then for any fixed $k > m$ there exists $1 \le i \le m$ with $u_k \wedge v - u_i \wedge v \in V$, and so

$$|u_k - u_i| \le \big(|u_k| - v\big)^+ + |u_k \wedge v - u_i \wedge v| + \big(|u_i| - v\big)^+ \in V + V + V\,,$$

which contradicts the choice of $\{u_n\}$, and the proof is finished. ■

From Theorems 6.59 and 6.49 the following result should be immediate.

COROLLARY 6.60. *Let (L, τ) be a Hausdorff τ-complete locally convex-solid Riesz space. If L has the countable sup property, then for a solid subset S of L the following statements are equivalent.*

(i) *The solid set S is relatively τ-compact.*
(ii) *The solid set S is relatively sequentially τ-compact.*

6.5. Semireflexive Riesz Spaces

Recall that for any given Hausdorff locally convex space (E, τ), the **strong topology** $\beta(E', E)$ on its topological dual E' is the locally convex topology of uniform convergence on the τ-bounded subsets of E. The **bidual** E'' of E is the topological dual of the Hausdorff locally convex space $\big(E', \beta(E', E)\big)$. As usual, each $x \in E$ gives rise to a $\beta(E', E)$-continuous linear functional $\widetilde{x}$ on E' via the formula

$$\widetilde{x}(\phi) = \phi(x)\,, \quad \phi \in E'\,.$$

The mapping $x \mapsto \widetilde{x}$ is a one-to-one operator, and so identifying $\widetilde{E}$ with E we can consider E as a vector subspace of E''.

DEFINITION 6.61. *A Hausdorff locally convex topological vector space (E, τ) is said to be* ***semireflexive*** *if E coincides with its bidual E''. A semireflexive Hausdorff locally convex space (E, τ) is said to be* ***reflexive*** *if $\tau = \beta(E'', E')$.*

This section studies semireflexive locally convex-solid Riesz spaces. The results in this section have appeared in many forms in the works of several authors; see for instance [**46**], [**58**], [**72**], [**82**], [**84, 85**], [**89**, Notes X, XI, XII, XIII], [**97, 98**], [**100**], [**105**], [**107**], [**123**, pp. 81–96], [**125**], and [**130**].

If a Hausdorff locally convex-solid Riesz space (L, τ) is semireflexive, then L must be an ideal (in fact a band) in its bidual L''. So, we begin by giving necessary and sufficient conditions for L to be an ideal (respectively a band) in L''.

THEOREM 6.62. *For a Hausdorff locally convex-solid Riesz space (L, τ) the following statements are equivalent.*

(i) *L is an ideal of L''.*
(ii) *L is Dedekind complete and τ is a Lebesgue topology.*
(iii) *Each order bounded disjoint sequence of L is τ-convergent to zero and each order interval of L is τ-complete.*
(iv) *Each order interval of L is $\sigma(L, L')$-compact.*

PROOF. (i) $\Longrightarrow$ (ii) Since L'' is Dedekind complete and L is an ideal in L'', it follows that L must be Dedekind complete. That τ is a Lebesgue topology now follows from Theorem 3.12(4).

(ii) $\Longrightarrow$ (iii) Since τ is a Lebesgue topology, each order bounded disjoint sequence is τ-convergent to zero; see Theorem 3.22. By Theorem 4.28 each order interval of L is τ-complete.

(iii) $\Longrightarrow$ (iv) It follows immediately from Theorem 6.46.

(iv) $\Longrightarrow$ (i) Again by Theorem 6.46 each order interval of L is a solid subset of L''. Thus L is a solid subset of L'' and so L is an ideal of L''. ■

Recall that a sequence $\{u_n\}$ in a topological vector space (E, τ) is said to have τ-**bounded partial sums** if the sequence $\big\{\sum_{i=1}^{n} u_i\big\}$ is τ-bounded.

THEOREM 6.63. *For a Hausdorff locally convex-solid Riesz space (L, τ) the following statements are equivalent.*

(i) *L is a band of L''.*
(ii) *The topology τ is Levi and Lebesgue.*
(iii) *L is τ-complete and every disjoint sequence of L^+ having a τ-bounded sequence of partial sums is τ-convergent to zero.*

(iv) *Each increasing* τ*-bounded net of* L^+ *is* τ*-convergent.*
(v) L *is* $|\sigma|(L,L')$*-complete.*
(vi) $L=(L')^\sim_{\mathrm{n}}$.

PROOF. (i) $\Longrightarrow$ (ii) That τ is a Lebesgue topology follows from Theorem 3.12. For the Levi property let $\{u_\alpha\}\subseteq L$ be a τ-bounded net with $0\le u_\alpha\uparrow$ in L. Then by Theorem 6.16 we have $u_\alpha\uparrow u$ in L'' and so $u\in L$ (since L is a band in L'') and the Levi property follows.

(ii) $\Longrightarrow$ (iii) The τ-completeness of (L,τ) follows from Theorem 4.37. Now let $\{u_n\}\subseteq L^+$ be a disjoint sequence whose sequence of partial sums $\{\sum_{i=1}^n u_i\}$ is τ-bounded. Since τ is a Levi topology, there exists some $u\in L^+$ such that $\sum_{i=1}^n u_i\uparrow u$ and so $u_n\xrightarrow{\tau}0$ since τ is also a Lebesgue topology.

(iii) $\Longrightarrow$ (iv) Let $\{u_\alpha\}\subseteq L$ be τ-bounded with $0\le u_\alpha\uparrow$ in L and let V be a solid τ-closed and convex τ-neighborhood of zero. By Theorem 6.52 there exists some $v\in L^+$ such that $|\phi|\big((u_\alpha-v)^+\big)<\frac13$ for all $\phi\in V^{\mathrm{o}}$ and all α. Now since $0\le u_\alpha\wedge v\uparrow\le v$ and the interval $[0,v]$ is $\sigma(L,L')$-compact by Theorem 6.46, it follows that $\{u_\alpha\wedge v\}$ is a τ-Cauchy net and so there exists an index α_0 such that for all $\phi\in V^{\mathrm{o}}$ and all $\alpha,\beta\succeq\alpha_0$ we have

$$|\phi|\big(|u_\alpha\wedge v-u_\beta\wedge v|\big)<\tfrac13\,.$$

Consequently, for all $\phi\in V^{\mathrm{o}}$ and all $\alpha,\beta\succeq\alpha_0$ we have

$$|\phi|\big(|u_\alpha-u_\beta|\big)\le|\phi|(u_\alpha-u_\alpha\wedge v)+|\phi|\big(|u_\alpha\wedge v-u_\beta\wedge v|\big)+|\phi|(u_\beta-u_\beta\wedge v)\le 1.$$

By the Bipolar Theorem 2.13, it follows that $u_\alpha-u_\beta\in V$ for all $\alpha,\beta\succeq\alpha_0$. This shows that $\{u_\alpha\}$ is a τ-Cauchy net and consequently τ-convergent.

(iv) $\Longrightarrow$ (v) According to Theorem 2.4(i) every $|\sigma|(L,L')$-bounded set is τ-bounded, so that if a net $\{u_\alpha\}$ in L is $|\sigma|(L,L')$-bounded and $0\le u_\alpha\uparrow$ holds, then there exists some $u\in L$ such that $u_\alpha\xrightarrow{\tau}u$. But then, by Theorem 2.21(c), we have $u_\alpha\uparrow u$ and hence $|\sigma|(L,L')$ is a Levi topology. Also since τ is a Lebesgue topology, $|\sigma|(L,L')$ is likewise so, and thus by Theorem 4.37 the Riesz space L is $|\sigma|(L,L')$-complete.

(v) $\Longrightarrow$ (vi) It follows from Theorem 6.10.

(vi) $\Longrightarrow$ (i) Since L'' is an ideal of $(L')^\sim$ and $(L')^\sim_{\mathrm{n}}$ is a band in $(L')^\sim$, the equality $L=(L')^\sim_{\mathrm{n}}$ shows that L is a band of L''. ■

The next theorem presents a number of conditions for the band generated by L in L'' to be all of L''.

THEOREM 6.64. *For a Hausdorff locally convex-solid Riesz space* (L,τ) *the following statements are equivalent.*

(i) $L^{\mathrm{d}}=\{0\}$ *holds in* L''.
(ii) L'' *coincides with the band generated by* L *in* L''.
(iii) $L''\subseteq(L')^\sim_{\mathrm{n}}$.
(iv) $\beta(L',L)$ *is a Lebesgue topology.*
(v) *Each order bounded disjoint sequence of* L' *is* $\beta(L',L)$*-convergent to zero.*

PROOF. (i) $\Longrightarrow$ (ii) If $L^{\mathrm{d}}=\{0\}$ holds in L'', then the band generated by L in L'' is precisely $L^{\mathrm{dd}}=L''$.

(ii) $\Longrightarrow$ (iii) Since $L \subseteq (L')^{\sim}_{\mathrm{n}}$ and L'' is an ideal in $(L')^{\sim}$, it follows that the band generated by L in L'' is included in the band generated by L in $(L')^{\sim}_{\mathrm{n}}$ and hence $L'' \subseteq (L')^{\sim}_{\mathrm{n}}$.

(iii) $\Longrightarrow$ (iv) Since L'' is the topological dual of $\big(L', \beta(L', L)\big)$, the conclusion follows from Theorem 3.12.

(iv) $\Longrightarrow$ (v) Since the Lebesgue property implies the pre-Lebesgue property, the conclusion follows from Theorem 3.22.

(v) $\Longrightarrow$ (i) According to Corollary 6.12 the topology $\beta(L', L)$ is a Fatou topology and by our hypothesis $\beta(L', L)$ is a pre-Lebesgue topology. Therefore, $\beta(L', L)$ is a Lebesgue topology by Theorem 4.8. Hence by Theorem 3.12 we have $L'' \subseteq (L')^{\sim}_{\mathrm{n}}$, and so by Theorem 6.9 the Riesz space L is $|\sigma|(L'', L')$-dense in L''. Now let $0 \le v \in L''$ satisfy $v \perp L$. Since L is $|\sigma|(L'', L')$-dense in L'', there exists a net $\{v_\alpha\} \subseteq L^+$ such that $\{v_\alpha\}$ is $|\sigma|(L'', L')$-convergent to the element $v \in L''$ and, in particular, the net $\{v_\alpha \wedge v\}$ is $|\sigma|(L'', L')$-convergent to v. But $v_\alpha \wedge v = 0$ for each α, and so $v = 0$, and the proof is finished. ■

Combining the conditions in Theorems 6.62, 6.63, and 6.64 we can characterize the semireflexive Hausdorff locally convex-solid Riesz spaces in terms of disjoint sequences as follows.

THEOREM 6.65. *For a Hausdorff locally convex-solid Riesz space (L, τ) the following statements are equivalent.*

(i) *L is semireflexive.*
(ii) *τ is a Lebesgue and Levi topology and $\beta(L', L)$ is Lebesgue.*
(iii) *L is τ-complete, each τ-bounded disjoint sequence of L^+ is $\sigma(L, L')$-convergent to zero, and each disjoint sequence of L^+ with τ-bounded sequence of partial sums is τ-convergent to zero.*
(iv) *L is $|\sigma|(L, L')$-complete and each τ-bounded disjoint sequence of L^+ is $\sigma(L, L')$-convergent to zero.*

PROOF. (i) $\Longrightarrow$ (ii) From Theorem 6.63 it follows that τ is a Levi Lebesgue topology. Also $\beta(L', L)$ is a Lebesgue topology by Theorem 3.12, since $L'' \subseteq (L')^{\sim}_{\mathrm{n}}$.

(ii) $\Longrightarrow$ (iii) That (L, τ) is τ-complete and each disjoint sequence of L^+ with τ-bounded partial sums is τ-convergent to zero follows from Theorem 6.63. Now let $\{u_n\} \subseteq L^+$ be a τ-bounded disjoint sequence and let S be the solid hull of $\{u_n\}$. Let $0 \le \phi \in L'$ and denote by P_n the projection of L' onto the band $C_{u_n} = N^{\mathrm{d}}_{u_n}$. For each n put $\psi_n = P_n(\phi)$ and observe that by Theorem 1.82 the sequence $\{\psi_n\}$ is disjoint and lies in the order interval $[0, \phi]$. Since $\beta(L', L)$ is a Lebesgue topology we have $\lim \rho_S(\psi_n) = 0$. Now use the inequalities $0 \le \phi(u_n) = \psi_n(u_n) \le \rho_S(\psi_n)$ to see that $\phi(u_n) \to 0$. Consequently, $\{u_n\}$ is $\sigma(L, L')$-convergent to zero.

(iii) $\Longrightarrow$ (iv) A glance at Theorem 6.63 guarantees that the Riesz space L is $|\sigma|(L, L')$-complete, and by our hypothesis each τ-bounded disjoint sequence of L^+ is $\sigma(L, L')$-convergent to zero.

(iv) $\Longrightarrow$ (i) Let S be a solid τ-bounded subset of L. Since each disjoint sequence of S is $\sigma(L, L')$-convergent to zero, it follows from Theorem 6.52 that $\lim \rho_S(\phi_n) = 0$ for each order bounded disjoint sequence $\{\phi_n\}$ of L'. Thus, each order bounded disjoint sequence of L' is $\beta(L', L)$-convergent to zero and so $L'' \subseteq (L')^{\sim}_{\mathrm{n}}$ by Theorem 6.64. But, by Theorem 6.63, we have $(L')^{\sim}_{\mathrm{n}} = L$. Therefore $L = L''$ and the proof is finished. ■

Since for Banach spaces the concepts of reflexivity and semireflexivity coincide, Theorem 6.65 applied to Banach lattices yields the classical characterization of reflexivity of T. Ogasawara [**105**]; see also [**89**, Note XIII, Theorem 40.1, p. 530].

THEOREM 6.66 (Ogasawara). *A Banach lattice L_ρ is reflexive if and only if it satisfies the following two properties:*

(i) *L_ρ and L'_ρ both have the Lebesgue property.*
(ii) *If $0 \le u_n \uparrow$ in L_ρ with $\sup_n \rho(u_n) < \infty$, then the supremum $\sup\{u_n\}$ exists in L_ρ.*

We continue with a useful separability property.

LEMMA 6.67. *Let (L, τ) be a Hausdorff locally convex-solid Riesz space. If L' is $|\sigma|(L', L)$-separable, then each principal ideal L_u of L is τ-separable.*

PROOF. Let $\{\phi_n\}$ be $|\sigma|(L', L)$-dense in L' and let $0 < u \in L$. For each n use Theorem 1.67 to choose u_n with $|u_n| \le u$ and such that $|\phi_n(u_n)| \ge \frac{1}{2}|\phi_n|(u)$. Let M be the set consisting of all (finite) rational linear combinations of $\{u_n\}$. We claim that the τ-closure $\overline{M}$ of M includes L_u.

Indeed, if L_u is not included in $\overline{M}$, then there exists some $v \in L_u$ with $v \notin \overline{M}$ and so, by the separation theorem, there exists some $\phi \in L'$ such that $\phi(v) = 1$ and $\phi(w) = 0$ for all $w \in M$. But then for every $\epsilon > 0$ there exists some $k \in \mathbb{N}$ (depending on ϵ) satisfying $|\phi - \phi_k|(u) < \epsilon$ and consequently

$$|\phi_k|(u) \le 2\big|\phi_k(u_k)\big| = 2\big|(\phi_k - \phi)(u_k)\big| \le 2\big|\phi_k - \phi\big|(u) < 2\epsilon\,.$$

This implies

$$|\phi|(u) \le |\phi_k|(u) + |\phi - \phi_k|(u) < 2\epsilon + \epsilon = 3\epsilon$$

for all $\epsilon > 0$ and hence $|\phi|(u) = 0$. Thus, $|\phi|(x) = 0$ for all $x \in L_u$, and therefore, $\phi(x) = 0$ for all $x \in L_u$, contradicting $\phi(v) = 1$ and $v \in L_u$.

Hence $L_u \subseteq \overline{M}$ and the conclusion follows by observing that M is a countable subset of L_u. ■

THEOREM 6.68. *Let (L, τ) be a topologically complete Hausdorff locally convex-solid Riesz space. If L'' is $\beta(L'', L')$-separable, then (L, τ) is semireflexive.*

PROOF. By Theorem 3.30 the topology $\beta(L'', L')$ is pre-Lebesgue and hence a Lebesgue topology by Theorem 4.8 and Corollary 6.12. Now let $\{u_n\} \subseteq L^+$ be a disjoint sequence with τ-bounded partial sums. By Theorem 6.16 we have $\sum_{i=1}^n u_i \uparrow u$ in L'' for some $u \in L''$ and hence $\{u_n\}$ is $\beta(L'', L')$-convergent to zero. From this it follows that $u_n \xrightarrow{\tau} 0$, for if V is a τ-neighborhood of zero then V° is $\beta(L', L)$-bounded and so $\{u_n\}$ converges to zero uniformly on V°. Therefore, by Theorem 6.63(iii) we conclude that $L = (L')^{\sim}_{\mathrm{n}}$.

Now, by Lemma 6.67, it follows that each principal ideal L'_ϕ of L' is separable for $\beta(L', L)$. So, according to Theorem 3.30, the topology $\beta(L', L)$ induces a pre-Lebesgue topology on L'_ϕ that is in fact a Lebesgue topology on L'_ϕ (in view of Theorem 4.8 and Corollary 6.12 again). So each $f \in L''$ is order continuous on L'_ϕ, from which it follows that $L'' \subseteq (L')^{\sim}_{\mathrm{n}}$. Therefore, $L = L''$ and the proof is finished. ■

Exercises

1. Let A be an ideal in a locally solid Riesz space (L,τ) and let P be the projection of L' onto $(A^{\circ})^{\mathrm{d}}$. Show that

$$P\phi(u) = \sup\{\phi(v)\colon\ v \in A \text{ and } 0 \le v \le u\}$$

holds for all $0 \le \phi \in L'$ and all $u \in L^+$.

2. Let (L,τ) be a Hausdorff locally convex-solid Riesz space. Assume that a convex solid subset S of L has the property that every positive increasing sequence of S has a supremum in L. Show that S is τ-bounded.

3. Let (L,τ) be a Hausdorff locally convex-solid Riesz space. Show that the topological dual of $\big(L', |\sigma|(L',L)\big)$ coincides with L if and only if L is Dedekind complete and τ is a Lebesgue topology.

4. Let (L,τ) be a Hausdorff locally convex-solid Riesz space with an order unit. Show that a convex solid and $\sigma(L,L')$-compact subset of L must be order bounded.

5. Let L_ρ be an arbitrary AL-space. If a subset A of L is relatively $\sigma\big(L_\rho, L'_\rho\big)$-compact, then show that the solid hull of A is also relatively $\sigma\big(L_\rho, L'_\rho\big)$-compact and that A is contained in a principal band of L.

6. Let (L,τ) be a Hausdorff locally convex-solid Riesz space and let I be the ideal generated by L in L''. If S is a convex and solid subset of L', show that S is $\sigma(L',I)$-compact if and only if S is order-equicontinuous on L and that the supremum of every increasing positive net of S exists in L' and belongs to S.

7. Assume that (L,τ) is a Hausdorff locally convex-solid Riesz space and that two nets $\{u_\alpha\}$ and $\{v_\alpha\}$ (with the same index set) satisfy $0 \le u_\alpha \le v_\alpha + w$ for all α and some fixed $w \in L^+$. If $v_\alpha \xrightarrow{\tau} 0$ and the order interval $[0,w]$ is weakly compact, then show that the net $\{u_\alpha\}$ has a weakly convergent subnet.

8. Let (L,τ) be a τ-complete Hausdorff locally convex-solid Riesz space and let L' be $\beta(L',L)$-separable. Show that (L,τ) is semireflexive if and only if (L,τ) has the σ-Levi property.

9. Let L be an Archimedean Riesz space with an order unit e. Define the Riesz norm $\rho(u) = \inf\{\lambda > 0\colon\ |u| \le \lambda e\}$. If L is an ideal of $(L^\sim)^\sim_{\mathrm{n}}$, then show that L_ρ is reflexive.

10. Let (L,τ) be a Hausdorff locally convex-solid Riesz space. An element $u \in L^+$ is called a **quasi-interior point** (or a **strictly positive vector**) if the principal ideal L_u generated by the element u is τ-dense.

Show that if $u \in L^+$ is a quasi-interior point and the order interval $[0,u]$ is $\sigma(L,L')$-compact, then each order interval of $\widehat{L}$ is $\sigma\big(\widehat{L}, \widehat{L}'\big)$-compact.

11. Let (L,τ) be a Hausdorff locally convex-solid Riesz space and for each $u \in L^+$ put $\langle 0,u\rangle = \big\{\widehat{u} \in \widehat{L}\colon\ 0 \le \widehat{u} \le u\big\}$. Show that $\langle 0,u\rangle$ is weakly compact if and only if for each disjoint sequence $\{v_n\} \subseteq \langle 0,u\rangle \cap L$ we have $v_n \xrightarrow{\tau} 0$.

12. Let (L, τ) be a Hausdorff locally convex-solid Riesz space such that $\sigma(L', L)$ is a locally solid topology—that is, $\sigma(L', L) = |\sigma|(L', L)$. Show that L is Dedekind complete, τ is a Lebesgue topology and that both L and L' are discrete spaces. [*Hint*: Use Exercise 3 above and the fact that each order interval of L' is $|\sigma|(L', L)$-compact.]

13. Show that every band in a semireflexive Riesz space with the induced topology is itself a semireflexive Riesz space.

14. Let (L, τ) be a semireflexive Riesz space and let A be a band of L. Show that $(L/A, \tau/A)$ is a semireflexive Riesz space.

15. Let L be a Riesz space with $L_n^\sim$ separating the points of L. Show that $(L, |\sigma|(L, L_n^\sim))$ is semireflexive if and only if L is a perfect Riesz space and $\beta(L_n^\sim, L)$ is a Lebesgue topology.

16. Let L be a Dedekind σ-complete Riesz space such that $L_n^\sim$ separates the points of L. Show that the Mackey topology $\tau(L, L_n^\sim)$ is locally solid and Lebesgue.

17. Assume that (X, Σ, μ) is an arbitrary σ-finite measure space and for simplicity put $L_\infty = L_\infty(X, \Sigma, \mu)$ and $L_1 = L_1(X, \Sigma, \mu)$. Establish the following.
(a) $(L_\infty)_n^\sim = (L_1')_n^\sim = L_1$.
(b) The Mackey topology $\tau(L_\infty, L_1)$ is a locally convex-solid Lebesgue topology on L_∞.
(c) The Mackey topology $\tau(\ell_\infty, \ell_1)$ is a locally convex-solid topology on ℓ_∞.

18. Let L be a Dedekind σ-complete Riesz space with $L_c^\sim$ separating the points of L. Show that the Mackey topology $\tau(L, L_c^\sim)$ is a locally solid topology on L.

19. (Grothendieck) Show that a subset A of a Banach space X is relatively weakly compact if and only if for each $\epsilon > 0$ there is some weakly compact set W such that $A \subseteq W + \epsilon U$, where U is the closed unit ball of X.

20. (Kaplan [**73**, Note II, p. 348]) Combining Theorems 6.9 and 6.10 it is not difficult to see that if $L = L_\infty[0, 1]$, then $L_1[0, 1]$ is a band in $L^\sim = L_1''[0, 1]$ and so $L^\sim = L_1[0, 1] \oplus L_1^d[0, 1]$. Show that $L_1^d[0, 1]$ separates the points of L.

CHAPTER 7

Lateral Completeness

This chapter deals with laterally and laterally σ-complete Riesz spaces, that is, with Riesz spaces having the property that the supremum of every disjoint subset or disjoint sequence of positive elements exists. The material is spread out in five sections.

The first section discusses the remarkable lattice properties of laterally complete Riesz spaces. For instance, it will be shown that Archimedean laterally σ-complete Riesz spaces satisfy the principal projection property, that Archimedean laterally complete Riesz spaces satisfy the projection property, and that in Archimedean Riesz spaces lateral and Dedekind completeness are independent properties.

The second section presents a comprehensive study of the universal completion of an Archimedean Riesz space. That is, given an Archimedean Riesz space L, it is shown that there exists a (uniquely) determined laterally complete and Dedekind complete Riesz space (called the universal completion of L) that "contains" L as an order dense Riesz subspace. The third section deals with some interesting special lattice properties of laterally complete Riesz spaces.

The fourth section discusses locally solid topologies on laterally or laterally σ-complete Riesz spaces. As we shall see, on an arbitrary laterally σ-complete Riesz space every Hausdorff locally solid topology must satisfy the pre-Lebesgue and σ-Lebesgue properties; a laterally σ-complete Riesz space can carry at most one Hausdorff Fatou topology (which must be necessarily Lebesgue) and that if it can carry this topology, then every other Hausdorff locally solid topology is finer and agrees with this Fatou topology on an order dense σ-ideal.

The fifth and final section of this chapter deals with minimal topologies on an Archimedean Riesz space L and makes (by means of the universal completion of L) a thorough investigation of their remarkable properties.

7.1. Laterally Complete Riesz Spaces

We start this section with the definition of laterally complete and laterally σ-complete Riesz spaces.

DEFINITION 7.1. *A Riesz space L is said to be:*

(1) ***laterally σ-complete**, if the supremum of every disjoint sequence of L^+ exists in L, and*
(2) ***laterally complete**, if the supremum of every disjoint subset of L^+ exists in L.*[1]

[1] In the terminology of W. A. J. Luxemburg and A. C. Zaanen a laterally complete Riesz space is referred to as a *universally complete Riesz space* [**90**, Definition 47.3, p. 323];

The lexicographic plane provides an example of a laterally complete Riesz space that is not Archimedean.

THEOREM 7.2. *For a laterally complete Riesz space L we have the following properties.*

(i) *Every band of L is a laterally complete Riesz space in its own right.*
(ii) *If L is Archimedean, then:*
   (a) *L contains a weak order unit.*
   (b) *Every band of L is a principal band.*

PROOF. (i) Let B be a band of L and let S be a disjoint subset of positive elements of B. Then $u = \sup S$ exists in L and u belongs to B, since B is a band of L. But then $u = \sup S$ also holds in B, and this shows that B is laterally complete.

(ii) Let $\{e_\alpha\}$ be a complete disjoint system of positive elements of L^+. Since L is laterally complete, $e = \sup\{e_\alpha\}$ exists in L and since L is Archimedean, e must be a weak order unit of L.

Now let B be a band of L. By part (i) the band B is laterally complete and so B contains a weak order unit e. But then, since L is Archimedean, $B = B_e$ (the band generated by e in L) holds, and the proof is finished. ■

The following result can be proven as in (i) above.

THEOREM 7.3. *Every σ-ideal of a laterally σ-complete Riesz space is a laterally σ-complete Riesz space in its own right.*

One of the most remarkable properties of the Archimedean laterally complete Riesz spaces is included in the next result that is due to A. I. Veksler and V. A. Geiler [**126**].

THEOREM 7.4 (Veksler–Geiler). *For an Archimedean Riesz space L we have the following.*

(i) *If L is laterally σ-complete, then L satisfies the principal projection property.*
(ii) *If L is laterally complete, then L satisfies the projection property.*

PROOF. The elegant proof below is due to S. Bernau [**37**, Theorem 1, p. 320]. Let $u, v \in L^+$ and for each $n = 0, 1, 2, \ldots$ let

$$w_n = \big[(n+2)u - v\big]^+ \wedge (nu - v)^- \,.$$

It is easy to check that $w_n \wedge w_{n+p} = 0$ holds for all $n \geq 0$ and all $p \geq 2$. Observe also that since $0 \leq w_n \leq (n+2)u$ for all $n \geq 0$, we have $w_n \in B_u$, the band generated by u in L. From the inequalities

$$\begin{aligned}
&\big\{n\big[(n+1)u - v\big]^+\big\} \wedge v + (n+1)w_n \\
&\quad \geq \big\{n\big[(n+1)u - v\big]^+ + (n+1)w_n\big\} \wedge v \\
&\quad = \big\{n\big[(n+1)u - v\big]^+ + (n+1)\big[\big((n+2)u - v\big)^+ \wedge \big(nu - v\big)^-\big]\big\} \wedge v \\
&\quad \geq \big\{(n+1)\big[(n+2)u - v\big]^+\big\} \wedge \big\{n\big[(n+1)u - v\big] + (n+1)\big[v - nu\big]\big\} \wedge v \\
&\quad = \big\{(n+1)\big[(n+2)u - v\big]^+\big\} \wedge v \,,
\end{aligned}$$

here a universally complete Riesz space will be defined to be a laterally and Dedekind complete Riesz space (see Definition 7.18). The term *laterally complete* is taken from the theory of lattice groups; see [**36**] and [**47**].

it follows that

$$(n+1)w_n \geq \{(n+1)[(n+2)u-v]^+\} \wedge v - \{n[(n+1)u-v]^+\} \wedge v$$

for all $n = 0, 1, 2, \ldots$. Consequently,

$$\sum_{r=0}^{n}(r+1)w_r \geq \{(n+1)[(n+2)u-v]^+\} \wedge v \geq (nu-v)^+ \wedge v\,. \qquad (\star)$$

Since L is laterally σ-complete and $w_n \wedge w_m = 0$ if $|m-n| \geq 2$, the suprema

$$f = \bigvee_{n=0}^{\infty} (2n+1)w_{2n} \quad \text{and} \quad g = \bigvee_{n=0}^{\infty} (2n+2)w_{2n+1}$$

both exist in L and obviously $f, g \in B_u$. In particular, $v \wedge (f+g) \in B_u$.

We shall show next that $v - v \wedge (f+g) \in B_u^{\mathrm{d}}$. To this end, observe that for every $k \geq 0$ it follows from $(\star)$ that

$$\begin{aligned} u \wedge [v - v \wedge (f+g)] &\leq u \wedge \Big[v - v \wedge \sum_{r=0}^{k}(r+1)w_r\Big] \\ &\leq u \wedge [v - v \wedge (ku-v)^+] \\ &\leq u \wedge [v - v \wedge (ku-v)] \\ &= u \wedge (2v - 2v \wedge ku) \\ &= (u+2v) \wedge (u+ku) \wedge 2v - 2v \wedge ku \\ &= 2v \wedge (k+1)u - 2v \wedge ku\,, \end{aligned}$$

and hence $n[u \wedge (v - v \wedge (f+g))] \leq \sum_{k=0}^{n-1}[2v \wedge (k+1)u - 2v \wedge ku] \leq 2v$ holds for $n = 1, 2, 3, \ldots$. Now, since the Riesz space L is Archimedean, it follows that $u \wedge [v - v \wedge (f+g)] = 0$ and so $v - v \wedge (f+g) \in B_u^{\mathrm{d}}$. Thus $L = B \oplus B_u^{\mathrm{d}}$ and this shows that L has the principal projection property.

(ii) This claim follows from Theorem 7.2(ii) and part (i) above. ■

The next four results are applications of the preceding theorem. The first theorem characterizes the Riesz spaces that are Dedekind and laterally complete.

THEOREM 7.5. *For an Archimedean Riesz space L the following statements are equivalent.*

(i) *L is Dedekind complete and laterally complete.*
(ii) *L is Dedekind σ-complete and laterally complete.*
(iii) *L is uniformly complete and laterally complete.*

PROOF. (i) $\Longrightarrow$ (ii) and (ii) $\Longrightarrow$ (iii) are obvious. To see that (iii) $\Longrightarrow$ (i) note first that by Theorem 7.4(ii) the Riesz space L has the projection property, and then that the uniform completeness and the projection property are equivalent to Dedekind completeness; see Theorem 1.59. ■

The proof of the next result is similar to the preceding one.

THEOREM 7.6. *An Archimedean laterally σ-complete Riesz space is Dedekind σ-complete if and only if it is uniformly complete.*

We continue with a useful approximation result.

LEMMA 7.7. *Let L be an Archimedean laterally σ-complete Riesz space and let $u_n \downarrow 0$ in L. Then for every $\epsilon > 0$ there exists some $u \in L^+$ (depending on ϵ) such that for each n we have*

$$0 \leq u_n \leq 2^{-n}u + \epsilon u_1 .$$

PROOF. Let $u_n \downarrow 0$ and fix $\epsilon > 0$. We denote by P_n the projection of L onto the band generated by $(u_n - \epsilon u_1)^+$ in L (the projection P_n exists by Theorem 7.4). Put $v_n = P_n(u_1)$ and $w_n = u_1 - v_n$ and observe that $w_n \wedge (u_n - \epsilon u_1)^+ = 0$. Since

$$0 \leq (u_n - \epsilon u_1)^+ = P_n(u_n - \epsilon u_1) = P_n(u_n) - \epsilon v_n \leq u_n - \epsilon v_n ,$$

we get $0 \leq \epsilon v_n \leq u_n$ and hence $v_n \downarrow 0$. Note next that each v_n is a component of u_1 and if $k > n$, then we have $v_{k+1} \leq v_k \leq v_{n+1} \leq v_n$. So, if $k > n$, then $(v_n - v_{n+1}) \wedge (v_k - v_{k+1}) = 0$ holds; see part (3) of Theorem 1.53. In particular, it follows that $\{2^{n+1}(v_n - v_{n+1})\}$ is a disjoint sequence of L^+, and consequently $0 \leq 2^{n+1}(v_n - v_{n+1}) \leq u$ must hold in L for all n and some $u \in L^+$. Also, from the inequalities $(u_n \wedge w_n - \epsilon u_1)^+ \leq (u_n - \epsilon u_1)^+$ and $(u_n \wedge w_n - \epsilon u_1)^+ \leq w_n$, we get

$$0 \leq (u_n \wedge w_n - \epsilon u_1)^+ \leq (u_n - \epsilon u_1)^+ \wedge w_n = 0 .$$

This implies $u_n \wedge w_n \leq \epsilon u_1$ for all n, and thus

$$\begin{aligned} 0 \leq u_n &= u_n \wedge u_1 = u_n \wedge (v_n + w_n) \leq u_n \wedge v_n + u_n \wedge w_n \\ &\leq v_n + \epsilon u_1 = (v_n - v_k) + \epsilon u_1 + v_k \\ &= \sum_{i=n}^{k-1} (v_i - v_{i+1}) + \epsilon u_1 + v_k \\ &\leq \Big[\sum_{i=n}^{\infty} 2^{-i-1}\Big] u + \epsilon u_1 + v_k \\ &= 2^{-n}u + \epsilon u_1 + v_k \end{aligned}$$

for all $k > n$. Since $v_n \downarrow 0$, we get $0 \leq u_n \leq 2^{-n}u + \epsilon u_1$ for all n. ■

Positive operators from Archimedean laterally σ-complete Riesz spaces to Archimedean Riesz spaces are always σ-order continuous.

THEOREM 7.8. *Let $T\colon L \to M$ be a positive operator between two Archimedean Riesz spaces. If L is laterally σ-complete, then the operator T is σ-order continuous.*

PROOF. Let $u_n \downarrow 0$ and let $\epsilon > 0$. According to Lemma 7.7 there exists some $u \in L^+$ such that $0 \leq u_n \leq 2^{-n}u + \epsilon u_1$ holds in L for all n.

Now if $0 \leq w \leq T(u_n)$ holds in M for all n, then it follows from the positivity of T that $0 \leq w \leq T(u_n) \leq 2^{-n}T(u) + \epsilon T(u_1)$ for all n. Now taking into account that M is an Archimedean Riesz space, we see that $0 \leq w \leq \epsilon T(u_1)$ holds for all $\epsilon > 0$. Using once more that M is Archimedean we get $w = 0$. Thus $T(u_n) \downarrow 0$ holds in M, so that T is σ-order continuous. ■

Note that a σ-order continuous operator between two Archimedean Riesz spaces need not be order continuous even if its domain is laterally complete. To see this, let $L = \mathbb{R}^X$ for some uncountable set X, and let A be the σ-ideal of L defined by

$$A = \big\{ f \in \mathbb{R}^X \colon \text{ The set } \{x \in X \mid f(x) \neq 0\} \text{ is at most countable} \big\} .$$

It follows that L/A is Archimedean. However, since A is not a band of L, the canonical projection of L onto L/A is not order continuous; see Lemma 1.32.

The next result is due to D. H. Fremlin [**59**, pp. 78–79].

THEOREM 7.9 (Fremlin). *For an Archimedean Riesz space we have the following.*

(i) *If L is laterally σ-complete and A is a σ-ideal of L, then L/A is an Archimedean laterally σ-complete Riesz space.*
(ii) *If L is laterally complete and A is a band of L, then L/A is an Archimedean laterally complete Riesz space.*

PROOF. (i) Let $\{\dot{u}_n\}$ be a disjoint sequence of positive elements of L/A. We can assume $\{u_n\} \subseteq L^+$. Let $v_1 = u_1$, and inductively if $v_1, \ldots, v_n$ have been selected, put $w_n = v_1 + \cdots + v_n$, and define

$$v_{n+1} = u_{n+1} - \sup_{m \in \mathbb{N}} u_{n+1} \wedge m w_n = u_{n+1} - P_{w_n}(u_{n+1}).$$

Observe that $\{v_n\}$ is a disjoint sequence. Indeed, for $m < n$ we have $v_m \leq w_{n-1}$ and so $P_{v_m}(u_n) \leq P_{w_{n-1}}(u_n)$, from which it follows that

$$\begin{aligned} 0 &\leq v_m \wedge v_n = v_m \wedge \big[u_n - P_{w_{n-1}}(u_n)\big] \\ &\leq v_m \wedge \big[u_n - P_{v_m}(u_n)\big] \in B_{v_m} \cap B_{v_m}^{\mathrm{d}} = \{0\}. \end{aligned}$$

Therefore $v_m \wedge v_n = 0$.

Now put $v = \sup\{v_n\} \in L$, and note that $\dot{v}_n = \dot{u}_n$ for all n and that the canonical projection of L onto L/A is a Riesz σ-homomorphism (since A is a σ-ideal). Hence $\dot{v} = \sup\{\dot{u}_n\}$ holds in L/A, so that L/A is laterally σ-complete. The Archimedeaness of L/A follows from the fact that A is a σ-ideal.

(ii) By Theorem 7.4(ii) the band A is a projection band of L and so L/A is Riesz isomorphic to A^{d}. ■

An important tool for the study of laterally complete Riesz spaces is the notion of a dominable set that was introduced by D. H. Fremlin [**59**, Definition 1.1(c), p. 72].

DEFINITION 7.10 (Fremlin). *Let L be a Riesz space and let A be a nonempty subset of L^+. Then A is said to be a* ***dominable set*** *if for every $0 < u \in L$ there exists some $0 < v \in L$ and a positive integer k satisfying $(ku - a)^+ \geq v$ for all $a \in A$.*

We shall see later (Theorem 7.37) that the dominable sets of an Archimedean Riesz space are precisely the subsets of L^+ that are order bounded in the universal completion of L.

We continue with a simple but very useful lemma.

LEMMA 7.11. *Let L be an order dense Riesz subspace of a Riesz space M and let A be a nonempty subset of L^+. Then A is a dominable subset of L if and only if A is a dominable subset of M.*

PROOF. Assume first that A is dominable in M. Let $0 < u \in L$. Then there exist $0 < u^* \in M$ and some $k \in \mathbb{N}$ such that $(ku - a)^+ \geq u^*$ for all $a \in A$. Now pick $0 < v \in L$ with $0 < v \leq u^*$, and note that $(ku - a)^+ \geq v$ for all $a \in A$, that is, A is a dominable subset of L.

On the other hand, if A is a dominable subset of L and $0 < u^* \in M$, pick $u \in L$ with $0 < u \leq u^*$. Now choose $0 < v \in L$ and a positive integer k satisfying

$(ku^* - a)^+ \geq (ku - a)^+ \geq v$ for all $a \in A$. Hence A is a dominable subset of M, and the proof is finished. ■

It should be clear that subsets of dominable sets are also dominable. More important properties of the dominable sets are included in the next few results.

LEMMA 7.12. *Let L be a Riesz space with the principal projection property and let A be a dominable subset of L^+. Then for each $u > 0$ there exist $0 < v \leq u$ and a positive integer n such that for each $a \in A$ we have:*

$$P_v(a) \leq nv\,.$$

PROOF. Let $u > 0$. Since A is a dominable set, there exists some $n \in \mathbb{N}$ and some $w > 0$ satisfying $nu - a \wedge nu = (nu - a)^+ \geq w$ for all $a \in A$ or

$$a \wedge nu \leq nu - w \quad \text{ for all } a \in A\,. \tag{$\star$}$$

Observe now that $0 < w \leq nu$ implies $w \wedge u > 0$ and so $v = P_w(u) > 0$. Applying P_w to $(\star)$ and using that P_w is a Riesz homomorphism, we get $P_w(a) \wedge nv \leq nv - w$. This implies

$$0 < w \leq nv - P_w(a) \wedge nv = \big[nv - P_w(a)\big]^+ = \big[P_w(a) - nv\big]^- \tag{$\star\star$}$$

for all $a \in A$.

Now if for any $a \in A$ we let $f = \big[P_w(a) - nv\big]^+$, then on one hand we have $f \in B_w$ and on the other hand, it follows from $(\star\star)$ that $f \perp w$ or $f \in B_w^{\mathrm{d}}$. Consequently, $f = 0$ or $P_w(a) \leq nv$ for all $a \in A$. Now since $v \in B_w$ implies $P_v(a) \in B_w$, it follows that

$$P_v(a) = P_w\big(P_v(a)\big) = P_v\big(P_w(a)\big) \leq nP_v(v) = nv$$

holds for all $a \in A$, and the proof is finished. ■

THEOREM 7.13. *Let L be a Riesz space with the principal projection property and let A be a dominable subset of L^+. Then there exists a complete disjoint system $\{e_i\colon\ i \in I\}$ of nonzero positive elements of L such that for each $i \in I$ there exists a positive integer n_i (depending on i) satisfying*

$$P_{e_i}(a) \leq n_i e_i$$

for all $a \in A$.

PROOF. Consider the collection $\mathcal{C}$ of all sets of the form $\{e_i\colon\ i \in I\}$, where $e_i > 0$ for each i, $e_i \wedge e_j = 0$ if $i \neq j$, and such that for each $i \in I$ there exists some $n_i \in \mathbb{N}$ (depending on i) satisfying $P_{e_i}(a) \leq n_i e_i$ for all $a \in A$. By Lemma 7.12 the collection $\mathcal{C}$ is nonempty and by Zorn's lemma there exists a maximal disjoint set with the above property, say $\{e_i\colon\ i \in I\}$. We claim now that $\{e_i\colon\ i \in I\}$ is a complete disjoint system of nonzero positive elements, i.e., $\{e_i\colon\ i \in I\}^{\mathrm{dd}} = L$. Indeed, if $0 < u \in L$ satisfies $u \wedge e_i = 0$ for all $i \in I$, then by Lemma 7.12 there exists $0 < v \leq u$ and a positive integer n satisfying $P_v(a) \leq nv$ for all $a \in A$. But then by adding v to the collection $\{e_i\colon\ i \in I\}$ we contradict the maximality property of the collection $\{e_i\colon\ i \in I\}$. This completes the proof. ■

We are now ready to characterize the dominable sets of the Archimedean laterally complete Riesz spaces.

THEOREM 7.14. *If L is an Archimedean laterally complete Riesz space, then a subset A of L^+ is dominable if and only if it is order bounded in L.*

PROOF. Assume first that A is a (nonempty) dominable set. According to Theorems 7.4 and 7.13 there exists a complete disjoint system $\{e_i\colon\ i \in I\}$ of nonzero positive elements with the property that for each $i \in I$ there exists a positive integer n_i satisfying $P_{e_i}(a) \le n_i e_i$ for all $a \in A$. Now since L is laterally complete, $u = \sup\{n_i e_i\colon\ i \in I\}$ exists in L and so, by Theorem 1.75, we have

$$a = \sup\{P_{e_i}(a)\colon\ i \in I\} \le \sup\{n_i e_i\colon\ i \in I\} = u$$

for all $a \in A$. Thus A is order bounded, and the proof is finished.

For the converse suppose that $0 \le a \le u$ holds in L for all $a \in A$ and let $w > 0$. Since L is Archimedean, there exists a positive integer k such that $(kw - u)^+ > 0$. But then $(kw - a)^+ \ge (kw - u)^+ = v$ holds for all $a \in A$, and hence A is dominable. (Note that the lateral completeness is not needed for the proof of this part.) ■

Among the most important properties of the laterally complete Riesz spaces are the ones included in the next theorem.

THEOREM 7.15. *For an order dense Riesz subspace L of an Archimedean Riesz space M we have the following.*

(i) *If L is laterally complete, then L majorizes M.*
(ii) *If L is laterally and Dedekind complete, then $L = M$.*

PROOF. (i) Let $u \in M$. Since L is order dense in M and M is Archimedean, there exists a net $\{u_\alpha\} \subseteq L^+$ with $u_\alpha \uparrow |u|$ in M; see Theorem 1.27. But then $\{u_\alpha\}$ is a dominable subset of M, and hence by Lemma 7.11 a dominable subset of L. Now according to Theorem 7.14 there exists some $w \in L^+$ satisfying $u_\alpha \le w$ for all α. It now follows that $|u| \le w$ holds, so that L majorizes M.

(ii) Use part (i) in conjunction with Theorem 1.40. ■

LEMMA 7.16. *Let L be an order dense Riesz subspace of an Archimedean Riesz space M. If L is Dedekind complete, then every element of M^+ is the supremum of a subset of pairwise disjoint vectors of L^+.*

PROOF. Note that L is an ideal of M; see Theorem 1.40. Now let $0 < u \in M$. Since M is Archimedean and L is order dense in M, there exists a net $\{u_\alpha\}$ of L with $0 \le u_\alpha \uparrow u$ in M. It follows that $\{u_\alpha\}$ is dominable in L^+ and so by Theorem 7.13 there exists a complete disjoint system $\{e_i\colon\ i \in I\}$ of L^+ such that for each $i \in I$ we have $P_{e_i}(u_\alpha) \le n_i e_i$ for all α and some fixed positive integer n_i. Now for each $i \in I$ define $w_i = \sup_\alpha P_{e_i}(u_\alpha) \le n_i e_i \in L^+$, and note that $\{w_i\colon\ i \in I\}$ is a disjoint subset of L^+. To complete the proof we shall show that $u = \sup_{i \in I} w_i$ holds in M.

Observe first that $P_{e_i}(u_\alpha) \le u_\alpha \le u$ for all α and $i \in I$ implies $w_i \le u$ for all $i \in I$. Now assume that $w_i \le v \in M$ holds for all $i \in I$. This implies $P_{e_i}(u_\alpha) \le v$ for all i and all α. But then, by Theorem 1.75, we get $u_\alpha = \sup_{i \in I} P_{e_i}(u_\alpha) \le v$ for all α, and hence $u \le v$. The proof is now complete. ■

We continue with an important extension theorem.

THEOREM 7.17. *Let $T\colon L \to K$ be a normal Riesz homomorphism from a Dedekind complete Riesz space L to an Archimedean laterally complete Riesz space K. If L is an order dense Riesz subspace of an Archimedean Riesz space M, then T has a unique normal Riesz homomorphic extension from M to K.*

In particular, if T is a normal Riesz isomorphism (into), then so is its extension.

PROOF. Let $0 < u \in M$. By Lemma 7.16 there exists a disjoint system $\{u_\alpha\}$ of L^+ with $u = \sup_\alpha u_\alpha$ in M. Then $\{T(u_\alpha)\}$ is a disjoint system of K^+ (since T is a Riesz homomorphism), and hence since K is laterally complete, $u^* = \sup_\alpha T(u_\alpha)$ exists in K. Now, if the net $\{v_\lambda\} \subseteq L^+$ satisfies $\sup_\lambda v_\lambda = u \in M$, then for each fixed λ we have $v_\lambda = \sup_\alpha v_\lambda \wedge u_\alpha$ in L and hence from the order continuity of T we get $T(v_\lambda) = \sup_\alpha T(v_\lambda \wedge u_\alpha) \leq u^*$. On the other hand, if $T(v_\lambda) \leq v^*$ holds in K for all λ, then from $u_\alpha = \sup_\lambda u_\alpha \wedge v_\lambda$ and the order continuity of T it follows that $T(u_\alpha) \leq v^*$ holds for all α and hence $u^* = \sup_\lambda T(v_\lambda)$ holds in K. Thus, the element u^* is independent of the system chosen from L^+. So, we extend T to M^+ by letting $T(u) = u^*$ for each $u \in M^+$.

This extension of T from M^+ to K^+ satisfies $T(u+v) = T(u) + T(v)$ for all $u, v \in M^+$. By Lemma 1.64 the mapping $T\colon M^+ \to K^+$ extends (uniquely) to a positive operator on all of M. The rest of the proof is left as an exercise for the reader. ■

We continue with the definitions of universally complete and universally σ-complete Riesz spaces.

DEFINITION 7.18. *A Riesz space that is both laterally and Dedekind complete is called a* ***universally complete Riesz space****.*

Similarly, a Dedekind σ-complete and laterally σ-complete Riesz space is called a ***universally σ-complete Riesz space****.*

Again we remind the reader that in the terminology of W. A. J. Luxemburg and A. C. Zaanen a laterally complete Riesz space is referred to as a universally complete Riesz space [**90**, Definition 47.3, p. 323]. D. H. Fremlin, inspired by Theorem 7.15(ii), calls a universally complete Riesz space an **inextensible Riesz space** [**59**, Definition 1.1(a), p. 72]. In the Russian literature a universally complete Riesz space is called an **extended Riesz space**; see [**109, 110**] and [**127**, pp. 140–144].

As an application of Theorem 7.17 we have the following result.

THEOREM 7.19. *Assume that two universally complete Riesz spaces L and M include two order dense Riesz subspaces L_1 and M_1, respectively, that are Riesz isomorphic, say via $\pi\colon L_1 \to M_1$. Then π extends uniquely to a Riesz isomorphism from L onto M.*

PROOF. Note first that since M_1 is order dense in M the operator $\pi\colon L_1 \to M$ is a normal Riesz homomorphism, and then that the ideal A generated by L_1 in L is the Dedekind completion of L_1. Next extend π to A^+ by

$$\pi(f) = \sup\{\pi(u)\colon\ u \in L^+ \text{ and } 0 \leq u \leq f\}$$

if $f \in A^+$ and then to all of A by $\pi(f) = \pi(f^+) - \pi(f^-)$. Clearly, this extension is a normal Riesz isomorphism (into). Now by Theorem 7.17 the operator π can be extended to a normal Riesz isomorphism from L into M. To complete the proof it suffices to show that the above extension of π is onto.

To this end, let $0 < u^* \in M$. Then there exists a net $\{u_\alpha\}$ of M_1^+ with $0 \leq u_\alpha \uparrow u^*$ in M. Pick a net $\{v_\alpha\}$ of L_1^+ with $u_\alpha = \pi(v_\alpha)$ for each α. Now observe that $\{u_\alpha\}$ is a dominable subset of M and hence of M_1. Thus, $\{v_\alpha\}$ is a dominable subset of L_1 and consequently of L. Theorem 7.14 implies now that

$0 \le v_\alpha \uparrow v^*$ holds in L for some $v^* \in L$, and therefore $u_\alpha = \pi(v_\alpha) \uparrow \pi(v^*) = u^*$, that is, π is onto. ■

7.2. The Universal Completion

We start this section with the definition of the universal completion and the universal σ-completion of a Riesz space.

DEFINITION 7.20. *A* ***universal completion*** *(resp. a* ***universal σ-completion****) of a Riesz space L is a universally complete (resp. a universally σ-complete) Riesz space K having an order dense (resp. a super order dense) Riesz subspace M that is Riesz isomorphic to L [and so identifying L with M we can consider L as an order dense Riesz subspace of K.]*

It should be clear that only Archimedean Riesz spaces can have universal completions and by Theorem 7.19 any two universal completions of a Riesz space are necessarily Riesz isomorphic. The objective of this section is to establish the following fundamental result, due to F. Maeda and T. Ogasawara [**91**], that guarantees the existence of universal completions.

THEOREM 7.21 (Maeda–Ogasawara). *Every Archimedean Riesz space has a unique (up to Riesz isomorphism) universal completion.*

As noted above, the uniqueness (up to Riesz isomorphism) of the universal completion follows immediately from Theorem 7.19. So, it makes sense to talk about "the universal completion" of an Archimedean Riesz space and we reserve a special symbol for it.

DEFINITION 7.22. *The universal completion of an Archimedean Riesz space L will be denoted by L^{u}.*

Note also that if L is an order dense Riesz subspace of an Archimedean Riesz space M, then the universal completion M^{u} of M serves equally well as the universal completion of L, since obviously L "sits" order densely in M^{u}. Thus, the universal completion L^{u} is Riesz isomorphic to M^{u} by Theorem 7.19. So, if L is order dense in M, then the universal completion of L^{u} "includes" M as an order dense Riesz subspace. More precisely, we have the following theorem.

THEOREM 7.23. *If an Archimedean Riesz space L is Riesz isomorphic to some order dense Riesz subspace of an Archimedean Riesz space K, then K is Riesz isomorphic to an order dense Riesz subspace of L^{u} (and hence L and K have the "same" universal completion).*

PROOF. Assume that π is a Riesz isomorphism between L and an order dense Riesz subspace, say K_1, of some Archimedean Riesz space K. Note that K^{u} is clearly the universal completion of K_1, and then use Theorem 7.19 to extend π to a Riesz isomorphism between L^{u} and K^{u}. The result now easily follows. ■

The rest of the discussion in this section is devoted to proving Theorem 7.21. To do this, we need to introduce some basic material. We start by recalling that the symbol $\mathbb{R}^*$ denotes the set of all **extended real numbers**, that is,

$$\mathbb{R}^* = [-\infty, \infty] = \mathbb{R} \cup \{-\infty, \infty\}.$$

The extended real numbers $\mathbb{R}^*$ are topologized in the usual manner so that $\mathbb{R}^*$ is a compact topological space that induces the standard topology on $\mathbb{R}$, where a base at the neighborhood system at ∞ is $\{(a,\infty]\colon\ a \in \mathbb{R}\}$ (and likewise a base at $-\infty$ is $\{[-\infty,a)\colon\ a \in \mathbb{R}\}$). The following simple result will be very useful for our discussion in this section.

LEMMA 7.24. *Let X be a topological space, let $f\colon X \to \mathbb{R}^*$ be a function and for each $a \in \mathbb{R}$ put*

$$V_a = \{x \in X\colon\ f(x) < a\} \quad \text{and} \quad F_a = \{x \in X\colon\ f(x) \le a\}.$$

Then the function f is continuous if and only if for each $a \in \mathbb{R}$ the set V_a is open and the set F_a is closed.[2]

PROOF. If f is continuous, then clearly V_a is open and F_a is closed for each $a \in \mathbb{R}$. For the converse, assume that V_a is open and F_a is closed for each $a \in \mathbb{R}$. If $-\infty < a < b < \infty$, then note that the sets

$$f^{-1}\big((a,b)\big) = V_b \setminus F_a\,, \quad f^{-1}\big([-\infty,a)\big) = V_a\,, \quad \text{and} \quad f^{-1}\big((b,\infty]\big) = X \setminus F_b$$

are all open subsets of X. Now observe that any open subset $\mathcal{O}$ of $\mathbb{R}^*$ can be written as union $\mathcal{O} = \bigcup_{i\in I} W_i$, where each W_i is of the form (a,b), $[-\infty,c)$, or $(d,\infty]$ with $a,b,c,d \in \mathbb{R}$. This implies that $f^{-1}(\mathcal{O}) = \bigcup_{i\in I} f^{-1}(W_i)$ is an open subset of X, and therefore the function f is continuous. ■

Recall that a topological space X is said to be **extremally disconnected** if the closure of every open set is open. An extremally disconnected Hausdorff compact topological space is referred to as a **Stonean space**. It is not difficult to see that every open subset of an extremally disconnected topological space with the induced topology is itself extremally disconnected. Also, it should be clear that in a Stonean space the clopen sets form a base for the topology.

The next result describes a basic extension property of continuous functions defined on open subsets of extremally disconnected topological spaces.

LEMMA 7.25. *Let $\mathcal{O}$ be an open subset of an extremally disconnected topological space X and assume that $f\colon \mathcal{O} \to \mathbb{R}$ is a continuous function. Then f has a unique continuous extension from $\overline{\mathcal{O}}$ (the closure of $\mathcal{O}$ in X) to $\mathbb{R}^*$.*

PROOF. Assume that $f\colon \mathcal{O} \to \mathbb{R}$ is a continuous function. For each $r \in \mathbb{R}$ consider the open set $V_r = \{x \in \mathcal{O}\colon\ f(x) < r\}$ and note that its closure $\overline{V}_r$ is open and satisfies $\overline{V}_r \subseteq \overline{\mathcal{O}}$. In particular, we have

$$\mathcal{O} \subseteq \bigcup_{r\in\mathbb{R}} \overline{V}_r \subseteq \overline{\mathcal{O}}\,.$$

Now define the function $\overline{f}\colon \overline{\mathcal{O}} \to \mathbb{R}^*$ via the formula

$$\overline{f}(x) = \begin{cases} \inf\{s \in \mathbb{R}\colon\ x \in \overline{V}_s\} & \text{if } \ x \in \bigcup_{r\in\mathbb{R}} \overline{V}_r \\ \infty & \text{if } \ x \in \overline{\mathcal{O}} \setminus \bigcup_{r\in\mathbb{R}} \overline{V}_r\,. \end{cases}$$

Now fix some $x \in \mathcal{O}$. From $f(x) < f(x) + \epsilon$ for each $\epsilon > 0$, it follows that $x \in V_{f(x)+\epsilon}$ and so $\overline{f}(x) \le f(x) + \epsilon$ for each $\epsilon > 0$. This shows that $\overline{f}(x) \le f(x)$. Now if $x \in \overline{V}_s$, then there exists a net $\{x_\alpha\} \subseteq V_s \subseteq \mathcal{O}$ such that $x_\alpha \to x$, and so by the continuity of f and $f(x_\alpha) < s$ for each α we get $f(x) = \lim f(x_\alpha) \le s$. By the definition of $\overline{f}$, we easily obtain $f(x) \le \overline{f}(x)$. Thus, $\overline{f}(x) = f(x)$ for all $x \in \mathcal{O}$, and so $\overline{f}$ is an extension of f to all of $\overline{\mathcal{O}}$.

[2] Incidentally, the same conditions characterize the continuity of a real-valued function $f\colon X \to \mathbb{R}$.

We shall finish the proof by establishing that $\overline{f}\colon \overline{\mathcal{O}} \to \mathbb{R}^*$ is continuous. To this end, let $a \in \mathbb{R}$. Observing that $-\infty < r_1 < r_2 < \infty$ implies $\overline{V}_{r_1} \subseteq \overline{V}_{r_2}$, an easy verification shows that

$$\{x \in \overline{\mathcal{O}}\colon \overline{f}(x) < a\} = \bigcup_{-\infty < s < a} \overline{V}_s \quad\text{and}\quad \{x \in \overline{\mathcal{O}}\colon \overline{f}(x) \le a\} = \bigcap_{a < s < \infty} \overline{V}_s .$$

It follows that the set $\{x \in \overline{\mathcal{O}}\colon \overline{f}(x) < a\}$ is an open set and that the set $\{x \in \overline{\mathcal{O}}\colon \overline{f}(x) \le a\}$ is a closed set. Now a glance at Lemma 7.24 guarantees that $\overline{f}\colon \overline{\mathcal{O}} \to \mathbb{R}^*$ is continuous. The uniqueness of the extension should be obvious. ■

Now let X be a fixed topological space. A continuous function $f\colon X \to \mathbb{R}^*$ is said to be an **extended continuous function** if the open set

$$\mathcal{O}_f = f^{-1}(\mathbb{R}) = \{x \in X\colon\ -\infty < f(x) < \infty\}$$

is dense in X.

DEFINITION 7.26. *The collection of all extended continuous functions on a topological space X is denoted by $C^\infty(X)$.*

Now assume that X is an extremally disconnected topological space. If f, g belong to $C^\infty(X)$, let $\mathcal{O}_f = f^{-1}(\mathbb{R})$ and $\mathcal{O}_g = g^{-1}(\mathbb{R})$ and note that $\mathcal{O}_f \cap \mathcal{O}_g$ is an open dense subset of X on which f and g are both real-valued and continuous. By Lemma 7.25 there exists a unique function $\phi \in C^\infty(X)$ such that $\phi(x) = f(x)+g(x)$ holds for all $x \in \mathcal{O}_f \cap \mathcal{O}_g$. We define the function ϕ to be the sum of f and g in $C^\infty(X)$, i.e., we let $\phi = f + g$ in $C^\infty(X)$. Similarly, if $\lambda \in \mathbb{R}$ is an arbitrary scalar, then the function $\lambda f\colon \mathcal{O}_f \to \mathbb{R}$ (defined by $(\lambda f)(x) = \lambda f(x)$ for each $x \in \mathcal{O}_f$) is continuous, and therefore there exists a unique function $\psi \in C^\infty(X)$ such that $\psi(x) = \lambda f(x)$ for all $x \in \mathcal{O}_f$. The function ψ is defined to be the scalar product of λ with f in $C^\infty(X)$, i.e., we let $\psi = \lambda f$. It is now a routine matter to verify that $C^\infty(X)$ under the above algebraic operations is a vector space.

Next, we equip $C^\infty(X)$ with the pointwise partial order $\ge$. That is, $f \ge g$ holds in $C^\infty(X)$ whenever $f(x) \ge g(x)$ holds in $\mathbb{R}^*$ for each $x \in X$. This is, of course, equivalent to saying that $f(x) \ge g(x)$ holds in $\mathbb{R}$ for all $x \in \mathcal{O}_f \cap \mathcal{O}_g$. Now, as above, consider two functions $f, g \in C^\infty(X)$. Then the function $f \vee g\colon \mathcal{O}_f \cap \mathcal{O}_g \to \mathbb{R}$, defined by $(f \vee g)(x) = \max\{f(x), g(x)\}$ for each $x \in \mathcal{O}_f \cap \mathcal{O}_g$, is continuous and so it extends to a unique continuous function $\overline{f \vee g}$ to all of X. It is easy to see that this function is the supremum of f and g in $C^\infty(X)$. Similar, statements can be made for the infimum $f \wedge g$ and the absolute value $|f|$. Thus, we have established the following.

- *If X is an extremally disconnected topological space, then $C^\infty(X)$ with the algebraic and lattice operations as defined above is a Riesz space.*

In actuality much more is true about the Riesz space $C^\infty(X)$.

THEOREM 7.27. *If X is an extremally disconnected topological space, then $C^\infty(X)$ is a universally complete Riesz space.*

PROOF. We shall establish first that $C^\infty(X)$ is Dedekind complete. To this end, let $\mathcal{F}$ be a nonempty subset of positive functions in $C^\infty(X)$ bounded above by a function $g \in C^\infty(X)$. If we let $\mathcal{O} = g^{-1}(\mathbb{R})$, then $0 \le f(x) \le g(x)$ holds for all $f \in \mathcal{F}$ and all $x \in \mathcal{O}$. Therefore, the set $\mathcal{F}_{\mathcal{O}} = \{f|_{\mathcal{O}}\colon\ f \in \mathcal{F}\}$ is a subset of $C(\mathcal{O})$ that is bounded above by the function $g|_{\mathcal{O}} \in C(\mathcal{O})$. Recalling that $\mathcal{O}$ (as an open set) is an extremally disconnected topological space in its own right, it follows from Theorem 1.50 that $h = \sup \mathcal{F}_{\mathcal{O}}$ exists in $C(\mathcal{O})$. If $\overline{h} \in C^\infty(X)$ is its unique continuous extension to all of X, then it is easy to see that $\overline{h} = \sup \mathcal{F}$ holds in $C^\infty(X)$. This shows that $C^\infty(X)$ is Dedekind complete.

To establish that $C^\infty(X)$ is laterally complete, let $\mathcal{F}$ be a nonempty pairwise disjoint subset of $C^\infty(X)$ consisting of positive functions. For each $f \in \mathcal{F}$, let $\mathcal{O}_f = \{x \in X\colon\ 0 < f(x) < \infty\}$ and note that the family of nonempty open sets $\{\mathcal{O}_f\colon\ f \in \mathcal{F}\}$ is pairwise disjoint. Now consider the open set $\mathcal{O} = \bigcup_{f\in\mathcal{F}} \mathcal{O}_f$ and define the function $g\colon \mathcal{O} \to \mathbb{R}$ via the formula $g(x) = f(x)$, where $f \in \mathcal{F}$ is the unique function for which $x \in \mathcal{O}_f$. Clearly, g is a continuous function and so by Lemma 7.25 it has a unique continuous extension $\overline{f}\colon \overline{\mathcal{O}} \to \mathbb{R}^*$. If we let $\overline{f}(x) = 0$ for all $x \in X \setminus \overline{\mathcal{O}}$, then $\overline{f} \in C^\infty(X)$ and an easy argument shows that $\overline{f} = \sup \mathcal{F}$ holds in $C^\infty(X)$. This shows that $C^\infty(X)$ is laterally complete and so $C^\infty(X)$ is a universally complete Riesz space. ■

The Stonean spaces (i.e., the Hausdorff compact extremally disconnected topological spaces) appear naturally in connection with Dedekind complete Boolean algebras. We shall recall briefly this relationship and we shall refer the reader for details to [**90**, Chapter 1]; see also Exercise 1 at the end of this chapter.

Recall that a **Boolean algebra** $\mathcal{B}$ is a complemented distributive lattice with smallest and largest elements. That is, $\mathcal{B}$ is a partially ordered set that is a distributive lattice with a smallest element 0 and a largest element e such that for each $a \in \mathcal{B}$ there exists a (necessarily unique) element $a' \in \mathcal{B}$ (called the *complement* of a) satisfying $a \wedge a' = 0$ and $a \vee a' = e$.

An **ideal** of a Boolean algebra $\mathcal{B}$ is a nonempty subset I of $\mathcal{B}$ that satisfies the following two properties:

(1) $u, v \in I$ imply $u \vee v \in I$.
(2) $w \leq u$ with $u \in I$ imply $w \in I$.

Clearly, $0 \in I$ holds true for each ideal I. An ideal I is called a **prime ideal** if $u \wedge v \in I$ implies that either $u \in I$ or $v \in I$.

Now let $\mathcal{B}$ be a Boolean algebra and let Ω denote the collection of all proper prime ideals of the Boolean algebra $\mathcal{B}$.[3] Then the collection of all subsets of Ω of the form $\Omega_u = \{\omega \in \Omega\colon\ u \notin \omega\}$ is a base for a Hausdorff compact topology on Ω, called the **hull-kernel topology**. Moreover, the sets of the form Ω_u are the only clopen sets of the hull-kernel topology; see part (iv) of Exercise 1 at the end of the chapter. Now if in addition $\mathcal{B}$ is Dedekind complete, that is, if every nonempty subset of $\mathcal{B}$ has a supremum, then the topological space Ω equipped with its hull-kernel topology is a Stonean space—called the **Stone space** of the Dedekind complete Boolean algebra $\mathcal{B}$.

If L is an Archimedean Riesz space, then the collection $\mathcal{B}_L$ of all bands (ordered by inclusion) is a Dedekind complete Boolean algebra; see Exercise 2 at the end of the chapter.

DEFINITION 7.28. *If L is an Archimedean Riesz space, then the Stone space of the Dedekind complete Boolean algebra $\mathcal{B}_L$ of all bands of L is called the* ***Stone space of*** *L and is denoted Ω_L (or by Ω if there is no ambiguity about the Riesz space L).*

---

[3] An ideal I of $\mathcal{B}$ is **proper** if $I \neq \mathcal{B}$. An ideal I of $\mathcal{B}$ is proper if and only if $e \notin I$.

The rest of the discussion in this section is devoted to proving Theorem 7.21 by establishing the following result of F. Maeda and T. Ogasawara [**91**].

THEOREM 7.29 (Maeda–Ogasawara). *If L is an Archimedean Riesz space with Stone space Ω, then $C^\infty(\Omega)$ is the universal completion of L.*

Moreover, Ω is the only (up to homeomorphism) Stonean space whose $C^\infty(\Omega)$-space is the universal completion of L.

Once we prove Theorem 7.29 the validity of Theorem 7.21 should be immediate. The proof of Theorem 7.29 is quite involved and it will be achieved by a series of steps in the form of lemmas. So, for our discussion below L will denote a fix Archimedean Riesz space. To simplify our arguments, we shall assume also that L has a weak unit $e > 0$. Moreover, to simplify our notation we shall denote the band generated by an element $f \in L$ by $[f]$. That is, we shall let $[f] = B_f$.

Start by recalling once more that the collection $\mathcal{B}_L$ of all bands of L is (under inclusion) a Dedekind complete Boolean algebra whose operations are given by

$$A \wedge B = A \cap B\,, \quad A \vee B = (A+B)^{\rm dd}\,, \quad \text{and} \quad A' = A^{\rm d}\,.$$

(See Exercise 2 at the end of the chapter.) So, if ω is an ideal of $\mathcal{B}_L$, then this means that ω is a nonempty subset of $\mathcal{B}_L$ such that:

(1) *If $A, B \in \omega$, then $(A+B)^{\rm dd} \in \omega$.*
(2) *If $A \in \omega$, then $B \in \omega$ for every $B \in \mathcal{B}_L$ that satisfies $B \subseteq A$.*

If ω is a proper ideal, then this means that

(3) $L \notin \omega$.

In case ω is a prime ideal, we have also the following:

(4) *If $A, B \in \mathcal{B}_L$ satisfy $A \cap B \in \omega$, then either $A \in \omega$ or $B \in \omega$.*

As in Theorem 7.29, we shall denote by Ω the Stone space of the Dedekind complete Boolean algebra $\mathcal{B}_L$ of all bands of L. So, an ideal of $\mathcal{B}_L$ belongs to Ω if and only if it satisfies properties (1)–(4) above. Notice that since e is a weak unit of L (and so $[e] = L$), it follows from property (3) that $[e] \notin \omega$ for each $\omega \in \Omega$.

Some basic properties of the ideals in Ω are included in the next result.

LEMMA 7.30. *For an ideal $\omega \in \Omega$ and any $u, v \in L$ we have:*

(a) *If $[u], [v] \in \omega$, then $[u+v] \in \omega$.*
(b) *$[u] \in \omega$ if and only if $[u^+], [u^-] \in \omega$.*
(c) *There is at most one $\alpha \in \mathbb{R}$ such that $[\alpha e - u] \in \omega$, and if there is such α, then:*
  (i) *For each $\beta < \alpha$ we have $[(\beta e - u)^+] \in \omega$ and $[(\beta e - u)^-] \notin \omega$.*
  (ii) *For each $\gamma > \alpha$ we have $[(\gamma e - u)^-] \in \omega$ and $[(\gamma e - u)^+] \notin \omega$.*

PROOF. (a) Assume $[u], [v] \in \omega$. From

$$u + v \in [u] + [v] \subseteq ([u] + [v])^{\rm dd} = [v] \vee [v]\,,$$

it follows that $[u+v] \subseteq [v] \vee [v] \in \omega$, and so $[u+v] \in \omega$.

(b) Clearly, $[u^+] \subseteq [u]$ and $[u^-] \subseteq [u]$. So, if $[u] \in \omega$, then $[u^+] \in \omega$ and $[u^-] \in \omega$ are both true. On the other hand if, $[u^+], [u^-] \in \omega$, then it follows from part (a) that $[u] = [|u|] = [u^+ + u^-] \in \omega$.

(c) Assume that for $\alpha_2 < \alpha_1$ we have $[(\alpha_1 e - u)^+], [(\alpha_1 e - vu)^-] \in \omega$ and $[(\alpha_2 e - u)^+], [(\alpha_2 e - u)^-] \in \omega$. From part (b) it easily follows that $[\alpha_1 e - u] \in \omega$

and $[\alpha_2 e - u] \in \omega$, and from part (a) we get

$$L = [e] = [(\alpha_1 - \alpha_2)e] = [(\alpha_1 e - u) - (\alpha_2 e - u)] \in \omega\,,$$

which is impossible. Hence there exists at most one $\alpha \in \mathbb{R}$ such that $[\alpha e - u] \in \omega$.

Now assume that for some $\alpha \in \mathbb{R}$ we have $[\alpha e - u] \in \omega$. From part (b), we easily obtain $[(\alpha e - u)^+] \in \omega$ and from the inclusion $[(\beta e - u)^+] \subseteq [(\alpha e - u)^+]$ for all $\beta < \alpha$, we see that $[(\beta e - u)^+] \in \omega$ for all $\beta < \alpha$. Now if for some $\beta < \alpha$ we have $[(\beta e - u)^-] \in \omega$, then $[(\beta e - u)] \in \omega$, contrary to the preceding conclusion, and the validity of (i) has been established. The validity of (ii) can be established in a similar manner. ■

We are now ready to associate with each vector $u \in L$ an extended function $\widehat{u}\colon \Omega \to \mathbb{R}^*$ that, as we shall later see, belongs to $C^\infty(\Omega)$.

DEFINITION 7.31. *If $u \in L$, then we define the function $\widehat{u}\colon \Omega \to \mathbb{R}^*$ via the formula*

$$\widehat{u}(\omega) = \sup\{\alpha \in \mathbb{R}\colon\ [(\alpha e - u)^+] \in \omega\}\,,$$

where the supremum is taken in $\mathbb{R}^*$ and we adhere to the convention $\sup \emptyset = -\infty$.

A glance at part (c) of Lemma 7.30 shows that $[(\alpha e - u)^+] \in \omega$ for all $\alpha < \widehat{u}(\omega)$ and $[(\beta e - u)^-] \in \omega$ for all $\beta > \widehat{u}(\omega)$. Consequently, we can describe the values of the functions $\widehat{u}$ in an another useful way.

LEMMA 7.32. *If $u \in L$, then for each $\omega \in \Omega$ we have*

$$\begin{aligned} \widehat{u}(\omega) &= \sup\{\alpha \in \mathbb{R}\colon\ [(\alpha e - u)^+] \in \omega\} \\ &= \inf\{\beta \in \mathbb{R}\colon\ [(\beta e - u)^-] \in \omega\}\,, \end{aligned}$$

subject to the conventions $\inf \emptyset = \infty$ *and* $\sup \emptyset = -\infty$.

Moreover, for any $\omega \in \Omega$ we have the following.

(i) *If some $\alpha \in \mathbb{R}$ satisfies $[\alpha e - u] \in \omega$, then $\widehat{u}(\omega) = \alpha$. In particular, $[u] \in \omega$ implies $\widehat{u}(\omega) = 0$.*
(ii) *If $\widehat{u}(\omega) = \infty$, then for each $\alpha \in \mathbb{R}$ we have $[(\alpha e - u)^+] \in \omega$ and $[(\alpha e - u)^-] \notin \omega$.*
(iii) *If $\widehat{u}(\omega) = -\infty$, then for each $\alpha \in \mathbb{R}$ we have $[(\alpha e - u)^+] \notin \omega$ and $[(\alpha e - u)^-] \in \omega$.*

PROOF. The first part in (i) follow immediately from Lemma 7.30. Next, we establish (ii). So, assume that $\widehat{u}(\omega) = \infty$. Clearly, $[(\alpha e - u)^+] \in \omega$ for all $\alpha \in \mathbb{R}$. Now, suppose by way of contradiction that $[(\alpha e - u)^-] \in \omega$ for some $\alpha \in \mathbb{R}$. But then, for any $\beta > \alpha$ we have $[(\beta e - u)^-] \subseteq [(\alpha e - u)^-]$ and so $[(\beta e - u)^-] \in \omega$. This implies that for any $\beta \in \mathbb{R}$ with $\beta > \alpha$ we have $[\alpha e - u], [\beta e - u] \in \omega$, which contradicts part (c) of Lemma 7.30. Hence, $[(\alpha e - u)^-] \notin \omega$ holds for each $\alpha \in \mathbb{R}$. Part (iii) can be established in a similar manner. ■

As an immediate consequence of Lemma 7.32, we have also the following handy observation.

LEMMA 7.33. *If $\alpha \in \mathbb{R}$ and $\omega \in \Omega$ satisfy $[(\alpha e - u)^+] \notin \omega$, then $\widehat{u}(\omega) \le \alpha$.*

The basic properties of the functions $\widehat{u}$ are included in the next lemma.

LEMMA 7.34. *We have the following properties.*

(1) *An element $u \in L$ satisfies $\widehat{u} = 0$ if and only if $u = 0$.*
(2) *The weak unit e satisfies $\widehat{e} = \mathbf{1}$, i.e., $\widehat{e}(\omega) = 1$ for all $\omega \in \Omega$.*
(3) *If $u \in L$ and $\lambda \in \mathbb{R}$, then $\widehat{\lambda u}(\omega) = \lambda\widehat{u}(\omega)$ holds for all $\omega \in \Omega$.*
(4) *If two elements $u, v \in L$ satisfy $\widehat{u}(\omega), \widehat{v}(\omega) \in \mathbb{R}$ for some $\omega \in \Omega$, then $(\widehat{u+v})(\omega) \in \mathbb{R}$ and $(\widehat{u+v})(\omega) = \widehat{u}(\omega) + \widehat{u}(\omega)$.*
(5) *If $u \in L$, then $|\widehat{u}(\omega)| = \widehat{|u|}(\omega)$ holds for all $\omega \in \Omega$.*

PROOF. (1) Assume first that $u = 0$. Then $[0] \in \omega$, and so by part (i) of Lemma 7.32 we get $\widehat{0}(\omega) = 0$ for all $\omega \in \Omega$.

For the converse, assume that $\widehat{u}(\omega) = 0$ holds for all $\omega \in \Omega$. Assume that for some $\alpha > 0$ we have

$$(u - \alpha e)^+ = (\alpha e - u)^- \neq 0\,.$$

Then $[(\alpha e - u)^-] \neq [0]$, and so there exists some $\omega_0 \in \Omega$ such that $[(\alpha e - u)^-] \notin \omega_0$; see part (ii) of Exercise 1 at the end of the chapter. From

$$[(\alpha e - u)^+] \wedge [(\alpha e - u)^-] = [0] \in \omega_0$$

and the fact that ω_0 is a prime ideal, we infer that $[(\alpha e - u)^+] \in \omega_0$. This implies $\widehat{u}(\omega_0) \geq \alpha > 0$, a contradiction. Hence, $(u - \alpha e)^+ = 0$ or $u \leq \alpha e$ holds for all $\alpha > 0$. Since L is Archimedean, the latter implies $u \leq 0$.

Next, we claim that for each $\alpha < 0$ we have $(\alpha e - u)^+ = 0$ or $\alpha e \leq u$. Indeed, if for some $\alpha < 0$ we have $(\alpha e - u)^+ \neq 0$, then (as above) there exists some $\omega_1 \in \Omega$ such that $[(\alpha e - u)^+] \notin \omega_1$, and so (by Lemma 7.33) we get $\widehat{u}(\omega_1) \leq \alpha < 0$, a contradiction. Now from $\alpha e \leq u$ for each $\alpha < 0$ and the Archimedean property of L, we get $0 \leq u$. Thus, $u = 0$.

(2) Note that $[1{\cdot}e - e] = [0] \in \omega$ holds for all $\omega \in \Omega$. So, by part (i) of Lemma 7.32, we get $\widehat{e}(\omega) = 1$ for all $\omega \in \Omega$.

(3) If $\lambda = 0$, then using part (1) (and that $0 \cdot x = 0$ for all $x \in \mathbb{R}^*$), we see that the claim is true for $\lambda = 0$. For $\lambda > 0$, note that

$$\begin{aligned}
(\widehat{\lambda u})(\omega) &= \sup\{\alpha \in \mathbb{R}\colon\ [(\alpha e - \lambda u)^+] \in \omega\} \\
&= \lambda \sup\{\tfrac{\alpha}{\lambda} \in \mathbb{R}\colon\ [(\tfrac{\alpha}{\lambda} e - u)^+] \in \omega\} \\
&= \lambda \sup\{\mu \in \mathbb{R}\colon\ [(\mu e - u)^+] \in \omega\} \\
&= \lambda \widehat{u}(\omega)\,.
\end{aligned}$$

Now assume that $\lambda < 0$. Without loss of generality, we can suppose that $\lambda = -1$. Then, we have

$$\begin{aligned}
(\widehat{-u})(\omega) &= \sup\{\alpha \in \mathbb{R}\colon\ [(\alpha e + u)^+] \in \omega\} \\
&= \sup\{\alpha \in \mathbb{R}\colon\ [(-\alpha e - u)^-] \in \omega\} \\
&= \sup\{-\beta \in \mathbb{R}\colon\ [(\beta e - u)^-] \in \omega\} \\
&= -\inf\{\beta \in \mathbb{R}\colon\ [(\beta e - u)^-] \in \omega\} \\
&= -\widehat{u}(\omega)\,.
\end{aligned}$$

(4) Assume that $\alpha < \widehat{u}(\omega)$ and $\beta < \widehat{v}(\omega)$. Clearly, $[(\alpha e - u)^+] \in \omega$ and $[(\beta e - u)^+] \in \omega$. From the lattice inequality

$$\big((\alpha + \beta)e - (u + v)\big)^+ \leq (\alpha e - u)^+ + (\beta e - v)^+\,,$$

we obtain

$$\begin{aligned}[((\alpha+\beta)e-(u+v))^+] &\subseteq [(\alpha e-u)^+ + (\beta e-v)^+]\\ &= [(\alpha e-u)^+]\vee[(\beta e-v)^+]\in\omega\,.\end{aligned}$$

This implies $\alpha+\beta\le\widehat{(u+v)}(\omega)$ for all $\alpha<\widehat{u}(\omega)$ and all $\beta<\widehat{v}(\omega)$. Consequently, $\widehat{u}(\omega)+\widehat{v}(\omega)\le\widehat{(u+v)}(\omega)$. Applying this conclusion to $-u$ and $-v$, and using part (3), we see that $-\widehat{u}(\omega)-\widehat{v}(\omega)\le-\widehat{(u+v)}(\omega)$ or $\widehat{u}(\omega)+\widehat{v}(\omega)\ge\widehat{(u+v)}(\omega)$. Therefore, $\widehat{(u+v)}(\omega)=\widehat{u}(\omega)+\widehat{v}(\omega)$.

(5) From the lattice inequality $\alpha e-(\pm u)\ge\alpha e-|u|$, we obtain the inclusion $[(\alpha e-(\pm u))^+]\supseteq[(\alpha e-|u|)^+]$, and from this we easily infer that

$$\big|\widehat{u}(\omega)\big|\le\widehat{|u|}(\omega)$$

for all $\omega\in\Omega$. If $\widehat{u}(\omega)=\pm\infty$, then clearly $\big|\widehat{u}(\omega)\big|=\widehat{|u|}(\omega)=\infty$.

Now assume that for some $\omega\in\Omega$ we have $\widehat{u}(\omega)\in\mathbb{R}$, and let $\alpha<\widehat{|u|}(\omega)$. From the lattice identity

$$\begin{aligned}(\alpha e-|u|)^+ &= \big(\alpha e-u\vee(-u)\big)^+\\ &= \big((\alpha e-u)\wedge(\alpha e+u)\big)\vee 0\\ &= \big((\alpha e-u)\vee 0\big)\wedge\big((\alpha e+u)\vee 0\big)\\ &= (\alpha e-u)^+\wedge(\alpha e+u)^+\,,\end{aligned}$$

it follows that $[(\alpha e-u)^+]\wedge[(\alpha e+u)^+]\in\omega$. This, in view of the fact that ω is a prime ideal, implies either $[(\alpha e-u)^+]\in\omega$ or $[(\alpha e+u)^+]\in\omega$. Therefore, either $\alpha\le\widehat{u}(\omega)\le|\widehat{u}(\omega)|$ or $\alpha\le\widehat{(-u)}(\omega)=-\widehat{u}(\omega)\le|\widehat{u}(\omega)|$. Thus, in either case we have $\alpha\le|\widehat{u}(\omega)|$, and from this we get $\widehat{|u|}(\omega)\le|\widehat{u}(\omega)|$. Consequently, $\widehat{|u|}(\omega)=|\widehat{u}(\omega)|$ holds for all $\omega\in\Omega$. ■

It turns out that each $\widehat{u}$ is an extended continuous function on Ω.

LEMMA 7.35. *For each $u\in L$ we have $\widehat{u}\in C^\infty(\Omega)$.*

PROOF. Recall that the clopen subsets of Ω are precisely the sets of the form $\Omega_B=\{\omega\in\Omega\colon\ B\notin\omega\}$, where $B\in\mathcal{B}_L$; see part (iv) of Exercise 1 at the end of the chapter.

Fix some $u\in L$. We shall verify first that $\widehat{u}$ is continuous. To this end, let $a\in\mathbb{R}$. From Lemma 7.33, it is easy to see that

$$\{\omega\in\Omega\colon\ \widehat{u}(\omega)<a\}=\bigcup_{\alpha<a}\Omega_{[(\alpha e-u)^+]}\,.$$

This implies that $\{\omega\in\Omega\colon\ \widehat{u}(\omega)<a\}$ is an open subset of Ω for each $a\in\mathbb{R}$. Similarly, it is easy to see that for each $a\in\mathbb{R}$ we have

$$\{\omega\in\Omega\colon\ \widehat{u}(\omega)\le a\}=\bigcap_{a<\alpha}\Omega_{[(\alpha e-u)^+]}\,,$$

and from this we infer that the set $\{\omega\in\Omega\colon\ \widehat{u}(\omega)\le a\}$ is closed for each $a\in\mathbb{R}$. By Lemma 7.24, the function $\widehat{u}\colon\Omega\to\mathbb{R}^*$ is continuous.

It remains to show that $\widehat{u}$ is an extended continuous function. Start by observing that (in view of part (5) of Lemma 7.34), we can assume that $u>0$. Now let $\mathcal{O}=\{\omega\in\Omega\colon\ 0\le\widehat{u}(\omega)<\infty\}$, and assume by way of contradiction that the clopen set $\overline{\mathcal{O}}$ is not equal to Ω. Pick some $0 < v \in L$ such that $\Omega_{[v]} = \Omega \setminus \overline{\mathcal{O}}$. From $[e] \notin \omega$ for each $\omega \in \Omega$, we see that $[v] \notin \omega$ holds for some $\omega \in \Omega$ if and only if $[v \wedge e] \notin \omega$. This implies $\Omega_{[v]} = \Omega_{[v \wedge e]}$, and so replacing v by $v \wedge e$, we can suppose that

$$0 < v \leq e\,. \tag{$\star$}$$

Clearly, $\widehat{u}(\omega) = \infty$ holds for each $\omega \in \Omega_{[v]}$. This implies that for all $\alpha > 0$ and all $\omega \in \Omega_{[v]}$ we have $[(\alpha e - u)^-] = [(u - \alpha e)^+] \notin \omega$; see part (ii) of Lemma 7.32. Therefore, $[(\alpha u - e)^+] \notin \omega$ for all $\alpha > 0$ and all $\omega \in \Omega_{[v]}$. Taking into account ($\star$), we infer that

$$[(\alpha u - v)^+] \notin \omega \quad \text{for all} \quad \alpha > 0 \quad \text{and all} \quad \omega \in \Omega_{[v]}\,. \tag{$\star\star$}$$

Now assume that for some $\beta > 0$ we have

$$(\beta u - v)^- = (v - \beta u)^+ > 0\,. \tag{$\star\star\star$}$$

Then there exists some $\omega_1 \in \Omega$ such that $[(\beta u - v)^-] \notin \omega_1$; see part (ii) of Exercise 1 at the end of the chapter. From ($\star\star\star$), it follows that $[v] \notin \omega_1$, and so $\omega_1 \in \Omega_{[v]}$. But then, from ($\star\star$), we get $[(\beta u - v)^+] \notin \omega_1$ which in view of $[(\beta u - v)^-] \notin \omega_1$ contradicts the fact that ω_1 is a prime ideal. Thus, $(\beta u - v)^- = (v - \beta u)^+ = 0$ or $0 < v \leq \beta u$ holds for all $\beta > 0$, contrary to the Archimedean property of L. This contradiction, shows that $\widehat{u} \in C^\infty(\Omega)$ and completes the proof. ■

The preceding results show that the mapping $u \mapsto \widehat{u}$, from L to $C^\infty(\Omega)$, is a Riesz isomorphism. So, identifying L with its image $\widehat{L}$, we can consider L as a Riesz subspace of $C^\infty(\Omega)$. It turns out that L is order dense in $C^\infty(\Omega)$.

LEMMA 7.36. *For each function $0 < f \in C^\infty(\Omega)$ there exists some element $0 < u \in L$ such that $0 < \widehat{u} \leq f$. That is, L is order dense in $C^\infty(\Omega)$ (and so $C^\infty(\Omega)$ is the universal completion of L).*

PROOF. Let $0 < f \in C^\infty(\Omega)$. Using the continuity of f, it easy to see that there exists some $\epsilon > 0$ and a clopen subset V of Ω such that $\epsilon\chi_V \leq f$. So, in order to establish the claim, it suffices to assume that $f = \chi_V$ for some nonempty clopen subset V of Ω.

To this end, pick some $0 < u \in L$ such that $V = \Omega_{[u]}$. Since $\Omega_{[u]} = \Omega_{[u \wedge e]}$ is true, we can assume that $0 < u \leq e$. Now notice that if $\omega \in V$, then we have $\widehat{u}(\omega) \leq \widehat{e}(\omega) = 1 = \chi_V(\omega)$, and if $\omega \notin V = \Omega_{[u]}$, then $[u] \in \omega$ so that $\widehat{u}(\omega) = 0 = \chi_V(\omega)$. Hence, $0 < \widehat{u} \leq \chi_V$, as desired. ■

The above discussion has established that if L is an Archimedean Riesz space with a weak unit, then $C^\infty(\Omega)$ is the universal completion of L. If now L is an Archimedean Riesz space without a weak unit, embed L order densely in a Dedekind complete Riesz space M with a weak unit (see Theorem 1.76) and then notice that the universal completion M^{u} of M is also the universal completion of L. This completes the proof of Theorem 7.21. Observe, also that if Ω_L and Ω_M are the Stone spaces of L and M, respectively, then Ω_L and Ω_M are homeomorphic. (This follows easily from the fact that $\mathcal{B}_L$ and $\mathcal{B}_M$ are isomorphic Boolean algebras; see Exercise 3 at the end of the chapter.) So, $L^{\mathrm{u}} = M^{\mathrm{u}} = C^\infty(\Omega_M) = C^\infty(\Omega_L)$.

Finally, we would like to mention that if $C^\infty(\Omega)$ is isomorphic to $C^\infty(Q)$, where Ω and Q are Stonean spaces, then Ω and Q are homeomorphic; see Exercise 4 at the end of the chapter. This implies that the universal completion L^u of an

Archimedean Riesz space L can be expressed in the form $C^\infty(\Omega)$ for a unique (up to homeomorphism) Stonean space Ω. This completes the proof of Theorem 7.29.

7.3. Lateral and Universal Completeness

In this section we shall collect a variety of results regarding laterally complete and universally complete Riesz spaces. We start by characterizing the dominable sets in Archimedean Riesz space.

THEOREM 7.37. *For an Archimedean Riesz space L and a nonempty subset A of L^+ the following statements are equivalent.*

(i) *A is a dominable subset of L^+.*
(ii) *A is order bounded in L^{u} (and hence $\sup A$ exists in L^{u}).*

PROOF. Apply Lemma 7.11 and Theorem 7.14. ■

The next three theorems are due to D. H. Fremlin [**59**, pp. 73–76]. The first one is a sequential analogue of the preceding result.

THEOREM 7.38 (Fremlin). *For a Dedekind σ-complete Riesz space L the following statements are equivalent.*

(i) *L is laterally σ-complete (and hence universally σ-complete).*
(ii) *Every countable dominable subset of L^+ is order bounded in L (and hence it has a supremum).*

PROOF. (i) $\Longrightarrow$ (ii) Assume first that L contains a weak order unit e. Let $\{u_1, u_2, \ldots\}$ be a countable dominable subset of L^+. By Theorem 7.37, the supremum $u^* = \sup_n u_n$ exists in L^{u}. Observe that for each $v \in L^+$ the supremum $w = \sup_n u_n \wedge v$ exists in L (since L is Dedekind σ-complete). Since the Riesz space L is order dense in L^{u} we have $u^* \wedge v = w \wedge v \in L$. Now if $w_n = [(n+2)e - u^*]^+ \wedge (ne - u^*)^-$, then by what was just noted, $\{w_n\} \subseteq L^+$. Next proceeding as in the proof of part (i) of Theorem 7.4 (and working in L^{u}), we see that the projection of u^* on e (which equals u^*) can be written as $u^* \wedge (f+g)$, where

$$f = \bigvee_{n=0}^{\infty} (2n+1)w_{2n} \quad \text{and} \quad g = \bigvee_{n=0}^{\infty} (2n+2)w_{2n+1}\,.$$

By the lateral σ-completeness of L both f and g belong to L. Consequently we have $u^* = u^* \wedge (f+g) \le f+g \in L$, so that $0 \le u_n \le f+g$ holds in L for all n, and the proof in this case is finished.

We shall show next that $\{u_n\}$ is included in a principal band B_e of L. We can then apply the above arguments to the universally σ-complete Riesz space B_e to show that $\{u_n\}$ is order bounded in B_e, and hence in L. To this end, define $v_1 = u_1$ and then inductively if $v_1, \ldots, v_n$ have been selected, put $w_n = v_1 + \cdots + v_n$, and let $v_{n+1} = u_{n+1} - P_{w_n}(u_{n+1})$. As in the first part of the proof of Theorem 7.9, it can be shown that $\{v_n\}$ is a disjoint sequence of L^+. Put $e = \bigvee_{n=1}^{\infty} v_n$ in L and note that since $0 \le v_n \le w_n = \bigvee_{i=1}^{n} v_i \le e$, we have $P_{w_n}(u_{n+1}), v_n \in B_e$ for all n and thus $u_{n+1} = P_{w_n}(u_{n+1}) + v_{n+1} \in B_e$ for all n. Hence $\{u_n\} \subseteq B_e$, and the desired conclusion follows.

(ii) $\Longrightarrow$ (i) Let $\{u_n\}$ be a disjoint sequence of L^+. Then $\{u_n\}$ is order bounded in L^{u}, and therefore $\{u_n\}$ is a dominable subset of L^{u}. It follows that $\{u_n\}$ is a dominable subset of L and thus order bounded by our hypothesis. Therefore, $\bigvee_{n=1}^{\infty} u_n$ exists in L and the proof is finished. ■

The previous theorem has a number of important consequences, some of which are included in the next results.

THEOREM 7.39 (Fremlin). *For a Dedekind complete Riesz space L the following statements are equivalent.*

(i) *L is universally complete, i.e., $L = L^{\mathrm{u}}$.*
(ii) *L is universally σ-complete and has a weak order unit.*

PROOF. (i) $\Longrightarrow$ (ii) Obvious.

(ii) $\Longrightarrow$ (i) Since L is Dedekind complete, L is an ideal of L^{u}; see Theorem 1.40. Now let $0 < e \in L$ be a weak order unit of L and let $0 < u^* \in L^{\mathrm{u}}$. Since e is also a weak order unit of the Riesz space L^{u}, we have $0 \le u^* \wedge ne \uparrow u^*$. From $\{u^* \wedge ne\} \subseteq L$, we see that $\{u^* \wedge ne\}$ is a dominable sequence of L. By Theorem 7.38 the sequence $\{u^* \wedge ne\}$ is order bounded in L and thus $u^* \in L$. Therefore, $L = L^{\mathrm{u}}$ and the proof is finished. ■

The next example shows that the Dedekind completeness cannot be dropped from Theorem 7.39.

EXAMPLE 7.40. Consider the Riesz space L consisting of all real-valued Lebesgue measurable functions defined on $[0,1]$ (with the pointwise ordering). Then L is universally σ-complete, has a weak order unit (the constant function $\mathbf{1}$), and it is neither laterally complete nor Dedekind complete. ■

Another example shows that the existence of the weak order unit in statement (ii) of Theorem 7.39 cannot be dropped either.

EXAMPLE 7.41. Consider an uncountable set X and let

$$L = \{f \in \mathbb{R}^X\colon \text{ The set } \{x \in X \mid f(x) \neq 0\} \text{ is at most countable}\},$$

with the pointwise ordering. It can be seen easily that L is Dedekind complete and universally σ-complete without weak order units. Note, however, that L is not universally complete—and that $L^{\mathrm{u}} = \mathbb{R}^X$. ■

We are now in the position to characterize the Riesz spaces that have universal σ-completions.

THEOREM 7.42 (Fremlin). *For a Riesz space L the following statements are equivalent.*

(i) *L is almost Dedekind σ-complete.*
(ii) *L has a universal σ-completion (determined uniquely up to a Riesz isomorphism).*

PROOF. (i) $\Longrightarrow$ (ii) Assume first that L is Dedekind σ-complete. Let

$$U(L^+) = \{u^* \in L^{\mathrm{u}}\colon \exists \{u_n\} \subseteq L^+ \text{ with } 0 \le u_n \uparrow u^* \text{ in } L^{\mathrm{u}}\},$$

and define $M = U(L^+) - U(L^+) = \{u^* - v^*\colon u^*, v^* \in U(L^+)\}$. Clearly, we have the following Riesz subspace inclusions $L \subseteq M \subseteq L^{\mathrm{u}}$. We shall show next that $M^+ = U(L^+)$. To this end, let $0 \le u^* - v^*$ with $u^*, v^* \in U(L^+)$. Pick two sequences $\{u_n\}$ and $\{v_n\}$ of L^+ with $0 \le u_n \uparrow u^*$ and $0 \le v_n \uparrow v^*$ in L^{u}. Note that since L is Dedekind σ-complete $v^* \wedge u$ belongs to L for each $u \in L^+$. Now

$$0 \le u_n - v^* \wedge u_n = u^* \wedge u_n - v^* \wedge u_n \le u^* - v^*$$

and $u_n - v^* \wedge u_n \xrightarrow{o} u^* - v^*$ in L^{u} imply that the sequence $\{w_n\}$ of L^+ defined by

$$w_n = \bigvee_{i=1}^{n} (u_i - v^* \wedge u_i)$$

satisfies $0 \le w_n \uparrow u^* - v^*$ in L^{u}. Consequently, $u^* - v^* \in U(L^+)$ and therefore $M^+ = U(L^+)$. It can be proven now easily that the Riesz space M is a universally σ-complete Riesz space that includes L as a super order dense Riesz subspace.

Now if L is almost Dedekind σ-complete, embed L super order densely in a Dedekind σ-complete Riesz space K and then embed K as above super order densely into a universally σ-complete Riesz space M. It easily follows that M is a universal σ-completion of L.

The uniqueness of the universal σ-completion can be derived easily by applying Theorem 7.38; the details are left to the reader as an exercise.

(ii) $\Longrightarrow$ (i) It follows immediately from the definition of the almost Dedekind σ-completeness. ■

By a **step function** s on a topological space X, we shall mean any function of $C^\infty(X)$ for which there exists a collection $\{V_i\colon\ i \in I\}$ of mutually disjoint clopen subsets of X whose union is dense in X and such that s is constant on each V_i. In other words, a function $s \in C^\infty(X)$ is a step function on X if it has a representation of the form $s = \sum_{i\in I} c_i \chi_{V_i}$, where each V_i is closed and open, $V_i \cap V_j = \emptyset$ if $i \ne j$, and the open set $\bigcup_{i\in I} V_i$ is dense in X. If the step function s has an at most countable representation, that is, if $s = \sum_{n=1}^{\infty} c_n \chi_{V_n}$, then it will be called a **σ-step function**. We shall denote by $S^\infty(X)$ (resp. by $S^\infty_\sigma(X)$) the collection of all step functions (resp. the collection of all σ-step functions) on a topological vector space X.

THEOREM 7.43. *If X is an extremally disconnected topological space, then we have:*

(i) *$S^\infty(X)$ is an Archimedean laterally complete Riesz space.*
(ii) *$S^\infty_\sigma(X)$ is an Archimedean laterally σ-complete Riesz space.*
(iii) *The following Riesz subspace inclusions hold:*

$$S^\infty_\sigma(X) \subseteq S^\infty(X) \subseteq C^\infty(X)\,,$$

with $S^\infty_\sigma(X)$ order dense in $C^\infty(X)$.

PROOF. (i) Note first that $S^\infty(X)$ is a vector subspace of $C^\infty(X)$. Indeed, if $s, t \in S^\infty(X)$ have the representations $s = \sum_{i\in I} c_i \chi_{V_i}$ and $t = \sum_{j\in J} b_j \chi_{U_j}$, then $\lambda s = \sum_{i\in I} \lambda c_i \chi_{V_i} \in S^\infty(X)$ for each $\lambda \in \mathbb{R}$ and

$$s + t = \sum_{i\in I} \sum_{j\in J} (c_i + b_j)\chi_{V_i \cap U_j} \in S^\infty(X)\,,$$

since the set

$$\bigcup\{V_i \cap U_j\colon\ i \in I \text{ and } j \in J\} = \Big[\bigcup_{i\in I} V_i\Big] \cap \Big[\bigcup_{j\in J} U_j\Big]$$

is dense in X. For the Riesz space structure of $S^\infty(X)$ note that:

$$s \vee t = \sum_{i\in I}\sum_{j\in J} \max\{c_i, b_j\}\chi_{V_i\cap U_j} \in S^\infty(X).$$

$$s \wedge t = \sum_{i\in I}\sum_{j\in J} \min\{c_i, b_j\}\chi_{V_i\cap U_j} \in S^\infty(X).$$

Since clearly $S^\infty(X)$ is Archimedean, it remains to be shown that $S^\infty(X)$ is a laterally complete Riesz space. To this end, let $\{s_\alpha\}$ be a disjoint system of positive elements in $S^\infty(X)$. For each α pick a representation for s_α. Then for each α collect all the clopen sets V appearing in the representation of s_α for which $s_\alpha(x) = c > 0$ for all $x \in V$. Now the collection of all those clopen sets forms a mutually disjoint family that we shall denote by $\{V_i\colon\ i \in I\}$. Note that for each V_i there exists some $c_i > 0$ and exactly one α such that $c_i\chi_{V_i}$ appears in the representation of s_α. Put $\mathcal{O} = \bigcup_{i\in I} V_i$ and $U = X \setminus \overline{\mathcal{O}}$ (a clopen set) and let $s = 0\chi_U + \sum_{i\in I} c_i\chi_{V_i}$ on the open dense set $\mathcal{O}\cup U$. By Lemma 7.25 the function s can be extended to a function of $C^\infty(X)$ and hence $s \in S^\infty(X)$. Now it should be clear that $s = \sup_\alpha s_\alpha$ holds in $S^\infty(X)$.

(ii) Proceed as in (i).

(iii) The proof is straightforward and is left to the reader. ■

The next theorem shows the relationship between Archimedean laterally complete Riesz spaces and the Riesz spaces of the form $S^\infty(X)$, with X extremally disconnected.

THEOREM 7.44. *Let L be an Archimedean laterally complete Riesz space with Stone space Ω. Then we have the following Riesz subspace inclusions:*

$$S^\infty(\Omega) \subseteq L \subseteq C^\infty(\Omega).$$

PROOF. Let $L^{\mathrm{u}} = C^\infty(\Omega)$ be the universal completion of L, where Ω is the Stone space of L. Since L has a weak order unit e, the embedding of L into $C^\infty(\Omega)$ can be taken to carry e to the constant function $\mathbf{1}$ on Ω. Now since L has the projection property [see Theorem 7.4(ii)], $\chi_V \in L$ for each clopen subset V of Ω; see Theorem 1.54. Now the lateral completeness of L easily guarantees that $S^\infty(\Omega) \subseteq L \subseteq C^\infty(\Omega)$. ■

We continue our discussion by presenting examples of Archimedean laterally complete and laterally σ-complete Riesz spaces.

EXAMPLE 7.45. (*An Archimedean laterally complete Riesz space that is not Dedekind complete.*) By a theorem of A. M. Gleason [**61**, Theorem 3.2, p. 486] there exists a Stonean space X and an onto continuous function $\phi\colon X \to [0,1]$ that does not carry any proper closed subset of X onto the interval $[0,1]$. Note that $\phi \in C(X) \subseteq C^\infty(X)$, and we shall show next that $\phi \notin S^\infty(X)$. Indeed, if $\phi \in S^\infty(X)$, then ϕ must be constant on some nonempty proper clopen subset V of X. But then ϕ maps the closed (actually compact) subset $X \setminus V$ onto $[0,1]$, in contradiction with the properties of ϕ. Now if $S^\infty(X)$ is Dedekind complete, then $S^\infty(X) = C^\infty(X)$ by Theorem 7.15(ii), in contradiction to $\phi \notin S^\infty(X)$. Thus $S^\infty(X)$ is the desired example.

Observe also that $C^\infty(X)$ is the universal completion of $C[0,1]$. Indeed, the mapping $T\colon C[0,1] \to C^\infty(X)$, defined by $T(f) = f \circ \phi$ for $f \in C[0,1]$, embeds $C[0,1]$ order densely into $C^\infty(X)$. ■

EXAMPLE 7.46. (*An Archimedean laterally σ-complete Riesz space that is neither Dedekind σ-complete nor laterally complete.*) If K is the Riesz space of Example 7.40 and $S^\infty(X)$ is as determined in Example 7.45, then the Riesz space $L = K \times S^\infty(X)$ has the desired properties. ■

The rest of the section deals with some characterizations of the laterally complete Riesz spaces of the form $\mathbb{R}^X$. Note that two such Riesz spaces $\mathbb{R}^X$ and $\mathbb{R}^Y$ are Riesz isomorphic if and only if X and Y have the same cardinality (i.e., if there exists a one-to-one mapping from X onto Y). The characterizations of the finite dimensional $\mathbb{R}^X$ spaces are given first.

THEOREM 7.47. *For an Archimedean laterally σ-complete Riesz space L the following statements are equivalent.*

(i) *L is Riesz isomorphic to some $\mathbb{R}^n$.*
(ii) *L admits an order unit.*
(iii) *L admits a Riesz norm.*

PROOF. (i) $\Longrightarrow$ (ii) Obvious.

(ii) $\Longrightarrow$ (iii) If e is an order unit of L, then the function $\rho\colon L \to \mathbb{R}$, defined by $\rho(u) = \inf\{\lambda > 0\colon\ |u| \le \lambda e\}$, is a Riesz norm on L.

(iii) $\Longrightarrow$ (i) Let ρ be a Riesz norm on L. If $\{u_n\}$ is a disjoint sequence of positive elements of norm one, then $\{nu_n\}$ is also a disjoint sequence of L^+ and so $u = \bigvee_{n=1}^\infty nu_n$ exists in L. But then we have $n = \rho(nu_n) \le \rho(u) < \infty$ for each n, which is impossible. Thus, every collection of disjoint elements of L is finite and hence L is Riesz isomorphic to some $\mathbb{R}^n$; see Exercise 13 of Chapter 1. ■

THEOREM 7.48. *For an Archimedean laterally complete Riesz space L the following statements are equivalent.*

(i) *L is Riesz isomorphic to some $\mathbb{R}^X$.*
(ii) *L admits a Hausdorff locally convex-solid Lebesgue topology.*
(iii) *$L_n^\sim$ separates the points of L.*
(iv) *L is a discrete Riesz space.*

PROOF. (i) $\Longrightarrow$ (ii) The topology of pointwise convergence on $\mathbb{R}^X$ is a Hausdorff locally convex-solid Lebesgue topology.

(ii) $\Longrightarrow$ (iii) Let τ be a Hausdorff Lebesgue topology on L that is also locally convex. By Theorem 3.12 the topological dual L' of (L,τ) is an ideal of $L_n^\sim$ and the conclusion follows from the Hahn–Banach theorem.

(iii) $\Longrightarrow$ (iv) Let $0 < u \in L$. Pick $0 < \phi \in L_n^\sim$ with $\phi(u) > 0$, and note that $N_\phi = \{v \in L\colon\ \phi(|v|) = 0\}$ is a band of L and so by Theorem 7.4(ii) we have $L = N_\phi \oplus C_\phi$; where $C_\phi = N_\phi^{\mathrm{d}}$. But C_ϕ admits a Riesz norm [namely $\rho(u) = \phi(|u|)$] and so by Theorem 7.47 the band C_ϕ is Riesz isomorphic to some $\mathbb{R}^n$. Now let v be the projection of u onto C_ϕ. Then there exists a discrete element w of C_ϕ, and hence of L, such that $0 < w \le v \le u$.

(iv) $\Longrightarrow$ (i) By Theorem 1.78 the Riesz space L is Riesz isomorphic to an order dense Riesz subspace of some $\mathbb{R}^X$. But then since L is laterally complete, this Riesz isomorphism must be onto and the proof is finished. ■

The equivalence (ii) $\iff$ (iv) in Theorem 7.48 was established by J. Quinn and R. Reichard [**113**, Theorem 3.1, p. 239].

7.4. Lateral Completeness and Local Solidness

In this section we shall study the locally solid topologies that a laterally σ-complete Riesz space can carry. Most of the results in this section have been obtained by D. H. Fremlin [**59**] for universally complete Riesz spaces. However, the generalizations and the presentation of the material follows the work of the authors [**18**].

THEOREM 7.49. *For a laterally σ-complete Riesz space L we have:*

(i) *Every disjoint sequence of L^+ converges to zero with respect to any locally solid topology on L. In particular, every locally solid topology on L satisfies the pre-Lebesgue property.*
(ii) *If in addition L is Archimedean, then every locally solid topology on L satisfies the σ-Lebesgue property.*

PROOF. (i) Note that if $\{u_n\}$ is a disjoint sequence of L^+, then so is $\{nu_n\}$ and hence $u = \bigvee_{n=1}^{\infty} nu_n$ exists in L. The conclusion now follows immediately by observing that $0 \leq u_n \leq \frac{1}{n}u$ holds for all n. For the validity of the last property use Theorem 3.22.

(ii) Let τ be a locally solid topology on L and let $u_n \downarrow 0$ in L. If V is a solid τ-neighborhood of zero, choose another τ-neighborhood U of zero with $U + U \subseteq V$ and then some $\epsilon > 0$ such that $\epsilon u_1 \in U$. Now, by Lemma 7.7 there exists some $u \in L^+$ such that $0 \leq u_n \leq 2^{-n}u + \epsilon u_1$ holds for all n. Pick some $k \in \mathbb{N}$ such that $2^{-n}u \in U$ for all $n \geq k$. But then we have $0 \leq u_n \leq 2^{-n}u + \epsilon u_1 \in U + U \subseteq V$ for all $n \geq k$, so that $u_n \xrightarrow{\tau} 0$, and the proof is finished. ■

We continue with more properties of the locally solid topologies on laterally σ-complete Riesz spaces.

THEOREM 7.50. *For a Hausdorff locally solid Riesz space (L, τ) with the Fatou property we have the following.*

(i) *Every increasing τ-bounded net of L^+ is dominable.*
(ii) *If L is laterally σ-complete, then τ is a Lebesgue topology.*
(iii) *If L is universally complete, then (L, τ) is τ-complete.*

PROOF. (i) Assume that $0 \leq u_\alpha \uparrow$ is τ-bounded and that $u > 0$. Pick a Fatou τ-neighborhood V of zero with $u \notin V$, and then choose some $n \in \mathbb{N}$ such that $\frac{1}{n}u_\alpha \in V$ for all α. Since V is solid and order closed, there exists $0 < w \in L$ such that $(\frac{1}{n}u_\alpha) \wedge u \leq u - w$ for all α (otherwise, if $0 \leq (\frac{1}{n}u_\alpha) \wedge u \uparrow u$, then $u \in V$, a contradiction). But then $(nu - u_\alpha)^+ = n\big[u - (\frac{1}{n}u_\alpha) \wedge u\big] \geq nw > 0$ for all α, so that $\{u_\alpha\}$ is dominable.

(ii) Note that by Theorem 7.49(i) the topology τ is pre-Lebesgue. The conclusion now follows immediately from Theorem 4.8.

(iii) By Theorem 4.31 the Riesz space L is order dense in its topological completion $\widehat{L}$ and by Theorem 7.15(ii) we get $\widehat{L} = L$, i.e., (L, τ) is τ-complete. ■

It is natural to ask when a Hausdorff Fatou topology on a Riesz space can be extended to a Hausdorff Fatou topology to its universal completion. Clearly, any such topology must be necessarily Lebesgue.

The following theorem tells us what Hausdorff Lebesgue topologies on a Riesz space have such an extension. This theorem will be the key result for the discussion in this section.

THEOREM 7.51. *For a Hausdorff locally solid Riesz space (L, τ) with the Lebesgue property the following statements are equivalent.*

(i) *τ extends to a Lebesgue topology on L^{u}.*
(ii) *τ extends to a locally solid topology on L^{u}.*
(iii) *τ is coarser than any Hausdorff σ-Fatou topology on L.*
(iv) *Every dominable subset of L^+ is τ-bounded.*
(v) *Every disjoint sequence of L^+ is τ-convergent to zero.*
(vi) *Every disjoint sequence of L^+ is τ-bounded.*
(vii) *The topological completion $\widehat{L}$ of (L, τ) is Riesz isomorphic to L^{u}, that is, $\widehat{L}$ is the universal completion of L.*

PROOF. (i) $\Longrightarrow$ (ii) Obvious.

(ii) $\Longrightarrow$ (iii) Let τ^* be a locally solid extension of τ to L^{u} and let τ' be a Hausdorff σ-Fatou topology on L. By Theorem 4.27 it is enough to show that $\tau \subseteq \tau'$ holds on every component N^{d} of the carrier C_τ. Consequently, let $N = \bigcap_{n=1}^{\infty} V_n$, where $\{V_n\}$ is a normal sequence of solid τ-neighborhoods of zero. Note that every disjoint subset of nonzero positive elements of N^{d} is at most countable. Indeed, if A is a disjoint subset of nonzero positive elements of N^{d}, then $A = \bigcup_{n=1}^{\infty} A_n$, where $A_n = \{u \in A\colon\ u \notin V_n\}$, and so if some A_k is infinite it must include a sequence $\{u_n\}$ that is τ^*-convergent to zero in L^{u} [see Theorem 7.49(i)] and hence is τ-convergent to zero in L, contradicting $u_n \notin V_k$ for all n; thus each A_n is finite and hence A is at most countable. By Zorn's lemma there exists an at most countable complete disjoint system.

Now since N^{d} admits a metrizable Lebesgue topology (its base is given by $\{V_n \cap N^{\mathrm{d}}\}$), it has the countable sup property and hence the restriction of τ' to N^{d}, say τ_1, has the Fatou property. By Theorem 4.18 there exists a metrizable Fatou topology, say τ_m, on N^{d} that is coarser than τ_1. Next consider $\big(\widehat{N^{\mathrm{d}}}, \widehat{\tau}_m\big)$, and note that since N^{d} is order dense in $\widehat{N^{\mathrm{d}}}$ (use Theorem 4.31 or 5.29 for this), and the universal completion of N^{d} is the band generated by N^{d} in L^{u}, we have the Riesz subspace inclusions $N^{\mathrm{d}} \subseteq \widehat{N^{\mathrm{d}}} \subseteq L^{\mathrm{u}}$.

Now since $\widehat{\tau}_m$ is a Fréchet topology, it follows from Theorem 5.20 that τ^* restricted to $\widehat{N^{\mathrm{d}}}$ is coarser that $\widehat{\tau}_m$. Hence τ is coarser than τ' on N^{d}.

(iii) $\Longrightarrow$ (iv) First use Theorem 4.12 to extend τ to the Hausdorff Lebesgue topology τ^δ on L^δ. Let A be a dominable subset of L. Then A is a dominable subset of L^δ. By Theorem 7.13 there exists a complete disjoint system $\{e_i\colon\ i \in I\}$ of nonzero positive elements of L^δ such that for each $i \in I$ we have $P_{e_i}(a) \le n_i e_i$ for all $a \in A$ and some $n_i \in \mathbb{N}$. Let B_i be the band generated by e_i in L^δ and then use Theorem 1.76 to embed L order densely in ΠB_i. But if τ_i denotes the Hausdorff Lebesgue topology of the restriction of τ^δ to B_i, then $\Pi\tau_i$ is a Hausdorff locally solid topology on ΠB_i and hence its restriction to L is also a Hausdorff Lebesgue topology. The conclusion now follows by observing that A (as embedded in ΠB_i) is $\Pi\tau_i$-bounded and that $\Pi\tau_i$ restricted to L is finer than τ by our hypothesis.

(iv) $\Longrightarrow$ (v) If $\{u_n\}$ is a disjoint sequence of L^+, then so is $\{nu_n\}$. Since $\{nu_n\}$ is order bounded in L^{u} it is dominable and hence by our assumption $\{u_n\}$ is τ-bounded. Now let V be a solid τ-neighborhood of zero. Pick some $k \in \mathbb{N}$ such that $nu_n \in kV$ holds for all n. But then $u_n \in V$ for $n \geq k$ and thus $u_n \xrightarrow{\tau} 0$.

(v) $\Longrightarrow$ (vi) Obvious.

(vi) $\Longrightarrow$ (vii) Let $(\widehat{L}, \widehat{\tau})$ be the topological completion of (L, τ). Then L is order dense in $\widehat{L}$ (see Theorem 4.31), $(\widehat{L}, \widehat{\tau})$ satisfies the Lebesgue property, and $\widehat{L}$ is Dedekind complete (see Theorem 3.27). Therefore, we have only to verify that $\widehat{L}$ is laterally complete. To this end, let $\{\widehat{u}_n\}$ be a disjoint sequence of $\widehat{L}^+$, and let V and W be solid τ-neighborhoods of zero such that $W + W \subseteq V$. Since L is order dense in $\widehat{L}$ and $\widehat{\tau}$ is a Lebesgue topology, for each n there exists $u_n \in L^+$ with $0 \leq u_n \leq \widehat{u}_n$ and $\widehat{u}_n - u_n \in \overline{W}$. Observe that $\{u_n\}$ is a disjoint sequence of L^+ and hence τ-bounded. Pick some $k \in \mathbb{N}$ with $\{u_n\} \subseteq kW$ and note that for each n we have

$$\widehat{u}_n = (\widehat{u}_n - u_n) + u_n \in k\overline{W} + k\overline{W} \subseteq k\overline{V}\,.$$

Thus, the sequence $\{\widehat{u}_n\}$ is $\widehat{\tau}$-bounded and consequently every disjoint sequence of $\widehat{L}^+$ is $\widehat{\tau}$-bounded.

Now assume that $\{\widehat{u}_\alpha\colon\ \alpha \in A\}$ is a disjoint family of $\widehat{L}^+$. Let Λ be the collection of all finite subsets of the index set A. For each $\lambda \in \Lambda$ put $\widehat{v}_\lambda = \sum_{\alpha\in\lambda} \widehat{u}_\alpha$. Then $0 \leq \widehat{v}_\lambda \uparrow$ holds in $\widehat{L}$, and we claim that $\{\widehat{v}_\lambda\}$ is in fact a $\widehat{\tau}$-Cauchy net. Indeed, if $\{\widehat{v}_\lambda\}$ is not a $\widehat{\tau}$-Cauchy net, then there exists (by Lemma 2.5) a solid τ-neighborhood V of zero and a sequence $\{\lambda_n\} \subseteq \Lambda$ satisfying $\lambda_n \subseteq \lambda_{n+1}$ and $\widehat{w}_n = \widehat{v}_{\lambda_{n+1}} - \widehat{v}_{\lambda_n} \notin \overline{V}$ for all n. But $\{\widehat{w}_n\}$ is a disjoint sequence of $\widehat{L}^+$ and hence so is $\{n\widehat{w}_n\}$. Thus, by what was shown above, $\{n\widehat{w}_n\}$ is $\widehat{\tau}$-bounded and so there exists some k with $\{n\widehat{w}_n\} \subseteq k\overline{V}$. In particular, it follows that $\widehat{w}_n \in \overline{V}$ for all sufficiently large n, a contradiction. Hence $\{\widehat{v}_\lambda\}$ is a $\widehat{\tau}$-Cauchy net and so $\widehat{v}_\lambda \xrightarrow{\widehat{\tau}} \widehat{u}$ for some $\widehat{u}$. But then $\widehat{v}_\lambda \uparrow \widehat{u}$ holds also in $\widehat{L}$. It now follows that $\widehat{u} = \bigvee_{\alpha\in A} \widehat{u}_\alpha$ in $\widehat{L}$, so that $\widehat{L}$ is laterally complete.

(vii) $\Longrightarrow$ (i) Observe that (according to Theorem 3.27) the locally solid topology $\widehat{\tau}$ is a Lebesgue extension of τ to $\widehat{L}$. ■

If a Hausdorff Lebesgue topology τ on a Riesz space L extends to a Lebesgue topology τ^{u} on its universal completion L^{u}, then any base $\mathcal{N}$ at zero for τ consisting of Fatou τ-neighborhoods of zero extends to a base $\{V^{\mathrm{u}}\colon\ V \in \mathcal{N}\}$ at zero for τ^{u} consisting of Fatou τ^{u}-neighborhoods of zero, where

$$V^{\mathrm{u}} = \big\{u^* \in L^{\mathrm{u}}\colon\ \exists\{u_\alpha\} \subseteq V \text{ with } 0 \leq u_\alpha \uparrow |u^*| \text{ in } L^{\mathrm{u}}\big\}\,.$$

Note, in particular, that τ^{u} is uniquely determined.

We are now in the position to state the main topological properties of laterally complete Riesz spaces.

THEOREM 7.52. *If a laterally σ-complete Riesz space L admits a Hausdorff Fatou topology τ, then any other Hausdorff locally solid topology on L is finer than τ.*

PROOF. Assume that τ is a Hausdorff Fatou topology and τ^* a Hausdorff locally solid topology on a laterally σ-complete Riesz space L. Then by Theorem 7.49(i) the topology τ satisfies the pre-Lebesgue property and hence the Lebesgue property; see Theorem 4.8. Likewise, by Theorem 7.49(ii) the topology τ^* satisfies the σ-Lebesgue property and hence the σ-Fatou property. But by Theorem 7.49 the topology τ satisfies condition (v) of Theorem 7.51 and hence condition (iii) of the same theorem is also true. Consequently, $\tau \subseteq \tau^*$ holds true and the proof is finished. ■

As an immediate consequence of the preceding result we obtain that the Hausdorff Fatou topologies on laterally σ-complete Riesz spaces are uniquely determined.

THEOREM 7.53. *A laterally σ-complete Riesz space admits at most one Hausdorff Fatou topology, which must be necessarily Lebesgue.*

The next result presents a uniqueness property of the Hausdorff Lebesgue topologies.

THEOREM 7.54. *For an Archimedean Riesz space L we have the following:*

(1) *L can admit at most one Hausdorff Lebesgue topology that extends to its universal completion as a locally solid topology.*
(2) *L admits a Hausdorff Lebesgue topology if and only if L^{u} does.*

PROOF. (1) This follows immediately from Theorem 7.53.

(2) Clearly, if L^{u} admits a Hausdorff (locally solid) Lebesgue topology, then so does the Riesz space L. For the converse, assume that L admits a Hausdorff locally solid Lebesgue topology τ. Let $\mathcal{N}$ be a base at zero for τ consisting of solid τ-neighborhoods. For each $u \in L^+$ and each $V \in \mathcal{N}$ let

$$V_u = \{v \in L\colon\ u \wedge |v| \in V\}\,.$$

Clearly, $V \subseteq V_u$.

It should be straightforward to verify that each V_u is a solid (and hence circled) and absorbing subset of L. Also, if $V, W \in \mathcal{N}$ satisfy $V \subseteq W$, then $V_u \subseteq W_u$. Moreover, if $V, W \in \mathcal{N}$ satisfy $V + V \subseteq W$, then $V_u + V_u \subseteq W_u$. Consequently, the family of solid sets

$$\{V_u\colon\ u \in L^+ \text{ and } V \in \mathcal{N}\}$$

is a filter base for the neighborhood system at zero for a locally solid topology τ^* on L. Now one more easy verification shows that τ^* is Hausdorff and Lebesgue.

Next, let $\{u_n\}$ be a disjoint sequence of L and fix some $u \in L^+$. Then $\{|u_n| \wedge u\}$ is an order bounded disjoint sequence of L. In particular, it follows from Theorem 3.22 that $|u_n| \wedge u \xrightarrow{\tau} 0$. The latter easily implies $u_n \xrightarrow{\tau^*} 0$. That is, every disjoint sequence in L is τ^*-convergent to zero. By Theorem 7.51, the Lebesgue topology τ^* extends to a Hausdorff Lebesgue topology to L^{u}, and so L^{u} admits a (unique) Hausdorff locally solid Lebesgue topology. ■

The following two theorems deal with metrizable locally solid topologies on laterally σ-complete Riesz spaces.

THEOREM 7.55. *If a laterally σ-complete Riesz space L admits a metrizable locally solid topology τ, then τ is the only Hausdorff locally solid topology that L can carry, and in addition τ is a Lebesgue topology.*

PROOF. Let τ be a metrizable locally solid topology on a laterally σ-complete Riesz space L. Then by combining Theorems 7.49 and 5.33, we infer that τ is a Lebesgue topology. In particular, it follows that L must satisfy the countable sup property. Thus every other Hausdorff locally solid topology on L must be a Lebesgue topology, and the desired conclusion follows from Theorem 7.53. ∎

THEOREM 7.56. *For a laterally complete Hausdorff locally solid Riesz space (L, τ) the following statements are equivalent.*

(i) *τ is metrizable—and hence uniquely determined.*
(ii) *L has the countable sup property.*

PROOF. (i) $\Longrightarrow$ (ii) By Theorem 7.55 the topology τ is a Lebesgue topology and so L must have the countable sup property.

(ii) $\Longrightarrow$ (i) Note first that τ is a Fatou topology. Hence by Theorem 4.18 (since L has a weak order unit), there exists a metrizable locally solid topology τ' on L, and by Theorem 7.55 we have $\tau = \tau'$. That is, τ is metrizable, and the proof is finished. ∎

The next result presents some conditions for the normal Riesz homomorphisms to be continuous.

THEOREM 7.57. *Assume that τ is a Hausdorff σ-Fatou topology on a Riesz space L, τ' is a Lebesgue topology on a laterally σ-complete Riesz space M, and that $T\colon (L,\tau) \to (M,\tau')$ is a normal Riesz homomorphism. Then T is continuous.*

PROOF. If $\mathcal{N}$ is a base at zero for τ' consisting of Fatou neighborhoods, note that since T is a normal Riesz homomorphism, $\{T^{-1}(V)\colon\ V \in \mathcal{N}\}$ is a base at zero for a Lebesgue topology, say τ_1, on L. Clearly $T\colon (L,\tau_1) \to (M,\tau')$ is continuous. Observe also that τ_1 (which is not necessarily Hausdorff) satisfies condition (v) of Theorem 7.51. Indeed, if $\{u_n\}$ is a disjoint sequence of L^+, then since T is a Riesz homomorphism $\{T(u_n)\}$ is a disjoint sequence of M^+ and hence by Theorem 7.49(i) we have $T(u_n) \xrightarrow{\tau'} 0$, so that $u_n \xrightarrow{\tau_1} 0$.

Now put $N = \bigcap\{T^{-1}(V)\colon\ V \in \mathcal{N}\}$, and note that N is a band of L and that τ_1 induces a Hausdorff Lebesgue topology on N^{d}. Thus by Theorem 7.51 the topology τ_1 must be coarser than τ if both are restricted to N^{d}. That is, T restricted to N^{d} is continuous. Obviously T restricted to N is continuous and hence T restricted to the order dense ideal $N \oplus N^{\mathrm{d}}$ is continuous.

Next let $\{u_\alpha\} \subseteq L^+$ satisfy $u_\alpha \xrightarrow{\tau} 0$, and let $V, U \in \mathcal{N}$ be such that $V + V \subseteq U$. Since $N \oplus N^{\mathrm{d}}$ is order dense in L, for each α there exists some $v_\alpha \in N \oplus N^{\mathrm{d}}$ with $0 \le v_\alpha \le u_\alpha$ and $T(u_\alpha) - T(v_\alpha) \in V$ (we use here that τ_1 is Lebesgue and that $T\colon (L,\tau_1) \to (M,\tau')$ is continuous). But then $v_\alpha \xrightarrow{\tau} 0$ holds in L and since $\{v_\alpha\} \subseteq N \oplus N^{\mathrm{d}}$, it follows $T(v_\alpha) \xrightarrow{\tau'} 0$ and thus $T(v_\alpha) \in V$ for all $\alpha \succeq \alpha_1$. Hence for all $\alpha \succeq \alpha_1$ we have

$$T(u_\alpha) = \big[T(u_\alpha) - T(v_\alpha)\big] + T(v_\alpha) \in V + V \subseteq U\,,$$

so that $T\colon (L,\tau) \to (M,\tau')$ is continuous, and the proof is finished. ∎

The next result describes a basic property of the unique Hausdorff Fatou topology on a laterally σ-complete Riesz space.

THEOREM 7.58. *If a laterally σ-complete Riesz space L admits a Hausdorff Fatou topology τ (that is necessarily unique and Lebesgue), then τ is coarser than any Hausdorff locally solid topology on L and finer than any Lebesgue topology on the Riesz space L.*

PROOF. The "coarser" part is just Theorem 7.52. To see the other part, let τ^* be a Lebesgue topology on L, and consider the identity operator $I\colon (L,\tau) \to (L,\tau^*)$. Now apply Theorem 7.57 to obtain the desired conclusion $\tau^* \subseteq \tau$. ■

We now state a theorem giving some conditions under which an order continuous operator is continuous.

THEOREM 7.59. *Let $T\colon (L,\tau) \to (M,\tau')$ be a positive order continuous operator, where (L,τ) is a laterally σ-complete Hausdorff locally solid Riesz space and (M,τ') a locally solid Riesz space with the Fatou property. Then the operator T is continuous.*

PROOF. Let $\mathcal{N}$ be a base at zero for the topology τ' consisting of Fatou τ'-neighborhoods. For each $V \in \mathcal{N}$ define

$$V^* = \{u \in L\colon\ T(|u|) \in V\},$$

and note that the collection $\{V^*\colon\ V \in \mathcal{N}\}$ defines a Fatou topology τ^* on L (and hence τ^* is Lebesgue) such that $T\colon (L,\tau^*) \to (M,\tau')$ is continuous. Now since τ is σ-Lebesgue, it follows from Theorem 7.57 that the identity operator $I\colon (L,\tau) \to (L,\tau^*)$ is continuous and hence $T\colon (L,\tau) \to (M,\tau')$ must be continuous too. ■

We can ask at this point whether there exist Archimedean laterally σ-complete Riesz spaces admitting no Hausdorff locally solid topologies. The following example gives an affirmative answer to this question.

EXAMPLE 7.60. Let L be the universal completion of $C[0,1]$. Assume that L admits a Hausdorff locally solid topology τ. Then by Theorem 7.49(ii) the topology τ is a σ-Lebesgue topology and hence τ restricted to $C[0,1]$ is also a Hausdorff σ-Lebesgue topology, contradicting the fact that $C[0,1]$ does not admit any Hausdorff σ-Lebesgue topology; see Example 3.2. This contradiction establishes that L does not admit any Hausdorff locally solid topology. ■

Regarding quotient Riesz spaces of laterally σ-complete Riesz spaces we have the following result.

THEOREM 7.61. *Let L be a laterally σ-complete Riesz space carrying a (necessarily unique) Hausdorff Lebesgue topology τ, and let A be a band of L. Then τ/A is the unique Hausdorff Lebesgue topology on the laterally σ-complete Riesz space L/A.*

PROOF. Note that L/A is laterally σ-complete (see Theorem 7.9), τ/A is Hausdorff (since every band is τ-closed) and τ/A is a Lebesgue topology by Theorem 3.10(b), and hence unique by Theorem 7.53. ■

For a laterally σ-complete Riesz space admitting the unique Hausdorff Fatou topology the following interesting result holds true.

THEOREM 7.62. *If a laterally σ-complete Riesz space L admits a Hausdorff Fatou topology τ (that is necessarily unique and Lebesgue), then there exists an order dense σ-ideal of L such that any other Hausdorff locally solid topology on L induces the same topology as τ on this σ-ideal.*

PROOF. By theorem 4.8 the topology τ is Lebesgue, and by Theorem 4.17 the carrier C_τ of τ is an order dense σ-ideal of L with the countable sup property. In particular, note that the carrier C_τ is a laterally σ-complete Riesz space in its own right.

Now if τ^* is any other Hausdorff locally solid topology on L, then τ^* satisfies the σ-Lebesgue property (see Theorem 7.49) and so, since C_τ has the countable sup property, τ^* restricted to C_τ must be a Lebesgue topology. The result now follows immediately from Theorem 7.53. ■

As an illustration of the preceding theorem we offer the following example.

EXAMPLE 7.63. Consider the universally complete Riesz space $L = \mathbb{R}^X$, where X is an uncountable set. Let τ be the Hausdorff Lebesgue topology of pointwise convergence on L. An easy verification shows that the carrier C_τ of τ consists of all functions of L whose support is at most countable. That is,

$$C_\tau = \left\{ f \in \mathbb{R}^X : \text{ The set } \{x \in X \mid f(x) \neq 0 \text{ is at most countable} \right\}.$$

According to Theorem 7.62 every Hausdorff locally solid topology on L induces the topology of pointwise convergence on C_τ. ■

7.5. Minimal Locally Solid Topologies

So far, we have discussed in many instances comparison properties of locally solid topologies. In this section we shall study minimal topologies, i.e., Hausdorff locally solid topologies that are coarser than any other Hausdorff locally solid topology. All results in this section are due to the authors [**20**].

We start with the definition of a minimal topology.

DEFINITION 7.64. *A Hausdorff locally solid topology ξ on a Riesz space L is said to be* ***minimal*** *if ξ is coarser than any other Hausdorff locally solid topology τ on L, i.e., $\xi \subseteq \tau$.*

It should be clear from the preceding definition that only Archimedean Riesz spaces can admit minimal topologies and that an Archimedean Riesz space can admit at most one minimal topology.

Let us present a class of Riesz spaces that admit minimal topologies.

THEOREM 7.65. *If a Riesz space L admits a complete metrizable Lebesgue topology (in particular, if L is a Banach lattice with order continuous norm), then L has a minimal topology.*

PROOF. Assume that a Riesz space L admits a complete metrizable Lebesgue topology τ. (According to Theorem 5.21 any such topology τ is, of course, uniquely determined.) A glance at part (2) of Theorem 7.54 guarantees that L also admits a (necessarily unique) Hausdorff Lebesgue topology τ^* that extends to a Hausdorff Lebesgue topology to L^{u}.

Now if τ' is any other Hausdorff locally solid topology on L, then by Theorem 5.20 we have $\tau' \subseteq \tau$. In particular, it follows that τ' has the Lebesgue property. But then, by condition (iii) of Theorem 7.51, we get $\tau^* \subseteq \tau'$. This shows that τ^* is a minimal topology on L. ■

In the terminology of minimal topologies Theorem 7.52 can be rephrased as follows.

THEOREM 7.66. *If a laterally σ-complete Riesz space L admits a Hausdorff Fatou topology τ, then τ is a minimal topology on L.*

We continue with two basic properties of minimal topologies.

THEOREM 7.67. *If ξ is a minimal topology on a Riesz space L, then:*

(i) *Every disjoint sequence of L is ξ-convergent to zero.*
(ii) *ξ is a Lebesgue topology.*

PROOF. We start the proof by observing that if ρ is a pseudonorm on a Riesz space L, then for each $v \in L^+$ the formula $\rho_v(u) = \rho(|u| \wedge v)$ also defines a Riesz pseudonorm satisfying $\rho_v(u) \leq \rho(u)$ for all $u \in L$.

Now let ξ be a minimal topology on a Riesz space L and let $\mathcal{P}$ denote the collection of all ξ-continuous Riesz pseudonorms on L.

(i) Let $\{u_n\}$ be a disjoint sequence of L^+ and define the set

$$A = \{v \in L^+ \colon \text{ There exists } k \text{ with } v \wedge u_n = 0 \text{ for all } n \geq k\}.$$

Clearly, $\{u_n\} \subseteq A$ and we claim that $A^{\mathrm{d}} = \{0\}$. Indeed, if $0 \leq v \in A^{\mathrm{d}}$, then $v \perp u_n$ for each n and so $v \in A$, from which it follows that $v = 0$.

Now consider the collection of Riesz pseudonorms

$$\mathcal{P}_A = \{\rho_v \colon \ \rho \in \mathcal{P} \text{ and } v \in A\},$$

and let τ be the topology generated by the family of Riesz pseudonorms $\mathcal{P}_A$. We claim that τ is a Hausdorff locally solid topology on L. To see this, assume that $\rho_v(u) = \rho(|u| \wedge v) = 0$ holds for all $\rho \in \mathcal{P}$ and all $v \in A$. Since ξ is a Hausdorff topology, we get $|u| \wedge v = 0$ for all $v \in A$, i.e., $u \in A^{\mathrm{d}} = \{0\}$. Thus, $u = 0$ and so τ is a Hausdorff locally solid topology.

The minimality of ξ implies $\xi \subseteq \tau$. Now note that $u_n \xrightarrow{\tau} 0$, and from this we easily infer that $u_n \xrightarrow{\xi} 0$.

(ii) Let $u_\alpha \downarrow 0$ in L and let V and U be two solid ξ-neighborhoods of zero such that $V + V \subseteq U$. Fix an index β and then pick some ϵ with $0 < \epsilon < 1$ such that $\epsilon u_\beta \in V$. Next, define the set

$$B = \{v \in L^+ \colon \ \exists \text{ some index } \alpha \text{ such that } v \wedge (u_\alpha - \epsilon u_\beta)^+ = 0\}.$$

The collection of Riesz pseudonorms

$$\mathcal{P}_B = \{\rho_v \colon \ \rho \in \mathcal{P} \text{ and } v \in B\}$$

defines a Hausdorff locally solid topology τ^* on L. To establish this, suppose that $\rho_v(u) = 0$ holds for all $\rho \in \mathcal{P}$ and all $v \in B$. Since $(u_\alpha - \epsilon u_\beta)^- \in B$ for each α, it follows that $|u| \wedge (u_\alpha - \epsilon u_\beta)^- = 0$ for each α. This, in conjunction with $u_\alpha \downarrow 0$, yields $|u| \wedge u_\beta = 0$ and so

$$|u| \wedge (u_\beta - \epsilon u_\beta)^+ = |u| \wedge [(1-\epsilon)u_\beta] = 0.$$

This implies that $|u| \in B$ and thus $\rho(|u|) = \rho(|u| \wedge |u|) = 0$ for all $\rho \in \mathcal{P}$ proving that $u = 0$. Therefore, τ^* is a Hausdorff locally solid topology and so $\xi \subseteq \tau^*$.

From the definition of τ^*, it easily follows that $(u_\alpha - \epsilon u_\beta)^+ \xrightarrow[\alpha]{\tau^*} 0$, and from this we infer that $(u_\alpha - \epsilon u_\beta)^+ \xrightarrow[\alpha]{\xi} 0$. Consequently, there exists an index α_0 such that $(u_\alpha - \epsilon u_\beta)^+ \in V$ for all $\alpha \succeq \alpha_0$. Now use the inequalities $0 \le u_\alpha \le (u_\alpha - \epsilon u_\beta)^+ + \epsilon u_\beta$ to infer that $u_\alpha \in V + V \subseteq U$ for all $\alpha \succeq \alpha_0$. This shows that $u_\alpha \xrightarrow{\xi} 0$, and the proof is finished. ■

A glance at conditions (ii) and (v) of Theorem 7.51 in conjunction with the preceding theorem yields the following result.

COROLLARY 7.68. *If ξ is a minimal topology on a Riesz space L, then ξ has a (necessarily unique) Lebesgue extension $\widehat{\xi}$ to the universal completion L^{u} of L and, moreover, $(L^{\mathrm{u}}, \widehat{\xi})$ is the topological completion of (L, ξ).*

Given that (according to Example 3.2) the Riesz space $C[0,1]$ has no Hausdorff σ-Lebesgue topology, also we have the following result.

COROLLARY 7.69. *The Riesz space $C[0,1]$ has no minimal topology and its universal completion does not admit any Hausdorff Lebesgue topology.*

Regarding minimal locally convex-solid topologies we have the following.

THEOREM 7.70. *An Archimedean Riesz space L admits a minimal topology that is also locally convex if and only if L is a discrete space—in which case the minimal topology is the topology of pointwise convergence.*

PROOF. Assume first that L is an Archimedean discrete Riesz space. Then, by Theorem 1.78, L is Riesz isomorphic to an order dense Riesz subspace M of some $\mathbb{R}^X$. Identifying L with M, we see that $\mathbb{R}^X$ is the universal completion of L. We claim that the Hausdorff locally-convex topology ξ of pointwise convergence on L is minimal. To see this, let τ be a Hausdorff locally solid topology on L, and assume that a net $\{u_\alpha\} \subseteq L^+$ satisfies $u_\alpha \xrightarrow{\tau} 0$. If $\lim u_\alpha(x) \neq 0$ for some $x \in X$, then by passing to a subnet we can assume that there exists some $\epsilon > 0$ such that $u_\alpha(x) \ge \epsilon$ holds for all α. But then the vector $v = \epsilon \chi_{\{x\}} \in L$ satisfies $0 < v \le u_\alpha$ for all α, contrary to $u_\alpha \xrightarrow{\tau} 0$. Therefore, ξ is a minimal topology.

For the converse, assume that L admits a minimal topology ξ that is in addition locally convex and let $0 < u \in L$. We must show that there exists a discrete element $v \in L^+$ satisfying $0 < v \le u$.

To see this, start by observing that by part (ii) of Theorem 7.67 the topology ξ is also a Lebesgue topology, and so the topological dual L' of (L, ξ) satisfies $L' \subseteq L_{\mathrm{n}}^{\sim}$. Choose some $0 < \phi \in L'$ with $\phi(u) > 0$. Since $\phi \in L_{\mathrm{n}}^{\sim}$, there exists some $0 < v \in C_\phi$ such that $0 < v \le u$. If $[0, v]$ does not contain any discrete element, then there exists a disjoint sequence $\{v_n\} \subseteq [0, v]$ of nonzero vectors. Since $v_n \in C_\phi$ holds for each n, we have $\phi(v_n) > 0$, and so if $\lambda_n = \frac{1}{\phi(v_n)} > 0$, then $\lambda_n \phi(v_n) = 1$. Now note that $\{\lambda_n v_n\}$ is a disjoint sequence, and so by part (1) of Theorem 7.67 we have $\lambda_n v_n \xrightarrow{\xi} 0$. But then we have $1 = \lambda_n \phi(v_n) = \phi(\lambda_n v_n) \to 0$, which is impossible. This contradiction completes the proof of the theorem. ■

Repeating almost verbatim the arguments of the second part of the previous proof, we can establish the following result.

THEOREM 7.71. *If an atomless Riesz space L admits a minimal topology ξ, then $(L, \xi)' = \{0\}$.*

Now we turn our discussion to minimal topologies on L_p-spaces. So, consider (X, Σ, μ), where X is a set, Σ is a σ-algebra of subsets of X, and μ is a σ-finite measure on Σ. A function $f\colon X \to \mathbb{R}$ is said to be **Σ-measurable** (or simply **measurable**) if $f^{-1}(B) \in \Sigma$ for each Borel subset B of $\mathbb{R}$. The collection of all measurable functions is a vector space. Two functions $f, g\colon X \to \mathbb{R}$ are said to be **μ-almost everywhere equal** (in brief μ-a.e. or simply a.e.) if $f(x) = g(x)$ holds for μ-almost all $x \in X$. That is, f and g are equal almost everywhere if there exists some $A \in \Sigma$ with $\mu(A) = 0$ such that $f(x) = g(x)$ for all $x \notin A$. Almost everywhere equality defines an equivalence relation on the vector space of all measurable function. It is easy to see that under the algebraic and lattice operations defined μ-a.e., the set of all equivalence classes is a Riesz space. This is the very important Riesz space $L_0(\mu)$.

DEFINITION 7.72. *The Riesz space of all equivalence classes of measurable real-valued functions on X is denoted by $L_0(\mu)$.*

As usual, we identify the equivalence classes with functions and agree that two arbitrary measurable functions $f, g\colon X \to \mathbb{R}$ are equal if and only if $f(x) = g(x)$ for μ-almost all $x \in X$. With this convention in mind, a measurable function f can be changed (or can even be left undefined) on a set of μ-measure zero and still represent the same equivalence class as f in $L_0(\mu)$.

The proof of the next result is left as an exercise; see also Exercise 23 at the end of the chapter.

THEOREM 7.73. *If the measure μ is σ-finite, then $L_0(\mu)$ is a Dedekind complete and laterally complete Riesz space with the countable sup property.*

In particular, for every $0 \le p \le \infty$ the Riesz space $L_0(\mu)$ is the universal completion of the Riesz space $L_p(\mu)$.

For the rest of the discussion in this section (X, Σ, μ) will denote a σ-finite measure space. For each $E \in \Sigma$ with $\mu(E) < \infty$ we define the Riesz pseudonorm $\rho_E\colon L_0(\mu) \to \mathbb{R}$ via the formula

$$\rho_E(u) = \int_E \tfrac{|u|}{1+|u|}\, d\mu\,.$$

Clearly, the family of Riesz pseudonorms $\{\rho_E\colon\; E \in \Sigma \text{ and } \mu(E) < \infty\}$ defines a Hausdorff locally solid topology τ_m on $L_0(\mu)$. This topology is called the **topology of convergence in measure** in $L_0(\mu)$ since a sequence $\{u_n\}$ in $L_0(\mu)$ satisfies $u_n \xrightarrow{\tau_m} u$ if and only if for each $\epsilon > 0$ and each $E \in \Sigma$ with $\mu(E) < \infty$ we have

$$\lim_{n\to\infty} \mu\bigl(\{x \in E\colon\; |u_n(x) - u(x)| \ge \epsilon\}\bigr) = 0\,.$$

Now if $u_n \downarrow 0$ holds in $L_0(\mu)$, then it follows from the Lebesgue Dominated Theorem that $\rho_E(u_n) \to 0$ for each $E \in \Sigma$ with $\mu(E) < \infty$. This shows that τ_m is a σ-Lebesgue topology and since $L_0(\mu)$ has the countable sup property, it follows that τ_m is a Hausdorff Lebesgue topology on $L_0(\mu)$. But then Theorem 7.53 guarantees that τ_m is the only Hausdorff locally solid topology that $L_0(\mu)$ can carry. In addition, it follows from Corollary 7.68 that the only possible minimal topology on any L_p-space is the restriction of τ_m on the space. It turns out that this is the case for all L_p-spaces except when $p = \infty$.

THEOREM 7.74. *If (X, Σ, μ) is an arbitrary σ-finite measure space, then for each $0 < p < \infty$ the topology of convergence in measure on $L_p(\mu)$ is a minimal topology. Moreover, if τ_p denotes the Hausdorff locally solid topology generated by the "L_p-norm," then every other Hausdorff locally solid topology τ on on $L_p(\mu)$ satisfies $\tau_m \subseteq \tau \subseteq \tau_p$.*

PROOF. For $0 < p < \infty$ the topology generated by the"L_p-norm" is a Lebesgue Fréchet topology. Thus, according to Theorem 5.20 the topology τ_p is finer than any other locally solid topology on $L_p(\mu)$. The rest of the proof can be completed by invoking Theorem 7.65 and Corollary 7.68. ■

For $p = \infty$ the situation is drastically different.

THEOREM 7.75. *If (X, Σ, μ) is a non-atomic σ-finite measure space, then $L_\infty(\mu)$ has no minimal topology.*

PROOF. Assume by way of contradiction that the Riesz space $L = L_\infty(\mu)$ (where μ is σ-finite and non-atomic) admits a minimal topology ξ. By Theorem 7.67 the Hausdorff topology ξ is Lebesgue, and so ξ is entire.

Now write $L^\sim = L_1(\mu) \oplus B$, where B is the disjoint complement of $L_1(\mu)$ in $L^\sim$. It follows that B separates the points of L (see Exercise 20 in Chapter 6), and so the absolute weak topology $\tau = |\sigma|(L, B)$ is a Hausdorff locally convex-solid topology on L. Moreover, by Theorem 2.33, we have $(L, \tau)' = B$.

Since ξ is a minimal topology it must be the case that $\xi \subseteq \tau$, and therefore $C_\xi \subseteq C_\tau$. This implies that τ is an entire topology. Now invoke Theorem 5.51 to conclude that the ideal $L'_{\rm n} = B \cap L^\sim_{\rm n} = \{0\}$ necessarily must be order dense in $L^\sim_{\rm n} = L_1(\mu)$, which is impossible. This contradiction establishes that $L_\infty(\mu)$ does admit a minimal topology. ■

The preceding theorem shows that not every Hausdorff Lebesgue topology on a Riesz space that has a Hausdorff Lebesgue extension to its universal completion is a minimal topology. Or, in other words, the restriction on L of the unique Hausdorff Lebesgue topology on $L^{\rm u}$ need not be a minimal topology on L. Our next result presents a condition for this induced topology to be minimal.

THEOREM 7.76. *Assume that ξ is a Hausdorff Lebesgue topology on a Riesz space L such that ξ has a (necessarily unique) Hausdorff Lebesgue extension to its universal completion $L^{\rm u}$. If every Hausdorff locally solid topology on L is entire, then ξ is minimal.*

PROOF. Let τ be a Hausdorff locally solid topology on L and let $\mathcal{N}$ be a base at zero for τ consisting of solid τ-neighborhoods. For each $V \in \mathcal{N}$, let V^* denote the ξ-closure of V in L. Clearly, each V^* is solid and order closed. Thus, the collection $\{V^*\colon\ V \in \mathcal{N}\}$ defines a Fatou topology τ^* on L such that $\tau^* \subseteq \tau$.

We next show that τ^* is Hausdorff. To see this, assume by way of contradiction that $0 < u \in V^*$ for all V^*. Since both C_ξ and C_τ are order dense, we can assume that $u \in C_\xi \cap C_\tau$. Pick a normal sequence $\{V_n\}$ of solid τ-neighborhoods of zero with $u \in N^{\rm d}$, where $N = \bigcap_{n=1}^\infty V_n$. Similarly, pick another normal sequence $\{W_n\}$ of solid ξ-neighborhoods of zero such that $u \in M^{\rm d}$, where $M = \bigcap_{n=1}^\infty W_n$. Clearly, $u \in V_n^*$ for all n. Now for each n choose some $u_n \in V_n$ with $0 \le u_n \le u$ and $u - u_n \in W_n$. Then $\{u_n\} \subseteq M^{\rm d}$ and $u_n \xrightarrow{\tau_1} u$ in $M^{\rm d}$, where τ_1 is the metrizable Lebesgue topology on $M^{\rm d}$ generated by $\{W_n \cap M^{\rm d}\}$. Now, by Theorem 5.10, there exists a subsequence $\{v_n\}$ of $\{u_n\}$ and a net $\{w_\alpha\}$ satisfying $0 \le w_\alpha \uparrow u$ in $M^{\rm d}$, and such that given any index α there exists some n_α satisfying $w_\alpha \le v_n$ for all $n \ge n_\alpha$. It easily follows that $\{w_\alpha\} \subseteq N$. Now since $\{w_\alpha\} \subseteq N^{\rm d}$, we get $w_\alpha = 0$ for all α. This implies $u = 0$, a contradiction.

Now invoke Theorem 7.51 to see that $\xi \subseteq \tau^*$. Hence $\xi \subseteq \tau$, and consequently the topology ξ is minimal. ■

COROLLARY 7.77. *Assume that L is an order dense Riesz subspace of an Archimedean Riesz space M. If L admits a minimal topology, then so does M.*

PROOF. Let ξ be the minimal topology of L. Since L is order dense in M, the universal completion M^{u} of M coincides with the universal completion of L. In particular, it follows from Theorem 7.51 that ξ extends (uniquely) to a Hausdorff Lebesgue topology τ^{u} on M^{u}. We claim that the restriction of τ^{u} to M is a minimal topology.

To this end, let τ be a Hausdorff locally solid topology on M. Then, τ restricted to L is entire, and hence (as is easily checked) τ has an order dense carrier in M, i.e., τ is entire. Now to complete the proof it remains to invoke Theorem 7.76. ■

Although a Hausdorff Lebesgue topology on a Riesz space L that extends to a (unique) Hausdorff Lebesgue topology on L^{u} need not be minimal, it does satisfy a nice minimality property. Let us call a Hausdorff locally solid topology τ on a Riesz space L **relatively minimal** if it follows from $\tau_1 \subseteq \tau$ and τ_1 a Hausdorff locally solid topology that $\tau_1 = \tau$.

THEOREM 7.78. *A Hausdorff Lebesgue topology τ on a Riesz space L is relatively minimal if and only if τ extends to a Hausdorff Lebesgue topology on L^{u}. In particular, a given Riesz space can have at most one Hausdorff Lebesgue topology that is relatively minimal.*

PROOF. Let τ be a Hausdorff Lebesgue topology on a Riesz space L. Assume first that τ extends to a Hausdorff Lebesgue topology on L^{u} and let a Hausdorff locally solid topology τ_1 satisfy $\tau_1 \subseteq \tau$. Clearly, τ_1 is a Lebesgue topology, and so by Theorem 7.51 we have $\tau \subseteq \tau_1$. Therefore, $\tau_1 = \tau$, and thus τ is relatively minimal.

For the converse, assume that τ is relatively minimal. According to part (2) of Theorem 7.54 there exists a Hausdorff Lebesgue topology τ^* on L that extends to a Hausdorff Lebesgue topology on L^{u}. It follows from Theorem 7.51 that $\tau^* \subseteq \tau$, and so by the relative minimality of τ we get $\tau = \tau^*$. Consequently, the topology τ extends to a Hausdorff Lebesgue topology on L^{u}. ■

Exercises

1. For a Boolean algebra $\mathcal{B}$ establish the following.
   (i) A proper ideal I of $\mathcal{B}$ is prime if and only if I is a maximal ideal.
   (ii) If an ideal I of $\mathcal{B}$ does not contain an element u, then there exists a prime ideal J of $\mathcal{B}$ such that $I \subseteq J$ and $u \notin J$.
   (iii) Let Ω denote the collection of all proper prime ideals of $\mathcal{B}$. For each $u \in \mathcal{B}$ let $\Omega_u = \{\omega \in \Omega\colon\ u \notin \omega\}$. Then, the collection $\{\Omega_u\colon\ u \in \mathcal{B}\}$ forms a base for a Hausdorff compact topology τ on Ω, called the **hull-kernel topology**.
   (iv) A subset A of Ω is clopen with respect to τ if and only if there exists some $u \in \mathcal{B}$ such that $A = \Omega_u$.

(v) The Hausdorff compact space (Ω, τ) is extremally disconnected if and only if $\mathcal{B}$ is a Dedekind complete Boolean algebra.
(vi) If X is Stonean, then there exists a unique (up to Boolean isomorphism) Dedekind complete Boolean algebra $\mathcal{B}$ whose Stone space (Ω, τ) is homeomorphic to X.

2. Show that if L is an Archimedean Riesz space, then the collection $\mathcal{B}_L$ of all bands of L ordered by inclusion is a Dedekind complete Boolean algebra whose Boolean operations for each $A, B \in \mathcal{B}_L$ are given by:

$$A \wedge B = A \cap B, \quad A \vee B = (A+B)^{\mathrm{dd}}, \quad \text{and} \quad A' = A^{\mathrm{d}}.$$

3. Show that if L is an order dense Riesz subspace of an Archimedean Riesz space M, then $\mathcal{B}_L$ and $\mathcal{B}_M$ are Boolean isomorphic.

4. Show that two Stonean spaces Ω and Q are homeomorphic if and only if the universally complete Riesz spaces $C^\infty(\Omega)$ and $C^\infty(Q)$ are Riesz isomorphic.

5. Show that $\mathbb{R}^X$ is Riesz isomorphic to $\mathbb{R}^Y$ if and only if the sets X and Y have the same cardinality.

6. Let L be the Riesz space consisting of all real-valued, bounded everywhere, Lebesgue measurable functions on $[0,1]$ with the pointwise ordering (everywhere). Determine the universal completion and the universal σ-completion of L.

7. Let L be an Archimedean laterally σ-complete Riesz space. Show that if $\phi \in L^\sim$, then C_ϕ is a projection band of L. [Compare this with part (ii) of Exercise 23 in Chapter 1.]

8. Let L be an Archimedean laterally σ-complete Riesz space. Show that L has no discrete elements if and only if $L_{\mathrm{n}}^\sim = \{0\}$.

9. Complete the details in the proof of Theorem 7.42.

10. (Bernau [**37**]) Let L be an Archimedean Riesz space. As usual, the **lateral completion** L^λ of L is defined as the intersection of all laterally complete Riesz subspaces between L and L^{u}. Show that $(L^\lambda)^\delta = (L^\delta)^\lambda = L^{\mathrm{u}}$.

11. Let $L = \mathbb{R}^X$ for some nonempty set X. Show that $L_{\mathrm{c}}^\sim = L^\sim$. [*Hint*: Use Exercise 26 of Chapter 1 or use Theorem 7.8.]

12. A set X is said to have a **non-measurable cardinal** if there is no probability measure μ defined on all subsets of X that vanishes on the finite subsets of X.

Show that if a set X has a non-measurable cardinal, then for every $\phi \in (\mathbb{R}^X)^\sim$ there exist points $x_1, \ldots, x_n \in X$ and real numbers $\lambda_1, \ldots, \lambda_n$ (all depending on ϕ) such that $\phi(f) = \sum_{i=1}^n \lambda_i f(x_i)$ for all $f \in \mathbb{R}^X$ and derive in this case that $(\mathbb{R}^X)_{\mathrm{n}}^\sim = (\mathbb{R}^X)^\sim$. (For more about this see [**86**].)

13. Let $L = \mathbb{R}^X$, where X is a nonempty set and let τ be the topology of pointwise convergence on L. For a sequence $\{u_n\}$ in L show the following:
(i) $u_n \xrightarrow{\tau} 0$ if and only if $u_n \xrightarrow{o} 0$ in L.
(ii) If τ_1 and τ_2 are two Hausdorff locally solid topologies on L, then $u_n \xrightarrow{\tau_1} 0$ if and only if $u_n \xrightarrow{\tau_2} 0$.

14. Let (L, τ) be a laterally σ-complete locally solid Riesz space, $\{V_n\}$ a normal sequence of solid τ-neighborhoods of zero and $N = \bigcap_{n=1}^\infty V_n$. Show that N^{d} is an Archimedean laterally complete Riesz space with the countable sup property.

15. Give an alternate proof of Theorem 7.57 by using Theorem 4.27.

16. Verify that the carrier of Example 7.63 is given by

$$C_\tau = \{ f \in \mathbb{R}^X : \ \{x \in X \mid f(x) \neq 0\} \text{ is at most countable} \}.$$

17. Let (X, Σ, μ) be an arbitrary finite measure space whose finite subsets have measure zero and let L be the Riesz space of all real-valued μ-measurable functions defined on X (with the everywhere pointwise ordering).

(i) Show that L is a universally σ-complete Riesz space whose unique Hausdorff Fatou topology τ is the topology of pointwise convergence.

(ii) Let $A = \{f \in L\colon\ f(x) = 0 \ a.e.\}$. Show that A is a σ-ideal of L and that L/A is a universally complete Riesz space.

(iii) For each n let

$$V_n = \Big\{\dot{u} \in L/A\colon\ \int \tfrac{|u|}{1+|u|}\, d\mu \le \tfrac{1}{2^n}\Big\}.$$

Show that $\{V_n\}$ is a base of solid order closed neighborhoods at zero for the unique Hausdorff Fatou topology on L/A.

(iv) What is τ/A? Compare your last answer with Theorems 7.61 and 4.44.

18. If τ is a locally solid topology on a laterally σ-complete Riesz space, show that C_τ is a laterally σ-complete Riesz space with the countable sup property and that τ restricted to C_τ is a Hausdorff Lebesgue topology.

19. Show that a Hausdorff locally solid topology τ on an arbitrary laterally σ-complete Riesz space is Lebesgue if and only if its carrier C_τ is τ-dense.

20. Let (L, τ) be a Hausdorff laterally σ-complete locally solid Riesz space. Show that L admits a Hausdorff Fatou topology (necessarily unique) if and only if C_τ is order dense in L.

21. Give an example of a universally σ-complete Riesz space with a Hausdorff locally solid topology that is not Lebesgue.

22. Let X be a nonempty set. If every Hausdorff locally solid topology on $\mathbb{R}^X$ is Lebesgue, show that X has a non-measurable cardinal. [*Hint*: If μ is a finite measure on all subsets of X that vanishes on the finite subsets of X, then the Hausdorff locally solid topology on $\mathbb{R}^X$ generated by the Riesz pseudonorms

$$\rho_x(u) = |u(x)| \text{ for } x \in X\,, \quad \text{and} \quad \rho(u) = \textstyle\int \frac{|u|}{1+|u|}\, d\mu\,,$$

is not a Lebesgue topology.]

23. If the measure μ is σ-finite, then show that $L_0(\mu)$ is a Dedekind complete and laterally complete Riesz space with the countable sup property.

In particular, conclude from this that for every $0 \le p \le \infty$ the Riesz space $L_0(\mu)$ is the universal completion of the Riesz space $L_p(\mu)$.

24. Use Theorem 7.54 in conjunction with Theorem 7.51 to provide an alternate proof of the Amemiya–Mori Theorem 4.22: *All Hausdorff Lebesgue topologies on a Riesz space induce the same topology on the order bounded sets.*

25. If a Riesz space L admits a minimal topology, then show that the collection of all Hausdorff locally solid topologies on L is (under the inclusion ordering) a Dedekind complete Boolean algebra.

CHAPTER 8

Market Economies

The purpose of this chapter is to present a brief but nevertheless comprehensive introduction to the theory of General Economic Equilibrium. The main emphasis of our exposition will be focused on the importance of the lattice structures in studying and analyzing properties related to efficiency.

It has become clear in the last couple of decades that in order to address real policy questions, one must employ economic models that are both stochastic and dynamic. These models lead naturally to infinite dimensional commodity spaces. The future is uncertain, and infinitely many resolutions of this uncertainty are conceivable. There is no natural terminal period, so time horizons must be open ended.

In this chapter, we shall discuss the problems of working with infinite dimensional spaces in connection with the existence and efficiency of equilibria in the Arrow–Debreu–McKenzie framework [**35, 96**]. In particular, the properties of the locally solid topologies will play a primary role in this study. To simplify the exposition, we only consider exchange economies.

Riesz spaces were introduced to economics for the first time in [**11**]. The present chapter is an expanded version of a paper by K. C. Border and the authors [**9**]. We express our gratitude to Academic Press Publishing Company for allowing us to reprint parts of this article. For details regarding the basic ideas of Microeconomic Theory we refer the reader to [**51**] and [**94**]. For an extensive treatment of the role of the lattice structures in General Economic Equilibrium see [**15**]. For a glimpse of the role of Riesz spaces in production economies see [**13**], [**111**], [**114**], and [**132**]. For a more general approach of General Equilibrium with non-linear prices see [**28**] and [**30**].

8.1. Preferences and Utility Functions

One of the basic primitive concepts in economics is that of a preference relation or simply a preference. It formalizes the intuitive idea of the taste of a consumer according to which in any decision making he chooses the most satisfactory to him bundle among many alternatives. The set of alternatives available to a consumer (or in general to a decision maker) is called the **opportunity** (or the **choice**) **set** of the consumer.

We start our discussion with the definition of a preference.

DEFINITION 8.1. *A* ***preference*** $\succeq$ *on a (choice) set* X *is a binary relation on* X *which is*

(1) ***reflexive:*** $x \succeq x$ *for all* $x \in X$,
(2) ***transitive:*** $x \succeq y$ *and* $y \succeq z$ *imply* $x \succeq z$, *and*

(3) ***total*** *or* ***complete:*** *for each* $x, y \in X$ *either* $x \succeq y$ *or* $y \succeq x$.[1]

As usual, the expression $x \succeq y$ means x **is as good as** y. When both $x \succeq y$ and $y \succeq x$, then x is **indifferent** to y, written $x \sim y$. In case both $x \succeq y$ and $y \not\succeq x$, we say that x **is strictly preferred to** y, denoted $x \succ y$.

If $\succeq$ is a preference relation on a set X and $x \in X$, then the **no worse-than-x** set is the set $\{y \in X\colon\ y \succeq x\}$. Similarly, for each $x \in X$ the **no better-than-x** set is the set $\{y \in X\colon\ x \succeq y\}$.

DEFINITION 8.2. *Let* $\succeq$ *be a preference on a subset* X *of a vector space. A vector* v*, not necessarily lying in* X*, is* ***desirable*** *(for* $\succeq$*) if for each* $x \in X$ *there exists some* $\delta > 0$ *such that for each* $0 < \alpha < \delta$ *we have* $x + \alpha v \in X$ *and* $x + \alpha v \succ x$.

We now introduce some important topological, algebraic, and order properties of preference relations. A preference relation $\succeq$ on a topological space X is:

- **upper semicontinuous**, whenever for each $x \in X$ the no worse-than-x set $\{y \in X\colon\ y \succeq x\}$ is closed.
- **lower semicontinuous**, whenever for each $x \in X$ the no better-than-x set $\{z \in X\colon\ x \succeq z\}$ is closed.
- **continuous**, if $\succeq$ is both upper and lower semicontinuous.

When X is a convex subset of a vector space, a preference relation $\succeq$ on X is said to be:

- **convex**, if the no worse-than-x set $\{y \in X\colon\ y \succeq x\}$ is convex for each $x \in X$.
- **strictly convex**, whenever $y \succeq x$, $z \succeq x$, $y \neq z$, and $0 < \alpha < 1$ imply $\alpha y + (1 - \alpha) z \succ x$.

In case the opportunity set X is also partially ordered by $\geq$, then we say that a preference relation $\succeq$ on X is:

- **monotone** (with respect to $\geq$), if $x \geq y$ implies $x \succeq y$.
- **strictly monotone**, if $x > y$ implies $x \succ y$.

An element $a \in X$ is **maximal** for a binary relation $\succeq$ on the set X if there is no other element $x \in X$ with $x \succ a$. When $\succeq$ is total and reflexive this is equivalent to a being a **greatest element**, i.e., $a \succeq x$ for all $x \in X$. The next theorem deals with the properties and existence of greatest elements for preference relations.

THEOREM 8.3. *For a preference* $\succeq$ *on a set* X*:*

(1) *Greatest elements are indifferent to each other.*
(2) *If* X *is a compact topological space and* $\succeq$ *is upper semicontinuous, then* $\succeq$ *has greatest elements.*
(3) *If* X *is convex and* $\succeq$ *is a convex preference, then the (possibly empty) set of greatest elements of* $\succeq$ *is convex.*
(4) *If* X *is convex and* $\succeq$ *is strictly convex, then* $\succeq$ *has at most one greatest element.*

PROOF. (1) If a and b are two greatest elements for $\succeq$, then $a \succeq b$ and $b \succeq a$ are obviously true, and so $a \sim b$.

---

[1] In its general formulation, a preference for a decision maker is simply a binary relation on his opportunity set. A preference as defined here is usually referred to as a **rational preference**; see [**94**, Chapter 1].

(2) For each $x \in X$ let $P(x) = \{y \in X\colon\ y \succeq x\}$, the no worse-than-$x$ set of x. The upper semicontinuity of $\succeq$ guarantees that each $P(x)$ is a closed set. Since $\succeq$ is total, it follows that the family of closed sets $\{P(x)\}_{x\in X}$ has the finite intersection property. Consequently, the compactness of X guarantees that the collection of all greatest elements $\bigcap_{x\in X} P(x)$ is nonempty (and compact).

(3) Let G be the collection of all greatest elements and let $a, b \in G$. Then $a \succeq b$ and so for each $0 \le \alpha \le 1$, it follows from the convexity of $\succeq$ that $\alpha a + (1-\alpha)b \succeq b$. Since b is a greatest element, the latter shows that $\alpha a + (1-\alpha)b$ is itself a greatest element, i.e., $\alpha a + (1-\alpha)b \in G$. Thus, G is a convex set.

(4) Assume that two distinct elements a and b are greatest elements. From $a \succeq b$, $a \neq b$ and the strict convexity of $\succeq$, we get $\frac{1}{2}a + \frac{1}{2}b \succ b$. This shows that b is not a greatest element, a contradiction. So, a strictly convex preference $\succeq$ can have at most one greatest element. ■

A **utility function** for a given preference relation $\succeq$ is a function $u\colon X \to \mathbb{R}$ such that $x \succeq y$ holds if and only if $u(x) \ge u(y)$. In this case, we say that the utility function u **represents** the preference $\succeq$ on X. Every real-valued function $u\colon X \to \mathbb{R}$ automatically gives rise to a preference relation $\succeq$ (called the **preference generated by** u) by simply letting $x \succeq y$ if $u(x) \ge u(y)$.

A utility function $u\colon X \to \mathbb{R}$ defined on a convex subset of a vector space is said to be **quasi-concave** whenever for each $x, y \in X$ and each $0 \le \alpha \le 1$ we have

$$u\big(\alpha x + (1-\alpha)y\big) \ge \min\big\{u(x), u(y)\big\}\,.$$

Note that a utility function is quasiconcave if and only it represents a convex preference; see Exercise 1 at the end of the chapter.

It is well known that even for finite dimensional commodity spaces, not every preference is representable by a utility function. For instance, the lexicographic ordering of the plane has no utility representation; see Exercise 2 at the end of the chapter. Additional assumptions (usually continuity assumptions) are needed for the existence of a utility representation. There is a vast literature dealing with the representation of preferences by utility functions. We refer the reader to the monograph by D. S. Bridges and G. B. Mehta [**43**] for details regarding the representation of preferences by utility functions.

8.2. Exchange Economies and Efficiency

One of the basic economic considerations associated with any market environment is the **commodity-price duality**. G. Debreu [**50**] was the first to express this in terms of a dual pair $\langle X, X'\rangle$.[2] The vector space X is the **commodity space** and its vectors are known as **commodity bundles** (or simply as **commodities** or **bundles**) and X' is the **price space** whose vectors are called **prices**. The real number $\langle x, x'\rangle$ is interpreted as the value of the bundle x at prices x'. It is customary in economic literature to designate the price vectors by p, q, etc., instead of using primes. Also, the evaluation $\langle x, p\rangle$ is denoted $p \cdot x$, i.e., $p \cdot x = \langle x, p\rangle$.

[2] Recall that a **dual pair** (or a **dual system**) is a pair of vector spaces $\langle X, X'\rangle$ together with a bilinear map $\langle\cdot,\cdot\rangle\colon X \times X' \to \mathbb{R}$ (called the **duality map** of the system) such that $\langle x, x'\rangle = 0$ for all $x \in X$ implies $x' = 0$ and $\langle x, x'\rangle = 0$ for all $x' \in X'$ implies $x = 0$.

An exchange economy consists of a set of consumers who exchange goods but are unable to produce anything other than their endowment. Formally, we have the following definition.

DEFINITION 8.4. *An* ***exchange economy*** *is a tuple*

$$\mathcal{E} = \big(\langle X, X'\rangle, (X_1, \succeq_1, \omega_1), \ldots, (X_m, \succeq_m, \omega_m)\big),$$

where:

- *The commodity-price duality is described by the dual pair $\langle X, X'\rangle$; X is the commodity space and X' is the price space.*
- *The economy has m consumers. Each consumer i has a* ***consumption set*** *$X_i \subseteq X$ and an* ***initial endowment*** *$\omega_i \in X_i$.*
- *The tastes of each consumer i are described by his preference relation $\succeq_i$ on the consumption set X_i.*

The vector $\omega = \sum_{i=1}^m \omega_i$ is known as the **total endowment** (or the **social**) **endowment**. An **allocation** for an exchange economy is simply an m-tuple $(x_1, \ldots, x_m)$ of vectors such that $x_i \in X_i$ for each i and $\sum_{i=1}^m x_i = \omega$. In other words, allocations are redistributions of the social endowment amongst the m consumers of the economy. When the commodity space X of the economy is an ordered vector space, we refer to the order interval $[0, \omega] = \{x \in X\colon\ 0 \le x \le \omega\}$ as the **Edgeworth box**.

For our discussion in this chapter we shall employ two kinds of dual pairs satisfying certain lattice properties. These are the Riesz pairs and the lattice pairs that are defined as follows.

DEFINITION 8.5. *A* ***Riesz pair*** *$\langle L, L'\rangle$ is a dual pair such that L is a Riesz space and L' is an ideal in the order dual $L^\sim$.*

Similarly, a ***lattice pair*** *$\langle L, L'\rangle$ is a dual pair such that L is a Riesz space and L' is a Riesz subspace of the order dual $L^\sim$.*

In particular, if $L^\sim$ separates the points of a Riesz space L, then $\langle L, L^\sim\rangle$ is automatically a Riesz pair.

DEFINITION 8.6. *A Riesz pair $\langle L, L'\rangle$ is said to be* ***symmetric*** *if L is an ideal in $(L')^\sim$ or, equivalently, if $\langle L', L\rangle$ is also a Riesz pair.*

Here are two useful properties of Riesz pairs.

- *If $\langle L, L'\rangle$ is Riesz pair, then $\langle L', L\rangle$ is a lattice pair and the order intervals of L' are $\sigma(L', L)$-compact.* (See Theorem 6.1.)
- *A Riesz pair $\langle L, L'\rangle$ is symmetric if and only if the order intervals of L are $\sigma(L, L')$-compact.* (See Theorem 6.62.)

It should be noticed that a lattice pair need not be a Riesz pair. For instance, $\langle ca[0,1], C[0,1]\rangle$ is a lattice pair but not a Riesz pair. Recall that if Ω is a Hausdorff compact topological space, then $\mathrm{ca}(\Omega)$ denotes the Banach lattice of all regular Borel measures on Ω with the total variation norm. According to the classical Riesz representation theorem, the Banach lattice $\mathrm{ca}(\Omega)$ is the norm dual of $C(\Omega)$ and so $\langle C(\Omega), \mathrm{ca}(\Omega)\rangle$ is a Riesz pair. For more about Riesz pairs see [**21**, Section 11] and [**8**, Sections 7.15 and 7.16].

If $\langle L, L'\rangle$ is a lattice pair, then an element $x \in L^+$ is said to be **strictly positive**, written $x \gg 0$, if $\langle x, x'\rangle > 0$ for each $0 < x' \in L'$. If $L = \mathbb{R}^\ell$, then a positive vector is strictly positive if and only if each component of x is strictly positive. Similarly, a vector $0 < x' \in L'$ is **strictly positive** if $0 < x \in L^+$ implies $\langle x, x' \rangle > 0$.

We are now ready to classify exchange economies according to the lattice properties of the commodity and price spaces.

DEFINITION 8.7. *An exchange economy is called:*

- *A* ***Riesz economy***, *if its commodity-price duality is described by a Riesz pair $\langle L, L' \rangle$ and each consumer's consumption set is L^+.*
- *A* ***lattice economy***, *if its commodity-price duality is described by a lattice pair $\langle L, L' \rangle$ and each consumer's consumption set is L^+.*
- *A* ***Riesz commodity economy***, *if its commodity space is a Riesz space (in which case we denote the commodity-price dual pair by $\langle L, X' \rangle$) and each consumer's consumption set is L^+.*
- *A* ***Riesz price economy***, *if its price space is a Riesz space (in which case we denote the commodity-price dual pair by $\langle X, L' \rangle$).*
- *A* ***convex economy***, *if all preferences and consumption sets are convex.*
- *A* ***finite dimensional economy***, *if it is a convex Riesz economy with Riesz pair $\langle \mathbb{R}^\ell, \mathbb{R}^\ell \rangle$ and the total endowment is strictly positive.*

We shall denote the collection of all allocations by $\mathcal{A}$, i.e.,

$$\mathcal{A} = \Big\{ (x_1, \ldots, x_m) \in X^m \colon \; x_i \in X_i \text{ for each } i \text{ and } \sum_{i=1}^m x_i = \omega \Big\}.$$

Clearly, $(\omega_1, \ldots, \omega_m) \in \mathcal{A}$, so $\mathcal{A}$ is always a nonempty set.

The fundamental value judgment made in neoclassical economics is that consumers' preferences should be used to evaluate their welfare. This leads naturally to the following notions.

DEFINITION 8.8. *An allocation $(x_1, \ldots, x_m)$ is:*

- ***Pareto efficient*** *(or* ***Pareto optimal****), if there is no other allocation $(z_1, \ldots, z_m)$ satisfying $z_i \succsim_i x_i$ for all i and $z_k \succ_k x_k$ for some k.*
- ***weakly Pareto efficient*** *(or* ***weakly Pareto optimal****), if there is no allocation $(z_1, \ldots, z_m)$ with $z_i \succ_i x_i$ for all i.*
- ***individually rational***, *if $x_i \succsim_i \omega_i$ for each i.*

The set of all individually rational allocations is denoted $\mathcal{A}_r$, i.e.,

$$\mathcal{A}_r = \big\{ (x_1, \ldots, x_m) \in \mathcal{A} \colon \; x_i \succsim_i \omega_i \text{ for each } i \big\}.$$

Likewise, the set of weakly Pareto efficient allocations is denoted $\mathcal{A}_{wp}$, i.e.,

$$\begin{aligned} \mathcal{A}_{wp} \;=\; & \big\{ (x_1, \ldots, x_m) \in \mathcal{A} \colon \text{There is no } (y_1, \ldots, y_m) \in \mathcal{A} \\ & \qquad\qquad \text{satisfying } y_i \succ_i x_i \text{ for each } i \big\}. \end{aligned}$$

Clearly, every Pareto efficient allocation is weakly Pareto efficient. But the converse is false; see Example 8.38 below. However, with strictly monotone and continuous preferences the notions of Pareto and weak Pareto efficiency coincide.

LEMMA 8.9. *In a Riesz commodity economy, if each preference is strictly monotone and lower semicontinuous for some linear topology, then every weakly Pareto efficient allocation is Pareto efficient.*

An interesting question is the following: *Do individually rational Pareto efficient allocations exist?* Before answering this question (affirmatively), we shall characterize the compactness of the set of all allocations $\mathcal{A}$.

THEOREM 8.10. *In a Riesz commodity economy with weakly closed Edgeworth box* $[0, \omega]$ *and at least two consumers, the set* $\mathcal{A}$ *of allocations is weakly compact in* L^m *(that is,* $\mathcal{A}$ *is* $\sigma(L^m, (X')^m)$*-compact) if and only if the Edgeworth box* $[0, \omega]$ *is weakly compact.*

PROOF. Notice that we have the inclusion $\mathcal{A} \subseteq [0, \omega]^m$. In addition, from the identity $\sigma(L^m, (X')^m) = \prod_{i=1}^m \sigma(L, X')$, it is easy to see that $\mathcal{A}$ is a $\sigma(L^m, (X')^m)$-closed subset of L^m. Now suppose that the Edgeworth box $[0, \omega]$ is $\sigma(L, X')$-compact. Then $[0, \omega]^m$ is $\sigma(L^m, (X')^m)$-compact, so the $\sigma(L^m, (X')^m)$-closed subset $\mathcal{A}$ of $[0, \omega]^m$ is likewise $\sigma(L^m, (X')^m)$-compact.

For the converse, assume that $\mathcal{A}$ is a $\sigma(L^m, (X')^m)$-compact subset of L^m and let $\{x_\alpha\}$ be a net in $[0, \omega]$. Put

$$y_\alpha = (x_\alpha, \omega - x_\alpha, 0, 0, \ldots, 0)$$

and note that $\{y_\alpha\}$ is a net of the weakly compact set $\mathcal{A}$. Hence, $\{y_\alpha\}$ has a $\sigma(L^m, (X')^m)$-convergent subnet in L^m, and from this it follows that $\{x_\alpha\}$ also has a $\sigma(L, X')$-convergent subnet in $[0, \omega]$. Consequently, the Edgeworth box $[0, \omega]$ is $\sigma(L, X')$-compact. ■

THEOREM 8.11. *In a Riesz commodity economy, if the Edgeworth box* $[0, \omega]$ *is weakly compact and preferences are weakly upper semicontinuous, then the nonempty sets* $\mathcal{A}_r$ *(of all individually rational allocations) and the set* $\mathcal{A}_{wp}$ *(of all weakly Pareto efficient allocations) are both weakly compact.*

PROOF. In this case, both $\mathcal{A}_r$ and $\mathcal{A}_{wp}$ are weakly closed subsets of the weakly compact set $\mathcal{A}$. ■

Relatively mild conditions guarantee the existence of individually rational Pareto efficient allocations.

THEOREM 8.12. *In a Riesz commodity economy, if the Edgeworth box* $[0, \omega]$ *is weakly compact and preferences are weakly upper semicontinuous, then individually rational Pareto efficient allocations exist.*

PROOF. Start by introducing on $\mathcal{A}_r$ (the weakly compact set of all individually rational allocations) the equivalence relation $\sim$ by letting

$$(x_1, \ldots, x_m) \sim (z_1, \ldots, z_m)$$

if $x_i \sim_i z_i$ for each i. Notice that $\sim$ is indeed an equivalence relation on $\mathcal{A}_r$. For simplicity, denote the set of all equivalence classes $\mathcal{A}_r/\sim$ by $\mathcal{A}_r$ again. Next, we define an ordering $\geq$ on (the equivalence classes) $\mathcal{A}_r$ by defining

$$x = (x_1, \ldots, x_m) \geq z = (z_1, \ldots, z_m)$$

whenever $x_i \succeq_i z_i$ for each i. Again, notice that $\geq$ is a partial order on $\mathcal{A}_r$ and that the maximal elements of $\geq$ are precisely the Pareto efficient allocations. So, in order to establish that Pareto efficient allocations exist, it suffices to show that $\geq$ has maximal elements. For this, we need to establish that the partially ordered set $(\mathcal{A}_r, \geq)$ satisfies the hypothesis of Zorn's lemma. That is, we must show that every infinite chain of $\mathcal{A}_r$ has an upper bound in $\mathcal{A}_r$.

To this end, let $\mathcal{C}$ be an infinite chain in $\mathcal{A}_r$, that is, for each pair $x, z \in \mathcal{C}$ either $x \geq z$ or $z \geq x$ is true. Put $x_\alpha = \alpha$ for each $\alpha \in \mathcal{C}$ and note that $\{x_\alpha\}_{\alpha \in \mathcal{C}}$ is a net in the weakly compact set $\mathcal{A}_r$. If $x \in \mathcal{A}_r$ is a limit point of the net $\{x_\alpha\}_{\alpha \in \mathcal{C}}$, then it follows from the weak upper semicontinuity of the preferences that $x \geq y$ for each $y \in \mathcal{C}$. Therefore, x is an upper bound of $\mathcal{C}$ in $\mathcal{A}_r$, as desired. ■

COROLLARY 8.13. *In a Riesz commodity economy, if the Edgeworth box* $[0, \omega]$ *is weakly compact and each preference is convex and continuous with respect to a consistent locally convex topology on* L*, then individually rational Pareto efficient allocations exist.*

PROOF. Notice that in this case every preference is weakly upper semicontinuous. ■

8.3. Efficiency, Prices, and the Welfare Theorems

We are now ready to discuss efficiency properties of allocations in the presence of prices. As we shall see, the Pareto efficiency of an allocation can be characterized in terms of its support by prices. We start with the definition of a price-supported allocation.

DEFINITION 8.14. *An allocation* $(x_1, \ldots, x_m)$ *in an exchange economy is said to be* ***supported by prices*** *if there exists a nonzero price* $p \in X'$ *(called a* ***supporting price****) such that for each consumer* i*:*

$$x \succeq_i x_i \ \textit{ implies } \ p \cdot x \geq p \cdot x_i.$$

In other words, an allocation $(x_1, \ldots, x_m)$ is supported by prices if there exists a nonzero price p such that for each i the price p attains its minimum value on the no worse-than-x_i set at x_i.

In Riesz commodity economies with monotone preferences, supporting prices are positive.

LEMMA 8.15. *In a Riesz commodity economy, if at least one consumer has a monotone preference, then every supporting price is positive.*

PROOF. Let the first consumer have a monotone preference and let $(x_1, \ldots, x_m)$ be an allocation supported by a nonzero price p. If $x \geq 0$, then $x + x_1 \succeq_1 x_1$, and since p is a supporting price, we infer that $p \cdot (x + x_1) \geq p \cdot x_1$. This inequality implies $p \cdot x \geq 0$, which means that $p \geq 0$. ■

The objective of this section is to establish versions of the following efficiency properties of allocations known as the first and second welfare theorems. These theorems were first proven in reasonable generality by K. J. Arrow [**34**] and G. Debreu [**49, 50**]. The two theorems can be stated (subject to appropriate conditions) as:

- **First Welfare Theorem:** *Every allocation supported by prices is Pareto efficient.*
- **Second Welfare Theorem:** *Every Pareto efficient allocation can by supported by prices.*

Relatively general conditions guarantee the validity of the First Welfare Theorem.

THEOREM 8.16 (First Welfare Theorem). *Assume that the consumption sets of an exchange economy are cones and that each preference is lower semicontinuous for some linear topology. Then every allocation which is supported by a price p satisfying $p{\cdot}\omega \neq 0$ is weakly Pareto efficient.*

PROOF. Given the hypotheses, let $(x_1, \ldots, x_m)$ be an allocation supported by a price p with $p{\cdot}\omega \neq 0$. Also, suppose by way of contradiction that $(x_1, \ldots, x_m)$ is not weakly Pareto efficient. So, there exists another allocation $(z_1, \ldots, z_m)$ such that $z_i \succ_i x_i$ holds for each i.

Since p is supporting, we infer that $p \cdot z_i \geq p \cdot x_i$ for each i, and from

$$\sum_{i=1}^{m} p \cdot z_i = p \cdot \Big[\sum_{i=1}^{m} z_i\Big] = p \cdot \omega = p \cdot \Big[\sum_{i=1}^{m} x_i\Big] = \sum_{i=1}^{m} p \cdot x_i,$$

we see that $p \cdot z_i = p \cdot x_i$ for each i. Now since each consumption set X_i is a cone and $\lim_{\delta \uparrow 1} \delta z_i = z_i$ holds true for each linear topology, it follows from the lower semicontinuity of the preferences that there exists some $0 < \epsilon < 1$ such that $1 - \epsilon \leq \delta \leq 1 + \epsilon$ implies $\delta z_i \succ x_i$. Using again that p is supporting, we get

$$\delta p \cdot x_i = \delta p \cdot z_i = p \cdot (\delta z_i) \geq p \cdot x_i,$$

or $(\delta - 1)p{\cdot}x_i \geq 0$ for all $1 - \epsilon \leq \delta \leq 1 + \epsilon$. Letting $\delta = 1 \pm \epsilon$, we see that $\pm \epsilon p{\cdot}x_i \geq 0$, or $p \cdot x_i = 0$ for each i. This implies $p \cdot \omega = \sum_{i=1}^{m} p \cdot x_i = 0$, which is impossible. Hence, $(x_1, \ldots, x_m)$ is weakly Pareto efficient. ■

In finite dimensional economies, it is also relatively easy to establish the second welfare theorem.

THEOREM 8.17 (Second Welfare Theorem). *Assume that a finite dimensional economy is convex and:*

(a) *The total endowment is desirable by each consumer.*
(b) *One consumer has a monotone preference.*

Then every weakly Pareto efficient (and hence every Pareto efficient) allocation is supported by a positive price p satisfying $p{\cdot}\omega > 0$.

PROOF. Let $(x_1, \ldots, x_m)$ be a weakly Pareto efficient allocation in a finite dimensional economy with convex preferences in which the total endowment ω is desirable by each consumer. Now for each consumer i define the convex set $F_i = \{x \in \mathbb{R}^\ell_+ \colon\ x \succeq_i x_i\}$ and let

$$F = F_1 + F_2 + \cdots + F_m - \omega\,.$$

Clearly, F is a convex set and we claim that $-\frac{1}{n}\omega \notin F$ for each n.

To see this, assume by way of contradiction that $-\frac{1}{n}\omega \in F$ for some n. Then for each $i = 1, \ldots, m$ there exists some $z_i \in F_i$ such that $-\frac{1}{n}\omega = z_1 + \cdots + z_m - \omega$. Rearranging terms, we get

$$\big(z_1 + \tfrac{1}{mn}\omega\big) + \big(z_2 + \tfrac{1}{mn}\omega\big) + \cdots + \big(z_m + \tfrac{1}{mn}\omega\big) = \omega\,.$$

So, if $y_i = z_i + \frac{1}{mn}\omega$ $(i = 1, \ldots, m)$, then $(y_1, \ldots, y_m)$ is an allocation and since ω is desirable by each consumer it satisfies $y_i \succ_i z_i \succeq_i x_i$ for each i. But this shows that $(x_1, \ldots, x_m)$ is not weakly Pareto efficient, which is impossible. So $-\frac{1}{n}\omega \notin F$ for each n.

Now, according to the finite dimensional separation theorem, for each n there exists some price $p_n \in \mathbb{R}^\ell$ with $\|p_n\| = 1$ satisfying

$$-\tfrac{1}{n} p_n \cdot \omega \le p_n \cdot z \quad \text{for all} \quad z \in F\,.$$

Next, let p be a limit point of the sequence $\{p_n\}$. Then p is a nonzero price (in fact, $\|p\| = 1$) and clearly $p \cdot z \ge 0$ for each $z \in F$.

Finally, we claim that the price p supports the allocation $(x_1, \ldots, x_m)$. To see this, assume $x \succeq_i x_i$. Then, from

$$x - x_i = x_1 + \cdots + x_{i-1} + x + x_{i+1} + \cdots + x_m - \omega \in F\,,$$

it follows that $p \cdot (x - x_i) = p \cdot x - p \cdot x_i \ge 0$ or $p \cdot x \ge p \cdot x_i$. Moreover, according to Lemma 8.15, we have $p > 0$ and from this and the fact that ω is strictly positive, we get $p \cdot \omega > 0$. Therefore, the price p supports the allocation $(x_1, \ldots, x_m)$. ■

The infinite dimensional analogue of the Second Welfare Theorem 8.17 may fail, as we can see from the following example.

EXAMPLE 8.18 (Aliprantis–Burkinshaw [**22**]). Consider a Riesz economy with two consumers having the following characteristics:

- Commodity-price duality the symmetric Riesz pair $\langle \ell_1, \ell_\infty \rangle$.
- Initial endowments $\omega_1 = \omega_2 = (\frac{1}{2^3}, \frac{1}{2^5}, \ldots, \frac{1}{2^{2n+1}}, \ldots)$.
- Preferences for both agents are represented on ℓ_1^+ by the utility function

$$u(x) = \sum_{n=1}^{\infty} \phi_n(x_n),$$

where the functions $\phi_n \colon \mathbb{R} \to \mathbb{R}$ are defined by

$$\phi_n(t) = \begin{cases} 2^n t & \text{if} \;\; t \le \frac{1}{2^{2n}}, \\ \frac{1}{2^n} - \frac{1}{2^{2n}} + t & \text{if} \;\; t > \frac{1}{2^{2n}}\,. \end{cases}$$

This utility function was introduced by A. Mas-Colell in [**92**]. Since $t \ge 0$ implies $0 \le \phi_n(t) \le \frac{1}{2^n} + t$, notice that the series defining the utility function u converges for each $x = (x_1, x_2, \ldots) \in \ell_1^+$.

A direct verification shows that the utility function u is concave, strictly monotone, and norm continuous. In addition, note that the total endowment $\omega = \omega_1 + \omega_2$ is strictly positive. We now claim that the allocation (ω_1, ω_2) is Pareto efficient (hence weakly Pareto efficient) but fails to be supported by prices.

We establish first that (ω_1, ω_2) is Pareto efficient. To see this, start by observing that if $a = (a_1, a_2, \ldots)$ and $b = (b_1, b_2, \ldots)$ are two arbitrary positive sequences satisfying $0 \le a_n + b_n \le 2^{-2n}$ for all n, then

$$u(a + b) = u(a) + u(b).$$

Next, let (x_1, x_2) be an allocation satisfying $x_1 \succeq \omega_1$ and $x_2 \succeq \omega_2$. From the identity $x_1 + x_2 = \omega_1 + \omega_2$ and the above observation, we see that

$$u(x_1) + u(x_2) = u(x_1 + x_2) = u(\omega_1 + \omega_2) = u(\omega_1) + u(\omega_2).$$

From $u(x_1) \ge u(\omega_1)$ and $u(x_2) \ge u(\omega_2)$, we easily infer that $u(x_1) = u(\omega_1)$ and $u(x_2) = u(\omega_2)$. This implies that the allocation (ω_1, ω_2) is Pareto efficient.

Finally, we establish that the allocation (ω_1, ω_2) cannot be supported by prices. To this end, let $p \in \ell_\infty$ be a price that supports (ω_1, ω_2). By the monotonicity of the utility function u, it follows that $p \ge 0$. Also, an easy computation shows that

$(1-2^{-k})\omega_1 + 2^{-2k}e_k \succ \omega_1$, where e_k denotes the sequence whose k^{th} component is one and every other is equal to zero. Since p is supporting, we get

$$(1-2^{-k})p\cdot\omega_1 + 2^{-2k}p\cdot e_k \geq p\cdot\omega_1.$$

Consequently, $0 \leq 2^k p\cdot\omega_1 \leq p\cdot e_k \leq \|p\|_\infty < \infty$ for all k. Thus, $p\cdot\omega_1 = 0$, and since all components of ω_1 are positive, we see that $p = 0$, which means that (ω_1, ω_2) cannot be supported by prices. ■

In order to establish the second welfare theorem in infinite dimensional spaces, we need to add an additional condition on preferences. This property is known in economics as properness and is the subject of discussion in the next section.

8.4. Properness

For the validity of the second welfare theorem, we need some additional condition on preferences. The new property needed is a "cone condition" that has appeared in the mathematical literature in many forms. It has been studied extensively by V. Klee [**76, 77**]. The importance of this condition in general equilibrium theory was established first by A. Mas-Colell in [**92**] under the name of "properness." Since then, properness conditions have been studied extensively in the economics literature. For an extensive discussion of the properness conditions in the economics literature see [**29**].

Let us call a nonempty open convex subset Γ of a vector space an **open cone** if for each $\lambda > 0$ and each $x \in \Gamma$ we have $\lambda x \in \Gamma$. Also, recall that a nonzero linear functional p on a vector space X **supports** a subset A of X at a point $a \in A$ if $p\cdot x \geq p\cdot a$ holds for each $x \in A$, i.e., if p attains its minimum value over A at a.

As mentioned above, there are several versions of the "properness condition" now available in the literature. But all of them are based really upon the following "cone condition" result.

THEOREM 8.19. *For a nonempty convex subset C of a locally convex space (X,τ) and a vector $c \in C$ the following statements are equivalent.*

1. *The set C is supported at c by a nonzero τ-continuous linear functional.*
2. *There is an open cone Γ such that $(c+\Gamma)\cap C = \emptyset$.*
3. *There exist a nonzero vector $v \in X$ and a τ-neighborhood V of zero such that:*

$$c - \alpha v + z \in C \quad \textit{with} \quad \alpha > 0 \quad \textit{implies} \quad z \notin \alpha V.$$

PROOF. $(1) \Longrightarrow (2)$ If p is a nonzero continuous linear functional supporting C at c, then the open cone $\Gamma = \{x \in X\colon\ p\cdot x < 0\}$ satisfies $(c+\Gamma)\cap C = \emptyset$.

$(2) \Longrightarrow (3)$ Fix some $w \in \Gamma$ and then choose an open and convex τ-neighborhood of zero with $w+V \subseteq \Gamma$. We claim that the vector $v = -w \neq 0$ and the neighborhood V satisfy the desired property. To see this, assume $c - \alpha v + z \in C$ with $\alpha > 0$. If $z \in \alpha V$, then we can write $z = \alpha x$ with $x \in V$ and obtain

$$c - \alpha v + z = c + \alpha(w + x) \in (c+\Gamma)\cap C = \emptyset\,,$$

which is impossible.

$(3) \Longrightarrow (1)$ We can suppose that V is an open convex τ-neighborhood at zero. Let

$$\Gamma = \{\alpha(-v+w)\colon\ \alpha > 0 \text{ and } w \in V\}.$$

Clearly, Γ is an open cone. Our condition guarantees $(c+\Gamma)\cap C=\emptyset$. By the separation theorem there exists some nonzero continuous linear functional p satisfying $p\cdot(c+\gamma)\le p\cdot x$ for all $\gamma\in\Gamma$ and $x\in C$. Since Γ is a cone, we have $p\cdot(c+\lambda\gamma)\le p\cdot x$ for all $\lambda>0$ and each $\gamma\in\Gamma$, and by letting $\lambda\downarrow 0$, we get $p\cdot c\le p\cdot x$ for all $x\in C$. That is, p supports C at c. ■

Inspired from Theorem 8.19, we shall say that a subset A of a locally convex space is **proper at a point** $a\in A$ if there exists an open cone Γ such that $(a+\Gamma)\cap A=\emptyset$. Any open cone Γ that satisfies $(a+\Gamma)\cap A=\emptyset$ is called a **cone of properness** for A at the point a. This leads us to the following definition.

DEFINITION 8.20. *A preference $\succeq$ defined on a subset C of a locally convex space is said to be:*

(1) ***proper at a point*** $c\in C$, *whenever the no worse-than-c set $\{x\in C\colon\ x\succeq c\}$ is proper at c.*
(2) ***proper***, *if the no worse-than-c set $\{x\in C\colon\ x\succeq c\}$ is proper at c for each $c\in C$.*
(3) ***uniformly proper***, *if it is proper and the cone of properness at the point c for the no worse-than-c set $\{x\in C\colon\ x\succeq c\}$ can be chosen independently of c, i.e., if there is an open cone Γ such that $(c+\Gamma)\cap\{x\in C\colon\ x\succeq c\}=\emptyset$ for each $c\in C$.*

If the set C is also convex, then Theorem 8.19 has several consequences.

THEOREM 8.21. *For a preference $\succeq$ defined on a convex subset C of a locally convex space X we have:.*

(1) *For a point $c\in C$ the following statements are equivalent.*
   (a) *The preference $\succeq$ is proper at c.*
   (b) *C is supported at c by a nonzero continuous linear functional.*
   (c) *There exist a nonzero vector v and a neighborhood V of zero such that*

$$c-\alpha v+z\succeq c\ \text{ with }\ \alpha>0\ \text{ implies }\ z\notin\alpha V.$$

(2) *The preference $\succeq$ is uniformly proper if and only if there exist a nonzero vector v and a neighborhood V of zero such that for any point $c\in C$:*

$$c-\alpha v+z\succeq c\ \text{ with }\ \alpha>0\ \text{ implies }\ z\notin\alpha V.$$

(3) *Assume X is a Riesz space, $C=X^+$ and $\succeq$ is monotone. If $\succeq$ is proper at some c, then the vector v in 1(c) above can be taken to be positive.*

PROOF. For (3) note that the nonzero vector in 1(c) must satisfy $v^+>0$. Indeed, if $v^+=0$, then $v=v^+-v^-=-v^-<0$ and so $c-v+0\ge c$, which (in view of the monotonicity of $\succeq$) yields $c-v+0\succeq c$. But then $0\notin V$, which is impossible. Hence $v^+>0$. Now assume $c-\alpha v^++z\succeq c$. From $c-\alpha v+z\ge c-\alpha v^++z$ and the monotonicity of $\succeq$, we get $c-\alpha v+z\succeq c$ and so $z\notin\alpha V$. ■

We now specialize the properness conditions to preferences defined on the positive cones of Riesz spaces.

DEFINITION 8.22 (Properness). *Let L be a Riesz space, τ a linear topology on L, $v > 0$, and V a τ-open neighborhood of zero. A preference relation $\succeq$ on L^+ is:*

(1) (v, V)***-proper*** *at a point $x \in L^+$, if $x - \alpha v + z \succeq x$ in L^+ with $\alpha > 0$ implies $z \notin \alpha V$, and*

(2) ***uniformly*** (v, V)***-proper***, *if $x - \alpha v + z \succeq x$ in L^+ with $\alpha > 0$ implies $z \notin \alpha V$.*

Any vector $v > 0$ that satisfies property (2) above for some neighborhood V is called a **vector of uniform properness** for $\succeq$. When the vector v and the neighborhood V are clear from the context, we simply say that $\succeq$ is uniformly proper. In case we want to emphasize the vector v only, we say that $\succeq$ is uniformly v-proper. If we wish to emphasize the topology τ, then we say that $\succeq$ is uniformly τ-proper.

LEMMA 8.23. *Vectors of uniform properness are desirable.*

PROOF. Let $\succeq$ be a uniformly (v, V)-proper preference. Fix $x \in L^+$ and let $\alpha > 0$. If $x \succeq x + \alpha v$, then from $(x + \alpha v) - \alpha v + 0 = x \succeq x + \alpha v$, it follows that $0 \notin \alpha V$, which is impossible. Hence, $x + \alpha v \succ x$ for all $x \in L^+$ and each $\alpha > 0$. Thus, v is a desirable vector. ■

THEOREM 8.24 (Mas-Colell [**92**]). *Let τ be a locally convex topology on the Riesz space L and let $\succeq$ be a preference on L^+. Then $\succeq$ is uniformly τ-proper if and only if there exists a nonempty τ-open convex cone Γ such that:*

(1) $\Gamma \cap (-L^+) \neq \emptyset$.

(2) $(x + \Gamma) \cap \{y \in L^+ : y \succeq x\} = \emptyset$ *for all* $x \in L^+$.

PROOF. Assume that (1) and (2) are true. Fix some $v \in L^+$ such that $-v \in \Gamma$ and then choose a τ-open neighborhood V of zero such that $-v + V \subseteq \Gamma$. It is now straightforward to verify that $\succeq$ is uniformly (v, V)-proper.

Conversely, if the preference $\succeq$ is uniformly(v, V)-proper, then consider the set $\Gamma = \{\alpha(-v + w)\colon\ \alpha > 0 \text{ and } w \in V\}$ (where V is an open and convex τ-neighborhood of zero) and note that Γ is an open convex cone that satisfies properties (1) and (2). ■

Monotone preferences on AM-spaces with units are uniformly proper.

THEOREM 8.25 (Mas-Colell [**92**]). *If a monotone preference on the positive cone of an AM-space L with unit has a desirable bundle, then it is uniformly norm proper. In particular, every monotone preference on the positive cone of some $\mathbb{R}^\ell$ space with a desirable bundle is uniformly proper.*

PROOF. Assume that $v > 0$ is a desirable bundle. Notice that if $w \in \operatorname{Int}(L^+)$, then $w - \alpha v \in \operatorname{Int}(L^+)$ for some $\alpha > 0$. In particular, for each $x \in L^+$, we have

$$x + w = x + \alpha v + (w - \alpha v) \succeq x + \alpha v \succ x\,.$$

Now let $\Gamma = -\operatorname{Int}(L^+)$. Clearly, $\Gamma \cap (-L^+) \neq \emptyset$.

We claim that $(x + \Gamma) \cap \{y \in L^+\colon\ y \succeq x\} = \emptyset$ for each $x \in L^+$. If this is not the case, then $(x + \Gamma) \cap \{y \in L^+\colon\ y \succeq x\} \neq \emptyset$ for some $x \in L^+$. Hence, there exists $w \in \operatorname{Int}(L^+)$ such that $x - w \geq 0$ and $x - w \succeq x$. But then we have

$$x = (x - w) + w \succ x - w \succeq x\,,$$

which is impossible. To complete the proof now invoke Theorem 8.24. ■

Strictly positive vectors are a source of vectors of uniform properness.

LEMMA 8.26. *Let $\langle L, L' \rangle$ be a Riesz pair and let τ be a consistent locally convex-solid topology on L. If $\succeq$ is a uniformly τ-proper preference on L^+, then every strictly positive element of L is a vector of uniform τ-properness for $\succeq$.*

PROOF. Let e be a strictly positive vector in L. Then the principal ideal L_e is weakly (and hence τ) dense in L. Since the preference $\succeq$ is uniformly τ-proper, there exists a τ-neighborhood V of zero and some $v > 0$ such that:

$$x - \alpha v + z \succeq x \text{ in } L^+ \text{ with } \alpha > 0 \text{ implies } z \notin \alpha V. \qquad (\star)$$

Next, choose a convex and solid τ-neighborhood W of zero satisfying $v \notin W$ and $W + W \subseteq V$. Pick some $y \in L_e$ with $v - y \in W$. Replacing y by $v \wedge y^+$ and taking into account the inequality $|v - v \wedge y^+| \leq |v - y|$, we can suppose without loss of generality that $0 < y \leq v$. Now note that if $x - \alpha y + z \succeq x$ in L^+ with $\alpha > 0$, then from $(\star)$ and the relation

$$x - \alpha v + [z - \alpha(y - v)] = x - \alpha y + z \succeq x\,,$$

it follows that $z - \alpha(y - v) \notin \alpha V$. So if $z \in \alpha W$, then

$$z - \alpha(y - v) \in \alpha W + \alpha W = \alpha(W + W) \subseteq \alpha V\,,$$

which is a contradiction. The above show that in $(\star)$, we can also assume (replacing v by y) that $0 < v \in L_e$. Now if we pick some $\lambda > 0$ such that $0 < y \leq \lambda e$, then it is easy to see that λe is a vector of uniform τ-properness for $\succeq$, and consequently e itself is also a vector of uniform τ-properness. ■

As an application of the preceding result, let us show that Mas-Colell's preference relation $\succeq$ introduced in Example 8.18 is not uniformly norm proper.

COROLLARY 8.27. *The preference relation $\succeq$ on ℓ_1^+ represented by the utility function defined via the formula*

$$u(x_1, x_2, \ldots) = \sum_{n=1}^{\infty} \phi_n(x_n)\,,$$

where

$$\phi_n(t) = \begin{cases} 2^n t & \text{if } t \leq \frac{1}{2^{2n}}\,, \\ \frac{1}{2^n} - \frac{1}{2^{2n}} + t & \text{if } t > \frac{1}{2^{2n}}\,, \end{cases}$$

is not uniformly norm proper.

PROOF. To see this, suppose by way of contradiction that the preference $\succeq$ is uniformly norm proper. Then, according to Lemma 8.26, the strictly positive vector $v = (\frac{1}{2^3}, \frac{1}{2^5}, \ldots, \frac{1}{2^{2n+1}}, \ldots)$ must be a vector of uniform norm properness for $\succeq$. That is, there exists some $\epsilon > 0$ such that $x - \alpha v + z \succeq x$ in ℓ_1^+ implies $\|z\|_1 \geq \alpha\epsilon$. Now note that $v - 2^{-n}v + 2^{-2n}e_n \succ v$ for each n, where e_n denotes the sequence having its nth component equal to one and every other equal to zero. Hence, $2^{-2n}\|e_n\|_1 \geq 2^{-n}\epsilon$ or $1 \geq 2^n\epsilon$ for each n, which is impossible. Hence $\succeq$ cannot be uniformly norm proper. ■

Let us close this section with a few remarks. In essence, the notion of properness replaces the interiority assumption of the positive cone. The notion of properness provides the means of studying general equilibrium problems in infinite dimensional Riesz spaces whose cones have empty interior. On the other hand, properness is not a panacea. The positive cone of an arbitrary Riesz space need not admit proper preferences that satisfy the other standard properties. For instance, for the Riesz space $\mathbb{R}_\infty = \mathbb{R}^{\mathbb{N}}$ (the Riesz space of real sequences equipped with the product topology) there is no proper continuous preference on $\mathbb{R}_\infty^+$ that can be represented by a quasiconcave and strictly monotone utility function. For details, see [**15**, Example 3.6.9, p. 174].

8.5. Properness and Efficiency

When preferences are uniformly proper the pathology of Example 8.18 cannot occur. A. Mas-Colell has established that under the additional assumption of uniform properness (and the other standard assumptions) the second welfare theorem is valid for Riesz economies.

THEOREM 8.28 (The Mas-Colell Second Welfare Theorem [**92**]). *In a Riesz economy, if preferences are convex, monotone, and uniformly (ω, V)-proper for a consistent locally convex-solid topology τ on L, then every weakly Pareto efficient allocation is supported by prices.*

Moreover, every weakly Pareto efficient allocation can be supported by a price $p > 0$ that satisfies

$$p \cdot \omega = 1 \quad \text{and} \quad |p \cdot z| \leq m \quad \text{for all} \quad z \in V\,,$$

i.e., the supporting price p is normalized at ω and belongs to the polar $(\frac{1}{m}V)^{\circ}$ of the τ-neighborhood $\frac{1}{m}V$.

PROOF. Let $(x_1, \ldots, x_m)$ be a weakly Pareto efficient allocation. As in the finite dimensional case (see the proof of Theorem 8.17), we define the nonempty convex sets $F_i = \{x \in L^+ \colon\ x \succeq_i x_i\}$ and consider the nonempty convex set

$$F = F_1 + F_2 + \cdots + F_m - \omega\,.$$

Since ω is a desirable bundle, the proof of Theorem 8.17 is valid and guarantees that $-\frac{1}{n}\omega \notin F$ for each n. However, since we are now in the infinite dimensional case, the finite dimensional version of the separation theorem is not applicable; the infinite dimensional versions require that F is either compact or that it has an interior point. To apply the infinite dimensional version of the separation theorem, we must employ the uniform properness. This will be done next.

We can assume that the neighborhood V is convex and solid (and hence also symmetric). Let $\Gamma = \{\alpha(\omega + \frac{1}{m}v) \colon\ \alpha > 0 \text{ and } v \in V\}$. Clearly, Γ is a convex set and since $\omega + \frac{1}{m}V \subseteq \Gamma$, we see that ω is an interior point of Γ, i.e., Γ has a nonempty interior. Next, we claim that $F \cap (-\Gamma) = \emptyset$. To see this, assume by way of contradiction that $F \cap (-\Gamma) \neq \emptyset$.

This means that there exist positive vectors $z_i \succeq_i x_i$ $(i = 1, \ldots, m)$, some $\alpha > 0$ and some $v \in V$ such that $\sum_{i=1}^m z_i - \omega = -\alpha(\omega + \frac{1}{m}v)$, or $\sum_{i=1}^m z_i - (\omega - \alpha\omega) \in \frac{\alpha}{m}V$. Letting $z = \sum_{i=1}^m z_i$ and $y = \omega - \alpha\omega$, we see that

$$z - y \in \tfrac{\alpha}{m}V\,. \qquad (\star)$$

From $z - y = \sum_{i=1}^m z_i - \omega + \alpha\omega \leq \sum_{i=1}^m z_i + \alpha\omega$, it follows that

$$(y - z)^- = (z - y)^+ \leq \sum_{i=1}^m z_i + \alpha\omega = \sum_{i=1}^m \big(z_i + \tfrac{\alpha}{m}\omega\big)\,.$$

Therefore, by the Riesz Decomposition Property (see Theorem 1.10), we can write $(y-z)^- = \sum_{i=1}^m s_i$ with $0 \le s_i \le z_i + \frac{\alpha}{m}\omega$ for each i. In particular, from the inequalities $0 \le s_i \le (y-z)^- \le |y-z|$, the solidness of $\frac{\alpha}{m}V$ and $(\star)$, we see that

$$s_i \in \tfrac{\alpha}{m}V \quad \text{for each } \ i\,. \tag{$\star\star$}$$

Next, let $y_i = z_i + \frac{\alpha}{m}\omega - s_i \ge 0$ for each i and note that $y_i \succ_i z_i \succeq x_i$ holds for each i. Indeed, if $z_i = y_i - \frac{\alpha}{m}\omega + s_i \succeq_i y_i$ holds for some i, then by the uniform (ω, V)-properness of $\succeq_i$, we see that $s_i \notin \frac{\alpha}{m}V$, contrary to $(\star\star)$. Now notice that

$$\begin{aligned}
\textstyle\sum_{i=1}^m y_i &= z + \alpha\omega - (y-z)^- \\
&\le z + \alpha\omega - (y-z)^- + (y-z)^+ \\
&= z + \alpha\omega + (y-z) \\
&= y + \alpha\omega = \omega
\end{aligned}$$

contradicts the weak Pareto efficiency of $(x_1, \ldots, x_m)$. So, we have $F \cap (-\Gamma) = \emptyset$.

Now by the infinite dimensional separation theorem, there exist a nonzero price $p \in L'$ and some scalar c such that

$$p \cdot (-\gamma) \le c \le p \cdot f \tag{$\star\star\star$}$$

for all $\gamma \in \Gamma$ and each $f \in F$. Since $\gamma \in \Gamma$ implies $\alpha\gamma \in \Gamma$ for each $\alpha > 0$, it easily follows that we can assume $c = 0$. In particular, we have $p \cdot \gamma \ge 0$ for each $\gamma \in \Gamma$. Letting $\gamma = \omega \pm \frac{1}{m}v$ with $v \in V$, we get $\pm\frac{1}{m}p \cdot v \le p \cdot \omega$. This implies $p \cdot \omega > 0$ and so replacing p by $\frac{p}{p\cdot\omega}$, we can assume that $p \cdot \omega = 1$. From $\pm\frac{1}{m}p \cdot v \le 1$ for each $v \in V$, we see that $p \in (\frac{1}{m}V)^{\circ}$.

Finally, let us show that p supports $(x_1, \ldots, x_m)$; and hence—in view of the monotonicity of the preferences—necessarily $p > 0$. If $x \succeq_i x_i$, then we have $x - x_i = x + \sum_{j \ne i} x_i - \omega \in F$, and so $p \cdot (x - x_i) \ge c = 0$, which implies $p \cdot x \ge p \cdot x_i$, and the proof is finished. ■

Notice that in Theorem 8.28 we did not assume that the preferences were continuous. The reader should also observe that a glance at Theorem 8.25 guarantees that Theorem 8.28 implies immediately the finite dimensional Second Welfare Theorem 8.17.

For lattice economies A. Mas-Colell and S. F. Richard have established another remarkable version of the second welfare theorem. To discuss this result, we need to set up some notation.

Assume that $\langle L, L' \rangle$ is a lattice pair. We let $\boldsymbol{L} = L^m$, $\boldsymbol{L'} = (L')^m$, and consider the lattice pair $\langle \boldsymbol{L}, \boldsymbol{L'} \rangle$ under the obvious duality map. The elements of $\boldsymbol{L}$ and $\boldsymbol{L'}$ are indicated by boldface letters. The preference relation $\succeq$ on $\boldsymbol{L}^+$ is defined by $\mathbf{x} \succeq \mathbf{y}$ if $x_i \succeq_i y_i$ for each i. As usual, a price $\mathbf{p} = (p_1, \ldots, p_m) \in \boldsymbol{L'}$ supports $\mathbf{x} \in \boldsymbol{L}^+$ if $\mathbf{y} \succeq \mathbf{x}$ implies $\mathbf{p} \cdot \mathbf{y} \ge \mathbf{p} \cdot \mathbf{x}$. (This is equivalent to the condition that if $x \succeq_i x_i$ for some i, then $p_i \cdot x \ge p_i \cdot x_i$.)

LEMMA 8.29 (Mas-Colell–Richard [**93**]). *Consider a lattice economy such that the preference of each consumer is convex, monotone, uniformly ω-proper and continuous for a consistent locally convex topology τ on L. Then there exists a nonempty convex and $\sigma(\boldsymbol{L'}, \boldsymbol{L})$-compact subset $\mathcal{K}$ of $\boldsymbol{L'_+}$ consisting of nonzero prices such that:*

(1) *For each weakly Pareto efficient allocation* $\boldsymbol{x} \in \boldsymbol{L}^+$ *the set*

$$\mathcal{K}(\boldsymbol{x}) = \{\boldsymbol{p} \in \mathcal{K}\colon\ \boldsymbol{x} \text{ maximizes } \boldsymbol{p} \text{ over the set of all allocations } \mathcal{A} \text{ and } \boldsymbol{p} \text{ supports } \boldsymbol{x}\},$$

is a nonempty, convex, and $\sigma(\boldsymbol{L}', \boldsymbol{L})$*-compact subset of* $\boldsymbol{L}'_+$.

(2) *If* $\boldsymbol{p} \in \mathcal{K}(\boldsymbol{x})$, *then* $\boldsymbol{p} \cdot \boldsymbol{x} > 0$.

Proof. Assume that each preference is uniformly (ω, V)-proper, where V is an open, convex and symmetric τ-neighborhood of zero. Let $\mathbf{V} = V^m$, $\boldsymbol{\omega} = (\omega, \ldots, \omega) \in \mathbf{L}$ and define the set

$$\begin{aligned}\mathcal{K} &= \{\mathbf{p} \in \boldsymbol{L}'_+\colon\ |\mathbf{p} \cdot \mathbf{y}| \le 1 \text{ for each } \mathbf{y} \in \mathbf{V} \text{ and } \mathbf{p} \cdot \boldsymbol{\omega} = 1\} \\ &= \mathbf{V}^{\circ} \cap \{\mathbf{p} \in \boldsymbol{L}'_+\colon\ \mathbf{p} \cdot \boldsymbol{\omega} = 1\}.\end{aligned}$$

By the Alaoglu–Bourbaki Theorem 2.12 the set $\mathcal{K}$ is a convex and $\sigma(\boldsymbol{L}', \boldsymbol{L})$-compact subset of $\boldsymbol{L}'$. We claim that $\mathcal{K}$ satisfies the desired properties. To see this, fix an arbitrary weakly Pareto efficient allocation $\mathbf{x} = (x_1, \ldots, x_m)$.

Start by defining the convex sets $F_i = \{x \in L^+\colon\ x \succeq_i x_i\}$, the open convex cone $\Gamma = \{\alpha(\omega + v)\colon\ \alpha > 0 \text{ and } v \in V\}$ of L and the sets $W_i = F_i + \Gamma$. We have the following properties.

(a) *Each* W_i *is an open and convex subset of* L.

(b) *Every* $0 \le z \in W_i$ *satisfies* $z \succ_i x_i$ (*and in particular* $x_i \notin W_i$ *for each* i).

To see (b), assume that some $z = x + \alpha(\omega + w) \in W_i \cap L^+$ satisfies $x_i \succeq_i z$. Then $x = z - \alpha\omega - \alpha w \succeq_i x_i \succeq_i z$, and so by the uniform (ω, V)-properness of $\succeq_i$, we see that $w \notin V$, a contradiction.

Now we introduce the open convex set $\mathbf{W} = W_1 \times W_2 \times \cdots \times W_m$ of $\boldsymbol{L}$ and we note that property (b) implies immediately $\mathcal{A} \cap \mathbf{W} = \emptyset$, where $\mathcal{A}$ denotes the convex set of all allocations of the economy. So, by the separation theorem, there is some nonzero price $\mathbf{p} = (p_1, \ldots, p_m) \in \boldsymbol{L}'$ satisfying

$$\mathbf{p} \cdot \mathbf{y} \le \mathbf{p} \cdot \mathbf{w} \qquad (\star)$$

for all $\mathbf{y} \in \mathcal{A}$ and $\mathbf{w} \in \mathbf{W}$. Since $\mathbf{x} + \epsilon\boldsymbol{\omega} = (x_1 + \epsilon\omega, \ldots, x_m + \epsilon\omega) \in \mathbf{W}$, we see that $\mathbf{p} \cdot \mathbf{y} \le \mathbf{p} \cdot \mathbf{x} + \epsilon\mathbf{p} \cdot \boldsymbol{\omega}$ for all $\mathbf{y} \in \mathcal{A}$ and each $\epsilon > 0$. This implies $\mathbf{p} \cdot \mathbf{y} \le \mathbf{p} \cdot \mathbf{x}$ for each $\mathbf{y} \in \mathcal{A}$, i.e., $\mathbf{x}$ maximizes $\mathbf{p}$ over $\mathcal{A}$.

Next, we show that p_i supports $\succeq_i$ at x_i. To see this, let $z \succeq_i x_i$. If $t_j = x_j + \epsilon\omega$ for $j \ne i$ and $t_i = z + \epsilon\omega$, then $\mathbf{t} = (t_1, \ldots, t_m) \in \mathbf{W}$. So, from $(\star)$, we see that

$$\mathbf{p} \cdot \mathbf{x} = \sum_{i=1}^{m} p_i \cdot x_i \le \mathbf{p} \cdot \mathbf{t} = \sum_{j \ne i} p_j \cdot x_j + p_i \cdot z + \epsilon\mathbf{p} \cdot \boldsymbol{\omega},$$

and hence $p_i \cdot z + \epsilon\mathbf{p} \cdot \boldsymbol{\omega} \ge p_i \cdot x_i$ for each $\epsilon > 0$. This implies $p_i \cdot z \ge p_i \cdot x_i$. The latter, coupled with the monotonicity of the preferences, implies that $p_i \ge 0$ for each i, i.e., $\mathbf{p} \ge 0$.

Now observe that $x_i + \omega + w_i \in W_i$ for each $w_i \in V$. Invoking $(\star)$ once more, we get $\mathbf{p} \cdot \mathbf{x} \le \mathbf{p} \cdot \mathbf{x} + \mathbf{p} \cdot \boldsymbol{\omega} + \mathbf{p} \cdot \mathbf{w}$ for all $\mathbf{w} \in \mathbf{V}$. In particular, $\pm\mathbf{p} \cdot \mathbf{w} \le \mathbf{p} \cdot \boldsymbol{\omega}$ for each $\mathbf{w} \in \mathbf{V}$, and so $\mathbf{p} \cdot \boldsymbol{\omega} > 0$. Replacing $\frac{\mathbf{p}}{\mathbf{p} \cdot \boldsymbol{\omega}}$, we can assume that $\mathbf{p} \cdot \boldsymbol{\omega} = 1$. This implies $\mathbf{p} \in \mathcal{K} \cap \mathcal{K}(\mathbf{x})$. Moreover, taking into consideration that $\frac{1}{m}\boldsymbol{\omega} \in \mathcal{A}$, we see that $\mathbf{p} \cdot \mathbf{x} \ge \frac{1}{m}\mathbf{p} \cdot \boldsymbol{\omega} = \frac{1}{m} > 0$, and the proof is finished. ■

And now we can state the Mas-Colell–Richard Second Welfare Theorem.

THEOREM 8.30 (The Mas-Colell–Richard Second Welfare Theorem [**93**]). *In a lattice economy, if the preference of each consumer is convex, monotone, uniformly ω-proper and continuous for a consistent locally convex topology τ on L, then every weakly efficient allocation can be supported by prices.*

PROOF. Let $\mathbf{x} = (x_1, \ldots, x_m)$ be an arbitrary weakly Pareto efficient allocation. According to Lemma 8.29 there exists some price $\mathbf{p} = (p_1, \ldots, p_m) \in \mathcal{K}(\mathbf{x})$. Put $\pi = \bigvee_{i=1}^{m} p_i > 0$. From the definition of the members of $\mathcal{K}(\mathbf{x})$ and the definition of the supremum of a finite set of linear functionals (see Theorem 1.67), it follows that $\pi \cdot \omega = \sum_{i=1}^{m} p_i \cdot x_i$. Moreover, $\pi \geq p_i$ implies $\pi \cdot x_i \geq p_i \cdot x_i$ for each i. Summing up, we get $\pi \cdot \omega = \sum_{i=1}^{m} \pi \cdot x_i \geq \sum_{i=1}^{m} p_i \cdot x_i = \pi \cdot \omega = \mathbf{p} \cdot \mathbf{x} > 0$, so $\pi \cdot x_i = p_i \cdot x_i$ for each i. Now, we claim that π supports $\mathbf{x}$. To see this, let $x \succeq_i x_i$. Then

$$\pi \cdot x \geq p_i \cdot x \geq p_i \cdot x_i = \pi \cdot x_i ,$$

and the proof is finished. ∎

For an illustration of Theorem 8.30 see the discussion after Example 8.63.

8.6. Equilibrium

In an exchange economy, the **budget set** of a consumer i at prices $p \in X'$ is the set

$$B_i(p) = \{x \in X_i \colon\ p \cdot x \leq p \cdot \omega_i\} .$$

At prices p consumer i chooses a bundle from $B_i(p)$ that is best under the preference relation $\succeq_i$. Any such choice bundle is known as a **demand bundle** for consumer i. When at some prices p we can select demand bundles for all consumers that clear the market, then we say that the economy is in equilibrium. Formally, we have the following definition.

DEFINITION 8.31. *An allocation $(x_1, \ldots, x_m)$ in an exchange economy is:*

- *A **competitive** (or a **Walrasian**) **equilibrium**, if for some nonzero price $p \in X'$, each x_i is a maximal element in consumer i's budget set $B_i(p)$.*
- *A **quasiequilibrium**, if there is some nonzero price $p \in X'$ such that $x \succeq_i x_i$ in X_i implies $p \cdot x \geq p \cdot \omega_i$.*

For simplicity, a competitive equilibrium is also called an **equilibrium**. Here are some elementary properties of equilibria.

LEMMA 8.32. *For an allocation $(x_1, \ldots, x_m)$ we have the following.*

1. *The allocation $(x_1, \ldots, x_m)$ is a competitive equilibrium with respect to some nonzero price $p \in X'$ if and only if:*
   (a) *$x_i \in B_i(p)$ for each i, and*
   (b) *$x \succ_i x_i$ in X_i implies $p \cdot x > p \cdot \omega_i$.*
2. *If each consumer has a desirable bundle, then every competitive equilibrium is a quasiequilibrium.*

PROOF. (1) Assume first that $(x_1, \ldots, x_m)$ is a competitive equilibrium with respect to a nonzero price p. So, $x_i \in B_i(p)$ for each i and $x_i \succeq_i y$ for each $y \in B_i(p)$. Now if some $x \in X_i$ satisfies $x \succ_i x_i$, the clearly $x \notin B_i(p)$, which means that $p \cdot x > p \cdot \omega_i$.

For the converse, suppose that for some nonzero price p we have $p \cdot x > p \cdot \omega_i$ whenever $x \in X_i$ satisfies $x \succ_i x_i$. But this readily says that each x_i is a best choice in the budget set $B_i(p)$, i.e., $(x_1, \ldots, x_m)$ is a competitive equilibrium with respect to the nonzero price p.

(2) Assume that $(x_1, \ldots, x_m)$ is a competitive equilibrium with respect to some nonzero price $p \in X'$ and let v_i be a desirable commodity for consumer i. So if $x \succeq_i x_i$ in X_i, then there exists some $\delta > 0$ such that $0 < \alpha < \delta$ implies $x + \alpha v_i \in X_i$ and $x + \alpha v_i \succ_i x_i$. Consequently, we have

$$p \cdot (x + \alpha v_i) = p \cdot x + \alpha p \cdot v_i > p \cdot \omega_i$$

for each $0 < \alpha < \delta$. Letting $\alpha \to 0^+$, we get $p \cdot x \geq p \cdot \omega_i$, so that the allocation $(x_1, \ldots, x_m)$ is a quasiequilibrium. ■

We continue with more properties of equilibria.

LEMMA 8.33. *If $(x_1, \ldots, x_m)$ is a quasiequilibrium in an exchange economy with respect to a nonzero price $p \in X'$, then $p \cdot x_i = p \cdot \omega_i$ for each i (so the price p supports the allocation).*

PROOF. From $x_i \succeq_i x_i$, it follows that $p \cdot x_i \geq p \cdot \omega_i$ for each consumer i. Now using the relation

$$p \cdot \omega = p \cdot \Big[\sum_{i=1}^m x_i\Big] = \sum_{i=1}^m p \cdot x_i \geq \sum_{i=1}^m p \cdot \omega_i = p \cdot \Big[\sum_{i=1}^m \omega_i\Big] = p \cdot \omega\,,$$

we see that $p \cdot x_i = p \cdot \omega_i$ holds for each i. ■

Because of the preceding result, any nonzero price that makes an allocation an equilibrium or a quasiequilibrium is referred to as a **supporting price**.

LEMMA 8.34. *In a Riesz commodity economy, if one consumer has a strictly monotone preference, then every price supporting an equilibrium is strictly positive.*

PROOF. Assume that the first consumer has a strictly monotone preference and that a nonzero price p supports an equilibrium $(x_1, \ldots, x_m)$. By Lemma 8.15, we know that the price p is positive. Now let $x > 0$. Then the strict monotonicity of $\succeq_1$ implies $x_1 + x \succ_1 x_1$. Consequently, we have $p \cdot (x_1 + x) = p \cdot x_1 + p \cdot x > p \cdot \omega_1 = p \cdot x_1$, so $p \cdot x > 0$. This establishes that p is strictly positive. ■

If consumers have positive income, then a quasiequilibrium is, in fact, an equilibrium.

THEOREM 8.35. *Assume that in a Riesz commodity economy preferences are continuous for some linear topology. If an allocation $(x_1, \ldots, x_m)$ is a quasiequilibrium with respect to a nonzero price $p \in L'$ and $p \cdot \omega_i > 0$ for some i, then x_i is a maximal element in $B_i(p)$.*

In particular, if $p \cdot \omega_i > 0$ for each i, then $(x_1, \ldots, x_m)$ is a competitive equilibrium.

PROOF. By Lemma 8.33, we know that $p \cdot x_i = p \cdot \omega_i$, so $x_i \in B_i(p)$ for each i. Now let $p \cdot \omega_i > 0$ for some i, and assume by way of contradiction that there exists some $x \in B_i(p)$ (i.e., $p \cdot x \leq p \cdot \omega_i$) such that $x \succ_i x_i$. Since the preference $\succeq_i$ is continuous for some linear topology, there exists some $0 < \delta < 1$ such that $\delta x \succ_i x_i$. It follows that

$$\delta p \cdot x = p \cdot (\delta x) \geq p \cdot \omega_i.$$

Now from $p \cdot x \leq p \cdot \omega_i$ and $p \cdot \omega_i > 0$, we see that $\delta p \cdot x \leq \delta p \cdot \omega_i < p \cdot \omega_i$, which is impossible, and the proof is finished. ■

COROLLARY 8.36. *Assume that in a lattice economy preferences are continuous for some linear topology. If a quasiequilibrium is supported by a price p with $p{\cdot}\omega > 0$, then the quasiequilibrium is weakly Pareto efficient.*

PROOF. This follows immediately from Theorem 8.16. ■

COROLLARY 8.37. *In a lattice economy, if preferences are strictly monotone and continuous for some linear topology and the total endowment is strictly positive, then an allocation is a competitive equilibrium if and only if it is a quasiequilibrium. Moreover, in this case, every nonzero price that supports a quasiequilibrium is strictly positive.*

PROOF. Let $(x_1, \ldots, x_m)$ be a quasiequilibrium supported by a nonzero price p. Clearly, $p > 0$. We claim that $p \gg 0$ is also true. To see this, start by noticing that $\omega \gg 0$ implies $p{\cdot}\omega = \sum_{i=1}^{m} p{\cdot}\omega_i > 0$. So $p{\cdot}\omega_i > 0$ for some i. By Theorem 8.35 the bundle x_i is a maximal element in $B_i(p)$. Now let $x > 0$. If $p \cdot x = 0$, then $p \cdot (x_i + x) = p \cdot \omega_i$ (i.e., $x + x_i \in B_i(p)$) and by the strict monotonicity of $\succeq_i$, we see that $x + x_i \succ_i x_i$, which contradicts the maximality of x_i. Thus, $p \cdot x > 0$, which means that p is strictly positive. Now to complete the proof notice that $p \cdot \omega_i > 0$ for each i and then invoke Theorem 8.35. ■

In general, a quasiequilibrium need not be a competitive equilibrium. Here is a simple example.

EXAMPLE 8.38. Consider a two consumer 2-dimensional exchange economy with the following characteristics:

- *Consumer* 1: Initial endowment $\omega_1 = (\frac{1}{2}, 0)$ and preference $\succeq_1$ represented by the utility function $u_1(x, y) = x + y$.
- *Consumer* 2: Initial endowment $\omega_2 = (\frac{1}{2}, 1)$ and preference $\succeq_2$ represented by the utility function $u_2(x, y) = y$.

A straightforward argument shows that the allocation $\big((1,0),(0,1)\big)$ is a quasiequilibrium with respect to the price $p = (0, 1)$ but fails to be a competitive equilibrium.

Now an easy computation shows that $(1, 0) \succ_1 \omega_1$ and $(0, 1) \succeq_2 \omega_2$. Consequently, (ω_1, ω_2) is not a Pareto efficient allocation. However, it is easy to see that (ω_1, ω_2) is weakly Pareto efficient. ■

8.7. Continuity Properties of Supporting Prices

Consider an allocation $(x_1, \ldots, x_m)$ in an exchange economy. We shall say that a linear functional p (called a price) on a vector subspace Y of the commodity space that contains each x_i **supports** the allocation $(x_1, \ldots, x_m)$ on Y if $x \in X_i \cap Y$ and $x \succeq_i x_i$ imply $p{\cdot}x \geq p{\cdot}x_i$. Our purpose here is to discuss two important continuity properties of supporting prices.

Recall that if L is a Riesz space and $e \in L^+$, then the **principal ideal** (or simply the ideal) L_e generated by e is the vector subspace

$$L_e = \big\{v \in L\colon \text{ There exists some } \lambda > 0 \text{ such that } |v| \leq \lambda e\big\}.$$

Clearly, the vector subspace L_e is an ideal of L. If we define the real-valued function $\|\cdot\|_\infty\colon L_e \to [0,\infty)$ by

$$\|v\|_\infty = \inf\{\lambda > 0\colon\ |v| \le \lambda e\},$$

then $\|\cdot\|_\infty$ is a Riesz seminorm on L_e (that is, $|v| \le |w|$ in L_e implies $\|v\|_\infty \le \|w\|_\infty$). If L is an Archimedean Riesz space, then the function $\|\cdot\|_\infty$ is also a norm, and in fact an M-norm. In case L is an Archimedean uniformly complete Riesz space (in particular a Banach lattice), then L_e is an AM-space with unit e, in which case $(L_e, \|\cdot\|_\infty)$ is Riesz isometric to some $C(\Omega)$-space (Ω Hausdorff and compact and uniquely determined) with the element e corresponding to the constant function one on Ω; see Theorem 3.40. The norm dual L'_e of $(L_e, \|\cdot\|_\infty)$ is an AL-space—which is lattice isometric to some concrete $L_1(\mu)$-space (see Theorem 3.33).

Returning now to our setting, we see that the Edgeworth box $[0,\omega]$ incorporates all the economic activities of a Riesz commodity economy. In fact, $[0,\omega]$ lies entirely in L_ω. Therefore, although the commodity-price duality of the economy is described in terms of the dual pair $\langle L, X'\rangle$, the underlying Riesz pair $\langle L_\omega, L'_\omega\rangle$ is a more realistic Riesz pair that describes the commodity-price duality of the economy. This Riesz pair appears over and over in the following discussion.

THEOREM 8.39 (Aliprantis–Brown–Burkinshaw [**12**]). *In a Riesz economy assume that:*

(a) *Preferences are lower semicontinuous for a consistent locally convex-solid topology τ on L.*
(b) *Every consumer desires the total endowment ω.*

Then every positive linear functional p on L_ω that supports an allocation satisfies the following properties.

(1) *The linear functional p is τ-continuous on the Edgeworth box $[0,\omega]$.*
(2) *If, in addition, $[0,\omega]$ is weakly compact, then p is also order continuous on L_ω, i.e., $v_\alpha \downarrow 0$ in L_ω implies $p\cdot v_\alpha \downarrow 0$.*

PROOF. Assume that a positive price p on the ideal L_ω supports an allocation $(x_1,\dots,x_m)$.

(1) Consider first the case of a net $\{v_\alpha\}$ that satisfies $0 \le v_\alpha \le x_i$ for some i and all α and $v_\alpha \xrightarrow{\tau} 0$. Fix $\epsilon > 0$. From $x_i - v_\alpha + \epsilon\omega \xrightarrow[\alpha]{\tau} x_i + \epsilon\omega$, $x_i + \epsilon\omega \succ_i x_i$, and the lower τ-semicontinuity of $\succeq_i$, we see that there exists some α_0 such that $x_i - v_\alpha + \epsilon\omega \succ_i x_i$ for all $\alpha \ge \alpha_0$. Therefore,

$$p\cdot(x_i - v_\alpha + \epsilon\omega) = p\cdot x_i - p\cdot v_\alpha + \epsilon p\cdot\omega \ge p\cdot x_i$$

for all $\alpha \ge \alpha_0$. Thus, $0 \le p\cdot v_\alpha \le \epsilon p\cdot\omega$ for all $\alpha \ge \alpha_0$, which means that $\lim_\alpha p\cdot v_\alpha = 0$.

Next, we consider a net $\{v_\alpha\}$ satisfying $0 \le v_\alpha \le \omega = \sum_{i=1}^m x_i$ for each α and $v_\alpha \xrightarrow{\tau} 0$. By the Riesz decomposition property (see Theorem 1.10), we can write $v_\alpha = \sum_{i=1}^m v_\alpha^i$ with $0 \le v_\alpha^i \le x_i$. Since $0 \le v_\alpha^i \le v_\alpha$ for all α and all i, it follows from the local solidness of τ that $v_\alpha^i \xrightarrow[\alpha]{\tau} 0$. By the preceding case, we get

$$\lim_\alpha p\cdot v_\alpha = \lim_\alpha \sum_{i=1}^m p\cdot v_\alpha^i = \sum_{i=1}^m \lim_\alpha p\cdot v_\alpha^i = 0.$$

Finally, we consider the general case. That is, assume that a net $\{v_\alpha\}$ in $[0,\omega]$ satisfies $v_\alpha \xrightarrow{\tau} v$. It follows that $|v_\alpha - v| \le \omega$ holds for each α and $|v_\alpha - v| \xrightarrow{\tau} 0$.

Now use the preceding case and the inequality $|p\cdot v_\alpha - p\cdot v| \le p\cdot|v_\alpha - v|$ to conclude that $\lim_\alpha p\cdot v_\alpha = p\cdot v$.

(2) Assume $\{v_\alpha\} \subseteq [0,\omega]$ for each α and $v_\alpha \downarrow 0$. Since $[0,\omega]$ is weakly compact, it follows that $\{v_\alpha\}$ has a weakly convergent subnet and from this, we infer that $v_\alpha \xrightarrow{\sigma(L,L')} 0$. Now a glance at Theorem 3.21 guarantees that $v_\alpha \xrightarrow{\tau} 0$. But then part (1) yields $p\cdot v_\alpha \downarrow 0$. ■

A positive linear functional that supports an allocation on L_ω (although it is τ-continuous on the Edgeworth box $[0,\omega]$) need not be τ-continuous on L_ω . An example follows.

EXAMPLE 8.40 (Aliprantis–Burkinshaw [**22**]). Consider the Riesz economy discussed in Example 8.18. Notice that

$$\omega = \omega_1 + \omega_2 = \left(\tfrac{1}{2^2}, \tfrac{1}{2^4}, \ldots, \tfrac{1}{2^{2n}}, \ldots\right).$$

Clearly, $\omega \gg 0$, so L_ω is norm dense in ℓ_1, i.e., $\overline{L}_\omega = \ell_1$.

Now consider the price $p = (2, 2^2, 2^3, \ldots, 2^n, \ldots)$. Then p defines a positive price on L_ω; in fact, notice that $p\cdot\omega = \sum_{n=1}^\infty \frac{1}{2^n} = 1$. We claim that p supports the allocation (ω_1,ω_2) on L_ω as a Walrasian equilibrium. This follows from the following two facts.

(1) $u(\omega_i) = p\cdot\omega_i = \frac{1}{2}$, and
(2) $p\cdot x \ge u(x)$ for each $x \in \ell_1^+ \cap L_\omega$.

To see (1), note that $\phi_n\left(\frac{1}{2^{2n+1}}\right) = 2^n \cdot \frac{1}{2^{2n+1}} = \frac{1}{2^{n+1}}$, so

$$u(\omega_i) = \sum_{n=1}^\infty \phi_n\left(\tfrac{1}{2^{2n+1}}\right) = \sum_{n=1}^\infty \tfrac{1}{2^{n+1}} = p\cdot\omega_i = \tfrac{1}{2}.$$

For (2) notice first $t \ge 0$ implies $2^n t \ge \phi_n(t)$ for each n. Indeed, if $t \le \frac{1}{2^{2n}}$, then $\phi_n(t) = 2^n t$. On the other hand, if $t > \frac{1}{2^{2n}}$, then

$$2^n t = (2^n - 1)t + t \ge (2^n - 1)\cdot\tfrac{1}{2^{2n}} + t = \tfrac{1}{2^n} - \tfrac{1}{2^{2n}} + t = \phi_n(t).$$

Consequently, for $x = (x_1, x_2, \ldots) \in \ell_1^+ \cap L_\omega$, we have

$$p\cdot x = \sum_{n=1}^\infty 2^n x_n \ge \sum_{n=1}^\infty \phi_n(x_n) = u(x).$$

Now observe that if $x \succ_i \omega_i$ in L_ω^+, then $p\cdot x \ge u(x) > u(\omega_i) = p\cdot\omega_i$, so $p\cdot x > p\cdot\omega_i$. In other words, p supports (ω_1,ω_2) on L_ω as a Walrasian equilibrium. Finally, note that the price p cannot be ℓ_1-norm continuous on the ideal L_ω. Indeed, if we let $x_n = (0, \ldots, 0, \frac{1}{2^n}, 0, 0, \ldots) \in L_\omega$, then $\|x_n\|_1 = \frac{1}{2^n} \to 0$ while $p\cdot x_n = 1$ for each n. ■

The pathology of Example 8.40 cannot occur in a Riesz economy with uniformly proper preferences. N. C. Yannelis and W. R. Zame [**131**] have shown that when preferences are uniformly proper, every price that supports an allocation on the ideal L_ω is topologically continuous on L_ω.

THEOREM 8.41 (Yannelis–Zame). *In a Riesz economy, if every preference is uniformly τ-proper on L_ω for some locally convex-solid consistent topology τ on L, then every positive price on L_ω that supports an allocation is τ-continuous on L_ω.*

PROOF. Consider a Riesz economy satisfying the properties of the theorem and let an allocation $(x_1, \dots, x_m)$ be supported by a price $0 < p \in L'_\omega$. Since every vector of L^+_ω is bounded by a multiple of ω, it follows that ω is a vector of uniform properness. So there exists a convex, solid τ-neighborhood V of zero such that $x - \alpha\omega + z \succeq_i x$ in L^+_ω and $\alpha > 0$ imply $z \notin \alpha V$.

Consider the Riesz seminorm $\rho\colon L \to \mathbb{R}$ defined by $\rho(y) = \inf\{\lambda > 0\colon\ y \in \lambda V\}$ for each $y \in L$. Clearly, ρ is a τ-continuous seminorm on L and, as we shall see, one of its multiples dominates p on L_ω.

Assume first $0 \le z \le \omega = \sum_{i=1}^m x_i$. Then by the Riesz decomposition property we can write $z = \sum_{i=1}^m z_i$ with $0 \le z_i \le x_i$ for each i. Put $\alpha_i = \rho(z_i)$ and let $\epsilon > 0$ be fixed. For each i let $y_i = x_i + (\alpha_i + \epsilon)\omega - z_i \ge 0$. Clearly, $x_i = y_i - (\alpha_i + \epsilon)\omega + z_i \ge 0$. If $y_i - (\alpha_i + \epsilon)\omega + z_i \succeq_i y_i$, then by the uniform τ-properness of $\succeq_i$ on L_ω, we see that $z_i \notin (\alpha_i + \epsilon)V$, which contradicts $\rho(z_i) = \alpha_i$. Therefore, $y_i \succ_i y_i - (\alpha_i + \epsilon)\omega + z_i = x_i$. Since p supports $(x_1, \dots, x_m)$ on L_ω, it follows that

$$p \cdot y_i \ge p \cdot x_i = p \cdot [y_i - (\alpha_i + \epsilon)\omega + z_i] = p \cdot y_i - (\alpha_i + \epsilon)p \cdot \omega + p \cdot z_i.$$

Hence, $p \cdot z_i \le (\alpha_i + \epsilon)p \cdot \omega$ for each i and all $\epsilon > 0$, from which it follows that $p \cdot z_i \le \alpha_i p \cdot \omega = (p \cdot \omega)\rho(z_i) \le (p \cdot \omega)\rho(z)$. This implies

$$p \cdot z = \sum_{i=1}^m p \cdot z_i \le \Big[\sum_{i=1}^m p \cdot \omega\Big]\rho(z) = m(p \cdot \omega)\rho(z)$$

for all z with $0 \le z \le \omega$. Now consider any $z \in L_\omega$. Pick some $\lambda > 0$ such that $|z| \le \lambda\omega$ and note that

$$|p \cdot z| \le p \cdot |z| = \lambda p \cdot \big(\tfrac{1}{\lambda}|z|\big) \le \lambda m(p \cdot \omega)\rho\big(\tfrac{1}{\lambda}|z|\big) = m(p \cdot \omega)\rho(z).$$

Thus, the price p is dominated by a τ-continuous seminorm on L_ω and therefore is itself τ-continuous on L_ω. ■

8.8. The Utility Space of an Economy and Efficiency

In this section, we discuss the utility space of an exchange economy. The set up is as follows.

- We consider a Riesz commodity economy with commodity-price dual pair $\langle L, X' \rangle$ in which preferences are represented by monotone quasiconcave and continuous utility functions.
- Unless otherwise stated, when we say that a utility function is continuous it should be understood that the utility function is continuous for some consistent locally convex topology on L.
- We shall denote by u_i a fixed utility function representing the preference of consumer i that is monotone quasiconcave and continuous.
- Without loss of generality, we further assume that $u_i(0) = 0$.

A **free disposal allocation** is an m-tuple $(x_1, \dots, x_m)$ such that $x_i \in X_i$ for each i and $\sum_{i=1}^m x_i \le \omega$. A **utility allocation** is any vector in $\mathbb{R}^m$ of the form $\big(u_1(x_1), \dots, u_m(x_m)\big)$, where $(x_1, \dots, x_m)$ is a free disposal allocation. If $x = (x_1, \dots, x_m)$ is a free disposal allocation, then for simplicity we shall also write $u(x)$ for the vector $\big(u_1(x_1), \dots, u_m(x_m)\big)$.

DEFINITION 8.42. *The set of all utility allocations is called the* ***utility space*** *of the economy, denoted* $\boldsymbol{U}$, *i.e.,*

$$\boldsymbol{U} = \big\{u(x) = \big(u_1(x_1), \ldots, u_m(x_m)\big) \colon \ x = (x_1, \ldots, x_m) \text{ is a free disposal allocation}\big\}.$$

Clearly, U is a subset of the Euclidean space $\mathbb{R}^m$. The utility space enjoys some simple but important topological properties.

LEMMA 8.43. *For a Riesz commodity economy we have the following.*

(1) *Its utility space a nonempty bounded subset of* $\mathbb{R}^m_+$.
(2) *The utility space* $\boldsymbol{U}$ *is a* ***comprehensive*** *set in the sense that if* $0 \le y \le z$ *holds in* $\mathbb{R}^m$ *and* $z \in \boldsymbol{U}$, *then* $y \in \boldsymbol{U}$.
(3) *If there is a common desirable bundle, then there exists some* $\epsilon > 0$ *such that* $0 \le z \in \mathbb{R}^m$ *with* $\|z\| < \epsilon$ *implies* $z \in \boldsymbol{U}$.

PROOF. (1) Clearly, $0 \in \mathbf{U}$ and the vector $\big(u_1(\omega), u_2(\omega), \ldots, u_m(\omega)\big)$ is an upper bound for $\mathbf{U}$.

(2) To see this, let $0 \le y = (y_1, \ldots, y_m) \le z = (z_1, \ldots, z_m) \in \mathbf{U}$. Pick a free disposal allocation $(x_1, \ldots, x_m)$ with $z_i = u_i(x_i)$ for each i. Since the real-valued function $f_i(t) = u_i(tx_i)$, $0 \le t \le 1$, is continuous, there exists (by the Intermediate Value Theorem) some $0 \le t_i \le 1$ such that $y_i = u_i(t_i x_i)$. Clearly, $(t_1 x_1, \ldots, t_m x_m)$ is a free disposal allocation, and from this we infer that $y \in \mathbf{U}$.

(3) If $v > 0$ is desirable by each consumer, then $u_i\big(\frac{1}{m}v\big) > 0$ for each i. Consequently, by part (2), the number $\epsilon = \min\big\{u_i\big(\frac{1}{m}v\big) : i = 1, \ldots, m\big\}$ satisfies the desired property. ■

When is $\boldsymbol{U}$ *compact?* Or, equivalently, *When is the utility space* $\boldsymbol{U}$ *closed?* The next result provides the answer.

LEMMA 8.44. *In a Riesz commodity economy whose consumer preferences are represented by monotone continuous utility functions, the following statements are equivalent.*

(1) *The utility space* $\boldsymbol{U}$ *is a closed subset of* $\mathbb{R}^m_+$.
(2) *The utility space* $\boldsymbol{U}$ *is a compact subset of* $\mathbb{R}^m_+$.
(3) *If a sequence* $\{(x_1^n, \ldots, x_m^n)\}$ *of free disposal allocations satisfies* $x_i^{n+1} \succeq_i x_i^n$ *for all* i *and all* n, *then there exists an allocation* $(x_1, \ldots, x_m)$ *such that* $x_i \succeq_i x_i^n$ *for all* i *and all* n.

DEFINITION 8.45. *An exchange economy is* ***utility closed*** *if its preferences are represented by continuous utility functions and its utility space* $\mathbf{U}$ *is a closed set (or equivalently, if its utility space* $\boldsymbol{U}$ *satisfies any one of the equivalent statements of Lemma* 8.44).

We emphasize once more that for an economy to have a utility space it is necessary that consumer preferences be represented by utility functions.

LEMMA 8.46. *In a Riesz commodity economy, if*

(1) *the Edgeworth box* $[0, \omega]$ *is weakly compact, and*
(2) *consumer preferences are represented by monotone, quasiconcave and continuous utility functions,*

then the economy is utility closed.

PROOF. Assume that $[0,\omega]$ is weakly compact. Let a sequence $\{z_n\}$ of $\mathbf{U}$ satisfy $z_n \to z$, where $z_n = (z_1^n, \dots, z_m^n)$ and $z = (z_1, \dots, z_m)$. For each n pick a free disposal allocation $(x_1^n, \dots, x_m^n)$ such that $z_i^n = u_i(x_i^n)$ for each i. Since $0 \le x_i^n \le \omega$ for each i and n and the order interval $[0,\omega]$ is weakly compact, there exists a subnet $\{n_\alpha\}$ of the sequence of natural numbers $\{n\}$ such that $x_i^{n_\alpha} \xrightarrow[\alpha]{\sigma(L,X')} x_i$ for each i. Clearly, $0 \le x_i \le \omega$ for each i and from $\sum_{i=1}^m x_i^{n_\alpha} \le \omega$, we get $\sum_{i=1}^m x_i \le \omega$. In other words, $(x_1, x_2, \dots, x_m)$ is a free disposal allocation.

Now since each $u_i\colon L^+ \to \mathbb{R}$ is Mackey-continuous and quasiconcave, we easily obtain

$$z_i = \lim_\alpha z_i^{n_\alpha} = \lim_\alpha u_i(x_i^{n_\alpha}) \le u_i(x_i)\,.$$

Next notice that for each i the real-valued function $f_i(t) = u_i(tx_i)$, $t \in [0,1]$, is continuous, and so by the Intermediate Value Theorem there exists for each i some $t_i \in [0,1]$ such that $u_i(t_i x_i) = z_i$. Since $(t_1x_1, t_2x_2, \dots, t_mx_m)$ is a free disposal allocation, we infer that $(z_1, \dots, z_m) \in \mathbf{U}$. This proves that $\mathbf{U}$ is a closed set. ■

An economy might be utility closed even without the Edgeworth box $[0,\omega]$ being weakly compact. Here is an example.

EXAMPLE 8.47. Consider a two consumer Riesz economy with the following characteristics.

- Riesz pair: $\langle C[0,1], ca[0,1]\rangle$, where $ca[0,1]$ is the norm dual of $C[0,1]$ consisting of all Borel measures on $[0,1]$.
- Total endowment: $\omega = \mathbf{1}$ (the constant function one).
- Consumer 1: Utility function $u_1(x) = \int_0^1 x(t)\,dt$.
- Consumer 2: Utility function $u_2(x) = \int_0^1 \sqrt{x(t)}\,dt$.

You can verify that:

(1) Each utility function is (sup) norm continuous, concave, and strictly monotone.
(2) The Edgeworth box $[0,\omega]$ is not weakly compact.
(3) The utility space of the economy is given by

$$\mathbf{U} = \big\{(x,y) \in \mathbb{R}_+^2\colon\ x + y^2 \le 1\big\}.$$

In particular, the economy is utility closed. ■

We shall denote by $\partial\mathbf{U}$ the boundary of the utility space $\mathbf{U}$ in $\mathbb{R}_+^m$. It turns out that the points of $\partial\mathbf{U}$ are closely related to the weakly Pareto efficient allocations. The next result clarifies the relationship.

LEMMA 8.48. *In a Riesz commodity economy whose consumer preferences are represented by monotone, quasiconcave and continuous utility functions we have*

$$\boldsymbol{U} \cap \partial\boldsymbol{U} = \big\{u(x)\colon\ x \text{ is a weakly Pareto efficient allocation}\big\}\,.$$

That is:

(1) *If $x = (x_1, \dots, x_m)$ is a weakly Pareto efficient allocation, then its utility allocation vector $u(x) = \big(u_1(x_1), \dots, u_m(x_m)\big)$ is a boundary point of the utility space $\boldsymbol{U}$ relative to $\mathbb{R}_+^m$.*
(2) *If $a = (a_1, \dots, a_m) \in \boldsymbol{U}$ is a boundary point of the utility space $\boldsymbol{U}$ relative to $\mathbb{R}_+^m$, then there exists a weakly Pareto efficient allocation $(y_1, \dots, y_m)$ such that $a = \big(u_1(y_1), \dots, u_m(y_m)\big)$.*

PROOF. (1) Let $x = (x_1, \ldots, x_m)$ be a weakly Pareto efficient allocation, whose utility allocation vector $u(x) = \big(u_1(x_1), \ldots, u_m(x_m)\big)$ does not belong to $\partial\mathbf{U}$. So, the there exists some $\epsilon > 0$ such that the open ball $B(u(x), 2\epsilon)$ (centered at $u(x)$ and having radius 2ϵ) satisfies $B(u(x), 2\epsilon) \cap \mathbb{R}^m_+ \subseteq \mathbf{U}$. In particular,

$$\big(u_1(x_1) + \epsilon, \ldots, u_m(x_m) + \epsilon\big) \in \mathbf{U},$$

and so there is a free disposal allocation $(y_1, \ldots, y_m)$ such that $u_i(y_i) = u_i(x_i) + \epsilon$. Letting $z_1 = y_1 + \big(\omega - \sum_{j=1}^n y_j\big)$ and $z_i = y_i$ for $i \neq 1$, we see that $(z_1, \ldots, z_n)$ is an allocation satisfying $u_i(z_i) > u_i(x_i)$ for each i. This contradicts the fact that $(x_1, \ldots, x_m)$ is weakly Pareto efficient, and the validity of (1) has been established.

(2) Pick a free disposal allocation $(x_1, x_2, \ldots, x_m)$ such that

$$a = \big(u_1(x_1), u_2(x_2), \ldots, u_m(x_m)\big).$$

If $\sum_{i=1}^m x_i = \omega$, then there is nothing to prove. So, assume that $\sum_{i=1}^m x_i < \omega$ holds and let $z = \omega - \sum_{i=1}^m x_i > 0$.

By the monotonicity of the preferences, we have $x_i + \frac{1}{m}z \succeq_i x_i$ for each i. On the other hand, if $x_i + \frac{1}{m}z \succ_i x_i$ for each i (i.e., if $u_i(x_i + \frac{1}{m}z) > u_i(x_i)$ for each i), then there exists some $\epsilon > 0$ such that

$$(a_1 + \epsilon, \ldots, a_m + \epsilon) \leq \big(u_1(x_1 + \tfrac{1}{m}z), u_2(x_2 + \tfrac{1}{m}z), \ldots, u_m(x_m + \tfrac{1}{m}z)\big) \in \mathbf{U},$$

which is a contradiction. Thus, $u_i(x_i + \frac{1}{m}z) = u_i(x_i)$ must hold for some i. Let $(x_1^1, x_2^1, \ldots, x_m^1)$ be a free disposal allocation with $x_i^1 = x_i$ for all but one i and for that exceptional i we have $x_i^1 = x_i + \frac{1}{m}z$ and $u_i(x_i + \frac{1}{m}z) = u_i(x_i)$. Clearly,

$$\omega - \sum_{i=1}^m x_i^1 = (1 - \tfrac{1}{m})z = z_1 > 0.$$

Now repeat the above argument with z_1 in place of z and continue on repeating the above process inductively. This guarantees the existence of a sequence $\{(x_{n,1}, x_{n,2}, \ldots, x_{n,m})\}$ of free disposal allocations with the following properties:

(a) $x_{n+1,i} = x_{n,i}$ for all but one i and for that exceptional i we have

$$x_{n+1,i} = x_{n,i} + \tfrac{1}{m}\big(1 - \tfrac{1}{m}\big)^{n-1} z.$$

(b) $u_i(x_{n,i}) = u_i(x_i)$ for all $i = 1, \ldots, m$ and all n.

(c) $\omega - \sum_{i=1}^m x_{n,i} = \big(1 - \frac{1}{m}\big)^n z$.

Clearly, $x_{n,i} \uparrow_n$ holds for each i and moreover (in view of the inequality $\sum_{n=1}^\infty (1 - \frac{1}{m})^n < \infty$), it follows that each sequence $\{x_{n,i}\}$ is τ-convergent in L. If $\lim_{n\to\infty} x_{n,i} = y_i$, then from (c), we see that $\sum_{i=1}^m y_i = \omega$, i.e., that $(y_1, \ldots, y_m)$ is an allocation. From (b) and the continuity of the utility functions, we infer that $u_i(y_i) = u_i(x_i) = a_i$ holds for each i, and consequently

$$a = \big(u_1(y_1), u_2(y_2), \ldots, u_m(y_m)\big).$$

Now observe that the allocation $(y_1, \ldots, y_m)$ is weakly Pareto efficient. ■

COROLLARY 8.49. *Assume that in a closed Riesz commodity economy consumer preferences are represented by monotone, quasiconcave and continuous utility functions. Then an allocation* $(x_1, \ldots, x_m)$ *is weakly Pareto efficient if and only if* $\big(u_1(x_1), \ldots, u_m(x_m)\big) \in \partial\mathbf{U}$.

The utility space will be used to establish the existence of quasiequilibria. Now we introduce the function $\rho\colon \Delta \to [0,\infty)$ via the formula

$$\rho(\mathbf{s}) = \sup\{\alpha \geq 0\colon\ \alpha\mathbf{s} \in \mathbf{U}\},$$

where as usual $\Delta = \{\mathbf{s} = (s_1,\dots,s_m) \in \mathbb{R}^m_+\colon\ \sum_{i=1}^m s_i = 1\}$.

LEMMA 8.50 (A. Mas-Colell [**92**]). *For a Riesz commodity economy, assume that preferences are representable by monotone and continuous utility functions and that the total endowment is desirable for each consumer. Then the function* $\rho\colon \Delta \to [0,\infty)$, *defined by*

$$\rho(\mathbf{s}) = \sup\{\alpha \geq 0 : \alpha\mathbf{s} \in \boldsymbol{U}\},$$

is continuous.

PROOF. Assume that $\mathbf{s}_n \to \mathbf{s}$ holds in Δ. We must show that $\rho(\mathbf{s}_n) \to \rho(\mathbf{s})$.

Let $\alpha > 0$ satisfy $\alpha\mathbf{s} \in \mathbf{U}$ and let $0 < \beta < \alpha$. Pick a free disposal allocation $(x_1,\dots,x_m)$ such that

$$\alpha\mathbf{s} = \big(u_1(x_1), u_2(x_2), \dots, u_m(x_m)\big) > \beta\mathbf{s}\,.$$

By the continuity of the utility functions there exists some $0 < \delta < 1$ such that

$$\big(u_1(\delta x_1), u_2(\delta x_2), \dots, u_m(\delta x_m)\big) > \beta\mathbf{s}\,.$$

Now note that $\big(\delta x_1 + \frac{1-\delta}{m}\omega, \delta x_2 + \frac{1-\delta}{m}\omega, \dots, \delta x_m + \frac{1-\delta}{m}\omega\big)$ is a free disposal allocation satisfying $u_i(\delta x_i + \frac{1-\delta}{m}\omega) > u_i(\delta x_i) \geq \beta s_i$ for each i. In view of $\mathbf{s}_n \to \mathbf{s}$, we see that

$$\big(u_1(\delta x_1 + \tfrac{1-\delta}{m}\omega), u_2(\delta x_2 + \tfrac{1-\delta}{m}\omega), \dots, u_m(\delta x_m + \tfrac{1-\delta}{m}\omega)\big) > \beta\mathbf{s}_n$$

holds for all sufficiently large n. This implies $\beta\mathbf{s}_n \in \mathbf{U}$ for all sufficiently large n, and so $\beta \leq \rho(\mathbf{s}_n)$ holds for all sufficiently large n. Consequently, $\beta \leq \liminf_{n\to\infty} \rho(\mathbf{s}_n)$ for all $0 < \beta < \alpha$ and thus $\alpha \leq \liminf_{n\to\infty} \rho(\mathbf{s}_n)$ for all $\alpha > 0$ with $\alpha\mathbf{s} \in \mathbf{U}$. We conclude that

$$\rho(\mathbf{s}) \leq \liminf_{n\to\infty} \rho(\mathbf{s}_n). \tag{$\star$}$$

Next, let $\rho(\mathbf{s}) < \beta$. Fix some γ with $\rho(\mathbf{s}) < \gamma < \beta$. From $\mathbf{s}_n \to \mathbf{s}$ and $\gamma s_i < \beta s_i$ for each i with $s_i > 0$, we infer that $\gamma\mathbf{s} \leq \beta\mathbf{s}_n$ holds for all sufficiently large n. Now note that if $\beta\mathbf{s}_n \in \mathbf{U}$, then (by Lemma 8.43) $\gamma\mathbf{s} \in \mathbf{U}$, contrary to $\rho(\mathbf{s}) < \gamma$. Thus, $\beta\mathbf{s}_n \notin \mathbf{U}$ likewise holds for all sufficiently large n, and so $\rho(\mathbf{s}_n) \leq \beta$ holds for all sufficiently large n. Therefore, $\limsup_{n\to\infty} \rho(\mathbf{s}_n) \leq \beta$ holds for all β with $\rho(\mathbf{s}) < \beta$, and consequently

$$\limsup_{n\to\infty} \rho(\mathbf{s}_n) \leq \rho(\mathbf{s})\,. \tag{$\star\star$}$$

Finally, from ($\star$) and ($\star\star$), we see that $\lim_{n\to\infty} \rho(\mathbf{s}_n) = \rho(\mathbf{s})$. ∎

8.9. Existence of Equilibria

Under appropriate hypotheses A. Mas-Colell [**92**] has established the following general result regarding the existence of quasiequilibria.

THEOREM 8.51 (Mas-Colell). *Assume that a Riesz economy satisfies the following properties.*

(1) *The preferences of each consumer are represented by a monotone and quasiconcave utility function.*

(2) *There is a consistent locally convex-solid topology τ on the commodity space such that each utility function is continuous.*
(3) *Each preference is ω-uniformly τ-proper.*
(4) *The economy is utility closed.*

Then the exchange economy has a quasiequilibrium.

PROOF. Pick a convex solid τ-neighborhood V of zero such that each preference is uniformly (ω, V)-proper. Next, for each $\mathbf{s} \in \Delta$ fix an allocation $x^{\mathbf{s}} = (x_1^{\mathbf{s}}, \ldots, x_m^{\mathbf{s}})$ such that

$$u(x^{\mathbf{s}}) = \big(u_1(x_1^{\mathbf{s}}), \ldots, u_m(x_m^{\mathbf{s}})\big) = \rho(\mathbf{s})\mathbf{s},$$

where ρ is the function defined in Lemma 8.50. (In view of Corollary 8.49 and Lemma 8.43 such allocation $x^{\mathbf{s}}$ always exists.) Let $W = \frac{1}{m}V$ and for each $\mathbf{s} \in \Delta$ we define

$$P(\mathbf{s}) = \big\{p \in W^{\circ}\colon\ p \cdot \omega = 1 \ \text{ and } \ p \ \text{ supports } \ x^{\mathbf{s}}\big\}.$$

Clearly, $P(\mathbf{s})$ is convex and (by Theorem 8.28) nonempty. Finally, for each $\mathbf{s} \in \Delta$ we define

$$\Phi(\mathbf{s}) = \big\{\big(p \cdot (\omega_1 - x_1^{\mathbf{s}}), \ldots, p \cdot (\omega_m - x_m^{\mathbf{s}})\big)\colon\ p \in P(\mathbf{s})\big\}.$$

Notice that $\Phi(\mathbf{s})$ is nonempty and convex and $z = (z_1, \ldots, z_m) \in \Phi(\mathbf{s})$ implies $\sum_{i=1}^m z_i = 0$. Moreover, notice the following crucial property:

$$0 \in \Phi(\mathbf{s}) \quad \Longleftrightarrow \quad x^{\mathbf{s}} \ \text{ is a quasiequilibrium}.$$

We shall complete the proof by showing that $0 \in \Phi(\mathbf{s})$ for some $\mathbf{s} \in \Delta$. This will be done in two steps. Let $T = \big\{z \in \mathbb{R}^m\colon\ \sum_{i=1}^m z_i = 0\big\}$.

STEP 1: *The correspondence $\Phi\colon \Delta \twoheadrightarrow T$ is upper hemicontinuous.*[3]

By the Alaoglu–Bourbaki Theorem 2.12, we know that the set W° is $\sigma(L', L)$-compact. This implies that the set $\{p \cdot \omega\colon\ p \in W^{\circ}\}$ is a bounded subset of $\mathbb{R}$. So, from $|p \cdot (\omega_i - x_i^{\mathbf{s}})| \le 2p \cdot \omega$, it easily follows that the range of Φ lies in a compact subset of $\mathbb{R}^m$. The latter guarantees that Φ is upper hemicontinuous if and only if it has a closed graph; see [**8**, Theorem 16.12, p. 529]. We shall establish next that Φ has indeed a closed graph.

To this end, let $\mathbf{s}_n \to \mathbf{s}$ in Δ and $z_n \to z$ in T, where $z^n \in \Phi(\mathbf{s}_n)$ for each n. We must show that $z \in \Phi(\mathbf{s})$. For each n fix some $p_n \in P(\mathbf{s}_n)$ such that

$$z_n = \big(p_n \cdot (\omega_1 - x_1^{\mathbf{s}_n}), \ldots, p_n \cdot (\omega_m - x_m^{\mathbf{s}_n})\big).$$

Since the set W° is $\sigma(L', L)$-compact, there exists a subnet $\{p_{n_\alpha}\}$ of the sequence $\{p_n\}$ such that $p_{n_\alpha} \xrightarrow{\sigma(L',L)} p$. Clearly, $p \in W^{\circ}$ and from $p_{n_\alpha} \cdot \omega = 1$, we see that $p \cdot \omega = 1$. Also, since p_n supports the allocation $x^{\mathbf{s}_n}$ and the preferences are monotone, we have $p_n \ge 0$ for each n; see Lemma 8.15. This implies $p > 0$.

Next, we claim that the price p supports the allocation $x^{\mathbf{s}}$. To see this, assume $x \succeq_i x_i^{\mathbf{s}}$ and let $\epsilon > 0$. The desirability of ω by each consumer (see Lemma 8.23) guarantees that $x + \epsilon\omega \succ_i x_i^{\mathbf{s}}$. Since $u(x^{\mathbf{s}_\alpha}) = \rho(\mathbf{s}_{n_\alpha})\mathbf{s}_{n_\alpha} \to \rho(\mathbf{s})\mathbf{s} = u(\mathbf{s})$, we see that there exists an index α_0 such that $x + \epsilon\omega \succ_i x_i^{\mathbf{s}_{n_\alpha}}$ for each $\alpha \succeq \alpha_0$. Since p_{n_α} is supporting $x^{\mathbf{s}_{n_\alpha}}$, we have

$$p_{n_\alpha} \cdot x + \epsilon p_{n_\alpha} \cdot \omega \ge p_{n_\alpha} \cdot x_i^{\mathbf{s}_{n_\alpha}} = p_{n_\alpha} \cdot \omega_i - z_i^{n_\alpha},$$

[3] For a comprehensive discussion of upper hemicontinuous correspondences see [**8**, Chapter 16].

for all $\alpha \succeq \alpha_0$. Taking the limits with respect to the indices $\alpha \succeq \alpha_0$, we obtain that $p \cdot x + \epsilon p \cdot \omega \geq p \cdot \omega_i - z_i$. Since $\epsilon > 0$ is arbitrary, $p \cdot x \geq p \cdot \omega_i - z_i$.

Now observing that $x_i^{\mathbf{s}} \succeq_i x_i^{\mathbf{s}}$ holds for each i, the preceding conclusion yields $p \cdot x_i^{\mathbf{s}} \geq p \cdot \omega_i - z_i$ for each i. Taking into account that

$$\sum_{i=1}^{m}(p \cdot \omega_i - z_i) = p \cdot \omega = \sum_{i=1}^{m} p \cdot x_i^{\mathbf{s}},$$

we infer that $p \cdot x_i^{\mathbf{s}} = p \cdot \omega_i - z_i$, or $z_i = p \cdot (\omega_i - x_i^{\mathbf{s}})$ for each i. This implies $z \in \Phi(\mathbf{s})$.

STEP 2: *The correspondence $\Psi\colon \Delta \twoheadrightarrow T$, defined by $\Psi(\mathbf{s}) = \mathbf{s} + \Phi(\mathbf{s})$, is inward pointing.*

Recall that a correspondence $\phi\colon A \twoheadrightarrow X$, where A is a subset of the vector space X, is said to be **inward pointing** if for each $x \in A$ there exist $y \in \phi(x)$ and $\lambda > 0$ such that $x + \lambda(y - x) \in A$.

Therefore, to prove that the correspondence Ψ is inward pointing, fix $\mathbf{s} \in \Delta$. Also, fix some $z \in \Phi(\mathbf{s})$ and pick some $p \in P(x^{\mathbf{s}})$ such that $z_i = p \cdot (\omega_i - x_i^{\mathbf{s}})$.

Notice that if $s_i = 0$, then $u_i(\omega_i) \geq 0 = \rho(\mathbf{s})s_i = u_i(x_i^{\mathbf{s}})$ (i.e., $\omega_i \succeq_i x_i^{\mathbf{s}}$), and so since p supports $x^{\mathbf{s}}$, we have $p \cdot \omega_i \geq p \cdot x_i^{\mathbf{s}}$. That is, $s_i = 0$ implies $z_i = p \cdot \omega_i - p \cdot x_i^{\mathbf{s}} \geq 0$. Now fix some $\lambda > 0$ such that $s_i + \lambda z_i > 0$ for each i with $s_i > 0$. But then, it is easy to see that the vector $\mathbf{y} = \mathbf{s} + z \in \Psi(\mathbf{s})$ satisfies $\mathbf{s} + \lambda(\mathbf{y} - \mathbf{s}) = \mathbf{s} + \lambda z \in \Delta$, proving that Ψ is indeed inward pointing.

The rest of the proof now goes as follows: By the above established properties, we see that $s \mapsto \Psi(\mathbf{s}) = \mathbf{s} + \Phi(\mathbf{s})$ is an upper hemicontinuous correspondence with nonempty, convex and compact values which is also inward pointing. It follows that Ψ has a fixed point by the Halpern–Bergman Fixed Point Theorem; see for instance [**8**, Theorem 16.50, p. 549]). If now $\mathbf{s} \in \Psi(\mathbf{s}) = \mathbf{s} + \Phi(\mathbf{s})$, then some $z \in \Phi(\mathbf{s})$ satisfies $z = 0$, and this guarantees that $x^{\mathbf{s}}$ is a quasiequilibrium. (For a different proof of this theorem see [**15**, Theorem 3.5.12, p. 161].) ■

The next important result regarding lattice economies is due to A. Mas-Colell and S. F. Richard [**93**].

THEOREM 8.52 (Mas-Colell–Richard). *Assume that in a lattice economy:*

(1) *Its Edgeworth box $[0, \omega]$ is weakly compact.*
(2) *The preference of each consumer i is represented by a monotone and quasiconcave utility function u_i.*
(3) *There is a consistent locally convex topology τ on the Riesz space L such that every utility function u_i is ω-uniformly τ-proper and τ-continuous.*

Then the economy has a quasiequilibrium.

PROOF. We sketch the basic steps of the Mas-Colell–Richard proof; see [**93**] for the details and [**7**] for a correction of the proof. We employ the notation used in Lemma 8.29.

For each $\mathbf{p} = (p_1, \ldots, p_m) \in \mathbf{L}'$, let $\pi_{\mathbf{p}}$ denote the supremum of the set $\{p_1, \ldots, p_m\}$, i.e., $\pi_{\mathbf{p}} = \bigvee_{i=1}^{m} p_i \in L'$. As in the proof of Theorem 8.51, for each $\mathbf{s} \in \Delta$ we fix an allocation $x^{\mathbf{s}} = (x_1^{\mathbf{s}}, \ldots, x_m^{\mathbf{s}})$ such that

$$u(x^{\mathbf{s}}) = \big(u_1(x_1^{\mathbf{s}}), \ldots, u_m(x_m^{\mathbf{s}})\big) = \rho(\mathbf{s})\mathbf{s},$$

where ρ is the function defined in Lemma 8.50.

Next we use the nonempty, convex, and weakly compact set $\mathcal{K}$ guaranteed by Lemma 8.29 and define the correspondence

$$\Phi\colon \Delta \twoheadrightarrow T = \Big\{(t_1,\dots,t_m)\in\mathbb{R}^m\colon \sum_{i=1}^m t_i = 0\Big\}$$

via the formula

$$\Phi(s) = \big\{(t_1,\dots,t_m)\in T\colon\ \exists\ \mathbf{p}\in\mathcal{K}(x^{\mathbf{s}}) \text{ such that } t_i = \pi_{\mathbf{p}}(\omega_i - x_i^{\mathbf{s}}) \text{ for each } i\big\}.$$

Since $\pi_{\mathbf{p}}$ supports $x^{\mathbf{s}}$, it suffices to show that $0\in\Phi(\mathbf{s})$ for some $\mathbf{s}\in\Delta$.

This is established by proving that the correspondence Φ is nonempty, convex, and compact valued and also upper hemicontinuous. It then follows that the correspondence $\Psi\colon \Delta\twoheadrightarrow\mathbb{R}^m$, defined via the formula $\Psi(\mathbf{s}) = \mathbf{s}+\Phi(\mathbf{s})$, satisfies the same properties.

The final argument shows that Ψ is "inward directed," which (as in the proof of Theorem 8.51) guarantees that Ψ has a fixed point. This easily implies that $0\in\Phi(\mathbf{s})$ for some $\mathbf{s}\in\Delta$, and hence $x^{\mathbf{s}}$ is a quasiequilibrium, as desired. ■

8.10. The Core of an Economy

In this section we discuss the core of an economy, the set of allocations that are efficient for every subeconomy. This notion was introduced by F. Y. Edgeworth [**56**].

A **coalition** is a nonempty set of consumers. The coalition of all consumers $\{1,\dots,m\}$ is called the **grand coalition**. We say that a coalition S **blocks** (or that it **can improve upon**) an allocation $(x_1,\dots,x_m)$ if there exists another allocation $(y_1,\dots,y_m)$ such that:

(1) $\sum_{i\in S} y_i = \sum_{i\in S}\omega_i$.
(2) $y_i \succ_i x_i$ for each $i\in S$.

In other words, a coalition can improve upon an allocation whenever there is another way of redistributing the resources so that every member of the coalition becomes better off—regardless of what is happening to the members of the complementary coalition.

DEFINITION 8.53 (Edgeworth). *A **core allocation** is an allocation that cannot be blocked by any coalition. The **core** of an economy $\mathcal{E}$ is the set of core allocations and is denoted* Core$(\mathcal{E})$.

Here are some well known properties of core allocations.

THEOREM 8.54. *In any economy:*

(1) *Every core allocation is weakly Pareto efficient and individually rational.*
(2) *Every Walrasian equilibrium is a core allocation.*

PROOF. (1) Notice that an allocation is weakly Pareto efficient if and only if it cannot be blocked by the grand coalition. This implies that every core allocation is Pareto efficient.

Now if $(x_1,\dots,x_m)$ is an allocation and $\omega_i\succ_i x_i$ holds for some i, then the coalition $\{i\}$ blocks (via $(\omega_1,\dots,\omega_m)$) the allocation $(x_1,\dots,x_m)$. So, every core allocation is individually rational.

(2) Let $(x_1,\dots,x_m)$ be a Walrasian equilibrium with respect to a nonzero price p. If $(x_1,\dots,x_m)$ is not a core allocation, then there exists another allocation

$(y_1, \ldots, y_m)$ such that $\sum_{i\in S} y_i = \sum_{i\in S} \omega_i$ and $y_i \succ_i x_i$ for each $i \in S$. From $y_i \succ_i x_i$, we see that $p \cdot y_i > p \cdot \omega_i$ for each $i \in S$, and so

$$p \cdot \Big(\sum_{i\in S} y_i\Big) = \sum_{i\in S} p \cdot y_i > \sum_{i\in S} p \cdot \omega_i = p \cdot \Big(\sum_{i\in S} \omega_i\Big),$$

which contradicts $\sum_{i\in S} y_i = \sum_{i\in S} \omega_i$. This contradiction shows that the allocation $(x_1, \ldots, x_m)$ is a core allocation. ∎

It is easy to see that a core allocation need not be a Walrasian equilibrium. This leads us to the following question: *Do core allocations exist?* By the above discussion, the answer is, of course, yes. Once we know that Walrasian equilibria exist, core allocations also exist. However, it is interesting to know that if we can establish the existence of core allocations in a different manner, then this can be used to present an alternate way of proving the existence of Walrasian equilibria. As we shall see later (Theorem 8.75), one can establish—in a class of convex Riesz economies—that the following two statements are equivalent.

(1) *Every economy has a Walrasian equilibrium.*
(2) *Every economy has a core allocation.*

For the time being, we remark that using game theoretic techniques one can establish the existence of core allocations independently of the existence of Walrasian equilibria. We state this result, which is essentially independently due to O. N. Bondareva [**42**] and H. E. Scarf [**119**]. (For a proof see [**15**, Theorem 3.3.6, p. 127].)

THEOREM 8.55 (Bondareva–Scarf). *In a Riesz economy assume that:*

(1) *The preference of each consumer are represented by a monotone and quasiconcave utility function.*
(2) *There exists a consistent locally convex-solid topology on the commodity space L such that each utility function is continuous.*
(3) *The Edgeworth box $[0, \omega]$ is weakly compact.*

Then the core of the economy is a nonempty weakly compact subset of L^m.

The following example shows that the weak compactness of the Edgeworth box $[0, \omega]$ cannot be dropped from the preceding theorem.

EXAMPLE 8.56. Consider a two consumer Riesz economy with the following characteristics.

- Riesz pair: $\langle C[0,1], ca[0,1]\rangle$.
- Consumer 1: Initial endowment $\omega_1 = \mathbf{1}$ (the constant function one) and preference given by the utility function

$$u_1(x) = \int_0^{\frac{1}{2}} \sqrt{x(t)}\, dt + \tfrac{1}{2} \int_{\frac{1}{2}}^1 \sqrt{x(t)}\, dt.$$

- Consumer 2: Initial endowment $\omega_2 = \mathbf{1}$ and preference given by the utility function

$$u_2(x) = \tfrac{1}{2} \int_0^{\frac{1}{2}} \sqrt{x(t)}\, dt + \int_{\frac{1}{2}}^1 \sqrt{x(t)}\, dt.$$

Both utility functions are strictly monotone, strictly concave, norm continuous, and norm proper. However, the Edgeworth box $[0, \omega]$ is not weakly compact. It can be shown that this exchange economy has an empty core; see Exercise 12 at the of the chapter. ■

8.11. Replication

Theorems which guarantee that the set of Walrasian equilibrium allocations coincides with the set of core allocations are known in the economic literature as **core equivalence theorems**. It was conjectured more than one hundred years ago by F. Y. Edgeworth [**56**] that when the number of consumers in an economy increases, the core of the economy "shrinks" to the set of Walrasian equilibria. The purpose of this section is to make precise this claim by establishing the core equivalence theorem of G. Debreu and H. E. Scarf [**52**]. To do this, we need to introduce the notion of a replica economy.

DEFINITION 8.57 (Debreu–Scarf). *The* ***r-fold replica*** *of an economy $\mathcal{E}$ is a new economy $\mathcal{E}_r$ with the same commodity-price dual pair as $\mathcal{E}$ and mr consumers indexed by (i,j) $(i=1,\dots,m;\ j=1,\dots,r)$ such that:*

- *The preference $\succeq_{ij}$ of consumer (i,j) is the same as the preference of consumer i, i.e., $\succeq_{ij}=\succeq_i$.*
- *The initial endowment ω_{ij} of consumer (i,j) coincides with the initial endowment of consumer i, i.e., $\omega_{ij}=\omega_i$.*

Clearly, the total endowment of the r-fold replica economy is precisely $r\omega$. The consumers of the form (i,j) for $j=1,\dots,r$ are all identical and are referred to as consumers of type i. An allocation

$$\big(x_{11},\dots,x_{1r},x_{21},\dots,x_{2r},\dots,x_{m1},\dots,x_{mr}\big)$$

in an r-fold replica economy has the **equal treatment property** whenever $x_{is}\sim_i x_{it}$ for all i, s, and t. That is, consumers of the same type receive basically "identical" bundles. Equal treatment allocations can be identified with allocations in the original economy. Indeed, if $(x_1,\dots,x_m)$ is an allocation of some economy, then by letting $x_{ij}=x_i$, we can consider $(x_1,\dots,x_m)$ as an allocation in every r-fold replica economy.

LEMMA 8.58. *In an exchange economy, every Walrasian equilibrium belongs to the core of each replica economy.*

PROOF. Let $(x_1,\dots,x_m)$ be a Walrasian equilibrium in an exchange economy supported by a price p and consider $(x_1,\dots,x_m)$ as an allocation in some r-fold replica. Clearly, $x_{ij}=x_i\in B_i(p)=B_{ij}(p)$ for all (i,j). This implies that the bundle x_{ij} is a maximal element in the budget set $B_{ij}(p)$ of consumer (i,j). So, $(x_1,\dots,x_m)$ is a Walrasian equilibrium in the r-fold replica economy. ■

For convex finite dimensional economies, the converse of the preceding result is the celebrated core equivalence theorem of G. Debreu and H. E. Scarf [**52**].

THEOREM 8.59 (The Debreu–Scarf Core Equivalence Theorem). *In a finite dimensional economy, if preferences are convex, continuous, and strictly monotone, then an allocation is a Walrasian equilibrium if and only if it belongs to the core of every replica economy.*

PROOF. Let $(x_1,\dots,x_m)$ be an allocation. If it is a Walrasian equilibrium, then (by Lemma 8.58) it belongs to the core of each replica economy. For the converse, assume that $(x_1, \ldots, x_m)$ belongs to the core of every replica economy. We must show that $(x_1, \ldots, x_m)$ is a Walrasian equilibrium.

For each i define the sets

$$F_i = \{x \in \mathbb{R}^\ell_+ \colon\ x \succ_i x_i\} \quad \text{and} \quad G_i = F_i - \omega_i\,,$$

where $\mathbb{R}^\ell$ is the commodity space of the economy. The strict monotonicity of the preferences guarantees that each F_i is nonempty and it is easy to see that each F_i is also convex. Therefore, each G_i is nonempty and convex. Now let G be the convex hull of $\bigcup_{i=1}^m G_i$, i.e.,

$$G = \operatorname{co}\Big(\bigcup_{i=1}^m G_i\Big) = \Big\{\sum_{i=1}^m \lambda_i y_i \colon\ y_i \in G_i,\ \lambda_i \geq 0\ \forall\, i \text{ and } \sum_{i=1}^m \lambda_i = 1\Big\}.$$

We claim that zero does not belong to the convex set G.

To see this, assume by way of contradiction that $0 \in G$. Then, there exist $y_i \in G$ and $\lambda_i \geq 0$ $(i = 1, \ldots, m)$ satisfying

$$\sum_{i=1}^m \lambda_i = 1 \qquad \text{and} \qquad \sum_{i=1}^m \lambda_i y_i = 0\,. \tag{$\star$}$$

Put $S = \{i\colon\ \lambda_i > 0\}$ and note that $S \neq \emptyset$. Next, for each i pick some $v_i \in \mathbb{R}^\ell_+$ such that $v_i \succ_i x_i$ and $y_i = v_i - \omega_i$. From $(\star)$, it follows that $\sum_{i=1}^m \lambda_i v_i = \sum_{i=1}^m \lambda_i \omega_i$ or

$$\sum_{i\in S} \lambda_i v_i = \sum_{i\in S} \lambda_i \omega_i\,. \tag{$\star\star$}$$

Now if n is a positive integer, denote by n_i the smallest integer greater than or equal to $n\lambda_i$, that is, $0 \leq n_i - n\lambda_i < 1$ (and, of course, $i \in S$ implies $n_i \leq n$). Since for each $i \in S$ we have $\lim_{n\to\infty} \frac{n\lambda_i}{n_i} = 1$ and $v_i \succ_i x_i$, there exists (by the continuity of the preferences) some n large enough satisfying

$$z_i = \tfrac{n\lambda_i}{n_i} v_i \succ_i x_i \quad \text{for each} \quad i \in S\,. \tag{$\star\star\star$}$$

Taking into account $(\star\star)$, we see that

$$\sum_{i\in S} n_i z_i = \sum_{i\in S} n\lambda_i v_i = \sum_{i\in S} n\lambda_i \omega_i \leq \sum_{i\in S} n_i \omega_i\,.$$

The preceding inequality, coupled with $(\star\star\star)$ and the monotonicity of preferences, shows that the allocation $(x_1, \ldots, x_m)$ can be improved upon in the n-fold replica of the economy $\mathcal{E}$, which is a contradiction. Hence, $0 \notin G$.

Now by the finite dimensional separation theorem (see, for instance [**6**, Problem 2.3.6, p. 102]), there exists some nonzero price $p \in \mathbb{R}^\ell_+$ such that $p \cdot g \geq 0$ holds for all $g \in G$. In particular, if $x \succ_i x_i$, then $x - \omega_i \in G$, and so $p \cdot x \geq p \cdot \omega_i$ holds. That is, $(x_1, \ldots, x_m)$ is a quasiequilibrium. By Corollary 8.37, $(x_1, \ldots, x_m)$ is a Walrasian equilibrium with respect to the price p, and the proof of the theorem is complete. ■

The next theorem takes care of the existence of allocations that belong to the core of every replica economy. In particular, it provides an alternate way of proving the existence of Walrasian equilibria in convex finite dimensional economies.

THEOREM 8.60. *Assume that in a Riesz economy $\mathcal{E}$:*

(1) *The preferences of each consumer are represented by a strictly monotone and quasiconcave utility function.*
(2) *There exists a consistent locally convex-solid topology τ on the commodity space L such that each utility function is continuous.*
(3) *The Edgeworth box $[0, \omega]$ is weakly compact.*

Then there exists an allocation of $\mathcal{E}$ that belongs to the core of every r-fold replica economy of $\mathcal{E}$ (and hence if the economy is finite dimensional, it is a Walrasian equilibrium).

PROOF. Let $\mathcal{E}$ be a Riesz economy satisfying the hypotheses stated in the theorem and let $\mathcal{E}_r$ denote the r-fold replica economy of $\mathcal{E}$. Let $\mathcal{C}_r$ be the set of allocations for $\mathcal{E}$ that belong to the core of the economy $\mathcal{E}_r$, i.e.,

$$\mathcal{C}_r = \big\{(x_1, \ldots, x_m) \in \mathcal{A} : (x_1, \ldots, x_m) \in \text{Core}(\mathcal{E}_r)\big\}.$$

Clearly, $\mathcal{C}_{r+1} \subseteq \mathcal{C}_r$ for each r and the intersection $\bigcap_{r=1}^{\infty} \mathcal{C}_r$ consists precisely of all allocations for $\mathcal{E}$ that belong to the core of every replica economy. If we establish that each $\mathcal{C}_r$ is nonempty and weakly closed, then it follows from the weak compactness of $\mathcal{A}$ (Theorem 8.10) and the finite intersection property that $\bigcap_{r=1}^{\infty} \mathcal{C}_r$ is nonempty. The fact that each $\mathcal{C}_r$ is a weakly closed set (and in fact a weakly compact set) is a matter of a straightforward verification. Next, we prove that each $\mathcal{C}_r$ is nonempty.

By Theorem 8.55 the r-fold economy has a nonempty core. Let

$$(x_{11}, \ldots, x_{1r}, x_{21}, \ldots, x_{2r}, \ldots, x_{m1}, \ldots, x_{mr})$$

be a core allocation for the economy $\mathcal{E}_r$. We claim that $x_{ij} \sim_i x_{ik}$ for $j, k = 1, \ldots, r$ and $i = 1, \ldots, m$, i.e., we claim that no consumer strictly prefers his bundle to that of another consumer of the same type. To see this, we can assume (by rearranging the consumers of each type) that $x_{ij} \succeq_i x_{i1}$ for each i and j. Let

$$y_{i1} = \tfrac{1}{n} \sum_{j=1}^{m} x_{ij}, \quad i = 1, \ldots, m,$$

and note (by the convexity of the preferences) that $y_{i1} \succeq_i x_{i1}$ for all i. If for some $1 \leq i \leq m$ and some $1 \leq s \leq r$ we have $x_{is} \succ_i x_{i1}$, then it follows that $y_{i1} \succ_i x_{i1}$. Now if each consumer $(i, 1)$ gets the bundle y_{i1}, then the coalition $\big\{(i, 1) : i = 1, \ldots, m\big\}$ blocks the original allocation (x_{ij}), which is a contradiction. To finish the proof note that $(y_{11}, y_{21}, \ldots, y_{m1})$ is an allocation of the economy $\mathcal{E}$ satisfying $y_{i1} \succeq_i x_{ij}$ for all i and j, so $(y_{11}, y_{21}, \ldots, y_{m1}) \in \mathcal{C}_r$. The parenthetical part follows immediately from Theorem 8.59. ■

8.12. Edgeworth Equilibria

We saw in the preceding section that in certain finite dimensional exchange economies the Walrasian equilibria coincide with the allocations that belong to the core of every replica economy. Also, Theorem 8.60 provides conditions that guarantee the existence of such allocations. Our purpose here is to isolate and study the allocations that belong to the core of every replica economy. In general, as we shall see, these allocations need not be Walrasian equilibria or even quasiequilibria.

DEFINITION 8.61 (Aliprantis–Brown–Burkinshaw [**12**]). *An allocation is said to be an* ***Edgeworth equilibrium*** *if it belongs to the core of every replication of the economy.*

Obviously, every Edgeworth equilibrium is a core allocation (and hence it is individually rational and weakly Pareto efficient). The notion of Edgeworth equilibrium is a price free notion in the following sense. If the commodity space L is a Riesz space, then all economic activity takes place inside the Edgeworth box $[0, \omega]$. So, in this case, if $(x_1, \dots, x_m)$ is an Edgeworth equilibrium, then the allocation $(x_1, \dots, x_m)$ remains an Edgeworth equilibrium in any exchange economy whose commodity space is a Riesz subspace of L containing the principal ideal L_ω (regardless of the nature of the price space L').

It is clear that every Walrasian equilibrium is automatically an Edgeworth equilibrium. As we shall see in the next section, the Edgeworth equilibria are precisely the allocations that are Walrasian equilibria with respect to the Riesz pair $\langle L_\omega, L'_\omega \rangle$. The fact that a Walrasian equilibrium with respect to a price of L'_ω is an Edgeworth equilibrium is easy to prove. We start by stating this result and leave its straightforward proof (which mimics the proof of Theorem 8.54(2)) for the reader.

LEMMA 8.62. *Let $(x_1, \dots, x_m)$ be an allocation in a Riesz commodity economy. If there exists a price $0 < p \in L'_\omega$ (not necessarily lying in X') such that $x \succ_i x_i$ in L^+_ω implies $p \cdot x > p \cdot \omega_i$, then the allocation $(x_1, \dots, x_m)$ is an Edgeworth equilibrium.*

As an illustration of the importance of Lemma 8.62, we present an example due to L. E. Jones [**66**].

EXAMPLE 8.63 (Jones). Consider a Riesz commodity economy with commodity space $L = L_1[0, 1]$ and two consumers with initial endowments $\omega_1 = \omega_2 = \frac{1}{2}\chi_{[0,1]}$ and preferences represented by the utility functions

$$u_1(x) = \int_0^1 t x(t)\, dt \quad \text{and} \quad u_2(x) = \int_0^1 (1 - t) x(t)\, dt.$$

Note that both utility functions are strictly monotone and convex (in fact, they are both linear). If the functions $z_1(t) = t$ and $z_2(t) = 1 - t$ belong to X', then u_1 and u_2 are also weakly continuous.

Clearly, $\omega = \mathbf{1}$ and if we put $x_1 = \chi_{(\frac{1}{2},1]}$ and $x_2 = \chi_{[0,\frac{1}{2}]}$, then (x_1, x_2) is an allocation. Let $p(t) = \max\{t, 1 - t\}$. We claim that $x \succ_i x_i$ implies $p \cdot x > p \cdot \omega_i$. To see this, note first that $p \cdot x_1 = p \cdot x_2 = p \cdot \omega_1 = p \cdot \omega_2 = \frac{3}{8}$. Now if $x \succ_1 x_1$, then

$$\begin{aligned} p \cdot x &= \int_0^1 p(t)x(t)\, dt \ge \int_0^1 t x(t)\, dt = u_1(x) \\ &> u_1(x_1) = \int_{\frac{1}{2}}^1 t\, dt = \tfrac{3}{8} = p \cdot \omega_1 . \end{aligned}$$

Similarly, $x \succ_2 x_2$ implies $p \cdot x > p \cdot \omega_2$. Hence, (x_1, x_2) is a Walrasian equilibrium with respect to the price p and so, by Lemma 8.62, (x_1, x_2) is an Edgeworth equilibrium. In particular, (x_1, x_2) is also a weakly Pareto efficient (and hence a Pareto efficient) allocation.

We now show that p is (up to a constant) the only price that supports the allocation (x_1, x_2). To this end, let q be a price supporting (x_1, x_2). By Lemma 8.15, we know that q is a positive linear functional. This implies that q belongs to the order dual of $L_1[0,1]$, i.e., $0 < q \in L_\infty[0,1]$.

Next, we claim that the function q is strictly positive. To see this, consider the set $A = \{t \in [0, \frac{1}{2}]\colon\ q(t) = 0\}$. If A has positive Lebesgue measure, then for some $k > 0$ we have $u_2(k\chi_A) = k \int_A (1-t)\,dt > u_2(x_2)$, so by the supportability of q, we get

$$0 = q \cdot (k\chi_A) \geq q \cdot x_1 = \int_{\frac{1}{2}}^1 q(t)\,dt \geq 0\,.$$

This implies that $q = 0$ a.e. on $[0, \frac{1}{2}]$. Now if we let $x = 8\chi_{[0,\frac{1}{2})}$, then we have $u_1(x) = 1 > \frac{3}{8} = u_1(x_1)$, i.e., $x \succeq_1 x_1$, and so since q is supporting we get

$$0 = \int_0^1 x(t)q(t)\,dt = q \cdot x \geq q \cdot x_1 = \tfrac{1}{2} \int_{\frac{1}{2}}^1 q(t)\,dt = \tfrac{1}{2} \int_0^1 q(t)\,dt > 0\,,$$

which is impossible. Hence A has Lebesgue measure zero so that q is strictly positive on $[0, \frac{1}{2}]$. Similarly, q is strictly positive on $[\frac{1}{2}, 1]$. So, $q(t) > 0$ for almost all t.

If A is a measurable subset of $[\frac{1}{2}, 1]$ of positive Lebesgue measure, then $\frac{u_1(x_1)}{u_1(\chi_A)}\chi_A \succeq_1 x_1$. Therefore, $\frac{u_1(x_1)}{u_1(\chi_A)} \int_A q(t)\,dt \geq \int_{\frac{1}{2}}^1 q(t)\,dt$, and consequently

$$\int_A q(t)\,dt \geq \Big[\tfrac{1}{u_1(x_1)} \int_{\frac{1}{2}}^1 q(t)\,dt\Big] \int_A t\,dt,$$

for each measurable subset A of $[\frac{1}{2}, 1]$. This implies $q(t) \geq ct$ for almost all t in $[\frac{1}{2}, 1]$, where $c = \frac{1}{u_1(x_1)} \int_{\frac{1}{2}}^1 q(t)\,dt$. We claim that $q(t) = ct$ for almost all $t \in [\frac{1}{2}, 1]$. To see this, assume that $\{t \in [\frac{1}{2}, 1]\colon\ q(t) > ct\}$ has positive Lebesgue measure. Then

$$\int_{\frac{1}{2}}^1 q(t)\,dt > c \int_{\frac{1}{2}}^1 t\,dt = cu_1(x_1) = \int_{\frac{1}{2}}^1 q(t)\,dt,$$

which is impossible. Hence $q(t) = ct$ for almost all $t \in [\frac{1}{2}, 1]$. Similarly, we get $q(t) = c^*(1-t)$ for almost all $t \in [0, \frac{1}{2}]$.

Next, we show that $c = c^*$. Recall that $u_1(x_1) = u_2(x_2) = \frac{3}{8}$. For $n \geq 2$ let $x = \frac{3}{4}\frac{n^2}{n-1}\chi_{[\frac{1}{2}-\frac{1}{n},\frac{1}{2}]}$ and note that $u_1(x) = \frac{3}{8}$. Thus, $x \succeq_1 x_1$, so support by q implies $q \cdot x \geq q \cdot x_1$. But $q \cdot x = \frac{3}{8}\frac{n+1}{n-1}c^*$ and $q \cdot x_1 = \frac{3}{8}c$ so that $\frac{3}{8}\frac{n+1}{n-1}c^* \geq \frac{3}{8}c$. Letting $n \to \infty$, we get $c^* \geq c$. Similarly, if $y = \frac{3}{4}\frac{n^2}{n-1}\chi_{[\frac{1}{2},\frac{1}{2}+\frac{1}{n}]}$, then $u_2(y) = \frac{3}{8}$ and so $y \succeq_2 x_2$. Support by q implies $q \cdot y = \frac{3}{8}\frac{n+1}{n-1}c \geq q \cdot x_2 = \frac{3}{8}c^*$ and by letting $n \to \infty$, we get $c \geq c^*$. Thus, $c = c^*$. Consequently, $q = cp$ proving that p is (essentially) the only price that supports (x_1, x_2).

If the price space X' does not contain the price p (for instance, let $X' = C^1[0,1]$, the vector subspace of $L_\infty[0,1]$ consisting of all continuously differentiable functions on $[0,1]$), then the allocation (x_1, x_2) is an Edgeworth equilibrium with respect to $\langle L, X' \rangle$ but not a Walrasian equilibrium. ■

The preceding example can be employed to illustrate Theorem 8.30, which asserts that in a lattice economy (with the standard hypotheses) every weakly efficient allocation can be supported by prices. Since a linear preference (i.e., a preference that is represented by a utility function of the form $u(x) = p \cdot x$ for some positive linear functional p on L) is proper if and only if it is continuous, we see that in the setting of the preceding example, the hypotheses of Theorem 8.30

guarantee that the linear functionals represented by the functions t and $1-t$ must lie in X'. In particular, if the price space X' is a Riesz subspace of the order dual $L_\infty[0,1]$ of $L_1[0,1]$, then X' must contain the Riesz subspace generated by $\mathbf{1}$ and t (which coincides with the vector space of all continuous piecewise linear functions). Therefore, if the hypotheses of Theorem 8.30 are satisfied, then X' contains all the continuous piecewise linear functions.

The next example illustrates the fact that Edgeworth equilibria need not be supported by prices in the given price space.

EXAMPLE 8.64. Consider a two consumer exchange economy with two identical consumers; each has an initial endowment $\omega > 0$ and a convex preference $\succeq$ defined on a common convex consumption set. Then, we claim that the allocation (ω, ω) is an Edgeworth equilibrium. To see this, assume in some n-fold replica economy there exist an allocation $\{x_{ij} : i = 1, \ldots, m;\ j = 1, \ldots, n\}$ and a coalition S such that:

(1) $\sum_{(i,j)\in S} x_{ij} = k\omega$ (where k is the number of consumers in S).
(2) $x_{ij} \succ \omega$ for each $(i,j) \in S$.

Hence, if $(i_0, j_0) \in S$ satisfies $x_{ij} \succeq x_{i_0 j_0}$ for each $(i,j) \in S$, then by the convexity of $\succeq$, we see that

$$\omega = \tfrac{1}{k} \sum_{(i,j)\in S} x_{ij} \succeq x_{i_0 j_0} \succ \omega,$$

which is impossible. Therefore, (ω, ω) is an Edgeworth equilibrium. However, such an equilibrium need not be supported by any price in X'; see Examples 8.18 and 8.40. ■

With any allocation $(x_1, \ldots, x_m)$ we define the two nonempty sets

$$F_i = \{x \in X_i : x \succeq_i x_i\} \quad \text{and} \quad G_i = F_i - \omega_i = \{x \in X : x + \omega_i \succeq_i x_i\}.$$

If preferences are convex, then the F_i and G_i are all convex sets, so the convex hull $\mathbf{G}$ of the union of the G_i is given by

$$\mathbf{G} = \operatorname{co}\Big(\bigcup_{i=1}^{m} G_i\Big) = \Big\{\sum_{i=1}^{m} \lambda_i g_i : g_i \in G_i,\ \lambda_i \geq 0\ \forall\, i \text{ and } \sum_{i=1}^{m} \lambda_i = 1\Big\}.$$

When the allocation is an Edgeworth equilibrium the convex set $\mathbf{G}$ satisfies an important approximation property. This approximation property is the key property in establishing several "core equivalence" type theorems. The next lemma was first proved by B. Peleg and M. Yaari [**106**] for a special case. In its present form the lemma was established by C. D. Aliprantis, D. J. Brown and O. Burkinshaw [**12**].

LEMMA 8.65. *Consider a Riesz economy such that:*

(1) *Preferences are convex and weakly upper semicontinuous.*
(2) *The Edgeworth box $[0, \omega]$ is weakly compact.*
(3) *The total endowment ω is desirable for each consumer.*

If $(x_1, \ldots, x_m)$ is an Edgeworth equilibrium, then for each $\epsilon > 0$ we have $0 \notin \epsilon\omega + \overline{\mathbf{G}}$ (where the bar denotes weak closure).

PROOF. Let $\langle L, L'\rangle$ denote the Riesz pair of the economy. Fix an Edgeworth equilibrium $(x_1, \ldots, x_m)$ and let $\epsilon > 0$. Also, assume by way of contradiction that $0 \in \epsilon\omega + \overline{\mathbf{G}}$ or $-\epsilon\omega \in \overline{\mathbf{G}}$.

Let τ denote the absolute weak topology on the Riesz space L, that is, assume that $\tau = |\sigma|(L, L')$. Since $\mathbf{G}$ is convex and τ is a consistent topology on L, it follows that the weak and τ closures of $\mathbf{G}$ coincide. Hence, there exists a net $\{g_\alpha\}$ in $\mathbf{G}$ such that $g_\alpha \xrightarrow{\tau} -\epsilon\omega$.

For each α choose $g_i^\alpha \in G_i$, $\lambda_i^\alpha \geq 0$ $(i = 1, \ldots, m)$ with $\sum_{i=1}^m \lambda_i^\alpha = 1$ and $g_\alpha = \sum_{i=1}^m \lambda_i^\alpha g_i^\alpha$. By passing to a subnet, we can assume that $\lambda_i^\alpha \to \lambda_i \geq 0$ for each i. From $\sum_{i=1}^m \lambda_i = 1$, we see that the set $S = \{n\colon \lambda_n > 0\}$ is nonempty. By passing to a subnet again, we can assume that there exists some $\delta > 0$ such that $\lambda_n^\alpha \geq \frac{1}{\delta}$ for all $n \in S$ and all α.

Next, for each i write $g_i^\alpha = f_i^\alpha - \omega_i$. We claim that for each $n \in S$ the net $\{f_n^\alpha\}$ has a weakly convergent subnet in L. To see this, fix $n \in S$. Then

$$\begin{aligned} 0 \leq \lambda_n^\alpha f_n^\alpha &\leq \sum_{i=1}^m \lambda_i^\alpha f_i^\alpha = \sum_{i=1}^m \lambda_i^\alpha (g_i^\alpha + \omega_i) = \sum_{i=1}^m \lambda_i^\alpha g_i^\alpha + \sum_{i=1}^m \lambda_i^\alpha \omega_i \\ &= g_\alpha + \sum_{i=1}^m \lambda_i^\alpha \omega_i \leq g_\alpha + \omega \leq (g_\alpha + \epsilon\omega) + \omega, \end{aligned}$$

and consequently

$$0 \leq f_n^\alpha \leq \tfrac{1}{\lambda_n^\alpha}(g_\alpha + \epsilon\omega) + \tfrac{1}{\lambda_n^\alpha}\omega \leq \delta(g_\alpha + \epsilon\omega) + \delta\omega.$$

Since $[0, \delta\omega] = \delta[0, \omega]$ is (by our hypothesis) a weakly compact order interval and $g_\alpha + \epsilon\omega \xrightarrow{\tau} 0$, it follows that $\{f_n^\alpha\}$ has a weakly convergent subnet; see Exercise 7 of Chapter 6. So, by passing to an appropriate subnet, we can assume that $f_n^\alpha \xrightarrow[\alpha]{\sigma(L,L')} f_n$ in L for each $n \in S$. By the weak upper semicontinuity of preferences each F_n is weakly closed, so $f_n \in F_n$ (i.e., $f_n \succeq_n x_n$) for each $n \in S$. From

$$g_\alpha = \sum_{i=1}^m \lambda_i^\alpha (f_i^\alpha - \omega_i) \geq \sum_{n\in S} \lambda_n^\alpha f_n^\alpha - \sum_{i=1}^m \lambda_n^\alpha \omega_n \xrightarrow[\alpha]{\sigma(L,L')} \sum_{n\in S} \lambda_n f_n - \sum_{n\in S} \lambda_n \omega_n$$

and the fact that L^+ is weakly closed, we see that

$$-\epsilon\omega = \sigma(L, L')\text{-}\lim_\alpha g_\alpha \geq \sum_{n\in S} \lambda_n f_n - \sum_{n\in S} \lambda_n \omega_n.$$

Rearranging terms, we get

$$\sum_{n\in S} \lambda_n (f_n + \epsilon\omega) \leq \sum_{n\in S} \lambda_n \omega_n. \tag{$\star$}$$

By the desirability of ω, we have $f_n + \epsilon\omega \succ_n f_n \succeq_n x_n$ for each $n \in S$. Finally, using $(\star)$, an easy argument shows that $(x_1, \ldots, x_m)$ can be blocked in some replica economy, which is a contradiction. The proof of the lemma is now complete. ■

8.13. Core Equivalence

Here we relate the Edgeworth equilibria to Walrasian equilibria and quasiequilibria. The results in this section are due to C. D. Aliprantis, D. J. Brown and O. Burkinshaw [**12**]. We start by introducing the notion of an approximate quasiequilibrium.

DEFINITION 8.66. *An allocation* $(x_1,\ldots,x_m)$ *in an economy is an* ***approximate quasiequilibrium*** *if for each* $\epsilon>0$ *there exists some nonzero price* $p\in X'$ *(depending upon* ϵ*) such that:*

(1) $p\cdot\omega=1$.
(2) $x\succeq_i x_i$ *implies* $p\cdot x\geq p\cdot\omega_i-\epsilon$.

In a Riesz commodity economy with at least one consumer having a monotone preference, say the first one, every price p that satisfies properties (1) and (2) of the above definition is necessarily a positive price. Indeed, if $x\geq 0$, then $x_1+\frac{1}{\delta}x\succeq_1 x_1$ for each $\delta>0$. Consequently, we have

$$p\cdot x_1+\tfrac{1}{\delta}p\cdot x\geq p\cdot\omega_1-\epsilon$$

or $p\cdot x\geq\delta\big(p\cdot\omega_1-p\cdot x_1-\epsilon\big)$ for each $\delta>0$ from which it follows that $p\cdot x\geq 0$.

THEOREM 8.67. *Consider a Riesz economy such that:*

(1) *Preferences are convex and weakly upper semicontinuous.*
(2) *The Edgeworth box* $[0,\omega]$ *is weakly compact.*
(3) *The total endowment* ω *is desirable by each consumer.*

Then every Edgeworth equilibrium is an approximate quasiequilibrium.

PROOF. Consider an exchange economy satisfying the stated hypotheses, let $(x_1,\ldots,x_m)$ be an Edgeworth equilibrium, and let $\epsilon>0$. By Lemma 8.65, we know that $0\notin\epsilon\omega+\overline{\mathbf{G}}$, where

$$\mathbf{G}=\operatorname{co}\Big(\bigcup_{i=1}^m G_i\Big)=\Big\{\sum_{i=1}^m\lambda_i g_i\colon\ g_i\in G_i,\ \lambda_i\geq 0\ \forall\, i \text{ and } \sum_{i=1}^m\lambda_i=1\Big\},$$

and $G_i=\{x\in L\colon\ x+\omega_i\succeq_i x_i\}$. So, by the separating hyperplane theorem, there exists some price $p\in L'$ satisfying $p\cdot(g+\epsilon\omega)>0$ for all $g\in\mathbf{G}$. Clearly $p\cdot\omega>0$. Indeed, $x_i-\omega_i\in G_i$, so $0=\sum_{i=1}^m\frac{1}{m}(x_i-\omega_i)\in\mathbf{G}$, and thus $p\cdot\omega=\frac{1}{\epsilon}p\cdot(0+\epsilon\omega)>0$.

Replacing p by $\frac{p}{p\cdot\omega}$, we can assume that $p\cdot\omega=1$. If $x\succeq_i x_i$, then $x-\omega_i\in\mathbf{G}$, so that $p\cdot(x-\omega_i+\epsilon\omega)>0$, from which we get $p\cdot x\geq p\cdot\omega_i-\epsilon$, as desired. ■

The approximate quasiequilibria are Walrasian equilibria for an appropriate subeconomy of the original economy.

LEMMA 8.68. *If* $(x_1,\ldots,x_m)$ *is an approximate quasiequilibrium for a Riesz economy, then there exists a positive linear functional* p *on* L_ω *(the ideal generated in* L *by the total endowment* ω*) such that:*

(1) $p\cdot\omega=1$.
(2) $x\succeq_i x_i$ *in* L_ω^+ *implies* $p\cdot x\geq p\cdot\omega_i$.

That is, every approximate quasiequilibrium is a quasiequilibrium with respect to another Riesz economy whose consumer characteristics are the same as the original ones and with Riesz pair $\langle L_\omega,L'_\omega\rangle$.

PROOF. Let $(x_1,\ldots,x_m)$ be an approximate quasiequilibrium for a Riesz economy. For each natural number n choose some $0\leq p_n\in L'$ with $p_n\cdot\omega=1$ and such that

$$x\succeq_i x_i \text{ in } L^+ \quad\text{implies}\quad p_n\cdot x\geq p_n\cdot\omega_i-\tfrac{1}{n}.$$

Now consider each p_n restricted to L_ω. The condition $p_n\cdot\omega=1$ implies that p_n as a functional on the normed space $(L_\omega,\|\cdot\|_\infty)$ has norm one. So, by the Alaoglu–Bourbaki Theorem 2.12, the sequence $\{p_n\}$ has a $\sigma(L'_\omega,L_\omega)$-accumulation point, say p. Then, p is a positive functional on L_ω (although it need not be defined on all of L), $p \cdot \omega = 1$, and $x \succeq_i x_i$ in L_ω^+ implies $p \cdot x \geq p \cdot \omega_i$, as claimed. ∎

And now we come to an important result that describes the relationships between the various equilibria notions introduced before.

THEOREM 8.69 (Aliprantis–Brown–Burkinshaw [**12**]). *Consider a Riesz economy with Riesz pair $\langle L, L' \rangle$ such that:*

(a) *Preferences are convex, continuous for some consistent locally convex-solid topology on L, and strictly monotone on L_ω^+.*
(b) *The Edgeworth box $[0, \omega]$ is weakly compact.*
(c) *The total endowment ω is desirable by each consumer.*

Then for an allocation $(x_1, \ldots, x_m)$ the following statements are equivalent.

(1) *The allocation $(x_1, \ldots, x_m)$ is an Edgeworth equilibrium.*
(2) *The allocation $(x_1, \ldots, x_m)$ is an approximate quasiequilibrium.*
(3) *The allocation $(x_1, \ldots, x_m)$ is a Walrasian equilibrium for the Riesz economy with the same preferences and initial endowments and Riesz pair $\langle L_\omega, L'_\omega \rangle$.*

PROOF. Consider an economy that satisfies the properties stated in the theorem and let $(x_1, \ldots, x_m)$ be an allocation.

(1) $\Longrightarrow$ (2) This is an immediate consequence of Theorem 8.67.

(2) $\Longrightarrow$ (3) By Lemma 8.68 the allocation $(x_1, \ldots, x_m)$ is a quasiequilibrium with respect to the Riesz pair $\langle L_\omega, L'_\omega \rangle$. Since preferences are strictly monotone on L_ω^+, it follows from Corollary 8.37 that $(x_1, \ldots, x_m)$ is a Walrasian equilibrium with respect to $\langle L_\omega, L'_\omega \rangle$.

(3) $\Longrightarrow$ (1) This is an immediate consequence of Lemma 8.62. ∎

The next example illustrates the preceding theorem.

EXAMPLE 8.70. Consider once more the Riesz economy introduced in Example 8.18. In Example 8.40 we established that (ω_1, ω_2) was a Walrasian equilibrium with respect to the Riesz pair $\langle L_\omega, L'_\omega \rangle$ (but not a Walrasian equilibrium with respect to the original Riesz pair $\langle \ell_1, \ell_\infty \rangle$). So, according to Theorem 8.69, the allocation (ω_1, ω_2) is an Edgeworth equilibrium or, equivalently, an approximate quasiequilibrium. We now demonstrate the existence of the prices that support (ω_1, ω_2) as an approximate quasiequilibrium. To this end, let $\epsilon > 0$.

Fix some $k \in \mathbb{N}$ such that $\sum_{n=k}^\infty \frac{1}{2^n} < \frac{\epsilon}{2}$ holds and consider the ℓ_∞-price $q = (2, 2^2, \ldots, 2^k, 1, 1, 1, \ldots)$. Then q satisfies the following properties.

- $u(\omega_1) \geq q \cdot \omega_1$.

To see this, note that

$$u(\omega_1) = \sum_{n=1}^{\infty} \tfrac{1}{2^{n+1}} \geq \sum_{n=1}^{k} \tfrac{1}{2^{n+1}} + \sum_{n=k+1}^{\infty} \tfrac{1}{2^{2n+1}} = q \cdot \omega_1 .$$

- $q \cdot x \geq u(x) - \frac{\epsilon}{2}$ *for each* $0 \leq x \in \ell_1$.

We proved in Example 8.40 that $2^n t \geq \phi_n(t)$ for each $t \geq 0$. Moreover, from the definition of ϕ_n, we see that $t + \frac{1}{2^n} \geq \phi_n(t)$ for all $t \geq 0$. So, if $x = (x_1, x_2, \ldots) \in \ell_1^+$, then

$$\begin{aligned} q \cdot x &= \sum_{n=1}^{k} 2^n x_n + \sum_{n=k+1}^{\infty} x_n = \sum_{n=1}^{k} 2^n x_n + \sum_{n=k+1}^{\infty} \left(x_n + \tfrac{1}{2^n}\right) - \sum_{n=k+1}^{\infty} \tfrac{1}{2^n} \\ &\geq \sum_{n=1}^{k} \phi_n(x_n) + \sum_{n=k+1}^{\infty} \phi_n(x_n) - \tfrac{\epsilon}{2} = u(x) - \tfrac{\epsilon}{2}. \end{aligned}$$

- *If $x \succeq_1 \omega_1$, then $q \cdot x \geq q \cdot \omega_1 - \frac{\epsilon}{2}$.*

Indeed, if $x \succeq_1 \omega_1$, then from the preceding two properties we see that

$$q \cdot x \geq u(x) - \tfrac{\epsilon}{2} \geq u(\omega_1) - \tfrac{\epsilon}{2} \geq q \cdot \omega_1 - \tfrac{\epsilon}{2}.$$

- *If $p = \frac{q}{q \cdot \omega} \in L'_\omega$, then $p \cdot \omega = 1$ and $x \succeq_1 \omega_1$ guarantees that $p \cdot x \geq p \cdot \omega_1 - \epsilon$.*

Clearly, $p \cdot \omega = 1$, where $\omega = \omega_1 + \omega_2 = \left(\frac{1}{2^2}, \frac{1}{2^4}, \ldots, \frac{1}{2^{2n}}, \ldots\right)$, and $q \cdot \omega > \frac{1}{2}$. Now if $x \succeq_1 \omega_1$, then from the last property we get

$$p \cdot x = \tfrac{q \cdot x}{q \cdot \omega} \geq \tfrac{q \cdot \omega_1}{q \cdot \omega} - \tfrac{\epsilon}{2 q \cdot \omega} \geq p \cdot \omega_1 - \epsilon.$$

Consequently, the price p is "ϵ-supporting" the Edgeworth equilibrium (ω_1, ω_2). ■

Statements (1) and (3) of Theorem 8.69 also remain equivalent without the assumption that the Edgeworth box $[0, \omega]$ is weakly compact. Specifically, we have the following companion result to Theorem 8.69.

THEOREM 8.71 (Aliprantis–Brown–Burkinshaw [**12**]). *In a Riesz economy, suppose that preferences are convex, continuous for some consistent linear topology on L, and strictly monotone on L_ω^+. Then for an allocation $(x_1, \ldots, x_m)$ the following statements are equivalent.*

1. *The allocation $(x_1, \ldots, x_m)$ is an Edgeworth equilibrium.*
2. *The allocation $(x_1, \ldots, x_m)$ is an approximate quasiequilibrium with respect to the Riesz pair $\langle L_\omega, L'_\omega \rangle$.*
3. *The allocation $(x_1, \ldots, x_m)$ is a Walrasian equilibrium with respect to the Riesz pair $\langle L_\omega, L'_\omega \rangle$.*

PROOF. (1) $\Longrightarrow$ (2) Fix an Edgeworth equilibrium $(x_1, \ldots, x_m)$ and note that it is also an Edgeworth equilibrium with respect to the Riesz pair $\langle L_\omega, L'_\omega \rangle$. Let $\epsilon > 0$ and keep in mind that preferences are automatically $\|\cdot\|_\infty$-continuous on L_ω^+. Put $F_i = \{x \in L_\omega^+ \colon x \succeq_i x_i\}$, $G_i = F_i - \omega_i$, and let G be the convex hull of $\bigcup_{i=1}^m G_i$. Pick some n with $n\omega > x_1$ and note that by the strong monotonicity of the preferences on L_ω^+, we have $n\omega \succ_1 x_1$. Since $n\omega$ is an $\|\cdot\|_\infty$-interior point of L_ω^+ and the preference $\succeq_1$ is $\|\cdot\|_\infty$-continuous, it follows that $n\omega$ is an $\|\cdot\|_\infty$-interior point of F_1. From $\epsilon\omega + F_1 - \omega_1 \subseteq \epsilon\omega + G$, we infer that $\epsilon\omega + G$ has $\|\cdot\|_\infty$-interior points.

Since the allocation $(x_1, \ldots, x_m)$ is an Edgeworth equilibrium with respect to L_ω, a repetition of the proof of Theorem 8.59 shows that $0 \notin \epsilon\omega + G$. Thus, by the classical separation theorem, there exists a nonzero price $p \in L'_\omega$ satisfying $p \cdot (\epsilon\omega + g) \geq 0$ for all $g \in G$. It follows that $p \geq 0$. On the other hand, $p \cdot \omega = 0$ implies $p = 0$ (why?), which is a contradiction. Thus $p \cdot \omega > 0$, and so replacing p by $\frac{p}{p \cdot \omega}$, we can assume that $p \cdot \omega = 1$. Now if $x \succeq_i x_i$ holds in L_ω^+, then $x - \omega_i \in G$, and so $p \cdot (\epsilon\omega + x - \omega_i) \geq 0$, from which it follows that

$$p \cdot x \geq p \cdot \omega_i - \epsilon\,.$$

Therefore, $(x_1, \ldots, x_m)$ is an approximate quasiequilibrium with respect to the Riesz pair $\langle L_\omega, L'_\omega \rangle$.

The implications (2) $\Longrightarrow$ (3) $\Longrightarrow$ (1) are straightforward and are left for the reader. ■

Next we shall establish the well-known equilibrium existence result of T. F. Bewley [**38**]. Consider the Riesz pair $\langle L_\infty, L_1 \rangle$, where $L_\infty = L_\infty(\mu)$ and $L_1 = L_1(\mu)$ for some σ-finite measure μ. We know that $\langle L_\infty, L_1 \rangle$ is a symmetric Riesz pair and that the Mackey topology $\tau(L_\infty, L_1)$ is locally solid and Lebesgue and that the only order continuous prices on L_∞ are the prices in L_1; see Exercise 17 of Chapter 6.

THEOREM 8.72 (Bewley). *Consider a Riesz economy having Riesz pair $\langle L_\infty, L_1 \rangle$ and whose preferences are represented by strictly monotone, quasiconcave, and Mackey continuous utility functions.*

If the total endowment ω is bounded away from zero (i.e., if there exists some $\epsilon > 0$ such that $\omega(x) \geq \epsilon$ for μ-almost all x), then the economy has a Walrasian equilibrium.

PROOF. Start by observing that since ω is bounded away from zero, we have $L_\omega = L_\infty$. Now, by Theorem 8.60, the economy has an Edgeworth equilibrium which, by Theorem 8.69, it is supported as a Walrasian equilibrium by a price $0 < p \in L'_\omega = L'_\infty$. Since the order interval $[0, \omega]$ is weakly compact, it follows from Theorem 8.39(2) that p is order continuous on L_∞, and thus $p \in L_1$. So, the economy has a Walrasian equilibrium. ■

We have seen so far that an Edgeworth equilibrium need not be a quasiequilibrium. However, if the economy has uniformly proper preferences, then every Edgeworth equilibrium is a quasiequilibrium.

THEOREM 8.73 (Aliprantis–Brown–Burkinshaw [**12**]). *Consider a Riesz economy such that each preference is monotone, convex, and uniformly proper for a consistent locally convex-solid topology τ on the commodity space. Then every Edgeworth equilibrium is a quasiequilibrium (with respect to the Riesz pair $\langle L, L' \rangle$).*

PROOF. Consider a Riesz economy having the properties stated in the theorem and let $(x_1, \ldots, x_m)$ be an Edgeworth equilibrium. As before, define

$$F_i = \{x \in L^+ \colon\ x \succeq_i x_i\} \quad \text{and} \quad G_i = F_i - \omega_i = \{x \in L \colon\ x + \omega_i \succeq_i x_i\},$$

and let $\mathbf{G} = \operatorname{co}\left(\bigcup_{i=1}^m G_i\right)$.

Since each preference is uniformly τ-proper, for each i there exist a convex and solid τ-neighborhood V_i of zero and some vector $v_i > 0$ such that $x - \alpha v_i + z \succeq_i x$ in L^+ with $\alpha > 0$ imply $z \notin \alpha V_i$. Put $V = \bigcap_{i=1}^m V_i$ and $v = \sum_{i=1}^m v_i$, and define the convex set

$$\Gamma = \left\{\alpha w \colon\ \alpha > 0 \text{ and } w \in L \text{ satisfies } w + v \in \tfrac{1}{2}V\right\}.$$

We claim that $\Gamma \cap \mathbf{G} = \emptyset$. If this is the case, then (since Γ has a nonempty interior) the separating hyperplane theorem guarantees the existence of some nonzero price $p \in L'$ satisfying $p \cdot g \geq p \cdot \gamma$ for all $g \in \mathbf{G}$ and all $\gamma \in \Gamma$. Since $\gamma \in \Gamma$ implies $\alpha\gamma \in \Gamma$ for all $\alpha > 0$, we see that $p \cdot g \geq 0$ for each $g \in \mathbf{G}$. Now, if $x \succeq_i x_i$, then $x - \omega_i \in \mathbf{G}$, so $p \cdot (x - \omega_i) = p \cdot x - p \cdot \omega_i \geq 0$ implies $p \cdot x \geq p \cdot \omega_i$. This means that $(x_1, \ldots, x_m)$ is a quasiequilibrium.

To finish the proof, we must establish that $\Gamma \cap \mathbf{G} = \emptyset$. To this end, assume by way of contradiction that $\Gamma \cap \mathbf{G} \neq \emptyset$. Fix $a \in \Gamma \cap \mathbf{G}$ and write

$$a = \sum_{i=1}^{m} \lambda_i (z_i - \omega_i) = \alpha w\,,$$

where $\lambda_i \geq 0$ and $z_i \succeq_i x_i$ for each i, $\sum_{i=1}^{m} \lambda_i = 1$, $\alpha > 0$, and $w \in L$ satisfies $w + v \in \frac{1}{2}V$. For each i pick a sequence $\{r_n^i\}$ of strictly positive rational numbers such that $r_n^i \downarrow_n \lambda_i$. So, there exist positive rational numbers $\frac{n_i}{n}$ (with the n_i and n positive integers) such that

$$\Big[\tfrac{1}{\alpha} \sum_{i=1}^{m} \tfrac{n_i}{n}(z_i - \omega_i) - w\Big] + (w + v) \in \tfrac{1}{2}V + \tfrac{1}{2}V = V\,.$$

Consequently,

$$\sum_{i=1}^{m} n_i z_i - \sum_{i=1}^{m} n_i \omega_i + \alpha n v \in \alpha n V\,. \qquad (\star)$$

Next, put

$$y = \sum_{i=1}^{m} n_i \omega_i - \alpha n v \quad \text{and} \quad z = \sum_{i=1}^{m} n_i z_i \geq 0\,.$$

From $z - y = z + \alpha n v - \sum_{i=1}^{m} n_i \omega_i \leq z + \alpha n v$, it follows that

$$(y - z)^- = (z - y)^+ \leq z + \alpha n v = \sum_{i=1}^{m} (n_i z_i + \alpha n v_i)\,.$$

From the Riesz decomposition property of the Riesz space L, there exist positive vectors $w_i \in L^+$ $(i = 1, \dots, m)$ satisfying the properties $0 \leq w_i \leq n_i z_i + \alpha n v_i$ and $\sum_{i=1}^{m} w_i = (y - z)^-$. Define

$$y_i = z_i + \tfrac{\alpha n}{n_i} v_i - \tfrac{1}{n_i} w_i = \tfrac{1}{n_i}(n_i z_i + \alpha n v_i - w_i) \geq 0\,,$$

and note that $y_i \succ_i z_i$ for each i. Indeed, if $z_i = y_i - \frac{\alpha n}{n_i} v_i + \frac{1}{n_i} w_i \succeq_i y_i$ is true, then the properness implies $\frac{1}{n_i} w_i \notin \frac{\alpha n}{n_i} V$, or $w_i \notin \alpha n V$. On the other hand, from

$$0 \leq w_i \leq (y - z)^- \leq |y - z| = \Big|\sum_{i=1}^{m} n_i (z_i - \omega_i) + \alpha n v\Big|$$

and $(\star)$, we see that $w_i \in \alpha n V$, which contradicts $w_i \notin \alpha n V$. So $y_i \succ_i z_i$ for all i. Finally, from $y_i \succ_i z_i \succeq_i x_i$ and

$$\begin{aligned} \sum_{i=1}^{m} n_i y_i &= \sum_{i=1}^{m} n_i z_i + \alpha n \sum_{i=1}^{m} v_i - \sum_{i=1}^{m} w_i = z + \alpha n v - (y - z)^- \\ &\leq z + \alpha n v + y - z = y + \alpha n v = \sum_{i=1}^{m} n_i \omega_i, \end{aligned}$$

it easily follows that the allocation $(x_1, \dots, x_m)$ can be blocked in some replica economy, which is impossible. Hence, $\Gamma \cap \mathbf{G} = \emptyset$ must be true, and the proof is finished. ■

Now the infinite dimensional analogue of the Debreu–Scarf Core Equivalence Theorem 8.59 can be formulated as follows. It was established by C. D. Aliprantis, D. J. Brown, and O. Burkinshaw [**12**].

THEOREM 8.74 (Core Equivalence Theorem). *In a convex Riesz economy whose preferences are strictly monotone, continuous, and uniformly proper for a consistent locally convex-solid topology and the total endowment is strictly positive, the sets of Walrasian and Edgeworth equilibria coincide.*

PROOF. Clearly, every Walrasian equilibrium is an Edgeworth equilibrium. On the other hand, any Edgeworth equilibrium in such an economy is (by Theorem 8.73) a quasiequilibrium which in turn (by Corollary 8.37) must be a Walrasian equilibrium. ■

THEOREM 8.75. *In the class of convex Riesz economies with strictly positive total endowments, weakly compact Edgeworth boxes, and strictly monotone, continuous and uniformly proper (for consistent locally convex-solid topologies) preferences, the following statements are equivalent.*

(1) *Every economy has a Walrasian equilibrium.*
(2) *Every economy has a core allocation.*

PROOF. To see that (2) implies (1), notice that a verbatim repetition of the proof of Theorem 8.60 guarantees that every economy from our class of economies has an Edgeworth equilibrium, which, by Theorem 8.74, must be a Walrasian equilibrium. ■

It is also interesting to mention the following converse-type result of Theorem 8.30. If (under the standard hypotheses) in a Riesz commodity economy every Edgeworth equilibrium is supported by a price as a quasiequilibrium, then the price space X' is necessarily a Riesz subspace of the order dual $L^{\sim}$ of L. For details see [**23**].

8.14. The Single Sector Growth Model

The standard formulation of the single sector optimal growth problem runs something like this.[4] There is a single commodity which is used as capital, along with labor, to produce more output. In the simplest formulation, labor is presumed to be supplied in fixed amounts and capital and consumption are interpreted in per capita terms. Raw output is either consumed or saved as capital until the next period. This process repeats *ad infinitum*.

Letting c_t denote the quantity of output consumed at time t and k_t denote the capital stock, where k_1 is given as a datum, the set of **feasible trajectories** of this economy is the set of all positive sequences

$$(c,k) = \big((c_1,c_2,\ldots),(k_1,k_2,\ldots)\big) = \big((c_1,k_1),(c_2,k_2),\ldots\big)$$

obeying

$$k_{t+1} = (1-\delta)k_t + f(k_t) - c_t \quad\text{and}\quad 0 \le c_t \le f(k_t)$$

for $t = 1, 2, \ldots$. Here the function $f\colon [0,\infty) \to [0,\infty)$ is the **production function** and $0 < \delta < 1$ is the **rate of physical depreciation** of capital. The production function f is assumed

[4] The presentation in this section follows [**10**]. For more details of the standard formulation see, for instance [**124**, Chapter 5].

- *to be continuous, strictly concave, strictly increasing, and*
- *to satisfy* $f(0) = 0$, $f'(0) = \infty$, *and* $\lim_{k\to\infty} \frac{f(k)}{k} = 0$.

The functions $f(k) = k^p$, where $0 < p < 1$, are all acceptable production functions.

The proof of the next result is left for the reader; see also Exercise 18 at the end of the Chapter.

LEMMA 8.76. *If f is a production function, then for each $m > 0$ there exists a unique $k_m > 0$ such that:*

(1) $f(k_m) = mk_m$.
(2) $f(k) > mk$ *for each* $0 < k < k_m$.
(3) $f(k) < mk$ *for each* $k > k_m$.

We continue with the notion of a free disposal trajectory.

DEFINITION 8.77. *The* ***free disposal feasible trajectory*** *with respect to a production function f, some $k_1 > 0$, and some $0 < \delta < 1$ is the set $Y \subseteq \ell_\infty^+ \times \ell_\infty^+$ defined by*

$$Y = \left\{ (c,k) \in \left(\ell_\infty^+\right)^2 \colon \begin{array}{l} 0 \le k_{t+1} \le (1-\delta)k_t + f(k_t) - c_t \\ 0 \le c_t \le f(k_t) \end{array}, \; t=1,2,\ldots \right\}.$$

We shall denote by $\mathbf{1}$ the constant sequence one of the space $\ell_\infty \times \ell_\infty$, that is, $\mathbf{1} = \big((1,1),(1,1),\ldots\big)$.

LEMMA 8.78. *Every feasible free disposal set of trajectories Y is a convex norm bounded subset of $\ell_\infty^+ \times \ell_\infty^+$ and satisfies $0 \in Y$ and $\eta\mathbf{1} \in Y$ for some $\eta > 0$.*

PROOF. Let Y be a free disposal feasible set of trajectories with respect to a production function f, some $0 < \delta < 1$ and some $k_1 > 0$. Clearly, Y is a convex set.

Next, we shall establish that Y is norm bounded. To this end, let (c,k) in Y. Since $\lim_{k\to\infty} \frac{f(k)}{k} = 0$, there exists n such that $f(k) < \delta k$ for each $k \ge n$. Put

$$M = \sup_{k\in[0,\infty)} \big[f(k) - \delta k\big] = \max_{k\in[0,n]} \big[f(k) - \delta k\big],$$

and let $b = \max\{n+M, k_1\} < \infty$. We claim that $k_t \le b$ holds for each t. The proof is by induction. For $t = 1$, the inequality is obvious. Now for the induction step, assume that $k_t \le b$. If $k_t > n$, then

$$k_{t+1} = (1-\delta)k_t + f(k_t) - c_t = k_t + \big[f(k_t) - \delta k_t\big] - c_t < k_t \le b.$$

If $k_t \le n$, then

$$k_{t+1} = k_t + \big[f(k_t) - \delta k_t\big] - c_t \le n + M \le b,$$

and the proof of our claim is complete. To see that the sequence $\{c_t\}$ is uniformly bounded, note that

$$0 \le c_t \le f(k_t) \le \max_{0\le k\le b} f(k)$$

holds for each t. Finally, note that $0 \in Y$ and if $f(\eta) = (1+\delta)\eta$, then $\eta\mathbf{1} \in Y$. ∎

It is also easy to see that since Y is defined pointwise by inequalities among continuous functions, that Y is pointwise closed. In particular, Y is closed for the $\sigma(\ell_\infty \times \ell_\infty, \ell_1 \times \ell_1)$-topology. This leads to the next lemma.

LEMMA 8.79. *Any set of feasible free disposal trajectories is $\sigma(\ell_\infty \times \ell_\infty, \ell_1 \times \ell_1)$-compact.*

PROOF. Recall that the norm dual of ℓ_1 is the Banach lattice ℓ_∞. Since the set Y is norm bounded and $\sigma(\ell_\infty \times \ell_\infty, \ell_1 \times \ell_1)$-closed, the Alaoglu–Bourbaki Theorem 2.12 guarantees that the set Y is also $\sigma(\ell_\infty \times \ell_\infty, \ell_1 \times \ell_1)$-compact. ■

The purpose of the optimal growth problem is to find the feasible plan (c, k) maximizing the objective function

$$U(c,k) = \sum_{t=1}^{\infty} \beta^t u(c_t),$$

where $u\colon [0,\infty) \to \mathbb{R}$ is continuous, concave, strictly increasing, and normalized so that $u(0) = 0$ and $0 < \beta < 1$ is a constant, called the **discount factor**. Note that in case $\{x_t\}$ is bounded so is $\{u(x_t)\}$ and so $\sum_{t=1}^{\infty} \beta^t u(x_t)$ is a real number. Therefore, $U\colon \ell_\infty^+ \times \ell_\infty^+ \to \mathbb{R}$ is a well defined function.

The objective function U may be interpreted either as the social welfare function of a central planner or as the utility function of a "representative agent" in the economy who has complete control of his own consumption and production. Since Y is $\sigma(\ell_\infty \times \ell_\infty, \ell_1 \times \ell_1)$-compact an optimal plan is guaranteed to exist if U is $\sigma(\ell_\infty \times \ell_\infty, \ell_1 \times \ell_1)$-upper semicontinuous; see part (2) of Theorem 8.3. Indeed, as we shall see later, U is $\sigma(\ell_\infty \times \ell_\infty, \ell_1 \times \ell_1)$-continuous on the free disposal feasible trajectory sets. Any free disposal trajectory that maximizes U over a free disposal feasible trajectory set Y is known as an **optimal plan**.

LEMMA 8.80. *If Y is a free disposal feasible trajectory set, then every optimal plan (c, k) is a feasible trajectory, i.e.,*

$$k_{t+1} = (1-\delta)k_t + f(k_t) - c_t \quad \text{and} \quad 0 \le c_t \le f(k_t)$$

hold for each t.

PROOF. Assume by way of contradiction that for some period s we have $k_{s+1} < (1-\delta)k_s + f(k_s) - c_s$. We distinguish two cases.

CASE 1. $c_s < f(k_s)$.

In this case, we can find some c'_s satisfying $c_s < c'_s < f(k_s)$ and such that $k_{s+1} < (1-\delta)k_s + f(k_s) - c'_s$. Define (c^*, k^*) by $k_t^* = k_t$ for all t, $c_t^* = c_t$ for all $t \neq s$ and $c_s^* = c'_s$. Then (c^*, k^*) is a free disposal feasible trajectory, i.e., $(c^*, k^*) \in Y$. However, $U(c^*, k^*) > U(c, k)$ holds true, which contradicts the optimality property of (c, k).

CASE 2. $c_s = f(k_s)$.

In this case, we have $k_{s+1} < (1-\delta)k_s$. Now define (c^*, k^*) by $c_t^* = c_t$ for each t, $k_t^* = k_t$ for $t \neq s+1$ and $k_{s+1}^* = (1-\delta)k_s$. Clearly, $(c^*, k^*) \in Y$ and $U(c^*, k^*) = U(c, k)$. However, (c^*, k^*) satisfies $c_{s+1}^* \le f(k_{s+1}) < f(k_{s+1}^*)$, which (as in the previous case) leads to a contradiction. ■

In sum: The single growth model represents an economy having a single commodity which is used as capital along with labor to produce more output. The goal in this economy is to establish the existence of a feasible trajectory that maximizes the social welfare function. To do this, we need one more technical result whose proof is left for the reader; see also Exercise 19 at the end of this chapter.

LEMMA 8.81. *For any concave nondecreasing function* $u\colon [0,\infty) \to [0,\infty)$ *satisfying* $u(0) = 0$ *the following statements hold.*

(1) *If* $0 < x \le y$, *then* $\frac{u(y)}{y} \le \frac{u(x)}{x}$.
(2) *If* $x > 0$ *and* $y \ge 0$, *then*

$$\big|u(x) - u(y)\big| \le \tfrac{u(x)}{x}|x-y|.$$

(3) *If* $0 \le x \le z \le y$ *and* $z > 0$, *then*

$$\big|u(x) - u(y)\big| \le \tfrac{u(z)}{z}|x-y|.$$

Now let $u\colon [0,\infty) \to [0,\infty)$ be a nondecreasing utility function. Then the **discounted utility** of u with respect to the discount factor $0 < \beta < 1$ is the function $U\colon \ell_\infty^+ \to [0,\infty)$ defined by

$$U(x) = \sum_{t=1}^{\infty} \beta^t u(x_t).$$

THEOREM 8.82. *Let* $u\colon [0,\infty) \to \mathbb{R}$ *be a nondecreasing concave function satisfying* $u(0) = 0$ *and let* U *be the discounted utility of* u *with respect to some* $0 < \beta < 1$. *Then:*

(1) U *is nondecreasing, concave, and* $|\sigma|(\ell_\infty, \ell_1)$*-continuous.*
(2) U *is* $\sigma(\ell_\infty, \ell_1)$*-continuous on the norm bounded subsets of* ℓ_∞^+.
(3) U *is norm continuous.*

PROOF. (1) To show that U is $|\sigma|(\ell_\infty, \ell_1)$-continuous, assume that $x^\alpha \xrightarrow{|\sigma|(\ell_\infty,\ell_1)} x$ holds in ℓ_∞^+. We wish to show that $\big|U(x^\alpha) - U(x)\big| \underset{\alpha}{\to} 0$. To this end, let $\epsilon > 0$.

Choose $\eta > 0$ such that $\sum_{t=1}^\infty \beta^t u(\eta) < \epsilon$ and let $B_\eta = \{t\colon\ x_t < \eta\}$. From $\sum_{t=1}^\infty \beta^t u(x_t) < \infty$, we get $\big(\beta^1 u(x_1), \beta^2 u(x_2), \beta^3 u(x_3), \ldots\big) \in \ell_1$. So, there exists (in view of $x^a \xrightarrow[\alpha]{|\sigma|(\ell_\infty,\ell_1)} 0$) some α_0 such that

$$\sum_{t=1}^{\infty} \beta^t \tfrac{u(x_t)}{\eta}|x_t^\alpha - x_t| < \epsilon \quad \text{and} \quad \sum_{t=1}^{\infty} \beta^t \tfrac{u(\eta)}{\eta}|x_t^\alpha - x_t| < \epsilon$$

for all $\alpha \ge \alpha_0$.

Now let $t \in B_\eta$. If $x_t^\alpha > \eta$ holds, then $|u(x_t^\alpha) - u(x_t)| \le \frac{u(\eta)}{\eta}|x_t^\alpha - x_t|$ (Lemma 8.81), and if $x_t^\alpha < \eta$, then $|u(x_t^\alpha) - u(x_t)| \le 2u(\eta)$. Thus,

$$\sum_{t\in B_\eta} \beta^t |u(x_t^\alpha) - u(x_t)| \le \sum_{t\in B_\eta} \beta^t \tfrac{u(\eta)}{\eta}|x_t^\alpha - x_t| + 2\sum_{t\in B_\eta} \beta^t u(\eta) < \epsilon + 2\epsilon = 3\epsilon.$$

Finally, using Lemma 8.81 once more, we see that

$$\begin{aligned}
\big|U(x^\alpha)-U(x)\big| &= \Big|\sum_{t=1}^{\infty}\beta^t u(x_t^\alpha)-\sum_{t=1}^{\infty}\beta^t u(x_t)\Big| \\
&\le \sum_{t=1}^{\infty}\beta^t|u(x_t^\alpha)-u(x_t)| \\
&= \sum_{t\in B_\eta}\beta^t|u(x_t^\alpha)-u(x_t)|+\sum_{t\in B_\eta^c}\beta^t|u(x_t^\alpha)-u(x_t)| \\
&\le \sum_{t\in B_\eta}\beta^t|u(x_t^\alpha)-u(x_t)|+\sum_{t\in B_\eta^c}\beta^t\tfrac{u(x_t)}{x_t}|x_t^\alpha-x_t| \\
&\le \sum_{t\in B_\eta}\beta^t|u(x_t^\alpha)-u(x_t)|+\sum_{t\in B_\eta^c}\beta^t\tfrac{u(x_t)}{\eta}|x_t^\alpha-x_t| \\
&< 3\epsilon+\epsilon=4\epsilon.
\end{aligned}$$

for all $\alpha \ge \alpha_0$, as desired.

(2) Assume $x^\alpha \xrightarrow{\sigma(\ell_\infty,\ell_1)} x$ and that $\|x\|_\infty \le b < \infty$ and $\|x^\alpha\|_\infty \le b$ for each index α. Fix $\epsilon > 0$ and then choose some $n \in \mathbb{N}$ such that $2\sum_{t=n+1}^{\infty}\beta^t u(b) < \epsilon$. Thus,

$$\big|U(x^\alpha)-U(x)\big| \le \sum_{t=1}^{n}\beta^t|u(x_t^\alpha)-u(x_t)|+\epsilon.$$

Taking into account that $x^\alpha \xrightarrow{\sigma(\ell_\infty,\ell_1)} x$ implies $x_t^\alpha \xrightarrow[\alpha]{} x_t$ for each t, the preceding inequality yields $\limsup_\alpha |U(x^\alpha)-U(x)| \le \epsilon$ for each $\epsilon > 0$. Therefore, $U(x^\alpha) \xrightarrow[\alpha]{} U(x)$.

(3) This follows from the fact that every norm convergent sequence is norm bounded and weakly convergent. ■

We now continue by introducing the characteristics of a single agent growth economy.[5] The **single agent growth economy** has the following characteristics.

(1) The commodity-price duality $\langle L, L'\rangle$ is given by the symmetric Riesz pair $\langle \ell_\infty \times \ell_\infty, \ell_1 \times \ell_1\rangle$.
(2) The production possibility set is a set $Y \subseteq L^+$ which is convex, weakly compact, and satisfies $0 \in Y$ and $\eta\mathbf{1} \in Y$ for some $\eta > 0$.
(3) There is a single agent who in his capacity as the manager of production chooses output to maximize the value of output treating prices parametrically. As a consumer, the agent maximizes a preference relation $\succeq$ defined on L^+ subject to a budget constraint, treating prices and income as parameters. The preference $\succeq$ is assumed to be strictly monotone, convex, and τ-continuous for a consistent locally convex linear topology τ on L.

We have already seen that the standard formulation of the single sector growth model with discounted utility fits this framework.

[5] A single agent economy is often called a "Robinson Crusoe" economy.

DEFINITION 8.83. *A* ***competitive equilibrium*** *for a single growth economy is a pair* $(x, p) \in L \times L'$ *such that:*

1. $x \in Y$ *and* $p \neq 0$.
2. *The bundle* x *maximizes the value of output at prices* p, *that is,* $p \cdot x \geq p \cdot y$ *for all* $y \in Y$.
3. *The bundle* x *also maximizes* $\succeq$ *subject to the budget constraint. That is, letting* $B = \{z \in L^+ \colon\ p \cdot z \leq p \cdot x\}$, *then* $x \succeq z$ *for all* $z \in B$.

And now we are ready to show that single agent growth economies have equilibria.

THEOREM 8.84. *Every single agent growth economy has a competitive equilibrium.*

PROOF. Since Y is $\sigma(L, L')$-compact and $\succeq$ is τ-continuous there is a $\succeq$-greatest element x of Y; see part (2) of Theorem 8.3.

By Theorem 8.25, we know that $\succeq$ is norm proper. Thus, $\succeq$ is uniformly (v, W)-proper for some norm open ball W centered at the origin. Therefore, v is a desirable bundle. Consider the open cone

$$\Gamma = \big\{\alpha(-v + w) \colon\ \alpha > 0 \text{ and } w \in W\big\}.$$

Now if F is the preferred set of x, i.e., if $F = \{z \in L \colon\ z \succeq x\}$, then we claim that $(F - Y) \cap \Gamma = \emptyset$.

To see this, assume by way of contradiction that $\big(F - Y\big) \cap \Gamma \neq \emptyset$. Hence, there exist $\alpha > 0$, $z \in F$, $y \in Y$, and $w \in W$ satisfying

$$z - y = \alpha(-v + w).$$

Then $z = y - \alpha v + \alpha w$, $z \succeq x$, and $x \succeq y$ since x is maximal in Y. Thus $y - \alpha v + \alpha w = z \succeq x$ and $x \succeq y$. Therefore, we have $y - \alpha v + \alpha w \succeq y$, and by properness $\alpha w \notin \alpha W$, a contradiction. Thus, $\big(F - Y\big) \cap \Gamma = \emptyset$ holds, as claimed.

Next, by the separation theorem there exist a nonzero price q in the norm dual of L (which may not lie in L') and some constant c such that

$$q \cdot (z - y) \geq c \geq q \cdot g$$

holds for all $z - y \in F - Y$ and all $g \in \Gamma$. Since Γ is a cone, we see that $c \geq 0$. Also $x \in Y$, so if $z \succeq x$ then $z - x \in F - Y$, and so $q \cdot z \geq q \cdot x$. Moreover, $x - y \in F - Y$ holds, and so $q \cdot x \geq q \cdot y$ for all $y \in Y$.

Since for each vector $z \geq 0$, we have $(x + z) - x \in F - Y$, it follows that $q \cdot (x + z) \geq q \cdot x$, or $q \cdot z \geq 0$. Therefore, $q > 0$. In particular, $q \cdot \mathbf{1} > 0$. The latter, in view of $\eta \mathbf{1} \in Y$, shows that $q \cdot x > 0$.

By the classical Yosida–Hewitt representation [**8**, p. 510], we can write $q = p + s$, where $0 \leq p \in \ell_1 \times \ell_1$, $s \geq 0$ and $s \cdot z = 0$ for all $z \in c_0 \times c_0$. To finish the proof, we shall show that (x, p) is an equilibrium.

Pick a sequence $\{x_n\} \in c_0 \times c_0$ such that $0 \leq x_n \uparrow x$ and note that $x_n \xrightarrow{\tau} x$. Fix $\epsilon > 0$. From $x + \epsilon v \succ x$, we see that there exists some n_0 such that $x_n + \epsilon v \succ x$ for all $n \geq n_0$. So, $q \cdot (x_n + \epsilon v) = q \cdot x_n + \epsilon q \cdot v \geq q \cdot x$ for all $n \geq n_0$. Consequently, $0 \leq q \cdot (x - x_n) \leq \epsilon q \cdot v$ for all $n \geq n_0$, which means that $\lim q \cdot x_n = q \cdot x$. Since the linear functional p is also $\sigma(L, L')$-continuous and $q \cdot x_n = p \cdot x_n + s \cdot x_n = p \cdot x_n$, we see that $p \cdot x = q \cdot x$. In particular, note that $p \neq 0$.

Now we are ready to show that (x, p) is an equilibrium. Let $y \in Y$. Then $y \geq 0$ and $q \geq p$, and so $p \cdot x = q \cdot x \geq q \cdot y \geq p \cdot y$. Next, let $z \succ x$. It easily follows that there exists some $0 \leq z_0 \leq z$ such that $z_0 \succ x$ and $z_0 \in c_0 \times c_0$. Choose some $0 < \delta < 1$ such that $\delta z_0 \succ x$. Then $p \cdot (\delta z_0) = (p+s) \cdot (\delta z_0) = q \cdot (\delta z_0) \geq q \cdot x = p \cdot x > 0$. Since $p \geq 0$ and $z \geq z_0$, we get $p \cdot z \geq \delta p \cdot z \geq \delta p \cdot z_0 \geq p \cdot x > 0$, which shows that $p \cdot z > p \cdot x$. In other words, x is the maximal element in the budget set, and the proof is finished. ■

Exercises

1. Show that a utility function $u \colon X \to \mathbb{R}$ defined on a convex subset of a vector space is quasiconcave if and only if it generates a convex preference.
2. Let $L = \mathbb{R}^2$ be the lexicographic plane and consider the preference relation $\succeq$ on L^+ defined by $x \succeq y$ whenever $x \geq y$. Show that the preference $\succeq$ cannot be represented by a utility function.
3. Establish the following properties.
   (a) If $\langle L, L' \rangle$ is Riesz pair, then $\langle L', L \rangle$ is a lattice pair and the order intervals of L' are $\sigma(L', L)$-compact.
   (b) A Riesz pair $\langle L, L' \rangle$ is symmetric if and only if the order intervals of L are $\sigma(L, L')$-compact.
4. Prove Lemma 8.9.
5. This exercise presents an alternate proof of the existence of individually rational Pareto efficient allocations; see Theorem 8.12.

   Assume that in Riesz commodity economy with m consumers the Edgeworth box $[0, \omega]$ is weakly compact and that each preference $\succeq_i$ is represented by a concave utility function u_i that is continuous with respect to a consistent locally convex topology on the commodity space L. Fix m positive real numbers $\lambda_1, \ldots, \lambda_m$ and consider the function $U \colon (L^+)^m \to \mathbb{R}$ defined by

$$U(x_1, \ldots, x_m) = \sum_{i=1}^{m} \lambda_i u_i(x_i).$$

   (a) Show that U attains a maximum value over $\mathcal{A}_r$, the nonempty weakly compact set of all individually rational allocations.
   (b) Show that every maximizer of U over $\mathcal{A}_r$ is an individually rational Pareto efficient allocation—and hence, in this case, individually rational Pareto efficient allocations exist.
6. This exercise demonstrates the existence of individually rational weakly Pareto efficient allocations. Consider a Riesz commodity economy with m consumers whose Edgeworth box $[0, \omega]$ is weakly compact and such that each preference $\succeq_i$ is convex and is represented by a utility function u_i that is continuous with respect to a consistent locally convex topology on the commodity space L. Define the function $V \colon (L^+)^m \to \mathbb{R}$ by

$$V(x_1, \ldots, x_m) = \min\{u_1(x_1), \ldots, u_m(x_m)\}.$$

   (a) Show that V is a Mackey-continuous quasiconcave function.
   (b) Show that V attains a maximum value on the nonempty compact set $\mathcal{A}_r$ of all individually rational allocations.
   (c) Show that every maximizer of V over $\mathcal{A}_r$ is an individually rational weakly Pareto efficient allocation—and hence, in this case, individually rational weakly Pareto efficient allocations exist.
7. Prove Lemma 8.44.

8. Consider a Riesz commodity economy with m consumers all having the same initial endowment ω and the same strictly convex preference $\succeq$. Show that $(\omega, \ldots, \omega)$ is the only individually rational and the only core allocation of the economy.

9. Assume that in a Riesz economy preferences are strictly monotone and lower semicontinuous for a consistent locally convex topology on L and that $\omega \gg 0$. Show that for an allocation $(x_1, \ldots, x_m)$ and a nonzero price $p \in L'$ the following statements are equivalent:

(a) For each consumer i the bundle x_i is a maximal element in the budget set $B_i(p) = \{x \in L^+ \colon\ p \cdot x \leq p \cdot \omega_i\}$.

(b) $p \gg 0$ and $x \succ_i x_i$ implies $p \cdot x > p \cdot \omega_i$.

(c) $x \succ_i x_i$ implies $p \cdot x \geq p \cdot \omega_i$.

(d) $x \succeq_i x_i$ implies $p \cdot x \geq p \cdot \omega_i$.

10. Assume that a Riesz economy with Riesz pair $\langle L_\infty[0,1], L_1[0,1]\rangle$ has m consumers with utility functions $u_i(x) = \int_0^1 \sqrt[i]{x(t)}\, dt$ for each $i = 1, \ldots, m$, and total endowment $\omega = \mathbf{1}$. Find the utility space of the economy.

11. This exercise completes the missing details in Example 8.47. Consider the Riesz economy having Riesz dual pair $\langle C[0,1], \mathrm{ca}[0,1]\rangle$ and two consumers with utility functions

$$u_1(x) = \int_0^1 x(t)\, dt \quad \text{and} \quad u_2(x) = \int_0^1 \sqrt{x(t)}\, dt$$

and initial endowments the functions ω_1 and ω_2 given by the formulas $\omega_1(t) = t$ and $\omega_2(t) = 1 - t$. Establish the following.

(a) Both utility functions are continuous, concave (in fact, the function u_2 is strictly concave), strictly monotone and uniformly $\|\cdot\|_\infty$-proper.

(b) The order interval $[0, \mathbf{1}]$ is not weakly compact.

(c) The utility space of the economy is given by

$$\mathbf{U} = \{(x, y) \in \mathbb{R}^2_+ \colon\ x + y^2 \leq 1\}.$$

(d) The economy has quasiequilibria which are, in fact, Walrasian equilibria.

(e) If $x_1 = x_2 = \frac{1}{2}\chi_{[0,1]}$, then the allocation (x_1, x_2) is a Walrasian equilibrium supported by the Lebesgue measure.

12. (Aliprantis–Brown–Burkinshaw [**14**]) Consider a two consumer Riesz economy whose Riesz pair is either $\langle C[0,1], \mathrm{ca}[0,1]\rangle$ or $\langle L_\infty[0,1], L_1[0,1]\rangle$. The consumers' initial endowments are given by $\omega_1 = \omega_2 = \mathbf{1}$, where $\mathbf{1}$ denotes the constant function one on $[0,1]$. The preferences are represented by the utility functions

$$u_1(x) = \int_0^{\frac{1}{2}} \sqrt{x(t)}\, dt + \tfrac{1}{2} \int_{\frac{1}{2}}^1 \sqrt{x(t)}\, dt$$

and

$$u_2(x) = \tfrac{1}{2} \int_0^{\frac{1}{2}} \sqrt{x(t)}\, dt + \int_{\frac{1}{2}}^1 \sqrt{x(t)}\, dt\,.$$

Establish the following.

(a) The Riesz economy has an empty core with respect to the Riesz pair $\langle C[0,1], \mathrm{ca}[0,1]\rangle$.

(b) The Riesz economy satisfies the closedness condition with respect to the Riesz pair $\langle L_\infty[0,1], L_1[0,1]\rangle$ (and so it has Walrasian equilibria).

(c) The Riesz economy fails to satisfy the closedness condition with respect to $\langle C[0,1], \mathrm{ca}[0,1]\rangle$.

13. Consider a Riesz economy having $\langle C[0,1], \mathrm{ca}[0,1]\rangle$ as its Riesz pair and two consumers with initial endowments $\omega_1 = \omega_2 = \mathbf{1}$ and utility functions given by

$$u_1(x) = \lambda \int_0^{\frac{1}{2}} \sqrt{x(t)}\, dt + (1-\lambda) \int_{\frac{1}{2}}^1 \sqrt{x(t)}\, dt$$

and

$$u_2(x) = (1-\lambda) \int_0^{\frac{1}{2}} \sqrt{x(t)}\, dt + \lambda \int_{\frac{1}{2}}^1 \sqrt{x(t)}\, dt\,,$$

where $0 < \lambda < 1$ is a fixed real number.

(a) Show that for $\lambda \neq \frac{1}{2}$ the economy does not have any core allocations.

(b) Show that for $\lambda = \frac{1}{2}$ the allocation (ω_1, ω_2) is a Walrasian equilibrium.

14. Consider a Riesz economy with Riesz pair $\langle L_p[0,1], L_q[0,1]\rangle$ (where, as usual, $1 \le p, q \le \infty$ and $\frac{1}{p} + \frac{1}{q} = 1$) and two consumers with the characteristics of Exercise 13 above. Show that the allocation (x_1, x_2) given by

$$x_1 = \tfrac{2\lambda^2}{\lambda^2+(1-\lambda)^2}\, \chi_{[0,\frac{1}{2}]} + \tfrac{2(1-\lambda)^2}{\lambda^2+(1-\lambda)^2}\, \chi_{(\frac{1}{2},1]}$$

and

$$x_2 = \tfrac{2(1-\lambda)^2}{\lambda^2+(1-\lambda)^2}\, \chi_{[0,\frac{1}{2}]} + \tfrac{2\lambda^2}{\lambda^2+(1-\lambda)^2}\, \chi_{(\frac{1}{2},1]}$$

is a Walrasian equilibrium supported by the price $\mathbf{1} \in L_q[0,1]$.

15. Consider a Riesz economy with Riesz pair $\langle L_p(T,\Sigma,\mu), L_q(T,\Sigma,\mu)\rangle$, where $1 \le p, q \le \infty$ satisfy $\frac{1}{p} + \frac{1}{q} = 1$ and (T,Σ,μ) a finite measure space, having m consumers and with total endowment $\omega = \mathbf{1}$. Assume that for each i there exists some function $f_i \in L_q^+(T,\Sigma,\mu)$ such that for each $x \in L_p^+(T,\Sigma,\mu)$ we have

$$u_i(x) = \int_T f_i(t) x(t)\, d\mu(t)\,.$$

Also, assume that $\omega_i = \chi_{A_i}$ for each i, where $A_1, \ldots, A_m$ are pairwise disjoint measurable sets with $T = \bigcup_{i=1}^m A_i$ and

$$\max\{f_1(t), f_2(t), \ldots, f_m(t)\} = f_i(t)$$

for each $t \in A_i$ $(i = 1, \ldots, m)$.

Show that the allocation $(\chi_{A_1}, \chi_{A_2}, \ldots, \chi_{A_m})$ is a Walrasian equilibrium supported by the price p defined by

$$p(t) = \max\{f_1(t), f_2(t), \ldots, f_m(t)\}\,, \quad t \in T\,.$$

16. (Araujo [**33**]) Consider a Riesz economy with Riesz pair $\langle \ell_\infty, \ell_1\rangle$ having two consumers with utility functions

$$u_1(x_1, x_2, \ldots) = \liminf_{n\to\infty} x_n \quad \text{and} \quad u_2(x_1, x_2, \ldots) = \sum_{n=1}^{\infty} \tfrac{x_n}{2^n}\,,$$

and total endowment $\omega = (1, 1, 1, \ldots)$. Establish the following.

(a) The utility function u_1 is monotone, concave, and has ω as a desirable bundle, but it fails to be Mackey continuous.

(b) The utility function u_2 is strictly monotone, concave and Mackey continuous.

(c) The utility space of this economy is $\mathbf{U} = [0,1] \times [0,1) \cup \{(0,1)\}$.

(d) If $\omega_1 > 0$ and $\omega_2 > 0$ hold, then the economy has no quasiequilibria.

17. (Jones [**66**]) Consider a Riesz economy with Riesz pair $\langle \ell_2, \ell_2 \rangle$ and one consumer with initial endowment $\omega = \left(1, \frac{1}{2^2}, \frac{1}{3^2}, \ldots\right)$ and utility function $u \colon \ell_2^+ \to \mathbb{R}$ defined by

$$u(x_1, x_2, \ldots) = \sum_{n=1}^{\infty} \tfrac{1-e^{-n^2 x_n}}{n^2} .$$

(a) Show that the utility function u is strictly monotone, strictly concave and weakly continuous.

(b) Show that the allocation (ω) is the only Edgeworth equilibrium of the economy.

(c) Show that the price $0 \ll p = (1, 1, 1, \ldots) \in L'_\omega$ supports the allocation (ω) on L_ω as a Walrasian equilibrium.

(d) Show that, aside from a scalar multiple, the only price that supports (ω) on L_ω as a Walrasian equilibrium is the strictly positive price $p = (1, 1, 1, \ldots)$.

(e) Show that the preference $\succeq$ represented by the utility function u is not uniformly ℓ_2-norm proper.

(f) Study this problem when the Riesz pair $\langle \ell_2, \ell_2 \rangle$ is replaced by the Riesz pair $\langle \ell_p, \ell_q \rangle$; $1 \le p,\ q \le \infty$, $\frac{1}{p} + \frac{1}{q} = 1$.

18. Prove Lemma 8.76

19. Prove Lemma 8.81

CHAPTER 9

Solutions to the Exercises

9.1. Chapter 1: The Lattice Structure of Riesz Spaces

1.1. Assume that there exists an element a in an ordered vector space E such that $u \vee a$ exists in E for each $u \in E$. If $x, y \in E$, then we claim that

$$x \vee y = [(x + a - y) \vee a] + (y - a)$$

holds in E. To see this, let $z = [(x + a - y) \vee a] + (y - a)$. This implies

$$z + a - y = (x + a - y) \vee a\,,$$

and so $x + a - y \leq z + a - y$ and $a \leq z + a - y$. Hence $x \leq z$ and $y \leq z$. Therefore z is an upper bound of the set $\{x, y\}$.

To see that z is the least upper bound of the set $\{x, y\}$ assume that an element $v \in E$ satisfies $x \leq v$ and $y \leq v$. It follows that

$$v + a - y \geq v + a - v = a \quad \text{and} \quad v + a - y \geq x + a - y\,,$$

and so $v + a - y \geq (x + a - y) \vee a$ or $v \geq [(x + a - y) \vee a] + (y - a) = z$. This shows that $z = x \vee y$ holds and so E is a Riesz space.

The converse is clearly true. As a matter of fact, if E is a Riesz space, then every $a \in E$ has the property that $x \vee a$ exists in E for each $x \in E$. ■

1.2. We shall prove only (a). Let $s = \sup A$. This implies $a \leq s$ or $-s \leq -a$ for all $a \in A$, that is, $-s$ is a lower bound of the set $-A$. To see that $-s$ is the greatest lower bound of $-A$ assume that some element t satisfies $t \leq -a$ for all $a \in A$. Therefore, $a \leq -t$ for all $a \in A$, and from this we infer that $s \leq -t$ or $t \leq -s$. Consequently, $\inf(-A) = -s$. ■

1.3. (a) We have:

$$\begin{aligned}(u - v)^+ + u \wedge v &= [u \vee v - v] + u \wedge v\\ &= [u \vee v + u \wedge v] - v\\ &= [u + v] - v\\ &= u\,.\end{aligned}$$

(b) Notice that

$$
\begin{aligned}
&|u+v| \wedge |u-v| \\
&\quad = \big[(u+v) \vee (-u-v)\big] \wedge \big[(u-v) \vee (v-u)\big] \\
&\quad = \big\{\big[(u+v) \vee (-u-v)\big] \wedge (u-v)\big\} \vee \big\{\big[(u+v) \vee (-u-v)\big] \wedge (v-u)\big\} \\
&\quad = \big[(u+v) \wedge (u-v)\big] \vee \big[(-u-v) \wedge (u-v)\big] \vee \cdots \\
&\qquad\quad \cdots \vee \big[(u+v) \wedge (v-u)\big] \vee \big[(-u-v) \wedge (v-u)\big] \\
&\quad = \big[u+v \wedge (-v)\big] \vee \big[-v+(-u) \wedge u\big] \vee \cdots \\
&\qquad\quad \cdots \vee \big[v+u \wedge (-u)\big] \vee \big[-u+(-v) \wedge v\big] \\
&\quad = \Big\{\big[u+v \wedge (-v)\big] \vee \big[-u+v \wedge (-v)\big]\Big\} \vee \cdots \\
&\qquad\quad \cdots \vee \Big\{\big[-v+(-u) \wedge u\big] \vee \big[v+u \wedge (-u)\big]\Big\} \\
&\quad = \big[u \vee (-u) + v \wedge (-v)\big] \vee \big[(-v) \vee v + u \wedge (-u)\big] \\
&\quad = \big[|u| - |v|\big] \vee \big[|v| - |u|\big] = \big||u| - |v|\big|,
\end{aligned}
$$

and we are finished. ■

1.4. Take $L = \mathbb{R}^2$ and as convex set the line segment with end points $A = (-2, 0)$ and $B = (1, 2)$.

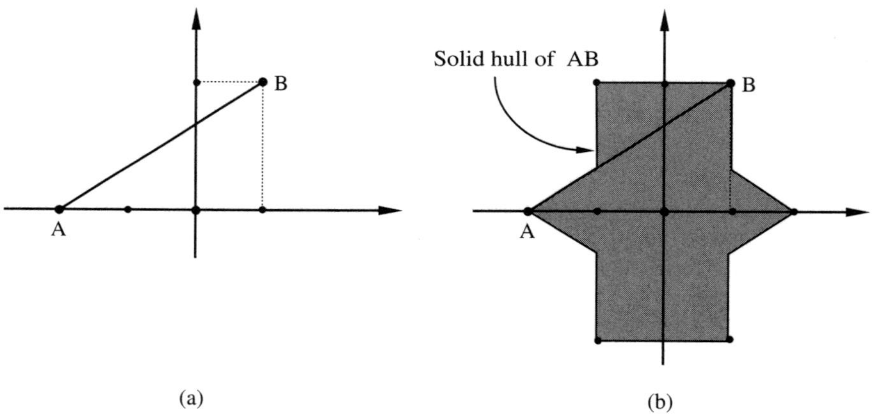

The line segment AB is shown in Figure (a). Its solid hull is shown in Figure (b). Clearly, the solid hull of the line segment AB (which is obviously a convex set) is not a convex subset of $\mathbb{R}^2$. ■

1.5. (i) Use the inequalities: $\big||u|-|v|\big| \le |u-v|$, $|u^+-v^+| \le |u-v|$, and $|u^- - v^-| \le |u-v|$.

(ii) Use the inequality: $|\lambda u + \mu v| \le |\lambda||u| + |\mu||v|$.

(iii) Use the inequalities:

$$|u \vee v - u_1 \vee v_1| \le |u - u_1| + |v - v_1| \quad \text{and} \quad |u \wedge v - u_1 \wedge v_1| \le |u - u_1| + |v - v_1|.$$

(iv) If $u_\alpha \uparrow u$, then let $v_\alpha = u - u_\alpha$ and note that $v_\alpha \downarrow 0$ and that $|u - u_\alpha| = v_\alpha$ holds for each α.

(v) Assume that $u_\alpha \xrightarrow{o} u$ and $u_\alpha \le w$ holds for all $\alpha \succeq \alpha_1$. Choose a net $\{v_\alpha\}$ with $|u_\alpha - u| \le v_\alpha \downarrow 0$. Now if $\alpha \succeq \alpha_1$, then we have

$$0 \le (u-w)^+ \le (u-u_\alpha)^+ \le |u-u_\alpha| \le v_\alpha .$$

Since α is arbitrary, it follows that $(u-w)^+ = 0$. This implies $u \le w$.

(vi) For each $\alpha \succeq \beta$ we have $u_\alpha - u_\beta \ge 0$ and $u_\alpha - u_\beta \xrightarrow[\alpha \succeq \beta]{o} u - u_\beta$. So, by (v), we get $u - u_\beta \ge 0$ or $u_\beta \le u$ for all β. That is, u is an upper bound of the net $\{u_\alpha\}$. Now if $u_\alpha \le v$ holds for all α, then it follows from (v) again that $u \le v$. Therefore, $u_\alpha \uparrow u$. ■

1.6. Note first that if $u, v \in U$, then $u \vee v \in U$. Indeed, if $\{u_\alpha\} \subseteq J$ and $\{v_\beta\} \subseteq J$ satisfy $0 \le u_\alpha \uparrow u$ and $0 \le v_\beta \uparrow v$, then the net $\{u_\alpha \vee v_\beta\} \subseteq J$ satisfies $u_\alpha \vee v_\beta \uparrow u \vee v$, so that $u \vee v \in U$. Also, if $v \in L$ satisfies $0 \le v \le u$ for some $u \in U$, then $v \in U$. To see this, pick a net $\{u_\alpha\} \subseteq J$ with $0 \le u_\alpha \uparrow u$. Then the net $\{u_\alpha \wedge v\} \subseteq J$ satisfies $0 \le u_\alpha \wedge v \uparrow u \wedge v = v$, and so $v \in U$.

Now let $C = U - U = \{w \in L\colon\ \exists\, u, v \in U \text{ with } w = u - v\}$. It should be clear that C is a vector subspace of L satisfying $J \subseteq C \subseteq B$. Now by virtue of $(u-v)^+ = u \vee v - v$ and the first remark, it follows that C is a Riesz subspace of L. Also if $u, v \in U$ satisfy $u - v \ge 0$, then from $0 \le u - v \le u \in U$ and the second remark, it follows that $u - v \in U$. Therefore, $C^+ = U$. Altogether the above show that C is an ideal of L.

Finally, we show that C is a band of L. Let $\{u_\alpha\} \subseteq C^+ = U$ be such that $0 \le u_\alpha \uparrow u$ in L. Each u_α is the supremum of a net from J^+. So, taking the net consisting of all finite suprema of the elements in these nets, we get a net $\{w_\lambda\}$ of J^+ satisfying $0 \le w_\lambda \uparrow u$ in L. Hence $u \in U$, and therefore C is a band. Thus $B = C$.

For the proof of the σ-ideal generated by J in L repeat the above arguments by letting $U = \big\{u \in L^+\colon\ \exists \text{ a sequence } \{u_n\} \subseteq J \text{ with } 0 \le u_n \uparrow u \text{ in } L\big\}$. ■

1.7. (i) $\Longrightarrow$ (ii) From the lattice identity $(u - u\wedge v) \wedge (v - u \wedge v) = 0$, we easily obtain $[\pi(u) - \pi(u \wedge v)] \wedge [\pi(v) - \pi(u\wedge v)] = 0$. This implies $\pi(u \wedge v) = \pi(u) \wedge \pi(v)$.

(ii) $\Longrightarrow$ (iii) Using the lattice identity $u \vee v = -[(-u)\wedge(-v)]$, we get

$$\pi(u\vee v) = -\big[(-\pi(u)) \wedge (-\pi(v))\big] = \pi(u) \vee \pi(v) .$$

(iii) $\Longrightarrow$ (iv) Obvious.

(iv) $\Longrightarrow$ (v) For this implication we need the following property.

- *If* $u = v - w$ *and* $v \wedge w = 0$, *then* $v = u^+$ *and* $w = u^-$.

To see this, note that $u^+ = (v-w)^+ = v \vee w - w = [v + w - v \wedge w] - w = v$.

Now fix some $u \in L$ and observe that $u^+ \wedge u^- = 0$. Consequently, we have

$$\begin{aligned} \pi(u^+) \vee \pi(u^-) &= \pi(u^+ \vee u^-) = \pi(|u|) = \pi(u^+ + u^-) = \pi(u^+) + \pi(u^-) \\ &= \pi(u^+) \vee \pi(u^-) + \pi(u^+) \wedge \pi(u^-) , \end{aligned}$$

and therefore $\pi(u^+)\wedge\pi(u^-) = 0$. Now note that $u = u^+ - u^-$ implies $\pi(u) = \pi(u^+) - \pi(u^-)$ and consequently, by the property already stated above, we get $[\pi(u)]^+ = \pi(u^+)$.

(v) $\Longrightarrow$ (vi) Clearly $\pi(|u|) = \pi(u^+) + \pi(u^-) = [\pi(u)]^+ + [\pi(u)]^- = |\pi(u)|$.

(vi) $\Longrightarrow$ (i) If $u \wedge v = 0$, then

$$\begin{aligned} \pi(u) \wedge \pi(v) &= \tfrac{1}{2}\big[\pi(u) + \pi(v) - |\pi(u) - \pi(v)|\big] \\ &= \tfrac{1}{2}\big[\pi(u) + \pi(v) - |\pi(u - v)|\big] \\ &= \tfrac{1}{2}\big[\pi(u) + \pi(v) - \pi(|u - v|)\big] \\ &= \pi\big(\tfrac{1}{2}[u + v - |u - v|]\big) = \pi(u \wedge v) = \pi(0) = 0 , \end{aligned}$$

and so π is a Riesz homomorphism. ■

1.8. Let J be an ideal in a Riesz space L. Keep in mind that J is a regular Riesz subspace of L.

(i) *L is super Dedekind complete.* Let $0 \leq u_\alpha \uparrow \leq u$ in J. There exists some $v \in L$ with $u_\alpha \uparrow v \leq u$ in L, and moreover, there exists an increasing sequence of indices $\{\alpha_n\}$ such that $u_{\alpha_n} \uparrow v$ in L. It follows that $v \in J$, $u_\alpha \uparrow v$ and $u_{\alpha_n} \uparrow v$ in J, so that J is super Dedekind complete.

(ii) *L is Dedekind complete or Dedekind σ-complete.* Repeat the preceding arguments.

(iii) *L is almost Dedekind σ-complete.* In this case L is super order dense in its Dedekind σ-completion L^σ. Let J_σ be the ideal generated by J in L^σ. Then J_σ is Dedekind σ-complete and J is super order dense in J_σ. To verify the last assertion, let $0 < u \in J_\sigma$; this means that there exists some $v \in J$ with $u \leq v$. Since L is super order dense in L^σ, there exists a sequence $\{u_n\} \subseteq L$ with $0 \leq u_n \uparrow u$ in L^σ. But since J is an ideal of L, we have $\{u_n\} \subseteq J$. Therefore $J_\sigma = J^\sigma$, and consequently J is almost Dedekind σ-complete.

(iv) *L has the projection property.* Let B be a band of J and denote by A the band generated by B in L. From Exercise **1.6**, it follows that $A \cap J = B$.

Now let $0 \leq u \in J$. Since $L = A \oplus A^{\mathrm{d}}$, we can write $u = u_1 + u_2$, where $0 \leq u_1 \in A$ and $0 \leq u_2 \in A^{\mathrm{d}}$. Since J is an ideal, it follows that $u_1, u_2 \in J$, and so $u_1 \in A \cap J = B$. Hence $u \in B \oplus B^{\mathrm{d}}$. That is, $J = B \oplus B^{\mathrm{d}}$ holds in J, where now B^{d} denotes the disjoint complement of B in J, and so J has the projection property.

(v) *L has the principal projection property.* Repeat the previous arguments.

(vi) *L has sufficiently many projections.* Let B be a nonzero band of J. If A is the band generated by B in L, then $A \cap J = B$. Pick a nonzero projection band C of L with $C \subseteq A$. Then $C \cap J$ is a nonzero projection band of J satisfying $C \cap J \subseteq B$. Hence J has sufficiently many projections.

(vii) *L is Archimedean.* In this case J is obviously Archimedean. ■

1.9. It should be clear that the Dedekind completion of L is $L^\delta = \ell_\infty(X)$. The Dedekind σ-completion of L consists of all functions $f \in \ell_\infty(X)$ for which there exists a constant $f(\infty)$ such that the set $\{x \in X\colon\ f(x) \neq f(\infty)\}$ is at most countable.

To see the last claim note first that L is order dense in

$$K = \bigl\{f \in \ell_\infty(X)\colon\ \exists\, f(\infty) \in \mathbb{R} \text{ such that } \{x \in X\colon\ f(x) \neq f(\infty)\} \text{ is at most countable}\bigr\},$$

and that K is Dedekind σ-complete. (Note also that order convergence in L implies pointwise convergence.) So, we have $L \subseteq L^\sigma \subseteq K \subseteq L^\delta = \ell_\infty(X)$. Now let $0 < u \in K$. Consider the function $w \in K$, defined by $w(x) = u(x)$ if $u(x) \geq u(\infty)$ and $w(x) = u(\infty)$ if $u(x) < u(\infty)$. Similarly, consider the function $v \in K$ defined by $v(x) = 0$ if $u(x) \geq u(\infty)$ and $v(x) = u(\infty) - u(x)$ if $u(x) < u(\infty)$. Clearly, $u = w - v$, and so if we show that $w, v \in L^\sigma$, then $u \in L^\sigma$, and so $L^\sigma = K$.

Let $E = \{x \in X\colon\ u(x) > u(\infty)\}$, and note that E is at most countable. If E is finite, then $w \in L \subseteq L^\sigma$. Now assume that $E = \{x_1, x_2, \ldots\}$ is a countable set. Define $u_n \in L$ by $u_n(x) = u(\infty)$ if $x \neq x_1, \ldots, x_n$ and $u_n(x_i) = u(x_i)$ for $i = 1, \ldots, n$. Then $0 \leq u_n \uparrow w$ in L^δ, and so $w \in L^\sigma$. Similarly, $v \in L^\sigma$.

Note also that L is not super order dense in L^σ. Since L has the principal projection property, this provides another example of a Riesz space with the principal projection property that is not almost Dedekind σ-complete. ■

1.10. Note that since L majorizes M, the Riesz space L also majorizes M/J. The desired conclusion now follows from Theorem 1.49. ■

1.11. Note first that order convergence in L implies pointwise convergence. Now assume by way of contradiction that L is almost Dedekind σ-complete; so that L is super order dense (as well as majorizing) in L^σ.

Let $\{r_1, r_2, \ldots\}$ be an enumeration of the rational numbers and let u_n be the characteristic function of the set $\{r_1, \ldots, r_n\}$. Then $0 \le u_n \uparrow \le \mathbf{1}$ in L, where $\mathbf{1}$ denotes the constant function one. There exists some $u^* \in L^\sigma$ with $u_n \uparrow u^* \le \mathbf{1}$ in L^σ. Choose a sequence $\{v_n\} \subseteq L$ such that $v_n \downarrow u^*$ in L^σ; there is no loss of generality in assuming that $v_n \le \mathbf{1}$ holds for all n. It easily follows that $v_n(r) = 1$ for all rational numbers r and all n and that $v_n - u_n \downarrow 0$ in L. Since each v_n (as an element of L) has at most countably many discontinuities, there exists an irrational number x where every v_n is continuous. But $u_n(x) = 0$ for all n implies that $v_n(x) = v_n(x) - u_n(x) \downarrow 0$ in $\mathbb{R}$, since order convergence in L implies pointwise convergence. In particular, for some sufficiently large $k \in \mathbb{N}$ we have $v_k(x) < \frac{1}{2}$ and so by the continuity of v_k at x we have $v_k(y) < \frac{1}{2}$ for all y in a neighborhood of x, contrary to $v_k(r) = 1$ for all rational numbers r. This contradiction shows that L cannot be almost Dedekind σ-complete. ■

1.12. (i) Fix a discrete element $u > 0$ and assume that $0 \le v \le u$ and $0 \le w \le u$ satisfy $v \wedge w = 0$. Since u is a discrete element, there exist two non-negative constants λ and μ such that $v = \lambda u$ and $w = \mu u$. But then

$$0 = v \wedge w = (\lambda u) \wedge (\mu u) = \min\{\lambda, \mu\} u$$

implies $\min\{\lambda, \mu\} = 0$, so that either $v = 0$ or $w = 0$. Therefore, u is an atom. (These arguments show that a discrete element is always an atom.)

To see that the converse is not true, let L be the lexicographic plane. Note that since L is totally ordered, every nonzero element of L is an atom. However, every nonzero positive element of L which does not lie on the y-axis is not a discrete element.

(ii) By the previous part every discrete element is an atom. Now let L be Archimedean and let $0 < u \in L$ be an atom. Assume that $v \in L$ satisfies $0 < v \le u$ and put $S = \{\mu \in \mathbb{R}^+ \colon\ \mu v \le u\}$. Since $1 \in S$, it follows that S is a nonempty. The fact that L is Archimedean implies that S is bounded from above in $\mathbb{R}^+$. Put $\lambda = \sup S > 0$ and note that the Archimedean property implies

$$\lambda v \le u\,. \tag{$\star$}$$

Now let $0 < \epsilon < 1$. Then $[(\lambda + \epsilon)v - u]^+ \le (\lambda + 1)v \le (\lambda + 1)u$ and $[(\lambda + \epsilon)v - u]^+ > 0$. Also, $[(\lambda + \epsilon)v - u]^- = [u - (\lambda + \epsilon)v]^+ \le u \le (\lambda + 1)u$ and

$$[(\lambda + \epsilon)v - u]^+ \wedge [(\lambda + \epsilon)v - u]^- = 0\,.$$

But then since $(\lambda + 1)u$ is also an atom, it follows that $[(\lambda + \epsilon)v - u]^- = 0$ or $u \le (\lambda + \epsilon)v$ for all $0 < \epsilon < 1$. Now let $\epsilon \to 0^+$ and use once more that L is Archimedean to get $u \le \lambda u$. The latter in conjunction with ($\star$) yields $u = \lambda v$ or $v = \frac{1}{\lambda} u$. This shows that u is a discrete element.

(iii) To show that $C[0, 1]$ does not contain any discrete elements, it is enough (according to the preceding part) to establish that $C[0, 1]$ does not have any atoms. To this end, let $0 < u \in C[0, 1]$. Then there exists $\epsilon > 0$ and an open interval $(a, b) \subseteq [0, 1]$ such that $u(x) \ge \epsilon$ holds for all $x \in (a, b)$. But then it is easy to construct two nonzero orthogonal functions v and w in $C[0, 1]$ which are less than $\epsilon \chi_{(a,b)}$ (and thus below u). Hence u cannot be an atom. ■

1.13. We show first that L is a discrete space. Let $0 < u \in L$. Assume that there exists no discrete element of L^+ below u. By the previous exercise, and an easy inductive argument, we can construct a disjoint sequence of nonzero positive elements of L below u, contradicting our hypothesis. Hence L is discrete.

By Theorem 1.78 the Riesz space L is Riesz isomorphic to an order dense Riesz subspace of some $\mathbb{R}^X$, where X has the same cardinality as any complete disjoint system of discrete elements of L. It now easily follows that X is finite and that L is Riesz isomorphic to $\mathbb{R}^X$. ■

1.14. Assume that $L \neq \{0\}$. Since L is totally ordered every disjoint subset of L has at most two elements (if two, then one must necessarily be the zero element). Hence, by the previous exercise, L is Riesz isomorphic to some $\mathbb{R}^X$, where X has the same cardinality as any complete disjoint system of discrete elements. In this case, X must be a singleton, and so L is Riesz isomorphic to $\mathbb{R}$. ■

1.15. (a) $\Longrightarrow$ (b) Let A be an arbitrary nonempty bounded above subset of L and put $u = \sup A$ in L^δ. We must show that there exists an at most countable subset B of A satisfying $u = \sup B$.

If we let $C = \{v \in L\colon\ v \geq u\}$, then we have $\sup(A - C) = 0$ or $\inf(C - A) = 0$. Pick an at most countable subset D of $A - C$ such that $\sup D = 0$ [or $\inf(-D) = 0$] and write $D = \{a_i - c_i\colon\ i \in I\}$, where the index set I is at most countable. Now from $0 \leq u - a_i \leq c_i - a_i$, it follows that the at most countable subset $B = \{a_i\colon\ i \in I\}$ of A satisfies $u = \sup B$.

(b) $\Longrightarrow$ (c) Let A be a nonempty bounded above disjoint subset of L^+. Put $u = \sup A$ in L^δ. Next, pick an at most countable subset B of the set A such that $\sup B = u$. We claim that $A = B$. If this is established, then A is clearly at most countable.

To see this, let $0 < a \in A$ and assume by way of contradiction that $a \notin B$. Now if C is a finite subset of B, then note that

$$a + \bigvee_{c \in C} c = a \vee \Big[\bigvee_{c \in C} c\Big] \leq \sup A = u\,. \tag{$\star$}$$

Since $u = \sup B = \sup\{\bigvee_{c \in C} c\colon\ C \text{ is a finite subset of } B\}$, it easily follows from ($\star$) that $a + u \leq u$ or $a \leq 0$, a contradiction. Hence $a \in B$ and so $A = B$.

(c) $\Longrightarrow$ (a) Let A be a nonempty subset of L such that $u = \sup A$ exists in L. Replacing A by the collection of all finite suprema of elements of A we can assume that $A \uparrow u$. Also, if $a_0 \in A$ is any fixed element and we consider the set $A_0 = \{a - a_0\colon\ a \in A \text{ and } a \geq a_0\}$, then $A_0 \subseteq L^+$ and $\sup A_0 = \sup A - a_0$. This shows that by replacing A by A_0 if necessary, we can assume from the outset that A is a subset of L^+ and that $A \uparrow u$.

Fix any $0 < \delta < 1$. We claim that there exists an at most countable subset B_δ of A such that

$$\delta u = \bigvee_{a \in B_\delta} a \wedge (\delta u) \tag{$\star\star$}$$

To establish the validity of ($\star\star$), consider the set $S = \bigcup_{a \in A} \big[0, (a - \delta u)^+\big]$. If we consider the collection of all disjoint subsets of S, then Zorn's lemma guarantees the existence of a maximal disjoint subset, say C, of S. Since $\big[0, (a - \delta u)^+\big] \subseteq [0, u]$ holds for each $a \in A$, it follows that C is bounded above. So, by our hypothesis, the set C is at most countable. For each $c \in C$ choose some $a_c \in A$ such that $0 \leq c \leq (a_c - \delta u)^+$ and note that the set $B_\delta = \{a_c\colon\ c \in C\}$ is at most countable. We shall show next that this set B_δ satisfies ($\star\star$).

If $(\star\star)$ is not valid for the above choice of B_δ, then there exists some $0 < v \leq \delta u$ such that $a_c \wedge \delta u \leq \delta u - v$. This implies

$$0 < v \leq \delta u - a_c \wedge \delta u = (a_c - \delta u)^-$$

for all $c \in C$. In particular, we have $v \wedge c = 0$ for all $c \in C$. Since C is a maximal disjoint subset of S, the latter implies $v \wedge (a - \delta u)^+ = 0$ for all $a \in A$. But then, in view of $A \uparrow u$, we get $v \wedge (1 - \delta)u = 0$. Since $1 - \delta > 0$ this yields $v \wedge u = 0$, from which it follows that $0 < v = v \wedge \delta u = 0$, a contradiction. This contradiction establishes the validity of $(\star\star)$.

To complete the solution, choose a sequence $\{\delta_n\} \subseteq (0,1)$ with $\delta_n \uparrow 1$ and consider the at most countable set $B = \bigcup_{n=1}^\infty B_{\delta_n}$. Then, working in L^δ, we see that

$$\begin{aligned} u &= \bigvee_{n=1}^\infty \delta_n u = \bigvee_{n=1}^\infty \bigvee_{a \in B_{\delta_n}} a \wedge \delta_n u \leq \bigvee_{n=1}^\infty \bigvee_{a \in B} a \wedge \delta_n u \\ &= \bigvee_{a \in B} \bigvee_{n=1}^\infty a \wedge \delta_n u = \bigvee_{a \in B} a \wedge u = \bigvee_{a \in B} a = u\,. \end{aligned}$$

Therefore, $\sup B = u = \sup A$, and the solution is finished. ■

1.16. Let $T\colon L \to M$ be an operator between two Riesz spaces that is one-to-one and onto. If T is a Riesz isomorphism, then clearly T and T^{-1} are both positive operators.

For the converse, assume that T and T^{-1} are both positive operators and let $u \wedge v = 0$. Put $w = T(u) \wedge T(v)$ and note that the positivity of T guarantees that $w \geq 0$. We must show that $w = 0$. From $0 \leq w \leq T(u)$, $0 \leq w \leq T(v)$, and the positivity of T^{-1} we get $0 \leq T^{-1}(w) \leq u$ and $0 \leq T^{-1}(w) \leq v$. This implies $T^{-1}(w) = 0$, and so $w = T(T^{-1}(w)) = 0$. ■

1.17. (a) $\Longrightarrow$ (b) Obvious.

(b) $\Longrightarrow$ (c) Let $u, v \in L^+$ and put $w = T(u) \wedge [v - T(v)] \geq 0$. From the inequality $w \leq [v - T(v)]$, we obtain

$$0 \leq T(w) \leq [T(I - T)](v) = (T - T^2)(v) = 0\,,$$

and thus $T(w) = 0$. Similarly, $[I - T](w) = 0$, and so we have $w = T(w) + [I - T](w) = 0$. This shows that the operators T and $I - T$ have disjoint ranges.

(c) $\Longrightarrow$ (a) Let A and B be the ideals generated by the ranges of T and $I - T$, respectively. From our hypothesis, it easily follows that $A \perp B$, and from the identity $u = T(u) + [I - T](u)$ we obtain that $L = A \oplus B$. The latter easily implies that $A = B^{\mathrm{d}}$ and $B = A^{\mathrm{d}}$, and hence both A and B are projection bands. Finally, note that the identity

$$P_A(u) - T(u) = P_A(u) - P_A(T(u)) = P_A\big(u - T(u)\big) = 0$$

shows that $T = P_A$ holds, i.e., that T is a band projection. ■

1.18. (a) We verify first that A^{d} is a projection band. Since A is a projection band, we have $L = A \oplus A^{\mathrm{d}}$. We claim that $A^{\mathrm{dd}} = A$, and this is enough to establish that A^{d} is a projection band. Clearly, $A \subseteq A^{\mathrm{dd}}$. Now let $u \in A^{\mathrm{dd}}$ and write $u = u_1 + u_2$ with $u_1 \in A$ and $u_2 \in A^{\mathrm{d}}$. It follows that $u_2 \in A^{\mathrm{d}} \cap A^{\mathrm{dd}} = \{0\}$, and hence $u = u_1 \in A$. Consequently, $A^{\mathrm{dd}} \subseteq A$ and so $A^{\mathrm{dd}} = A$. The identity $P_{A^{\mathrm{d}}} = I - P_A$ should be obvious.

To see that $A \cap B$ is a projection band, let $u \in L^+$. From $B \cap [0, u] = [0, P_B(u)]$, it follows that $A \cap B \cap [0, u] = A \cap [0, P_B(u)]$ and that

$$[P_A P_B](u) = \sup A \cap [0, P_B(u)] = \sup A \cap B \cap [0, u]\,.$$

This establishes (according to Theorem 1.46) that the band $A \cap B$ is a projection band and that $P_{A \cap B}(u) = [P_A P_B](u)$ holds for all $u \in L^+$; and so $P_{A \cap B} = P_A P_B = P_B P_A$.

Next we shall prove that $A+B$ is a projection band. Assume first that $A\cap B=\{0\}$. Let $u\in L^+$. If $0\le a+b\in A\oplus B$ satisfies $a+b\le u$, then $a\in A\cap[0,u]$ and $b\in B\cap[0,u]$ and so $a+b\le P_A(u)+P_B(u)\in A\oplus B$. This implies

$$\sup(A+B)\cap[0,u]=P_A(u)+P_B(u)\,.$$

Now a glance at Theorem 1.46 guarantees that $A+B$ is a projection band satisfying $P_{A+B}=P_A+P_B$.

The general case can be established by observing first that $A\cap B^{\mathrm d}$ and B are projection bands and then that $A+B=(A\cap B^{\mathrm d})+B$. By the preceding case, we have

$$\begin{aligned} P_{A+B} &= P_{(A\cap B^{\mathrm d})+B}=P_{A\cap B^{\mathrm d}}+P_B=P_AP_{B^{\mathrm d}}+P_B=P_A(I-P_B)+P_B\\ &= P_A-P_AP_B+P_B=P_A+P_B-P_{A\cap B}\,. \end{aligned}$$

(b) Let $u,v\in L^+$. Clearly, $P_A(u)\wedge P_B(u)\in A\cap B$ and if $w\in A\cap B$ satisfies $w\le u$, then $w\le P_A(u)\wedge P_B(u)$. This shows that

$$P_{A\cap B}(u)=\sup\{w\in A\cap B\colon\ 0\le w\le u\}=P_A(u)\wedge P_B(u)\,.$$

Similarly, we have

$$P_A(u)\wedge P_B(v)=\sup\{w\in A\cap B\colon\ 0\le w\le u\wedge v\}=P_{A\cap B}(u\wedge v)\,,$$

and the solution is finished. ∎

1.19. It is not difficult to check that

$$\mathcal P(T+S)=(T+S)_A=T_A+S_A=\mathcal P(T)+\mathcal P(S)$$

holds for all $S,T\in\mathcal L_{\mathrm b}^+(L,M)$. So, the mapping $\mathcal P\colon\mathcal L_{\mathrm b}^+(L,M)\to\mathcal L_{\mathrm b}^+(L,M)$ is additive, and consequently it extends (uniquely) via the formula

$$\mathcal P(T)=\mathcal P(T^+)-\mathcal P(T^-)$$

to a positive operator from $\mathcal L_{\mathrm b}(L,M)$ to $\mathcal L_{\mathrm b}(L,M)$; see Lemma 1.64. Now notice that for each $T\in\mathcal L_{\mathrm b}^+(L,M)$ we have $T=T_A$ on A, and from this it is easy to see that $\mathcal P^2(T)=(T_A)_A=T_A=\mathcal P(T)$. That is, $\mathcal P$ is a positive projection on $\mathcal L_{\mathrm b}(L,M)$. Next note that for each $T\in\mathcal L_{\mathrm b}^+(L,M)$ we have $0\le\mathcal P(T)=T_A\le T=\mathcal I(T)$, where $\mathcal I$ denotes the identity operator on $\mathcal L_{\mathrm b}(L,M)$. Now a glance at Exercise **1.17** guarantees that $\mathcal P$ is a band projection on $\mathcal L_{\mathrm b}(L,M)$. ∎

1.20. (a)$\Longrightarrow$(b) If $u_\alpha\downarrow 0$, then $u_\alpha\xrightarrow{o}0$, and so $T(u_\alpha)\xrightarrow{o}0$.

(b)$\Longrightarrow$(c) Assume that $u_\alpha\downarrow 0$ in L and let some $y\in M$ satisfy $y\le|T(u_\alpha)|$ for each α. Since $T(u_\alpha)\xrightarrow{o}0$, there is a net $\{v_\alpha\}$ satisfying $|T(u_\alpha)|\le v_\alpha\downarrow 0$. This implies $y\le v_\alpha$ for each α, and so $y\le 0$. This shows that $\inf_\alpha|T(u_\alpha)|=0$ in M.

(c)$\Longrightarrow$(d) It suffices to establish that T^+ is order continuous. To this end, assume that $u_\alpha\downarrow 0$ in L. Let $0\le w\le T^+(u_\alpha)$ for each α. To show that T^+ is order continuous, we must prove that $w=0$.

Fix some index α_0, and let $u=u_{\alpha_0}$. Clearly, for each $0\le v\le u$ and all $\alpha\succeq\alpha_0$ we have $0\le v-v\wedge u_{\alpha_0}=v\wedge u-v\wedge u_\alpha\le u-u_\alpha$. This implies

$$T(v)-T(v\wedge u_\alpha)=T(v-v\wedge u_\alpha)\le T^+(u-u_\alpha)=T^+(u)-T^+(u_\alpha)\,,$$

and consequently

$$0\le w\le T^+(u_\alpha)\le T^+(u)+|T(v\wedge u_\alpha)|-T(v) \qquad (\star)$$

for each $\alpha\succeq\alpha_0$ and all $0\le v\le u$. Since for each $0\le v\le u$ we have $v\wedge u_\alpha\downarrow_{\alpha\succeq\alpha_0}0$, it follows from our hypothesis that $\inf_{\alpha\succeq\alpha_0}|T(v\wedge u_\alpha)|=0$. Now a glance at $(\star)$ yields $0\le w\le T^+(u)-T(v)$ for each $0\le v\le u$. Taking into consideration the Riesz–Kantorovich formula $T^+(u)=\sup_{0\le v\le u}T(v)$, the latter shows that $w=0$, as desired.

(d) $\Longrightarrow$ (e) Obvious.

(e) $\Longrightarrow$ (a) It follows from the inequality $|T(u)| \le |T|(|u|)$. ■

1.21. (i) Straightforward.

(ii) Assume $u \wedge v = 0$ in M. Then for each $w \in L^+$ we have

$$0 \le (T_u \wedge T_v)(w) \le T_u(w) \wedge T_v(w) = \big[\phi(w)u\big] \wedge \big[\phi(w)v\big] = \phi(w)(u \wedge v) = 0\,,$$

and hence $T_u \wedge T_v = 0$. Thus the operator $T\colon M \to \mathcal{L}_\mathrm{b}(L,M)$ is a Riesz isomorphism (into), which is obviously order continuous.

(iii) Assume that $\mathcal{L}_\mathrm{b}(L,M)$ is a Dedekind complete Riesz space, and suppose that $0 \le u_\alpha \uparrow\le u$ in M. Then $0 \le T_{u_\alpha} \uparrow\le T_u$ holds in $\mathcal{L}_\mathrm{b}(L,M)$ and so since $\mathcal{L}_\mathrm{b}(L,M)$ is Dedekind complete, there exists $T \in \mathcal{L}_\mathrm{b}(L,M)$ with $T_{u_\alpha} \uparrow T$ in $\mathcal{L}_\mathrm{b}(L,M)$. Choose $v \in L^+$ with $\phi(v) = 1$. We claim that $u_\alpha \uparrow T(v)$ holds in M. Indeed, note first that $u_\alpha = T_{u_\alpha}(v) \le T(v)$ holds for all α, and then that if $u_\alpha \le w$ for all α, the relation $T_{u_\alpha} \le T_w$ implies $T \le T_w$ and so $T(v) \le T_w(v) = w$. Thus M is Dedekind complete. ■

1.22. (i) Let $e > 0$ be an order unit of L, and let $T_\alpha \downarrow 0$ in $\mathcal{L}_\mathrm{b}(L,M)$. Then $T_\alpha(e) \downarrow 0$ holds in M. Since M is super Dedekind complete, there exists a sequence $\{T_{\alpha_n}\} \subseteq \{T_\alpha\}$ with $T_{\alpha_n}(e) \downarrow 0$ in M; we can assume without loss of generality that $0 \le T_{\alpha_n}\!\downarrow$ in $\mathcal{L}_\mathrm{b}(L,M)$. We claim that $T_{\alpha_n} \downarrow 0$ in $\mathcal{L}_\mathrm{b}(L,M)$. To see this, assume $0 \le T \le T_{\alpha_n}$ for all n. Now if $u \in L^+$, choose $\lambda > 0$ with $u \le \lambda e$ and note that $0 \le T(u) \le \lambda T(e) \le \lambda T_{\alpha_n}(e) \downarrow 0$ implies $T(u) = 0$, i.e., $T = 0$. Thus, $T_{\alpha_n} \downarrow 0$ holds in $\mathcal{L}_\mathrm{b}(L,M)$, and hence $\mathcal{L}_\mathrm{b}(L,M)$ is super Dedekind complete.

(ii) Let $e > 0$ be a weak order unit of L and let $T_\alpha \downarrow 0$ in $\mathcal{L}_\mathrm{n}(L,M)$. As in the previous part, there exists a sequence $\{T_{\alpha_n}\} \subseteq \{T_\alpha\}$ satisfying $T_{\alpha_n}\!\downarrow$ in $\mathcal{L}_\mathrm{n}(L,M)$ and $T_{\alpha_n}(e) \downarrow 0$ in M. Also (as above) if $0 \le T \le T_{\alpha_n}$ holds for all n, then $T(v) = 0$ for all $v \in L_e$. But then since L_e is order dense in the Archimedean Riesz space L (e is a weak unit of L) and $T \in \mathcal{L}_\mathrm{n}(L,M)$, it follows that $T = 0$. Hence $T_{\alpha_n} \downarrow 0$ in $\mathcal{L}_\mathrm{n}(L,M)$. ■

1.23. (i) We can assume that $T \ge 0$, otherwise we replace T by $|T|$. Since N_T^d (where $N_T = \{u \in L\colon\ T(|u|) = 0\}$) is an ideal of L and L is Dedekind σ-complete, N_T^d is likewise Dedekind σ-complete. Also, since the operator $T\colon N_T^\mathrm{d} \to M$ is strictly positive and order continuous, it follows from Theorem 1.45 that N_T^d has the countable sup property, and hence N_T^d is super Dedekind complete.

(ii) Let $0 < u \in L$ and consider the set $V = \{v \in N_T^\mathrm{d}\colon\ 0 \le v \le u\}$. Then the subset $T(V)$ of M is order bounded from above by $T(u)$, and hence $h = \sup T(V)$ exists in M. Since M is super Dedekind complete, there exists a sequence $\{v_n\} \subseteq V$ with $T(v_n) \uparrow h$; we can assume $0 \le v_n \uparrow\le u$ in L. Now since L is Dedekind σ-complete, there exists some $w \in L$ such that $v_n \uparrow w$. Clearly, $w \in N_T^\mathrm{d}$. Since $0 \le T \in \mathcal{L}_\mathrm{c}(L,M)$, we have $T(v_n) \uparrow T(w)$ in M, and so $h = T(w)$. We claim that $w = \sup V$ holds in L. For if $v \in V$, then $v \vee v_n \uparrow v \vee w$, and so it follows from the relation $T(v_n) \uparrow\le T(v \vee v_n) \uparrow\le h$, that $T(v \vee w) = T(w) = h$. Thus, $T(w \vee v - w) = 0$ which (in view of $0 \le w \vee v - w \in N_T^\mathrm{d}$) implies $w \vee v - w = 0$. Hence $v \le w \vee v = w$ holds, and so $w = \sup V$. Theorem 1.46 now shows that N_T^d is a projection band of L.

(iii) If $T \in \mathcal{L}_\mathrm{n}(L,M)$, then clearly N_T is a band of L. For the converse assume that N_T is a band of L, and that $u_\alpha \downarrow 0$ in L. By part (ii) we have $L = N_T \oplus N_T^\mathrm{d}$, and so $u_\alpha = v_\alpha + w_\alpha$ with $0 \le v_\alpha \in N_T$ and $0 \le w_\alpha \in N_T^\mathrm{d}$ for all α. In particular, we have $w_\alpha \downarrow 0$. Therefore, by part (i), we have $T(u_\alpha) = T(v_\alpha) + T(w_\alpha) = T(w_\alpha) \downarrow 0$. This shows that $T \in \mathcal{L}_\mathrm{n}(L,M)$. ■

1.24. Assume $u_n \downarrow 0$ in L. Then $u_n(x) \downarrow 0$ holds in $\mathbb{R}$ for all $x \in X$. For each n there exists an at most countable subset X_n of X such that $u_n(x) = u_n(\infty)$ for each $x \notin X_n$. Choose some $x_0 \in X \setminus \bigcup_{n=1}^{\infty} X_n$. Then $u_n(\infty) = u_n(x_0) \downarrow 0$, thus $u_n(x) \downarrow 0$ for all $x \in X_\infty$. Now by Dini's classical convergence theorem, it follows that $\{u_n\}$ converges uniformly to zero and from this we easily infer that $\phi(u_n) \downarrow 0$. Therefore, $\phi \in L_c^{\sim}$.

An easy verification shows that $N_\phi = \{u \in L\colon\ u(x_n) = 0 \text{ for all } n\}$, which is obviously a band of L. We claim that ϕ is not order continuous.

To see that $\phi \notin L_n^{\sim}$, let $\{\chi_F\}$ be the net of the characteristic functions of the finite subsets of X. Then $0 \leq \chi_F \uparrow \mathbf{1}$ (where $\mathbf{1}(x) = 1$ for all $x \in X$) holds in L, but $\phi(\chi_F) \leq 1$ for all finite subsets F and $\phi(\mathbf{1}) = 2$. Thus, $\phi \notin L_n^{\sim}$. This, however, does not contradict part (iii) of the previous exercise since the Riesz space L is not Dedekind σ-complete. ■

1.25. The solution below is due to W. A. J. Luxemburg and A. C. Zaanen [**89**, Note VII, Example 24.5(ii), p. 674]. Let $L = C[0,1]$, $0 \leq \phi \in L_c^{\sim}$, and let $\{r_1, r_2, \ldots,\}$ be an enumeration of the rational numbers of $[0,1]$. For each n choose some function $u_n \in C[0,1]$ such that:

(1) $0 \leq u_n(x) \leq 1$ for all $x \in [0,1]$.
(2) $u_n(r_n) = 1$.
(3) $u_n(x) = 0$ for all $x \in [0,1]$ with $|x - r_n| > 2^{-n}$.

Next, for each n and k let $v_{n,k} = \bigvee_{i=n}^{n+k} u_i$. Then $v_{n,k} \uparrow_k$ holds for each n, and since $v_{n,k}(r_k) = 1$ for all $k \geq n$, it follows that $v_{n,k} \uparrow_k \mathbf{1}$, where $\mathbf{1}(x) = 1$ for each $x \in [0,1]$.

Now let $\epsilon > 0$. Since $0 \leq \phi \in L_c^{\sim}$, we have $\phi(v_{n,k}) \uparrow_k \phi(\mathbf{1})$. For each n choose some k_n such that $\phi(\mathbf{1} - v_{n,k_n}) < \epsilon 2^{-n}$. Observe that by condition (3) above, the set $\{x \in [0,1]\colon\ v_{n,k}(x) > 0\}$ is at most of Lebesgue measure 2^{1-n}. For each n put $w_n = \bigwedge_{m=1}^{n} v_{m,k_m}$. Then $w_n \downarrow$ in L and the set $\{x \in [0,1]\colon\ w_n(x) > 0\}$ is at most of Lebesgue measure 2^{1-n}. This implies that $w_n \downarrow 0$ holds in L, and so $\phi(w_n) \downarrow 0$. From

$$0 \leq \mathbf{1} - w_n = \bigvee_{m=1}^{n} (\mathbf{1} - v_{m,k_m}) \leq \sum_{m=1}^{n} (\mathbf{1} - v_{m,k_m}),$$

it follows that $0 \leq \phi(\mathbf{1} - w_n) \leq \sum_{m=1}^{n} \phi(\mathbf{1} - v_{m,k_m}) < \sum_{m=1}^{n} \epsilon 2^{-m} < \epsilon$ holds for all n, and therefore $0 \leq \phi(\mathbf{1}) \leq \epsilon$ for all $\epsilon > 0$. Thus $\phi(\mathbf{1}) = 0$, and since $\mathbf{1}$ is an order unit of L, we get $\phi = 0$. Therefore, $L_n^{\sim} = L_c^{\sim} = \{0\}$.

To determine $L^{\sim}$ note first that $L^{\sim} = L'$, where L is equipped with the sup norm (see also Theorem 5.19). Now, by the classical Riesz Representation Theorem, it can be easily seen that L' can be realized as the Riesz space of all Borel measures on $[0,1]$. (For this Riesz space see [**90**, Example 11.2(viii), p. 50].) ■

1.26. The solution below is due to W. A. J. Luxemburg and A. C. Zaanen [**89**, Note VI, Example 20.8(ii), p. 665]. Let $\{u_n\} \subseteq L = C(X)$ satisfy $u_n(x) \downarrow 0$ for all $x \in X$, and let $0 \leq \phi \in L^{\sim}$. Assume that there exists some $\epsilon > 0$ satisfying $\phi(u_n) \geq \epsilon$ for all n. Without loss of generality we can assume that $u_1(x) > 0$ for all $x \in X$; otherwise, replace u_1 by $\mathbf{1} + u_1$. Note next that for each $x \in X$ we have $u_n(x) < \frac{\epsilon u_1(x)}{2\phi(u_1)}$ for all sufficiently large n, and so the series $\sum_{n=1}^{\infty} \big[u_n(x) - \frac{\epsilon u_1(x)}{2\phi(u_1)}\big]^+$ is a finite sum. This implies that $w = \sum_{n=1}^{\infty} \big[u_n - \frac{\epsilon u_1}{2\phi(u_1)}\big]^+ \in \mathbb{R}^X$.

We claim that $w \in C(X)$. To see this, let $x_0 \in X$. Choose some k satisfying $u_k(x_0) < \frac{\epsilon u_1(x_0)}{2\phi(u_1)}$. Since u_k and u_1 belong to $C(X)$, there exists a neighborhood V of x_0

such that $u_k(x) < \frac{\epsilon u_1(x)}{2\phi(u_1)}$ holds for all $x \in V$. Hence for every $x \in V$ we have

$$w(x) = \sum_{n=1}^{k} \left[u_n(x) - \tfrac{\epsilon u_1(x)}{2\phi(u_1)}\right]^+ ,$$

and this shows that w is continuous at x_0, and hence $w \in C(X)$.

It now follows from the positivity of ϕ that

$$\phi(w) \geq \phi\Big(\sum_{i=1}^{n} \left[u_i(x) - \tfrac{\epsilon u_1(x)}{2\phi(u_1)}\right]^+\Big) \geq \sum_{i=1}^{n} \left[\phi(u_i) - \tfrac{\epsilon}{2}\right] \geq \sum_{i=1}^{n} \left(\epsilon - \tfrac{\epsilon}{2}\right) = \tfrac{n\epsilon}{2}$$

for all n. This implies $\phi(w) = \infty$, which is impossible. Thus, $\phi(u_n) \downarrow 0$. ■

1.27. (a) Assume that $0 \leq nu \leq v$ holds for some $u, v \in L^+$ and all n. Then for each $0 \leq \phi \in L_n^\sim$ we have $0 \leq n\phi(u) \leq \phi(v)$ for all n, and so $\phi(u) = 0$. This implies $\phi(u) = 0$ for all $\phi \in L_n^\sim$. Consequently, the separation of the points of L by $L_n^\sim$ yields $u = 0$ so that L is Archimedean.

(b) Let $\psi \in (L_n^\sim)^d$. Then, according to Theorem 1.81, we have $C_\psi \subseteq N_\phi$ for all $\phi \in L_n^\sim$. Form this and the separation of points of L by $L_n^\sim$ we get $C_\psi = N_\psi^d = \{0\}$. Hence $N_\psi^{dd} = C_\psi^d = L$. Now a glance at Theorem 1.24 guarantees that N_ψ is order dense in L. ■

1.28. Consider the operator $\pi\colon \mathbb{R}^2 \to \mathbb{R}^2$ defined by $\pi(x,y) = (x, 2y)$ for all $(x,y) \in \mathbb{R}^2$. An easy verification shows that π is an onto Riesz isomorphism. Now if $I\colon \mathbb{R}^2 \to \mathbb{R}^2$ denotes the identity operator, then $0 \leq I \leq \pi$ holds in $\mathcal{L}_b(\mathbb{R}^2)$, but $I \neq \lambda\pi$ for each $\lambda \in \mathbb{R}$. This shows that π is not a discrete element of $\mathcal{L}_b(\mathbb{R}^2)$. ■

1.29. (i) $\Longrightarrow$ (ii) Embed the Riesz space L super order densely in some Dedekind σ-complete Riesz space K and let $0 \leq u_n \downarrow$ in L. Then there exists some $u^* \in K^+$ with $u_n \downarrow u^*$ in K. Choose a sequence $\{v_n\} \subseteq L^+$ satisfying $0 \leq v_n \uparrow u^*$ in K. Now an easy verification shows that the sequence $\{v_n\}$ satisfies the properties stated in (ii).

(ii) $\Longrightarrow$ (i) Note first that L must be an Archimedean Riesz space. Indeed, if we assume that $0 \leq nu \leq v$ holds in L for all n and some $u, v \in L^+$, then $0 \leq v - nu \downarrow$, and so there exists a sequence $\{v_n\} \subseteq L^+$ with $0 \leq v_n \uparrow$, $v_k \leq v - nu$ for all n and k and such that $v - nu - v_n \downarrow 0$. In particular, $v_n \leq v - (n+1)u$ for all n implies $0 \leq u \leq v - nu - v_n$, from which it follows that $u = 0$.

Now consider the collection L^* of all $u^* \in L^\delta$ for which there exist two countable subsets E and F of L such that $u^* = \sup E = \inf F$ holds in L^δ. Clearly, L^* is a Riesz subspace of L^δ including L as a super order dense Riesz subspace.

To complete the solution, we shall show that L^* is Dedekind σ-complete. To this end, let $0 \leq u_n^* \downarrow$ in L^*. Then $u_n^* \downarrow v^*$ in L^δ. We shall show that $v^* \in L^*$. Since each u_n^* is the infimum of some appropriate countable subset of L, there exists a sequence $\{w_n\} \subseteq L^+$ with $w_n \downarrow v^*$ in L^δ. Now choose a sequence $\{v_n\} \subseteq L$ with $0 \leq v_n\uparrow$, $v_k \leq w_n$ for all n and k and with $w_n - v_n \downarrow 0$. This implies $v_n \uparrow v^*$ in L^δ, so that $v^* \in L^*$, and the solution is finished. ■

1.30. (a) Note that for each n we have

$$\left|\int_0^1 u(x)\sin(nx)\,dx\right| \leq \int_0^1 |u(x)|\,dx = \|u\|_1 .$$

Consequently, for every $u \in L_1[0,1]$ we have

$$\|T(u)\|_\infty = \sup_{n\in\mathbb{N}} \left|\int_0^1 u(x)\sin(nx)\,dx\right| \le \|u\|_1 .$$

This establishes that T is norm continuous.

(b) If $\mathbf{1}$ denotes the constant function one on $[0,1]$, we shall show that $T([-\mathbf{1},\mathbf{1}])$ is not an order bounded subset of c_0. To see this, assume by way of contradiction that there exists some $0 \le a = (a_1, a_2, \ldots) \in c_0$ such that $|T(u)| \le a$ holds for all $u \in [-\mathbf{1},\mathbf{1}]$. Let $\delta = \frac{1}{\pi} - \frac{1}{4} > 0$. Now for each n, let s_n denote the sign function of $\sin(nx)$. That is, $s_n(x) = 1$ if $\sin(nx) \ge 0$ and $s_n(x) = -1$ if $\sin(nx) < 0$. Clearly, each s_n is measurable and $|s_n(x)| = 1$ holds or all x, and so $s_n \in [-\mathbf{1},\mathbf{1}]$.

Fix any $n \in \mathbb{N}$ with $n \ge 4$ and let k be the unique natural number satisfying $k\pi < n < (k+1)\pi$. In particular, note that $\frac{k}{n} > \frac{1}{\pi} - \frac{1}{n} \ge \delta > 0$. Hence, we have

$$\begin{aligned} a_n &\ge \int_0^1 s_n(x)\sin(nx)\,dx = \int_0^1 |\sin(nx)|\,dx \\ &= \tfrac{1}{n}\int_0^n |\sin(t)|\,dt \ge \tfrac{1}{n}\int_0^{k\pi} |\sin(t)|\,dt = \tfrac{1}{n}(2k) > 2\delta , \end{aligned}$$

and so $a_n \not\to 0$, which is impossible. This contradiction shows that T is not order bounded.

(c) We shall verify next that T is σ-order continuous. To see this, assume that $u_n \xrightarrow{o} 0$ holds in $L_1[0,1]$. Since $L_1[0,1]$ has order continuous norm, it follows that $\|u_n\|_1 \to 0$, and so the set $\Omega = \{0, u_1, u_2, \ldots\}$ is a norm compact subset of $L_1[0,1]$. Now note that the sequence of functions $\{\sin(nx)\}$ is norm bounded in $L_\infty[0,1]$, the norm dual of $L_1[0,1]$. In particular, $\{\sin(nx)\}$ can be viewed as a norm bounded equicontinuous sequence on $L_1[0,1]$. We can also consider the sequence $\{\sin(nx)\}$ as a norm bounded equicontinuous sequence on $C(\Omega)$, which satisfies $\lim \int_0^1 u(x)\sin(nx)\,dx = 0$ for each $u \in \Omega$. By the classical Ascoli–Arzelà theorem, the sequence $\{\sin(nx)\}$ converges uniformly to zero on Ω. Now for each n and k let

$$v_n^k = \sup_{i\ge k} \left|\int_0^1 u_i(x)\sin(nx)\,dx\right| ,$$

and for each n consider the sequence $v_n = (v_n^1, v_n^2, \ldots) \in c_0$. From $u_n \xrightarrow{o} 0$, it is easy to see that $v_n \downarrow 0$ holds in c_0. Now an easy argument shows that $|T(u_n)| \le v_n$ for all n, and therefore $T(u_n) \xrightarrow{o} 0$ holds in c_0. Consequently, T is σ-order continuous.

Finally, notice that T cannot be order continuous since it is not order bounded; see Lemma 1.72. ■

1.31. (a) Assume that L is a norm dense Riesz subspace of the Riesz space $C(X)$ and let $0 < u \in C(X)$. If $\mathbf{2} \in C(X)$ denotes the function defined by $\mathbf{2}(x) = 2$ for all $x \in X$, then choose some function $v \in L^+$ satisfying $\|v - \mathbf{2}\| < 1$ and note that $v(x) \ge 1$ holds for each $x \in X$. Now for each $n \in \mathbb{N}$ pick some $v_n \in L^+$ such that $\|v_n - u\|_\infty < \frac{1}{n}$. It follows that $v_n - u \le \frac{1}{n}v$ and from this we see that the element $w_n = \left(v_n - \frac{1}{n}v\right)^+$ satisfies $0 \le w_n \le u$ and

$$0 \le u - w_n \le u - \left(v_n - \tfrac{1}{n}v\right) = u - v_n + \tfrac{1}{n}v \le \tfrac{2}{n}v$$

for each n. Consequently, if for each n we let $x_n = \bigvee_{k=1}^n w_k$, then from $0 \le x_n\!\uparrow \le u$ and $0 \le u - x_n \le u - w_n \le \frac{2}{n}v$, we see that $x_n \uparrow u$. So, L is super order dense in $C(X)$.

(b) If π is of the described form, then π is clearly a Riesz homomorphism. For the converse, assume that $\pi\colon C(X) \to \mathbb{R}$ is a lattice homomorphism. Clearly, π is a positive linear functional. Hence, by the Riesz representation theorem, there exists a unique regular Borel measure μ on X satisfying $\pi(u) = \int u\,d\mu$ for each $u \in C(X)$. We claim that the support of μ is a singleton. To see this, assume that two distinct points, say x and y, belong to the support of μ. Pick two functions $u, v \in C(X)$ satisfying $u \wedge v = 0$ and $u(x) = v(y) = 1$. This implies

$$0 = \pi(u \wedge v) = \min\{\pi(u), \pi(v)\} = \min\{\textstyle\int u\,d\mu, \int v\,d\mu\} > 0\,,$$

which is a contradiction. Hence, the support of μ consists of a single point, say ξ. So, for each $u \in C(X)$ we have

$$\pi(u) = \int u\,d\mu = \int_{\{\xi\}} u\,d\mu = u(\xi)\mu(\{\xi\}) = \alpha u(\xi)\,,$$

where $\alpha = \mu(\{\xi\}) > 0$. ■

1.32. Assume that L is an Archimedean Riesz space and let $\{L_i\}_{i \in I}$ be a family of Riesz subspaces of L^δ between L and L^δ such that each L_i is a Dedekind σ-complete Riesz space in its own right.

It should be clear that every L_i includes L as a Riesz subspace. In particular, since L is order dense in L^δ, we infer that each L_i is order dense in L^δ. A consequence of this and Theorem 1.23 is the following: If $x_\alpha \uparrow x$ holds in some L_i, then automatically $x_\alpha \uparrow x$ also holds in L^δ.

Now let $K = \bigcap_{i \in I} L_i$ and note that K is a Riesz subspace between L and L^δ. To see that K is Dedekind σ-complete, assume that $0 \leq x_n \uparrow \leq x$ holds in K. Since L^δ is Dedekind complete, there exists some $x^* \in L^\delta$ such that $x_n \uparrow x^*$ holds in L^δ. On the other hand, since each L_i is Dedekind σ-complete, for each $i \in I$ there exists some $u_i \in L_i$ such that $x_n \uparrow u_i$ holds in L_i. By the remark at the beginning of the solution, it follows that $x_n \uparrow u_i$ also holds in L^δ for each $i \in I$. Consequently, it must be the case that $x^* = u_i \in L_i$ for each $i \in I$, and so $x^* \in K$. This implies that $x_n \uparrow x^*$ holds in K, and so K is Dedekind σ-complete. ■

9.2. Chapter 2: Locally Solid Topologies

2.1. (a) The square in $\mathbb{R}^2$ with vertices $(0,0)$, $(1,0)$, $(1,1)$, and $(0,1)$ is a full set that is not a solid set.

(b) The solid hull in $\mathbb{R}^2$ of the line segment AB, where $A = (-2,0)$ and $B = (1,2)$, is a solid set that is not full. For instance, the order interval $[A, B]$ is not included in the solid hull of AB; see Exercise **1.4**.

(c) Let us introduce some notation here. If u, v are two arbitrary elements in a Riesz space, let us put

$$[u, v] = \begin{cases} \emptyset & \text{if } u \not\leq v \\ \{w\colon u \leq w \leq v\} & \text{if } u \leq v\,. \end{cases}$$

If A is a subset of a Riesz space, then its **full hull** is the set $F(A) = \bigcup\{[u, v]\colon u, v \in A\}$. Clearly, $F(A)$ is the smallest (with respect to inclusion) full set that includes A.

Now let τ be a linear topology on a Riesz space L. Assume first that τ is a locally solid topology. Then, by condition (6) of Theorem 2.17, the mapping $u \mapsto u^+$ (from (L, τ) to (L, τ)), is uniformly continuous. In particular, given any τ-neighborhood U of zero there exists another τ-neighborhood V of zero such that $v \in V$ implies $v^+ \in U$.

To see that τ is also locally full, let U be a τ-neighborhood of zero and then select a solid τ-neighborhood V of zero with $V + V \subseteq U$. Now pick another solid τ-neighborhood W of zero such that $w \in W$ implies $w^+ \in V$. Next, assume that $u \leq x \leq v$ holds with $u, v \in W$. From $0 \leq x^+ \leq v^+ \in V$ and $0 \leq x^- \leq u^- \in V$, it follows that $x = x^+ - x^- \in V + V \subseteq U$. This implies $W \subseteq F(W) \subseteq U$, and from this we infer that τ is locally full.

For the converse assume that τ is locally full and that for each τ-neighborhood N of zero there exists another τ-neighborhood N_1 of zero such that $x \in N_1$ implies $x^+ \in N$. Pick a τ-neighborhood U of zero and then choose a full τ-neighborhood V of zero with $V + V \subseteq U$. Next pick a symmetric τ-neighborhood W of zero such that $w \in W$ implies $w^+ \in V$. To finish the solution, we shall establish that the solid hull of W satisfies $\operatorname{Sol}(W) \subseteq U$.

To see this, assume that $|u| \leq |w| = |w^+ - w^-|$ holds with $w \in W$. Using the Riesz decomposition property, we can write $u = u_1 + u_2$, where $-w^+ \leq u_1 \leq w^+$ and $-w^- \leq u_2 \leq w^-$. Since $\pm w^+$ and $\pm w^-$belong to the full set V, it follows that $u_1, u_2 \in V$ and so $u = u_1 + u_2 \in V + V \subseteq U$. Consequently, $\operatorname{Sol}(W) \subseteq U$. ■

2.2. Observe first that K_τ is τ-closed, since each neighborhood of zero includes a τ-closed neighborhood of zero. Similarly, since each neighborhood of zero includes a solid τ-neighborhood of zero, K_τ is a solid set. So, it remains to be shown that K_τ is a vector subspace of L.

To this end, let $u, v \in K_\tau$ and let V be an arbitrary τ-neighborhood of zero. Choose another τ-neighborhood W of zero with $W + W \subseteq V$. Then $u, v \in W$ and consequently $u + v \in W + W \subseteq V$, from which get $u + v \in V$, and thus $u + v \in K_\tau$. Similarly, if $c \neq 0$, then $u \in c^{-1}V$ and so $cu \in V$, and this implies that $cu \in K_\tau$. Therefore, K_τ is a vector subspace of L, and thus K_τ is a τ-closed ideal.

Now if τ is a Hausdorff topology and $u \neq 0$, then there exists a τ-neighborhood U of zero with $u \notin U$. So $u \notin K_\tau$ and hence $K_\tau = \{0\}$. Conversely, if $K_\tau = \{0\}$ and $u \neq 0$, then $u \notin K_\tau$ and so $u \notin V$ for some τ-neighborhood V of zero, proving that τ is a Hausdorff topology. ■

2.3. We need to note first that the cone L^+ is τ-closed. Indeed, since τ is Hausdorff, the set $\{0\}$ is τ-closed, and the desired conclusion follows from continuity of the mapping $u \mapsto u^-$ and the fact that $L^+ = \{u \in L\colon\ u^- = 0\}$.

Next let $\{u_{\alpha_\lambda}\}$ be a subnet of the net $\{u_\alpha\}$ such that $u_{\alpha_\lambda} \xrightarrow{\tau} u$. Now fix any index α and then select an index λ_0 such that $\lambda \succeq \lambda_0$ implies $\alpha_\lambda \succeq \alpha$. It follows that $u_{\alpha_\lambda} - u_\alpha \geq 0$ for all $\lambda \succeq \lambda_0$ and from $u_{\alpha_\lambda} - u_\alpha \xrightarrow[\lambda \succeq \lambda_0]{\tau} u - u_\alpha$ and the τ-closedness of L^+, we see that $u - u_\alpha \geq 0$. Therefore, $u \geq u_\alpha$ for each α so that u is an upper bound of the net $\{u_\alpha\}$.

To see that u is the least upper bound of the net $\{u_\alpha\}$, assume that a vector $v \in L$ satisfies $u_\alpha \leq v$ for all α. It follows that $u_{\alpha_\lambda} \leq v$ (or $u_{\alpha_\lambda} - v \geq 0$) for all λ. Using the τ-closedness of L^+ once more, we see that $v - u \geq 0$ or $v \geq u$. Therefore, $u = \sup\{u_\alpha\}$ holds true, as desired. ■

2.4. We show the first formula only; the proof of the other formula is similar. Clearly $u \leq u \vee v$ for each $v \in A$. Choose a net $\{v_\alpha\} \subseteq A$ with $v_\alpha \xrightarrow{\tau} u$. Since τ is a locally solid topology we have $u \vee v_\alpha \xrightarrow{\tau} u \vee u = u$. Hence if $w \leq u \vee v$ holds for all $v \in A$, then in particular we have $w \leq u \vee v_\alpha$ for all α, and so (by the previous exercise) $w \leq u$. This establishes that $u = \inf\{u \vee v\colon\ v \in A\}$. ■

2.5. Let τ^* be a locally solid extension of τ to M, and let $0 < w \in M$. Choose some $u \in L$ with $0 < u \leq w$ and then pick a τ-neighborhood V of zero such that $u \notin V$. Now select a solid τ^*-neighborhood V^* of zero such that $L \cap V^* \subseteq V$. Then $w \notin V^*$, for if

$w \in V^*$, then $u \in V^*$ and so $u \in L \cap V^* \subseteq V$, which is impossible. Hence the linear topology τ^* must be Hausdorff. ■

2.6. Note that if $e \in W$ and $u \in L^+$, then $e + u \in W$. Assume that $W \neq \emptyset$. Choose $e \in W$. Also let $u \in L^+$ and pick a τ-neighborhood V of zero. Pick some $\lambda > 0$ such that $\lambda e \in V$. Then $v = \lambda e + u \in W$ and $v - u = \lambda e \in V$, proving that W is τ-dense in L^+. ■

2.7. Let J be a τ-dense ideal in L. Since τ is a Hausdorff topology, the band $J^{\rm dd}$ is, according to part (d) of Theorem 2.21, τ-closed. Therefore, the relation $J \subseteq J^{\rm dd}$ implies $L = \overline{J} \subseteq J^{\rm dd} \subseteq L$, so that $J^{\rm dd} = L$. The conclusion now follows from Theorem 1.24. ■

2.8. Let $\mathcal{N}$ be the base of (L, τ) consisting of all solid τ-neighborhoods of zero. For each $V \in \mathcal{N}$ put

$$V^* = \big\{u^* \in L^\delta \colon \text{ There exists } u \in V \text{ with } |u^*| \leq u\big\}.$$

Clearly V^* is the solid hull of V in L^δ. The collection $\{V^* \colon V \in \mathcal{N}\}$ has the following properties:

(i) Each V^* is a solid (and hence a balanced) subset of L^δ.
(ii) Each V^* is an absorbing subset of L^δ.
(iii) If $W + W \subseteq V$ holds in L, then $W^* + W^* \subseteq V^*$ holds in L^δ.
(iv) For every pair $V, W \in \mathcal{N}$ we have $(V \cap W)^* \subseteq V^* \cap W^*$.
(v) For each $V \in \mathcal{N}$ we have $V^* \cap L = V$.

It follows that the family $\{V^* \colon V \in \mathcal{N}\}$ is a base at zero for a locally solid topology τ^* on L^δ that induces τ on L. ■

2.9. Assume that (E, τ) is an arbitrary Hausdorff locally convex space and let $\rho \colon E \to \mathbb{R}$ be a τ-continuous seminorm. Since ρ is τ-uniformly continuous, it follows from Theorem 2.6 that ρ has a unique $\widehat{\tau}$-uniformly continuous extension $\widehat{\rho} \colon \widehat{E} \to \mathbb{R}$. It should be also obvious that $\widehat{\rho}$ is a seminorm on $\widehat{E}$. For each $\epsilon > 0$ let

$$U_{\widehat{\rho}}(\epsilon) = \big\{\widehat{x} \in \widehat{E} \colon \widehat{\rho}(\widehat{x}) \leq \epsilon\big\} \quad \text{and} \quad U_\rho(\epsilon) = \big\{x \in E \colon \rho(x) \leq \epsilon\big\}.$$

We claim that

$$U_{\widehat{\rho}}(\epsilon) = \overline{U_\rho(\epsilon)} \quad \text{and} \quad U_{\widehat{\rho}}(\epsilon) \cap E = U_\rho(\epsilon). \tag{$\star$}$$

Clearly, $U_\rho(\epsilon) \subseteq U_{\widehat{\rho}}(\epsilon)$ and so $\overline{U_\rho(\epsilon)} \subseteq U_{\widehat{\rho}}(\epsilon)$. For the converse inclusion, let $\widehat{u} \in U_{\widehat{\rho}}(\epsilon)$, i.e., $\widehat{\rho}(\widehat{u}) \leq \epsilon$. Pick a net $\{u_\alpha\} \subseteq E$ such that $u_\alpha \xrightarrow{\widehat{\tau}} \widehat{u}$. In particular, it follows that $\rho(u_\alpha) \to \widehat{\rho}(\widehat{u})$ and so for each $\delta > 1$ there is an index α_δ such that $\rho(u_\alpha) < \epsilon\delta$ for all $\alpha \succeq \alpha_\delta$. From $\big\{\frac{u_\alpha}{\delta} \colon \alpha \succeq \alpha_\delta\big\} \subseteq U_\rho(\epsilon)$ and $\frac{u_\alpha}{\delta} \xrightarrow[\alpha \succeq \alpha_\delta]{} \frac{\widehat{u}}{\delta}$, we infer that $\frac{\widehat{u}}{\delta} \in \overline{U_\rho(\epsilon)}$. Now letting $\delta \downarrow 1$, we obtain $\widehat{u} \in \overline{U_\rho(\epsilon)}$. The identity $U_{\widehat{\rho}}(\epsilon) \cap E = U_\rho(\epsilon)$ should be obvious.

From ($\star$) it should be clear that if a family of τ-continuous seminorms $\{p_\alpha\}_{\alpha \in A}$ generates the topology τ, then the family $\{\widehat{p}_\alpha\}_{\alpha \in A}$ of $\widehat{\tau}$-continuous seminorms on $\widehat{E}$ generates the locally convex topology $\widehat{\tau}$.

Finally, let (E, τ) be a Hausdorff locally convex-solid Riesz space and let $\rho \colon E \to \mathbb{R}$ be a τ-continuous Riesz seminorm. We must show that $\widehat{\rho} \colon \widehat{E} \to \mathbb{R}$ is a Riesz seminorm. To see this, assume $|\widehat{u}| \leq |\widehat{v}|$ in $\widehat{E}$. Pick two nets $\{u_\alpha\}$ and $\{v_\alpha\}$ of E such that $u_\alpha \xrightarrow{\widehat{\tau}} \widehat{u}$ and $v_\alpha \xrightarrow{\widehat{\tau}} \widehat{v}$. Letting

$$w_\alpha = \big[(-|v_\alpha|) \vee u_\alpha\big] \wedge |v_\alpha|,$$

we see that $|w_\alpha| \leq |v_\alpha|$ for each α and that $w_\alpha \xrightarrow{\widehat{\tau}} \widehat{u}$. It follows that $|w_\alpha| \xrightarrow{\widehat{\tau}} |\widehat{u}|$ and $|v_\alpha| \xrightarrow{\widehat{\tau}} |\widehat{v}|$. Using the inequality

$$\rho(w_\alpha) = \rho(|w_\alpha|) \leq \rho(|v_\alpha|) = \rho(v_\alpha)$$

and taking limits, we get $\widehat{\rho}(\widehat{u}) = \widehat{\rho}(|\widehat{u}|) \le \widehat{\rho}(|\widehat{v}|) = \widehat{\rho}(\widehat{v})$. This shows $\widehat{\rho}\colon \widehat{E} \to \mathbb{R}$ is a Riesz seminorm, and the solution is finished. ■

2.10. The verification that $\phi \mapsto \widehat{\phi}$ is a one-to-one linear mapping from L' onto $\widehat{L}'$ is straightforward. We shall show that $\phi \mapsto \widehat{\phi}$ is a Riesz isomorphism by establishing that $\widehat{\phi^+} = (\widehat{\phi})^+$. To see this, note first that if $0 \le \widehat{u} \le u \in L$ holds in $\widehat{L}$, then there exists a net $\{u_\alpha\} \subseteq [0,u] \cap L$ satisfying $u_\alpha \xrightarrow{\widehat{\tau}} \widehat{u}$, and so $\lim \phi(u_\alpha) = \widehat{\phi}(\widehat{u})$. Thus for each $u \in L^+$ we have

$$\begin{aligned}(\widehat{\phi})^+(u) &= \sup\{\widehat{\phi}(\widehat{v})\colon\ \widehat{v} \in \widehat{L} \text{ and } 0 \le \widehat{v} \le u\}\\ &= \sup\{\phi(v)\colon\ v \in L \text{ and } 0 \le v \le u\}\\ &= \phi^+(u) = \widehat{\phi^+}(u)\,.\end{aligned}$$

Hence $(\widehat{\phi})^+(u) = \widehat{\phi^+}(u)$ holds for all $u \in L$ and therefore (since L is $\widehat{\tau}$-dense in $\widehat{L}$) we have $(\widehat{\phi})^+(\widehat{u}) = \widehat{\phi^+}(\widehat{u})$ for all $\widehat{u} \in \widehat{L}$. Consequently $(\widehat{\phi})^+ = \widehat{\phi^+}$. ■

2.11. If L is sequentially τ-complete, then clearly every order interval of L is sequentially τ-complete and every increasing τ-Cauchy sequence of L^+ is τ-convergent.

Conversely, assume that (L,τ) satisfies the above two properties. Note first that L is sequentially τ-complete if and only if $L = L_\sigma$, where

$$L_\sigma = \{\widehat{u} \in \widehat{L}\colon\ \text{There exists } \{u_n\} \subseteq L \text{ with } u_n \xrightarrow{\widehat{\tau}} \widehat{u}\}\,.$$

Let $0 \le \widehat{u} \in L_\sigma$. Choose a sequence $\{u_n\} \subseteq L^+$ with $u_n \xrightarrow{\widehat{\tau}} \widehat{u}$; clearly $u_n \wedge \widehat{u} \xrightarrow{\widehat{\tau}} \widehat{u}$.

Now for each fixed k we have $u_n \wedge u_k \xrightarrow[n\to\infty]{\widehat{\tau}} \widehat{u} \wedge u_k$ and $u_n \wedge u_k \in [0,u_k] \cap L$ for each n and k. It follows from our hypothesis that $\{u_k \wedge \widehat{u}\} \subseteq L$. For each n put $w_n = \bigvee_{i=1}^n u_i \wedge \widehat{u}$, and note that $0 \le w_n \uparrow$ in L and $w_n \xrightarrow{\widehat{\tau}} \widehat{u}$. Thus, using our hypothesis, we see that $\widehat{u} \in L$. This shows that $L = L_\sigma$. ■

2.12. Let $\{u_n\}$ be a disjoint sequence of L^+, let $u \in L^+$, and fix $\epsilon > 0$. In order to show that $\lim \rho(u - u \wedge u_n) = \rho(u)$ it suffices to establish that the set

$$I = \{k \in \mathbb{N}\colon\ \rho(u - u \wedge u_k) \le \rho(u) - \epsilon\}$$

is a finite set.

To this end, let $\{k_1, \ldots, k_n\} \subseteq I$. From the relation

$$\begin{aligned}\sum_{m=1}^n (u - u \wedge u_{k_m}) &= nu - \sum_{m=1}^n u \wedge u_{k_m} = nu - \bigvee_{m=1}^n u \wedge u_{k_m}\\ &\ge nu - u = (n-1)u \ge 0\,,\end{aligned}$$

it follows that

$$\begin{aligned}(n-1)\rho(u) &\le \rho\Big(\sum_{m=1}^n (u - u \wedge u_{k_m})\Big) \le \sum_{m=1}^n \rho(u - u \wedge u_{k_m})\\ &\le n[\rho(u) - \epsilon] = n\rho(u) - n\epsilon\,.\end{aligned}$$

This implies $n\epsilon \le \rho(u)$, and consequently I must be a finite set. ■

2.13. Let $L = \mathbb{R}^n$ for some n, and let $\rho\colon \mathbb{R}^n \to \mathbb{R}$ be a seminorm on L such that $\rho(u) = \rho(|u|)$ for all $u \in L$. Assume that $0 \le v \le u = (u_1, \ldots, u_n)$. Consider the collection of all vectors w of L whose i^{th} coordinate is either u_i or $-u_i$, where $i = 1, \ldots, n$. There are, of course, finitely many vectors with this property and all of them satisfy $|w| = |u|$, and hence by our hypothesis $\rho(w) = \rho(u)$. Thus, all these vectors lie in the closed ball centered at zero with radius $\rho(u)+\epsilon$, for any $\epsilon>0$. In particular, the convex hull of these vectors lies in the above balls, and it clearly contains v. This implies that $\rho(v)\le\rho(u)+\epsilon$ for all $\epsilon>0$, that is, $\rho(v)\le\rho(u)$. Hence ρ is a Riesz seminorm. ■

2.14. The space L is norm dense in c_0 and from this we get $L'=c_0'=\ell_1$. The linear functional $\phi\colon L\to\mathbb{R}$, defined via the formula $\phi(u)=\sum_{i=1}^{\infty}u_i$ for all $u=(u_1,u_2,\dots)\in L$, satisfies $\phi\in L_{\mathrm{n}}^{\sim}$ and $\phi\notin L'$. ■

2.15. It should be clear that if $0<e\in L'$ is an order unit of L', then $|\sigma|(L,L')$ is generated by the Riesz norm $\rho(u)=e(|u|)$ for all $u\in L$. In particular, it follows that $|\sigma|(L,L')$ is a metrizable topology.

We shall show next that $\tau=|\sigma|(L,L')$. Since $|\sigma|(L,L')\subseteq\tau$, we only need to establish $\tau\subseteq|\sigma|(L,L')$. To this end, let $u_n\xrightarrow{|\sigma|(L,L')}0$ and suppose by way of contradiction that $\{u_n\}$ is not τ-convergent to zero. By passing to a subsequence if necessary, we can assume without loss of generality that there exists a τ-closed, convex, and solid τ-neighborhood V of zero satisfying $u_n\notin V$ and $\rho(u_n)<2^{-2n}$ for all n. Since $u_n\notin V$, there exists $0\le\phi_n\in V^{\circ}$ $(=\{\psi\in L'\colon |\psi(u)|\le 1 \text{ for all } u\in V\})$ satisfying $\phi_n(u_n)\ge 1$ for each n. From the convexity and solidness of V°, it follows that if we let $\psi_n=\sum_{i=1}^{n}2^{-i}\phi_i\ge 0$, then $\{\psi_n\}\subseteq V^{\circ}$ and $0\le\psi_n=\sum_{i=1}^{n}2^{-i}\phi_i\uparrow$. Also from the $\sigma(L',L)$-compactness of V°, it follows that the sequence $\{\psi_n\}$ has a $\sigma(L',L)$-accumulation point $\psi\in V^{\circ}$. Now an easy argument (similar to the one in the solution of Exercise **2.3**) shows that $\psi_n\uparrow\psi$ in L'.

Now choose some $\lambda>0$ with $\psi\le\lambda e$, and note that

$$2^{-n}\le 2^{-n}\phi_n(u_n)\le\psi(u_n)\le\lambda e(u_n)=\lambda\rho(u_n)\le\lambda 2^{-2n}.$$

This implies $2^n\le\lambda$ for all n, which is a contradiction. Thus, $u_n\xrightarrow{\tau}0$ and therefore $|\sigma|(L,L')=\tau$. ■

2.16. Consider the Riesz space $\mathbb{R}^{[0,1]}$ equipped with the topology of pointwise convergence and $L_1[0,1]$ with the topology generated by the L_1-norm. Put $M=\mathbb{R}^{[0,1]}\times L_1[0,1]$ and let τ^* be the product topology on M. Note that (M,τ^*) is τ^*-complete. We claim that $(\widehat{L},\widehat{\tau})=(M,\tau^*)$.

To see this, consider the mapping $u\mapsto\phi(u)=(u,[u])$ from L to M, where $[u]$ denotes the equivalence class determined by u in $L_1[0,1]$. A straightforward verification shows that ϕ is a Riesz isomorphism (into), which is also a homeomorphism between $L=C[0,1]$ and $\phi(L)$. Subject to this Riesz isomorphism, we identify (L,τ) with $(\phi(L),\tau^*)$. To complete the proof, we shall show that $\phi(L)$ is τ^*-dense in M.

To this end, let $(f,g)\in M$, let $\{x_1,\dots,x_n\}\subseteq[0,1]$, and let $\epsilon>0$. Choose some $u\in C[0,1]$ satisfying $u(x_i)=f(x_i)$ for each $i=1,\dots,n$ and $\int_0^1|g(x)-u(x)|\,dx<\epsilon$ (note that indeed such a function u exists). But then $(u,[u])\in\phi(L)$ satisfies $\rho_{x_i}(u-f)=0$ for $i=1,\dots,n$ and $\rho(f-g)<\epsilon$. This shows that $\phi(L)$ is τ^*-dense in M, as required. ■

2.17. Let $\{u_1,u_2,\dots\}$ be a countable τ-dense subset of L. We only need to show that L' has the countable sup property (since it is already Dedekind complete by Theorem 2.22). To this end, let $\phi_\alpha\downarrow 0$ in L'. For each n choose an index α_n such that $\phi_{\alpha_n}(|u_k|)<\frac{1}{n}$ for each $k=1,\dots,n$; we can assume that $\alpha_{n+1}\succeq\alpha_n$ holds for each n. We claim that $\phi_{\alpha_n}\downarrow 0$ in L'.

Indeed, if $0\le\phi\le\phi_{\alpha_n}$ holds for all $n\in\mathbb{N}$, then given k and any $n>k$ we have $0\le|\phi(u_k)|\le\phi(|u_k|)\le\phi_{\alpha_n}(|u_k|)<\frac{1}{n}$. This implies $\phi(u_k)=0$ for all k, and since $\{u_1,u_2,\dots\}$ is τ-dense in L and $\phi\in L'$, it follows that $\phi=0$. Therefore, $\phi_{\alpha_n}\downarrow 0$ holds in L', so that L' is super Dedekind complete. ■

2.18. It should be clear that $\phi_n \xrightarrow{\sigma(L_n^\sim, L^\delta)} \phi$ implies $\phi_n \xrightarrow{\sigma(L_n^\sim, L)} \phi$. Thus in order to complete the solution, we must establish the converse.

To this end, let $\phi_n \xrightarrow{\sigma(L_n^\sim, L)} \phi$, let $u^* \in L^\delta$, and fix $\epsilon > 0$. We must show that $\lim \phi_n(u^*) = \phi(u^*)$. Start by choosing $u, v \in L$ with $u \le u^* \le v$ and such that $\phi(v-u) < \epsilon$ (this is possible since $\phi \in L_n^\sim$). Then we have

$$\begin{aligned} \phi_n(u^*) - \phi(u^*) &\le \phi_n(v) - \phi(u) = (\phi_n - \phi)(v) + \phi(v-u) \\ &\le (\phi_n - \phi)(v) + \epsilon \le |(\phi_n - \phi)(v)| + |(\phi_n - \phi)(u)| + \epsilon \,. \end{aligned}$$

Similarly, $\phi(u^*) - \phi_n(u^*) \le |(\phi_n - \phi)(v)| + |(\phi_n - \phi)(u)| + \epsilon$. So, for all n we have

$$\big|\phi_n(u^*) - \phi(u^*)\big| \le \big|(\phi_n - \phi)(v)\big| + \big|(\phi_n - \phi)(u)\big| + \epsilon \,.$$

Now our hypothesis yields $\limsup |\phi_n(u^*) - \phi(u^*)| \le \epsilon$. Since $\epsilon > 0$ is arbitrary, we get $\lim |\phi_n(u^*) - \phi(u^*)| = 0$, and thus $\lim \phi_n(u^*) = \phi(u^*)$. ■

2.19. Let u be an interior point of L^+. Pick a balanced τ-neighborhood V of zero with $u + V \subseteq L^+$. Thus, if $v \in V$, then $-v \in V$ and so $u + v$ and $u - v$ belong to L^+. Hence $v \in V$ implies $v \in [-u, u]$, that is, $V \subseteq [-u, u]$. It now follows that $[-u, u]$ is a τ-neighborhood of zero.

Now if $\phi \in L^\sim$, then ϕ is bounded on the τ-neighborhood $[-u, u]$ of zero and therefore is τ-continuous, that is, $\phi \in L'$. Hence $L' = L^\sim$. ■

2.20. Fix $u \in L^+$. By part (6) of Theorem 1.53 we know that the extreme points of the order interval $[0, u]$ are precisely the components of u. Let C denote the τ-closed convex hull of the extreme points of $[0, u]$, and assume by way of contradiction that $C \ne [0, u]$.

Pick some $0 < v < u$ with $v \notin C$. Since $\{v\}$ is convex and τ-compact, C is convex and τ-closed, and $\{v\} \cap C = \emptyset$, it follows from the separation theorem that there exist some $c \in \mathbb{R}$ and some $\phi \in L'$ such that $\phi(w) \le c$ holds for all $w \in C$ and $\phi(v) > c$. Now by Theorem 1.68 we have

$$\phi^+(u) = \sup\{\phi(x)\colon\ x \wedge (u - x) = 0\} \,.$$

So, since the components of u belong to C, we infer that $\phi^+(u) \le c$. In particular, it follows that

$$c < \phi(v) \le \phi^+(v) \le \phi^+(u) \le c \,,$$

which is a contradiction. This implies that $C = [0, u]$, as claimed. ■

9.3. Chapter 3: Lebesgue Topologies

3.1. Let K_τ be the intersection of all τ-neighborhoods of zero. Then K_τ is a τ-closed ideal of L, and τ is a Hausdorff topology if and only if $K_\tau = \{0\}$; see Exercise **2.2**. Now if $0 \le u \in K_\tau$, then $nu \in K_\tau$ for each $n \in \mathbb{N}$, so that the sequence $\{nu\}$ is τ-bounded. But then since τ is a σ-Levi topology and $0 \le nu \uparrow$, we see that $v = \sup\{nu\}$ exists in L. Now from $2v = 2\sup\{nu\} = \sup\{2nu\} = v$, we get $v = 0$. This implies that $u = 0$, and thus $K_\tau = \{0\}$. ■

3.2. The inclusion $|\sigma|(L, L') \subseteq \tau$ guarantees that if τ is a Lebesgue topology, then so is $|\sigma|(L, L')$.

On the other hand, if $|\sigma|(L, L')$ is a Lebesgue topology and $u_\alpha \downarrow 0$ holds in L, then the inclusion $\sigma(L, L') \subseteq |\sigma|(L, L')$ guarantees that $u_\alpha \xrightarrow{\sigma(L,L')} 0$. But then Theorem 3.21 implies $u_\alpha \xrightarrow{\tau} 0$, and so τ is a Lebesgue topology. ■

3.3. Note first that if τ is a Hausdorff Lebesgue topology and $0 < \phi \in L'$ satisfies $C_\phi = N_\phi^{\mathrm{d}} = \{0\}$, then ϕ vanishes on the order dense ideal $N_\phi \oplus C_\phi$, and therefore, ϕ must vanish everywhere (by virtue of $L' \subseteq L_{\mathrm{n}}^{\sim}$), contradicting $0 < \phi \in L'$. Thus $C_\phi \neq \{0\}$ for every $0 < \phi \in L'$.

On the other hand, let the Hausdorff locally convex-solid topology τ on L satisfy $C_\phi \neq \{0\}$ for every $0 < \phi \in L'$, and let J be an order dense ideal of L. If $\overline{J} \neq L$, then by the separation theorem there exists some nonzero $\phi \in L'$ such that $\phi(u) = 0$ for all $u \in J$. It now follows from Theorem 1.67 that $J \subseteq N_\psi \subseteq N_\psi^{\mathrm{dd}}$, where $\psi = |\phi|$. But then by the order denseness of J we obtain $N_\psi^{\mathrm{dd}} = L$. Therefore, $C_\psi = N_\psi^{\mathrm{d}} = \{0\}$, contrary to $0 < |\phi| = \psi \in L'$. Thus $\overline{J} = L$, and the desired conclusion follows from Theorem 3.12. ■

3.4. Let $u_n \downarrow 0$ in L. Since τ is a σ-Lebesgue topology, we have $u_n \xrightarrow{\tau} 0$ and so $u_n \xrightarrow{\sigma(L,L')} 0$. Consequently, the continuity of T implies that $T(u_n) \xrightarrow{\sigma(M,M')} 0$. Now observe first that by the positivity of T we have $0 \leq T(u_n) \downarrow$ in M, and then use the arguments in the proof of Theorem 3.21 to obtain $T(u_n) \xrightarrow{\tau^*} 0$, and from this that $T(u_n) \downarrow 0$. ■

3.5. Let $u > 0$ be an interior point. As in the solution of Exercise **2.19**, we see that the order interval $[-u, u]$ is a τ-neighborhood of zero. In particular, $[-u, u]$ is an absorbing set, and so u is a strong order unit.

Now assume that there exists a disjoint sequence $\{w_n\}$ of nonzero positive elements. Replacing $\{w_n\}$ by $\{u \wedge w_n\}$, we can assume without loss of generality that $0 \leq w_n \leq u$ holds for all n. Next for each n let $\lambda_n = \sup\{\lambda > 0\colon\ \lambda w_n \leq u\}$, and note that (since L is Archimedean) $\lambda_n \in \mathbb{R}$ and $\lambda_n w_n \leq u$ for all n. Put $v_n = 2\lambda_n w_n$ and note that $v_n \notin [-u, u]$ and $0 < v_n \leq 2u$ for each n. Also $\{v_n\}$ is a disjoint sequence of L^+, and hence from the pre-Lebesgue property it follows that $v_n \xrightarrow{\tau} 0$. In particular, we have $v_n \in [-u, u]$ for all sufficiently large n, a contradiction. Thus every disjoint subset of L is finite and hence by Exercise **1.13** the Riesz space L is finite dimensional, contrary to our hypothesis. This contradiction establishes that L^+ does not have interior points. ■

3.6. (i) Let $L = C[0, 1]$ and let τ be the Hausdorff locally solid topology generated on L by the Riesz norm $\rho(u) = \int_0^1 |u(x)|\, dx$. Let L^* be the Riesz space consisting of all real-valued (everywhere) Lebesgue integrable functions on $[0, 1]$, and let τ^* be the locally solid topology generated on L^* by the Riesz seminorm $\rho(u) = \int_0^1 |u(x)|\, dx$. Then (L^*, τ^*) is a non-Hausdorff topological completion of (L, τ).

(ii) Repeat the proof of Theorem 3.26. ■

3.7. Consider $L = C[0, 1]$ and let τ be the Hausdorff locally solid topology generated by the sup norm. Note that (L, τ) is τ-separable, but it does not satisfy the pre-Lebesgue property. (It should be remarked here that if X is a Hausdorff compact topological space, then $C(X)$ is separable if and only if X is metrizable; see [**8**, Theorem 8.48, p. 323].) ■

3.8. For any Banach lattice $L = L_\rho$ we have $L' = L^\sim$. So, if in addition the Banach lattice L_ρ satisfies the Lebesgue property, then from $L' \subseteq L_n^\sim \subseteq L^\sim$ it follows that $L' = L_n^\sim = L^\sim$.

Now note that $A = c_0$ (with the sup norm) is a Banach lattice with the Lebesgue property whose norm dual is ℓ_1. Consequently $A_n^\sim = c_0' = \ell_1$. Similarly, ℓ_1 is a Banach lattice with the Lebesgue property, and therefore we have $A_{nn} = (A_n^\sim)_n^\sim = \ell_1' = \ell_\infty$. However, c_0 embedded in its natural way in ℓ_∞ is a proper ideal. Also, since $\ell_1^\sim = \ell_\infty$, it follows immediately that ℓ_∞ is a perfect Riesz space. ■

3.9. If a Riesz space L is perfect, then $|\sigma|(L, L_n^\sim)$ is a Lebesgue locally convex-solid topology on L with the Levi property; see Theorem 3.17.

For the converse assume that a Riesz space L admits a Lebesgue locally convex-solid topology τ with the Levi property—and so by Exercise **3.1** the topology τ is Hausdorff. Now according to Theorem 3.17 we must show that $|\sigma|(L, L_n^\sim)$ is a Levi topology. So, assume that a $|\sigma|(L, L_n^\sim)$-bounded net $\{u_\alpha\}$ satisfy $0 \le u_\alpha \uparrow$. Now since τ is a Lebesgue topology, $L' \subseteq L_n^\sim$ holds, and therefore $\{u_\alpha\}$ is $|\sigma|(L, L')$-bounded, and consequently a τ-bounded net. But then $u = \sup\{u_\alpha\}$ exists in L (since τ is a Levi topology), and the solution is finished. ■

3.10. Put $K = \mathcal{L}_b(L, M)$. According to Theorem 3.17 we must show that $|\sigma|(K, K_n^\sim)$ is a Levi topology. To this end, assume $0 \le T_\alpha \uparrow$ holds in K with the net $\{T_\alpha\}$ being $|\sigma|(K, K_n^\sim)$-bounded. We must prove that there exists some $T \in K$ satisfying $T_\alpha \uparrow T$.

Note that for each $0 \le \phi \in M_n^\sim$ and each $u \in L^+$, the linear functional $F_{u,\phi}\colon K \to \mathbb{R}$, defined by $F_{u,\phi}(T) = \phi(T(u))$, belongs to $K_n^\sim$. Now taking into account that $\{T_\alpha\}$ is $|\sigma|(K, K_n^\sim)$-bounded, we see that for each $u \in L^+$ the net $\{T_\alpha(u)\}$ of M is $|\sigma|(M, M_n^\sim)$-bounded. But then since M is perfect and $0 \le T_\alpha(u) \uparrow$ holds in M for each $u \in L^+$, we get $T_\alpha(u) \uparrow T(u)$ in M for each $u \in L^+$. It should be clear that $T\colon L^+ \to M^+$ is additive and so by Lemma 1.64 it extends uniquely to a positive operator from L to M. Now note that $T \in K$ and that $T_\alpha \uparrow T$ holds in K. ■

3.11. Assume first that the norm of a normed Riesz space L satisfies the identity $\|u \vee v\| = \max\{\|u\|, \|v\|\}$ for all $u, v \in L^+$. In particular, if $u, v \in L^+$ satisfy $u \wedge v = 0$, then we have $\|u + v\| = \|u \vee v\| = \max\{\|u\|, \|v\|\}$, and so L is an M-space.

For the converse, assume that L is an M-space. Then its double norm dual L'' is an AM-space with unit, and hence by the Kakutani–Bohnenblust–Krein Representation Theorem 3.40 we know that L'' is lattice isometric to some $C(X)$-space. So, if $u, v \in L^+$, then [considering $v, u \in C(X)$] we have

$$\begin{aligned} \|u \vee v\| &= \max_{x \in X} \max\{u(x), v(x)\} = \max\{\max_{x \in X} u(x), \max_{x \in X} v(x)\} \\ &= \max\{\|u\|, \|v\|\}, \end{aligned}$$

as desired. ■

3.12. Let ρ be a p-additive Riesz seminorm on L, and let A be an ideal of L. Assume that $0 \le \dot{u}, \dot{v} \in L/A$ satisfy $\dot{u} \wedge \dot{v} = 0$. Replacing u and v by $|u| - |u| \wedge |v|$ and $|v| - |u| \wedge |v|$, respectively, we can assume without loss of generality that $u \wedge v = 0$ holds in L.

Now let $0 \le w \in \dot{u} + \dot{v}$. Then from

$$\begin{aligned} |u - w \wedge u| + |v - w \wedge v| &= u + v - (w \wedge u + w \wedge v) \\ &= u + v - w \wedge (u + v) \\ &\le |u + v - w| \in A, \end{aligned}$$

it follows that $w \wedge u - u \in A$ (i.e., $w \wedge u \in \dot{u}$) and $w \wedge v \in \dot{v}$. Moreover, we have $(w \wedge u) \wedge (w \wedge v) = 0$, and $0 \leq w \wedge u + w \wedge v = w \wedge (u+v) \leq w$. So, by applying the p-additivity of the Riesz seminorm ρ, we get

$$\begin{aligned}[\dot{\rho}(\dot{u})]^p + [\dot{\rho}(\dot{v})]^p &\leq [\rho(w \wedge u)]^p + [\rho(w \wedge v)]^p = [\rho(w \wedge u + w \wedge v)]^p \\ &= [\rho(w \wedge (u+v))]^p \leq [\rho(w)]^p ,\end{aligned}$$

and consequently

$$[\dot{\rho}(\dot{u})]^p + [\dot{\rho}(\dot{v})]^p \leq [\dot{\rho}(\dot{u} + \dot{v})]^p . \qquad (\star)$$

On the other hand, given $\epsilon > 0$, there exist $u_1 \in \dot{u}$ with $0 \leq u_1 \leq u$ and $v_1 \in \dot{v}$ with $0 \leq v_1 \leq v$ satisfying $\rho(u_1) \leq \dot{\rho}(\dot{u}) + \epsilon$ and $\rho(v_1) \leq \dot{\rho}(\dot{v}) + \epsilon$. In particular, note that $u_1 \wedge v_1 = 0$. So, from $u_1 + v_1 \in \dot{u} + \dot{v}$ we get

$$[\dot{\rho}(\dot{u} + \dot{v})]^p \leq [\rho(u_1 + v_1)]^p = [\rho(u_1)]^p + [\rho(v_1)]^p \leq [\dot{\rho}(\dot{u}) + \epsilon]^p + [\dot{\rho}(\dot{v}) + \epsilon]^p$$

for all $\epsilon > 0$. Hence

$$[\dot{\rho}(\dot{u} + \dot{v})]^p \leq [\dot{\rho}(\dot{u})]^p + [\dot{\rho}(\dot{v})]^p . \qquad (\star\star)$$

Now $(\star)$ and $(\star\star)$ combined together show that $\dot{\rho}$ is p-additive on L/A. The proof for the M-seminorm is similar. ∎

3.13. Clearly, every AM-space of the form $c_0(X)$ (with the sup norm) satisfies the Lebesgue property.

Now assume that an AM-space L_ρ satisfies the Lebesgue property. We shall establish first that L is a discrete space. To this end, let $0 < u \in L$, and consider the nonempty set

$$C = \{w \in L\colon\ 0 < w \leq u \text{ and } \rho(w) = \rho(u)\} .$$

Observe that C is partially ordered by $\leq$. Let $\{w_\alpha\colon\ \alpha \in A\}$ be a chain of C. If A is finite, then $w = \inf_{\alpha \in A} w_\alpha$ is a lower bound for the chain in C. If A is infinite, then $\{w_\alpha\}$ considered as a net, satisfies $0 \leq w_\alpha \downarrow$, and hence by virtue of the Lebesgue property, $\{w_\alpha\}$ is a ρ-Cauchy net and thus $\rho(w_\alpha - w) \to 0$ for some $w \in L$. But then $w_\alpha \downarrow w$ and $\rho(w) = \rho(u)$ hold, that is, $w \in C$ and w is a lower element for the chain. Now according to Zorn's lemma, there exists a minimal element $v \in C$. We claim that v is a discrete element. Indeed, if v is not a discrete element, then there exist $0 < v_1 < v$ and $0 < v_2 < v$ with $v_1 \wedge v_2 = 0$ (see Exercise **1.12**), and so $(v - v_1) \vee (v - v_2) = v - v_1 \wedge v_2 = v$. But then since ρ is an M-norm, we must have $\rho(u) = \rho(v) = \max\{\rho(v - v_1), \rho(v - v_2)\}$ (see Exercise **3.11** above) with $0 < v - v_1 < v \leq u$ and $0 < v - v_2 < v \leq u$, contradicting the minimality property of v. Hence v is a discrete element satisfying $0 < v \leq u$, and therefore L is a discrete space.

Now consider a complete disjoint system $\{e_x\colon\ x \in X\}$ consisting of discrete elements satisfying $\rho(e_x) = 1$ for each $x \in X$. Let L_0 be the order dense (and hence norm dense) Riesz subspace of L consisting of the finite linear combinations of the elements of the set $\{e_x\colon\ x \in X\}$, and let L_1 be the order dense (and hence norm dense) Riesz subspace of $c_0(X)$ consisting of the functions vanishing off finite subsets of X. It can be easily seen that the mapping

$$u = \sum_{n=1}^{k} \lambda_n e_{x_n} \longmapsto \phi(u) = \sum_{n=1}^{k} \lambda_n \chi_{\{x_n\}}$$

is a norm preserving Riesz isomorphism from L_0 onto L_1. Now using that L_0 and L_1 are norm dense in L and $c_0(X)$, respectively, we see that ϕ can be extended to a lattice isometry from L onto $c_0(X)$. ∎

3.14. Assume that a norm bounded net $\{u_\alpha\}$ in some abstract L_p-space L satisfies $0 \le u_\alpha \uparrow$. Define $F(\phi) = \sup\{\phi(u_\alpha)\}$ for each $0 \le \phi \in L'$ and extend F to all of L' in the obvious manner; see Lemma 1.64. Clearly, $F \in L''$ and $0 \le u_\alpha \uparrow F$ in L''. Since L'' is an abstract L_p-space (since L is so), the net $\{u_\alpha\}$ is a Cauchy net of $L''$and hence of L. But then $\{u_\alpha\}$ is norm convergent to some u in L. So, $u_\alpha \uparrow u$ holds in L. It now follows from Exercise **3.9** that L is a perfect Riesz space. ■

3.15. Let τ be a σ-Lebesgue topology on an almost Dedekind σ-complete Riesz space L. Assume that $0 \le u_n \uparrow \le u$ holds in L and let V be some solid τ-neighborhood of zero. There exists a sequence $\{v_n\} \subseteq L^+$ satisfying $u_n \le v_k$ for all n and k and such that $v_n - u_n \downarrow 0$; see Exercise **1.29**.

Now choose some $m \in \mathbb{N}$ such that $v_n - u_n \in V$ for all $n \ge m$, and note that $0 \le u_{n+p} - u_n \le v_n - u_n$ implies $u_{n+p} - u_n \in V$ for all $n \ge m$ and all $p \in \mathbb{N}$. This shows that $\{u_n\}$ is a τ-Cauchy sequence and so τ is a pre-Lebesgue topology. ■

3.16. If M is order dense in L, then clearly M is also τ-dense in L. For the converse assume that M is τ-dense in L and let $0 < u \in L$. Note that every $\phi \in M'$ has a unique continuous linear extension to L (which we shall denote by ϕ again).

The element u defines a linear functional on M' via the standard formula $u(\phi) = \phi(u)$ and, as we have seen before, $u \in (M')^\sim_{\rm n}$. Since M is a regular Riesz subspace of L, the topology τ restricted to M is also a Lebesgue topology, and so according to Theorem 3.12 the Riesz space M is order dense in $(M')^\sim_{\rm n}$. Choose some $0 < v \in M$ such that $0 \le v \le u$ in $(M')^\sim_{\rm n}$. Now since τ is a Hausdorff locally convex topology, an easy application of the Hahn–Banach–Kantorovich Theorem 1.69 shows that $0 < v \le u$ also holds in L. Therefore M is order dense in L. ■

3.17. The verification of the fact that the formula defining $\|\cdot\|_\infty$ is a Riesz norm is straightforward. From $|u| \le \lambda e$ it follows that $\|u\| \le \lambda\|e\|$, and so

$$\|u\| \le \|e\| \cdot \|u\|_\infty \tag{$\star$}$$

holds for each $u \in L$.

Now assume by way of contradiction that $\|\cdot\|$ is not equivalent to $\|\cdot\|_\infty$. That is, assume that the identity operator $I\colon (L, \|\cdot\|) \to (L, \|\cdot\|_\infty)$ is not continuous. So, there exists a sequence $\{u_n\}$ satisfying $\|u_n\| < n^{-3}$ and $\|u_n\|_\infty \ge 1$ for each n. Now let $v_n = \sum_{i=1}^n n|u_n| \in L^+$ and note that the sequence $\{v_n\}$ is $\|\cdot\|$-Cauchy. Since L is a Banach lattice, it follows that $\{v_n\}$ is $\|\cdot\|$-convergent to some $v \in L$. From $0 \le v_n \uparrow$, we get $v_n \uparrow v$, and so $0 \le n|u_n| \le v$ holds for all n. But then we have $n \le \|nu_n\|_\infty \le \|v\|_\infty < \infty$ for all n, which is impossible. Hence the identity operator $I\colon (L, \|\cdot\|) \to (L, \|\cdot\|_\infty)$ is continuous and thus the Riesz norm $\|\cdot\|_\infty$ is equivalent to $\|\cdot\|$. ■

9.4. Chapter 4: Fatou Topologies

4.1. Note that L is super order dense in its Dedekind σ-completion L^σ. Let $\mathcal{N}$ be the base at zero for τ consisting of all solid and σ-order closed neighborhoods. For each $V \in \mathcal{N}$ put

$$V^\sigma = \{u^* \in L^\sigma\colon \text{ There exists } \{u_n\} \subseteq V \text{ with } 0 \le u_n \uparrow |u^*| \text{ in } L^\sigma\}.$$

To complete the rest of the solution now follow almost verbatim the arguments in the proof of Theorem 4.12. ■

4.2. In view of Theorem 4.13, we can assume without loss of generality that L is Dedekind σ-complete. Let $\mathcal{N}$ denote the collection of all solid σ-order closed τ-neighborhoods of zero. For each $V \in \mathcal{N}$ put

$$V^* = \{u^* \in L^\delta\colon \text{ There exists } v \in V \text{ with } |u^*| \le v\}.$$

That is, V^* is the solid hull of V in L^δ. The family $\{V^*\colon\ V \in \mathcal{N}\}$ satisfies the following properties:

(i) Each V^* is solid (and hence balanced) and absorbing.
(ii) If $V + V \subseteq W$, then $V^* + V^* \subseteq W^*$.
(iii) For every pair $V, W \in \mathcal{N}$ we have $(V \cap W)^* = V^* \cap W^*$.
(iv) For each $V \in \mathcal{N}$ we have $V^* \cap L = V$.
(v) Each V^* is σ-order closed in L^δ.
(vi) $\bigcap\{V^*\colon\ V \in \mathcal{N}\} = \{0\}$.

To see (v) let $0 \le u_n^* \uparrow u^*$ in L^δ with $\{u_n^*\} \subseteq V^*$. Choose some $v \in L$ with $u^* \le v$, and then for each n pick some $v_n \in V$ such that $u_n^* \le v_n \le v$. Since L is Dedekind σ-complete, the element $w_n = \bigwedge_{k=n}^\infty v_k$ exists in L for each n. Clearly, $\{w_n\} \subseteq V$. Also $0 \le w_n \uparrow \le v$ holds in L and so, using the Dedekind σ-completeness of L once more, we get $w_n \uparrow w$ in L. The σ-order closedness of V implies $w \in V$. It now follows that $u^* \le w$, and hence $u^* \in V^*$. So, V^* is σ-order closed.

The above properties establish that the collection $\{V^*\colon\ V \in \mathcal{N}\}$ is a base for a Hausdorff σ-Fatou topology τ^* on L^δ that induces τ on L. ■

4.3. Let $\mathcal{N}$ be a base at zero for τ' consisting of solid sets. For each $V \in \mathcal{N}$ put $V^* = \{u \in L\colon\ T(|u|) \in V\}$. The family of sets $\mathcal{N}^* = \{V^*\colon\ V \in \mathcal{N}\}$ has the following properties:

(i) Each V^* is solid (and hence balanced) and absorbing.
(ii) If $V + V \subseteq W$, then $V^* + V^* \subseteq W^*$.
(iii) For every pair $V, W \in \mathcal{N}$ we have $(V \cap W)^* = V^* \cap W^*$.

Thus $\mathcal{N}^*$ generates a locally solid topology τ^* on L which is also a Lebesgue topology. Indeed, if $u_\alpha \downarrow 0$ in L, then $T(u_\alpha) \downarrow 0$ in M (since T is order continuous), and so $T(u_\alpha) \xrightarrow{\tau'} 0$, since τ' is a Lebesgue topology. Hence, if $V \in \mathcal{N}$ is given, then $T(u_\alpha) \in V$ for all $\alpha \succeq \alpha_0$, so that $u_\alpha \in V^*$ for all $\alpha \succeq \alpha_0$. Hence $u_\alpha \xrightarrow{\tau^*} 0$.

Clearly, $T\colon (L, \tau^*) \to (M, \tau')$ is continuous. Now let A be an order bounded subset of L. By Theorem 4.21 we have $\tau^* \subseteq \tau$ on A, and so the identity mapping $I\colon (A, \tau) \to (A, \tau^*)$ is continuous. Now, the scheme of continuous functions $(A, \tau) \xrightarrow{I} (A, \tau^*) \xrightarrow{T} (M, \tau')$ implies that the mapping $T\colon (A, \tau) \to (M, \tau')$ is continuous. ■

4.4. Recall that an ideal J of a Riesz space L is order dense if and only if $J^{\mathrm{d}} = \{0\}$.

Now note that $C_\tau = \bigcup_{\alpha \in A} N_\alpha^{\mathrm{d}}$ implies $C_\tau^{\mathrm{d}} = \bigcap_{\alpha \in A} N_\alpha^{\mathrm{dd}}$. Thus C_τ is order dense in L if and only if $\bigcap_{\alpha \in A} N_\alpha^{\mathrm{dd}} = \{0\}$. ■

4.5. Let $\mathcal{N}$ be the base at zero for τ consisting of Fatou τ-neighborhoods of zero. Then a base at zero for τ^δ is $\{V^\delta\colon\ V\in\mathcal{N}\}$; see Theorem 4.12. Let $\{V_n^\delta\}$ be a normal sequence from these basic τ^δ-neighborhoods of L^δ and let $M=\bigcap_{n=1}^\infty V_n^\delta$. Note that $M\cap L=\bigcap_{n=1}^\infty(V_n^\delta\cap L)=\bigcap_{n=1}^\infty V_n=N$. Now since N is order dense in M, it follows that the disjoint complement M^{d} of M in L^δ satisfies $M^{\mathrm{d}}\cap L=N^{\mathrm{d}}$, where now N^{d} is the disjoint complement of N in L. From this equation, we easily get $C_{\tau^\delta}\cap L=C_\tau$. ■

4.6. Let $u\in L$ and let V be a solid τ-neighborhood of zero. Choose a normal sequence $\{V_n\}$ of solid and order closed τ-neighborhoods of zero with $V_1\subseteq V$. Then $N=\bigcap_{n=1}^\infty V_n$ is a projection band and so $L=N\oplus N^{\mathrm{d}}$. Write $u=u_1+u_2$ with $u_1\in N$ and $u_2\in N^{\mathrm{d}}\subseteq C_\tau$. Thus $u_2=u-u_1\in u+V$ and $u_2\in C_\tau$. This implies $(u+V)\cap C_\tau\neq\emptyset$ and thus C_τ is τ-dense in L. ■

4.7. Consider the set

$$A=\big\{f\in L\colon\ \exists \text{ an at most countable set } E \text{ with } f(x)=0 \text{ for all } x\notin E\big\}\,.$$

We claim that $C_\tau=A$.

To see this, let $\{V_n\}$ be any normal sequence consisting of solid basic τ-neighborhoods of zero. For each n there exist some $k_n\in\mathbb{N}$, some $\epsilon_n>0$, and a finite subset F_n of $[0,1]$ such that $V_n=W_{F_n,k_n,\epsilon_n}$; see Example 3.6 for the definition of the set W_{F_n,k_n,ϵ_n}. Put $E=\bigcup_{n=1}^\infty F_n$ and note that E is at most countable. Let

$$B=\Big\{f\in L\colon\ \int_0^1|f(x)|^p\,dx=0 \text{ and } f(x)=0 \text{ for all } x\in E\Big\}\,,$$

and note that $B\subseteq N=\bigcap_{n=1}^\infty V_n$. This implies

$$N^{\mathrm{d}}\subseteq B^{\mathrm{d}}=\big\{f\in L\colon\ f(x)=0 \text{ for all } x\notin E\big\}\subseteq A\,.$$

Therefore, $C_\tau\subseteq A$.

On the other hand, if $f\in A$, then there exists an at most countable set E with $f(x)=0$ for all $x\notin E$. If E is finite put $V_n=W_{E,2^n,2^{-n}}$ and if $E=\{x_1,x_2,\dots\}$ is a countable set put $V_n=W_{E_n,2^n,2^{-n}}$, where $E_n=\{x_1,\dots,x_n\}$. Then $\{V_n\}$ is a normal sequence of solid τ-neighborhoods of zero and $f\in N^{\mathrm{d}}$ (where $N=\bigcap_{n=1}^\infty V_n$), so that $f\in C_\tau$. Hence $A\subseteq C_\tau$, and our claim has been established.

The carrier C_τ is order dense in L since the characteristic functions of the points belong to C_τ. However, C_τ is not τ-dense in L. Indeed, if $\mathbf{1}$ denotes the constant function one on $[0,1]$, then $\mathbf{1}$ is not in the τ-closure of C_τ. For if, there exists a net $\{f_\alpha\}\subseteq C_\tau$ with $f_\alpha\xrightarrow{\tau}\mathbf{1}$, then

$$1=\int_0^1[\mathbf{1}(x)]^p\,dx=\int_0^1\big|1-f_\alpha(x)\big|^p\,dx\longrightarrow 0\,,$$

which is impossible.

The topology τ is not a Fatou topology. To see this, assume by way of contradiction that τ is a Fatou topology. Let $E=\{0\}$. Choose a Fatou τ-neighborhood V of zero with $V\subseteq W_{E,1,1}$, and then pick another solid τ-neighborhood $W_{G,n,\epsilon}$ such that $W_{G,n,\epsilon}\subseteq V$. Now the net $\{\chi_F\}$ of the characteristic functions of the finite subsets of the set $[0,1]\setminus G$ satisfies $\{\chi_F\}\subseteq W_{G,n,\epsilon}$ and $0\le\chi_F\uparrow u=\chi_{[0,1]\setminus G}$ in L. Therefore, $u\in V$ and thus $u\in W_{E,1,1}$. But $\int_0^1|u(x)|^p\,dx=1$ and so $u\notin W_{E,1,1}$, a contradiction. Thus, τ is not a Fatou topology.

Another way of proving that τ is not a Fatou topology is by observing that L is not order dense in its topological completion (see part (4) of Example 3.6) and then by applying Theorem 4.31. ■

4.8. By Theorem 4.17 the carrier C_τ is an order dense σ-ideal of L. Now let $0 \le \phi \in L^\sim$. Since ϕ acts on C_ϕ as a strictly positive linear functional, it follows from Theorem 1.45 that C_ϕ has the countable sup property.

Now let $0 \le u \in C_\phi$. Pick a net $\{u_\alpha\} \subseteq C_\tau$ with $0 \le u_\alpha \uparrow u$ in L; clearly $\{u_\alpha\} \subseteq C_\phi$. Choose a sequence $\{u_{\alpha_n}\} \subseteq \{u_\alpha\}$ with $0 \le u_{\alpha_n} \uparrow u$ in L. But then since C_τ is a σ-ideal of L, it follows that $u \in C_\tau$. So, $C_\phi \subseteq C_\tau$. ■

4.9. (i) Note that K_τ, as the intersection of all σ-Fatou τ-neighborhoods of zero, is a σ-ideal. It is also a τ-closed ideal since it can also be taken to be the intersection of all τ-closed neighborhoods of zero.

(ii) Apply Theorem 4.48. ■

4.10. Here K_τ is a band of L, and therefore, K_τ is a projection band. The conclusion now follows from Theorem 4.45. ■

4.11. The solution below is due to D. H. Fremlin. We start the solution with three remarks clarifying a few points.

Remark 1: *For each $f \in K_1$ and each $\epsilon > 0$ the set $\mathcal{O}_\epsilon$ is at most countable.*

Note first that each subset of $[0,1]$ with no strictly decreasing sequence is at most countable. Indeed, if A is such a subset of $[0,1]$ consider the set

$$S = \big\{s \in [0,1]\colon \ \{x \in A\colon \ x \le s\} \text{ is at most countable}\big\},$$

and let $x = \sup S$. Clearly, $x \in S$, and we claim that $x = 1$. For if $x < 1$, then for each $\delta > 0$ with $x + \delta < 1$, the interval $[x, x+\delta]$ contains uncountably many elements of A. In particular, there exists a strictly decreasing sequence of A, which is impossible. Thus $x = 1$, and so the set $A = \{x \in A\colon \ x \le 1\}$ is at most countable.

To complete the proof of Remark 1, we shall show that $\mathcal{O}_\epsilon$ does not contain strictly decreasing sequences. To see this, assume by way of contradiction that a sequence $\{x_n\}$ of $\mathcal{O}_\epsilon$ is strictly decreasing; let $x_n \downarrow x_0$ in $\mathbb{R}$. Choose a disjoint sequence $\{I_n\}$ of open intervals with $x_n \in I_n$ for each n. Subsequently, for each n pick $y_n, z_n \in I_n$ such that $\big|f(y_n) - f(x_n)\big| \ge \epsilon$. Note that $y_n \downarrow x_0$ and $z_n \downarrow x_0$. Hence $\lim f(y_n) = \lim f(z_n)$ (since $f \in K_1$) contradicting $\big|f(y_n) - f(z_n)\big| \ge \epsilon$ for all n. Therefore, $\mathcal{O}_\epsilon$ does not contain strictly decreasing sequences.

Remark 2: *Order convergence in K_1 and K_2 implies pointwise convergence.*

To see this, let $u_\alpha \downarrow 0$ in K_1. If there is some $x \in [0,1]$ with $u_\alpha(x) \ge \epsilon > 0$ for all α, then $v = \epsilon\chi_{\{x\}} \in K_1$ satisfies $0 < v \le u_\alpha$ for all α, contrary to $u_\alpha \downarrow 0$ in K_1.

Remark 3: *The vector subspace L is a Riesz subspace of K.*

Let $(u, v) \in L$. Then $(u,v)^+ = (u^+, v^+)$ holds in K. We must show that $(u^+, v^+) \in L$. To this end, let t be a point in $[0,1]$ where u^+ and v^+ are both continuous. There exists a sequence $\{t_n\} \subseteq [0,1]$ satisfying $t_n \to t$ such that u and v are both continuous at each t_n (this is possible by Remark 1). Since $(u,v) \in L$, we have $u(t_n) = v(t_n)$ for all n, and so

$$u^+(t_n) = \max\big\{u(t_n), 0\big\} = \max\big\{v(t_n), 0\big\} = v^+(t_n)$$

for all n. The continuity of u^+ and v^+ at t implies $u^+(t) = v^+(t)$, so that $(u^+, v^+) \in L$.

We now proceed by steps with the solution of the exercise.

1. *(L, τ) satisfies the Monotone Completeness Property.*

Let $\{(u_\alpha, v_\alpha)\}$ be a τ-Cauchy net of L satisfying $0 \le (u_\alpha, v_\alpha) \uparrow$ in L. This means that $0 \le u_\alpha \uparrow$ and $\{u_\alpha\}$ is a τ_1-Cauchy net of K_1, and similarly for $\{v_\alpha\}$. Since K_1 is τ_1-complete, there exists some $u \in K_1$ such that $u_\alpha \xrightarrow{\tau_1} u$; and similarly there exists some

$v \in K_2$ such that $v_\alpha \xrightarrow{\tau_2} v$. To complete the proof of this part, we need to show that $(u,v) \in L$. To this end, let t_0 be a point of continuity of both u and v. Assume that $u(t_0) \neq v(t_0)$; we can suppose without loss of generality that $u(t_0) < v(t_0)$. Then there exists a closed interval I of $[0,1]$ and constants a, b, and c such that:

$$u(t) \leq a < b < c \leq v(t) \text{ for all } t \in I. \tag{$\dagger$}$$

Choose a strictly increasing sequence $\{x_n\} \subseteq I$ with $x_n \uparrow x_0$. Also pick an index α_0 such that $0 \leq v(x_n) - v_{\alpha_0}(x_n) < c - b$ for all n (this is possible since $v_\alpha \xrightarrow{\tau_2} v$). In particular, we have $v_{\alpha_0}(x_n) \geq v(x_n) - c + b \geq b$ for all n. Hence

$$\lim_{t \uparrow x_0} v_{\alpha_0}(t) = \lim v_{\alpha_0}(x_n) \geq b > a. \tag{$\dagger\dagger$}$$

Next for each n choose some $y_n \in (x_n, x_{n+1})$ such that both u_{α_0} and v_{α_0} are continuous—this is possible since the v_{α_0} and u_{α_0} have at most countably many discontinuities. Thus $u_{\alpha_0}(y_n) = v_{\alpha_0}(y_n)$ holds for all n. It now follows from ($\dagger$) that $v_{\alpha_0}(y_n) = u_{\alpha_0}(y_n) \leq u(y_n) \leq a$ for all n (note that we have used here the monotonicity of the net $\{u_\alpha\}$), and so

$$\lim_{t \uparrow x_0} v_{\alpha_0}(t) = \lim v_{\alpha_0}(y_n) \leq a. \tag{$\dagger\dagger\dagger$}$$

Now ($\dagger\dagger\dagger$) contradicts ($\dagger\dagger$), so that $u(t_0) = v(t_0)$. This implies $(u,v) \in L$, and therefore (L,τ) has the Monotone Completeness Property.

2. *L is τ_3-dense in K, that is $(\widehat{L}, \widehat{\tau}) = (K, \tau_3)$.*

If this is established, then it will be clear that the locally solid Riesz space (L,τ) cannot be τ-complete. To this end, let $(f,g) \in K_1 \times K_2$ and let m sequences $s_i = (t_0^i, t_1^i, t_2^i, t_3^i, \ldots)$ (where $i = 1, \ldots, k, k+1, \ldots, m$) be such that $t_n^i \uparrow t_0^i$ holds for $i = 1, \ldots, k$ and $t_n^i \downarrow t_0^i$ for $i = k+1, \ldots, m$. We shall show that there exists $(u,v) \in L$ satisfying $\rho_{s_i}(f-u) = 0$ for $i = 1, \ldots, k$ and $\rho_{s_i}(g - v) = 0$ for $i = k+1, \ldots, m$. If this is established, then it will be immediate that $(\widehat{L}, \widehat{\tau}) = (K, \tau_3)$.

Let A be the collection of all real numbers appearing in the sequences $s_1, \ldots, s_k$ and likewise let B be the collection of all real numbers appearing in the sequences $s_{k+1}, \ldots, s_m$. Note that both A and B are compact sets. Clearly, A does not contain strictly increasing sequences and B does not contain strictly decreasing sequences. This implies that $A \cap B$ is a finite (possibly empty) subset of $[0,1]$. We can also assume without loss of generality that both A and B contain 0 and 1. In particular, observe that we can write $[0,1] \setminus A \cup B = \bigcup_{\lambda \in \Lambda} (a_\lambda, b_\lambda)$, where Λ is at most countable such that $\lambda \neq \mu$ implies $(a_\lambda, b_\lambda) \cap (a_\mu, b_\mu) = \emptyset$. Now define the function $u\colon [0,1] \to \mathbb{R}$ by

(i) $u(x) = f(x)$ if $x \in A$.
(ii) $u(x) = g(x)$ if $x \in B \setminus A$.
(iii) If $x \in [0,1] \setminus A \cup B$, then there exists exactly one λ with $x \in (a_\lambda, b_\lambda)$; clearly a_λ and b_λ belong to $A \cup B$. Put:
 (a) $u(x) = \frac{b_\lambda - x}{b_\lambda - a_\lambda} u(a_\lambda) + \frac{x - a_\lambda}{b_\lambda - a_\lambda} u(b_\lambda)$ if neither a_λ nor b_λ belongs to $A \cap B$,
 (b) $u(x) = u(b_\lambda)$, if $a_\lambda \in A \cap B$ but $b_\lambda \notin A \cap B$.
 (c) $u(x) = u(a_\lambda)$, if $a_\lambda \notin A \cap B$ but $b_\lambda \in A \cap B$.
 (d) $u(x) = 0$, if a_λ and b_λ both belong to $A \cap B$.

We see that u is continuous at every point of $[0,1] \setminus A \cup B$, so that $\lim_{t \downarrow x} u(t)$ exists for every $x \in [0,1] \setminus A \cup B$. On the other hand, take any $x \in A \cup B$ with $x \neq 1$. As B contains no strictly decreasing sequences, there exists some $\eta > 0$ with $(x, x+\eta) \cap B = \emptyset$. If there is $0 < \eta_1 < \eta$ with $(x, x+\eta_1) \cap A = \emptyset$, then $(x, x+\eta_1) \subseteq A \cup B$; so $(x, x+\eta_1) \subseteq (a_\lambda, b_\lambda)$ for some λ, and hence $\lim_{t \downarrow x} u(t)$ exists. Assume that $(x, x+\delta) \cap A \neq \emptyset$ for every $0 < \delta < \eta$.

Now given $\epsilon > 0$, choose $0 < \delta < \eta$ with $x + \delta \in A$ and $|f(t) - s| < \epsilon$ for all $x < t < x+\delta$, where $s = \lim_{t\downarrow x} f(t)$. Now for each $x < t < x+\delta$ we have two possibilities:

(1) $t \in A$; in this case $|u(t) - s| = |f(t) - s| < \epsilon$.

(2) $t \notin A$; in this case there exists some λ with $t \in (a_\lambda, b_\lambda)$. Since $x \in A \cup B$ and $x + \delta \in A \cup B$, we have $(a_\lambda, b_\lambda) \subseteq (x, x+\delta)$. Thus, a_λ and b_λ do not belong to B, and since $a_\lambda, b_\lambda \in A \cup B$ we have a_λ and b_λ in A. Now case (iii)(a) is applicable, and so by using that $u(a_\lambda) = f(a_\lambda)$ and $u(b_\lambda) = f(b_\lambda)$ we get

$$\begin{aligned} \big|u(t) - s\big| &= \big|\tfrac{b_\lambda - t}{b_\lambda - a_\lambda} f(a_\lambda) + \tfrac{t - a_\lambda}{b_\lambda - a_\lambda} f(b_\lambda) - s\big| \\ &\le \tfrac{b_\lambda - t}{b_\lambda - a_\lambda}\big|f(a_\lambda) - s\big| + \tfrac{t - a_\lambda}{b_\lambda - a_\lambda}\big|f(b_\lambda) - s\big| \\ &< \tfrac{b_\lambda - t}{b_\lambda - a_\lambda}\epsilon + \tfrac{t - a_\lambda}{b_\lambda - a_\lambda}\epsilon = \epsilon\,. \end{aligned}$$

Therefore, $\lim_{t\downarrow x} u(t) = s$, and thus $u \in K_1$.

Next define v in the same way, but with the roles of A and B in (i) and (ii) interchanged. Then we have $v \in K_2$. Since $u = v$ except possibly on the finite set $A \cap B$, we have $u = v$ on a dense set, so certainly $u(x) = v(x)$ at any point x where u and v are both continuous. Thus $(u, v) \in L$ is an element satisfying the requirements stated at the beginning of this part.

3. *A special subset of L.*

Consider the subset of L defined as follows:

$$S = \big\{(u,0) \in L\colon\ u(1) = 0 \text{ and } |u(x)| \le 1 \text{ for all } x \in [0,1]\big\}\,.$$

Then S has the following properties.

(α) *S is a solid subset of L.*

If $|(v,0)| \le |(u,0)|$ holds in L with $(u,0) \in S$ and $(v,0) \in L$, then $|v| \le |u|$ holds. Therefore, $v(1) = 0$ and $|v(x)| \le 1$ for all $x \in [0,1]$, that is, $(v,0) \in S$.

(β) *S is order closed in L.*

If $\{(u_\alpha, 0)\} \subseteq S$ satisfies $(u_\alpha, 0) \xrightarrow{o} (u,0)$ in L, then $u_\alpha(x) \to u(x)$ for all $x \in [0,1]$. Thus $u(1) = 0$ and $|u(x)| \le 1$ for all $x \in [0,1]$, i.e., $(u,0) \in S$.

(γ) *S as a subset of K is τ_3-closed.*

If $\{(u_\alpha, 0)\} \subseteq S$ satisfies $(u_\alpha, 0) \xrightarrow{\tau_3} (f,g)$ in K, then from $u_\alpha(x) \to f(x)$ for all $x \in [0,1]$ it follows that $|f(x)| \le 1$ for all $x \in [0,1]$, and clearly $f(1) = 0$. Similarly, $g(x) = 0$ for all $x \in [0,1]$. To see that $(f,0) \in L$, let f be continuous at x_0. If $x_0 = 1$, then $f(x_0) = f(1) = 0$; so assume $x_0 < 1$. Choose a strictly decreasing sequence $\{x_n\} \subseteq [0,1]$ with $x_n \downarrow x_0$.

Given $\epsilon > 0$, select some α such that $|u_\alpha(x_n) - f(x_n)| < \epsilon$ for all n. Then for each $n = 1, 2, \ldots$ choose some $y_n \in (x_{n+1}, x_n)$ such that the function u_α is continuous at y_n. It follows that $u_\alpha(y_n) = 0$ for all n, and consequently $\lim u_\alpha(x_n) = \lim u_\alpha(y_n) = 0$. Next, choose m so that $|u_\alpha(x_m)| < \epsilon$ and $|f(x_0) - f(x_m)| < \epsilon$ (the last inequality is possible since the function f is continuous at x_0). Then we have

$$|f(x_0)| \le \big|f(x_0) - f(x_m)\big| + \big|f(x_m) - u_\alpha(x_m)\big| + |u_\alpha(x_m)| < 3\epsilon$$

for all $\epsilon > 0$. Hence $f(x_0) = 0$, as desired.

(δ) *S is not order closed in K.*

Let $\{\chi_F\}$ be the net of characteristic functions of the finite subsets of $[0,1)$. Then we have $\{(\chi_F, 0)\} \subseteq S$ and $(\chi_F, 0) \uparrow (\chi_{[0,1)}, 0)$ in K. Since $(\chi_{[0,1)}, 0) \notin S$, it follows that S is not order closed in K. ■

4.12. Apply Theorem 4.32. ■

4.13. Assume that $0 \le \dot{u}_\alpha \uparrow \dot{u}$ in L/I_ρ. We must show that $\dot{\rho}(\dot{u}_\alpha) \uparrow \dot{\rho}(\dot{u})$. Note first that without loss of generality, we can assume that $0 \le u_\alpha \uparrow \le u$ holds in L. Since L is Archimedean, there exists a net $\{v_\lambda\} \subseteq L^+$ such that $v_\lambda \downarrow \le u$ and $v_\lambda - u_\alpha \downarrow_{(\lambda,\alpha)} 0$. In particular, we have $0 \le u + u_\alpha - v_\lambda \uparrow_{(\lambda,\alpha)} u$, and therefore $\rho(u + u_\alpha - v_\lambda) \uparrow_{(\lambda,\alpha)} \rho(u)$, by the Fatou property of ρ.

Now note that $u_\alpha \le v_\lambda \le u$ for all α and λ implies $\dot{u}_\alpha \le \dot{v}_\lambda \le \dot{u}$ in L/I_ρ and so, since $\dot{u}_\alpha \uparrow \dot{u}$, we get $\dot{v}_\lambda = \dot{u}$ for all λ. Thus $\rho(u - v_\lambda) = 0$, i.e., $u - v_\lambda \in I_\rho$ for all λ, and so $\dot{\rho}(\dot{u}_\alpha) = \rho(u_\alpha) = \rho(u + u_\alpha - v_\lambda) \uparrow \rho(u) = \dot{\rho}(\dot{u})$. ■

4.14. Repeat the arguments of the previous exercise. ■

4.15. (i) The solution is similar to that of Exercise **1.24**.

(ii) It should be clear that $u \in I_\rho$ implies $u(x_i) = 0$ for $i = 1, 2, \ldots$. On the other hand, if $u \in L$ satisfies $u(x_i) = 0$ for every $i = 1, 2, \ldots$, then $u(\infty) = \lim u(x_n) = 0$, so that $\rho(u) = 0$. Now since order convergence implies pointwise convergence, it easily follows that I_ρ is a band.

(iii) Let $u_n = \chi_{\{x_1,\ldots,x_n\}}$ for $n = 1, 2, \ldots$, and note that $0 \le u_n \le e$ holds in L, where $e(x) = 1$ for all $x \in X$. We claim that $0 \le \dot{u}_n \uparrow \dot{e}$ holds in L/I_ρ, although $\sup\{u_n\}$ does not exist in L. Indeed, if $w \in L$ satisfies $u_n \le w \le e$ for all n, then $w(x_i) = 1$ for all i, and so $w(\infty) = \lim w(x_n) = 1$. Thus $\rho(e - w) = 0$, that is, $\dot{w} = \dot{e}$. Since L is Archimedean, there exists a net $\{w_\lambda\} \subseteq L$ satisfying $w_\lambda \downarrow \le e$ and $w_\lambda - u_n \downarrow_{(\lambda,n)} 0$ in L. Now since I_ρ is a band of L, we have $\dot{w}_\lambda - \dot{u}_n \downarrow_{(\lambda,n)} 0$ in L/I_ρ, that is, $\dot{e} - \dot{u}_n \downarrow 0$ in L/I_ρ. Hence $0 \le \dot{u}_n \uparrow \dot{e}$ holds in L/I_ρ.

Now note that $\dot{\rho}(\dot{e}) = \rho(e) = 2$ and $\dot{\rho}(\dot{u}_n) = \rho(u_n) = 1$ hold true for all n, so that $\lim \dot{\rho}(\dot{u}_n) \ne \dot{\rho}(\dot{e})$. Therefore, $\dot{\rho}$ is not a σ-Fatou norm.

The above conclusion does not contradict the previous exercise since L is not almost Dedekind σ-complete. ■

4.16. (a) Clearly, I_ρ is a solid set which (by the triangle inequality) is also closed under addition. The only thing missing is the closedness of I_ρ under scalar multiplication. To see this, let $u \in I_\rho$ and $\lambda \in \mathbb{R}$. Pick some $n \in \mathbb{N}$ with $|\lambda| \le n$ and note that from $|\lambda u| \le n|u|$, we get

$$0 \le \rho(\lambda u) \le \rho(nu) \le n\rho(u) = 0\,.$$

This implies $\rho(\lambda u) = 0$, i.e., $\lambda u \in I_\rho$, so that I_ρ is an ideal.

(b) To see that ρ is a Riesz pseudonorm notice first that if $r\colon L \to \mathbb{R}$ is any Riesz pseudonorm on a Riesz space L, then the function $r^*\colon L \to \mathbb{R}$, defined by $r^*(u) = \frac{r(u)}{1+r(u)}$, is also a Riesz pseudonorm—which is also τ-continuous if r is also τ-continuous for some locally solid topology τ on L.

The τ-continuity of ρ follows easily from the fact that for each k we have

$$\rho(u) = \sum_{n=1}^{\infty} \tfrac{1}{2^n} \cdot \tfrac{\rho_n(u)}{1+\rho_n(u)} \le \tfrac{1}{2^k} + \sum_{n=1}^{k} \tfrac{1}{2^n} \cdot \tfrac{\rho_n(u)}{1+\rho_n(u)}\,.$$

The identity $I_\rho = \bigcap_{n=1}^{\infty} I_{\rho_n}$ is a consequence of the property that $\rho(u) = 0$ if and only if $\rho_n(u) = 0$ for each n.

(c) Let τ be a locally solid topology on a Riesz space L. Assume that $\rho \in \mathcal{F}$, i.e., ρ is a τ-continuous Riesz pseudonorm. For each $n \in \mathbb{N}$ let $V_n = \{u \in L\colon\ \rho(u) < \frac{1}{2^n}\}$ and note that $\{V_n\}$ is a normal sequence of solid τ-neighborhoods of zero satisfying $I_\rho = \bigcap_{n=1}^{\infty} V_n$. This implies $I_\rho^{\mathrm{d}} \subseteq C_\tau$ and so $\bigcup_{\rho \in \mathcal{F}} I_\rho^{\mathrm{d}} \subseteq C_\tau$.

For the converse inclusion, assume that $\{V_n\}$ is a normal sequence of τ-neighborhoods of zero and let $N=\bigcap_{n=1}^{\infty} V_n$. Then, as in the proof of Theorem 2.28, for each n there is some $\epsilon_n>0$ and a τ-continuous Riesz pseudonorm ρ_n such that $\{u \in L\colon\ \rho_n(u)<\epsilon_n\} \subseteq V_n$. Now consider the τ-continuous Riesz pseudonorm ρ defined by

$$\rho(u)=\sum_{n=1}^{\infty} \tfrac{1}{2^n} \cdot \tfrac{\rho_n(u)}{1+\rho_n(u)}\,.$$

Clearly, $I_\rho \subseteq N$. This implies $N^{\mathrm{d}} \subseteq I_\rho^{\mathrm{d}}$, and from this we infer that $C_\tau \subseteq \bigcup_{\rho\in\mathcal{F}} I_\rho^{\mathrm{d}}$. Consequently, $C_\tau=\bigcup_{\rho\in\mathcal{F}} I_\rho^{\mathrm{d}}$. ■

4.17. Let τ be a Fatou topology on a Riesz space L. If τ is a Hausdorff topology, then C_τ is order dense in L by part (ii) of Theorem 4.17.

Now assume that C_τ is order dense in L and let $u \neq 0$. Choose $w \in C_\tau$ with $0<w \leq |u|$, and then select a normal sequence $\{V_n\}$ of solid τ-neighborhoods of zero with $w \in N^{\mathrm{d}}$, where $N=\bigcap_{n=1}^{\infty} V_n$. Clearly, there exists some n such that $w \notin V_n$, and consequently $u \notin V_n$. Thus τ is a Hausdorff topology. (The last argument actually shows that if τ is a locally solid topology and C_τ is order dense in L, then τ is a Hausdorff topology.) ■

9.5. Chapter 5: Metrizability

5.1. Let $\mathcal{N}=\{V_1,V_2,\ldots\}$ be a countable base at zero for τ, and let D be an order bounded disjoint subset of L. Put $D_n=\{u\in D\colon\ u\notin V_n\}$ and note that $D=\bigcup_{n=1}^{\infty} D_n$. We claim that each D_n is finite. Indeed, if $\{u_n\}$ is a sequence of distinct elements of some D_k, then by the pre-Lebesgue property, we must have $u_n \xrightarrow{\tau} 0$ (see Theorem 3.22), and therefore $u_n \in V_k$ holds true for all sufficiently large n, a contradiction. Thus, each D_k is finite and so D is at most countable. ■

5.2. Let $u_\alpha \downarrow 0$ in L. Then $\{u_\alpha\}$ is a τ-Cauchy net, and so $u_\alpha \xrightarrow{\widehat{\tau}} \widehat{u}$ in $\widehat{L}$. Choose a sequence $\{u_{\alpha_n}\} \subseteq \{u_\alpha\}$ with $u_{\alpha_n}\!\downarrow$ and $u_{\alpha_n} \xrightarrow{\widehat{\tau}} \widehat{u}$. Now if $0\leq w\leq u_{\alpha_n}$ holds in L for all n, then $0 \leq w \leq \widehat{u} \leq u_\alpha$ for all α, and so $w=0$. Thus $u_{\alpha_n}\!\downarrow 0$ in L, and hence L has the countable sup property. ■

5.3. If (L,τ) satisfies the Lebesgue property, then it obviously satisfies the pseudo-Lebesgue property and by Theorem 3.23 it also satisfies the pre-Lebesgue property.

Conversely, if (L,τ) satisfies both the pseudo-Lebesgue and pre-Lebesgue properties and $u_\alpha \downarrow 0$, then $\{u_\alpha\}$ is a τ-Cauchy net, by the pre-Lebesgue property, and hence $u_\alpha \xrightarrow{\tau} 0$ by the pseudo-Lebesgue property. ■

5.4. Fix a normal base $\{V_n\}$ at zero for τ, and let V be a balanced τ-neighborhood of zero. We claim that there exists some n such that the solid hull of V_n is included in V. To see this, assume that the above claim is false. Then $\mathrm{Sol}\left(\frac{1}{n}V_n\right) \not\subseteq V$ for each n. Choose a sequence $\{u_n\}$ such that $u_n \in \mathrm{Sol}\left(\frac{1}{n}V_n\right)$ and $u_n \notin V$ for all n. Next for each n pick $v_n \in V_n$ with $|u_n| \leq \frac{1}{n}|v_n|$. But then the sequence $\{nu_n\}$ lies in the solid hull of the sequence $\{v_n\}$, which is τ-bounded as being τ-convergent to zero—due to the fact that

$v_n \in V_n$ for each n. It now follows from our hypothesis that $\{nu_n\}$ is τ-bounded, and so, in particular, $nu_n \in kV$ for some k. This implies that $u_n \in V$ eventually, contrary to the choice of $\{u_n\}$. Therefore, τ is a locally solid topology. ■

5.5. The σ-Lebesgue property implies that every τ-closed ideal is a σ-ideal. Now let A be a σ-ideal and let $u \in \overline{A}$. Pick a sequence $\{u_n\} \subseteq A$ with $u_n \xrightarrow{\tau} u$. Put $w_n = |u| \wedge \bigvee_{i=1}^{n} |u_i|$ for each n, and note that $\{w_n\} \subseteq A$. Now $0 \leq |u| - w_n \leq |u - u_n|$ implies $0 \leq w_n \uparrow |u|$ in L. Hence $|u| \in A$ and so $u \in A$. This shows that A is τ-closed. ■

5.6. It should be clear that $\widehat{T}$ is positive. Let $\widehat{f}_n \downarrow 0$ in $\widehat{L}$. As in the first part of the proof of Theorem 5.36, there exists a sequence $\{\widehat{u}_n\}$ of upper elements of L with $\widehat{f}_n \leq \widehat{u}_n$ for each n and with $\widehat{u}_n \downarrow 0$. It is enough to show that $\widehat{T}(\widehat{u}_n) \downarrow 0$ in $\widehat{M}$. To this end, assume $0 \leq \widehat{h} \leq \widehat{T}(\widehat{u}_n)$ for all n and some $\widehat{h} \in \widehat{M}$. We must show that $\widehat{h} = 0$.

Let W be a solid $\widehat{\tau^*}$-neighborhood of zero in $\widehat{M}$. Choose another solid $\widehat{\tau^*}$-neighborhood W_1 of zero of $\widehat{M}$ satisfying $W_1 + W_1 \subseteq W$. Pick a solid τ-neighborhood V of zero of $\widehat{L}$ such that $\widehat{T}(\widehat{u}) \in W_1$ whenever $\widehat{u} \in V$. Let $\{V_n\}$ be a normal base at zero for $\widehat{L}$ with $V_1 + V_1 \subseteq V$. Now for each n choose $u_n \in L^+$ with $0 \leq u_n \leq \widehat{u}_n$ and $\widehat{u}_n - u_n \in V_n$. Put $w_n = \bigwedge_{i=1}^{n} u_i$ and note that $w_n \downarrow 0$ in L, and so $T(w_n) \downarrow 0$ in M. Therefore, $T(w_n) \xrightarrow{\tau^*} 0$ in M and, in particular, $T(w_n) \in W_1$ for all $n \geq n_1$. Now from

$$\begin{aligned} 0 \leq \widehat{u}_n - w_n &= \widehat{u}_n - \bigwedge_{i=1}^{n} u_i = \bigvee_{i=1}^{n} (\widehat{u}_n - u_i) \leq \bigvee_{i=1}^{n} (\widehat{u}_i - u_i) \\ &\leq \sum_{i=1}^{n} (\widehat{u}_i - u_i) \in V_1 + \cdots + V_n \subseteq V_1 + V_1 \subseteq V, \end{aligned}$$

it follows that $\widehat{u}_n - w_n \in V$ for all n. Hence

$$0 \leq \widehat{h} \leq \widehat{T}(\widehat{u}_n) = \widehat{T}(\widehat{u}_n - w_n) + T(w_n) \in W_1 + W_1 \subseteq W$$

for all $n \geq n_1$. So, $\widehat{h} \in W$ for all $\widehat{\tau^*}$-neighborhoods W of zero of $\widehat{M}$. Since $\widehat{\tau^*}$ is a Hausdorff topology, we infer that $\widehat{h} = 0$. ■

5.7. Let (L, τ) be a separable Fréchet space. Choose a countable dense subset $\{u_1, u_2, \ldots\}$ of L and let $\{V_n\}$ be a normal base at zero for τ consisting of solid sets. For each n pick some $\lambda_n > 0$ such that $\lambda_n |u_n| \in V_n$. Put $w_n = \sum_{i=1}^{n} \lambda_i |u_i|$ for each n. Then $w_n \uparrow$ and since

$$0 \leq w_{n+p} - w_n = \sum_{i=n+1}^{n+p} \lambda_i |u_i| \in V_{n+1} + \cdots + V_{n+p} \subseteq V_n ,$$

we get that $\{w_n\}$ is a τ-Cauchy sequence. So, there exists some $w \in L$ with $w_n \xrightarrow{\tau} w$ and, of course, $0 \leq w_n \uparrow w$ holds in L.

We claim that w is a weak order unit of L. Indeed, if $w \wedge u = 0$, then it should be clear that $(\lambda_n |u_n|) \wedge u = 0$ for all n and so $|u_n| \wedge u = 0$ for all n. Since $\{u_1, u_2, \ldots\}$ is τ-dense in L, it follows that $u = u \wedge u = 0$. This shows that w is indeed a weak order unit of the Riesz space L. ■

5.8. Let $\{V_n\}$ be a normal base at zero for τ consisting of solid sets. Choose a subsequence $\{v_n\}$ of the sequence $\{u_n\}$ such that $|v_n - u| \in \frac{1}{n} V_n$ for all n and put $w_n = \sum_{k=1}^{n} k|v_k - u|$ for each n. Then $w_n \uparrow$ and since

$$0 \leq w_{n+p} - w_n = \sum_{k=n+1}^{n+p} k|v_k - u| \in V_{n+1} + \cdots + V_{n+p} \subseteq V_n ,$$

it follows that $\{w_n\}$ is a τ-Cauchy sequence. Hence $w_n \xrightarrow{\tau} v$ holds for some $v \in L$, and thus $w_n \uparrow v$. In particular, we have $n|v_n - u| \leq w_n \leq v$ and so $|v_n - u| \leq \frac{1}{n}v$ for all n. ■

5.9. Let A be a maximal ideal in a Fréchet space (L, τ) and let $u \in \overline{A}$. Choose a sequence $\{u_n\} \subseteq A$ such that $u_n \xrightarrow{\tau} u$. By the previous exercise, we can assume without loss of generality that there exists some $v \in L^+$ such that $|u_n - u| \leq \frac{1}{n}v$ for all n.

If $v \in A$, then $|u| \leq |u_1| + v \in A$ and so $u \in A$. If $v \notin A$, then define

$$B = \{w \in L\colon\ (n|w| - v)^+ \in A \text{ for all } n\}.$$

An easy verification shows that B is an ideal of L satisfying $A \subseteq B$ and $v \notin B$. By the maximality property of A we obtain $B = A$. Now from $|u_n - u| \leq \frac{1}{n}v$ we get $|u| - \frac{1}{n}v \leq |u_n|$ and so $(n|u| - v)^+ \leq n|u_n| \in A$, from which it follows that $(n|u| - v)^+ \in A$ for all n. Consequently $u \in B = A$, and thus A is τ-closed. ■

5.10. (i) Hölder's inequality yields

$$\begin{aligned} \rho_1(u) &\geq \int_0^{\frac{1}{2}} |u(x)|\,dx + \sqrt{2}\int_{\frac{1}{2}}^1 |u(x)|\,dx \\ &\geq \int_0^{\frac{1}{2}} |u(x)|\,dx + \int_{\frac{1}{2}}^1 |u(x)|\,dx = \int_0^1 |u(x)|\,dx\,, \end{aligned}$$

similarly $\rho_2(u) \geq \int_0^1 |u(x)|\,dx$. On the other hand, the relation

$$[\rho_1 \wedge \rho_2](u) \leq \rho_1\big(\chi_{[0,\frac{1}{2}]}u\big) + \rho_2\big(\chi_{[\frac{1}{2},1]}u\big) = \int_0^1 |u(x)|\,dx\,,$$

implies that $[\rho_1 \wedge \rho_2](u) = \int_0^1 |u(x)|\,dx$ for all $u \in L$.

(ii) Note that:

$$\big[(\rho_1 \wedge \rho_2) \vee \rho_3\big](\mathbf{1}) = \max\Big\{\int_0^1 \mathbf{1}(x)\,dx, \rho_3(\mathbf{1})\Big\} = \max\{1, 1\} = 1\,.$$

(iii) Put $\rho^* = (\rho_1 \vee \rho_3) \wedge (\rho_2 \vee \rho_3)$ and note that given $\epsilon > 0$ there exists some $v \in L$ satisfying $0 \leq v \leq \mathbf{1}$ and

$$\max\{\rho_1(v), \rho_3(v)\} + \max\{\rho_2(\mathbf{1} - v), \rho_3(\mathbf{1} - v)\} \leq \rho^*(\mathbf{1}) + \epsilon\,. \qquad (\star)$$

If $m = \operatorname{ess\,inf}|v|$, then $\rho_3(v) + 1 - m = \rho_3(v) + \rho_3(\mathbf{1} - v) \leq \rho^*(\mathbf{1}) + \epsilon$ and so

$$c := 1 + \rho_3(v) - \rho^*(\mathbf{1}) - \epsilon \leq m\,. \qquad (\star\star)$$

Also since $0 \leq m\mathbf{1} \leq v$ and $0 \leq [1 - \rho_3(v)]\mathbf{1} \leq \mathbf{1} - v$ hold in L, and ρ_1 and ρ_2 are Riesz seminorms, we have

$$m\rho_1(\mathbf{1}) = \rho_1(m\mathbf{1}) \leq \rho_1(v) \quad \text{and} \quad \big[1 - \rho_3(v)\big]\rho_2(\mathbf{1}) \leq \rho_2(\mathbf{1} - v)\,. \qquad (\star\star\star)$$

Thus, using $(\star)$, $(\star\star)$ and $(\star\star\star)$, we obtain

$$c\rho_1(\mathbf{1}) + \big[1 - \rho_3(v)\big]\rho_2(\mathbf{1}) \leq m\rho_1(\mathbf{1}) + \rho_2(\mathbf{1} - v) \leq \rho_1(v) + \rho_2(\mathbf{1} - v) \leq \rho^*(\mathbf{1}) + \epsilon\,.$$

Now, taking into account that $\rho_1(\mathbf{1}) = \rho_2(\mathbf{1}) = \frac{1+\sqrt{2}}{2}$ and the definition of c from $(\star\star)$, we get from the last inequality that

$$c\rho_1(\mathbf{1}) + \big[1 - \rho_3(v)\big]\rho_2(\mathbf{1}) = \big[2 - \rho^*(\mathbf{1}) - \epsilon\big]\rho_1(\mathbf{1}) \leq \rho^*(\mathbf{1}) + \epsilon$$

for all $\epsilon > 0$. Therefore, $\big[2 - \rho^*(\mathbf{1})\big]\rho_1(\mathbf{1}) \leq \rho^*(\mathbf{1})$. Thus

$$\rho^*(\mathbf{1}) \geq \tfrac{2\rho_1(\mathbf{1})}{1+\rho_1(\mathbf{1})} = \tfrac{2(1+\sqrt{2})}{3+\sqrt{2}} > 1\,,$$

and so

$$\big[(\rho_1 \wedge \rho_2) \vee \rho_3\big](\mathbf{1}) < \big[(\rho_1 \vee \rho_3) \wedge (\rho_2 \vee \rho_3)\big](\mathbf{1})\,.$$

Consequently, $\mathcal{S}_L$ is not a distributive lattice. ■

5.11. (a) Since $u_\alpha \downarrow 0$ in L implies $u_\alpha(x) \downarrow 0$ for all $x \in X$, it easily follows that for every subset E of X the set $B_E = \{f \in L\colon\ f(x) = 0 \text{ for all } x \in E\}$ is a projection band of L, whose disjoint complement is the band B_{E^c} (where $E^c = X \setminus E$).

Let A be a band of L and put $E = \{x \in X\colon\ \exists f \in A \text{ with } f(x) \neq 0\}$. We claim that $A^d = B_E$. Clearly, $f \in A^d$ implies $f(x) = 0$ for all $x \in E$, i.e., $f \in B_E$. On the other hand, if $f \in B_E$ and $g \in A$, then for each $x \in X$ we have $f(x) = 0$ if $x \in E$ and $g(x) = 0$ if $x \notin E$, i.e., $|f| \wedge |g| = 0$, and thus $f \in A^d$.

Now observe that $A = A^{dd} = B_{E^c}$, and so A is a projection band. This shows that L has the projection property.

(b) Let $0 \le u_\alpha \uparrow u$ hold in L. Assume that there exists some $\epsilon > 0$ such that $\rho(u_\alpha) \le \rho(u) - \epsilon$ for all α. Choose some $x \in X$ such that $u(x) > \rho(u) - \epsilon$. But then, from $u_\alpha(x) \le \rho(u_\alpha) \le \rho(u) - \epsilon < u(x)$ for all α, we see that $u_\alpha(x) \not\to u(x)$, a contradiction. Hence $\rho(u_\alpha) \uparrow \rho(u)$, and so ρ is a Fatou norm.

(c) We claim that the norm completion of L_ρ is $\ell_\infty(X)$ (with the sup norm). To see this, pick some $u \in \ell_\infty(X)$ with $u \notin L$ and let $\epsilon > 0$. Put $S = \sup_{x\in X} u(x)$, $s = \inf_{x\in X} u(x)$ and $d = S - s$. Also choose some $n > 2$ such that $0 < \frac{d}{n} < \epsilon$. Now for each $k = 1, \ldots, n-1$ define the set

$$X_k = \left\{x \in X\colon\ s + (k-1)\tfrac{d}{n} \le u(x) < s + k\tfrac{d}{n}\right\},$$

and also let $X_n = \left\{x \in X\colon\ s + (n-1)\frac{d}{n} \le u(x) \le S\right\}$.

Observe that $X = \bigcup_{k=1}^n X_k$ and that $X_i \cap X_j = \emptyset$ if $i \neq j$. Now define $f\colon X \to \mathbb{R}$ via the formula $f(x) = s + (k-1)\frac{d}{n}$ if $x \in X_k$, and note that $f \in L$. Moreover, we have $|f(x) - u(x)| \le \frac{d}{n} < \epsilon$ for all x, and consequently $\|f - u\|_\infty < \epsilon$. This shows that the norm completion of L_ρ is $\ell_\infty(X)$. ■

5.12. Recall Tietze's Extension Theorem: *If C is a closed subset of a normal Hausdorff topological space X and $f\colon C \to [a,b]$ is a continuous function, where C is equipped with the induced topology and $[a,b]$ is a closed interval of $\mathbb{R}$, then there exists a continuous function $\phi\colon X \to [a,b]$ with $\phi(x) = f(x)$ for all $x \in C$.*

Let M be the Riesz space of all real-valued functions on $[0,1]$ whose restrictions to the rational and irrational numbers of $[0,1]$ are both continuous. Clearly, M is a Riesz space including $C[0,1]$ as a Riesz subspace. Let τ^* be the Hausdorff locally solid topology on M generated by the family of Riesz seminorms $\{\rho_{K_1,K_2}\}$ described in Example 5.30. We claim that τ^* is a complete topology. To see this, let $\{u_\alpha\}$ be a τ^*-Cauchy net of M. Then there exists a function $u \in \mathbb{R}^{[0,1]}$ such that $\{u_\alpha\}$ converges uniformly to u whenever restricted to any compact subset of rational or irrational numbers of $[0,1]$. We shall show that $u \in M$.

Indeed, since each u_α is continuous on every compact subset of rational numbers of $[0,1]$, u is continuous on any such set; hence if $r_n \to r$ (with all numbers rational), then $\{r_1, r_2, \ldots\} \cup \{r\}$ is a compact subset of rational numbers, and so $u(r_n) \to u(r)$. Hence, u restricted to the rational numbers of $[0,1]$ is continuous and, similarly, for the irrational numbers. So, $u \in M$ and the τ^*-completeness of M follows.

It remains to be shown that $C[0,1]$ is τ^*-dense in M. Observe first that the family $\{\rho_{K_1,K_2}\}$ is closed under finite suprema. Now let $f \in M$ and K_1 and K_2 be two compact subsets of rational and irrational numbers of $[0,1]$, respectively. Put $C = K_1 \cup K_2$, and note that f restricted to C is continuous. By Tietze's Extension Theorem, there exists some $u \in C[0,1]$ with $u(x) = f(x)$ for all $x \in C$. So, $\rho_{K_1,K_2}(f - u) = 0$, and therefore $(\widehat{L}, \widehat{\tau}) = (M, \tau^*)$. ■

5.13. The solution is identical to that of Theorem 5.24. We only need to observe that if E is a measurable set satisfying $\mu(E) > 0$, then for each $0 \le \epsilon \le \mu(E)$ the fact that μ is atomless guarantees the existence of a measurable subset F of E such that $\mu(F) = \epsilon$. ■

5.14. Assume $0 \le u_n \uparrow \le u$ in $C[0,1]$. Then $\int_0^1 u_n(x)\,dx \uparrow s < \infty$ in $\mathbb{R}$, and so

$$\rho(u_{n+p} - u_n) = \int_0^1 u_{n+p}(x)\,dx - \int_0^1 u_n(x)\,dx \xrightarrow[n,p\to\infty]{} s - s = 0\,.$$

Hence $\{u_n\}$ is a ρ-Cauchy sequence, and the pre-Lebesgue property follows.

Since $C[0,1]$ is not order dense in its topological completion $L_1[0,1]$, the Riesz seminorm ρ cannot satisfy the pseudo-Lebesgue property; see Theorem 5.29. Alternatively, pick a sequence $\{u_n\}$ of $C[0,1]$ with $u_n \downarrow 0$ and $\int_0^1 u_n(x) \ge \epsilon > 0$ for all n and some $\epsilon > 0$. Then $\{u_n\}$ is ρ-Cauchy but $\rho(u_n) \ge \epsilon$ for all n. (For the construction of such a sequence $\{u_n\}$ see [**90**, Exercise 18.14, p. 104].) ■

5.15. The solution will be based upon the following two properties.

(1) *If an Archimedean Riesz space admits a strictly positive linear functional, then its Dedekind completion also admits a strictly positive linear functional.*

To see this, let $\phi\colon L \to \mathbb{R}$ be a strictly positive linear functional on an Archimedean Riesz space. Define $p\colon L^\delta \to \mathbb{R}^+$ by

$$p(u) = \inf\{\phi(v)\colon\ v \in L \text{ and } v \ge |u|\}\,.$$

Then p is a sublinear mapping and satisfies $\phi(u) \le p(u)$ for all $u \in L$. By Theorem 1.69 there exists an extension ψ of ϕ to all of L^δ satisfying $\psi(u) \le p(u)$ for all $u \in L^\delta$. Clearly, $\psi \in (L^\delta)^\sim$. Now, by Theorem 1.67, for each $0 < u \in L^\delta$ we have

$$\begin{aligned}\psi^+(u) &= \sup\{\psi(v^*)\colon\ v^* \in L^\delta \text{ and } 0 \le v^* \le u\}\\ &\ge \sup\{\psi(v)\colon\ v \in L \text{ and } 0 \le v \le u\}\\ &= \sup\{\phi(v)\colon\ v \in L \text{ and } 0 \le v \le u\} > 0\,.\end{aligned}$$

Hence ψ^+ is a desired strictly positive linear functional on L^δ.

(2) *If a Riesz space L satisfies $L^\sim = \{0\}$ and a Dedekind complete Riesz space M admits a strictly positive linear functional, then $\mathcal{L}_b(L,M) = \{0\}$.*

Let $0 \le T \in \mathcal{L}_b(L,M)$ and let ϕ be a strictly positive linear functional on M. Then the formula $\psi(f) = \phi\big(T(f)\big)$ for all $f \in L$ defines a positive linear functional on L, and hence by our hypothesis $\psi = 0$. Therefore, $\phi\big(T(f)\big) = 0$ for all $f \in L^+$, and since ϕ is strictly positive, we get $T(f) = 0$ for all $f \in L^+$, i.e., $T = 0$. Thus, $\mathcal{L}_b(L,M) = \{0\}$.

In our case we have $L^\sim = \{0\}$; see Theorem 5.24. Also M admits a strictly positive linear functional, namely, $\phi(u) = \int_0^1 u(x)\,dx$, and hence by Property (1) the Dedekind completion M^δ of M likewise does. Now by Property (2) we obtain $\mathcal{L}_b(L,M^\delta) = \{0\}$. Finally, the desired conclusion follows from the inclusion $\mathcal{L}_b(L,M) \subseteq \mathcal{L}_b(L,M^\delta)$. This example does not contradict part (ii) of Exercise **1.21** since $L^\sim = \{0\}$. ■

5.16. Assume that a sequence $\{\Psi_n\} \subseteq L''$ satisfies $0 \le \Psi_n \uparrow \Psi$ in the Riesz space $(L')^\sim$. From $\Psi_n(\phi) \to \Psi(\phi)$ for all $\phi \in L'$, it follows that $\{\Psi_n\}$ is a $\sigma(L'',L')$-Cauchy sequence of L''. Since L'' is sequentially $\sigma(L'',L')$-complete (see [**78**, p. 396]), it follows that $\Psi_n \xrightarrow{\sigma(L'',L')} F$ for some $F \in L''$. So, $\Psi_n(\phi) \to F(\phi)$ for all $\phi \in L'$. This implies $F(\phi) = \Psi(\phi)$ for all $\phi \in L'$, i.e., $F = \Psi$. Hence L'' is a σ-ideal of $(L')^\sim$. ■

5.17. For every $\widehat{u} \in \widehat{L}_\rho$ put

$$\rho^*(\widehat{u}) = \inf\Big\{ \lim_{n\to\infty} \rho(u_n)\colon \ \{u_n\} \subseteq L^+, \ u_n\uparrow \ \text{and} \ u_n \wedge |\widehat{u}| \uparrow |\widehat{u}| \ \text{in} \ \widehat{L}_\rho \Big\}.$$

From Theorem 5.13 we know that $\widehat{\rho}$ is a σ-Fatou norm on $\widehat{L}_\rho$. Consequently, if $\{u_n\} \subseteq L^+$ satisfies $u_n\uparrow$ and $u_n \wedge |\widehat{u}| \uparrow |\widehat{u}|$, then it follows from $\widehat{\rho}(u_n \wedge |\widehat{u}|) \le \rho(u_n)$ that $\widehat{\rho}(\widehat{u}) = \lim \rho(u_n \wedge |\widehat{u}|) \le \lim \rho(u_n)$. Hence, for all $\widehat{u} \in \widehat{L}_\rho$ we have

$$\widehat{\rho}(\widehat{u}) \le \rho^*(\widehat{u}). \tag{$\star$}$$

On the other hand, if $\widehat{u} \in \widehat{L}_\rho$, then according to Theorem 5.4 there exists a sequence of upper elements $\{\widehat{u}_n\}$ with $\widehat{u}_n \downarrow |\widehat{u}|$ and $\lim \widehat{\rho}(\widehat{u}_n - |\widehat{u}|) = 0$. In particular, we have $\widehat{\rho}(\widehat{u}_n) \downarrow \widehat{\rho}(\widehat{u})$. Now for each n there exists a sequence $\{u_{n,k}\} \subseteq L^+$ with $u_{n,k}\uparrow_k$ and $u_{n,k}\uparrow_k \widehat{u}_n$. Therefore, $u_{n,k} \wedge |\widehat{u}| \uparrow_k |\widehat{u}|$ and $\lim_{k\to\infty} \rho(u_{n,k}) = \widehat{\rho}(\widehat{u}_n)$ for each n. Consequently, $\rho^*(\widehat{u}) \le \widehat{\rho}(\widehat{u})$ for all $\widehat{u} \in \widehat{L}_\rho$, and the desired conclusion follows from ($\star$). ■

5.18. Let (L, τ) be a topologically complete locally convex-solid Riesz space. Assume first that (L, τ) satisfies the B-property. If $0 \le u_n \uparrow\le u$ holds in L, then $\{u_n\}$ is a τ-bounded sequence, and so $\{u_n\}$ is a τ-Cauchy sequence. This implies that (L, τ) satisfies the pre-Lebesgue property and hence (by Theorem 3.24) L is Dedekind complete and τ is a Lebesgue topology. But then, by Theorem 3.13, the Riesz space L is an ideal of $(L')^\sim_{\rm n}$.

To see that L is also a band in $(L')^\sim_{\rm n}$ assume that $0 \le u_\alpha \uparrow F$ holds in $(L')^\sim_{\rm n}$ with $\{u_\alpha\} \subseteq L$. From $\phi(u_\alpha) = u_\alpha(\phi) \uparrow F(\phi) < \infty$, it follows that the net $\{u_\alpha\}$ is $\sigma(L, L')$-bounded. Consequently, $\{u_\alpha\}$ is also τ-bounded, and hence by the B-property (see Theorem 5.42) the net $\{u_\alpha\}$ is a τ-Cauchy net. If $u_\alpha \xrightarrow{\tau} u$ holds in L, then $u_\alpha \uparrow u$ in L, and since L is an ideal of $(L')^\sim_{\rm n}$ we infer that $F = u \in L$. This shows that L is a band in $(L')^\sim_{\rm n}$.

For the converse, assume that L is a band of $(L')^\sim_{\rm n}$ and let a sequence $\{u_n\}$ of L be τ-bounded and satisfying $0 \le u_n \uparrow$ in L. It follows that the formula $F(\phi) = \lim \phi(u_n)$ for $\phi \in L'$ defines a positive linear functional on L' such that $0 \le u_n \uparrow F$ holds in $(L')^\sim$. This implies that $F \in (L')^\sim_{\rm n}$ and consequently (since L is a band in $(L')^\sim_{\rm n}$), we see that $F \in L$ and that $u_n \xrightarrow{\sigma(L,L')} F$. Now a glance at Theorem 3.21 shows that $u_n \xrightarrow{\tau} F$, and thus $\{u_n\}$ is a τ-Cauchy sequence. Hence, (L, τ) satisfies the B-property. ■

5.19. Note that $\tau = |\sigma|(L, B)$ is a Hausdorff σ-Lebesgue locally convex-solid topology on L. By Theorem 2.33 we know that $(L, \tau)' = B$. Now a glance at Theorem 5.51 guarantees that the band $B \cap L^\sim_{\rm n} = \{0\}$ is order dense in $L^\sim_{\rm n}$, and this implies that $L^\sim_{\rm n} = \{0\}$. ■

5.20. Let X be a metric space and let τ be a Hausdorff locally solid topology on $C_c(X)$. Fix some function $0 < f \in C_c(X)$. We must show that there exists some $0 < g \in C_\tau$ such that $0 < g \le f$.

Start by observing that the support $S = \overline{\{x \in X\colon \ f(x) \ne 0\}}$ is a compact subset. Since X is a metric space, it follows that S is a separable metric space in its own right. Let $\{x_1, x_2, \ldots\}$ be a countable dense subset of S. We can assume without loss of generality that $f(x_n) > 0$ holds for each n. Next, for each pair $k, n \in \mathbb{N}$ pick a function $f_{k,n} \in C(X)$ such that:

(1) $0 \le f_{k,n}(x) \le \frac{1}{n}$ for all $x \in X$,
(2) $f_{k,n}(x) = 0$ if $d(x_k, x) \ge \frac{1}{n}$, where d denotes the metric, and
(3) $f_{k,n}(x_k) = \frac{1}{n}$.

Letting $g_{k,n} = f_{k,n} \wedge f$, we see that $g_{k,n} \in C_c(X)$ for each k and n and that the functions $g_{k,n}$ satisfy the following properties:

(a) $0 \le g_{k,n}(x) \le \frac{1}{n}$ for all $x \in X$,
(b) $g_{k,n}(x) = 0$ if $d(x_k, x) \ge \frac{1}{n}$, and
(c) $0 < g_{k,n}(x_k) \le \frac{1}{n}$.

Let $\{u_1, u_2, \ldots\}$ be an enumeration of the double sequence $\{g_{k,n}\}$.

Now, using induction, it is easy to see that we can pick a normal sequence $\{V_n\}$ of solid τ-neighborhoods of zero such that $u_n \notin V_n$ holds for each n. Put $N = \bigcap_{n=1}^{\infty} V_n$ and we claim that $N \cap [0, f] = \{0\}$. To see this, assume by way of contradiction that there exists some $0 < g \le f$ such that $g \in N$. Pick some n with $g(x_n) > 0$ and then note that some u_k must satisfy $0 \le u_k \le g$. This implies $u_k \in N$ and so $u_k \in V_k$, which is impossible. Hence, $N \cap [0, f] = \{0\}$.

Since $N \oplus N^{\mathrm{d}}$ is order in $C_c(X)$, there exists some $g_1 + g_2 \in N \oplus N^{\mathrm{d}}$ such that $0 < g_1 + g_2 \le f$. By the above discussion, we have $g_1 = 0$, and hence $0 < g_2 \le f$. This shows that C_τ is order dense in $C_c(X)$, and so τ is an entire topology. ■

5.21. Without loss of generality we can suppose that $g > 0$. Now assume that τ is a Hausdorff locally solid topology on B and let $0 < h \in B$. Since $h \wedge ng \uparrow h$ holds true in $C_c(X)$, there exists some k such that $0 < h \wedge kg \le h$. Put $f = h \wedge kg$.

As in the solution of the previous exercise, we can construct the sequence $\{u_n\}$ that satisfies the properties (a), (b), and (c). Also, we can choose a normal sequence $\{V_n\}$ of solid τ-neighborhoods of zero in B such that $u_n \notin V_n$ for each n. As in the solution of the previous exercise, there exists some $v \in N^{\mathrm{d}}$ such that $0 < v \le f$. This implies $0 < v \le f \le h$, and this shows that C_τ is order dense in B so that τ an entire topology on B. ■

9.6. Chapter 6: Weak Compactness in Riesz Spaces

6.1. Let $0 \le \phi \in L'$. Put $\psi(u) = \sup\{\phi(v)\colon\ v \in A \text{ and } 0 \le v \le u\}$ for each $u \in L^+$, and note that ψ is additive on L^+. So, by Lemma 1.64, the formula $\psi(u) = \psi(u^+) - \psi(u^-)$ extends ψ to a positive linear functional on L.

Now $\psi(u) = \phi(u)$ if $u \in A$, and so $\phi - \psi \in A^{\mathrm{o}}$. Also, $\phi - P(\phi) \in A^{\mathrm{o}}$ implies $\phi(u) = [P(\phi)](u)$ for all $u \in A$, and hence

$$\begin{aligned} 0 \le \psi(u) &= \sup\{\phi(v)\colon\ v \in A \text{ and } 0 \le v \le u\} \\ &= \sup\{[P(\phi)](v)\colon\ v \in A \text{ and } 0 \le v \le u\} \\ &\le [P(\phi)](u) \end{aligned}$$

for all $u \in L^+$, i.e., $0 \le \psi \le P(\phi) \in (A^{\mathrm{o}})^{\mathrm{d}}$. It follows that $\psi \in (A^{\mathrm{o}})^{\mathrm{d}}$, and so from $\phi = \psi + (\phi - \psi)$ we see that $\psi = P(\phi)$. ■

6.2. Let S be a convex and solid subset of (L, τ) such that every increasing positive sequence of S has a supremum in L. Assume that the set S is not τ-bounded. Then there exists a τ-neighborhood V of zero such that $S \not\subseteq \lambda V$ for each $\lambda > 0$.

For each n choose some $u_n \in S$ such that $0 \le u_n \notin n2^n V$. Now for every n let $w_n = \sum_{i=1}^{n} 2^{-i} u_i$ and note that $0 \le w_n \uparrow$ in L. Also from

$$0 \le w_n \le \sum_{i=1}^{n} 2^{-i} u_i + \Big(1 - \sum_{i=1}^{n} 2^{-i}\Big) u_n \in S\,,$$

we see that $\{w_n\} \subseteq S$ and so by our hypothesis $0 \le w_n \uparrow w$ holds in L. Pick some k such that $w \in kV$ and note that $0 \le 2^{-k}u_k \le w$ implies $2^{-k}u_k \in kV$. Hence $u_k \in k2^kV$, a contradiction. Therefore S is τ-bounded. ■

6.3. Assume that the topological dual of $(L', |\sigma|(L', L))$ equals L, that is, for every $F \in (L', |\sigma|(L', L))'$ there exists a (unique) $u \in L$ satisfying $F(\phi) = \phi(u)$ for all $\phi \in L'$. Since $(L', |\sigma|(L', L))$ is a locally solid topology, the topological dual of $(L', |\sigma|(L', L))$ is an ideal of $(L')^\sim$. Hence, L is an ideal of $(L')^\sim$, and so L is Dedekind complete and τ is a Lebesgue topology by Theorem 3.12.

On the other hand, if L is Dedekind complete and τ is a Lebesgue topology, then by Theorem 3.12 the Riesz space L is an ideal of $(L')^\sim_{\rm n}$. But then, by Theorem 2.33, the topological dual of $(L', |\sigma|(L', L))$ is precisely the ideal generated by L in $(L')^\sim_{\rm n}$, which is exactly L. ■

6.4. Let $e > 0$ be an order unit of L and let S be a convex, solid and $\sigma(L', L)$-compact subset of L. If S is not order bounded, then $S \not\subseteq [-\lambda e, \lambda e]$ for each $\lambda > 0$.

For each n choose $0 \le u_n \in S$ such that $u_n \notin [0, n2^n e]$, and then define the element $w_n = \sum_{k=1}^{n} 2^{-k}u_k$ for each n. Observe that $\{w_n\} \subseteq S$. Now by Theorem 6.47 we have $0 \le w_n \uparrow w$ in L, and therefore $w \le ke$ for some k. But then $0 \le 2^{-k}u_k \le w \le ke$ implies $u_k \in [0, k2^k e]$, which is impossible. ■

6.5. Let A be a relatively $\sigma(L_\rho, L'_\rho)$-compact subset of L_ρ. Note that L_ρ is a perfect Riesz space and that $L_\rho = (L'_\rho)^\sim_{\rm n}$. Hence, A is relatively $\sigma\big((L'_\rho)^\sim_{\rm n}, L'_\rho\big)$-compact and therefore, by Corollary 6.31, its solid hull S is also relatively $\sigma\big((L'_\rho)^\sim_{\rm n}, L'_\rho\big)$-compact. This implies that the set S is relatively $\sigma(L_\rho, L'_\rho)$-compact.

Now, by Theorem 6.30, S is order-equicontinuous on L'_ρ. So, by condition (iii) of Theorem 6.25, for each n there exists some $u_n \in L^+$ such that $\phi\big((|u| - u_n)^+\big) < \frac{1}{n}$ for all $u \in S$, where $0 \le \phi \in L'$ is defined by $\phi(v) = \rho(v^+) - \rho(v^-)$ for all $v \in L_\rho$. Put $w = \sum_{n=1}^{\infty} \frac{2^{-n}}{1+\rho(u_n)} u_n$, and we claim that $S \subseteq B_w$.

To see this, let $u \in S$ with $|u| \wedge w = 0$. Then $|u| \wedge u_n = 0$ for all n and therefore $|u| = (|u| - u_n)^+ + |u| \wedge u_n = (|u| - u_n)^+$ for all n. Consequently, for each n we have $0 \le \phi(|u|) = \phi\big((|u| - u_n)^+\big) < \frac{1}{n}$. This implies $\phi(|u|) = 0$ and so $u = 0$. Since B_w is a projection band, it easily follows that $S \subseteq B_w$. ■

6.6. Assume that S is a convex and solid subset of L'. If the set S is $\sigma(L', I)$-compact, then S is order-equicontinuous on L by Theorem 6.34. Also, if a net $\{\phi_\alpha\} \subseteq S$ satisfies $0 \le \phi_\alpha \uparrow$ in L', then the net $\{\phi_\alpha\}$ has a $\sigma(L', I)$-accumulation point $\phi \in S$ which, by the monotonicity of the net, it satisfies $0 \le \phi_\alpha \uparrow \phi$ in L'.

For the converse, assume that S is order-equicontinuous on L and that every increasing positive net of S has a supremum in L' which belongs to S. Consider $(L', |\sigma|(L', L))$. Then, by Theorem 2.33, the topological dual of this space is precisely I. Also, since S is order-equicontinuous on L, it follows from Theorem 6.25 that every disjoint sequence of the set S is $|\sigma|(L', L)$-convergent to zero, and therefore, $|\sigma|(L, I)$-convergent to zero. In particular, every disjoint sequence of S is $\sigma(L, I)$-convergent to zero. Next notice that if $\{\phi_\alpha\} \subseteq S$ satisfies $0 \le \phi_\alpha \uparrow$, then $\phi_\alpha \uparrow \phi$ holds in L' by our hypothesis, and hence $\phi_\alpha \xrightarrow{\sigma(L', L)} \phi$ holds. It now easily follows that S satisfies condition (ii) of Theorem 6.47 and so S is $\sigma(L', I)$-compact. ■

6.7. Since $v_\alpha \xrightarrow{\tau} 0$ implies $v_\alpha^+ \xrightarrow{\tau} 0$, replacing $\{v_\alpha\}$ by $\{v_\alpha^+\}$, we can assume without loss of generality that $v_\alpha \geq 0$ holds for all α. Now use the Riesz decomposition property to write $u_\alpha = x_\alpha + y_\alpha$ with $0 \leq x_\alpha \leq v_\alpha$ and $0 \leq y_\alpha \leq w$ for all α. Clearly, $x_\alpha \xrightarrow{\tau} 0$ and since the order interval $[0, w]$ is weakly compact, it follows that the net $\{y_\alpha\}$ has a weakly convergent subnet, say $y_{\lambda_\kappa} \xrightarrow{\sigma(L,L')} y$. It is now easy to see that $u_{\lambda_\kappa} = x_{\lambda_\kappa} + y_{\lambda_\kappa} \xrightarrow[\kappa]{\sigma(L,L')} y$. Therefore, the net $\{u_\alpha\}$ has a weakly convergent subnet. ■

6.8. If (L, τ) is semireflexive, then τ is necessarily a Levi topology by Theorem 6.65.

Now assume that (L, τ) is τ-complete, that L' is $\beta(L', L)$-separable and that τ is a σ-Levi topology. First note that L is Dedekind σ-complete, and then that L' is $|\sigma|(L', L)$-separable. Thus, by Lemma 6.67, each principal ideal of L is τ-separable. But then it follows from Theorem 3.30 that τ induces a pre-Lebesgue topology on each principal ideal. Therefore, τ is a pre-Lebesgue topology and so, by Theorem 3.24, the topology τ is a Lebesgue topology. Now, according to Theorem 3.30, the topology $\beta(L', L)$ satisfies the pre-Lebesgue property and since $\beta(L', L)$ is always a Fatou topology (see Corollary 6.12), it follows from Theorem 4.8 that $\beta(L', L)$ is a Lebesgue topology. So, we have the inclusions $L \subseteq L'' \subseteq (L')_{\mathrm{n}}^{\sim}$, with L being an order dense ideal in L'', by Theorem 3.12. But then the $\beta(L', L)$-separability of L' implies that L is super order dense in L''; see Exercise **2.17**.

Now let $0 < u \in L''$. Then there exists a sequence $\{u_n\} \subseteq L$ with $0 \leq u_n \uparrow u$ in L''. But $\{u_n\}$ is $\sigma(L, L')$-bounded and so τ-bounded. Thus, by the σ-Levi property, there exists some $v \in L$ with $u_n \uparrow v$ in L. Since L is an ideal of L'', we have $u_n \uparrow v$ in L'', and thus $u = v \in L$. Hence $L = L''$, and (L, τ) is semireflexive. ■

6.9. Note that the closed unit ball of L_ρ is $[-e, e]$. We must show that the order interval $[-e, e]$ is $\sigma(L_\rho, L_\rho')$-compact.

Since L is an ideal of $(L^\sim)_{\mathrm{n}}^{\sim}$, it follows that $[-e, e]$ is also an interval in $(L^\sim)_{\mathrm{n}}^{\sim}$. But then, by Theorem 6.1, the order interval $[-e, e]$ is $\sigma\big((L^\sim)_{\mathrm{n}}^{\sim}, L'\big)$-compact. Hence $[-e, e]$ is $\sigma(L, L')$-compact. ■

6.10. Let $u > 0$ be a quasi-interior point of L and assume that $[0, u]$ is $\sigma(L, L')$-compact. By Theorem 6.46 each order interval of L_u (the ideal generated by u in L) is τ-complete, and therefore, L_u is an ideal of $\widehat{L}$. Since L_u is τ-dense in L and L is $\widehat{\tau}$-dense in $\widehat{L}$, it follows that L_u is $\widehat{\tau}$-dense in $\widehat{L}$.

Now consider an interval $[0, \widehat{v}]$ of $\widehat{L}$. To see that $[0, \widehat{v}]$ is $\sigma(\widehat{L}, \widehat{L}')$-compact, we need only to show that each disjoint sequence of $[0, \widehat{v}]$ is $\widehat{\tau}$-convergent to zero; see Theorem 6.46. So, let $\{\widehat{v}_n\} \subseteq [0, \widehat{v}]$ be a disjoint sequence, and let V and V_1 be two solid $\widehat{\tau}$-neighborhoods of zero of $\widehat{L}$ with $V_1 + V_1 \subseteq V$. Choose some $0 \leq w \in L_u$ with $\widehat{v} - w \in V_1$, and then note that $\{\widehat{v}_n \wedge w\}$ is a disjoint sequence of L, lying in some order interval of L_u. Therefore, since the order intervals of L_u are $\sigma(L, L')$-compact, it follows that $\widehat{v}_n \wedge w \xrightarrow{\tau} 0$. Hence, $0 \leq \widehat{v}_n = \widehat{v}_n \wedge w + (\widehat{v}_n - w)^+ \in V_1 + V_1 \subseteq V$ holds true for all sufficiently large n, and so $\widehat{v}_n \xrightarrow{\widehat{\tau}} 0$. ■

6.11. Fix $u \in L^+$. If $\langle 0, u\rangle$ is weakly compact, then it follows from Theorem 6.46 (applied to $(\widehat{L}, \widehat{\tau})$) that each disjoint sequence of $\langle 0, u\rangle$ is $\widehat{\tau}$-convergent to zero. In particular, every disjoint sequence of L in $\langle 0, u\rangle$ is τ-convergent to zero.

For the converse, assume that every disjoint sequence of the set $\langle 0, u\rangle \cap L$ is τ-convergent to zero. To see that $\langle 0, u\rangle$ is weakly compact, we need only to show that every disjoint sequence of $\langle 0, u\rangle$ is $\widehat{\tau}$-convergent to zero; see Theorem 6.46. So, let $\{v_n\} \subseteq \langle 0, u\rangle$ be a disjoint sequence, and let the solid $\widehat{\tau}$-neighborhoods V and V_1 of zero satisfy $V_1 + V_1 \subseteq V$.

Also let L_u be the ideal generated by u in L. It now follows from Theorem 3.22 that τ restricted to L_u is a pre-Lebesgue topology and so, according to Theorem 6.17, there exist $\delta > 0$ and $0 \le \phi \in L'$ such that:

$$\textit{If } v \in L \textit{ satisfies } |v| \le u \textit{ and } \phi(|v|) < 2\delta, \textit{ then } v \in V_1 . \qquad (\star)$$

Since $\{v_n\}$ is $|\sigma|(\widehat{L}, \widehat{L}')$-convergent to zero (because $\sum_{i=1}^n v_i \le u$ holds for all n), there exists some k such that $\phi(v_n) < \delta$ for all $n > k$. Next, for each n choose some $w_n \in L$ satisfying $0 \le w_n \le u$ and $v_n - w_n \in V_1$ and $\phi(|v_n - w_n|) < \delta$. Thus $\phi(w_n) < 2\delta$ for $n \ge k$, and so from $(\star)$ we see that $w_n \in V_1$ for all $n \ge k$. So,

$$0 \le v_n = (v_n - w_n) + w_n \in V_1 + V_1 \subseteq V$$

for $n \ge k$. This implies $v_n \in V$ for all $n \ge k$, and consequently $v_n \xrightarrow{\tau} 0$. ■

6.12. From Exercise **6.3** it follows that L is Dedekind complete and that τ is a Lebesgue topology. To see that L' is a discrete space note that by Theorem 6.1 each interval of L' is $\sigma(L', L)$-compact, and hence $|\sigma|(L', L)$-compact. Now the discreteness of L' follows immediately from Corollary 6.57.

For the discreteness of L we apply Theorem 6.56 to $\big(L, |\sigma|(L, L')\big)$. We shall verify that statement (iii) holds, and the conclusion will follow. So, let $0 < u \in L$. Since $|\sigma|(L, L')$ is also a Lebesgue topology, it follows from Nakano's Theorem 4.28 that $[0, u]$ is τ-complete. Now let $\{u_n\} \subseteq [0, u]$ satisfy $u_n \xrightarrow{\sigma(L,L')} v$. We need to show that $u_n \xrightarrow{|\sigma|(L,L')} v$.

Since $u_n \xrightarrow{\sigma(L'',L')} v$, it follows from Theorem 6.39 (applied to (L', L'')) that $\{v - u_n\}$ is order-equicontinuous on L', and hence the solid hull of the sequence $\{|v - u_n|\}$ is also order-equicontinuous on L'. Let $\mathcal{A}$ be the set of all atoms of L', and note that (since L' is discrete) the ideal generated by $\mathcal{A}$ is order dense in L'. Also $|\phi(w)| = \phi(|w|)$ holds for all $0 \le \phi \in \mathcal{A}$ and $w \in L$. Now if $0 \le \psi \in L'$, then there exists a net $\{\phi_\alpha\}$ with $0 \le \phi_\alpha \uparrow \psi$, where each ϕ_α is a linear combination of elements of $\mathcal{A}$. From the order-equicontinuity of $\{|v - u_n|\}$, it follows that $\{\phi_\alpha\}$ converges uniformly to ψ on $\{|v - u_n|\}$. Thus, for each $\epsilon > 0$ there exists some index β such that $(\psi - \phi_\beta)(|v - u_n|) < \epsilon$ for all n. Write $\phi_\beta = \sum_{i=1}^k \lambda_i \psi_i$, where each $\psi_i \in \mathcal{A}$, and then choose some m such that $\sum_{i=1}^k |\lambda_i||\psi_i(v - u_n)| < \epsilon$ holds for all $n \ge m$. Thus, for all $n \ge m$ we have

$$\begin{aligned} \psi(|v - u_n|) &= (\psi - \phi_\beta)(|v - u_n|) + \phi_\beta(|v - u_n|) \\ &\le \epsilon + \sum_{i=1}^k |\lambda_i| |\psi_i(|v - u_n|)| \\ &= \epsilon + \sum_{i=1}^k |\lambda_i| |\psi_i(v - u_n)| < 2\epsilon . \end{aligned}$$

Consequently, $u_n \xrightarrow{|\sigma|(L,L')} v$, as desired. ■

6.13. Let B be a band of a semireflexive Riesz space (L, τ). We shall show that τ restricted to B satisfies condition (iii) of Theorem 6.65.

Note that B is τ-complete since it is τ-closed; see Theorem 2.21. Also each disjoint sequence of B^+ whose sequence of partial sums is τ-bounded, is τ-convergent to zero. (This is so simply because L has this property.) Consequently, it remains to be shown that each τ-bounded disjoint sequence of B^+ is $\sigma(B, B')$-convergent to zero. But this is also immediate by observing that any such sequence is $\sigma(L, L')$-convergent to zero (since (L, τ) is semireflexive), and that every continuous linear functional on B is the restriction of some continuous linear functional on L by the Hahn–Banach theorem. ■

6.14. Since L is Dedekind complete, A is a projection band. Consequently, $(L/A, \tau/A)$ is Riesz isomorphic to (A^d, τ), and the conclusion follows from the preceding exercise. ∎

6.15. According to Theorem 2.33 we have $\big(L, |\sigma|(L, L_n^\sim)\big)' = L_n^\sim$. Therefore, if $\big(L, |\sigma|(L, L_n^\sim)\big)$ is semireflexive, then $\beta(L_n^\sim, L)$ is a Lebesgue topology and $|\sigma|(L, L_n^\sim)$ is a Levi topology by Theorem 6.65. Therefore, by Theorem 3.17, the Riesz space L is perfect.

Conversely, assume that L is a perfect Riesz space and that $\beta(L_n^\sim, L)$ is a Lebesgue topology. Then $|\sigma|(L, L_n^\sim)$ is a Levi topology (see Theorem 3.17), and it is obviously a Lebesgue topology. Hence, by Theorem 6.65, the Hausdorff locally convex-solid Riesz space $\big(L, |\sigma|(L, L_n^\sim)\big)$ is semireflexive. ∎

6.16. Recall first the definition of the Mackey topology $\tau(X, X')$ on X for a dual system $\langle X, X'\rangle$. Let $\mathfrak{S}$ denote the collection of all circled, convex and $\sigma(X', X)$-compact subsets of X'. Then a base at zero for the Mackey topology $\tau(X, X')$ is the collection of sets of the form

$$V = (A_1)^\circ \cap (A_2)^\circ \cap \cdots \cap (A_k)^\circ\,, \tag{$\star$}$$

where $A_i \in \mathfrak{S}$ for each i and the polar of any subset A of X' is defined as usual by $A^\circ = \{x \in X\colon\ |\langle x, x'\rangle| \le 1 \text{ for all } x' \in A\}$.

We now consider the dual system $\langle L, L_n^\sim\rangle$. (Since $L_n^\sim$ separates the points of L notice that $\langle L, L_n^\sim\rangle$ is indeed a dual system.) Moreover, it is easy to see that the polar of any solid subset of $L_n^\sim$ is also a solid subset of L; see the discussion at the beginning of Section 6.1.

Let $\mathfrak{S}$ be the collection of all circled, convex and $\sigma(L_n^\sim, L)$-compact subsets of $L_n^\sim$. Also, let V be a typical $\tau(L, L_n^\sim)$-neighborhood of L as in ($\star$). According to Corollary 6.31, the $\sigma(L_n^\sim, L)$-closed convex-solid hull B_i of each A_i is also $\sigma(L_n^\sim, L)$-compact—and of course, as A_i being circled, each B_i is also circled. That is, $B_i \in \mathfrak{S}$ for each i and clearly

$$(B_1)^\circ \cap (B_2)^\circ \cap \cdots \cap (B_k)^\circ \subseteq (A_1)^\circ \cap (A_2)^\circ \cap \cdots \cap (A_k)^\circ = V\,.$$

This shows that the solid $\tau(L, L_n^\sim)$-neighborhoods at zero form a base for the Mackey topology $\tau(L, L_n^\sim)$, and so $\tau(L, L_n^\sim)$ is locally solid.

To see that $\tau(L, L_n^\sim)$ is a Lebesgue topology, let $u_\alpha \downarrow 0$ in L. Then $\phi(u_\alpha) \downarrow 0$ holds in $\mathbb{R}$ for each $0 \le \phi \in L_n^\sim$. This implies $u_\alpha \xrightarrow{\sigma(L,L_n^\sim)} 0$. Now a glance at Theorem 3.21 guarantees that $u_\alpha \xrightarrow{\tau(L,L_n^\sim)} 0$. This shows that the Mackey topology $\tau(L, L_n^\sim)$ is a Lebesgue topology. ∎

6.17. We claim that L_1 is a band of $(L_\infty)^\sim$. To see this, let us simplify further the notation by letting $L = L_1 = L_1(X, \Sigma, \mu)$. Clearly, L is an AL-space. This implies that $L' = L_\infty$ is an AM-space and consequently $L'' = L'_\infty = L_\infty^\sim$ is an AL-space. In particular, L'' also has order continuous norm; see Theorem 3.38 and 3.31.

Since L has order continuous norm, it follows that L is an ideal of L''; see Theorem 6.62. Now assume that $0 \le u_\alpha \uparrow F$ holds in L'' with $\{u_\alpha\} \subseteq L$. Since L'' has order continuous norm, there is an increasing sequence of indices $\{\alpha_n\}$ such that $\|u_{\alpha_n} - F\| \to 0$, and of course $u_{\alpha_n} \uparrow F$. Now by the classical theorem of B. Levi from the theory of integration, we know that there exists some $u \in L$ such that $u_{\alpha_n} \uparrow u$ in L. Since L is an ideal of L'', it follows that $u_{\alpha_n} \uparrow u$ also holds in L''. This implies $F = u \in L$, and so L is a band of L''.

Now a glance at Theorem 6.9 guarantees that L is $|\sigma|\big((L')_n^\sim, L'\big)$-dense in $(L')_n^\sim$. But by part (d) of Theorem 2.21, we know that L, as being a band of $(L')^\sim$ and hence of $(L')_n^\sim$, is $|\sigma|\big((L')_n^\sim, L'\big)$-closed. So, $L = (L')_n^\sim = (L_\infty)_n^\sim$.

Finally, since L_∞ is Dedekind complete and $(L_\infty)_n^\sim = L_1$ separates the points of L_∞, it follows from the preceding exercise that the Mackey topology $\tau(L_\infty, L_1)$ is locally solid and Lebesgue.

For the last part of this exercise, notice that $\ell_\infty = L_\infty(\mathbb{N}, \mathcal{P}(\mathbb{N}), \mu)$ and that $\ell_1 = L_1(\mathbb{N}, \mathcal{P}(\mathbb{N}), \mu)$, where μ is the counting measure (which is a σ-finite measure) on the σ-algebra $\mathcal{P}(\mathbb{N})$ of all subsets of $\mathbb{N}$. ∎

6.18. Let $B \subseteq L_c^\sim$ be convex, balanced and $\sigma(L_c^\sim, L)$-compact. By Corollary 6.29 the $\sigma(L_c^\sim, L)$-closed solid hull S of B is also $\sigma(L_c^\sim, L)$-compact—and also convex by Theorem 1.11. It follows from $B \subseteq S$ that $S^\circ \subseteq B^\circ$. But the polar S° is a solid $\tau(L, L_c^\sim)$-neighborhood of zero and since B° is a basic $\tau(L, L_c^\sim)$-neighborhood of zero, it follows that $\tau(L, L_c^\sim)$ is a locally solid topology. ∎

6.19. If A is relatively weakly compact, then A automatically satisfies the condition. For the converse, let A satisfy the condition and fix $\epsilon > 0$. Pick a weakly compact subset W of X such that $A \subseteq W + \epsilon U \subseteq W + \epsilon U''$, where U'' denotes the closed unit ball of X''. Clearly, A is a norm bounded set, and so its $\sigma(X'', X')$-closure $\overline{A}$ in X'' is $\sigma(X'', X')$-compact. On the other hand, since $W + \epsilon U''$ is $\sigma(X'', X')$-compact, it follows that $\overline{A} \subseteq W + \epsilon U'' \subseteq X + \epsilon U''$ holds for all $\epsilon > 0$. Since X is norm closed in X'', we get $\overline{A} \subseteq X$. Taking into consideration that $\sigma(X'', X')$ induces $\sigma(X, X')$ on X, we conclude that A is a relatively weakly compact subset of X. ∎

6.20. Let $0 < u \in L_\infty[0,1]$. It suffices to show that there exists some $0 < \psi \in L_1^{\mathrm{d}}[0,1]$ such that $\psi(u) > 0$. Once this is established, the solution can be completed as follows. Take any $v \in L_\infty[0,1]$ with $v \neq 0$. Then either $v^+ > 0$ or $v^- > 0$; we can assume that $v^+ > 0$. Pick some $0 < \psi_1 \in L_1^{\mathrm{d}}[0,1]$ such that $\psi_1(v^+) > 0$. Now viewing v as an order bounded linear functional on $L_1^{\mathrm{d}}[0,1]$, we see that

$$0 < \psi_1(v^+) = v^+(\psi_1) = \sup\{\phi(v)\colon\ \phi \in L_1^{\mathrm{d}}[0,1] \text{ and } 0 \leq \phi \leq \psi_1\}.$$

Hence $\phi(v) \neq 0$ must hold for some $\phi \in L_1^{\mathrm{d}}[0,1]$.

To complete the solution, pick some $\epsilon > 0$ and then choose a Lebesgue measurable set $A \subseteq [0,1]$ of positive Lebesgue measure such that $u \geq \epsilon\chi_A$. Note that

$$L_\infty[0,1] = L_\infty(A) \oplus L_\infty(A^c).$$

Also, fix a disjoint sequence $\{A_n\}$ of Lebesgue measurable sets each of positive measure such that $\bigcup_{n=1}^\infty A_n = A$ (this is possible since the Lebesgue measure on $[0,1]$ is non-atomic). Now observe that the mapping $T\colon \ell_\infty \to L_\infty(A)$, defined by

$$T(a_1, a_2, \ldots) = \sum_{n=1}^\infty a_n \chi_{A_n},$$

is a norm preserving (into) Riesz isomorphism such that $T(e) = \chi_A$, where as usual $e = (1, 1, \ldots)$.

Next, pick some $0 < \psi \in \ell_1^{\mathrm{d}}$ with $\psi(e) = 1$, and consider it as a positive linear functional on $T(\ell_\infty)$ in $L_\infty(A)$. In view of $T(e) = \chi_A$, the Riesz subspace $T(\ell_\infty)$ is majorizing $L_\infty(A)$. This implies that ψ has a positive linear extension (denoted by ψ again) to all of $L_\infty(A)$. By assigning zero values on $L_\infty(A^c)$, we can assume without loss of generality that $0 < \psi \in L^\sim$.

We claim that $\psi \in L_1^{\mathrm{d}}[0,1]$. To see this, let $0 \leq w \in L_1[0,1]$. Then w acts on $T(\ell_\infty)$ as the ℓ_1-sequence $\left(\int_{A_1} w(x)\,dx, \int_{A_2} w(x)\,dx, \ldots\right)$. Consequently, from $[w \wedge \psi](e) = 0$, we infer that $[w \wedge \psi](\mathbf{1}) = 0$, where $\mathbf{1}$ is the constant function one on the interval $[0,1]$. That is, $\psi \perp w$ for all $w \in L_1[0,1]$, which means that $\psi \in L_1^{\mathrm{d}}[0,1]$. Now note that $\psi(u) \geq \psi(\epsilon\chi_A) = \epsilon > 0$. ∎

9.7. Chapter 7: Lateral Completeness

7.1. (i) Let I be a proper prime ideal of $\mathcal{B}$ and let J be an ideal of $\mathcal{B}$ such that $I \subseteq J$ and $I \neq J$. Pick $a \in J$ with $a \notin I$. Since $a \wedge a' = 0 \in I$ and I is a prime ideal, it follows that $a' \in I$ and consequently $a' \in J$. But then $e = a \vee a' \in J$ and so $J = \mathcal{B}$, and this shows that I is a maximal ideal.

Now assume that I is a (proper) maximal ideal of $\mathcal{B}$ and that $a \wedge b \in I$. If neither a nor b belong to I, then the ideals generated by $I \cup \{a\}$ and $I \cup \{b\}$ must equal $\mathcal{B}$, due to the maximality of I. Therefore, if $x \in \mathcal{B}$, then there exists some $y \in I$ such that $x \leq y \vee a$ and $x \leq y \vee b$ and consequently $x \leq (y \vee a) \wedge (y \vee b) = y \vee (a \wedge b) \in I$. Hence $x \in I$ and thus $I = \mathcal{B}$, a contradiction. This contradiction establishes that I is a prime ideal of $\mathcal{B}$.

(ii) Consider the collection

$$\mathcal{C} = \{J\colon\ J \text{ is an ideal of } \mathcal{B},\ I \subseteq J, \text{ and } u \notin J\}.$$

Clearly, $\mathcal{C}$ is partially ordered by the inclusion $\supseteq$ and since $I \in \mathcal{C}$, the partially ordered set $\mathcal{C}$ is nonempty. Now if $\{J_\alpha\colon\ \alpha \in A\}$ is a chain of $\mathcal{C}$, then it is easy to check that $J = \bigcup_{\alpha \in A} J_\alpha$ is an ideal of $\mathcal{B}$ which belongs to $\mathcal{C}$. By Zorn's lemma there exists a maximal element I_0 of $\mathcal{C}$, and we claim that I_0 is a prime ideal of $\mathcal{B}$.

To see this, assume that I_0 is not a prime ideal of $\mathcal{B}$. Then there exist $a, b \in \mathcal{B}$ with $a \wedge b \in I_0$ and with $a \notin I_0$ and $b \notin I_0$. It now follows from the maximality property of I_0 that the ideals generated by $I_0 \cup \{a\}$ and $I_0 \cup \{b\}$ both contain u, and thus there exists some $x \in I_0$ with $u \leq x \vee a$ and $u \leq x \vee b$. Consequently, $u \leq (x \vee a) \wedge (x \vee b) = x \vee (a \wedge b) \in I_0$ and therefore $u \in I_0$, a contradiction. This establishes that I_0 is a prime ideal.

(iii) Observe first that $\Omega_u \cup \Omega_v = \Omega_{u \vee v}$ and $\Omega_u \cap \Omega_v = \Omega_{u \wedge v}$ hold for all $u, v \in \mathcal{B}$. These formulas can be verified as follows:

$$(1) \qquad \omega \in \Omega_u \cup \Omega_v \iff \omega \in \Omega_u \text{ or } \omega \in \Omega_v \iff u \notin \omega \text{ or } v \notin \omega$$
$$\iff u \vee v \notin \omega \iff \omega \in \Omega_{u \vee v}$$
$$(2) \qquad \omega \in \Omega_u \cap \Omega_v \iff \omega \in \Omega_u \text{ and } \omega \in \Omega_v \iff u \notin \omega \text{ and } v \notin \omega$$
$$\iff u \wedge v \notin \omega \iff \omega \in \Omega_{u \wedge v}\,.$$

From (1) and (2) it also follows that the collection $\{\Omega_u\colon\ u \in \mathcal{B}\}$ is a base for a topology τ on Ω (called the **hull-kernel topology** on Ω). We shall establish that (Ω, τ) is a Hausdorff compact topological space.

- *τ is a Hausdorff topology.*

Let $\omega_1 \neq \omega_2$. Since by (i), both ω_1 and ω_2 are maximal ideals of $\mathcal{B}$ neither ω_1 is included in ω_2 nor ω_2 is included in ω_1. Choose $u_1 \in \omega_1$ with $u_1 \notin \omega_2$ and $u_2 \in \omega_2$ with $u_2 \notin \omega_1$. Put $v_1 = u_1 \wedge (u_1 \wedge u_2)'$ and $v_2 = u_2 \wedge (u_1 \wedge u_2)'$. Since $v_1 \wedge v_2 = 0$, it follows from (2) above that $\Omega_{v_1} \cap \Omega_{v_2} = \Omega_{v_1 \wedge v_2} = \Omega_0 = \emptyset$. On the other hand, note that $v_1 \notin \omega_2$. Indeed, if $v_1 \in \omega_2$, then $(u_1 \wedge u_2)' \in \omega_2$ and so $e = (u_1 \wedge u_2) \vee (u_1 \wedge u_2)' \in \omega_2$ from which it follows that $\omega_2 = \mathcal{B}$, a contradiction. Hence $v_1 \notin \omega_2$ and thus $\omega_2 \in \Omega_{v_1}$. Similarly, $\omega_1 \in \Omega_{v_2}$. These conclusions imply that τ is a Hausdorff topology.

- *τ is a compact topology.*

Since $\Omega = \Omega_e$ (where, of course, the vector e is the unit of $\mathcal{B}$), it is enough to establish that each Ω_u is compact. To this end, let $u \in \mathcal{B}$ and assume that $\Omega_u \subseteq \bigcup_{\alpha \in A} \Omega_{u_\alpha}$. Let I be the ideal generated by the set $\{u_\alpha\colon\ \alpha \in A\}$ in $\mathcal{B}$. If $u \notin I$, then by part (ii) there exists a prime ideal ω satisfying $I \subseteq \omega$ and $u \notin \omega$, so that $\omega \in \Omega_u$. This implies that $u_\alpha \notin \omega$ for some α, which is impossible. Thus, $u \in I$ and so there exist indices $\alpha_1, \ldots, \alpha_n$ with $u \leq \bigvee_{i=1}^n u_{\alpha_i}$. But then we have $\Omega_u \subseteq \bigcup_{i=1}^n \Omega_{u_{\alpha_i}}$, so that Ω_u is τ-compact.

(iv) From part (iii) it should be clear that every set of the form Ω_u is τ-clopen. Now let $\mathcal{O}$ be a τ-clopen subset of Ω. Since $\mathcal{O}$ is open it must be of the form $\mathcal{O} = \bigcup_{\alpha \in A} \Omega_{u_\alpha}$ and since $\mathcal{O}$ is closed and (Ω, τ) is Hausdorff and compact, $\mathcal{O}$ must be also compact. Thus there exist indices $\alpha_1, \ldots, \alpha_n$ such that $\mathcal{O} = \bigcup_{i=1}^n \Omega_{u_{\alpha_i}}$. Now put $u = \bigvee_{i=1}^n u_{\alpha_i}$ and use part (ii) to obtain $\mathcal{O} = \Omega_u$.

(v) Assume that (Ω, τ) is extremally disconnected and that S is a nonempty subset of $\mathcal{B}$. Put $\mathcal{O} = \bigcup_{u \in S} \Omega_u$ and note that $\overline{\mathcal{O}}$ is a τ-clopen subset of Ω. From part (iv) it follows immediately that $\overline{\mathcal{O}} = \Omega_w$ for precisely one $w \in \mathcal{B}$. (Using part (ii) it can be established easily that $\Omega_u = \Omega_v$ if and only if $u = v$.) Clearly, $u \leq w$ holds in $\mathcal{B}$ for all $u \in S$. On the other hand, if $u \leq v$ holds in $\mathcal{B}$ for all $u \in S$, then $\mathcal{O} = \bigcup_{u \in S} \Omega_u \subseteq \Omega_v$ must be the case, and consequently $\Omega_w \subseteq \Omega_v$, from which it follows that $w \leq v$. Thus $w = \sup S$ in $\mathcal{B}$, and so $\mathcal{B}$ is a Dedekind complete Boolean algebra.

Now assume that $\mathcal{B}$ is an arbitrary Dedekind complete Boolean algebra, and let $\mathcal{O} = \bigcup_{u \in S} \Omega_u$ be an open subset of Ω. We can assume that $\mathcal{O} \neq \emptyset$. Put $w = \sup S$ in $\mathcal{B}$ and note that $\mathcal{O} \subseteq \Omega_w$ implies $\overline{\mathcal{O}} \subseteq \Omega_w$. To establish this claim, we shall show that $\overline{\mathcal{O}} = \Omega_w$.

To this end, assume by way of contradiction that the open set $\Omega_w \setminus \overline{\mathcal{O}}$ is nonempty. Then there exists a nonempty basic open set Ω_v such that $\Omega_v = \Omega_w \setminus \overline{\mathcal{O}}$. It follows that $\Omega_v \cap \Omega_u = \emptyset$ for all $u \in S$ and therefore $v \wedge u = 0$ for all $u \in S$. It now follows from the infinite distributive law of the Boolean algebra $\mathcal{B}$ that $v \wedge w = 0$ and hence (in view of $v \leq w$) that $v = 0$. The latter implies $\Omega_v = \emptyset$, a contradiction, and the desired conclusion follows.

(vi) Let $\mathcal{B}$ be the Dedekind complete Boolean algebra consisting of all clopen subsets of X. For each $x \in X$ put $\omega_x = \{\mathcal{O} \in \mathcal{B} \colon\ x \notin \mathcal{O}\}$ and note that ω_x is a proper prime ideal of $\mathcal{B}$. Next note that every proper prime ideal of $\mathcal{B}$ must be of the above form. Indeed, if ω is a proper prime ideal of $\mathcal{B}$ such that $\omega \neq \omega_x$ for all $x \in X$, then for every $x \in X$ there exists some $\mathcal{O}_x \in \omega$ with $x \in \mathcal{O}_x$. But then $X = \bigcup_{x \in X} \mathcal{O}_x$ and the compactness of X guarantees that $X = \bigcup_{i=1}^n \mathcal{O}_{x_i} \in \omega$ and so $\omega = \mathcal{B}$, a contradiction. Also note that $\omega_x = \omega_y$ if and only if $x = y$. (If $x \neq y$, then there exists a clopen set $\mathcal{O}$ with $x \in \mathcal{O}$ and $y \notin \mathcal{O}$ and so $\omega_x \neq \omega_y$.)

Thus, we have established a mapping $x \mapsto \omega_x$ from X onto the Stone space Ω of $\mathcal{B}$. Now let $x_\alpha \to x$ in X and let $\Omega_{\mathcal{O}} = \{\omega \in \Omega \colon\ \mathcal{O} \notin \omega\}$ be a basic neighborhood of ω_x. Then $x \in \mathcal{O}$ and therefore $x_\alpha \in \mathcal{O}$ for all $\alpha \succeq \alpha_1$. So $\mathcal{O} \notin \omega_{x_\alpha}$ for all $\alpha \succeq \alpha_1$ and hence $\omega_{x_\alpha} \in \Omega_{\mathcal{O}}$ for all $\alpha \succeq \alpha_1$, which shows that the mapping $x \mapsto \omega_x$ is continuous, and therefore a homeomorphism.

Finally, assume that $\phi\colon \Omega_1 \to \Omega_2$ is a homeomorphism between the Stone spaces of the Dedekind complete Boolean algebras $\mathcal{B}_1$ and $\mathcal{B}_2$. If Ω_u is a clopen set of Ω_1, then $\phi(\Omega_u)$ is a clopen subset of Ω_2, and hence there exists a unique element $\pi(u)$ of $\mathcal{B}_2$ with $\phi(\Omega_u) = \Omega_{\pi(u)}$. So, we can establish a mapping $\pi\colon \mathcal{B}_1 \to \mathcal{B}_2$ which as it can be easily seen is a Boolean isomorphism from $\mathcal{B}_1$ onto $\mathcal{B}_2$. ■

7.2. Clearly $\mathcal{B}_L$ is a partially ordered set under inclusion which is also a lattice under the operations: $A \wedge B = A \cap B$ and $A \vee B = (A + B)^{\mathrm{dd}}$. Also $0 = \{0\}$ is the smallest element of $\mathcal{B}_L$ and $E = L$ is the largest. Now since $A \cap A^{\mathrm{d}} = \{0\}$ and $(A + A^{\mathrm{d}})^{\mathrm{dd}} = E$, it follows that $\mathcal{B}_L$ is complemented by $A' = A^{\mathrm{d}}$. It remains only to be shown that $\mathcal{B}_L$ is distributive.

To this end, let $A, B, C \in \mathcal{B}_L$. We must show that

$$A \wedge (B \vee C) \subseteq (A \wedge B) \vee (A \wedge C)$$

or, equivalently, that

$$A \cap (B + C)^{\mathrm{dd}} \subseteq (A \cap B + A \cap C)^{\mathrm{dd}}.$$

If $0 \le u \in A \cap (B+C)^{\rm dd}$, then since L is Archimedean, there exists a net $\{u_\alpha\}$ in $B+C$ with $0 \le u_\alpha \uparrow u$. Write $u_\alpha = v_\alpha + w_\alpha$ with $0 \le v_\alpha \in B$ and $0 \le w_\alpha \in C$ for all α, and note that $\{v_\alpha\} \subseteq A \cap B$ and $\{w_\alpha\} \subseteq A \cap C$ imply that $\{u_\alpha\} \subseteq A \cap B + A \cap C$, from which it follows that $u \in (A \cap B + A \cap C)^{\rm dd}$. The distributivity of $\mathcal{B}_L$ is now immediate.

Remark. The converse of this exercise is also true. That is: *If the collection $\mathcal{B}_L$ of all bands of a Riesz space L partially ordered by inclusion forms a Boolean algebra, then L is Archimedean.*

To see this, let us denote by B_D the band generated in L by any subset D of L. Note first that if $A, C \in \mathcal{B}_L$, then

$$A \wedge C = A \cap C \quad \text{and} \quad A \vee C = B_{A+C}$$

must be the case. Now let $A \in \mathcal{B}_L$ and $0 \le u \in A^{\rm dd}$. Pick $C \in \mathcal{B}_L$ such that $A \cap C = \{0\}$ and $B_{A+C} = L$, and then choose a net $\{u_\alpha\} \subseteq A+C$ with $0 \le u_\alpha \uparrow u$. Write $u_\alpha = v_\alpha + w_\alpha$ with $0 \le v_\alpha \in A$ and $0 \le w_\alpha \in C$ for all α. Now it can be seen easily that $C \subseteq A^{\rm d}$, and so $w_\alpha \in A^{\rm d} \cap A^{\rm dd}$ implies $w_\alpha = 0$ for all α. Hence $0 \le v_\alpha \uparrow u$ holds and so $u \in A$, since A is a band. Thus $A = A^{\rm dd}$ holds for all $A \in \mathcal{B}_L$, and the Archimedeaness of L follows from Theorem 1.28. ■

7.3. Consider the mapping $\phi\colon \mathcal{B}_L \to \mathcal{B}_M$ by defining $\phi(B)$ to be the band generated by B in M. Since L is order dense, we see that if $C \in \mathcal{B}_M$, then $C \cap L \in \mathcal{B}_L$ and $\phi(C \cap L) = C$. This shows that ϕ is onto. An easy argument shows that ϕ is also one-to-one. Finally, notice that $B \subseteq A$ holds in $\mathcal{B}_L$ if and only if $\phi(B) \subseteq \phi(A)$ holds in $\mathcal{B}_L$. This shows that ϕ is a Boolean isomorphism from $\mathcal{B}_L$ onto $\mathcal{B}_M$. ■

7.4. For a given topological space X, we shall denote by Clop(X) the collection of all clopen subsets of X. Clearly, Clop(X) ordered with the inclusion relation is a Boolean algebra, which is Dedekind complete if X is extremally disconnected.

We shall sketch the basic steps of the solution below and we shall leave the routine verification of the details for the reader. So, let Ω and Q be two Stonean spaces. If a mapping $\phi\colon Q \to \Omega$ is a homeomorphism, then it is easy to see that the operator $\pi\colon C^\infty(\Omega) \to C^\infty(Q)$, defined by $\pi(f) = f \circ \phi$ for each $f \in C^\infty(\Omega)$, is an onto Riesz isomorphism.

For the converse, assume that $\pi\colon C^\infty(\Omega) \to C^\infty(Q)$ is an onto Riesz isomorphism. First, we shall describe the bands in $C^\infty(\Omega)$. For each clopen subset V of Ω, let

$$\begin{aligned} B_V &= \big\{f \in C^\infty(\Omega)\colon\ f(\omega) = 0 \text{ for all } \omega \notin V\big\} \\ &= \big\{f\chi_V\colon\ f \in C^\infty(\Omega\big\} = C^\infty(V)\,. \end{aligned}$$

It should be clear that B_V is a band of $C^\infty(\Omega)$. We claim that every band is of the above form. To see this, let B be a band of $C^\infty(\Omega)$. Then, B is a projection band (see Theorem 7.4), and if we let u be the projection of $\mathbf{1}_\Omega$ on B, then a straightforward verification shows that $u = \chi_V$ for some clopen subset V of Ω and that $B = B_V$. Moreover, it should be noticed that the clopen set V is uniquely determined and that two clopen sets U and W of Ω satisfy $U \subseteq W$ if and only if $B_U \subseteq B_W$. Therefore, the mapping $V \mapsto B_V$ is a Boolean isomorphism from Clop(Ω) onto $\mathcal{B}_{C^\infty(\Omega)}$.

Now let $V \in$ Clop(Ω). Since $\pi\colon C^\infty(\Omega) \to C^\infty(Q)$ is an onto Riesz isomorphism, it follows that $\pi(B_V)$ is a band of $C^\infty(Q)$. Consequently, there exists a unique clopen subset $\phi(V)$ of Q such that $\pi(B_V) = B_{\phi(V)}$. Clearly, $V, W \in$ Clop(Ω) satisfy $V \subseteq W$ if and only if $\phi(V) \subseteq \phi(W)$. This implies that $\phi\colon$ Clop$(\Omega) \to$ Clop(Q) is an onto Boolean isomorphism.

For each $\omega \in \Omega$ let $\mathcal{I}_\omega = \{\mathcal{O} \in \text{Clop}(\Omega)\colon\ \omega \notin \mathcal{O}\}$. As in part (vi) of Exercise **7.1**, it follows that the proper prime ideals of the Boolean algebra Clop(Ω) are of the form $\mathcal{I}_\omega$. Now if $\mathcal{I}_\omega$ is a prime ideal of the Boolean algebra Clop(Ω), then $\phi(\mathcal{I}_\omega) = \{\phi(\mathcal{O})\colon\ \mathcal{O} \in \mathcal{I}_\omega\}$ is also a proper prime ideal of the Boolean algebra Clop(Q). Consequently, there exists a unique point $\psi(\omega) \in Q$ such that $\phi(\mathcal{I}_\omega) = \mathcal{I}_{\psi(\omega)}$. This shows that we have established a mapping $\psi\colon \Omega \to Q$ which is clearly one-to-one and onto. We shall finish the solution by proving that ψ is a homeomorphism.

To this end, start by observing that since the clopen subsets of Q is a base for the topology of Q, it suffices to show that $\psi^{-1}(U)$ is an open subset of Ω for all clopen subsets U of Q. So, let U be a clopen subset of Q and choose the uniquely determined clopen subset V of Ω such that $\phi(V) = U$. Now note that:

$$\begin{aligned}\psi^{-1}(U) &= \{\omega \in \Omega\colon\ \psi(\omega) \in U\} = \{\omega \in \Omega\colon\ U \notin \mathcal{I}_{\psi(\omega)}\}\\ &= \{\omega \in \Omega\colon\ \phi(V) = U \notin \mathcal{I}_{\psi(\omega)} = \phi(\mathcal{I}_\omega)\}\\ &= \{\omega \in \Omega\colon\ V \notin \mathcal{I}_\omega\} = \{\omega \in \Omega\colon\ \omega \in V\} = V\,.\end{aligned}$$

This shows that ψ is a homeomorphism and completes the solution. ■

7.5. Assume first that $\pi\colon \mathbb{R}^X \to \mathbb{R}^Y$ is an onto Riesz isomorphism. Notice that the atoms of $\mathbb{R}^X$ are precisely the elements $\chi_{\{x\}}$ for $x \in X$. Similarly, the atoms of $\mathbb{R}^Y$ are the elements $\chi_{\{y\}}$ for $y \in Y$. For simplicity, let us identify the element $\chi_{\{x\}}$ with x. Now note that since π is a Riesz isomorphism $\pi(x)$ is an atom of $\mathbb{R}^Y$, i.e., $\pi(x) \in Y$. Since $\pi\colon \mathbb{R}^X \to \mathbb{R}^Y$ is an onto Riesz isomorphism, it follows that $\pi\colon X \to Y$ is also an onto one-to-one function. This shows that the sets X and Y have the same cardinality.

For the converse, assume that X and Y have the same cardinality and let $\pi\colon X \to Y$ be an onto one-to-one function. Let L_1 be the order dense Riesz subspace of $\mathbb{R}^X$ consisting of all functions of the form $f = \sum_{a\in F} c_a\chi_{\{a\}}$, where F is a finite subset of X. The order dense Riesz subspace M_1 in $\mathbb{R}^Y$ is defined similarly. Observe that $\mathbb{R}^X$ and $\mathbb{R}^Y$ are the universal completions of L_1 and M_1, respectively. Now consider the mapping $T\colon L_1 \to M_1$, defined by

$$T(f) = \sum_{a\in F} c_a\chi_{\{\pi(a)\}}$$

for all $f = \sum_{a\in F} c_a\chi_{\{a\}} \in L_1$, and note that it is an onto Riesz isomorphism. By Theorem 7.19, π extends to a (unique) Riesz isomorphism from $\mathbb{R}^X$ onto $\mathbb{R}^Y$. ■

7.6. Since the characteristic functions of the points of $[0,1]$ belong to L, it follows that L is order dense in $\mathbb{R}^{[0,1]}$, and thus $L^{\text{u}} = \mathbb{R}^{[0,1]}$ (since $\mathbb{R}^{[0,1]}$ is universally complete.)

Now let $M = \{f \in \mathbb{R}^{[0,1]}\colon\ f \text{ is Lebesgue measurable}\}$. Then M (with the pointwise ordering everywhere) is a universally σ-complete Riesz space including L as a super order dense Riesz subspace. (If $0 \le f \in M$ and $\mathbf{1}$ denotes the constant function one on the interval $[0,1]$, then $\{f \wedge n\mathbf{1}\} \subseteq L$ and $0 \le f \wedge n\mathbf{1} \uparrow f$ holds in the Riesz space M.) Consequently M is the universal σ-completion of L. ■

7.7. If $\phi = 0$ the result is trivial. So, assume that $\phi \ne 0$. Note that C_ϕ is a laterally σ-complete Riesz space in its own right which admits a Riesz norm (namely, $\rho(u) = |\phi|(|u|)$ and so, by Theorem 7.47, C_ϕ admits an order unit. Hence, C_ϕ must be a principal band which by Theorem 7.4 must be a projection band. ■

7.8. If $u > 0$ is a discrete element in an Archimedean Riesz space L, then L_u is a projection band of L (see Theorem 1.77), and hence every element $v \in L$ has a unique representation of the form $v = \lambda u + w$ with $w \perp u$. It easily follows that the mapping $\phi\colon L \to \mathbb{R}$, defined by $\phi(v) = \lambda$, is a nonzero order continuous linear functional, and hence $L_n^\sim \neq \{0\}$. Thus, if $L_n^\sim = \{0\}$, then L does not contain any discrete elements.

On the other hand, if the laterally σ-complete Riesz space L has no discrete elements, then $L_n^\sim = \{0\}$ must be the case. Indeed, if $0 < \phi \in L_n^\sim$, then $C_\phi \neq \{0\}$, and (as shown in the previous exercise) C_ϕ admits an order unit. This conclusion and Theorem 7.47 imply that C_ϕ is Riesz isomorphic to some $\mathbb{R}^n$. But as $\mathbb{R}^n$ has plenty of discrete elements, C_ϕ has likewise plenty of discrete elements and hence L also has discrete elements, contrary to our hypothesis. ■

7.9. We complete the details of the implication (i) $\Longrightarrow$ (ii). We show that the Riesz space $M = U(L^+) - U(L^+)$ is universally σ-complete. For the Dedekind σ-completeness of M assume that $0 \leq u_n^* \uparrow \leq v^*$ in M. Then $0 \leq u_n^* \uparrow u^*$ holds in L^{u}. For each k choose $\{u_n^k\colon\ n = 1, 2, \ldots\} \subseteq L^+$ with $0 \leq u_n^k \uparrow_n u_k^*$ and note that if we let $w_n = \bigvee_{i=1}^n \bigvee_{k=1}^n u_i^k$, then $\{w_n\} \subseteq L^+$ and $0 \leq w_n \uparrow u^*$ in L^{u}. Therefore, $u^* \in M$ and hence M is Dedekind σ-complete.

For the lateral σ-completeness of M let $\{x_n^*\}$ be a disjoint sequence of $U(L^+)$. Put $x^* = \bigvee_{n=1}^\infty x_n^*$ in L^{u} and then for each k choose a sequence $\{y_n^k\colon\ n = 1, 2, \ldots\}$ of L^+ with $0 \leq y_n^k \uparrow_n x_k^*$. So, if $z_n = \bigvee_{i=1}^n \bigvee_{k=1}^n y_i^k$, then $\{z_n\} \subseteq L^+$ and $0 \leq z_n \uparrow x^*$ in L^{u}, so that $x^* \in M$.

For the uniqueness of the universal σ-completion, assume that there exist two universally σ-complete Riesz spaces M_1 and M_2 and Riesz isomorphisms (into) $\pi_1\colon L \to M_1$ and $\pi_2\colon L \to M_2$ so that $\pi_1(L)$ and $\pi_2(L)$ are super order dense in M_1 and M_2, respectively. Now if $0 \leq u \in M_1$, then choose a sequence $\{u_n\} \subseteq L^+$ with $0 \leq \pi_1(u_n) \uparrow u$ in M_1. Then $\{\pi_1(u_n)\}$ is a dominable subset of M_1 and therefore $\{u_n\}$ is a dominable sequence of L. But then $\{\pi_2(u_n)\}$ is a dominable sequence of $\pi_2(L)$, and hence of M_2. By Theorem 7.38 there exists some $w \in M_2$ such that $0 \leq \pi_2(u_n) \uparrow w$ holds in M_2. To see that w is independent of the sequence $\{u_n\}$, let another sequence $\{v_n\} \subseteq L^+$ satisfy $0 \leq \pi_1(v_n) \uparrow u$. Then, as above, $\pi_2(v_n) \uparrow w_1$ in M_2. But for each fixed k we have

$$\pi_1(v_n) \wedge \pi_1(u_k) = \pi_1(v_n \wedge u_k) \uparrow_n \pi_1(u_k)$$

in M_1. This implies $v_n \wedge u_k \uparrow_n u_k$ in L, and so

$$\pi_2(v_n) \wedge \pi_2(u_k) = \pi_2(v_n \wedge u_k) \uparrow_n \pi_2(u_k)$$

in M_2. It follows that $w \leq w_1$. By the symmetry of the situation, we also have $w_1 \leq w$ and so $w = w_1$.

The preceding discussion shows that the formula $\pi(u) = w$ defines a mapping $\pi\colon M_1^+ \to M_2^+$. This mapping is additive and therefore, by means of the formula $\pi(u) = \pi(u^+) - \pi(u^-)$, it extends to a unique linear operator from M_1 to M_2. We leave it for the reader to verify that π is a Riesz isomorphism from M_1 onto M_2. This establishes the uniqueness of the universal σ-completion. ■

7.10. Clearly the Riesz space L^λ must be laterally complete and the desired equality $(L^\lambda)^\delta = L^{\mathrm{u}}$ follows immediately from Theorem 7.15. The equality $(L^\delta)^\lambda = L^{\mathrm{u}}$ follows from Lemma 7.16. ■

7.11. We present two solutions. For the first solution start by observing that $u_\alpha \xrightarrow{o} 0$ in L implies $u_\alpha(x) \to 0$ for all $x \in X$. Indeed, if $u_\alpha \downarrow 0$ in $\mathbb{R}^X$ and if $u_\alpha(x) \geq \epsilon > 0$ holds for all α and some x and $\epsilon > 0$, then $u_\alpha \geq \epsilon\chi_{\{x\}} > 0$ holds in L for all α, violating $u_\alpha \downarrow 0$ in L. Note next that $\mathbb{R}^X = C(X)$, where X is topologized with the discrete topology. Now use Exercise **1.26** to establish that $L_c^\sim = L^\sim$.

For a second solution, notice that according to Theorem 7.8 every positive linear functional on L is order continuous, and hence $L_c^\sim = L^\sim$. ■

7.12. Assume that X has a non-measurable cardinal. Let $0 < \phi \in (\mathbb{R}^X)^\sim$, and assume by way of contradiction that ϕ is not of the described form.

Observe first that $\phi(\chi_{\{x\}}) \neq 0$ for finitely many x. Indeed, if there exists a countable subset $\{x_1, x_2, \ldots\}$ of X for which $\phi(\chi_{\{x_n\}}) = \lambda_n > 0$, then the function $f \in \mathbb{R}^X$ defined to be zero outside the set $\{x_1, x_2, \ldots\}$ and $f(x_n) = \frac{1}{\lambda_n}$ for each $n = 1, 2, \ldots$ satisfies $\phi(f) \geq \sum_{n=1}^k f(x_n)\phi(\chi_{\{x_n\}}) = k$ for all k, which is impossible. Thus, the expression $\sum_{x\in E} \phi(\chi_{\{x\}})$ makes sense as a real number for every subset E of X, and its meaning is obvious.

Now define the set function $\mu\colon \mathcal{P}(X) \to \mathbb{R}^+$, where $\mathcal{P}(X)$ is the power set of the set X, by $\mu(E) = \phi(\chi_E) - \sum_{x\in E} \phi(\chi_{\{x\}})$ for all $E \in \mathcal{P}(X)$ and note that:

(i) $\mu(E) = 0$, if E is finite.
(ii) $\mu\big(\bigcup_{n=1}^\infty E_n\big) = \sum_{n=1}^\infty \mu(E_n)$ for every pairwise disjoint sequence $\{E_n\}$ of subsets of X.

To see (ii), let F be the finite set of points for which $\phi(\chi_{\{x\}}) > 0$. Put $E = \bigcup_{n=1}^\infty E_n$. Using the fact that ϕ is σ-order continuous (see the previous exercise), we get

$$\sum_{n=1}^\infty \mu(E_n) = \sum_{n=1}^\infty \phi(\chi_{E_n}) - \sum_{x\in E\cap F} \phi(\chi_{\{x\}}) = \phi(\chi_E) - \sum_{x\in E\cap F} \phi(\chi_{\{x\}}) = \mu(E)\,.$$

Consequently μ is a finite measure on $\mathcal{P}(X)$, which is actually nonzero. To establish the last claim assume that $\mu(X) = 0$. If $F \neq \emptyset$, then it follows from the previous exercise that $\phi = 0$, which is a contradiction. So, let $F = \{x_1, \ldots, x_n\}$ and for each $i = 1, \ldots, n$ put $\lambda_i = \phi(\chi_{\{x_i\}}) > 0$. Observe that $\phi(\chi_X) = \sum_{i=1}^n \lambda_i$. Put $E = X \setminus F$ and note that in view of $\phi(\chi_F) = \sum_{i=1}^n \lambda_i$, it easily follows that $\phi(\chi_E) = 0$. Now if $0 \leq f \in \mathbb{R}^X$, then $0 \leq f \wedge n\chi_E \uparrow f\chi_E$, and thus by the σ-order continuity of ϕ we obtain $\phi(f\chi_E) = 0$. Hence

$$\phi(f) = \phi(f\chi_F) + \phi(f\chi_E) = \phi(f\chi_F) = \sum_{i=1}^n \lambda_i f(x_i)\,,$$

which is impossible, since ϕ is not of the above type.

It now easily follows (via $\nu = [\mu(X)]^{-1}\mu$) that X has a measurable cardinal contradicting our hypothesis, and the desired conclusion follows.

Remark. It should be noted that the converse also holds true. Namely: *If every $\phi \in (\mathbb{R}^X)^\sim$ is of the described form, then X has a non-measurable cardinal.* (For details see [**86**].) ■

7.13. (i) Assume that $\{u_n\} \subseteq L$ satisfies $u_n \xrightarrow{\tau} 0$. This implies $u_n(x) \to 0$ for all $x \in X$. So, $\limsup u_n(x) = \liminf u_n(x) = 0$ holds for all $x \in X$, and therefore both $\limsup u_n$ and $\liminf u_n$ exist in L, and they satisfy $\limsup u_n = \liminf u_n = 0$. Hence $u_n \xrightarrow{o} 0$ holds in L. On the other hand, if $u_n \xrightarrow{o} 0$ holds in L, then (as in the solution of Exercise **7.11**) $u_n(x) \to 0$ for all x, which means, of course, that $u_n \xrightarrow{\tau} 0$.

(ii) Let τ_1 and τ_2 be two Hausdorff locally solid topologies on L. Assume that $u_n \xrightarrow{\tau_1} 0$. By Theorem 7.52 we have $\tau \subseteq \tau_1$, and therefore, $u_n \xrightarrow{\tau} 0$. Now by part (i) above $u_n \xrightarrow{o} 0$, and by part (ii) of Theorem 7.49 we get $u_n \xrightarrow{\tau_2} 0$. Similarly, $u_n \xrightarrow{\tau_2} 0$ implies $u_n \xrightarrow{\tau_1} 0$. ■

7.14. Observe that N^{d} is a laterally σ-complete Riesz space which admits a metrizable locally solid topology τ^* whose base at zero is $\{V_n \cap N^{\mathrm{d}}\}$. Thus N^{d} is Archimedean.

To see that N^{d} is laterally complete, let D be a disjoint subset of N^{d} consisting of nonzero positive vectors. Clearly, $D = \bigcup_{n=1}^{\infty} D_n$ holds, where $D_n = \{u \in D\colon\ u \notin V_n\}$. Since by part (i) of Theorem 7.49 every disjoint sequence of N^{d} converges to zero with respect to τ^*, it easily follows that each D_n is finite, and hence D is at most countable. Thus $\sup D$ exists in L, and therefore $\sup D \in N^{\mathrm{d}}$. This shows that N^{d} is laterally complete. The rest of the solution can be completed by invoking Theorem 7.56. ■

7.15. As in the first part of the proof of Theorem 7.57, the topology τ_1 (as defined there) is a Lebesgue topology and $T\colon (L,\tau_1) \to (M,\tau')$ is continuous. From the scheme of operators $(L,\tau) \xrightarrow{I} (L,\tau_1) \xrightarrow{T} (M,\tau')$, it follows that in order to show that the operator $T\colon (L,\tau) \to (M,\tau')$ is continuous, it suffices to establish that τ_1 is coarser than τ, i.e., $\tau_1 \subseteq \tau$.

To prove this, it is enough (in view of Theorem 4.27) to show that $\tau_1 \subseteq \tau$ holds whenever both topologies are restricted to the components of C_{τ_1}. To see this, let N^{d} be a component of C_{τ_1}. Then τ_1 restricted to N^{d} is a Hausdorff Lebesgue topology which (as its is easily verified) satisfies condition (v) of Theorem 7.51. But then since τ restricted to N^{d} is a also a σ-Fatou topology, the conclusion follows immediately from statement (iii) of Theorem 7.51. ■

7.16. Start by observing that the typical basic τ-neighborhood V of zero is of the form $V = \{f \in \mathbb{R}^X\colon\ |f(x_i)| \le \epsilon \text{ for } i = 1,\dots,n\}$, where $\{x_1,\dots,x_n\}$ is a given finite subset of X and ϵ is a given positive number. This easily implies that if $\{V_n\}$ is a normal sequence of τ-neighborhoods of zero, then there exists an at most countable subset E of X such that:

$$\{f \in \mathbb{R}^X\colon\ f(x) = 0 \text{ for all } x \in E\} \subseteq N = \bigcap_{n=1}^{\infty} V_n\,.$$

On the other hand, it is not difficult to observe that every set of the form

$$\{f \in \mathbb{R}^X\colon\ f(x) = 0 \text{ for all } x \in E\}\,,$$

with E at most countable, is the null ideal of an appropriate normal sequence of τ-neighborhoods of zero. Indeed, if the set $E = \{x_1, x_2, \dots\}$ is countable, let

$$V_n = \{f \in \mathbb{R}^X\colon\ |f(x_i)| \le 2^{-n} \text{ for } i = 1,\dots,n\}$$

and if $E = \{x_1,\dots,x_k\}$ is finite, let

$$V_n = \{f \in \mathbb{R}^X\colon\ |f(x_i)| \le 2^{-n} \text{ for } 1 \le i \le k\}\,.$$

Now the desired conclusion should be immediate. ■

7.17. (i) If $\{f_n\}$ is a disjoint sequence of L^+, then the formula $f(x) = \sum_{n=1}^{\infty} f_n(x)$ defines a real-valued function on the set X, that is also μ-measurable since it is the pointwise limit of the μ-measurable sequence of its partial sums $\{\sum_{k=1}^{n} f_k\}$. Clearly, $f = \sup_{n\in\mathbb{N}} f_n$ holds in L, and so L is laterally σ-complete. Now if $0 \le f_n \uparrow\le g$ holds in the Riesz space L, then $f(x) = \sup_{n\in\mathbb{N}} f_n(x) \in \mathbb{R}$ for each $x \in X$. Obviously, $f \in L$ and $0 \le f_n \uparrow f$ in L. So, the Riesz space L is Dedekind σ-complete and hence L is universally σ-complete.

Next observe that since the finite subsets of X are measurable, the characteristic functions of the points belong to L, and so L is order dense in $\mathbb{R}^X$. Therefore, the topology of pointwise convergence is the unique Hausdorff Lebesgue topology on L.

(ii) That A is a σ-ideal follows immediately from the fact that $u_\alpha \xrightarrow{o} u$ in L implies $u_\alpha(x) \to 0$ for each $x \in X$. For the universal σ-completeness of L/A use part (i) of Theorem 7.9.

(iii) Since $0 \leq \xi \leq \eta$ is equivalent to $\frac{\xi}{1+\xi} \leq \frac{\eta}{1+\eta}$, it easily follows that each V_n is solid—and so also balanced. Given $u \in L^+$ and any $k \in \mathbb{N}$, the relations

$$\left| \tfrac{n^{-1}u}{1+n^{-1}u} \right| \leq 1, \quad \mu(X) < \infty, \quad \text{and} \quad \tfrac{n^{-1}u}{1+n^{-1}u} \to 0$$

combined with the Lebesgue dominated convergence theorem, guarantees that $\frac{1}{n}u \in V_k$ for all sufficiently large n. That is, each V_k is also absorbing. From the inequalities

$$\tfrac{|\xi+\eta|}{1+|\xi+\eta|} \leq \tfrac{|\xi|+|\eta|}{1+|\xi|+|\eta|} \leq \tfrac{|\xi|}{1+|\xi|} + \tfrac{|\eta|}{1+|\eta|},$$

we get $V_{n+1} + V_{n+1} \subseteq V_n$ for all n. Finally, by observing that $\bigcap_{n=1}^{\infty} V_n = \{0\}$, we obtain that $\{V_n\}$ defines a metrizable locally solid topology on L/A.

Now the proof of this part can be completed either by a direct verification or by using Theorem 7.55.

(iv) Since τ is generated by the family of Riesz seminorms $\{\rho_x\colon\ x \in X\}$, where $\rho_x(u) = |u(x)|$, the topology τ/A is generated by the family of Riesz seminorms $\{\dot\rho_x\colon\ x \in X\}$. Now, since each finite set has measure zero, it easily follows that $\dot\rho_x(\dot u) = \inf\{|v(x)|\colon\ v - u \in A\} = 0$ for all $x \in X$, and so $\tau/A = \{\emptyset, L/A\}$. Thus τ/A is the indiscrete topology. ■

7.18. Repeating the proof of part (i) of Theorem 4.17, we see that C_τ is a σ-ideal of L, and hence C_τ is laterally σ-complete. For the countable sup property let $u_\alpha \downarrow 0$ in C_τ. We can assume without loss of generality that $\{u_\alpha\} \subseteq N^{\mathrm{d}}$ for some component N^{d} of C_τ, and the countable sup property follows from Exercise **7.14**. To see that τ restricted to C_τ is a Hausdorff topology, let $0 < u \in C_\tau$. Then $u \in N^{\mathrm{d}}$ for some component N^{d} of C_τ, where $N = \bigcap_{n=1}^{\infty} V_n$ for some normal sequence $\{V_n\}$ of τ-neighborhoods of zero. If $u \in V_n$ for each n, then $u \in N$ and hence $u = 0$, contrary to $u > 0$. Thus $u \notin V_n$ holds true for some n, and therefore τ restricted to C_τ is a Hausdorff topology.

That τ has the Lebesgue property when restricted to C_τ, follows immediately from Theorem 7.49 in conjunction with the countable sup property of C_τ. ■

7.19. If τ is a Hausdorff Lebesgue topology on L, then C_τ is order dense in L, and therefore (since L is also Archimedean) C_τ must be τ-dense.

For the converse, assume that C_τ is τ-dense in L and let $u_\alpha \downarrow 0$ in L. Let V be a solid τ-neighborhood of zero. Choose another solid τ-neighborhood W of zero with $W+W \subseteq V$ and fix an index α_1. Pick $0 \leq w \in C_\tau$ with $u_{\alpha_1} - w \in W$. Now $u_\alpha \wedge w \downarrow 0$ holds in C_τ, and hence $u_\alpha \wedge w \xrightarrow{\tau} 0$ (since τ is a Lebesgue topology on C_τ by the previous exercise). Choose some index $\beta \succeq \alpha_1$ such that $u_\alpha \wedge w \in W$ holds for all $\alpha \succeq \beta$. Then, for all $\alpha \succeq \beta$, we have

$$0 \leq u_\alpha = (u_\alpha - w)^+ + u_\alpha \wedge w \leq (u_{\alpha_1} - w)^+ + u_\alpha \wedge w \in W + W \subseteq V.$$

This implies that $u_\alpha \in V$ holds for all $\alpha \succeq \beta$. Therefore, $u_\alpha \xrightarrow{\tau} 0$, and thus τ is a Lebesgue topology. ■

7.20. If L admits a Hausdorff Lebesgue topology τ^*, then by Theorem 7.52 we have $\tau^* \subseteq \tau$. So, $C_{\tau^*} \subseteq C_\tau$, and the order denseness of C_τ follows from the order denseness of C_{τ^*}.

Now if C_τ is order dense in L, then the universal completion of C_τ is precisely L^{u}. But the restriction of τ to C_τ is (according to Exercise **7.18**) a Hausdorff Lebesgue topology that satisfies condition (v) of Theorem 7.51. Therefore, the restriction of τ to the carrier C_τ has a Hausdorff Lebesgue extension to the universal completion L^{u} of C_τ. Hence L admits a Hausdorff Lebesgue topology. ■

7.21. Let $L = \{f \in \mathbb{R}^{[0,1]}\colon\ f \text{ is Lebesgue measurable}\}$. Then L with the pointwise ordering (everywhere) is a universally σ-complete Riesz space; but it fails to be laterally complete.

Let τ be the Hausdorff locally solid topology generated by the family of Riesz seminorms $\{\rho_x\colon\ x \in [0,1]\}$, where $\rho_x(u) = |u(x)|$ for all $u \in L$ and all $x \in [0,1]$, together with the Riesz pseudonorm ρ defined by $\rho(u) = \int_0^1 \frac{|u(x)|}{1+|u(x)|}\,dx$ for $u \in L$. Then (L,τ) satisfies the σ-Lebesgue and pre-Lebesgue properties, but it fails to satisfy the Lebesgue property. To see the latter, consider the net $\{\chi_F\}$ with F running over the finite subsets of $[0,1]$. Then $0 \le \chi_F \uparrow \mathbf{1}$ (where $\mathbf{1}(x) = 1$ for all $x \in [0,1]$) in L, $\rho(\chi_F) = 0$ for all finite subsets F of $[0,1]$ and $\rho(\mathbf{1}) = \frac{1}{2}$, which shows that the net $\{\chi_F\}$ is not τ-convergent to $\mathbf{1}$. ■

7.22. If X has a measurable cardinal, then the Hausdorff locally solid topology generated by the indicated (in the hint) Riesz pseudonorms is not a Lebesgue topology. To see this, use the net of the previous exercise. The conclusion now should be obvious. ■

7.23. Fix a sequence $\{X_n\} \subseteq \Sigma$ satisfying $0 < \mu(X_n) < \infty$ for each $n \in \mathbb{N}$ and $X = \bigcup_{n=1}^\infty X_n$. Also, define $\rho\colon L_0^+(\mu) \to [0,\infty)$ by

$$\rho(u) = \sum_{n=1}^\infty \tfrac{1}{2^n} \cdot \tfrac{1}{\mu(X_n)} \int_{X_n} \tfrac{u}{1+u}\,d\mu\,.$$

It is easy to see that ρ possesses the following properties:

(1) A function $u \in L_0^+(\mu)$ satisfies $\rho(u) = 0$ if and only if $u = 0$.
(2) If $u, v \in L_0^+(\mu)$ satisfy $0 \le u < v$, then $\rho(u) < \rho(v)$, i.e., ρ is strictly monotone.
(3) If $0 \le u_n \uparrow u$ holds in $L_0^+(\mu)$, then $\rho(u_n) \uparrow \rho(u)$.

Now let D be a nonempty subset of $L_0(\mu)$ that is bounded from above in $L_0(\mu)$. Since the collection of all finite suprema of D has the same upper bounds as D, we can assume that D is directed upward. Now fix some $u_0 \in D$ and consider the set $D_1 = \{u \in D\colon\ u \ge u_0\} - u_0$. Notice that $\sup D$ exists if and only $\sup D_1$ exists and $\sup D_1 = (\sup D) - u_0$. So, replacing D with the set consisting of all finite suprema of D_1, we can assume without loss of generality that D is an upward directed subset of $L_0^+(\mu)$.

Put $m = \sup_{u\in D} \rho(u) < \infty$, and then choose a sequence $\{u_n\} \subseteq D$ such that $u_n \uparrow$ and $\rho(u_n) \uparrow m$. Since D is bounded from above, it follows that there exists some $u \in L_0^+(\mu)$ such that $u_n \uparrow u$. This implies $\rho(u_n) \uparrow \rho(u)$. We claim that $u = \sup D$.

First, we shall show that u is an upper bound of D. To see this, let $v \in D$. Then $v \vee u_n \in D$ and $v \vee u_n \uparrow v \vee u$. From $u_n \le v \vee u_n$ and $\rho(v \vee u_n) \le m$, it follows that $\rho(v \vee u) = m$. Now from $0 \le u \le v \vee u$, $\rho(v \vee u) = \rho(u) = m$, and the strict monotonicity of ρ, we get $u = v \vee u$. This implies $v \le u$, i.e., u is an upper bound of D. To see that u is the least upper bound of D, let $w \in L_0(\mu)$ be an upper bound of D, i.e., $v \le w$ for each $v \in D$. Then $u_n \le w$ also holds in $L_0(\mu)$ for each n. Since u is the μ-a.e. pointwise limit of the sequence $\{u_n\}$, we see that $u \le w$. Thus, $u = \sup D$.

The above arguments show that $L_0(\mu)$ is a Dedekind complete Riesz space with the countable sup property. Finally, we shall show that $L_0(\mu)$ is universally complete. To this end, notice first that the Riesz space $L_0(\mu)$ is laterally σ-complete. We shall finish the proof by showing that every disjoint subset of $L_0(\mu)$ is at most countable.

To see this, let D be a disjoint subset of L^+ consisting of nonzero positive functions. Clearly, $D = \bigcup_{n=1}^{\infty} D_n$, where $D_n = \{f \in D\colon\ f \wedge \chi_{X_n} \neq 0\}$. The desired conclusion will be finished if we show that each D_n is at most countable. To see this, for each $g \in D_n$ let $S_g = \{x \in X_n\colon\ g(x) > 0\}$ and note that $S_g \in \Sigma$ and $\mu(S_g) > 0$. Thus, we can write $D_n = \bigcup_{k=1}^{\infty} A_k$, where $A_k = \{g \in D_n\colon\ \mu(S_g) > \frac{1}{k}\}$. Since the collection of measurable sets $\{S_g\colon\ g \in A_k\}$ is pairwise disjoint and $\mu(X_n) < \infty$, it follows that each A_k is a finite set. This implies that each D_n is at most countable, and the solution is finished. ■

7.24. Let τ be a Hausdorff locally solid Lebesgue topology on a (necessarily Archimedean) Riesz space L. By part (2) of Theorem 7.54 the topology τ extends to a (unique) Hausdorff locally solid topology τ^* on L^{u}. By part (iii) of Theorem 7.51 we have $\tau^* \subseteq \tau$ on L. Moreover, as in the proof of Theorem 7.54, we know that a base at zero for the restriction of τ^* on L is the collection of all solid sets of the form $V_u = \{v \in L\colon\ u \wedge |v| \in V\}$, where V is a solid τ-neighborhood of zero and $u \in L^+$.

Now assume that a net $\{u_\alpha\}$ of L satisfies $0 \leq u_\alpha \leq w$ for all α and some $w \in L^+$ and that $u_\alpha \xrightarrow{\tau^*} 0$. If V is a solid τ-neighborhood of zero, then choose some index α_0 such that $u_\alpha \in V_w$ (or $u_\alpha = w \wedge u_\alpha \in V$) for all $\alpha \succeq \alpha_0$. This shows that $u_\alpha \xrightarrow{\tau} 0$, and thus τ^* and τ agree on $[0, u]$. Consequently, all Hausdorff locally solid Lebesgue topologies on L induce the same topology on any order bounded subset A of L (and this topology is the restriction on A of the unique Hausdorff locally solid Lebesgue topology of L^{u}). ■

7.25. Let as denote by $\mathcal{H}$ the collection of all Hausdorff locally solid topologies on a Riesz space and assume that L admits a minimal locally solid topology ξ. Also, let $\mathcal{S}$ denotes the collection of all pseudonorms on L. For any $\tau \in \mathcal{H}$ we shall denote the collection of all τ-continuous pseudonorms by $\mathcal{S}_\tau$.

If τ_m denotes the locally solid topology generated by $\mathcal{S}$, then every $\tau \in \mathcal{H}$ satisfies $\xi \subseteq \tau \subseteq \tau_m$. Now let us describe the lattice operations in $\mathcal{H}$. Let $\tau_1, \tau_2 \in \mathcal{H}$. Clearly, the infimum is given by

$$\begin{aligned} \tau_1 \wedge \tau_2 &= \tau_1 \cap \tau_2 \\ &= \text{The Hausdorff locally solid topology generated by } \mathcal{S}_{\tau_1} \cap \mathcal{S}_{\tau_2}\,. \end{aligned}$$

For the supremum operation note that

$$\tau_1 \vee \tau_2 = \text{The Hausdorff locally solid topology generated by } \mathcal{S}_{\tau_1} \cup \mathcal{S}_{\tau_2}\,.$$

Also, it should be clear that if $\tau \in \mathcal{H}$, then the complement of τ is given by:

$$\tau' = \text{The Hausdorff locally solid topology generated by } [\mathcal{S} \setminus \mathcal{S}_\tau] \cup \mathcal{S}_\xi\,.$$

Finally, to see that $\mathcal{H}$ is a Dedekind complete Boolean algebra, let $\mathcal{F}$ be a nonempty subset of $\mathcal{H}$. Then, a direct verification shows that

$$\inf \mathcal{F} = \bigcap_{\tau \in \mathcal{F}} \tau\,,$$

and that

$$\sup \mathcal{F} = \text{The Hausdorff locally solid topology generated by } \bigcup_{\tau \in \mathcal{F}} \mathcal{S}_\tau\,.$$

Therefore, $\mathcal{H}$ under inclusion is a Dedekind complete Boolean algebra with smallest element ξ and largest element τ_m. ■

9.8. Chapter 8: Market Economies

8.1. Let $u: X \to \mathbb{R}$ be a function defined on a convex subset of a vector space and denote by $\succeq$ the preference generated by u. Assume first that $\succeq$ is a convex preference and fix $x, y \in X$ and $0 < \alpha < 1$. We also can assume without loss of generality that $x \succeq y$. But then convexity of $\succeq$ implies that $\alpha x + (1-\alpha)y \succeq y$ or

$$u(\alpha x + (1-\alpha)y) \geq u(y) = \min\{u(x), u(y)\}.$$

This shows that the function u is quasiconcave.

For the converse assume that u is quasiconcave. Fix $y, z \in X$ such that $y \succeq x$ and $z \succeq x$ or $u(y) \geq u(x)$ and $u(z) \geq u(x)$. Now if $0 \leq \alpha \leq 1$, then it follows from the quasiconcavity of u that

$$u(\alpha y + (1-\alpha)z) \geq \min\{u(y), u(z)\} \geq u(x).$$

That is, $\alpha y + (1-\alpha)z \succeq x$ so that $\succeq$ is a convex preference. ■

8.2. Assume by way of contradiction that there exists a function $u: L^+ \to \mathbb{R}$ satisfying $u(x) \geq u(y)$ if and only if $x \geq y$ holds, where $\geq$ denotes the lexicographic order of $\mathbb{R}^2$. For each $\alpha \geq 0$ we have $(\alpha, 1), (\alpha, 0) \in L^+$ and clearly $(\alpha, 1) > (\alpha, 0)$. This implies $u(\alpha, 1) > u(\alpha, 0)$ for all $\alpha \geq 0$. Now for each $\alpha \geq 0$ pick a rational number r_α such that $u(\alpha, 0) < r_\alpha < u(\alpha, 1)$.

We claim that the mapping $\alpha \mapsto r_\alpha$, from $[0, \infty)$ to $\mathbb{Q}$, is one-to-one. Indeed, if $\alpha > \beta \geq 0$, then $(\alpha, 0) > (\beta, 1)$ and so $u(\alpha, 0) > u(\beta, 1)$. Consequently, we have

$$u(\beta, 0) < r_\beta < u(\beta, 1) < u(\alpha, 0) < r_\alpha < u(\alpha, 1),$$

and hence $r_\beta < r_\alpha$. Thus, $\alpha \mapsto r_\alpha$ is one-to-one and so the interval $[0, \infty)$ is at most countable, a contradiction. Therefore, the preference generated by the lexicographic order does not have a utility representation. ■

8.3. Assume that $\langle L, L' \rangle$ is Riesz pair. That is, L' is an ideal of the order dual $L^\sim$ that separates the points of L and the duality map is given by $\langle x, x' \rangle = x'(x)$.

(a) According to part (2) of Nakano's Theorem 1.83 the Riesz space L can be considered as a Riesz subspace of $(L')^\sim_n$ and hence it can be considered as a Riesz subspace of $(L')^\sim$. This shows that $\langle L', L \rangle$ is a lattice pair.

To see that every order interval of L' is $\sigma(L', L)$-compact, let $0 \leq x' \in L'$. Then, it is not difficult to see that the order interval $[0, x']$ is a closed and bounded subset of $\mathbb{R}^L$, and hence a compact subset of $\mathbb{R}^L$. Since the product topology of $\mathbb{R}^L$ induces the topology $\sigma(L', L)$ on L', it follows that $[0, x']$ is $\sigma(L', L)$-compact.

(b) Assume that the Riesz pair $\langle L, L' \rangle$ is symmetric. This means that L embedded in $(L')^\sim$ (as in Theorem 1.83) is an ideal of $(L')^\sim$. In particular, by part (a) the order intervals of L are $\sigma(L, L')$-compact.

To establish the converse, assume that the order intervals of L are $\sigma(L, L')$-compact. Since L' is the topological dual of L equipped with the absolute weak topology $|\sigma|(L, L')$ (see Theorem 2.33), an easy application of Theorem 6.62 shows that L is an ideal in L'', the topological dual of $\big(L', \beta(L', L)\big)$. Since (according to Theorem 2.22) L'' is an ideal of $(L')^\sim$, it follows that L is also an ideal of $(L')^\sim$. This implies that $\langle L', L \rangle$ is a Riesz pair and so $\langle L, L' \rangle$ is a symmetric Riesz pair. (The proof of this part also follows immediately from Theorem 6.62.) ■

8.4. Assume that each preference is strictly monotone and lower semicontinuous for some linear topology. For $m = 1$ the result is trivial. So, assume $m > 1$. Let $(x_1, \ldots, x_m)$ be a weakly Pareto efficient allocation.

Now suppose that an allocation $(y_1, \ldots, y_m)$ satisfies $y_i \succeq_i x_i$ for each i and $y_k \succ_k x_k$ for some k. By the lower semicontinuity of the preference $\succeq_k$, there exists some $0 < \alpha < 1$ such that $(1-\alpha)y_k \succ_k x_k$. Now if we let $z_i = y_i + \frac{\alpha}{m-1} y_k$ whenever $i \neq k$ and $z_k = (1-\alpha)y_k$, then $(z_1, \ldots, z_m)$ is an allocation. Moreover, by the strict monotonicity of the preferences, we see that $z_i \succ_i x_i$ holds for all i, contradicting the weak Pareto efficiency of $(x_1, \ldots, x_m)$. This implies that $(x_1, \ldots, x_m)$ is Pareto efficient. ■

8.5. (a) Since each function is continuous for a consistent locally convex topology on the commodity space L, it follows that each utility function is also Mackey-continuous, i.e., $\tau(L, X')$-continuous. This implies that U is a Mackey-continuous function. Clearly, U is also a concave function.

From the Mackey-continuity of U it is easy to see that U is also weakly upper semicontinuous. In particular, it follows from part (2) of Theorem 8.3 that U attains a maximum value over the weakly compact set $\mathcal{A}_r$.

(b) Let $(x_1, \ldots, x_m) \in \mathcal{A}_r$ be a maximizer of U over $\mathcal{A}_r$. That is,

$$U(y_1, \ldots, y_m) \leq U(x_1, \ldots, x_m) \qquad (\star)$$

holds for each $(y_1, \ldots, y_m) \in \mathcal{A}_r$. If $(x_1, \ldots, x_m)$ is not Pareto efficient, then there exists an allocation $(z_1, \ldots, z_m)$ satisfying $z_i \succeq_i x_i$ for all i and $z_i \succ_i x_i$ for at least one i. But then $(z_1, \ldots, z_m) \in \mathcal{A}_r$ and

$$U(z_1, \ldots, z_m) > U(x_1, \ldots, x_m),$$

contrary to $(\star)$. Thus, every maximizer of U over $\mathcal{A}_r$ is an individually rational Pareto efficient allocation. ■

8.6. (a) For the Mackey-continuity of V notice first that for each pair of real numbers α_1 and α_2 we have

$$\min\{\alpha_1, \alpha_2\} = \tfrac{1}{2}(\alpha_1 + \alpha_2 - |\alpha_1 - \alpha_2|).$$

This implies that the function $(\alpha_1, \alpha_2) \mapsto \min\{\alpha_1, \alpha_2\}$ is continuous. It is now easy to verify by an inductive argument that the real-valued function

$$(\alpha_1, \ldots, \alpha_m) \longmapsto \min\{\alpha_1, \ldots, \alpha_m\}$$

is also continuous on $\mathbb{R}^m$. From this conclusion and the fact that each utility function u_i is Mackey-continuous, it follows that V is also Mackey-continuous.

Next, we shall prove that V is quasiconcave. From Exercise **8.1** we know that each u_i is quasiconcave. This implies that V is also a quasiconcave function. To see this, let $x = (x_1, \ldots, x_m)$ and $y = (y_1, \ldots, y_m)$ be two vectors in $(L^+)^m$ and let $0 < \alpha < 1$. Clearly, we have $u_i(x_i) \geq V(x)$ and $u_i(y_i) \geq V(y)$ for all i. Now the quasiconcavity of each u_i implies

$$u_i(\alpha x_i + (1-\alpha)y_i) \geq \min\{u_i(x_i), u_i(y_i)\} \geq \min\{V(x), V(y)\}.$$

Consequently,

$$\begin{aligned} V(\alpha x + (1-\alpha)y) &= \min\{u_i(\alpha x_i + (1-\alpha)y_i)\colon\ i = 1, \ldots, m\} \\ &\geq \min\{V(x), V(y)\}, \end{aligned}$$

so that V is a quasiconcave function.

(b) It follows that V is a weakly upper semicontinuous quasiconcave function on $(L^+)^m$. By part (2) of Theorem 8.3 the function V attains a maximum value over the weakly compact set $\mathcal{A}_r$.

(c) Assume that $(x_1,\dots,x_m)\in\mathcal{A}_r$ is a maximizer of V over $\mathcal{A}_r$, i.e.,

$$V(y_1,\dots,y_m)\le V(x_1,\dots,x_m) \qquad (\star\star)$$

holds for each $(y_1,\dots,y_m)\in\mathcal{A}_r$. Now assume by way of contradiction that the individually rational allocation $(x_1,\dots,x_m)$ is not weakly Pareto efficient. This means that there exists another allocation $(z_1,\dots,z_m)$ satisfying $u_i(z_i)>u_i(x_i)$ for each i. In particular, $(z_1,\dots,z_m)\in\mathcal{A}_r$. If k satisfies $V(z_1,\dots,z_m)=u_k(z_k)$, then we have

$$V(z_1,\dots,z_m)=u_k(z_k)>u_k(x_k)\ge V(x_1,\dots,x_m),$$

which contradicts $(\star\star)$. This contradiction guarantees that $(x_1,\dots,x_m)$ is an individually rational weakly Pareto efficient allocation. ■

8.7. (1)$\Longrightarrow$(2) Since the utility space $\mathbf{U}$ is bounded, it follows that $\mathbf{U}$ is compact if and only if it is closed.

(2)$\Longrightarrow$(3) Consider a sequence $\{(x_1^n,\dots,x_m^n)\}$ of free disposal allocations such that $x_i^{n+1}\succeq_i x_i^n$, i.e., $u_i(x_i^{n+1})\ge u_i(x_i^n)$, holds for all i and all n. This implies that the sequence $\{(u_1(x_1^n),\dots,u_m(x_m^n))\}$ of $\mathbf{U}$ is increasing. By the compactness of $\mathbf{U}$, the sequence converges to an element of $\mathbf{U}$, say $(u_1(x_1),\dots,u_m(x_m))$. Now an easy argument shows that $x_i\succeq_i x_i^n$ holds for each i and each n.

(3)$\Longrightarrow$(1) Let $\{(z_1^n,\dots,z_m^n)\}$ be a sequence of $\mathbf{U}$ satisfying

$$\mathbf{z}_n=(z_1^n,\dots,z_m^n)\longrightarrow\mathbf{z}=(z_1,\dots,z_m)$$

in $\mathbb{R}^m$. By passing to a subsequence, we can suppose that $|z_i^n-z_i|\le\frac{1}{n}$ holds for each n and all i. Thus, $\mathbf{z}-\mathbf{z}_n\le|\mathbf{z}_n-\mathbf{z}|\le\frac{1}{n}\mathbf{e}$ holds for each n, where $\mathbf{e}=(1,1,\dots,1)$. From $\mathbf{0}\le(\mathbf{z}-\frac{1}{n}\mathbf{e})^+\le\mathbf{z}_n$ and part (2) of Lemma 8.43, we see that $(\mathbf{z}-\frac{1}{n}\mathbf{e})^+\in\mathbf{U}$ holds for all n, and from $(\mathbf{z}-\frac{1}{n}\mathbf{e})^+\uparrow\mathbf{z}$ and our hypothesis, we easily infer that $\mathbf{z}\in\mathbf{U}$. ■

8.8. For the solution to this exercise, we need the following simple property of strictly convex preferences:

($\bullet$) *Let $\succeq$ be a strictly convex preference on a convex set C. If $x,x_1,x_2,\dots,x_k$ in C satisfy $x_i\succeq x$ for each i and $x_i\succ x$ for some i, then $\sum_{i=1}^k\alpha_ix_i\succ x$ for every proper convex combination of $x_1,\dots,x_k$.*

The proof is by induction. To see this, recall first that a preference $\succeq$ on a convex set is said to be **strictly convex** if $y\succeq x$, $z\succeq x$ and $y\ne z$ imply $\alpha y+(1-\alpha)z\succ x$ for all $0<\alpha<1$. So, let $\succeq$ be a strictly convex preference on a convex set C. For $k=1$ the claim is obvious. For $k=2$ assume $x_1\succ x$, $x_2\succeq x$ and let $0<\alpha<1$. If $x_1=x_2$, then $\alpha x_1+(1-\alpha)x_2=x_1\succ x$, and in case $x_1\ne x_2$ the strict convexity of $\succeq$ implies $\alpha x_1+(1-\alpha)x_2\succ x$. For the induction step assume our claim to be true for some k and assume that $x_i\succeq x$ for all $i=1,\dots,k,k+1$, and let $\alpha_i>0$ $(i=1,\dots,k+1)$ satisfy $\sum_{i=1}^{k+1}\alpha_i=1$. Put $\lambda=\sum_{i=1}^k\alpha_i$ and note that $1-\lambda=\alpha_{k+1}>0$. If $\beta_i=\frac{\alpha_i}{\lambda}$ $(i=1,\dots,k)$, then $\sum_{i=1}^k\beta_i=1$ and our induction hypothesis yields $y=\sum_{i=1}^k\beta_ix_i\succ x$. But then, from the case $k=2$, we obtain $\sum_{i=1}^{k+1}\alpha_ix_i=\lambda y+(1-\lambda)x_{k+1}\succ x$, and the proof of our claim is complete.

Now let $(\mathbf{x}_1,\dots,\mathbf{x}_m)$ be an individually rational allocation in a Riesz commodity economy where each consumer has the same initial endowment ω and the same strictly convex preference $\succeq$. Since $(\mathbf{x}_1,\dots,\mathbf{x}_m)$ is an allocation, $\sum_{i=1}^m\mathbf{x}_i=m\omega$ or $\sum_{i=1}^m\frac{1}{m}\mathbf{x}_i=\omega$. Next, notice that the individual rationality of $(\mathbf{x}_1,\dots,\mathbf{x}_m)$ implies $\mathbf{x}_i\succeq\omega$ for each i. Now assume by way of contradiction that $\mathbf{x}_j\ne\omega$ holds for some j. Then the strict convexity of $\succeq$ implies $\frac12\mathbf{x}_j+\frac12\omega\succ_j\omega$. Since $\frac12\mathbf{x}_i+\frac12\omega\succeq_i\omega$ is true for all i, the above property ($\bullet$) yields

$$\omega=\textstyle\sum_{i=1}^m\frac{1}{m}\mathbf{x}_i=\sum_{i=1}^m\frac{1}{m}\big(\frac12\mathbf{x}_i+\frac12\omega\big)\succ\omega,$$

which is impossible. Hence, $\mathbf{x}_i = \omega$ for each i, proving that $(\omega,\dots,\omega)$ is the only individually rational allocation in the economy.

Since every core allocation is individually rational, the allocation $(\omega,\dots,\omega)$ is the only possible candidate for a core allocation. To see that $(\omega,\dots,\omega)$ is indeed a core allocation, assume by way of contradiction that there exist a coalition S consisting of k consumers and an allocation $(\mathbf{y}_1,\dots,\mathbf{y}_m)$ with $\mathbf{y}_i \succ \omega$ for each $i \in S$ and $\sum_{i\in S}\mathbf{y}_i = k\omega$. But then from $(\bullet)$, we see that $\omega = \sum_{i\in S}\frac{1}{k}\mathbf{y}_i \succ \omega$, which is impossible. Hence, $(\omega,\dots,\omega)$ is also the only core allocation of this economy. ■

8.9. (a) $\Longrightarrow$ (b) Let $x \succ_i x_i$. Since x_i is a maximal element in the budget set $B_i(p)$, it follows that $x \notin B_i(p)$ or, equivalently, $p\cdot x > p\cdot\omega_i$. Since $x_i + \epsilon\omega \succ_i x_i$ for each $\epsilon > 0$, we see that $p\cdot x_i + \epsilon p\cdot\omega > p\cdot\omega_i$ for all $\epsilon > 0$ so that $p\cdot x_i \geq p\cdot\omega_i$ for each i. Next, from

$$p\cdot\omega = p\cdot\Big[\sum_{i=1}^m x_i\Big] = \sum_{i=1}^m p\cdot x_i \geq \sum_{i=1}^m p\cdot\omega_i = p\cdot\Big[\sum_{i=1}^m \omega_i\Big] = p\cdot\omega\,,$$

we easily get $p\cdot x_i = p\cdot\omega_i$ for each consumer i.

Now if $x > 0$, then $x + x_1 \succ_1 x_1$, and so $p\cdot(x + x_1) > p\cdot\omega_1 = p\cdot x_1$. This implies $p\cdot x > 0$ and thus p is strictly positive.

(b) $\Longrightarrow$ (c) Obvious.

(c) $\Longrightarrow$ (d) Let $x \succeq_i x_i$. Since the preference of each consumer i is strictly monotone, we see that $x + \epsilon\omega \succ_i x_i$ holds for each $\epsilon > 0$. Therefore,

$$p\cdot x + \epsilon p\cdot\omega = p\cdot(x + \epsilon\omega) > p\cdot\omega_i$$

holds for each $\epsilon > 0$, and so $p\cdot x \geq p\cdot\omega_i$.

(d) $\Longrightarrow$ (a) Observe that $x_i \succeq_i x_i$ implies $p\cdot x_i \geq p\cdot\omega_i$ for each i. From

$$p\cdot\omega = p\cdot\Big[\sum_{i=1}^m x_i\Big] = \sum_{i=1}^m p\cdot x_i \geq \sum_{i=1}^m p\cdot\omega_i = p\cdot\Big[\sum_{i=1}^m \omega_i\Big] = p\cdot\omega\,,$$

we see that $p\cdot x_i = p\cdot\omega_i$ must hold for each i. Now if $x > 0$, then $x + x_1 \succ_1 x_1$ implies $p\cdot x + p\cdot x_1 \geq p\cdot\omega_1 = p\cdot x_1$, and so $p\cdot x \geq 0$. This shows that $p > 0$, and consequently $p\cdot\omega > 0$.

Now fix some i such that $p\cdot\omega_i > 0$. Then, we claim that x_i is a maximal element in the budget set $B_i(p)$. To see this, assume by way of contradiction that there exists some $x \in B_i(p)$ (i.e., $p\cdot x \leq p\cdot\omega_i$) satisfying $x \succ_i x_i$. Since the set $\{y \in L^+:\ y \succ_i x_i\}$ is open in L^+ (by the lower semicontinuity of $\succeq_i$), $x \succ_i x_i$ and $\lim_{\epsilon\uparrow 1}\epsilon x = x$, there exists some $0 < \epsilon < 1$ such that $\epsilon x \succ_i x_i$. Therefore, $p\cdot(\epsilon x) \geq p\cdot\omega_i$. On the other hand, from $p\cdot\omega_i > 0$ and $p\cdot x \leq p\cdot\omega_i$, we see that

$$p\cdot\omega_i > \epsilon(p\cdot\omega_i) \geq \epsilon(p\cdot x) = p\cdot(\epsilon x) \geq p\cdot\omega_i\,,$$

which is impossible. Thus, x_i is maximal in $B_i(p)$.

The latter conclusion implies that $p \gg 0$ must be the case. Otherwise, if for some $x > 0$ we have $p\cdot x = 0$, then also we have $x + x_i \in B_i(p)$ and $x + x_i \succ_i x_i$ which proves that x_i is not a maximal element in $B_i(p)$.

Finally, notice that $p\cdot\omega_i > 0$ for each i in conjunction with the above conclusion guarantees that each x_i is a maximal element in $B_i(p)$. ■

8.10. We shall establish that

$$\mathbf{U} = \big\{(x_1,\dots,x_m) \in \mathbb{R}^m_+:\ x_1 + (x_2)^2 + \cdots + (x_m)^m \leq 1\big\}\,. \qquad (\star)$$

To see this, let $(\mathbf{x}_1,\dots,\mathbf{x}_m)$ be a free disposal allocation. This means $\mathbf{x}_i \geq 0$ for each i and $\sum_{i=1}^m \mathbf{x}_i \leq \omega = \mathbf{1}$, and so $\sum_{i=1}^m \int_0^1 \mathbf{x}_i(t)\,dt \leq 1$. Now if $x_i = u_i(\mathbf{x}_i) = \int_0^1 \sqrt[i]{\mathbf{x}_i(t)}\,dt$, then from Hölder's inequality,[1] we see that

$$x_i = \int_0^1 \sqrt[i]{\mathbf{x}_i(t)}\, dt \le \Big[\int_0^1 \mathbf{x}_i(t)\, dt\Big]^{\frac{1}{i}},$$

or $(x_i)^i \le \int_0^1 \mathbf{x}_i(t)\, dt$ for each i. This implies $\sum_{i=1}^m (x_i)^i \le \sum_{i=1}^m \int_0^1 \mathbf{x}_i(t)\, dt \le 1$. Therefore,

$$\mathbf{U} \subseteq \big\{(x_1, \ldots, x_m) \in \mathbb{R}^m_+ :\ x_1 + (x_2)^2 + \cdots + (x_m)^m \le 1\big\}.$$

Now let $(x_1, \ldots, x_m) \in \mathbb{R}^m_+$ satisfy $\sum_{i=1}^m (x_i)^i \le 1$. Put $\mathbf{x}_i = (x_i)^i \cdot \mathbf{1} \ge 0$, $i = 1, \ldots, m$. From $\sum_{i=1}^m \mathbf{x}_i = \big[\sum_{i=1}^m (x_i)^i\big] \cdot \mathbf{1} \le \mathbf{1}$, we see that $(\mathbf{x}_1, \ldots, \mathbf{x}_m)$ is a free disposal allocation. But then we have

$$u_i(\mathbf{x}_i) = \int_0^1 \sqrt[i]{(x_i)^i}\, dt = x_i$$

for each i, and so $(x_1, \ldots, x_m) \in \mathbf{U}$. Therefore,

$$\big\{(x_1, \ldots, x_m) \in \mathbb{R}^m_+ :\ x_1 + (x_2)^2 + \cdots + (x_m)^m \le 1\big\} \subseteq \mathbf{U},$$

and the validity of $(\star)$ follows. ■

8.11. (a) The utility function u_1 is linear (in fact, it represents the Lebesgue measure). So, it is concave and, of course, $\|\cdot\|_\infty$-continuous. The strict monotonicity of u_1 follows from the strict monotonicity of the Riemann integral. Since the square root function is strictly concave, it is easy to see that the utility function u_2 is likewise strictly concave. The strict monotonicity and the $\|\cdot\|_\infty$-continuity of u_2 should be obvious. The uniform $\|\cdot\|_\infty$-norm properness of the preferences is a consequence of Theorem 8.25.

(b) Since the sup norm of $C[0,1]$ is not order continuous and $\mathbf{1}$ is an order unit, it follows that order interval $[0, \mathbf{1}]$ is not weakly compact.

(c) Repeat the arguments of the solution of Exercise **8.10**. In particular, notice that $\mathbf{U}$ is a closed set.

(d) Our exchange economy satisfies all the hypotheses of Mas-Colell's Theorem 8.51. Hence, it has a quasiequilibrium. Since $\omega = \mathbf{1} \gg 0$, it follows from Exercise **8.9** that in this case the concepts of Walrasian equilibrium and quasiequilibrium coincide.

(e) Now let $x_1 = x_2 = \frac{1}{2} \cdot \mathbf{1}$. We shall establish that the allocation (x_1, x_2) is a Walrasian equilibrium supported by the price p defined by the Lebesgue measure, i.e., $p \cdot x = \int_0^1 x(t)\, dt$. Note first that

$$p \cdot x_1 = p \cdot x_2 = p \cdot \omega_1 = p \cdot \omega_2 = \tfrac{1}{2}.$$

If $x \succ_1 x_1$, then

$$p \cdot x = \int_0^1 x(t)\, dt = u_1(x) > u_1(x_1) = \int_0^1 x_1(t)\, dt = p \cdot x_1 = p \cdot \omega_1.$$

In addition, if $x \succ_2 x_2$, then from

$$\tfrac{1}{\sqrt{2}} = u_2(x_2) < u_2(x) = \int_0^1 \sqrt{x(t)}\, dt \le \Big[\int_0^1 x(t)\, dt\Big]^{\frac{1}{2}},$$

we see that $p \cdot x = \int_0^1 x(t)\, dt > \frac{1}{2} = p \cdot \omega_2$. ■

8.12. For simplicity we shall write L_p instead of $L_p[0,1]$. It is easy to see that the utility functions are strictly monotone and strictly concave—and consequently, by Theorem 8.25, they are also uniformly proper with respect to the Riesz dual system $\langle C[0,1], \mathrm{ca}[0,1]\rangle$. The

---

[1] **Hölder's inequality** asserts that if $f \in L_p(\mu)$ and $g \in L_q(\mu)$, where $1 \le p, q \le \infty$ satisfy $\frac{1}{p} + \frac{1}{q} = 1$, then the function fg belongs to $L_1(\mu)$ and $\int |fg|\, d\mu \le \|f\|_p \cdot \|g\|_q$; see [**24**, p. 256].

order interval $[0,\omega]$ (where $\omega=\omega_1+\omega_2=\mathbf{2}\gg 0$) is $\sigma(L_p,L_q)$-compact but (as easily checked) it is not $\sigma(C[0,1],\mathrm{ca}[0,1])$-compact.

We claim that the Riesz economy with Riesz pair $\langle C[0,1],\mathrm{ca}[0,1]\rangle$ has an empty core—and so, since in this case the core allocations coincide with the individually rational Pareto optimal allocations (why?), the economy does not have any individually rational Pareto optimal allocation and, of course, has no Walrasian equilibrium either. The proof of this claim will be accomplished with a number of steps.

1. *The utility functions are continuous for both the $\|\cdot\|_1$-norm and the Mackey topology $\tau(L_\infty,L_1)$. In particular, they are continuous for the sup norm and all the L_p-norms.*

Let a net $\{x_\alpha\}\subseteq L_\infty^+$ satisfy $x_\alpha \xrightarrow{\tau(L_\infty,L_1)} x$. Since $[-\mathbf{1},\mathbf{1}]$ is a convex, circled and $\sigma(L_1,L_\infty)$-compact subset of L_1, it follows that

$$\begin{aligned} V=[-\mathbf{1},\mathbf{1}]^\circ &= \Big\{x\in L_\infty\colon\ \Big|\int_0^1 x(t)y(t)\,dt\Big|\le 1 \quad\text{for all}\quad y\in[-\mathbf{1},\mathbf{1}]\Big\}\\ &= \Big\{x\in L_\infty\colon\ \int_0^1 |x(t)|\,dt\le 1\Big\}\end{aligned}$$

is a $\tau(L_\infty,L_1)$-neighborhood of zero. From this, we see that

$$\|x_\alpha-x\|_1=\int_0^1|x_\alpha(t)-x(t)|\,dt\longrightarrow 0\,.$$

Thus, if a net $\{x_\alpha\}$ converges to x for the Mackey topology $\tau(L_\infty,L_1)$, then it also converges to x for the L_1-norm.

Next, let us show that the utility functions are continuous for the L_1-norm. So, assume that $\|x_\alpha-x\|_1\longrightarrow 0$ holds in L_1^+. If $f=\chi_{[0,\frac12]}+\frac12\chi_{(\frac12,1]}$, then we have

$$\begin{aligned} \big|u_1(x_\alpha)-u_1(x)\big| &\le \int_0^{\frac12}\big|\sqrt{x_\alpha(t)}-\sqrt{x(t)}\big|\,dt+\tfrac12\int_{\frac12}^1\big|\sqrt{x_\alpha(t)}-\sqrt{x(t)}\big|\,dt\\ &= \int_0^1 f(t)\big|\sqrt{x_\alpha(t)}-\sqrt{x(t)}\big|\,dt\\ &\le \int_0^1 f(t)\sqrt{|x_\alpha(t)-x(t)|}\,dt\\ &\le \sqrt{\tfrac58}\cdot\Big[\int_0^1|x_\alpha(t)-x(t)|\,dt\Big]^{\frac12}\longrightarrow 0\,,\end{aligned}$$

where the last inequality holds by virtue of Hölder's inequality. Hence, $u_1(x_\alpha)\to u_1(x)$, and similarly $u_2(x_\alpha)\longrightarrow u_2(x)$.

2. *Let (x_1,x_2) be an allocation with respect to L_1 satisfying $x_1>0$ and $x_2>0$. Then there exist two constants $0\le a\le 2$ and $0\le b\le 2$ with $a\ne b$ such that the allocation (x_1^*,x_2^*), given by*

$$x_1^*=a\chi_{[0,\frac12]}+b\chi_{(\frac12,1]}\quad\textit{and}\quad x_2^*=(2-a)\chi_{[0,\frac12]}+(2-b)\chi_{(\frac12,1]}\,,$$

satisfies $x_1^*\succeq_1 x_1$ and $x_2^*\succeq_2 x_2$.

To establish this claim, let (x_1,x_2) be an allocation with respect to L_1 with $x_1>0$ and $x_2>0$. Put $a=2\int_0^{\frac12}x_1(t)\,dt$ and $b=2\int_{\frac12}^1 x_1(t)\,dt$, and note that $0\le a\le 2$ and $0\le b\le 2$ both hold. Let x_1^* and x_2^* be defined as above. Now using Hölder's inequality, we obtain

$$\int_0^{\frac12}\sqrt{x_1(t)}\,dt\le\tfrac{1}{\sqrt2}\Big[\int_0^{\frac12}x_1(t)\,dt\Big]^{\frac12}=\tfrac{\sqrt a}{2}\quad\text{and}\quad\int_{\frac12}^1\sqrt{x_1(t)}\,dt\le\tfrac{\sqrt b}{2}\,.$$

Thus, $u_1(x_1) = \int_0^{\frac{1}{2}} \sqrt{x_1(t)}\,dt + \frac{1}{2}\int_{\frac{1}{2}}^1 \sqrt{x_1(t)}\,dt \le \frac{\sqrt{a}}{2} + \frac{\sqrt{b}}{4} = u_1(x_1^*)$, i.e., $x_1^* \succeq_1 x_1$. Also,

$$\begin{aligned} u_2(x_2) &= \tfrac{1}{2}\int_0^{\frac{1}{2}} \sqrt{2-x_1(t)}\,dt + \int_{\frac{1}{2}}^1 \sqrt{2-x_1(t)}\,dt \\ &\le \tfrac{1}{2\sqrt{2}}\Big[\int_0^{\frac{1}{2}} [2-x_1(t)]\,dt\Big]^{\frac{1}{2}} + \tfrac{1}{\sqrt{2}}\Big[\int_{\frac{1}{2}}^1 [2-x_1(t)]\,dt\Big]^{\frac{1}{2}} \\ &= \tfrac{1}{2\sqrt{2}}\big(1-\tfrac{a}{2}\big)^{\frac{1}{2}} + \tfrac{1}{\sqrt{2}}\big(1-\tfrac{b}{2}\big)^{\frac{1}{2}} = \tfrac{1}{4}\sqrt{2-a} + \tfrac{1}{2}\sqrt{2-b} = u_2(x_2^*)\,, \end{aligned}$$

and so $x_2^* \succeq_2 x_2$.

Next, we will verify that we can choose a and b with $a \neq b$. So, assume that $a = b$. In this case, we have $x_1^* = a$ and $x_2^* = 2 - a$. From $x_1^* \succeq_1 x_1 > 0$ and $x_2^* \succeq_2 x_2 > 0$, we see that $0 < a < 2$. By the symmetry of the situation, we can also assume that $0 < a \le 1$. Then we claim that the allocation (y_1, y_2), given by

$$y_1 = \tfrac{3}{2}a\chi_{[0,\frac{1}{2}]} + \tfrac{1}{2}a\chi_{(\frac{1}{2},1]} \qquad \text{and} \qquad y_2 = (2-\tfrac{3}{2}a)\chi_{[0,\frac{1}{2}]} + (2-\tfrac{1}{2}a)\chi_{(\frac{1}{2},1]}\,,$$

satisfies $y_1 \succ_1 x_1^*$ and $y_2 \succ_2 x_2^*$ (and hence, $y_1 \succ_1 x_1$ and $y_2 \succ_2 x_2$). Indeed, note that

$$u_1(y_1) = \tfrac{1}{2}\sqrt{\tfrac{3a}{2}} + \tfrac{1}{4}\sqrt{\tfrac{a}{2}} = \tfrac{2\sqrt{3}+1}{4\sqrt{2}}\sqrt{a} > \tfrac{3}{4}\sqrt{a} = u_1(x_1^*)\,.$$

On the other hand, we see that

$$u_2(y_2) = \tfrac{1}{4}\sqrt{2-\tfrac{3a}{2}} + \tfrac{1}{2}\sqrt{2-\tfrac{a}{2}} = \tfrac{1}{4\sqrt{2}}\big(\sqrt{4-3a} + 2\sqrt{4-a}\big)\,,$$

and a direct calculation shows that $u_2(y_2) > \frac{3}{4}\sqrt{2-a} = u_2(x_2^*)$. The proof of step 2 is now complete.

3. *Assume that (x_1, x_2) is an allocation with $0 < x_i \in C[0,1]$ for each $i = 1, 2$. Then there exists an allocation (y_1, y_2) with $y_i \succ_i x_i$ and $y_i \in C[0,1]$ for $i = 1, 2$. In other words, the economy with Riesz pair $\langle C[0,1], \mathrm{ca}[0,1]\rangle$ has no core allocations.*

To see this, let (x_1, x_2) be an allocation with $0 < x_i \in C[0,1]\,(i = 1,2)$. According to part (2) there exist two constants $0 \le a \le 2$ and $0 \le b \le 2$ with $a \neq b$ such that the allocation (x_1^*, x_2^*), given by

$$x_1^* = a\chi_{[0,\frac{1}{2}]} + b\chi_{(\frac{1}{2},1]} \quad \text{and} \quad x_2^* = (2-a)\chi_{[0,\frac{1}{2}]} + (2-b)\chi_{(\frac{1}{2},1]}\,,$$

satisfies $x_1^* \succeq_1 x_1$ and $x_2^* \succeq_2 x_2$. Since x_1^* and x_2^* are not continuous functions, we see that $x_1^* \neq x_1$ and $x_2^* \neq x_2$. Thus, by the strict concavity of the utility functions, we obtain that the allocation (ψ_1, ψ_2), given by $\psi_1 = \frac{1}{2}(x_1 + x_1^*)$ and $\psi_2 = \frac{1}{2}(x_2 + x_2^*)$, satisfies $\psi_1 \succ_1 x_1$ and $\psi_2 \succ_2 x_2$.

Now since $C[0,1]$ is $\|\cdot\|_1$-dense in L_1, there exists a sequence $\{z_n\}$ of $C[0,1]$ satisfying $0 \le z_n \le \mathbf{2}$ for each n and $\lim_{n\to\infty} \|\psi_1 - z_n\|_1 = 0$ (and so $\lim_{n\to\infty} \|\psi_2 - (\mathbf{2} - z_n)\|_1 = 0$). By virtue of the $\|\cdot\|_1$-continuity of the utility functions (see step 1), there exists some n so that the allocation $(z_n, \mathbf{2} - z_n)$ satisfies $z_n \succ_1 x_1$ and $\mathbf{2} - z_n \succ_2 x_2$. Therefore, the economy has no core allocations. ■

8.13. (a) The solution of this part is due to Mike Maxwell. Let (x_1, x_2) be a core allocation with x_1 and x_2 continuous functions. Since (x_1, x_2) is individually rational, we have $x_1 \succeq_1 \omega_1$ and $x_2 \succeq_2 \omega_2$. If x_1 is a constant, then it is easy to see that $x_1 = x_2 = \mathbf{1}$, and so (ω_1, ω_2) is the only individually rational allocation of constant functions. Now consider the allocation (y_1, y_2) defined by

$$y_1 = 2\lambda\chi_{[0,\frac{1}{2})} + (2-2\lambda)\chi_{[\frac{1}{2},1]} \quad \text{and} \quad y_2 = (2-2\lambda)\chi_{[0,\frac{1}{2})} + 2\lambda\chi_{[\frac{1}{2},1]}\,.$$

Since $\lambda \neq \frac{1}{2}$, it follows that y_1 and y_2 are not continuous functions. Moreover, from

$$\begin{aligned} u_1(y_1) &= u_2(y_2) = \sqrt{2}\Big[\tfrac{\lambda^{\frac{3}{2}}+(1-\lambda)^{\frac{3}{2}}}{2}\Big] \\ &\geq \sqrt{2}\Big[\tfrac{\lambda+(1-\lambda)}{2}\Big]^{\frac{3}{2}} = \tfrac{1}{2} = u_1(\omega_1) = u_2(\omega_2) \end{aligned}$$

and the strict concavity of u_1 and u_2, we get $\frac{y_1+\omega_1}{2} \succ_1 \omega_1$ and $\frac{y_2+\omega_2}{2} \succ_2 \omega_2$. Taking into account that preferences are $\|\cdot\|_1$-continuous and $C[0,1]$ is $\|\cdot\|_1$-dense in $L_1[0,1]$, it easily follows that an allocation of continuous functions approximating $\big(\frac{y_1+\omega_1}{2}, \frac{y_2+\omega_2}{2}\big)$ can improve upon (ω_1, ω_2). So, (ω_1, ω_2) cannot be a core allocation.

Next, assume that x_1 and x_2 are non-constant. Without loss of generality, we can suppose that there exist $0 < t_1 < t_2 < \frac{1}{2}$ with $x_1(t_1) \neq x_2(t_2)$. Now consider the allocation (y_1, y_2) of discontinuous functions defined by

$$y_1(t) = \begin{cases} x_1(t) & \text{if } t \notin [t_1, t_2] \\ x_1(t_1+t_2-t) & \text{if } t \in [t_1, t_2] \end{cases} \qquad \text{and} \qquad y_2(t) = 2 - y_1(t)\,.$$

An easy computation shows that $u_1(y_1) = u_1(x_1)$ and $u_2(y_2) = u_2(x_2)$. Consequently, we have $\frac{x_1+y_1}{2} \succ_1 x_1$ and $\frac{x_2+y_2}{2} \succ_2 x_2$. Thus, an allocation of continuous functions approximating $\big(\frac{x_1+y_1}{2}, \frac{x_2+y_2}{2}\big)$ can improve upon (x_1, x_2). The above show that the core of the economy is empty.

(b) If $\lambda = \frac{1}{2}$, then $u_1(x) = u_2(x) = \frac{1}{2}\int_0^1 \sqrt{x(t)}\,dt$ and $u_1(\omega_1) = u_2(\omega_2) = \frac{1}{2}$. So, if $x \succ_i \omega_i$, i.e., if $u_i(x) > u_i(\omega_i)$, then from Hölder's inequality we get

$$\tfrac{1}{2} = u_i(\omega_i) < u_i(x) = \tfrac{1}{2}\int_0^1 \sqrt{x(t)}\,dt \leq \tfrac{1}{2}\Big[\int_0^1 1^2\,dt\Big]^{\frac{1}{2}} \cdot \Big[\int_0^1 x(t)\,dt\Big]^{\frac{1}{2}}.$$

This implies $\int_0^1 x(t)\,dt > 1 = \int_0^1 \omega_i(t)\,dt$, and this shows that the Lebesgue measure on the interval $[0,1]$ supports the allocation (ω_1, ω_2) as a Walrasian equilibrium. ■

8.14. Let x_1 and x_2 be defined as in the statement of the exercise. Clearly,

$$\mathbf{1} \cdot g = \int_0^1 g(t)\,dt\,, \quad g \in L_p[0,1]\,.$$

Direct computations show that $\mathbf{1} \cdot x_i = \mathbf{1} \cdot \omega_i = 1$ and $u_i(x_i) = \sqrt{\frac{\lambda^2+(1-\lambda)^2}{2}}$ for all i.

Now if $f = \lambda\chi_{[0,\frac{1}{2}]} + (1-\lambda)\chi_{(\frac{1}{2},1]}$, then note that $u_1(x) = \int_0^1 f(t)\sqrt{x(t)}\,dt$, and so, from Hölder's inequality, we get

$$u_1(x) \leq \Big[\int_0^1 f^2(t)\,dt\Big]^{\frac{1}{2}} \cdot \Big[\int_0^1 x(t)\,dt\Big]^{\frac{1}{2}} = \sqrt{\tfrac{\lambda^2+(1-\lambda)^2}{2}} \cdot \Big[\int_0^1 x(t)\,dt\Big]^{\frac{1}{2}}.$$

In particular, if $x \succ_1 x_1$, then $u_1(x) > u_1(x_1)$, and so from

$$\sqrt{\tfrac{\lambda^2+(1-\lambda)^2}{2}} = u_1(x_1) < u_1(x) \leq \sqrt{\tfrac{\lambda^2+(1-\lambda)^2}{2}}\Big[\int_0^1 x(t)\,dt\Big]^{\frac{1}{2}},$$

we get $\int_0^1 x(t)\,dt > 1 = \int_0^1 \omega_1(t)\,dt$, or $\mathbf{1} \cdot x > \mathbf{1} \cdot \omega_1$.

Similarly, $x \succ_2 x_2$ implies $\mathbf{1} \cdot x > \mathbf{1} \cdot \omega_2$. This shows that (x_1, x_2) is a Walrasian equilibrium supported by the price $\mathbf{1}$. ■

8.15. Let $y \succ_i x_{A_i}$ in $L_p^+(T,\Sigma,\mu)$. This means $u_i(y) > u_i(\chi_{A_i})$. Hence,

$$\begin{aligned}\int_T f_i(t)y(t)\,d\mu(t) &> \int_T f_i(t)\chi_{A_i}(t)\,d\mu(t) = \int_{A_i} f_i(t)\,d\mu(t)\\ &= \int_{A_i} p(t)\,d\mu(t) = \int_T p(t)\chi_{A_i}(t)\,d\mu(t)\\ &= p\cdot\omega_i\end{aligned}$$

Now, from $p(t) \geq f_i(t)$ for each $t \in T$, we get

$$p\cdot y = \int_T p(t)y(t)\,d\mu(t) \geq \int_T f_i(t)y(t)\,d\mu(t) > p\cdot\omega_i\,.$$

This shows that the allocation $(\chi_{A_1}, \chi_{A_2}, \dots, \chi_{A_m})$ is a Walrasian equilibrium supported by the price p. ∎

8.16. (a) The monotonicity of u_1 follows immediately from the fact that the lim inf function preserves inequalities. That is, if two sequences $\{x_n\}$ and $\{y_n\}$ of real numbers satisfy $x_n \leq y_n$ for each n (in fact, we need only $x_n \leq y_n$ for all $n \geq n_0$), then $\liminf_{n\to\infty} x_n \leq \liminf_{n\to\infty} y_n$. (One way of proving this is to use the familiar formula: $\liminf_{n\to\infty} z_n = \sup_n \inf_{k\geq n} z_k$.)

To see that u_1 has ω as a desirable bundle, we shall use the following property of the lim inf function.

- *If $\{x_n\}$ and $\{y_n\}$ are two bounded real sequences, then*

$$\liminf_{n\to\infty}(x_n + y_n) \geq \liminf_{n\to\infty} x_n + \liminf_{n\to\infty} y_n\,.$$

The above inequality is equality if one of the sequences converges.

So, if $x = (x_1, x_2, \dots) \in \ell_\infty$ and $\epsilon > 0$, then

$$u_1(x + \epsilon\omega) = \liminf_{n\to\infty} x_n + \epsilon > \liminf_{n\to\infty} x_n = u_1(x)\,,$$

which shows that ω is desirable for consumer 1.

For the concavity of u_1, let $x = (x_1, x_2, \dots)$, $y = (y_1, y_2, \dots) \in \ell_\infty$ and $0 \leq \alpha \leq 1$. Then we have

$$\begin{aligned}u_1\big(\alpha x + (1-\alpha)y\big) &= u_1\big(\alpha x_1 + (1-\alpha)y_1, \alpha x_2 + (1-\alpha)y_2, \dots\big)\\ &= \liminf_{n\to\infty}\,[\alpha x_n + (1-\alpha)y_n]\\ &\geq \liminf_{n\to\infty}\,[\alpha x_n] + \liminf_{n\to\infty}\,[(1-\alpha)y_n]\\ &= \alpha\liminf_{n\to\infty} x_n + (1-\alpha)\liminf_{n\to\infty} y_n\\ &= \alpha u_1(x) + (1-\alpha)u_1(y)\,.\end{aligned}$$

In order to establish that u_1 is not Mackey continuous, we need to recall the definition of the Mackey topology. If $\langle X, X'\rangle$ is a dual system, then the **Mackey topology** $\tau(X,X')$ is the largest Hausdorff locally convex topology on X which is consistent with $\langle X, X'\rangle$, i.e., $(X, \tau(X,X'))' = X'$. A base at zero for the Mackey topology $\tau(X,X')$ consists of the polars of all convex, circled and $\sigma(X',X)$-compact subsets of X'. In convergence terms, a net $\{x_\alpha\}$ of X satisfies $x_\alpha \xrightarrow{\tau(X,X')} x$ if and only if the net $\{x_\alpha\}$ converges to x uniformly on every convex, circled and $\sigma(X',X)$-compact subset of X'. That is, $x_\alpha \xrightarrow{\tau(X,X')} x$ if and only if for each convex, circled and $\sigma(X',X)$-compact subset A of X' we have $\sup_{x'\in A}\big|\langle x_\alpha - x, x'\rangle\big| \to 0$. For a complete discussion of the Mackey topology see [**8**, pp. 219–221].

We now return to the dual system $\langle \ell_\infty, \ell_1 \rangle$. We know that ℓ_∞ is the norm dual of ℓ_1. Moreover, it is known that a subset of ℓ_1 is $\sigma(\ell_1, \ell_\infty)$-compact if and only if it is norm compact; see [**21**, pp. 200–201].

Let $\mathbf{x}_n = (0, \ldots, 0, 1, 1, 1, \ldots)$, where the zeros occupy the first n positions, and note that the sequence $\{\mathbf{x}_n\}$ belongs to ℓ_∞^+. Also, let A be a $\sigma(\ell_1, \ell_\infty)$-compact subset of ℓ_1. By the above discussion, we know that the set A is a norm compact subset of ℓ_1. This means that given any $\epsilon > 0$, there exists some n_0 such that $\sum_{i=n_0}^\infty |x_i| < \epsilon$ for all $x = (x_1, x_2, \ldots) \in A$; see [**21**, Theorem 13.2, p. 201]. In particular, if $x \in A$ and $n \geq n_0$, then

$$\big|\langle \mathbf{x}_n, x\rangle\big| = \sum_{i=n}^{\infty} |x_i| \leq \sum_{i=n_0}^{\infty} |x_i| < \epsilon\,,$$

and this shows that the sequence $\{\mathbf{x}_n\}$ converges to zero uniformly on every $\sigma(\ell_1, \ell_\infty)$-compact subset of ℓ_1. So, by the definition of the Mackey topology, we see that $\mathbf{x}_n \xrightarrow{\tau(\ell_\infty, \ell_1)} 0$. Finally, notice that $u_1(\mathbf{x}_n) = 1$ for each n while $u_1(0) = 0$. This shows that u_1 is not Mackey continuous on ℓ_∞^+.

(b) Let $p = (\frac{1}{2}, \frac{1}{2^2}, \frac{1}{2^3}, \ldots) \in \ell_1$ and notice that

$$u_2(x) = p \cdot x$$

for each x. In addition, p is strictly positive on ℓ_∞^+, from which it easily follows that u_2 is strictly positive on ℓ_∞^+. The linearity of u_2 shows that u_2 is concave and Mackey continuous.

(c) To verify that $\mathbf{U} = [0,1] \times [0,1) \cup \{(0,1)\}$ is the utility space of this economy, note first that $(x,y) \in \mathbf{U}$ implies $0 \leq x \leq 1$ and $0 \leq y \leq 1$. Now assume that $(x,y) \in \mathbb{R}_+^2$ satisfies $0 \leq x \leq 1$ and $0 \leq y < 1$.

Since $\lim_{n\to\infty} \frac{2^n}{2^n - 1} = 1$, there exists some n such that $\frac{2^n}{2^n-1} y = z < 1$. Next, consider the vectors

$$\mathbf{x}_1 = (\underbrace{0, 0, \ldots, 0}_{n}, x, x, x, \ldots) \quad \text{and} \quad \mathbf{x}_2 = (\underbrace{z, z, \ldots, z}_{n}, 0, 0, 0, \ldots)\,,$$

and notice that $(\mathbf{x}_1, \mathbf{x}_2)$ is a free disposable allocation. Moreover, we have

$$u_1(\mathbf{x}_1) = x \quad \text{and} \quad u_2(\mathbf{x}_2) = \Big(\sum_{i=1}^{n} \tfrac{1}{2^i}\Big) z = \big(1 - \tfrac{1}{2^n}\big) z = y\,,$$

which shows that $(x,y) \in \mathbf{U}$.

Now, assume that $(x,1) \in \mathbf{U}$ for some $0 \leq x \leq 1$. Pick some free disposable allocation $(\mathbf{x}_1, \mathbf{x}_2)$ such that $u_1(\mathbf{x}_1) = x$ and $u_2(\mathbf{x}_2) = 1$. Also, write $\mathbf{x}_2 = (x_1, x_2, \ldots)$ with $0 \leq x_i \leq 1$ for each i. If $0 \leq x_k < 1$ holds for some k, then

$$1 = u_2(\mathbf{x}_2) = \sum_{n=1}^{\infty} \tfrac{x_n}{2^n} \leq \sum_{n \neq k} \tfrac{1}{2^n} + \tfrac{x_k}{2^k} \leq \sum_{n=1}^{\infty} \tfrac{1}{2^n} + \tfrac{x_k}{2^k} - \tfrac{1}{2^k} = 1 - \tfrac{1 - x_k}{2^k} < 1\,,$$

which is a contradiction. Hence $x_i = 1$ for each i. This implies $\mathbf{x}_1 = 0$ and so $(x,1) = (0,1)$. In other words, we have shown that

$$\mathbf{U} = [0,1] \times [0,1) \cup \{(0,1)\}\,.$$

(d) Assume $(\mathbf{x}_1, \mathbf{x}_2)$ is a quasiequilibrium supported by a price $0 < \mathbf{p} \in \ell_1$. Now note that if $\mathbf{x}_1 = (x_1, x_2, x_3, \ldots) \in \ell_\infty$ and

$$\mathbf{z}_n = (0, 0, \ldots, 0, x_{n+1}, x_{n+2}, \ldots)\,,$$

where the zeros occupy the first n positions, then $\mathbf{z}_n \sim_1 \mathbf{x}_1$ for each n. This implies $\mathbf{p} \cdot \mathbf{z}_n \geq \mathbf{p} \cdot \omega_1 \geq 0$ for all n. Thus, $\mathbf{p} \cdot \omega_1 = \lim_{n\to\infty} \mathbf{p} \cdot \mathbf{z}_n = 0$. Since $\mathbf{p} \cdot \omega > 0$, the latter guarantees that $\mathbf{p} \cdot \omega_2 > 0$.

Now notice that the strict monotonicity of $\succeq_2$ implies $\omega_2 + \omega_1 \succ_2 \omega_2$. So, by the Mackey continuity of u_2, there is some $0 < \epsilon < 1$ such that $\epsilon(\omega_2 + \omega_1) \succ_2 \omega_2$. The supportability from $\mathbf{p}$ yields $\mathbf{p} \cdot \epsilon(\omega_2 + \omega_1) \geq \mathbf{p} \cdot \omega_2$, or $\epsilon \mathbf{p} \cdot \omega_1 \geq (1-\epsilon)\mathbf{p} \cdot \omega_2 > 0$. Therefore, $\mathbf{p} \cdot \omega_1 > 0$, a contradiction. Hence, if $\omega_1 > 0$ and $\omega_2 > 0$, then the economy has no quasiequilibria. ■

8.17. (a) Notice that the function $f_n \colon \mathbb{R} \to \mathbb{R}$ defined by $f_n(t) = \frac{1-e^{-n^2 t}}{n^2}$ satisfies $f_n'(t) = e^{-n^2 t} > 0$ and $f_n''(t) = -n^2 e^{-n^2 t} < 0$ for each t. So, each f_n is a strictly increasing and strictly concave function on $\mathbb{R}$. This easily implies that the utility function $u \colon \ell_2^+ \to \mathbb{R}$ defined by $u(x_1, x_2, \ldots) = \sum_{n=1}^{\infty} f_n(x_n)$ is also strictly monotone and strictly concave.

For the weak continuity of u, let $\mathbf{x}^\alpha = (x_1^\alpha, x_2^\alpha, \ldots) \xrightarrow{w} \mathbf{x} = (x_1, x_2, \ldots)$ in ℓ_2^+ and let $\epsilon > 0$. Fix some k such that $\sum_{n=k}^{\infty} \frac{2}{n^2} < \epsilon$ and then select some index α_0 such that $\sum_{n=1}^{k-1} \frac{|e^{-n^2 x_n^\alpha} - e^{-n^2 x_n}|}{n^2} < \epsilon$ for all $\alpha \succeq \alpha_0$. (Such an index α_0 exists since $\mathbf{x}^\alpha \xrightarrow{w} \mathbf{x}$ implies $x_i^\alpha \xrightarrow[\alpha]{} x_i$ in $\mathbb{R}$ for each i.) But then, for $\alpha \succeq \alpha_0$ we have

$$\big|u(\mathbf{x}^\alpha) - u(\mathbf{x})\big| \leq \sum_{n=1}^{k-1} \frac{|e^{-n^2 x_n^\alpha} - e^{-n^2 x_n}|}{n^2} + \sum_{n=k}^{\infty} \frac{2}{n^2} < \epsilon + \epsilon = 2\epsilon\,,$$

which shows that $u(\mathbf{x}^\alpha) \to u(\mathbf{x})$, i.e., u is weakly continuous.

(b) We must show that (ω) belongs to the core of every r-fold replica economy. To see this, assume by way of contradiction that there exists some allocation $(x_1, \ldots, x_r)$ in an r-fold replica economy and a coalition S (i.e., a nonempty subset of $\{1, 2, \ldots, r\}$) of k consumers such that

(a) $\sum_{i \in S} x_i = \sum_{i \in S} \omega = k\omega$, and
(b) $x_i \succ \omega$ for each $i \in S$.

Since $\succeq$ is strictly convex, it follows from the statement at the beginning of the solution to Exercise **8.8** that $\omega = \frac{1}{k} \sum_{i \in S} x_i \succ \omega$, which is impossible. Hence (ω) is an Edgeworth equilibrium.

(c) Consider the function $f \colon \mathbb{R} \to \mathbb{R}$ defined via the formula $f(x) = e^{1-x} + x - 2$. From $f'(x) = 1 - e^{1-x}$ and $f''(x) = e^{1-x}$, we see that f attains an absolute minimum at $x = 1$. So, $f(x) = e^{1-x} + x - 2 \geq f(1) = 0$, or $x \geq 2 - e^{1-x}$ for each $x \in \mathbb{R}$.

Now assume that $x = (x_1, x_2, \ldots) \succ \omega$ holds in L_ω^+. That is, we assume that $\sum_{n=1}^{\infty} \frac{1-e^{-n^2 x_n}}{n^2} > \sum_{n=1}^{\infty} \frac{1-e^{-1}}{n^2}$. Then note that

$$\begin{aligned} p \cdot x &= \sum_{n=1}^{\infty} x_n = \sum_{n=1}^{\infty} \frac{n^2 x_n}{n^2} \\ &\geq \sum_{n=1}^{\infty} \frac{2-e^{1-n^2 x_n}}{n^2} = \sum_{n=1}^{\infty} \frac{2-e+e(1-e^{-n^2 x_n})}{n^2} \\ &> \sum_{n=1}^{\infty} \frac{2-e+e(1-e^{-1})}{n^2} = \sum_{n=1}^{\infty} \frac{1}{n^2} = p \cdot \omega\,. \end{aligned}$$

So, the allocation (ω) is a Walrasian equilibrium relative to the Riesz dual system $\langle L_\omega, L_\omega' \rangle$ supported by the price p.

(d) Assume that $p \in L'_\omega$ is a nonzero price that supports (ω) as a Walrasian equilibrium. Since $\succeq$ is strictly monotone and $\langle \ell_2, \ell_2 \rangle$ is a symmetric Riesz dual system, it follows that p is positive and order continuous on L_ω; see Theorem 8.39. Now let $p_i = p \cdot \mathbf{e}_i \geq 0$, where $\mathbf{e}_i$ is the usual unit vector. If $0 \leq x \in L_\omega$, then from $\sum_{i=1}^n x_i \mathbf{e}_i \uparrow x$ and the order continuity of p, we get

$$\sum_{i=1}^{\infty} p_i x_i = \lim_{n \to \infty} p \cdot \Big(\sum_{i=1}^{n} x_i \mathbf{e}_i \Big) = p \cdot x\,.$$

This shows that $p \cdot x = \sum_{n=1}^{\infty} p_n x_n$ for each $x = (x_1, x_2, \ldots) \in L_\omega$, and thus the price p can be identified with the non-negative sequence $(p_1, p_2, \ldots)$.

Next, we assert that the price $\mathbf{p}_\ell = (p_1, p_2, \ldots, p_\ell)$ supports the allocation $\omega_\ell = \big(1, \frac{1}{2^2}, \frac{1}{3^2}, \ldots, \frac{1}{\ell^2}\big)$ on $\mathbb{R}^\ell$, where the utility function is now

$$U_\ell(x_1, x_2, \ldots, x_\ell) = \sum_{n=1}^{\ell} \tfrac{1 - e^{-n^2 x_n}}{n^2}\,.$$

To see this, assume $\mathbf{x} = (x_1, \ldots, x_\ell) \succ_\ell \omega_\ell$ in $\mathbb{R}^\ell_+$, i.e., $U_\ell(\mathbf{x}) > U_\ell(\omega_\ell)$. This implies

$$\begin{aligned} u\big(x_1, \ldots, x_\ell, \tfrac{1}{(\ell+1)^2}, \tfrac{1}{(\ell+2)^2}, \ldots\big) &= U_\ell(x_1, \ldots, x_\ell) + \sum_{n=\ell+1}^{\infty} \tfrac{1-e^{-1}}{n^2} \\ &> U_\ell(\omega_\ell) + \sum_{n=\ell+1}^{\infty} \tfrac{1-e^{-1}}{n^2} = u(\omega) \end{aligned}$$

in L^+_ω. The supportability from p yields $\sum_{i=1}^{\ell} p_i x_i + \sum_{i=\ell+1}^{\infty} \frac{p_i}{i^2} > \sum_{i=1}^{\infty} \frac{p_i}{i^2}$, which implies $\mathbf{p}_\ell \cdot \mathbf{x} > \mathbf{p} \cdot \omega_\ell$.

Now since $\nabla U_\ell(x_1, x_2, \ldots, x_\ell) = \big(e^{-x_1}, e^{-2^2 x_2}, \ldots, e^{-\ell^2 x_\ell}\big)$, it is easy to obtain (see for instance [**6**, Problem 1.1.20] that

$$(p_1, p_2, \ldots, p_\ell) = \lambda \nabla U_\ell\big(1, \tfrac{1}{2^2}, \tfrac{1}{3^2}, \ldots, \tfrac{1}{\ell^2}\big) = \lambda e^{-1}(1, 1, \ldots, 1)\,.$$

Hence, $p_1 = p_2 = \cdots = p_\ell$ for each ℓ. This shows that the price p must be a scalar multiple of the price $(1, 1, \ldots)$.

(e) Assume by way of contradiction that $\succeq$ is uniformly ℓ_2-norm proper. Then by Theorem 8.41 the price $p = (1, 1, \ldots)$ must be ℓ_2-norm continuous on L_ω. Since L_ω is ℓ_2-norm dense in ℓ_2, it follows that p extends to an ℓ_2-norm continuous linear functional on ℓ_2. This means that there exists some $q = (q_1, q_2, \ldots) \in \ell_2$ satisfying $q \cdot x = p \cdot x$ for all $x \in L_\omega$. Since $\mathbf{e}_n \in L_\omega$, we have $q \cdot \mathbf{e}_n = p \cdot \mathbf{e}_n$, or $q_n = 1$ for each n. In other words, $q = (1, 1, \ldots) \in \ell_2$, which is impossible. Therefore, $\succeq$ is not a uniformly ℓ_2-norm proper preference.

(f) With "ℓ_p" replacing "ℓ_2", the solutions to (a), (b), (c) and (d) are clearly valid for any of the Riesz dual systems $\langle \ell_p, \ell_q \rangle$. The only really nontrivial observation that must be made for parts (c) and (d) is that the price $p = (1, 1, \ldots)$ is in L'_ω. This is the case since $\omega \in \ell_p$ for all $1 \leq p \leq \infty$. (The reader should also notice that since $p \in \ell_\infty$, the allocation (ω) is a Walrasian equilibrium also with respect to $\langle \ell_1, \ell_\infty \rangle$.)

Since L_ω is ℓ_p-norm dense in ℓ_p for $1 \leq p < \infty$, a close look at the solution shows that statement (e) holds for any $1 \leq p < \infty$. Now L_ω is not ℓ_∞-norm dense in ℓ_∞. In fact, the result is not true for $p = \infty$; see Theorem 8.25.

However, for the interested reader the following is true: the preference $\succeq$ represented by u is not uniformly Mackey-proper. The strategy of the solution to (e) holds provided that L_ω is Mackey-dense in ℓ_∞. To see this, it suffices to show that $x_n = (0, \ldots, 0, 1, 1, \ldots) \xrightarrow{\tau(\ell_\infty, \ell_1)} 0$. This is shown in the last paragraph of the solution to Exercise **8.16**(a). ■

8.18. (1) We shall establish first that there exists at most one k such that $f(k) = mk$. To this end, assume by way of contradiction that there exist $0 < k_1 < k_2$ satisfying $f(k_1) = mk_1$ and $f(k_2) = mk_2$. If $\alpha = \frac{k_1}{k_2} < 1$, then we have

$$\begin{aligned} f(k_1) &= f(\alpha k_2) = f(\alpha k_2 + (1-\alpha)0) \\ &> \alpha f(k_2) + (1-\alpha)f(0) = \alpha f(k_2) \\ &= \alpha m k_2 = mk_1, \end{aligned}$$

which is a contradiction.

For the existence of k_m, note that $\lim_{k\to\infty} \frac{f(k)}{k} = 0$ implies $\frac{f(k)}{k} < m$ for all sufficiently large k, while $\frac{f(k)}{k} = \frac{f(k)-f(0)}{k-0} \xrightarrow[k\to 0^+]{} \infty$ implies $\frac{f(k)}{k} > m$ for all k sufficently large. The intermediate value theorem guarantees the existence of some $k_m > 0$ with $f(k_m) = mk_m$.

(2) Let $0 < k < k_m$ and put $\alpha = \frac{k}{k_m} < 1$. Then

$$f(k) = f(\alpha k_m) > \alpha f(k_m) = \alpha m k_m = mk.$$

(3) Let $k > k_m$. If $f(k) \geq mk$, then (in view of $f(k) < mk$ for all eventually large k), we must have $f(k) = mk$ for some $k > k_m$, a contradiction. Hence, $f(k) < mk$ holds for all $k > k_m$, and the solution is finished. ■

8.19. (1) Assume that $0 < x \leq y$. From the convex combination $x = \left(\frac{x}{y}\right)y + \left(\frac{y-x}{y}\right)0$ and the concavity of u, we get

$$u(x) \geq \tfrac{x}{y}u(y) + \tfrac{y-x}{y}u(y) = \tfrac{x}{y}u(y)\,,$$

and the desired inequality follows.

(2) If $x = y$ or $y = 0$, the inequality is trivial. So, we consider only the cases $0 < y < x$ and $0 < x < y$. Assume first that $0 < y < x$. Then from part (1), we have

$$0 \leq u(x) - u(y) = \tfrac{u(x)}{x}x - \tfrac{u(y)}{y}y \leq \tfrac{u(x)}{x}x - \tfrac{u(x)}{x}y = \tfrac{u(x)}{x}(x-y).$$

On the other hand, when $0 < x < y$, the preceding case (by exchanging x and y) yields

$$0 \leq u(y) - u(x) \leq \tfrac{u(y)}{y}(y-x) \leq \tfrac{u(x)}{x}(y-x).$$

(3) Let $0 \leq x \leq z \leq y$ with $z > 0$. If $z = x$ or $z = y$, the inequality follows from part (2). So, we can assume $0 \leq x < z < y$. If $x = 0$, then the inequality follows from part (1). Therefore, we can also suppose that $x > 0$. Consequently, from (2), we get

$$\big|u(x) - u(y)\big| \leq \tfrac{u(y)}{y}|x-y| \leq \tfrac{u(z)}{z}|x-y|,$$

and the solution is finished. ■

Bibliography

1. Y. A. Abramovich, Injective envelopes of normed lattices, *Soviet Math. Dokl.* **12** (1971), 511–514.
2. C. D. Aliprantis, Some order and topological properties of locally solid linear topological Riesz spaces, *Proc. Amer. Math. Soc.* **40** (1973), 443–447.
3. C. D. Aliprantis, On the completion of Hausdorff locally solid Riesz spaces, *Trans. Amer. Math. Soc.* **196** (1974), 105–125.
4. C. D. Aliprantis, Riesz seminorms with Fatou properties, *Proc. Amer. Math. Soc.* **45** (1974), 383–388.
5. C. D. Aliprantis, On order properties of order bounded transformations, *Canadian J. Math.* **27** (1975), 666–678.
6. C. D. Aliprantis, *Problems in Equilibrium Theory*, Springer–Verlag, New York and Heidelberg, 1996.
7. C. D. Aliprantis, On the Mas-Colell–Richard equilibrium theorem, *J. Econom. Theory* **74** (1997), 414–424.
8. C. D. Aliprantis and K. C. Border, *Infinite Dimensional Analysis: A Hitchhikers Guide*, 2nd Edition, Springer–Verlag, Heidelberg and New York, 1999.
9. C. D. Aliprantis, K. C. Border, and O. Burkinshaw, Economies with many commodities, *J. Econom. Theory* **74** (1997), 62–105.
10. C. D. Aliprantis, K. C. Border, and O. Burkinshaw, New proof of the existence of equilibrium in a single sector growth model, *Macroeconomic Dynamics* **1** (1997), 669–679.
11. C. D. Aliprantis and D. J. Brown, Equilibria in markets with a Riesz space of commodities, *J. Math. Econom.* **11** (1983), 189–207.
12. C. D. Aliprantis, D. J. Brown, and O. Burkinshaw, Edgeworth equilibria, *Econometrica* **55** (1987), 1109–1137.
13. C. D. Aliprantis, D. J. Brown, and O. Burkinshaw, Edgeworth equilibria in production economies, *J. Econom. Theory* **43** (1987), 252–291.
14. C. D. Aliprantis, D. J. Brown, and O. Burkinshaw, An economy with infinite dimensional commodity space and empty core, *Econom. Lett.* **23** (1987), 1–4.
15. C. D. Aliprantis, D. J. Brown, and O. Burkinshaw, *Existence and Optimality of Competitive Equilibria*, Springer–Verlag, Heidelberg and New York, 1989.
16. C. D. Aliprantis and O. Burkinshaw, A new proof of Nakano's theorem in locally solid Riesz spaces, *Math. Z.* **144** (1975), 25–33.
17. C. D. Aliprantis and O. Burkinshaw, Nakano's theorem revisited, *Michigan Math. J.* **23** (1976), 173–176.
18. C. D. Aliprantis and O. Burkinshaw, On universally complete Riesz spaces, *Pacific J. Math.* **71** (1977), 1–12.
19. C. D. Aliprantis and O. Burkinshaw, On the structure of locally solid topologies, *Canad. Math. Bull.* **23** (1980), 185–191.
20. C. D. Aliprantis and O. Burkinshaw, Minimal topologies and L_p-spaces, *Illinois J. Math.* **24** (1980), 164–172.
21. C. D. Aliprantis and O. Burkinshaw, *Positive Operators*, Academic Press, New York and London, 1985.
22. C. D. Aliprantis and O. Burkinshaw, The fundamental theorems of welfare economics without proper preferences, *J. Math. Econom.* **17** (1988), 41–54.
23. C. D. Aliprantis and O. Burkinshaw, When is the core equivalence theorem valid?, *Economic Theory* **1** (1991), 169–18.

24. C. D. Aliprantis and O. Burkinshaw, *Principles of Real Analysis*, 3rd Edition, Academic Press, New York and London, 1998.
25. C. D. Aliprantis and O. Burkinshaw, *Problems in Real Analysis*, 2rd Edition, Academic Press, New York and London, 1998.
26. C. D. Aliprantis and E. Langford, Almost σ-Dedekind complete Riesz spaces and the main inclusion theorem, *Proc. Amer. Math. Soc.* **44** (1974), 421–426.
27. C. D. Aliprantis and E. Langford, Order completions of Archimedean Riesz spaces and ℓ-groups, *Algebra Universalis* **19** (1984), 151–159.
28. C. D. Aliprantis and R. Tourky, The super order dual of an ordered vector space and the Riesz–Kantorovich formula, *Trans. Amer. Math. Soc.* **354** (2002), 2055–2077.
29. C. D. Aliprantis, R. Tourky, and N. C. Yannelis, Cone conditions in general equilibrium, *J. Econom. Theory* **92** (2000), 96–121.
30. C. D. Aliprantis, R. Tourky, and N. C. Yannelis, A theory of value with non-linear prices: equilibrium analysis beyond vector lattices, *J. Econom. Theory* **100** (2001), 22–72.
31. I. Amemiya, On ordered topological linear spaces. In *Proc. Internat. Symp. Linear Spaces*, Jerusalem, 1960, pp. 14–23. Israel Acad. Sci. Humanities, Jerusalem, 1961.
32. I. Amemiya and T. Mori, Topological structures in ordered linear spaces, *J. Math. Soc. Japan* **9** (1957), 131–142.
33. A. Araujo, Lack of Pareto optimality in economies with infinitely many commodities: The need for impatience, *Econometrica* **53** (1985), 455–461.
34. K. J. Arrow, An extension of the basic theorems of classical welfare economics, in: J. Neyman ed., *Proceedings of the Second Berkeley Symposium on Mathematical Statistics and Probability II, Part I* (Berkeley, California, 1951), University of California Press, Berkeley, CA, 1951, 507–532.
35. K. J. Arrow and G. Debreu, Existence of an equilibrium for a competitive economy, *Econometrica* **22** (1954), 265–290.
36. S. J. Bernau, Orthocompletion of lattice groups, *J. London Math. Soc.* (3) **16** (1966), 107–130.
37. S. J. Bernau, Lateral and Dedekind completion of Archimedean lattice groups, *J. London Math. Soc.* (2) **12** (1976), 320–322.
38. T. F. Bewley, Existence of equilibria in economies with infinitely many commodities, *J. Econom. Theory* **24** (1981), 265–295.
39. G. Birkhoff, *Lattice Theory*, 3rd Edition, Amer. Math. Soc. Colloq. Publ., # **25**, Providence, RI, 1967.
40. H. F. Bohnenblust, On axiomatic characterization of L_p-spaces, *Duke Math. J.* **6** (1940), 627–640.
41. H. F. Bohnenblust and S. Kakutani, Concrete representations of (M)-spaces, *Ann. of Math.* **42** (1941), 1025–1028.
42. O. N. Bondareva, Kernel theory in n-person games, *Vestnik Leningrad. Univ. Mat. Mekh. Astronom.* **17** (1962), 141–142.
43. D. S. Bridges and G. B. Mehta, *Representation of Preference Orderings*, Lecture Notes in Economics and Mathematical Systems, # **422**, Springer–Verlag, Berlin and New York, 1995.
44. O. Burkinshaw, Weak compactness in the order dual of a vector lattice, *Trans. Amer. Math. Soc.* **187** (1974), 183–201.
45. O. Burkinshaw and P. G. Dodds, Weak sequential compactness in Riesz spaces, *Canadian J. Math.* **28** (1976), 1332–1339.
46. O. Burkinshaw and P. G. Dodds, Disjoint sequences, compactness and semireflexivity in locally convex Riesz spaces, *Illinois J. Math.* **21** (1977), 759–775.
47. P. Conrad, The lateral completion of a lattice-ordered group, *Proc. London Math. Soc.* **19** (1969), 444–480.
48. M. M. Day, The spaces L^p with $0 < p < 1$, *Bull. Amer. Math. Soc.* **46** (1940), 816–823.
49. G. Debreu, The coefficient of resource utilization, *Econometrica* **19** (1951), 257–273.
50. G. Debreu, Valuation equilibrium and Pareto optimum, *Proc. Nat. Acad. Sci. U.S.A.* **40** (1954), 588–592.
51. G. Debreu, *Theory of Value*, Yale University Press, New Haven, CT, 1959.
52. G. Debreu and H. E. Scarf, A limit theorem on the core of an economy, *Internat. Econom. Rev.* **4** (1963), 235–246.
53. P. G. Dodds, Sequential convergence in the order dual of certain classes of Riesz spaces, *Trans. Amer. Math. Soc.* **203** (1975), 391–403.

54. M, Duhoux, Le complété d'un treillis localement solide, *Bull. Soc. Math. Belg.* **24** (1972), 133–151.
55. N. Dunford and J. T. Schwartz, *Linear Operators*, Vol. I, Wiley, New York, 1958.
56. F. Y. Edgeworth, *Mathematical Psychics*, Kegan Paul, London, 1881.
57. D. H. Fremlin, On the completion of locally solid vector lattices, *Pacific J. Math.* **43** (1972), 341–347.
58. D. H. Fremlin, *Topological Riesz Spaces and Measure Theory*, Cambridge Univ. Press, London and New York, 1974.
59. D. H. Fremlin, Inextensible Riesz spaces, *Math. Proc. Cambridge Phil. Soc.* **77** (1975), 71–89.
60. H. Freudenthal, Teilweise geordnete Moduln, *Nederl. Akad. Wetensch. Proc. Ser. A* **39** (1936), 641–651.
61. A. M. Gleason, Projective topological spaces, *Illinois J. Math.* **2** (1958), 482–489.
62. C. Goffman, Compatible seminorms in a vector lattice, *Proc. Nat. Acad. Sci. U.S.A.* **42** (1956), 536–538.
63. C. Goffman, Completeness in topological vector lattices, *Amer. Math. Monthly* **66** (1959), 87–92.
64. J. Horváth, *Topological Vector Spaces and Distributions*, I, Addison-Wesley, Reading, Massachusetts, 1966.
65. G. Jameson, *Ordered Linear Spaces*, Lecture Notes in Mathematics, **141**, Springer–Verlag, Berlin and New York, 1970.
66. L. E. Jones, Special problems arising in the study of economies with infinitely many commodities, in: H. F. Sonnenschein, ed., *Models of Economic Dynamics* (Lecture Notes in Economics and Mathematical Systems, # **264**), Springer–Verlag, New York and Berlin, 1986, 184–205.
67. S. Kakutani, Concrete representations of (L)-spaces and the mean ergodic theorem, *Ann. of Math.* **42** (1941), 523–537.
68. S. Kakutani, Concrete representations of (M)-spaces, *Ann. of Math.* **42** (1941), 994–1024.
69. L. V. Kantorovich, Sur les propriétés des espaces semi-ordonnés linéaires, *C. R. Acad. Sci. Paris Ser. A-B* **202** (1936), 813–816.
70. L. V. Kantorovich, Concerning the general theory of operations in partially ordered spaces, *DAN SSSR* **1** (1936), 271–274. (In Russian.)
71. L. V. Kantorovich, Lineare halbgeordnete Räume, *Math. Sbornik* **44** (1937), 121–168.
72. L. V. Kantorovich, B. Z. Vulikh, and A. G. Pinsker, *Functional Analysis in Partially Ordered Spaces*, Moscow and Leningrad, 1950.
73. S. Kaplan, On the second dual of the space of continuous functions, *Trans. Amer. Math. Soc.*, Note I **86** (1957), 70–90; Note II **93** (1959), 329–350; Note III **101** (1961), 34–51; Note IV **113** (1964), 512–546.
74. I. Kawai, Locally convex lattices, *J. Math. Soc. Japan* **9** (1957), 281–314.
75. J. L. Kelley and I. Namioka, *Linear Topological Spaces*, Van Nostrand, New York and London, 1963.
76. V. L. Klee, The support property of a convex set, *Duke Math. J.* **15** (1948), 767–772.
77. V. L. Klee, Convex sets in linear spaces, *Duke Math. J.* **18** (1951), 443–466.
78. G. Köthe, *Topological Vector Spaces*, I, Springer–Verlag, New York and Heidelberg, 1969. (English translation from the German.)
79. M. G. Krein and S. G. Krein, One internal characterization of the space of all continuous functions defined on a Hausdorff bicompact set, *Dokl. USSR* **27** (1940), 427–431.
80. K. K. Kutty and J. Quinn, Some characterizations of the projection property in Archimedean Riesz spaces, *Canadian J. Math.* **24** (1972), 306–311.
81. E. Langford and C. D. Aliprantis, Regularity properties of quotient Riesz seminorms, *Nederl. Akad. Wetensch. Proc. Ser. A* **78** (1975), 199–212.
82. H. P. Lotz, Minimal and reflexive Banach lattices, *Math. Ann.* **209** (1974), 117–126.
83. G. Ya. Lozanovsky, Two remarks concerning operators in partially ordered spaces, *Vestnik Leningrad Univ. Mat. Mekh. Astronom.* **19** (1965), 159–160. (Russian). **MR 32** # 6227; **Zbl 154**, 158.
84. G. Ya. Lozanovsky, On Banach lattices and bases, *Funktsional. Anal. i Prilozhen* **1** (1967), 92.
85. G. Ya. Lozanovsky, Some topological properties of Banach lattices and reflexivity conditions on them, *Soviet Math. Dokl.* **9** (1968), 1415–1418.
86. W. A. J. Luxemburg, Is every integral normal?, *Bull. Amer. Math. Soc.* **73** (1967), 685–688.

87. W. A. J. Luxemburg, Notes on Banach function spaces, *Nederl. Akad. Wetensch. Proc. Ser. A*, Note XIV **68** (1965), 229–248; Note XV **68** (1965), 415–446; Note XVI **68** (1965), 646–667.
88. W. A. J. Luxemburg, Integration with respect to finitely additive measures, in: C. D. Aliprantis, K. C. Border and W. A. J. Luxemburg, eds., *Positive Operators, Riesz Spaces, and Economics* (Springer–Verlag, Berlin and New York, 1990), 109–150.
89. W. A. J. Luxemburg and A. C. Zaanen, Notes on Banach function spaces, *Nederl. Akad. Wetensch. Proc. Ser. A*, Note I **66** (1963), 135–147; Note II **66** (1963), 148–153; Note III **66** (1963), 239–250; Note IV **66** (1963), 251–263; Note V **66** (1963), 496–504; Note VI **66** (1963), 655–668; Note VII **66** (1963), 669–681; Note VIII **67** (1964), 104–119; Note IX **67** (1964), 360–376; Note X **67** (1964), 493–506; Note XI **67** (1964), 507–518; Note XII **67** (1964), 519–529; Note XIII **67** (1964), 530–543.
90. W. A. J. Luxemburg and A. C. Zaanen, *Riesz Spaces*, I, North-Holland, Amsterdam, 1971.
91. F. Maeda and T. Ogasawara, Representation of vector lattices, *J. Sci. Hiroshima Univ. A* **12** (1942), 17–35.
92. A. Mas-Colell, The price equilibrium existence problem in topological vector lattices, *Econometrica* **54** (1986), 1039–1053.
93. A. Mas-Colell and S. F. Richard, A new approach to the existence of equilibria in vector lattices, *J. Econom. Theory* **53** (1991), 1–11.
94. A. Mas-Colell, M. D. Whinston, and J. R. Green, *Microeconomic Theory*, Oxford University Press, New York and Oxford, 1995.
95. W. Maxey, *The Dedekind completion of $C(X)$ and its second dual*, Ph.D. Dissertation, Purdue University, West Lafayette, Indiana, 1973.
96. L. W. McKenzie, On equilibrium in Graham's model of world trade and other competitive systems, *Econometrica* **22** (1954), 147–161.
97. P. Meyer-Nieberg, Zur schwachen Kompaktheit in Banachverbänden, *Math. Z.* **134** (1973), 303–315.
98. P. Meyer-Nieberg, Charakterisierung einiger topologischer und ordnungstheretischer Eigenschaften von Banachverbänden mit Hilfe disjunkter Folgen, *Arch. Math. (Basel)* **24** (1973), 640–647.
99. H. Nakano, Über das System aller stetigen Funktionen auf einen Topologischen Räum, *Proc. Imp. Acad. (Tokyo)* **17** (1941), 308–310.
100. H. Nakano, *Modulared Semi-Ordered Linear Spaces*, Maruzen Co., Tokyo, 1950.
101. H. Nakano, Linear topologies on semi-ordered linear spaces, *J. Fac. Sci. Hokkaido Univ. Ser. I* **12** (1953), 87–104.
102. H. Nakano, *Semi-Ordered Linear Spaces*, Japan Society for the Promotion of Science, Tokyo, 1955.
103. H. Nakano, *Linear Lattices*, Wayne State Univ. Press, Detroit, Michigan, 1966.
104. I. Namioka, Partially ordered linear topological spaces, *Mem. Amer. Math. Soc.*, # **24**, Providence, RI,1957.
105. T. Ogasawara, *Vector Lattices I and II*, *J. Sci. Hiroshima Univ. Ser. A* **12** (1942), 37–100 and **12** (1944), 41–161 (Japanese). MR **10**, 545
106. B. Peleg and M. E. Yaari, Markets with countably many commodities, *Internat. Econom. Rev.* **11** (1970), 369–377.
107. A. L. Peressini, On topologies in ordered vector spaces, *Math. Ann.* **144** (1961), 199–223.
108. A. L. Peressini, *Ordered Topological Vector Spaces*, Harper and Row, New York and London, 1967.
109. A. G. Pinsker, On the extension of partially ordered spaces, *DAN SSSR* **21** (1938), 6–10. (In Russian.)
110. A. G. Pinsker, Extension of partially ordered groups and spaces, *Leningrad Gos. Ped. Inst. Uchen. Zap.* **86** (1949), 285–315. (In Russian.)
111. K. Podczeck, Equilibria in vector lattices without ordered preferences and without uniform properness, *J. Math. Economics* **25** (1996), 465–484.
112. J. Quinn, Intermediate Riesz spaces, *Pacific J. Math.* **56** (1975), 225–263.
113. J. Quinn and R. Reichard, Characterizing discrete vector lattices, *Colloq. Math.* **31** (1974), 235–242.
114. S. F. Richard, A new approach to production equilibria in vector lattices, *J. Math. Econom.* **18** (1989), 41–56.

115. F. Riesz, Sur la décomposition des opérations linéaires, *Atti. Congr. Internaz. Mat. Bologna* **3** (1930), 143–148.

116. F. Riesz, Sur quelques notions fondamentals dans la theorie générale des opérations linéaires, *Ann. of Math.* **41** (1940), 174–206. (This work was published first in 1937 in Hungarian.)

117. G. T. Roberts, Topologies in vector lattices, *Math. Proc. Cambridge Phil. Soc.* **48** (1952), 533–546.

118. A. Robertson and W. Robertson, *Topological Vector Spaces*, 2nd Edition, Cambridge Univ. Press, London, 1973.

119. H. E. Scarf, The core of an N person game, *Econometrica* **35** 91967), 50–69.

120. H. H. Schaefer, Halbgeordnete lokalkonvexe Vektorräume, *Math. Ann.*, Note I **135** (1958), 115–141; Note II **138** (1959), 259–286; Note III **141** (1960), 113–142.

121. H. H. Schaefer, On the completeness of topological vector lattices, *Michigan Math. J.* **7** (1960), 303–309.

122. H. H. Schaefer, *Topological Vector Spaces*, Springer–Verlag, Heidelberg and New York, 1971.

123. H. H. Schaefer, *Banach Lattices and Positive Operators*, Springer–Verlag, Berlin and New York, 1974.

124. N. Stokey, R. E. Lucas, Jr., and E. C. Prescott, *Recursive Methods in Economic Dynamics*, Harvard University Press, Cambridge, MA, 1989.

125. L. Tzafriri, Reflexivity in Banach lattices and their subspaces, *J. Funct. Anal.* **10** (1972), 1–18.

126. A. I. Veksler and V. A. Geiler, Order and disjoint completeness of linear partially ordered spaces, *Siberian Math. J.* **13** (1972), 30–35.

127. B. Z. Vulikh, *Introduction to the Theory of Partially Ordered Spaces*, Wolter–Hoordhoff, Groningen, Netherlands, 1967. (English translation from the Russian.)

128. W. Wnuk, *Banach Lattices with Order Continuous Norms*, Advanced Topics in Mathematics, Polish Scientific Publishers PWN, Warszawa, 1999.

129. W. Wnuk, On regular Riesz subspaces, *Positivity*, to appear.

130. Y. C. Wong and K. F. Ng, *Partially Ordered Topological Vector Spaces*, Clarendon Press, Oxford, 1973.

131. N. C. Yannelis and W. R. Zame, Equilibria in Banach lattices without ordered preferences, *J. Math. Econom.* **15** (1986), 85–110.

132. W. R. Zame, Competitive equilibria in production economies with an infinite-dimensional commodity space, *Econometrica* **55** (1987), 1075–1108.

Index

Titles in This Series

106 **Hiroyuki Yoshida,** Absolute CM-periods, 2003

105 **Charalambos D. Aliprantis and Owen Burkinshaw,** Locally solid Riesz spaces with applications to economics, second edition, 2003

104 **Graham Everest, Alf van der Poorten, Igor Shparlinski, and Thomas Ward,** Recurrence sequences, 2003

103 **Octav Cornea, Gregory Lupton, John Oprea, and Daniel Tanré,** Lusternik-Schnirelmann category, 2003

102 **Linda Rass and John Radcliffe,** Spatial deterministic epidemics, 2003

101 **Eli Glasner,** Ergodic theory via joinings, 2003

99 **Philip S. Hirschhorn,** Model categories and their localizations, 2003

98 **Victor Guillemin, Viktor Ginzburg, and Yael Karshon,** Moment maps, cobordisms, and Hamiltonian group actions, 2002

97 **V. A. Vassiliev,** Applied Picard-Lefschetz theory, 2002

96 **Martin Markl, Steve Shnider, and Jim Stasheff,** Operads in algebra, topology and physics, 2002

95 **Seiichi Kamada,** Braid and knot theory in dimension four, 2002

94 **Mara D. Neusel and Larry Smith,** Invariant theory of finite groups, 2002

93 **Nikolai K. Nikolski,** Operators, functions, and systems: An easy reading. Volume 2: Model operators and systems, 2002

92 **Nikolai K. Nikolski,** Operators, functions, and systems: An easy reading. Volume 1: Hardy, Hankel, and Toeplitz, 2002

91 **Richard Montgomery,** A tour of subriemannian geometries, their geodesics and applications, 2002

90 **Christian Gérard and Izabella Łaba,** Multiparticle quantum scattering in constant magnetic fields, 2002

89 **Michel Ledoux,** The concentration of measure phenomenon, 2001

88 **Edward Frenkel and David Ben-Zvi,** Vertex algebras and algebraic curves, 2001

87 **Bruno Poizat,** Stable groups, 2001

86 **Stanley N. Burris,** Number theoretic density and logical limit laws, 2001

85 **V. A. Kozlov, V. G. Maz′ya, and J. Rossmann,** Spectral problems associated with corner singularities of solutions to elliptic equations, 2001

84 **László Fuchs and Luigi Salce,** Modules over non-Noetherian domains, 2001

83 **Sigurdur Helgason,** Groups and geometric analysis: Integral geometry, invariant differential operators, and spherical functions, 2000

82 **Goro Shimura,** Arithmeticity in the theory of automorphic forms, 2000

81 **Michael E. Taylor,** Tools for PDE: Pseudodifferential operators, paradifferential operators, and layer potentials, 2000

80 **Lindsay N. Childs,** Taming wild extensions: Hopf algebras and local Galois module theory, 2000

79 **Joseph A. Cima and William T. Ross,** The backward shift on the Hardy space, 2000

78 **Boris A. Kupershmidt,** KP or mKP: Noncommutative mathematics of Lagrangian, Hamiltonian, and integrable systems, 2000

77 **Fumio Hiai and Dénes Petz,** The semicircle law, free random variables and entropy, 2000

76 **Frederick P. Gardiner and Nikola Lakic,** Quasiconformal Teichmüller theory, 2000

75 **Greg Hjorth,** Classification and orbit equivalence relations, 2000

74 **Daniel W. Stroock,** An introduction to the analysis of paths on a Riemannian manifold, 2000

73 **John Locker,** Spectral theory of non-self-adjoint two-point differential operators, 2000

For a complete list of titles in this series, visit the
AMS Bookstore at **www.ams.org/bookstore/**.